Elastoplasticity Theory

Mechanical Engineering Series

Frank Kreith - Series Editor

Published Titles

Distributed Generation: The Power Paradigm for the New Millennium
Anne-Marie Borbely & Jan F. Kreider
Elastoplasticity Theory
Vlado A. Lubarda
Energy Audit of Building Systems: An Engineering Approach
Moncef Krarti
Entropy Generation Minimization
Adrian Bejan
Finite Element Method Using MATLAB, 2nd Edition
Young W. Kwon & Hyochoong Bang
Fluid Power Circuits and Controls: Fundamentals and Applications
John S. Cundiff
Fundamentals of Environmental Discharge Modeling
Lorin R. Davis
Introductory Finite Element Method
Chandrakant S. Desai & Tribikram Kundu
Intelligent Transportation Systems: New Principles and Architectures
Sumit Ghosh & Tony Lee
Mathematical & Physical Modeling of Materials Processing Operations
Olusegun Johnson Ileghus, Manabu Iguchi & Walter E. Wahnsiedler
Mechanics of Composite Materials
Autar K. Kaw
Mechanics of Fatigue
Vladimir V. Bolotin
Mechanism Design: Enumeration of Kinematic Structures According to Function
Lung-Wen Tsai
Nonlinear Analysis of Structures
M. Sathyamoorthy
Practical Inverse Analysis in Engineering
David M. Trujillo & Henry R. Busby
Principles of Solid Mechanics
Rowland Richards, Jr.
Thermodynamics for Engineers
Kau-Fui Wong
Viscoelastic Solids
Roderic S. Lakes

Forthcoming Titles

Engineering Experimentation
Euan Somerscales
Heat Transfer in Single and Multiphase Systems
Greg F. Naterer
Mechanics of Solids & Shells
Gerald Wempner & Demosthenes Talaslidis

Elastoplasticity Theory

Vlado A. Lubarda

CRC Press
Taylor & Francis Group
Boca Raton London New York

CRC Press is an imprint of the
Taylor & Francis Group, an **informa** business

CRC Press
Taylor & Francis Group
6000 Broken Sound Parkway NW, Suite 300
Boca Raton, FL 33487-2742

First issued in paperback 2019

ISBN-13: 978-0-8493-1138-3 (hbk)
ISBN-13: 978-0-367-39712-8 (pbk)

Library of Congress Cataloging-in-Publication Data

Lubarda, Vlado A.
Elastoplasticity theory/ Vlado A. Lubarda
p. cm. -- (Mechanical engineering series)
Includes bibliographical references and index.
ISBN 0-8493-1138-1 (alk. paper)
1. Elastoplasticiy.
I. Title. II. Advanced topics in mechanical engineering series.
QA931 .L9386 2001
620.1'1232—dc21 2001025780
CIP

Library of Congress Card Number 2001025780

Visit the Taylor & Francis Web site at
http://www.taylorandfrancis.com

and the CRC Press Web site at
http://www.crcpress.com

Contents

Preface

This book grew out of my lecture notes for graduate courses on the theory of plasticity and nonlinear continuum mechanics that I taught at several universities in the USA and former Yugoslavia during the past two decades. The book consists of three parts. The first part is an introduction to nonlinear continuum mechanics. After tensor preliminaries in Chapter 1, selected topics of kinematics and kinetics of deformation are presented in Chapters 2 and 3. Hill's theory of conjugate stress and strain measures is used. Chapter 4 is a brief treatment of the thermodynamics of deformation, with an accent given to formulation with internal state variables. Part 2 of the book is devoted to nonlinear elasticity. Constitutive theory of finite strain elasticity is presented in Chapter 5, and its rate-type formulation in Chapter 6. An analysis of elastic stability at finite strain is given in Chapter 7. Nonlinear elasticity is included in the book because it illustrates an application of many general concepts from Part 1, and because it is combined in Part 3 with finite deformation plasticity to derive general constitutive structure of finite deformation elastoplasticity. Part 3 is the largest part of the book, consisting of seven chapters on plasticity. Chapter 8 is an analysis of the constitutive framework for rate-independent and rate-dependent plasticity. The postulates of Drucker and Ilyushin are discussed in the context of finite strain. Derivation of elastoplastic constitutive equations for various phenomenological models of material response is presented in Chapter 9. Formulations in stress and strain space, using the yield surfaces with and without vertices, are given. Isotropic, kinematic, combined isotropic–kinematic and multisurface hardening models are introduced. Pressure-dependent plasticity and non-associative flow rules are then discussed. Fundamental aspects of thermoplasticity, rate-dependent plasticity and deformation theory of plasticity are also included. Hill's theory of uniqueness and plastic stability is presented in Chapter 10, together with an analysis of eigenmodal deformations and acceleration waves in elastoplastic solids. Rice's treatment of plastic flow localization in pressure-insensitive and pressure-sensitive materials is then given. Chapter 11 is devoted to formulation of the constitutive theory of elastoplasticity in the framework of Lee's multiplicative decomposition of deformation gradient into its elastic and plastic parts. Isotropic and orthotropic materials are considered, with an introductory treatment of damage-elastoplasticity. The theory of monocrystalline plasticity is presented in Chapter 12. Crystallographic slip is assumed to be the only mechanism of plastic deformation. Hardening rules and uniqueness of slip rates are examined. Specific forms of constitutive equations for rate-independent and rate-dependent crystals are derived. Chapter 13 covers some fundamental topics of micro-to-macro transition in the constitutive description. The analysis is aimed toward the derivation of constitutive equations for a polycrystalline aggregate from known constitutive equations of single crystals. The fourteenth, and final chapter of the book is devoted to approximate models of polycrystalline plasticity. The classical model of Taylor and the analysis of Bishop and Hill are presented. The main theme is the self-consistent method, introduced in polycrystalline plasticity by Kröner, Budiansky and Wu. Hill's formulation of the method is used in the finite deformation presentation. Calculations of the polycrystalline stress-strain curve and polycrystalline yield surface, development of the crystallographic texture, and effects of the grain-size on the aggregate response are discussed.

This book is an advanced treatment of finite deformation elastoplasticity and is intended for graduate students and other interested readers who are familiar with an introductory treatment of plasticity. Such treatment is usually given in an infinitesimal strain context and with a focus on the geometry of admissible yield surfaces, von Mises and Tresca yield conditions, derivation of the Levy-Mises and Prandtl-Reuss equations, and the analysis of some elementary elastoplastic problems. Familiarity with basic concepts of crystallography and the dislocation theory from an undergraduate course in materials science is also assumed. Important topics of the slip-line theory and limit analysis are not discussed, since they have been repeatedly well covered in a number of existing plasticity books. Numerical treatments of boundary value problems and experimental techniques are not included either, as they require books on their own. A recent text by Simo and Hughes can be consulted as a reference to computational plasticity.

I began to study plasticity as a graduate student of Professor Erastus Lee at Stanford University in the late seventies. His research work and teaching of plasticity was a great inspiration to all his students. I am indebted to him for his guidance during our research on the rate-type constitutive theory of elastoplasticity based on the multiplicative decomposition of deformation gradient. The influence of Rodney Hill's development of the theory of plasticity on my writing is evident from the contents of this book. Large parts of all chapters are based on his research papers from 1948 to 1993. Communications with Professor Hill in 1994 were most inspirational. Two years spent in the solid mechanics group at Brown University in the late eighties and collaborations with Alan Needleman and Fong Shih were rewarding to my understanding of plasticity. Much of the first two parts of this book I wrote in the mid-nineties while teaching and conducting research in the Mechanical and Aerospace Engineering Department of Arizona State University. Collaboration with Dusan Krajcinovic on damage-elastoplasticity was a beneficial experience. A major part of the book was written while I was an Adjunct Professor in the Department of Applied Mechanics and Engineering Sciences of the University of California in San Diego. Professors Xanthippi Markenscoff and Marc Meyers repeatedly encouraged me to write a book on plasticity, and I express my gratitude to them for their support. Collaboration with David Benson on viscoplasticity and dynamic plasticity is also acknowledged. The books by Ray Ogden and Kerry Havner were in many aspects exemplary to my writing in chapters devoted to nonlinear elasticity and crystalline plasticity. I am indebted to Dr. Owen Richmond from Alcoa Laboratories for his continuing support of my research work at Brown, ASU and UCSD. The research support from NSF and the US Army is also acknowledged. Several chapters of this book were written while I was visiting the University of Montenegro during summers of the last two years. Docent Borko Vujičić from the Physics Department was always available to help with Latex related issues in the preparation of the manuscript. I thank him for that. Computer specialists Todd Porteous and Andres Burgos from UCSD were also of help. My appreciation finally extends to Cindy Renee Carelli, acquisitions editor, and Bill Heyward, project editor from CRC Press, for their assistance in publishing this book.

Vlado A. Lubarda
San Diego, April 2001

Professor Vlado A. Lubarda received his Ph.D. degree from Stanford University in 1980. He has been a Docent and an Associate Professor at the University of Montenegro, and a Fulbright Fellow and a Visiting Associate Professor at Brown University and the Arizona State University. Currently, he is an Adjunct Professor of Applied Mechanics in the Department of Mechanical and Aerospace Engineering at the University of California, San Diego. Dr. Lubarda has done extensive research in the constitutive theory of large deformation elastoplasticity, damage mechanics, and dislocation theory. He is the author of 75 journal and conference publications and the textbook *Strength of Materials* (in Serbo-Croatian). He has served as a reviewer to numerous international journals, and was elected in 2000 to the Montenegrin Academy of Sciences and Arts.

Part 1

ELEMENTS OF CONTINUUM MECHANICS

CHAPTER 1

TENSOR PRELIMINARIES

1.1. Vectors

An orthonormal basis for the three-dimensional Euclidean vector space is a set of three orthogonal unit vectors. The scalar product of any two of these vectors is

$$\mathbf{e}_i \cdot \mathbf{e}_j = \delta_{ij} = \begin{cases} 1, & \text{if } i = j, \\ 0, & \text{if } i \neq j, \end{cases} \tag{1.1.1}$$

δ_{ij} being the Kronecker delta symbol. An arbitrary vector $\mathbf{a}$ can be decomposed in the introduced basis as

$$\mathbf{a} = a_i \mathbf{e}_i, \quad a_i = \mathbf{a} \cdot \mathbf{e}_i. \tag{1.1.2}$$

The summation convention is assumed over the repeated indices. The scalar product of the vectors $\mathbf{a}$ and $\mathbf{b}$ is

$$\mathbf{a} \cdot \mathbf{b} = a_i b_i. \tag{1.1.3}$$

The vector product of two base vectors is defined by

$$\mathbf{e}_i \times \mathbf{e}_j = \epsilon_{ijk} \mathbf{e}_k, \tag{1.1.4}$$

where ϵ_{ijk} is the permutation symbol

$$\epsilon_{ijk} = \begin{cases} 1, & \text{if } ijk \text{ is an even permutation of 123}, \\ -1, & \text{if } ijk \text{ is an odd permutation of 123}, \\ 0, & \text{otherwise}. \end{cases} \tag{1.1.5}$$

The vector product of the vectors $\mathbf{a}$ and $\mathbf{b}$ can consequently be written as

$$\mathbf{a} \times \mathbf{b} = \epsilon_{ijk} a_i b_j \mathbf{e}_k. \tag{1.1.6}$$

The triple scalar product of the base vectors is

$$(\mathbf{e}_i \times \mathbf{e}_j) \cdot \mathbf{e}_k = \epsilon_{ijk}, \tag{1.1.7}$$

so that

$$(\mathbf{a} \times \mathbf{b}) \cdot \mathbf{c} = \epsilon_{ijk} a_i b_j c_k = \begin{vmatrix} a_1 & b_1 & c_1 \\ a_2 & b_2 & c_2 \\ a_3 & b_3 & c_3 \end{vmatrix}. \tag{1.1.8}$$

In view of the vector relationship

$$(\mathbf{e}_i \times \mathbf{e}_j) \cdot (\mathbf{e}_k \times \mathbf{e}_l) = (\mathbf{e}_i \cdot \mathbf{e}_k)(\mathbf{e}_j \cdot \mathbf{e}_l) - (\mathbf{e}_i \cdot \mathbf{e}_l)(\mathbf{e}_j \cdot \mathbf{e}_k), \tag{1.1.9}$$

there is an $\epsilon - \delta$ identity

$$\epsilon_{ijm}\epsilon_{klm} = \delta_{ik}\delta_{jl} - \delta_{il}\delta_{jk}. \tag{1.1.10}$$

In particular,

$$\epsilon_{ikl}\epsilon_{jkl} = 2\delta_{ij}, \quad \epsilon_{ijk}\epsilon_{ijk} = 6. \tag{1.1.11}$$

The triple vector product of the base vectors is

$$(\mathbf{e}_i \times \mathbf{e}_j) \times \mathbf{e}_k = \epsilon_{ijm}\epsilon_{klm}\mathbf{e}_l = \delta_{ik}\mathbf{e}_j - \delta_{jk}\mathbf{e}_i. \tag{1.1.12}$$

Thus,

$$(\mathbf{a} \times \mathbf{b}) \times \mathbf{c} = a_i b_j (c_i \mathbf{e}_j - c_j \mathbf{e}_i), \tag{1.1.13}$$

which confirms the vector identity

$$(\mathbf{a} \times \mathbf{b}) \times \mathbf{c} = (\mathbf{a} \cdot \mathbf{c})\mathbf{b} - (\mathbf{b} \cdot \mathbf{c})\mathbf{a}. \tag{1.1.14}$$

1.2. Second-Order Tensors

A dyadic product of two base vectors is the second-order tensor $\mathbf{e}_i \otimes \mathbf{e}_j$, such that

$$(\mathbf{e}_i \otimes \mathbf{e}_j) \cdot \mathbf{e}_k = \mathbf{e}_k \cdot (\mathbf{e}_j \otimes \mathbf{e}_i) = \delta_{jk}\mathbf{e}_i. \tag{1.2.1}$$

For arbitrary vectors **a**, **b** and **c**, it follows that

$$(\mathbf{a} \otimes \mathbf{b}) \cdot \mathbf{e}_k = b_k \mathbf{a}, \quad (\mathbf{a} \otimes \mathbf{b}) \cdot \mathbf{c} = (\mathbf{b} \cdot \mathbf{c})\mathbf{a}. \tag{1.2.2}$$

The tensors $\mathbf{e}_i \otimes \mathbf{e}_j$ serve as base tensors for the representation of an arbitrary second-order tensor,

$$\mathbf{A} = A_{ij}\mathbf{e}_i \otimes \mathbf{e}_j, \quad A_{ij} = \mathbf{e}_i \cdot \mathbf{A} \cdot \mathbf{e}_j. \tag{1.2.3}$$

A dot product of the second-order tensor **A** and the vector **a** is the vector

$$\mathbf{b} = \mathbf{A} \cdot \mathbf{a} = b_i \mathbf{e}_i, \quad b_i = A_{ij} a_j. \tag{1.2.4}$$

Similarly, a dot product of two second-order tensors **A** and **B** is the second-order tensor

$$\mathbf{C} = \mathbf{A} \cdot \mathbf{B} = C_{ij}\mathbf{e}_i \otimes \mathbf{e}_j, \quad C_{ij} = A_{ik}B_{kj}. \tag{1.2.5}$$

The unit (identity) second-order tensor is

$$\mathbf{I} = \delta_{ij}\mathbf{e}_i \otimes \mathbf{e}_j, \tag{1.2.6}$$

which satisfies

$$\mathbf{A} \cdot \mathbf{I} = \mathbf{I} \cdot \mathbf{A} = \mathbf{A}, \quad \mathbf{I} \cdot \mathbf{a} = \mathbf{a}. \tag{1.2.7}$$

The transpose of the tensor **A** is the tensor $\mathbf{A}^T$, which, for any vectors **a** and **b**, meets

$$\mathbf{A} \cdot \mathbf{a} = \mathbf{a} \cdot \mathbf{A}^T, \quad \mathbf{b} \cdot \mathbf{A} \cdot \mathbf{a} = \mathbf{a} \cdot \mathbf{A}^T \cdot \mathbf{b}. \tag{1.2.8}$$

Thus, if $\mathbf{A} = A_{ij}\mathbf{e}_i \otimes \mathbf{e}_j$, then

$$\mathbf{A}^T = A_{ji}\mathbf{e}_i \otimes \mathbf{e}_j. \tag{1.2.9}$$

The tensor $\mathbf{A}$ is symmetric if $\mathbf{A}^T = \mathbf{A}$; it is antisymmetric (or skew-symmetric) if $\mathbf{A}^T = -\mathbf{A}$. If $\mathbf{A}$ is nonsingular ($\det \mathbf{A} \neq 0$), there is a unique inverse tensor $\mathbf{A}^{-1}$ such that

$$\mathbf{A} \cdot \mathbf{A}^{-1} = \mathbf{A}^{-1} \cdot \mathbf{A} = \mathbf{I}. \tag{1.2.10}$$

In this case, $\mathbf{b} = \mathbf{A} \cdot \mathbf{a}$ implies $\mathbf{a} = \mathbf{A}^{-1} \cdot \mathbf{b}$. For an orthogonal tensor $\mathbf{A}^T = \mathbf{A}^{-1}$, so that $\det \mathbf{A} = \pm 1$. The plus sign corresponds to proper and minus to improper orthogonal tensors.

The trace of the tensor $\mathbf{A}$ is a scalar obtained by the contraction ($i = j$) operation

$$\operatorname{tr} \mathbf{A} = A_{ii}. \tag{1.2.11}$$

For a three-dimensional identity tensor, $\operatorname{tr} \mathbf{I} = 3$. Two inner (scalar or double-dot) products of two second-order tensors are defined by

$$\mathbf{A} \cdot\cdot\, \mathbf{B} = \operatorname{tr} (\mathbf{A} \cdot \mathbf{B}) = A_{ij} B_{ji}, \tag{1.2.12}$$

$$\mathbf{A} : \mathbf{B} = \operatorname{tr} \left(\mathbf{A} \cdot \mathbf{B}^T\right) = \operatorname{tr} \left(\mathbf{A}^T \cdot \mathbf{B}\right) = A_{ij} B_{ij}. \tag{1.2.13}$$

The connections are

$$\mathbf{A} \cdot\cdot\, \mathbf{B} = \mathbf{A}^T : \mathbf{B} = \mathbf{A} : \mathbf{B}^T. \tag{1.2.14}$$

If either $\mathbf{A}$ or $\mathbf{B}$ is symmetric, $\mathbf{A} \cdot\cdot\, \mathbf{B} = \mathbf{A} : \mathbf{B}$. Also,

$$\operatorname{tr} \mathbf{A} = \mathbf{A} : \mathbf{I}, \quad \operatorname{tr} (\mathbf{a} \otimes \mathbf{b}) = \mathbf{a} \cdot \mathbf{b}. \tag{1.2.15}$$

Since the trace product is unaltered by any cyclic rearrangement of the factors, we have

$$\mathbf{A} \cdot\cdot\, (\mathbf{B} \cdot \mathbf{C}) = (\mathbf{A} \cdot \mathbf{B}) \cdot\cdot\, \mathbf{C} = (\mathbf{C} \cdot \mathbf{A}) \cdot\cdot\, \mathbf{B}, \tag{1.2.16}$$

$$\mathbf{A} : (\mathbf{B} \cdot \mathbf{C}) = \left(\mathbf{B}^T \cdot \mathbf{A}\right) : \mathbf{C} = \left(\mathbf{A} \cdot \mathbf{C}^T\right) : \mathbf{B}. \tag{1.2.17}$$

A deviatoric part of $\mathbf{A}$ is defined by

$$\mathbf{A}' = \mathbf{A} - \frac{1}{3} (\operatorname{tr} \mathbf{A}) \mathbf{I}, \tag{1.2.18}$$

with the property $\operatorname{tr} \mathbf{A}' = 0$. It is easily verified that $\mathbf{A}' : \mathbf{A} = \mathbf{A}' : \mathbf{A}'$ and $\mathbf{A}' \cdot\cdot\, \mathbf{A} = \mathbf{A}' \cdot\cdot\, \mathbf{A}'$. A nonsymmetric tensor $\mathbf{A}$ can be decomposed into its symmetric and antisymmetric parts, $\mathbf{A} = \mathbf{A}_s + \mathbf{A}_a$, such that

$$\mathbf{A}_s = \frac{1}{2} \left(\mathbf{A} + \mathbf{A}^T\right), \quad \mathbf{A}_a = \frac{1}{2} \left(\mathbf{A} - \mathbf{A}^T\right). \tag{1.2.19}$$

If $\mathbf{A}$ is symmetric and $\mathbf{W}$ is antisymmetric, the trace of their dot product is equal to zero, $\operatorname{tr} (\mathbf{A} \cdot \mathbf{W}) = 0$. The axial vector $\boldsymbol{\omega}$ of an antisymmetric tensor $\mathbf{W}$ is defined by

$$\mathbf{W} \cdot \mathbf{a} = \boldsymbol{\omega} \times \mathbf{a}, \tag{1.2.20}$$

for every vector $\mathbf{a}$. This gives the component relationships

$$W_{ij} = -\epsilon_{ijk} \omega_k, \quad \omega_i = -\frac{1}{2} \epsilon_{ijk} W_{jk}. \tag{1.2.21}$$

Since $\mathbf{A} \cdot \mathbf{e}_i = A_{ji}\mathbf{e}_j$, the determinant of $\mathbf{A}$ can be calculated from Eq. (1.1.8) as

$$\det \mathbf{A} = [(\mathbf{A} \cdot \mathbf{e}_1) \times (\mathbf{A} \cdot \mathbf{e}_2)] \cdot (\mathbf{A} \cdot \mathbf{e}_3) = \epsilon_{ijk} A_{i1} A_{j2} A_{k3}. \tag{1.2.22}$$

Thus,

$$\epsilon_{\alpha\beta\gamma}(\det \mathbf{A}) = \epsilon_{ijk} A_{i\alpha} A_{j\beta} A_{k\gamma}, \tag{1.2.23}$$

and by second of Eq. (1.1.11)

$$\det \mathbf{A} = \frac{1}{6} \epsilon_{ijk} \epsilon_{\alpha\beta\gamma} A_{i\alpha} A_{j\beta} A_{k\gamma}. \tag{1.2.24}$$

For further details, standard texts such as Brillouin (1964) can be consulted.

1.3. Eigenvalues and Eigenvectors

The vector $\mathbf{n}$ is an eigenvector of the second-order tensor $\mathbf{A}$ if there is a scalar λ such that $\mathbf{A} \cdot \mathbf{n} = \lambda \mathbf{n}$, i.e.,

$$(\mathbf{A} - \lambda \mathbf{I}) \cdot \mathbf{n} = \mathbf{0}. \tag{1.3.1}$$

A scalar λ is called an eigenvalue of $\mathbf{A}$ corresponding to the eigenvector $\mathbf{n}$. Nontrivial solutions for $\mathbf{n}$ exist if $\det(\mathbf{A} - \lambda \mathbf{I}) = 0$, which gives the characteristic equation for $\mathbf{A}$,

$$\lambda^3 - J_1 \lambda^2 - J_2 \lambda - J_3 = 0. \tag{1.3.2}$$

The scalars J_1, J_2 and J_3 are the principal invariants of $\mathbf{A}$, which remain unchanged under any orthogonal transformation of the orthonormal basis of $\mathbf{A}$. These are

$$J_1 = \operatorname{tr} \mathbf{A}, \tag{1.3.3}$$

$$J_2 = \frac{1}{2} \left[\operatorname{tr} \left(\mathbf{A}^2 \right) - \left(\operatorname{tr} \mathbf{A} \right)^2 \right], \tag{1.3.4}$$

$$J_3 = \det \mathbf{A} = \frac{1}{6} \left[2 \operatorname{tr} \left(\mathbf{A}^3 \right) - 3 \left(\operatorname{tr} \mathbf{A} \right) \operatorname{tr} \left(\mathbf{A}^2 \right) + \left(\operatorname{tr} \mathbf{A} \right)^3 \right]. \tag{1.3.5}$$

If $\lambda_1 \neq \lambda_2 \neq \lambda_3 \neq \lambda_1$, there are three mutually orthogonal eigenvectors $\mathbf{n}_1$, $\mathbf{n}_2$, $\mathbf{n}_3$, so that $\mathbf{A}$ has a spectral representation

$$\mathbf{A} = \sum_{i=1}^{3} \lambda_i \mathbf{n}_i \otimes \mathbf{n}_i. \tag{1.3.6}$$

If $\lambda_1 \neq \lambda_2 = \lambda_3$,

$$\mathbf{A} = (\lambda_1 - \lambda_2) \mathbf{n}_1 \otimes \mathbf{n}_1 + \lambda_2 \mathbf{I}, \tag{1.3.7}$$

while $\mathbf{A} = \lambda \mathbf{I}$, if $\lambda_1 = \lambda_2 = \lambda_3 = \lambda$.

A symmetric real tensor has all real eigenvalues. An antisymmetric tensor has only one real eigenvalue, which is equal to zero. The corresponding eigendirection is parallel to the axial vector of the antisymmetric tensor. A proper orthogonal (rotation) tensor has also one real eigenvalue, which is equal to one. The corresponding eigendirection is parallel to the axis of rotation.

1.4. Cayley–Hamilton Theorem

A second-order tensor satisfies its own characteristic equation

$$\mathbf{A}^3 - J_1\mathbf{A}^2 - J_2\mathbf{A} - J_3\mathbf{I} = \mathbf{0}. \tag{1.4.1}$$

This is a Cayley–Hamilton theorem. Thus, if $\mathbf{A}^{-1}$ exists, it can be expressed as

$$J_3\mathbf{A}^{-1} = \mathbf{A}^2 - J_1\mathbf{A} - J_2\mathbf{I}, \tag{1.4.2}$$

which shows that eigendirections of $\mathbf{A}^{-1}$ are parallel to those of $\mathbf{A}$. A number of useful results can be extracted from the Cayley–Hamilton theorem. An expression for $(\det \mathbf{F})$ in terms of traces of $\mathbf{A}$, $\mathbf{A}^2$, $\mathbf{A}^3$, given in Eq. (1.3.5), is obtained by taking the trace of Eq. (1.4.1). Similarly,

$$\det(\mathbf{I} + \mathbf{A}) - \det \mathbf{A} = 1 + J_1 - J_2. \tag{1.4.3}$$

If $\mathbf{X}^2 = \mathbf{A}$, an application of Eq. (1.4.1) to $\mathbf{X}$ gives

$$\mathbf{A} \cdot \mathbf{X} - I_1\mathbf{A} - I_2\mathbf{X} - I_3\mathbf{I} = \mathbf{0}, \tag{1.4.4}$$

where I_i are the principal invariants of $\mathbf{X}$. Multiplying this with I_1 and $\mathbf{X}$, and summing up the resulting two equations yields

$$\mathbf{X} = \frac{1}{I_1 I_2 + I_3}\left[\mathbf{A}^2 - \left(I_1^2 + I_2\right)\mathbf{A} - I_1 I_3 \mathbf{I}\right]. \tag{1.4.5}$$

The invariants I_i can be calculated from the principal invariants of $\mathbf{A}$, or from the eigenvalues of $\mathbf{A}$. Alternative route to solve $\mathbf{X}^2 = \mathbf{A}$ is via eigendirections and spectral representation (diagonalization) of $\mathbf{A}$.

1.5. Change of Basis

Under a rotational change of basis, the new base vectors are $\mathbf{e}_i^* = \mathbf{Q}\cdot\mathbf{e}_i$, where $\mathbf{Q}$ is a proper orthogonal tensor. An arbitrary vector $\mathbf{a}$ can be decomposed in the two bases as

$$\mathbf{a} = a_i\mathbf{e}_i = a_i^*\mathbf{e}_i^*, \quad a_i^* = Q_{ji}a_j. \tag{1.5.1}$$

If the vector $\mathbf{a}^*$ is introduced, with components a_i^* in the original basis ($\mathbf{a}^* = a_i^*\mathbf{e}_i$), then $\mathbf{a}^* = \mathbf{Q}^T \cdot \mathbf{a}$.

Under an arbitrary orthogonal transformation $\mathbf{Q}$ ($\mathbf{Q} \cdot \mathbf{Q}^T = \mathbf{Q}^T \cdot \mathbf{Q} = \mathbf{I}$, $\det \mathbf{Q} = \pm 1$), the components of so-called axial vectors transform according to $\omega_i^* = (\det \mathbf{Q})Q_{ji}\omega_j$. On the other hand, the components of absolute vectors transform as $a_i^* = Q_{ji}a_j$. If attention is confined to proper orthogonal transformations, i.e., the rotations of the basis only ($\det \mathbf{Q} = 1$), no distinction is made between axial and absolute vectors.

An invariant of $\mathbf{a}$ is $\mathbf{a} \cdot \mathbf{a}$. A scalar product of two vectors $\mathbf{a}$ and $\mathbf{b}$ is an even invariant of vectors $\mathbf{a}$ and $\mathbf{b}$, since it remains unchanged under both proper and improper orthogonal transformation of the basis (rotation and reflection). A triple scalar product of three vectors is an odd invariant of those vectors, since it remains unchanged under all proper orthogonal transformations ($\det \mathbf{Q} = 1$), but changes the sign under improper orthogonal transformations ($\det \mathbf{Q} = -1$).

A second-order tensor $\mathbf{A}$ can be decomposed in the considered bases as

$$\mathbf{A} = A_{ij}\mathbf{e}_i \otimes \mathbf{e}_j = A^*_{ij}\mathbf{e}^*_i \otimes \mathbf{e}^*_j, \quad A^*_{ij} = Q_{ki}A_{kl}Q_{lj}. \tag{1.5.2}$$

If the tensor $\mathbf{A}^* = A^*_{ij}\mathbf{e}_i \otimes \mathbf{e}_j$ is introduced, it is related to $\mathbf{A}$ by $\mathbf{A}^* = \mathbf{Q}^T \cdot \mathbf{A} \cdot \mathbf{Q}$. The two tensors share the same eigenvalues, which are thus invariants of $\mathbf{A}$ under rotation of the basis. Invariants are also symmetric functions of the eigenvalues, such as

$$\operatorname{tr}\mathbf{A} = \lambda_1 + \lambda_2 + \lambda_3, \quad \operatorname{tr}\left(\mathbf{A}^2\right) = \lambda_1^2 + \lambda_2^2 + \lambda_3^2, \quad \operatorname{tr}\left(\mathbf{A}^3\right) = \lambda_1^3 + \lambda_2^3 + \lambda_3^3, \tag{1.5.3}$$

or the principal invariants of Eqs. (1.3.3)–(1.3.5),

$$J_1 = \lambda_1 + \lambda_2 + \lambda_3, \quad J_2 = -\left(\lambda_1\lambda_2 + \lambda_2\lambda_3 + \lambda_3\lambda_1\right), \quad J_3 = \lambda_1\lambda_2\lambda_3. \tag{1.5.4}$$

All invariants of the second-order tensors under orthogonal transformations are even invariants.

1.6. Higher-Order Tensors

Triadic and tetradic products of the base vectors are

$$\mathbf{e}_i \otimes \mathbf{e}_j \otimes \mathbf{e}_k, \quad \mathbf{e}_i \otimes \mathbf{e}_j \otimes \mathbf{e}_k \otimes \mathbf{e}_l, \tag{1.6.1}$$

with obvious extension to higher-order polyadic products. These tensors serve as base tensors for the representation of higher-order tensors. For example, the permutation tensor is

$$\boldsymbol{\epsilon} = \epsilon_{ijk}\mathbf{e}_i \otimes \mathbf{e}_j \otimes \mathbf{e}_k, \tag{1.6.2}$$

where ϵ_{ijk} is defined by Eq. (1.1.5). If $\mathbf{A}$ is a symmetric second-order tensor,

$$\boldsymbol{\epsilon} : \mathbf{A} = \epsilon_{ijk}A_{jk}\mathbf{e}_i = \mathbf{0}. \tag{1.6.3}$$

The fourth-order tensor $\boldsymbol{\mathcal{L}}$ can be expressed as

$$\boldsymbol{\mathcal{L}} = \mathcal{L}_{ijkl}\mathbf{e}_i \otimes \mathbf{e}_j \otimes \mathbf{e}_k \otimes \mathbf{e}_l. \tag{1.6.4}$$

A dot product of $\boldsymbol{\mathcal{L}}$ with a vector $\mathbf{a}$ is

$$\boldsymbol{\mathcal{L}} \cdot \mathbf{a} = \mathcal{L}_{ijkl}a_l\mathbf{e}_i \otimes \mathbf{e}_j \otimes \mathbf{e}_k. \tag{1.6.5}$$

Two inner products of the fourth- and second-order tensors can be defined by

$$\boldsymbol{\mathcal{L}} \cdot\cdot\, \mathbf{A} = \mathcal{L}_{ijkl}A_{lk}\mathbf{e}_i \otimes \mathbf{e}_j, \quad \boldsymbol{\mathcal{L}} : \mathbf{A} = \mathcal{L}_{ijkl}A_{kl}\mathbf{e}_i \otimes \mathbf{e}_j. \tag{1.6.6}$$

If $\mathbf{W}$ is antisymmetric and $\boldsymbol{\mathcal{L}}$ has the symmetry in its last two indices,

$$\boldsymbol{\mathcal{L}} : \mathbf{W} = \mathbf{0}. \tag{1.6.7}$$

The symmetries of the form $\mathcal{L}_{ijkl} = \mathcal{L}_{jikl} = \mathcal{L}_{ijlk}$ will frequently, but not always, hold for the fourth-order tensors considered in this book. We also introduce the scalar products

$$\boldsymbol{\mathcal{L}} :: (\mathbf{A} \otimes \mathbf{B}) = \mathbf{B} : \boldsymbol{\mathcal{L}} : \mathbf{A} = B_{ij}\mathcal{L}_{ijkl}A_{kl}, \tag{1.6.8}$$

and

$$\boldsymbol{\mathcal{L}} \cdot\cdot\cdot\cdot\, (\mathbf{A} \otimes \mathbf{B}) = \mathbf{B} \cdot\cdot\, \boldsymbol{\mathcal{L}} \cdot\cdot\, \mathbf{A} = B_{ji}\mathcal{L}_{ijkl}A_{lk}. \tag{1.6.9}$$

The transpose of $\mathcal{L}$ satisfies

$$\mathcal{L} : \mathbf{A} = \mathbf{A} : \mathcal{L}^T, \quad \mathbf{B} : \mathcal{L} : \mathbf{A} = \mathbf{A} : \mathcal{L}^T : \mathbf{B}, \tag{1.6.10}$$

hence, $\mathcal{L}^T_{ijkl} = \mathcal{L}_{klij}$. The tensor $\mathcal{L}$ is symmetric if $\mathcal{L}^T = \mathcal{L}$, i.e., $\mathcal{L}_{ijkl} = \mathcal{L}_{klij}$ (reciprocal symmetry).

The symmetric fourth-order unit tensor $\boldsymbol{I}$ is

$$\boldsymbol{I} = I_{ijkl}\mathbf{e}_i \otimes \mathbf{e}_j \otimes \mathbf{e}_k \otimes \mathbf{e}_l, \quad I_{ijkl} = \frac{1}{2}\left(\delta_{ik}\delta_{jl} + \delta_{il}\delta_{jk}\right). \tag{1.6.11}$$

If $\mathcal{L}$ possesses the symmetry in its leading and terminal pair of indices ($\mathcal{L}_{ijkl} = \mathcal{L}_{jikl}$ and $\mathcal{L}_{ijkl} = \mathcal{L}_{ijlk}$) and if $\mathbf{A}$ is symmetric ($A_{ij} = A_{ji}$), then

$$\mathcal{L} : \boldsymbol{I} = \boldsymbol{I} : \mathcal{L} = \mathcal{L}, \quad \boldsymbol{I} : \mathbf{A} = \mathbf{A} : \boldsymbol{I} = \mathbf{A}. \tag{1.6.12}$$

For an arbitrary nonsymmetric second-order tensor $\mathbf{A}$,

$$\boldsymbol{I} : \mathbf{A} = \mathbf{A}_{\mathrm{s}} = \frac{1}{2}(\mathbf{A} + \mathbf{A}^T). \tag{1.6.13}$$

The fourth-order tensor with rectangular components

$$\hat{I}_{ijkl} = \frac{1}{2}\left(\delta_{ik}\delta_{jl} - \delta_{il}\delta_{jk}\right) \tag{1.6.14}$$

can also be introduced, such that

$$\hat{\boldsymbol{I}} : \mathbf{A} = \mathbf{A}_{\mathrm{a}} = \frac{1}{2}(\mathbf{A} - \mathbf{A}^T). \tag{1.6.15}$$

Note the symmetry properties

$$\hat{I}_{ijkl} = \hat{I}_{klij}, \quad \hat{I}_{jikl} = \hat{I}_{ijlk} = -\hat{I}_{ijkl}. \tag{1.6.16}$$

A fourth-order tensor $\mathcal{L}$ is invertible if there exists another such tensor $\mathcal{L}^{-1}$ which obeys

$$\mathcal{L} : \mathcal{L}^{-1} = \mathcal{L}^{-1} : \mathcal{L} = \boldsymbol{I}. \tag{1.6.17}$$

In this case, $\mathbf{B} = \mathcal{L} : \mathbf{A}$ implies $\mathbf{A} = \mathcal{L}^{-1} : \mathbf{B}$, and *vice versa.* The inner product of two fourth-order tensors $\mathcal{L}$ and $\mathcal{M}$ is defined by

$$\mathcal{L} : \mathcal{M} = \mathcal{L}_{ijmn}\mathcal{M}_{mnkl}\mathbf{e}_i \otimes \mathbf{e}_j \otimes \mathbf{e}_k \otimes \mathbf{e}_l. \tag{1.6.18}$$

The trace of the fourth-order tensor $\mathcal{L}$ is

$$\operatorname{tr}\mathcal{L} = \mathcal{L} :: \boldsymbol{I} = \mathcal{L}_{ijij}. \tag{1.6.19}$$

In particular, $\operatorname{tr}\boldsymbol{I} = 6$. A fourth-order tensor defined by

$$\mathcal{L}^{\mathrm{d}} = \mathcal{L} - \frac{1}{6}(\operatorname{tr}\mathcal{L})\boldsymbol{I}, \tag{1.6.20}$$

satisfies

$$\operatorname{tr}\mathcal{L}^{\mathrm{d}} = 0, \quad \mathcal{L}^{\mathrm{d}} :: \mathcal{L} = \mathcal{L}^{\mathrm{d}} :: \mathcal{L}^{\mathrm{d}}. \tag{1.6.21}$$

The tensor

$$\hat{\mathcal{L}}^{\mathrm{d}} = \mathcal{L} - \frac{1}{3}(\operatorname{tr}\mathcal{L})\mathbf{I} \otimes \mathbf{I} \tag{1.6.22}$$

also has the property $\operatorname{tr}\hat{\mathcal{L}}^{\mathrm{d}} = 0$.

Under rotational change of the basis specified by a proper orthogonal tensor $\mathbf{Q}$, the components of the fourth-order tensor change according to

$$\mathcal{L}^*_{ijkl} = Q_{\alpha i} Q_{\beta j} \mathcal{L}_{\alpha\beta\gamma\delta} Q_{\gamma k} Q_{\delta l}. \tag{1.6.23}$$

The trace of the fourth-order tensor is one of its invariants under rotational change of basis. Other invariants are discussed in the paper by Betten (1987).

1.6.1. Traceless Tensors

A traceless part of the symmetric second-order tensor $\mathbf{A}$ has the rectangular components

$$A'_{ij} = A_{ij} - \frac{1}{3} A_{kk} \delta_{ij}, \tag{1.6.24}$$

such that $A'_{ii} = 0$. For a symmetric third-order tensor $\mathbf{Z}$ ($Z_{ijk} = Z_{jik} = Z_{jki}$), the traceless part is

$$Z'_{ijk} = Z_{ijk} - \frac{1}{5} \left(Z_{mmi} \delta_{jk} + Z_{mmj} \delta_{ki} + Z_{mmk} \delta_{ij} \right), \tag{1.6.25}$$

which is defined so that the contraction of any two of its indices gives a zero vector, e.g.,

$$Z'_{iij} = Z'_{jii} = Z'_{iji} = 0. \tag{1.6.26}$$

A traceless part of the symmetric fourth-order tensor ($\mathcal{L}_{ijkl} = \mathcal{L}_{jikl} = \mathcal{L}_{ijlk} = \mathcal{L}_{klij}$) is defined by

$$\begin{aligned} \mathcal{L}'_{ijkl} = \mathcal{L}_{ijkl} &- \frac{1}{7} \left(\mathcal{L}_{mmij} \delta_{kl} + \mathcal{L}_{mmkl} \delta_{ij} + \mathcal{L}_{mmjk} \delta_{il} + \mathcal{L}_{mmil} \delta_{jk} \right. \\ &\left. + \mathcal{L}_{mmik} \delta_{jl} + \mathcal{L}_{mmjl} \delta_{ik} \right) + \frac{1}{35} \mathcal{L}_{mmnn} \left(\delta_{ij} \delta_{kl} + \delta_{ik} \delta_{jl} + \delta_{il} \delta_{jk} \right). \end{aligned} \tag{1.6.27}$$

A contraction of any two of its indices also yields a zero tensor, e.g.,

$$\mathcal{L}'_{iikl} = \mathcal{L}'_{kiil} = \mathcal{L}'_{ikli} = 0. \tag{1.6.28}$$

For further details see the papers by Spencer (1970), Kanatani (1984), and Lubarda and Krajcinovic (1993).

1.7. Covariant and Contravariant Components

1.7.1. Vectors

A pair of vector bases, $\mathbf{e}_1$, $\mathbf{e}_2$, $\mathbf{e}_3$ and $\mathbf{e}^1$, $\mathbf{e}^2$, $\mathbf{e}^3$, are said to be reciprocal if

$$\mathbf{e}_i \cdot \mathbf{e}^j = \delta_i^{\ j}, \tag{1.7.1}$$

where $\delta_i^{\ j}$ is the Kronecker delta symbol (Fig. 1.1). The base vectors of each basis are neither unit nor mutually orthogonal vectors, so that

$$2D\,\mathbf{e}^i = \epsilon_{ijk} (\mathbf{e}_j \times \mathbf{e}_k), \quad D = \mathbf{e}_1 \cdot (\mathbf{e}_2 \times \mathbf{e}_3). \tag{1.7.2}$$

Any vector $\mathbf{a}$ can be decomposed in the primary basis as

$$\mathbf{a} = a^i \mathbf{e}_i, \quad a^i = \mathbf{a} \cdot \mathbf{e}^i, \tag{1.7.3}$$

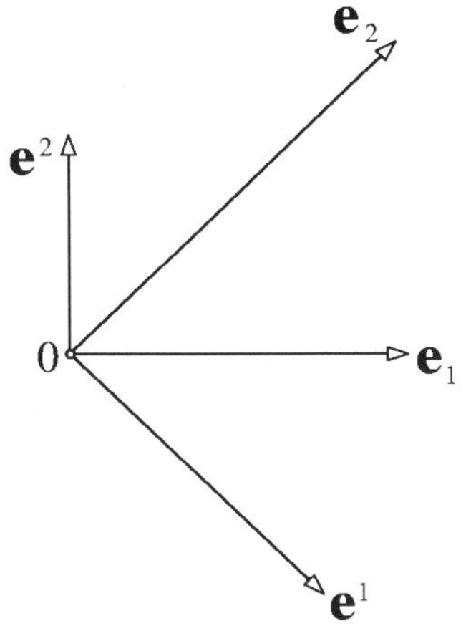

FIGURE 1.1. Primary and reciprocal bases in two dimensions ($\mathbf{e}_1 \cdot \mathbf{e}^2 = \mathbf{e}_2 \cdot \mathbf{e}^1 = 0$).

and in the reciprocal basis as

$$\mathbf{a} = a_i \mathbf{e}^i, \quad a_i = \mathbf{a} \cdot \mathbf{e}_i. \tag{1.7.4}$$

The components a^i are called contravariant, and a_i covariant components of the vector $\mathbf{a}$.

1.7.2. Second-Order Tensors

Denoting the scalar products of the base vectors by

$$g^{ij} = \mathbf{e}^i \cdot \mathbf{e}^j = g^{ji}, \quad g_{ij} = \mathbf{e}_i \cdot \mathbf{e}_j = g_{ji}, \tag{1.7.5}$$

there follows

$$a^i = g^{ij} a_j, \quad a_i = g_{ij} a^j, \tag{1.7.6}$$

$$\mathbf{e}^i = g^{ik} \mathbf{e}_k, \quad \mathbf{e}_i = g_{ik} \mathbf{e}^k. \tag{1.7.7}$$

This shows that the matrices of g^{ij} and g_{ij} are mutual inverses. The components g^{ij} and g_{ij} are contravariant and covariant components of the second-order unit (metric) tensor

$$\mathbf{I} = g^{ij} \mathbf{e}_i \otimes \mathbf{e}_j = g_{ij} \mathbf{e}^i \otimes \mathbf{e}^j = \mathbf{e}^j \otimes \mathbf{e}_j = \mathbf{e}_j \otimes \mathbf{e}^j. \tag{1.7.8}$$

Note that $g^i{}_j = \delta^i{}_j$ and $g_i{}^j = \delta_i{}^j$, both being the Kronecker delta. The scalar product of two vectors $\mathbf{a}$ and $\mathbf{b}$ can be calculated from

$$\mathbf{a} \cdot \mathbf{b} = g^{ij} a_i b_j = g_{ij} a^i b^j = a^i b_i = a_i b^i. \tag{1.7.9}$$

The second-order tensor has four types of decompositions

$$\mathbf{A} = A^{ij} \mathbf{e}_i \otimes \mathbf{e}_j = A_{ij} \mathbf{e}^i \otimes \mathbf{e}^j = A^i{}_j \mathbf{e}_i \otimes \mathbf{e}^j = A_i{}^j \mathbf{e}^i \otimes \mathbf{e}_j. \tag{1.7.10}$$

These are, respectively, contravariant, covariant, and two kinds of mixed components of **A**, such that

$$A^{ij} = \mathbf{e}^i \cdot \mathbf{A} \cdot \mathbf{e}^j, \quad A_{ij} = \mathbf{e}_i \cdot \mathbf{A} \cdot \mathbf{e}_j, \quad A^i{}_j = \mathbf{e}^i \cdot \mathbf{A} \cdot \mathbf{e}_j, \quad A_i{}^j = \mathbf{e}_i \cdot \mathbf{A} \cdot \mathbf{e}^j. \tag{1.7.11}$$

The relationships between different components are easily established by using Eq. (1.7.7). For example,

$$A_{ij} = g_{ik} A^k{}_j = A_i{}^k g_{kj} = g_{ik} A^{kl} g_{lj}. \tag{1.7.12}$$

The transpose of **A** can be decomposed as

$$\mathbf{A}^T = A^{ji} \mathbf{e}_i \otimes \mathbf{e}_j = A_{ji} \mathbf{e}^i \otimes \mathbf{e}^j = A_j{}^i \mathbf{e}_i \otimes \mathbf{e}^j = A^j{}_i \mathbf{e}^i \otimes \mathbf{e}_j. \tag{1.7.13}$$

If **A** is symmetric ($\mathbf{A} \cdot \mathbf{a} = \mathbf{a} \cdot \mathbf{A}$), one has

$$A^{ij} = A^{ji}, \quad A_{ij} = A_{ji}, \quad A^i{}_j = A_j{}^i, \tag{1.7.14}$$

although $A^i{}_j \neq A_i{}^j$.

A dot product of a second-order tensor **A** and a vector **a** is the vector

$$\mathbf{b} = \mathbf{A} \cdot \mathbf{a} = b^i \mathbf{e}_i = b_i \mathbf{e}^i. \tag{1.7.15}$$

The contravariant and covariant components of **b** are

$$b^i = A^{ij} a_j = A^i{}_j a^j, \quad b_i = A_{ij} a^j = A_i{}^j a_j. \tag{1.7.16}$$

A dot product of two second-order tensors **A** and **B** is the second-order tensor **C**, such that

$$\mathbf{C} \cdot \mathbf{a} = \mathbf{A} \cdot (\mathbf{B} \cdot \mathbf{a}), \tag{1.7.17}$$

for any vector **a**. Each type of components of **C** has two possible representations. For example,

$$C^{ij} = A^{ik} B_k{}^j = A^i{}_k B^{kj}, \quad C^i{}_j = A^{ik} B_{kj} = A^i{}_k B^k{}_j. \tag{1.7.18}$$

The trace of a tensor **A** is the scalar obtained by contraction of the subscript and superscript in the mixed component tensor representation. Thus,

$$\operatorname{tr} \mathbf{A} = A^i{}_i = A_i{}^i = g_{ij} A^{ij} = g^{ij} A_{ij}. \tag{1.7.19}$$

Two kinds of inner products are defined by

$$\mathbf{A} \cdot\cdot \mathbf{B} = \operatorname{tr}(\mathbf{A} \cdot \mathbf{B}) = A^{ij} B_{ji} = A_{ij} B^{ji} = A^i{}_j B^j{}_i = A_i{}^j B_j{}^i, \tag{1.7.20}$$

$$\mathbf{A} : \mathbf{B} = \operatorname{tr}\left(\mathbf{A} \cdot \mathbf{B}^T\right) = A^{ij} B_{ij} = A_{ij} B^{ij} = A^i{}_j B_i{}^j = A_i{}^j B^i{}_j. \tag{1.7.21}$$

If either **A** or **B** is symmetric, $\mathbf{A} \cdot\cdot \mathbf{B} = \mathbf{A} : \mathbf{B}$. The trace of **A** in Eq. (1.7.19) can be written as $\operatorname{tr} \mathbf{A} = \mathbf{A} : \mathbf{I}$, where **I** is defined by (1.7.8).

1.7.3. Higher-Order Tensors

An n-th order tensor has one completely contravariant, one completely covariant, and $(2^n - 2)$ kinds of mixed component representations. For a third-order tensor $\mathbf{\Gamma}$, for example, these are respectively Γ^{ijk}, Γ_{ijk}, and

$$\Gamma^{ij}{}_k, \quad \Gamma^{i}{}_j{}^k, \quad \Gamma_i{}^{jk}, \quad \Gamma^i{}_{jk}, \quad \Gamma_i{}^j{}_k, \quad \Gamma_{ij}{}^k. \tag{1.7.22}$$

As an illustration,

$$\mathbf{\Gamma} = \Gamma^{ijk}\mathbf{e}_i \otimes \mathbf{e}_j \otimes \mathbf{e}_k = \Gamma^{ij}{}_k \mathbf{e}_i \otimes \mathbf{e}_j \otimes \mathbf{e}^k. \tag{1.7.23}$$

The relationships between various components are analogous to those in Eq. (1.7.12), e.g.,

$$\Gamma^i{}_{jk} = \Gamma^i{}_j{}^m g_{mk} = \Gamma_m{}^n{}_k g^{mi} g_{nj} = \Gamma_m{}^{np} g^{mi} g_{nj} g_{pk}. \tag{1.7.24}$$

Four types of components of the inner product of the fourth- and second-order tensors, $\mathbf{C} = \boldsymbol{\mathcal{L}} : \mathbf{A}$, can all be expressed in terms of the components of $\boldsymbol{\mathcal{L}}$ and $\mathbf{A}$. For example, contravariant and mixed (right-covariant) components are

$$C^{ij} = \mathcal{L}^{ijkl} A_{kl} = \mathcal{L}^{ij}{}_{kl} A^{kl} = \mathcal{L}^{ijk}{}_l A_k{}^l = \mathcal{L}^{ij}{}_k{}^l A^k{}_l, \tag{1.7.25}$$

$$C^i{}_j = \mathcal{L}^i{}_j{}^{kl} A_{kl} = \mathcal{L}^i{}_{jkl} A^{kl} = \mathcal{L}^i{}_j{}^k{}_l A_k{}^l = \mathcal{L}^i{}_{jk}{}^l A^k{}_l. \tag{1.7.26}$$

1.8. Induced Tensors

Let $\{\mathbf{e}_i\}$ and $\{\mathbf{e}^i\}$ be a pair of reciprocal bases, and let $\mathbf{F}$ be a nonsingular mapping that transforms the base vectors $\mathbf{e}_i$ into

$$\hat{\mathbf{e}}_i = \mathbf{F} \cdot \mathbf{e}_i = F^j{}_i \mathbf{e}_j, \tag{1.8.1}$$

and the vectors $\mathbf{e}^i$ into

$$\hat{\mathbf{e}}^i = \mathbf{e}^i \cdot \mathbf{F}^{-1} = (F^{-1})^i{}_j \mathbf{e}^j, \tag{1.8.2}$$

such that $\hat{\mathbf{e}}_i \cdot \hat{\mathbf{e}}^j = \delta_i{}^j$ (Fig. 1.2). Then, in view of Eqs. (1.7.10) and (1.7.13) applied to $\mathbf{F}$ and $\mathbf{F}^T$, we have

$$\mathbf{F}^T \cdot \mathbf{F} = \hat{g}_{ij} \mathbf{e}^i \otimes \mathbf{e}^j, \quad \mathbf{F}^{-1} \cdot \mathbf{F}^{-T} = \hat{g}^{ij} \mathbf{e}_i \otimes \mathbf{e}_j, \tag{1.8.3}$$

where $\hat{g}_{ij} = \hat{\mathbf{e}}_i \cdot \hat{\mathbf{e}}_j$ and $\hat{g}^{ij} = \hat{\mathbf{e}}^i \cdot \hat{\mathbf{e}}^j$. Thus, covariant components of $\mathbf{F}^T \cdot \mathbf{F}$ and contravariant components of $\mathbf{F}^{-1} \cdot \mathbf{F}^{-T}$ in the original bases are equal to covariant and contravariant components of the metric tensor in the transformed bases ($\mathbf{I} = \hat{g}_{ij}\hat{\mathbf{e}}^i \otimes \hat{\mathbf{e}}^j = \hat{g}^{ij}\hat{\mathbf{e}}_i \otimes \hat{\mathbf{e}}_j$).

An arbitrary vector $\mathbf{a}$ can be decomposed in the original and transformed bases as

$$\mathbf{a} = a^i \mathbf{e}_i = a_i \mathbf{e}^i = \hat{a}^i \hat{\mathbf{e}}_i = \hat{a}_i \hat{\mathbf{e}}^i. \tag{1.8.4}$$

Evidently,

$$\hat{a}^i = (F^{-1})^i{}_j a^j, \quad \hat{a}_i = F^j{}_i a_j. \tag{1.8.5}$$

Introducing the vectors

$$\mathbf{a}^* = \hat{a}^i \mathbf{e}_i, \quad \mathbf{a}_* = \hat{a}_i \mathbf{e}^i, \tag{1.8.6}$$

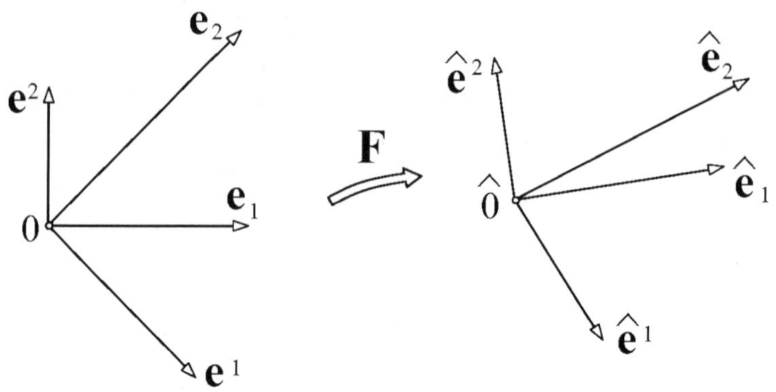

FIGURE 1.2. Upon mapping $\mathbf{F}$ the pair of reciprocal bases $\mathbf{e}_i$ and $\mathbf{e}^j$ transform into reciprocal bases $\hat{\mathbf{e}}_i$ and $\hat{\mathbf{e}}^j$.

it follows that

$$\hat{\mathbf{a}}^* = \mathbf{F}^{-1} \cdot \mathbf{a}, \quad \hat{\mathbf{a}}_* = \mathbf{F}^T \cdot \mathbf{a}. \tag{1.8.7}$$

The vectors $\mathbf{a}^*$ and $\mathbf{a}_*$ are induced from $\mathbf{a}$ by the transformation of bases. The contravariant components of $\mathbf{F}^{-1} \cdot \mathbf{a}$ in the original basis are numerically equal to contravariant components of $\mathbf{a}$ in the transformed basis. Analogous statement applies to covariant components.

Let $\mathbf{A}$ be a second-order tensor with components in the original basis given by Eq. (1.7.10), and in the transformed basis by

$$\mathbf{A} = \hat{A}^{ij} \hat{\mathbf{e}}_i \otimes \hat{\mathbf{e}}_j = \hat{A}_{ij} \hat{\mathbf{e}}^i \otimes \hat{\mathbf{e}}^j = \hat{A}^i{}_j \hat{\mathbf{e}}_i \otimes \hat{\mathbf{e}}^j = \hat{A}_i{}^j \hat{\mathbf{e}}^i \otimes \hat{\mathbf{e}}_j. \tag{1.8.8}$$

The components are related through

$$\hat{A}^{ij} = (F^{-1})^i{}_k A^{kl} (F^{-1})^j{}_l, \quad \hat{A}_{ij} = F^k_{\ i} A_{kl} F^l{}_j, \tag{1.8.9}$$

$$\hat{A}^i{}_j = (F^{-1})^i{}_k A^k{}_l F^l{}_j, \quad \hat{A}_i{}^j = F^k_{\ i} A_k{}^l (F^{-1})^j{}_l. \tag{1.8.10}$$

Introducing the tensors

$$\mathbf{A}^* = \hat{A}^{ij} \mathbf{e}_i \otimes \mathbf{e}_j, \quad \mathbf{A}_* = \hat{A}_{ij} \mathbf{e}^i \otimes \mathbf{e}^j, \tag{1.8.11}$$

$$\mathbf{A}^\star = \hat{A}^i{}_j \mathbf{e}_i \otimes \mathbf{e}^j, \quad \mathbf{A}_\star = \hat{A}_i{}^j \mathbf{e}^i \otimes \mathbf{e}_j, \tag{1.8.12}$$

we recognize from Eqs. (1.8.9) and (1.8.10) that

$$\mathbf{A}^* = \mathbf{F}^{-1} \cdot \mathbf{A} \cdot \mathbf{F}^{-T}, \quad \mathbf{A}_* = \mathbf{F}^T \cdot \mathbf{A} \cdot \mathbf{F}, \tag{1.8.13}$$

$$\mathbf{A}^\star = \mathbf{F}^{-1} \cdot \mathbf{A} \cdot \mathbf{F}, \quad \mathbf{A}_\star = \mathbf{F}^T \cdot \mathbf{A} \cdot \mathbf{F}^{-T}. \tag{1.8.14}$$

These four tensors are said to be induced from $\mathbf{A}$ by transformation of the bases (Hill, 1978). The contravariant components of the tensor $\mathbf{F}^{-1} \cdot \mathbf{A} \cdot \mathbf{F}^{-T}$ in the original basis are numerically equal to the contravariant components of the tensor $\mathbf{A}$ in the transformed basis. Analogous statements apply to covariant and mixed components.

1.9. Gradient of Tensor Functions

Let $f = f(\mathbf{A})$ be a scalar function of the second-order tensor argument $\mathbf{A}$. The change of f associated with an infinitesimal change of $\mathbf{A}$ can be determined from

$$\mathrm{d}f = \mathrm{tr}\left(\frac{\partial f}{\partial \mathbf{A}} \cdot \mathrm{d}\mathbf{A}\right). \tag{1.9.1}$$

If $\mathrm{d}\mathbf{A}$ is decomposed on the fixed primary and reciprocal bases as

$$\mathrm{d}\mathbf{A} = \mathrm{d}A^{ij}\mathbf{e}_i \otimes \mathbf{e}_j = \mathrm{d}A_{ij}\mathbf{e}^i \otimes \mathbf{e}^j = \mathrm{d}A^i{}_j\mathbf{e}_i \otimes \mathbf{e}^j = \mathrm{d}A_i{}^j\mathbf{e}^i \otimes \mathbf{e}_j, \tag{1.9.2}$$

the gradient of f with respect to $\mathbf{A}$ is the second-order tensor with decompositions

$$\frac{\partial f}{\partial \mathbf{A}} = \frac{\partial f}{\partial A^{ji}}\,\mathbf{e}^i \otimes \mathbf{e}^j = \frac{\partial f}{\partial A_{ji}}\,\mathbf{e}_i \otimes \mathbf{e}_j = \frac{\partial f}{\partial A^j{}_i}\,\mathbf{e}_i \otimes \mathbf{e}^j = \frac{\partial f}{\partial A_j{}^i}\,\mathbf{e}^i \otimes \mathbf{e}_j, \tag{1.9.3}$$

since then (Ogden, 1984)

$$\mathrm{d}f = \frac{\partial f}{\partial A^{ij}}\,\mathrm{d}A^{ij} = \frac{\partial f}{\partial A_{ij}}\,\mathrm{d}A_{ij} = \frac{\partial f}{\partial A^i{}_j}\,\mathrm{d}A^i{}_j = \frac{\partial f}{\partial A_i{}^j}\,\mathrm{d}A_i{}^j. \tag{1.9.4}$$

Let $\mathbf{F} = \mathbf{F}(\mathbf{A})$ be a second-order tensor function of the second-order tensor argument $\mathbf{A}$. The change of $\mathbf{F}$ associated with an infinitesimal change of $\mathbf{A}$ can be determined from

$$\mathrm{d}\mathbf{F} = \frac{\partial \mathbf{F}}{\partial \mathbf{A}} \cdot\cdot\, \mathrm{d}\mathbf{A}. \tag{1.9.5}$$

If $\mathrm{d}\mathbf{A}$ is decomposed on the fixed primary and reciprocal bases as in Eq. (1.9.2), the gradient of $\mathbf{F}$ with respect to $\mathbf{A}$ is the fourth-order tensor, such that

$$\frac{\partial \mathbf{F}}{\partial \mathbf{A}} = \frac{\partial \mathbf{F}}{\partial A^{ji}}\,\mathbf{e}^i \otimes \mathbf{e}^j = \frac{\partial \mathbf{F}}{\partial A_{ji}}\,\mathbf{e}_i \otimes \mathbf{e}_j = \frac{\partial \mathbf{F}}{\partial A^j{}_i}\,\mathbf{e}_i \otimes \mathbf{e}^j = \frac{\partial \mathbf{F}}{\partial A_j{}^i}\,\mathbf{e}^i \otimes \mathbf{e}_j, \tag{1.9.6}$$

for then

$$\mathrm{d}\mathbf{F} = \frac{\partial \mathbf{F}}{\partial A^{ij}}\,\mathrm{d}A^{ij} = \frac{\partial \mathbf{F}}{\partial A_{ij}}\,\mathrm{d}A_{ij} = \frac{\partial \mathbf{F}}{\partial A^i{}_j}\,\mathrm{d}A^i{}_j = \frac{\partial \mathbf{F}}{\partial A_i{}^j}\,\mathrm{d}A_i{}^j. \tag{1.9.7}$$

For example,

$$\frac{\partial \mathbf{F}}{\partial \mathbf{A}} = \frac{\partial F_{ij}}{\partial A^{lk}}\,\mathbf{e}^i \otimes \mathbf{e}^j \otimes \mathbf{e}^k \otimes \mathbf{e}^l. \tag{1.9.8}$$

As an illustration, if $\mathbf{A}$ is symmetric and invertible second-order tensor, by taking a gradient of $\mathbf{A} \cdot \mathbf{A}^{-1} = \mathbf{I}$ with respect to $\mathbf{A}$, it readily follows that

$$\frac{\partial A_{ij}^{-1}}{\partial A_{kl}} = -\frac{1}{2}\left(A_{ik}^{-1}A_{jl}^{-1} + A_{il}^{-1}A_{jk}^{-1}\right). \tag{1.9.9}$$

The gradients of the three invariants of $\mathbf{A}$ in Eqs. (1.3.3)–(1.3.5) are

$$\frac{\partial J_1}{\partial \mathbf{A}} = \mathbf{I}, \quad \frac{\partial J_2}{\partial \mathbf{A}} = \mathbf{A} - J_1\mathbf{I}, \quad \frac{\partial J_3}{\partial \mathbf{A}} = \mathbf{A}^2 - J_1\mathbf{A} - J_2\mathbf{I}. \tag{1.9.10}$$

Since $\mathbf{A}^2$ has the same principal directions as $\mathbf{A}$, the gradients in Eq. (1.9.10) also have the same principal directions as $\mathbf{A}$. It is also noted that by the Cayley–Hamilton theorem (1.4.1), the last of Eq. (1.9.10) can be rewritten as

$$\frac{\partial J_3}{\partial \mathbf{A}} = J_3 \mathbf{A}^{-1}, \quad \text{i.e.,} \quad \frac{\partial (\det \mathbf{A})}{\partial \mathbf{A}} = (\det \mathbf{A})\, \mathbf{A}^{-1}. \tag{1.9.11}$$

Furthermore, if $\mathbf{F} = \mathbf{A} \cdot \mathbf{A}^T$, then with respect to an orthonormal basis

$$\frac{\partial A_{ij}}{\partial A_{kl}} = \delta_{ik}\delta_{jl}, \quad \frac{\partial F_{ij}}{\partial A_{kl}} = \delta_{ik}A_{jl} + \delta_{jk}A_{il}. \tag{1.9.12}$$

The gradients of the principal invariants $\bar{J}_i$ of $\mathbf{A} \cdot \mathbf{A}^T$ with respect to $\mathbf{A}$ are consequently

$$\frac{\partial \bar{J}_1}{\partial \mathbf{A}} = 2\mathbf{A}^T, \quad \frac{\partial \bar{J}_2}{\partial \mathbf{A}} = 2\left(\mathbf{A}^T \cdot \mathbf{A} \cdot \mathbf{A}^T - \bar{J}_1 \mathbf{A}^T\right), \quad \frac{\partial \bar{J}_3}{\partial \mathbf{A}} = 2\bar{J}_3 \mathbf{A}^{-1}. \tag{1.9.13}$$

1.10. Isotropic Tensors

An isotropic tensor is one whose components in an orthonormal basis remain unchanged by any proper orthogonal transformation (rotation) of the basis. All scalars are isotropic zero-order tensors. There are no isotropic first-order tensors (vectors), except the zero-vector. The only isotropic second-order tensors are scalar multiples of the second-order unit tensor δ_{ij}. The scalar multiples of the permutation tensor ϵ_{ijk} are the only isotropic third-order tensors. The most general isotropic fourth-order tensor has the components

$$\mathcal{L}_{ijkl} = a\,\delta_{ij}\delta_{kl} + b\,\delta_{ik}\delta_{jl} + c\,\delta_{il}\delta_{jk}, \tag{1.10.1}$$

where a, b, c are scalars. If $\boldsymbol{\mathcal{L}}$ is symmetric, $b = c$ and

$$\mathcal{L}_{ijkl} = a\,\delta_{ij}\delta_{kl} + 2b\,I_{ijkl}. \tag{1.10.2}$$

Isotropic tensors of even order can be expressed as a linear combination of outer products of the Kronecker deltas only; those of odd order can be expressed as a linear combination of outer products of the Kronecker deltas and permutation tensors. Since the outer product of two permutation tensors,

$$\epsilon_{ijk}\epsilon_{\alpha\beta\gamma} = \begin{vmatrix} \delta_{i\alpha} & \delta_{i\beta} & \delta_{i\gamma} \\ \delta_{j\alpha} & \delta_{j\beta} & \delta_{j\gamma} \\ \delta_{k\alpha} & \delta_{k\beta} & \delta_{k\gamma} \end{vmatrix}, \tag{1.10.3}$$

is expressed solely in terms of the Kronecker deltas, each term of an isotropic tensor of odd order contains at most one permutation tensor. Such tensors change sign under improper orthogonal transformation. Isotropic tensors of even order are unchanged under both proper and improper orthogonal transformations. For example, the components of an isotropic symmetric sixth-order tensor are

$$S_{ijklmn} = a\,\delta_{ij}\delta_{kl}\delta_{mn} + b\,\delta_{(ij}I_{klmn)} + c\,\delta_{(ik}\delta_{lm}\delta_{nj)}, \tag{1.10.4}$$

where the notation such as $\delta_{(ij}I_{klmn)}$ designates the symmetrization with respect to i and j, k and l, m and n, and ij, kl and mn (Eringen, 1971). Specifically,

$$\begin{aligned}\delta_{(ij}I_{klmn)} &= \frac{1}{3}\left(\delta_{ij}I_{klmn} + \delta_{kl}I_{mnij} + \delta_{mn}I_{ijkl}\right),\\ \delta_{(ik}\delta_{lm}\delta_{nj)} &= \frac{1}{4}\left(\delta_{ik}I_{jlmn} + \delta_{il}I_{jkmn} + \delta_{im}I_{klnj} + \delta_{in}I_{klmj}\right).\end{aligned} \tag{1.10.5}$$

In some applications it may be convenient to introduce the fourth-order base tensors (Hill, 1965; Walpole, 1981)

$$\boldsymbol{K} = \frac{1}{3}\mathbf{I}\otimes\mathbf{I}, \quad \boldsymbol{J} = \boldsymbol{I} - \boldsymbol{K}. \tag{1.10.6}$$

These tensors are such that $\operatorname{tr}\boldsymbol{K} = K_{ijij} = 1$, $\operatorname{tr}\boldsymbol{J} = J_{ijij} = 5$, and

$$\boldsymbol{J}:\boldsymbol{J} = \boldsymbol{J}, \quad \boldsymbol{K}:\boldsymbol{K} = \boldsymbol{K}, \quad \boldsymbol{J}:\boldsymbol{K} = \boldsymbol{K}:\boldsymbol{J} = \mathbf{0}. \tag{1.10.7}$$

Consequently,

$$(a_1\,\boldsymbol{J} + b_1\,\boldsymbol{K}) : (a_2\,\boldsymbol{J} + b_2\,\boldsymbol{K}) = a_1a_2\,\boldsymbol{J} + b_1b_2\,\boldsymbol{K}, \tag{1.10.8}$$

$$(a_1\,\boldsymbol{J} + b_1\,\boldsymbol{K})^{-1} = a_1^{-1}\,\boldsymbol{J} + b_1^{-1}\,\boldsymbol{K}. \tag{1.10.9}$$

An isotropic fourth-order tensor $\boldsymbol{\mathcal{L}}$ can be decomposed in this basis as

$$\boldsymbol{\mathcal{L}} = \mathcal{L}_J\,\boldsymbol{J} + \mathcal{L}_K\,\boldsymbol{K}, \tag{1.10.10}$$

where

$$\mathcal{L}_K = \operatorname{tr}(\boldsymbol{\mathcal{L}}:\boldsymbol{K}), \quad \mathcal{L}_K + 5\,\mathcal{L}_J = \operatorname{tr}\boldsymbol{\mathcal{L}}. \tag{1.10.11}$$

Product of any pair of isotropic fourth-order tensors is isotropic and commutative. The base tensors $\boldsymbol{K}$ and $\boldsymbol{J}$ partition the second-order tensor $\mathbf{A}$ into its spherical and deviatoric parts, such that

$$\mathbf{A}_{\text{sph}} = \boldsymbol{K}:\mathbf{A} = \frac{1}{3}(\operatorname{tr}\mathbf{A})\,\mathbf{I}, \quad \mathbf{A}_{\text{dev}} = \boldsymbol{J}:\mathbf{A} = \mathbf{A} - \mathbf{A}_{\text{sph}}. \tag{1.10.12}$$

1.11. Isotropic Functions

1.11.1. Isotropic Scalar Functions

A scalar function of the second-order symmetric tensor argument is said to be an isotropic function if

$$f\left(\mathbf{Q}\cdot\mathbf{A}\cdot\mathbf{Q}^T\right) = f(\mathbf{A}), \tag{1.11.1}$$

where $\mathbf{Q}$ is an arbitrary proper orthogonal (rotation) tensor. Such a function depends on $\mathbf{A}$ only through its three invariants, $f = f(J_1, J_2, J_3)$. For isotropic $f(\mathbf{A})$, the principal directions of the gradient $\partial f/\partial\mathbf{A}$ are parallel to those of $\mathbf{A}$. This follows because the gradients $\partial J_i/\partial\mathbf{A}$ are all parallel to $\mathbf{A}$, by Eq. (1.9.10).

A scalar function of two symmetric second-order tensors $\mathbf{A}$ and $\mathbf{B}$ is said to be an isotropic function of both $\mathbf{A}$ and $\mathbf{B}$, if

$$f\left(\mathbf{Q}\cdot\mathbf{A}\cdot\mathbf{Q}^T, \mathbf{Q}\cdot\mathbf{B}\cdot\mathbf{Q}^T\right) = f(\mathbf{A}, \mathbf{B}). \tag{1.11.2}$$

Such a function can be represented as a polynomial of its irreducible integrity basis consisting of the individual and joint invariants of **A** and **B**. The independent joint invariants are the traces of the following products

$$(\mathbf{A}\cdot\mathbf{B}), \quad (\mathbf{A}\cdot\mathbf{B}^2)^*, \quad (\mathbf{A}^2\cdot\mathbf{B}^2)\,. \tag{1.11.3}$$

The joint invariants of three symmetric second-order tensors are the traces of

$$(\mathbf{A}\cdot\mathbf{B}\cdot\mathbf{C}), \quad (\mathbf{A}^2\cdot\mathbf{B}\cdot\mathbf{C})^*, \quad (\mathbf{A}^2\cdot\mathbf{B}^2\cdot\mathbf{C})^*\,. \tag{1.11.4}$$

A superposed asterisk (*) indicates that the integrity basis also includes invariants formed by cyclic permutation of symmetric tensors involved. The integrity basis can be written for any finite set of second-order tensors. Spencer (1971) provides a list of invariants and integrity bases for a polynomial scalar function dependent on one up to six second-order symmetric tensors. An integrity basis for an arbitrary number of tensors is obtained by taking the bases for the tensors six at a time, in all possible combinations. For invariants of second-order tensors alone, it is not necessary to distinguish between the full and the proper orthogonal groups.

The trace of an antisymmetric tensor, or any power of it, is equal to zero, so that the integrity basis for the antisymmetric tensor **X** is $\mathrm{tr}\,(\mathbf{X}^2)$. A joint invariant of two antisymmetric tensors **X** and **Y** is $\mathrm{tr}\,(\mathbf{X}\cdot\mathbf{Y})$. The independent joint invariants of a symmetric tensor **A** and an antisymmetric tensor **X** are the traces of the products

$$(\mathbf{X}^2\cdot\mathbf{A}), \quad (\mathbf{X}^2\cdot\mathbf{A}^2), \quad (\mathbf{X}^2\cdot\mathbf{A}^2\cdot\mathbf{X}\cdot\mathbf{A}^2)\,. \tag{1.11.5}$$

In the case of two symmetric and one antisymmetric tensor, the joint invariants include the traces of

$$\begin{array}{lll} (\mathbf{X}\cdot\mathbf{A}\cdot\mathbf{B}), & (\mathbf{X}\cdot\mathbf{A}^2\cdot\mathbf{B})^*, & (\mathbf{X}\cdot\mathbf{A}^2\cdot\mathbf{B}^2), \\ (\mathbf{X}\cdot\mathbf{A}^2\cdot\mathbf{B}\cdot\mathbf{A})^*, & (\mathbf{X}\cdot\mathbf{A}^2\cdot\mathbf{B}^2\cdot\mathbf{A})^*, & (\mathbf{X}^2\cdot\mathbf{A}\cdot\mathbf{B}), \\ (\mathbf{X}^2\cdot\mathbf{A}^2\cdot\mathbf{B})^*, & (\mathbf{X}^2\cdot\mathbf{A}\cdot\mathbf{X}\cdot\mathbf{B}), & (\mathbf{X}^2\cdot\mathbf{A}\cdot\mathbf{X}\cdot\mathbf{B}^2)^*. \end{array} \tag{1.11.6}$$

1.11.2. Isotropic Tensor Functions

A second-order tensor function is said to be an isotropic function of its second-order tensor argument if

$$\mathbf{F}\left(\mathbf{Q}\cdot\mathbf{A}\cdot\mathbf{Q}^T\right) = \mathbf{Q}\cdot\mathbf{F}(\mathbf{A})\cdot\mathbf{Q}^T. \tag{1.11.7}$$

An isotropic symmetric function of a symmetric tensor **A** can be expressed as

$$\mathbf{F}(\mathbf{A}) = a_0\mathbf{I} + a_1\mathbf{A} + a_2\mathbf{A}^2, \tag{1.11.8}$$

where a_i are scalar functions of the principal invariants of **A**.

A second-order tensor function is said to be an isotropic function of its two second-order tensor arguments if

$$\mathbf{F}\left(\mathbf{Q}\cdot\mathbf{A}\cdot\mathbf{Q}^T,\mathbf{Q}\cdot\mathbf{B}\cdot\mathbf{Q}^T\right) = \mathbf{Q}\cdot\mathbf{F}(\mathbf{A},\mathbf{B})\cdot\mathbf{Q}^T. \tag{1.11.9}$$

An isotropic symmetric tensor function which is a polynomial of two symmetric tensors **A** and **B** can be expressed in terms of nine tensors, such that

$$
\begin{aligned}
\mathbf{F}(\mathbf{A},\mathbf{B}) &= a_1\mathbf{I} + a_2\mathbf{A} + a_3\mathbf{A}^2 + a_4\mathbf{B} + a_5\mathbf{B}^2 \\
&+ a_6\left(\mathbf{A}\cdot\mathbf{B} + \mathbf{B}\cdot\mathbf{A}\right) + a_7\left(\mathbf{A}^2\cdot\mathbf{B} + \mathbf{B}\cdot\mathbf{A}^2\right) \\
&+ a_8\left(\mathbf{A}\cdot\mathbf{B}^2 + \mathbf{B}^2\cdot\mathbf{A}\right) + a_9\left(\mathbf{A}^2\cdot\mathbf{B}^2 + \mathbf{B}^2\cdot\mathbf{A}^2\right).
\end{aligned} \tag{1.11.10}
$$

The scalars a_i are scalar functions of ten individual and joint invariants of $\mathbf{A}$ and $\mathbf{B}$. An antisymmetric tensor polynomial function of two symmetric tensors allows a representation

$$
\begin{aligned}
\mathbf{F}(\mathbf{A},\mathbf{B}) &= a_1(\mathbf{A}\cdot\mathbf{B} - \mathbf{B}\cdot\mathbf{A}) + a_2\left(\mathbf{A}^2\cdot\mathbf{B} - \mathbf{B}\cdot\mathbf{A}^2\right) \\
&+ a_3\left(\mathbf{B}^2\cdot\mathbf{A} - \mathbf{A}\cdot\mathbf{B}^2\right) + a_4\left(\mathbf{A}^2\cdot\mathbf{B}^2 - \mathbf{B}^2\cdot\mathbf{A}^2\right) \\
&+ a_5\left(\mathbf{A}^2\cdot\mathbf{B}\cdot\mathbf{A} - \mathbf{A}\cdot\mathbf{B}\cdot\mathbf{A}^2\right) + a_6\left(\mathbf{B}^2\cdot\mathbf{A}\cdot\mathbf{B} - \mathbf{B}\cdot\mathbf{A}\cdot\mathbf{B}^2\right) \\
&+ a_7\left(\mathbf{A}^2\cdot\mathbf{B}^2\cdot\mathbf{A} - \mathbf{A}\cdot\mathbf{B}^2\cdot\mathbf{A}^2\right) + a_8\left(\mathbf{B}^2\cdot\mathbf{A}^2\cdot\mathbf{B} - \mathbf{B}\cdot\mathbf{A}^2\cdot\mathbf{B}^2\right).
\end{aligned} \tag{1.11.11}
$$

A derivation of Eq. (1.11.11) is instructive. The most general scalar invariant of two symmetric and one antisymmetric tensor $\mathbf{X}$, linear in $\mathbf{X}$, can be written from Eq. (1.11.6) as

$$
\begin{aligned}
g(\mathbf{A},\mathbf{B},\mathbf{X}) &= a_1\,\mathrm{tr}\left[(\mathbf{A}\cdot\mathbf{B} - \mathbf{B}\cdot\mathbf{A})\cdot\mathbf{X}\right] + a_2\,\mathrm{tr}\left[\left(\mathbf{A}^2\cdot\mathbf{B} - \mathbf{B}\cdot\mathbf{A}^2\right)\cdot\mathbf{X}\right] \\
&+ a_3\,\mathrm{tr}\left[\left(\mathbf{B}^2\cdot\mathbf{A} - \mathbf{A}\cdot\mathbf{B}^2\right)\cdot\mathbf{X}\right] + a_4\,\mathrm{tr}\left[\left(\mathbf{A}^2\cdot\mathbf{B}^2 - \mathbf{B}^2\cdot\mathbf{A}^2\right)\cdot\mathbf{X}\right] \\
&+ a_5\,\mathrm{tr}\left[\left(\mathbf{A}^2\cdot\mathbf{B}\cdot\mathbf{A} - \mathbf{A}\cdot\mathbf{B}\cdot\mathbf{A}^2\right)\cdot\mathbf{X}\right] + a_6\,\mathrm{tr}\left[\left(\mathbf{B}^2\cdot\mathbf{A}\cdot\mathbf{B}\right.\right. \\
&- \left.\left.\mathbf{B}\cdot\mathbf{A}\cdot\mathbf{B}^2\right)\cdot\mathbf{X}\right] + a_7\,\mathrm{tr}\left[\left(\mathbf{A}^2\cdot\mathbf{B}^2\cdot\mathbf{A} - \mathbf{A}\cdot\mathbf{B}^2\cdot\mathbf{A}^2\right)\cdot\mathbf{X}\right] \\
&+ a_8\,\mathrm{tr}\left[\left(\mathbf{B}^2\cdot\mathbf{A}^2\cdot\mathbf{B} - \mathbf{B}\cdot\mathbf{A}^2\cdot\mathbf{B}^2\right)\cdot\mathbf{X}\right].
\end{aligned} \tag{1.11.12}
$$

The coefficients a_i depend on the invariants of $\mathbf{A}$ and $\mathbf{B}$. Recall that the trace of the product of symmetric and antisymmetric matrix, such as $(\mathbf{A}\cdot\mathbf{B}+\mathbf{B}\cdot\mathbf{A})\cdot\mathbf{X}$, is equal to zero. The antisymmetric function $\mathbf{F}(\mathbf{A},\mathbf{B})$ is obtained from Eq. (1.11.12) as the gradient $\partial g/\partial\mathbf{X}$, which yields Eq. (1.11.11).

1.12. Rivlin's Identities

Applying the Cayley–Hamilton theorem to a second-order tensor $a\mathbf{A} + b\mathbf{B}$, where a and b are arbitrary scalars, and equating to zero the coefficient of a^2b, gives

$$
\begin{aligned}
&\mathbf{A}^2\cdot\mathbf{B} + \mathbf{B}\cdot\mathbf{A}^2 + \mathbf{A}\cdot\mathbf{B}\cdot\mathbf{A} - I_A(\mathbf{A}\cdot\mathbf{B} + \mathbf{B}\cdot\mathbf{A}) - I_B\mathbf{A}^2 - II_A\mathbf{B} \\
&\quad - \left[\mathrm{tr}\,(\mathbf{A}\cdot\mathbf{B}) - I_AI_B\right]\mathbf{A} - \left[III_A\,\mathrm{tr}\left(\mathbf{A}^{-1}\cdot\mathbf{B}\right)\right]\mathbf{I} = \mathbf{0}.
\end{aligned} \tag{1.12.1}
$$

The principal invariants of $\mathbf{A}$ and $\mathbf{B}$ are denoted by I_A, I_B, etc. Identity (1.12.1) is known as the Rivlin's identity (Rivlin, 1955). If $\mathbf{B} = \mathbf{A}$, the original Cayley–Hamilton theorem of Eq. (1.4.1) is recovered. In addition, from the Cayley–Hamilton theorem we have

$$
III_A\,\mathrm{tr}\left(\mathbf{A}^{-1}\cdot\mathbf{B}\right) = \mathrm{tr}\left(\mathbf{A}^2\cdot\mathbf{B}\right) - I_A\,\mathrm{tr}\,(\mathbf{A}\cdot\mathbf{B}) - I_BII_A. \tag{1.12.2}
$$

An identity among three tensors is obtained by applying the Cayley–Hamilton theorem to a second-order tensor $a\mathbf{A} + b\mathbf{B} + c\mathbf{C}$, and by equating to zero the coefficient of abc.

Suppose that $\mathbf{A}$ is symmetric, and $\mathbf{B}$ is antisymmetric. Equations (1.12.1) and (1.12.2) can then be rewritten as

$$\begin{aligned}\mathbf{A}\cdot(\mathbf{A}\cdot\mathbf{B}+\mathbf{B}\cdot\mathbf{A}) + (\mathbf{A}\cdot\mathbf{B}+\mathbf{B}\cdot\mathbf{A})\cdot\mathbf{A} - I_A(\mathbf{A}\cdot\mathbf{B}+\mathbf{B}\cdot\mathbf{A}) \\ - II_A\mathbf{B} - \mathbf{A}\cdot\mathbf{B}\cdot\mathbf{A} = \mathbf{0}.\end{aligned} \tag{1.12.3}$$

Postmultiplying Eq. (1.12.3) with $\mathbf{A}$ and using the Cayley–Hamilton theorem yields another identity

$$\mathbf{A}\cdot(\mathbf{A}\cdot\mathbf{B}+\mathbf{B}\cdot\mathbf{A})\cdot\mathbf{A} + III_A\mathbf{B} - \mathbf{A}\cdot\mathbf{B}\cdot\mathbf{A} = \mathbf{0}. \tag{1.12.4}$$

If A is invertible, Eq. (1.12.4) is equivalent to

$$III_A\mathbf{A}^{-1}\cdot\mathbf{B}\cdot\mathbf{A}^{-1} = I_A\mathbf{B} - (\mathbf{A}\cdot\mathbf{B}+\mathbf{B}\cdot\mathbf{A}). \tag{1.12.5}$$

1.12.1. Matrix Equation $\mathbf{A}\cdot\mathbf{X}+\mathbf{X}\cdot\mathbf{A}=\mathbf{B}$

The matrix equation

$$\mathbf{A}\cdot\mathbf{X}+\mathbf{X}\cdot\mathbf{A}=\mathbf{B} \tag{1.12.6}$$

can be solved by using Rivlin's identities. Suppose $\mathbf{A}$ is symmetric and $\mathbf{B}$ is antisymmetric. The solution $\mathbf{X}$ of Eq. (1.12.6) is then an antisymmetric matrix, and the Rivlin identities (1.12.3) and (1.12.4) become

$$\mathbf{A}\cdot\mathbf{B}+\mathbf{B}\cdot\mathbf{A} - I_A\mathbf{B} - II_A\mathbf{X} - \mathbf{A}\cdot\mathbf{X}\cdot\mathbf{A} = \mathbf{0}, \tag{1.12.7}$$

$$\mathbf{A}\cdot\mathbf{B}\cdot\mathbf{A} + III_A\mathbf{X} - I_A\mathbf{A}\cdot\mathbf{X}\cdot\mathbf{A} = \mathbf{0}. \tag{1.12.8}$$

Upon eliminating $\mathbf{A}\cdot\mathbf{X}\cdot\mathbf{A}$, we obtain the solution for $\mathbf{X}$

$$(I_A II_A + III_A)\mathbf{X} = I_A(\mathbf{A}\cdot\mathbf{B}+\mathbf{B}\cdot\mathbf{A}) - I_A^2\mathbf{B} - \mathbf{A}\cdot\mathbf{B}\cdot\mathbf{A}, \tag{1.12.9}$$

which can be rewritten as

$$(I_A II_A + III_A)\mathbf{X} = -(I_A\mathbf{I}-\mathbf{A})\cdot\mathbf{B}\cdot(I_A\mathbf{I}-\mathbf{A}). \tag{1.12.10}$$

Since

$$I_A II_A + III_A = -\det(I_A\mathbf{I}-\mathbf{A}), \tag{1.12.11}$$

and having in mind Eq. (1.12.5), the solution for $\mathbf{X}$ in Eq. (1.12.10) can be expressed in an alternative form

$$\mathbf{X} = [\operatorname{tr}(I_A\mathbf{I}-\mathbf{A})^{-1}]\,\mathbf{B} - (I_A\mathbf{I}-\mathbf{A})^{-1}\cdot\mathbf{B} - \mathbf{B}\cdot(I_A\mathbf{I}-\mathbf{A})^{-1}, \tag{1.12.12}$$

provided that $I_A\mathbf{I}-\mathbf{A}$ is not a singular matrix.

Consider now the solution of Eq. (1.12.6) when both $\mathbf{A}$ and $\mathbf{B}$ are symmetric, and so is $\mathbf{X}$. If Eq. (1.12.6) is premultiplied by $\mathbf{A}$, it can be recast in the form

$$\mathbf{A}\cdot\left(\mathbf{A}\cdot\mathbf{X}-\frac{1}{2}\mathbf{B}\right) + \left(\mathbf{A}\cdot\mathbf{X}-\frac{1}{2}\mathbf{B}\right)\cdot\mathbf{A} = \frac{1}{2}(\mathbf{A}\cdot\mathbf{B}-\mathbf{B}\cdot\mathbf{A}). \tag{1.12.13}$$

Since the right-hand side of this equation is an antisymmetric matrix, it follows that

$$\frac{\mathbf{Y}}{2} = \mathbf{A} \cdot \mathbf{X} - \frac{1}{2}\mathbf{B} = \frac{1}{2}\mathbf{B} - \mathbf{X} \cdot \mathbf{A} \tag{1.12.14}$$

is also antisymmetric, and Eq. (1.12.13) has the solution for $\mathbf{Y}$ according to Eq. (1.12.10) or (1.12.12), e.g.,

$$(I_A II_A + III_A)\mathbf{Y} = -(I_A\mathbf{I} - \mathbf{A}) \cdot (\mathbf{A} \cdot \mathbf{B} - \mathbf{B} \cdot \mathbf{A}) \cdot (I_A\mathbf{I} - \mathbf{A}). \tag{1.12.15}$$

Thus, from Eq. (1.12.14), the solution for $\mathbf{X}$ is

$$\mathbf{X} = \frac{1}{4}\left[\mathbf{A}^{-1}(\mathbf{B} + \mathbf{Y}) + (\mathbf{B} - \mathbf{Y}) \cdot \mathbf{A}^{-1}\right]. \tag{1.12.16}$$

For further analysis the papers by Sidoroff (1978), Guo (1984), and Scheidler (1994) can be consulted.

1.13. Tensor Fields

Tensors fields are comprised by tensors whose values depend on the position in space. For simplicity, consider the rectangular Cartesian coordinates. The position vector of an arbitrary point of three-dimensional space is $\mathbf{x} = x_i\mathbf{e}_i$, where $\mathbf{e}_i$ are the unit vectors in the coordinate directions. The tensor field is denoted by $\mathbf{T}(\mathbf{x})$. This can represent a scalar field $f(\mathbf{x})$, a vector field $\mathbf{a}(\mathbf{x})$, a second-order tensor field $\mathbf{A}(\mathbf{x})$, or any higher-order tensor field. It is assumed that $\mathbf{T}(\mathbf{x})$ is differentiable at a point $\mathbf{x}$ of the considered domain.

1.13.1. Differential Operators

The gradient of a scalar field $f = f(\mathbf{x})$ is the operator which gives a directional derivative of f, such that

$$\mathrm{d}f = \boldsymbol{\nabla} f \cdot \mathrm{d}\mathbf{x}. \tag{1.13.1}$$

Thus, with respect to rectangular Cartesian coordinates,

$$\boldsymbol{\nabla} f = \frac{\partial f}{\partial x_i}\mathbf{e}_i, \quad \boldsymbol{\nabla} = \frac{\partial}{\partial x_i}\mathbf{e}_i. \tag{1.13.2}$$

In particular, if $\mathrm{d}\mathbf{x}$ is taken to be parallel to the level surface $f(\mathbf{x}) = \text{const.}$, it follows that $\boldsymbol{\nabla} f$ is normal to the level surface at the considered point (Fig. 1.3).

The gradient of a vector field $\mathbf{a} = \mathbf{a}(\mathbf{x})$, and its transpose, are the second-order tensors

$$\boldsymbol{\nabla}\mathbf{a} = \boldsymbol{\nabla} \otimes \mathbf{a} = \frac{\partial a_j}{\partial x_i}\mathbf{e}_i \otimes \mathbf{e}_j, \quad \mathbf{a}\boldsymbol{\nabla} = \mathbf{a} \otimes \boldsymbol{\nabla} = \frac{\partial a_i}{\partial x_j}\mathbf{e}_i \otimes \mathbf{e}_j. \tag{1.13.3}$$

They are introduced such that

$$\mathrm{d}\mathbf{a} = (\mathbf{a}\boldsymbol{\nabla}) \cdot \mathrm{d}\mathbf{x} = \mathrm{d}\mathbf{x} \cdot (\boldsymbol{\nabla}\mathbf{a}). \tag{1.13.4}$$

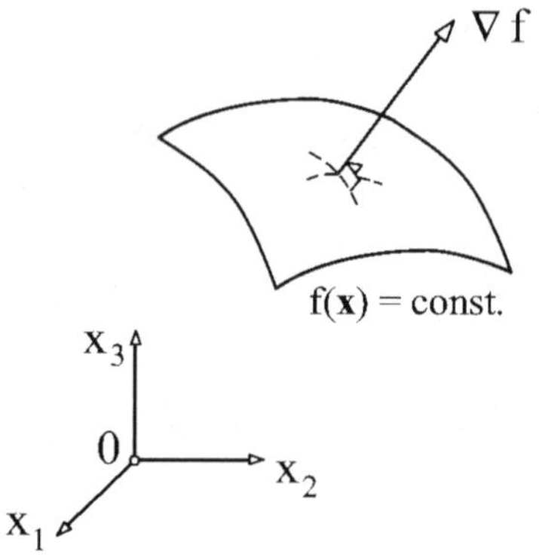

FIGURE 1.3. The gradient $\boldsymbol{\nabla} f$ is perpendicular to the level surface $f(\mathbf{x}) = \text{const}$.

The gradient of a second-order tensor field $\mathbf{A} = \mathbf{A}(\mathbf{x})$ is similarly

$$\boldsymbol{\nabla}\mathbf{A} = \boldsymbol{\nabla} \otimes \mathbf{A} = \frac{\partial A_{ij}}{\partial x_k}\,\mathbf{e}_k \otimes \mathbf{e}_i \otimes \mathbf{e}_j, \quad \mathbf{A}\boldsymbol{\nabla} = \mathbf{A} \otimes \boldsymbol{\nabla} = \frac{\partial A_{ij}}{\partial x_k}\,\mathbf{e}_i \otimes \mathbf{e}_j \otimes \mathbf{e}_k, \tag{1.13.5}$$

so that

$$\mathrm{d}\mathbf{A} = (\mathbf{A}\boldsymbol{\nabla}) \cdot \mathrm{d}\mathbf{x} = \mathrm{d}\mathbf{x} \cdot (\boldsymbol{\nabla}\mathbf{A}). \tag{1.13.6}$$

The divergence of a vector field is the scalar

$$\boldsymbol{\nabla} \cdot \mathbf{a} = \operatorname{tr}(\boldsymbol{\nabla}\mathbf{a}) = \frac{\partial a_i}{\partial x_i}. \tag{1.13.7}$$

The divergence of the gradient of a scalar field is

$$\boldsymbol{\nabla} \cdot (\boldsymbol{\nabla} f) = \nabla^2 f = \frac{\partial^2 f}{\partial x_i \partial x_i}, \quad \nabla^2 = \frac{\partial^2}{\partial x_i \partial x_i}. \tag{1.13.8}$$

The operator ∇^2 is the Laplacian operator. The divergence of the gradient of a vector field can be written as

$$\boldsymbol{\nabla} \cdot (\boldsymbol{\nabla}\mathbf{a}) = \nabla^2 \mathbf{a} = \frac{\partial^2 a_i}{\partial x_j \partial x_j}\,\mathbf{e}_i. \tag{1.13.9}$$

The divergence of a second-order tensor field is defined by

$$\boldsymbol{\nabla} \cdot \mathbf{A} = \frac{\partial A_{ij}}{\partial x_i}\,\mathbf{e}_j, \quad \mathbf{A} \cdot \boldsymbol{\nabla} = \frac{\partial A_{ij}}{\partial x_j}\,\mathbf{e}_i. \tag{1.13.10}$$

The curl of a vector field is the vector

$$\boldsymbol{\nabla} \times \mathbf{a} = \epsilon_{ijk}\,\frac{\partial a_j}{\partial x_i}\,\mathbf{e}_k. \tag{1.13.11}$$

It can be shown that the vector field $\boldsymbol{\nabla} \times \mathbf{a}$ is an axial vector field of the antisymmetric tensor field $(\mathbf{a}\boldsymbol{\nabla} - \boldsymbol{\nabla}\mathbf{a})$. The curl of a second-order tensor

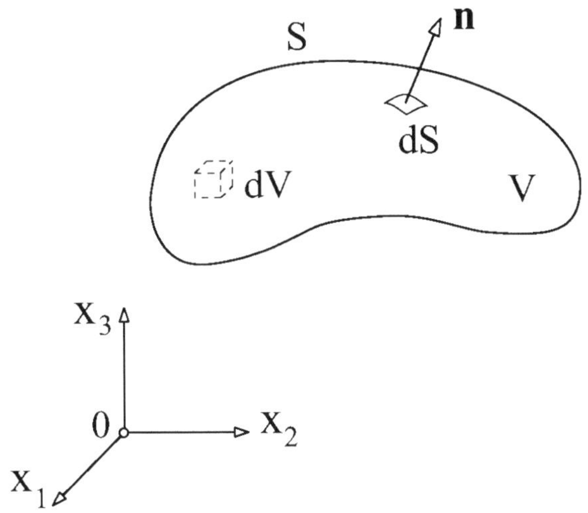

FIGURE 1.4. Three-dimensional domain V bounded by a closed surface S with unit outward normal $\mathbf{n}$.

field is similarly

$$\boldsymbol{\nabla} \times \mathbf{A} = \epsilon_{ijk} \frac{\partial A_{jl}}{\partial x_i} \mathbf{e}_k \otimes \mathbf{e}_l. \tag{1.13.12}$$

It is noted that $\mathbf{A} \times \boldsymbol{\nabla} = -\left(\boldsymbol{\nabla} \times \mathbf{A}^T\right)^T$, while $\mathbf{a} \times \boldsymbol{\nabla} = -\boldsymbol{\nabla} \times \mathbf{a}$.

We list bellow three formulas used later in the book. If $\mathbf{a}$ is an arbitrary vector, $\mathbf{x}$ is a position vector, and if $\mathbf{A}$ and $\mathbf{B}$ are two second-order tensors, then

$$\boldsymbol{\nabla} \cdot (\mathbf{A} \cdot \mathbf{a}) = (\boldsymbol{\nabla} \cdot \mathbf{A}) \cdot \mathbf{a} + \mathbf{A} : (\boldsymbol{\nabla} \otimes \mathbf{a}), \tag{1.13.13}$$

$$\boldsymbol{\nabla} \cdot (\mathbf{A} \cdot \mathbf{B}) = (\boldsymbol{\nabla} \cdot \mathbf{A}) \cdot \mathbf{B} + \left(\mathbf{A}^T \cdot \boldsymbol{\nabla}\right) \cdot \mathbf{B}, \tag{1.13.14}$$

$$\boldsymbol{\nabla} \cdot (\mathbf{A} \times \mathbf{x}) = (\boldsymbol{\nabla} \cdot \mathbf{A}) \times \mathbf{x} - \boldsymbol{\epsilon} : \mathbf{A}. \tag{1.13.15}$$

The permutation tensor is $\boldsymbol{\epsilon}$, and : designates the inner product, defined by Eq. (1.2.13). The nabla operator in Eqs. (1.13.13)–(1.13.15) acts on the quantity to the right of it. The formulas can be easily proven by using the component tensor representations. A comprehensive treatment of tensor fields can be found in Truesdell and Toupin (1960), and Ericksen (1960).

1.13.2. Integral Transformation Theorems

Let V be a three dimensional domain bounded by a closed surface S with unit outward normal $\mathbf{n}$ (Fig. 1.4). For a tensor field $\mathbf{T} = \mathbf{T}(\mathbf{x})$, continuously

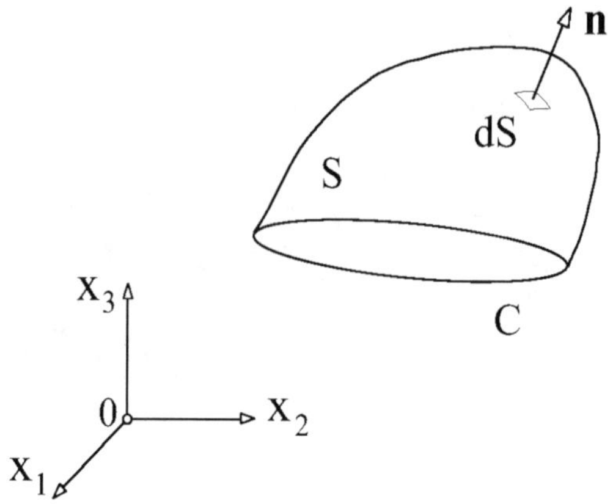

FIGURE 1.5. An open surface S with unit outward normal $\mathbf{n}$ and a bounding edge C.

differentiable in V and continuous on S, the generalized Gauss theorem asserts that

$$\int_V (\boldsymbol{\nabla} * \mathbf{T})\, \mathrm{d}V = \int_S \mathbf{n} * \mathbf{T}\, \mathrm{d}S. \tag{1.13.16}$$

The asterisk $(*)$ product can be either a dot $(\cdot)$ or cross $(\times)$ product, and $\mathbf{T}$ represents a scalar, vector, second- or higher-order tensor field (Malvern, 1969). For example, for a second-order tensor field $\mathbf{A}$, expressed in rectangular Cartesian coordinates,

$$\int_V \frac{\partial A_{ij}}{\partial x_i}\, \mathrm{d}V = \int_S n_i A_{ij}\, \mathrm{d}S. \tag{1.13.17}$$

Let S be a portion of an oriented surface with unit outward normal $\mathbf{n}$. The bounding edge of the surface is a closed curve C (Fig. 1.5). For tensors fields that are continuously differentiable in S and continuous on C, the generalized Stokes theorem asserts that

$$\int_S (\mathbf{n} \times \boldsymbol{\nabla}) * \mathbf{T}\, \mathrm{d}S = \int_C \mathrm{d}\mathbf{C} * \mathbf{T}. \tag{1.13.18}$$

For example, for a second-order tensor $\mathbf{A}$ this becomes, in the rectangular Cartesian coordinates,

$$\int_S \epsilon_{ijk} n_i \frac{\partial A_{kl}}{\partial x_j}\, \mathrm{d}S = \int_C A_{kl}\, \mathrm{d}C_k. \tag{1.13.19}$$

References

Betten, J. (1987), Invariants of fourth-order tensors, in *Application of Tensor Functions in Solid Mechanics*, ed. J. P. Boehler, pp. 203–226, Springer, Wien.

Brillouin, L. (1964), *Tensors in Mechanics and Elasticity*, Academic Press, New York.

Ericksen, J. L. (1960), Tensor fields, in *Handbuch der Physik*, ed. S. Flügge, Band III/1, pp. 794–858, Springer-Verlag, Berlin.

Eringen, A. C. (1971), Tensor analysis, in *Continuum Physics*, ed. A. C. Eringen, Vol. 1, pp. 1–155, Academic Press, New York.

Guo, Z.-H. (1984), Rates of stretch tensors, *J. Elasticity*, Vol. 14, pp. 263–267.

Hill, R. (1965), Continuum micro-mechanics of elastoplastic polycrystals, *J. Mech. Phys. Solids*, Vol. 13, pp. 89–101.

Hill, R. (1978), Aspects of invariance in solid mechanics, *Adv. Appl. Mech.*, Vol. 18, pp. 1–75.

Kanatani, K.-I. (1984), Distribution of directional data and fabric tensors, *Int. J. Engng. Sci.*, Vol. 22, pp. 149–164.

Lubarda, V. A. and Krajcinovic, D. (1993), Damage tensors and the crack density distribution, *Int. J. Solids Struct.*, Vol. 30, pp. 2859–2877.

Malvern, L. E. (1969), *Introduction to the Mechanics of a Continuous Medium*, Prentice-Hall, Englewood Cliffs, New Jersey.

Ogden, R. W. (1984), *Non-Linear Elastic Deformations*, Ellis Horwood Ltd., Chichester, England (2nd ed., Dover, 1997).

Rivlin, R. S. (1955), Further remarks on the stress-deformation relations for isotropic materials, *J. Rat. Mech. Anal.*, Vol. 4, pp. 681–701.

Scheidler, M. (1994), The tensor equation $AX + XA = \Phi(A, H)$, with applications to kinematics of continua, *J. Elasticity*, Vol. 36, pp. 117–153.

Sidoroff, F. (1978), Tensor equation $AX + XA = H$, *Comp. Acad. Sci. A Math.*, Vol. 286, pp. 71–73.

Spencer, A. J. M. (1970), A note on the decomposition of tensors into traceless symmetric tensors, *Int. J. Engng. Sci.*, Vol. 8, pp. 475–481.

Spencer, A. J. M. (1971), Theory of invariants, in *Continuum Physics*, ed. A. C. Eringen, Vol. 1, pp. 240–353, Academic Press, New York.

Truesdell, C. and Toupin, R. (1960), The classical field theories, in *Handbuch der Physik*, ed. S. Flügge, Band III/1, pp. 226–793, Springer-Verlag, Berlin.

Walpole, L. J. (1981), Elastic behavior of composite materials: Theoretical foundations, *Adv. Appl. Mech.*, Vol. 21, pp. 169–242.

CHAPTER 2

KINEMATICS OF DEFORMATION

2.1. Material and Spatial Description of Motion

The locations of material points of a three-dimensional body in its initial, undeformed configuration are specified by vectors $\mathbf{X}$. Their locations in deformed configuration at time t are specified by vectors $\mathbf{x}$, such that

$$\mathbf{x} = \mathbf{x}(\mathbf{X}, t). \tag{2.1.1}$$

The one-to-one deformation mapping from $\mathbf{X}$ to $\mathbf{x}$ is assumed to be twice continuously differentiable. The components of $\mathbf{X}$ are the material coordinates of the particle, while those of $\mathbf{x}$ are the spatial coordinates. They can be referred to the same or different bases. For example, if the orthonormal base vectors in the undeformed configuration are $\mathbf{e}_J^0$, and those in the deformed configuration are $\mathbf{e}_i$, then $\mathbf{X} = X_J\mathbf{e}_J^0$ and $\mathbf{x} = x_i\mathbf{e}_i$. Often, the same basis is used for both configurations (common frame).

If a tensor field $\mathbf{T}$ is expressed as a function of the material coordinates,

$$\mathbf{T} = \mathbf{T}(\mathbf{X}, t), \tag{2.1.2}$$

the description is referred to as the material or Lagrangian description. If the changes of $\mathbf{T}$ are observed at fixed points in space,

$$\mathbf{T} = \mathbf{T}(\mathbf{x}, t), \tag{2.1.3}$$

the description is spatial or Eulerian. The time derivative of $\mathbf{T}$ can be calculated as

$$\dot{\mathbf{T}} = \frac{\partial \mathbf{T}(\mathbf{X}, t)}{\partial t} = \frac{\partial \mathbf{T}(\mathbf{x}, t)}{\partial t} + \mathbf{v} \cdot (\boldsymbol{\nabla} \otimes \mathbf{T}). \tag{2.1.4}$$

The $\boldsymbol{\nabla}$ operator in Eq. (2.1.4) is defined with respect to spatial coordinates $\mathbf{x}$, and

$$\mathbf{v} = \frac{\partial \mathbf{x}(\mathbf{X}, t)}{\partial t} \tag{2.1.5}$$

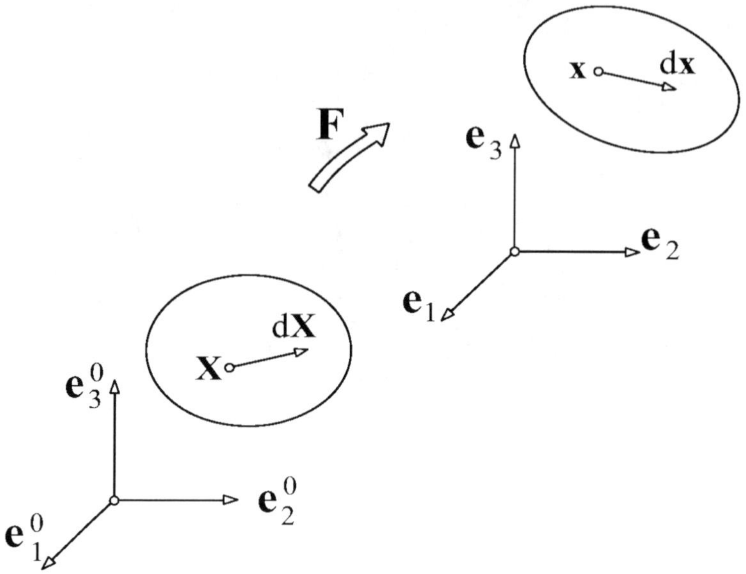

FIGURE 2.1. An infinitesimal material element $\mathbf{dX}$ from the initial configuration becomes $\mathbf{dx} = \mathbf{F} \cdot \mathbf{dX}$ in the deformed configuration, where $\mathbf{F}$ is the deformation gradient. The orthonormal base vectors in the undeformed and deformed configurations are $\mathbf{e}_J^0$ and $\mathbf{e}_i$.

is the velocity of a considered material particle at time t. The first term on the right-hand side of Eq. (2.1.4) is the local rate of change of $\mathbf{T}$, while the second term represents the convective rate of change (e.g., Eringen, 1967; Chadwick, 1976).

2.2. Deformation Gradient

An infinitesimal material element $\mathbf{dX}$ from the initial configuration becomes

$$\mathbf{dx} = \mathbf{F} \cdot \mathbf{dX}, \quad \mathbf{F} = \mathbf{x} \otimes \boldsymbol{\nabla}^0 = \frac{\partial \mathbf{x}}{\partial \mathbf{X}} \tag{2.2.1}$$

in the deformed configuration at time t (Fig. 2.1). The gradient operator $\boldsymbol{\nabla}^0$ is defined with respect to material coordinates. The tensor $\mathbf{F}$ is called the deformation gradient. If the orthonormal base vectors in the undeformed and deformed configurations are $\mathbf{e}_J^0$ and $\mathbf{e}_i$, then

$$\mathbf{F} = F_{iJ} \mathbf{e}_i \otimes \mathbf{e}_J^0, \quad F_{iJ} = \frac{\partial x_i}{\partial X_J}. \tag{2.2.2}$$

This represents a two-point tensor: when the base vectors in the deformed configuration are rotated by $\mathbf{Q}$, and those in the undeformed configuration by $\mathbf{Q}^0$, the components F_{iJ} change into $Q_{ki}F_{kL}Q^0_{LJ}$. If $\mathbf{Q}^0$ is the unit tensor, the components of $\mathbf{F}$ transform like those of a vector. Physically possible deformation mappings have the positive Jacobian determinant,

$$\det \mathbf{F} > 0. \tag{2.2.3}$$

Hence, $\mathbf{F}$ is an invertible tensor and $d\mathbf{X}$ can be recovered from $\mathbf{dx}$ by the inverse operation

$$\mathrm{d}\mathbf{X} = \mathbf{F}^{-1} \cdot \mathrm{d}\mathbf{x}. \tag{2.2.4}$$

The transpose and the inverse of $\mathbf{F}$ have the rectangular representations

$$\mathbf{F}^T = F_{iJ}\mathbf{e}^0_J \otimes \mathbf{e}_i, \quad \mathbf{F}^{-1} = F^{-1}_{Ji}\mathbf{e}^0_J \otimes \mathbf{e}_i. \tag{2.2.5}$$

2.2.1. Polar Decomposition

By the polar decomposition theorem, $\mathbf{F}$ can be decomposed into the product of a proper orthogonal tensor and a positive-definite symmetric tensor, such that (Truesdell and Noll, 1965; Malvern, 1969)

$$\mathbf{F} = \mathbf{R} \cdot \mathbf{U} = \mathbf{V} \cdot \mathbf{R}. \tag{2.2.6}$$

The symmetric tensor $\mathbf{U}$ is the right stretch tensor, $\mathbf{V}$ is the left stretch tensor, and $\mathbf{R}$ is the rotation tensor (Fig. 2.2). Evidently,

$$\mathbf{V} = \mathbf{R} \cdot \mathbf{U} \cdot \mathbf{R}^T, \tag{2.2.7}$$

so that $\mathbf{V}$ and $\mathbf{U}$ share the same eigenvalues (principal stretches λ_i), while their eigenvectors are related by

$$\mathbf{n}_i = \mathbf{R} \cdot \mathbf{N}_i. \tag{2.2.8}$$

The right and left Cauchy–Green deformation tensors are

$$\mathbf{C} = \mathbf{F}^T \cdot \mathbf{F} = \mathbf{U}^2, \quad \mathbf{B} = \mathbf{F} \cdot \mathbf{F}^T = \mathbf{V}^2. \tag{2.2.9}$$

The inverse of the left Cauchy–Green deformation tensor, $\mathbf{B}^{-1}$, is often referred to as the Finger deformation tensor. If there are three distinct principal stretches, $\mathbf{C}$ and $\mathbf{B}$ have the spectral representations

$$\mathbf{C} = \sum_{i=1}^{3} \lambda_i^2 \mathbf{N}_i \otimes \mathbf{N}_i, \quad \mathbf{B} = \sum_{i=1}^{3} \lambda_i^2 \mathbf{n}_i \otimes \mathbf{n}_i. \tag{2.2.10}$$

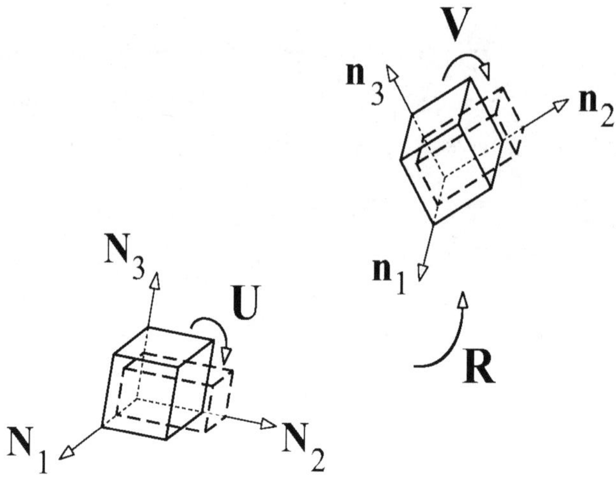

FIGURE 2.2. Schematic representation of the polar decomposition of deformation gradient. Material element is first stretched by **U** and then rotated by **R**, or first rotated by **R** and then stretched by **V**. The principal directions of **U** are $\mathbf{N}_i$, and those of **V** are $\mathbf{n}_i = \mathbf{R} \cdot \mathbf{N}_i$.

Furthermore,

$$\mathbf{U} = \sum_{i=1}^{3} \lambda_i \, \mathbf{N}_i \otimes \mathbf{N}_i, \quad \mathbf{V} = \sum_{i=1}^{3} \lambda_i \, \mathbf{n}_i \otimes \mathbf{n}_i, \quad \mathbf{R} = \sum_{i=1}^{3} \mathbf{n}_i \otimes \mathbf{N}_i, \tag{2.2.11}$$

and

$$\mathbf{F} = \sum_{i=1}^{3} \lambda_i \, \mathbf{n}_i \otimes \mathbf{N}_i. \tag{2.2.12}$$

If

$$j_1 = \lambda_1 + \lambda_2 + \lambda_3, \quad j_2 = -(\lambda_1\lambda_2 + \lambda_2\lambda_3 + \lambda_3\lambda_1), \quad j_3 = \lambda_1\lambda_2\lambda_3 \tag{2.2.13}$$

are the principal invariants of **U**, then (Hoger and Carlson, 1984; Simo and Hughes, 1998)

$$\mathbf{U} = \frac{1}{j_1 j_2 + j_3}\left[\mathbf{C}^2 - (j_1^2 + j_2)\mathbf{C} - j_3 j_1 \mathbf{I}^0\right], \quad \mathbf{U}^{-1} = \frac{1}{j_3}\left(\mathbf{C} - j_1\mathbf{U} - j_2\mathbf{I}^0\right). \tag{2.2.14}$$

The unit second-order tensors $\mathbf{I}^0$ is defined by

$$\mathbf{I}^0 = \sum_{i=1}^{3} \mathbf{N}_i \otimes \mathbf{N}_i. \tag{2.2.15}$$

2.2.2. Nanson's Relation

An infinitesimal volume element $\mathrm{d}V^0$ from the undeformed configuration becomes

$$\mathrm{d}V = (\det \mathbf{F})\mathrm{d}V^0 \tag{2.2.16}$$

in the deformed configuration. An infinitesimal area $\mathrm{d}S^0$ with unit normal $\mathbf{n}^0$ in the undeformed configuration becomes the area $\mathrm{d}S$ with unit normal $\mathbf{n}$ in the deformed configuration, such that (Nanson's relation)

$$\mathbf{n}\,\mathrm{d}S = (\det \mathbf{F})\mathbf{F}^{-T} \cdot \mathbf{n}^0\,\mathrm{d}S^0. \tag{2.2.17}$$

The following is a proof of (2.2.17). Consider a triad of vectors in the undeformed configuration $\mathbf{e}_J^0$, and its reciprocal triad $\mathbf{e}_0^J$. Then, the vector area

$$\mathrm{d}\mathbf{S}^0 = \mathbf{e}_1^0 \times \mathbf{e}_2^0 = D^0\,\mathbf{e}_0^3, \quad D^0 = (\mathbf{e}_1^0 \times \mathbf{e}_2^0) \cdot \mathbf{e}_3^0, \tag{2.2.18}$$

by definition of the reciprocal vectors (Hill, 1978). If the primary vectors are embedded in the material, they become in the deformed configuration $\mathbf{e}_i = \mathbf{F} \cdot \mathbf{e}_i^0$. Their reciprocal vectors are $\mathbf{e}^i = \mathbf{F}^{-T} \cdot \mathbf{e}_0^i$ (Fig. 2.3). Thus, the vector area corresponding to (2.2.18) is in the deformed configuration

$$\mathrm{d}\mathbf{S} = \mathbf{e}_1 \times \mathbf{e}_2 = D\,\mathbf{e}^3 = (\det \mathbf{F})\mathbf{F}^{-T} \cdot \mathrm{d}\mathbf{S}^0, \tag{2.2.19}$$

because

$$\mathbf{e}^3 = \mathbf{F}^{-T} \cdot \mathbf{e}_0^3, \quad D = (\mathbf{e}_1 \times \mathbf{e}_2) \cdot \mathbf{e}_3 = (\det \mathbf{F})D^0. \tag{2.2.20}$$

Equation (2.2.19) is the Nanson's relation.

By Eq. (1.13.16) the integral of $\mathbf{n}\,\mathrm{d}S$ over any closed surface S is equal to zero. Therefore, by applying the Gauss theorem to the integral of the right-hand side of Eq. (2.2.17) over the corresponding surface S^0 in the undeformed configuration gives

$$\boldsymbol{\nabla}^0 \cdot \left[(\det \mathbf{F})\mathbf{F}^{-1}\right] = \mathbf{0}. \tag{2.2.21}$$

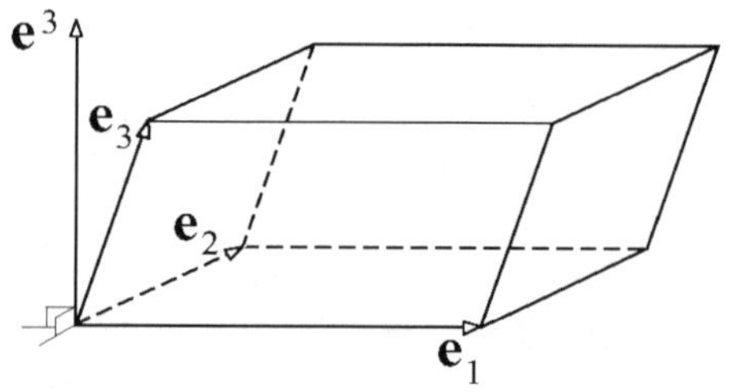

FIGURE 2.3. Deformed primary base vectors define an infinitesimal volume element dV in the deformed configuration. The reciprocal vector $\mathbf{e}^3 = D^{-1}(\mathbf{e}_1 \times \mathbf{e}_2)$, where $D = dV = (\mathbf{e}_1 \times \mathbf{e}_2) \cdot \mathbf{e}_3$.

2.2.3. Simple Shear

This is an isochoric plane deformation in which the planes with unit normal $\mathbf{N}$ slide relative to each other in the direction $\mathbf{M}$ (Fig. 2.4), such that

$$\mathbf{x} = \mathbf{X} + \gamma\left[\mathbf{N} \cdot (\mathbf{X} - \mathbf{X}_0)\right] \mathbf{m}. \tag{2.2.22}$$

The point $\mathbf{X}_0$ is fixed during the deformation, as are all other points within the plane for which $\mathbf{X} - \mathbf{X}_0$ is perpendicular to $\mathbf{N}$. The amount of shear is specified by $\gamma = \tan\varphi$, where φ is the shear angle. The vectors embedded in the planes of shearing preserve their length and orientation, so that $\mathbf{m} = \mathbf{M}$. The deformation gradient corresponding to Eq. (2.2.22), and its inverse are

$$\mathbf{F} = \mathbf{I} + \gamma(\mathbf{m} \otimes \mathbf{N}), \quad \mathbf{F}^{-1} = \mathbf{I} - \gamma(\mathbf{m} \otimes \mathbf{N}). \tag{2.2.23}$$

It is assumed that the same basis is used in both undeformed and deformed configurations. Clearly,

$$\mathbf{m} = \mathbf{F} \cdot \mathbf{M} = \mathbf{M}, \quad \mathbf{n} = \mathbf{N} \cdot \mathbf{F}^{-1} = \mathbf{N}, \tag{2.2.24}$$

where $\mathbf{n}$ is the unit normal to shear plane in the deformed configuration.

If different orthogonal bases are used in the undeformed and deformed configurations, we have

$$\mathbf{F} = g_{iJ}\left(\mathbf{e}_i \otimes \mathbf{e}_J^0\right) + \gamma(\mathbf{m} \otimes \mathbf{N}), \quad \mathbf{F}^{-1} = g_{Ji}\left(\mathbf{e}_J^0 \otimes \mathbf{e}_I\right) - \gamma(\mathbf{M} \otimes \mathbf{n}), \tag{2.2.25}$$

where

$$g_{iJ} = \mathbf{e}_i \cdot \mathbf{e}_J^0. \tag{2.2.26}$$

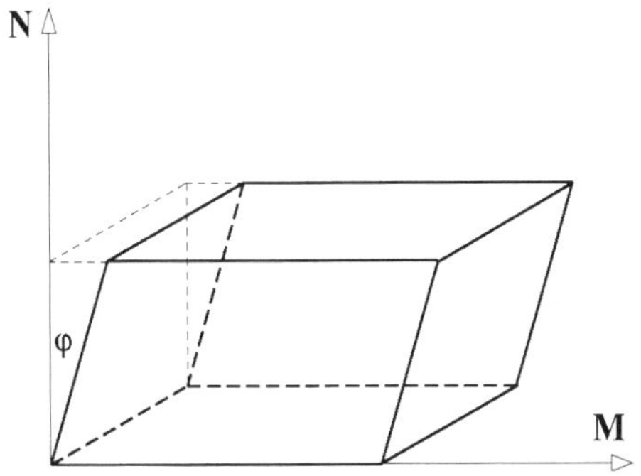

FIGURE 2.4. Simple shear of a rectangular block in the direction **M**, parallel to the plane with normal **N**. The shear angle is φ.

These are components of orthogonal matrices such that $g_{Ij}g_{jK} = \delta_{IK}$ and $g_{iJ}g_{Jk} = \delta_{ik}$ represent the components of unit tensors in the undeformed and deformed configurations, respectively, i.e.,

$$\mathbf{I}^0 = \delta_{IK}\mathbf{e}_I^0 \otimes \mathbf{e}_K^0, \quad \mathbf{I} = \delta_{ik}\mathbf{e}_i \otimes \mathbf{e}_k. \tag{2.2.27}$$

The corresponding right and left Cauchy–Green deformation tensors are accordingly

$$\mathbf{C} = \mathbf{I}^0 + \gamma(\mathbf{M} \otimes \mathbf{N} + \mathbf{N} \otimes \mathbf{M}) + \gamma^2(\mathbf{N} \otimes \mathbf{N}), \tag{2.2.28}$$

$$\mathbf{B} = \mathbf{I} + \gamma(\mathbf{m} \otimes \mathbf{n} + \mathbf{n} \otimes \mathbf{m}) + \gamma^2(\mathbf{m} \otimes \mathbf{m}). \tag{2.2.29}$$

2.3. Strain Tensors

2.3.1. Material Strain Tensors

Various tensor measures of strain can be defined. A fairly general definition of material strain measures, reckoned relative to the initial configuration, was introduced by Seth (1964,1966) and Hill (1968,1978). This is

$$\mathbf{E}_{(n)} = \frac{1}{2n}\left(\mathbf{U}^{2n} - \mathbf{I}^0\right) = \sum_{i=1}^{3} \frac{1}{2n}\left(\lambda_i^{2n} - 1\right)\mathbf{N}_i \otimes \mathbf{N}_i, \tag{2.3.1}$$

where $2n$ is a positive or negative integer, and λ_i and $\mathbf{N}_i$ are the principal values and directions of the right stretch tensor $\mathbf{U}$. The unit tensor in the initial configuration is $\mathbf{I}^0$. For $n = 1$, Eq. (2.3.1) gives the Lagrangian or

Green strain

$$\mathbf{E}_{(1)} = \frac{1}{2}(\mathbf{U}^2 - \mathbf{I}^0), \tag{2.3.2}$$

for $n = -1$ the Almansi strain

$$\mathbf{E}_{(-1)} = \frac{1}{2}(\mathbf{I}^0 - \mathbf{U}^{-2}), \tag{2.3.3}$$

and for $n = 1/2$ the Biot strain

$$\mathbf{E}_{(1/2)} = (\mathbf{U} - \mathbf{I}^0). \tag{2.3.4}$$

There is a general connection

$$\mathbf{E}_{(-n)} = \mathbf{U}^{-n} \cdot \mathbf{E}_{(n)} \cdot \mathbf{U}^{-n}. \tag{2.3.5}$$

The logarithmic or Hencky strain is obtained from (2.3.1) in the limit $n \to 0$, and is given by

$$\mathbf{E}_{(0)} = \ln \mathbf{U} = \sum_{i=1}^{3} \ln \lambda_i \, \mathbf{N}_i \otimes \mathbf{N}_i. \tag{2.3.6}$$

For isochoric deformation ($\lambda_1 \lambda_2 \lambda_3 = 1$), $\mathbf{E}_{(0)}$ is a traceless tensor.

Since,

$$\ln \lambda = (\lambda - 1) - \frac{1}{2}(\lambda - 1)^2 + \frac{1}{3}(\lambda - 1)^3 - \cdots, \quad 0 < \lambda \le 2, \tag{2.3.7}$$

$$\begin{aligned} \frac{1}{2n}\left(\lambda^{2n} - 1\right) &= (\lambda - 1) + \frac{1}{2}(2n - 1)(\lambda - 1)^2 \\ &+ \frac{1}{3}(n - 1)(2n - 1)(\lambda - 1)^3 + \cdots, \quad \lambda > 0, \end{aligned} \tag{2.3.8}$$

there follows

$$\mathbf{E}_{(0)} = \mathbf{E}_{(1/2)} - \frac{1}{2}\mathbf{E}^2_{(1/2)} + \frac{1}{3}\mathbf{E}^3_{(1/2)} + \mathcal{O}\left(\mathbf{E}^4_{(1/2)}\right), \tag{2.3.9}$$

$$\mathbf{E}_{(n)} = \mathbf{E}_{(1/2)} + \frac{1}{2}(2n - 1)\mathbf{E}^2_{(1/2)} + \frac{1}{3}(n - 1)(2n - 1)\mathbf{E}^3_{(1/2)} + \mathcal{O}\left(\mathbf{E}^4_{(1/2)}\right). \tag{2.3.10}$$

From this we can deduce the following useful connections

$$\mathbf{E}_{(0)} = \mathbf{E}_{(n)} - n\mathbf{E}^2_{(n)} + \frac{4}{3}n^2\mathbf{E}^3_{(n)} + \mathcal{O}\left(\mathbf{E}^4_{(n)}\right), \tag{2.3.11}$$

$$\mathbf{E}_{(n)} = \mathbf{E}_{(0)} + n\mathbf{E}^2_{(0)} + \frac{2}{3}n^2\mathbf{E}^3_{(0)} + \mathcal{O}\left(\mathbf{E}^4_{(0)}\right). \tag{2.3.12}$$

For the later purposes it is also noted that

$$\mathbf{E}^2_{(0)} = \mathbf{E}^2_{(n)} + \mathcal{O}\left(\mathbf{E}^3_{(n)}\right). \tag{2.3.13}$$

2.3.2. Spatial Strain Tensors

A family of spatial strain measures, reckoned relative to the deformed configuration and corresponding to material strain measures of Eqs. (2.3.1) and (2.3.6), is defined by

$$\boldsymbol{\mathcal{E}}_{(n)} = \frac{1}{2n}\left(\mathbf{V}^{2n} - \mathbf{I}\right) = \sum_{i=1}^{3} \frac{1}{2n}\left(\lambda_i^{2n} - 1\right) \mathbf{n}_i \otimes \mathbf{n}_i, \tag{2.3.14}$$

$$\boldsymbol{\mathcal{E}}_{(0)} = \ln \mathbf{V} = \sum_{i=1}^{3} \ln \lambda_i \, \mathbf{n}_i \otimes \mathbf{n}_i. \tag{2.3.15}$$

The unit tensor in the deformed configuration is $\mathbf{I}$, and $\mathbf{n}_i$ are the principal directions of the left stretch tensor $\mathbf{V}$. For example,

$$\boldsymbol{\mathcal{E}}_{(1)} = \frac{1}{2}(\mathbf{V}^2 - \mathbf{I}), \tag{2.3.16}$$

and

$$\boldsymbol{\mathcal{E}}_{(-1)} = \frac{1}{2}(\mathbf{I} - \mathbf{V}^{-2}), \tag{2.3.17}$$

the latter being known as the Eulerian strain tensor. Since

$$\mathbf{U}^{2n} = \mathbf{R}^T \cdot \mathbf{V}^{2n} \cdot \mathbf{R}, \tag{2.3.18}$$

and $\mathbf{n}_i = \mathbf{R} \cdot \mathbf{N}_i$, the material and spatial strain measures are related by

$$\mathbf{E}_{(n)} = \mathbf{R}^T \cdot \boldsymbol{\mathcal{E}}_{(n)} \cdot \mathbf{R}, \quad \mathbf{E}_{(0)} = \mathbf{R}^T \cdot \boldsymbol{\mathcal{E}}_{(0)} \cdot \mathbf{R}, \tag{2.3.19}$$

i.e., the former are induced from the latter by the rotation $\mathbf{R}$. Also, for any integer m, $\mathbf{E}_{(n)}^m$ is induced from $\boldsymbol{\mathcal{E}}_{(n)}^m$ by the rotation $\mathbf{R}$.

If $\mathrm{d}\mathbf{X}$ and $\delta\mathbf{X}$ are two material line elements in the undeformed configuration, and $\mathrm{d}\mathbf{x}$ and $\delta\mathbf{x}$ are the corresponding elements in the deformed configuration, it follows that

$$\mathrm{d}\mathbf{x} \cdot \delta\mathbf{x} - \mathrm{d}\mathbf{X} \cdot \delta\mathbf{X} = 2\,\mathrm{d}\mathbf{X} \cdot \mathbf{E}_{(1)} \cdot \delta\mathbf{X} = 2\,\mathrm{d}\mathbf{x} \cdot \boldsymbol{\mathcal{E}}_{(-1)} \cdot \delta\mathbf{x}. \tag{2.3.20}$$

Evidently, the Lagrangian and Eulerian strains are related by

$$\mathbf{E}_{(1)} = \mathbf{F}^T \cdot \boldsymbol{\mathcal{E}}_{(-1)} \cdot \mathbf{F}, \tag{2.3.21}$$

so that $\mathbf{E}_{(1)}$ is one of the induced tensors from $\boldsymbol{\mathcal{E}}_{(-1)}$ by the deformation $\mathbf{F}$ (Section 1.8). In the component form, the material and spatial strain tensors can be expressed as

$$\mathbf{E}_{(n)} = E_{(n)}^{IJ} \mathbf{e}_I^0 \otimes \mathbf{e}_J^0, \quad \boldsymbol{\mathcal{E}}_{(n)} = \mathcal{E}_{(n)}^{ij} \mathbf{e}_i \otimes \mathbf{e}_j, \tag{2.3.22}$$

relative to primary bases in the undeformed and deformed configuration, respectively. Covariant and two mixed representations are similarly written.

2.3.3. Infinitesimal Strain and Rotation Tensors

Introducing the displacement vector $\mathbf{u} = \mathbf{u}(\mathbf{X}, t)$ such that

$$\mathbf{x} = \mathbf{X} + \mathbf{u}, \tag{2.3.23}$$

the deformation gradient can be written as

$$\mathbf{F} = \mathbf{x} \otimes \boldsymbol{\nabla}^0 = \mathbf{I} + \mathbf{u} \otimes \boldsymbol{\nabla}^0. \tag{2.3.24}$$

The tensor $\mathbf{u} \otimes \boldsymbol{\nabla}^0$ is called the displacement gradient tensor. The right Cauchy–Green deformation tensor is expressed in terms of the displacement gradient tensor as

$$\mathbf{C} = \mathbf{U}^2 = \mathbf{F}^T \cdot \mathbf{F} = \mathbf{I} + \mathbf{u} \otimes \boldsymbol{\nabla}^0 + \boldsymbol{\nabla}^0 \otimes \mathbf{u} + (\boldsymbol{\nabla}^0 \otimes \mathbf{u}) \cdot (\mathbf{u} \otimes \boldsymbol{\nabla}^0). \tag{2.3.25}$$

If each component of the displacement gradient tensor is small compared with unity, Eq. (2.3.25) becomes

$$\mathbf{U}^2 \approx \mathbf{I} + \mathbf{u} \otimes \boldsymbol{\nabla}^0 + \boldsymbol{\nabla}^0 \otimes \mathbf{u}, \tag{2.3.26}$$

upon neglecting quadratic terms in the displacement gradient. Consequently,

$$\mathbf{U} \approx \mathbf{I} + \boldsymbol{\varepsilon}, \quad \mathbf{U}^{2n} \approx \mathbf{I} + 2n\boldsymbol{\varepsilon}, \tag{2.3.27}$$

where

$$\boldsymbol{\varepsilon} = \frac{1}{2}\left(\mathbf{u} \otimes \boldsymbol{\nabla}^0 + \boldsymbol{\nabla}^0 \otimes \mathbf{u}\right). \tag{2.3.28}$$

The material strain tensors are, therefore,

$$\mathbf{E}_{(n)} = \frac{1}{2n}\left(\mathbf{U}^{2n} - \mathbf{I}\right) \approx \boldsymbol{\varepsilon}, \quad \mathbf{E}_{(0)} = \ln \mathbf{U} \approx \boldsymbol{\varepsilon}, \tag{2.3.29}$$

all being approximately equal to $\boldsymbol{\varepsilon}$. The tensor $\boldsymbol{\varepsilon}$ defined by (2.3.28) is called the infinitesimal strain tensor. This tensor can also be expressed as (Hunter, 1976)

$$\boldsymbol{\varepsilon} = \frac{1}{2}\left(\mathbf{F} + \mathbf{F}^T\right) - \mathbf{I}. \tag{2.3.30}$$

If the displacement gradient is decomposed into its symmetric and antisymmetric parts,

$$\mathbf{u} \otimes \boldsymbol{\nabla}^0 = \boldsymbol{\varepsilon} + \boldsymbol{\omega}, \tag{2.3.31}$$

we have

$$\boldsymbol{\omega} = \frac{1}{2}\left(\mathbf{u} \otimes \boldsymbol{\nabla}^0 - \boldsymbol{\nabla}^0 \otimes \mathbf{u}\right) = \frac{1}{2}\left(\mathbf{F} - \mathbf{F}^T\right). \tag{2.3.32}$$

The tensor $\boldsymbol{\omega}$ is the infinitesimal rotation tensor. Its corresponding axial vector is $(1/2)(\boldsymbol{\nabla}^0 \times \mathbf{u})$. When the deformation gradient is decomposed by polar decomposition as $\mathbf{F} = \mathbf{V} \cdot \mathbf{R} = \mathbf{R} \cdot \mathbf{U}$, it follows that

$$\mathbf{V} \approx \mathbf{U} \approx \mathbf{I} + \boldsymbol{\varepsilon}, \quad \mathbf{R} \approx \mathbf{I} + \boldsymbol{\omega}, \tag{2.3.33}$$

again neglecting quadratic terms in the displacement gradient. Note also that, within the same order of approximation,

$$\det \mathbf{F} \approx 1 + \operatorname{tr} \boldsymbol{\varepsilon}. \tag{2.3.34}$$

If an infinitesimal strain tensor is defined by

$$\hat{\boldsymbol{\varepsilon}} = \frac{1}{2}\left(\mathbf{u} \otimes \boldsymbol{\nabla} + \boldsymbol{\nabla} \otimes \mathbf{u}\right), \tag{2.3.35}$$

then

$$\hat{\boldsymbol{\varepsilon}} = \mathbf{I} - \frac{1}{2}\left(\mathbf{F}^{-1} + \mathbf{F}^{-T}\right). \tag{2.3.36}$$

Since,

$$\mathbf{F}^{-1} = \left[\mathbf{I} + (\mathbf{F} - \mathbf{I})\right]^{-1} = \mathbf{I} - (\mathbf{F} - \mathbf{I}) + (\mathbf{F} - \mathbf{I})^2 - \cdots, \tag{2.3.37}$$

it follows that $\hat{\boldsymbol{\varepsilon}} = \boldsymbol{\varepsilon}$, provided that quadratic and higher-order terms in $(\mathbf{F} - \mathbf{I})$ are neglected. Indeed, in infinitesimal deformation (displacement gradient) theory, no distinction is made between the Lagrangian and Eulerian coordinates. For further details, the texts by Jaunzemis (1967), Spencer (1971), and Chung (1996) can be reviewed.

2.4. Velocity Gradient, Velocity Strain, and Spin Tensors

Consider a material line element $d\mathbf{x}$ in the deformed configuration at time t. If the velocity field is

$$\mathbf{v} = \mathbf{v}(\mathbf{x}, t), \tag{2.4.1}$$

the velocities of the end points of $d\mathbf{x}$ differ by

$$d\mathbf{v} = (\mathbf{v} \otimes \boldsymbol{\nabla}) \cdot d\mathbf{x} = \mathbf{L} \cdot d\mathbf{x}, \tag{2.4.2}$$

where $\boldsymbol{\nabla}$ represents the gradient operator with respect to spatial coordinates (Fig. 2.5). The tensor

$$\mathbf{L} = \mathbf{v} \otimes \boldsymbol{\nabla} \tag{2.4.3}$$

is called the velocity gradient. Its rectangular Cartesian components are

$$L_{ij} = \frac{\partial v_i}{\partial x_j}. \tag{2.4.4}$$

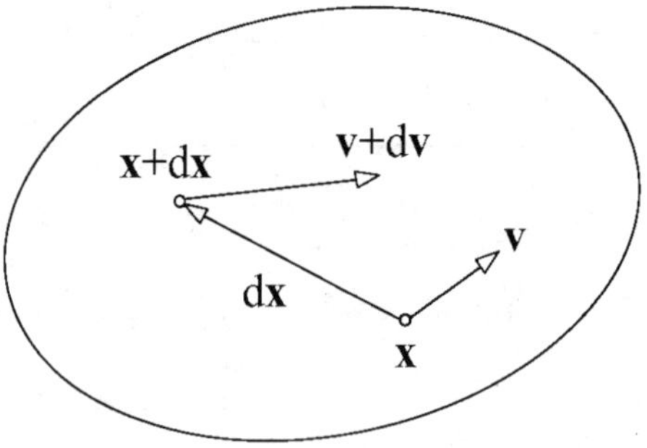

FIGURE 2.5. The velocity vectors of two nearby material points in deformed configuration at time t. The velocity gradient $\mathbf{L}$ is defined such that $\mathbf{dv} = \mathbf{L} \cdot \mathbf{dx}$.

The gradient operators with respect to material and spatial coordinates are related by

$$\overleftarrow{\nabla} = \overleftarrow{\nabla}^0 \cdot \mathbf{F}^{-1}, \quad \overrightarrow{\nabla} = \mathbf{F}^{-T} \cdot \overrightarrow{\nabla}^0. \tag{2.4.5}$$

For clarity, the arrows above the nabla operators are attached to indicate the direction in which the operators apply. Since from Eq. (2.2.1), the rate of deformation gradient is

$$\dot{\mathbf{F}} = \mathbf{v} \otimes \overleftarrow{\nabla}^0, \tag{2.4.6}$$

the substitution into Eq. (2.4.3) gives the relationship

$$\mathbf{L} = \dot{\mathbf{F}} \cdot \mathbf{F}^{-1}. \tag{2.4.7}$$

The symmetric and antisymmetric parts of $\mathbf{L}$ are the velocity strain or rate of deformation tensor, and the spin tensor, i.e.,

$$\mathbf{D} = \frac{1}{2}\left(\mathbf{L} + \mathbf{L}^T\right), \quad \mathbf{W} = \frac{1}{2}\left(\mathbf{L} - \mathbf{L}^T\right). \tag{2.4.8}$$

For example, the rate of change of the length $\mathrm{d}s$ of the material element $\mathbf{dx}$ can be calculated from

$$\frac{\mathrm{d}}{\mathrm{d}t}(\mathrm{d}s)^2 = 2\,\mathbf{dx} \cdot \mathbf{D} \cdot \mathbf{dx}, \quad \frac{\mathrm{d}}{\mathrm{d}t}(\mathrm{d}s) = (\mathbf{m} \cdot \mathbf{D} \cdot \mathbf{m})\,\mathrm{d}s, \tag{2.4.9}$$

where $\mathbf{m} = \mathbf{dx}/\mathrm{d}s$. By differentiating $\mathbf{dx}/\mathrm{d}s$ it also follows that the rate of unit vector $\mathbf{m}$ along the material direction $\mathbf{dx}$ is

$$\frac{\mathbf{dm}}{\mathrm{d}t} = \mathbf{L} \cdot \mathbf{m} - (\mathbf{m} \cdot \mathbf{D} \cdot \mathbf{m})\mathbf{m}. \tag{2.4.10}$$

If $\mathbf{m}$ is an eigenvector of $\mathbf{D}$, then

$$\frac{\mathbf{dm}}{\mathrm{d}t} = \mathbf{W} \cdot \mathbf{m}. \tag{2.4.11}$$

Thus, we can interpret $\mathbf{W}$ as the spin of the triad of line elements directed, at the considered instant of deformation, along the principal axes of the rate of deformation $\mathbf{D}$.

The rate of the inverse $\mathbf{F}^{-1}$ and the rate of the Jacobian determinant are

$$\left(\mathbf{F}^{-1}\right)^{\cdot} = -\mathbf{F}^{-1} \cdot \dot{\mathbf{F}} \cdot \mathbf{F}^{-1}, \quad \frac{\mathrm{d}}{\mathrm{d}t}(\det \mathbf{F}) = (\det \mathbf{F}) \operatorname{tr} \mathbf{D}. \tag{2.4.12}$$

The first expression follows by differentiating $\mathbf{F} \cdot \mathbf{F}^{-1} = \mathbf{I}$, and the second from

$$\frac{\mathrm{d}}{\mathrm{d}t}(\det \mathbf{F}) = \operatorname{tr}\left[\frac{\partial(\det \mathbf{F})}{\partial \mathbf{F}} \cdot \dot{\mathbf{F}}\right] = \operatorname{tr}\left[(\det \mathbf{F})\, \mathbf{F}^{-1} \cdot \dot{\mathbf{F}}\right] = (\det \mathbf{F}) \operatorname{tr} \mathbf{D}, \tag{2.4.13}$$

because $\operatorname{tr} \mathbf{W} = 0$. Furthermore, since $\mathrm{d}V = (\det \mathbf{F})\mathrm{d}V^0$, the rate of volume change is

$$\frac{\mathrm{d}}{\mathrm{d}t}(\mathrm{d}V) = (\operatorname{tr} \mathbf{D})\mathrm{d}V. \tag{2.4.14}$$

By differentiating Nanson's relation (2.2.17), we have

$$\frac{\mathrm{d}}{\mathrm{d}t}(\mathrm{d}\mathbf{S}) = \frac{\mathrm{d}}{\mathrm{d}t}(\mathrm{d}S\mathbf{n}) = [(\operatorname{tr} \mathbf{D})\mathbf{n} - (\mathbf{n} \cdot \mathbf{L})]\, \mathrm{d}S. \tag{2.4.15}$$

Since $\dot{\mathbf{n}} \cdot \mathbf{n} = 0$, $\mathbf{n}$ being the unit vector normal to $\mathrm{d}S$, and having in mind that

$$\frac{\mathrm{d}}{\mathrm{d}t}(\mathrm{d}\mathbf{S}) = \frac{\mathrm{d}}{\mathrm{d}t}(\mathrm{d}S\mathbf{n}) = \frac{\mathrm{d}}{\mathrm{d}t}(\mathrm{d}S)\, \mathbf{n} + \mathrm{d}S\, \frac{\mathrm{d}}{\mathrm{d}t}(\mathbf{n}), \tag{2.4.16}$$

there follows

$$\frac{\mathrm{d}}{\mathrm{d}t}(\mathrm{d}S) = (\operatorname{tr} \mathbf{D} - \mathbf{n} \cdot \mathbf{D} \cdot \mathbf{n})\, \mathrm{d}S. \tag{2.4.17}$$

$$\frac{\mathrm{d}}{\mathrm{d}t}(\mathbf{n}) = (\mathbf{n} \cdot \mathbf{D} \cdot \mathbf{n}) \cdot \mathbf{n} - \mathbf{n} \cdot \mathbf{L}. \tag{2.4.18}$$

In the case of simple shearing deformation considered in Subsection 2.2.2, the velocity gradient can be written as

$$\mathbf{L} = \dot{\gamma}(\mathbf{m} \otimes \mathbf{n}). \tag{2.4.19}$$

2.5. Convected Derivatives

Consider the primary and reciprocal bases in the undeformed configuration, $\mathbf{e}_I^0$ and $\mathbf{e}_0^I$. If the primary basis is embedded in the material, its base vectors in the deformed configuration become $\mathbf{e}_i = \mathbf{F} \cdot \mathbf{e}_I^0$. The associated reciprocal (non-embedded) base vectors are $\mathbf{e}^i = \mathbf{e}_0^I \cdot \mathbf{F}^{-1}$. Thus, by differentiation it follows that

$$\dot{\mathbf{e}}_i = \mathbf{L} \cdot \mathbf{e}_i, \quad \dot{\mathbf{e}}^i = -\mathbf{L}^T \cdot \mathbf{e}^i. \tag{2.5.1}$$

In view of Eq. (1.7.8), the velocity gradient can be expressed as

$$\mathbf{L} = \dot{\mathbf{e}}_i \otimes \mathbf{e}^i. \tag{2.5.2}$$

The rate of change of an arbitrary vector in the deformed configuration, $\mathbf{a} = a^i \mathbf{e}_i = a_i \mathbf{e}^i$, is

$$\dot{\mathbf{a}} = \dot{a}^i \mathbf{e}_i + \mathbf{L} \cdot \mathbf{a} = \dot{a}_i \mathbf{e}^i - \mathbf{L}^T \cdot \mathbf{a}. \tag{2.5.3}$$

The two derivatives,

$$\overset{\triangle}{\mathbf{a}} = \dot{a}^i \mathbf{e}_i = \dot{\mathbf{a}} - \mathbf{L} \cdot \mathbf{a}, \quad \overset{\triangledown}{\mathbf{a}} = \dot{a}_i \mathbf{e}^i = \dot{\mathbf{a}} + \mathbf{L}^T \cdot \mathbf{a}, \tag{2.5.4}$$

are the two convected-type derivatives of the vector $\mathbf{a}$. The first gives the rate of change observed in the embedded basis $\mathbf{e}_i$, which is convected with the deforming material. The second is the rate of change observed in the basis $\mathbf{e}^i$, reciprocal to the embedded basis $\mathbf{e}_i$.

The corotational or Jaumann derivative of $\mathbf{a}$ is

$$\overset{\circ}{\mathbf{a}} = \dot{\mathbf{a}} - \mathbf{W} \cdot \mathbf{a}, \tag{2.5.5}$$

which represents the rate of change observed in the basis that momentarily rotates with the material spin $\mathbf{W}$. Two types of convected, and the Jaumann derivative of a two-point deformation gradient tensor are likewise

$$\overset{\triangle}{\mathbf{F}} = \dot{\mathbf{F}} - \mathbf{L} \cdot \mathbf{F} = \mathbf{0}, \quad \overset{\triangledown}{\mathbf{F}} = \dot{\mathbf{F}} + \mathbf{L}^T \cdot \mathbf{F}, \quad \overset{\circ}{\mathbf{F}} = \dot{\mathbf{F}} - \mathbf{W} \cdot \mathbf{F}. \tag{2.5.6}$$

Therefore,

$$\overset{\triangle}{\mathbf{F}} \cdot \mathbf{F}^{-1} = \mathbf{0}, \quad \overset{\triangledown}{\mathbf{F}} \cdot \mathbf{F}^{-1} = 2\mathbf{D}, \quad \overset{\circ}{\mathbf{F}} \cdot \mathbf{F}^{-1} = \mathbf{D}. \tag{2.5.7}$$

Four kinds of convected derivatives of a second-order tensor $\mathbf{A}$ in the deformed configuration can be similarly introduced. They are given by the following formulas

$$\overset{\triangle}{\mathbf{A}} = \dot{A}^{ij} \mathbf{e}_i \otimes \mathbf{e}_j = \dot{\mathbf{A}} - \mathbf{L} \cdot \mathbf{A} - \mathbf{A} \cdot \mathbf{L}^T, \tag{2.5.8}$$

$$\overset{\triangledown}{\mathbf{A}} = \dot{A}_{ij} \mathbf{e}^i \otimes \mathbf{e}^j = \dot{\mathbf{A}} + \mathbf{L}^T \cdot \mathbf{A} + \mathbf{A} \cdot \mathbf{L}, \tag{2.5.9}$$

$$\overset{\triangleleft}{\mathbf{A}} = \dot{A}^i{}_j \mathbf{e}_i \otimes \mathbf{e}^j = \dot{\mathbf{A}} - \mathbf{L} \cdot \mathbf{A} + \mathbf{A} \cdot \mathbf{L}, \tag{2.5.10}$$

$$\overset{\triangleright}{\mathbf{A}} = \dot{A}_i{}^j \mathbf{e}^i \otimes \mathbf{e}_j = \dot{\mathbf{A}} + \mathbf{L}^T \cdot \mathbf{A} - \mathbf{A} \cdot \mathbf{L}^T. \tag{2.5.11}$$

The rate $\overset{\triangle}{\mathbf{A}}$ is often referred to as the Oldroyd, and $\overset{\triangledown}{\mathbf{A}}$ as the Cotter–Rivlin convected rate. Additional discussion can be found in Prager (1961), Truesdell and Noll (1965), Sedov (1966), and Hill (1978). Convected derivatives of the second-order tensors can also be interpreted as the Lie derivatives (Marsden and Hughes, 1983). Note that convected derivatives of the unit tensor in the deformed configuration are

$$\overset{\triangledown}{\mathbf{I}} = -\overset{\triangle}{\mathbf{I}} = 2\mathbf{D}, \quad \overset{\triangleleft}{\mathbf{I}} = \overset{\triangleright}{\mathbf{I}} = \mathbf{0}. \tag{2.5.12}$$

The Jaumann (or Jaumann–Zaremba) derivative of a second-order tensor $\mathbf{A}$ is

$$\overset{\circ}{\mathbf{A}} = \dot{\mathbf{A}} - \mathbf{W} \cdot \mathbf{A} + \mathbf{A} \cdot \mathbf{W}. \tag{2.5.13}$$

The relationships hold

$$\overset{\circ}{\mathbf{A}} = \frac{1}{2}\left(\overset{\triangle}{\mathbf{A}} + \overset{\triangledown}{\mathbf{A}}\right) = \frac{1}{2}\left(\overset{\triangleleft}{\mathbf{A}} + \overset{\triangleright}{\mathbf{A}}\right). \tag{2.5.14}$$

It is easily verified that

$$\left(\mathbf{F}^{-1}\right)^{\triangle} = -\mathbf{F}^{-1} \cdot \overset{\triangledown}{\mathbf{F}} \cdot \mathbf{F}^{-1} = -2\mathbf{F}^{-1} \cdot \mathbf{D}, \quad \left(\mathbf{F}^{-1}\right)^{\triangledown} = -\mathbf{F}^{-1} \cdot \overset{\triangle}{\mathbf{F}} \cdot \mathbf{F}^{-1} = \mathbf{0}. \tag{2.5.15}$$

Convected derivatives of the higher-order tensors can be introduced analogously.

2.5.1. Convected Derivatives of Tensor Products

Let $\mathbf{F}$ be a two-point tensor such that

$$\mathbf{F} = F^{iJ}\mathbf{e}_i \otimes \mathbf{e}_J^0, \tag{2.5.16}$$

and similarly for the other three decompositions. Its convected and corotational derivatives are

$$\overset{\triangle}{\mathbf{F}} = \overset{\triangleleft}{\mathbf{F}} = \dot{\mathbf{F}} - \mathbf{L} \cdot \mathbf{F}, \quad \overset{\triangledown}{\mathbf{F}} = \overset{\triangleright}{\mathbf{F}} = \dot{\mathbf{F}} + \mathbf{L}^T \cdot \mathbf{F}, \quad \overset{\circ}{\mathbf{F}} = \dot{\mathbf{F}} - \mathbf{W} \cdot \mathbf{F}. \tag{2.5.17}$$

Introduce a two-point tensor $\mathbf{G}$ such that

$$\mathbf{G} = G^{Ji}\mathbf{e}_J^0 \otimes \mathbf{e}_i, \tag{2.5.18}$$

and similarly for the other three decompositions. Its convected and corotational derivatives are

$$\overset{\triangle}{\mathbf{G}} = \overset{\triangleright}{\mathbf{G}} = \dot{\mathbf{G}} - \mathbf{G} \cdot \mathbf{L}^T, \quad \overset{\triangledown}{\mathbf{G}} = \overset{\triangleleft}{\mathbf{G}} = \dot{\mathbf{G}} + \mathbf{G} \cdot \mathbf{L}, \quad \overset{\circ}{\mathbf{G}} = \dot{\mathbf{G}} + \mathbf{G} \cdot \mathbf{W}. \tag{2.5.19}$$

The tensor $\mathbf{B} = \mathbf{F} \cdot \mathbf{G}$ is a spatial tensor, whose convected derivatives are defined by Eqs. (2.5.8)–(2.5.11). The following connections hold

$$\overset{\triangle}{\mathbf{B}} = \overset{\triangle}{\mathbf{F}} \cdot \mathbf{G} + \mathbf{F} \cdot \overset{\triangle}{\mathbf{G}}, \quad \overset{\triangledown}{\mathbf{B}} = \overset{\triangledown}{\mathbf{F}} \cdot \mathbf{G} + \mathbf{F} \cdot \overset{\triangledown}{\mathbf{G}}, \quad \overset{\circ}{\mathbf{B}} = \overset{\circ}{\mathbf{F}} \cdot \mathbf{G} + \mathbf{F} \cdot \overset{\circ}{\mathbf{G}}. \tag{2.5.20}$$

The same type of chain rule applies to $\overset{\triangleleft}{\mathbf{B}}$ and $\overset{\triangleright}{\mathbf{B}}$. Two additional identities exist, which are

$$\overset{\triangleleft}{\mathbf{B}} = \overset{\triangle}{\mathbf{F}} \cdot \mathbf{G} + \mathbf{F} \cdot \overset{\triangledown}{\mathbf{G}}, \quad \overset{\triangleright}{\mathbf{B}} = \overset{\triangledown}{\mathbf{F}} \cdot \mathbf{G} + \mathbf{F} \cdot \overset{\triangle}{\mathbf{G}}. \tag{2.5.21}$$

On the other hand, the tensor $\mathbf{C} = \mathbf{G} \cdot \mathbf{F}$ is a material tensor, unaffected by convected operations in the deformed configuration, so that

$$\overset{\triangle}{\mathbf{C}} = \overset{\triangledown}{\mathbf{C}} = \overset{\circ}{\mathbf{C}} = \dot{\mathbf{C}}. \tag{2.5.22}$$

The following identities are easily verified

$$\dot{\mathbf{C}} = \overset{\triangleleft}{\mathbf{G}} \cdot \mathbf{F} + \mathbf{G} \cdot \overset{\triangleleft}{\mathbf{F}} = \overset{\triangleright}{\mathbf{G}} \cdot \mathbf{F} + \mathbf{G} \cdot \overset{\triangleright}{\mathbf{F}} = \overset{\circ}{\mathbf{G}} \cdot \mathbf{F} + \mathbf{G} \cdot \overset{\circ}{\mathbf{F}}. \tag{2.5.23}$$

Furthermore,

$$\dot{\mathbf{C}} = \overset{\triangle}{\mathbf{G}} \cdot \mathbf{F} + \mathbf{G} \cdot \overset{\triangle}{\mathbf{F}} + 2\mathbf{G} \cdot \mathbf{D} \cdot \mathbf{F} = \overset{\triangledown}{\mathbf{G}} \cdot \mathbf{F} + \mathbf{G} \cdot \overset{\triangledown}{\mathbf{F}} - 2\mathbf{G} \cdot \mathbf{D} \cdot \mathbf{F}, \tag{2.5.24}$$

and

$$\dot{\mathbf{C}} = \overset{\triangle}{\mathbf{G}} \cdot \mathbf{F} + \mathbf{G} \cdot \overset{\triangledown}{\mathbf{F}} = \overset{\triangledown}{\mathbf{G}} \cdot \mathbf{F} + \mathbf{G} \cdot \overset{\triangle}{\mathbf{F}}. \tag{2.5.25}$$

If both $\mathbf{A}$ and $\mathbf{B}$ are spatial tensors, then $\mathbf{K} = \mathbf{A} \cdot \mathbf{B}$ is as well. Its convected derivatives are defined by Eqs. (2.5.8)–(2.5.11). It can be shown that

$$\overset{\triangleleft}{\mathbf{K}} = \overset{\triangleleft}{\mathbf{A}} \cdot \mathbf{B} + \mathbf{A} \cdot \overset{\triangleleft}{\mathbf{B}}, \quad \overset{\triangleright}{\mathbf{K}} = \overset{\triangleright}{\mathbf{A}} \cdot \mathbf{B} + \mathbf{A} \cdot \overset{\triangleright}{\mathbf{B}}, \tag{2.5.26}$$

$$\overset{\circ}{\mathbf{K}} = \overset{\circ}{\mathbf{A}} \cdot \mathbf{B} + \mathbf{A} \cdot \overset{\circ}{\mathbf{B}}, \tag{2.5.27}$$

$$\overset{\triangle}{\mathbf{K}} = \overset{\triangle}{\mathbf{A}} \cdot \mathbf{B} + \mathbf{A} \cdot \overset{\triangle}{\mathbf{B}} + 2\mathbf{A} \cdot \mathbf{D} \cdot \mathbf{B}, \tag{2.5.28}$$

$$\overset{\triangledown}{\mathbf{K}} = \overset{\triangledown}{\mathbf{A}} \cdot \mathbf{B} + \mathbf{A} \cdot \overset{\triangledown}{\mathbf{B}} - 2\mathbf{A} \cdot \mathbf{D} \cdot \mathbf{B}, \tag{2.5.29}$$

$$\overset{\triangleleft}{\mathbf{K}} = \overset{\triangle}{\mathbf{A}} \cdot \mathbf{B} + \mathbf{A} \cdot \overset{\triangledown}{\mathbf{B}}, \quad \overset{\triangleright}{\mathbf{K}} = \overset{\triangledown}{\mathbf{A}} \cdot \mathbf{B} + \mathbf{A} \cdot \overset{\triangle}{\mathbf{B}}. \tag{2.5.30}$$

2.6. Rates of Strain

2.6.1. Rates of Material Strains

The rate of the Lagrangian strain is expressed in terms of the rate of deformation tensor as

$$\dot{\mathbf{E}}_{(1)} = \mathbf{F}^T \cdot \mathbf{D} \cdot \mathbf{F} = \mathbf{U} \cdot \hat{\mathbf{D}} \cdot \mathbf{U}, \tag{2.6.1}$$

where

$$\hat{\mathbf{D}} = \mathbf{R}^T \cdot \mathbf{D} \cdot \mathbf{R}. \tag{2.6.2}$$

The rate of the Almansi strain is similarly

$$\dot{\mathbf{E}}_{(-1)} = \mathbf{F}^{-1} \cdot \mathbf{D} \cdot \mathbf{F}^{-T} = \mathbf{U}^{-1} \cdot \hat{\mathbf{D}} \cdot \mathbf{U}^{-1}. \tag{2.6.3}$$

Evidently, the two strain rates are related by

$$\dot{\mathbf{E}}_{(-1)} = \mathbf{U}^{-2} \cdot \dot{\mathbf{E}}_{(1)} \cdot \mathbf{U}^{-2}. \tag{2.6.4}$$

This is a particular case of the general relationship (Ogden, 1984)

$$\dot{\mathbf{E}}_{(-n)} = \mathbf{U}^{-2n} \cdot \dot{\mathbf{E}}_{(n)} \cdot \mathbf{U}^{-2n}, \quad n \neq 0, \tag{2.6.5}$$

which holds because

$$\left(\mathbf{U}^{-n}\right)^{\cdot} = -\mathbf{U}^{-n} \cdot (\mathbf{U}^{n})^{\cdot} \cdot \mathbf{U}^{-n}. \tag{2.6.6}$$

An expression for the rate of the logarithmic strain can be derived as follows. From Eq. (2.3.11), we have

$$\dot{\mathbf{E}}_{(0)} = \dot{\mathbf{E}}_{(n)} - n\left(\mathbf{E}_{(n)} \cdot \dot{\mathbf{E}}_{(n)} + \dot{\mathbf{E}}_{(n)} \cdot \mathbf{E}_{(n)}\right) + \mathcal{O}\left(\mathbf{E}_{(n)}^2 \cdot \dot{\mathbf{E}}_{(n)}\right). \tag{2.6.7}$$

To evaluate $\dot{\mathbf{E}}_{(0)}$, any $\mathbf{E}_{(n)}$ can be used. For example, if $\mathbf{E}_{(1)}$ is used, from Eq. (2.6.1) we have

$$\dot{\mathbf{E}}_{(1)} = \hat{\mathbf{D}} + \mathbf{E}_{(1)} \cdot \hat{\mathbf{D}} + \hat{\mathbf{D}} \cdot \mathbf{E}_{(1)} + \mathcal{O}\left(\mathbf{E}_{(1)}^2 \cdot \hat{\mathbf{D}}\right). \tag{2.6.8}$$

Substitution of Eq. (2.6.8) into Eq. (2.6.7), therefore, gives

$$\dot{\mathbf{E}}_{(0)} = \hat{\mathbf{D}} + \mathcal{O}\left(\mathbf{E}_{(n)}^2 \cdot \hat{\mathbf{D}}\right). \tag{2.6.9}$$

Recall from Eqs. (2.3.11) and (2.3.12) that $\mathbf{E}_{(1)}^2 = \mathbf{E}_{(n)}^2$, neglecting cubic are higher-order terms in strain. If principal directions of $\mathbf{U}$ remain fixed ($\dot{\mathbf{N}}_i = \mathbf{0}$), we have

$$\dot{\mathbf{E}}_{(0)} = \hat{\mathbf{D}}, \tag{2.6.10}$$

exactly. Further analysis can be found in the papers by Fitzgerald (1980), Hoger (1986), and Dui, Ren, and Shen (1999).

2.6.2. Rates of Spatial Strains

The following relationships hold for convected rates of the strains $\boldsymbol{\mathcal{E}}_{(1)}$ and $\boldsymbol{\mathcal{E}}_{(-1)}$,

$$\overset{\triangle}{\boldsymbol{\mathcal{E}}}_{(1)} = \mathbf{D}, \quad \overset{\triangledown}{\boldsymbol{\mathcal{E}}}_{(1)} = \mathbf{D} + 2\left(\boldsymbol{\mathcal{E}}_{(1)} \cdot \mathbf{D} + \mathbf{D} \cdot \boldsymbol{\mathcal{E}}_{(1)}\right), \tag{2.6.11}$$

$$\overset{\triangledown}{\boldsymbol{\mathcal{E}}}_{(-1)} = \mathbf{D}, \quad \overset{\triangle}{\boldsymbol{\mathcal{E}}}_{(-1)} = \mathbf{D} - 2\left(\boldsymbol{\mathcal{E}}_{(-1)} \cdot \mathbf{D} + \mathbf{D} \cdot \boldsymbol{\mathcal{E}}_{(-1)}\right). \tag{2.6.12}$$

The rate of the deformation tensor $\mathbf{B} = \mathbf{F} \cdot \mathbf{F}^T$ is

$$\dot{\mathbf{B}} = \mathbf{L} \cdot \mathbf{B} + \mathbf{B} \cdot \mathbf{L}^T, \tag{2.6.13}$$

so that

$$\overset{\triangle}{\mathbf{B}} = \mathbf{0}, \quad \overset{\triangledown}{\mathbf{B}} = 2(\mathbf{B} \cdot \mathbf{D} + \mathbf{D} \cdot \mathbf{B}), \quad \overset{\triangleleft}{\mathbf{B}} = 2\mathbf{B} \cdot \mathbf{D}, \quad \overset{\triangleright}{\mathbf{B}} = 2\mathbf{D} \cdot \mathbf{B}, \tag{2.6.14}$$

and

$$\left(\mathbf{B}^{-1}\right)^{\triangle} = -\mathbf{B}^{-1} \cdot \overset{\triangledown}{\mathbf{B}} \cdot \mathbf{B}^{-1}, \quad \left(\mathbf{B}^{-1}\right)^{\triangledown} = -\mathbf{B}^{-1} \cdot \overset{\triangle}{\mathbf{B}} \cdot \mathbf{B}^{-1}, \tag{2.6.15}$$

$$\left(\mathbf{B}^{-1}\right)^{\triangleleft} = -\mathbf{B}^{-1} \cdot \overset{\triangleleft}{\mathbf{B}} \cdot \mathbf{B}^{-1}, \quad \left(\mathbf{B}^{-1}\right)^{\triangleright} = -\mathbf{B}^{-1} \cdot \overset{\triangleright}{\mathbf{B}} \cdot \mathbf{B}^{-1}. \tag{2.6.16}$$

Furthermore,

$$\overset{\circ}{\mathbf{B}} = \mathbf{B} \cdot \mathbf{D} + \mathbf{D} \cdot \mathbf{B}, \quad \overset{\bullet}{\mathbf{B}} = 2\mathbf{V} \cdot \mathbf{D} \cdot \mathbf{V}, \tag{2.6.17}$$

where

$$\overset{\circ}{\mathbf{B}} = \dot{\mathbf{B}} - \mathbf{W} \cdot \mathbf{B} + \mathbf{B} \cdot \mathbf{W}, \quad \overset{\bullet}{\mathbf{B}} = \dot{\mathbf{B}} - \boldsymbol{\omega} \cdot \mathbf{B} + \mathbf{B} \cdot \boldsymbol{\omega}. \tag{2.6.18}$$

The corotational rate with respect to $\boldsymbol{\omega} = \dot{\mathbf{R}} \cdot \mathbf{R}^{-1}$ is sometimes referred to as the Green–Naghdi–McInnis corotational rate.

The expressions for the rates of other strain measures in terms of $\mathbf{D}$ are more involved. Since $\boldsymbol{\mathcal{E}}_{(n)} = \mathbf{R} \cdot \mathbf{E}_{(n)} \cdot \mathbf{R}^T$, there is a general connection

$$\overset{\bullet}{\boldsymbol{\mathcal{E}}}_{(n)} = \mathbf{R} \cdot \dot{\mathbf{E}}_{(n)} \cdot \mathbf{R}^T, \quad \overset{\bullet}{\boldsymbol{\mathcal{E}}}_{(n)} = \dot{\boldsymbol{\mathcal{E}}}_{(n)} - \boldsymbol{\omega} \cdot \boldsymbol{\mathcal{E}}_{(n)} + \boldsymbol{\mathcal{E}}_{(n)} \cdot \boldsymbol{\omega}. \tag{2.6.19}$$

Higher rates of strain can be investigated along similar lines. For example, it can be shown that

$$\ddot{\mathbf{E}}_{(1)} = \mathbf{F}^T \cdot \overset{\triangledown}{\mathbf{D}} \cdot \mathbf{F}, \quad \overset{\triangledown}{\mathbf{D}} = \dot{\mathbf{D}} + \mathbf{L}^T \cdot \mathbf{D} + \mathbf{D} \cdot \mathbf{L}. \tag{2.6.20}$$

2.7. Relationship between Spins W and ω

The velocity gradient $\mathbf{L}$ can be written, in terms of the constituents of the polar decomposition of deformation gradient $\mathbf{F} = \mathbf{V} \cdot \mathbf{R}$, as

$$\mathbf{L} = \dot{\mathbf{V}} \cdot \mathbf{V}^{-1} + \mathbf{V} \cdot \boldsymbol{\omega} \cdot \mathbf{V}^{-1} = \boldsymbol{\omega} + \overset{\bullet}{\mathbf{V}} \cdot \mathbf{V}^{-1}, \tag{2.7.1}$$

where

$$\overset{\bullet}{\mathbf{V}} = \dot{\mathbf{V}} - \boldsymbol{\omega} \cdot \mathbf{V} + \mathbf{V} \cdot \boldsymbol{\omega}, \quad \boldsymbol{\omega} = \dot{\mathbf{R}} \cdot \mathbf{R}^{-1}. \tag{2.7.2}$$

By taking symmetric and antisymmetric parts of Eq. (2.7.1), there follows

$$\mathbf{D} = \left(\overset{\bullet}{\mathbf{V}} \cdot \mathbf{V}^{-1}\right)_s, \quad \mathbf{W} = \boldsymbol{\omega} + \left(\overset{\bullet}{\mathbf{V}} \cdot \mathbf{V}^{-1}\right)_a. \tag{2.7.3}$$

Similarly, if the decomposition $\mathbf{F} = \mathbf{R} \cdot \mathbf{U}$ is used, we obtain

$$\mathbf{L} = \boldsymbol{\omega} + \mathbf{R} \cdot \left(\dot{\mathbf{U}} \cdot \mathbf{U}^{-1}\right) \cdot \mathbf{R}^T. \tag{2.7.4}$$

This can be rewritten as

$$\hat{\mathbf{L}} = \hat{\boldsymbol{\omega}} + \dot{\mathbf{U}} \cdot \mathbf{U}^{-1}, \tag{2.7.5}$$

where

$$\hat{\mathbf{L}} = \mathbf{R}^T \cdot \mathbf{L} \cdot \mathbf{R}, \quad \hat{\boldsymbol{\omega}} = \mathbf{R}^T \cdot \boldsymbol{\omega} \cdot \mathbf{R} \tag{2.7.6}$$

are the tensors induced from $\mathbf{L}$ and $\boldsymbol{\omega}$ by the rotation $\mathbf{R}$. Upon taking symmetric and antisymmetric parts of Eq. (2.7.5),

$$\hat{\mathbf{D}} = \left(\dot{\mathbf{U}} \cdot \mathbf{U}^{-1}\right)_s, \quad \hat{\mathbf{W}} = \hat{\boldsymbol{\omega}} + \left(\dot{\mathbf{U}} \cdot \mathbf{U}^{-1}\right)_a. \tag{2.7.7}$$

Since $\mathbf{V} = \mathbf{R} \cdot \mathbf{U} \cdot \mathbf{R}^T$, we also have

$$\overset{\bullet}{\mathbf{V}} = \mathbf{R} \cdot \dot{\mathbf{U}} \cdot \mathbf{R}^T. \tag{2.7.8}$$

In particular, if $\dot{\mathbf{U}} = \mathbf{0}$, then $\overset{\bullet}{\mathbf{V}} = \mathbf{0}$ and

$$\dot{\mathbf{V}} = \boldsymbol{\omega} \cdot \mathbf{V} - \mathbf{V} \cdot \boldsymbol{\omega}, \quad \boldsymbol{\omega} = \dot{\mathbf{R}} \cdot \mathbf{R}^{-1}. \tag{2.7.9}$$

With these preliminaries, we now derive a relationship between $\mathbf{W}$ and $\boldsymbol{\omega}$ (or $\hat{\mathbf{W}}$ and $\hat{\boldsymbol{\omega}}$). First, observe the identity

$$\mathbf{V}^{-1} \cdot \left(\overset{\bullet}{\mathbf{V}} \cdot \mathbf{V}^{-1}\right) = \left(\overset{\bullet}{\mathbf{V}} \cdot \mathbf{V}^{-1}\right)^T \cdot \mathbf{V}^{-1}, \tag{2.7.10}$$

which can be rewritten as

$$\mathbf{V}^{-1} \cdot \left(\overset{\bullet}{\mathbf{V}} \cdot \mathbf{V}^{-1}\right)_a + \left(\overset{\bullet}{\mathbf{V}} \cdot \mathbf{V}^{-1}\right)_a \cdot \mathbf{V}^{-1} = \mathbf{D} \cdot \mathbf{V}^{-1} - \mathbf{V}^{-1} \cdot \mathbf{D}. \tag{2.7.11}$$

This can be solved for $\left(\overset{\bullet}{\mathbf{V}} \cdot \mathbf{V}^{-1}\right)_a$ by using the procedure described in Subsection 1.12.1. The result is

$$\begin{aligned}\left(\overset{\bullet}{\mathbf{V}} \cdot \mathbf{V}^{-1}\right)_a &= K_1 \left(\mathbf{D} \cdot \mathbf{V}^{-1} - \mathbf{V}^{-1} \cdot \mathbf{D}\right) \\ &- \left[\left(J_1 \mathbf{I} - \mathbf{V}^{-1}\right)^{-1} \cdot \left(\mathbf{D} \cdot \mathbf{V}^{-1} - \mathbf{V}^{-1} \cdot \mathbf{D}\right) \right. \\ &+ \left. \left(\mathbf{D} \cdot \mathbf{V}^{-1} - \mathbf{V}^{-1} \cdot \mathbf{D}\right) \cdot \left(J_1 \mathbf{I} - \mathbf{V}^{-1}\right)^{-1}\right],\end{aligned} \tag{2.7.12}$$

where

$$J_1 = \operatorname{tr} \mathbf{V}^{-1}, \quad K_1 = \operatorname{tr} \left(J_1 \mathbf{I} - \mathbf{V}^{-1}\right)^{-1}. \tag{2.7.13}$$

Substitution of Eq. (2.7.12) into the second of Eq. (2.7.3) gives

$$\begin{aligned}\boldsymbol{\omega} = \mathbf{W} &- K_1 \left(\mathbf{D} \cdot \mathbf{V}^{-1} - \mathbf{V}^{-1} \cdot \mathbf{D}\right) \\ &+ \left[\left(J_1 \mathbf{I} - \mathbf{V}^{-1}\right)^{-1} \cdot \left(\mathbf{D} \cdot \mathbf{V}^{-1} - \mathbf{V}^{-1} \cdot \mathbf{D}\right) \right. \\ &+ \left. \left(\mathbf{D} \cdot \mathbf{V}^{-1} - \mathbf{V}^{-1} \cdot \mathbf{D}\right) \cdot \left(J_1 \mathbf{I} - \mathbf{V}^{-1}\right)^{-1}\right],\end{aligned} \tag{2.7.14}$$

which shows that the spin $\boldsymbol{\omega}$ can be determined at each stage of deformation solely in terms of $\mathbf{V}$, $\mathbf{D}$, and $\mathbf{W}$.

Analogous derivation proceeds to find

$$\begin{aligned}\left(\dot{\mathbf{U}} \cdot \mathbf{U}^{-1}\right)_a &= K_1 \left(\hat{\mathbf{D}} \cdot \mathbf{U}^{-1} - \mathbf{U}^{-1} \cdot \hat{\mathbf{D}}\right) \\ &- \left[\left(J_1 \mathbf{I} - \mathbf{U}^{-1}\right)^{-1} \cdot \left(\hat{\mathbf{D}} \cdot \mathbf{U}^{-1} - \mathbf{U}^{-1} \cdot \hat{\mathbf{D}}\right) \right. \\ &+ \left. \left(\hat{\mathbf{D}} \cdot \mathbf{U}^{-1} - \mathbf{U}^{-1} \cdot \hat{\mathbf{D}}\right) \cdot \left(J_1 \mathbf{I} - \mathbf{U}^{-1}\right)^{-1}\right].\end{aligned} \tag{2.7.15}$$

Substitution into second of Eq. (2.7.7) gives

$$\begin{aligned}\hat{\boldsymbol{\omega}} = \hat{\mathbf{W}} &- K_1 \left(\hat{\mathbf{D}} \cdot \mathbf{U}^{-1} - \mathbf{U}^{-1} \cdot \hat{\mathbf{D}}\right) \\ &+ \left[\left(J_1 \mathbf{I} - \mathbf{U}^{-1}\right)^{-1} \cdot \left(\hat{\mathbf{D}} \cdot \mathbf{U}^{-1} - \mathbf{U}^{-1} \cdot \hat{\mathbf{D}}\right) \right. \\ &+ \left. \left(\hat{\mathbf{D}} \cdot \mathbf{U}^{-1} - \mathbf{U}^{-1} \cdot \hat{\mathbf{D}}\right) \cdot \left(J_1 \mathbf{I} - \mathbf{U}^{-1}\right)^{-1}\right],\end{aligned} \tag{2.7.16}$$

as anticipated at the outset from its duality with Eq. (2.7.14). Additional kinematic analysis is provided by Mehrabadi and Nemat-Nasser (1987), and Reinhardt and Dubey (1996).

2.8. Rate of F in Terms of Principal Stretches

From Eq. (2.2.11) the right stretch tensor can be expressed in terms of its eigenvalues – principal stretches λ_i (assumed here to be different), and corresponding eigendirections $\mathbf{N}_i$ as

$$\mathbf{U} = \sum_{i=1}^{3} \lambda_i \, \mathbf{N}_i \otimes \mathbf{N}_i. \tag{2.8.1}$$

The rate of $\mathbf{U}$ is then

$$\dot{\mathbf{U}} = \sum_{i=1}^{3} \left[\dot{\lambda}_i \, \mathbf{N}_i \otimes \mathbf{N}_i + \lambda_i \left(\dot{\mathbf{N}}_i \otimes \mathbf{N}_i + \mathbf{N}_i \otimes \dot{\mathbf{N}}_i \right) \right]. \tag{2.8.2}$$

If $\mathbf{e}_i^0$ $(i = 1, 2, 3)$ are the fixed reference unit vectors, the unit vectors $\mathbf{N}_i$ of the principal directions of $\mathbf{U}$ can be expressed as

$$\mathbf{N}_i = \boldsymbol{\mathcal{R}}_0 \cdot \mathbf{e}_i^0, \tag{2.8.3}$$

where $\boldsymbol{\mathcal{R}}_0$ is the rotation that carries the orthogonal triad $\{\mathbf{e}_i^0\}$ into the Lagrangian triad $\{\mathbf{N}_i\}$. Defining the spin of the Lagrangian triad by

$$\boldsymbol{\Omega}_0 = \dot{\boldsymbol{\mathcal{R}}}_0 \cdot \boldsymbol{\mathcal{R}}_0^{-1}, \tag{2.8.4}$$

it follows that

$$\dot{\mathbf{N}}_i = \dot{\boldsymbol{\mathcal{R}}}_0 \cdot \mathbf{e}_i^0 = \boldsymbol{\Omega}_0 \cdot \mathbf{N}_i = -\mathbf{N}_i \cdot \boldsymbol{\Omega}_0, \tag{2.8.5}$$

and the substitution into Eq. (2.8.2) gives

$$\dot{\mathbf{U}} = \sum_{i=1}^{3} \dot{\lambda}_i \, \mathbf{N}_i \otimes \mathbf{N}_i + \boldsymbol{\Omega}_0 \cdot \mathbf{U} - \mathbf{U} \cdot \boldsymbol{\Omega}_0. \tag{2.8.6}$$

If the spin tensor $\boldsymbol{\Omega}_0$ is expressed on the axes of the Lagrangian triad as

$$\boldsymbol{\Omega}_0 = \sum_{i \neq j} \Omega_{ij}^0 \, \mathbf{N}_i \otimes \mathbf{N}_j, \tag{2.8.7}$$

it is readily found that

$$\begin{aligned} \boldsymbol{\Omega}_0 \cdot \mathbf{U} = \Omega_{12}^0 (\lambda_2 - \lambda_1) \, \mathbf{N}_1 \otimes \mathbf{N}_2 + \Omega_{23}^0 (\lambda_3 - \lambda_2) \, \mathbf{N}_2 \otimes \mathbf{N}_3 \\ + \Omega_{31}^0 (\lambda_1 - \lambda_3) \, \mathbf{N}_3 \otimes \mathbf{N}_1. \end{aligned} \tag{2.8.8}$$

Consequently,

$$\boldsymbol{\Omega}_0 \cdot \mathbf{U} - \mathbf{U} \cdot \boldsymbol{\Omega}_0 = \boldsymbol{\Omega}_0 \cdot \mathbf{U} + (\boldsymbol{\Omega}_0 \cdot \mathbf{U})^T = \sum_{i \neq j} \Omega_{ij}^0 \, (\lambda_j - \lambda_i) \, \mathbf{N}_i \otimes \mathbf{N}_j. \tag{2.8.9}$$

The substitution into Eq. (2.8.6) yields

$$\dot{\mathbf{U}} = \sum_{i=1}^{3} \dot{\lambda}_i \, \mathbf{N}_i \otimes \mathbf{N}_i + \sum_{i \neq j} \Omega_{ij}^0 \, (\lambda_j - \lambda_i) \, \mathbf{N}_i \otimes \mathbf{N}_j. \tag{2.8.10}$$

Similarly, the rate of the material strain tensor of Eq. (2.3.1) is

$$\dot{\mathbf{E}}_{(n)} = \sum_{i=1}^{3} \lambda_i^{2n-1} \dot{\lambda}_i \, \mathbf{N}_i \otimes \mathbf{N}_i + \sum_{i \neq j} \Omega_{ij}^0 \, \frac{\lambda_j^{2n} - \lambda_i^{2n}}{2n} \, \mathbf{N}_i \otimes \mathbf{N}_j. \tag{2.8.11}$$

The principal directions of the left stretch tensor $\mathbf{V}$, appearing in the spectral representation

$$\mathbf{V} = \sum_{i=1}^{3} \lambda_i \, \mathbf{n}_i \otimes \mathbf{n}_i, \tag{2.8.12}$$

are related to principal directions $\mathbf{N}_i$ of the right stretch tensor $\mathbf{U}$ by

$$\mathbf{n}_i = \mathbf{R} \cdot \mathbf{N}_i = \boldsymbol{\mathcal{R}} \cdot \mathbf{e}_i^0, \quad \boldsymbol{\mathcal{R}} = \mathbf{R} \cdot \boldsymbol{\mathcal{R}}_0. \tag{2.8.13}$$

The rotation tensor $\mathbf{R}$ is from the polar decomposition of the the deformation gradient $\mathbf{F} = \mathbf{V} \cdot \mathbf{R} = \mathbf{R} \cdot \mathbf{U}$. By differentiating Eq. (2.8.13) there follows

$$\dot{\mathbf{n}}_i = \boldsymbol{\Omega} \cdot \mathbf{n}_i, \tag{2.8.14}$$

where the spin of the Eulerian triad $\{\mathbf{n}_i\}$ is defined by

$$\boldsymbol{\Omega} = \dot{\boldsymbol{\mathcal{R}}} \cdot \boldsymbol{\mathcal{R}}^{-1} = \boldsymbol{\omega} + \mathbf{R} \cdot \boldsymbol{\Omega}_0 \cdot \mathbf{R}^T, \quad \boldsymbol{\omega} = \dot{\mathbf{R}} \cdot \mathbf{R}^{-1}. \tag{2.8.15}$$

On the axes $\mathbf{n}_i$, the spin $\boldsymbol{\Omega}$ can be decomposed as

$$\boldsymbol{\Omega} = \sum_{i \neq j} \Omega_{ij} \, \mathbf{n}_i \otimes \mathbf{n}_j. \tag{2.8.16}$$

By an analogous derivation as used to obtain the rate $\dot{\mathbf{U}}$ it follows that

$$\dot{\mathbf{V}} = \sum_{i=1}^{3} \dot{\lambda}_i \, \mathbf{n}_i \otimes \mathbf{n}_i + \sum_{i \neq j} \Omega_{ij} \, (\lambda_j - \lambda_i) \, \mathbf{n}_i \otimes \mathbf{n}_j. \tag{2.8.17}$$

The rate of the rotation tensor

$$\mathbf{R} = \sum_{i=1}^{3} \mathbf{n}_i \otimes \mathbf{N}_i \tag{2.8.18}$$

is

$$\dot{\mathbf{R}} = \sum_{i=1}^{3} \left(\dot{\mathbf{n}}_i \otimes \mathbf{N}_i + \mathbf{n}_i \otimes \dot{\mathbf{N}}_i \right) = \boldsymbol{\Omega} \cdot \mathbf{R} - \mathbf{R} \cdot \boldsymbol{\Omega}_0, \tag{2.8.19}$$

or

$$\dot{\mathbf{R}} = \sum_{i \neq j} \left(\Omega_{ij} - \Omega_{ij}^0 \right) \mathbf{n}_i \otimes \mathbf{N}_j. \tag{2.8.20}$$

Finally, the rate of the deformation gradient

$$\mathbf{F} = \sum_{i=1}^{3} \lambda_i \, \mathbf{n}_i \otimes \mathbf{N}_i \tag{2.8.21}$$

is

$$\dot{\mathbf{F}} = \sum_{i=1}^{3} \left[\dot{\lambda}_i \, \mathbf{n}_i \otimes \mathbf{N}_i + \lambda_i \left(\dot{\mathbf{n}}_i \otimes \mathbf{N}_i + \mathbf{n}_i \otimes \dot{\mathbf{N}}_i \right) \right]. \tag{2.8.22}$$

Since $\dot{\mathbf{n}}_i = \boldsymbol{\Omega} \cdot \mathbf{n}_i$ and $\dot{\mathbf{N}}_i = \boldsymbol{\Omega}_0 \cdot \mathbf{N}_i$, it follows that

$$\dot{\mathbf{F}} = \sum_{i=1}^{3} \dot{\lambda}_i \, \mathbf{n}_i \otimes \mathbf{N}_i + \boldsymbol{\Omega} \cdot \mathbf{F} - \mathbf{F} \cdot \boldsymbol{\Omega}_0, \tag{2.8.23}$$

and

$$\dot{\mathbf{F}} = \sum_{i=1}^{3} \dot{\lambda}_i \, \mathbf{n}_i \otimes \mathbf{N}_i + \sum_{i \neq j} \left(\lambda_j \Omega_{ij} - \lambda_i \Omega_{ij}^0 \right) \mathbf{n}_i \otimes \mathbf{N}_j. \tag{2.8.24}$$

2.8.1. Spins of Lagrangian and Eulerian Triads

The inverse of the deformation gradient can be written in terms of the principal stretches as

$$\mathbf{F}^{-1} = \sum_{i=1}^{3} \frac{1}{\lambda_i} \mathbf{N}_i \otimes \mathbf{n}_i. \tag{2.8.25}$$

Using this and Eq. (2.8.24) we obtain an expression for the velocity gradient

$$\mathbf{L} = \dot{\mathbf{F}} \cdot \mathbf{F}^{-1} = \sum_{i=1}^{3} \frac{\dot{\lambda}_i}{\lambda_i} \mathbf{n}_i \otimes \mathbf{n}_i + \sum_{i \neq j} \left(\Omega_{ij} - \frac{\lambda_i}{\lambda_j} \Omega_{ij}^0 \right) \mathbf{n}_i \otimes \mathbf{n}_j. \tag{2.8.26}$$

The symmetric part of this is the rate of deformation tensor,

$$\mathbf{D} = \sum_{i=1}^{3} \frac{\dot{\lambda}_i}{\lambda_i} \mathbf{n}_i \otimes \mathbf{n}_i + \sum_{i \neq j} \frac{\lambda_j^2 - \lambda_i^2}{2\lambda_i \lambda_j} \Omega_{ij}^0 \, \mathbf{n}_i \otimes \mathbf{n}_j, \tag{2.8.27}$$

while the antisymmetric part is the spin tensor

$$\mathbf{W} = \sum_{i \neq j} \left(\Omega_{ij} - \frac{\lambda_i^2 + \lambda_j^2}{2\lambda_i \lambda_j} \Omega_{ij}^0 \right) \mathbf{n}_i \otimes \mathbf{n}_j. \tag{2.8.28}$$

Evidently, for $i \neq j$ from Eq. (2.8.27) we have

$$\Omega_{ij}^0 = \frac{2\lambda_i \lambda_j}{\lambda_j^2 - \lambda_i^2} D_{ij}, \quad \lambda_i \neq \lambda_j, \tag{2.8.29}$$

which is an expression for the components of the Lagrangian spin $\boldsymbol{\Omega}_0$ in terms of the stretch ratios and the components of the rate of deformation tensor. Substituting (2.8.29) into (2.8.28) we obtain an expression for the components of the Eulerian spin $\boldsymbol{\Omega}$ in terms of the stretch ratios and the components of the rate of deformation and spin tensors, i.e.,

$$\Omega_{ij} = W_{ij} + \frac{\lambda_i^2 + \lambda_j^2}{\lambda_j^2 - \lambda_i^2} D_{ij}, \quad \lambda_i \neq \lambda_j. \tag{2.8.30}$$

Lastly, we note that the inverse of the rotation tensor $\mathbf{R}$ is

$$\mathbf{R}^{-1} = \sum_{i=1}^{3} \mathbf{N}_i \otimes \mathbf{n}_i, \tag{2.8.31}$$

so that, by virtue of Eq. (2.8.20), the spin $\boldsymbol{\omega}$ can be expressed as

$$\boldsymbol{\omega} = \dot{\mathbf{R}} \cdot \mathbf{R}^{-1} = \sum_{i \neq j} \left(\Omega_{ij} - \Omega_{ij}^0\right) \mathbf{n}_i \otimes \mathbf{n}_j. \tag{2.8.32}$$

Thus,

$$\omega_{ij} = \Omega_{ij} - \Omega_{ij}^0, \tag{2.8.33}$$

where Ω_{ij}^0 are the components of $\boldsymbol{\Omega}_0$ on the Lagrangian triad $\{\mathbf{N}_i\}$, while Ω_{ij} are the components of $\boldsymbol{\Omega}$ on the Eulerian triad $\{\mathbf{n}_i\}$. When Eqs. (2.8.29) and (2.8.30) are substituted into Eq. (2.8.33), we obtain an expression for the spin components ω_{ij} in terms of the stretch ratios and the components of the rate of deformation and spin tensors, which is

$$\omega_{ij} = W_{ij} + \frac{\lambda_j - \lambda_i}{\lambda_i + \lambda_j} D_{ij}. \tag{2.8.34}$$

This complements the previously derived expression for the spin $\boldsymbol{\omega}$ in terms of $\mathbf{V}$, $\mathbf{D}$, and $\mathbf{W}$, given by Eq. (2.7.14). Further analysis can be found in Biot (1965) and Hill (1970,1978).

2.9. Behavior under Superimposed Rotation

If a time-dependent rotation $\mathbf{Q}$ is superimposed to the deformed configuration at time t, an infinitesimal material line element $\mathrm{d}\mathbf{x}$ becomes (Fig. 2.6)

$$\mathrm{d}\mathbf{x}^* = \mathbf{Q} \cdot \mathrm{d}\mathbf{x}, \tag{2.9.1}$$

while in the undeformed configuration

$$\mathrm{d}\mathbf{X}^* = \mathrm{d}\mathbf{X}. \tag{2.9.2}$$

Consequently, since $\mathrm{d}\mathbf{x} = \mathbf{F} \cdot \mathrm{d}\mathbf{X}$, we have

$$\mathbf{F}^* = \mathbf{Q} \cdot \mathbf{F}. \tag{2.9.3}$$

This implies that

$$\mathbf{U}^* = \mathbf{U}, \quad \mathbf{C}^* = \mathbf{C}, \quad \mathbf{E}_{(n)}^* = \mathbf{E}_{(n)}, \tag{2.9.4}$$

and

$$\mathbf{V}^* = \mathbf{Q} \cdot \mathbf{V} \cdot \mathbf{Q}^T, \quad \mathbf{B}^* = \mathbf{Q} \cdot \mathbf{B} \cdot \mathbf{Q}^T, \quad \boldsymbol{\mathcal{E}}_{(n)}^* = \mathbf{Q} \cdot \boldsymbol{\mathcal{E}}_{(n)} \cdot \mathbf{Q}^T. \tag{2.9.5}$$

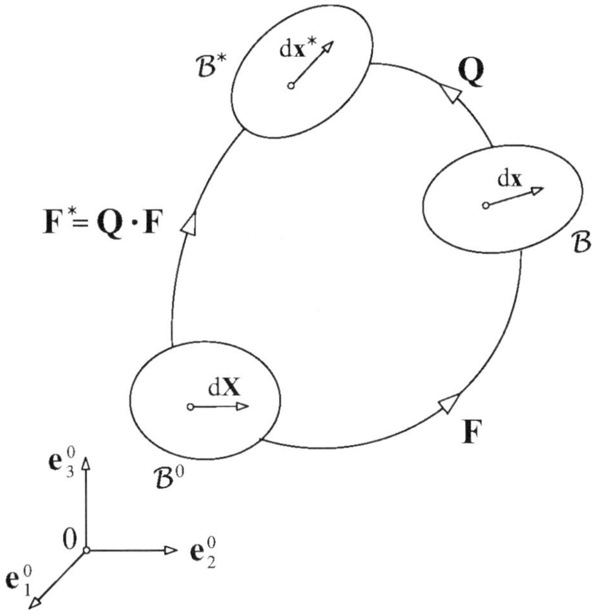

FIGURE 2.6. The material element $\mathbf{dX}$ from the undeformed configuration $\mathcal{B}^0$ becomes $\mathbf{dx} = \mathbf{F} \cdot \mathbf{dX}$ in the deformed configuration $\mathcal{B}$, and $\mathbf{dx}^* = \mathbf{Q} \cdot \mathbf{dx}$ in the rotated deformed configuration $\mathcal{B}^*$.

The objective rates of the spatial vector $\mathbf{a}$ transform according to

$$\overset{\triangle}{\mathbf{a}}{}^* = \mathbf{Q} \cdot \overset{\triangle}{\mathbf{a}}, \quad \overset{\triangledown}{\mathbf{a}}{}^* = \mathbf{Q} \cdot \overset{\triangledown}{\mathbf{a}}, \quad \overset{\circ}{\mathbf{a}}{}^* = \mathbf{Q} \cdot \overset{\circ}{\mathbf{a}}, \tag{2.9.6}$$

as do the objective rates of the deformation gradient $\mathbf{F}$. The rotation $\mathbf{R}$ becomes

$$\mathbf{R}^* = \mathbf{Q} \cdot \mathbf{R}. \tag{2.9.7}$$

The spin $\boldsymbol{\omega} = \dot{\mathbf{R}} \cdot \mathbf{R}^{-1}$ changes to

$$\boldsymbol{\omega}^* = \boldsymbol{\Omega} + \mathbf{Q} \cdot \boldsymbol{\omega} \cdot \mathbf{Q}^T, \quad \boldsymbol{\Omega} = \dot{\mathbf{Q}} \cdot \mathbf{Q}^{-1}. \tag{2.9.8}$$

The velocity gradient transforms as

$$\mathbf{L}^* = \boldsymbol{\Omega} + \mathbf{Q} \cdot \mathbf{L} \cdot \mathbf{Q}^T, \tag{2.9.9}$$

while the velocity strain and the spin tensors become

$$\mathbf{D}^* = \mathbf{Q} \cdot \mathbf{D} \cdot \mathbf{Q}^T, \tag{2.9.10}$$

$$\mathbf{W}^* = \boldsymbol{\Omega} + \mathbf{Q} \cdot \mathbf{W} \cdot \mathbf{Q}^T. \tag{2.9.11}$$

The rates of the material and spatial strain tensors change according to

$$\dot{\mathbf{E}}^*_{(n)} = \dot{\mathbf{E}}_{(n)}, \tag{2.9.12}$$

$$\dot{\boldsymbol{\mathcal{E}}}^*_{(n)} = \mathbf{Q} \cdot \left(\dot{\boldsymbol{\mathcal{E}}}_{(n)} + \hat{\boldsymbol{\Omega}} \cdot \boldsymbol{\mathcal{E}}_{(n)} - \boldsymbol{\mathcal{E}}_{(n)} \cdot \hat{\boldsymbol{\Omega}} \right) \cdot \mathbf{Q}^T, \tag{2.9.13}$$

where

$$\hat{\boldsymbol{\Omega}} = \mathbf{Q}^T \cdot \boldsymbol{\Omega} \cdot \mathbf{Q}, \quad \boldsymbol{\Omega} = \dot{\mathbf{Q}} \cdot \mathbf{Q}^{-1}. \tag{2.9.14}$$

The transformation formulas for the convected rates of spatial strain tensors are

$$\overset{\triangle}{\boldsymbol{\mathcal{E}}}{}^*_{(1)} = \mathbf{Q} \cdot \overset{\triangle}{\boldsymbol{\mathcal{E}}}_{(1)} \cdot \mathbf{Q}^T, \quad \overset{\triangledown}{\boldsymbol{\mathcal{E}}}{}^*_{(-1)} = \mathbf{Q} \cdot \overset{\triangledown}{\boldsymbol{\mathcal{E}}}_{(-1)} \cdot \mathbf{Q}^T. \tag{2.9.15}$$

Since $\overset{\triangle}{\boldsymbol{\mathcal{E}}}_{(1)} = \overset{\triangledown}{\boldsymbol{\mathcal{E}}}_{(-1)}$ by Eqs. (2.6.11) and (2.6.12), it follows that

$$\overset{\triangle}{\boldsymbol{\mathcal{E}}}{}^*_{(1)} = \overset{\triangledown}{\boldsymbol{\mathcal{E}}}{}^*_{(-1)}, \tag{2.9.16}$$

as expected. The same transformation, as in Eq. (2.9.15), applies to other objective rates of spatial tensors, such as $\overset{\triangledown}{\boldsymbol{\mathcal{E}}}_{(1)}$ and $\overset{\triangle}{\boldsymbol{\mathcal{E}}}_{(-1)}$, or $\overset{\circ}{\mathbf{B}}$ and $\overset{\bullet}{\mathbf{B}}$. Furthermore,

$$\overset{\bullet}{\boldsymbol{\mathcal{E}}}{}^*_{(n)} = \mathbf{Q} \cdot \overset{\bullet}{\boldsymbol{\mathcal{E}}}_{(n)} \cdot \mathbf{Q}^T, \tag{2.9.17}$$

where $\overset{\bullet}{\boldsymbol{\mathcal{E}}}_{(n)}$ is defined in Eq. (2.6.19).

In summary, while objective material tensors remain unchanged by the rotation of the deformed configuration, e.g., Eqs. (2.9.4) and (2.9.12), the objective spatial tensors change according to transformation rules specified by equations such as (2.9.5) and (2.9.10).

References

Biot, M. A. (1965), *Mechanics of Incremental Deformations*, John Wiley, New York.

Chadwick, P. (1976), *Continuum Mechanics, Concise Theory and Problems*, George Allen and Unwin, London.

Chung, T. J. (1996), *Applied Continuum Mechanics*, Cambridge University Press, Cambridge.

Dui, G.-S., Ren, Q.-W., and Shen, Z.-J. (1999), Time rates of Hill's strain tensors, *J. Elasticity*, Vol. 54, pp. 129–140.

Eringen, A. C. (1967), *Mechanics of Continua*, John Wiley, New York.

Fitzgerald, J. E. (1980), Tensorial Hencky measure of strain and strain rate for finite deformations, *J. Appl. Phys.*, Vol. 51, pp. 5111–5115.

Hill, R. (1968), On constitutive inequalities for simple materials–I, *J. Mech. Phys. Solids*, Vol. 16, pp. 229–242.

Hill, R. (1970), Constitutive inequalities for isotropic elastic solids under finite strain, *Proc. Roy. Soc. London A*, Vol. 314, pp. 457–472.

Hill, R. (1978), Aspects of invariance in solid mechanics, *Adv. Appl. Mech.*, Vol. 18, pp. 1–75.

Hoger, A. (1986), The material time derivative of logarithmic strain, *Int. J. Solids Struct.*, Vol. 22, pp. 1019–1032.

Hoger, A. and Carlson, D. E. (1984), Determination of the stretch and rotation in the polar decomposition of the deformation gradient, *Quart. Appl. Math.*, Vol. 42, pp. 113–117.

Hunter, S. C. (1983), *Mechanics of Continuous Media*, Ellis Horwood, Chichester, England.

Jaunzemis, W. (1967), *Continuum Mechanics*, The Macmillan, New York.

Malvern, L. E. (1969), *Introduction to the Mechanics of a Continuous Medium*, Prentice-Hall, Englewood Cliffs, New Jersey.

Marsden, J. E. and Hughes, T. J. R. (1983), *Mathematical Foundations of Elasticity*, Prentice Hall, Englewood Cliffs, New Jersey.

Mehrabadi, M. M. and Nemat-Nasser, S. (1987), Some basic kinematical relations for finite deformations of continua, *Mech. Mater.*, Vol. 6, pp. 127–138.

Ogden, R. W. (1984), *Non-Linear Elastic Deformations*, Ellis Horwood Ltd., Chichester, England (2nd ed., Dover, 1997).

Prager, W. (1961), *Introduction of Mechanics of Continua*, Ginn and Company, Boston.

Reinhardt, W. D. and Dubey, R. N. (1996), Application of objective rates in mechanical modeling of solids, *J. Appl. Mech.*, Vol. 118, pp. 692–698.

Sedov, L. I. (1966), *Foundations of the Non-Linear Mechanics of Continua*, Pergamon Press, Oxford.

Seth, B. R. (1964), Generalized strain measure with applications to physical problems, in *Second-Order Effects in Elasticity, Plasticity and Fluid*

Dynamics (Haifa 1962), eds. M. Reiner and D. Abir, pp. 162–172, Pergamon Press, Oxford.

Seth, B. R. (1966), Generalized strain and transition concepts for elastic-plastic deformation – creep and relaxation, in *Applied Mechanics: Proc. 11th Int. Congr. Appl. Mech. (Munich 1964)*, eds. H. Görtler and P. Sorger, pp. 383–389, Springer-Verlag, Berlin.

Simo, J. C. and Hughes, T. J. R. (1998), *Computational Inelasticity*, Springer-Verlag, New York.

Spencer, A. J. M. (1992), *Continuum Mechanics*, Longman Scientific & Technical, London.

Truesdell, C. and Noll, W. (1965), The nonlinear field theories of mechanics, in *Handbuch der Physik*, ed. S. Flügge, Band III/3, Springer-Verlag, Berlin (2nd ed., 1992).

CHAPTER 3

KINETICS OF DEFORMATION

3.1. Cauchy Stress

Consider an internal surface S within a loaded deformable body. If the resultant force across an infinitesimal surface element $\mathrm{d}S$ with unit normal $\mathbf{n}$ is $\mathrm{d}\mathbf{f}_n$, the corresponding traction vector is (Fig. 3.1)

$$\mathbf{t}_n = \frac{\mathrm{d}\mathbf{f}_n}{\mathrm{d}S}. \tag{3.1.1}$$

The Cauchy or true stress is the second-order tensor $\boldsymbol{\sigma}$ related to the traction vector $\mathbf{t}_n$ by

$$\mathbf{t}_n = \mathbf{n} \cdot \boldsymbol{\sigma}. \tag{3.1.2}$$

When $\boldsymbol{\sigma}$ is decomposed on an orthonormal basis in the deformed configuration as

$$\boldsymbol{\sigma} = \sigma_{ij}\mathbf{e}_i \otimes \mathbf{e}_j, \tag{3.1.3}$$

the traction vector over the area with the normal in the coordinate direction $\mathbf{e}_i$ can be written as

$$\mathbf{t}_i = \mathbf{e}_i \cdot \boldsymbol{\sigma} = \sigma_{ij}\mathbf{e}_j. \tag{3.1.4}$$

From Eqs. (3.1.2) and (3.1.4) we conclude that the traction vector over the surface element with unit normal $\mathbf{n} = n_i\mathbf{e}_i$ can be expressed in terms of the traction vectors $\mathbf{t}_i$ as

$$\mathbf{t}_n = n_i\mathbf{t}_i. \tag{3.1.5}$$

Equation (3.1.5), known as the Cauchy relation, can also be derived directly by applying the balance law of linear momentum to an infinitesimal tetrahedron around a point of the stressed body (e.g., Prager, 1961; Fung, 1965). In Section 3.3 it will be shown that the Cauchy stress $\boldsymbol{\sigma}$ is a symmetric tensor, provided that there are no distributed surface or body couples acting within the body.

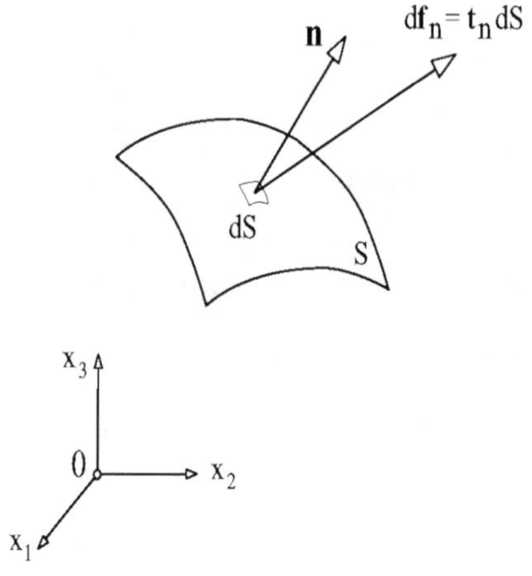

FIGURE 3.1. The traction vector $\mathbf{t}_n$ over the surface element with outward normal $\mathbf{n}$. The total force over dS is $\mathbf{df}_n = \mathbf{t}_n \mathrm{d}S$.

A spherical part of the Cauchy stress is equal to $(\mathrm{tr}\,\boldsymbol{\sigma})\mathbf{I}/3$. The remainder is the deviatoric part,

$$\boldsymbol{\sigma}' = \boldsymbol{\sigma} - \frac{1}{3}(\mathrm{tr}\,\boldsymbol{\sigma})\mathbf{I}. \tag{3.1.6}$$

Since $\boldsymbol{\sigma}'$ is a traceless tensor $(\mathrm{tr}\,\boldsymbol{\sigma}' = 0)$, there are in general only two nonvanishing invariants of $\boldsymbol{\sigma}'$. These are, from Eqs. (1.3.4) and (1.3.5),

$$J_2 = \frac{1}{2}\,\mathrm{tr}\,(\boldsymbol{\sigma}'^2), \quad J_3 = \frac{1}{3}\,\mathrm{tr}\,(\boldsymbol{\sigma}'^3). \tag{3.1.7}$$

If I_1, I_2 and I_3 are the invariants of $\boldsymbol{\sigma}$, we have the relationships

$$J_2 = I_2 + \frac{1}{3}\,I_1^2, \quad J_3 = I_3 + \frac{1}{3}\,I_1 I_2 + \frac{2}{27}\,I_1^3. \tag{3.1.8}$$

Physically, J_2 can be related to shear stress on the octahedral plane ($n_i = \pm 1/\sqrt{3}$ with respect to principal stress directions), since $J_2 = (3/2)\tau_{\mathrm{oct}}^2$. The octahedral planes are shown in Fig. 3.2. The normal stress on the octahedral plane is $\sigma_{\mathrm{oct}} = I_1/3$. In two-dimensional plane stress problems, the third invariant of the stress tensor $I_3 = 0$, so that in three-dimensional problems I_3 can be viewed as a measure of the stress state triaxiality. For later use, it is also noted that

$$\frac{\partial J_2}{\partial \boldsymbol{\sigma}} = \boldsymbol{\sigma}', \quad \frac{\partial J_3}{\partial \boldsymbol{\sigma}} = \boldsymbol{\sigma}'^2 - \frac{2}{3}\,J_2\,\mathbf{I}, \quad \frac{\partial \boldsymbol{\sigma}'}{\partial \boldsymbol{\sigma}} = \boldsymbol{I} - \frac{1}{3}\,\mathbf{I} \otimes \mathbf{I}, \tag{3.1.9}$$

where $\mathbf{I}$ is the second-order, and $\boldsymbol{I}$ is the fourth-order unit tensor.

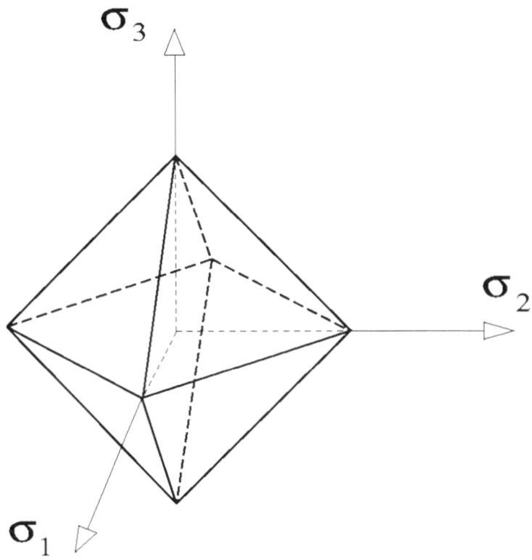

FIGURE 3.2. Octahedral planes in the coordinate system of principal stresses.

3.2. Continuity Equation

If $\rho = \rho(\mathbf{x}, t)$ is a continuous mass density function, the conservation of mass requires that $\mathrm{d}m = \rho\, \mathrm{d}V$ is constant during the deformation process. Since $\mathrm{d}V = (\det \mathbf{F})\, \mathrm{d}V^0$, this implies that

$$\rho\, (\det \mathbf{F}) = \text{const.} \tag{3.2.1}$$

By differentiating we obtain the continuity equation

$$\frac{\mathrm{d}\rho}{\mathrm{d}t} + \rho\, (\boldsymbol{\nabla} \cdot \mathbf{v}) = 0, \tag{3.2.2}$$

where $\mathbf{v}$ is the velocity of the particle in the position $\mathbf{x}$ at time t. Recall from Eq. (2.4.12) that the time rate

$$(\det \mathbf{F})^{\cdot} = (\det \mathbf{F})(\boldsymbol{\nabla} \cdot \mathbf{v}). \tag{3.2.3}$$

In view of Eq. (2.1.4) for the total time rate of a spatial field, Eq. (3.2.2) can be rewritten as

$$\frac{\partial \rho}{\partial t} + \boldsymbol{\nabla} \cdot (\rho\, \mathbf{v}) = 0. \tag{3.2.4}$$

If the deformation process is volume preserving (isochoric), so that $\det \mathbf{F} = 1$ and $\rho =$ const., the continuity equation reduces to

$$\boldsymbol{\nabla} \cdot \mathbf{v} = 0, \tag{3.2.5}$$

i.e., the velocity field is a divergence free vector field.

The Reynolds transport theorem states that for any continuously differentiable tensor field $\mathbf{T} = \mathbf{T}(\mathbf{x}, t)$ within the volume V bounded by surface S,

$$\frac{\mathrm{d}}{\mathrm{d}t}\int_V \rho\,\mathbf{T}\,\mathrm{d}V = \int_V \frac{\partial}{\partial t}(\rho\,\mathbf{T})\,\mathrm{d}V + \int_S \rho\,\mathbf{T}(\mathbf{v}\cdot\mathbf{n})\,\mathrm{d}S, \tag{3.2.6}$$

where ρ is the mass density, and $\mathbf{n}$ is the unit normal to S (e.g., Malvern, 1969; Gurtin, 1981). By applying the Gauss theorem, Eq. (1.13.16), to convert the surface integral in Eq. (3.2.6) to volume integral, and having in mind the continuity equation (3.2.4), there follows

$$\frac{\mathrm{d}}{\mathrm{d}t}\int_V \rho\,\mathbf{T}\,\mathrm{d}V = \int_V \rho\,\frac{\mathrm{d}\mathbf{T}}{\mathrm{d}t}\,\mathrm{d}V. \tag{3.2.7}$$

This important formula of continuum mechanics will be frequently utilized in subsequent derivations. For example, by taking $\mathbf{T}$ to be ρ^{-1}, and by using (3.2.2), Eq. (3.2.7) gives

$$\frac{\mathrm{d}}{\mathrm{d}t}\int_V \mathrm{d}V = \int_V (\boldsymbol{\nabla}\cdot\mathbf{v})\,\mathrm{d}V. \tag{3.2.8}$$

3.3. Equations of Motion

Consider an arbitrary portion of a continuous body in the deformed configuration. Denote its volume by V and its bounding surface by S (Fig. 3.3). The rate of change of the linear mass momentum within V is equal to the sum of all surface forces acting on S and all body forces acting in V (first Euler's law of motion), i.e.,

$$\int_S \mathbf{t}_n\,\mathrm{d}S + \int_V \rho\,\mathbf{b}\,\mathrm{d}V = \frac{\mathrm{d}}{\mathrm{d}t}\int_V \rho\,\mathbf{v}\,\mathrm{d}V. \tag{3.3.1}$$

The body force per unit mass is

$$\mathbf{b} = \frac{\mathrm{d}\mathbf{f}_b}{\mathrm{d}m}, \tag{3.3.2}$$

and $\mathbf{v} = \mathbf{v}(\mathbf{x}, t)$ is the velocity field. Applying the Gauss theorem to convert the surface into volume integral, and incorporating Eq. (3.2.7) in the right-hand side of Eq. (3.3.1), we obtain

$$\int_V \left(\boldsymbol{\nabla}\cdot\boldsymbol{\sigma} + \rho\,\mathbf{b} - \rho\,\frac{\mathrm{d}\mathbf{v}}{\mathrm{d}t}\right)\mathrm{d}V = 0. \tag{3.3.3}$$

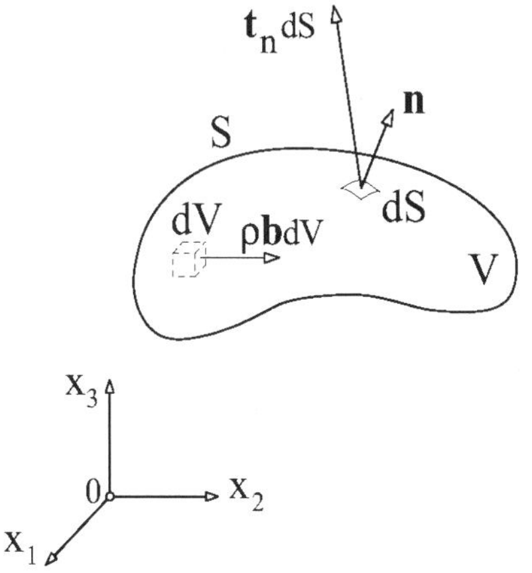

FIGURE 3.3. The volume V of the body bounded by closed surface S. The body force per unit mass is $\mathbf{b}$ and the surface traction over S is $\mathbf{t}_n$.

Since this holds for an arbitrary volume V, the integrand must vanish at each point of the deforming body,

$$\boldsymbol{\nabla}\cdot\boldsymbol{\sigma}+\rho\,\mathbf{b}=\rho\,\frac{\mathrm{d}\mathbf{v}}{\mathrm{d}t}\,. \tag{3.3.4}$$

These are the Cauchy equations of motion for continuous media that apply at any point $\mathbf{x}$ in the deformed configuration. Equilibrium equations are obtained by setting the acceleration $\mathrm{d}\mathbf{v}/\mathrm{d}t$ equal to zero.

The transition to corresponding equations at points $\mathbf{X}$ in the undeformed configuration is straightforward. We only need to multiply Eq. (3.3.4) with $(\det\mathbf{F})$. Since

$$\rho(\det\mathbf{F})=\rho^0 \tag{3.3.5}$$

is the density in the undeformed configuration, $\rho^0=\rho^0(\mathbf{X})$, and since

$$(\det\mathbf{F})\boldsymbol{\nabla}\cdot\boldsymbol{\sigma}=\boldsymbol{\nabla}^0\cdot\left(\mathbf{F}^{-1}\cdot\boldsymbol{\tau}\right), \tag{3.3.6}$$

where

$$\boldsymbol{\tau}=(\det\mathbf{F})\boldsymbol{\sigma} \tag{3.3.7}$$

is the Kirchhoff stress, Eq. (3.3.4) becomes

$$\boldsymbol{\nabla}^0\cdot\left(\mathbf{F}^{-1}\cdot\boldsymbol{\tau}\right)+\rho^0\,\mathbf{b}=\rho^0\,\frac{\mathrm{d}\mathbf{v}}{\mathrm{d}t}\,. \tag{3.3.8}$$

The stress tensor

$$\mathbf{P} = \mathbf{F}^{-1} \cdot \boldsymbol{\tau} \tag{3.3.9}$$

is a nonsymmetric nominal stress, and will be considered in more detail later in Section 3.7.

It is left to prove the identity in Eq. (3.3.6). First, by Eq. (1.13.14),

$$\boldsymbol{\nabla}^0 \cdot \left[(\det \mathbf{F})\mathbf{F}^{-1} \cdot \boldsymbol{\sigma}\right] = \left\{\boldsymbol{\nabla}^0 \cdot \left[(\det \mathbf{F})\mathbf{F}^{-1}\right]\right\} \cdot \boldsymbol{\sigma} + \left[(\det \mathbf{F})\mathbf{F}^{-T} \cdot \boldsymbol{\nabla}^0\right] \cdot \boldsymbol{\sigma}. \tag{3.3.10}$$

The first term on the right-hand side is equal to zero, in view of Eq. (2.2.21). Equation (3.3.8) follows because $\mathbf{F}^{-T} \cdot \boldsymbol{\nabla}^0 = \boldsymbol{\nabla}$, by Eq. (2.4.5).

3.4. Symmetry of Cauchy Stress

The balance law of angular momentum requires that the Cauchy stress is symmetric, if there are no distributed surface or body couples acting on the body (nonpolar case). This is now proven. The rate of change of angular momentum of the mass within V is equal to the sum of the moments of all forces acting on V and S (second Euler's law of motion), i.e.,

$$\int_S (\mathbf{x} \times \mathbf{t}_n)\, \mathrm{d}S + \int_V (\mathbf{x} \times \rho\, \mathbf{b})\, \mathrm{d}V = \frac{\mathrm{d}}{\mathrm{d}t} \int_V (\mathbf{x} \times \rho\, \mathbf{v})\, \mathrm{d}V. \tag{3.4.1}$$

Applying the Gauss theorem to convert the surface integral into the volume integral, we obtain

$$\int_S (\mathbf{x} \times \mathbf{t}_n)\, \mathrm{d}S = -\int_V \boldsymbol{\nabla} \cdot (\boldsymbol{\sigma} \times \mathbf{x})\, \mathrm{d}V. \tag{3.4.2}$$

The integrand on the right-hand side can be expanded as

$$\boldsymbol{\nabla} \cdot (\boldsymbol{\sigma} \times \mathbf{x}) = (\boldsymbol{\nabla} \cdot \boldsymbol{\sigma}) \times \mathbf{x} - \boldsymbol{\epsilon} : \boldsymbol{\sigma}, \tag{3.4.3}$$

where $\boldsymbol{\epsilon}$ is the permutation tensor, and : designates the trace product; see Eq. (1.13.15). Thus, Eq. (3.4.1) becomes

$$\int_V \mathbf{x} \times \left(\boldsymbol{\nabla} \cdot \boldsymbol{\sigma} + \rho\, \mathbf{b} - \rho \frac{\mathrm{d}\mathbf{v}}{\mathrm{d}t}\right) \mathrm{d}V + \int_V (\boldsymbol{\epsilon} : \boldsymbol{\sigma})\, \mathrm{d}V = 0. \tag{3.4.4}$$

The integrand of the first integral in Eq. (3.4.4) vanishes by equations of motion (3.3.4). The second integral has to vanish for all choices of V (the whole body or any part of it), hence

$$\boldsymbol{\epsilon} : \boldsymbol{\sigma} = \mathbf{0} \tag{3.4.5}$$

at each point of the body. Since the permutation tensor $\boldsymbol{\epsilon}$ is antisymmetric with respect to its last two indices, Eq. (3.4.5) requires the Cauchy stress $\boldsymbol{\sigma}$ to be symmetric,

$$\boldsymbol{\sigma} = \boldsymbol{\sigma}^T. \tag{3.4.6}$$

3.5. Stress Power

The rate at which external surface and body forces are doing work on the mass instantaneously occupying the volume V bounded by S is the power input

$$\mathcal{P} = \int_S \mathbf{t}_n \cdot \mathbf{v}\, \mathrm{d}S + \int_V \rho\, \mathbf{b} \cdot \mathbf{v}\, \mathrm{d}V. \tag{3.5.1}$$

Converting the surface integral into the volume integral, this becomes

$$\mathcal{P} = \int_V \left[(\boldsymbol{\nabla} \cdot \boldsymbol{\sigma} + \rho\, \mathbf{b}) \cdot \mathbf{v} + \boldsymbol{\sigma} : \mathbf{D} \right] \mathrm{d}V. \tag{3.5.2}$$

The formula (1.13.13) was used, giving

$$\boldsymbol{\nabla} \cdot (\boldsymbol{\sigma} \cdot \mathbf{v}) = (\boldsymbol{\nabla} \cdot \boldsymbol{\sigma}) \cdot \mathbf{v} + \boldsymbol{\sigma} : \mathbf{L}^T. \tag{3.5.3}$$

The symmetry of the Cauchy stress makes

$$\boldsymbol{\sigma} : \mathbf{L}^T = \boldsymbol{\sigma} : \mathbf{D}. \tag{3.5.4}$$

The deformation gradient is

$$\mathbf{L} = \mathbf{v} \otimes \boldsymbol{\nabla}, \tag{3.5.5}$$

and its symmetric part $\mathbf{D}$ is the rate of deformation tensor. Using the Cauchy equations of motion (3.3.4) and Eq. (3.2.7), the rate at which external forces do work is, from Eq. (3.5.2),

$$\mathcal{P} = \frac{\mathrm{d}}{\mathrm{d}t} \int_V \frac{1}{2} \rho\, \mathbf{v} \cdot \mathbf{v}\, \mathrm{d}V + \int_V \boldsymbol{\sigma} : \mathbf{D}\, \mathrm{d}V. \tag{3.5.6}$$

The first term represents the rate of macroscopic kinetic energy of the total mass. The second term is the total stress power expended at the considered instant to deform the material. This contributes to internal energy of the material, and, depending on the nature of deformation, part of it may be dissipated in the form of heat. The scalar quantity $\boldsymbol{\sigma} : \mathbf{D}$ is called the stress power per unit current volume. If it is reckoned with respect to unit initial volume, it becomes $\boldsymbol{\tau} : \mathbf{D}$.

3.6. Conjugate Stress Tensors

3.6.1. Material Stress Tensors

A systematic construction of stress tensors as work conjugates to strain tensors was introduced by Hill (1968). For any material strain $\mathbf{E}_{(n)}$ of Eq. (2.3.1), its work conjugate stress $\mathbf{T}_{(n)}$ is defined such that the stress power per unit reference volume is

$$\mathbf{T}_{(n)} : \dot{\mathbf{E}}_{(n)} = \boldsymbol{\tau} : \mathbf{D}, \tag{3.6.1}$$

where $\boldsymbol{\tau} = (\det \mathbf{F})\boldsymbol{\sigma}$ is the Kirchhoff stress. For $n = 1$, Eq. (3.6.1) gives

$$\mathbf{T}_{(1)} = \mathbf{F}^{-1} \cdot \boldsymbol{\tau} \cdot \mathbf{F}^{-T} = \mathbf{U}^{-1} \cdot \hat{\boldsymbol{\tau}} \cdot \mathbf{U}^{-1} \quad \Leftrightarrow \quad \mathbf{E}_{(1)} = \frac{1}{2}\left(\mathbf{U}^2 - \mathbf{I}^0\right). \tag{3.6.2}$$

For $n = 1/2$ it follows that

$$\mathbf{T}_{(1/2)} = \frac{1}{2}\left(\mathbf{U}^{-1} \cdot \hat{\boldsymbol{\tau}} + \hat{\boldsymbol{\tau}} \cdot \mathbf{U}^{-1}\right) \quad \Leftrightarrow \quad \mathbf{E}_{(1/2)} = \mathbf{U} - \mathbf{I}^0. \tag{3.6.3}$$

The symbol $\Leftrightarrow$ stands for "conjugate to" and the stress

$$\hat{\boldsymbol{\tau}} = \mathbf{R}^T \cdot \boldsymbol{\tau} \cdot \mathbf{R} \tag{3.6.4}$$

is induced from $\boldsymbol{\tau}$ by the rotation $\mathbf{R}$. Similarly,

$$\mathbf{T}_{(-1)} = \mathbf{F}^T \cdot \boldsymbol{\tau} \cdot \mathbf{F} = \mathbf{U} \cdot \hat{\boldsymbol{\tau}} \cdot \mathbf{U} \quad \Leftrightarrow \quad \mathbf{E}_{(-1)} = \frac{1}{2}\left(\mathbf{I}^0 - \mathbf{U}^{-2}\right), \tag{3.6.5}$$

$$\mathbf{T}_{(-1/2)} = \frac{1}{2}\left(\mathbf{U} \cdot \hat{\boldsymbol{\tau}} + \hat{\boldsymbol{\tau}} \cdot \mathbf{U}\right) \quad \Leftrightarrow \quad \mathbf{E}_{(-1/2)} = \mathbf{I}^0 - \mathbf{U}^{-1}. \tag{3.6.6}$$

In view of Eq. (2.6.5), there is a general relationship

$$\mathbf{T}_{(-n)} = \mathbf{U}^{2n} \cdot \mathbf{T}_{(n)} \cdot \mathbf{U}^{2n}. \tag{3.6.7}$$

Furthermore, for positive n we have

$$\begin{aligned}\dot{\mathbf{E}}_{(n)} = \frac{1}{2n}\Big(&\dot{\mathbf{U}} \cdot \mathbf{U}^{2n-1} + \mathbf{U} \cdot \dot{\mathbf{U}} \cdot \mathbf{U}^{2n-2} + \cdots \\ &+ \mathbf{U}^{2n-2} \cdot \dot{\mathbf{U}} \cdot \mathbf{U} + \mathbf{U}^{2n-1} \cdot \dot{\mathbf{U}}\Big).\end{aligned} \tag{3.6.8}$$

Thus, since

$$\mathbf{T}_{(n)} : \dot{\mathbf{E}}_{(n)} = \mathbf{T}_{(1/2)} : \dot{\mathbf{E}}_{(1/2)}, \tag{3.6.9}$$

it follows that (Ogden, 1984)

$$\begin{aligned}\mathbf{T}_{(1/2)} = \frac{1}{2n}\big(&\mathbf{U}^{2n-1} \cdot \mathbf{T}_{(n)} + \mathbf{U}^{2n-2} \cdot \mathbf{T}_{(n)} \cdot \mathbf{U} + \cdots \\ &+ \mathbf{U} \cdot \mathbf{T}_{(n)} \cdot \mathbf{U}^{2n-2} + \mathbf{T}_{(n)} \cdot \mathbf{U}^{2n-1}\big), \quad n > 0.\end{aligned} \tag{3.6.10}$$

Similarly,

$$\begin{aligned}\mathbf{T}_{(-1/2)} = \frac{1}{2n}\big(&\mathbf{U}^{1-2n}\cdot\mathbf{T}_{(-n)} + \mathbf{U}^{2-2n}\cdot\mathbf{T}_{(-n)}\cdot\mathbf{U}^{-1} + \cdots \\ &+ \mathbf{U}^{-1}\cdot\mathbf{T}_{(-n)}\cdot\mathbf{U}^{2-2n} + \mathbf{T}_{(-n)}\cdot\mathbf{U}^{1-2n}\big), \quad n > 0.\end{aligned} \tag{3.6.11}$$

If $\mathbf{T}_{(n)}$ and $\mathbf{U}$ are commutative,

$$\mathbf{T}_{(1/2)} = \mathbf{U}^{2n-1}\cdot\mathbf{T}_{(n)}, \quad \mathbf{T}_{(-1/2)} = \mathbf{U}^{1-2n}\cdot\mathbf{T}_{(-n)}. \tag{3.6.12}$$

A derivation of an explicit expression for the stress tensor conjugate to logarithmic strain $\mathbf{E}_{(0)}$ is more involved. The approximate expression can be obtained as follows. From Eq. (2.3.12), by differentiation,

$$\begin{aligned}\dot{\mathbf{E}}_{(n)} &= \dot{\mathbf{E}}_{(0)} + 2n\left(\mathbf{E}_{(0)}\cdot\dot{\mathbf{E}}_{(0)} + \dot{\mathbf{E}}_{(0)}\cdot\mathbf{E}_{(0)}\right) \\ &+ \frac{2}{3}n^2\left(\mathbf{E}_{(0)}^2\cdot\dot{\mathbf{E}}_{(0)} + \dot{\mathbf{E}}_{(0)}\cdot\mathbf{E}_{(0)}^2 + \mathbf{E}_{(0)}\cdot\dot{\mathbf{E}}_{(0)}\cdot\mathbf{E}_{(0)}\right) \\ &+ \mathcal{O}\left(\mathbf{E}_{(0)}^3\cdot\dot{\mathbf{E}}_{(0)}\right).\end{aligned} \tag{3.6.13}$$

Substitution of this into

$$\mathbf{T}_{(n)} : \dot{\mathbf{E}}_{(n)} = \mathbf{T}_{(0)} : \dot{\mathbf{E}}_{(0)} \tag{3.6.14}$$

gives

$$\begin{aligned}\mathbf{T}_{(0)} &= \mathbf{T}_{(n)} + n\left(\mathbf{E}_{(n)}\cdot\mathbf{T}_{(n)} + \mathbf{T}_{(n)}\cdot\mathbf{E}_{(n)}\right) \\ &- \frac{1}{3}n^2\left(\mathbf{E}_{(n)}^2\cdot\mathbf{T}_{(n)} + \mathbf{T}_{(n)}\cdot\mathbf{E}_{(n)}^2 - 2\mathbf{E}_{(n)}\cdot\mathbf{T}_{(n)}\cdot\mathbf{E}_{(n)}\right) \\ &+ \mathcal{O}\left(\mathbf{E}_{(n)}^3\cdot\mathbf{T}_{(n)}\right).\end{aligned} \tag{3.6.15}$$

Furthermore, from any of Eqs. (3.6.2)–(3.6.6) for the stress $\mathbf{T}_{(n)}$, it can be shown that

$$\mathbf{T}_{(n)} = \hat{\boldsymbol{\tau}} - n\left(\mathbf{E}_{(n)}\cdot\hat{\boldsymbol{\tau}} + \hat{\boldsymbol{\tau}}\cdot\mathbf{E}_{(n)}\right) + \mathcal{O}\left(\mathbf{E}_{(n)}^2\cdot\hat{\boldsymbol{\tau}}\right). \tag{3.6.16}$$

The substitution into Eq. (3.6.15) then yields

$$\mathbf{T}_{(0)} = \hat{\boldsymbol{\tau}} + \mathcal{O}\left(\mathbf{E}_{(n)}^2\cdot\hat{\boldsymbol{\tau}}\right) \quad \Leftrightarrow \quad \mathbf{E}_{(0)} = \ln\mathbf{U}. \tag{3.6.17}$$

The approximation $\mathbf{T}_{(0)} \approx \hat{\boldsymbol{\tau}}$ may be acceptable at moderate strains (Hill, 1978). If deformation is such that the principal directions of $\mathbf{V}$ and $\boldsymbol{\tau}$ are parallel (as in the deformation of isotropic elastic materials), the matrices $\mathbf{E}_{(n)}$ and $\mathbf{T}_{(n)}$ commute, and the term proportional to n^2 in Eq. (3.6.15) vanishes, as well as all other higher-order terms. In that case, therefore,

$\mathbf{T}_{(0)} = \hat{\boldsymbol{\tau}}$ exactly. Also, if principal directions of $\mathbf{U}$ remain fixed during deformation,

$$\dot{\mathbf{E}}_{(0)} = \dot{\mathbf{U}} \cdot \mathbf{U}^{-1} = \hat{\mathbf{D}}, \quad \mathbf{T}_{(0)} = \hat{\boldsymbol{\tau}}. \tag{3.6.18}$$

Additional analysis can be found in the articles by Hoger (1987), Guo and Man (1992), Lehmann and Liang (1993), Heiduschke (1995), and Xiao (1995).

3.6.2. Spatial Stress Tensors

The spatial strain tensors $\boldsymbol{\mathcal{E}}_{(n)}$ in general do not have their conjugate stress tensors $\boldsymbol{\mathcal{T}}_{(n)}$ such that $\mathbf{T}_{(n)} : \dot{\mathbf{E}}_{(n)} = \boldsymbol{\mathcal{T}}_{(n)} : \dot{\boldsymbol{\mathcal{E}}}_{(n)}$. This is clear at the outset, because a spatial stress tensor should be objective ($\boldsymbol{\mathcal{T}}^*_{(n)} = \mathbf{Q} \cdot \boldsymbol{\mathcal{T}}_{(n)} \cdot \mathbf{Q}^T$). Since $\dot{\boldsymbol{\mathcal{E}}}_{(n)}$ is not objective, as seen from Eq. (2.9.13), their trace product cannot in general be equal to an invariant quantity $\mathbf{T}_{(n)} : \dot{\mathbf{E}}_{(n)}$ (which is independent of the rotation $\mathbf{Q}$ superimposed to the deformed configuration). However, the spatial stress tensors conjugate to strain tensors $\boldsymbol{\mathcal{E}}_{(n)}$ can be introduced by requiring that

$$\mathbf{T}_{(n)} : \dot{\mathbf{E}}_{(n)} = \boldsymbol{\mathcal{T}}_{(n)} : \overset{\bullet}{\boldsymbol{\mathcal{E}}}_{(n)}, \tag{3.6.19}$$

where the objective, corotational rate of strain $\overset{\bullet}{\boldsymbol{\mathcal{E}}}_{(n)}$ is defined by Eq. (2.6.19), i.e.,

$$\overset{\bullet}{\boldsymbol{\mathcal{E}}}_{(n)} = \dot{\boldsymbol{\mathcal{E}}}_{(n)} - \boldsymbol{\omega} \cdot \boldsymbol{\mathcal{E}}_{(n)} + \boldsymbol{\mathcal{E}}_{(n)} \cdot \boldsymbol{\omega}, \quad \boldsymbol{\omega} = \dot{\mathbf{R}} \cdot \mathbf{R}^{-1}. \tag{3.6.20}$$

In view of the relationship

$$\overset{\bullet}{\boldsymbol{\mathcal{E}}}_{(n)} = \mathbf{R} \cdot \dot{\mathbf{E}}_{(n)} \cdot \mathbf{R}^T, \tag{3.6.21}$$

it follows that

$$\boldsymbol{\mathcal{T}}_{(n)} = \mathbf{R} \cdot \mathbf{T}_{(n)} \cdot \mathbf{R}^T. \tag{3.6.22}$$

This is the conjugate stress to spatial strains $\boldsymbol{\mathcal{E}}_{(n)}$ according to Eq. (3.6.19). Therefore, in this sense we consider

$$\boldsymbol{\mathcal{T}}_{(1)} = \mathbf{F}^{-T} \cdot \hat{\boldsymbol{\tau}} \cdot \mathbf{F}^{-1} = \mathbf{V}^{-1} \cdot \boldsymbol{\tau} \cdot \mathbf{V}^{-1} \quad \Leftrightarrow \quad \boldsymbol{\mathcal{E}}_{(1)} = \frac{1}{2}\left(\mathbf{V}^2 - \mathbf{I}\right), \tag{3.6.23}$$

$$\boldsymbol{\mathcal{T}}_{(-1)} = \mathbf{F} \cdot \hat{\boldsymbol{\tau}} \cdot \mathbf{F}^{T} = \mathbf{V} \cdot \boldsymbol{\tau} \cdot \mathbf{V} \quad \Leftrightarrow \quad \boldsymbol{\mathcal{E}}_{(-1)} = \frac{1}{2}\left(\mathbf{I} - \mathbf{V}^{-2}\right), \tag{3.6.24}$$

$$\boldsymbol{\mathcal{T}}_{(1/2)} = \frac{1}{2}\left(\mathbf{V}^{-1} \cdot \boldsymbol{\tau} + \boldsymbol{\tau} \cdot \mathbf{V}^{-1}\right) \quad \Leftrightarrow \quad \boldsymbol{\mathcal{E}}_{(1/2)} = \mathbf{V} - \mathbf{I}, \tag{3.6.25}$$

$$\mathcal{T}_{(-1/2)} = \frac{1}{2}(\mathbf{V}\cdot\boldsymbol{\tau}+\boldsymbol{\tau}\cdot\mathbf{V}) \quad \Leftrightarrow \quad \mathcal{E}_{(-1/2)} = \mathbf{I}-\mathbf{V}^{-1}. \tag{3.6.26}$$

It is easy to derive equations dual to Eqs. (3.6.8)–(3.6.12). For example, if $\mathcal{T}$ and $\mathbf{V}$ are coaxial tensors,

$$\mathcal{T}_{(1/2)} = \mathbf{V}^{2n-1}\cdot\mathcal{T}_{(n)}, \quad \mathcal{T}_{(-1/2)} = \mathbf{V}^{1-2n}\cdot\mathcal{T}_{(-n)}. \tag{3.6.27}$$

If the principal directions of $\mathbf{T}_{(n)}$ and $\mathbf{E}_{(n)}$ are parallel (as in the deformation of elastically isotropic materials), so are the principal directions of $\mathcal{T}_{(n)}$ and $\mathcal{E}_{(n)}$. In this case

$$\mathcal{T}_{(n)} : \overset{\bullet}{\mathcal{E}}_{(n)} = \mathcal{T}_{(n)} : \dot{\mathcal{E}}_{(n)}, \tag{3.6.28}$$

because the tensor $\left(\boldsymbol{\Omega}\cdot\mathcal{E}_{(n)} - \mathcal{E}_{(n)}\cdot\boldsymbol{\Omega}\right)$ is orthogonal to $\mathcal{E}_{(n)}$ and thus to $\mathcal{T}_{(n)}$, so that

$$\mathcal{T}_{(n)} : \left(\boldsymbol{\omega}\cdot\mathcal{E}_{(n)} - \mathcal{E}_{(n)}\cdot\boldsymbol{\omega}\right) = 0. \tag{3.6.29}$$

Note that $\mathbf{R}\cdot\boldsymbol{\tau}\cdot\mathbf{R}^T$ is not the work conjugate to any strain measure, since the material stress tensor $\mathbf{T}_{(n)}$ in Eq. (3.6.22) cannot be equal to spatial stress tensor $\boldsymbol{\tau}$. Likewise, although $\hat{\boldsymbol{\tau}} : \hat{\mathbf{D}} = \boldsymbol{\tau} : \mathbf{D}$, the stress tensor $\hat{\boldsymbol{\tau}} = \mathbf{R}^T\cdot\boldsymbol{\tau}\cdot\mathbf{R}$ is not the work conjugate to any strain measure, because $\hat{\mathbf{D}} = \mathbf{R}^T\cdot\mathbf{D}\cdot\mathbf{R}$ is not the rate of any strain. Of course, $\boldsymbol{\tau}$ itself is not the work conjugate to any strain, because $\mathbf{D}$ is not the rate of any strain, either.

If deformation is uniform extension or compression ($\mathbf{F} = \lambda\mathbf{I}$), it can be shown that

$$\dot{\mathbf{E}}_{(n)} = \lambda^{2n}\mathbf{D}, \quad \dot{\mathbf{E}}_{(0)} = \mathbf{D} = \frac{\dot{\lambda}}{\lambda}\mathbf{I}, \tag{3.6.30}$$

and in this case

$$\mathbf{T}_{(n)} = \lambda^{-2n}\boldsymbol{\tau}, \quad \mathbf{T}_{(0)} = \boldsymbol{\tau}. \tag{3.6.31}$$

3.7. Nominal Stress

If the element of area $\mathrm{d}\mathbf{S} = \mathrm{d}S\mathbf{n}$ in the deformed configuration carries the force $d\mathbf{f}_n$, the corresponding traction vector is $\mathbf{t}_n = \mathrm{d}\mathbf{f}_n/\mathrm{d}S$. It is related to Cauchy stress by $\mathbf{t}_n = \mathbf{n}\cdot\boldsymbol{\sigma}$. Let $\mathrm{d}\mathbf{S}^0 = \mathrm{d}S^0\,\mathbf{n}^0$ be the element of area in the undeformed configuration, corresponding to $\mathrm{d}\mathbf{S}$ in the deformed configuration. The nominal traction vector is defined as the actual force in the

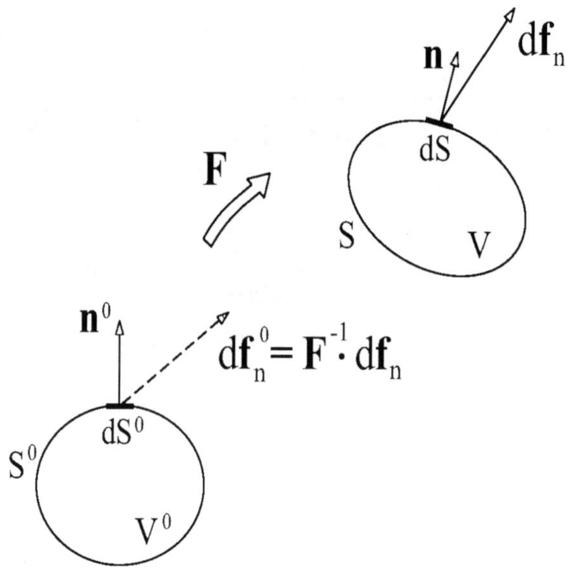

FIGURE 3.4. An infinitesimal surface element $\mathrm{d}S$ with unit normal $\mathbf{n}$ in deformed configuration carries the force $\mathbf{df}_n$. The nominal traction vector with respect to undeformed configuration is $\mathbf{df}_n/\mathrm{d}S^0$. A pseudo-force vector is $\mathbf{df}_n^0$.

deformed configuration divided by the area in the undeformed configuration, i.e.,

$$\mathbf{p}_n = \frac{\mathbf{df}_n}{\mathrm{d}S^0}, \tag{3.7.1}$$

so that (Fig. 3.4)

$$\mathbf{p}_n \, \mathrm{d}S^0 = \mathbf{t}_n \, \mathrm{d}S. \tag{3.7.2}$$

The nominal stress tensor $\mathbf{P}$ is introduced by

$$\mathbf{p}_n = \mathbf{n}^0 \cdot \mathbf{P}. \tag{3.7.3}$$

In view of Nanson's relation (2.2.17), it follows that

$$\mathbf{P} = \mathbf{F}^{-1} \cdot \boldsymbol{\tau}. \tag{3.7.4}$$

The nominal stress is a nonsymmetric two-point tensor. Its transpose

$$\mathbf{P}^T = \boldsymbol{\tau} \cdot \mathbf{F}^{-T} \tag{3.7.5}$$

is often referred to as the first or nonsymmetric Piola–Kirchhoff stress tensor, $\mathbf{n}^0 \cdot \mathbf{P} = \mathbf{P}^T \cdot \mathbf{n}^0$; Truesdell and Noll (1965).

Observe that the rate of work can be expressed in terms of the nominal stress as

$$\mathbf{P} \cdot\cdot \dot{\mathbf{F}} = \boldsymbol{\tau} : \mathbf{D}. \tag{3.7.6}$$

This, in turn, can serve as a starting point to define $\mathbf{P}$, since

$$\mathbf{P} \cdot\cdot \dot{\mathbf{F}} = (\mathbf{F} \cdot \mathbf{P}) \cdot\cdot \left(\dot{\mathbf{F}} \cdot \mathbf{F}^{-1}\right), \quad \mathbf{F} \cdot \mathbf{P} = \boldsymbol{\tau}. \tag{3.7.7}$$

The balance law of linear momentum can be written with respect to undeformed geometry as

$$\int_{S^0} \mathbf{p}_n \, \mathrm{d}S^0 + \int_{V^0} \rho^0 \, \mathbf{b} \, \mathrm{d}V^0 = \frac{\mathrm{d}}{\mathrm{d}t} \int_{V^0} \rho^0 \, \mathbf{v} \, \mathrm{d}V^0, \tag{3.7.8}$$

which, in view of Eq. (3.7.3) and the Gauss theorem, reproduces the equations of motion (3.3.8), written at points of the undeformed configuration.

3.7.1. Piola–Kirchhoff Stress

The second or symmetric Piola–Kirchhoff stress tensor is the stress tensor $\mathbf{T}_{(1)}$, introduced previously as the work conjugate to the Lagrangian strain $\mathbf{E}_{(1)}$. An alternative construction of this stress tensor is as follows. A pseudo-force vector $\mathrm{d}\mathbf{f}_n^0$ in the undeformed configuration is introduced such that

$$\mathrm{d}\mathbf{f}_n = \mathbf{F} \cdot \mathrm{d}\mathbf{f}_n^0. \tag{3.7.9}$$

The associated pseudo-traction is (Fig. 3.4)

$$\mathbf{t}_n^0 = \frac{\mathrm{d}\mathbf{f}_n^0}{\mathrm{d}S^0}. \tag{3.7.10}$$

The second Piola–Kirchhoff stress tensor satisfies

$$\mathbf{t}_n^0 = \mathbf{n}^0 \cdot \mathbf{T}_{(1)}. \tag{3.7.11}$$

This gives

$$\mathbf{T}_{(1)} = \mathbf{F}^{-1} \cdot \boldsymbol{\tau} \cdot \mathbf{F}^{-T}, \tag{3.7.12}$$

which is symmetric whenever $\boldsymbol{\tau}$ is symmetric (nonpolar case). The connection with the nominal stress is

$$\mathbf{P} = \mathbf{T}_{(1)} \cdot \mathbf{F}^T, \tag{3.7.13}$$

so that $\mathbf{F} \cdot \mathbf{P}$ is symmetric. It is also noted that

$$\mathbf{T}_{(1/2)} = (\mathbf{P} \cdot \mathbf{R})_s, \tag{3.7.14}$$

which is referred to as the Biot stress (Biot, 1965; Hill, 1968; Ogden, 1984).

Returning to expression (3.7.6) for the rate of work, it is noted that in the case when there is momentarily no rate of stretching tensor $\mathbf{U}$ (from the polar decomposition of deformation gradient $\mathbf{F} = \mathbf{R} \cdot \mathbf{U}$), i.e., when $\dot{\mathbf{U}} = \mathbf{0}$, we have

$$\mathbf{P} \cdot\cdot\, \dot{\mathbf{F}} = \mathbf{P} \cdot\cdot\, \dot{\mathbf{R}} \cdot \mathbf{U} = \mathbf{T}_{(1)} \cdot \mathbf{F}^T \cdot\cdot\, \dot{\mathbf{R}} \cdot \mathbf{U} = (\mathbf{F} \cdot \mathbf{T}_{(1)} \cdot \mathbf{F}^T) : (\dot{\mathbf{R}} \cdot \mathbf{R}^{-1}) = 0. \quad (3.7.15)$$

The last trace product vanishes because $\mathbf{F} \cdot \mathbf{T}_{(1)} \cdot \mathbf{F}^T$ is symmetric, while $\dot{\mathbf{R}} \cdot \mathbf{R}^{-1}$ is antisymmetric tensor. The result was expected because there can not be any rate of work associated with instantaneous rigid-body spin of an already stretched body.

3.8. Stress Rates

The material stress tensors $\mathbf{T}_{(n)}$ can be decomposed in four different ways on the primary and reciprocal bases in the undeformed configuration. Likewise, the spatial stress tensors $\boldsymbol{\mathcal{T}}_{(n)}$ can be decomposed on the bases in the deformed configuration. For example, the contravariant decompositions are

$$\mathbf{T}_{(n)} = T_{(n)}^{IJ} \mathbf{e}_I^0 \otimes \mathbf{e}_J^0, \quad \boldsymbol{\mathcal{T}}_{(n)} = \mathcal{T}_{(n)}^{ij} \mathbf{e}_i \otimes \mathbf{e}_j. \quad (3.8.1)$$

Since

$$\dot{\mathbf{e}}_i = \mathbf{L} \cdot \mathbf{e}_i, \quad \dot{\mathbf{e}}^i = -\mathbf{L}^T \cdot \mathbf{e}^i, \quad (3.8.2)$$

by Eq. (2.5.1), there are four types of the convected derivatives of the spatial stress tensors. They are given by Eqs. (2.5.8)–(2.5.11), if $\mathbf{A}$ is there replaced by $\boldsymbol{\mathcal{T}}_{(n)}$. In view of Eq. (3.6.22), there is a connection between the rates of material and spatial stress tensors,

$$\overset{\bullet}{\boldsymbol{\mathcal{T}}}_{(n)} = \mathbf{R} \cdot \dot{\mathbf{T}}_{(n)} \cdot \mathbf{R}^T, \quad \overset{\bullet}{\boldsymbol{\mathcal{T}}}_{(n)} = \dot{\boldsymbol{\mathcal{T}}}_{(n)} - \boldsymbol{\omega} \cdot \boldsymbol{\mathcal{T}}_{(n)} + \boldsymbol{\mathcal{T}}_{(n)} \cdot \boldsymbol{\omega}. \quad (3.8.3)$$

Here, $\boldsymbol{\omega} = \dot{\mathbf{R}} \cdot \mathbf{R}^{-1}$ is the spin due to $\mathbf{R}$.

The rates of material stress tensors $\mathbf{T}_{(1)}$ and $\mathbf{T}_{(-1)}$ are related to convected rates of the Kirchhoff stress by

$$\dot{\mathbf{T}}_{(1)} = \mathbf{F}^{-1} \cdot \overset{\triangle}{\boldsymbol{\tau}} \cdot \mathbf{F}^{-T}, \quad \overset{\triangle}{\boldsymbol{\tau}} = \dot{\boldsymbol{\tau}} - \mathbf{L} \cdot \boldsymbol{\tau} - \boldsymbol{\tau} \cdot \mathbf{L}^T, \quad (3.8.4)$$

$$\dot{\mathbf{T}}_{(-1)} = \mathbf{F}^T \cdot \overset{\triangledown}{\boldsymbol{\tau}} \cdot \mathbf{F}, \quad \overset{\triangledown}{\boldsymbol{\tau}} = \dot{\boldsymbol{\tau}} + \mathbf{L}^T \cdot \boldsymbol{\tau} + \boldsymbol{\tau} \cdot \mathbf{L}. \quad (3.8.5)$$

The rate of stress conjugate to logarithmic strain is obtained from Eq. (3.6.15) by differentiation, and is given by

$$\dot{\mathbf{T}}_{(0)} = \dot{\mathbf{T}}_{(n)} + n\left(\dot{\mathbf{E}}_{(n)} \cdot \mathbf{T}_{(n)} + \mathbf{T}_{(n)} \cdot \dot{\mathbf{E}}_{(n)}\right) + \mathcal{O}\left(\mathbf{E}_{(n)}\right). \tag{3.8.6}$$

3.8.1. Rate of Nominal Stress

The nominal stress tensor, being a two-point tensor, has four kinds of decompositions

$$\mathbf{P} = P^{Ji}\mathbf{e}_J^0 \otimes \mathbf{e}_i = P_{Ji}\mathbf{e}_0^J \otimes \mathbf{e}^i = P^J{}_i\mathbf{e}_J^0 \otimes \mathbf{e}^i = P_J{}^i\mathbf{e}_0^J \otimes \mathbf{e}_i, \tag{3.8.7}$$

but only two different convected derivatives result. They are

$$\overset{\triangle}{\mathbf{P}} = \overset{\triangleright}{\mathbf{P}} = \dot{\mathbf{P}} - \mathbf{P} \cdot \mathbf{L}^T, \quad \overset{\nabla}{\mathbf{P}} = \overset{\triangleleft}{\mathbf{P}} = \dot{\mathbf{P}} + \mathbf{P} \cdot \mathbf{L}. \tag{3.8.8}$$

The Jaumann derivative of the nominal stress is

$$\overset{\circ}{\mathbf{P}} = \dot{\mathbf{P}} + \mathbf{P} \cdot \mathbf{W}. \tag{3.8.9}$$

Observe the difference in the structure of the expressions (3.8.8) and (3.8.9) for the convected and Jaumann derivatives of a two-point nominal stress tensor, and the corresponding expressions (2.5.6) for objective derivatives of a two-point deformation gradient tensor. This is because, for example,

$$\mathbf{P} = P^{Ji}\mathbf{e}_J^0 \otimes \mathbf{e}_i, \quad \text{while} \quad \mathbf{F} = F^{iJ}\mathbf{e}_i \otimes \mathbf{e}_J^0. \tag{3.8.10}$$

The transpose tensor $\mathbf{P}^T$ has the convected and the Jaumann derivatives defined according to Eqs. (2.5.6).

The rate of the nominal stress is

$$\dot{\mathbf{P}} = \mathbf{F}^{-1} \cdot (\dot{\boldsymbol{\tau}} - \mathbf{L} \cdot \boldsymbol{\tau}). \tag{3.8.11}$$

The following relationships are easily established between the objective rates of the nominal and Kirchhoff stress

$$\overset{\triangle}{\mathbf{P}} = \mathbf{F}^{-1} \cdot \overset{\triangle}{\boldsymbol{\tau}}, \quad \overset{\nabla}{\mathbf{P}} = \mathbf{F}^{-1} \cdot \overset{\triangleleft}{\boldsymbol{\tau}}, \tag{3.8.12}$$

and

$$\overset{\circ}{\mathbf{P}} = \mathbf{F}^{-1} \cdot \left(\overset{\circ}{\boldsymbol{\tau}} - \mathbf{D} \cdot \boldsymbol{\tau}\right), \quad \overset{\circ}{\boldsymbol{\tau}} = \dot{\boldsymbol{\tau}} - \mathbf{W} \cdot \boldsymbol{\tau} + \boldsymbol{\tau} \cdot \mathbf{W}. \tag{3.8.13}$$

Furthermore, the rates of the material stress tensors can be expressed as

$$\dot{\mathbf{T}}_{(1)} = \overset{\triangle}{\mathbf{P}} \cdot \mathbf{F}^{-T}, \quad \dot{\mathbf{T}}_{(1/2)} = \left(\overset{\bullet}{\mathbf{P}} \cdot \mathbf{R}\right)_s, \tag{3.8.14}$$

where

$$\overset{\bullet}{\mathbf{P}} = \dot{\mathbf{P}} + \mathbf{P} \cdot \boldsymbol{\omega}. \tag{3.8.15}$$

Finally, the rates of nominal and true tractions are related by

$$\dot{\mathbf{p}}_n \, \mathrm{d}S^0 = \left[\dot{\mathbf{t}}_n + (\operatorname{tr} \mathbf{D} - \mathbf{n} \cdot \mathbf{D} \cdot \mathbf{n}) \, \mathbf{t}_\mathrm{n} \right] \mathrm{d}S. \tag{3.8.16}$$

This follows by differentiation of $\mathbf{p}_n \, \mathrm{d}S^0 = \mathbf{t}_n \, \mathrm{d}S$, having in mind the connection (2.4.17).

Higher rates of stress can be investigated similarly, but will not be needed in this book. They are used in modeling certain viscoelastic-type materials. A paper by Prager (1962) and a treatise by Truesdell and Noll (1965) can be consulted in this respect.

3.9. Stress Rates with Current Configuration as Reference

If the current configuration is chosen as the reference configuration ($\mathbf{F} = \mathbf{I}$), all strain measures vanish, and all corresponding stresses are equal to Cauchy stress. All material strain rates are equal to the rate of deformation tensor,

$$\underline{\dot{\mathbf{E}}}_{(n)} = \mathbf{D}. \tag{3.9.1}$$

Since

$$\mathbf{D} = \underline{\dot{\mathbf{U}}}, \quad \mathbf{W} = \underline{\dot{\mathbf{R}}} = \underline{\boldsymbol{\omega}}, \tag{3.9.2}$$

from Eq. (2.6.19) it follows that

$$\underline{\overset{\circ}{\boldsymbol{\mathcal{E}}}}_{(n)} = \underline{\overset{\bullet}{\boldsymbol{\mathcal{E}}}}_{(n)} = \underline{\dot{\mathbf{E}}}_{(n)} = \mathbf{D}. \tag{3.9.3}$$

The underline indicates that the current configuration is used as the reference configuration.

The rate of stress $\underline{\mathbf{T}}_{(0)}$ is, from Eq. (8.5),

$$\underline{\dot{\mathbf{T}}}_{(0)} = \underline{\dot{\mathbf{T}}}_{(n)} + n(\mathbf{D} \cdot \boldsymbol{\sigma} + \boldsymbol{\sigma} \cdot \mathbf{D}). \tag{3.9.4}$$

Any $\mathbf{T}_{(n)}$ can be used in Eq. (3.9.4) to evaluate $\underline{\dot{\mathbf{T}}}_{(0)}$. For example, for $n = 1$ we have from Eq. (3.8.4)

$$\underline{\dot{\mathbf{T}}}_{(1)} = \overset{\triangle}{\boldsymbol{\sigma}} + \boldsymbol{\sigma} \operatorname{tr} \mathbf{D}, \quad \overset{\triangle}{\boldsymbol{\sigma}} = \dot{\boldsymbol{\sigma}} - \mathbf{L} \cdot \boldsymbol{\sigma} - \boldsymbol{\sigma} \cdot \mathbf{L}^T. \tag{3.9.5}$$

Substitution into Eq. (3.9.4) gives

$$\underline{\dot{\mathbf{T}}}_{(0)} = \overset{\circ}{\boldsymbol{\sigma}} + \boldsymbol{\sigma} \operatorname{tr} \mathbf{D}, \quad \overset{\circ}{\boldsymbol{\sigma}} = \dot{\boldsymbol{\sigma}} - \mathbf{W} \cdot \boldsymbol{\sigma} + \boldsymbol{\sigma} \cdot \mathbf{W}. \tag{3.9.6}$$

The rate of stress $\dot{\underline{\mathbf{T}}}_{(n)}$ for an arbitrary n can be deduced from Eq. (3.9.4) by inserting $\dot{\underline{\mathbf{T}}}_{(0)}$ from Eq. (3.9.6). The result is

$$\dot{\underline{\mathbf{T}}}_{(n)} = \overset{\circ}{\boldsymbol{\sigma}} + \boldsymbol{\sigma}\,\mathrm{tr}\,\mathbf{D} - n(\mathbf{D}\cdot\boldsymbol{\sigma} + \boldsymbol{\sigma}\cdot\mathbf{D}). \tag{3.9.7}$$

This is also equal to the Jaumann rate of the spatial stress

$$\overset{\circ}{\underline{\mathcal{T}}}_{(n)} = \dot{\underline{\mathbf{T}}}_{(n)}, \tag{3.9.8}$$

again, of course, with the current configuration taken as the reference configuration. Recall that

$$\overset{\bullet}{\underline{\mathcal{T}}}_{(n)} = \overset{\circ}{\underline{\mathcal{T}}}_{(n)}, \tag{3.9.9}$$

since $\underline{\boldsymbol{\omega}} = \mathbf{W}$. Finally, the rate of nominal stress, momentarily equal to $\boldsymbol{\sigma}$, is

$$\dot{\underline{\mathbf{P}}} = \dot{\boldsymbol{\sigma}} + \boldsymbol{\sigma}\,\mathrm{tr}\,\mathbf{D} - \mathbf{L}\cdot\boldsymbol{\sigma}, \tag{3.9.10}$$

which can be rewritten as either of

$$\overset{\triangle}{\underline{\mathbf{P}}} = \overset{\triangle}{\boldsymbol{\sigma}} + \boldsymbol{\sigma}\,\mathrm{tr}\,\mathbf{D}, \quad \overset{\circ}{\underline{\mathbf{P}}} = \overset{\circ}{\boldsymbol{\sigma}} + \boldsymbol{\sigma}\,\mathrm{tr}\,\mathbf{D} - \mathbf{D}\cdot\boldsymbol{\sigma}. \tag{3.9.11}$$

The stress rate

$$\overset{\circ}{\underline{\boldsymbol{\tau}}} = \overset{\circ}{\boldsymbol{\sigma}} + \boldsymbol{\sigma}\,\mathrm{tr}\,\mathbf{D} \tag{3.9.12}$$

repeatedly appears in the above equations. It is the rate of Kirchhoff stress when the current configuration is taken for the reference configuration. Similarly,

$$\overset{\triangle}{\underline{\boldsymbol{\tau}}} = \overset{\triangle}{\boldsymbol{\sigma}} + \boldsymbol{\sigma}\,\mathrm{tr}\,\mathbf{D}, \quad \overset{\triangledown}{\underline{\boldsymbol{\tau}}} = \overset{\triangledown}{\boldsymbol{\sigma}} + \boldsymbol{\sigma}\,\mathrm{tr}\,\mathbf{D}, \tag{3.9.13}$$

with the connections

$$\overset{\circ}{\boldsymbol{\tau}} = (\det\mathbf{F})\overset{\circ}{\underline{\boldsymbol{\tau}}}, \quad \overset{\triangle}{\boldsymbol{\tau}} = (\det\mathbf{F})\overset{\triangle}{\underline{\boldsymbol{\tau}}}, \quad \overset{\triangledown}{\boldsymbol{\tau}} = (\det\mathbf{F})\overset{\triangledown}{\underline{\boldsymbol{\tau}}}. \tag{3.9.14}$$

The stress rate $\overset{\triangle}{\underline{\boldsymbol{\tau}}}$ is also known as the Truesdell rate of the Cauchy stress $\boldsymbol{\sigma}$ (the Oldroyd rate of the Cauchy stress plus $\boldsymbol{\sigma}\,\mathrm{tr}\,\mathbf{D}$). Evidently,

$$\dot{\underline{\mathbf{T}}}_{(0)} = \overset{\circ}{\underline{\boldsymbol{\tau}}}, \quad \dot{\underline{\mathbf{T}}}_{(1)} = \overset{\triangle}{\underline{\boldsymbol{\tau}}}, \quad \dot{\underline{\mathbf{T}}}_{(-1)} = \overset{\triangledown}{\underline{\boldsymbol{\tau}}}, \quad \dot{\underline{\mathbf{P}}} = \dot{\underline{\boldsymbol{\tau}}} - \mathbf{L}\cdot\boldsymbol{\sigma}, \tag{3.9.15}$$

and

$$\dot{\mathbf{T}}_{(1)} = (\det\mathbf{F})\mathbf{F}^{-1}\cdot\dot{\underline{\mathbf{T}}}_{(1)}\cdot\mathbf{F}^{-T}, \quad \dot{\mathbf{T}}_{(-1)} = (\det\mathbf{F})\mathbf{F}^{T}\cdot\dot{\underline{\mathbf{T}}}_{(-1)}\cdot\mathbf{F}, \tag{3.9.16}$$

$$\dot{\mathbf{P}} = (\det\mathbf{F})\mathbf{F}^{-1}\cdot\dot{\underline{\mathbf{P}}}. \tag{3.9.17}$$

Lastly, it is noted that at the current state as reference, the rates of nominal and true tractions are related by

$$\dot{\underline{\mathbf{p}}}_n = \dot{\mathbf{t}}_n + (\operatorname{tr}\mathbf{D} - \mathbf{n}\cdot\mathbf{D}\cdot\mathbf{n})\,\mathbf{t}_n. \tag{3.9.18}$$

This follows directly from Eq. (3.8.16), since $\mathrm{d}S^0 = \mathrm{d}S$ at the current state as reference.

3.10. Behavior under Superimposed Rotation

If a time-dependent rotation $\mathbf{Q}$ is superimposed to the deformed configuration at time t, the material stress tensors $\mathbf{T}_{(n)}$ do not change,

$$\mathbf{T}^*_{(n)} = \mathbf{T}_{(n)}, \tag{3.10.1}$$

because the strain rates $\dot{\mathbf{E}}_{(n)}$ remain unchanged ($\mathbf{E}^*_{(n)} = \mathbf{E}_{(n)}$), and

$$\dot{w} = \mathbf{T}^*_{(n)} : \dot{\mathbf{E}}^*_{(n)} = \mathbf{T}_{(n)} : \dot{\mathbf{E}}_{(n)}. \tag{3.10.2}$$

In view of Eq. (3.6.22), the spatial stress tensors change into

$$\boldsymbol{\mathcal{T}}^*_{(n)} = \mathbf{Q}\cdot\boldsymbol{\mathcal{T}}_{(n)}\cdot\mathbf{Q}^T. \tag{3.10.3}$$

The same transformation rule applies to Cauchy and Kirchhoff stress. Since the nominal stress is defined by $\mathbf{P} = \mathbf{F}^{-1}\cdot\boldsymbol{\tau}$, it becomes

$$\mathbf{P}^* = \mathbf{P}\cdot\mathbf{Q}^T. \tag{3.10.4}$$

The transformation rule for the Cauchy stress can be independently deduced from the basic relation $\mathbf{t}_n = \mathbf{n}\cdot\boldsymbol{\sigma}$. Under rotation $\mathbf{Q}$ of the deformed configuration, the traction vector changes into

$$\mathbf{t}^*_n = \mathbf{Q}\cdot\mathbf{t}_n, \tag{3.10.5}$$

and the unit normal becomes $\mathbf{n}^* = \mathbf{Q}\cdot\mathbf{n}$. Hence, the transformation

$$\boldsymbol{\sigma}^* = \mathbf{Q}\cdot\boldsymbol{\sigma}\cdot\mathbf{Q}^T. \tag{3.10.6}$$

Likewise,

$$\boldsymbol{\tau}^* = \mathbf{Q}\cdot\boldsymbol{\tau}\cdot\mathbf{Q}^T. \tag{3.10.7}$$

On the other hand,

$$\hat{\boldsymbol{\tau}}^* = \hat{\boldsymbol{\tau}}, \quad \hat{\boldsymbol{\tau}} = \mathbf{R}^T\cdot\boldsymbol{\tau}\cdot\mathbf{R}. \tag{3.10.8}$$

The following transformation rules apply for the rates of material and spatial stress tensors

$$\dot{\mathbf{T}}^*_{(n)} = \dot{\mathbf{T}}_{(n)}, \quad \dot{\boldsymbol{\mathcal{T}}}^*_{(n)} = \mathbf{Q} \cdot \left(\dot{\boldsymbol{\mathcal{T}}}_{(n)} + \hat{\boldsymbol{\Omega}} \cdot \boldsymbol{\mathcal{T}}_{(n)} - \boldsymbol{\mathcal{T}}_{(n)} \cdot \hat{\boldsymbol{\Omega}} \right) \cdot \mathbf{Q}^T, \tag{3.10.9}$$

where $\hat{\boldsymbol{\Omega}} = \mathbf{Q}^T \cdot \boldsymbol{\Omega} \cdot \mathbf{Q}$ and $\boldsymbol{\Omega} = \dot{\mathbf{Q}} \cdot \mathbf{Q}^{-1}$. The rate of nominal stress becomes

$$\dot{\mathbf{P}}^* = \left(\dot{\mathbf{P}} - \mathbf{P} \cdot \hat{\boldsymbol{\Omega}} \right) \cdot \mathbf{Q}^T. \tag{3.10.10}$$

The objective spatial stress rates change according to

$$\overset{\bullet}{\boldsymbol{\mathcal{T}}}{}^*_{(n)} = \mathbf{Q} \cdot \overset{\bullet}{\boldsymbol{\mathcal{T}}}_{(n)} \cdot \mathbf{Q}^T, \quad \overset{\triangle}{\boldsymbol{\tau}}{}^* = \mathbf{Q} \cdot \overset{\triangle}{\boldsymbol{\tau}} \cdot \mathbf{Q}^T, \quad \overset{\circ}{\boldsymbol{\tau}}{}^* = \mathbf{Q} \cdot \overset{\circ}{\boldsymbol{\tau}} \cdot \mathbf{Q}^T, \tag{3.10.11}$$

while objective rates of the nominal stress transform as

$$\overset{\triangle}{\mathbf{P}}{}^* = \overset{\triangle}{\mathbf{P}} \cdot \mathbf{Q}^T, \quad \overset{\circ}{\mathbf{P}}{}^* = \overset{\circ}{\mathbf{P}} \cdot \mathbf{Q}^T. \tag{3.10.12}$$

3.11. Principle of Virtual Velocities

Kinematically admissible velocity field is one possessing continuous first partial derivatives in the interior of the body (analytically admissible), and satisfying prescribed kinematic (velocity) boundary conditions. Kinetically admissible stress and acceleration fields satisfy equations of motion and prescribed kinetic (traction) boundary conditions. Statically admissible stress field satisfies equations of equilibrium and prescribed traction boundary conditions.

Principle of virtual velocities: If the stress and acceleration fields are kinetically admissible, the rate of work of external and inertial forces on any kinematically admissible virtual velocity field is equal to

$$\int_V \boldsymbol{\sigma} : \delta\mathbf{D} \, \mathrm{d}V. \tag{3.11.1}$$

Conversely, if the rate of work of external and inertial forces is equal to (3.11.1), for the assumed stress and acceleration fields and for every kinematically admissible virtual velocity field, then the stress and acceleration fields are kinetically admissible.

Proof: The rate of work of the surface traction $\mathbf{t}_n$ on an analytically admissible virtual velocity field $\delta\mathbf{v}$ vanishing on S_v is

$$\int_S \mathbf{t}_n \cdot \delta\mathbf{v} \, \mathrm{d}S = \int_V \boldsymbol{\nabla} \cdot (\boldsymbol{\sigma} \cdot \delta\mathbf{v}) \, \mathrm{d}V. \tag{3.11.2}$$

If the traction is applied only on the S_t part of S, while velocity is prescribed on the remainder S_v of the boundary, then $\delta\mathbf{v} = \mathbf{0}$ on S_v by definition of the kinematically admissible virtual velocity field. Thus, the integral on the left-hand side of Eq. (3.11.2) can always be taken over the total S. Applying Eq. (1.13.13) to the integrand on the right-hand side of Eq. (3.11.2), and by the symmetry of $\boldsymbol{\sigma}$, we obtain

$$\int_S \mathbf{t}_n \cdot \delta\mathbf{v}\,\mathrm{d}S - \int_V (\boldsymbol{\nabla}\cdot\boldsymbol{\sigma})\cdot\delta\mathbf{v}\,\mathrm{d}V = \int_V \boldsymbol{\sigma} : \delta\mathbf{D}\,\mathrm{d}V. \tag{3.11.3}$$

If $\boldsymbol{\sigma}$ and $\mathrm{d}\mathbf{v}/\mathrm{d}t$ are kinetically admissible, from equations of motion (3.3.4) it follows that

$$\boldsymbol{\nabla}\cdot\boldsymbol{\sigma} = \rho\left(\frac{\mathrm{d}\mathbf{v}}{\mathrm{d}t} - \mathbf{b}\right). \tag{3.11.4}$$

Substitution into Eq. (3.11.3) gives the desired expression

$$\int_S \mathbf{t}_n \cdot \delta\mathbf{v}\,\mathrm{d}S + \int_V \rho\left(\mathbf{b} - \frac{\mathrm{d}\mathbf{v}}{\mathrm{d}t}\right)\cdot\delta\mathbf{v}\,\mathrm{d}V = \int_V \boldsymbol{\sigma} : \delta\mathbf{D}\,\mathrm{d}V. \tag{3.11.5}$$

Conversely, assume that Eq. (3.11.5) holds for a prescribed traction on S_t, given body forces in V, and for assumed stress and acceleration fields. Subtracting from both sides of Eq. (3.11.5) the integral of $(\mathbf{n}\cdot\boldsymbol{\sigma})\cdot\delta\mathbf{v}$ over the surface S, we have

$$\int_S (\mathbf{t}_n - \mathbf{n}\cdot\boldsymbol{\sigma})\cdot\delta\mathbf{v}\,\mathrm{d}S + \int_V \left[\boldsymbol{\nabla}\cdot\boldsymbol{\sigma} + \rho\left(\mathbf{b} - \frac{\mathrm{d}\mathbf{v}}{\mathrm{d}t}\right)\right]\cdot\delta\mathbf{v}\,\mathrm{d}V = 0. \tag{3.11.6}$$

This is identically satisfied if $\boldsymbol{\sigma}$ and $\mathrm{d}\mathbf{v}/\mathrm{d}t$ are kinetically admissible, satisfying equations of motion (3.3.4) and the boundary conditions $\mathbf{n}\cdot\boldsymbol{\sigma} = \mathbf{t}_n$ on S_t.

If integrals are written with respect to undeformed geometry, Eq. (3.11.5) is replaced with

$$\int_{S^0} \mathbf{p}_n \cdot \delta\mathbf{v}\,\mathrm{d}S^0 + \int_{V^0} \rho^0\left(\mathbf{b} - \frac{\mathrm{d}\mathbf{v}}{\mathrm{d}t}\right)\cdot\delta\mathbf{v}\,\mathrm{d}V^0 = \int_{V^0} \mathbf{P}\cdot\cdot\,\delta\dot{\mathbf{F}}\,\mathrm{d}V^0, \tag{3.11.7}$$

where $\delta\dot{\mathbf{F}} = \delta\mathbf{v}\otimes\boldsymbol{\nabla}^0$. If Eq. (3.11.7) holds, the nominal stress $\mathbf{P}$ and the acceleration field satisfy equations of motion (3.3.8), and the boundary conditions $\mathbf{n}^0\cdot\mathbf{P} = \mathbf{p}_n$ on S_t^0.

A straightforward extension of the previous result is obtained by using the rates of nominal stress and traction. Indeed, if

$$\int_{S^0} \dot{\mathbf{p}}_n \cdot \delta\mathbf{v}\,\mathrm{d}S^0 + \int_{V^0} \rho^0\left(\dot{\mathbf{b}} - \frac{\mathrm{d}^2\mathbf{v}}{\mathrm{d}t^2}\right)\cdot\delta\mathbf{v}\,\mathrm{d}V^0 = \int_{V^0} \dot{\mathbf{P}}\cdot\cdot\,\delta\dot{\mathbf{F}}\,\mathrm{d}V^0, \tag{3.11.8}$$

for all analytically admissible $\delta\mathbf{v}$ vanishing on S_v^0, the rates of nominal stress $\dot{\mathbf{P}}$ and the rate of acceleration field satisfy the rate-type equations

$$\boldsymbol{\nabla}^0 \cdot \dot{\mathbf{P}} + \rho^0 \dot{\mathbf{b}} = \rho^0 \frac{\mathrm{d}^2\mathbf{v}}{\mathrm{d}t^2}, \tag{3.11.9}$$

and the rate-type boundary conditions

$$\mathbf{n}^0 \cdot \dot{\mathbf{P}} = \dot{\mathbf{p}}_n \quad \text{on} \quad S_t^0. \tag{3.11.10}$$

The rate-type equations (3.11.9) also follow from equations of motion (3.3.8) by differentiation.

For static problems, $\mathrm{d}\mathbf{v}/\mathrm{d}t$ and $\mathrm{d}^2\mathbf{v}/\mathrm{d}t^2$ are equal to zero in Eqs. (3.11.4)–(3.11.9), so that

$$\boldsymbol{\nabla}^0 \cdot \dot{\mathbf{P}} + \rho^0 \dot{\mathbf{b}} = \mathbf{0}. \tag{3.11.11}$$

If $\dot{\mathbf{P}}$ satisfies Eq. (3.11.11), by Gauss divergence theorem it also follows that

$$\int_{V^0} \dot{\mathbf{P}} \cdot\cdot\, \dot{\mathbf{F}}' \,\mathrm{d}V^0 = \int_{V^0} \rho^0 \dot{\mathbf{b}} \cdot \mathbf{v}' \,\mathrm{d}V^0 + \int_{S^0} \mathbf{n}^0 \cdot \dot{\mathbf{P}} \cdot \mathbf{v}' \,\mathrm{d}S^0, \tag{3.11.12}$$

for any analytically admissible velocity field $\mathbf{v}'$. A direct consequence is a Kirchhoff type identity

$$\begin{aligned} \int_{V^0} (\dot{\mathbf{P}} - \dot{\mathbf{P}}') \cdot\cdot\, (\dot{\mathbf{F}} - \dot{\mathbf{F}}') \,\mathrm{d}V^0 = \int_{V^0} \rho^0 (\dot{\mathbf{b}} - \dot{\mathbf{b}}') \cdot (\mathbf{v} - \mathbf{v}') \,\mathrm{d}V^0 \\ + \int_{S^0} \mathbf{n}^0 \cdot (\dot{\mathbf{P}} - \dot{\mathbf{P}}') \cdot (\mathbf{v} - \mathbf{v}') \,\mathrm{d}S^0, \end{aligned} \tag{3.11.13}$$

where $\dot{\mathbf{P}}'$ and $\dot{\mathbf{b}}'$ are related by Eq. (3.11.11).

If $\mathbf{v}' = \mathbf{v}$, the surface integral in Eq. (3.11.12) is

$$\int_{S^0} \mathbf{n}^0 \cdot \dot{\mathbf{P}} \cdot \mathbf{v} \,\mathrm{d}S^0 = \int_{S_t^0} \dot{\mathbf{p}}_n \cdot \mathbf{v} \,\mathrm{d}S_t^0 + \int_{S_v^0} \mathbf{n}^0 \cdot \dot{\mathbf{P}} \cdot \mathbf{v} \,\mathrm{d}S_v^0, \tag{3.11.14}$$

with $\mathbf{v}$ prescribed on S_v^0, and $\mathbf{n}^0 \cdot \dot{\mathbf{P}} = \dot{\mathbf{p}}_n$ prescribed on S_t^0. If $\mathbf{v} \neq \mathbf{v}'$ in Eq. (3.11.13), but both correspond to the same data ($\dot{\mathbf{b}}$ in V^0, $\dot{\mathbf{p}}_n$ on S_t^0, and $\mathbf{v} = \mathbf{v}'$ on S_v^0), the right-hand side of Eq. (3.11.13) vanishes.

3.12. Principle of Virtual Work

If displacement rather than velocity field is used, we arrive at the principle of virtual displacement (or virtual work). Displacement field is $\mathbf{u} = \mathbf{x} - \mathbf{X}$ (with the same coordinate origin for both $\mathbf{x}$ and $\mathbf{X}$). Geometrically admissible displacement field is one possessing continuous first partial derivatives in the interior of the body, and satisfying prescribed geometric (displacement) boundary conditions. Statically admissible stress field satisfies equations of equilibrium and prescribed static (traction) boundary conditions. Thus, if

$$\int_{S^0} \mathbf{p}_n \cdot \delta\mathbf{u}\, \mathrm{d}S^0 + \int_{V^0} \rho^0\, \mathbf{b} \cdot \delta\mathbf{u}\, \mathrm{d}V^0 = \int_{V^0} \mathbf{P} \cdot\cdot\, \delta\mathbf{F}\, \mathrm{d}V^0, \tag{3.12.1}$$

for all analytically admissible virtual displacements $\delta\mathbf{u}$ vanishing on S_u^0, the nominal stress $\mathbf{P}$ satisfies the equilibrium equations

$$\boldsymbol{\nabla}^0 \cdot \mathbf{P} + \rho^0\, \mathbf{b} = \mathbf{0}, \tag{3.12.2}$$

and the traction boundary conditions

$$\mathbf{n}^0 \cdot \mathbf{P} = \mathbf{p}_n \quad \text{on} \quad S_t^0. \tag{3.12.3}$$

In general, the nominal traction $\mathbf{p}_n$ applied at $\mathbf{X}$ depends on the deformation $\mathbf{x}$ and its gradient $\mathbf{F}$. A particular type of loading for which $\mathbf{p}_n$ depends only on $\mathbf{X}$ is known as dead loading. During dead loading an increase in load deforms the body, but the resulting changes in surface geometry do not modify the load.

If $\mathbf{P}$ satisfies Eq. (3.12.2), by Gauss divergence theorem it follows that

$$\int_{V^0} \mathbf{P} \cdot\cdot\, \mathbf{F}'\, \mathrm{d}V^0 = \int_{V^0} \rho^0\, \mathbf{b} \cdot \mathbf{x}'\, \mathrm{d}V^0 + \int_{S^0} \mathbf{n}^0 \cdot \mathbf{P} \cdot \mathbf{x}'\, \mathrm{d}S^0, \tag{3.12.4}$$

for any analytically admissible deformation field $\mathbf{x}'$. A direct consequence is the Kirchhoff identity

$$\begin{aligned} \int_{V^0} (\mathbf{P} - \mathbf{P}') \cdot\cdot\, (\mathbf{F} - \mathbf{F}')\, \mathrm{d}V^0 &= \int_{V^0} \rho^0 (\mathbf{b} - \mathbf{b}') \cdot (\mathbf{x} - \mathbf{x}')\, \mathrm{d}V^0 \\ &+ \int_{S^0} \mathbf{n}^0 \cdot (\mathbf{P} - \mathbf{P}') \cdot (\mathbf{x} - \mathbf{x}')\, \mathrm{d}S^0, \end{aligned} \tag{3.12.5}$$

where $\mathbf{P}'$ and $\mathbf{b}'$ are related by Eq. (3.12.2).

If $\mathbf{x}' = \mathbf{x}$, the surface integral in Eq. (3.12.4) becomes

$$\int_{S^0} \mathbf{n}^0 \cdot \mathbf{P} \cdot \mathbf{x}\, \mathrm{d}S^0 = \int_{S_t^0} \mathbf{p}_n \cdot \mathbf{x}\, \mathrm{d}S_t^0 + \int_{S_u^0} \mathbf{n}^0 \cdot \mathbf{P} \cdot \mathbf{x}\, \mathrm{d}S_u^0, \tag{3.12.6}$$

with $\mathbf{x}$ prescribed on S_u^0, and $\mathbf{n}^0 \cdot \mathbf{P} = \mathbf{p}_n$ prescribed on S_t^0.

References

Biot, M. A. (1965), *Mechanics of Incremental Deformations*, John Wiley, New York.

Fung, Y. C. (1965), *Foundations of Solid Mechanics*, Prentice-Hall, Englewood Cliffs, New Jersey.

Guo, Z.-H. and Man, C.-S. (1992), Conjugate stress and tensor equation $\sum_{r=1}^{m} U^{m-r} X U^{r-1} = C$, *Int. J. Solids Struct.*, Vol. 29, pp. 2063–2076.

Gurtin, M. E. (1981), *An Introduction to Continuum Mechanics*, Academic Press, New York.

Heiduschke, K. (1995), The logarithmic strain space description, *Int. J. Solids Struct.*, Vol. 32, pp. 1047–1062.

Hill, R. (1968), On constitutive inequalities for simple materials–I, *J. Mech. Phys. Solids*, Vol. 16, pp. 229–242.

Hill, R. (1978), Aspects of invariance in solid mechanics, *Adv. Appl. Mech.*, Vol. 18, pp. 1–75.

Hoger, A. (1987), The stress conjugate to logarithmic strain, *Int. J. Solids Struct.*, Vol. 23, pp. 1645–1656.

Lehmann, Th. and Liang, H. Y. (1993), The stress conjugate to the logarithmic strain, *Z. angew. Math. Mech.*, Vol. 73, pp. 357–363.

Malvern, L. E. (1969), *Introduction to the Mechanics of a Continuous Medium*, Prentice-Hall, Englewood Cliffs, New Jersey.

Ogden, R. W. (1984), *Non-Linear Elastic Deformations*, Ellis Horwood Ltd., Chichester, England (2nd ed., Dover, 1997).

Prager, W. (1961), *Introduction of Mechanics of Continua*, Ginn and Company, Boston.

Prager, W. (1962), On higher rates of stress and deformation, *J. Mech. Phys. Solids*, Vol. 10, pp. 133–138.

Truesdell, C. and Noll, W. (1965), The nonlinear field theories of mechanics, in *Handbuch der Physik*, ed. S. Flügge, Band III/3, Springer-Verlag, Berlin (2nd ed., 1992).

Xiao, H. (1995), Unified explicit basis-free expressions for time rate and conjugate stress of an arbitrary Hill's strain, *Int. J. Solids Struct.*, Vol. 32, pp. 3327–3340.

CHAPTER 4

THERMODYNAMICS OF DEFORMATION

4.1. Energy Equation

A deforming body, or a given portion of it, can be considered to be a thermodynamic system in continuum mechanics. The first law of thermodynamics relates the mechanical work done on the system and the heat transferred into the system to the change in total energy of the system. The rate at which external surface and body forces are doing work on a body currently occupying the volume V bounded by the surface S is given by Eq. (3.5.6), i.e.,

$$\mathcal{P} = \frac{\mathrm{d}}{\mathrm{d}t}\int_V \frac{1}{2}\rho\,\mathbf{v}\cdot\mathbf{v}\,\mathrm{d}V + \int_V \boldsymbol{\sigma} : \mathbf{D}\,\mathrm{d}V. \tag{4.1.1}$$

Let $\mathbf{q}$ be a vector whose magnitude gives the rate of heat flow by conduction across a unit area normal to $\mathbf{q}$. The direction of $\mathbf{q}$ is the direction of heat flow, so that in time $\mathrm{d}t$ the heat amount $\mathbf{q}\,\mathrm{d}t$ would flow through a unit area normal to $\mathbf{q}$. If the area $\mathrm{d}S$ is oriented so that its normal $\mathbf{n}$ is not in the direction of $\mathbf{q}$, the rate of outward heat flow through $\mathrm{d}S$ is $\mathbf{q}\cdot\mathbf{n}\,\mathrm{d}S$ (Fig. 4.1). Let a scalar r be the rate of heat input per unit mass due to distributed internal heat sources. The total heat input rate into the system is then

$$\mathcal{Q} = -\int_S \mathbf{q}\cdot\mathbf{n}\,\mathrm{d}S + \int_V \rho\, r\,\mathrm{d}V = \int_V (-\boldsymbol{\nabla}\cdot\mathbf{q} + \rho\, r)\,\mathrm{d}V. \tag{4.1.2}$$

According to the first law of thermodynamics there exists a state function of a thermodynamic system, called the total energy of the system $\mathcal{E}_{\mathrm{tot}}$, such that its rate of change is

$$\dot{\mathcal{E}}_{\mathrm{tot}} = \mathcal{P} + \mathcal{Q}. \tag{4.1.3}$$

Neither $\mathcal{P}$ nor $\mathcal{Q}$ is in general the rate of any state function, but their sum is. The total energy of the system consists of the macroscopic kinetic energy

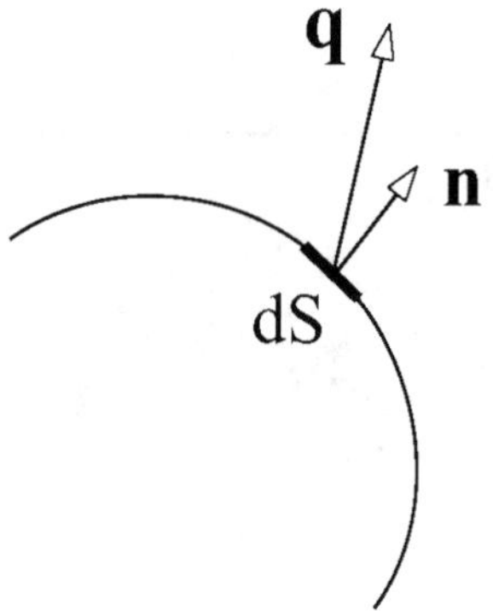

FIGURE 4.1. The heat flow vector $\mathbf{q}$ through the surface element $\mathrm{d}S$ with a unit normal $\mathbf{n}$.

and the internal energy of the system,

$$\mathcal{E}_{\mathrm{tot}} = \frac{\mathrm{d}}{\mathrm{d}t} \int_V \frac{1}{2} \rho \, \mathbf{v} \cdot \mathbf{v} \, \mathrm{d}V + \int_V \rho \, u \, \mathrm{d}V. \tag{4.1.4}$$

The specific internal energy (internal energy per unit mass) is denoted by u. It includes the elastic strain energy and all other forms of energy that do not contribute to macroscopic kinetic energy (e.g., latent strain energy around dislocations, phase-transition energy, energy of random thermal motion of atoms, etc.).

Substituting Eqs. (4.1.1), (4.1.2), and (4.1.4) into Eq. (4.1.3), and having in mind Eq. (3.2.7), gives

$$\int_V (\rho \, \dot{u} - \boldsymbol{\sigma} : \mathbf{D} + \boldsymbol{\nabla} \cdot \mathbf{q} - \rho \, r) \, \mathrm{d}V = 0. \tag{4.1.5}$$

This holds for the whole body and for any part of it, so that locally, at each point, we can write

$$\rho \, \dot{u} = \boldsymbol{\sigma} : \mathbf{D} - \boldsymbol{\nabla} \cdot \mathbf{q} + \rho \, r. \tag{4.1.6}$$

This is the energy equation in the deformed configuration (spatial form of the energy equation).

4.1.1. Material Form of Energy Equation

The corresponding equation written relative to the undeformed configuration is obtained by multiplying Eq. (4.1.6) with $(\det \mathbf{F})$. Since $\rho(\det \mathbf{F}) = \rho^0$,

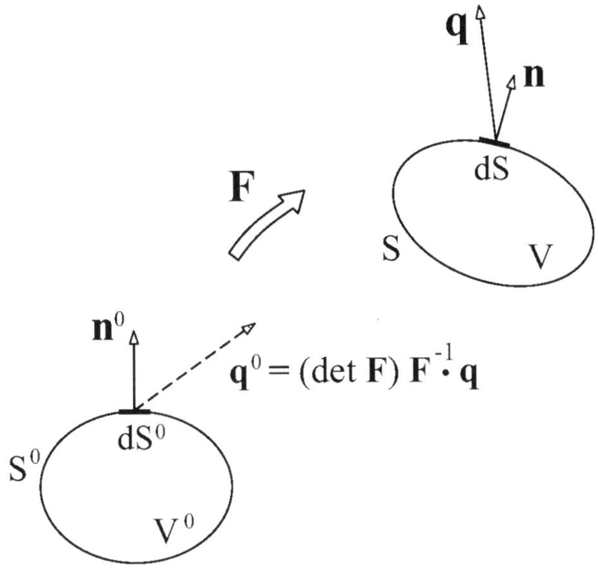

FIGURE 4.2. The nominal rate of the heat flow vector $\mathbf{q}^0$ is related to the heat flow vector $\mathbf{q}$ in the deformed configuration by $\mathbf{q}^0 = (\det \mathbf{F})\mathbf{F}^{-1} \cdot \mathbf{q}$.

and since

$$(\det \mathbf{F})\boldsymbol{\nabla} \cdot \mathbf{q} = \boldsymbol{\nabla}^0 \cdot \left[(\det \mathbf{F})\mathbf{F}^{-1} \cdot \mathbf{q}\right], \tag{4.1.7}$$

by an equation such as (3.3.10), Eq. (4.1.6) becomes

$$\rho^0 \dot{u} = \mathbf{P} \cdot\cdot \dot{\mathbf{F}} - \boldsymbol{\nabla}^0 \cdot \mathbf{q}^0 + \rho^0 r. \tag{4.1.8}$$

The nominal stress $\mathbf{P}$ is defined by Eq. (3.7.4) ($\mathbf{P} \cdot\cdot \dot{\mathbf{F}} = \boldsymbol{\tau} : \mathbf{D}$), and

$$\mathbf{q}^0 = (\det \mathbf{F})\mathbf{F}^{-1} \cdot \mathbf{q} \tag{4.1.9}$$

is the nominal rate of the heat flow vector $\left(\mathbf{q}^0 \cdot \mathbf{n}^0 \, dS^0 = \mathbf{q} \cdot \mathbf{n} \, dS\right)$; see Fig. 4.2. Equation (4.1.8) is a material form of the energy equation.

The rate of specific internal energy can consequently be written as either of

$$\dot{u} = \frac{1}{\rho}\boldsymbol{\sigma} : \mathbf{D} - \frac{1}{\rho}\boldsymbol{\nabla} \cdot \mathbf{q} + r = \frac{1}{\rho^0}\mathbf{P} \cdot\cdot \dot{\mathbf{F}} - \frac{1}{\rho^0}\boldsymbol{\nabla}^0 \cdot \mathbf{q}^0 + r. \tag{4.1.10}$$

The stress dependent term,

$$\frac{1}{\rho}\boldsymbol{\sigma} : \mathbf{D} = \frac{1}{\rho^0}\mathbf{P} \cdot\cdot \dot{\mathbf{F}} = \frac{1}{\rho^0}\mathbf{T}_{(n)} : \dot{\mathbf{E}}_{(n)} \tag{4.1.11}$$

is the contribution to the change of internal energy due to the rate of mechanical work, while the remaining terms in Eq. (4.1.10) represent the rate of heat input per unit mass. The stress $\mathbf{T}_{(n)}$ is conjugate to strain $\mathbf{E}_{(n)}$ in the spirit of Eq. (4.1.11), as discussed in Section 3.6.

4.2. Clausius–Duhem Inequality

The first law of thermodynamics is a statement of the energy balance, which applies regardless of the direction in which the energy conversion between work and heat is assumed to occur. The second law of thermodynamics imposes restrictions on possible directions of thermodynamic processes. A state function, called the entropy of the system, is introduced as a measure of microstructural disorder of the system. The entropy can change by interaction of the system with its surroundings through the heat transfer, and by irreversible changes that take place inside the system due to local rearrangements of microstructure caused by deformation. The entropy input rate due to heat transfer is (Truesdell and Noll, 1965; Malvern, 1969)

$$-\int_S \frac{\mathbf{q}\cdot\mathbf{n}}{\theta}\,\mathrm{d}S + \int_V \rho\frac{r}{\theta}\,\mathrm{d}V = \int_V \left[-\frac{1}{\rho}\boldsymbol{\nabla}\cdot\left(\frac{\mathbf{q}}{\theta}\right) + \frac{r}{\theta}\right]\rho\,\mathrm{d}V, \tag{4.2.1}$$

where $\theta > 0$ is the absolute temperature. The temperature is defined as a measure of the coldness or hotness. It appears in the denominators of the above integrands, because a given heat input causes more disorder (higher entropy change) at lower than at higher temperature (state at lower temperature being less disordered and thus more sensitive to the heat input).

An explicit expression for the rate of entropy change caused by irreversible microstructural changes inside the system depends on the type of deformation and constitution of the material. Denote this part of the rate of entropy change (per unit mass) by γ. The total rate of entropy change of the whole system is then

$$\int_V \rho\frac{\mathrm{d}\eta}{\mathrm{d}t}\,\mathrm{d}V = \int_V \left[-\frac{1}{\rho}\boldsymbol{\nabla}\cdot\left(\frac{\mathbf{q}}{\theta}\right) + \frac{r}{\theta} + \gamma\right]\rho\,\mathrm{d}V. \tag{4.2.2}$$

Locally, at each point of a deformed body, the rate of specific entropy is

$$\dot{\eta} = -\frac{1}{\rho}\boldsymbol{\nabla}\cdot\left(\frac{\mathbf{q}}{\theta}\right) + \frac{r}{\theta} + \gamma. \tag{4.2.3}$$

Since irreversible microstructural changes increase a disorder, they always contribute to an increase of the entropy. Thus, γ is always positive, and is referred to as the entropy production rate. The inequality

$$\gamma > 0 \tag{4.2.4}$$

is a statement of the second law of thermodynamics for irreversible processes. Therefore, from Eq. (4.2.3) we can write

$$\dot{\eta} \geq -\frac{1}{\rho}\boldsymbol{\nabla}\cdot\left(\frac{\mathbf{q}}{\theta}\right) + \frac{r}{\theta}. \tag{4.2.5}$$

The equality sign applies only to reversible processes ($\gamma = 0$). Inequality (4.2.5) is known as the Clausius–Duhem inequality (e.g., Müller, 1985; Ericksen, 1991).

Since

$$\boldsymbol{\nabla}\cdot\left(\frac{\mathbf{q}}{\theta}\right) = \frac{1}{\theta}\boldsymbol{\nabla}\cdot\mathbf{q} - \frac{1}{\theta^2}\mathbf{q}\cdot\boldsymbol{\nabla}\theta, \tag{4.2.6}$$

the inequality (4.2.5) can be rewritten as

$$\dot{\eta} \geq -\frac{1}{\rho\theta}\boldsymbol{\nabla}\cdot\mathbf{q} + \frac{r}{\theta} + \frac{1}{\rho\,\theta^2}\mathbf{q}\cdot\boldsymbol{\nabla}\theta. \tag{4.2.7}$$

The heat spontaneously flows in the direction from the hot to cold part of the body, so that $\mathbf{q}\cdot\boldsymbol{\nabla}\theta \leq 0$. Since $\theta > 0$, it follows that

$$\frac{1}{\rho\,\theta^2}\mathbf{q}\cdot\boldsymbol{\nabla}\theta \leq 0. \tag{4.2.8}$$

Thus, a stronger (more restrictive) form of the Clausius–Duhem inequality is

$$\dot{\eta} \geq -\frac{1}{\rho\theta}\boldsymbol{\nabla}\cdot\mathbf{q} + \frac{r}{\theta}. \tag{4.2.9}$$

Inequality (4.2.9) can alternatively be adopted if the temperature gradients are negligible or equal to zero.

The material forms of the inequalities (4.2.8) and (4.2.9) are

$$\dot{\eta} \geq -\frac{1}{\rho^0\,\theta}\boldsymbol{\nabla}^0\cdot\mathbf{q}^0 + \frac{r}{\theta}, \tag{4.2.10}$$

and

$$\frac{1}{\rho^0\,\theta^2}\mathbf{q}^0\cdot\boldsymbol{\nabla}^0\theta \leq 0. \tag{4.2.11}$$

4.3. Reversible Thermodynamics

If deformation is such that there are no permanent microstructural rearrangements within the material (e.g., thermoelastic deformation), the entropy production rate γ is equal to zero. The rate of entropy change is due to heat transfer only, and

$$\theta\dot{\eta} = -\frac{1}{\rho}\boldsymbol{\nabla}\cdot\mathbf{q} + r. \tag{4.3.1}$$

The energy equation (4.1.10) in this case becomes

$$\dot{u} = \frac{1}{\rho^0}\,\mathbf{T}_{(n)} : \dot{\mathbf{E}}_{(n)} + \theta\,\dot{\eta}. \tag{4.3.2}$$

Equation (4.3.2) shows that the internal energy is a thermodynamic potential for determining $\mathbf{T}_{(n)}$ and θ, when $\mathbf{E}_{(n)}$ and η are considered to be independent state variables. Indeed, by partial differentiation of

$$u = u\left(\mathbf{E}_{(n)}, \eta\right), \tag{4.3.3}$$

we have

$$\dot{u} = \frac{\partial u}{\partial \mathbf{E}_{(n)}} : \dot{\mathbf{E}}_{(n)} + \frac{\partial u}{\partial \eta}\,\dot{\eta}, \tag{4.3.4}$$

and comparison with Eq. (4.3.2) gives

$$\mathbf{T}_{(n)} = \rho^0\,\frac{\partial u}{\partial \mathbf{E}_{(n)}}, \quad \theta = \frac{\partial u}{\partial \eta}. \tag{4.3.5}$$

4.3.1. Thermodynamic Potentials

The Helmholtz free energy is related to internal energy by

$$\psi = u - \theta\,\eta. \tag{4.3.6}$$

By differentiating and incorporating Eq. (4.3.2), the rate of the Helmholtz free energy is

$$\dot{\psi} = \frac{1}{\rho^0}\,\mathbf{T}_{(n)} : \dot{\mathbf{E}}_{(n)} - \eta\,\dot{\theta}. \tag{4.3.7}$$

This indicates that ψ is the portion of internal energy u available for doing work at constant temperature ($\dot{\theta} = 0$). The Helmholtz free energy is a thermodynamic potential for $\mathbf{T}_{(n)}$ and η, when $\mathbf{E}_{(n)}$ and θ are considered to be independent state variables. Indeed, by partial differentiation of

$$\psi = \psi\left(\mathbf{E}_{(n)}, \theta\right), \tag{4.3.8}$$

we have

$$\dot{\psi} = \frac{\partial \psi}{\partial \mathbf{E}_{(n)}} : \dot{\mathbf{E}}_{(n)} + \frac{\partial \psi}{\partial \theta}\,\dot{\theta}, \tag{4.3.9}$$

and comparison with Eq. (4.3.7) gives

$$\mathbf{T}_{(n)} = \rho^0\,\frac{\partial \psi}{\partial \mathbf{E}_{(n)}}, \quad \eta = -\frac{\partial \psi}{\partial \theta}. \tag{4.3.10}$$

The Gibbs energy can be defined as a Legendre transform of the Helmholtz free energy, i.e.,

$$\Phi_{(n)}\left(\mathbf{T}_{(n)},\theta\right)=\frac{1}{\rho^0}\,\mathbf{T}_{(n)}:\mathbf{E}_{(n)}-\psi\left(\mathbf{E}_{(n)},\theta\right). \tag{4.3.11}$$

Note that $\Phi_{(n)}$ is not measure invariant, although ψ is, because for a given geometry change, the quantity $\mathbf{T}_{(n)}:\mathbf{E}_{(n)}$ in general depends on the selected strain and stress measures $\mathbf{E}_{(n)}$ and $\mathbf{T}_{(n)}$. Recall that these are conjugate in the sense that $\mathbf{T}_{(n)}:\mathrm{d}\mathbf{E}_{(n)}$ is measure invariant.

By differentiating Eq. (4.3.11) and using (4.3.7), it follows that

$$\dot{\phi}_{(n)}=\frac{\partial\Phi_{(n)}}{\partial\mathbf{T}_{(n)}}:\dot{\mathbf{T}}_{(n)}+\frac{\partial\Phi_{(n)}}{\partial\theta}\,\dot{\theta}=\frac{1}{\rho^0}\,\mathbf{E}_{(n)}:\dot{\mathbf{T}}_{(n)}+\eta\,\dot{\theta}, \tag{4.3.12}$$

so that

$$\mathbf{E}_{(n)}=\rho^0\,\frac{\partial\Phi_{(n)}}{\partial\mathbf{T}_{(n)}},\quad \eta=\frac{\partial\Phi_{(n)}}{\partial\theta}. \tag{4.3.13}$$

Finally, the enthalpy function is introduced by

$$h_{(n)}\left(\mathbf{T}_{(n)},\eta\right)=\frac{1}{\rho^0}\,\mathbf{T}_{(n)}:\mathbf{E}_{(n)}-u_{(n)}\left(\mathbf{E}_{(n)},\eta\right)=\Phi_{(n)}\left(\mathbf{T}_{(n)},\theta\right)-\theta\,\eta. \tag{4.3.14}$$

By either Eq. (4.3.2) or Eq. (4.3.12), the rate of enthalpy is

$$\dot{h}_{(n)}=\frac{\partial h_{(n)}}{\partial\mathbf{T}_{(n)}}:\dot{\mathbf{T}}_{(n)}+\frac{\partial h_{(n)}}{\partial\eta}\,\dot{\eta}=\frac{1}{\rho^0}\,\mathbf{E}_{(n)}:\dot{\mathbf{T}}_{(n)}-\theta\,\dot{\eta}. \tag{4.3.15}$$

This demonstrates that the enthalpy is a portion of the internal energy that can be released as heat when stress $\mathbf{T}_{(n)}$ is held constant. Furthermore, Eq. (4.3.15) yields

$$\mathbf{E}_{(n)}=\rho^0\,\frac{\partial h_{(n)}}{\partial\mathbf{T}_{(n)}},\quad \theta=-\frac{\partial h_{(n)}}{\partial\eta}. \tag{4.3.16}$$

The fourth-order tensors

$$\mathbf{\Lambda}_{(n)}=\frac{\partial\mathbf{T}_{(n)}}{\partial\mathbf{E}_{(n)}}=\frac{\partial^2\left(\rho^0\,\psi\right)}{\partial\mathbf{E}_{(n)}\otimes\partial\mathbf{E}_{(n)}}, \tag{4.3.17}$$

$$\mathbf{M}_{(n)}=\frac{\partial\mathbf{E}_{(n)}}{\partial\mathbf{T}_{(n)}}=\frac{\partial^2\left(\rho^0\,\Phi_{(n)}\right)}{\partial\mathbf{T}_{(n)}\otimes\partial\mathbf{T}_{(n)}} \tag{4.3.18}$$

are the isothermal elastic stiffness and compliance tensors corresponding to the selected pair $\left(\mathbf{E}_{(n)},\mathbf{T}_{(n)}\right)$ of conjugate stress and strain tensors. The two fourth-order tensors are the inverse of each other $\left(\mathbf{M}_{(n)}=\mathbf{\Lambda}_{(n)}^{-1}\right)$, since

$$\frac{\partial\mathbf{T}_{(n)}}{\partial\mathbf{E}_{(n)}}:\frac{\partial\mathbf{E}_{(n)}}{\partial\mathbf{T}_{(n)}}=\boldsymbol{I}^0. \tag{4.3.19}$$

Being defined as the Hessians of $\rho^0\,\psi$ and $\rho^0\,\Phi_{(n)}$ with respect to $\mathbf{E}_{(n)}$ and $\mathbf{T}_{(n)}$, respectively, the tensors $\mathbf{\Lambda}_{(n)}$ and $\mathbf{M}_{(n)}$ possess reciprocal symmetries

$$\Lambda^{(n)}_{ijkl} = \Lambda^{(n)}_{klij}, \quad M^{(n)}_{ijkl} = M^{(n)}_{klij}. \tag{4.3.20}$$

The adiabatic elastic stiffness and compliance tensors are defined as the Hessians of $\rho^0 u$ and $\rho^0 h_{(n)}$ with respect to $\mathbf{E}_{(n)}$ and $\mathbf{T}_{(n)}$, respectively. The relationship with their isothermal counterparts has been discussed by Truesdell and Toupin (1960), McLellan (1980), and Hill (1981).

4.3.2. Specific and Latent Heats

Specific heats at constant strain and stress are defined by

$$C_{E_n} = \theta\,\frac{\partial \bar{\eta}}{\partial \theta}, \quad C_{T_n} = \theta\,\frac{\partial \hat{\eta}}{\partial \theta}, \tag{4.3.21}$$

where

$$\eta = \bar{\eta}\left(\mathbf{E}_{(n)}, \theta\right) = \hat{\eta}\left(\mathbf{T}_{(n)}, \theta\right). \tag{4.3.22}$$

The latent heats of change of strain and stress are the second-order tensors (e.g., Callen, 1960; Fung, 1965; Kestin, 1979)

$$\boldsymbol{\ell}_{E_n} = \theta\,\frac{\partial \bar{\eta}}{\partial \mathbf{E}_{(n)}}, \quad \boldsymbol{\ell}_{T_n} = \theta\,\frac{\partial \hat{\eta}}{\partial \mathbf{T}_{(n)}}. \tag{4.3.23}$$

In view of the reciprocal relations

$$\rho^0\,\frac{\partial \bar{\eta}}{\partial \mathbf{E}_{(n)}} = -\frac{\partial \mathbf{T}_{(n)}}{\partial \theta}, \quad \rho^0\,\frac{\partial \hat{\eta}}{\partial \mathbf{T}_{(n)}} = -\frac{\partial \mathbf{E}_{(n)}}{\partial \theta}, \tag{4.3.24}$$

the latent heats can also be expressed as

$$\boldsymbol{\ell}_{E_n} = -\frac{1}{\rho^0}\,\theta\,\frac{\partial \mathbf{T}_{(n)}}{\partial \theta}, \quad \boldsymbol{\ell}_{T_n} = \frac{1}{\rho^0}\,\theta\,\frac{\partial \mathbf{E}_{(n)}}{\partial \theta}. \tag{4.3.25}$$

The physical interpretation of the specific and latent heats follows from

$$\mathrm{d}\eta = \frac{\partial \bar{\eta}}{\partial \mathbf{E}_{(n)}} : \mathrm{d}\mathbf{E}_{(n)} + \frac{\partial \bar{\eta}}{\partial \theta}\,\mathrm{d}\theta = \frac{1}{\theta}\left(\boldsymbol{\ell}_{E_n} : \mathrm{d}\mathbf{E}_{(n)} + C_{E_n}\mathrm{d}\theta\right), \tag{4.3.26}$$

$$\mathrm{d}\eta = \frac{\partial \hat{\eta}}{\partial \mathbf{T}_{(n)}} : \mathrm{d}\mathbf{T}_{(n)} + \frac{\partial \bar{\eta}}{\partial \theta}\,\mathrm{d}\theta = \frac{1}{\theta}\left(\boldsymbol{\ell}_{T_n} : \mathrm{d}\mathbf{T}_{(n)} + C_{T_n}\mathrm{d}\theta\right). \tag{4.3.27}$$

Thus, the specific heat at constant strain C_{E_n} (often denoted by C_V) is the heat amount $(\theta\,\mathrm{d}\eta)$ required to increase the temperature of a unit mass for the amount $\mathrm{d}\theta$ at constant strain ($\mathrm{d}\mathbf{E}_{(n)} = 0$). Similar interpretation holds for C_{T_n} (often denoted by C_P). The latent heat $\boldsymbol{\ell}_{E_n}$ is the second-order tensor whose ij component represents the heat amount associated with a change of the corresponding strain component by $\mathrm{d}E_{ij}^{(n)}$, at fixed temperature and fixed values of the remaining five strain components. Analogous interpretation applies to $\boldsymbol{\ell}_{T_n}$.

By partial differentiation, we have from Eq. (4.3.22)

$$\frac{\partial \hat{\eta}}{\partial \theta} = \frac{\partial \bar{\eta}}{\partial \theta} + \frac{\partial \bar{\eta}}{\partial \mathbf{E}_{(n)}} : \frac{\partial \mathbf{E}_{(n)}}{\partial \theta}. \tag{4.3.28}$$

The multiplication by θ and incorporation of Eqs. (4.3.21)–(4.3.25) gives the relationship

$$C_{T_n} - C_{E_n} = \frac{\rho^0}{\theta} \boldsymbol{\ell}_{T_n} : \boldsymbol{\ell}_{E_n}. \tag{4.3.29}$$

Furthermore, since

$$\frac{\partial \hat{\eta}}{\partial \mathbf{T}_{(n)}} = \frac{\partial \bar{\eta}}{\partial \mathbf{E}_{(n)}} : \mathbf{M}_{(n)}, \tag{4.3.30}$$

it follows that

$$\boldsymbol{\ell}_{T_n} = \mathbf{M}_{(n)} : \boldsymbol{\ell}_{E_n}. \tag{4.3.31}$$

When this is inserted into Eq. (4.3.29), we obtain

$$C_{T_n} - C_{E_n} = \frac{\rho^0}{\theta} \mathbf{M}_{(n)} : (\boldsymbol{\ell}_{E_n} \otimes \boldsymbol{\ell}_{E_n}). \tag{4.3.32}$$

For positive definite elastic compliance $\mathbf{M}_{(n)}$, it follows that

$$C_{T_n} > C_{E_n}. \tag{4.3.33}$$

The change in temperature caused by adiabatic straining $\mathrm{d}\mathbf{E}_{(n)}$, or adiabatic stressing $\mathrm{d}\mathbf{T}_{(n)}$, is obtained by setting $\mathrm{d}\eta = 0$ in Eqs. (4.3.26) and (4.3.27). This gives

$$\mathrm{d}\theta = -\frac{1}{C_{E_n}} \boldsymbol{\ell}_{E_n} : \mathrm{d}\mathbf{E}_{(n)}, \quad \mathrm{d}\theta = -\frac{1}{C_{T_n}} \boldsymbol{\ell}_{T_n} : \mathrm{d}\mathbf{T}_{(n)}. \tag{4.3.34}$$

4.4. Irreversible Thermodynamics

For irreversible thermodynamic processes (e.g., processes involving plastic deformation) we shall adopt a thermodynamics with internal state variables (Coleman and Gurtin, 1967; Shapery, 1968; Kestin and Rice, 1970; Rice, 1971, 1975). A set of internal (structural) variables is introduced to describe, in some average sense, the essential features of microstructural changes that occurred at the considered place during the deformation process. These variables are denoted by ξ_j ($j = 1, 2, \ldots, n$). For simplicity, they are assumed to be scalars (extension to include tensorial internal variables is straightforward). Inelastic deformation is considered to be a sequence of constrained equilibrium states. These states are created by a conceptual constraining of internal variables at their current values through imposed thermodynamic forces f_j. The thermodynamic forces or constraints are defined such that the power dissipation (temperature times the entropy production rate) due to structural rearrangements can be expressed as

$$\theta\,\gamma = f_j\,\dot{\xi}_j. \tag{4.4.1}$$

The rates of internal variables $\dot{\xi}_j$ are called the fluxes, and the forces f_j are their affinities.

If various equilibrium states are considered, each corresponding to the same set of values of internal variables ξ_j, the neighboring states are related by the usual laws of reversible thermodynamics (thermoelasticity), such as Eqs. (4.3.1) and (4.3.2). If neighboring constrained equilibrium states correspond to different values of internal variables, then

$$\theta\,\dot{\eta} = -\frac{1}{\rho}\,\boldsymbol{\nabla}\cdot\mathbf{q} + r + f_j\,\dot{\xi}_j. \tag{4.4.2}$$

Combining this with the energy equation (4.1.10) gives

$$\dot{u} = \frac{1}{\rho^0}\,\mathbf{T}_{(n)} : \dot{\mathbf{E}}_{(n)} + \theta\,\dot{\eta} - f_j\,\dot{\xi}_j. \tag{4.4.3}$$

Thus, the internal energy is a thermodynamic potential for determining $\mathbf{T}_{(n)}$, θ and f_j, when $\mathbf{E}_{(n)}$, η and ξ_j are considered to be independent state variables. Indeed, after partial differentiation of

$$u = u\left(\mathbf{E}_{(n)}, \eta, \boldsymbol{\xi}\right), \tag{4.4.4}$$

the comparison with Eq. (4.4.3) gives

$$\mathbf{T}_{(n)} = \rho^0\,\frac{\partial u}{\partial \mathbf{E}_{(n)}}, \quad \theta = \frac{\partial u}{\partial \eta}, \quad f_j = \frac{\partial u}{\partial \xi_j}. \tag{4.4.5}$$

The internal variables are collectively denoted by $\boldsymbol{\xi}$. The Helmholtz free energy

$$\psi = \psi\left(\mathbf{E}_{(n)}, \theta, \boldsymbol{\xi}\right) \tag{4.4.6}$$

is a thermodynamic potential for determining $\mathbf{T}_{(n)}$, η and f_j, such that

$$\mathbf{T}_{(n)} = \rho^0\,\frac{\partial \psi}{\partial \mathbf{E}_{(n)}}, \quad \eta = -\frac{\partial \psi}{\partial \theta}, \quad f_j = -\frac{\partial \psi}{\partial \xi_j}. \tag{4.4.7}$$

If the Gibbs energy

$$\phi_{(n)} = \phi_{(n)}\left(\mathbf{T}_{(n)}, \theta, \boldsymbol{\xi}\right) \tag{4.4.8}$$

is used, we have

$$\mathbf{E}_{(n)} = \rho^0 \frac{\partial \Phi_{(n)}}{\partial \mathbf{T}_{(n)}}, \quad \eta = \frac{\partial \Phi_{(n)}}{\partial \theta}, \quad f_j = \frac{\partial \Phi_{(n)}}{\partial \xi_j}. \tag{4.4.9}$$

Note that in Eq. (4.4.7),

$$f_j = \bar{f}_j \left(\mathbf{E}_{(n)}, \theta, \boldsymbol{\xi}\right), \tag{4.4.10}$$

while in Eq. (4.4.9),

$$f_j = \hat{f}_j \left(\mathbf{T}_{(n)}, \theta, \boldsymbol{\xi}\right), \tag{4.4.11}$$

indicating different functional dependences of the respective arguments. Finally, with the enthalpy

$$h_{(n)} = h_{(n)} \left(\mathbf{T}_{(n)}, \eta, \boldsymbol{\xi}\right) \tag{4.4.12}$$

used as a thermodynamic potential, one has

$$\mathbf{E}_{(n)} = \rho^0 \frac{\partial h_{(n)}}{\partial \mathbf{T}_{(n)}}, \quad \theta = -\frac{\partial h_{(n)}}{\partial \eta}, \quad f_j = \frac{\partial h_{(n)}}{\partial \xi_j}. \tag{4.4.13}$$

By taking appropriate cross-derivatives of the previous expressions, we obtain the Maxwell relations. For example,

$$\begin{aligned} \frac{\partial \mathbf{E}_{(n)} \left(\mathbf{T}_{(n)}, \theta, \boldsymbol{\xi}\right)}{\partial \theta} &= \rho^0 \frac{\partial \hat{\eta} \left(\mathbf{T}_{(n)}, \theta, \boldsymbol{\xi}\right)}{\partial \mathbf{T}_{(n)}}, \\ \frac{\partial \mathbf{T}_{(n)} \left(\mathbf{E}_{(n)}, \theta, \boldsymbol{\xi}\right)}{\partial \theta} &= -\rho^0 \frac{\partial \bar{\eta} \left(\mathbf{E}_{(n)}, \theta, \boldsymbol{\xi}\right)}{\partial \mathbf{E}_{(n)}}, \end{aligned} \tag{4.4.14}$$

and

$$\begin{aligned} \frac{\partial \mathbf{E}_{(n)} \left(\mathbf{T}_{(n)}, \theta, \boldsymbol{\xi}\right)}{\partial \xi_j} &= \rho^0 \frac{\partial \hat{f}_j \left(\mathbf{T}_{(n)}, \theta, \boldsymbol{\xi}\right)}{\partial \mathbf{T}_{(n)}}, \\ \frac{\partial \mathbf{T}_{(n)} \left(\mathbf{E}_{(n)}, \theta, \boldsymbol{\xi}\right)}{\partial \xi_j} &= -\rho^0 \frac{\partial \bar{f}_j \left(\mathbf{E}_{(n)}, \theta, \boldsymbol{\xi}\right)}{\partial \mathbf{E}_{(n)}}. \end{aligned} \tag{4.4.15}$$

4.4.1. Evolution of Internal Variables

The selection of appropriate internal variables is a difficult task, which depends on the material constitution and the type of deformation. Once internal variables are selected, it is necessary to construct evolution equations that govern their change during the deformation. For example, if the fluxes are assumed to be linearly dependent on the affinities, we may write

$$\dot{\xi}_j = \Lambda_{ij} \, f_j. \tag{4.4.16}$$

The coefficients obey the Onsager reciprocity relations if $\Lambda_{ij} = \Lambda_{ji}$ (e.g., Ziegler, 1983; Germain, Nguyen, and Suquet, 1983).

For some materials and the range of deformation, it may be appropriate to assume that at a given temperature θ and the pattern of internal rearrangements $\boldsymbol{\xi}$, each flux depends only on its own affinity, i.e.,

$$\dot{\xi}_j = \text{function}\,(f_j, \theta, \boldsymbol{\xi})\,. \tag{4.4.17}$$

The flux dependence on the stress $\mathbf{T}_{(n)}$ comes only through the fact that $f_j = \hat{f}_j\left(\mathbf{T}_{(n)}, \theta, \boldsymbol{\xi}\right)$. This type of evolution equation is often adopted in metal plasticity, where it is assumed that the crystallographic slip on each slip system is governed by the resolved shear stress on that system (or, at the dislocation level, the motion of each dislocation segment is governed by the Peach–Koehler force on that segment; Rice, 1971).

4.4.2. Gibbs Conditions of Thermodynamic Equilibrium

The system is in a thermodynamic equilibrium if its state variables do not spontaneously change with time. Thus, among all neighboring states with the same internal energy (in the sense of variational calculus), the equilibrium state is one with the highest entropy. This follows from the laws of thermodynamics. If no external work was done on the system nor heat was transferred to the system, so that its internal energy is constant, any spontaneous change from equilibrium would be accompanied by an increase in the entropy (by the second law). Since there is no spontaneous change from the equilibrium, among all neighboring states with the same internal energy, entropy is at maximum in the state of thermodynamic equilibrium (Fung, 1965).

Alternatively, among all neighboring states with the same entropy, the equilibrium state is one with the lowest internal energy. This again follows from the laws of thermodynamics. With no external work done, the system can change its internal energy only by the heat exchange, and from Eq. (4.4.3) and the second law, $du = -f_j\, d\xi_j < 0$, where $d\xi_j$ designates a virtual change of ξ_j between the two considered neighboring states at the same entropy. Thus, any disturbance from the thermodynamic equilibrium by a spontaneous heat transfer would decrease the internal energy. Since there is no spontaneous heat exchange from the equilibrium, among all neighboring states with the same entropy, internal energy is at minimum in the state of thermodynamic equilibrium. It also follows that among all neighboring states with the same temperature, the Helmholtz free energy $\psi = u - \theta\,\eta$ is at minimum in the state of thermodynamic equilibrium.

4.5. Internal Rearrangements without Explicit State Variables

For some inelastic deformation processes it may be more appropriate to assume that there is a set of variables ξ_j that describe internal rearrangements of the material, but that these are not state variables (in the sense that thermodynamic potentials are not point functions of ξ_j), but instead depend on their path history (Rice, 1971). Denoting symbolically by $\mathcal{H}$ the pattern of internal rearrangements, i.e., the set of internal variables ξ_j including the path history by which they were achieved, the Helmholtz free energy can be written as

$$\psi = \psi\left(\mathbf{E}_{(n)}, \theta, \mathcal{H}\right). \tag{4.5.1}$$

At any given state of deformation, an infinitesimal change of $\mathcal{H}$ is assumed to be fully described by a set of scalar infinitesimals $d\xi_j$, such that the change in ψ due to $\mathrm{d}\mathbf{E}_{(n)}$, $\mathrm{d}\theta$ and $d\xi_j$ is, to first order,

$$\mathrm{d}\psi = \frac{\partial\psi}{\partial\mathbf{E}_{(n)}} : \mathrm{d}\mathbf{E}_{(n)} + \frac{\partial\psi}{\partial\theta}\,\mathrm{d}\theta - f_j\,d\xi_j. \tag{4.5.2}$$

It is not necessary that any variable ξ_j exists such that $d\xi_j$ represents an infinitesimal change of ξ_j (the use of an italic d in $d\xi_j$ is meant to indicate this). The stress response and the entropy are

$$\mathbf{T}_{(n)} = \rho^0\,\frac{\partial\psi}{\partial\mathbf{E}_{(n)}}, \quad \eta = -\frac{\partial\psi}{\partial\theta}, \tag{4.5.3}$$

evaluated from ψ at fixed values of $\mathcal{H}$. The thermodynamic forces f_j are associated with infinitesimals $d\xi_j$, so that irreversible (inelastic) change of the free energy, due to change in $\mathcal{H}$ alone, is given by

$$\begin{aligned}\mathrm{d}^{\mathrm{i}}\psi &= \psi\left(\mathbf{E}_{(n)}, \theta, \mathcal{H} + d\mathcal{H}\right) - \psi\left(\mathbf{E}_{(n)}, \theta, \mathcal{H}\right)\\ &= -f_j\,d\xi_j = -\bar{f}_j\left(\mathbf{E}_{(n)}, \theta, \mathcal{H}\right) d\xi_j.\end{aligned} \tag{4.5.4}$$

Higher-order terms, such as $(1/2)df_j\,d\xi_j$, associated with an infinitesimal change of f_j during the variations $d\xi_j$, are neglected.

From Eqs. (4.5.3) and (4.5.4), the inelastic part of the stress increment can be defined by (Hill and Rice, 1973)

$$\begin{aligned} \mathrm{d}^{\mathrm{i}}\mathbf{T}_{(n)} &= \mathbf{T}_{(n)}\left(\mathbf{E}_{(n)}, \theta, \mathcal{H} + d\mathcal{H}\right) - \mathbf{T}_{(n)}\left(\mathbf{E}_{(n)}, \theta, \mathcal{H}\right) \\ &= \rho^0 \frac{\partial}{\partial \mathbf{E}_{(n)}} \left(\mathrm{d}^{\mathrm{i}}\psi\right) = -\rho^0 \frac{\partial \bar{f}_j\left(\mathbf{E}_{(n)}, \theta, \mathcal{H}\right)}{\partial \mathbf{E}_{(n)}} d\xi_j. \end{aligned} \tag{4.5.5}$$

The gradient of $\mathrm{d}^{\mathrm{i}}\psi$ with respect to $\mathbf{E}_{(n)}$ is evaluated at fixed values of θ, $\mathcal{H}$ and $d\mathcal{H}$. The entropy change due to infinitesimal change of $\mathcal{H}$ alone is determined from

$$\bar{\mathrm{d}}^{\mathrm{i}}\eta = -\frac{\partial}{\partial \theta} \left(\mathrm{d}^{\mathrm{i}}\psi\right) = \frac{\partial \bar{f}_j\left(\mathbf{E}_{(n)}, \theta, \mathcal{H}\right)}{\partial \theta} d\xi_j. \tag{4.5.6}$$

Considering the functions $\mathbf{T}_{(n)}\left(\mathbf{E}_{(n)}, \theta, \mathcal{H}\right)$ and $\bar{\eta}\left(\mathbf{E}_{(n)}, \theta, \mathcal{H}\right)$, we can also write

$$\mathrm{d}^{\mathrm{i}}\mathbf{T}_{(n)} = \mathrm{d}\mathbf{T}_{(n)} - \frac{\partial \mathbf{T}_{(n)}}{\partial \mathbf{E}_{(n)}} : \mathrm{d}\mathbf{E}_{(n)} - \frac{\partial \mathbf{T}_{(n)}}{\partial \theta} \mathrm{d}\theta, \tag{4.5.7}$$

$$\bar{\mathrm{d}}^{\mathrm{i}}\eta = \mathrm{d}\eta - \frac{\partial \bar{\eta}}{\partial \mathbf{E}_{(n)}} : \mathrm{d}\mathbf{E}_{(n)} - \frac{\partial \bar{\eta}}{\partial \theta} \mathrm{d}\theta. \tag{4.5.8}$$

Dually, the change of Gibbs energy due to $\mathrm{d}\mathbf{T}_{(n)}$, $\mathrm{d}\theta$ and $d\xi_j$ is

$$\mathrm{d}\Phi_{(n)} = \frac{\partial \Phi_{(n)}}{\partial \mathbf{T}_{(n)}} : \mathrm{d}\mathbf{T}_{(n)} + \frac{\partial \Phi_{(n)}}{\partial \theta} \mathrm{d}\theta + f_j\, d\xi_j. \tag{4.5.9}$$

The strain response and the entropy are

$$\mathbf{E}_{(n)} = \rho^0 \frac{\partial \Phi_{(n)}}{\partial \mathbf{T}_{(n)}}, \quad \eta = \frac{\partial \Phi_{(n)}}{\partial \theta}, \tag{4.5.10}$$

evaluated from $\Phi_{(n)}$ at fixed values of $\mathcal{H}$. The inelastic change of Gibbs energy, due to change in $\mathcal{H}$ alone, is

$$\begin{aligned} \mathrm{d}^{\mathrm{i}}\Phi_{(n)} &= \Phi_{(n)}\left(\mathbf{T}_{(n)}, \theta, \mathcal{H} + d\mathcal{H}\right) - \Phi_{(n)}\left(\mathbf{T}_{(n)}, \theta, \mathcal{H}\right) \\ &= f_j\, d\xi_j = \hat{f}_j\left(\mathbf{T}_{(n)}, \theta, \mathcal{H}\right) d\xi_j. \end{aligned} \tag{4.5.11}$$

Equations (4.5.4) and (4.5.11) show that

$$\mathrm{d}^{\mathrm{i}}\psi + \mathrm{d}^{\mathrm{i}}\Phi_{(n)} = 0, \tag{4.5.12}$$

within the order of accuracy used in Eqs. (4.5.4) and (4.5.11). The inelastic part of strain increment is

$$\begin{aligned} \mathrm{d}^{\mathrm{i}}\mathbf{E}_{(n)} &= \mathbf{E}_{(n)}\left(\mathbf{T}_{(n)}, \theta, \mathcal{H} + d\mathcal{H}\right) - \mathbf{E}_{(n)}\left(\mathbf{T}_{(n)}, \theta, \mathcal{H}\right) \\ &= \rho^0 \frac{\partial}{\partial \mathbf{T}_{(n)}} \left(\mathrm{d}^{\mathrm{i}}\Phi_{(n)}\right) = \rho^0 \frac{\partial \hat{f}_j\left(\mathbf{T}_{(n)}, \theta, \mathcal{H}\right)}{\partial \mathbf{T}_{(n)}} d\xi_j. \end{aligned} \tag{4.5.13}$$

The change of entropy associated with $d\mathcal{H}$ alone is

$$\hat{\mathrm{d}}^{\mathrm{i}}\eta = \frac{\partial}{\partial\theta}\left(\mathrm{d}^{\mathrm{i}}\Phi_{(n)}\right) = \frac{\partial \hat{f}_j\left(\mathbf{T}_{(n)},\theta,\mathcal{H}\right)}{\partial\theta}\,d\xi_j, \tag{4.5.14}$$

which is different from the entropy change in Eq. (4.5.6). The difference is discussed in the next section. If the functions $\mathbf{E}_{(n)}\left(\mathbf{T}_{(n)},\theta,\mathcal{H}\right)$ and $\hat{\eta}\left(\mathbf{E}_{(n)},\theta,\mathcal{H}\right)$ are considered, we can also write

$$\mathrm{d}^{\mathrm{i}}\mathbf{E}_{(n)} = \mathrm{d}\mathbf{E}_{(n)} - \frac{\partial\mathbf{E}_{(n)}}{\partial\mathbf{T}_{(n)}} : \mathrm{d}\mathbf{T}_{(n)} - \frac{\partial\mathbf{E}_{(n)}}{\partial\theta}\,\mathrm{d}\theta, \tag{4.5.15}$$

$$\hat{\mathrm{d}}^{\mathrm{i}}\eta = \mathrm{d}\eta - \frac{\partial\hat{\eta}}{\partial\mathbf{T}_{(n)}} : \mathrm{d}\mathbf{T}_{(n)} - \frac{\partial\hat{\eta}}{\partial\theta}\,\mathrm{d}\theta. \tag{4.5.16}$$

4.6. Relationship between Inelastic Increments

The relationship between the inelastic increments of stress $\mathrm{d}^{\mathrm{i}}\mathbf{T}_{(n)}$ and strain $\mathrm{d}^{\mathrm{i}}\mathbf{E}_{(n)}$ is easily established from Eqs. (4.5.5) and (4.5.13). Since

$$\mathrm{d}^{\mathrm{i}}\mathbf{T}_{(n)} = -\rho^0\,\frac{\partial \bar{f}_j\left(\mathbf{E}_{(n)},\theta,\mathcal{H}\right)}{\partial\mathbf{E}_{(n)}}\,d\xi_j = -\rho^0\left[\frac{\partial \hat{f}_j\left(\mathbf{T}_{(n)},\theta,\mathcal{H}\right)}{\partial\mathbf{T}_{(n)}} : \frac{\partial\mathbf{T}_{(n)}}{\partial\mathbf{E}_{(n)}}\right]d\xi_j, \tag{4.6.1}$$

we have

$$\mathrm{d}^{\mathrm{i}}\mathbf{T}_{(n)} = -\frac{\partial\mathbf{T}_{(n)}}{\partial\mathbf{E}_{(n)}} : \mathrm{d}^{\mathrm{i}}\mathbf{E}_{(n)}. \tag{4.6.2}$$

Therefore,

$$\mathrm{d}^{\mathrm{i}}\mathbf{T}_{(n)} = -\mathbf{\Lambda}_{(n)} : \mathrm{d}^{\mathrm{i}}\mathbf{E}_{(n)}, \quad \mathrm{d}^{\mathrm{i}}\mathbf{E}_{(n)} = -\mathbf{M}_{(n)} : \mathrm{d}^{\mathrm{i}}\mathbf{T}_{(n)}, \tag{4.6.3}$$

where

$$\mathbf{\Lambda}_{(n)} = \frac{\partial\mathbf{T}_{(n)}}{\partial\mathbf{E}_{(n)}} = \rho^0\,\frac{\partial^2\psi\left(\mathbf{E}_{(n)},\theta,\mathcal{H}\right)}{\partial\mathbf{E}_{(n)}\otimes\partial\mathbf{E}_{(n)}}, \tag{4.6.4}$$

$$\mathbf{M}_{(n)} = \frac{\partial\mathbf{E}_{(n)}}{\partial\mathbf{T}_{(n)}} = \rho^0\,\frac{\partial^2\Phi_{(n)}\left(\mathbf{T}_{(n)},\theta,\mathcal{H}\right)}{\partial\mathbf{T}_{(n)}\otimes\partial\mathbf{T}_{(n)}} \tag{4.6.5}$$

are the instantaneous elastic stiffness and compliance tensors of the material at a given state of deformation and internal structure.

An alternative proof of Eq. (4.6.3) is instructive. In view of the reciprocal relations such as given by Eqs. (4.4.14), we can rewrite Eqs. (4.5.7) and (4.5.15) as

$$\mathrm{d}^{\mathrm{i}}\mathbf{T}_{(n)} = \mathrm{d}\mathbf{T}_{(n)} - \left(\mathbf{\Lambda}_{(n)} : \mathrm{d}\mathbf{E}_{(n)} - \rho^0\,\frac{\partial\bar{\eta}}{\partial\mathbf{E}_{(n)}}\,\mathrm{d}\theta\right), \tag{4.6.6}$$

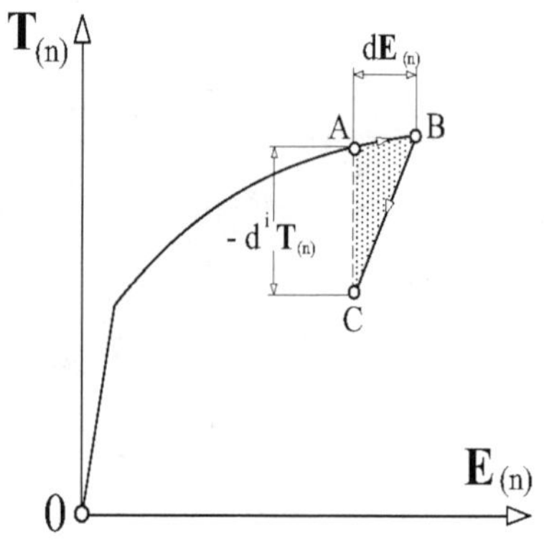

FIGURE 4.3. Schematic representation of an infinitesimal cycle of strain and temperature that involves a change of the pattern of internal rearrangements due to plastic deformation along the segment AB.

$$\mathrm{d}^{\mathrm{i}}\mathbf{E}_{(n)} = \mathrm{d}\mathbf{E}_{(n)} - \left(\mathbf{M}_{(n)} : \mathrm{d}\mathbf{T}_{(n)} + \rho^0 \frac{\partial \hat{\eta}}{\partial \mathbf{T}_{(n)}} \,\mathrm{d}\theta\right). \tag{4.6.7}$$

Taking the inner product of $\mathrm{d}^{\mathrm{i}}\mathbf{E}_{(n)}$ in Eq. (4.6.7) with $\mathbf{\Lambda}_{(n)}$, and having in mind that

$$\mathbf{\Lambda}_{(n)} : \frac{\partial \hat{\eta}}{\partial \mathbf{T}_{(n)}} = \frac{\partial \bar{\eta}}{\partial \mathbf{E}_{(n)}}, \tag{4.6.8}$$

yields Eq. (4.6.3).

The relationship between $\hat{\mathrm{d}}^{\mathrm{i}}\eta$ and $\bar{\mathrm{d}}^{\mathrm{i}}\eta$ can also be established. Since by partial differentiation

$$\frac{\partial \hat{f}_j}{\partial \theta} = \frac{\partial \bar{f}_j}{\partial \theta} + \frac{\partial \bar{f}_j}{\partial \mathbf{E}_{(n)}} : \frac{\partial \mathbf{E}_{(n)}}{\partial \theta}, \quad \frac{\partial \bar{f}_j}{\partial \theta} = \frac{\partial \hat{f}_j}{\partial \theta} + \frac{\partial \hat{f}_j}{\partial \mathbf{T}_{(n)}} : \frac{\partial \mathbf{T}_{(n)}}{\partial \theta}, \tag{4.6.9}$$

and in view of reciprocal relations, Eqs. (4.5.6) and (4.5.14) give

$$\hat{\mathrm{d}}^{\mathrm{i}}\eta = \bar{\mathrm{d}}^{\mathrm{i}}\eta + \frac{\partial \bar{\eta}}{\partial \mathbf{E}_{(n)}} \,\mathrm{d}^{\mathrm{i}}\mathbf{E}_{(n)}, \quad \bar{\mathrm{d}}^{\mathrm{i}}\eta = \hat{\mathrm{d}}^{\mathrm{i}}\eta + \frac{\partial \hat{\eta}}{\partial \mathbf{T}_{(n)}} \,\mathrm{d}^{\mathrm{i}}\mathbf{T}_{(n)}. \tag{4.6.10}$$

Alternatively, one can use Eqs. (4.5.8) and (4.5.16), and the connections

$$\frac{\partial \hat{\eta}}{\partial \theta} = \frac{\partial \bar{\eta}}{\partial \theta} + \frac{\partial \bar{\eta}}{\partial \mathbf{E}_{(n)}} : \frac{\partial \mathbf{E}_{(n)}}{\partial \theta}, \quad \frac{\partial \bar{\eta}}{\partial \theta} = \frac{\partial \hat{\eta}}{\partial \theta} + \frac{\partial \hat{\eta}}{\partial \mathbf{T}_{(n)}} : \frac{\partial \mathbf{T}_{(n)}}{\partial \theta}. \tag{4.6.11}$$

In a rate-independent elastoplastic material, the only way to vary $\mathcal{H}$ but not $\mathbf{E}_{(n)}$ and θ is to perform a cycle of $\mathbf{E}_{(n)}$ and θ that includes $d\mathcal{H}$. Consider a cycle that starts at the state $A\left(\mathbf{E}_{(n)}, \theta, \mathcal{H}\right)$, goes through the state $B\left(\mathbf{E}_{(n)} + \mathrm{d}\mathbf{E}_{(n)}, \theta + \mathrm{d}\theta, \mathcal{H} + d\mathcal{H}\right)$, and ends at the state $C\left(\mathbf{E}_{(n)}, \theta, \mathcal{H} + d\mathcal{H}\right)$.

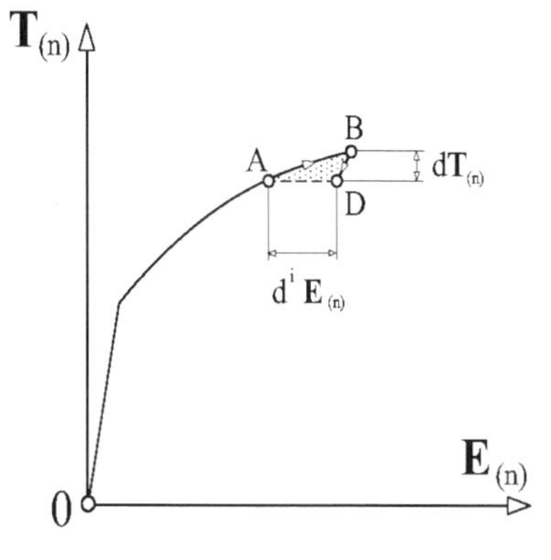

FIGURE 4.4. Schematic representation of an infinitesimal cycle of stress and temperature that involves a change of the pattern of internal rearrangements due to plastic deformation along the segment AB.

The cycle is shown in Fig. 4.3. If the stress and entropy at A were $\mathbf{T}_{(n)}$ and η, in the state B they are $\mathbf{T}_{(n)} + \mathrm{d}\mathbf{T}_{(n)}$ and $\eta + \mathrm{d}\eta$. The change of entropy during the loading from A to B caused by $d\mathcal{H}$ is such that

$$\theta(\mathrm{d}\eta)^{\mathrm{i}} = f_j \, d\xi_j, \tag{4.6.12}$$

by Eq. (4.4.1) for the entropy production rate. After strain and temperature are returned to their values at the beginning of the cycle by elastic unloading, the state C is reached. The stress there is $\mathbf{T}_{(n)} + \mathrm{d}^{\mathrm{i}}\mathbf{T}_{(n)}$, and the entropy is $\eta + \bar{\mathrm{d}}^{\mathrm{i}}\eta$. The stress difference $\mathrm{d}^{\mathrm{i}}\mathbf{T}_{(n)}$ is the stress decrement after the cycle of strain and temperature that includes $d\mathcal{H}$. The entropy difference $\bar{\mathrm{d}}^{\mathrm{i}}\eta$ is different from $(\mathrm{d}\eta)^{\mathrm{i}}$ in Eq. (4.6.12), because the heat input during the unloading from B to C, required to return the temperature to its value before the cycle, is in general different than the heat input during the loading path from A to B.

Alternatively, consider a stress/temperature cycle $A \rightarrow B \rightarrow D$ (Fig. 4.4). In the state D the stress and temperature are returned to their values before the cycle, so that $A\left(\mathbf{T}_{(n)}, \theta, \mathcal{H}\right)$, $B\left(\mathbf{T}_{(n)} + \mathrm{d}\mathbf{T}_{(n)}, \theta + \mathrm{d}\theta, \mathcal{H} + d\mathcal{H}\right)$, and $D\left(\mathbf{T}_{(n)}, \theta, \mathcal{H} + d\mathcal{H}\right)$. The strain and entropy in the state A are $\mathbf{E}_{(n)}$ and η. In the state B they are $\mathbf{E}_{(n)} + \mathrm{d}\mathbf{E}_{(n)}$ and $\eta + \mathrm{d}\eta$. The entropy change from A to B caused by $d\mathcal{H}$ is as in Eq. (4.6.12). After stress and temperature are returned to their values before the cycle by elastic unloading, the state

D is reached, where the strain is $\mathbf{E}_{(n)} + \mathrm{d}^{\mathrm{i}}\mathbf{E}_{(n)}$, and the entropy $\eta + \hat{\mathrm{d}}^{\mathrm{i}}\eta$. The strain difference $\mathrm{d}^{\mathrm{i}}\mathbf{E}_{(n)}$ is the strain increment after the cycle of stress and temperature that includes $d\mathcal{H}$. The entropy difference $\hat{\mathrm{d}}^{\mathrm{i}}\eta$ is different from $(\mathrm{d}\eta)^{\mathrm{i}}$, because the heat input along the unloading path from B to D is in general different than along the loading path from A to B. The entropy differences $\hat{\mathrm{d}}^{\mathrm{i}}\eta$ and $\bar{\mathrm{d}}^{\mathrm{i}}\eta$ are also different because there is a heat exchange along the unloading portion of the path between D and C, which makes the entropies in the states C and D in general different.

References

Callen, H. B. (1960), *Thermodynamics*, John Wiley, New York.

Coleman, B. D. and M. Gurtin, M. (1967), Thermodynamics with internal variables, *J. Chem. Phys.*, Vol. 47, pp. 597–613.

Ericksen, J. L. (1991), *Introduction to the Thermodynamics of Solids*, Chapman and Hall, London.

Fung, Y. C. (1965), *Foundations of Solid Mechanics*, Prentice-Hall, Englewood Cliffs, New Jersey.

Germain, P., Nguyen, Q. S., and Suquet, P. (1983), Continuum thermodynamics, *J. Appl. Mech.*, Vol. 50, pp. 1010–1020.

Hill, R. (1981), Invariance relations in thermoelasticity with generalized variables, *Math. Proc. Camb. Phil. Soc.*, Vol. 90, pp. 373–384.

Hill, R. and Rice, J. R. (1973), Elastic potentials and the structure of inelastic constitutive laws, *SIAM J. Appl. Math.*, Vol. 25, pp. 448–461.

Kestin, J. (1979), *A Course in Thermodynamics*, McGraw-Hill, New York.

Kestin, J. and Rice, J. R. (1970), Paradoxes in the application of thermodynamics to strained solids, in *A Critical Review of Thermodynamics*, eds. E. B. Stuart, B. Gal-Or, and A. J. Brainard, pp. 275–298, Mono-Book, Baltimore.

Malvern, L. E. (1969), *Introduction to the Mechanics of a Continuous Medium*, Prentice-Hall, Englewood Cliffs, New Jersey.

McLellan, A. G. (1980), *The Classical Thermodynamics of Deformable Materials*, Cambridge University Press, Cambridge.

Müller, I. (1985), *Thermodynamics*, Pitman Publishing Inc., Boston.

Rice, J. R. (1971), Inelastic constitutive relations for solids: An internal variable theory and its application to metal plasticity, *J. Mech. Phys. Solids*, Vol. 19, pp. 433–455.

Rice, J. R. (1975), Continuum mechanics and thermodynamics of plasticity in relation to micro-scale deformation mechanisms, in *Constitutive Equations in Plasticity*, ed. A. S. Argon, pp. 23–75, MIT Press, Cambridge, Massachusetts.

Shapery, R. A. (1968), On a thermodynamic constitutive theory and its application to various nonlinear materials, in *Irreversible Aspects of Continuum Mechanics*, eds. H. Parkus and L. I. Sedov, pp. 259–285, Springer-Verlag, Berlin.

Truesdell, C. and Noll, W. (1965), The nonlinear field theories of mechanics, in *Handbuch der Physik*, ed. S. Flügge, Band III/3, Springer-Verlag, Berlin (2nd ed., 1992).

Truesdell, C. and Toupin, R. (1960), The Classical Field Theories, in *Handbuch der Physik*, ed. S. Flügge, Band III/1, pp. 226–793, Springer-Verlag, Berlin.

Ziegler, H. (1983), *An Introduction to Thermomechanics*, 2nd revised ed., North-Holland, Amsterdam.

Part 2

THEORY OF ELASTICITY

CHAPTER 5

FINITE STRAIN ELASTICITY

5.1. Green-Elasticity

Elastic deformation does not cause irreversible rearrangement of internal structure, and the corresponding Helmholtz free energy is a function of stress and temperature only. Restricting consideration to isothermal elastic deformation ($\dot{\theta} = 0$), Eqs. (4.3.7) and (4.3.9) give

$$\dot{\psi} = \frac{\partial \psi}{\partial \mathbf{E}_{(n)}} : \dot{\mathbf{E}}_{(n)} = \frac{1}{\rho^0} \mathbf{T}_{(n)} : \dot{\mathbf{E}}_{(n)}, \tag{5.1.1}$$

i.e.,

$$\mathbf{T}_{(n)} = \frac{\partial \Psi}{\partial \mathbf{E}_{(n)}}, \quad \Psi = \rho^0 \psi \left(\mathbf{E}_{(n)}\right). \tag{5.1.2}$$

Alternatively, Eq. (5.1.2) can be deduced by adopting an experimentally observed property that there is no net work left in a body upon any closed cycle of elastic strain, i.e.,

$$\oint \mathbf{T}_{(n)} : \mathrm{d}\mathbf{E}_{(n)} = 0. \tag{5.1.3}$$

This implies that

$$\mathbf{T}_{(n)} : \mathrm{d}\mathbf{E}_{(n)} = \mathrm{d}\Psi \tag{5.1.4}$$

is a total differential, which leads to Eq. (5.1.2). The function $\Psi = \Psi\left(\mathbf{E}_{(n)}\right)$ is the strain energy function per unit initial volume. It represents the work done to isothermally deform a unit of initial volume to the state of strain $\mathbf{E}_{(n)}$. The explicit representation of the function $\Psi\left(\mathbf{E}_{(n)}\right)$ depends on the selected strain measure $\mathbf{E}_{(n)}$ and the material properties.

Since the material and spatial strain tensors (see Section 2.3) are related by

$$\mathbf{E}_{(n)} = \hat{\boldsymbol{\mathcal{E}}}_{(n)} = \mathbf{R}^T \cdot \boldsymbol{\mathcal{E}}_{(n)} \cdot \mathbf{R}, \tag{5.1.5}$$

the strain energy per unit mass can be written as

$$\psi = \psi\left(\mathbf{E}_{(n)}\right) = \psi\left(\hat{\boldsymbol{\mathcal{E}}}_{(n)}\right). \tag{5.1.6}$$

It can be easily verified that

$$\frac{\partial \psi}{\partial \boldsymbol{\mathcal{E}}_{(n)}} = \mathbf{R} \cdot \frac{\partial \psi}{\partial \mathbf{E}_{(n)}} \cdot \mathbf{R}^T, \quad \dot{\hat{\boldsymbol{\mathcal{E}}}}_{(n)} = \mathbf{R}^T \cdot \overset{\bullet}{\boldsymbol{\mathcal{E}}}_{(n)} \cdot \mathbf{R}, \tag{5.1.7}$$

and the rate of ψ becomes

$$\dot{\psi} = \frac{\partial \psi}{\partial \mathbf{E}_{(n)}} : \dot{\mathbf{E}}_{(n)} = \frac{\partial \psi}{\partial \boldsymbol{\mathcal{E}}_{(n)}} : \overset{\bullet}{\boldsymbol{\mathcal{E}}}_{(n)}. \tag{5.1.8}$$

The stress tensor $\boldsymbol{\mathcal{T}}_{(n)}$ conjugate to spatial strain tensor $\boldsymbol{\mathcal{E}}_{(n)}$ is defined in Section 3.6 by

$$\mathbf{T}_{(n)} : \dot{\mathbf{E}}_{(n)} = \boldsymbol{\mathcal{T}}_{(n)} : \overset{\bullet}{\boldsymbol{\mathcal{E}}}_{(n)}, \quad \boldsymbol{\mathcal{T}}_{(n)} = \mathbf{R} \cdot \mathbf{T}_{(n)} \cdot \mathbf{R}^T. \tag{5.1.9}$$

Consequently, in addition to (5.1.2), from Eq. (5.1.7) we deduce that

$$\boldsymbol{\mathcal{T}}_{(n)} = \frac{\partial \Psi}{\partial \boldsymbol{\mathcal{E}}_{(n)}}, \quad \Psi = \rho^0 \psi \left(\hat{\boldsymbol{\mathcal{E}}}_{(n)} \right). \tag{5.1.10}$$

In view of the expressions for the conjugate stress and strain tensors corresponding to $n = \pm 1$, given in Section 3.6, the following expressions for the Kirchhoff stress $\boldsymbol{\tau} = (\det \mathbf{F})\boldsymbol{\sigma}$ are obtained from Eqs. (5.1.2) and (5.1.10)

$$\boldsymbol{\tau} = \mathbf{F} \cdot \frac{\partial \Psi}{\partial \mathbf{E}_{(1)}} \cdot \mathbf{F}^T = \mathbf{F}^{-T} \cdot \frac{\partial \Psi}{\partial \mathbf{E}_{(-1)}} \cdot \mathbf{F}^{-1}, \tag{5.1.11}$$

$$\boldsymbol{\tau} = \mathbf{V} \cdot \frac{\partial \Psi}{\partial \boldsymbol{\mathcal{E}}_{(1)}} \cdot \mathbf{V} = \mathbf{V}^{-1} \cdot \frac{\partial \Psi}{\partial \boldsymbol{\mathcal{E}}_{(-1)}} \cdot \mathbf{V}^{-1}. \tag{5.1.12}$$

If the conjugate pair associated with $n = 1/2$ is used, from Eq. (3.6.3) and Eq. (5.1.2) there follows

$$\hat{\boldsymbol{\tau}} = \frac{1}{2} \left(\mathbf{U} \cdot \frac{\partial \Psi}{\partial \mathbf{E}_{(1/2)}} + \frac{\partial \Psi}{\partial \mathbf{E}_{(1/2)}} \cdot \mathbf{U} \right) - \frac{1}{2} \left(\mathbf{U} \cdot \hat{\mathbf{K}} - \hat{\mathbf{K}} \cdot \mathbf{U} \right). \tag{5.1.13}$$

Here, $\hat{\boldsymbol{\tau}} = \mathbf{R}^T \cdot \boldsymbol{\tau} \cdot \mathbf{R}$ and

$$\begin{aligned} \hat{\mathbf{K}} = \frac{1}{J_1 J_2 + J_3} \left(J_1 \mathbf{I}^0 - \mathbf{U}^{-1} \right) \cdot \left(\mathbf{U}^{-1} \cdot \frac{\partial \Psi}{\partial \mathbf{E}_{(1/2)}} - \frac{\partial \Psi}{\partial \mathbf{E}_{(1/2)}} \cdot \mathbf{U}^{-1} \right) \\ \cdot \left(J_1 \mathbf{I}^0 - \mathbf{U}^{-1} \right). \end{aligned} \tag{5.1.14}$$

The invariants of $\mathbf{U}^{-1}$ are denoted by J_i. In the derivation, the results from Subsection 1.12.1 were used to solve the matrix equation of the type $\mathbf{A} \cdot \mathbf{X} + \mathbf{X} \cdot \mathbf{A} = \mathbf{B}$.

If Eq. (3.6.25) is used, Eq. (5.1.10) gives

$$\boldsymbol{\tau} = \frac{1}{2} \left(\mathbf{V} \cdot \frac{\partial \Psi}{\partial \boldsymbol{\mathcal{E}}_{(1/2)}} + \frac{\partial \Psi}{\partial \boldsymbol{\mathcal{E}}_{(1/2)}} \cdot \mathbf{V} \right) - \frac{1}{2} \left(\mathbf{V} \cdot \mathbf{K} - \mathbf{K} \cdot \mathbf{V} \right), \tag{5.1.15}$$

where

$$\mathbf{K} = \frac{1}{J_1 J_2 + J_3} \left(J_1 \mathbf{I} - \mathbf{V}^{-1}\right) \cdot \left(\mathbf{V}^{-1} \cdot \frac{\partial \Psi}{\partial \boldsymbol{\mathcal{E}}_{(1/2)}} - \frac{\partial \Psi}{\partial \boldsymbol{\mathcal{E}}_{(1/2)}} \cdot \mathbf{V}^{-1}\right) \cdot \left(J_1 \mathbf{I} - \mathbf{V}^{-1}\right). \tag{5.1.16}$$

The invariants of $\mathbf{V}^{-1}$ are equal to those of $\mathbf{U}^{-1}$ and are again denoted by J_i. The transition from Eq. (5.1.13) to (5.1.15) is straightforward by noting that

$$\hat{\mathbf{K}} = \mathbf{R}^T \cdot \mathbf{K} \cdot \mathbf{R}. \tag{5.1.17}$$

For elastically isotropic materials, considered in the next section, the tensors $\mathbf{V}^{-1}$ and $\partial \Psi / \partial \boldsymbol{\mathcal{E}}_{(1/2)}$ are coaxial, hence commutative, and $\mathbf{K} = 0$. Similar expressions are obtained when Eqs. (3.6.6) and (3.6.26) are used to specify the conjugate stress and strain measures corresponding to $n = -1/2$.

With a properly specified strain energy function $\Psi\left(\mathbf{E}_{(n)}\right)$ for a given material, Eqs. (5.1.11) and (5.1.12), or (5.1.13) and (5.1.15), define the stress response at any state of finite elastic deformation. Since stress is derived from the strain energy function, the equations are referred to as the constitutive equations of hyperelasticity or Green-elasticity (Doyle and Ericksen, 1956; Truesdell and Noll, 1965).

The nominal stress is

$$\mathbf{P} = \frac{\partial \Psi}{\partial \mathbf{F}}, \tag{5.1.18}$$

which follows from

$$\dot{\Psi} = \mathbf{P} \cdot \cdot \dot{\mathbf{F}}, \tag{5.1.19}$$

and $\Psi = \Psi(\mathbf{F})$. Since Ψ is unaffected by rotation of the deformed configuration,

$$\Psi(\mathbf{F}) = \Psi(\mathbf{Q} \cdot \mathbf{F}). \tag{5.1.20}$$

By choosing $\mathbf{Q} = \mathbf{R}^T$, it follows that Ψ depends on $\mathbf{F}$ only through $\mathbf{U}$, or $\mathbf{C} = \mathbf{U}^2$, i.e.,

$$\Psi = \Psi(\mathbf{C}), \quad \mathbf{C} = \mathbf{F}^T \cdot \mathbf{F}. \tag{5.1.21}$$

The functional dependences of Ψ on different tensor arguments such as $\mathbf{F}$, $\mathbf{U}$ or $\mathbf{C}$ are, of course, different.

5.2. Cauchy-Elasticity

Constitutive equations of finite elasticity can be derived without assuming the existence of the strain energy function. Suppose that at any state of elastic deformation, the stress is a single-valued function of strain, regardless of the history or deformation path along which the state has been reached. Since no strain energy is assumed to exist, the work done by the stress could in general be different for different deformation paths. This type of elasticity is known as Cauchy-elasticity, although experimental evidence does not indicate existence of any Cauchy-elastic material that is also not Green-elastic. In any case, we write

$$\mathbf{T}_{(n)} = \mathbf{f}\left(\mathbf{E}_{(n)}\right), \tag{5.2.1}$$

where $\mathbf{f}$ is a second-order tensor function, whose representation depends on the selected strain measure $\mathbf{E}_{(n)}$ (relative to an undeformed configuration and its orientation), and on elastic properties of the material. In terms of the spatial stress and strain measures, Eq. (5.2.1) can be rewritten as

$$\hat{\boldsymbol{\mathcal{T}}}_{(n)} = \mathbf{f}(\hat{\boldsymbol{\mathcal{E}}}_{(n)}), \quad \hat{\boldsymbol{\mathcal{T}}}_{(n)} = \mathbf{R}^T \cdot \boldsymbol{\mathcal{T}}_{(n)} \cdot \mathbf{R}. \tag{5.2.2}$$

The rotated Kirchhoff stress can be expressed from these equations by using any of the conjugate stress and strain measures. For example,

$$\begin{gathered}\hat{\boldsymbol{\tau}} = \mathbf{g}\left(\mathbf{E}_{(1)}\right), \\ \mathbf{g}\left(\mathbf{E}_{(1)}\right) = \left(\mathbf{I}^0 + 2\mathbf{E}_{(1)}\right)^{1/2} \cdot \mathbf{f}\left(\mathbf{E}_{(1)}\right) \cdot \left(\mathbf{I}^0 + 2\mathbf{E}_{(1)}\right)^{1/2},\end{gathered} \tag{5.2.3}$$

or

$$\begin{gathered}\hat{\boldsymbol{\tau}} = \mathbf{g}\left(\mathbf{E}_{(-1)}\right), \\ \mathbf{g}\left(\mathbf{E}_{(-1)}\right) = \left(\mathbf{I}^0 - 2\mathbf{E}_{(-1)}\right)^{1/2} \cdot \mathbf{f}\left(\mathbf{E}_{(-1)}\right) \cdot \left(\mathbf{I}^0 - 2\mathbf{E}_{(-1)}\right)^{1/2}.\end{gathered} \tag{5.2.4}$$

Note that $(\det \mathbf{F})$ can be cast in terms of the invariants of $\mathbf{E}_{(n)}$, since

$$(\det \mathbf{F})^{2n} = 1 + 2nI_E - 4n^2 II_E + 8n^3 III_E. \tag{5.2.5}$$

Thus, Eqs. (5.2.3) and (5.2.4) also define $\hat{\boldsymbol{\sigma}} = \mathbf{R}^T \cdot \boldsymbol{\sigma} \cdot \mathbf{R}$ in terms of $\mathbf{E}_{(1)}$ and $\mathbf{E}_{(-1)}$.

All constitutive equations given in this section are objective under rigid-body rotation of the deformed configuration. The material tensors are unaffected by the transformation $\mathbf{F}^* = \mathbf{Q} \cdot \mathbf{F}$, since $\mathbf{E}^*_{(n)} = \mathbf{E}_{(n)}$ and $\mathbf{T}^*_{(n)} = \mathbf{T}_{(n)}$. The spatial tensors transform according to $\boldsymbol{\mathcal{E}}^*_{(n)} = \mathbf{Q} \cdot \boldsymbol{\mathcal{E}}_{(n)} \cdot \mathbf{Q}^T$ and

$\mathcal{T}^*_{(n)} = \mathbf{Q} \cdot \mathcal{T}_{(n)} \cdot \mathbf{Q}^T$, preserving the physical structure of the constitutive equations such as Eq. (5.2.2).

5.3. Isotropic Green-Elasticity

If the strain energy does not depend along which material directions the principal strains are applied, so that

$$\Psi\left(\mathbf{Q}_0 \cdot \mathbf{E}_{(n)} \cdot \mathbf{Q}_0^T\right) = \Psi\left(\mathbf{E}_{(n)}\right) \tag{5.3.1}$$

for any rotation tensor $\mathbf{Q}_0$, the material is elastically isotropic. A scalar function which satisfies Eq. (5.3.1) is said to be an isotropic function of its second-order tensor argument. Such a function can be expressed in terms of the principal invariants of the strain tensor $\mathbf{E}_{(n)}$, defined according to Eqs. (1.3.3)–(1.3.5), i.e.,

$$\Psi = \Psi\left(I_E, II_E, III_E\right). \tag{5.3.2}$$

Since

$$\frac{\partial I_E}{\partial \mathbf{E}_{(n)}} = \mathbf{I}^0, \quad \frac{\partial II_E}{\partial \mathbf{E}_{(n)}} = \mathbf{E}_{(n)} - I_E \mathbf{I}^0,$$
$$\frac{\partial III_E}{\partial \mathbf{E}_{(n)}} = \mathbf{E}^2_{(n)} - I_E \mathbf{E}_{(n)} - II_E \mathbf{I}^0, \tag{5.3.3}$$

Equation (5.1.2) yields, by partial differentiation,

$$\mathbf{T}_{(n)} = c_0 \mathbf{I}^0 + c_1 \mathbf{E}_{(n)} + c_2 \mathbf{E}^2_{(n)}. \tag{5.3.4}$$

The parameters are

$$c_0 = \frac{\partial \Psi}{\partial I_E} - I_E \frac{\partial \Psi}{\partial II_E} - II_E \frac{\partial \Psi}{\partial III_E}, \quad c_1 = \frac{\partial \Psi}{\partial II_E} - I_E \frac{\partial \Psi}{\partial III_E},$$
$$c_2 = \frac{\partial \Psi}{\partial III_E}. \tag{5.3.5}$$

For example, if it is assumed that (Saint-Venant–Kirchhoff assumption)

$$\Psi = \frac{1}{2}(\lambda + 2\mu) I_E^2 + 2\mu II_E, \tag{5.3.6}$$

a generalized Hooke's law for finite strain is obtained as

$$\mathbf{T}_{(n)} = \lambda I_E \mathbf{I}^0 + 2\mu \mathbf{E}_{(n)}. \tag{5.3.7}$$

The Lamé material constants λ and μ should be specified for each selected strain measure $\mathbf{E}_{(n)}$. If a cubic representation of Ψ is assumed (Murnaghan,

1951), i.e.,

$$\Psi = \frac{1}{2}(\lambda + 2\mu)I_E^2 + 2\mu II_E + \frac{l + 2m}{3} I_E^3 + 2m I_E II_E + n III_E, \tag{5.3.8}$$

the stress response is

$$\begin{aligned} \mathbf{T}_{(n)} &= [\lambda I_E + l I_E^2 + (2m - n) II_E]\mathbf{I}^0 \\ &+ [2\mu + (2m - n) I_E]\mathbf{E}_{(n)} + n\mathbf{E}_{(n)}^2. \end{aligned} \tag{5.3.9}$$

The constants l, m, and n are the Murnaghan's constants.

By choosing $\mathbf{Q}_0 = \mathbf{R}$, Eq. (5.3.1) gives

$$\Psi\left(\boldsymbol{\mathcal{E}}_{(n)}\right) = \Psi\left(\mathbf{E}_{(n)}\right), \tag{5.3.10}$$

so that Ψ is also an isotropic function of $\boldsymbol{\mathcal{E}}_{(n)}$. Since $\boldsymbol{\mathcal{E}}_{(n)}$ and $\mathbf{E}_{(n)} = \hat{\boldsymbol{\mathcal{E}}}_{(n)}$ share the same invariants, from Eqs. (5.1.10) and (5.3.10) it follows that

$$\boldsymbol{\mathcal{T}}_{(n)} = c_0\mathbf{I} + c_1\boldsymbol{\mathcal{E}}_{(n)} + c_2\boldsymbol{\mathcal{E}}_{(n)}^2. \tag{5.3.11}$$

The parameters c_i are defined by Eq. (5.3.5), with $I_E = I_{\mathcal{E}}$, $II_E = II_{\mathcal{E}}$, and $III_E = III_{\mathcal{E}}$. Equation (5.3.11) shows that, for elastic deformation of isotropic materials, the tensors $\boldsymbol{\mathcal{T}}_{(n)}$ and $\boldsymbol{\mathcal{E}}_{(n)}$ have principal directions parallel. Likewise, $\mathbf{T}_{(n)}$ and $\mathbf{E}_{(n)}$ have parallel their principal directions.

The conjugate stress to logarithmic strain $\mathbf{E}_{(0)}$ for an elastically isotropic material is $\mathbf{T}_{(0)} = \hat{\boldsymbol{\tau}}$. The corresponding constitutive structures are

$$\begin{aligned} \hat{\boldsymbol{\tau}} &= \frac{\partial \Psi}{\partial \mathbf{E}_{(0)}} = c_0\mathbf{I}^0 + c_1\mathbf{E}_{(0)} + c_2\mathbf{E}_{(0)}^2, \\ \boldsymbol{\tau} &= \frac{\partial \Psi}{\partial \boldsymbol{\mathcal{E}}_{(0)}} = c_0\mathbf{I} + c_1\boldsymbol{\mathcal{E}}_{(0)} + c_2\boldsymbol{\mathcal{E}}_{(0)}^2, \end{aligned} \tag{5.3.12}$$

where c_i are given by Eq. (5.3.5), in which the invariants of the logarithmic strain are appropriately used. Recall that the invariants of $\boldsymbol{\mathcal{E}}_{(0)} = \ln \mathbf{V}$ are equal to those of $\mathbf{E}_{(0)} = \ln \mathbf{U}$.

5.4. Further Expressions for Isotropic Green-Elasticity

Using Eq. (3.6.12) to express $\mathbf{T}_{(n)}$ in terms of $\mathbf{T}_{(1/2)}$, we have

$$\mathbf{T}_{(n)} = \mathbf{U}^{1-2n} \cdot \mathbf{T}_{(1/2)} = \mathbf{U}^{-2n} \cdot \hat{\boldsymbol{\tau}}. \tag{5.4.1}$$

Substituting this into Eq. (5.1.2), carrying in mind that $\mathbf{U}^{2n} = \mathbf{I}^0 + 2n\mathbf{E}_{(n)}$, gives

$$\hat{\boldsymbol{\tau}} = \frac{\partial \Psi}{\partial \mathbf{E}_{(n)}} + n\left(\mathbf{E}_{(n)} \cdot \frac{\partial \Psi}{\partial \mathbf{E}_{(n)}} + \frac{\partial \Psi}{\partial \mathbf{E}_{(n)}} \cdot \mathbf{E}_{(n)}\right), \tag{5.4.2}$$

written in a symmetrized form. Equation (5.4.2) applies for either positive or negative n. A dual representation, employing the spatial stress and strain tensors, is

$$\boldsymbol{\tau} = \frac{\partial \Psi}{\partial \boldsymbol{\mathcal{E}}_{(n)}} + n\left(\boldsymbol{\mathcal{E}}_{(n)} \cdot \frac{\partial \Psi}{\partial \boldsymbol{\mathcal{E}}_{(n)}} + \frac{\partial \Psi}{\partial \boldsymbol{\mathcal{E}}_{(n)}} \cdot \boldsymbol{\mathcal{E}}_{(n)}\right). \tag{5.4.3}$$

Since Ψ is an isotropic function, it follows that all material strain tensors $\mathbf{E}_{(n)}$ are coaxial with $\hat{\boldsymbol{\tau}}$, and all spatial strain tensors $\boldsymbol{\mathcal{E}}_{(n)}$ are coaxial with $\boldsymbol{\tau}$.

When the strain energy Ψ is represented in terms of the strain invariants, Eqs. (5.4.2) and (5.4.3) give, upon partial differentiation,

$$\hat{\boldsymbol{\tau}} = b_0\mathbf{I}^0 + b_1\mathbf{E}_{(n)} + b_2\mathbf{E}_{(n)}^2, \tag{5.4.4}$$

$$\boldsymbol{\tau} = b_0\mathbf{I} + b_1\boldsymbol{\mathcal{E}}_{(n)} + b_2\boldsymbol{\mathcal{E}}_{(n)}^2, \tag{5.4.5}$$

with the parameters

$$\begin{gathered} b_0 = c_0 + 2nc_2 III_E, \quad b_1 = c_1 + 2n\left(c_0 + c_2 II_E\right), \\ b_2 = c_2 + 2n\left(c_1 + c_2 I_E\right). \end{gathered} \tag{5.4.6}$$

More specifically, these are

$$b_0 = \frac{\partial \Psi}{\partial I_E} - I_E\frac{\partial \Psi}{\partial II_E} - \left(II_E - 2nIII_E\right)\frac{\partial \Psi}{\partial III_E}, \tag{5.4.7}$$

$$b_1 = 2n\frac{\partial \Psi}{\partial I_E} + \left(1 - 2nI_E\right)\frac{\partial \Psi}{\partial II_E} - I_E\frac{\partial \Psi}{\partial III_E}, \tag{5.4.8}$$

$$b_2 = 2n\frac{\partial \Psi}{\partial II_E} + \frac{\partial \Psi}{\partial III_E}. \tag{5.4.9}$$

5.5. Constitutive Equations in Terms of B

The finite strain constitutive equations of isotropic elasticity are often expressed in terms of the left Cauchy–Green deformation tensor $\mathbf{B} = \mathbf{V}^2$. Since $\boldsymbol{\mathcal{E}}_{(1)} = (\mathbf{B} - \mathbf{I})/2$, from Eq. (5.4.3) it follows that

$$\boldsymbol{\tau} = \mathbf{B} \cdot \frac{\partial \Psi}{\partial \mathbf{B}} + \frac{\partial \Psi}{\partial \mathbf{B}} \cdot \mathbf{B}, \tag{5.5.1}$$

written in a symmetrized form. Alternatively, this follows directly from

$$\dot{\Psi} = \frac{\partial \Psi}{\partial \mathbf{B}} : \overset{\circ}{\mathbf{B}} = \boldsymbol{\tau} : \mathbf{D}, \tag{5.5.2}$$

and the connection

$$\overset{\circ}{\mathbf{B}} = \mathbf{B} \cdot \mathbf{D} + \mathbf{D} \cdot \mathbf{B}. \tag{5.5.3}$$

The function $\Psi(\mathbf{B})$ is an isotropic function of $\mathbf{B}$. Introducing the strain energy representation

$$\Psi = \Psi\left(I_B, II_B, III_B\right), \tag{5.5.4}$$

Equation (5.5.1) gives (Rivlin, 1960)

$$\boldsymbol{\tau} = 2\left[\left(III_B \frac{\partial \Psi}{\partial III_B}\right)\mathbf{I} + \left(\frac{\partial \Psi}{\partial I_B} - I_B \frac{\partial \Psi}{\partial II_B}\right)\mathbf{B} + \left(\frac{\partial \Psi}{\partial II_B}\right)\mathbf{B}^2\right]. \tag{5.5.5}$$

If $\mathbf{B}^2$ is eliminated by using the Cayley–Hamilton theorem, Eq. (5.5.5) can be restructured as

$$\boldsymbol{\tau} = 2\left[\left(III_B \frac{\partial \Psi}{\partial III_B} + II_B \frac{\partial \Psi}{\partial II_B}\right)\mathbf{I} + \left(\frac{\partial \Psi}{\partial I_B}\right)\mathbf{B} + \left(III_B \frac{\partial \Psi}{\partial II_B}\right)\mathbf{B}^{-1}\right]. \tag{5.5.6}$$

These are in accord with Eq. (5.4.5), which can be verified by inspection. In the transition, the following relationships between the invariants of $\mathbf{E}_{(1)}$ or $\mathcal{E}_{(1)}$, and $\mathbf{B}$ are noted

$$\begin{aligned} I_E =& \frac{1}{2}\left(I_B - 3\right), \quad II_E = \frac{1}{4} II_B + \frac{1}{2} I_B - \frac{3}{4}, \\ III_E &= \frac{1}{8}\left(III_B + II_B + I_B - 1\right), \end{aligned} \tag{5.5.7}$$

$$\begin{aligned} I_B =& 2I_E + 3, \quad II_B = 4II_E - 4I_E - 3, \\ III_B &= 8III_E - 4II_E + 2I_E + 1. \end{aligned} \tag{5.5.8}$$

The constitutive equation of isotropic elastic material in terms of the nominal stress is

$$\mathbf{P} = \mathbf{F}^{-1} \cdot \boldsymbol{\tau} = \mathbf{F}^T \cdot \left(\frac{\partial \Psi}{\partial \mathbf{B}} + \mathbf{B}^{-1} \cdot \frac{\partial \Psi}{\partial \mathbf{B}} \cdot \mathbf{B}\right). \tag{5.5.9}$$

By using the strain energy representation of Eq. (5.5.4), this becomes

$$\mathbf{P} = 2\mathbf{F}^T \cdot \left[\left(\frac{\partial \Psi}{\partial I_B} - I_B \frac{\partial \Psi}{\partial II_B}\right)\mathbf{I} + \left(\frac{\partial \Psi}{\partial II_B}\right)\mathbf{B} + \left(III_B \frac{\partial \Psi}{\partial III_B}\right)\mathbf{B}^{-1}\right]. \tag{5.5.10}$$

Different specific forms of the strain energy function were used in the literature. For example, Ogden (1984) constructed a strain energy function

$$\Psi = \frac{a}{2}\left(I_B - 3 - \ln III_B\right) + c\left(III_B^{1/2} - 1\right)^2, \tag{5.5.11}$$

where a and c are the material parameters. Based on their theoretical analysis and experimental data Blatz and Ko (1962) proposed an expression for the strain energy for compressible foamed elastomers. Other representations can be found in Blatz, Sharda, and Tschoegl (1974), Morman (1986), Ciarlet (1988), Beatty (1996), and Holzapfel (2000).

5.6. Constitutive Equations in Terms of Principal Stretches

The strain energy of an isotropic material can be often conveniently expressed in terms of the principal stretches λ_i (the eigenvalues of $\mathbf{U}$ and $\mathbf{V}$, which are invariant quantities), i.e.,

$$\Psi = \Psi(\lambda_1, \lambda_2, \lambda_3). \tag{5.6.1}$$

Suppose that all principal stretches are different, and that $\mathbf{N}_i$ and $\mathbf{n}_i$ are the principal directions of the right and left stretch tensors $\mathbf{U}$ and $\mathbf{V}$, respectively, so that

$$\mathbf{U} = \sum_{i=1}^{3} \lambda_i \mathbf{N}_i \otimes \mathbf{N}_i, \quad \mathbf{E}_{(n)} = \sum_{i=1}^{3} \frac{1}{2n} \left(\lambda_i^{2n} - 1\right) \mathbf{N}_i \otimes \mathbf{N}_i, \tag{5.6.2}$$

and

$$\mathbf{V} = \sum_{i=1}^{3} \lambda_i \mathbf{n}_i \otimes \mathbf{n}_i, \quad \mathbf{F} = \sum_{i=1}^{3} \lambda_i \mathbf{n}_i \otimes \mathbf{N}_i. \tag{5.6.3}$$

For an isotropic elastic material, the principal directions of the strain tensor $\mathbf{E}_{(n)}$ are parallel to those of its conjugate stress tensor $\mathbf{T}_{(n)}$, and we can write

$$\mathbf{T}_{(n)} = \sum_{i=1}^{3} T_i^{(n)} \mathbf{N}_i \otimes \mathbf{N}_i. \tag{5.6.4}$$

The principal stresses are here

$$T_i^{(n)} = \frac{\partial \Psi}{\partial E_i^{(n)}} = \lambda_i^{1-2n} \frac{\partial \Psi}{\partial \lambda_i}, \tag{5.6.5}$$

with no sum on i. Recall that $\lambda_i^{2n} = 1 + 2nE_i^{(n)}$. For example, for $n = 1$ we obtain the principal components of the symmetric Piola–Kirchhoff stress,

$$T_i^{(1)} = \frac{\partial \Psi}{\partial E_i^{(1)}} = \frac{1}{\lambda_i} \frac{\partial \Psi}{\partial \lambda_i}. \tag{5.6.6}$$

The principal directions of the Kirchhoff stress $\boldsymbol{\tau}$ of an isotropic elastic material are parallel to those of $\mathbf{V}$, so that

$$\boldsymbol{\tau} = \sum_{i=1}^{3} \tau_i \, \mathbf{n}_i \otimes \mathbf{n}_i. \tag{5.6.7}$$

The corresponding principal components are

$$\tau_i = \lambda_i^2 T_i^{(1)} = \lambda_i \frac{\partial \Psi}{\partial \lambda_i}. \tag{5.6.8}$$

Finally, decomposing the nominal stress as

$$\mathbf{P} = \sum_{i=1}^{3} P_i \, \mathbf{n}_i \otimes \mathbf{N}_i, \tag{5.6.9}$$

we have

$$P_i = \lambda_i T_i^{(1)} = \frac{\partial \Psi}{\partial \lambda_i}. \tag{5.6.10}$$

5.7. Incompressible Isotropic Elastic Materials

For an incompressible material the deformation is necessarily isochoric, so that $\det \mathbf{F} = 1$. Only two invariants of $\mathbf{E}_{(n)}$ are independent, since

$$III_E = -\frac{1}{4n^2} \left(I_E - 2n II_E \right). \tag{5.7.1}$$

Thus, the strain energy can be expressed as

$$\Psi = \Psi \left(I_E, II_E \right), \tag{5.7.2}$$

and we obtain

$$\boldsymbol{\sigma} = (b_0 - p)\mathbf{I} + b_1 \boldsymbol{\mathcal{E}}_{(n)} + b_2 \boldsymbol{\mathcal{E}}_{(n)}^2. \tag{5.7.3}$$

Here, p is an arbitrary pressure, and b_i are defined by Eqs. (5.4.7)–(5.4.9), without terms proportional to $\partial \Psi / \partial III_E$. Alternatively, if Eqs. (5.5.5) and (5.5.6) are specialized to incompressible materials, there follows

$$\boldsymbol{\sigma} = -p\mathbf{I} + 2 \left[\left(\frac{\partial \Psi}{\partial I_B} - I_B \frac{\partial \Psi}{\partial II_B} \right) \mathbf{B} + \left(\frac{\partial \Psi}{\partial II_B} \right) \mathbf{B}^2 \right], \tag{5.7.4}$$

and

$$\boldsymbol{\sigma} = -p_0 \mathbf{I} + 2 \left[\left(\frac{\partial \Psi}{\partial I_B} \right) \mathbf{B} + \left(\frac{\partial \Psi}{\partial II_B} \right) \mathbf{B}^{-1} \right]. \tag{5.7.5}$$

In Eq. (5.7.5), all terms proportional to $\mathbf{I}$ are absorbed in p_0.

Equation (5.7.4) can also be derived by viewing an incompressible material as a material with internal constraint

$$III_B - 1 = 0. \tag{5.7.6}$$

A Lagrangian multiplier $-p/2$ can then be introduced, such that

$$\Psi = \Psi\left(I_B, II_B\right) - \frac{p}{2}\left(III_B - 1\right), \tag{5.7.7}$$

and Eq. (5.5.1) directly leads to Eq. (5.7.4).

For the Mooney–Rivlin material (rubber model; see Treloar, 1975), the strain energy is

$$\Psi = aI_E + bII_E = \frac{a+b}{2}\left(I_B - 3\right) + \frac{b}{4}\left(II_B + 3\right), \tag{5.7.8}$$

and for the neo-Hookean material

$$\Psi = aI_E = \frac{a}{2}\left(I_B - 3\right). \tag{5.7.9}$$

The strain energy representation, suggested by Ogden (1972,1982),

$$\Psi = \sum_{n=1}^{N} a_n \operatorname{tr} \mathbf{E}_{(n)} = \sum_{n=1}^{N} \frac{a_n}{\alpha_n}\left(\lambda_1^{\alpha_n} + \lambda_2^{\alpha_n} + \lambda_3^{\alpha_n} - 3\right) \tag{5.7.10}$$

may be used in some applications, where N is positive integer, but α_n need not be integers (the tensors $\mathbf{E}_{(n)}$ are here defined by Eq. (2.3.1) with α_n replacing $2n$; Hill, 1978). The material parameters are a_n and α_n. Incompressibility constraint is $\lambda_1\lambda_2\lambda_3 = 1$. Other representations in terms of principal stretches λ_i have also been explored (Valanis and Landel, 1967; Rivlin and Sawyers, 1976; Anand, 1986; Arruda and Boyce, 1993).

5.8. Isotropic Cauchy-Elasticity

For isotropic elastic material the tensor function $\mathbf{f}$ in Eq. (5.2.1) is an isotropic function of strain,

$$\mathbf{f}\left(\mathbf{Q}_0 \cdot \mathbf{E}_{(n)} \cdot \mathbf{Q}_0^T\right) = \mathbf{Q}_0 \cdot \mathbf{f}\left(\mathbf{E}_{(n)}\right) \cdot \mathbf{Q}_0^T, \tag{5.8.1}$$

and, by the representation theorem from Section 1.11, the stress response can be written as

$$\mathbf{T}_{(n)} = c_0 \mathbf{I}^0 + c_1 \mathbf{E}_{(n)} + c_2 \mathbf{E}_{(n)}^2. \tag{5.8.2}$$

The parameters c_i are scalar functions of the invariants of $\mathbf{E}_{(n)}$. Similarly, from Eq. (5.2.2) it follows that

$$\boldsymbol{\mathcal{T}}_{(n)} = c_0 \mathbf{I} + c_1 \boldsymbol{\mathcal{E}}_{(n)} + c_2 \boldsymbol{\mathcal{E}}_{(n)}^2. \tag{5.8.3}$$

In view of the isotropic elasticity relationships

$$\hat{\boldsymbol{\tau}} = \mathbf{U}^{2n} \cdot \mathbf{T}_{(n)}, \quad \boldsymbol{\tau} = \mathbf{V}^{2n} \cdot \boldsymbol{\mathcal{T}}_{(n)}, \tag{5.8.4}$$

equations (5.8.2) and (5.8.3) can be rephrased as

$$\hat{\boldsymbol{\tau}} = b_0\mathbf{I}^0 + b_1\mathbf{E}_{(n)} + b_2\mathbf{E}^2_{(n)}, \quad \boldsymbol{\tau} = b_0\mathbf{I} + b_1\boldsymbol{\mathcal{E}}_{(n)} + b_2\boldsymbol{\mathcal{E}}^2_{(n)}, \tag{5.8.5}$$

where b_i are given by Eq. (5.4.6). The constitutive equations of Green-elasticity are recovered if the strain energy function exists, so that the constants c_i in Eq. (5.4.6) are specified by Eq. (5.3.5).

Finally, it is noted that Eqs. (5.8.5) can be recast in terms of $\mathbf{C} = \mathbf{U}^2$ and $\mathbf{B} = \mathbf{V}^2$, with the results

$$\hat{\boldsymbol{\tau}} = a_0\mathbf{I}^0 + a_1\mathbf{C} + a_2\mathbf{C}^2, \quad \boldsymbol{\tau} = a_0\mathbf{I} + a_1\mathbf{B} + a_2\mathbf{B}^2. \tag{5.8.6}$$

The scalar parameters a_i depend on the invariants of $\mathbf{C}$ or $\mathbf{B}$. The last expression can also be deduced directly from $\mathbf{T}_{(1)} = \mathbf{f}\left(\mathbf{E}_{(1)}\right)$ by the representation theorem for the isotropic function $\mathbf{f}$, dependent on the Lagrangian strain $\mathbf{E}_{(1)} = \left(\mathbf{C} - \mathbf{I}^0\right)/2$. Furthermore, since $(\det \mathbf{F}) = III_C^{1/2}$, Eqs. (5.8.6) define the stress tensors $\hat{\boldsymbol{\sigma}}$ and $\boldsymbol{\sigma}$, as well ($\boldsymbol{\sigma}$ being the Cauchy stress). For incompressible materials

$$\boldsymbol{\sigma} = -p_1\mathbf{I} + b_1\boldsymbol{\mathcal{E}}_{(n)} + b_2\boldsymbol{\mathcal{E}}^2_{(n)} = -p_2\mathbf{I} + a_1\mathbf{B} + a_2\mathbf{B}^2, \tag{5.8.7}$$

where p_1 and p_2 are arbitrary pressures. Additional discussion can be found in the books by Leigh (1968) and Malvern (1969).

5.9. Transversely Isotropic Materials

For an elastically isotropic material, elastic properties are equal in all directions. Any rotation of the undeformed reference configuration before the application of a given stress has no effect on the subsequent stress-deformation response. The material symmetry group is the full orthogonal group. If the symmetry group of the material is less than the full orthogonal group, the material is anisotropic (aelotropic). For the most general anisotropy, the isotropy group consists only of identity transformation 1 and the central inversion transformation $\bar{1}$. Any rotation of the reference configuration prior to application of stress will change the elastic response of such a material.

The material is said to have a plane of elastic symmetry if the reference configuration obtained from the undeformed configuration by reflection in the plane of symmetry is indistinguishable from the undeformed configuration (in the sense of elastic response).

Transversely isotropic material has a single preferred direction (axis of isotropy). Its symmetry group consists of arbitrary rotations about the axis of isotropy, say $\mathbf{m}^0$, and rotations that carry $\mathbf{m}^0$ into $-\mathbf{m}^0$. Every plane containing $\mathbf{m}^0$ is a plane of elastic symmetry, so that reflections in these planes also belong to the symmetry group. The elastic strain energy function can be consequently written as

$$\Psi = \Psi\left(I_E, II_E, III_E, E_{33}, E_{31}^2 + E_{32}^2\right), \tag{5.9.1}$$

provided that the coordinate axes are selected so that $\mathbf{m}^0$ is in the coordinate direction $\mathbf{e}_3$. The arguments in Eq. (5.9.1) are invariant under the transformations from the symmetry group of transverse isotropy. This can be derived as follows. For transversely isotropic material, the strain energy is a scalar function of the strain tensor $\mathbf{E}_{(n)}$ and the unit vector $\mathbf{m}^0$,

$$\Psi = \Psi\left(\mathbf{E}_{(n)}, \mathbf{m}^0\right). \tag{5.9.2}$$

The function Ψ is invariant under all orthogonal transformations of the reference configuration that carry both $\mathbf{E}_{(n)}$ and $\mathbf{m}^0$, i.e.,

$$\Psi\left(\mathbf{Q}_0 \cdot \mathbf{E}_{(n)} \cdot \mathbf{Q}_0^T, \mathbf{Q}_0 \cdot \mathbf{m}^0\right) = \Psi\left(\mathbf{E}_{(n)}, \mathbf{m}^0\right). \tag{5.9.3}$$

Such a function Ψ is said to be an isotropic function of both $\mathbf{E}_{(n)}$ and $\mathbf{m}^0$, simultaneously. Physically, the rotated strain $\mathbf{Q}_0 \cdot \mathbf{E}_{(n)} \cdot \mathbf{Q}_0^T$, applied relative to the rotated axis of isotropy $\mathbf{Q}_0 \cdot \mathbf{m}^0$, gives the same strain energy as the strain $\mathbf{E}_{(n)}$ applied relative to the original axis of isotropy $\mathbf{m}^0$. Of course, Ψ is not an isotropic function of the strain alone, i.e.,

$$\Psi\left(\mathbf{Q}_0 \cdot \mathbf{E}_{(n)} \cdot \mathbf{Q}_0^T, \mathbf{m}^0\right) \neq \Psi\left(\mathbf{E}_{(n)}, \mathbf{m}^0\right) \tag{5.9.4}$$

in general, although the equality sign holds for those $\mathbf{Q}_0$ that belong to the symmetry group of transverse isotropy.

Representation of isotropic scalar functions of second-order tensors and vectors is well-known (e.g., Boehler, 1987). The function $\Psi\left(\mathbf{E}_{(n)}, \mathbf{m}^0\right)$ can be expressed in terms of individual and joint invariants of $\mathbf{E}_{(n)}$ and $\mathbf{m}^0$, i.e.,

$$\Psi = \Psi\left(I_E, II_E, III_E, \mathbf{m}^0 \cdot \mathbf{E}_{(n)} \cdot \mathbf{m}^0, \mathbf{m}^0 \cdot \mathbf{E}_{(n)}^2 \cdot \mathbf{m}^0\right). \tag{5.9.5}$$

It is convenient to introduce the second-order tensor

$$\mathbf{M}^0 = \mathbf{m}^0 \otimes \mathbf{m}^0. \tag{5.9.6}$$

This is an idempotent tensor, for which

$$\mathbf{M}^0 \cdot \mathbf{M}^0 = \mathbf{M}^0, \quad I_M = 1, \quad II_M = III_M = 0. \tag{5.9.7}$$

When applied to an arbitrary vector $\mathbf{a}_0$, the tensor $\mathbf{M}^0$ projects it on $\mathbf{m}^0$,

$$\mathbf{M}^0 \cdot \mathbf{a}_0 = (\mathbf{m}^0 \cdot \mathbf{a}_0)\mathbf{m}^0. \tag{5.9.8}$$

The joint invariants of $\mathbf{E}_{(n)}$ and $\mathbf{m}^0$ in Eq. (5.9.5) can thus be written as

$$\begin{aligned} K_1 &= \mathbf{m}^0 \cdot \mathbf{E}_{(n)} \cdot \mathbf{m}^0 = \operatorname{tr}\left(\mathbf{M}^0 \cdot \mathbf{E}_{(n)}\right), \\ K_2 &= \mathbf{m}^0 \cdot \mathbf{E}_{(n)}^2 \cdot \mathbf{m}^0 = \operatorname{tr}\left(\mathbf{M}^0 \cdot \mathbf{E}_{(n)}^2\right), \end{aligned} \tag{5.9.9}$$

and the strain energy becomes

$$\Psi = \Psi\left(I_E, II_E, III_E, K_1, K_2\right). \tag{5.9.10}$$

The stress response is accordingly

$$\mathbf{T}_{(n)} = c_0\mathbf{I}^0 + c_1\mathbf{E}_{(n)} + c_2\mathbf{E}_{(n)}^2 + c_3\mathbf{M}^0 + c_4\left(\mathbf{M}^0 \cdot \mathbf{E}_{(n)} + \mathbf{E}_{(n)} \cdot \mathbf{M}^0\right). \tag{5.9.11}$$

The parameters c_0, c_1 and c_3 are defined by Eqs. (5.3.5), and

$$c_3 = \frac{\partial \Psi}{\partial K_1}, \quad c_4 = \frac{1}{2}\frac{\partial \Psi}{\partial K_2}. \tag{5.9.12}$$

If we choose $\mathbf{Q}_0 = \mathbf{R}$ (rotation tensor from the polar decomposition of deformation gradient), from Eq. (5.9.3) it follows that

$$\Psi\left(\boldsymbol{\mathcal{E}}_{(n)}, \overline{\mathbf{m}}\right) = \Psi\left(\mathbf{E}_{(n)}, \mathbf{m}^0\right), \tag{5.9.13}$$

where

$$\overline{\mathbf{m}} = \mathbf{R} \cdot \mathbf{m}^0. \tag{5.9.14}$$

Thus, Ψ is also an isotropic function of the spatial strain $\boldsymbol{\mathcal{E}}_{(n)}$ and the vector $\overline{\mathbf{m}}$. A dual equation to Eq. (5.9.11), expressed relative to the deformed configuration, is consequently

$$\boldsymbol{\mathcal{T}}_{(n)} = c_0\mathbf{I} + c_1\boldsymbol{\mathcal{E}}_{(n)} + c_2\boldsymbol{\mathcal{E}}_{(n)}^2 + c_3\overline{\mathbf{M}} + c_4\left(\overline{\mathbf{M}} \cdot \boldsymbol{\mathcal{E}}_{(n)} + \boldsymbol{\mathcal{E}}_{(n)} \cdot \overline{\mathbf{M}}\right). \tag{5.9.15}$$

The tensor $\overline{\mathbf{M}}$ is defined by

$$\overline{\mathbf{M}} = \overline{\mathbf{m}} \otimes \overline{\mathbf{m}} = \mathbf{R} \cdot \mathbf{M}^0 \cdot \mathbf{R}^T. \tag{5.9.16}$$

For example, if $n = 1$, Eq. (5.9.15) gives the Kirchhoff stress

$$\boldsymbol{\tau} = b_0\mathbf{I} + b_1\boldsymbol{\mathcal{E}}_{(1)} + b_2\boldsymbol{\mathcal{E}}_{(1)}^2 + c_3\mathbf{M} + c_4\left(\mathbf{M} \cdot \boldsymbol{\mathcal{E}}_{(1)} + \boldsymbol{\mathcal{E}}_{(1)} \cdot \mathbf{M}\right). \tag{5.9.17}$$

The coefficients b_i are written in terms of c_i by Eqs. (5.4.6), and

$$\mathbf{M} = \mathbf{m} \otimes \mathbf{m} = \mathbf{F} \cdot \mathbf{M}^0 \cdot \mathbf{F}^T, \quad \mathbf{m} = \mathbf{V} \cdot \overline{\mathbf{m}} = \mathbf{F} \cdot \mathbf{m}^0. \tag{5.9.18}$$

The vector $\mathbf{m}$ in the deformed configuration is obtained by deformation $\mathbf{F}$ from the vector $\mathbf{m}^0$ in the undeformed configuration. However, while $\mathbf{m}^0$ and $\overline{\mathbf{m}}$ are the unit vectors, the (embedded) vector $\mathbf{m}$ is not. The tensor $\mathbf{M}^0 = \mathbf{F}^{-1} \cdot \mathbf{M} \cdot \mathbf{F}^{-T}$ is induced from $\mathbf{M}$ by a transformation of the contravariant type.

If transversely isotropic material is inextensible in the direction of the axis of isotropy, so that there exists a deformation constraint

$$\mathbf{m}^0 \cdot \mathbf{C} \cdot \mathbf{m}^0 = \overline{\mathbf{m}} \cdot \mathbf{B} \cdot \overline{\mathbf{m}} = 1, \quad \text{or} \quad \mathbf{m}^0 \cdot \mathbf{E}_{(1)} \cdot \mathbf{m}^0 = \overline{\mathbf{m}} \cdot \boldsymbol{\mathcal{E}}_{(1)} \cdot \overline{\mathbf{m}} = 0, \tag{5.9.19}$$

the strain energy can be written by using the Lagrangian multiplier as

$$\Psi = \Psi\left(I_E, II_E, III_E, K_1, K_2\right) + (\det \mathbf{F})\sigma_m\, \mathbf{m}^0 \cdot \mathbf{E}_{(1)} \cdot \mathbf{m}^0. \tag{5.9.20}$$

Thus, we add to the right-hand side of Eq. (5.9.11) the term $(\det \mathbf{F})\sigma_m \mathbf{M}^0$, and to the right-hand side of Eq. (5.9.17) the term $(\det \mathbf{F})\sigma_m \mathbf{M}$, where the Lagrangian multiplier σ_m is an arbitrary tension in the direction $\mathbf{m}$.

5.9.1. Transversely Isotropic Cauchy-Elasticity

In this case, the stress is assumed to be a function of $\mathbf{E}_{(n)}$ and $\mathbf{M}^0$ at the outset,

$$\mathbf{T}_{(n)} = \mathbf{f}\left(\mathbf{E}_{(n)}, \mathbf{M}^0\right). \tag{5.9.21}$$

This must be an isotropic tensor function of both $\mathbf{E}_{(n)}$ and $\mathbf{M}^0$, so that

$$\mathbf{Q}_0 \cdot \mathbf{T}_{(n)} \cdot \mathbf{Q}_0^T = \mathbf{f}\left(\mathbf{Q}_0 \cdot \mathbf{E}_{(n)} \cdot \mathbf{Q}_0^T, \mathbf{Q}_0 \cdot \mathbf{M}^0 \cdot \mathbf{Q}_0^T\right). \tag{5.9.22}$$

Representation of isotropic second-order tensor functions of two symmetric second-order tensor arguments is well-known. The set of generating tensors is given in Eq. (1.11.10). Indeed, consider the most general isotropic invariant of $\mathbf{E}_{(n)}$, $\mathbf{M}^0$ and a symmetric tensor $\mathbf{H}$, which is linear in $\mathbf{H}$. Since $\mathbf{M}^0$ is idempotent, this invariant is

$$\begin{aligned} g = c_0 \operatorname{tr}\mathbf{H} &+ c_1 \operatorname{tr}\left(\mathbf{E}_{(n)} \cdot \mathbf{H}\right) + c_2 \operatorname{tr}\left(\mathbf{E}_{(n)}^2 \cdot \mathbf{H}\right) + c_3 \operatorname{tr}\left(\mathbf{M}^0 \cdot \mathbf{H}\right) \\ &+ c_4 \operatorname{tr}\left[\left(\mathbf{M}^0 \cdot \mathbf{E}_{(n)} + \mathbf{E}_{(n)} \cdot \mathbf{M}^0\right) \cdot \mathbf{H}\right] \\ &+ c_5 \operatorname{tr}\left[\left(\mathbf{M}^0 \cdot \mathbf{E}_{(n)}^2 + \mathbf{E}_{(n)}^2 \cdot \mathbf{M}^0\right) \cdot \mathbf{H}\right]. \end{aligned} \tag{5.9.23}$$

The parameters c_i are scalar invariants of $\mathbf{E}_{(n)}$ and $\mathbf{M}^0$. The stress tensor is derived as the gradient of g with respect to $\mathbf{H}$, which gives

$$\mathbf{T}_{(n)} = c_0\mathbf{I}^0 + c_1\mathbf{E}_{(n)} + c_2\mathbf{E}^2_{(n)} + c_3\mathbf{M}^0 + c_4\left(\mathbf{M}^0\cdot\mathbf{E}_{(n)} + \mathbf{E}_{(n)}\cdot\mathbf{M}^0\right) + c_5\left(\mathbf{M}^0\cdot\mathbf{E}^2_{(n)} + \mathbf{E}^2_{(n)}\cdot\mathbf{M}^0\right). \tag{5.9.24}$$

The term proportional to c_5 in Eq. (5.9.24) for transversely isotropic Cauchy-elasticity is absent in the case of transversely isotropic Green-elasticity, cf. Eq. (5.9.11). Also, it is noted that in the transition to linearized theory (retaining linear terms in strain $\mathbf{E}_{(n)}$ only), the Cauchy-elasticity of transversely isotropic materials involves six independent material parameters, while the Green-elasticity involves only five of them.

5.10. Orthotropic Materials

Elastic material is orthotropic in its reference configuration if it possesses three mutually orthogonal planes of elastic symmetry. Its symmetry group consists of reflections in these planes. Therefore, we introduce two second -order tensors

$$\mathbf{M}^0 = \mathbf{m}^0\otimes\mathbf{m}^0, \quad \mathbf{N}^0 = \mathbf{n}^0\otimes\mathbf{n}^0, \tag{5.10.1}$$

which are associated with the unit vectors $\mathbf{m}^0$ and $\mathbf{n}^0$, normal to two of the planes of elastic symmetry in the undeformed configuration. The tensor associated with the third plane of symmetry is $\mathbf{I}^0 - \mathbf{M}^0 - \mathbf{N}^0$, and need not be considered. The strain energy is then

$$\Psi = \Psi\left(\mathbf{E}_{(n)}, \mathbf{M}^0, \mathbf{N}^0\right). \tag{5.10.2}$$

This must be an isotropic function of all three tensor arguments,

$$\Psi\left(\mathbf{Q}_0\cdot\mathbf{E}_{(n)}\cdot\mathbf{Q}_0^T, \mathbf{Q}_0\cdot\mathbf{M}^0\cdot\mathbf{Q}_0^T, \mathbf{Q}_0\cdot\mathbf{N}^0\cdot\mathbf{Q}_0^T\right) = \Psi\left(\mathbf{E}_{(n)}, \mathbf{M}^0, \mathbf{N}^0\right), \tag{5.10.3}$$

and thus dependent on individual and joint invariants of its tensor arguments. Since $\mathbf{M}^0\cdot\mathbf{N}^0 = 0$, by the orthogonality of $\mathbf{m}^0$ and $\mathbf{n}^0$, it follows that

$$\Psi = \Psi\left(I_E, II_E, III_E, K_1, K_2, K_3, K_4\right). \tag{5.10.4}$$

The invariants K_1 and K_2 are defined by Eq. (5.9.9), and K_3 and K_4 by the corresponding expressions in which $\mathbf{M}^0$ is replaced with $\mathbf{N}^0$. The stress response is

$$\mathbf{T}_{(n)} = c_0\mathbf{I}^0 + c_1\mathbf{E}_{(n)} + c_2\mathbf{E}_{(n)}^2 + c_3\mathbf{M}^0 + c_4\left(\mathbf{M}^0 \cdot \mathbf{E}_{(n)} + \mathbf{E}_{(n)} \cdot \mathbf{M}^0\right)$$
$$+ c_5\mathbf{N}^0 + c_6\left(\mathbf{N}^0 \cdot \mathbf{E}_{(n)} + \mathbf{E}_{(n)} \cdot \mathbf{N}^0\right). \tag{5.10.5}$$

The coefficients c_0 to c_4 are specified by Eqs. (5.3.5) and (5.9.12), and c_5 and c_6 by equations (5.9.12) in which the derivatives are taken with respect to K_3 and K_4.

Equation (5.10.5) has a dual equation in the deformed configuration

$$\begin{aligned}\mathcal{T}_{(n)} = c_0\mathbf{I} + c_1\boldsymbol{\mathcal{E}}_{(n)} + c_2\boldsymbol{\mathcal{E}}_{(n)}^2 + c_3\overline{\mathbf{M}} + c_4\left(\overline{\mathbf{M}} \cdot \boldsymbol{\mathcal{E}}_{(n)} + \boldsymbol{\mathcal{E}}_{(n)} \cdot \overline{\mathbf{M}}\right)\\ + c_5\overline{\mathbf{N}} + c_6\left(\overline{\mathbf{N}} \cdot \boldsymbol{\mathcal{E}}_{(n)} + \boldsymbol{\mathcal{E}}_{(n)} \cdot \overline{\mathbf{N}}\right),\end{aligned} \tag{5.10.6}$$

where

$$\overline{\mathbf{M}} = \overline{\mathbf{m}} \otimes \overline{\mathbf{m}}, \quad \overline{\mathbf{N}} = \overline{\mathbf{n}} \otimes \overline{\mathbf{n}}, \tag{5.10.7}$$

and

$$\overline{\mathbf{m}} = \mathbf{R} \cdot \mathbf{m}^0, \quad \overline{\mathbf{n}} = \mathbf{R} \cdot \mathbf{n}^0. \tag{5.10.8}$$

In particular, for $n = 1$, Eq. (5.10.6) gives

$$\begin{aligned}\boldsymbol{\tau} = b_0\mathbf{I} + b_1\boldsymbol{\mathcal{E}}_{(1)} + b_2\boldsymbol{\mathcal{E}}_{(1)}^2 + c_3\mathbf{M} + c_4\left(\mathbf{M} \cdot \boldsymbol{\mathcal{E}}_{(1)} + \boldsymbol{\mathcal{E}}_{(1)} \cdot \mathbf{M}\right)\\ + c_5\mathbf{N} + c_6\left(\mathbf{N} \cdot \boldsymbol{\mathcal{E}}_{(1)} + \boldsymbol{\mathcal{E}}_{(1)} \cdot \mathbf{N}\right).\end{aligned} \tag{5.10.9}$$

The coefficients b_i are expressed in terms of c_i by Eqs. (5.4.6), and

$$\mathbf{M} = \mathbf{m} \otimes \mathbf{m}, \quad \mathbf{N} = \mathbf{n} \otimes \mathbf{n}. \tag{5.10.10}$$

The vectors $\mathbf{m}$ and $\mathbf{n}$ are

$$\mathbf{m} = \mathbf{V} \cdot \overline{\mathbf{m}} = \mathbf{F} \cdot \mathbf{m}^0, \quad \mathbf{n} = \mathbf{V} \cdot \overline{\mathbf{n}} = \mathbf{F} \cdot \mathbf{n}^0. \tag{5.10.11}$$

5.10.1. Orthotropic Cauchy-Elasticity

The stress is here assumed to be a function of three tensor arguments, such that

$$\mathbf{T}_{(n)} = \mathbf{f}\left(\mathbf{E}_{(n)}, \mathbf{M}^0, \mathbf{N}^0\right). \tag{5.10.12}$$

If the undeformed configuration is rotated by $\mathbf{Q}_0$, we have

$$\mathbf{Q}_0 \cdot \mathbf{T}_{(n)} \cdot \mathbf{Q}_0^T = \mathbf{f}\left(\mathbf{Q}_0 \cdot \mathbf{E}_{(n)} \cdot \mathbf{Q}_0^T, \mathbf{Q}_0 \cdot \mathbf{M}^0 \cdot \mathbf{Q}_0^T, \mathbf{Q}_0 \cdot \mathbf{N}^0 \cdot \mathbf{Q}_0^T\right), \tag{5.10.13}$$

which implies that $\mathbf{f}$ must be an isotropic tensor function of all three of its tensor arguments. The most general form of this function is

$$\begin{aligned}\mathbf{T}_{(n)} &= c_0\mathbf{I}^0 + c_1\mathbf{E}_{(n)} + c_2\mathbf{E}^2_{(n)} + c_3\mathbf{M}^0 + c_6\mathbf{N}^0 \\ &+ c_4\left(\mathbf{M}^0\cdot\mathbf{E}_{(n)} + \mathbf{E}_{(n)}\cdot\mathbf{M}^0\right) + c_5\left(\mathbf{M}^0\cdot\mathbf{E}^2_{(n)} + \mathbf{E}^2_{(n)}\cdot\mathbf{M}^0\right) \\ &+ c_7\left(\mathbf{N}^0\cdot\mathbf{E}_{(n)} + \mathbf{E}_{(n)}\cdot\mathbf{N}^0\right) + c_8\left(\mathbf{N}^0\cdot\mathbf{E}^2_{(n)} + \mathbf{E}^2_{(n)}\cdot\mathbf{N}^0\right).\end{aligned} \tag{5.10.14}$$

The terms proportional to c_5 and c_8 in Eq. (5.10.14) are absent in the case of orthotropic Green-elasticity, cf. Eq. (5.10.5). In the transition to linearized theory (retaining linear terms in strain $\mathbf{E}_{(n)}$ only), the Cauchy-elasticity of orthotropic materials involves twelve independent material parameters, while the Green-elasticity involves only nine of them.

5.11. Crystal Elasticity

5.11.1. Crystal Classes

Anisotropic materials known as crystal classes possess three preferred directions, defined by unit vectors $\mathbf{a}_1$, $\mathbf{a}_2$, and $\mathbf{a}_3$. There are thirty two crystal classes (point groups). Each class is characterized by a group of orthogonal transformations which carry the reference undeformed configuration into an equivalent configuration, indistinguishable from the original configuration. Since elastic properties of crystals are centrosymmetric, the eleven Laue groups can be identified. All point groups belonging to the same Laue group have common polynomial representation of the strain energy function in terms of the corresponding polynomial strain invariants. Crystal classes are grouped into seven crystal systems. In describing them, the following convention will be used. By $\frac{n}{m}$ is meant the rotation by an angle $2\pi/n$, followed by a reflection in the plane normal to the axis of rotation. By $\bar{n}$ is meant the rotation by an angle $2\pi/n$, followed by an inversion.

i) Triclinic System (Laue group N). For this crystal system there is no restriction on the orientation of the vectors $\mathbf{a}_i$. Two point groups of this system are $(1, \bar{1})$. Since components of the strain tensor $\mathbf{E}_{(n)}$ are unaltered by identity and central inversion transformations, no restriction is placed on the form of the polynomial representation of the strain energy in terms of the six strain components, i.e.,

$$\Psi = \Psi\left(E_{11}, E_{22}, E_{33}, E_{12}, E_{31}, E_{32}\right). \tag{5.11.1}$$

Any rectangular Cartesian coordinate system may be chosen as a reference system.

ii) Monoclinic System (Laue group M). The preferred directions $\mathbf{a}_1$ and $\mathbf{a}_2$ are not orthogonal, and the direction $\mathbf{a}_3$ is perpendicular to the plane $(\mathbf{a}_1, \mathbf{a}_2)$. There are three point groups of the monoclinic system. They are $\left(2,\, m,\, \frac{2}{m}\right)$. The symmetry transformation of the first point group is the rotation $\mathcal{Q}_3$ about X_3 axis through 180°, for the second it is reflection $\mathcal{R}_3$ in the plane normal to X_3 axis, and for the third it is the rotation $\mathcal{Q}_3$ followed by the reflection $\mathcal{R}_3$. For each point group, the strain energy is a polynomial of the seven polynomial strain invariants of this system, i.e.,

$$\Psi = \Psi\left(E_{11}, E_{22}, E_{33}, E_{12}, E_{31}^2, E_{32}^2, E_{31}E_{32}\right). \tag{5.11.2}$$

The rectangular Cartesian system is used with the axis X_3 parallel to $\mathbf{a}_3$, and with the axes X_1 and X_2 in any two orthogonal directions within $(\mathbf{a}_1, \mathbf{a}_2)$ plane.

iii) Orthorombic System (Laue group O). The preferred directions $\mathbf{a}_i$ are mutually perpendicular. There are three point groups of this system. They are $\left(222,\, mm2,\, \frac{2}{m}\frac{2}{m}\frac{2}{m}\right)$. For each point group, the strain energy is a polynomial of the seven polynomial strain invariants,

$$\Psi = \Psi\left(E_{11}, E_{22}, E_{33}, E_{12}^2, E_{31}^2, E_{32}^2, E_{12}E_{31}E_{32}\right). \tag{5.11.3}$$

The axes of the reference coordinate system are parallel to $\mathbf{a}_i$.

iv) Tetragonal System (Laue groups TII and TI). The vectors $\mathbf{a}_i$ are mutually perpendicular, but the direction $\mathbf{a}_3$ has a special significance and is called the principal axis of symmetry. The Laue group TII contains three point groups $\left(4,\, \bar{4},\, \frac{4}{m}\right)$. The corresponding strain energy is expressible as a polynomial in the twelve polynomial strain invariants. These are

$$\begin{aligned}
&E_{11}+E_{22}, \quad E_{33}, \quad E_{31}^2+E_{32}^2, \quad E_{12}^2, \quad E_{11}E_{22},\\
&E_{12}(E_{11}-E_{22}), \quad E_{31}E_{32}(E_{11}-E_{22}), \quad E_{12}E_{31}E_{32},\\
&E_{12}\left(E_{31}^2-E_{32}^2\right), \quad E_{11}E_{32}^2+E_{22}E_{31}^2, \quad E_{31}^2E_{32}^2,\\
&E_{31}E_{32}\left(E_{31}^2-E_{32}^2\right).
\end{aligned} \tag{5.11.4}$$

The Laue group TI contains four point groups $\left(422,\ 4mm,\ \bar{4}2m,\ \frac{4}{m}\frac{2}{m}\frac{2}{m}\right)$. The corresponding strain energy can be expressed as a polynomial in the eight polynomial strain invariants,

$$\begin{aligned} &E_{11}+E_{22}, \quad E_{33}, \quad E_{31}^2+E_{32}^2, \quad E_{12}^2, \quad E_{11}E_{22}, \\ &E_{12}E_{31}E_{32}, \quad E_{11}E_{32}^2+E_{22}E_{31}^2, \quad E_{31}^2E_{32}^2. \end{aligned} \tag{5.11.5}$$

The axes of the reference coordinate system are parallel to $\mathbf{a}_i$.

v) Cubic System (Laue groups CII and CI). The vectors $\mathbf{a}_i$ are mutually perpendicular. The Laue group CII contains two point groups $\left(23,\ \frac{2}{m}\bar{3}\right)$. The corresponding strain energy is a polynomial in the fourteen polynomial strain invariants. They are listed by Green and Adkins (1960), Eq. (1.11.2). The Laue group CI contains three point groups $\left(432,\ \bar{4}3m,\ \frac{4}{m}\bar{3}\frac{2}{m}\right)$. The corresponding strain energy is a polynomial in the nine polynomial strain invariants, which are listed in *op. cit.*, Eq. (1.11.4).

vi) Rhombohedral System (Laue groups RII and RI). The vector $\mathbf{a}_3$ is perpendicular to the basal plane defined by vectors $\mathbf{a}_1$ and $\mathbf{a}_2$, where $\mathbf{a}_2$ is at 120° from $\mathbf{a}_1$. The Laue group RII contains two point groups $(3,\ \bar{3})$. The corresponding strain energy is a polynomial in the fourteen polynomial strain invariants. They are listed in *op. cit.*, Eq. (1.12.5). The Laue group RI contains three point groups $\left(32,\ 3m,\ \bar{3}\frac{2}{m}\right)$. The corresponding strain energy is a polynomial in the nine polynomial strain invariants, listed by Green and Adkins (1960) in Eq. (1.12.8) (rhombohedral system is there considered to be hexagonal).

vii) Hexagonal System (Laue groups HII and HI). The vector $\mathbf{a}_3$ is perpendicular to the basal plane defined by vectors $\mathbf{a}_1$ and $\mathbf{a}_2$, where $\mathbf{a}_2$ is at 120° from $\mathbf{a}_1$. The Laue group HII contains three point groups $\left(6,\ \bar{6},\ \frac{6}{m}\right)$. The corresponding strain energy is a polynomial in the fourteen polynomial strain invariants; Eq. (1.12.11) of *op. cit.* The Laue group HI contains four point groups $\left(622,\ 6mm,\ \bar{6}m2,\ \frac{6}{m}\frac{2}{m}\frac{2}{m}\right)$. The corresponding strain energy is a polynomial in the nine polynomial strain invariants. These are given by Eq. (1.12.13) of *op. cit.*

In the remaining two subsections we consider the general strain energy representation, with a particular attention given to cubic crystals and their elastic constants.

5.11.2. Strain Energy Representation

For each Laue group, the strain energy can be expanded in a Taylor series about the state of zero strain and stress as

$$\Psi = \frac{1}{2!}\, C_{ijkl}E_{ij}E_{kl} + \frac{1}{3!}\, C_{ijklmn}E_{ij}E_{kl}E_{mn} + \cdots . \tag{5.11.6}$$

The E_{ij} are the rectangular Cartesian components of the strain tensor $\mathbf{E}_{(n)}$, and $C_{ijklmn\ldots}$ are the corresponding elastic stiffness constants or elastic moduli. For simplicity, we omit the label (n). The components of the conjugate stress are

$$T_{ij} = \frac{\partial \Psi}{\partial E_{ij}} = C_{ijkl}E_{kl} + \frac{1}{2}\, C_{ijklmn}E_{kl}E_{mn} + \cdots . \tag{5.11.7}$$

Elastic constants of the kth order are the components of the tensor of the order $2k$. Since they are the strain gradients of Ψ evaluated at zero strain,

$$C_{ijkl} = \left(\frac{\partial^2 \Psi}{\partial E_{ij}\partial E_{kl}}\right)_0, \quad C_{ijklmn} = \left(\frac{\partial^3 \Psi}{\partial E_{ij}\partial E_{kl}\partial E_{mn}}\right)_0, \ldots, \tag{5.11.8}$$

they possess the obvious basic symmetries. For example, the third-order elastic constants satisfy

$$C_{ijklmn} = C_{jiklmn}, \quad C_{ijklmn} = C_{klijmn} = C_{mnklij}. \tag{5.11.9}$$

Following the Voigt notation

$$11 \sim 1, \quad 22 \sim 2, \quad 33 \sim 3, \quad 23 \sim 4, \quad 13 \sim 5, \quad 12 \sim 6, \tag{5.11.10}$$

and the recipe

$$E_{ij} = \frac{1}{2}(1 + \delta_{ij})\eta_\vartheta, \quad \vartheta = 1, 2, \ldots, 6, \tag{5.11.11}$$

Equation (5.11.6) can be rewritten as (Brugger, 1964)

$$\begin{aligned}\Psi = {} & \frac{1}{2}\sum_i c_{ii}\eta_i^2 + \sum_{i<j} c_{ij}\eta_i\eta_j + \frac{1}{6}\sum_i c_{iii}\eta_i^3 \\ & + \frac{1}{2}\sum_{i\neq j} c_{iij}\eta_i^2\eta_j + \sum_{i<j<k} c_{ijk}\eta_i\eta_j\eta_k + \cdots .\end{aligned} \tag{5.11.12}$$

For triclinic crystals, whose symmetry group consists solely of the identity transformation, there are $\binom{5+k}{k}$ independent k^{th} order elastic constants (Toupin and Bernstein, 1961), i.e., there are at most 21 independent second-order elastic constants c_{ij}, and at most 56 independent third-order elastic constants c_{ijk}. For other crystal systems, fewer independent constants are involved, since they must be invariant under the group of transformations defining the material symmetry. This requires certain constants to vanish and supplies relations among some of the remaining ones. The tables for the second- and third-order independent elastic constants in crystals for all crystallographic groups can be found in Brugger (1965) and Thurston (1984). An analysis of eigenvalues and eigentensors of the elastic constants C_{ijkl} of an anisotropic material is given by Ting (1987), Mehrabadi and Cowin (1990), and Sutcliffe (1992).

5.11.3. Elastic Constants of Cubic Crystals

For cubic crystals belonging to the Laue group CI, there are at most three independent second-order and six independent third-order elastic constants. Written with respect to principal cubic axes, the strain energy can be expressed as (Birch, 1947)

$$
\begin{aligned}
\Psi = & \frac{1}{2} c_{11}\left(\eta_1^2+\eta_2^2+\eta_3^2\right)+\frac{1}{2} c_{44}\left(\eta_4^2+\eta_5^2+\eta_6^2\right) \\
& +c_{12}\left(\eta_1 \eta_2+\eta_2 \eta_3+\eta_3 \eta_1\right)+\frac{1}{6} c_{111}\left(\eta_1^3+\eta_2^3+\eta_3^3\right) \\
& +\frac{1}{2} c_{112}\left[\eta_1^2\left(\eta_2+\eta_3\right)+\eta_2^2\left(\eta_3+\eta_1\right)+\eta_3^2\left(\eta_1+\eta_2\right)\right] \\
& +\frac{1}{2} c_{144}\left(\eta_4^2 \eta_1+\eta_5^2 \eta_2+\eta_6^2 \eta_3\right)+\frac{1}{2} c_{244}\left[\eta_4^2\left(\eta_2+\eta_3\right)\right. \\
& \left.+\eta_5^2\left(\eta_3+\eta_1\right)+\eta_6^2\left(\eta_1+\eta_2\right)\right]+c_{123} \eta_1 \eta_2 \eta_3+c_{456} \eta_4 \eta_5 \eta_6,
\end{aligned} \tag{5.11.13}
$$

to third-order terms in strain. The corresponding components of the fourth-order tensor of the second-order elastic moduli, written with respect to an arbitrary rectangular Cartesian basis, are

$$
C_{ijkl} = c_{12}\delta_{ij}\delta_{kl} + 2c_{44}I_{ijkl} + (c_{11} - c_{12} - 2c_{44})\, A_{ijkl}. \tag{5.11.14}
$$

The components of the symmetric fourth-order unit tensor are again denoted by I_{ijkl}, and

$$
A_{ijkl} = a_i a_j a_k a_l + b_i b_j b_k b_l + c_i c_j c_k c_l. \tag{5.11.15}
$$

The vectors $\mathbf{a}$, $\mathbf{b}$, and $\mathbf{c}$ are the orthogonal unit vectors along the principal cubic axes (previously denoted by $\mathbf{a}_1$, $\mathbf{a}_2$, and $\mathbf{a}_3$).

Two independent linear invariants of the elastic moduli tensor C_{ijkl} are

$$
C_{iijj} = 3\,(c_{11} + 2c_{12}), \quad C_{ijij} = 3\,(c_{11} + 2c_{44}). \tag{5.11.16}
$$

In the case when $c_{11} - c_{12} = 2c_{44}$, the components C_{ijkl} are the components of an isotropic fourth-order tensor,

$$C_{ijkl} = c_{12}\delta_{ij}\delta_{kl} + 2c_{44}I_{ijkl}. \tag{5.11.17}$$

If the Cauchy symmetry

$$C_{ijkl} = C_{ikjl} \tag{5.11.18}$$

applies, then $c_{12} = c_{44}$. For example, in atomistic calculations the Cauchy symmetry is an inevitable consequence whenever the atomic interactions are modeled by pairwise central forces.

The sixth-order tensor of the third-order elastic moduli has the Cartesian components

$$\begin{aligned} C_{ijklmn} &= c_1\delta_{ij}\delta_{kl}\delta_{mn} + c_2\delta_{(ij}I_{klmn)} + c_3\delta_{(ik}\delta_{lm}\delta_{nj)} \\ &\quad + c_4\delta_{(ij}A_{klmn)} + c_5 a_{(i}a_j b_k b_l c_m c_{n)} + c_6 a_{(i}b_j c_k a_l b_m c_{n)}. \end{aligned} \tag{5.11.19}$$

The following constants are conveniently introduced

$$\begin{aligned} c_1 &= -\frac{1}{2}\left(c_{111} - 3c_{112} + 4c_{144} - 4c_{244}\right), \\ c_2 &= 6c_{144}, \quad c_3 = 4\left(c_{244} - c_{144}\right), \\ c_4 &= -\frac{3}{2}\left(c_{112} - c_{111} + 4c_{244}\right), \quad c_5 = 6\left(c_{123} - c_1\right), \\ c_6 &= 24\left(c_{144} - c_{244} + 2c_{456}\right). \end{aligned} \tag{5.11.20}$$

The notation such as $\delta_{(ij}A_{klmn)}$ designates the symmetrization. For example, we have

$$\begin{aligned} \delta_{(ij}I_{klmn)} &= \frac{1}{3}\left(\delta_{ij}I_{klmn} + \delta_{kl}I_{mnij} + \delta_{mn}I_{ijkl}\right), \\ \delta_{(ik}\delta_{lm}\delta_{nj)} &= \frac{1}{4}\left(\delta_{ik}I_{jlmn} + \delta_{il}I_{jkmn} + \delta_{im}I_{klnj} + \delta_{in}I_{klmj}\right). \end{aligned} \tag{5.11.21}$$

The tensors $\delta_{ij}\delta_{kl}\delta_{mn}$, $\delta_{(ij}I_{klmn)}$, and $\delta_{(ik}\delta_{lm}\delta_{nj)}$ constitute an integrity basis for the sixth-order isotropic tensors (Spencer, 1982). The tensors appearing on the right-hand side of Eq. (5.11.19) are the base tensors for the sixth-order elastic stiffness tensor with cubic symmetry. Other base tensors could also be constructed. The tensor representations of the second- and third-order elastic compliances are given by Lubarda (1997,1999).

Three independent linear invariants of the sixth-order tensor in Eq. (5.11.19) are

$$\begin{aligned} C_{iijjkk} &= 3\left(c_{111} + 6c_{112} + 2c_{123}\right), \\ C_{iiklkl} &= 3\left(c_{111} + 2c_{112} + 2c_{144} + 4c_{244}\right), \\ C_{ijjkki} &= 3\left(c_{111} + 6c_{244} + 2c_{456}\right). \end{aligned} \tag{5.11.22}$$

For isotropic materials

$$\begin{aligned} c_{111} &= c_{123} + 6c_{144} + 8c_{456}, \\ c_{112} &= c_{123} + 2c_{144}, \\ c_{244} &= c_{144} + 2c_{456}, \end{aligned} \tag{5.11.23}$$

so that C_{ijklmn} becomes an isotropic sixth-order tensor

$$C_{ijklmn} = c_{123}\delta_{ij}\delta_{kl}\delta_{mn} + 6c_{144}\delta_{(ij}I_{klmn)} + 8c_{456}\delta_{(ik}\delta_{lm}\delta_{nj)}. \tag{5.11.24}$$

If the Milder symmetry

$$C_{ijklmn} = C_{ikjlmn} \tag{5.11.25}$$

applies, then $c_{123} = c_{144} = c_{456}$.

The three independent third-order elastic constants of an isotropic material (c_{123}, c_{144}, and c_{456}) are related to Murnaghan's constants l, m, and n, which appear in the strain energy representation (5.3.8), by

$$l = c_{144} + \frac{1}{2}c_{123}, \quad m = c_{144} + 2c_{456}, \quad n = 4c_{456}. \tag{5.11.26}$$

Toupin and Bernstein (1961) used the notation $\nu_1 = c_{123}$, $\nu_2 = c_{144}$, and $\nu_3 = c_{456}$, referring to them as the third-order Lamé constants.

References

Anand, L. (1986), Moderate deformations in extension-torsion of incompressible isotropic elastic materials, *J. Mech. Phys. Solids*, Vol. 34, pp. 293–304.

Arruda, E. M. and Boyce, M. C. (1993), A three-dimensional constitutive model for the large stretch behavior of rubber elastic materials, *J. Mech. Phys. Solids*, Vol. 41, pp. 389–412.

Beatty, M. F. (1996), Introduction to nonlinear elasticity, in *Nonlinear Effects in Fluids and Solids*, eds. M. M. Carroll and M. A. Hayes, pp. 13–112, Plenum Press, New York.

Birch, F. (1947), Finite elastic strain of cubic crystals, *Phys. Rev.*, Vol. 71, pp. 809–824.

Blatz, P. J. and Ko, W. L. (1962), Application of finite elasticity theory to the deformation of rubbery materials, *Trans. Soc. Rheol.*, Vol. 6, pp. 223–251.

Blatz, P. J., Sharda, S. C., and Tschoegl, N. W. (1974), Strain energy function for rubberlike materials based on a generalized measure of strain, *Trans. Soc. Rheol.*, Vol. 18, pp. 145–161.

Boehler, J. P. (1987), Representations for isotropic and anisotropic nonpolynomial tensor functions, in *Applications of Tensor Functions in Solid Mechanics*, ed. J. P. Boehler, pp. 31–53, CISM Courses and Lectures No. 292, Springer, Wien.

Brugger, K. (1964), Thermodynamic definition of higher order elastic coefficients, *Phys. Rev.*, Vol. 133, pp. A1611–A1612.

Brugger, K. (1965), Pure modes for elastic waves in crystals, *J. Appl. Phys.*, Vol. 36, pp. 759–768.

Ciarlet, P. G. (1988), *Mathematical Elasticity, Volume I: Three-Dimensional Elasticity*, North-Holland, Amsterdam.

Doyle, T. C. and Ericksen, J. L. (1956), Nonlinear elasticity, *Adv. Appl. Mech.*, Vol. 4, pp. 53–115.

Green, A. E. and Adkins, J. E. (1960), *Large Elastic Deformations*, Oxford University Press, Oxford.

Hill, R. (1978), Aspects of invariance in solid mechanics, *Adv. Appl. Mech.*, Vol. 18, pp. 1–75.

Holzapfel, G. A. (2000), *Nonlinear Solid Mechanics*, John Wiley & Sons, Ltd, Chichester, England.

Leigh, D. C. (1968), *Nonlinear Continuum Mechanics*, McGraw Hill, New York.

Lubarda, V. A. (1997), New estimates of the third-order elastic constants for isotropic aggregates of cubic crystals, *J. Mech. Phys. Solids*, Vol. 45, pp. 471–490.

Lubarda, V. A. (1999), Apparent elastic constants of cubic crystals and their pressure derivatives, *Int. J. Nonlin. Mech.*, Vol. 34, pp. 5–11.

Malvern, L. E. (1969), *Introduction to the Mechanics of a Continuous Medium*, Prentice-Hall, Englewood Cliffs, New Jersey.

Mehrabadi, M. M. and Cowin, S. C. (1990), Eigentensors of linear anisotropic elastic materials, *Quart. J. Mech. Appl. Math.*, Vol. 43, pp. 15–41.

Morman, K. N. (1986), The generalized strain measure with application to nonhomogeneous deformations in rubber-like solids, *J. Appl. Mech.*, Vol. 53, pp. 726–728.

Murnaghan, F. D. (1951), *Finite Deformation of an Elastic Solid*, John-Wiley & Sons, New York.

Ogden, R. W. (1972), Large deformation isotropic elasticity: On the correlation of theory and experiment for incompressible rubberlike solids, *Proc. Roy. Soc. Lond. A*, Vol. 326, pp. 565–584.

Ogden, R. W. (1982), Elastic deformations of rubberlike solids, in *Mechanics of Solids: The Rodney Hill 60th Anniversary Volume*, eds. H. G. Hopkins and M. J. Sewell, pp. 499–537, Pergamon Press, Oxford.

Ogden, R. W. (1984), *Non-Linear Elastic Deformations*, Ellis Horwood Ltd., Chichester, England (2nd ed., Dover, 1997).

Rivlin, R. S. (1960), Some topics in finite elasticity, in *Structural Mechanics*, eds. J. N. Goodier and N. Hoff, pp. 169–198, Pergamon Press, New York.

Rivlin, R. S. and Sawyers, K. N. (1976), The strain-energy function for elastomers, *Trans. Soc. Rheol.*, Vol. 20, pp. 545–557.

Spencer, A. J. M. (1982), The formulation of constitutive equation for anisotropic solids, in *Mechanical Behavior of Anisotropic Solids*, ed. J. P. Boehler, pp. 2–26, Martinus Nijhoff Publishers, The Hague.

Sutcliffe, S. (1992), Spectral decomposition of the elasticity tensor, *J. Appl. Mech.*, Vol. 59, pp. 762–773.

Thurston, R. N. (1984), Waves in solids, in *Mechanics of Solids*, Vol. IV, ed. C. Truesdell, Springer-Verlag, Berlin.

Ting, T. C. T. (1987), Invariants of anisotropic elastic constants, *Quart. J. Mech. Appl. Math.*, Vol. 40, pp. 431–448.

Toupin, R. A. and Bernstein, B. (1961), Sound waves in deformed perfectly elastic materials. Acoustoelastic effect, *J. Acoust. Soc. Amer.*, Vol. 33, pp. 216–225.

Treloar, L. R. G. (1975), *The Physics of Rubber Elasticity*, Clarendon Press, Oxford.

Truesdell, C. and Noll, W. (1965), The nonlinear field theories of mechanics, in *Handbuch der Physik*, ed. S. Flügge, Band III/3, Springer-Verlag, Berlin (2nd ed., 1992).

Valanis, K. C. and Landel, R. F. (1967), The strain-energy function of a hyperelastic material in terms of the extension ratios, *J. Appl. Phys.*, Vol. 38, pp. 2997–3002.

Suggested Reading

Adkins, J. E. (1961), Large elastic deformations, in *Progress in Solid Mechanics*, eds. R. Hill and I. N. Sneddon, Vol. 2, pp. 1–60, North-Holland, Amsterdam.

Antman, S. S. (1995), *Nonlinear Problems of Elasticity*, Springer-Verlag, New York.

Beatty, M. F. (1987), Topics in finite elasticity: Hyperelasticity of rubber, elastomers, and biological tissues – with examples, *Appl. Mech. Rev.*, Vol. 40, pp. 1699–1734.

Carlson, D. E. and Shield, R. T., eds. (1982), *Finite Elasticity*, Martinus Nijhoff Publishers, The Hague.

Ericksen, J. L. (1977), Special topics in elastostatics, *Adv. Appl. Mech.*, Vol. 17, pp. 189–244.

Green, A. E. and Zerna, W. (1968), *Theoretical Elasticity*, Oxford University Press, Oxford.

Gurtin, M. E. (1981), *Topics in Finite Elasticity*, SIAM, Philadelphia.

Hanyga, A. (1985), *Mathematical Theory of Non-Linear Elasticity*, Ellis Horwood, Chichester, England, and PWN–Polish Scientific Publishers, Warsaw, Poland.

Marsden, J. E. and Hughes, T. J. R. (1983), *Mathematical Foundations of Elasticity*, Prentice Hall, Englewood Cliffs, New Jersey.

Rivlin, R. S. (1955), Further remarks on stress-deformation relations for isotropic materials, *J. Rat. Mech. Anal.*, Vol. 4, pp. 681–701.

Rivlin, R. S., ed. (1977), *Finite Elasticity*, ASME, AMD, Vol. 27, New York.

Smith, G. F. and Rivlin, R. S. (1958), The strain energy function for anisotropic elastic materials, *Trans. Am. Math. Soc.*, Vol. 88, pp. 175–193.

Ting, T. C. T. (1996), *Anisotropic Elasticity: Theory and Applications*, Oxford University Press, New York.

Truesdell, C. (1985), *The Elements of Continuum Mechanics*, Springer-Verlag, New York.

Wang, C.-C. and Truesdell, C. (1973), *Introduction to Rational Elasticity*, Noordhoff International Publishing, Leyden, The Netherlands.

CHAPTER 6

RATE-TYPE ELASTICITY

6.1. Elastic Moduli Tensors

The rate-type constitutive equation for finite deformation elasticity is obtained by differentiating Eq. (5.1.2) with respect to a time-like monotonically increasing parameter t. This gives

$$\dot{\mathbf{T}}_{(n)} = \boldsymbol{\Lambda}_{(n)} : \dot{\mathbf{E}}_{(n)}, \quad \boldsymbol{\Lambda}_{(n)} = \frac{\partial^2 \Psi\left(\mathbf{E}_{(n)}\right)}{\partial \mathbf{E}_{(n)} \otimes \partial \mathbf{E}_{(n)}}. \tag{6.1.1}$$

The fourth-order tensor $\boldsymbol{\Lambda}_{(n)}$ is the tensor of elastic moduli (or tensor of elasticities) associated with a conjugate pair of material tensors $\left(\mathbf{E}_{(n)}, \mathbf{T}_{(n)}\right)$. Its representation in an orthonormal basis in the undeformed configuration is

$$\boldsymbol{\Lambda}_{(n)} = \Lambda^{(n)}_{IJKL} \mathbf{e}^0_I \otimes \mathbf{e}^0_J \otimes \mathbf{e}^0_K \otimes \mathbf{e}^0_L. \tag{6.1.2}$$

Similarly, by applying to Eq. (5.1.10) the Jaumann derivative with respect to spin $\boldsymbol{\omega} = \dot{\mathbf{R}} \cdot \mathbf{R}^{-1}$, we obtain the rate-type constitutive equation

$$\overset{\bullet}{\boldsymbol{\mathcal{T}}}_{(n)} = \bar{\boldsymbol{\Lambda}}_{(n)} : \overset{\bullet}{\boldsymbol{\mathcal{E}}}_{(n)}, \quad \bar{\boldsymbol{\Lambda}}_{(n)} = \frac{\partial^2 \Psi\left(\hat{\boldsymbol{\mathcal{E}}}_{(n)}\right)}{\partial \boldsymbol{\mathcal{E}}_{(n)} \otimes \partial \boldsymbol{\mathcal{E}}_{(n)}}. \tag{6.1.3}$$

The fourth-order tensor $\bar{\boldsymbol{\Lambda}}_{(n)}$ is the tensor of elastic moduli associated with a conjugate pair of spatial tensors $\left(\boldsymbol{\mathcal{E}}_{(n)}, \boldsymbol{\mathcal{T}}_{(n)}\right)$. This can be represented in an orthonormal basis in the deformed configuration as

$$\bar{\boldsymbol{\Lambda}}_{(n)} = \bar{\Lambda}^{(n)}_{ijkl} \mathbf{e}_i \otimes \mathbf{e}_j \otimes \mathbf{e}_k \otimes \mathbf{e}_l. \tag{6.1.4}$$

The relationship between the tensors $\boldsymbol{\Lambda}_{(n)}$ and $\bar{\boldsymbol{\Lambda}}_{(n)}$ follows by recalling that

$$\hat{\boldsymbol{\mathcal{E}}}_{(n)} = \mathbf{E}_{(n)}, \quad \dot{\mathbf{T}}_{(n)} = \mathbf{R}^T \cdot \overset{\bullet}{\boldsymbol{\mathcal{T}}}_{(n)} \cdot \mathbf{R}, \quad \dot{\mathbf{E}}_{(n)} = \mathbf{R}^T \cdot \overset{\bullet}{\boldsymbol{\mathcal{E}}}_{(n)} \cdot \mathbf{R}, \tag{6.1.5}$$

which gives

$$\bar{\boldsymbol{\Lambda}}_{(n)} = \mathbf{R}\,\mathbf{R}\,\boldsymbol{\Lambda}_{(n)}\,\mathbf{R}^T\,\mathbf{R}^T. \tag{6.1.6}$$

The tensor products in Eq. (6.1.6) are defined so that the Cartesian components are related by

$$\bar{\Lambda}^{(n)}_{ijkl} = R_{iM} R_{jN} \Lambda^{(n)}_{MNPQ} R^T_{Pk} R^T_{Ql}. \tag{6.1.7}$$

In performing the Jaumann derivation of Eq. (5.1.10) it should be kept in mind that

$$\overset{\bullet}{\hat{\boldsymbol{\mathcal{E}}}}_{(n)} = \dot{\hat{\boldsymbol{\mathcal{E}}}}_{(n)} = \dot{\mathbf{E}}_{(n)}, \tag{6.1.8}$$

since corotational (and convected) derivatives of the material tensors are equal to ordinary material derivatives (material tensors not being affected by the transformation of the base tensors in the deformed configuration). It is instructive to discuss this point a little further. To be more specific, consider a transversely isotropic material from Section 5.9, for which the strain energy is

$$\Psi = \Psi\left(\mathbf{E}_{(n)}, \mathbf{M}^0\right) = \Psi\left(\boldsymbol{\mathcal{E}}_{(n)}, \overline{\mathbf{M}}\right), \tag{6.1.9}$$

with the spatial stress tensor

$$\boldsymbol{\mathcal{T}}_{(n)} = \frac{\partial \Psi}{\partial \boldsymbol{\mathcal{E}}_{(n)}}. \tag{6.1.10}$$

The application of the Jaumann derivative with respect to spin $\boldsymbol{\omega}$ to Eq. (6.1.10) gives

$$\begin{aligned}\overset{\bullet}{\boldsymbol{\mathcal{T}}}_{(n)} &= \frac{\partial^2 \Psi}{\partial \boldsymbol{\mathcal{E}}_{(n)} \otimes \partial \boldsymbol{\mathcal{E}}_{(n)}} : \overset{\bullet}{\boldsymbol{\mathcal{E}}}_{(n)} + \frac{\partial^2 \Psi}{\partial \boldsymbol{\mathcal{E}}_{(n)} \otimes \partial \overline{\mathbf{M}}} : \overset{\bullet}{\overline{\mathbf{M}}} \\ &= \frac{\partial^2 \Psi}{\partial \boldsymbol{\mathcal{E}}_{(n)} \otimes \partial \boldsymbol{\mathcal{E}}_{(n)}} : \overset{\bullet}{\boldsymbol{\mathcal{E}}}_{(n)},\end{aligned} \tag{6.1.11}$$

because $\overset{\bullet}{\overline{\mathbf{M}}} = \mathbf{0}$. Recall that $\overline{\mathbf{M}} = \mathbf{R} \cdot \mathbf{M}^0 \cdot \mathbf{R}^T$, so that

$$\dot{\overline{\mathbf{M}}} = \boldsymbol{\omega} \cdot \overline{\mathbf{M}} - \overline{\mathbf{M}} \cdot \boldsymbol{\omega}. \tag{6.1.12}$$

If the ordinary material derivative of Eq. (6.1.10) is taken, we have

$$\dot{\boldsymbol{\mathcal{T}}}_{(n)} = \frac{\partial^2 \Psi}{\partial \boldsymbol{\mathcal{E}}_{(n)} \otimes \partial \boldsymbol{\mathcal{E}}_{(n)}} : \dot{\boldsymbol{\mathcal{E}}}_{(n)} + \frac{\partial^2 \Psi}{\partial \boldsymbol{\mathcal{E}}_{(n)} \otimes \partial \overline{\mathbf{M}}} : \dot{\overline{\mathbf{M}}}. \tag{6.1.13}$$

This is in accord with Eq. (6.1.11) because the identity holds

$$\begin{aligned}&\frac{\partial^2 \Psi}{\partial \boldsymbol{\mathcal{E}}_{(n)} \otimes \partial \boldsymbol{\mathcal{E}}_{(n)}} : \left(\boldsymbol{\mathcal{E}}_{(n)} \cdot \boldsymbol{\omega} - \boldsymbol{\omega} \cdot \boldsymbol{\mathcal{E}}_{(n)}\right) + \frac{\partial^2 \Psi}{\partial \boldsymbol{\mathcal{E}}_{(n)} \otimes \partial \overline{\mathbf{M}}} \\ &\quad : \left(\overline{\mathbf{M}} \cdot \boldsymbol{\omega} - \boldsymbol{\omega} \cdot \overline{\mathbf{M}}\right) = \boldsymbol{\mathcal{T}}_{(n)} \cdot \boldsymbol{\omega} - \boldsymbol{\omega} \cdot \boldsymbol{\mathcal{T}}_{(n)}.\end{aligned} \tag{6.1.14}$$

To verify Eq. (6.1.14), we can differentiate both sides of Eq. (6.1.9) to obtain

$$\frac{\partial \Psi}{\partial \mathbf{E}_{(n)}} : \dot{\mathbf{E}}_{(n)} = \frac{\partial \Psi}{\partial \boldsymbol{\mathcal{E}}_{(n)}} : \overset{\bullet}{\boldsymbol{\mathcal{E}}}_{(n)} = \frac{\partial \Psi}{\partial \boldsymbol{\mathcal{E}}_{(n)}} : \dot{\boldsymbol{\mathcal{E}}}_{(n)} + \frac{\partial \Psi}{\partial \overline{\mathbf{M}}} : \dot{\overline{\mathbf{M}}}, \tag{6.1.15}$$

which establishes the identity

$$\frac{\partial \Psi}{\partial \boldsymbol{\mathcal{E}}_{(n)}} : \left(\boldsymbol{\mathcal{E}}_{(n)} \cdot \boldsymbol{\omega} - \boldsymbol{\omega} \cdot \boldsymbol{\mathcal{E}}_{(n)}\right) = \frac{\partial \Psi}{\partial \overline{\mathbf{M}}} : \left(\boldsymbol{\omega} \cdot \overline{\mathbf{M}} - \overline{\mathbf{M}} \cdot \boldsymbol{\omega}\right). \tag{6.1.16}$$

Differentiation of Eq. (6.1.16) with respect to $\boldsymbol{\mathcal{E}}_{(n)}$ gives Eq. (6.1.14).

6.2. Elastic Moduli for Conjugate Measures with $\mathbf{n} = \pm 1$

The rates of the conjugate tensors $\mathbf{E}_{(1)}$ and $\mathbf{T}_{(1)}$ are, from Eqs. (2.6.1) and (3.8.4),

$$\dot{\mathbf{E}}_{(1)} = \mathbf{F}^T \cdot \mathbf{D} \cdot \mathbf{F}, \quad \dot{\mathbf{T}}_{(1)} = \mathbf{F}^{-1} \cdot \overset{\triangle}{\boldsymbol{\tau}} \cdot \mathbf{F}^{-T}. \tag{6.2.1}$$

Substitution into Eq. (6.1.1) gives the Oldroyd rate of the Kirchhoff stress $\boldsymbol{\tau}$ in terms of the rate of deformation $\mathbf{D}$,

$$\overset{\triangle}{\boldsymbol{\tau}} = \boldsymbol{\mathcal{L}}_{(1)} : \mathbf{D}. \tag{6.2.2}$$

The corresponding elastic moduli tensor is

$$\boldsymbol{\mathcal{L}}_{(1)} = \mathbf{F}\,\mathbf{F}\,\boldsymbol{\Lambda}_{(1)}\,\mathbf{F}^T\,\mathbf{F}^T. \tag{6.2.3}$$

The products in Eq. (6.2.3) are such that the Cartesian components of the two tensors of elasticities are related by

$$\mathcal{L}^{(1)}_{ijkl} = F_{iM} F_{jN} \Lambda^{(1)}_{MNPQ} F^T_{Pk} F^T_{Ql}. \tag{6.2.4}$$

Equation (6.2.2) can also be derived from the first of Eq. (5.1.11) by applying to it, for example, the convected derivative $\overset{\triangle}{(\;)}$, and by recalling that

$$\overset{\triangle}{\mathbf{F}} = \mathbf{0}, \quad \overset{\triangle}{\mathbf{E}}_{(1)} = \dot{\mathbf{E}}_{(1)}. \tag{6.2.5}$$

See also Truesdell and Noll (1965), and Marsden and Hughes (1983).

Similarly, from Eqs. (2.6.3) and (3.8.5), the rates of the conjugate measures $\mathbf{E}_{(-1)}$ and $\mathbf{T}_{(-1)}$ are

$$\dot{\mathbf{E}}_{(-1)} = \mathbf{F}^{-1} \cdot \mathbf{D} \cdot \mathbf{F}^{-T}, \quad \dot{\mathbf{T}}_{(-1)} = \mathbf{F}^T \cdot \overset{\triangledown}{\boldsymbol{\tau}} \cdot \mathbf{F}. \tag{6.2.6}$$

Substitution into Eq. (6.1.1) gives

$$\overset{\triangledown}{\boldsymbol{\tau}} = \boldsymbol{\mathcal{L}}_{(-1)} : \mathbf{D}, \tag{6.2.7}$$

where

$$\boldsymbol{\mathcal{L}}_{(-1)} = \mathbf{F}^{-T}\,\mathbf{F}^{-T}\,\boldsymbol{\Lambda}_{(-1)}\,\mathbf{F}^{-1}\,\mathbf{F}^{-1}. \tag{6.2.8}$$

This can be alternatively derived by applying the convected derivative $\overset{\triangledown}{(\ )}$ to the second of Eq. (5.1.11), and by recalling that

$$(\mathbf{F}^{-1})^{\triangledown} = \mathbf{0}, \quad \overset{\triangledown}{\mathbf{E}}_{(-1)} = \dot{\mathbf{E}}_{(-1)}. \tag{6.2.9}$$

In view of the connection (6.1.6) between the moduli $\boldsymbol{\Lambda}_{(n)}$ and $\bar{\boldsymbol{\Lambda}}_{(n)}$, Eqs. (6.2.3) and (6.2.8) can be rewritten as

$$\boldsymbol{\mathcal{L}}_{(1)} = \mathbf{V}\,\mathbf{V}\,\bar{\boldsymbol{\Lambda}}_{(1)}\,\mathbf{V}\,\mathbf{V}, \quad \boldsymbol{\mathcal{L}}_{(-1)} = \mathbf{V}^{-1}\,\mathbf{V}^{-1}\,\bar{\boldsymbol{\Lambda}}_{(-1)}\,\mathbf{V}^{-1}\,\mathbf{V}^{-1}. \tag{6.2.10}$$

Another route to derive Eq. (6.2.2) is by differentiation of Eqs. (5.1.12). For example, by applying the Jaumann derivative $\overset{\bullet}{(\ )}$ to the first of Eqs. (5.1.12) gives

$$\begin{aligned}\overset{\bullet}{\boldsymbol{\tau}} = {} & \left(\overset{\bullet}{\mathbf{V}}\cdot\mathbf{V}^{-1}\right)\cdot\boldsymbol{\tau} + \boldsymbol{\tau}\cdot\left(\mathbf{V}^{-1}\cdot\overset{\bullet}{\mathbf{V}}\right)\\ & + \mathbf{V}\left(\frac{\partial^2\Psi}{\partial\boldsymbol{\mathcal{E}}_{(1)}\otimes\partial\hat{\boldsymbol{\mathcal{E}}}_{(1)}} : \dot{\hat{\boldsymbol{\mathcal{E}}}}_{(1)}\right)\mathbf{V}.\end{aligned} \tag{6.2.11}$$

Since

$$\overset{\bullet}{\mathbf{V}}\cdot\mathbf{V}^{-1} = \mathbf{L} - \boldsymbol{\omega}, \quad \dot{\hat{\boldsymbol{\mathcal{E}}}}_{(1)} = \mathbf{R}^T\cdot\overset{\bullet}{\boldsymbol{\mathcal{E}}}_{(1)}\cdot\mathbf{R}, \quad \overset{\bullet}{\boldsymbol{\mathcal{E}}}_{(1)} = \mathbf{V}\cdot\mathbf{D}\cdot\mathbf{V}, \tag{6.2.12}$$

Equation (6.2.11) becomes

$$\overset{\triangle}{\boldsymbol{\tau}} = \left(\mathbf{V}\mathbf{V}\,\frac{\partial^2\Psi}{\partial\boldsymbol{\mathcal{E}}_{(1)}\otimes\partial\boldsymbol{\mathcal{E}}_{(1)}}\,\mathbf{V}\mathbf{V}\right) : \mathbf{D} = \boldsymbol{\mathcal{L}}_{(1)} : \mathbf{D}. \tag{6.2.13}$$

The rate-type constitutive Eqs. (6.2.2) and (6.2.7) can be rewritten in terms of the Jaumann rate $\overset{\circ}{\boldsymbol{\tau}}$ as

$$\overset{\circ}{\boldsymbol{\tau}} = \boldsymbol{\mathcal{L}}_{(0)} : \mathbf{D}, \tag{6.2.14}$$

where

$$\boldsymbol{\mathcal{L}}_{(0)} = \boldsymbol{\mathcal{L}}_{(1)} + 2\boldsymbol{\mathcal{S}} = \boldsymbol{\mathcal{L}}_{(-1)} - 2\boldsymbol{\mathcal{S}}. \tag{6.2.15}$$

This follows because of the relationships (see Section 3.8)

$$\overset{\circ}{\boldsymbol{\tau}} = \overset{\triangle}{\boldsymbol{\tau}} + \mathbf{D}\cdot\boldsymbol{\tau} + \boldsymbol{\tau}\cdot\mathbf{D} = \overset{\triangledown}{\boldsymbol{\tau}} - \mathbf{D}\cdot\boldsymbol{\tau} - \boldsymbol{\tau}\cdot\mathbf{D}. \tag{6.2.16}$$

The Cartesian components of the fourth-order tensor $\boldsymbol{\mathcal{S}}$ are

$$\mathcal{S}_{ijkl} = \tau_{(ik}\delta_{jl)} = \frac{1}{4}\left(\tau_{ik}\delta_{jl} + \tau_{jk}\delta_{il} + \tau_{il}\delta_{jk} + \tau_{jl}\delta_{ik}\right). \tag{6.2.17}$$

The elastic moduli tensors $\boldsymbol{\Lambda}_{(n)}$, $\bar{\boldsymbol{\Lambda}}_{(n)}$ and $\boldsymbol{\mathcal{L}}_{(n)}$ all possess the basic and reciprocal (major) symmetries, e.g.,

$$\mathcal{L}_{ijkl}^{(n)} = \mathcal{L}_{jikl}^{(n)} = \mathcal{L}_{ijlk}^{(n)}, \quad \mathcal{L}_{ijkl}^{(n)} = \mathcal{L}_{klij}^{(n)}. \tag{6.2.18}$$

Further analysis of elastic moduli tensors can be found in Truesdell and Toupin (1960), Ogden (1984), and Holzapfel (2000).

6.3. Instantaneous Elastic Moduli

The instantaneous elastic moduli relate the rates of conjugate stress and strain tensors, when these are evaluated at the current configuration as the reference. Thus, since $\dot{\underline{\mathbf{E}}}_{(n)} = \mathbf{D}$, we write

$$\dot{\underline{\mathbf{T}}}_{(n)} = \underline{\boldsymbol{\Lambda}}_{(n)} : \dot{\underline{\mathbf{E}}}_{(n)} = \underline{\boldsymbol{\Lambda}}_{(n)} : \mathbf{D}. \tag{6.3.1}$$

The tensor of instantaneous elastic moduli $\underline{\boldsymbol{\Lambda}}_{(n)}$ can be related to the corresponding tensor of elastic moduli $\boldsymbol{\Lambda}_{(n)}$ by using the relationship between $\dot{\mathbf{E}}_{(n)}$ and $\dot{\underline{\mathbf{E}}}_{(n)}$. For example, for $n = 1$, from Eq. (3.9.16) we obtain

$$\dot{\mathbf{T}}_{(1)} = (\det \mathbf{F})\, \mathbf{F}^{-1} \cdot \dot{\underline{\mathbf{T}}}_{(1)} \cdot \mathbf{F}^{-T}, \quad \dot{\mathbf{E}}_{(1)} = \mathbf{F}^T \cdot \mathbf{D} \cdot \mathbf{F}. \tag{6.3.2}$$

The substitution into Eq. (6.1.1) gives

$$\begin{gathered} \dot{\underline{\mathbf{T}}}_{(1)} = \underline{\boldsymbol{\Lambda}}_{(1)} : \mathbf{D}, \\ \underline{\boldsymbol{\Lambda}}_{(1)} = (\det \mathbf{F})^{-1}\, \mathbf{F}\, \mathbf{F}\, \boldsymbol{\Lambda}_{(1)}\, \mathbf{F}^T\, \mathbf{F}^T = (\det \mathbf{F})^{-1}\, \boldsymbol{\mathcal{L}}_{(1)}. \end{gathered} \tag{6.3.3}$$

Recalling from Eq. (3.9.15) that $\dot{\underline{\mathbf{T}}}_{(1)} = \overset{\triangle}{\underline{\boldsymbol{\tau}}}$, Eq. (6.3.3) becomes

$$\overset{\triangle}{\underline{\boldsymbol{\tau}}} = \underline{\boldsymbol{\mathcal{L}}}_{(1)} : \mathbf{D}, \quad \underline{\boldsymbol{\mathcal{L}}}_{(1)} = \underline{\boldsymbol{\Lambda}}_{(1)}. \tag{6.3.4}$$

Similarly,

$$\overset{\triangledown}{\underline{\boldsymbol{\tau}}} = \underline{\boldsymbol{\mathcal{L}}}_{(-1)} : \mathbf{D}, \quad \underline{\boldsymbol{\mathcal{L}}}_{(-1)} = (\det \mathbf{F})^{-1}\, \boldsymbol{\mathcal{L}}_{(-1)}. \tag{6.3.5}$$

Furthermore, from Eq. (3.9.7) we have

$$\dot{\underline{\mathbf{T}}}_{(n)} = \overset{\circ}{\underline{\boldsymbol{\tau}}} - n(\mathbf{D} \cdot \boldsymbol{\sigma} + \boldsymbol{\sigma} \cdot \mathbf{D}) = \overset{\triangle}{\underline{\boldsymbol{\tau}}} - (n-1)(\mathbf{D} \cdot \boldsymbol{\sigma} + \boldsymbol{\sigma} \cdot \mathbf{D}). \tag{6.3.6}$$

Thus, Eq. (6.3.1) can be recast in the form

$$\overset{\triangle}{\underline{\boldsymbol{\tau}}} - (n-1)(\mathbf{D} \cdot \boldsymbol{\sigma} + \boldsymbol{\sigma} \cdot \mathbf{D}) = \underline{\boldsymbol{\mathcal{L}}}_{(n)} : \mathbf{D}, \tag{6.3.7}$$

since, in general,

$$\underline{\boldsymbol{\mathcal{L}}}_{(n)} = \underline{\boldsymbol{\Lambda}}_{(n)} = \underline{\bar{\boldsymbol{\Lambda}}}_{(n)}. \tag{6.3.8}$$

Substituting the expression (6.3.4) for $\overset{\triangle}{\underline{\boldsymbol{\tau}}}$ into Eq. (6.3.7) gives

$$\underline{\boldsymbol{\mathcal{L}}}_{(1)} : \mathbf{D} - (n-1)(\mathbf{D}\cdot\boldsymbol{\sigma} + \boldsymbol{\sigma}\cdot\mathbf{D}) = \underline{\boldsymbol{\mathcal{L}}}_{(n)} : \mathbf{D}. \tag{6.3.9}$$

This establishes the relationship between the instantaneous elastic moduli $\underline{\boldsymbol{\mathcal{L}}}_{(n)}$ and $\underline{\boldsymbol{\mathcal{L}}}_{(1)}$,

$$\underline{\boldsymbol{\mathcal{L}}}_{(n)} = \underline{\boldsymbol{\mathcal{L}}}_{(1)} - 2(n-1)\underline{\boldsymbol{\mathcal{S}}}. \tag{6.3.10}$$

The Cartesian components of the tensor $\underline{\boldsymbol{\mathcal{S}}}$ are

$$\underline{\mathcal{S}}_{ijkl} = \sigma_{(ik}\delta_{jl)} = \frac{1}{4}\left(\sigma_{ik}\delta_{jl} + \sigma_{jk}\delta_{il} + \sigma_{il}\delta_{jk} + \sigma_{jl}\delta_{ik}\right). \tag{6.3.11}$$

Thus, the difference between the various instantaneous elastic moduli in Eq. (6.3.10) is of the order of the Cauchy stress.

If the logarithmic strain is used, we have

$$\dot{\underline{\mathbf{T}}}_{(0)} = \overset{\circ}{\underline{\boldsymbol{\tau}}} = \underline{\boldsymbol{\mathcal{L}}}_{(0)} : \mathbf{D}, \tag{6.3.12}$$

and comparison with Eq. (6.3.7) gives

$$\underline{\boldsymbol{\mathcal{L}}}_{(n)} = \underline{\boldsymbol{\mathcal{L}}}_{(0)} - 2n\underline{\boldsymbol{\mathcal{S}}}. \tag{6.3.13}$$

In particular,

$$\underline{\boldsymbol{\mathcal{L}}}_{(0)} = \underline{\boldsymbol{\mathcal{L}}}_{(1)} + 2\underline{\boldsymbol{\mathcal{S}}} = \underline{\boldsymbol{\mathcal{L}}}_{(-1)} - 2\underline{\boldsymbol{\mathcal{S}}}, \tag{6.3.14}$$

as expected from Eq. (6.2.15). Further details are available in Hill (1978) and Ogden (1984).

6.4. Elastic Pseudomoduli

The nonsymmetric nominal stress $\mathbf{P}$ is derived from the strain energy function as its gradient with respect to deformation gradient $\mathbf{F}$, such that

$$\mathbf{P} = \frac{\partial\Psi}{\partial\mathbf{F}}, \quad P_{Ji} = \frac{\partial\Psi}{\partial F_{iJ}}. \tag{6.4.1}$$

The rate of the nominal stress is, therefore,

$$\dot{\mathbf{P}} = \boldsymbol{\Lambda}\cdot\cdot\,\dot{\mathbf{F}} = \boldsymbol{\Lambda}\cdot\cdot\,(\mathbf{L}\cdot\mathbf{F}), \quad \boldsymbol{\Lambda} = \frac{\partial^2\Psi}{\partial\mathbf{F}\otimes\partial\mathbf{F}}. \tag{6.4.2}$$

A two-point tensor of elastic pseudomoduli is denoted by $\boldsymbol{\Lambda}$. The Cartesian component representation of Eq. (6.4.2) is

$$\dot{P}_{Ji} = \Lambda_{JiLk}\dot{F}_{kL}, \quad \Lambda_{JiLk} = \frac{\partial^2\Psi}{\partial F_{iJ}\partial F_{kL}}. \tag{6.4.3}$$

The elastic pseudomoduli Λ_{JiLk} are not true moduli since they are partly associated with the material spin. They clearly possess the reciprocal symmetry

$$\Lambda_{JiLk} = \Lambda_{LkJi}. \tag{6.4.4}$$

In view of the connection

$$\mathbf{P} = \mathbf{T}_{(1)} \cdot \mathbf{F}^T, \tag{6.4.5}$$

the differentiation gives

$$\boldsymbol{\Lambda} \cdot\cdot \dot{\mathbf{F}} = \left(\boldsymbol{\Lambda}_{(1)} : \dot{\mathbf{E}}_{(1)}\right) \cdot \mathbf{F}^T + \mathbf{T}_{(1)} \cdot \dot{\mathbf{F}}^T. \tag{6.4.6}$$

Upon using

$$\dot{\mathbf{E}}_{(1)} = \frac{1}{2}\left(\dot{\mathbf{F}}^T \cdot \mathbf{F} + \mathbf{F}^T \cdot \dot{\mathbf{F}}\right), \tag{6.4.7}$$

Equation (6.4.6) yields the connection between the elastic moduli $\boldsymbol{\Lambda}$ and $\boldsymbol{\Lambda}_{(1)}$. Their Cartesian components are related by

$$\Lambda_{JiLk} = \Lambda^{(1)}_{JMLN} F_{iM} F_{kN} + T^{(1)}_{JL} \delta_{ik}. \tag{6.4.8}$$

Since $\mathbf{F} \cdot \mathbf{P}$ is a symmetric tensor, i.e.,

$$F_{iK} P_{Kj} = F_{jK} P_{Ki}, \tag{6.4.9}$$

by differentiation and incorporation of Eq. (6.4.3) it follows that

$$F_{jM} \Lambda_{MiLk} - F_{iM} \Lambda_{MjLk} = \delta_{ik} P_{Lj} - \delta_{jk} P_{Li}. \tag{6.4.10}$$

This corresponds to the symmetry in the leading pair of indices of the true elastic moduli

$$\Lambda^{(1)}_{IJKL} = \Lambda^{(1)}_{JIKL}. \tag{6.4.11}$$

The tensor of elastic pseudomoduli $\boldsymbol{\Lambda}$ can be related to the tensor of instantaneous elastic moduli, appearing in the expression

$$\underline{\dot{\mathbf{P}}} = \underline{\boldsymbol{\Lambda}} \cdot\cdot \mathbf{L}, \tag{6.4.12}$$

by recalling the relationship

$$\dot{\mathbf{P}} = (\det \mathbf{F}) \mathbf{F}^{-1} \cdot \underline{\dot{\mathbf{P}}}, \tag{6.4.13}$$

from Section 3.9. This gives

$$\underline{\boldsymbol{\Lambda}} = (\det \mathbf{F})^{-1} \mathbf{F} \, \boldsymbol{\Lambda} \, \mathbf{F}^T, \tag{6.4.14}$$

with the Cartesian component representation

$$\underline{\Lambda}_{ijkl} = (\det \mathbf{F})^{-1} F_{iM} \Lambda_{MjNk} F^T_{Nl}. \tag{6.4.15}$$

In addition, from Eq. (6.4.8), we have

$$\underline{\Lambda}_{jilk} = \underline{\Lambda}^{(1)}_{jilk} + \sigma_{jl}\delta_{ik}. \tag{6.4.16}$$

6.5. Elastic Moduli of Isotropic Elasticity

For isotropic elasticity, the strain energy function is an isotropic function of strain, so that

$$\Psi = \Psi\left(\hat{\mathcal{E}}_{(n)}\right) = \Psi\left(\mathcal{E}_{(n)}\right), \tag{6.5.1}$$

and

$$\mathcal{T}_{(n)} = \frac{\partial \Psi\left(\mathcal{E}_{(n)}\right)}{\partial \mathcal{E}_{(n)}} = c_0\mathbf{I} + c_1\mathcal{E}_{(n)} + c_2\mathcal{E}^2_{(n)}. \tag{6.5.2}$$

By definition of the Jaumann derivative, we have

$$\left(\frac{\partial \Psi}{\partial \mathcal{E}_{(n)}}\right)^{\circ} = \left(\frac{\partial \Psi}{\partial \mathcal{E}_{(n)}}\right)^{\cdot} - \mathbf{W}\cdot\frac{\partial \Psi}{\partial \mathcal{E}_{(n)}} + \frac{\partial \Psi}{\partial \mathcal{E}_{(n)}}\cdot\mathbf{W}. \tag{6.5.3}$$

Since Ψ is an isotropic function of $\mathcal{E}_{(n)}$, there is an identity

$$\frac{\partial^2 \Psi}{\partial \mathcal{E}_{(n)}\otimes\partial \mathcal{E}_{(n)}} : \left(\mathbf{W}\cdot\mathcal{E}_{(n)} - \mathcal{E}_{(n)}\cdot\mathbf{W}\right) = \mathbf{W}\cdot\frac{\partial \Psi}{\partial \mathcal{E}_{(n)}} - \frac{\partial \Psi}{\partial \mathcal{E}_{(n)}}\cdot\mathbf{W}, \tag{6.5.4}$$

which is easily verified by using Eq. (6.5.2). Thus, we can write

$$\overset{\circ}{\mathcal{T}}_{(n)} = \frac{\partial^2 \Psi}{\partial \mathcal{E}_{(n)}\otimes\partial \mathcal{E}_{(n)}} : \overset{\circ}{\mathcal{E}}_{(n)}. \tag{6.5.5}$$

This is one of the constitutive structures of the rate-type isotropic elasticity. It is pointed out that Eq. (6.5.5) also applies if $\overset{\circ}{(\;)}$ is replaced by the material derivative, or the Jaumann derivative with respect to spin $\boldsymbol{\omega}$, or any other spin associated with the deformed configuration.

An appealing rate-type constitutive structure of isotropic elasticity is obtained by using Eq. (5.5.5) to express the Kirchhoff stress in terms of the left Cauchy–Green deformation tensor $\mathbf{B}$. The application of the Jaumann derivative $\overset{\circ}{(\;)}$ gives (e.g., Lubarda, 1986)

$$\begin{aligned}\overset{\circ}{\boldsymbol{\tau}} = {} & \frac{1}{2}\left(\mathbf{D}\cdot\boldsymbol{\tau} + \boldsymbol{\tau}\cdot\mathbf{D}\right) + \frac{1}{2}\left[\mathbf{B}\cdot\left(\mathbf{D}\cdot\boldsymbol{\tau}\right)\cdot\mathbf{B}^{-1} + \mathbf{B}^{-1}\cdot\left(\boldsymbol{\tau}\cdot\mathbf{D}\right)\cdot\mathbf{B}\right] \\ & + 4\left(\mathbf{B}\,\frac{\partial^2\Psi}{\partial\mathbf{B}\otimes\partial\mathbf{B}}\,\mathbf{B}\right) : \mathbf{D} = \mathcal{L}_{(0)} : \mathbf{D}.\end{aligned} \tag{6.5.6}$$

Recall that

$$\overset{\circ}{\mathbf{B}} = \mathbf{B}\cdot\mathbf{D} + \mathbf{D}\cdot\mathbf{B}, \tag{6.5.7}$$

and that Ψ is an isotropic function of $\mathbf{B}$, which allows us to write

$$\left(\frac{\partial \Psi}{\partial \mathbf{B}}\right)^{\circ} = \frac{\partial^2 \Psi}{\partial \mathbf{B} \otimes \partial \mathbf{B}} : \overset{\circ}{\mathbf{B}}. \tag{6.5.8}$$

The Cartesian components of the elastic moduli tensor $\boldsymbol{\mathcal{L}}_{(0)}$ are

$$\mathcal{L}^{(0)}_{ijkl} = \tau_{(ik}\delta_{jl)} + B_{(ik}\tau_{lm}B^{-1}_{mj)} + B_{(im}\bar{\Lambda}^{(1)}_{mjkn}B_{nl)}, \tag{6.5.9}$$

where

$$\bar{\Lambda}^{(1)}_{mjkn} = \frac{\partial^2 \Psi}{\partial \mathcal{E}^{(1)}_{mj} \partial \mathcal{E}^{(1)}_{kn}} = 4\,\frac{\partial^2 \Psi}{\partial B_{mj}\partial B_{kn}}\,. \tag{6.5.10}$$

The symmetry in i and j, k and l, and ij and kl is ensured by Eq. (6.2.17), and by the symmetrization

$$\begin{aligned} B_{(ik}\tau_{lm}B^{-1}_{mj)} = \frac{1}{4}\left(B_{ik}\tau_{lm}B^{-1}_{mj} + B_{jk}\tau_{lm}B^{-1}_{mi}\right. \\ \left. + B_{il}\tau_{km}B^{-1}_{mj} + B_{jl}\tau_{km}B^{-1}_{mi}\right), \end{aligned} \tag{6.5.11}$$

and

$$\begin{aligned} B_{(im}\bar{\Lambda}^{(1)}_{mjkn}B_{nl)} = \frac{1}{4}\left(B_{im}\bar{\Lambda}^{(1)}_{mjkn}B_{nl} + B_{jm}\bar{\Lambda}^{(1)}_{mikn}B_{nl}\right. \\ \left. + B_{im}\bar{\Lambda}^{(1)}_{mjln}B_{nk} + B_{jm}\bar{\Lambda}^{(1)}_{miln}B_{nk}\right). \end{aligned} \tag{6.5.12}$$

Equation (6.5.6) can be recast in terms of the convected derivatives of the Kirchhoff stress as

$$\begin{aligned} \overset{\triangle}{\boldsymbol{\tau}} &= \boldsymbol{\mathcal{L}}_{(0)} : \mathbf{D} - \mathbf{D}\cdot\boldsymbol{\tau} - \boldsymbol{\tau}\cdot\mathbf{D} = \boldsymbol{\mathcal{L}}_{(1)} : \mathbf{D}, \\ \overset{\triangledown}{\boldsymbol{\tau}} &= \boldsymbol{\mathcal{L}}_{(0)} : \mathbf{D} + \mathbf{D}\cdot\boldsymbol{\tau} + \boldsymbol{\tau}\cdot\mathbf{D} = \boldsymbol{\mathcal{L}}_{(-1)} : \mathbf{D}. \end{aligned} \tag{6.5.13}$$

By using the instantaneous elastic moduli, these become

$$\begin{aligned} \overset{\triangle}{\underline{\boldsymbol{\tau}}} &= (\underline{\boldsymbol{\mathcal{L}}}_{(0)} - 2\underline{\boldsymbol{S}}) : \mathbf{D} = \underline{\boldsymbol{\mathcal{L}}}_{(1)} : \mathbf{D}, \\ \overset{\triangledown}{\underline{\boldsymbol{\tau}}} &= (\underline{\boldsymbol{\mathcal{L}}}_{(0)} + 2\underline{\boldsymbol{S}}) : \mathbf{D} = \underline{\boldsymbol{\mathcal{L}}}_{(-1)} : \mathbf{D}. \end{aligned} \tag{6.5.14}$$

The tensor $\underline{\boldsymbol{S}}$ is defined by Eq. (6.3.11), and

$$\underline{\boldsymbol{\mathcal{L}}}_{(0)} = (\det \mathbf{F})^{-1}\,\boldsymbol{\mathcal{L}}_{(0)}, \quad \underline{\boldsymbol{\mathcal{L}}}_{(\pm 1)} = (\det \mathbf{F})^{-1}\,\boldsymbol{\mathcal{L}}_{(\pm 1)}. \tag{6.5.15}$$

To obtain the elastic pseudomoduli we can proceed from the general expressions given in Section 3.4, or alternatively use Eq. (3.8.12) to express the rate of nominal stress as

$$\dot{\mathbf{P}} = \overset{\triangle}{\mathbf{P}} + \mathbf{P}\cdot\mathbf{L}^T = \mathbf{F}^{-1}\cdot\overset{\triangle}{\boldsymbol{\tau}} + \mathbf{P}\cdot\mathbf{L}^T. \tag{6.5.16}$$

Since, from Eq. (6.5.13),

$$\overset{\triangle}{\boldsymbol{\tau}} = \boldsymbol{\mathcal{L}}_{(1)} : \mathbf{D} = \boldsymbol{\mathcal{L}}_{(1)} : \mathbf{L}, \tag{6.5.17}$$

by the reciprocal symmetry of $\mathcal{L}_{(1)}$, the substitution into Eq. (6.5.16) gives

$$\dot{P}_{Ji} = \Lambda_{JiLk}\dot{F}_{kL}, \quad \Lambda_{JiLk} = F_{Jm}^{-1}\mathcal{L}_{mikn}^{(1)}F_{nL}^{-T} + P_{Jm}F_{mL}^{-T}\delta_{ik}. \tag{6.5.18}$$

The instantaneous elastic pseudomoduli $\underline{\Lambda}_{jilk}$ follow from Eq. (6.5.18) by setting $\mathbf{F} = \mathbf{I}$,

$$\underline{\Lambda}_{jilk} = \underline{\mathcal{L}}_{jilk}^{(1)} + \sigma_{jl}\delta_{ik}. \tag{6.5.19}$$

This is in agreement with Eq. (6.4.16), because $\underline{\mathcal{L}}_{(1)} = \underline{\mathbf{\Lambda}}_{(1)}$.

6.5.1. Components of Elastic Moduli in Terms of C

When the Lagrangian strain and its conjugate Piola–Kirchhoff stress are used, the rate-type constitutive structure of isotropic elasticity is

$$\begin{aligned}\mathbf{T}_{(1)} = \frac{\partial\Psi}{\partial\mathbf{E}_{(1)}} = 2\frac{\partial\Psi}{\partial\mathbf{C}} = 2\bigg[&\left(\frac{\partial\Psi}{\partial I_C} - I_C\frac{\partial\Psi}{\partial II_C}\right)\mathbf{I}^0 + \left(\frac{\partial\Psi}{\partial II_C}\right)\mathbf{C} \\ &+ \left(III_C\frac{\partial\Psi}{\partial III_C}\right)\mathbf{C}^{-1}\bigg].\end{aligned} \tag{6.5.20}$$

The strain energy function $\Psi = \Psi(I_C, II_C, III_C)$ is here expressed in terms of the principal invariants of the right Cauchy–Green deformation tensor $\mathbf{C} = \mathbf{F}^T\cdot\mathbf{F} = \mathbf{I}^0 + 2\mathbf{E}_{(1)}$. The corresponding elastic moduli tensor is

$$\mathbf{\Lambda}_{(1)} = \frac{\partial\mathbf{T}_{(1)}}{\partial\mathbf{E}_{(1)}} = \frac{\partial^2\Psi}{\partial\mathbf{E}_{(1)}\otimes\partial\mathbf{E}_{(1)}} = 4\frac{\partial^2\Psi}{\partial\mathbf{C}\otimes\partial\mathbf{C}}, \tag{6.5.21}$$

which is thus defined by the fully symmetric tensor $\partial^2\Psi/(\partial\mathbf{C}\otimes\partial\mathbf{C})$. Since

$$\begin{aligned}&\frac{\partial I_C}{\partial\mathbf{C}} = \mathbf{I}^0, \quad \frac{\partial II_C}{\partial\mathbf{C}} = \mathbf{C} - I_C\mathbf{I}^0, \\ &\frac{\partial III_C}{\partial\mathbf{C}} = \mathbf{C}^2 - I_C\mathbf{C} - II_C\mathbf{I}^0 = III_C\mathbf{C}^{-1},\end{aligned} \tag{6.5.22}$$

and in view of the symmetry $C_{ij} = C_{ji}$, we obtain

$$\begin{aligned}\frac{\partial^2\Psi}{\partial C_{ij}\partial C_{kl}} &= c_1\delta_{ij}\delta_{kl} + c_2(\delta_{ij}C_{kl} + C_{ij}\delta_{kl}) + c_3C_{ij}C_{kl} \\ &+ c_4\left(\delta_{ij}C_{kl}^{-1} + C_{ij}^{-1}\delta_{kl}\right) + c_5\left(C_{ij}C_{kl}^{-1} + C_{ij}^{-1}C_{kl}\right) \\ &+ c_6C_{ij}^{-1}C_{kl}^{-1} + c_7\left(C_{ik}^{-1}C_{jl}^{-1} + C_{il}^{-1}C_{jk}^{-1}\right) \\ &+ c_8(\delta_{ik}\delta_{jl} + \delta_{il}\delta_{jk}).\end{aligned} \tag{6.5.23}$$

The parameters c_i $(i = 1, 2, \ldots, 8)$ are (e.g., Lubarda and Lee, 1981)

$$c_1 = \frac{\partial^2\Psi}{\partial I_C^2} - 2I_C\frac{\partial^2\Psi}{\partial I_C\partial II_C} + I_C^2\frac{\partial^2\Psi}{\partial II_C^2} - \frac{\partial\Psi}{\partial II_C}, \tag{6.5.24}$$

$$c_2 = \frac{\partial^2 \Psi}{\partial I_C \partial II_C} - I_C \frac{\partial^2 \Psi}{\partial II_C^2}, \tag{6.5.25}$$

$$c_3 = \frac{\partial^2 \Psi}{\partial II_C^2}, \quad c_5 = III_C \frac{\partial^2 \Psi}{\partial II_C \partial III_C}, \tag{6.5.26}$$

$$c_4 = III_C \frac{\partial^2 \Psi}{\partial III_C \partial I_C} - III_C I_C \frac{\partial^2 \Psi}{\partial II_C \partial III_C}, \tag{6.5.27}$$

$$c_6 = III_C^2 \frac{\partial^2 \Psi}{\partial III_C^2} + III_C \frac{\partial \Psi}{\partial III_C}, \tag{6.5.28}$$

$$c_7 = -\frac{1}{2} III_C \frac{\partial \Psi}{\partial III_C}, \quad c_8 = \frac{1}{2} \frac{\partial \Psi}{\partial II_C}. \tag{6.5.29}$$

6.5.2. Elastic Moduli in Terms of Principal Stretches

For isotropic elastic material the principal directions $\mathbf{N}_i$ of the right Cauchy–Green deformation tensor

$$\mathbf{C} = \sum_{i=1}^{3} \lambda_i^2 \, \mathbf{N}_i \otimes \mathbf{N}_i, \quad C_i = \lambda_i^2, \tag{6.5.30}$$

where λ_i are the principal stretches, are parallel to those of the symmetric Piola–Kirchhoff stress $\mathbf{T}_{(1)}$. Thus, the spectral representation of $\mathbf{T}_{(1)}$ is

$$\mathbf{T}_{(1)} = \sum_{i=1}^{3} T_i^{(1)} \, \mathbf{N}_i \otimes \mathbf{N}_i. \tag{6.5.31}$$

From the analysis presented in Section 2.8 it readily follows that

$$\dot{\mathbf{C}} = \sum_{i=1}^{3} 2\lambda_i \dot{\lambda}_i \, \mathbf{N}_i \otimes \mathbf{N}_i + \sum_{i \neq j} \Omega_{ij}^0 \left(\lambda_j^2 - \lambda_i^2\right) \mathbf{N}_i \otimes \mathbf{N}_j, \tag{6.5.32}$$

and

$$\dot{\mathbf{T}}_{(1)} = \sum_{i=1}^{3} \dot{T}_i^{(1)} \, \mathbf{N}_i \otimes \mathbf{N}_i + \sum_{i \neq j} \Omega_{ij}^0 \left(T_j^{(1)} - T_i^{(1)}\right) \mathbf{N}_i \otimes \mathbf{N}_j. \tag{6.5.33}$$

The components of the spin tensor $\boldsymbol{\Omega}_0 = \dot{\boldsymbol{\mathcal{R}}}_0 \cdot \boldsymbol{\mathcal{R}}_0^{-1}$ on the axes $\mathbf{N}_i$ are denoted by Ω_{ij}^0. The rotation tensor $\boldsymbol{\mathcal{R}}_0$ maps the reference triad of unit vectors $\mathbf{e}_i$ into the Lagrangian triad $\mathbf{N}_i = \boldsymbol{\mathcal{R}}_0 \cdot \mathbf{e}_i^0$. For elastically isotropic material the strain energy can be expressed as a function of the principal stretches, $\Psi = \Psi(\lambda_1, \lambda_2, \lambda_3)$, so that

$$T_i^{(1)} = \frac{\partial \Psi}{\partial E_i^{(1)}} = \frac{1}{\lambda_i} \frac{\partial \Psi}{\partial \lambda_i}. \tag{6.5.34}$$

$$\dot{T}_i^{(1)} = \sum_{j=1}^{3} \frac{\partial T_i^{(1)}}{\partial \lambda_j}\, \dot{\lambda}_j, \qquad \frac{\partial T_i^{(1)}}{\partial \lambda_j} = -\delta_{ij}\, \frac{1}{\lambda_i^2}\frac{\partial \Psi}{\partial \lambda_i} + \frac{1}{\lambda_i}\frac{\partial^2 \Psi}{\partial \lambda_i \partial \lambda_j}. \tag{6.5.35}$$

Thus, Eq. (6.5.33) can be rewritten as

$$\dot{\mathbf{T}}_{(1)} = \sum_{i,j=1}^{3} \frac{\partial T_i^{(1)}}{\partial \lambda_j}\, \dot{\lambda}_j\, \mathbf{N}_i \otimes \mathbf{N}_i + \sum_{i \neq j} \Omega_{ij}^0 \left(\lambda_j^2 - \lambda_i^2\right) \frac{T_j^{(1)} - T_i^{(1)}}{\lambda_j^2 - \lambda_i^2}\, \mathbf{N}_i \otimes \mathbf{N}_j. \tag{6.5.36}$$

Since

$$\dot{\mathbf{T}}_{(1)} = \boldsymbol{\Lambda}_{(1)} : \dot{\mathbf{E}}_{(1)} = \frac{1}{2}\, \boldsymbol{\Lambda}_{(1)} : \dot{\mathbf{C}}, \tag{6.5.37}$$

we recognize from Eqs. (6.5.32) and (6.5.36) by inspection (Chadwick and Ogden, 1971; Ogden, 1984) that

$$\begin{aligned} \boldsymbol{\Lambda}_{(1)} = & \sum_{i,j=1}^{3} \frac{1}{\lambda_j}\frac{\partial T_i^{(1)}}{\partial \lambda_j}\, \mathbf{N}_i \otimes \mathbf{N}_i \otimes \mathbf{N}_j \otimes \mathbf{N}_j \\ & + \sum_{i \neq j} \frac{T_j^{(1)} - T_i^{(1)}}{\lambda_j^2 - \lambda_i^2}\, \mathbf{N}_i \otimes \mathbf{N}_j \otimes \left(\mathbf{N}_i \otimes \mathbf{N}_j + \mathbf{N}_j \otimes \mathbf{N}_i\right). \end{aligned} \tag{6.5.38}$$

Note also

$$\frac{\partial T_i^{(1)}}{\partial E_j^{(1)}} = \frac{1}{\lambda_j}\frac{\partial T_i^{(1)}}{\partial \lambda_j}, \qquad \frac{T_j^{(1)} - T_i^{(1)}}{E_j^{(1)} - E_i^{(1)}} = 2\, \frac{T_j^{(1)} - T_i^{(1)}}{\lambda_j^2 - \lambda_i^2}. \tag{6.5.39}$$

If $\lambda_j \to \lambda_i$, i.e., $E_j^{(1)} \to E_i^{(1)}$, then by the l'Hopital rule

$$\lim_{E_j \to E_i} \frac{T_j^{(1)} - T_i^{(1)}}{E_j^{(1)} - E_i^{(1)}} = \frac{\partial (T_j^{(1)} - T_i^{(1)})}{\partial E_j^{(1)}}, \tag{6.5.40}$$

so that the representation of the elastic moduli tensor in Eq. (6.5.38) holds regardless of the relative magnitude of the principal stretches.

6.6. Hypoelasticity

The material is hypoelastic if its rate-type constitutive equation can be expressed in the form (Truesdell, 1955; Truesdell and Noll, 1965)

$$\overset{\circ}{\boldsymbol{\sigma}} = \mathbf{f}(\boldsymbol{\sigma}, \mathbf{D}). \tag{6.6.1}$$

Under rigid-body rotation $\mathbf{Q}$ of the deformed configuration, Eq. (6.6.1) transforms according to

$$\mathbf{Q} \cdot \overset{\circ}{\boldsymbol{\sigma}} \cdot \mathbf{Q}^T = \mathbf{f}\left(\mathbf{Q} \cdot \boldsymbol{\sigma} \cdot \mathbf{Q}^T, \mathbf{Q} \cdot \mathbf{D} \cdot \mathbf{Q}^T\right), \tag{6.6.2}$$

which requires the second-order tensor function $\mathbf{f}$ to be an isotropic function of both of its arguments. Such a function can be expressed by Eq. (1.11.10) as

$$\begin{aligned}\overset{\circ}{\boldsymbol{\sigma}} &= a_1\mathbf{I} + a_2\boldsymbol{\sigma} + a_3\boldsymbol{\sigma}^2 + a_4\mathbf{D} + a_5\mathbf{D}^2 \\ &+ a_6(\boldsymbol{\sigma}\cdot\mathbf{D} + \mathbf{D}\cdot\boldsymbol{\sigma}) + a_7\left(\boldsymbol{\sigma}^2\cdot\mathbf{D} + \mathbf{D}\cdot\boldsymbol{\sigma}^2\right) \\ &+ a_8\left(\boldsymbol{\sigma}\cdot\mathbf{D}^2 + \mathbf{D}^2\cdot\boldsymbol{\sigma}\right) + a_9\left(\boldsymbol{\sigma}^2\cdot\mathbf{D}^2 + \mathbf{D}^2\cdot\boldsymbol{\sigma}^2\right).\end{aligned} \tag{6.6.3}$$

The coefficients a_i are the scalar functions of ten individual and joint invariants of $\boldsymbol{\sigma}$ and $\mathbf{D}$. These are

$$\begin{gathered}\operatorname{tr}(\boldsymbol{\sigma}),\quad \operatorname{tr}\left(\boldsymbol{\sigma}^2\right),\quad \operatorname{tr}\left(\boldsymbol{\sigma}^3\right),\quad \operatorname{tr}(\mathbf{D}),\quad \operatorname{tr}\left(\mathbf{D}^2\right),\quad \operatorname{tr}\left(\mathbf{D}^3\right), \\ \operatorname{tr}(\boldsymbol{\sigma}\cdot\mathbf{D}),\quad \operatorname{tr}\left(\boldsymbol{\sigma}\cdot\mathbf{D}^2\right),\quad \operatorname{tr}\left(\boldsymbol{\sigma}^2\cdot\mathbf{D}\right),\quad \operatorname{tr}\left(\boldsymbol{\sigma}^2\cdot\mathbf{D}^2\right).\end{gathered} \tag{6.6.4}$$

Suppose that the material behavior is time independent, in the sense that any monotonically increasing parameter can serve as a time scale (materials without a natural time; Hill, 1959). The function $\mathbf{f}$ is then a homogeneous function of degree one in the rate of deformation tensor $\mathbf{D}$. Indeed, if two different time scales are used (t and $t' = kt$, $k = \text{const.}$), we have

$$\overset{\circ}{\boldsymbol{\sigma}}_t = k\overset{\circ}{\boldsymbol{\sigma}}_{t'},\quad \mathbf{D}_t = k\mathbf{D}_{t'}, \tag{6.6.5}$$

and

$$\mathbf{f}(\boldsymbol{\sigma}, k\mathbf{D}_{t'}) = k\mathbf{f}(\boldsymbol{\sigma}, \mathbf{D}_{t'}). \tag{6.6.6}$$

Consequently, in this case, the constitutive structure of Eq. (6.6.3) does not contain quadratic and higher order terms in $\mathbf{D}$, so that

$$\overset{\circ}{\boldsymbol{\sigma}} = a_1\mathbf{I} + a_2\boldsymbol{\sigma} + a_3\boldsymbol{\sigma}^2 + a_4\mathbf{D} + a_6(\boldsymbol{\sigma}\cdot\mathbf{D} + \mathbf{D}\cdot\boldsymbol{\sigma}) + a_7\left(\boldsymbol{\sigma}^2\cdot\mathbf{D} + \mathbf{D}\cdot\boldsymbol{\sigma}^2\right), \tag{6.6.7}$$

where

$$\begin{aligned} a_1 &= c_1\operatorname{tr}(\mathbf{D}) + c_2\operatorname{tr}(\boldsymbol{\sigma}\cdot\mathbf{D}) + c_3\operatorname{tr}\left(\boldsymbol{\sigma}^2\cdot\mathbf{D}\right), \\ a_2 &= c_4\operatorname{tr}(\mathbf{D}) + c_5\operatorname{tr}(\boldsymbol{\sigma}\cdot\mathbf{D}) + c_6\operatorname{tr}\left(\boldsymbol{\sigma}^2\cdot\mathbf{D}\right), \\ a_3 &= c_7\operatorname{tr}(\mathbf{D}) + c_8\operatorname{tr}(\boldsymbol{\sigma}\cdot\mathbf{D}) + c_9\operatorname{tr}\left(\boldsymbol{\sigma}^2\cdot\mathbf{D}\right),\end{aligned} \tag{6.6.8}$$

and

$$a_4 = c_{10},\quad a_6 = c_{11},\quad a_7 = c_{12}. \tag{6.6.9}$$

The coefficients c_i ($i = 1, 2, \ldots, 12$) are the scalar functions of the invariants of $\boldsymbol{\sigma}$ (e.g., $I_\sigma, II_\sigma, III_\sigma$). The structure of the expressions for a_i in Eq. (6.6.8) ensures that $\overset{\circ}{\boldsymbol{\sigma}}$ in Eq. (6.6.7) is linearly dependent on $\mathbf{D}$, i.e.,

$$\overset{\circ}{\boldsymbol{\sigma}} = \boldsymbol{\mathcal{L}} : \mathbf{D}. \tag{6.6.10}$$

The fourth-order tensor $\boldsymbol{\mathcal{L}}$ has the Cartesian components

$$
\begin{aligned}
\mathcal{L}_{ijkl} &= c_1\delta_{ij}\delta_{kl} + c_2\delta_{ij}\sigma_{kl} + c_3\delta_{ij}\sigma^2_{kl} + c_4\sigma_{ij}\delta_{kl} \\
&+ c_5\sigma_{ij}\sigma_{kl} + c_6\sigma_{ij}\sigma^2_{kl} + c_7\sigma^2_{ij}\delta_{kl} + c_8\sigma^2_{ij}\sigma_{kl} \\
&+ c_9\sigma^2_{ij}\sigma^2_{kl} + c_{10}\delta_{(ik}\delta_{jl)} + c_{11}\sigma_{(ik}\delta_{jl)} + c_{12}\sigma^2_{(ik}\delta_{jl)}.
\end{aligned} \tag{6.6.11}
$$

If $c_2 = c_4$, $c_3 = c_7$ and $c_6 = c_8$, the tensor $\boldsymbol{\mathcal{L}}$ obeys the reciprocal symmetry $\mathcal{L}_{ijkl} = \mathcal{L}_{klij}$.

A hypoelastic material is of degree N if $\mathbf{f}$ is a polynomial of degree N in the components of $\boldsymbol{\sigma}$. For example, for hypoelastic material of degree one,

$$
\begin{aligned}
&c_1 = \alpha_1 + \alpha_2 \operatorname{tr}(\boldsymbol{\sigma}), \quad c_{10} = \alpha_3 + \alpha_4 \operatorname{tr}(\boldsymbol{\sigma}), \\
&c_2 = \alpha_5, \quad c_4 = \alpha_6, \quad c_{11} = \alpha_7, \\
&c_3 = c_5 = c_6 = c_7 = c_8 = c_9 = c_{12} = 0,
\end{aligned} \tag{6.6.12}
$$

where α_i $(i = 1, 2, \ldots, 7)$ are seven constants available as material parameters.

In general, elasticity and hypoelasticity are different concepts, although under infinitesimal deformation from an arbitrary stressed configuration, Eq. (6.6.10), with anisotropic tensor $\boldsymbol{\mathcal{L}}$ given by Eq. (6.6.11), corresponds to some type of anisotropic elastic response. However, a hypoelastic constitutive equation cannot describe an anisotropic elastic material in infinitesimal deformation from the unstressed configuration, because the tensor $\boldsymbol{\mathcal{L}}$ becomes an isotropic fourth-order tensor in the unstressed state ($\boldsymbol{\sigma} = \mathbf{0}$).

Furthermore, a general rate-type constitutive equation of anisotropic elasticity, e.g., Eq. (6.2.14), is not of the hypoelastic type, because the anisotropic elastic moduli $\boldsymbol{\mathcal{L}}_{(0)}$ depend on the nine components of the deformation gradient $\mathbf{F}$, which cannot be expressed in terms of the six components of the stress tensor $\boldsymbol{\sigma}$, as required by the hypoelastic constitutive structure. However, a rate-type constitutive equation of finite strain isotropic elasticity (with invertible stress-strain relation) is of hypoelastic type. This follows because $\boldsymbol{\mathcal{L}}_{(0)}$ in Eq. (6.5.9) depends on $\mathbf{V}$, and for isotropic elasticity the six components of $\mathbf{V}$ can be expressed in terms of the six components of $\boldsymbol{\sigma}$, from an invertible type of Eq. (5.5.1). For additional discussion and comparison between elasticity and hypoelasticity, the papers by Pinsky, Ortiz, and Pister (1983), Simo and Pister (1984), and Simo and Ortiz (1985) can be consulted. A majority of hypoelastic solids are inelastic, in the sense that the stress state is generally not recovered upon an arbitrary closed cycle of strain (Hill, 1959). Illustrative examples can be found in Kojić and Bathe (1987), Weber and Anand (1990), Christoffersen (1991), and Bruhns, Xiao, and Meyers (1999). For instance, there is no truly hyperelastic material corresponding to hypoelastic constitutive equation

$$\overset{\circ}{\boldsymbol{\sigma}} = (\lambda \mathbf{I} \otimes \mathbf{I} + 2\mu \boldsymbol{I}) : \mathbf{D}, \tag{6.6.13}$$

where λ and μ are the Lamé type elasticity constants. Integration of Eq. (6.6.13) over a closed cycle of strain gives rise to a small net work left upon a cycle and the hysteresis effects. This is a consequence of the fact that Eq. (6.6.13) is not exactly an integrable equation. As pointed out by Simo and Ortiz (1985), a hypoelastic response with constant components of the fourth-order tensor in Eq. (6.6.13) cannot integrate into a truly hyperelastic response. Further discussion of hypoelastic constitutive equations, particularly regarding the use of different objective stress rates, is given by Dienes (1979), Atluri (1984), Johnson and Bammann (1984), Sowerby and Chu (1984), Metzger and Dubey (1987), and Szabó and Balla (1989).

References

Atluri, S. N. (1984), On constitutive relations at finite strain: Hypoelasticity and elastoplasticity with isotropic or kinematic hardening, *Comput. Meth. Appl. Mech. Engrg.*, Vol. 43, pp. 137–171.

Bruhns, O. T., Xiao, H., and Meyers, A. (1999), Self-consistent Eulerian rate type elasto-plasticity models based upon the logarithmic stress rate, *Int. J. Plasticity*, Vol. 15, pp. 479–520.

Chadwick, P. and Ogden, R. W. (1971), On the definition of elastic moduli, *Arch. Rat. Mech. Anal.*, Vol. 44, pp. 41–53.

Christoffersen, J. (1991), Hyperelastic relations with isotropic rate forms appropriate for elastoplasticity, *Eur. J. Mech., A/Solids*, Vol. 10, pp. 91–99.

Dienes, J. K. (1979), On the analysis of rotation and stress rate in deforming bodies, *Acta Mech.*, Vol. 32, pp. 217–232.

Hill, R. (1959), Some basic principles in the mechanics of solids without natural time, *J. Mech. Phys. Solids*, Vol. 7, pp. 209–225.

Hill, R. (1978), Aspects of invariance in solid mechanics, *Adv. Appl. Mech.*, Vol. 18, pp. 1–75.

Holzapfel, G. A. (2000), *Nonlinear Solid Mechanics*, John Wiley & Sons, Ltd, Chichester, England.

Johnson, G. C. and Bammann, D. J. (1984), A discussion of stress rates in finite deformation problems, *Int. J. Solids Struct.*, Vol. 20, pp. 725–737.

Kojić, M. and Bathe, K. (1987), Studies of finite-element procedures – Stress solution of a closed elastic strain path with stretching and shearing using the updated Lagrangian Jaumann formulation, *Comp. Struct.*, Vol. 26, pp. 175–179.

Lubarda, V. A. (1986), On the rate-type finite elasticity constitutive law, *Z. angew. Math. Mech.*, Vol. 66, pp. 631–632.

Lubarda, V. A. and Lee, E. H. (1981), A correct definition of elastic and plastic deformation and its computational significance, *J. Appl. Mech.*, Vol. 48, pp. 35–40.

Marsden, J. E. and Hughes, T. J. R. (1983), *Mathematical Foundations of Elasticity*, Prentice Hall, Englewood Cliffs, New Jersey.

Metzger, D. R. and Dubey, R. N. (1987), Corotational rates in constitutive modeling of elastic-plastic deformation, *Int. J. Plasticity*, Vol. 4, pp. 341–368.

Ogden, R. W. (1984), *Non-Linear Elastic Deformations*, Ellis Horwood Ltd., Chichester, England (2nd ed., Dover, 1997).

Pinsky, P. M., Ortiz, M., and Pister, K. S. (1983), Numerical integration of rate constitutive equations in finite deformation analysis, *Comput. Meth. Appl. Mech. Engrg.*, Vol. 40, pp. 137–158.

Simo, J. C. and Ortiz, M. (1985), A unified approach to finite deformation elastoplastic analysis based on the use of hyperelastic constitutive equations, *Comput. Meth. Appl. Mech. Engrg.*, Vol. 49, pp. 221–245.

Simo, J. C. and Pister, K. S. (1984), Remarks on rate constitutive equations for finite deformation problems: Computational implications, *Comput. Meth. Appl. Mech. Engrg.*, Vol. 46, pp. 201–215.

Sowerby, R. and Chu, E. (1984), Rotations, stress rates and strain measures in homogeneous deformation processes, *Int. J. Solids Struct.*, Vol. 20, pp. 1037–1048.

Szabó, L. and Balla, M. (1989), Comparison of some stress rates, *Int. J. Solids Struct.*, Vol. 25, pp. 279–297.

Truesdell, C. (1955), Hypo-elasticity, *J. Rat. Mech. Anal.*, Vol. 4, pp. 83–133.

Truesdell, C. and Noll, W. (1965), The nonlinear field theories of mechanics, in *Handbuch der Physik*, ed. S. Flügge, Band III/3, Springer-Verlag, Berlin (2nd ed., 1992).

Truesdell, C. and Toupin, R. (1960), The classical field theories, in *Handbuch der Physik*, ed. S. Flügge, Band III/1, pp. 226–793, Springer-Verlag, Berlin.

Weber, G. and Anand, L. (1990), Finite deformation constitutive equations and a time integration procedure for isotropic, hyperelastic-viscoplastic solids, *Comput. Meth. Appl. Mech. Engrg.*, Vol. 79, pp. 173–202.

CHAPTER 7

ELASTIC STABILITY

7.1. Principle of Stationary Potential Energy

Denote by $\delta\mathbf{F}$ the variation of the deformation gradient $\mathbf{F}$. Since for Green elasticity $\mathbf{P} = \partial\Psi/\partial\mathbf{F}$, where $\Psi = \Psi(\mathbf{F})$ is the strain energy per unit initial volume, we can write

$$\mathbf{P} \cdot\cdot\, \delta\mathbf{F} = \frac{\partial\Psi}{\partial\mathbf{F}} \cdot\cdot\, \delta\mathbf{F} = \delta\Psi, \tag{7.1.1}$$

and the principle of virtual work of Eq. (3.12.1) becomes

$$\int_{V^0} \delta\Psi \,\mathrm{d}V^0 = \int_{V^0} \rho^0\, \mathbf{b}\cdot\delta\mathbf{u}\,\mathrm{d}V^0 + \int_{S_t^0} \mathbf{p}_n \cdot \delta\mathbf{u}\,\mathrm{d}S_t^0. \tag{7.1.2}$$

In general, for arbitrary loading there is no true variational principle associated with Eq. (7.1.2), because the variation δ affects the applied body force $\mathbf{b}$ and the surface traction $\mathbf{p}_{(n)}$. However, if the loading is conservative, as in the case of dead loading, then

$$\mathbf{b}\cdot\delta\mathbf{u} = \delta(\mathbf{b}\cdot\mathbf{u}), \quad \mathbf{p}_n \cdot \delta\mathbf{u} = \delta(\mathbf{p}_n\cdot\mathbf{u}), \tag{7.1.3}$$

and Eq. (7.1.2) can be recast in the variational form

$$\delta\mathcal{P} = 0, \tag{7.1.4}$$

where

$$\mathcal{P} = \int_{V^0} \Psi\,\mathrm{d}V^0 - \int_{V^0} \rho^0\,\mathbf{b}\cdot\mathbf{u}\,\mathrm{d}V^0 - \int_{S_t^0} \mathbf{p}_n\cdot\mathbf{u}\,\mathrm{d}S_t^0. \tag{7.1.5}$$

Among all geometrically admissible displacement fields, the actual displacement field (whether unique or not) of the considered boundary-value problem makes stationary the potential energy functional $\mathcal{P}(\mathbf{u})$ given by Eq. (7.1.5). See also Nemat-Nasser (1974) and Washizu (1982).

7.2. Uniqueness of Solution

Consider a finite elasticity problem described by the equilibrium equations

$$\nabla^0 \cdot \mathbf{P} + \rho^0 \mathbf{b} = \mathbf{0}, \tag{7.2.1}$$

and the mixed boundary conditions

$$\mathbf{u} = \mathbf{u}(\mathbf{X}) \quad \text{on} \quad S_u^0, \qquad \mathbf{n}^0 \cdot \mathbf{P} = \mathbf{p}_n(\mathbf{X}) \quad \text{on} \quad S_t^0. \tag{7.2.2}$$

For simplicity, restrict attention to dead loading on S_t^0, and dead body forces $\mathbf{b} = \mathbf{b}(\mathbf{X})$ in V^0. Suppose that there are two different solutions of Eqs. (7.2.1) and (7.2.2), $\mathbf{u}$ and $\mathbf{u}^*$ (i.e., $\mathbf{x}$ and $\mathbf{x}^*$). The corresponding deformation gradients are $\mathbf{F}$ and $\mathbf{F}^*$, and the nominal stresses $\mathbf{P}$ and $\mathbf{P}^*$. The equilibrium fields $(\mathbf{P}, \mathbf{F})$ and $(\mathbf{P}^*, \mathbf{F}^*)$ necessarily satisfy the condition

$$\int_{V^0} (\mathbf{P}^* - \mathbf{P}) \cdot\cdot (\mathbf{F}^* - \mathbf{F})\, \mathrm{d}V^0 = 0, \tag{7.2.3}$$

which follows from Eq. (3.12.5). Consequently, the solution $\mathbf{x} = \mathbf{x}(\mathbf{X})$ is unique if

$$\int_{V^0} (\mathbf{P}^* - \mathbf{P}) \cdot\cdot (\mathbf{F}^* - \mathbf{F})\, \mathrm{d}V^0 \neq 0, \tag{7.2.4}$$

for all geometrically admissible $\mathbf{x}^*$ giving rise to

$$\mathbf{F}^* = \frac{\partial \mathbf{x}^*}{\partial \mathbf{X}}, \quad \mathbf{P}^* = \frac{\partial \Psi}{\partial \mathbf{F}^*}. \tag{7.2.5}$$

The stress field $\mathbf{P}^*$ in (7.2.4) need not be statically admissible, so even if equality sign applies in (7.2.4) for some $\mathbf{x}^*$, the uniqueness is not lost unless that $\mathbf{x}^*$ gives rise to statically admissible stress field $\mathbf{P}^*$. Therefore, a sufficient condition for $\mathbf{x}$ to be unique solution is that for all geometrically admissible deformation fields $\mathbf{x}^*$,

$$\int_{V^0} (\mathbf{P}^* - \mathbf{P}) \cdot\cdot (\mathbf{F}^* - \mathbf{F})\, \mathrm{d}V^0 > 0. \tag{7.2.6}$$

The reversed inequality could also serve as a sufficient condition for uniqueness. The solution $\mathbf{x}$ which obeys such inequality for all geometrically admissible $\mathbf{x}^*$ would define unique, but unstable equilibrium configuration, as will be discussed in Section 7.3.

A stronger (more restrictive) condition for uniqueness is

$$(\mathbf{P}^* - \mathbf{P}) \cdot\cdot (\mathbf{F}^* - \mathbf{F}) > 0, \tag{7.2.7}$$

which clearly implies (7.2.6). However, unique solution in finite elasticity is not expected in general (particularly under dead loading), so that inequalities such as (7.2.6) and (7.2.7) are too strong restrictions on elastic constitutive relation. In fact, a nonuniqueness in finite elasticity is certainly anticipated whenever the stress-deformation relation $\mathbf{P} = \partial\Psi/\partial\mathbf{F}$ is not uniquely invertible. For example, Ogden (1984) provides examples in which two, four or more possible states of deformation correspond to a given state of nominal stress. See also Antman (1995). A study of the existence of solutions to boundary-value problems in finite strain elasticity is more difficult, with only few results presently available (e.g., Ball, 1977; Hanyga, 1985; Ciarlet, 1988). A comprehensive account of the uniqueness theorems in linear elasticity is given by Knops and Payne (1971).

7.3. Stability of Equilibrium

Consider the inequality

$$\Psi\left(\mathbf{F}^*\right) - \Psi\left(\mathbf{F}\right) - \mathbf{P} \cdot\cdot \left(\mathbf{F}^* - \mathbf{F}\right) > 0, \tag{7.3.1}$$

where $(\mathbf{P}, \mathbf{F})$ correspond to equilibrium configuration $\mathbf{x}$, and $(\mathbf{P}^*, \mathbf{F}^*)$ to any geometrically admissible configuration $\mathbf{x}^*$ (Coleman and Noll, 1959). This inequality implies (7.2.7), so that (7.3.1) also represents a sufficient condition for uniqueness. (To see that (7.3.1) implies (7.2.7), write another inequality by reversing the role of $\mathbf{F}$ and $\mathbf{F}^*$ in (7.3.1), and add the results; Ogden, *op. cit.*). Inequality (7.3.1) is particularly appealing because it directly leads to stability criterion. To that goal, integrate (7.3.1) to obtain

$$\int_{V^0} \left[\Psi\left(\mathbf{F}^*\right) - \Psi(\mathbf{F})\right] \mathrm{d}V^0 > \int_{V^0} \mathbf{P} \cdot\cdot \left(\mathbf{F}^* - \mathbf{F}\right) \mathrm{d}V^0. \tag{7.3.2}$$

Using Eq. (3.12.4) to express the integral on the right-hand side gives

$$\begin{aligned} \int_{V^0} \left[\Psi\left(\mathbf{F}^*\right) - \Psi(\mathbf{F})\right] \mathrm{d}V^0 > & \int_{V^0} \rho^0 \mathbf{b} \cdot \left(\mathbf{x}^* - \mathbf{x}\right) \mathrm{d}V^0 \\ & + \int_{S_t^0} \mathbf{p}_n \cdot \left(\mathbf{x}^* - \mathbf{x}\right) \mathrm{d}S_t^0. \end{aligned} \tag{7.3.3}$$

This means that the increase of the strain energy in moving from the configuration $\mathbf{x}$ to $\mathbf{x}^*$ exceeds the work done by the prescribed dead loading on that transition. According to the classical energy criterion of stability this means that $\mathbf{x}$ is a stable equilibrium configuration (Hill, 1957; Pearson, 1959).

Recalling the expression for the potential energy from Eq. (7.1.5), and the identity

$$\mathbf{x}^* - \mathbf{x} = \mathbf{u}^* - \mathbf{u}, \tag{7.3.4}$$

the inequality (7.3.3) can be rewritten as

$$\mathcal{P}(\mathbf{u}^*) > \mathcal{P}(\mathbf{u}). \tag{7.3.5}$$

Consequently, among all geometrically admissible configurations the potential energy is minimized in the configuration of stable equilibrium.

In a broader sense, stability of equilibrium at $\mathbf{x}$ is stable if for some geometrically admissible $\mathbf{x}^*$, $\mathcal{P}(\mathbf{u}^*) = \mathcal{P}(\mathbf{u})$, while for all others $\mathcal{P}(\mathbf{u}^*) > \mathcal{P}(\mathbf{u})$. In this situation, however, equilibrium configuration $\mathbf{x}$ is not necessarily unique, because $\mathbf{x}^*$ for which $\mathcal{P}(\mathbf{u}^*) = \mathcal{P}(\mathbf{u})$ may give rise to statically admissible stress field (in which case $\mathbf{x}^*$ is also an equilibrium configuration). Therefore, stability in the sense $\mathcal{P}(\mathbf{u}^*) \geq \mathcal{P}(\mathbf{u})$ does not in general imply uniqueness. Conversely, unique configuration need not be stable. It is unstable if $\mathcal{P}(\mathbf{u}^*) < \mathcal{P}(\mathbf{u})$ for at least one $\mathbf{u}^*$, and $\mathcal{P}(\mathbf{u}^*) > \mathcal{P}(\mathbf{u})$ for all other geometrically admissible $\mathbf{x}^*$.

In summary, the inequality

$$\mathcal{P}(\mathbf{u}^*) \geq \mathcal{P}(\mathbf{u}) \tag{7.3.6}$$

is a global sufficient condition for stability of equilibrium configuration $\mathbf{x}$. It is, however, too restrictive criterion, because it is formulated relative to all geometrically admissible configurations around $\mathbf{x}$.

7.4. Incremental Uniqueness and Stability

Physically more appealing stability criterion is obtained if $\mathbf{x}^*$ is confined to adjacent configurations, in the neighborhood of $\mathbf{x}$. In that case we talk about local or incremental (infinitesimal) stability (Truesdell and Noll, 1965). We start from the inequality (7.3.1). If $\mathbf{F}^*$ is near $\mathbf{F}$ (corresponding to an equilibrium configuration), so that

$$\mathbf{F}^* = \mathbf{F} + \delta\mathbf{F}, \tag{7.4.1}$$

the Taylor expansion gives

$$\Psi(\mathbf{F} + \delta\mathbf{F}) = \Psi(\mathbf{F}) + \mathbf{P} \cdot\cdot\, \delta\mathbf{F} + \frac{1}{2}\,\boldsymbol{\Lambda} \cdot\cdot\cdot\cdot\, (\delta\mathbf{F} \otimes \delta\mathbf{F}) + \cdots\,. \tag{7.4.2}$$

Consequently, to second-order terms, the inequality (7.3.1) becomes

$$\frac{1}{2}\boldsymbol{\Lambda}\cdots(\delta\mathbf{F}\otimes\delta\mathbf{F})>0. \tag{7.4.3}$$

This is a sufficient condition for incremental (infinitesimal) uniqueness, or uniqueness in the small neighborhood of $\mathbf{F}$. An integration over the volume V^0 yields

$$\frac{1}{2}\int_{V^0}\boldsymbol{\Lambda}\cdots(\delta\mathbf{F}\otimes\delta\mathbf{F})\,\mathrm{d}V^0>0. \tag{7.4.4}$$

Using (7.4.2), Eq. (7.1.5) gives in the case of dead loading

$$\begin{aligned}\mathcal{P}(\mathbf{u}+\delta\mathbf{u})-\mathcal{P}(\mathbf{u})=&\frac{1}{2}\int_{V^0}\boldsymbol{\Lambda}\cdots(\delta\mathbf{F}\otimes\delta\mathbf{F})\,\mathrm{d}V^0\\+\int_{V^0}\mathbf{P}\cdot\cdot\,\delta\mathbf{F}\,\mathrm{d}V^0-&\int_{V^0}\rho^0\,\mathbf{b}\cdot\delta\mathbf{u}\,\mathrm{d}V^0-\int_{S_t^0}\mathbf{p}_n\cdot\delta\mathbf{u}\,\mathrm{d}S^0,\end{aligned} \tag{7.4.5}$$

where $\delta\mathbf{u}=\delta\mathbf{x}$. The last three integrals on the right-hand side of Eq. (7.4.5) cancel each other by Gauss theorem, equilibrium equations, and the condition $\delta\mathbf{u}=\mathbf{0}$ on S_u^0; see Eq. (3.12.1). Thus,

$$\mathcal{P}(\mathbf{u}+\delta\mathbf{u})-\mathcal{P}(\mathbf{u})=\frac{1}{2}\int_{V^0}\boldsymbol{\Lambda}\cdots(\delta\mathbf{F}\otimes\delta\mathbf{F})\,\mathrm{d}V^0. \tag{7.4.6}$$

If equilibrium configuration $\mathbf{x}$ is incrementally unique, so that (7.4.3) applies, then from (7.4.6) it follows that

$$\mathcal{P}(\mathbf{u}+\delta\mathbf{u})>\mathcal{P}(\mathbf{u}), \tag{7.4.7}$$

which means that equilibrium configuration $\mathbf{x}$ is locally or incrementally stable. If for some $\delta\mathbf{u}$, $\mathcal{P}(\mathbf{u}+\delta\mathbf{u})=\mathcal{P}(\mathbf{u})$, while for other $\delta\mathbf{u}$, $\mathcal{P}(\mathbf{u}+\delta\mathbf{u})>\mathcal{P}(\mathbf{u})$, the configuration $\mathbf{x}$ is a state of neutral incremental stability, although the configuration may not be incrementally unique. The strict inequality (7.4.7) is sometimes referred to as the criterion of local (incremental) superstability. See also Knops and Wilkes (1973), and Gurtin (1982).

7.5. Rate-Potentials and Variational Principle

In this section we examine the existence of the variational principle, and the uniqueness and stability of the boundary-value problem of the rate-type elasticity considered in Chapter 6. First, we recall that from Eq. (6.4.2) the rate of nominal stress is

$$\dot{\mathbf{P}}=\boldsymbol{\Lambda}\cdot\cdot\,\dot{\mathbf{F}},\quad \boldsymbol{\Lambda}=\frac{\partial^2\Psi}{\partial\mathbf{F}\otimes\partial\mathbf{F}}\,. \tag{7.5.1}$$

Since the tensor of elastic pseudomoduli $\boldsymbol{\Lambda}$ obeys the reciprocal symmetry, Eq. (7.5.1) can be rephrased by introducing the rate-potential function χ as

$$\dot{\mathbf{P}} = \frac{\partial \chi}{\partial \dot{\mathbf{F}}}, \quad \chi = \frac{1}{2}\boldsymbol{\Lambda} \cdots (\dot{\mathbf{F}} \otimes \dot{\mathbf{F}}). \tag{7.5.2}$$

Its Cartesian component representation is

$$\dot{P}_{Ji} = \frac{\partial \chi}{\partial \dot{F}_{iJ}}, \quad \chi = \frac{1}{2}\Lambda_{JiLk}\dot{F}_{iJ}\dot{F}_{kL}. \tag{7.5.3}$$

Consequently, we have

$$\dot{\mathbf{P}} \cdot\cdot\, \delta\dot{\mathbf{F}} = \frac{\partial \chi}{\partial \dot{\mathbf{F}}} \cdot\cdot\, \delta\dot{\mathbf{F}} = \delta\chi, \tag{7.5.4}$$

and the principle of virtual velocity from Eq. (3.11.8) becomes, for static problems,

$$\int_{V^0} \delta\chi \, \mathrm{d}V^0 = \int_{V^0} \rho^0 \, \dot{\mathbf{b}} \cdot \delta\mathbf{v} \, \mathrm{d}V^0 + \int_{S_t^0} \dot{\mathbf{p}}_n \cdot \delta\mathbf{v} \, \mathrm{d}S_t^0, \tag{7.5.5}$$

for any analytically admissible virtual velocity field $\delta\mathbf{v}$ vanishing on S_v^0.

For general, nonconservative loading there is no true variational principle associated with Eq. (7.5.5), because the variation δ affects $\dot{\mathbf{b}}$ and $\dot{\mathbf{p}}_{(n)}$. However, if the rates of loading are deformation insensitive (remain unaltered during the variation $\delta\mathbf{v}$), there is a variational principle

$$\delta\Xi = 0, \tag{7.5.6}$$

with

$$\Xi = \int_{V^0} \chi \, \mathrm{d}V^0 - \int_{V^0} \rho^0 \, \dot{\mathbf{b}} \cdot \mathbf{v} \, \mathrm{d}V^0 - \int_{S_t^0} \dot{\mathbf{p}}_n \cdot \mathbf{v} \, \mathrm{d}S_t^0. \tag{7.5.7}$$

Among all kinematically admissible velocity fields, the actual velocity field (whether unique or not) of the considered rate boundary-value problem renders stationary the functional $\Xi(\mathbf{v})$.

There is also a variational principle associated with Eq. (7.5.5) if the rates of prescribed tractions and body forces are self-adjoint in the sense that (Hill, 1978)

$$\int_{S_t^0} (\dot{\mathbf{p}}_n \cdot \delta\mathbf{v} - \mathbf{v} \cdot \delta\dot{\mathbf{p}}_n) \, \mathrm{d}S_t^0 = 0, \tag{7.5.8}$$

and similarly for the body forces, since then

$$\delta \int_{S_t^0} (\dot{\mathbf{p}}_n \cdot \mathbf{v})\, \mathrm{d}S_t^0 = 2 \int_{S_t^0} (\dot{\mathbf{p}}_n \cdot \delta\mathbf{v})\, \mathrm{d}S_t^0,$$
$$\delta \int_{V^0} \left(\dot{\mathbf{b}} \cdot \mathbf{v}\right) \mathrm{d}V^0 = 2 \int_{V^0} \left(\dot{\mathbf{b}} \cdot \delta\mathbf{v}\right) \mathrm{d}V^0. \tag{7.5.9}$$

In this case the variational integral is

$$\Xi = \int_{V^0} \chi\, \mathrm{d}V^0 - \frac{1}{2} \int_{V^0} \rho^0\, \dot{\mathbf{b}} \cdot \mathbf{v}\, \mathrm{d}V^0 - \frac{1}{2} \int_{S_t^0} \dot{\mathbf{p}}_n \cdot \mathbf{v}\, \mathrm{d}S_t^0. \tag{7.5.10}$$

A loading that is partly controllable (independent of $\mathbf{v}$), and partly deformation sensitive but self-adjoint in the above sense also allows the variational principle. Detailed analysis is available in Hill (*op. cit.*).

7.5.1. Betti's Theorem and Clapeyron's Formula

Let

$$\mathbf{v} = \dot{\mathbf{x}}, \quad \dot{\mathbf{F}} = \frac{\partial \mathbf{v}}{\partial \mathbf{X}}, \quad \dot{\mathbf{P}} = \mathbf{\Lambda} : \dot{\mathbf{F}} \tag{7.5.11}$$

be a solution of the boundary-value problem associated with the prescribed rates of body forces $\dot{\mathbf{b}}$ in V^0, surface tractions $\dot{\mathbf{p}}_n$ on S_t^0, and velocities $\mathbf{v}$ on S_v^0. Similarly, let

$$\mathbf{v}^* = \dot{\mathbf{x}}^*, \quad \dot{\mathbf{F}}^* = \frac{\partial \mathbf{v}^*}{\partial \mathbf{X}}, \quad \dot{\mathbf{P}}^* = \mathbf{\Lambda} : \dot{\mathbf{F}}^* \tag{7.5.12}$$

be a solution of the boundary-value problem associated with the prescribed rates of body forces $\dot{\mathbf{b}}^*$ in V^0, surface tractions $\dot{\mathbf{p}}_n^*$ on S_t^0, and velocities $\mathbf{v}^*$ on S_v^0. By reciprocal symmetry of pseudomoduli $\mathbf{\Lambda}$ we have the reciprocal relation

$$\dot{\mathbf{P}} \cdot\cdot\, \dot{\mathbf{F}}^* = \dot{\mathbf{P}}^* \cdot\cdot\, \dot{\mathbf{F}}. \tag{7.5.13}$$

Upon integration over the volume V^0, and by using Eq. (3.11.12), it follows that

$$\int_{V^0} \rho^0\, \dot{\mathbf{b}} \cdot \mathbf{v}^*\, \mathrm{d}V^0 + \int_{S^0} \mathbf{n}^0 \cdot \dot{\mathbf{P}} \cdot \mathbf{v}^*\, \mathrm{d}S^0$$
$$= \int_{V^0} \rho^0\, \dot{\mathbf{b}}^* \cdot \mathbf{v}\, \mathrm{d}V^0 + \int_{S^0} \mathbf{n}^0 \cdot \dot{\mathbf{P}}^* \cdot \mathbf{v}\, \mathrm{d}S^0. \tag{7.5.14}$$

This is analogous to Betti's reciprocal theorem of classical elasticity. Also, by incorporating $\dot{\mathbf{P}} = \mathbf{\Lambda} : \dot{\mathbf{F}}$ in the integral on the left-hand side of Eq. (3.11.12), there follows

$$\int_{V^0} \chi\, \mathrm{d}V^0 = \frac{1}{2} \int_{V^0} \rho^0 \dot{\mathbf{b}} \cdot \mathbf{v}\, \mathrm{d}V^0 + \frac{1}{2} \int_{S^0} \mathbf{n}^0 \cdot \dot{\mathbf{P}} \cdot \mathbf{v}\, \mathrm{d}S^0, \tag{7.5.15}$$

which is analogous to Clapeyron's formula from linear elasticity (Hill, 1978).

7.5.2. Other Rate-Potentials

The rate potential χ was introduced in Eq. (7.5.2) for the rate of nominal stress $\dot{\mathbf{P}}$. We can also introduce the rate-potentials for the rates of material and spatial stress tensors, such that

$$\dot{\mathbf{T}}_{(n)} = \frac{\partial \chi_{(n)}}{\partial \dot{\mathbf{E}}_{(n)}}, \quad \chi_{(n)} = \frac{1}{2}\boldsymbol{\Lambda}_{(n)} :: \left(\dot{\mathbf{E}}_{(n)} \otimes \dot{\mathbf{E}}_{(n)}\right), \tag{7.5.16}$$

$$\overset{\bullet}{\mathcal{T}}_{(n)} = \frac{\partial \bar{\chi}_{(n)}}{\partial \overset{\bullet}{\mathcal{E}}_{(n)}}, \quad \bar{\chi}_{(n)} = \frac{1}{2}\bar{\boldsymbol{\Lambda}}_{(n)} :: \left(\overset{\bullet}{\mathcal{E}}_{(n)} \otimes \overset{\bullet}{\mathcal{E}}_{(n)}\right). \tag{7.5.17}$$

7.5.3. Current Configuration as Reference

If the current configuration is taken as the reference configuration, we have

$$\dot{\underline{\mathbf{P}}} = \frac{\partial \underline{\chi}}{\partial \mathbf{L}}, \quad \underline{\chi} = \frac{1}{2}\underline{\boldsymbol{\Lambda}} \cdots (\mathbf{L} \otimes \mathbf{L}), \tag{7.5.18}$$

since $\dot{\underline{\mathbf{F}}} = \mathbf{L}$ (see Section 6.4). Substituting Eq. (6.4.16) for $\underline{\boldsymbol{\Lambda}}$, there follows

$$\underline{\chi} = \frac{1}{2}\underline{\boldsymbol{\mathcal{L}}}_{(1)} :: (\mathbf{D} \otimes \mathbf{D}) + \frac{1}{2}\boldsymbol{\sigma} : \left(\mathbf{L}^T \cdot \mathbf{L}\right). \tag{7.5.19}$$

Alternatively, in view of Eq. (6.3.14),

$$\underline{\chi} = \frac{1}{2}\underline{\boldsymbol{\mathcal{L}}}_{(0)} :: (\mathbf{D} \otimes \mathbf{D}) + \frac{1}{2}\boldsymbol{\sigma} : \left(\mathbf{L}^T \cdot \mathbf{L} - 2\mathbf{D}^2\right). \tag{7.5.20}$$

The symmetry of the instantaneous elastic moduli $\underline{\boldsymbol{\mathcal{L}}}_{(1)}$ was used in arriving at Eq. (7.5.19). With the current configuration as the reference, the variational integral of Eq. (7.5.7) becomes

$$\underline{\Xi} = \int_V \underline{\chi}\,\mathrm{d}V - \int_V \rho\,\dot{\mathbf{b}} \cdot \mathbf{v}\,\mathrm{d}V - \int_{S_t} \dot{\underline{\mathbf{p}}}_n \cdot \delta\mathbf{v}\,\mathrm{d}S_t, \tag{7.5.21}$$

where $\mathbf{n} \cdot \dot{\underline{\mathbf{P}}} = \dot{\underline{\mathbf{p}}}_n$ on S_t. The traction rate $\dot{\underline{\mathbf{p}}}_n$ is related to the rate of Cauchy traction $\dot{\mathbf{t}}_n$ by Eq. (3.9.18).

The rate potentials $\underline{\chi}_{(n)}$ are introduced such that

$$\dot{\underline{\mathbf{T}}}_{(n)} = \frac{\partial \underline{\chi}_{(n)}}{\partial \mathbf{D}}, \quad \underline{\chi}_{(n)} = \frac{1}{2}\underline{\boldsymbol{\mathcal{L}}}_{(n)} :: (\mathbf{D} \otimes \mathbf{D}). \tag{7.5.22}$$

In view of Eqs. (6.3.10) and (6.3.13), the various rate potentials are related by

$$\underline{\chi}_{(n)} = \underline{\chi}_{(0)} - n\boldsymbol{\sigma} : \mathbf{D}^2 = \underline{\chi}_{(1)} + (1-n)\boldsymbol{\sigma} : \mathbf{D}^2, \tag{7.5.23}$$

and

$$\underline{\chi} = \underline{\chi}_{(n)} + \frac{1}{2}\boldsymbol{\sigma} : \left[\mathbf{L}^T \cdot \mathbf{L} - 2(1-n)\mathbf{D}^2\right]. \tag{7.5.24}$$

Using the results from Section 3.9 for the rates $\dot{\underline{\mathbf{T}}}_{(n)}$, Eq. (7.5.22) gives, for $n = 0$ and $n = \pm 1$,

$$\overset{\circ}{\underline{\boldsymbol{\tau}}} = \frac{\partial \underline{\chi}_{(0)}}{\partial \mathbf{D}}, \quad \overset{\triangle}{\underline{\boldsymbol{\tau}}} = \frac{\partial \underline{\chi}_{(1)}}{\partial \mathbf{D}}, \quad \overset{\triangledown}{\underline{\boldsymbol{\tau}}} = \frac{\partial \underline{\chi}_{(-1)}}{\partial \mathbf{D}}. \tag{7.5.25}$$

7.6. Uniqueness of Solution to Rate Problem

We examine now the uniqueness of solution to the boundary-value problem described by the rate equilibrium equations

$$\boldsymbol{\nabla}^0 \cdot \dot{\mathbf{P}} + \rho^0 \dot{\mathbf{b}} = \mathbf{0}, \tag{7.6.1}$$

and the boundary conditions

$$\mathbf{v} = \mathbf{v}_0 \quad \text{on} \quad S_v^0, \qquad \mathbf{n}^0 \cdot \dot{\mathbf{P}} = \dot{\mathbf{p}}_n \quad \text{on} \quad S_t^0. \tag{7.6.2}$$

It is assumed that incremental loading is deformation insensitive, so that $\dot{\mathbf{b}}$ in V^0 and $\dot{\mathbf{p}}_n$ on S_t^0 do not depend on the velocity.

Suppose that there are two different solutions of Eqs. (7.6.1) and (7.6.2), $\mathbf{v}$ and $\mathbf{v}^*$. The corresponding rates of deformation gradients are $\dot{\mathbf{F}}$ and $\dot{\mathbf{F}}^*$, with the rates of nominal stresses $\dot{\mathbf{P}}$ and $\dot{\mathbf{P}}^*$. The equilibrium fields $\left(\dot{\mathbf{P}}, \dot{\mathbf{F}}\right)$ and $\left(\dot{\mathbf{P}}^*, \dot{\mathbf{F}}^*\right)$ necessarily satisfy the condition

$$\int_{V^0} (\dot{\mathbf{P}}^* - \dot{\mathbf{P}}) \cdot\cdot (\dot{\mathbf{F}}^* - \dot{\mathbf{F}})\, \mathrm{d}V^0 = 0, \tag{7.6.3}$$

which follows from Eq. (3.11.13). Consequently, from Eq. (7.6.3), the velocity field $\mathbf{v}$ is unique if

$$\begin{aligned} &\int_{V^0} \left(\dot{\mathbf{P}}^* - \dot{\mathbf{P}}\right) \cdot\cdot \left(\dot{\mathbf{F}}^* - \dot{\mathbf{F}}\right) \mathrm{d}V^0 \\ &\quad = \int_{V^0} \boldsymbol{\Lambda} \cdot\cdot\cdot\cdot \left(\dot{\mathbf{F}}^* - \dot{\mathbf{F}}\right) \otimes \left(\dot{\mathbf{F}}^* - \dot{\mathbf{F}}\right) \mathrm{d}V^0 \neq 0, \end{aligned} \tag{7.6.4}$$

for all kinematically admissible $\mathbf{v}^*$ giving rise to

$$\dot{\mathbf{F}}^* = \frac{\partial \mathbf{v}^*}{\partial \mathbf{X}}, \quad \dot{\mathbf{P}}^* = \boldsymbol{\Lambda} : \dot{\mathbf{F}}^*. \tag{7.6.5}$$

The stress rate $\dot{\mathbf{P}}^*$ in (7.6.4) need not be statically admissible, so even if the equality sign applies in (7.6.4) for some $\mathbf{v}^*$, the uniqueness is lost only if $\mathbf{v}^*$ gives rise to statically admissible stress-rate field $\dot{\mathbf{P}}^*$. Therefore, a sufficient condition for $\mathbf{v}$ to be unique solution is that for all kinematically admissible velocity fields $\mathbf{v}^*$,

$$\int_{V^0} \mathbf{\Lambda} \cdots \left(\dot{\mathbf{F}}^* - \dot{\mathbf{F}}\right) \otimes \left(\dot{\mathbf{F}}^* - \dot{\mathbf{F}}\right) \mathrm{d}V^0 > 0. \tag{7.6.6}$$

The reversed inequality could also serve as a sufficient condition for uniqueness. The solution $\mathbf{v}$ which obeys such inequality for all kinematically admissible $\mathbf{v}^*$ would define unique, but unstable equilibrium configuration, analogous to the consideration in Section 7.3.

A more restrictive condition for uniqueness is evidently

$$\mathbf{\Lambda} \cdots \left(\dot{\mathbf{F}} - \dot{\mathbf{F}}^*\right) \otimes \left(\dot{\mathbf{F}} - \dot{\mathbf{F}}^*\right) > 0, \tag{7.6.7}$$

which implies (7.6.6), and which states that $\mathbf{\Lambda}$ is positive definite. However, since unique solution to a finite elasticity rate problem cannot be expected in general, the inequality (7.6.7) may fail at certain states of deformation. A nonuniqueness of the rate problem is certainly a possibility if the state of deformation is reached when $\mathbf{\Lambda}$ becomes singular, so that $\mathbf{\Lambda} \cdot\cdot \dot{\mathbf{F}} = \mathbf{0}$ has nontrivial solutions for $\dot{\mathbf{F}}$. Details of the calculations for isotropic materials can be found in Ogden (1984).

If a sufficient condition for uniqueness (7.6.6) applies, then

$$\Xi(\mathbf{v}^*) > \Xi(\mathbf{v}), \tag{7.6.8}$$

and the variational principle is strengthened to a minimum principle: among all kinematically admissible velocity fields, the actual field renders Ξ the minimum. Indeed, from Eq. (7.5.7) it follows that

$$\Xi(\mathbf{v}^*) - \Xi(\mathbf{v}) = \frac{1}{2}\int_{V^0} \left(\dot{\mathbf{P}}^* - \dot{\mathbf{P}}\right) \cdot\cdot \left(\dot{\mathbf{F}}^* - \dot{\mathbf{F}}\right) \mathrm{d}V^0. \tag{7.6.9}$$

In the derivation, Eq. (3.11.12) was used, and the reciprocity relation

$$\dot{\mathbf{P}} \cdot\cdot \dot{\mathbf{F}}^* = \dot{\mathbf{P}}^* \cdot\cdot \dot{\mathbf{F}}. \tag{7.6.10}$$

A useful identity, resulting from the reciprocity of $\mathbf{\Lambda}$, is

$$\dot{\mathbf{P}}^* \cdot\cdot \dot{\mathbf{F}}^* - \dot{\mathbf{P}} \cdot\cdot \dot{\mathbf{F}} = \left(\dot{\mathbf{P}}^* - \dot{\mathbf{P}}\right) \cdot\cdot \left(\dot{\mathbf{F}}^* - \dot{\mathbf{F}}\right) + 2\,\dot{\mathbf{P}} \cdot\cdot \left(\dot{\mathbf{F}}^* - \dot{\mathbf{F}}\right). \tag{7.6.11}$$

7.7. Bifurcation Analysis

It was shown in the previous section, if displacement fields $\mathbf{v}$ and $\mathbf{v}^*$ are both solutions of incrementally linear inhomogeneous rate problem described by

Eqs. (7.6.1) and (7.6.2), then

$$\frac{1}{2}\int_{V^0}\left(\Delta\dot{\mathbf{P}}\cdot\cdot\Delta\dot{\mathbf{F}}\right)\mathrm{d}V^0=\frac{1}{2}\int_{V^0}\boldsymbol{\Lambda}\cdot\cdot\cdot\cdot\left(\Delta\dot{\mathbf{F}}\otimes\Delta\dot{\mathbf{F}}\right)\mathrm{d}V^0=0, \tag{7.7.1}$$

with

$$\Delta\dot{\mathbf{F}}=\dot{\mathbf{F}}-\dot{\mathbf{F}}^*,\quad \Delta\dot{\mathbf{P}}=\dot{\mathbf{P}}-\dot{\mathbf{P}}^*. \tag{7.7.2}$$

Consider the associated homogeneous rate problem, described by

$$\boldsymbol{\nabla}^0\cdot\dot{\boldsymbol{P}}=\mathbf{0}, \tag{7.7.3}$$

and the boundary conditions

$$\mathbf{w}=\mathbf{0}\quad\text{on}\quad S_v^0,\qquad \mathbf{n}^0\cdot\dot{\boldsymbol{P}}=\mathbf{0}\quad\text{on}\quad S_t^0, \tag{7.7.4}$$

where

$$\dot{\boldsymbol{F}}=\frac{\partial\mathbf{w}}{\partial\mathbf{X}},\quad \dot{\boldsymbol{P}}=\boldsymbol{\Lambda}\cdot\cdot\dot{\boldsymbol{F}}. \tag{7.7.5}$$

The bold face italic notation is used for the fields associated with the displacement field $\mathbf{w}$. The rate problem described by (7.7.3) and (7.7.4) has always a nul solution $\mathbf{w}=\mathbf{0}$. If the homogeneous problem also has a nontrivial solution

$$\mathbf{w}\neq\mathbf{0}, \tag{7.7.6}$$

then by Eq. (7.7.1)

$$\frac{1}{2}\int_{V^0}\left(\dot{\boldsymbol{P}}\cdot\cdot\dot{\boldsymbol{F}}\right)\mathrm{d}V^0=\frac{1}{2}\int_{V^0}\boldsymbol{\Lambda}\cdot\cdot\cdot\cdot\left(\dot{\boldsymbol{F}}\otimes\dot{\boldsymbol{F}}\right)\mathrm{d}V^0=0. \tag{7.7.7}$$

This condition places the same restrictions on the moduli $\boldsymbol{\Lambda}$ as does (7.7.1), as expected, since (7.7.7) follows directly from (7.7.1) by taking

$$\mathbf{w}=\mathbf{v}-\mathbf{v}^*. \tag{7.7.8}$$

The examination of the uniqueness of solution to incrementally linear inhomogeneous rate problem (7.6.1) and (7.6.2) is thus equivalent to the examination of the uniqueness of solution to the associated homogeneous rate problem (7.7.3) and (7.7.4).

7.7.1. Exclusion Functional

If for all kinematically admissible $\mathbf{w}$ giving rise to $\dot{\boldsymbol{F}}=\partial\mathbf{w}/\partial\mathbf{X}$,

$$\int_{V^0}\chi(\mathbf{w})\,\mathrm{d}V^0=\frac{1}{2}\int_{V^0}\boldsymbol{\Lambda}\cdot\cdot\cdot\cdot\left(\dot{\boldsymbol{F}}\otimes\dot{\boldsymbol{F}}\right)\mathrm{d}V^0>0, \tag{7.7.9}$$

from Eq. (7.6.6) it follows that $\mathbf{w}=\mathbf{0}$ is the only solution of the homogeneous rate problem. Furthermore, by Eq. (7.4.4) it follows that the underlying equilibrium configuration $\mathbf{x}$ is incrementally stable (and thus incrementally unique), under a considered dead loading. At some states of deformation, however, there may exist a nontrivial solution $\mathbf{w} \neq \mathbf{0}$ to the homogeneous rate problem. This $\mathbf{w}$ then satisfies Eq. (7.7.7), implying nonuniqueness of the homogeneous rate problem, and from Section 7.4 nonuniqueness and neutral incremental stability of the underlying equilibrium configuration $\mathbf{x}$. The deformation state at which this happens is called an eigenstate. A nontrivial solution to the homogeneous rate problem is called an eigenmode (Hill, 1978). Therefore, since inhomogeneous rate problem with an incrementally linear stress-deformation response is linear, its solution is unique if and only if the current configuration is not an eigenstate for the associated homogeneous rate problem. If the current configuration is an eigenstate, than any multiple of an eigenmode ($k\mathbf{w}$) could be added to one solution of inhomogeneous rate problem ($\mathbf{v}$) to generate others ($\mathbf{v} + k\mathbf{w}$). Thus, to guarantee uniqueness it is enough to exclude the possibility of eigenmodes. Consequently, following Hill (1978), introduce the exclusion functional

$$\mathcal{F} = \int_{V^0} \chi(\mathbf{w})\,\mathrm{d}V^0, \quad \chi(\mathbf{w}) = \frac{1}{2}\mathbf{\Lambda} \cdots \left(\dot{\boldsymbol{F}} \otimes \dot{\boldsymbol{F}}\right), \tag{7.7.10}$$

for any kinematically admissible $\mathbf{w}$ giving rise to $\dot{\boldsymbol{F}} = \partial\mathbf{w}/\partial\mathbf{X}$. Starting the deformation from a stable reference configuration, a state is reached where the exclusion functional becomes positive semidefinite ($\mathcal{F} \geq 0$), vanishing for some kinematically admissible $\mathbf{w}$. The state at which

$$\mathcal{F} = 0 \tag{7.7.11}$$

is first reached for some $\mathbf{w}$ is called a primary eigenstate. In this state the uniqueness fails, and the deformation path branches (usually by infinitely many eigenmodes). The phenomenon is referred to as bifurcation. (Beyond the region $\mathcal{F} \geq 0$, the exclusion functional is indefinite. If a kinematically admissible $\mathbf{w}$ makes $\mathcal{F} = 0$ for some configuration in this region, the configuration is an eigenstate, but $\mathbf{w}$ is not an eigenmode unless it gives rise to statically admissible stress rate field $\dot{\boldsymbol{F}}$. Since this region is unstable, it will not be considered further).

In any eigenstate at the boundary $\mathcal{F} \geq 0$, an eigenmode $\mathbf{w}$ makes the exclusion functional stationary within the class of kinematically admissible variations $\delta\mathbf{w}$. Indeed, for homogeneous data

$$\frac{1}{2}\int_{V^0} \mathbf{\Lambda} \cdot\cdot\cdot\cdot \left(\dot{\boldsymbol{F}} \otimes \delta\dot{\boldsymbol{F}}\right) \mathrm{d}V^0 = \frac{1}{2}\int_{V^0} \dot{\boldsymbol{P}} \cdot\cdot\, \delta\dot{\boldsymbol{F}}\, \mathrm{d}V^0 = 0, \tag{7.7.12}$$

by Eq. (3.11.12), since the stress rate $\dot{\boldsymbol{P}}$, associated with an eigenmode $\mathbf{w}$, is statically admissible field for the homogeneous rate problem. Since $\mathbf{\Lambda}$ possesses reciprocal symmetry, Eq. (7.7.12) implies

$$\delta\mathcal{F} = 0. \tag{7.7.13}$$

Conversely, any kinematically admissible velocity field $\mathbf{w}$ that makes $\mathcal{F}$ stationary is an eigenmode. This is so because for homogeneous problem the variational integral of Eq. (7.5.7) is equal to the exclusion functional $(\Xi = \mathcal{F})$.

As previously indicated, from Eq (7.4.6) it follows that

$$\mathcal{P}(\mathbf{u} + \delta\mathbf{u}) = \mathcal{P}(\mathbf{u}) \tag{7.7.14}$$

for any eigenmode $\mathbf{w}$ giving rise to displacement increment $\delta\mathbf{u} = \mathbf{w}\,\delta t$. Thus, the potential energies are equal in any two adjacent equilibrium states differing under dead load by an eigenmode deformation. These states are neutrally stable, within the second-order approximations used in deriving Eq. (7.4.6). To assess stability of an eigenmode more accurately, higher order terms in the expansion (7.4.2), leading to (7.4.6), would have to be retained.

The criticality of the exclusion functional is independent of the incepient loading rates (inhomogeneous data) in the current configuration. However, inhomogeneous data cannot be prescribed freely in an eigenstate, if the inhomogeneous rate problem is to admit a solution. Indeed, when the reciprocal theorem (7.5.14) is applied to the fields $\left(\mathbf{v}, \dot{\mathbf{P}}\right)$ and $(\mathbf{w}, \mathbf{0})$, it follows that

$$\int_{V^0} \rho^0\, \dot{\mathbf{b}} \cdot \mathbf{w}\, \mathrm{d}V^0 + \int_{S_t^0} \dot{\mathbf{p}}_n \cdot \mathbf{w}\, \mathrm{d}S_t^0 = 0, \tag{7.7.15}$$

for every distinct eigenmode. This may be regarded as a generalized orthogonality between the rates of loading (inhomogeneous data) and the eigenmodes (Hill, 1978; Ogden, 1984).

In the case of homogeneous material and homogeneous deformation, Eq. (7.7.9) implies that $\mathbf{\Lambda}$ is positive definite. A primary eigenstate is characterized by positive semidefinite $\mathbf{\Lambda}$, i.e.,

$$\chi(\mathbf{w}) = \frac{1}{2}\mathbf{\Lambda} \cdots\cdot (\dot{\boldsymbol{F}} \otimes \dot{\boldsymbol{F}}) \geq 0, \tag{7.7.16}$$

with equality sign for some $\dot{\boldsymbol{F}}$ (uniform throughout the body). The corresponding eigenmode is subject to stationary condition $\delta\mathcal{F} = 0$, which gives $\dot{\boldsymbol{P}} \cdot\cdot\, \delta\dot{\boldsymbol{F}} = 0$ for all $\delta\dot{\boldsymbol{F}}$ from kinematically admissible $\delta\mathbf{w}$. Thus,

$$\dot{\boldsymbol{P}} = \mathbf{\Lambda} \cdot\cdot\, \dot{\boldsymbol{F}} = \mathbf{0} \tag{7.7.17}$$

in a primary (uniformly deformed) eigenstate, as anticipated since $\mathbf{\Lambda}$ becomes singular in this state.

In the case of deformation sensitive loading rates, the exclusion condition is

$$\mathcal{F} > 0 \tag{7.7.18}$$

for all kinematically admissible fields $\mathbf{w}$, where

$$\mathcal{F} = \int_{V^0} \chi(\mathbf{w})\, \mathrm{d}V^0 - \frac{1}{2}\int_{V^0} \rho^0\, \dot{\mathbf{b}} \cdot \mathbf{w}\, \mathrm{d}V^0 - \frac{1}{2}\int_{S_t^0} \dot{\mathbf{p}}_n \cdot \mathbf{w}\, \mathrm{d}S_t^0. \tag{7.7.19}$$

If the loading rates are self-adjoint in the sense of Eq. (7.5.8), both the exclusion functional and its first variation vanish for an eigenmode. Detailed analysis is given by Hill (1978).

7.8. Localization Bifurcation

Consider a homogeneous elastic body in the state of uniform deformation. For prescribed velocities on the boundary which give rise to uniform $\dot{\mathbf{F}}$ throughout the body, conditions are sought under which bifurcation by localization of deformation within a planar band can occur. This is associated with a primary eigenmode

$$\mathbf{w} = f(\mathbf{N} \cdot \mathbf{X})\, \boldsymbol{\eta}, \quad \dot{\boldsymbol{F}} = f'\, \boldsymbol{\eta} \otimes \mathbf{N}. \tag{7.8.1}$$

For $\dot{\boldsymbol{F}}$ to be discontinuous across the band, the gradient f' is piecewise constant across the band, whose unit normal in the undeformed configuration is $\mathbf{N}$. The localization vector is $\boldsymbol{\eta}$. For example, in the case of shear band, $\mathbf{n}\cdot\boldsymbol{\eta} = 0$, where $\mathbf{n} = \mathbf{N}\cdot\boldsymbol{F}^{-1}$ is the band normal in the deformed configuration (Fig. 7.1). (Although shear and necking instabilities are usually associated

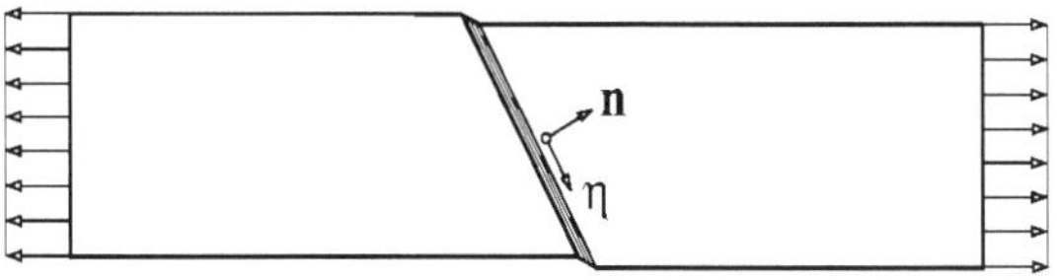

FIGURE 7.1. A shear band with normal $\mathbf{n}$ and localization vector $\boldsymbol{\eta}$ in a homogeneously deformed specimen under plane strain tension.

with plastic response, they can also occur in certain nonlinearly elastic materials; Silling, 1988; Antman, 1974,1995). The stress rate associated with Eq. (7.8.1) is

$$\dot{\boldsymbol{P}} = f'\,\boldsymbol{\Lambda}\cdot\cdot(\boldsymbol{\eta}\otimes\mathbf{N}) = f'\,\boldsymbol{\Lambda}:(\mathbf{N}\otimes\boldsymbol{\eta}). \tag{7.8.2}$$

Substituting this into equilibrium equation (7.7.3) gives

$$f''\,\mathbf{N}\cdot\boldsymbol{\Lambda}:(\mathbf{N}\otimes\boldsymbol{\eta}) = \mathbf{0}. \tag{7.8.3}$$

Thus,

$$\mathbf{N}\cdot\boldsymbol{\Lambda}:(\mathbf{N}\otimes\boldsymbol{\eta}) = \mathbf{A}(\mathbf{N})\cdot\boldsymbol{\eta} = \mathbf{0}. \tag{7.8.4}$$

The second-order tensor

$$\mathbf{A}(\mathbf{N}) = \boldsymbol{\Lambda}:(\mathbf{N}\otimes\mathbf{N}), \quad A_{ij}(\mathbf{N}) = \Lambda_{KiLj}N_K N_L \tag{7.8.5}$$

is a symmetric tensor, obeying the symmetry $\Lambda_{KiLj} = \Lambda_{LjKi}$. For a nontrivial $\boldsymbol{\eta}$ to be determined from the condition

$$\mathbf{A}(\mathbf{N})\cdot\boldsymbol{\eta} = \mathbf{0}, \tag{7.8.6}$$

the matrix $\mathbf{A}(\mathbf{N})$ has to be singular, i.e.,

$$\det\mathbf{A}(\mathbf{N}) = 0. \tag{7.8.7}$$

Note that Eq. (7.8.4) implies

$$\dot{\boldsymbol{p}}_n = \mathbf{N}\cdot\dot{\boldsymbol{P}} = \mathbf{0}, \tag{7.8.8}$$

which is obtained by multiplying Eq. (7.8.2) with $\mathbf{N}$. This means that the rate of nominal traction across the localization band vanishes.

Constitutive law and equilibrium equations are said to be elliptic in any state where

$$\det\mathbf{A}(\mathbf{N}) \neq 0, \quad \text{for all } \mathbf{N}. \tag{7.8.9}$$

Thus, if uniform deformation bifurcates by a band localization eigenmode, the constitutive law and governing equilibrium equations loose their ellipticity. Since there is a correspondence between the conditions for a localization bifurcation and the occurrence of stationary body waves (waves with vanishing wave speeds), the latter is briefly discussed in the next section.

7.9. Acoustic Tensor

Consider a homogeneous elastic body in a state of homogeneous deformation. Its response to small amplitude wave disturbances is examined. Solutions to the rate equations

$$\boldsymbol{\nabla}^0 \cdot \dot{\mathbf{P}} = \rho^0 \, \frac{\mathrm{d}^2 \mathbf{v}}{\mathrm{d}t^2} \tag{7.9.1}$$

are sought in the form of a plane wave propagating with a speed c in the direction $\mathbf{N}$,

$$\mathbf{v} = \boldsymbol{\eta} f(\mathbf{N} \cdot \mathbf{X} - ct). \tag{7.9.2}$$

The unit vector $\boldsymbol{\eta}$ defines the polarization of the wave. On substituting (7.9.2) into (7.9.1), the propagation condition is found to be

$$\mathbf{A}(\mathbf{N}) \cdot \boldsymbol{\eta} = \rho^0 \, c^2 \, \boldsymbol{\eta}. \tag{7.9.3}$$

The second-order tensor $\mathbf{A}(\mathbf{N})$ is referred to as the acoustic tensor. It is explicitly defined by Eq. (7.8.5). From Eq. (7.9.3) we conclude that $\rho^0 c^2$ is an eigenvalue and $\boldsymbol{\eta}$ is an eigenvector of the acoustic tensor $\mathbf{A}(\mathbf{N})$. Since $\mathbf{A}(\mathbf{N})$ is real and symmetric, c^2 must be real. If $c^2 > 0$, there is a stability with respect to propagation of small disturbances. For stationary waves (stationary discontinuity) $c = 0$, which signifies the transition from stability to instability. The instability is associated with $c^2 < 0$, and a divergent growth of an initial disturbance.

Taking a scalar product of Eq. (7.9.3) with $\boldsymbol{\eta}$ gives

$$\boldsymbol{\eta} \cdot \mathbf{A}(\mathbf{N}) \cdot \boldsymbol{\eta} = \rho^0 \, c^2. \tag{7.9.4}$$

Therefore, if $\mathbf{A}(\mathbf{N})$ is positive definite,

$$\boldsymbol{\eta} \cdot \mathbf{A}(\mathbf{N}) \cdot \boldsymbol{\eta} > 0 \tag{7.9.5}$$

for all $\boldsymbol{\eta}$, we have $c^2 > 0$, and Eq. (7.9.1) admits three linearly independent plane progressive waves for each direction of propagation $\mathbf{N}$. In this case, small amplitude elastic plane waves can propagate along a given direction in three distinct, mutually orthogonal modes. These modes are generally neither longitudinal nor transverse. We say that the wave is longitudinal if $\boldsymbol{\eta}$ and $\mathbf{n} = \mathbf{N} \cdot \mathbf{F}^{-1}$ are parallel, and transverse if $\boldsymbol{\eta}$ and $\mathbf{n}$ are perpendicular. Brugger (1965) calculated directions of propagation of pure mode, longitudinal and transverse waves for most anisotropic crystal classes in their undeformed state. See also Hill (1975) and Milstein (1982).

7.9.1. Strong Ellipticity Condition

If the condition holds

$$\boldsymbol{\eta} \cdot \mathbf{A}(\mathbf{N}) \cdot \boldsymbol{\eta} = \boldsymbol{\Lambda} :: [(\mathbf{N} \otimes \boldsymbol{\eta}) \otimes (\mathbf{N} \otimes \boldsymbol{\eta})] > 0 \tag{7.9.6}$$

for each $\mathbf{N} \otimes \boldsymbol{\eta}$, the system of equations (7.9.1) with zero acceleration is said to be strongly elliptic. Clearly, strong ellipticity implies ellipticity, since for positive definite acoustic tensor

$$\det \mathbf{A}(\mathbf{N}) > 0. \tag{7.9.7}$$

Not every strain energy function will yield an acoustic tensor satisfying the conditions of strong ellipticity in every configuration. For example, in the case of undeformed isotropic elastic material, the strong ellipticity requires that the Lamé constants satisfy

$$\lambda + 2\mu > 0, \quad \mu > 0. \tag{7.9.8}$$

This does not imply that the corresponding Ψ is positive definite. The conditions for the latter are

$$\lambda + \frac{2}{3}\mu > 0, \quad \mu > 0. \tag{7.9.9}$$

Thus, while the strong ellipticity condition is strong enough to preclude occurrence of shear band localization, it is not strong enough to ensure the physically observed behavior with necessarily positive value of the elastic bulk modulus ($\kappa = \lambda + 2\mu/3$).

A weaker inequality

$$\boldsymbol{\eta} \cdot \mathbf{A}(\mathbf{N}) \cdot \boldsymbol{\eta} = \boldsymbol{\Lambda} :: [(\mathbf{N} \otimes \boldsymbol{\eta}) \otimes (\mathbf{N} \otimes \boldsymbol{\eta})] \geq 0 \tag{7.9.10}$$

for all $\mathbf{N} \otimes \boldsymbol{\eta}$, is known as the Hadamard condition of stability. This condition does not exclude nonpropagating or stationary waves (discontinuities, singular surfaces). The condition is further discussed by Truesdell and Noll (1965), and Marsden and Hughes (1983).

If the current configuration is taken as the reference, Eq. (7.9.1) becomes

$$\nabla \cdot \dot{\underline{\mathbf{P}}} = \rho \frac{\mathrm{d}^2 \mathbf{v}}{\mathrm{d}t^2}, \tag{7.9.11}$$

where

$$\dot{\underline{\mathbf{P}}} = \underline{\mathbf{\Lambda}} \cdot\cdot \mathbf{L}, \quad \underline{\Lambda}_{jilk} = \underline{\mathcal{L}}^{(1)}_{jilk} + \sigma_{jl}\delta_{ik}. \tag{7.9.12}$$

The propagation condition is

$$\underline{\mathbf{A}}(\mathbf{n}) \cdot \boldsymbol{\eta} = \rho c^2 \boldsymbol{\eta}, \quad \underline{A}_{ij}(\mathbf{n}) = \underline{\Lambda}_{kilj} n_k n_l, \tag{7.9.13}$$

while the strong ellipticity requires that

$$\boldsymbol{\eta} \cdot \underline{\mathbf{A}}(\mathbf{n}) \cdot \boldsymbol{\eta} = \underline{\mathbf{\Lambda}} :: [(\mathbf{n} \otimes \boldsymbol{\eta}) \otimes (\mathbf{n} \otimes \boldsymbol{\eta})] > 0. \tag{7.9.14}$$

Since the moduli $\mathbf{\Lambda}$ and $\underline{\mathbf{\Lambda}}$ are related by Eq. (6.4.14), and since $\mathbf{n} = \mathbf{N} \cdot \mathbf{F}^{-1}$, there is a connection

$$\mathbf{A}(\mathbf{N}) = (\det \mathbf{F})\, \underline{\mathbf{A}}(\mathbf{n}). \tag{7.9.15}$$

7.10. Constitutive Inequalities

A significant amount of research was devoted to find a constitutive inequality for elastic materials under finite deformation that would hold irrespective of the geometry of the boundary value problem, or prescribed displacement and traction boundary conditions. For example, in the range of infinitesimal deformation such an inequality is $\boldsymbol{\sigma} : \boldsymbol{\varepsilon} > 0$, where $\boldsymbol{\varepsilon}$ is an infinitesimal strain. This is a consequence of positive definiteness of the strain energy function $\Psi = (1/2)\,\boldsymbol{\sigma} : \boldsymbol{\varepsilon}$. For finite elastic deformation, Caprioli (1955) proposed that the elastic work is non-negative on any path, open or closed, from the ground state. This implies the existence of Ψ, which must have an absolute minimum in the ground (unstressed) state.

Constitutive inequalities must be objective, i.e., independent of a superimposed rotation to the deformed configuration. For example, the inequality

$$\dot{\mathbf{P}} \cdot\cdot \dot{\mathbf{F}} = \mathbf{\Lambda} \cdot\cdot\cdot\cdot \left(\dot{\mathbf{F}} \otimes \dot{\mathbf{F}}\right) > 0, \tag{7.10.1}$$

derived from the considerations of uniqueness and stability of the rate boundary value problem, is not objective, since under the rotation $\mathbf{Q}$,

$$\dot{\mathbf{P}}^* \cdot\cdot \dot{\mathbf{F}}^* = \left(\dot{\mathbf{P}} - \mathbf{P} \cdot \hat{\mathbf{\Omega}}\right) \cdot\cdot \left(\dot{\mathbf{F}} + \hat{\mathbf{\Omega}} \cdot \mathbf{F}\right) \neq \dot{\mathbf{P}} \cdot\cdot \dot{\mathbf{F}}. \tag{7.10.2}$$

There is no universal constitutive inequality applicable to all types of finite elastic deformation. Instead, various inequalities have been proposed to hold in certain domains of deformation around the reference state, and for particular types of elastic materials (e.g., Truesdell and Noll, 1965; Hill, 1968,1970; Ogden, 1970). Such an inequality is

$$\left(\mathbf{T}^*_{(n)} - \mathbf{T}_{(n)}\right) : \left(\mathbf{E}^*_{(n)} - \mathbf{E}_{(n)}\right) = \left(\frac{\partial \Psi}{\partial \mathbf{E}^*_{(n)}} - \frac{\partial \Psi}{\partial \mathbf{E}_{(n)}}\right) : \left(\mathbf{E}^*_{(n)} - \mathbf{E}_{(n)}\right) > 0, \tag{7.10.3}$$

for all $\mathbf{E}_{(n)} \neq \mathbf{E}^*_{(n)}$. If the strain domain in which (7.10.3) holds is convex, the inequality implies that $\Psi(\mathbf{E}_{(n)})$ is globally strictly convex in that domain. It also implies that $\partial\Psi/\partial\mathbf{E}_{(n)}$ is one-to-one in that domain. For different n, (7.10.3) represents different physical requirements, so that inequality may hold for some n, and fail for others.

Another inequality is obtained by requiring that

$$\dot{\mathbf{T}}_{(n)} : \dot{\mathbf{E}}_{(n)} = \mathbf{\Lambda}_{(n)} :: (\dot{\mathbf{E}}_{(n)} \otimes \dot{\mathbf{E}}_{(n)}) > 0, \quad \mathbf{\Lambda}_{(n)} = \frac{\partial^2 \Psi}{\partial \mathbf{E}_{(n)} \otimes \partial \mathbf{E}_{(n)}}. \tag{7.10.4}$$

This means that $\mathbf{\Lambda}_{(n)}$, the Hessian of Ψ with respect to $\mathbf{E}_{(n)}$, is positive definite, i.e., that the strain energy Ψ is locally strictly convex in a considered strain domain. It can be shown that in a convex strain domain local convexity implies global convexity, and *vice versa*. To demonstrate former, for instance, we can choose the strain rate in (7.10.4) to be directed along the line from $\mathbf{E}_{(n)}$ to $\mathbf{E}^*_{(n)}$; integration from $\mathbf{E}_{(n)}$ to $\mathbf{E}^*_{(n)}$ leads (7.10.3). As in the case of (7.10.3), the inequality (7.10.4) represents different physical requirements for different choices of n. Convexity of Ψ is not an invariant property, so that convexity in the space of one strain measure may be lost in the space of another strain measure.

If the current configuration is taken as the reference, (7.10.4) becomes

$$\dot{\underline{\mathbf{T}}}_{(n)} : \mathbf{D} = \underline{\boldsymbol{\mathcal{L}}}_{(n)} :: (\mathbf{D} \otimes \mathbf{D}) = 2\,\underline{\chi}_{(n)} > 0. \tag{7.10.5}$$

This in general imposes different restrictions on the constitutive law than (7.10.4) does. In view of Eqs. (6.3.12) and (6.3.13), we can rewrite (7.10.5) as

$$\overset{\circ}{\underline{\boldsymbol{\tau}}} : \mathbf{D} > 2n\left(\boldsymbol{\sigma} : \mathbf{D}^2\right), \quad \overset{\circ}{\underline{\boldsymbol{\tau}}} = \overset{\circ}{\boldsymbol{\sigma}} + \boldsymbol{\sigma}\,\mathrm{tr}\,\mathbf{D}. \tag{7.10.6}$$

Hill (1968) proposed that the most appealing inequality is obtained from (7.10.6) for $n = 0$, so that

$$\overset{\circ}{\underline{\mathcal{T}}} : \mathbf{D} > 0. \tag{7.10.7}$$

This inequality is found to be in best agreement with the anticipated features of elastic response. See also Leblond (1992).

Alternative representation of the inequalities (7.10.3) and (7.10.4) is obtained by using spatial tensor measures. They are

$$\left(\boldsymbol{\mathcal{T}}^*_{(n)} - \boldsymbol{\mathcal{T}}_{(n)}\right) : \left(\boldsymbol{\mathcal{E}}^*_{(n)} - \boldsymbol{\mathcal{E}}_{(n)}\right) = \left(\frac{\partial \Psi}{\partial \boldsymbol{\mathcal{E}}^*_{(n)}} - \frac{\partial \Psi}{\partial \boldsymbol{\mathcal{E}}_{(n)}}\right) : \left(\boldsymbol{\mathcal{E}}^*_{(n)} - \boldsymbol{\mathcal{E}}_{(n)}\right) > 0, \tag{7.10.8}$$

$$\overset{\bullet}{\boldsymbol{\mathcal{T}}}_{(n)} : \overset{\bullet}{\boldsymbol{\mathcal{E}}}_{(n)} = \bar{\mathbf{\Lambda}}_{(n)} :: \left(\overset{\bullet}{\boldsymbol{\mathcal{E}}}_{(n)} \otimes \overset{\bullet}{\boldsymbol{\mathcal{E}}}_{(n)}\right) > 0, \quad \bar{\mathbf{\Lambda}}_{(n)} = \frac{\partial^2 \Psi}{\partial \boldsymbol{\mathcal{E}}_{(n)} \otimes \partial \boldsymbol{\mathcal{E}}_{(n)}}. \tag{7.10.9}$$

Inequality (7.10.5) remains the same, because $\overset{\bullet}{\underline{\boldsymbol{\mathcal{T}}}}_{(n)} = \dot{\underline{\mathbf{T}}}_{(n)}$ and $\bar{\underline{\mathbf{\Lambda}}}_{(n)} = \underline{\boldsymbol{\mathcal{L}}}_{(n)}$.

If $\mathbf{E}^*_{(n)}$ is nearby $\mathbf{E}_{(n)}$, so that

$$\mathbf{E}^*_{(n)} = \mathbf{E}_{(n)} + \delta\mathbf{E}_{(n)}, \tag{7.10.10}$$

by Taylor expansion of $\partial\Psi/\partial\mathbf{E}^*_{(n)}$ we obtain

$$\delta\mathbf{T}_{(n)} = \mathbf{\Lambda}_{(n)} : \delta\mathbf{E}_{(n)} + \frac{1}{2}\frac{\partial \mathbf{\Lambda}_{(n)}}{\partial \mathbf{E}_{(n)}} :: \left(\delta\mathbf{E}_{(n)} \otimes \delta\mathbf{E}_{(n)}\right) + \cdots . \tag{7.10.11}$$

Thus,

$$\begin{aligned} \delta\mathbf{T}_{(n)} : \delta\mathbf{E}_{(n)} &= \mathbf{\Lambda}_{(n)} :: \left(\delta\mathbf{E}_{(n)} \otimes \delta\mathbf{E}_{(n)}\right) \\ &+ \frac{1}{2}\frac{\partial \mathbf{\Lambda}_{(n)}}{\partial \mathbf{E}_{(n)}} ::: \left(\delta\mathbf{E}_{(n)} \otimes \delta\mathbf{E}_{(n)} \otimes \delta\mathbf{E}_{(n)}\right) + \cdots . \end{aligned} \tag{7.10.12}$$

The sixth-order tensor

$$\frac{\partial \mathbf{\Lambda}_{(n)}}{\partial \mathbf{E}_{(n)}} = \frac{\partial^2 \mathbf{T}_{(n)}}{\partial \mathbf{E}_{(n)} \otimes \partial \mathbf{E}_{(n)}} = \frac{\partial^3 \Psi}{\partial \mathbf{E}_{(n)} \otimes \partial \mathbf{E}_{(n)} \otimes \partial \mathbf{E}_{(n)}} \tag{7.10.13}$$

is a tensor of the third-order elastic moduli, previously encountered in Section 5.11 within the context of higher-order elastic constants of cubic crystals. The third-order pseudomoduli are similarly defined as $\partial\mathbf{\Lambda}/\partial\mathbf{F}$. These tensors play an important role in assessing the true nature of stability of equilibrium in the cases when the second-order expansions, such as those used in Section 7.4, lead to an assessment of neutral stability. Details are available in Hill (1982) and Ogden (1984).

References

Antman, S. S. (1974), Qualitative theory of the ordinary differential equations of nonlinear elasticity, in *Mechanics Today*, Vol. 1, ed. S. Nemat-Nasser, pp. 58–101, Pergamon Press, New York.

Antman, S. S. (1995), *Nonlinear Problems of Elasticity*, Springer-Verlag, New York.

Ball, J. M. (1977), Convexity conditions and existence theorems in non-linear elasticity, *Arch. Rat. Mech. Anal.*, Vol. 63, pp. 337–403.

Beatty, M. F. (1996), Introduction to nonlinear elasticity, in *Nonlinear Effects in Fluids and Solids*, eds. M. M. Carroll and M. A. Hayes, pp. 13–112, Plenum Press, New York.

Brugger, K. (1965), Pure modes for elastic waves in crystals, *J. Appl. Phys.*, Vol. 36, pp. 759–768.

Caprioli, L. (1965), Su un criterio per l'esistenza dell'energia di deformazione, *Boll. Un. Mat. Ital.*, Vol. 10, pp. 481–483 (1955); English translation in *Foundations of Elasticity Theory*, Intl. Sci. Rev. Ser., Gordon & Breach, New York.

Ciarlet, P. G. (1988), *Mathematical Elasticity, Volume I: Three-Dimensional Elasticity*, North-Holland, Amsterdam.

Coleman, B. and Noll, W. (1959), On the thermostatics of continuous media, *Arch. Rat. Mech. Anal.*, Vol. 4, pp. 97–128.

Ericksen, J. L. (1977), Special topics in elastostatics, *Adv. Appl. Mech.*, Vol. 17, pp. 189–244.

Gurtin, M. E. (1982), On uniquensess in finite elasticity, in *Finite Elasticity*, eds. D. E. Carlson and R. T. Shield, pp. 191–199, Martinus Nijhoff Publishers, The Hague.

Hanyga, A. (1985), *Mathematical Theory of Non-Linear Elasticity*, Ellis Horwood, Chichester, England, and PWN–Polish Scientific Publishers, Warsaw, Poland.

Hill, R. (1957), On uniqueness and stability in the theory of finite elastic strain, *J. Mech. Phys. Solids*, Vol. 5, pp. 229–241.

Hill, R. (1968), On constitutive inequalities for simple materials – I, *J. Mech. Phys. Solids*, Vol. 16, pp. 229–242.

Hill, R. (1970), Constitutive inequalities for isotropic elastic solids under finite strain, *Proc. Roy. Soc. London A*, Vol. 314, pp. 457–472.

Hill, R. (1975), On the elasticity and stability of perfect crystals at finite strain, *Math. Proc. Camb. Phil. Soc.*, Vol. 77, pp. 225–240.

Hill, R. (1978), Aspects of invariance in solid mechanics, *Adv. Appl. Mech.*, Vol. 18, pp. 1–75.

Hill, R. (1982), Constitutive branching in elastic materials, *Math. Proc. Cambridge Philos. Soc.*, Vol. 92, pp. 167–181.

Knops, R. J. and Payne, L. E. (1971), *Uniqueness Theorems in Linear Elasticity*, Springer-Verlag, New York.

Knops, R. J. and Wilkes, E. W. (1973), Theory of elastic stability, in *Handbuch der Physik*, ed. C. Truesdell, Band VIa/3, pp. 125–302, Springer-Verlag, Berlin.

Leblond, J. B. (1992), A constitutive inequality for hyperelastic materials in finite strain, *Eur. J. Mech., A/Solids*, Vol. 11, pp. 447–466.

Marsden, J. E. and Hughes, T. J. R. (1983), *Mathematical Foundations of Elasticity*, Prentice Hall, Englewood Cliffs, New Jersey.

Milstein, F. (1982), Crystal elasticity, in *Mechanics of Solids – The Rodney Hill 60th Anniversary Volume*, eds. H. G. Hopkins and M. J. Sewell, pp. 417–452, Pergamon Press, Oxford.

Nemat-Nasser, S. (1974), General variational principles in nonlinear and linear elasticity with applications, in *Mechanics Today*, Vol. 1, ed. S. Nemat-Nasser, pp. 214–261, Pergamon Press, New York.

Ogden, R. W. (1970), Compressible isotropic elastic solids under finite strain: Constitutive inequalities, *Quart. J. Mech. Appl. Math.*, Vol. 23, pp. 457–468.

Ogden, R. W. (1984), *Non-Linear Elastic Deformations*, Ellis Horwood Ltd., Chichester, England (2nd ed., Dover, 1997).

Person, C. E. (1959), *Theoretical Elasticity*, Harvard University Press, Cambridge, Massachusetts.

Silling, S. A. (1988), Two-dimensional effects in the necking of elastic bars, *J. Appl. Mech.*, Vol. 55, pp. 530–535.

Truesdell, C. (1985), *The Elements of Continuum Mechanics*, Springer-Verlag, New York.

Truesdell, C. and Noll, W. (1965), The nonlinear field theories of mechanics, in *Handbuch der Physik*, ed. S. Flügge, Band III/3, Springer-Verlag, Berlin (2nd ed. 1992).

Washizu, K. (1982), *Variational Methods in Elasticity and Plasticity*, 3rd ed., Pergamon Press, Oxford.

Part 3

THEORY OF PLASTICITY

CHAPTER 8

ELASTOPLASTIC CONSTITUTIVE FRAMEWORK

This chapter provides a basic framework for the constitutive analysis of elastoplastic materials. Such materials are capable of exhibiting, under certain loadings, purely elastic response at any stage of deformation. The development is originally due to Hill and Rice (1973). Rate-independent and rate-dependent plastic materials are both encompassed by this framework. For rate-independent materials, purely elastic response results when stress variations are directed within the current yield surface, which is introduced for such materials. For rate-dependent materials, the response may be purely elastic only in the limit, when stress variations are sufficiently rapid compared to fastest rates at which inelastic processes can take place. We start the analysis by defining elastic and plastic increments of stress and strain tensors. Normality properties are then discussed for rate-independent plastic materials which admit the yield surface. Formulations in both stress and strain space are given. Plasticity postulates of Ilyushin and Drucker are studied in detail. Conditions for the existence of flow potential for rate-dependent materials are also examined.

8.1. Elastic and Plastic Increments

An introductory thermodynamic analysis of inelastic deformation process within the framework of thermodynamics with internal state variables was presented in Sections 4.4–4.6. We proceed in this chapter with the analysis of elastoplastic deformation under isothermal conditions only. Basic physical mechanisms of such deformation are described in standard texts, such as Cottrell (1961,1964) and Honeycombe (1984). We shall assume that there is a set of variables ξ_j that, in some approximate sense, represent internal rearrangements of the material due to plastic deformation. These variables are not necessarily state variables in the sense that the free or complementary energy is not a point function of ξ_j but, instead, depends on their path history (Rice, 1971). Denoting the pattern of internal rearrangements symbolically by $\mathcal{H}$ (the set of internal variables ξ_j together with the path history by which they were achieved), the free energy per unit reference volume can be expressed as

$$\Psi = \Psi\left(\mathbf{E}_{(n)}, \mathcal{H}\right). \tag{8.1.1}$$

At any given state of deformation, an infinitesimal change of $\mathcal{H}$ is assumed to be fully described by a set of scalar infinitesimals $d\xi_j$, such that the change in Ψ due to $\mathrm{d}\mathbf{E}_{(n)}$ and $d\xi_j$ is, to first order,

$$\mathrm{d}\Psi = \frac{\partial \Psi}{\partial \mathbf{E}_{(n)}} : \mathrm{d}\mathbf{E}_{(n)} - \rho^0 f_j\, d\xi_j = \mathbf{T}_{(n)} : \mathrm{d}\mathbf{E}_{(n)} - \rho^0 f_j\, d\xi_j. \tag{8.1.2}$$

The reference density is ρ^0, and $f_j\, d\xi_j$ is an increment of dissipative work per unit mass. It is not necessary that any variable ξ_j exists such that $d\xi_j$ represents an infinitesimal change of ξ_j. The use of an italic d in $d\xi_j$ is intended to indicate this. The stress response is

$$\mathbf{T}_{(n)} = \frac{\partial \Psi}{\partial \mathbf{E}_{(n)}}, \tag{8.1.3}$$

evaluated from Ψ at fixed values of $\mathcal{H}$. The energetic forces f_j are associated with the infinitesimals $d\xi_j$, so that plastic change of the free energy, due to change of $\mathcal{H}$ alone,

$$\mathrm{d}^{\mathrm{p}}\Psi = \Psi\left(\mathbf{E}_{(n)}, \mathcal{H} + d\mathcal{H}\right) - \Psi\left(\mathbf{E}_{(n)}, \mathcal{H}\right), \tag{8.1.4}$$

is equal to

$$\mathrm{d}^{\mathrm{p}}\Psi = -\rho^0 f_j\, d\xi_j = -\rho^0 \bar{f}_j\left(\mathbf{E}_{(n)}, \mathcal{H}\right) d\xi_j. \tag{8.1.5}$$

Higher-order terms, such as $(1/2) df_j\, d\xi_j$, associated with infinitesimal changes of f_j during the variations $d\xi_j$, are neglected.

8.1.1. Plastic Stress Increment

The plastic part of stress increment is defined by Hill and Rice (1973) as

$$\mathrm{d}^{\mathrm{p}}\mathbf{T}_{(n)} = \mathbf{T}_{(n)}\left(\mathbf{E}_{(n)}, \mathcal{H} + d\mathcal{H}\right) - \mathbf{T}_{(n)}\left(\mathbf{E}_{(n)}, \mathcal{H}\right). \tag{8.1.6}$$

In view of Eqs. (8.1.3) and (8.1.4), this gives

$$\mathrm{d}^{\mathrm{p}}\mathbf{T}_{(n)} = \frac{\partial}{\partial \mathbf{E}_{(n)}}\left(\mathrm{d}^{\mathrm{p}}\Psi\right). \tag{8.1.7}$$

Thus, the plastic increment of free energy $\mathrm{d}^{\mathrm{p}}\Psi$ can be viewed as a potential for the plastic part of stress increment $\mathrm{d}^{\mathrm{p}}\mathbf{T}_{(n)}$. From Eqs. (8.1.5) and (8.1.7), we also have

$$\mathrm{d}^{\mathrm{p}}\mathbf{T}_{(n)} = -\rho^0 \frac{\partial \bar{f}_j}{\partial \mathbf{E}_{(n)}}\, d\xi_j. \tag{8.1.8}$$

Furthermore, by considering the function

$$\mathbf{T}_{(n)} = \mathbf{T}_{(n)}\left(\mathbf{E}_{(n)}, \mathcal{H}\right), \tag{8.1.9}$$

we deduce from Eq. (8.1.6) that

$$\mathrm{d}^{\mathrm{p}}\mathbf{T}_{(n)} = \mathrm{d}\mathbf{T}_{(n)} - \frac{\partial \mathbf{T}_{(n)}}{\partial \mathbf{E}_{(n)}} : \mathrm{d}\mathbf{E}_{(n)} = \mathrm{d}\mathbf{T}_{(n)} - \boldsymbol{\Lambda}_{(n)} : \mathrm{d}\mathbf{E}_{(n)}. \tag{8.1.10}$$

The fourth-order tensor

$$\boldsymbol{\Lambda}_{(n)} = \frac{\partial \mathbf{T}_{(n)}}{\partial \mathbf{E}_{(n)}} = \frac{\partial^2 \Psi}{\partial \mathbf{E}_{(n)} \otimes \partial \mathbf{E}_{(n)}} \tag{8.1.11}$$

is the tensor of elastic moduli corresponding to the selected strain measure $\mathbf{E}_{(n)}$.

In a rate-independent elastoplastic material, the only way to vary $\mathcal{H}$ but not $\mathbf{E}_{(n)}$ is to consider a cycle of strain $\mathbf{E}_{(n)}$ that involves $d\mathcal{H}$. Suppose that the cycle emanates from the state $A\left(\mathbf{E}_{(n)}, \mathcal{H}\right)$, it goes through $B\left(\mathbf{E}_{(n)} + \mathrm{d}\mathbf{E}_{(n)}, \mathcal{H} + d\mathcal{H}\right)$, and ends at the state $C\left(\mathbf{E}_{(n)}, \mathcal{H} + d\mathcal{H}\right)$, as shown in Fig. 8.1. If the stress at A was $\mathbf{T}_{(n)}$, in the state B it is $\mathbf{T}_{(n)} + \mathrm{d}\mathbf{T}_{(n)}$. After the strain is returned to its value at the beginning of the cycle by elastic unloading, the state C is reached. The stress there is $\mathbf{T}_{(n)} + \mathrm{d}^{\mathrm{p}}\mathbf{T}_{(n)}$. The stress difference $\mathrm{d}^{\mathrm{p}}\mathbf{T}_{(n)}$ is then the stress decrement left after the cycle of strain that involves $\mathrm{d}\mathcal{H}$.

8.1.2. Plastic Strain Increment

Dually, consider a complementary energy defined by the Legendre transform of the free energy as

$$\Phi_{(n)}\left(\mathbf{T}_{(n)}, \mathcal{H}\right) = \mathbf{T}_{(n)} : \mathbf{E}_{(n)} - \Psi\left(\mathbf{E}_{(n)}, \mathcal{H}\right). \tag{8.1.12}$$

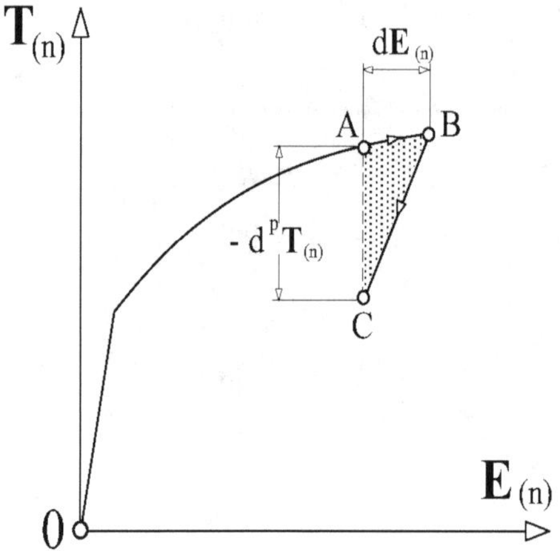

FIGURE 8.1. Strain cycle ABC involving plastic deformation along an infinitesimal segment AB.

The change of complementary energy due to $\mathrm{d}\mathbf{T}_{(n)}$ and $d\xi_j$ is

$$\mathrm{d}\Phi_{(n)} = \frac{\partial \Phi_{(n)}}{\partial \mathbf{T}_{(n)}} : \mathrm{d}\mathbf{T}_{(n)} + \rho^0\, f_j\, d\xi_j = \mathbf{E}_{(n)} : \mathrm{d}\mathbf{T}_{(n)} + \rho^0\, f_j\, d\xi_j. \tag{8.1.13}$$

The strain response is accordingly

$$\mathbf{E}_{(n)} = \frac{\partial \Phi_{(n)}}{\partial \mathbf{T}_{(n)}}, \tag{8.1.14}$$

evaluated from $\Phi_{(n)}$ at fixed values of $\mathcal{H}$. The plastic change of complementary energy, due to change of $\mathcal{H}$ alone,

$$\mathrm{d}^{\mathrm{p}}\Phi_{(n)} = \Phi_{(n)}\left(\mathbf{T}_{(n)}, \mathcal{H} + d\mathcal{H}\right) - \Phi_{(n)}\left(\mathbf{T}_{(n)}, \mathcal{H}\right), \tag{8.1.15}$$

is equal to

$$\mathrm{d}^{\mathrm{p}}\Phi_{(n)} = \rho^0\, f_j\, d\xi_j = \rho^0\, \hat{f}_j\left(\mathbf{T}_{(n)}, \mathcal{H}\right) d\xi_j. \tag{8.1.16}$$

The plastic part of strain increment is defined by

$$\mathrm{d}^{\mathrm{p}}\mathbf{E}_{(n)} = \mathbf{E}_{(n)}\left(\mathbf{T}_{(n)}, \mathcal{H} + d\mathcal{H}\right) - \mathbf{E}_{(n)}\left(\mathbf{T}_{(n)}, \mathcal{H}\right). \tag{8.1.17}$$

In view of Eqs. (8.1.14) and (8.1.15), this gives

$$\mathrm{d}^{\mathrm{p}}\mathbf{E}_{(n)} = \frac{\partial}{\partial \mathbf{T}_{(n)}}\left(\mathrm{d}^{\mathrm{p}}\Phi_{(n)}\right). \tag{8.1.18}$$

Thus, the plastic increment of complementary energy $\mathrm{d}^{\mathrm{p}}\Phi_{(n)}$ can be viewed as a potential for the plastic part of strain increment $\mathrm{d}^{\mathrm{p}}\mathbf{E}_{(n)}$. From Eqs.

(8.1.16) and (8.1.18), we also have

$$\mathrm{d}^{\mathrm{p}}\mathbf{E}_{(n)} = \rho^0 \frac{\partial \hat{f}_j}{\partial \mathbf{T}_{(n)}} d\xi_j. \tag{8.1.19}$$

Furthermore, by taking a differential of the function

$$\mathbf{E}_{(n)} = \mathbf{E}_{(n)}\left(\mathbf{T}_{(n)}, \mathcal{H}\right), \tag{8.1.20}$$

and by employing Eq. (8.1.17), we have

$$\mathrm{d}^{\mathrm{p}}\mathbf{E}_{(n)} = \mathrm{d}\mathbf{E}_{(n)} - \frac{\partial \mathbf{E}_{(n)}}{\partial \mathbf{T}_{(n)}} : \mathrm{d}\mathbf{T}_{(n)} = \mathrm{d}\mathbf{E}_{(n)} - \mathbf{M}_{(n)} : \mathrm{d}\mathbf{T}_{(n)}. \tag{8.1.21}$$

The fourth-order tensor

$$\mathbf{M}_{(n)} = \frac{\partial \mathbf{E}_{(n)}}{\partial \mathbf{T}_{(n)}} = \frac{\partial^2 \Phi}{\partial \mathbf{T}_{(n)} \otimes \partial \mathbf{T}_{(n)}} \tag{8.1.22}$$

is the tensor of elastic compliances corresponding to selected stress measure $\mathbf{T}_{(n)}$.

In a rate-independent elastoplastic material, the only way to vary $\mathcal{H}$ but not $\mathbf{T}_{(n)}$ is to consider a cycle of stress $\mathbf{T}_{(n)}$ that involves $d\mathcal{H}$. Consider a cycle $A \to B \to D$; see Fig. 8.2. In state D the stress is returned to its value before the cycle, i.e., $A\left(\mathbf{T}_{(n)}, \mathcal{H}\right)$, $B\left(\mathbf{T}_{(n)} + \mathrm{d}\mathbf{T}_{(n)}, \mathcal{H} + d\mathcal{H}\right)$ and $D\left(\mathbf{T}_{(n)}, \mathcal{H} + d\mathcal{H}\right)$. The strains in the states A and B are $\mathbf{E}_{(n)}$ and $\mathbf{E}_{(n)} + \mathrm{d}\mathbf{E}_{(n)}$, respectively. After stress is returned to its value before the cycle by elastic unloading, the state D is reached, where the strain is $\mathbf{E}_{(n)} + \mathrm{d}^{\mathrm{p}}\mathbf{E}_{(n)}$. The strain difference $\mathrm{d}^{\mathrm{p}}\mathbf{E}_{(n)}$ is the strain increment left after the cycle of stress that involves $d\mathcal{H}$.

For a rate-dependent material, $\mathrm{d}^{\mathrm{p}}\mathbf{E}_{(n)}$ is the difference between the strains when $\mathbf{T}_{(n)}$ is instantaneously applied after inelastic histories $\mathcal{H}$ and $\mathcal{H} + d\mathcal{H}$, respectively.

8.1.3. Relationship between Plastic Increments

Equations (8.1.4) and (8.1.16) show that

$$\mathrm{d}^{\mathrm{p}}\Psi + \mathrm{d}^{\mathrm{p}}\Phi_{(n)} = 0, \tag{8.1.23}$$

within the order of accuracy used in Eqs. (8.1.4) and (8.1.16). The relationship between the plastic increments $\mathrm{d}^{\mathrm{p}}\mathbf{E}_{(n)}$ and $\mathrm{d}^{\mathrm{p}}\mathbf{T}_{(n)}$ is easily established

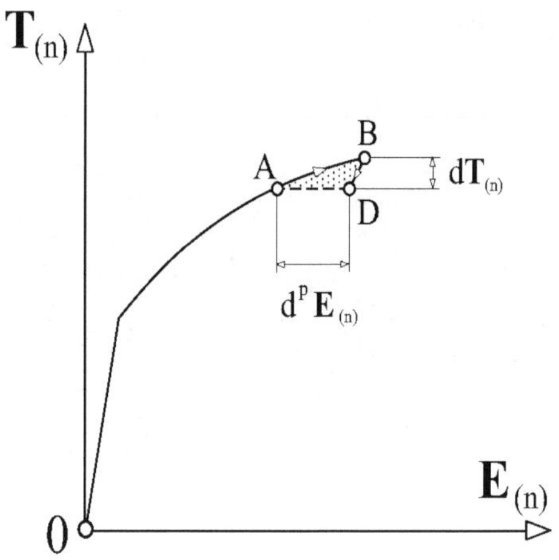

FIGURE 8.2. Stress cycle ABD involving plastic deformation along an infinitesimal segment AB.

from Eqs. (8.1.8) and (8.1.19). This is

$$\mathrm{d}^{\mathrm{p}}\mathbf{T}_{(n)} = -\rho^0 \frac{\partial \bar{f}_j}{\partial \mathbf{E}_{(n)}} d\xi_j = -\rho^0 \left(\frac{\partial \hat{f}_j}{\partial \mathbf{T}_{(n)}} : \frac{\partial \mathbf{T}_{(n)}}{\partial \mathbf{E}_{(n)}} \right) d\xi_j = -\frac{\partial \mathbf{T}_{(n)}}{\partial \mathbf{E}_{(n)}} : \mathrm{d}^{\mathrm{p}}\mathbf{E}_{(n)}. \tag{8.1.24}$$

Similarly,

$$\mathrm{d}^{\mathrm{p}}\mathbf{E}_{(n)} = \rho^0 \frac{\partial \hat{f}_j}{\partial \mathbf{T}_{(n)}} d\xi_j = \rho^0 \left(\frac{\partial \bar{f}_j}{\partial \mathbf{E}_{(n)}} : \frac{\partial \mathbf{E}_{(n)}}{\partial \mathbf{T}_{(n)}} \right) d\xi_j = -\frac{\partial \mathbf{E}_{(n)}}{\partial \mathbf{T}_{(n)}} : \mathrm{d}^{\mathrm{p}}\mathbf{T}_{(n)}. \tag{8.1.25}$$

Therefore, the plastic increments are related by

$$\mathrm{d}^{\mathrm{p}}\mathbf{T}_{(n)} = -\mathbf{\Lambda}_{(n)} : \mathrm{d}^{\mathrm{p}}\mathbf{E}_{(n)}, \quad \mathrm{d}^{\mathrm{p}}\mathbf{E}_{(n)} = -\mathbf{M}_{(n)} : \mathrm{d}^{\mathrm{p}}\mathbf{T}_{(n)}. \tag{8.1.26}$$

These expressions also follow directly from Eqs. (8.1.10) and (8.1.21), since $\mathbf{\Lambda}_{(n)}$ and $\mathbf{M}_{(n)}$ are mutual inverses.

Note that purely elastic increment of strain is related to the corresponding increment of stress by

$$\delta\mathbf{T}_{(n)} = \mathbf{\Lambda}_{(n)} : \delta\mathbf{E}_{(n)}, \quad \delta\mathbf{E}_{(n)} = \mathbf{M}_{(n)} : \delta\mathbf{T}_{(n)}. \tag{8.1.27}$$

The variation of the free energy associated with $\delta\mathbf{E}_{(n)}$ is

$$\delta\Psi = \frac{\partial \Psi}{\partial \mathbf{E}_{(n)}} : \delta\mathbf{E}_{(n)} = \mathbf{T}_{(n)} : \delta\mathbf{E}_{(n)}. \tag{8.1.28}$$

8.2. Yield Surface for Rate-Independent Materials

Rate-independent plastic materials have an elastic range within which they respond in a purely elastic manner. The boundary of this range, in either stress or strain space, is called the yield surface. The shape of the yield surface depends on the entire history of deformation from the reference state. During plastic deformation the states of stress or strain remain on the subsequent yield surfaces. The yield surfaces for actual materials are experimentally found to be mainly smooth, although they may develop pyramidal or conical vertices, or regions of high curvature (Hill, 1978). If elasticity within the yield surface is linear and unaffected by plastic flow, the yield surfaces for metals are convex in the Cauchy stress space. General discussion regarding the geometry and experimental determination of the yield surfaces can be found in Drucker (1960), Naghdi (1960), and Hecker (1976).

8.2.1. Yield Surface in Strain Space

Consider the yield surface in strain space defined by

$$g_{(n)}\left(\mathbf{E}_{(n)}, \mathcal{H}\right) = 0, \tag{8.2.1}$$

where $\mathcal{H}$ represents the pattern of internal rearrangements due to plastic deformation. The strain $\mathbf{E}_{(n)}$ is defined relative to an arbitrary reference state. The shape of the yield surface at each stage of deformation is different for different choices of $\mathbf{E}_{(n)}$, so that different functions $g_{(n)}$ correspond to different n. It is assumed that elastic response within the yield surface is Green-elastic, associated with the strain energy

$$\Psi = \Psi\left(\mathbf{E}_{(n)}, \mathcal{H}\right) \tag{8.2.2}$$

per unit reference volume, such that

$$\mathbf{T}_{(n)} = \frac{\partial \Psi}{\partial \mathbf{E}_{(n)}}. \tag{8.2.3}$$

Let the state of strain $\mathbf{E}_{(n)}$ be on the current yield surface. An increment of strain $\mathrm{d}\mathbf{E}_{(n)}$ directed inside the yield surface constitutes an elastic unloading. The corresponding incremental elastic response is governed by the rate-type equation

$$\dot{\mathbf{T}}_{(n)} = \mathbf{\Lambda}_{(n)} : \dot{\mathbf{E}}_{(n)}, \quad \mathbf{\Lambda}_{(n)} = \frac{\partial^2 \Psi}{\partial \mathbf{E}_{(n)} \otimes \partial \mathbf{E}_{(n)}}, \tag{8.2.4}$$

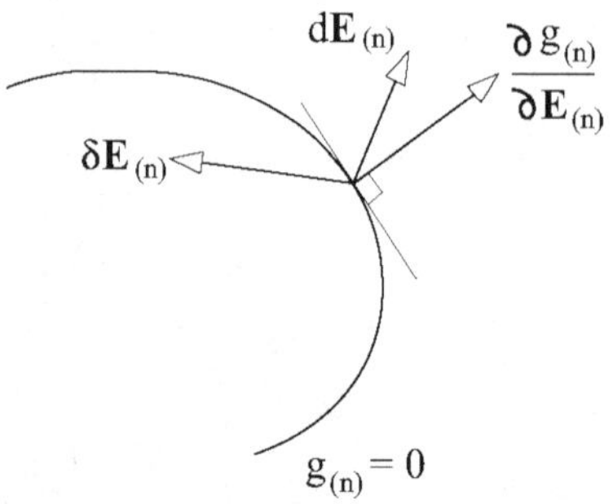

FIGURE 8.3. Strain increment associated with plastic loading $\mathrm{d}\mathbf{E}_{(n)}$ is directed outside the current yield surface in strain space. Strain increment of elastic unloading $\delta\mathbf{E}_{(n)}$ is directed inside the current yield surface.

where $\mathbf{\Lambda}_{(n)} = \mathbf{\Lambda}_{(n)}\left(\mathbf{E}_{(n)}, \mathcal{H}\right)$ is the tensor of instantaneous elastic moduli of the material at the considered state of strain and internal structure.

An increment of strain directed outside the current yield surface constitutes plastic loading. The resulting increment of stress consists of elastic and plastic parts, such that

$$\dot{\mathbf{T}}_{(n)} = \dot{\mathbf{T}}^{\mathrm{e}}_{(n)} + \dot{\mathbf{T}}^{\mathrm{p}}_{(n)} = \mathbf{\Lambda}_{(n)} : \dot{\mathbf{E}}_{(n)} + \dot{\mathbf{T}}^{\mathrm{p}}_{(n)}. \tag{8.2.5}$$

During plastic loading increment, the yield surface locally expands, while the strain state remains on the yield surface. The consistency condition assuring this is

$$g_{(n)}\left(\mathbf{E}_{(n)} + \mathrm{d}\mathbf{E}_{(n)}, \mathcal{H} + d\mathcal{H}\right) = 0. \tag{8.2.6}$$

The elastic stress decrement $\mathrm{d}^{\mathrm{e}}\mathbf{T}_{(n)}$ is associated with the elastic removal of the strain increment $\mathrm{d}\mathbf{E}_{(n)}$ from the state of strain $\mathbf{E}_{(n)}+\mathrm{d}\mathbf{E}_{(n)}$, where the elastic moduli are $\mathbf{\Lambda}_{(n)} + \mathrm{d}\mathbf{\Lambda}_{(n)}$. This is $\mathrm{d}^{\mathrm{e}}\mathbf{T}_{(n)} = \left(\mathbf{\Lambda}_{(n)} + \mathrm{d}\mathbf{\Lambda}_{(n)}\right) : \mathrm{d}\mathbf{E}_{(n)}$, which is, to first order, equal to $\mathbf{\Lambda}_{(n)} : \mathrm{d}\mathbf{E}_{(n)}$. Thus, in the limit

$$\dot{\mathbf{T}}^{\mathrm{e}}_{(n)} = \mathbf{\Lambda}_{(n)} : \dot{\mathbf{E}}_{(n)}. \tag{8.2.7}$$

The plastic part of the stress rate $\dot{\mathbf{T}}^{\mathrm{p}}_{(n)}$ corresponds to residual stress decrement $\mathrm{d}^{\mathrm{p}}\mathbf{T}_{(n)}$ in the considered infinitesimal strain cycle (Fig. 8.3). A transition between elastic unloading and plastic loading is a neutral loading. In this case an infinitesimal strain increment is tangential to the yield surface and represents purely elastic deformation. Therefore,

$$\frac{\partial g_{(n)}}{\partial \mathbf{E}_{(n)}} : \dot{\mathbf{E}}_{(n)} \begin{cases} > 0, & \text{for plastic loading,} \\ = 0, & \text{for neutral loading,} \\ < 0, & \text{for elastic unloading.} \end{cases} \tag{8.2.8}$$

The gradient $\partial g_{(n)}/\partial \mathbf{E}_{(n)}$ is codirectional with the outward normal to a locally smooth yield surface $g_{(n)} = 0$ at the state of strain $\mathbf{E}_{(n)}$. For incrementally linear response, all infinitesimal increments $\mathrm{d}\mathbf{E}_{(n)}$, which have equal projections on the normal $\partial g_{(n)}/\partial \mathbf{E}_{(n)}$ (thus forming a cone around $\partial g_{(n)}/\partial \mathbf{E}_{(n)}$), produce the same plastic increment of stress $\mathrm{d}^{\mathrm{p}}\mathbf{T}_{(n)}$. The components obtained by projecting $\mathrm{d}\mathbf{E}_{(n)}$ on the plane tangential to the yield surface represent elastic deformation only (Fig. 8.4).

8.2.2. Yield Surface in Stress Space

If the yield surface is introduced in stress space, it can be generally expressed as

$$f_{(n)}\left(\mathbf{T}_{(n)}, \mathcal{H}\right) = 0. \tag{8.2.9}$$

The stress $\mathbf{T}_{(n)}$ is conjugate to strain $\mathbf{E}_{(n)}$, and the function $f_{(n)}$ corresponds to $g_{(n)}$ such that

$$f_{(n)}\left[\mathbf{T}_{(n)}\left(\mathbf{E}_{(n)}, \mathcal{H}\right), \mathcal{H}\right] = g_{(n)}\left(\mathbf{E}_{(n)}, \mathcal{H}\right) = 0. \tag{8.2.10}$$

This implies that physically identical yield conditions are imposed in both stress and strain spaces. The shape of the yield surface is at each stage of deformation different for different choices of $\mathbf{T}_{(n)}$, so that different functions $f_{(n)}$ correspond to different n. It will be assumed that elastic response within the yield surface is Green-elastic, associated with the complementary strain energy

$$\Phi_{(n)} = \Phi_{(n)}\left(\mathbf{T}_{(n)}, \mathcal{H}\right) \tag{8.2.11}$$

per unit reference volume. Since $\Phi_{(n)}$ is not measure invariant (see Section 4.3), the index (n) is attached to Φ. We assume here that at any given $\mathcal{H}$ there is a one-to-one relationship between $\mathbf{T}_{(n)}$ and $\mathbf{E}_{(n)}$, such that

$$\mathbf{E}_{(n)} = \frac{\partial \Phi_{(n)}}{\partial \mathbf{T}_{(n)}}. \tag{8.2.12}$$

Let the stress state $\mathbf{T}_{(n)}$ be on the current yield surface. If material is in the hardening range relative to the pair $\mathbf{E}_{(n)}$ and $\mathbf{T}_{(n)}$ (precise definition

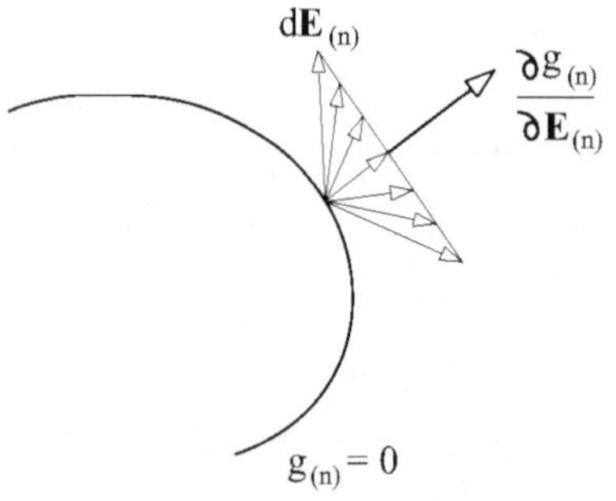

FIGURE 8.4. All strain increments $\mathbf{dE}_{(n)}$ within a cone around the yield surface normal in strain space, which have the same projection on the axis of the cone, give rise to the same plastic stress increment $\mathrm{d}^{\mathrm{p}}\mathbf{T}_{(n)}$.

of hardening is given in Sections 8.8 and 9.2), an increment of stress $\mathrm{d}\mathbf{T}_{(n)}$ directed inside the yield surface will cause purely elastic deformation ($d\mathcal{H} = 0$). This constitutes an elastic unloading from the current yield surface. The corresponding incremental elastic response is governed by the rate-type equation

$$\dot{\mathbf{E}}_{(n)} = \mathbf{M}_{(n)} : \dot{\mathbf{T}}_{(n)}, \quad \mathbf{M}_{(n)} = \frac{\partial^2 \Phi_{(n)}}{\partial \mathbf{T}_{(n)} \otimes \partial \mathbf{T}_{(n)}}. \tag{8.2.13}$$

The tensor $\mathbf{M}_{(n)} = \mathbf{M}_{(n)}\left(\mathbf{T}_{(n)}, \mathcal{H}\right)$ is the tensor of instantaneous elastic compliance of the material at the considered state of stress and internal structure.

An increment of stress directed outside the current yield surface constitutes plastic loading in the hardening range of the material response. The resulting increment of strain consists of elastic and plastic parts, such that

$$\dot{\mathbf{E}}_{(n)} = \dot{\mathbf{E}}^{\mathrm{e}}_{(n)} + \dot{\mathbf{E}}^{\mathrm{p}}_{(n)} = \mathbf{M}_{(n)} : \dot{\mathbf{T}}_{(n)} + \dot{\mathbf{E}}^{\mathrm{p}}_{(n)}. \tag{8.2.14}$$

During plastic loading, the yield surface of a hardening material locally expands, while the stress state remains on it. The consistency condition that assures this is

$$f\left(\mathbf{T}_{(n)} + \mathrm{d}\mathbf{T}_{(n)}, \mathcal{H} + d\mathcal{H}\right) = 0. \tag{8.2.15}$$

The elastic increment of strain $\mathrm{d}^{\mathrm{e}}\mathbf{E}_{(n)}$ is recovered upon elastic unloading of the stress increment $\mathrm{d}\mathbf{T}_{(n)}$. Since elastic unloading takes place from the state of stress $\mathbf{T}_{(n)} + \mathrm{d}\mathbf{T}_{(n)}$, where the elastic compliance is $\mathbf{M}_{(n)} + \mathrm{d}\mathbf{M}_{(n)}$,

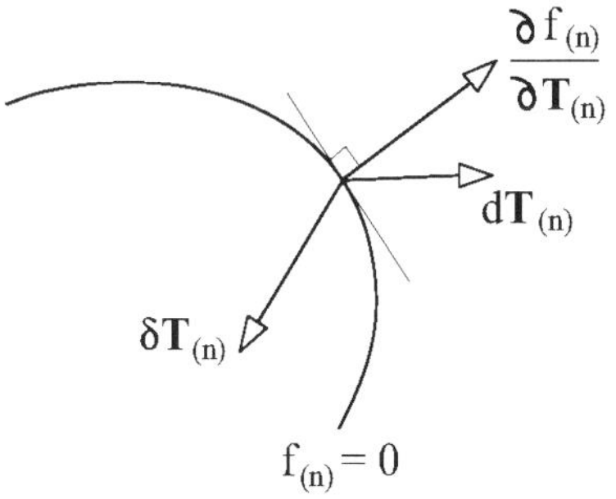

FIGURE 8.5. Stress increment associated with plastic loading $\mathbf{dT}_{(n)}$ is directed outside the current yield surface in stress space. Stress increment of elastic unloading $\delta\mathbf{T}_{(n)}$ is directed inside the current yield surface.

the removal of the stress increment $\mathbf{dT}_{(n)}$ recovers the elastic deformation $\mathrm{d^e}\mathbf{E}_{(n)} = \left(\mathbf{M}_{(n)} + \mathrm{d}\mathbf{M}_{(n)}\right) : \mathrm{d}\mathbf{T}_{(n)}$. To first order this is equal to $\mathbf{M}_{(n)} : \mathrm{d}\mathbf{T}_{(n)}$, and in the limit we have

$$\dot{\mathbf{E}}^{\mathrm{e}}_{(n)} = \mathbf{M}_{(n)} : \dot{\mathbf{T}}_{(n)}, \tag{8.2.16}$$

as used in Eq. (8.2.14). The plastic part of the strain rate $\dot{\mathbf{E}}^{\mathrm{p}}_{(n)}$ corresponds to residual increment of strain $\mathrm{d^p}\mathbf{E}_{(n)}$, left upon removal of the stress increment $\mathrm{d}\mathbf{T}_{(n)}$ (Fig. 8.5).

A transition between elastic unloading and plastic loading is a neutral loading. Here, an infinitesimal stress increment is tangential to the yield surface and produces only elastic deformation. Thus, we have in the hardening range

$$\frac{\partial f_{(n)}}{\partial \mathbf{T}_{(n)}} : \dot{\mathbf{T}}_{(n)} \begin{cases} > 0, & \text{for plastic loading,} \\ = 0, & \text{for neutral loading,} \\ < 0, & \text{for elastic unloading.} \end{cases} \tag{8.2.17}$$

The gradient $\partial f_{(n)}/\partial \mathbf{T}_{(n)}$ is codirectional with the outward normal to a locally smooth yield surface $f_{(n)} = 0$ at the state of stress $\mathbf{T}_{(n)}$. Assuming incrementally linear response, it follows that all infinitesimal increments $\mathrm{d}\mathbf{T}_{(n)}$, which have equal projection on $\partial f_{(n)}/\partial \mathbf{T}_{(n)}$, thus forming a cone around $\partial f_{(n)}/\partial \mathbf{T}_{(n)}$, produce the same plastic increment of deformation $\mathrm{d^p}\mathbf{E}_{(n)}$. The components obtained by projecting $\mathrm{d}\mathbf{T}_{(n)}$ on the plane tangential to the yield surface give rise to elastic deformation only. This is schematically depicted in Fig. 8.6.

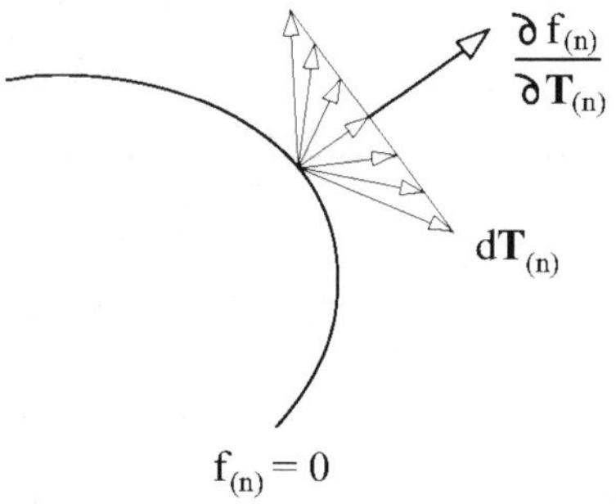

FIGURE 8.6. All stress increments $d\mathbf{T}_{(n)}$ within a cone around the yield surface normal in stress space, which have the same projection on the axis of the cone, give rise to the same plastic strain increment $d^p\mathbf{E}_{(n)}$.

In the softening range of material response, Eq. (8.2.14) still holds, although elastic and plastic parts of the strain rate have purely formal significance, because in the softening range it is not physically possible to perform an infinitesimal cycle of stress starting from the stress point on the yield surface. It should be noted, however, that the hardening is a relative term: the material may be in the hardening range relative to one pair of stress and strain measures, and in the softening range relative to another pair (Hill, 1978).

There are theories of plasticity proposed for rate-independent response which do not use the concept of the yield surface, such as the endochronic theory of Valanis (1971,1975), and a generalized theory of plasticity by Lubliner (1974,1984,1991). They are not discussed in this book, but we refer to original papers, and to Bažant (1978), Murakami and Read (1987), and Hüttel and Matzenmiller (1999). Gurtin (1983) developed a hypoelastic formulation of plasticity in which the existence of the yield surface is a consequence rather than an initial assumption of the theory. Pipkin and Rivlin (1965) earlier proposed a functional-type theory for rate-independent plasticity in which the strain history was defined as a function of the arc length along the strain path. See also Ilyushin (1954) for his geometric theory of plasticity, and Mróz (1966) for his nonlinear formulation of the rate-type theory. The so-called deformation theory of plasticity for proportional or nearly proportional loading paths is presented separately in Section 9.11.

8.3. Normality Rules

Let $d^p\mathbf{E}_{(n)}$ be the plastic increment of strain produced by the stress increment $d\mathbf{T}_{(n)}$ applied from the state of stress $\mathbf{T}_{(n)}$ on the current yield surface. Denote by $\delta\mathbf{T}_{(n)}$ an arbitrary stress variation emanating from the same $\mathbf{T}_{(n)}$ and directed inside the yield surface. If

$$\delta\mathbf{T}_{(n)} : d^p\mathbf{E}_{(n)} < 0, \tag{8.3.1}$$

for every such $\delta\mathbf{T}_{(n)}$, the material obeys the normality rule: the plastic strain increment must be codirectional with the outward normal to a locally smooth yield surface in stress space (Fig. 8.7), whereas at the vertex it must lie within or on the cone of limiting outward normals (Hill and Rice, 1973).

Since

$$\delta\mathbf{T}_{(n)} : d^p\mathbf{E}_{(n)} = -\delta\mathbf{E}_{(n)} : d^p\mathbf{T}_{(n)}, \tag{8.3.2}$$

Equation (8.3.1) implies

$$\delta\mathbf{E}_{(n)} : d^p\mathbf{T}_{(n)} > 0, \tag{8.3.3}$$

for all strain variations $\delta\mathbf{E}_{(n)}$ emanating from the same $\mathbf{E}_{(n)}$ on the yield surface in strain space and directed inside the yield surface. This expresses a dual normality, requiring that $d^p\mathbf{T}_{(n)}$ must be codirectional with the inward normal to a locally smooth yield surface in strain space (Fig. 8.8), with an appropriate generalization at a vertex. Further discussion of normality rules for rate-independent plastic materials is presented in Sections 8.5 and 8.6.

8.3.1. Invariance of Normality Rules

The normality rules (8.3.1) and (8.3.3) are invariant to reference configuration and strain measure, i.e., they apply for every choice of reference configuration and strain measure, or for none. In proof, we first observe that from Eqs. (8.1.7) and (8.1.18),

$$\delta\mathbf{E}_{(n)} : d^p\mathbf{T}_{(n)} = \delta\mathbf{E}_{(n)} : \frac{\partial}{\partial\mathbf{E}_{(n)}}(d^p\Psi) = \delta(d^p\Psi), \tag{8.3.4}$$

$$\delta\mathbf{T}_{(n)} : d^p\mathbf{E}_{(n)} = \delta\mathbf{T}_{(n)} : \frac{\partial}{\partial\mathbf{T}_{(n)}}(d^p\Phi) = \delta(d^p\Phi). \tag{8.3.5}$$

For example, $\delta(d^p\Psi)$ represents the difference between the values of $d^p\Psi$ evaluated at $\mathbf{E}_{(n)} + \delta\mathbf{E}_{(n)}$ and $\mathbf{E}_{(n)}$, for the same $\mathcal{H}$ and $\mathcal{H} + d\mathcal{H}$. Thus,

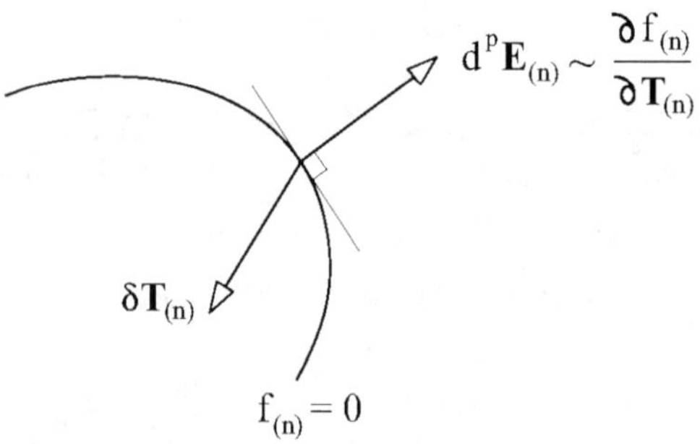

FIGURE 8.7. Normality rule in stress space. The plastic strain increment $d^p\mathbf{E}_{(n)}$ is codirectional with the outward normal to a locally smooth yield surface, so that $\delta\mathbf{T}_{(n)} : d^p\mathbf{E}_{(n)} < 0$, where $\delta\mathbf{T}_{(n)}$ is a stress increment associated with elastic unloading.

either from (8.3.2) or (8.1.23), we have

$$\delta(d^p\Psi) = -\delta(d^p\Phi). \tag{8.3.6}$$

On the other hand,

$$\delta(d^p\Psi) = (d^p\Psi)_{\mathbf{E}_{(n)}+d\mathbf{E}_{(n)}} - (d^p\Psi)_{\mathbf{E}_{(n)}} = (\delta\Psi)_{\mathcal{H}+d\mathcal{H}} - (\delta\Psi)_{\mathcal{H}} = d^p(\delta\Psi). \tag{8.3.7}$$

Since elastic work per unit mass (at fixed $\mathcal{H}$),

$$\frac{1}{\rho}\,\delta\Psi = \frac{1}{\rho}\,\mathbf{T}_{(n)} : \delta\mathbf{E}_{(n)}, \tag{8.3.8}$$

is invariant to choice of reference state and strain measure (provided that all strains define the same geometry change), it follows that

$$\frac{1}{\rho}\,d^p(\delta\Psi) = \frac{1}{\rho}\,\delta\mathbf{E}_{(n)} : d^p\mathbf{T}_{(n)} \tag{8.3.9}$$

is also the reference and strain measure invariant. Therefore, since the mass density of the reference state is positive ($\rho > 0$), we conclude that both normality rules (8.3.1) and (8.3.3) are invariant to choice of reference configuration and strain measure.

It is noted that

$$\begin{aligned}\frac{1}{\rho}\left(\delta\mathbf{T}_{(n)} : d^p\mathbf{E}_{(n)}\right) &= \frac{1}{\rho}\,\delta\mathbf{T}_{(n)} : \left(d\mathbf{E} - \mathbf{M}_{(n)} : d\mathbf{T}_{(n)}\right)\\ &= \frac{1}{\rho}\left(\delta\mathbf{T}_{(n)} : d\mathbf{E}_{(n)} - \delta\mathbf{E}_{(n)} : d\mathbf{T}_{(n)}\right),\end{aligned} \tag{8.3.10}$$

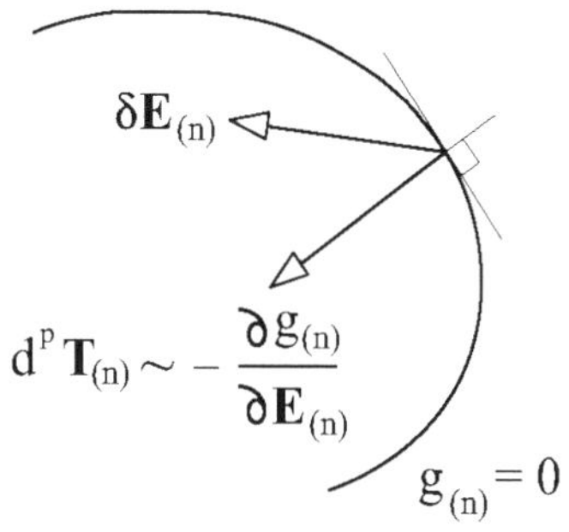

FIGURE 8.8. Normality rule in strain space. The plastic stress increment $\mathrm{d}^{\mathrm{p}}\mathbf{T}_{(n)}$ is codirectional with the inward normal to a locally smooth yield surface, so that $\delta\mathbf{E}_{(n)} : \mathrm{d}^{\mathrm{p}}\mathbf{T}_{(n)} > 0$, where $\delta\mathbf{E}_{(n)}$ is a strain increment associated with elastic unloading.

which demonstrates that the combination on the far right-hand side is invariant. This is a particular type of Hill's (1972) invariant bilinear form.

Normality rules can be expressed in terms of internal variables and conjugate energetic forces by recalling that, from Eq. (8.1.5),

$$\frac{1}{\rho}\,\mathrm{d}^{\mathrm{p}}\Psi = -f_j\, d\xi_j. \tag{8.3.11}$$

This implies that

$$-\frac{1}{\rho}\,\delta\mathbf{T}_{(n)} : \mathrm{d}^{\mathrm{p}}\mathbf{E}_{(n)} = \frac{1}{\rho}\,\delta(\mathrm{d}^{\mathrm{p}}\Psi) = -\delta f_j\, d\xi_j. \tag{8.3.12}$$

Thus, the normality rule (8.3.1) is obeyed if

$$\delta f_j\, d\xi_j < 0. \tag{8.3.13}$$

The inequality is, for example, guaranteed if each increment $d\xi_j$ is, at given $\mathcal{H}$, governed only by its own energetic force f_j. Indeed, the yield criterion for the j-th variable is solely expressed in terms of f_j as

$$f_j^{\mathrm{L}} < f_j < f_j^{\mathrm{U}}. \tag{8.3.14}$$

The yield emanating from the lower bound f_j^{L} involves $d\xi_j < 0$, while any elastic variation δf_j must be positive. The yield emanating from the upper bound f_j^{U} involves $d\xi_j > 0$, while any elastic variation δf_j must be negative. Thus, for each j the product $\delta f_j\, d\xi_j$ is negative, and so is the sum over all j (Rice, 1971).

8.4. Flow Potential for Rate-Dependent Materials

The constitutive framework of Sections 8.2 and 8.3 applies to rate-dependent plastic materials which exhibit elastic response to sufficiently rapid loading or straining (instantaneous elasticity). For the plastic part of strain rate we take

$$\frac{\mathrm{d}^{\mathrm{p}}\mathbf{E}_{(n)}}{\mathrm{d}t} = \frac{\mathrm{d}\mathbf{E}_{(n)}}{\mathrm{d}t} - \mathbf{M}_{(n)} : \frac{\mathrm{d}\mathbf{T}_{(n)}}{\mathrm{d}t}\,, \tag{8.4.1}$$

where t is the physical time. The plastic part of strain rate is a function of the current stress and accumulated inelastic history $\mathcal{H}$,

$$\frac{\mathrm{d}^{\mathrm{p}}\mathbf{E}_{(n)}}{\mathrm{d}t} = \frac{\mathrm{d}^{\mathrm{p}}\mathbf{E}_{(n)}}{\mathrm{d}t}\left(\mathbf{T}_{(n)}, \mathcal{H}\right). \tag{8.4.2}$$

Therefore, an instantaneous change of stress $\delta\mathbf{T}_{(n)}$ causes an instantaneous change of the plastic part of strain rate, but not the change of $\mathcal{H}$ or the plastic strain itself. We can thus examine the functional dependence of $\mathrm{d}^{\mathrm{p}}\mathbf{E}_{(n)}/\mathrm{d}t$ on $\mathbf{T}_{(n)}$ at any fixed $\mathcal{H}$. If this is such that $\delta\mathbf{T}_{(n)} : \mathrm{d}^{\mathrm{p}}\mathbf{E}_{(n)}/\mathrm{d}t$ is a perfect differential at fixed $\mathcal{H}$, i.e., if

$$\delta\mathbf{T}_{(n)} : \frac{\mathrm{d}^{\mathrm{p}}\mathbf{E}_{(n)}}{\mathrm{d}t} = \delta\hat{\Omega}\left(\mathbf{T}_{(n)}, \mathcal{H}\right), \tag{8.4.3}$$

then (Hill and Rice, 1973)

$$\frac{\mathrm{d}^{\mathrm{p}}\mathbf{E}_{(n)}}{\mathrm{d}t} = \frac{\partial\hat{\Omega}\left(\mathbf{T}_{(n)}, \mathcal{H}\right)}{\partial\mathbf{T}_{(n)}}. \tag{8.4.4}$$

This establishes the existence of a scalar flow potential for the plastic part of strain rate in rate-dependent materials,

$$\Omega = \hat{\Omega}\left(\mathbf{T}_{(n)}, \mathcal{H}\right). \tag{8.4.5}$$

Since

$$\delta\mathbf{T}_{(n)} : \frac{\mathrm{d}^{\mathrm{p}}\mathbf{E}_{(n)}}{\mathrm{d}t} = -\delta\mathbf{E}_{(n)} : \frac{\mathrm{d}^{\mathrm{p}}\mathbf{T}_{(n)}}{\mathrm{d}t}\,, \tag{8.4.6}$$

there follows

$$\frac{\mathrm{d}^{\mathrm{p}}\mathbf{T}_{(n)}}{\mathrm{d}t} = -\frac{\partial\bar{\Omega}\left(\mathbf{E}_{(n)}, \mathcal{H}\right)}{\partial\mathbf{E}_{(n)}}. \tag{8.4.7}$$

This shows that Ω, when expressed in terms of strain and inelastic history,

$$\Omega = \bar{\Omega}\left(\mathbf{E}_{(n)}, \mathcal{H}\right), \tag{8.4.8}$$

is also a flow potential for the plastic part of stress rate.

The normality rules (8.4.4) and (8.4.7) are clearly invariant to choice of reference configuration and strain measure. Deduction of the normality rules for rate-independent materials as singular limits of the normality rules for rate-dependent materials has been demonstrated by Rice (1970, 1971).

If it is assumed that, at a given $\mathcal{H}$, each $d\xi_j/\mathrm{d}t$ depends only on its own energetic force,

$$\frac{d\xi_j}{\mathrm{d}t} = \mathrm{function}(f_j, \mathcal{H}), \tag{8.4.9}$$

then

$$\delta \mathbf{T}_{(n)} : \frac{\mathrm{d}^{\mathrm{p}}\mathbf{E}_{(n)}}{\mathrm{d}t} = \delta f_j \frac{d\xi_j}{\mathrm{d}t} \tag{8.4.10}$$

is a perfect differential, because each term in the sum on the right-hand side is a perfect differential. This, for example, establishes the existence of flow potential in rate-dependent crystal plasticity, in which it is assumed that the crystallographic slip on each slip system is governed by the resolved shear stress on that system. A study of crystal plasticity is presented in Chapter 12.

8.5. Ilyushin's Postulate

The remaining sections in this chapter deal with the so-called plasticity postulates of rate-independent plasticity. These postulates are in the form of constitutive inequalities, proposed for certain types of materials undergoing plastic deformation. The two most well-known are by Drucker (1951) and Ilyushin (1961). They are discussed here within the framework of conjugate stress and strain measures, following the presentations by Hill (1968), and Hill and Rice (1973). Particular attention is given to the relationship between these postulates and the plastic normality rules. We begin with the Ilyushin postulate, and consider the Drucker postulate in Section 8.6. Other postulates are discussed in Section 8.9.

Ilyushin (1961) proposed that the net work in an isothermal cycle of strain must be positive,

$$\oint_E \mathbf{T}_{(n)} : \mathrm{d}\mathbf{E}_{(n)} > 0, \tag{8.5.1}$$

if a cycle involves plastic deformation at some stage. The integral in (8.5.1) over an elastic strain cycle is equal to zero, which implies the existence of elastic potential, such that $\mathbf{T}_{(n)} = \partial\Psi/\partial\mathbf{E}_{(n)}$. Since the cycle of strain that includes plastic deformation in general does not return the material to its state at the beginning of the cycle, the inequality (8.5.1) is not a law of

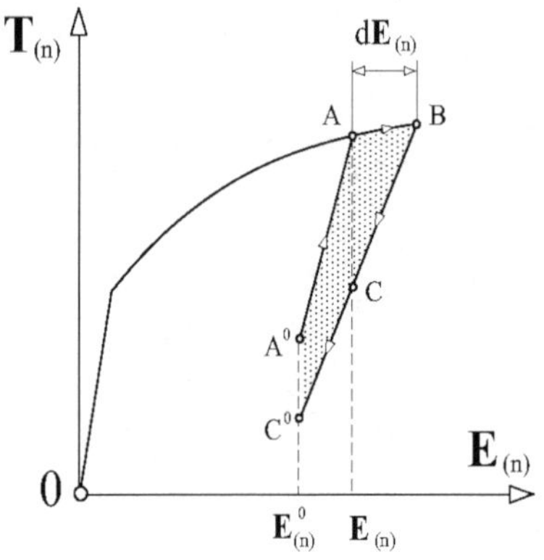

FIGURE 8.9. A strain cycle A^0ABCC^0 involving plastic deformation along an infinitesimal segment AB.

thermodynamics. For example, it does not apply to materials which dissipate energy by friction (Drucker, 1964; Rice, 1971; Dafalias, 1977; Chandler, 1985).

The inequality (8.5.1) is invariant to change of the reference configuration and strain measure, because it is based on an invariant work quantity. The value of the integral over a strain cycle that involves plastic deformation, and that begins and ends at the state of identical geometry, is independent of n and the reference state used to define $\mathbf{E}_{(n)}$. This has been examined in detail by Hill (1968).

The Ilyushin postulate imposes constitutive restrictions on the materials to which it applies. To elaborate, let $A^0\left(\mathbf{E}^0_{(n)}, \mathcal{H}\right)$ be an arbitrary state within the yield surface in strain space. Consider a strain cycle that starts from A^0, includes an elastic segment from A^0 to the state $A\left(\mathbf{E}_{(n)}, \mathcal{H}\right)$ on the current yield surface, followed by an infinitesimal elastoplastic segment from A to $B\left(\mathbf{E}_{(n)} + \mathrm{d}\mathbf{E}_{(n)}, \mathcal{H} + d\mathcal{H}\right)$, and elastic unloading segments from B to $C(\mathbf{E}_{(n)}, \mathcal{H}+d\mathcal{H})$, and from C to $C^0\left(\mathbf{E}^0_{(n)}, \mathcal{H} + d\mathcal{H}\right)$, as shown in Fig. 8.9. By using Eq. (8.1.3), the work done along the segment A^0A is readily evaluated to be

$$\begin{aligned}\int_{A^0}^{A} \mathbf{T}_{(n)} : \mathrm{d}\mathbf{E}_{(n)} &= \int_{A^0}^{A} \frac{\partial \Psi}{\partial \mathbf{E}_{(n)}} : \mathrm{d}\mathbf{E}_{(n)} \\ &= \Psi\left(\mathbf{E}_{(n)}, \mathcal{H}\right) - \Psi\left(\mathbf{E}^0_{(n)}, \mathcal{H}\right),\end{aligned} \tag{8.5.2}$$

while along the segment CC^0,

$$\int_C^{C^0} \mathbf{T}_{(n)} : \mathrm{d}\mathbf{E}_{(n)} = \int_C^{C^0} \frac{\partial \Psi}{\partial \mathbf{E}_{(n)}} : \mathrm{d}\mathbf{E}_{(n)} \\ = \Psi\left(\mathbf{E}^0_{(n)}, \mathcal{H} + d\mathcal{H}\right) - \Psi\left(\mathbf{E}_{(n)}, \mathcal{H} + d\mathcal{H}\right). \tag{8.5.3}$$

The work done along the segments AB and BC is, by the trapezoidal rule of quadrature,

$$\int_A^B \mathbf{T}_{(n)} : \mathrm{d}\mathbf{E}_{(n)} = \mathbf{T}_{(n)} : \mathrm{d}\mathbf{E}_{(n)} + \frac{1}{2}\mathrm{d}\mathbf{T}_{(n)} : \mathrm{d}\mathbf{E}_{(n)}, \tag{8.5.4}$$

$$\int_B^C \mathbf{T}_{(n)} : \mathrm{d}\mathbf{E}_{(n)} = -\mathbf{T}_{(n)} : \mathrm{d}\mathbf{E}_{(n)} - \frac{1}{2}\left(\mathrm{d}\mathbf{T}_{(n)} + \mathrm{d}^{\mathrm{p}}\mathbf{T}_{(n)}\right) : \mathrm{d}\mathbf{E}_{(n)}, \tag{8.5.5}$$

accurate to second-order terms. The plastic stress increment $\mathrm{d}^{\mathrm{p}}\mathbf{T}_{(n)}$ is introduced following Eq. (8.2.7), and is indicated schematically in Fig. 8.1. Consequently, the net work in the considered strain cycle is

$$\oint_E \mathbf{T}_{(n)} : \mathrm{d}\mathbf{E}_{(n)} = -\frac{1}{2}\mathrm{d}^{\mathrm{p}}\mathbf{T}_{(n)} : \mathrm{d}\mathbf{E}_{(n)} + (\mathrm{d}^{\mathrm{p}}\Psi)^0 - \mathrm{d}^{\mathrm{p}}\Psi, \tag{8.5.6}$$

where

$$\mathrm{d}^{\mathrm{p}}\Psi = \Psi\left(\mathbf{E}_{(n)}, \mathcal{H} + d\mathcal{H}\right) - \Psi\left(\mathbf{E}_{(n)}, \mathcal{H}\right), \tag{8.5.7}$$

$$(\mathrm{d}^{\mathrm{p}}\Psi)^0 = \Psi\left(\mathbf{E}^0_{(n)}, \mathcal{H} + d\mathcal{H}\right) - \Psi\left(\mathbf{E}^0_{(n)}, \mathcal{H}\right). \tag{8.5.8}$$

8.5.1. Normality Rule in Strain Space

If the strain cycle emanates from the state on the yield surface, i.e., if $A^0 = A$ and $\mathbf{E}^0_{(n)} = \mathbf{E}_{(n)}$, Eq. (8.5.6) reduces to

$$\oint_E \mathbf{T}_{(n)} : \mathrm{d}\mathbf{E}_{(n)} = -\frac{1}{2}\mathrm{d}^{\mathrm{p}}\mathbf{T}_{(n)} : \mathrm{d}\mathbf{E}_{(n)}. \tag{8.5.9}$$

By Ilyushin's postulate this must be positive, so that

$$\mathrm{d}^{\mathrm{p}}\mathbf{T}_{(n)} : \mathrm{d}\mathbf{E}_{(n)} < 0. \tag{8.5.10}$$

Since during plastic loading the strain increment $\mathrm{d}\mathbf{E}_{(n)}$ is directed outward from the yield surface, and since the same $\mathrm{d}^{\mathrm{p}}\mathbf{T}_{(n)}$ is associated with a fan of infinitely many $\mathrm{d}\mathbf{E}_{(n)}$ around the normal $\partial g_{(n)}/\partial \mathbf{E}_{(n)}$, all having the same projection on that normal, the inequality (8.5.10) requires that $\mathrm{d}^{\mathrm{p}}\mathbf{T}_{(n)}$ is codirectional with the inward normal to a locally smooth yield surface in strain $\mathbf{E}_{(n)}$ space, i.e.,

$$\mathrm{d}^{\mathrm{p}}\mathbf{T}_{(n)} = -\mathrm{d}\gamma_{(n)} \frac{\partial g_{(n)}}{\partial \mathbf{E}_{(n)}} \,. \tag{8.5.11}$$

The scalar multiplier

$$\mathrm{d}\gamma_{(n)} > 0 \tag{8.5.12}$$

is referred to as the loading index. At the vertex of the yield surface, $\mathrm{d}^{\mathrm{p}}\mathbf{T}_{(n)}$ must lie within the cone of limiting inward normals.

The inequality (8.5.10) and the normality rule (8.5.11) hold for all pairs of conjugate stress and strain measures, irrespective of the nature of elastic changes caused by plastic deformation, or possible elastic nonlinearities within the yield surface. Also, (8.5.11) applies regardless of whether the material is in the hardening or softening range.

8.5.2. Convexity of the Yield Surface in Strain Space

If elastic response is nonlinear, we can not conclude from (8.5.6) that the yield surface is necessarily convex. Consider, instead, a linear elastic response within the yield surface, for which the strain energy can be expressed as

$$\Psi(\mathbf{E}_{(n)}, \mathcal{H}) = \frac{1}{2}\,\boldsymbol{\Lambda}_{(n)}(\mathcal{H}) :: \left[\left(\mathbf{E}_{(n)} - \mathbf{E}^{\mathrm{p}}_{(n)}(\mathcal{H})\right) \otimes \left(\mathbf{E}_{(n)} - \mathbf{E}^{\mathrm{p}}_{(n)}(\mathcal{H})\right)\right], \tag{8.5.13}$$

so that

$$\mathbf{T}_{(n)} = \frac{\partial \Psi}{\partial \mathbf{E}_{(n)}} = \boldsymbol{\Lambda}_{(n)}(\mathcal{H}) : \left(\mathbf{E}_{(n)} - \mathbf{E}^{\mathrm{p}}_{(n)}(\mathcal{H})\right). \tag{8.5.14}$$

The tensor $\mathbf{E}^{\mathrm{p}}_{(n)}(\mathcal{H})$ represents a residual or plastic strain that is left upon (actual or conceptual) unloading to zero stress, at fixed values of the internal structure $\mathcal{H}$. Incorporating (8.5.13) and (8.5.14) into (8.5.6) gives

$$\begin{aligned} \oint_E \mathbf{T}_{(n)} : \mathrm{d}\mathbf{E}_{(n)} &= -\frac{1}{2}\,\mathrm{d}^{\mathrm{p}}\mathbf{T}_{(n)} : \mathrm{d}\mathbf{E}_{(n)} + \left(\mathbf{E}^{0}_{(n)} - \mathbf{E}_{(n)}\right) : \mathrm{d}^{\mathrm{p}}\mathbf{T}_{(n)} \\ &+ \frac{1}{2}\,\mathrm{d}\boldsymbol{\Lambda}_{(n)} :: \left[\left(\mathbf{E}_{(n)} - \mathbf{E}^{0}_{(n)}\right) \otimes \left(\mathbf{E}_{(n)} - \mathbf{E}^{0}_{(n)}\right)\right], \end{aligned} \tag{8.5.15}$$

where

$$\mathrm{d}\boldsymbol{\Lambda}_{(n)} = \boldsymbol{\Lambda}_{(n)}(\mathbf{E}_{(n)}, \mathcal{H} + d\mathcal{H}) - \boldsymbol{\Lambda}_{(n)}(\mathbf{E}_{(n)}, \mathcal{H}). \tag{8.5.16}$$

In the derivation, the following relationship was used

$$\begin{aligned}\boldsymbol{\Lambda}_{(n)}(\mathcal{H}+d\mathcal{H}) &: \mathbf{E}^{\mathrm{p}}_{(n)}(\mathcal{H}+d\mathcal{H}) - \boldsymbol{\Lambda}_{(n)}(\mathcal{H}) : \mathbf{E}^{\mathrm{p}}_{(n)}(\mathcal{H}) \\ &= \mathrm{d}\boldsymbol{\Lambda}_{(n)} : \mathbf{E}_{(n)} - \mathrm{d}^{\mathrm{p}}\mathbf{T}_{(n)}.\end{aligned} \tag{8.5.17}$$

By taking the strain cycle with a sufficiently small $\mathrm{d}\mathbf{E}_{(n)}$ comparing to $\mathbf{E}_{(n)} - \mathbf{E}^0_{(n)}$, the first term in Eq. (8.5.15) can be neglected, and for such cycles

$$\begin{aligned}\oint_E \mathbf{T}_{(n)} : \mathrm{d}\mathbf{E}_{(n)} &= \left(\mathbf{E}^0_{(n)} - \mathbf{E}_{(n)}\right) : \mathrm{d}^{\mathrm{p}}\mathbf{T}_{(n)} \\ &+ \frac{1}{2}\,\mathrm{d}\boldsymbol{\Lambda}_{(n)} :: \left[\left(\mathbf{E}_{(n)} - \mathbf{E}^0_{(n)}\right) \otimes \left(\mathbf{E}_{(n)} - \mathbf{E}^0_{(n)}\right)\right] > 0,\end{aligned} \tag{8.5.18}$$

i.e.,

$$\left(\mathbf{E}^0_{(n)} - \mathbf{E}_{(n)}\right) : \mathrm{d}^{\mathrm{p}}\mathbf{T}_{(n)} > -\frac{1}{2}\,\mathrm{d}\boldsymbol{\Lambda}_{(n)} :: \left[\left(\mathbf{E}_{(n)} - \mathbf{E}^0_{(n)}\right) \otimes \left(\mathbf{E}_{(n)} - \mathbf{E}^0_{(n)}\right)\right]. \tag{8.5.19}$$

Thus, if the change of elastic stiffness caused by plastic deformation is such that $\mathrm{d}\boldsymbol{\Lambda}_{(n)}$ is negative semi-definite, or if there is no change in elastic stiffness, from (8.5.19) it follows that (Fig. 8.10)

$$\left(\mathbf{E}^0_{(n)} - \mathbf{E}_{(n)}\right) : \mathrm{d}^{\mathrm{p}}\mathbf{T}_{(n)} > 0. \tag{8.5.20}$$

Since $\mathrm{d}^{\mathrm{p}}\mathbf{T}_{(n)}$ is codirectional with the inward normal to a locally smooth yield surface in strain $\mathbf{E}_{(n)}$ space, (8.5.20) implies that the yield surface is convex. It should be observed, however, that for some $\mathbf{E}_{(n)}$ and $\mathbf{T}_{(n)}$ the stiffness change $\mathrm{d}\boldsymbol{\Lambda}_{(n)}$ can be negative definite, but not for others, so that convexity of the yield surface is not invariant to change of stress and strain measures.

Returning to Eq. (8.5.15), we can write

$$\begin{aligned}\oint_E \mathbf{T}_{(n)} : \mathrm{d}\mathbf{E}_{(n)} &= -\frac{1}{2}\,\mathrm{d}^{\mathrm{p}}\mathbf{T}_{(n)} : \mathrm{d}\mathbf{E}_{(n)} \\ &+ \frac{1}{2}\left(\mathbf{E}^0_{(n)} - \mathbf{E}_{(n)}\right) : \left[\mathrm{d}^{\mathrm{p}}\mathbf{T}_{(n)} + \left(\mathrm{d}^{\mathrm{p}}\mathbf{T}_{(n)}\right)^0\right],\end{aligned} \tag{8.5.21}$$

where

$$\mathrm{d}^{\mathrm{p}}\mathbf{T}_{(n)} = \mathbf{T}_{(n)}\left(\mathbf{E}_{(n)}, \mathcal{H}+d\mathcal{H}\right) - \mathbf{T}_{(n)}\left(\mathbf{E}_{(n)}, \mathcal{H}\right), \tag{8.5.22}$$

$$\left(\mathrm{d}^{\mathrm{p}}\mathbf{T}_{(n)}\right)^0 = \mathbf{T}_{(n)}\left(\mathbf{E}^0_{(n)}, \mathcal{H}+d\mathcal{H}\right) - \mathbf{T}_{(n)}\left(\mathbf{E}^0_{(n)}, \mathcal{H}\right). \tag{8.5.23}$$

If there is no change in elastic stiffness,

$$\left(\mathrm{d}^{\mathrm{p}}\mathbf{T}_{(n)}\right)^0 = \mathrm{d}^{\mathrm{p}}\mathbf{T}_{(n)}. \tag{8.5.24}$$

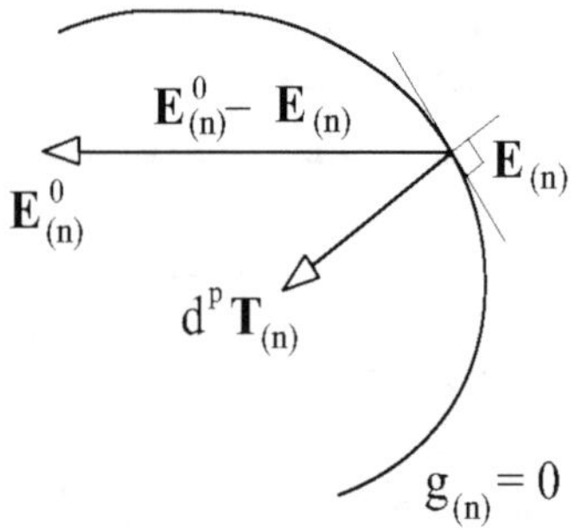

FIGURE 8.10. The plastic stress increment $\mathrm{d}^{\mathrm{p}}\mathbf{T}_{(n)}$ is codirectional with the inward normal to locally smooth yield surface in strain space, so that $(\mathbf{E}^0_{(n)} - \mathbf{E}_{(n)}) : \mathrm{d}^{\mathrm{p}}\mathbf{T}_{(n)} > 0$, where $\mathbf{E}_{(n)}$ is the strain state on the current yield surface and $\mathbf{E}^0_{(n)}$ is the strain state within the yield surface.

8.5.3. Normality Rule in Stress Space

By taking a trace product of Eq. (8.2.14) with $\mathbf{\Lambda}_{(n)} = \mathbf{M}^{-1}_{(n)}$, we obtain

$$\mathbf{\Lambda}_{(n)} : \dot{\mathbf{E}}_{(n)} = \dot{\mathbf{T}}_{(n)} + \mathbf{\Lambda}_{(n)} : \dot{\mathbf{E}}^{\mathrm{p}}_{(n)}, \tag{8.5.25}$$

and comparison with Eq. (8.2.13) establishes

$$\dot{\mathbf{T}}^{\mathrm{p}}_{(n)} = -\mathbf{\Lambda}_{(n)} : \dot{\mathbf{E}}^{\mathrm{p}}_{(n)}. \tag{8.5.26}$$

Thus, to first order,

$$\mathrm{d}^{\mathrm{p}}\mathbf{T}_{(n)} = -\mathbf{\Lambda}_{(n)} : \mathrm{d}^{\mathrm{p}}\mathbf{E}_{(n)}. \tag{8.5.27}$$

Since for any elastic strain increment $\delta\mathbf{E}_{(n)}$, emanating from a point on the yield surface in strain space and directed inside of it,

$$\mathrm{d}^{\mathrm{p}}\mathbf{T}_{(n)} : \delta\mathbf{E}_{(n)} > 0, \tag{8.5.28}$$

the substitution of (8.5.27) into (8.5.28) gives

$$\mathrm{d}^{\mathrm{p}}\mathbf{E}_{(n)} : \mathbf{\Lambda}_{(n)} : \delta\mathbf{E}_{(n)} = \mathrm{d}^{\mathrm{p}}\mathbf{E}_{(n)} : \delta\mathbf{T}_{(n)} < 0. \tag{8.5.29}$$

Here,

$$\delta\mathbf{T}_{(n)} = \mathbf{\Lambda}_{(n)} : \delta\mathbf{E}_{(n)} \tag{8.5.30}$$

is the stress increment from a point on the yield surface in stress space, directed inside of the yield surface (elastic unloading increment associated with elastic strain increment $\delta\mathbf{E}_{(n)}$). Inequality (8.5.29) holds for any such

$\delta\mathbf{T}_{(n)}$ and, consequently, $\mathrm{d}^{\mathrm{p}}\mathbf{E}_{(n)}$ must be codirectional with the outward normal to a locally smooth yield surface in stress $\mathbf{T}_{(n)}$ space, i.e.,

$$\mathrm{d}^{\mathrm{p}}\mathbf{E}_{(n)} = \mathrm{d}\gamma_{(n)} \frac{\partial f_{(n)}}{\partial \mathbf{T}_{(n)}}, \quad \mathrm{d}\gamma_{(n)} > 0. \tag{8.5.31}$$

At a vertex of the yield surface, $\mathrm{d}^{\mathrm{p}}\mathbf{E}_{(n)}$ must lie within the cone of limiting outward normals. Inequality (8.5.29) and the normality rule (8.5.31) hold for all pairs of conjugate stress and strain measures.

If material is in the hardening range relative to $\mathbf{E}_{(n)}$ and $\mathbf{T}_{(n)}$, the stress increment $\mathrm{d}\mathbf{T}_{(n)}$ producing plastic deformation $\mathrm{d}^{\mathrm{p}}\mathbf{E}_{(n)}$ is directed outside the yield surface, and satisfies the condition

$$\mathrm{d}^{\mathrm{p}}\mathbf{E}_{(n)} : \mathrm{d}\mathbf{T}_{(n)} > 0. \tag{8.5.32}$$

If material is in the softening range, the stress increment $\mathrm{d}\mathbf{T}_{(n)}$ producing plastic deformation $\mathrm{d}^{\mathrm{p}}\mathbf{E}_{(n)}$ is directed inside the yield surface, and satisfies the reversed inequality in (8.5.32). The normality rule (8.5.31) applies to both hardening and softening. Inequality (8.5.32) is not measure invariant, since the material may be in the hardening range relative to one pair of conjugate stress and strain measures, but in the softening range relative to another pair.

In view of (8.5.11), (8.5.27), and (8.5.31), the yield surface normals in stress and strain space are related by

$$\frac{\partial g_{(n)}}{\partial \mathbf{E}_{(n)}} = \mathbf{\Lambda}_{(n)} : \frac{\partial f_{(n)}}{\partial \mathbf{T}_{(n)}}. \tag{8.5.33}$$

This also follows directly from Eq. (8.2.10) by partial differentiation.

8.5.4. Additional Inequalities for Strain Cycles

Additional inequalities can be derived as follows. First, by partial differentiation we have

$$\mathbf{T}_{(n)} : \mathrm{d}\mathbf{E}_{(n)} = \mathrm{d}\left(\mathbf{T}_{(n)} : \mathbf{E}_{(n)}\right) - \mathbf{E}_{(n)} : \mathrm{d}\mathbf{T}_{(n)}. \tag{8.5.34}$$

The substitution of Eq. (8.5.34) into the integral of (8.5.1) gives, for the strain cycle A^0ABCC^0,

$$\begin{aligned}\oint_E \mathbf{T}_{(n)} : \mathrm{d}\mathbf{E}_{(n)} = &\left[\mathbf{T}_{(n)}\left(\mathbf{E}^0_{(n)}, \mathcal{H} + d\mathcal{H}\right) - \mathbf{T}_{(n)}\left(\mathbf{E}^0_{(n)}, \mathcal{H}\right)\right] : \mathbf{E}^0_{(n)} \\ &- \oint \mathbf{E}_{(n)} : \mathrm{d}\mathbf{T}_{(n)}.\end{aligned} \tag{8.5.35}$$

This must be positive by Ilyushin's postulate, so that

$$\oint_E \mathbf{E}_{(n)} : \mathrm{d}\mathbf{T}_{(n)} < \left[\mathbf{T}_{(n)}\left(\mathbf{E}^0_{(n)}, \mathcal{H} + d\mathcal{H}\right) - \mathbf{T}_{(n)}\left(\mathbf{E}^0_{(n)}, \mathcal{H}\right)\right] : \mathbf{E}^0_{(n)}. \tag{8.5.36}$$

Alternatively, the inequality (8.5.36) can be written as

$$\oint_E \left(\mathbf{E}_{(n)} - \mathbf{E}^0_{(n)}\right) : \mathrm{d}\mathbf{T}_{(n)} < 0, \tag{8.5.37}$$

for all strain cycles that at some stage involve plastic deformation (not necessarily infinitesimal). Since (8.5.1) is invariant, the inequality (8.5.37) holds irrespective of the reference state and strain measure. In particular, if we choose a reference state for strain measure $\mathbf{E}_{(n)}$ to be the state A^0, the strain $\mathbf{E}^0_{(n)}$ vanishes and (8.5.37) gives

$$\oint_E \mathbf{E}_{(n)} : \mathrm{d}\mathbf{T}_{(n)} < 0. \tag{8.5.38}$$

This applies for all strain measures defined relative to A^0, and for all strain cycles that involve plastic deformation at some stage. Further discussion can be found in Hill (1968, 1978) and Nemat-Nasser (1983).

8.6. Drucker's Postulate

Drucker (1951) introduced a postulate by considering the work done in stress cycles. His original formulation was in the context of infinitesimal strain and is presented in Subsection 8.6.3. We consider here a (noninvariant) dual inequality to (8.5.1), which is

$$\oint_T \mathbf{E}_{(n)} : \mathrm{d}\mathbf{T}_{(n)} < 0. \tag{8.6.1}$$

This means that a net complementary work (relative to measures $\mathbf{E}_{(n)}$ and $\mathbf{T}_{(n)}$) in an isothermal cycle of stress is negative, if a cycle involves plastic deformation at some stage. Inequality (8.6.1) is noninvariant because the value of the integral in (8.6.1) depends on the selected measures $\mathbf{E}_{(n)}$ and $\mathbf{T}_{(n)}$, and the reference state with respect to which they are defined. This is so because $\mathbf{T}_{(n)}$ is introduced as a conjugate stress to $\mathbf{E}_{(n)}$ such that, for the same geometry change, $\mathbf{T}_{(n)} : \mathrm{d}\mathbf{E}_{(n)}$, and not $\mathbf{E}_{(n)} : \mathrm{d}\mathbf{T}_{(n)}$, is measure invariant. Physically, cycling one stress measure does not necessarily imply cycling of another stress measure. Thus, for different n the integral in (8.6.1) corresponds to different physical cycles, and has different values.

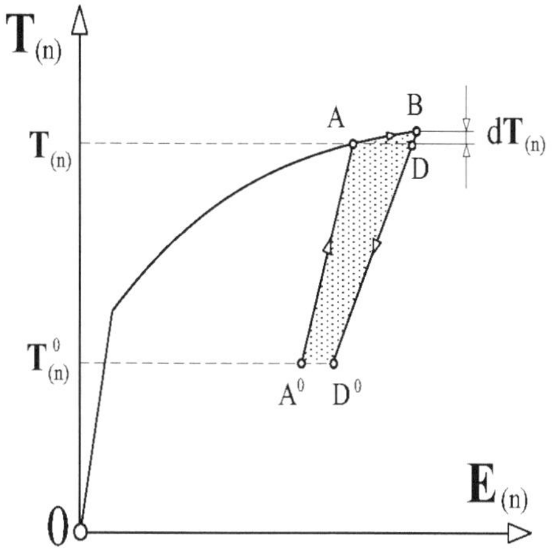

FIGURE 8.11. A stress cycle A^0ABDD^0 involving plastic deformation along an infinitesimal segment AB.

Constitutive inequalities which depend on the choice of reference configuration are not well suited for plastically deforming materials, for which no preferred state can be single out (Hill, 1968). Nevertheless, we proceed with the analysis of (8.6.1) and examine its consequences for different choices of strain measure and reference state.

First, since the cycle of stress that involves plastic deformation in general does not return the material to its state at the beginning of the cycle, the inequality (8.6.1) does not represent a law of thermodynamics for any n. If the integral in (8.6.1) vanishes for stress cycles that give rise to elastic deformation only (for a selected pair $\mathbf{E}_{(n)}$ and $\mathbf{T}_{(n)}$), the material admits a complementary strain energy $\Phi_{(n)} = \Phi_{(n)}\left(\mathbf{T}_{(n)}, \mathcal{H}\right)$, such that $\mathbf{E}_{(n)} = \partial\Phi_{(n)}/\partial\mathbf{T}_{(n)}$. In contrast to measure invariant strain energy Ψ, the complementary energy is in general not measure invariant. However, if the integral in (8.6.1) over an elastic cycle vanishes for some n, it vanishes for other n, as well.

Consider a yield surface in stress $\mathbf{T}_{(n)}$ space. Assume that within the yield surface there is one-to-one relationship between the stress $\mathbf{T}_{(n)}$ and strain $\mathbf{E}_{(n)}$, at a given state of internal structure $\mathcal{H}$. Let $A^0\left(\mathbf{T}^0_{(n)}, \mathcal{H}\right)$ be an arbitrary state within the yield surface. Consider a stress cycle that starts from A^0, includes an elastic segment from A^0 to $A\left(\mathbf{T}_{(n)}, \mathcal{H}\right)$ on the current yield surface, followed by an infinitesimal elastoplastic segment from A to $B\left(\mathbf{T}_{(n)}+\mathrm{d}\mathbf{T}_{(n)}, \mathcal{H}+d\mathcal{H}\right)$, and elastic unloading segments from B to $D\left(\mathbf{T}_{(n)}, \mathcal{H}+d\mathcal{H}\right)$, and from D to $D^0\left(\mathbf{T}^0_{(n)}, \mathcal{H}+d\mathcal{H}\right)$; see Fig. 8.11. The complementary work along the segment A^0A is

$$\begin{aligned}\int_{A^0}^{A} \mathbf{E}_{(n)} : \mathrm{d}\mathbf{T}_{(n)} &= \int_{A^0}^{A} \frac{\partial \Phi_{(n)}}{\partial \mathbf{T}_{(n)}} : \mathrm{d}\mathbf{T}_{(n)} \\ &= \Phi_{(n)}\left(\mathbf{T}_{(n)}, \mathcal{H}\right) - \Phi_{(n)}\left(\mathbf{T}^0_{(n)}, \mathcal{H}\right),\end{aligned} \tag{8.6.2}$$

while along the segment DD^0,

$$\begin{aligned}\int_{D}^{D^0} \mathbf{E}_{(n)} : \mathrm{d}\mathbf{T}_{(n)} &= \int_{D}^{D^0} \frac{\partial \Phi_{(n)}}{\partial \mathbf{T}_{(n)}} : \mathrm{d}\mathbf{T}_{(n)} \\ &= \Phi_{(n)}\left(\mathbf{T}^0_{(n)}, \mathcal{H}+d\mathcal{H}\right) - \Phi_{(n)}\left(\mathbf{T}_{(n)}, \mathcal{H}+d\mathcal{H}\right).\end{aligned} \tag{8.6.3}$$

The complementary work along the segments AB and BD is, by the trapezoidal rule of quadrature,

$$\int_{A}^{B} \mathbf{E}_{(n)} : \mathrm{d}\mathbf{T}_{(n)} = \mathbf{E}_{(n)} : \mathrm{d}\mathbf{T}_{(n)} + \frac{1}{2}\mathrm{d}\mathbf{E}_{(n)} : \mathrm{d}\mathbf{T}_{(n)}, \tag{8.6.4}$$

$$\int_{B}^{C} \mathbf{E}_{(n)} : \mathrm{d}\mathbf{T}_{(n)} = -\mathbf{E}_{(n)} : \mathrm{d}\mathbf{T}_{(n)} - \frac{1}{2}\left(\mathrm{d}\mathbf{E}_{(n)} + \mathrm{d}^{\mathrm{p}}\mathbf{E}_{(n)}\right) : \mathrm{d}\mathbf{T}_{(n)}, \tag{8.6.5}$$

accurate to second-order terms. The plastic strain increment $\mathrm{d}^{\mathrm{p}}\mathbf{E}_{(n)}$ is defined following Eq. (8.2.13), and is indicated schematically in Fig. 8.2. Consequently,

$$\oint_T \mathbf{E}_{(n)} : \mathrm{d}\mathbf{T}_{(n)} = -\frac{1}{2}\mathrm{d}^{\mathrm{p}}\mathbf{E}_{(n)} : \mathrm{d}\mathbf{T}_{(n)} + \left(\mathrm{d}^{\mathrm{p}}\Phi_{(n)}\right)^0 - \mathrm{d}^{\mathrm{p}}\Phi_{(n)}, \tag{8.6.6}$$

where

$$\mathrm{d}^{\mathrm{p}}\Phi_{(n)} = \Phi_{(n)}\left(\mathbf{T}_{(n)}, \mathcal{H}+d\mathcal{H}\right) - \Phi_{(n)}\left(\mathbf{T}_{(n)}, \mathcal{H}\right), \tag{8.6.7}$$

$$\left(\mathrm{d}^{\mathrm{p}}\Phi_{(n)}\right)^0 = \Psi\left(\mathbf{T}^0_{(n)}, \mathcal{H}+d\mathcal{H}\right) - \Phi_{(n)}\left(\mathbf{T}^0_{(n)}, \mathcal{H}\right). \tag{8.6.8}$$

8.6.1. Normality Rule in Stress Space

Assume that material is in the hardening range relative to $\mathbf{E}_{(n)}$ and $\mathbf{T}_{(n)}$. An infinitesimal stress cycle can be performed starting from the point on the yield surface. Thus, taking $A^0 = A$ and $\mathbf{T}^0_{(n)} = \mathbf{T}_{(n)}$, Eq. (8.6.6) reduces to

$$\oint_T \mathbf{E}_{(n)} : \mathrm{d}\mathbf{T}_{(n)} = -\frac{1}{2}\mathrm{d}^{\mathrm{p}}\mathbf{E}_{(n)} : \mathrm{d}\mathbf{T}_{(n)}. \tag{8.6.9}$$

If the inequality (8.6.1) applies to conjugate pair $\mathbf{E}_{(n)}$, $\mathbf{T}_{(n)}$, the integral in (8.6.9) must be negative, so that

$$\mathrm{d}^{\mathrm{p}}\mathbf{E}_{(n)} : \mathrm{d}\mathbf{T}_{(n)} > 0. \tag{8.6.10}$$

During plastic loading in the hardening range relative to $\mathbf{E}_{(n)}$ and $\mathbf{T}_{(n)}$, the stress increment $\mathrm{d}\mathbf{T}_{(n)}$ is directed outward from the yield surface. Since one $\mathrm{d}^{\mathrm{p}}\mathbf{E}_{(n)}$ is associated with a fan of infinitely many $\mathrm{d}\mathbf{T}_{(n)}$ around the normal $\partial f_{(n)}/\partial\mathbf{T}_{(n)}$ (all having the same projection on the normal), the inequality (8.6.10) requires that $\mathrm{d}^{\mathrm{p}}\mathbf{E}_{(n)}$ is codirectional with the outward normal to a locally smooth yield surface in stress $\mathbf{T}_{(n)}$ space, i.e.,

$$\mathrm{d}^{\mathrm{p}}\mathbf{E}_{(n)} = \mathrm{d}\gamma_{(n)}\,\frac{\partial f_{(n)}}{\partial\mathbf{T}_{(n)}}, \quad \mathrm{d}\gamma_{(n)} > 0. \tag{8.6.11}$$

At the vertex of the yield surface, $\mathrm{d}^{\mathrm{p}}\mathbf{E}_{(n)}$ must lie within the cone of limiting outward normals.

The inequality (8.6.10) and the normality rule (8.6.11) apply to a conjugate pair of stress and strain which obey (8.6.1), irrespective of the nature of elastic changes caused by plastic deformation, or possible elastic nonlinearities within the yield surface. If inequality (8.6.1) holds for all pairs of conjugate stress and strain measures, then (8.6.10) and (8.6.11) also hold with respect to all conjugate stress and strain measures.

When material is in the softening range, relative to a considered pair of stress and strain measures, it is physically impossible to perform a cycle of stress starting from a point on the yield surface. In this case, however, we can choose an infinitesimal stress cycle A^0AB, where $A^0\left(\mathbf{T}_{(n)} + \mathrm{d}\mathbf{T}_{(n)}, \mathcal{H}\right)$ is inside the yield surface, while $A\left(\mathbf{T}_{(n)}, \mathcal{H}\right)$ and $B\left(\mathbf{T}_{(n)} + \mathrm{d}\mathbf{T}_{(n)}, \mathcal{H} + d\mathcal{H}\right)$ are on the current and subsequent yield surfaces. Then,

$$\oint_T \mathbf{E}_{(n)} : \mathrm{d}\mathbf{T}_{(n)} = \frac{1}{2}\,\mathrm{d}^{\mathrm{p}}\mathbf{E}_{(n)} : \mathrm{d}\mathbf{T}_{(n)} < 0. \tag{8.6.12}$$

Since in the softening range $\mathrm{d}\mathbf{T}_{(n)}$ is directed inside the current yield surface, (8.6.12) requires that $\mathrm{d}^{\mathrm{p}}\mathbf{E}_{(n)}$ is codirectional with the outward normal to a locally smooth yield surface in stress $\mathbf{T}_{(n)}$ space.

8.6.2. Convexity of the Yield Surface in Stress Space

Returning to (8.6.6), if elastic response is nonlinear we can not conclude from it that the yield surface in stress space is necessarily convex. In fact, a concavity of the yield surface in the Cauchy stress space in the presence of nonlinear elasticity has been demonstrated for a particular material model by Palmer, Maier, and Drucker (1967). Consider, instead, a linear elastic response within the yield surface, for which the complementary energy can be expressed as

$$\Phi_{(n)}\left(\mathbf{T}_{(n)},\mathcal{H}\right)=\mathbf{E}_{(n)}(0,\mathcal{H}):\mathbf{T}_{(n)}+\frac{1}{2}\mathbf{M}_{(n)}(\mathcal{H})::\left(\mathbf{T}_{(n)}\otimes\mathbf{T}_{(n)}\right),\quad(8.6.13)$$

so that

$$\mathbf{E}_{(n)}=\frac{\partial\Phi_{(n)}}{\partial\mathbf{T}_{(n)}}=\mathbf{E}_{(n)}(0,\mathcal{H})+\mathbf{M}_{(n)}(\mathcal{H}):\mathbf{T}_{(n)}.\quad(8.6.14)$$

The tensor $\mathbf{E}_{(n)}(0,\mathcal{H})$, which is equal to $\mathbf{E}^{\mathrm{p}}_{(n)}(\mathcal{H})$ in the notation of Section 8.5, represents a residual or plastic strain, left upon elastic unloading to zero stress at the fixed values of internal structure $\mathcal{H}$. Incorporating (8.6.13) and (8.6.14) into (8.6.6) gives

$$\begin{aligned}\oint_T \mathbf{E}_{(n)}:\mathrm{d}\mathbf{T}_{(n)}=&-\frac{1}{2}\,\mathrm{d}^{\mathrm{p}}\mathbf{E}_{(n)}:\mathrm{d}\mathbf{T}_{(n)}-\left(\mathbf{T}_{(n)}-\mathbf{T}^0_{(n)}\right):\mathrm{d}^{\mathrm{p}}\mathbf{E}_{(n)}\\&+\frac{1}{2}\,\mathrm{d}\mathbf{M}_{(n)}::\left[\left(\mathbf{T}_{(n)}-\mathbf{T}^0_{(n)}\right)\otimes\left(\mathbf{T}_{(n)}-\mathbf{T}^0_{(n)}\right)\right],\end{aligned}\quad(8.6.15)$$

where

$$\mathrm{d}\mathbf{M}_{(n)}=\mathbf{M}_{(n)}(\mathbf{T}_{(n)},\mathcal{H}+d\mathcal{H})-\mathbf{M}_{(n)}(\mathbf{T}_{(n)},\mathcal{H}).\quad(8.6.16)$$

In the derivation, the following expression was used

$$\mathbf{E}_{(n)}(0,\mathcal{H}+d\mathcal{H})-\mathbf{E}_{(n)}(0,\mathcal{H})=-\mathrm{d}\mathbf{M}_{(n)}:\mathbf{T}_{(n)}+\mathrm{d}^{\mathrm{p}}\mathbf{E}_{(n)}.\quad(8.6.17)$$

By taking the stress cycle with a sufficiently small $\mathrm{d}\mathbf{T}_{(n)}$ comparing to $\mathbf{T}_{(n)}-\mathbf{T}^0_{(n)}$, the first term in Eq. (8.6.15) can be neglected, and for such cycles

$$\begin{aligned}\oint_T \mathbf{E}_{(n)}:\mathrm{d}\mathbf{T}_{(n)}=&-\left(\mathbf{T}_{(n)}-\mathbf{T}^0_{(n)}\right):\mathrm{d}^{\mathrm{p}}\mathbf{E}_{(n)}\\&+\frac{1}{2}\,\mathrm{d}\mathbf{M}_{(n)}::\left[\left(\mathbf{T}_{(n)}-\mathbf{T}^0_{(n)}\right)\otimes\left(\mathbf{T}_{(n)}-\mathbf{T}^0_{(n)}\right)\right]<0.\end{aligned}\quad(8.6.18)$$

This gives

$$\left(\mathbf{T}_{(n)}-\mathbf{T}^0_{(n)}\right):\mathrm{d}^{\mathrm{p}}\mathbf{E}_{(n)}>\frac{1}{2}\,\mathrm{d}\mathbf{M}_{(n)}::\left[\left(\mathbf{T}_{(n)}-\mathbf{T}^0_{(n)}\right)\otimes\left(\mathbf{T}_{(n)}-\mathbf{T}^0_{(n)}\right)\right].\quad(8.6.19)$$

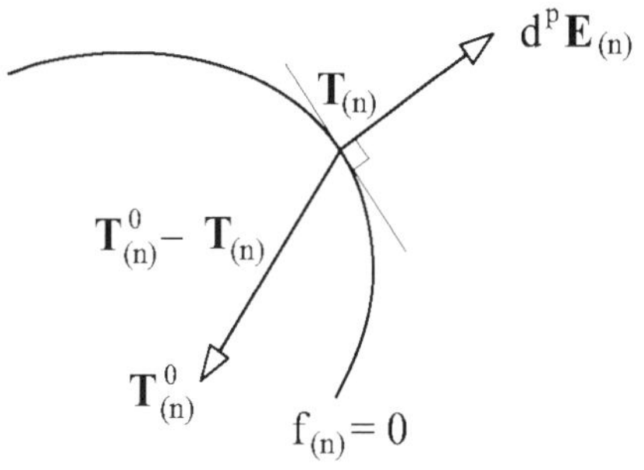

FIGURE 8.12. The plastic strain increment $d^p\mathbf{E}_{(n)}$ is codirectional with the outward normal to locally smooth yield surface in stress space, so that $(\mathbf{T}^0_{(n)} - \mathbf{T}_{(n)}) : d^p\mathbf{E}_{(n)} < 0$, where $\mathbf{T}_{(n)}$ is the stress state on the current yield surface and $\mathbf{T}^0_{(n)}$ is the stress state within the yield surface.

Thus, if the change of elastic stiffness caused by plastic deformation is such that $d\mathbf{M}_{(n)}$ is positive semi-definite, or if there is no change in $\mathbf{M}_{(n)}$, from (8.6.19) it follows that (Fig. 8.12)

$$\left(\mathbf{T}_{(n)} - \mathbf{T}^0_{(n)}\right) : d^p\mathbf{E}_{(n)} > 0. \tag{8.6.20}$$

Since $d^p\mathbf{E}_{(n)}$ is codirectional with the outward normal to a locally smooth yield surface in stress $\mathbf{T}_{(n)}$ space, (8.6.20) implies that the yield surface is convex in the considered stress space.

Returning to (8.6.15), it is noted that it can be rewritten as

$$\begin{aligned}\oint_T \mathbf{E}_{(n)} : d\mathbf{T}_{(n)} &= -\frac{1}{2}\, d^p\mathbf{E}_{(n)} : d\mathbf{T}_{(n)} \\ &\quad - \frac{1}{2}\left(\mathbf{T}_{(n)} - \mathbf{T}^0_{(n)}\right) : \left[d^p\mathbf{E}_{(n)} + \left(d^p\mathbf{E}_{(n)}\right)^0\right],\end{aligned} \tag{8.6.21}$$

where

$$d^p\mathbf{E}_{(n)} = \mathbf{E}_{(n)}(\mathbf{T}_{(n)}, \mathcal{H} + d\mathcal{H}) - \mathbf{E}_{(n)}(\mathbf{T}_{(n)}, \mathcal{H}), \tag{8.6.22}$$

$$\left(d^p\mathbf{E}_{(n)}\right)^0 = \mathbf{E}_{(n)}\left(\mathbf{T}^0_{(n)}, \mathcal{H} + d\mathcal{H}\right) - \mathbf{E}_{(n)}\left(\mathbf{T}^0_{(n)}, \mathcal{H}\right). \tag{8.6.23}$$

8.6.3. Normality Rule in Strain Space

The normality rule for the yield surface in strain space can be deduced from the results based on the inequality (8.6.1) in stress space. By taking a trace product of Eq. (8.2.5) with $\mathbf{M}_{(n)} = \boldsymbol{\Lambda}_{(n)}^{-1}$, we obtain

$$\mathbf{M}_{(n)} : \dot{\mathbf{T}}_{(n)} = \dot{\mathbf{E}}_{(n)} + \mathbf{M}_{(n)} : \dot{\mathbf{T}}_{(n)}^{\mathrm{p}}, \tag{8.6.24}$$

and comparison with Eq. (8.2.14) yields

$$\dot{\mathbf{E}}_{(n)}^{\mathrm{p}} = -\mathbf{M}_{(n)} : \dot{\mathbf{T}}_{(n)}^{\mathrm{p}}, \tag{8.6.25}$$

in accord with Eq. (8.5.26). Thus, to first order,

$$\mathrm{d}^{\mathrm{p}}\mathbf{E}_{(n)} = -\mathbf{M}_{(n)} : \mathrm{d}^{\mathrm{p}}\mathbf{T}_{(n)}. \tag{8.6.26}$$

Since for any elastic strain increment $\delta\mathbf{T}_{(n)}$, emanating from a point on the yield surface in stress space and directed inside of it,

$$\mathrm{d}^{\mathrm{p}}\mathbf{E}_{(n)} : \delta\mathbf{T}_{(n)} < 0, \tag{8.6.27}$$

substitution of (8.6.26) into (8.6.27) gives

$$\mathrm{d}^{\mathrm{p}}\mathbf{T}_{(n)} : \mathbf{M}_{(n)} : \delta\mathbf{T}_{(n)} = \mathrm{d}^{\mathrm{p}}\mathbf{T}_{(n)} : \delta\mathbf{E}_{(n)} > 0. \tag{8.6.28}$$

Here,

$$\delta\mathbf{E}_{(n)} = \mathbf{M}_{(n)} : \delta\mathbf{T}_{(n)} \tag{8.6.29}$$

is the elastic strain increment from a point on the yield surface in strain space, associated with the stress increment $\delta\mathbf{T}_{(n)}$, and directed inside the yield surface. Inequality (8.6.28) holds for any such $\delta\mathbf{E}_{(n)}$ and, therefore, $\mathrm{d}^{\mathrm{p}}\mathbf{T}_{(n)}$ must be codirectional with the inward normal to a locally smooth yield surface in strain $\mathbf{E}_{(n)}$ space,

$$\mathrm{d}^{\mathrm{p}}\mathbf{T}_{(n)} = -\mathrm{d}\gamma_{(n)} \frac{\partial g_{(n)}}{\partial \mathbf{E}_{(n)}}, \quad \mathrm{d}\gamma_{(n)} > 0. \tag{8.6.30}$$

At the vertex of the yield surface, $\mathrm{d}^{\mathrm{p}}\mathbf{T}_{(n)}$ must lie within the cone of limiting inward normals.

In view of (8.6.11), (8.5.28), and (8.6.30), the yield surface normals in stress and strain space are related by

$$\frac{\partial f_{(n)}}{\partial \mathbf{T}_{(n)}} = \mathbf{M}_{(n)} : \frac{\partial f_{(n)}}{\partial \mathbf{E}_{(n)}}, \tag{8.6.31}$$

in agreement with Eq. (8.5.33).

8.6.4. Additional Inequalities for Stress Cycles

Dually to the analysis from Subsection 8.5.2, we can write

$$\mathbf{E}_{(n)} : \mathrm{d}\mathbf{T}_{(n)} = \mathrm{d}\left(\mathbf{E}_{(n)} : \mathbf{T}_{(n)}\right) - \mathbf{T}_{(n)} : \mathrm{d}\mathbf{E}_{(n)}, \tag{8.6.32}$$

and substitution into (8.6.1) gives, for the stress cycle A^0ABDD^0,

$$\begin{aligned} \oint_T \mathbf{E}_{(n)} : \mathrm{d}\mathbf{T}_{(n)} = &\left[\mathbf{E}_{(n)}\left(\mathbf{T}^0_{(n)}, \mathcal{H} + d\mathcal{H}\right) - \mathbf{E}_{(n)}\left(\mathbf{T}^0_{(n)}, \mathcal{H}\right)\right] : \mathbf{T}^0_{(n)} \\ &- \oint_T \mathbf{T}_{(n)} : \mathrm{d}\mathbf{E}_{(n)}. \end{aligned} \tag{8.6.33}$$

If this is assumed to be negative by (8.6.1), there follows

$$\oint_T \mathbf{T}_{(n)} : \mathrm{d}\mathbf{E}_{(n)} > \left[\mathbf{E}_{(n)}\left(\mathbf{T}^0_{(n)}, \mathcal{H} + d\mathcal{H}\right) - \mathbf{E}_{(n)}\left(\mathbf{T}^0_{(n)}, \mathcal{H}\right)\right] : \mathbf{T}^0_{(n)}. \tag{8.6.34}$$

Alternatively, (8.6.34) can be written as

$$\oint_T \left(\mathbf{T}_{(n)} - \mathbf{T}^0_{(n)}\right) : \mathrm{d}\mathbf{E}_{(n)} > 0. \tag{8.6.35}$$

Since (8.6.1) is not invariant, neither is (8.6.35). For example, if we choose a reference state for the strain measure $\mathbf{E}_{(n)}$ to be the state A^0, we have

$$\mathbf{E}_{(n)}\left(\mathbf{T}^0_{(n)}, \mathcal{H}\right) = \mathbf{0}, \quad \mathbf{T}^0_{(n)} = \boldsymbol{\sigma}^0, \tag{8.6.36}$$

where $\boldsymbol{\sigma}^0$ is the Cauchy stress at A^0. Thus, (8.6.35) gives

$$\oint_T \mathbf{T}_{(n)} : \mathrm{d}\mathbf{E}_{(n)} > \boldsymbol{\sigma}^0 : \mathbf{E}_{(n)}\left(\boldsymbol{\sigma}^0, \mathcal{H} + d\mathcal{H}\right). \tag{8.6.37}$$

This shows that the bound on the work done in a stress cycle that involves plastic deformation (the right-hand side of the above inequality) depends on the selected strain measure. This was expected on physical grounds, because cycling one stress measure does not necessarily cycle another stress measure, and different amounts of work are done in cycles of different stress measures. These cycles are different cycles; they involve the same plastic, but not elastic deformation of the material.

8.6.5. Infinitesimal Strain Formulation

In the infinitesimal strain theory all stress measures reduce to the Cauchy stress $\boldsymbol{\sigma}$, and (8.6.35) becomes

$$\oint_\sigma \left(\boldsymbol{\sigma} - \boldsymbol{\sigma}^0\right) : \mathrm{d}\boldsymbol{\varepsilon} > 0. \tag{8.6.38}$$

This is the original postulate of Drucker (1951, 1959). The net work of added stresses in all physically possible stress cycles originating and terminating at some initial stress state $\boldsymbol{\sigma}^0$ within the yield surface is positive, if plastic deformation occurred at some stage of the cycle. In the hardening range $\boldsymbol{\sigma}^0$ can be inside or on the current yield surface, while in the softening range $\boldsymbol{\sigma}^0$ must be inside the current yield surface. If Drucker's postulate is restricted to stress cycles that involve only infinitesimal increment of plastic deformation, (8.6.38) becomes

$$\frac{1}{2}\,\mathrm{d}\boldsymbol{\sigma} : \mathrm{d}^{\mathrm{p}}\boldsymbol{\varepsilon} + (\boldsymbol{\sigma} - \boldsymbol{\sigma}^0) : \mathrm{d}^{\mathrm{p}}\boldsymbol{\varepsilon} > 0, \tag{8.6.39}$$

to terms of second order (assuming that there is no change in elastic properties due to plastic deformation). If the stress state $\boldsymbol{\sigma}^0$ is well inside the current yield surface, or on the yield surface far from the state of stress $\boldsymbol{\sigma}$, the first term in (8.6.39) can be neglected, and

$$(\boldsymbol{\sigma} - \boldsymbol{\sigma}^0) : \mathrm{d}^{\mathrm{p}}\boldsymbol{\varepsilon} > 0. \tag{8.6.40}$$

The inequality is referred to as the principle of maximum plastic work. It was introduced in continuum plasticity by Hill (1948), and in crystalline plasticity by Bishop and Hill (1951) (see Chapter 12). Detailed discussion of the inequality can be found in Hill (1950), Johnson and Mellor (1973), Martin (1975), and Lubliner (1990). It assures both normality and convexity. Its other implications in mathematical theory of plasticity are examined by Duvaut and Lions (1976), Temam (1985), and Han and Reddy (1998).

In the hardening range, the initial state can be chosen to be on the yield surface, so that $\boldsymbol{\sigma}^0 = \boldsymbol{\sigma}$ and (8.6.39) gives

$$\mathrm{d}\boldsymbol{\sigma} : \mathrm{d}^{\mathrm{p}}\boldsymbol{\varepsilon} > 0. \tag{8.6.41}$$

In the softening range, the initial state

$$\boldsymbol{\sigma}^0 = \boldsymbol{\sigma} + \mathrm{d}\boldsymbol{\sigma} \tag{8.6.42}$$

is chosen to be inside the yield surface, and (8.6.39) gives

$$\mathrm{d}\boldsymbol{\sigma} : \mathrm{d}^{\mathrm{p}}\boldsymbol{\varepsilon} < 0. \tag{8.6.43}$$

Both, (8.6.41) and (8.6.43), imply that $\mathrm{d}^{\mathrm{p}}\boldsymbol{\varepsilon}$ is codirectional with the outward normal to a locally smooth yield surface in the Cauchy stress space. Further discussion is given in the paper by Palgen and Drucker (1983).

8.7. Relationship between Work in Stress and Strain Cycles

The Ilyushin work in the cycle of strain $A^0ABD^0C^0$ can be written as

$$W_I = \oint_E \mathbf{T}_{(n)} : \mathrm{d}\mathbf{E}_{(n)} = \oint_T \mathbf{T}_{(n)} : \mathrm{d}\mathbf{E}_{(n)} + \int_{D^0}^{C^0} \mathbf{T}_{(n)} : \mathrm{d}\mathbf{E}_{(n)}. \tag{8.7.1}$$

Denoting the work of added stresses in the cycle of stress A^0ABD^0 as

$$W_D = \oint_T \left(\mathbf{T}_{(n)} - \mathbf{T}^0_{(n)}\right) : \mathrm{d}\mathbf{E}_{(n)}, \tag{8.7.2}$$

and recalling that $\mathbf{T}_{(n)} = \partial\Psi/\partial\mathbf{E}_{(n)}$, we rewrite Eq. (8.7.1) as

$$\begin{aligned} W_I - W_D &= \mathbf{T}^0_{(n)} : \left(\mathrm{d}^{\mathrm{p}}\mathbf{E}_{(n)}\right)^0 \\ &+ \Psi\left(\mathbf{E}^0_{(n)}, \mathcal{H} + \mathrm{d}\mathcal{H}\right) - \Psi\left[\mathbf{E}^0_{(n)} + \left(\mathrm{d}^{\mathrm{p}}\mathbf{E}_{(n)}\right)^0, \mathcal{H} + d\mathcal{H}\right]. \end{aligned} \tag{8.7.3}$$

Furthermore,

$$\begin{aligned} &\Psi\left[\mathbf{E}^0_{(n)} + \left(\mathrm{d}^{\mathrm{p}}\mathbf{E}_{(n)}\right)^0, \mathcal{H} + d\mathcal{H}\right] - \Psi\left(\mathbf{E}^0_{(n)}, \mathcal{H} + d\mathcal{H}\right) \\ &= \left(\frac{\partial\Psi}{\partial\mathbf{E}_{(n)}}\right)_{C^0} : \left(\mathrm{d}^{\mathrm{p}}\mathbf{E}_{(n)}\right)^0 \\ &+ \frac{1}{2}\left(\frac{\partial^2\Psi}{\partial\mathbf{E}_{(n)} \otimes \partial\mathbf{E}_{(n)}}\right)_{C^0} : \left[\left(\mathrm{d}^{\mathrm{p}}\mathbf{E}_{(n)}\right)^0 \otimes \left(\mathrm{d}^{\mathrm{p}}\mathbf{E}_{(n)}\right)^0\right] \\ &= \left[\mathbf{T}^0_{(n)} + \frac{1}{2}\left(\mathrm{d}^{\mathrm{p}}\mathbf{T}_{(n)}\right)^0\right] : \left(\mathrm{d}^{\mathrm{p}}\mathbf{E}_{(n)}\right)^0, \end{aligned} \tag{8.7.4}$$

neglecting the higher-order infinitesimals. The subscript C^0 in Eq. (8.7.4) indicates that partial derivatives are evaluated in the state C^0, where the stress is $\mathbf{T}_{(n)} + \left(\mathrm{d}^{\mathrm{p}}\mathbf{T}_{(n)}\right)^0$. Substitution of (8.7.4) into (8.7.3) gives

$$W_I - W_D = -\frac{1}{2}\left(\mathrm{d}^{\mathrm{p}}\mathbf{T}_{(n)}\right)^0 : \left(\mathrm{d}^{\mathrm{p}}\mathbf{E}_{(n)}\right)^0. \tag{8.7.5}$$

Here,

$$\left(\mathrm{d}^{\mathrm{p}}\mathbf{T}_{(n)}\right)^0 = -\boldsymbol{\Lambda}_{(n)} : \left(\mathrm{d}^{\mathrm{p}}\mathbf{E}_{(n)}\right)^0 \tag{8.7.6}$$

is the stress decrement from A^0 to C^0 caused by infinitesimal plastic deformation along AB (Fig. 8.13). Therefore, if elastic stiffness tensor $\boldsymbol{\Lambda}_{(n)}$ is positive definite, (8.7.5) implies that

$$W_I > W_D. \tag{8.7.7}$$

It is recalled that W_I is independent of the reference state and strain measure, while W_D is not. Thus, the right-hand side of (8.7.5) is dependent on the reference state and measure. However, if $\boldsymbol{\Lambda}_{(n)}$ is positive definite in each

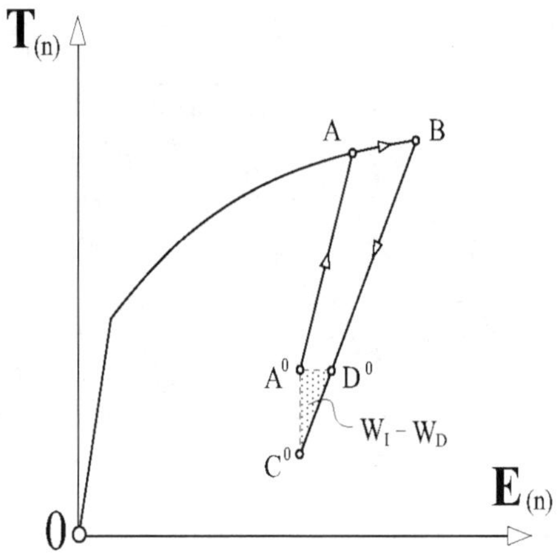

FIGURE 8.13. The dotted area represents the difference between the work done in the Ilyushin and Drucker closed cycles of strain and stress, indicating that $W_I > W_D$.

case, the inequality (8.7.7) holds for all pairs of conjugate stress and strain measures, and for any reference state.

Since $W_I > W_D$, the class of materials obeying inequality (8.5.1) is broader than that obeying (8.6.1). For example, it may happen that material behavior is such that over some stress cycles $W_D < 0$, while $W_I > 0$ for every strain cycle. Since Ilyushin's postulate (8.5.1) is a sufficient condition for the normality rule, it follows that plastic part of strain increment can be normal to a locally smooth yield surface in stress space, although the material does not satisfy (8.6.1) for some stress cycles. Thus, although sufficient, (8.6.1) is not a necessary condition for the normality. This was anticipated, because (8.6.1) places strong restrictions on material behavior, when imposed on all cycles of stress, involving infinitesimal or large plastic deformation. Weaker restrictions on material response are placed by requiring (8.6.1) to hold for stress cycles that involve only infinitesimal plastic deformation, such as cycle A^0ABD^0 considered in Section 8.6.

Returning to Ilyushin's postulate (8.5.1), although it imposes less restrictions than (8.6.1), it is not a necessary condition for the normality rule, either. For example, Palmer, Maier, and Drucker (1967) provide an example of negative work in certain strain cycles for materials that have experienced enormous cyclic work-softening. Yet, normality rule can be used to describe behavior of such materials in a satisfactory manner. For an analysis of plasticity postulates and nonassociative flow rules, considered in Chapter 9, the papers by Nicholson (1987), Lade, Bopp, and Peters (1993), and Lubarda, Mastilovic, and Knap (1996) can be consulted. See also Dougill (1975) and Lee (1994).

8.8. Further Inequalities

If the material obeys Ilyushin's postulate, we have from (8.5.10) a measure-invariant inequality

$$\dot{\mathbf{T}}^{\mathrm{p}}_{(n)} : \dot{\mathbf{E}}_{(n)} < 0. \tag{8.8.1}$$

Since

$$\dot{\mathbf{T}}^{\mathrm{p}}_{(n)} = -\mathbf{\Lambda}_{(n)} : \dot{\mathbf{E}}^{\mathrm{p}}_{(n)}, \quad \dot{\mathbf{T}}^{\mathrm{e}}_{(n)} = \mathbf{\Lambda}_{(n)} : \dot{\mathbf{E}}_{(n)}, \tag{8.8.2}$$

the inequality (8.8.1) is equivalent to

$$\dot{\mathbf{T}}^{\mathrm{e}}_{(n)} : \dot{\mathbf{E}}^{\mathrm{p}}_{(n)} > 0. \tag{8.8.3}$$

By taking a trace product of the first of (8.8.2) with $\dot{\mathbf{E}}^{\mathrm{p}}_{(n)}$, and of the second with $\dot{\mathbf{E}}_{(n)}$, it follows that

$$\dot{\mathbf{T}}^{\mathrm{p}}_{(n)} : \dot{\mathbf{E}}^{\mathrm{p}}_{(n)} < 0, \quad \dot{\mathbf{T}}^{\mathrm{e}}_{(n)} : \dot{\mathbf{E}}_{(n)} > 0, \tag{8.8.4}$$

provided that $\mathbf{\Lambda}_{(n)}$ is positive definite. Both inequalities in (8.8.4) are measure-invariant. Furthermore, since

$$\dot{\mathbf{E}}^{\mathrm{e}}_{(n)} = \mathbf{M}_{(n)} : \dot{\mathbf{T}}_{(n)}, \tag{8.8.5}$$

a trace product with $\dot{\mathbf{T}}_{(n)}$ yields another measure-invariant inequality

$$\dot{\mathbf{T}}_{(n)} : \dot{\mathbf{E}}^{\mathrm{e}}_{(n)} > 0. \tag{8.8.6}$$

In view of (8.8.2) and (8.8.3), there is an identity

$$\dot{\mathbf{T}}^{\mathrm{p}}_{(n)} : \dot{\mathbf{E}}^{\mathrm{e}}_{(n)} = -\dot{\mathbf{T}}_{(n)} : \dot{\mathbf{E}}^{\mathrm{p}}_{(n)}. \tag{8.8.7}$$

If material is in the hardening range, relative to a particular pair of stress and strain measures, then for that pair

$$\dot{\mathbf{T}}_{(n)} : \dot{\mathbf{E}}^{\mathrm{p}}_{(n)} > 0, \quad \dot{\mathbf{T}}^{\mathrm{p}}_{(n)} : \dot{\mathbf{E}}^{\mathrm{e}}_{(n)} < 0. \tag{8.8.8}$$

These are not measure-invariant inequalities, so that hardening with respect to one pair of measures may appear as softening relative to another pair.

If (8.8.8) holds for a particular pair of stress and strain measures, we have for that pair

$$\dot{\mathbf{T}}^{\mathrm{e}}_{(n)} : \dot{\mathbf{E}}^{\mathrm{e}}_{(n)} = \dot{\mathbf{T}}_{(n)} : \dot{\mathbf{E}}^{\mathrm{e}}_{(n)} - \dot{\mathbf{T}}^{\mathrm{p}}_{(n)} : \dot{\mathbf{E}}^{\mathrm{e}}_{(n)} > 0, \tag{8.8.9}$$

in view of (8.8.6) and (8.8.8). Since, by (8.8.2) and (8.8.5),

$$\dot{\mathbf{T}}^{\mathrm{e}}_{(n)} : \dot{\mathbf{E}}^{\mathrm{e}}_{(n)} = \dot{\mathbf{T}}_{(n)} : \dot{\mathbf{E}}_{(n)}, \tag{8.8.10}$$

the inequality (8.8.9) gives

$$\dot{\mathbf{T}}_{(n)} : \dot{\mathbf{E}}_{(n)} > 0, \tag{8.8.11}$$

for the same conjugate pair. Neither (8.8.9) nor (8.8.11) is measure-invariant.

In the softening range the directions of inequalities in (8.8.8) are reversed. Since the first term on the right-hand side of the equality sign in (8.8.9) is always positive, by measure-invariant (8.8.6), the direction of inequalities in (8.8.9) and (8.8.11) is uncertain. Thus, in the softening range, corresponding to given n, $\dot{\mathbf{T}}_{(n)} : \dot{\mathbf{E}}_{(n)}$ can be either positive or negative. As a result, (8.8.11) is not a criterion of hardening. A necessary and sufficient condition for hardening, relative to selected stress and strain measures, is given by (8.8.8).

8.8.1. Inequalities with Current State as Reference

If current state is taken as the reference, we have from Section 3.9

$$\underline{\dot{\mathbf{E}}}_{(n)} = \mathbf{D}, \quad \underline{\dot{\mathbf{T}}}_{(n)} = \overset{\circ}{\underline{\boldsymbol{\tau}}} - n\,(\mathbf{D}\cdot\boldsymbol{\sigma} + \boldsymbol{\sigma}\cdot\mathbf{D}). \tag{8.8.12}$$

Equation (8.2.5) consequently becomes

$$\underline{\dot{\mathbf{T}}}_{(n)} = \underline{\dot{\mathbf{T}}}^{\mathrm{e}}_{(n)} + \underline{\dot{\mathbf{T}}}^{\mathrm{p}}_{(n)}, \quad \underline{\dot{\mathbf{T}}}^{\mathrm{e}}_{(n)} = \underline{\boldsymbol{\Lambda}}_{(n)} : \mathbf{D}, \tag{8.8.13}$$

while Eq. (8.2.13) gives

$$\mathbf{D} = \mathbf{D}^{\mathrm{e}}_{(n)} + \mathbf{D}^{\mathrm{p}}_{(n)}, \quad \mathbf{D}^{\mathrm{e}}_{(n)} = \underline{\mathbf{M}}_{(n)} : \underline{\dot{\mathbf{T}}}_{(n)}. \tag{8.8.14}$$

Inequalities (8.8.1) and (8.8.6) yield

$$\underline{\dot{\mathbf{T}}}^{\mathrm{p}}_{(n)} : \mathbf{D} < 0, \quad \underline{\dot{\mathbf{T}}}_{(n)} : \mathbf{D}^{\mathrm{e}}_{(n)} > 0. \tag{8.8.15}$$

In addition, the inequalities in (8.8.4) reduce to

$$\underline{\dot{\mathbf{T}}}^{\mathrm{p}}_{(n)} : \mathbf{D}^{\mathrm{p}}_{(n)} < 0, \quad \underline{\dot{\mathbf{T}}}^{\mathrm{e}}_{(n)} : \mathbf{D} > 0. \tag{8.8.16}$$

For example, for $n = 0, \pm 1$, the inequalities in (8.8.15) give

$$\overset{\circ}{\underline{\tau}}{}^{\mathrm{p}} : \mathbf{D} < 0, \quad \overset{\triangle}{\underline{\tau}}{}^{\mathrm{p}} : \mathbf{D} < 0, \quad \overset{\nabla}{\underline{\tau}}{}^{\mathrm{p}} : \mathbf{D} < 0, \tag{8.8.17}$$

$$\overset{\circ}{\underline{\tau}} : \mathbf{D}^{\mathrm{e}}_{(0)} > 0, \quad \overset{\triangle}{\underline{\tau}} : \mathbf{D}^{\mathrm{e}}_{(1)} < 0, \quad \overset{\nabla}{\underline{\tau}} : \mathbf{D}^{\mathrm{e}}_{(-1)} < 0. \tag{8.8.18}$$

Similarly, from (8.8.16), we obtain

$$\overset{\circ}{\underline{\tau}}{}^{\mathrm{p}} : \mathbf{D}^{\mathrm{p}}_{(0)} < 0, \quad \overset{\triangle}{\underline{\tau}}{}^{\mathrm{p}} : \mathbf{D}^{\mathrm{p}}_{(1)} < 0, \quad \overset{\nabla}{\underline{\tau}}{}^{\mathrm{p}} : \mathbf{D}^{\mathrm{p}}_{(-1)} < 0, \tag{8.8.19}$$

$$\overset{\circ}{\underline{\tau}}{}^{\mathrm{e}} : \mathbf{D} > 0, \quad \overset{\triangle}{\underline{\tau}}{}^{\mathrm{e}} : \mathbf{D} > 0, \quad \overset{\nabla}{\underline{\tau}}{}^{\mathrm{e}} : \mathbf{D} > 0. \tag{8.8.20}$$

It is observed that

$$\dot{\underline{\mathbf{T}}}{}^{\mathrm{e}}_{(n)} = \mathbf{\Lambda}_{(n)} : \mathbf{D} = (\underline{\mathbf{\Lambda}}_{(0)} - 2n\,\underline{\mathbf{S}}) : \mathbf{D}, \tag{8.8.21}$$

and

$$\dot{\underline{\mathbf{T}}}{}^{\mathrm{p}}_{(n)} = \dot{\underline{\mathbf{T}}}_{(n)} - \dot{\underline{\mathbf{T}}}{}^{\mathrm{e}}_{(n)} = \overset{\circ}{\underline{\tau}} - \overset{\circ}{\underline{\tau}}{}^{\mathrm{e}} = \overset{\circ}{\underline{\tau}}{}^{\mathrm{p}}, \tag{8.8.22}$$

for all n. In particular,

$$\overset{\circ}{\underline{\tau}}{}^{\mathrm{p}} = \overset{\triangle}{\underline{\tau}}{}^{\mathrm{p}} = \overset{\nabla}{\underline{\tau}}{}^{\mathrm{p}}. \tag{8.8.23}$$

Furthermore,

$$\dot{\underline{\mathbf{T}}}_{(n)} : \dot{\underline{\mathbf{E}}}_{(n)} = \dot{\underline{\mathbf{T}}}_{(n)} : \mathbf{D} = \overset{\circ}{\underline{\tau}} : \mathbf{D} - 2n\,(\boldsymbol{\sigma} : \mathbf{D}^2). \tag{8.8.24}$$

To illustrate that $\dot{\underline{\mathbf{T}}}_{(n)} : \dot{\underline{\mathbf{E}}}_{(n)}$ can have a different sign for different n, consider a tensile test under superposed hydrostatic pressure. The corresponding stress and rate of deformation tensors are

$$\boldsymbol{\sigma} = \sigma\,\mathbf{e}_3 \otimes \mathbf{e}_3 - p\mathbf{I}, \quad \mathbf{D} = \frac{3}{2}\frac{\dot{l}}{l}\,\mathbf{e}_3 \otimes \mathbf{e}_3 - \frac{1}{2}\frac{\dot{l}}{l}\,\mathbf{I}, \tag{8.8.25}$$

where l is a current length of the specimen under tensile stress σ (in the direction $\mathbf{e}_3$), and under constant superposed pressure p. Substitution into Eq. (8.8.24) yields

$$\dot{\underline{\mathbf{T}}}_{(n)} : \mathbf{D} = \frac{\dot{\sigma}}{\sigma}\frac{\dot{l}}{l}\left[\sigma - 2n\left(1 - \frac{3}{2}\frac{p}{\sigma}\right)\frac{\dot{l}}{l}\right]. \tag{8.8.26}$$

For $n = 0$ this gives

$$\dot{\underline{\mathbf{T}}}_{(0)} : \mathbf{D} = \dot{\sigma}\,\frac{\dot{l}}{l}\,. \tag{8.8.27}$$

If this is positive, from (8.8.26) it follows that for other n the trace product $\dot{\underline{\mathbf{T}}}_{(n)} : \mathbf{D}$ can be either positive or negative, depending on the magnitude of the superposed pressure p (Hill, 1968).

8.9. Related Postulates

Consider again the measure-invariant inequality (8.8.1), i.e.,

$$\dot{\mathbf{E}}_{(n)} : \dot{\mathbf{T}}^{\mathrm{p}}_{(n)} < 0. \tag{8.9.1}$$

Since plastic parts of the stress and strain rates are related by

$$\dot{\mathbf{T}}^{\mathrm{p}}_{(n)} = -\mathbf{\Lambda}_{(n)} : \dot{\mathbf{E}}^{\mathrm{p}}_{(n)}, \tag{8.9.2}$$

there follows

$$\dot{\mathbf{E}}_{(n)} : \mathbf{\Lambda}_{(n)} : \dot{\mathbf{E}}^{\mathrm{p}}_{(n)} > 0. \tag{8.9.3}$$

Thus, recalling that

$$\dot{\mathbf{E}}_{(n)} = \dot{\mathbf{T}}_{(n)} : \mathbf{M}_{(n)} + \dot{\mathbf{E}}^{\mathrm{p}}_{(n)}, \tag{8.9.4}$$

the substitution into (8.9.3) gives

$$\dot{\mathbf{T}}_{(n)} : \dot{\mathbf{E}}^{\mathrm{p}}_{(n)} > -\dot{\mathbf{E}}^{\mathrm{p}}_{(n)} : \mathbf{\Lambda}_{(n)} : \dot{\mathbf{E}}^{\mathrm{p}}_{(n)}. \tag{8.9.5}$$

An inequality of this type was originally proposed by Nguyen and Bui (1974). See also Lubliner (1986). In particular, with the current state as the reference, and with the logarithmic strain measure, we obtain

$$\overset{\circ}{\underline{\tau}} : \mathbf{D}^{\mathrm{p}}_{(0)} > -\mathbf{D}^{\mathrm{p}}_{(0)} : \underline{\mathbf{\Lambda}}_{(0)} : \mathbf{D}^{\mathrm{p}}_{(0)}. \tag{8.9.6}$$

Naghdi and Trapp (1975a,b) proposed that the external work done on the body by surface tractions and body forces in any smooth spatially homogeneous closed cycle is non-negative, i.e.,

$$\int_{t_1}^{t_2} \mathcal{P}\,\mathrm{d}t \geq 0, \tag{8.9.7}$$

where

$$\mathcal{P} = \int_{S^0} \mathbf{p}_n \cdot \mathbf{v}\,\mathrm{d}S^0 + \int_{V^0} \rho^0\,\mathbf{b} \cdot \mathbf{v}\,\mathrm{d}V^0. \tag{8.9.8}$$

A smooth closed cycle is defined as a closed cycle of deformation which also restores the velocity and thus the kinetic energy,

$$\mathbf{E}_{(n)}(t_2) = \mathbf{E}_{(n)}(t_1), \quad \mathbf{v}(t_2) = \mathbf{v}(t_1). \tag{8.9.9}$$

By rewriting Eq. (8.9.8) as (see Section 3.5)

$$\mathcal{P} = \frac{\mathrm{d}}{\mathrm{d}t} \int_{V^0} \frac{1}{2} \rho^0\,\mathbf{v} \cdot \mathbf{v}\,\mathrm{d}V^0 + \int_{V^0} \mathbf{T}_{(n)} : \dot{\mathbf{E}}_{(n)}\,\mathrm{d}V^0, \tag{8.9.10}$$

substitution into (8.9.7) gives

$$\int_{t_1}^{t_2} \left(\int_{V^0} \mathbf{T}_{(n)} : \dot{\mathbf{E}}_{(n)} \, \mathrm{d}V^0 \right) \mathrm{d}t \geq 0. \tag{8.9.11}$$

Since deformation is assumed to be spatially uniform, this reduces to

$$\int_{t_1}^{t_2} \mathbf{T}_{(n)} : \dot{\mathbf{E}}_{(n)} \, \mathrm{d}t \geq 0, \tag{8.9.12}$$

or

$$\oint_{E} \mathbf{T}_{(n)} : \dot{\mathbf{E}}_{(n)} \, \mathrm{d}t \geq 0. \tag{8.9.13}$$

This, in fact, is Ilyushin's postulate in the form presented by Hill (1968). Additional discussion can be found in Carroll (1987), Hill and Rice (1987), and Rajagopal and Srinivasa (1998). The work inequalities in plastic fracturing materials were discussed by Bažant (1980), among others, and for elastic-viscoplastic materials by Naghdi (1984).

References

Bažant, Z. P. (1978), Endochronic inelasticity and incremental plasticity, *Int. J. Solids Struct.*, Vol. 14, pp. 691–714.

Bažant, Z. P. (1980), Work inequalities for plastic fracturing materials, *Int. J. Solids Struct.*, Vol. 16, pp. 873–901.

Bishop, J. F. W. and Hill, R. (1951), A theory of plastic distortion of a polycrystalline aggregate under combined stresses, *Phil. Mag.*, Vol. 42, pp. 414–427.

Carroll, M. M. (1987), A rate-independent constitutive theory for finite inelastic deformation, *J. Appl. Mech.*, Vol. 54, pp. 15–21.

Chandler, H. W. (1985), A plasticity theory without Drucker's postulate, suitable for granular materials, *J. Mech. Phys. Solids*, Vol. 33, pp. 215–226.

Cottrell, A. H. (1961), *Dislocations and Plastic Flow in Crystals*, Oxford University Press, London.

Cottrell, A. H. (1964), *The Mechanical Properties of Matter*, John Wiley & Sons, New York.

Dafalias, Y. F. (1977), Ilyushin's postulate and resulting thermodynamic conditions on elasto-plastic coupling, *Int. J. Solids Struct.*, Vol. 13, pp. 239–251.

Dougill, Y. W. (1975), Some remarks on path independence in the small in plasticity, *Q. Appl. Math.*, Vol. 33, pp. 233–243.

Drucker, D. C. (1951), A more fundamental approach to plastic stress-strain relations, in *Proc. 1st U.S. Natl. Congr. Appl. Mech.*, ed. E. Sternberg, pp. 487–491, ASME, New York.

Drucker, D. C. (1959), A definition of stable inelastic material, *J. Appl. Mech.*, Vol. 26, pp. 101–106.

Drucker, D. C. (1960), Plasticity, in *Structural Mechanics – Proc. 1st Symp. Naval Struct. Mechanics*, eds. J. N. Goodier and N. J. Hoff, pp. 407–455, Pergamon Press, New York.

Drucker, D. C. (1964), Stress-strain-time relations and irreversible thermodynamics, in *Second-Order Effects in Elasticity, Plasticity and Fluid Dynamics*, eds. M. Reiner and D. Abir, pp. 331–351, Pergamon Press, Oxford.

Duvaut, G. and Lions, J.-L. (1976), *Inequalities in Mechanics and Physics*, Springer-Verlag, Berlin.

Gurtin, M. E. (1983), On the hypoelastic formulation of plasticity using the past maximum of stress, *J. Appl. Mech.*, Vol. 50, pp. 894–896.

Han, W. and Reddy, B. D. (1999), *Plasticity – Mathematical Theory and Numerical Analysis*, Springer-Verlag, New York.

Hecker, S. S. (1976), Experimental studies of yield phenomena in biaxially loaded metals, in *Constitutive Equations in Viscoplasticity: Computational and Engineering Aspects*, AMD Vol. 20, eds. J. A. Stricklin and K. J. Saczalski, pp. 1–33, ASME, New York.

Hill, R. (1948), A variational principle of maximum plastic work in classical plasticity, *Quart. J. Mech. Appl. Math.*, Vol. 1, pp. 18–28.

Hill, R. (1950), *The Mathematical Theory of Plasticity*, Oxford University Press, London.

Hill, R. (1967), The essential structure of constitutive laws for metal composites and polycrystals, *J. Mech. Phys. Solids*, Vol. 15, pp. 79–95.

Hill, R. (1968), On constitutive inequalities for simple materials – II, *J. Mech. Phys. Solids*, Vol. 16, pp. 315–322.

Hill, R. (1972), On constitutive macro-variables for heterogeneous solids at finite strain, *Proc. Roy. Soc. Lond. A*, Vol. 326, pp. 131–147.

Hill, R. (1978), Aspects of invariance in solid mechanics, *Adv. Appl. Mech.*, Vol. 18, pp. 1–75.

Hill, R. and Rice, J. R. (1973), Elastic potentials and the structure of inelastic constitutive laws, *SIAM J. Appl. Math.*, Vol. 25, pp. 448–461.

Hill, R. and Rice, J. R. (1987), Discussion of "A rate-independent constitutive theory for finite inelastic deformation," *J. Appl. Mech.*, Vol. 54, pp. 745–747.

Honeycombe, R. W. K. (1984), *The Plastic Deformation of Metals*, 2nd ed., Edward Arnold Publ., London.

Hüttel, C. and Matzenmiller, A. (1999), Extension of generalized plasticity to finite deformations and non-linear hardening, *Int. J. Solids Struct.*, Vol. 36, pp. 5255–5276.

Ilyushin, A. A. (1954), On the relation between stresses and small strains in the mechanics of continua, *Prikl. Mat. Mekh.*, Vol. 18, pp. 641–666 (in Russian).

Ilyushin, A. A. (1961), On the postulate of plasticity, *Prikl. Math. Mekh.*, Vol. 25, pp. 503–507.

Johnson, W. and Mellor, P. B. (1973), *Engineering Plasticity*, Van Nostrand Reinhold, London.

Koiter, W. (1960), General theorems for elastic-plastic solids, in *Progress in Solid Mechanics*, eds. I. N. Sneddon and R. Hill, Vol. 1, pp. 165–221, North-Holland, Amsterdam.

Lade, P. V., Bopp, P. A., and Peters, J. F. (1993), Instability of dilating sand, *Mech. Mater.*, Vol. 16, pp. 249–264.

Lee, Y. K. (1994), A unified framework for compressible plasticity, *Int. J. Plasticity*, Vol. 10, pp. 695–717.

Lubarda, V. A., Mastilovic, S., and Knap, J. (1996), Some comments on plasticity postulates and non-associative flow rules, *Int. J. Mech. Sci.*, Vol. 38, pp. 247–258.

Lubliner, J. (1974), A simple theory of plasticity, *Int. J. Solids Struct.*, Vol. 10, pp. 313–319.

Lubliner, J. (1984), A maximum dissipation principle in generalized plasticity, *Acta Mech.*, Vol. 52, pp. 225–237.

Lubliner, J. (1986), Normality rules in large-deformation plasticity, *Mech. Mater.*, Vol. 5, pp. 29–34.

Lubliner, J. (1990), *Plasticity Theory*, Macmillan Publishing Comp., New York.

Lubliner, J. (1991), A simple model of generalized plasticity, *Int. J. Solids Struct.*, Vol. 28, pp. 769–778.

Martin, J. B. (1975), *Plasticity: Fundamentals and General Results*, MIT Press, Cambridge, Massachusetts.

Mróz, Z. (1966), On forms of constitutive laws for elastic-plastic solids, *Arch. Mech. Stosow.*, Vol. 18, pp. 3–35.

Murakami, H. and Read, H. E. (1987), Endochronic plasticity: Some basic properties of plastic flow and fracture, *Int. J. Solids Struct.*, Vol. 23, pp. 133–151.

Naghdi, P. M. (1984), Constitutive restrictions for idealized elastic-viscoplastic materials, *J. Appl. Mech.*, Vol. 51, pp. 93–101.

Naghdi, P. M. and Trapp, J. A. (1975a), Restrictions on constitutive equations of finitely deformed elastic-plastic materials, *Quart. J. Mech. Appl. Math.*, Vol. 28, pp. 25–46.

Naghdi, P. M. and Trapp, J. A. (1975b), On the nature of normality of plastic strain rate and convexity of yield surfaces in plasticity, *J. Appl. Mech.*, Vol. 42, pp. 61–66.

Nemat-Nasser, S. (1983), On finite plastic flow of crystalline solids and geomaterials, *J. Appl. Mech.*, Vol. 50, pp. 1114–1126.

Nguyen, Q. S. and Bui, H. D. (1974), Sur les matereériaux élastoplastiques à écrouissage positif ou négatif, *J. de Mécanique*, Vol. 13, pp. 321–342.

Nicholson, D. W. (1987), Research note on application of a constitutive inequality in nonassociated plasticity, *Int. J. Plasticity*, Vol. 3, pp. 295–301.

Palgen, L. and Drucker, D. C. (1983), The structure of stress-strain relations in finite elasto-plasticity, *Int. J. Solids Struct.*, Vol. 19, pp. 519–531.

Palmer, A. C., Maier, G., and Drucker, D. C. (1967), Normality relations and convexity of yield surfaces for unstable materials and structures, *J. Appl. Mech.*, Vol. 34, pp. 464–470.

Pipkin, A. C. and Rivlin, R. S. (1965), Mechanics of rate-independent materials, *Z. angew. Math. Phys.*, Vol. 16, pp. 313–327.

Rajagopal, K. R. and Srinivasa, A. R. (1998), Mechanics of the inelastic behavior of materials. Part II: Inelastic response, *Int. J. Plasticity*, Vol. 14, pp. 969–995.

Rice, J. R. (1970), On the structure of stress-strain relations for time-dependent plastic deformation in metals, *J. Appl. Mech.*, Vol. 37, pp. 728–737.

Rice, J. R. (1971), Inelastic constitutive relations for solids: An internal variable theory and its application to metal plasticity, *J. Mech. Phys. Solids*, Vol. 19, pp. 433–455.

Temam, R. (1985), *Mathematical Problems in Plasticity*, Gauthier-Villars, Paris (translation of French 1983 edition).

Valanis, K. C. (1971), A theory of viscoplasticity without a yield surface, *Arch. Mech.*, Vol. 23, pp. 517–551.

Valanis, K. C. (1975), On the foundations of the endochronic theory of viscoplasticity, *Arch. Mech.*, Vol. 27, pp. 857–868.

CHAPTER 9

PHENOMENOLOGICAL PLASTICITY

This chapter contains a detailed analysis of phenomenological constitutive equations for large deformation elastoplasticity. First eight sections are devoted to rate-independent models of isothermal elastoplastic behavior. Formulations in stress and strain space are both given. Different hardening models of metal plasticity are discussed, including isotropic, kinematic, combined and multisurface hardening models. Constitutive equations accounting for the yield vertices are also included. Pressure-dependent and nonassociative flow rules are then analyzed, with an application to rock mechanics. Constitutive theories of thermoplasticity, rate-dependent plasticity and viscoplasticity are considered in Sections 9.9 and 9.10. The final section of the chapter deals with the deformation theory of plasticity.

9.1. Formulation in Strain Space

In the rate-independent elastoplastic theory with the yield surface in strain space, the stress rate is decomposed into elastic and plastic parts, such that

$$\dot{\mathbf{T}}_{(n)} = \dot{\mathbf{T}}^{\mathrm{e}}_{(n)} + \dot{\mathbf{T}}^{\mathrm{p}}_{(n)} = \mathbf{\Lambda}_{(n)} : \dot{\mathbf{E}}_{(n)} - \dot{\gamma}_{(n)}\, \frac{\partial g_{(n)}}{\partial \mathbf{E}_{(n)}}\,. \tag{9.1.1}$$

The function $g_{(n)}\left(\mathbf{E}_{(n)}, \mathcal{H}\right)$ is the yield function, and

$$\dot{\gamma}_{(n)} > 0 \tag{9.1.2}$$

is the loading index, both corresponding to selected strain measure and reference state. The yield surface is defined by

$$g_{(n)}\left(\mathbf{E}_{(n)}, \mathcal{H}\right) = 0. \tag{9.1.3}$$

Assuming an incrementally linear response and a continuity of the response between loading and unloading, defined by Eq. (8.2.8), the loading index can be written as

$$\dot{\gamma}_{(n)} = \frac{1}{h_{(n)}} \left(\frac{\partial g_{(n)}}{\partial \mathbf{E}_{(n)}} : \dot{\mathbf{E}}_{(n)} \right), \quad \frac{\partial g_{(n)}}{\partial \mathbf{E}_{(n)}} : \dot{\mathbf{E}}_{(n)} > 0. \tag{9.1.4}$$

The parameter

$$h_{(n)} > 0 \tag{9.1.5}$$

is a scalar function of the plastic state on the yield surface, to be determined from the consistency condition and a given representation of the yield function. If the strain rate is such that

$$\frac{\partial g_{(n)}}{\partial \mathbf{E}_{(n)}} : \dot{\mathbf{E}}_{(n)} \leq 0, \tag{9.1.6}$$

only elastic deformation takes place, and

$$\dot{\gamma}_{(n)} = 0. \tag{9.1.7}$$

An alternative derivation of (9.1.4) is based on the consistency condition for continuing plastic deformation. This can be expressed as

$$\mathrm{d}g_{(n)} = \frac{\partial g_{(n)}}{\partial \mathbf{E}_{(n)}} : \mathrm{d}\mathbf{E}_{(n)} + \mathrm{d}^{\mathrm{p}} g_{(n)} = 0, \tag{9.1.8}$$

where

$$\mathrm{d}^{\mathrm{p}} g_{(n)} = g_{(n)} \left(\mathbf{E}_{(n)}, \mathcal{H} + d\mathcal{H} \right) - g_{(n)} \left(\mathbf{E}_{(n)}, \mathcal{H} \right) \tag{9.1.9}$$

is the plastic part of the increment of $\mathrm{d}^{\mathrm{p}} g_{(n)}$, due to change of the internal structure. Writing

$$\mathrm{d}^{\mathrm{p}} g_{(n)} = -h_{(n)}\, \mathrm{d}\gamma_{(n)}, \tag{9.1.10}$$

Equation (9.1.8) yields Eq. (9.1.4).

When Eq. (9.1.4) is substituted into Eq. (9.1.1), the constitutive equation for elastoplastic loading becomes (Hill, 1967a, 1978)

$$\dot{\mathbf{T}}_{(n)} = \left[\mathbf{\Lambda}_{(n)} - \frac{1}{h_{(n)}} \left(\frac{\partial g_{(n)}}{\partial \mathbf{E}_{(n)}} \otimes \frac{\partial g_{(n)}}{\partial \mathbf{E}_{(n)}} \right) \right] : \dot{\mathbf{E}}_{(n)}. \tag{9.1.11}$$

The fourth-order tensor within the square brackets is the elastoplastic stiffness tensor, associated with the considered stress and strain measures and the reference state. Within the employed framework based on the Green-elasticity and normality rule, the elastoplastic stiffness tensor obeys the reciprocal or self-adjoint symmetry (with respect to first and second pair of indices), in addition to symmetries in the first two and last two indices (minor symmetry), associated with the symmetry of the stress and strain tensors. The formulation of elastoplasticity theory based on the yield surface in strain space was also studied by Naghdi and Trapp (1975), Casey and Naghdi (1981,1983), Yoder and Iwan (1981), Klisinski, Mróz, and Runesson (1992), and Negahban (1995). A review by Naghdi (1990) contains additional related references.

It is of interest to invert the constitutive structure (9.1.11), and express the strain rate in terms of the stress rate. By taking a trace product of Eq. (9.1.11) with $\mathbf{M}_{(n)} = \boldsymbol{\Lambda}_{(n)}^{-1}$, there follows

$$\mathbf{M}_{(n)} : \dot{\mathbf{T}}_{(n)} = \dot{\mathbf{E}}_{(n)} - \frac{1}{h_{(n)}} \mathbf{M}_{(n)} : \frac{\partial g_{(n)}}{\partial \mathbf{E}_{(n)}} \left(\frac{\partial g_{(n)}}{\partial \mathbf{E}_{(n)}} : \dot{\mathbf{E}}_{(n)} \right). \tag{9.1.12}$$

A trace product of (9.1.12) with $\partial g_{(n)} / \partial \mathbf{E}_{(n)}$ gives

$$\frac{\partial g_{(n)}}{\partial \mathbf{E}_{(n)}} : \dot{\mathbf{E}}_{(n)} = \frac{h_{(n)}}{H_{(n)}} \frac{\partial g_{(n)}}{\partial \mathbf{E}_{(n)}} : \mathbf{M}_{(n)} : \dot{\mathbf{T}}_{(n)}, \tag{9.1.13}$$

where

$$H_{(n)} = h_{(n)} - \frac{\partial g_{(n)}}{\partial \mathbf{E}_{(n)}} : \mathbf{M}_{(n)} : \frac{\partial g_{(n)}}{\partial \mathbf{E}_{(n)}}. \tag{9.1.14}$$

For plastic loading the quantity in Eq. (9.1.13) must be positive. The substitution of Eq. (9.1.13) into Eq. (9.1.12) yields a desired inverted form

$$\dot{\mathbf{E}}_{(n)} = \left[\mathbf{M}_{(n)} + \frac{1}{H_{(n)}} \left(\mathbf{M}_{(n)} : \frac{\partial g_{(n)}}{\partial \mathbf{E}_{(n)}} \right) \otimes \left(\frac{\partial g_{(n)}}{\partial \mathbf{E}_{(n)}} : \mathbf{M}_{(n)} \right) \right] : \dot{\mathbf{T}}_{(n)}. \tag{9.1.15}$$

If current state is taken as the reference state, Eq. (9.1.11) becomes

$$\dot{\underline{\mathbf{T}}}_{(n)} = \left[\underline{\boldsymbol{\Lambda}}_{(n)} - \frac{1}{\underline{h}_{(n)}} \left(\frac{\partial \underline{g}_{(n)}}{\partial \underline{\mathbf{E}}_{(n)}} \otimes \frac{\partial \underline{g}_{(n)}}{\partial \underline{\mathbf{E}}_{(n)}} \right) \right] : \mathbf{D}. \tag{9.1.16}$$

Incorporating Eq. (6.3.6) for $\dot{\underline{\mathbf{T}}}_{(n)}$, and Eq. (6.3.13) for $\underline{\boldsymbol{\Lambda}}_{(n)} = \underline{\boldsymbol{\mathcal{L}}}_{(n)}$, gives

$$\overset{\circ}{\underline{\boldsymbol{\tau}}} = \left[\underline{\boldsymbol{\mathcal{L}}}_{(0)} - \frac{1}{\underline{h}_{(n)}} \left(\frac{\partial \underline{g}_{(n)}}{\partial \underline{\mathbf{E}}_{(n)}} \otimes \frac{\partial \underline{g}_{(n)}}{\partial \underline{\mathbf{E}}_{(n)}} \right) \right] : \mathbf{D}. \tag{9.1.17}$$

It is noted that

$$\dot{\underline{\mathbf{T}}}^{\mathrm{p}}_{(n)} = -\frac{1}{\underline{h}_{(n)}} \left(\frac{\partial \underline{g}_{(n)}}{\partial \underline{\mathbf{E}}_{(n)}} \otimes \frac{\partial \underline{g}_{(n)}}{\partial \underline{\mathbf{E}}_{(n)}} \right) : \mathbf{D} = \overset{\circ}{\underline{\boldsymbol{\tau}}}{}^{\mathrm{p}}, \tag{9.1.18}$$

for all n. In particular, the gradient $\partial \underline{g}_{(n)} / \partial \underline{\mathbf{E}}_{(n)}$ at the yield point is in the same direction for all n.

9.1.1. Translation and Expansion of the Yield Surface

Let the yield surface in strain space be defined by

$$g_{(n)}\left(\mathbf{E}_{(n)} - \mathbf{E}^{\mathrm{p}}_{(n)}, k_{(n)}\right) = 0, \tag{9.1.19}$$

where $g_{(n)}$ is an isotropic function of its tensor argument, and $\mathbf{E}^{\mathrm{p}}_{(n)}$ represents the center of the current yield surface (Fig. 9.1). This yield surface translates and expands in strain space, although it physically corresponds to isotropic hardening in stress space. The current center of the yield surface is determined by integration from an appropriate evolution equation, along a given deformation path. For instance, the evolution of $\mathbf{E}^{\mathrm{p}}_{(n)}$ can be described by

$$\dot{\mathbf{E}}^{\mathrm{p}}_{(n)} = -\mathbf{M}_{(n)} : \dot{\mathbf{T}}^{\mathrm{p}}_{(n)} = \dot{\gamma}_{(n)}\, \mathbf{M}_{(n)} : \frac{\partial g_{(n)}}{\partial \mathbf{E}_{(n)}}. \tag{9.1.20}$$

The scalar function

$$k_{(n)} = k_{(n)}\left(\varphi_{(n)}\right) \tag{9.1.21}$$

in Eq. (9.1.19) specifies the size of the current yield surface. The parameter $\varphi_{(n)}$ accounts for the history of plastic deformation, and can be taken as

$$\varphi_{(n)} = -\int_0^t \left(\frac{1}{2}\,\dot{\mathbf{T}}^{\mathrm{p}}_{(n)} : \dot{\mathbf{T}}^{\mathrm{p}}_{(n)}\right)^{1/2} \mathrm{d}t. \tag{9.1.22}$$

The consistency condition for continuing plastic deformation is

$$\frac{\partial g_{(n)}}{\partial \mathbf{E}_{(n)}} : \left(\dot{\mathbf{E}}_{(n)} - \dot{\mathbf{E}}^{\mathrm{p}}_{(n)}\right) + \frac{\partial g_{(n)}}{\partial k_{(n)}} \frac{\mathrm{d}k_{(n)}}{\mathrm{d}\varphi_{(n)}}\, \dot{\varphi}_{(n)} = 0. \tag{9.1.23}$$

Substitution of Eqs. (9.1.20) and (9.1.22) into Eq. (9.1.23) gives the loading index as in Eq. (9.1.4), with

$$h_{(n)} = \frac{\partial g_{(n)}}{\partial \mathbf{E}_{(n)}} : \mathbf{M}_{(n)} : \frac{\partial g_{(n)}}{\partial \mathbf{E}_{(n)}} + \frac{\partial g_{(n)}}{\partial k_{(n)}} \frac{\mathrm{d}k_{(n)}}{\mathrm{d}\varphi_{(n)}} \left(\frac{1}{2} \frac{\partial g_{(n)}}{\partial \mathbf{E}_{(n)}} : \frac{\partial g_{(n)}}{\partial \mathbf{E}_{(n)}}\right)^{1/2}. \tag{9.1.24}$$

Suppose that current state on the yield surface is taken as the reference state for the strain measure, so that $\underline{\mathbf{E}}_{(n)} = \mathbf{0}$. Then,

$$-\underline{\mathbf{E}}^{\mathrm{p}}_{(n)} = \boldsymbol{\mathcal{E}}^{\mathrm{e}}_{(-n)}, \tag{9.1.25}$$

where $\boldsymbol{\mathcal{E}}^{\mathrm{e}}_{(-n)}$ is a spatial measure of elastic strain at the current yield state, relative to the state at the center of the yield surface. To recognize this,

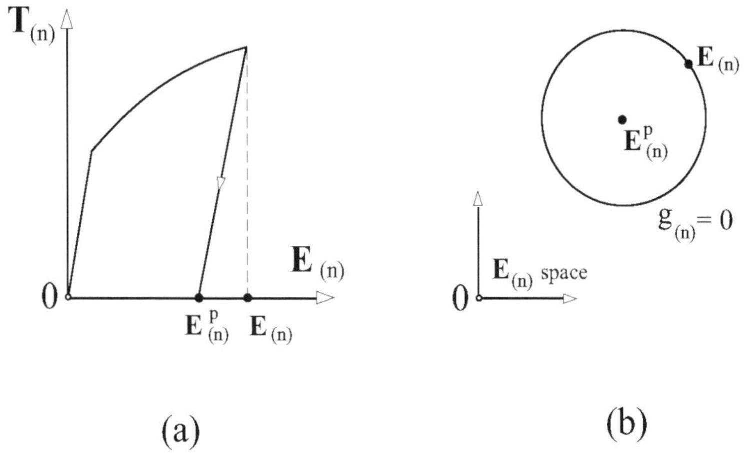

FIGURE 9.1. (a) Uniaxial stress-strain curve. (b) Yield surface in strain space corresponding to isotropic hardening. The center of the yield surface is at the plastic state of strain $\mathbf{E}^{\mathrm{p}}_{(n)}$, corresponding to zero state of stress.

denote by $\mathbf{F}^{\mathrm{e}} = \mathbf{V}^{\mathrm{e}} \cdot \mathbf{R}^{\mathrm{e}}$ the deformation gradient from the state at the center of the yield surface to the current state on the yield surface. It follows that

$$\underline{\mathbf{F}}^{\mathrm{p}} = (\mathbf{F}^{\mathrm{e}})^{-1}, \quad \underline{\mathbf{U}}^{\mathrm{p}} = (\mathbf{V}^{\mathrm{e}})^{-1}, \tag{9.1.26}$$

and

$$\underline{\mathbf{E}}^{\mathrm{p}}_{(n)} = \frac{1}{2n}\left[(\underline{\mathbf{U}}^{\mathrm{p}})^{2n} - \mathbf{I}\right] = -\boldsymbol{\mathcal{E}}^{\mathrm{e}}_{(-n)}. \tag{9.1.27}$$

Thus, Eqs. (9.1.4) and (9.1.24) give

$$\dot{\underline{\gamma}}_{(n)} = \frac{1}{\underline{h}_{(n)}}\left(\frac{\partial \underline{g}_{(n)}}{\partial \boldsymbol{\mathcal{E}}^{\mathrm{e}}_{(-n)}} : \mathbf{D}\right), \tag{9.1.28}$$

where

$$\underline{h}_{(n)} = \frac{\partial \underline{g}_{(n)}}{\partial \boldsymbol{\mathcal{E}}^{\mathrm{e}}_{(-n)}} : \underline{\mathbf{M}}_{(n)} : \frac{\partial \underline{g}_{(n)}}{\partial \boldsymbol{\mathcal{E}}^{\mathrm{e}}_{(-n)}} + \frac{\partial g_{(n)}}{\partial k_{(n)}} \frac{\mathrm{d}\underline{k}_{(n)}}{\mathrm{d}\underline{\varphi}_{(n)}} \left(\frac{1}{2} \frac{\partial \underline{g}_{(n)}}{\partial \boldsymbol{\mathcal{E}}^{\mathrm{e}}_{(-n)}} : \frac{\partial \underline{g}_{(n)}}{\partial \boldsymbol{\mathcal{E}}^{\mathrm{e}}_{(-n)}}\right)^{1/2}. \tag{9.1.29}$$

It is recalled from Eq. (9.1.18) that all stress rates $\dot{\underline{\mathbf{T}}}^{\mathrm{p}}_{(n)}$ are equal to each other, and thus all the history parameters $\underline{\varphi}_{(n)}$ are also equal (independent of n); see Eq. (9.1.22). These general expressions are next specialized by assuming that the elastic component of strain is infinitesimally small.

Infinitesimal Elasticity

If elastic deformation within the yield surface is infinitesimal, all strain measures $\boldsymbol{\mathcal{E}}^{\mathrm{e}}_{(-n)}$ reduce to infinitesimal elastic strain $\boldsymbol{\varepsilon}^{\mathrm{e}}$, whose deviatoric part is related to Cauchy stress by

$$\boldsymbol{\varepsilon}^{\mathrm{e}\,\prime} = \frac{1}{2\mu}\,\boldsymbol{\sigma}^{\prime}. \tag{9.1.30}$$

For example, let the yield surface be specified by

$$g = 4\mu^2\left[\frac{1}{2}\,\boldsymbol{\varepsilon}^{\mathrm{e}\,\prime} : \boldsymbol{\varepsilon}^{\mathrm{e}\,\prime} - k^2(\varphi)\right] = 0\,. \tag{9.1.31}$$

The factor $4\mu^2$ is introduced for the sake of comparison with the corresponding yield surface in stress space, considered later in Subsection 9.2.1. From Eqs. (9.1.18) and (9.1.31), we have

$$\frac{\partial g}{\partial \boldsymbol{\varepsilon}^{\mathrm{e}}} = 4\mu^2\boldsymbol{\varepsilon}^{\mathrm{e}\,\prime}, \quad \overset{\circ}{\underline{\boldsymbol{\tau}}}{}^{\mathrm{p}} = -4\mu^2\dot{\gamma}\,\boldsymbol{\varepsilon}^{\mathrm{e}\,\prime}, \tag{9.1.32}$$

while Eqs. (9.1.22), (9.1.24), and (9.1.29) give

$$\dot{\varphi} = -4\mu^2 k\,\dot{\gamma}, \quad \dot{\gamma} = \frac{4\mu^2}{h}\left(\boldsymbol{\varepsilon}^{\mathrm{e}\,\prime} : \mathbf{D}\right), \quad h = 16\mu^3 k^2\left(1 - 2\mu\,\frac{\mathrm{d}k}{\mathrm{d}\varphi}\right). \tag{9.1.33}$$

Consequently,

$$\overset{\circ}{\underline{\boldsymbol{\tau}}} = \left(\underline{\boldsymbol{\mathcal{L}}}_{(0)} - \frac{2\mu}{1 - 2\mu\,\mathrm{d}k/\mathrm{d}\varphi}\,\frac{\boldsymbol{\varepsilon}^{\mathrm{e}\,\prime} \otimes \boldsymbol{\varepsilon}^{\mathrm{e}\,\prime}}{\boldsymbol{\varepsilon}^{\mathrm{e}\,\prime} : \boldsymbol{\varepsilon}^{\mathrm{e}\,\prime}}\right) : \mathbf{D}. \tag{9.1.34}$$

The elastic stiffness or moduli tensor is taken as

$$\underline{\boldsymbol{\mathcal{L}}}_{(0)} = \lambda\,\mathbf{I} \otimes \mathbf{I} + 2\mu\,\boldsymbol{I}. \tag{9.1.35}$$

A similar approach to derive elastoplastic constitutive equations with the yield surface in strain space was used, within infinitesimal strain context, by Yoder and Iwan (1981).

It is convenient to express the elastic stiffness tensor (9.1.35) in an alternative form as

$$\underline{\boldsymbol{\mathcal{L}}}_{(0)} = 2\mu\,\boldsymbol{J} + 3\kappa\,\boldsymbol{K}, \tag{9.1.36}$$

where

$$\kappa = \lambda + \frac{2}{3}\,\mu \tag{9.1.37}$$

is the elastic bulk modulus. The base tensors $\boldsymbol{J}$ and $\boldsymbol{K}$ sum to give the fourth-order unit tensor, $\boldsymbol{J} + \boldsymbol{K} = \boldsymbol{I}$. The rectangular components of $\boldsymbol{I}$ and

$\boldsymbol{K}$ are

$$I_{ijkl} = \frac{1}{2}\left(\delta_{ik}\,\delta_{jl} + \delta_{il}\,\delta_{jk}\right), \quad K_{ijkl} = \frac{1}{3}\,\delta_{ij}\,\delta_{kl}\,. \tag{9.1.38}$$

These are convenient base tensors, because $\boldsymbol{J} : \boldsymbol{J} = \boldsymbol{J}$ and $\boldsymbol{K} : \boldsymbol{K} = \boldsymbol{K}$, as well as $\boldsymbol{J} : \boldsymbol{K} = \boldsymbol{K} : \boldsymbol{J} = \mathbf{0}$ (Hill, 1965; Walpole, 1981). In the trace operation with any second-order tensor $\mathbf{A}$, the tensor $\boldsymbol{J}$ extracts its deviatoric part, while the tensor $\boldsymbol{K}$ extracts its spherical part ($\boldsymbol{J} : \mathbf{A} = \mathbf{A}'$ and $\boldsymbol{K} : \mathbf{A} = \mathbf{A} - \mathbf{A}'$). It is then easily verified that the inverse of (9.1.36) is simply

$$\underline{\boldsymbol{\mathcal{L}}}_{(0)}^{-1} = \frac{1}{2\mu}\,\boldsymbol{J} + \frac{1}{3\kappa}\,\boldsymbol{K}. \tag{9.1.39}$$

9.2. Formulation in Stress Space

In the rate-independent elastoplastic theory with the yield surface in stress space, the strain rate is decomposed as the sum of elastic and plastic parts, such that

$$\dot{\mathbf{E}}_{(n)} = \dot{\mathbf{E}}^{\mathrm{e}}_{(n)} + \dot{\mathbf{E}}^{\mathrm{p}}_{(n)} = \mathbf{M}_{(n)} : \dot{\mathbf{T}}_{(n)} + \dot{\gamma}_{(n)}\,\frac{\partial f_{(n)}}{\partial \mathbf{T}_{(n)}}\,. \tag{9.2.1}$$

The function $f_{(n)}\left(\mathbf{T}_{(n)}, \mathcal{H}\right)$ is the yield function, and $\dot{\gamma}_{(n)} > 0$ is the loading index, both corresponding to selected measure and reference state. The yield surface is

$$f_{(n)}\left(\mathbf{T}_{(n)}, \mathcal{H}\right) = 0. \tag{9.2.2}$$

Assuming an incrementally linear response and a continuity of the response, the loading index can be expressed as

$$\dot{\gamma}_{(n)} = \frac{1}{H_{(n)}}\left(\frac{\partial f_{(n)}}{\partial \mathbf{T}_{(n)}} : \dot{\mathbf{T}}_{(n)}\right). \tag{9.2.3}$$

The scalar function $H_{(n)}$ is determined from the consistency condition and a given representation of the yield function. Substitution of Eq. (9.2.3) into Eq. (9.2.1) gives

$$\dot{\mathbf{E}}_{(n)} = \left[\mathbf{M}_{(n)} + \frac{1}{H_{(n)}}\left(\frac{\partial f_{(n)}}{\partial \mathbf{T}_{(n)}} \otimes \frac{\partial f_{(n)}}{\partial \mathbf{T}_{(n)}}\right)\right] : \dot{\mathbf{T}}_{(n)}. \tag{9.2.4}$$

The fourth-order tensor within the square brackets is the elastoplastic compliance tensor associated with the considered stress and strain measures and the reference state.

The relationship between $h_{(n)}$ in Eq. (9.1.11) and $H_{(n)}$ in Eq. (9.2.4) can be obtained by equating Eqs. (9.1.4) and (9.2.3), i.e.,

$$\frac{1}{h_{(n)}}\left(\frac{\partial g_{(n)}}{\partial \mathbf{E}_{(n)}}:\dot{\mathbf{E}}_{(n)}\right)=\frac{1}{H_{(n)}}\left(\frac{\partial f_{(n)}}{\partial \mathbf{T}_{(n)}}:\dot{\mathbf{T}}_{(n)}\right). \tag{9.2.5}$$

Substituting Eq. (9.2.4) for $\dot{\mathbf{E}}_{(n)}$ and by using the relationship between the yield surface normals in stress and strain space,

$$\frac{\partial f_{(n)}}{\partial \mathbf{T}_{(n)}}=\mathbf{M}_{(n)}:\frac{\partial g_{(n)}}{\partial \mathbf{E}_{(n)}}, \tag{9.2.6}$$

there follows

$$H_{(n)}=h_{(n)}-\frac{\partial g_{(n)}}{\partial \mathbf{E}_{(n)}}:\frac{\partial f_{(n)}}{\partial \mathbf{T}_{(n)}}, \tag{9.2.7}$$

in agreement with Eq. (9.1.14). Consequently, Eq. (9.2.4) is equivalent to Eq. (9.1.15).

The scalar parameter $H_{(n)}$ can be positive, negative or equal to zero. Three types of response are thus possible within this constitutive framework. They are

$$\begin{aligned}
&H_{(n)}>0, \quad \frac{\partial f_{(n)}}{\partial \mathbf{T}_{(n)}}:\dot{\mathbf{T}}_{(n)}>0 \quad \text{hardening},\\
&H_{(n)}<0, \quad \frac{\partial f_{(n)}}{\partial \mathbf{T}_{(n)}}:\dot{\mathbf{T}}_{(n)}<0 \quad \text{softening},\\
&H_{(n)}=0, \quad \frac{\partial f_{(n)}}{\partial \mathbf{T}_{(n)}}:\dot{\mathbf{T}}_{(n)}=0 \quad \text{ideally plastic.}
\end{aligned} \tag{9.2.8}$$

Starting from the current yield surface in stress space, the stress point moves outward in the case of hardening, inward in the case of softening, and tangentially to the yield surface in the case of ideally plastic response. In the case of softening, $\dot{\mathbf{E}}_{(n)}$ is not uniquely determined by the prescribed stress rate $\dot{\mathbf{T}}_{(n)}$, since either Eq. (9.2.4) applies, or the elastic unloading expression

$$\dot{\mathbf{E}}_{(n)}=\mathbf{M}_{(n)}:\dot{\mathbf{T}}_{(n)}. \tag{9.2.9}$$

In the case of ideally plastic response, the plastic part of the strain rate is indeterminate to the extent of an arbitrary positive multiple, since $\dot{\gamma}_{(n)}$ in Eq. (9.2.3) is indeterminate.

Inverted form of Eq. (9.2.4) can be obtained along similar lines as used to invert Eq. (9.1.11). The result is

$$\dot{\mathbf{T}}_{(n)}=\left[\mathbf{\Lambda}_{(n)}-\frac{1}{h_{(n)}}\left(\mathbf{\Lambda}_{(n)}:\frac{\partial f_{(n)}}{\partial \mathbf{T}_{(n)}}\right)\otimes\left(\frac{\partial f_{(n)}}{\partial \mathbf{T}_{(n)}}:\mathbf{\Lambda}_{(n)}\right)\right]:\dot{\mathbf{E}}_{(n)}, \tag{9.2.10}$$

where

$$h_{(n)} = H_{(n)} + \frac{\partial f_{(n)}}{\partial \mathbf{T}_{(n)}} : \mathbf{\Lambda}_{(n)} : \frac{\partial f_{(n)}}{\partial \mathbf{T}_{(n)}}, \tag{9.2.11}$$

which is in agreement with Eq. (9.1.14). If current state is taken as the reference state, Eq. (9.2.4) becomes

$$\mathbf{D} = \left[\underline{\mathbf{M}}_{(n)} + \frac{1}{\underline{H}_{(n)}} \left(\frac{\partial \underline{f}_{(n)}}{\partial \underline{\mathbf{T}}_{(n)}} \otimes \frac{\partial \underline{f}_{(n)}}{\partial \underline{\mathbf{T}}_{(n)}} \right) \right] : \dot{\underline{\mathbf{T}}}_{(n)}. \tag{9.2.12}$$

9.2.1. Yield Surface in Cauchy Stress Space

It is most convenient to apply Eq. (9.2.12) for $n = 0$. In the near neighborhood of the current stress state on the yield surface, the conjugate stress to logarithmic strain (relative to the state on the yield surface) is, from Eq. (3.6.17),

$$\mathbf{T}_{(0)} = \mathbf{R} \cdot \boldsymbol{\tau} \cdot \mathbf{R}^T + \mathcal{O}(\boldsymbol{\tau} \cdot \mathbf{E}_{(n)}^2), \tag{9.2.13}$$

where $\boldsymbol{\tau} = (\det \mathbf{F})\boldsymbol{\sigma}$ is the Kirchhoff stress. On the other hand, for $n \neq 0$,

$$\mathbf{T}_{(n)} = \mathbf{R} \cdot \boldsymbol{\tau} \cdot \mathbf{R}^T + \mathcal{O}(\boldsymbol{\tau} \cdot \mathbf{E}_{(n)}), \tag{9.2.14}$$

by Eq. (3.6.16). In the last two equations, the deformation gradient $\mathbf{F}$ and the rotation $\mathbf{R}$ are measured from the current, deformed configuration as the reference. Thus,

$$f_{(0)}(\mathbf{T}_{(0)}, \mathcal{H}) \approx f(\boldsymbol{\sigma}, \mathcal{H}) \tag{9.2.15}$$

in the near neighborhood of the current yield state, where

$$f(\boldsymbol{\sigma}, \mathcal{H}) = 0 \tag{9.2.16}$$

represents the yield surface in the Cauchy stress space. Equation (9.2.12) consequently becomes

$$\mathbf{D} = \left[\underline{\boldsymbol{\mathcal{M}}}_{(0)} + \frac{1}{H} \left(\frac{\partial f}{\partial \boldsymbol{\sigma}} \otimes \frac{\partial f}{\partial \boldsymbol{\sigma}} \right) \right] : \overset{\circ}{\underline{\boldsymbol{\tau}}}. \tag{9.2.17}$$

The tensor

$$\underline{\mathbf{M}}_{(0)} = \underline{\boldsymbol{\mathcal{M}}}_{(0)} = \underline{\boldsymbol{\mathcal{L}}}_{(0)}^{-1} \tag{9.2.18}$$

is the corresponding instantaneous compliance tensor, and H is an appropriate scalar function of the deformation history.

The elastic and plastic parts of the rate of deformation tensor $\mathbf{D}$, corresponding to $\overset{\circ}{\underline{\boldsymbol{\tau}}}$, are

$$\mathbf{D}^{\mathrm{e}}_{(0)} = \underline{\boldsymbol{\mathcal{M}}}_{(0)} : \overset{\circ}{\underline{\boldsymbol{\tau}}}, \quad \mathbf{D}^{\mathrm{p}}_{(0)} = \frac{1}{H}\left(\frac{\partial f}{\partial \boldsymbol{\sigma}} \otimes \frac{\partial f}{\partial \boldsymbol{\sigma}}\right) : \overset{\circ}{\underline{\boldsymbol{\tau}}}. \tag{9.2.19}$$

If elastic component of strain is neglected, a model of rigid-plasticity is obtained. The rate of deformation is due to plastic deformation only, so that

$$\mathbf{D} = \frac{1}{H}\left(\frac{\partial f}{\partial \boldsymbol{\sigma}} \otimes \frac{\partial f}{\partial \boldsymbol{\sigma}}\right) : \overset{\circ}{\underline{\boldsymbol{\tau}}}. \tag{9.2.20}$$

9.3. Nonuniqueness of the Rate of Deformation Partition

Within the considered framework of conjugate stress and strain tensors, there are infinitely many partitions of the rate of deformation tensor, one associated with each n. Thus, we can write (Lubarda, 1994)

$$\mathbf{D} = \mathbf{D}^{\mathrm{e}}_{(0)} + \mathbf{D}^{\mathrm{p}}_{(0)} = \mathbf{D}^{\mathrm{e}}_{(n)} + \mathbf{D}^{\mathrm{p}}_{(n)}. \tag{9.3.1}$$

The elastic parts of $\mathbf{D}$ are defined by

$$\mathbf{D}^{\mathrm{e}}_{(0)} = \underline{\boldsymbol{\mathcal{M}}}_{(0)} : \overset{\circ}{\underline{\boldsymbol{\tau}}}, \quad \mathbf{D}^{\mathrm{e}}_{(n)} = \underline{\boldsymbol{\mathcal{M}}}_{(n)} : \underline{\dot{\mathbf{T}}}_{(n)}, \tag{9.3.2}$$

where

$$\underline{\dot{\mathbf{T}}}_{(n)} = \overset{\circ}{\underline{\boldsymbol{\tau}}} - 2n\,\underline{\boldsymbol{\mathcal{S}}} : \mathbf{D}, \quad \underline{\boldsymbol{\mathcal{L}}}_{(n)} = \underline{\boldsymbol{\mathcal{L}}}_{(0)} - 2n\,\underline{\boldsymbol{\mathcal{S}}}. \tag{9.3.3}$$

The fourth-order tensor $\underline{\boldsymbol{\mathcal{S}}}$ is defined in Eq. (6.3.11) as

$$\underline{\mathcal{S}}_{ijkl} = \frac{1}{4}\left(\sigma_{ik}\delta_{jl} + \sigma_{jk}\delta_{il} + \sigma_{il}\delta_{jk} + \sigma_{jl}\delta_{ik}\right). \tag{9.3.4}$$

Since, from Eq. (8.8.22),

$$\underline{\dot{\mathbf{T}}}^{\mathrm{p}}_{(0)} = \underline{\dot{\mathbf{T}}}^{\mathrm{p}}_{(n)}, \tag{9.3.5}$$

and since

$$\underline{\dot{\mathbf{T}}}^{\mathrm{p}}_{(n)} = -\underline{\boldsymbol{\mathcal{L}}}_{(n)} : \mathbf{D}^{\mathrm{p}}_{(n)} = -\left(\underline{\boldsymbol{\mathcal{L}}}_{(0)} - 2n\,\underline{\boldsymbol{\mathcal{S}}}\right) : \mathbf{D}^{\mathrm{p}}_{(n)}, \tag{9.3.6}$$

the following relationships hold

$$\mathbf{D}^{\mathrm{p}}_{(0)} = \mathbf{D}^{\mathrm{p}}_{(n)} - 2n\,\underline{\boldsymbol{\mathcal{M}}}_{(0)} : \underline{\boldsymbol{\mathcal{S}}} : \mathbf{D}^{\mathrm{p}}_{(n)}, \tag{9.3.7}$$

$$\mathbf{D}^{\mathrm{e}}_{(0)} = \mathbf{D}^{\mathrm{e}}_{(n)} + 2n\,\underline{\boldsymbol{\mathcal{M}}}_{(0)} : \underline{\mathbf{S}} : \mathbf{D}^{\mathrm{p}}_{(n)}. \tag{9.3.8}$$

Alternatively, these can be expressed as

$$\mathbf{D}^{\mathrm{p}}_{(n)} = \mathbf{D}^{\mathrm{p}}_{(0)} + 2n\,\underline{\boldsymbol{\mathcal{M}}}_{(n)} : \underline{\boldsymbol{\mathcal{S}}} : \mathbf{D}^{\mathrm{p}}_{(0)}, \tag{9.3.9}$$

$$\mathbf{D}^{\mathrm{e}}_{(n)} = \mathbf{D}^{\mathrm{e}}_{(0)} - 2n\,\underline{\boldsymbol{\mathcal{M}}}_{(n)} : \underline{\boldsymbol{S}} : \mathbf{D}^{\mathrm{p}}_{(0)}. \tag{9.3.10}$$

The relative difference between the components of elastic (and plastic) rate of deformation tensors for various n are thus of the order of Cauchy stress over elastic modulus. In the sequel, the elastic and plastic parts of the rate of deformation tensor corresponding to $\overset{\circ}{\boldsymbol{\tau}}$ will be designated simply by $\mathbf{D}^{\mathrm{e}}$ and $\mathbf{D}^{\mathrm{p}}$, i.e.,

$$\mathbf{D}^{\mathrm{e}}_{(0)} = \mathbf{D}^{\mathrm{e}}, \quad \mathbf{D}^{\mathrm{p}}_{(0)} = \mathbf{D}^{\mathrm{p}}. \tag{9.3.11}$$

9.4. Hardening Models in Stress Space

9.4.1. Isotropic Hardening

The experimental determination of the yield surface is commonly done with respect to Cauchy stress. Suppose that this is given by

$$f(\sigma, K) = 0, \tag{9.4.1}$$

where f is an isotropic function of $\boldsymbol{\sigma}$, and

$$K = K(\vartheta) \tag{9.4.2}$$

is a scalar function which defines the size of the yield surface. The hardening model in which the yield surface expands during plastic deformation, preserving its shape, is known as the isotropic hardening model. Since f is taken to be an isotropic function of stress, the material is assumed to be isotropic. The history parameter ϑ is the effective (generalized) plastic strain, defined by

$$\vartheta = \int_0^t \left(2\,\mathbf{D}^{\mathrm{p}} : \mathbf{D}^{\mathrm{p}}\right)^{1/2} \mathrm{d}t. \tag{9.4.3}$$

In view of the isotropy of the function f, we may write

$$f(\boldsymbol{\sigma}, K) = f\left(\mathbf{R}^T \cdot \boldsymbol{\sigma} \cdot \mathbf{R}, K\right) \approx f(\mathbf{T}_{(0)}, K). \tag{9.4.4}$$

The approximation holds in the near neighborhood of the current state, relative to which $\mathbf{R}$ and $\mathbf{T}_{(0)}$ are measured. The consistency condition for continuing plastic deformation,

$$\dot{f} = 0, \tag{9.4.5}$$

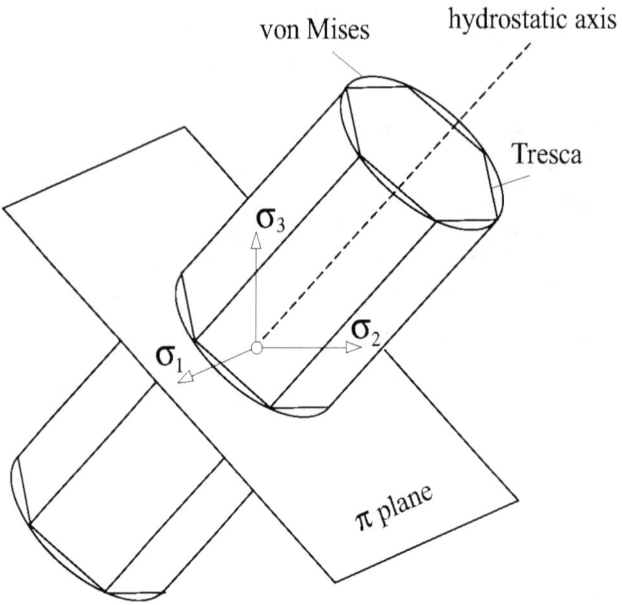

FIGURE 9.2. Von Mises and Tresca yield surfaces in principal stress space. The yield cylinder and the yield prism have their axis parallel to the hydrostatic axis, which is perpendicular to the π plane ($\sigma_1 + \sigma_2 + \sigma_3 = 0$).

gives

$$\frac{\partial f}{\partial \underline{\mathbf{T}}_{(0)}} : \dot{\underline{\mathbf{T}}}_{(0)} = \frac{\partial f}{\partial \boldsymbol{\sigma}} : \overset{\circ}{\underline{\boldsymbol{\tau}}} = -\frac{\partial f}{\partial K} \frac{\mathrm{d}K}{\mathrm{d}\vartheta} \dot{\vartheta}. \tag{9.4.6}$$

Upon substitution of Eqs. (9.4.3), the loading index becomes

$$\dot{\gamma} = \frac{1}{H} \left(\frac{\partial f}{\partial \boldsymbol{\sigma}} : \overset{\circ}{\underline{\boldsymbol{\tau}}} \right), \quad H = -\frac{\partial f}{\partial K} \frac{\mathrm{d}K}{\mathrm{d}\vartheta} \left(2 \frac{\partial f}{\partial \boldsymbol{\sigma}} : \frac{\partial f}{\partial \boldsymbol{\sigma}} \right)^{1/2}. \tag{9.4.7}$$

J_2 Flow Theory of Plasticity

For nonporous metals the onset of plastic deformation and plastic yielding is unaffected by a moderate superimposed pressure. The yield condition for such materials can consequently be written as an isotropic function of the deviatoric part of Cauchy stress, i.e., as a function of its second and third invariant,

$$f(J_2, J_3, K) = 0. \tag{9.4.8}$$

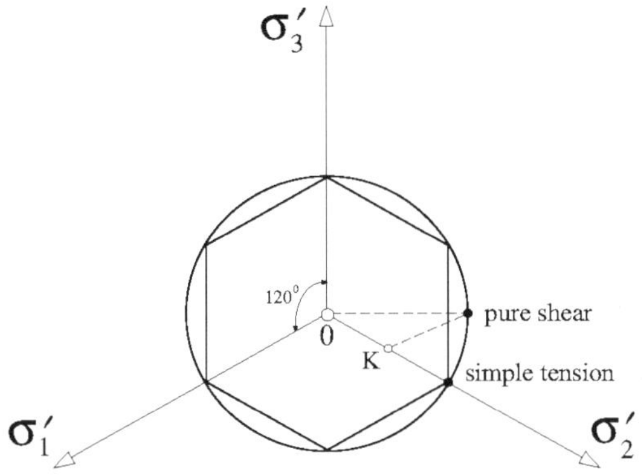

FIGURE 9.3. The trace of the von Mises and Tresca yield surfaces in the π plane. The states of simple tension and pure shear are indicated.

The classical examples are the Tresca maximum shear stress criterion and the von Mises yield criterion (Fig. 9.2). In the latter case,

$$f = J_2 - K^2(\vartheta) = 0, \quad J_2 = \frac{1}{2}\boldsymbol{\sigma}' : \boldsymbol{\sigma}'. \tag{9.4.9}$$

The corresponding plasticity theory is known as the J_2 flow theory of plasticity. If the yield stress in uniaxial tension is σ_Y, and in shear loading τ_Y, we have (Fig. 9.3)

$$K = \frac{1}{\sqrt{3}}\sigma_Y = \tau_Y. \tag{9.4.10}$$

For the J_2 plasticity,

$$\frac{\partial f}{\partial \boldsymbol{\sigma}} = \boldsymbol{\sigma}', \quad \mathbf{D}^{\mathrm{p}} = \dot{\gamma}\boldsymbol{\sigma}', \tag{9.4.11}$$

and

$$\dot{\vartheta} = 2K\dot{\gamma}, \quad H = 4K^2 h_{\mathrm{t}}^{\mathrm{p}}, \quad \dot{\gamma} = \frac{1}{4K^2 h_{\mathrm{t}}^{\mathrm{p}}}\left(\boldsymbol{\sigma}' : \overset{\circ}{\underline{\boldsymbol{\tau}}}\right). \tag{9.4.12}$$

The plastic tangent modulus in shear test is

$$h_{\mathrm{t}}^{\mathrm{p}} = \frac{\mathrm{d}K}{\mathrm{d}\vartheta}. \tag{9.4.13}$$

Equation (9.4.11) implies that plastic deformation is isochoric

$$\operatorname{tr}\mathbf{D}^{\mathrm{p}} = 0. \tag{9.4.14}$$

The total rate of deformation is

$$\mathbf{D} = \left[\underline{\mathcal{M}}_{(0)} + \frac{1}{4K^2 h_{\rm t}^{\rm p}} \left(\boldsymbol{\sigma}' \otimes \boldsymbol{\sigma}'\right) \right] : \overset{\circ}{\underline{\boldsymbol{\tau}}}. \tag{9.4.15}$$

For infinitesimal elasticity the elastic compliance tensor can be taken as

$$\underline{\mathcal{M}}_{(0)} = \frac{1}{2\mu} \left(\boldsymbol{I} - \frac{\lambda}{2\mu + 3\lambda} \mathbf{I} \otimes \mathbf{I} \right) = \frac{1}{2\mu} \boldsymbol{J} + \frac{1}{3\kappa} \boldsymbol{K}. \tag{9.4.16}$$

By using Eq. (9.4.9) to express K in terms of stress, Eq. (9.4.15) is rewritten as

$$\mathbf{D} = \left(\underline{\mathcal{M}}_{(0)} + \frac{1}{2h_{\rm t}^{\rm p}} \frac{\boldsymbol{\sigma}' \otimes \boldsymbol{\sigma}'}{\boldsymbol{\sigma}' : \boldsymbol{\sigma}'} \right) : \overset{\circ}{\underline{\boldsymbol{\tau}}}. \tag{9.4.17}$$

The plastic loading condition in the hardening range is

$$\boldsymbol{\sigma}' : \overset{\circ}{\underline{\boldsymbol{\tau}}} > 0. \tag{9.4.18}$$

The inverse equation is

$$\overset{\circ}{\underline{\boldsymbol{\tau}}} = \left(\underline{\mathcal{L}}_{(0)} - \frac{2\mu}{1 + h_{\rm t}^{\rm p}/\mu} \frac{\boldsymbol{\sigma}' \otimes \boldsymbol{\sigma}'}{\boldsymbol{\sigma}' : \boldsymbol{\sigma}'} \right) : \mathbf{D}, \tag{9.4.19}$$

which applies for

$$\boldsymbol{\sigma}' : \mathbf{D} > 0. \tag{9.4.20}$$

Note that

$$\underline{\mathcal{L}}_{(0)} : \boldsymbol{\sigma}' = 2\mu\, \boldsymbol{\sigma}'. \tag{9.4.21}$$

In retrospect, the plastic rate of deformation can be expressed either in terms of stress rate or total rate of deformation as

$$\mathbf{D}^{\rm p} = \frac{1}{2h_{\rm t}^{\rm p}} \frac{\boldsymbol{\sigma}' \otimes \boldsymbol{\sigma}'}{\boldsymbol{\sigma}' : \boldsymbol{\sigma}'} : \overset{\circ}{\underline{\boldsymbol{\tau}}} = \frac{1}{1 + h_{\rm t}^{\rm p}/\mu} \frac{\boldsymbol{\sigma}' \otimes \boldsymbol{\sigma}'}{\boldsymbol{\sigma}' : \boldsymbol{\sigma}'} : \mathbf{D}. \tag{9.4.22}$$

An often utilized expression for $K = K(\vartheta)$ corresponds to nonlinear hardening that saturates to linear hardening at large ϑ (Fig. 9.4), i.e.,

$$K = K_0 + h_1\vartheta + (K_1 - K_0) \left[1 - \exp\left(-\frac{h_0 - h_1}{K_1 - K_0}\, \vartheta \right) \right]. \tag{9.4.23}$$

The corresponding plastic tangent modulus is

$$h_{\rm t}^{\rm p} = h_1 + (h_0 - h_1) \exp\left(-\frac{h_0 - h_1}{K_1 - K_0}\, \vartheta \right). \tag{9.4.24}$$

In the case of linear hardening, $K = K_0 + h_{\rm t}^{\rm p}\vartheta$, where $h_{\rm t}^{\rm p}$ is a constant. For ideal (perfect) plasticity, $h_{\rm t}^{\rm p} = 0$ can be substituted in the expression on the far right-hand side of Eq. (9.4.22), since $\boldsymbol{\sigma}' : \boldsymbol{\sigma}' = 2K_0^2$, where K_0 is the constant radius of the yield surface.

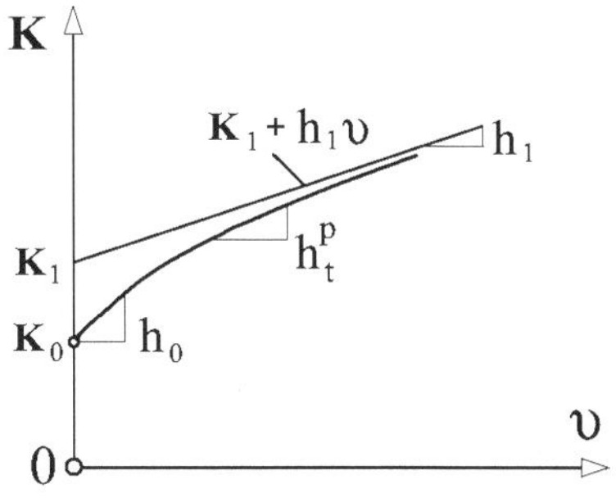

FIGURE 9.4. Nonlinear hardening that saturates to linear hardening with the rate h_1 at large ϑ, according to Eq. (9.4.17). The initial yield stress is K_0 and the initial hardening rate is h_0. The plastic tangent modulus at an arbitrary ϑ is h_t^p.

Constitutive structures (9.4.17) and (9.4.19) have been used in analytical and numerical treatments of various plastic deformation problems (e.g., Hutchinson, 1973; McMeeking and Rice, 1975; Neale, 1981; Needleman, 1982). More generally, when f is defined by Eq. (9.4.8), we can write

$$\mathbf{D} = \left(\underline{\boldsymbol{\mathcal{M}}}_{(0)} + \frac{1}{2h^{\mathrm{p}}}\, \mathbb{M} \otimes \mathbb{M} \right) : \overset{\circ}{\underline{\boldsymbol{\tau}}}, \tag{9.4.25}$$

$$\overset{\circ}{\underline{\boldsymbol{\tau}}} = \left(\underline{\boldsymbol{\mathcal{L}}}_{(0)} - \frac{2\mu}{1 + h^{\mathrm{p}}/\mu}\, \mathbb{M} \otimes \mathbb{M} \right) : \mathbf{D}. \tag{9.4.26}$$

The normalized tensor $\mathbb{M}$ is in the direction of outward normal to the yield surface, and h^{p} is the hardening parameter. They are defined by

$$\mathbb{M} = \frac{\frac{\partial f}{\partial \boldsymbol{\sigma}}}{\left(\frac{\partial f}{\partial \boldsymbol{\sigma}} : \frac{\partial f}{\partial \boldsymbol{\sigma}} \right)^{1/2}}, \quad h^{\mathrm{p}} = -\frac{\frac{\partial f}{\partial K}\, h_{\mathrm{t}}^{\mathrm{p}}}{\left(2\, \frac{\partial f}{\partial \boldsymbol{\sigma}} : \frac{\partial f}{\partial \boldsymbol{\sigma}} \right)^{1/2}}. \tag{9.4.27}$$

If f is given by Eq. (9.4.9), then

$$h^{\mathrm{p}} = h_{\mathrm{t}}^{\mathrm{p}} = \frac{\mathrm{d}K}{\mathrm{d}\vartheta}. \tag{9.4.28}$$

Derived equations are in accord with the constitutive structure (9.1.34), obtained within formulation based on the yield surface in strain space. This can be easily verified by observing that

$$K = 2\mu k, \quad \frac{\mathrm{d}K}{\mathrm{d}\vartheta} = 2\mu\, \frac{\mathrm{d}k}{\mathrm{d}\varphi}\, \frac{\dot{\varphi}}{\dot{\vartheta}}, \quad \frac{\dot{\varphi}}{\dot{\vartheta}} = -\mu, \tag{9.4.29}$$

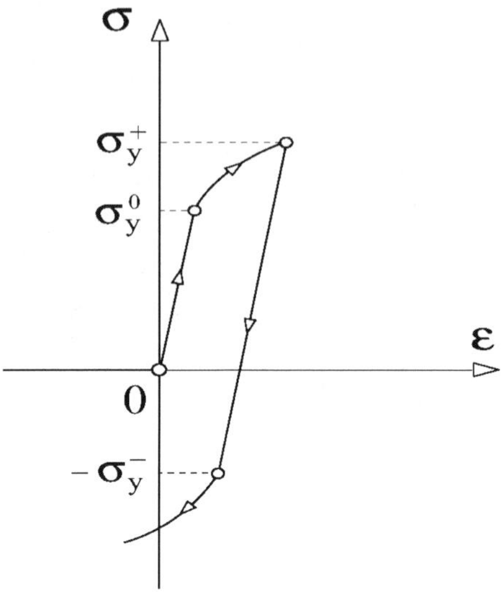

FIGURE 9.5. Illustration of the Bauschinger effect ($|\sigma_y^-| < \sigma_y^+$) in uniaxial tension. The Cauchy stress is σ and the logarithmic strain is ϵ.

and

$$H - h = -4\mu K^2. \tag{9.4.30}$$

The formulation of the constitutive equations for isotropic hardening plasticity within the framework of infinitesimal strain is presented in standard texts or review papers, such as Hill (1950), Drucker (1960), and Naghdi (1960). Derivation of classical Prandtl–Reuss equations for elastic-ideally plastic, and Levy–Mises equations for rigid-ideally plastic material models is also there given. The effects of the third invariant of the stress deviator on plastic deformation are discussed by Novozhilov (1952), Ohashi, Tokuda, and Yamashita (1975), and Gupta and Meyers (1992, 1994). The book by Zyczkowski (1981) contains a comprehensive list of references to various other topics of classical plasticity.

9.4.2. Kinematic Hardening

To account for the Bauschinger effect (Fig. 9.5) and anisotropic hardening, and thus provide better description of material response under cyclic loading, a simple model of kinematic hardening was introduced by Melan (1938) and

Prager (1955,1956). According to this model, the initial yield surface does not change its size and shape during plastic deformation, but translates in the stress space according to some prescribed rule. If the yield condition is pressure-independent, it is assumed that

$$f\,(\boldsymbol{\sigma}' - \boldsymbol{\alpha}, K_0) = 0, \quad K_0 = \text{const.}, \tag{9.4.31}$$

where $\boldsymbol{\alpha}$ represents the current center of the yield locus in the deviatoric plane $\operatorname{tr}\boldsymbol{\sigma} = 0$ (back stress), and f is an isotropic function of the stress difference $\boldsymbol{\sigma}' - \boldsymbol{\alpha}$. The back stress in the plane $\operatorname{tr}\boldsymbol{\sigma} = \text{const.}$ would be $\boldsymbol{\alpha} + (\operatorname{tr}\boldsymbol{\sigma}/3)\mathbf{I}$. The size of the yield locus is specified by the constant K_0. By an analysis similar to that used in the previous subsection, the consistency condition for continuing plastic deformation can be written as

$$\frac{\partial f}{\partial \boldsymbol{\sigma}} : \left(\overset{\circ}{\underline{\boldsymbol{\tau}}} - \overset{\circ}{\underline{\boldsymbol{\alpha}}}\right) = 0, \tag{9.4.32}$$

where $\partial f/\partial \boldsymbol{\sigma} = \partial f/\partial \boldsymbol{\sigma}'$. Suppose that the yield surface instantaneously translates so that the evolution of back stress is governed by

$$\overset{\circ}{\underline{\boldsymbol{\alpha}}} = c(\boldsymbol{\alpha}, \vartheta)\, \mathbf{D}^{\mathrm{p}} + \mathbf{C}(\boldsymbol{\alpha}, \vartheta)\, (\mathbf{D}^{\mathrm{p}} : \mathbf{D}^{\mathrm{p}})^{1/2}, \tag{9.4.33}$$

where c and $\mathbf{C}$ are the appropriate scalar and tensor functions of $\boldsymbol{\alpha}$ and ϑ. This representation is in accord with assumed time-independence of plastic deformation, which requires Eq. (9.4.33) to be homogeneous function of degree one in the components of plastic rate of deformation. Since the plastic rate of deformation is

$$\mathbf{D}^{\mathrm{p}} = \dot{\gamma}\, \frac{\partial f}{\partial \boldsymbol{\sigma}}, \tag{9.4.34}$$

the substitution of Eq. (9.4.33) into Eq. (9.4.32) gives the loading index

$$\dot{\gamma} = \frac{1}{H}\left(\frac{\partial f}{\partial \boldsymbol{\sigma}} : \overset{\circ}{\underline{\boldsymbol{\tau}}}\right), \quad H = c\left(\frac{\partial f}{\partial \boldsymbol{\sigma}} : \frac{\partial f}{\partial \boldsymbol{\sigma}}\right) + \left(\frac{\partial f}{\partial \boldsymbol{\sigma}} : \frac{\partial f}{\partial \boldsymbol{\sigma}}\right)^{1/2} \left(\mathbf{C} : \frac{\partial f}{\partial \boldsymbol{\sigma}}\right). \tag{9.4.35}$$

If the yield condition is specified by

$$f = \frac{1}{2}(\boldsymbol{\sigma}' - \boldsymbol{\alpha}) : (\boldsymbol{\sigma}' - \boldsymbol{\alpha}) - K_0^2 = 0, \tag{9.4.36}$$

then

$$\frac{\partial f}{\partial \boldsymbol{\sigma}} = \boldsymbol{\sigma}' - \boldsymbol{\alpha}, \quad \mathbf{D}^{\mathrm{p}} = \dot{\gamma}\,(\boldsymbol{\sigma}' - \boldsymbol{\alpha}), \tag{9.4.37}$$

and

$$\dot{\gamma} = \frac{1}{H}\,(\boldsymbol{\sigma}' - \boldsymbol{\alpha}) : \overset{\circ}{\underline{\boldsymbol{\tau}}}, \quad H = 2K_0\left[cK_0 + \frac{1}{\sqrt{2}}\mathbf{C} : (\boldsymbol{\sigma}' - \boldsymbol{\alpha})\right]. \tag{9.4.38}$$

Consequently,

$$\mathbf{D} = \left[\underline{\mathcal{M}}_{(0)} + \frac{1}{H} (\boldsymbol{\sigma}' - \boldsymbol{\alpha}) \otimes (\boldsymbol{\sigma}' - \boldsymbol{\alpha}) \right] : \overset{\circ}{\underline{\boldsymbol{\tau}}}. \tag{9.4.39}$$

Linear and Nonlinear Kinematic Hardening

When $\mathbf{C} = \mathbf{0}$ and c is taken to be a constant, the model with evolution equation (9.4.33) reduces to Prager's linear kinematic hardening (Fig. 9.6). The plastic tangent modulus $h_{\mathrm{t}}^{\mathrm{p}}$ from the shear test is constant, and related to c by

$$c = 2h_{\mathrm{t}}^{\mathrm{p}}. \tag{9.4.40}$$

In this case, Eq. (9.4.39) becomes

$$\mathbf{D} = \left[\underline{\mathcal{M}}_{(0)} + \frac{1}{2h_{\mathrm{t}}^{\mathrm{p}}} \frac{(\boldsymbol{\sigma}' - \boldsymbol{\alpha}) \otimes (\boldsymbol{\sigma}' - \boldsymbol{\alpha})}{(\boldsymbol{\sigma}' - \boldsymbol{\alpha}) : (\boldsymbol{\sigma}' - \boldsymbol{\alpha})} \right] : \overset{\circ}{\underline{\boldsymbol{\tau}}}, \tag{9.4.41}$$

with plastic loading condition in the hardening range

$$(\boldsymbol{\sigma}' - \boldsymbol{\alpha}) : \overset{\circ}{\underline{\boldsymbol{\tau}}} > 0. \tag{9.4.42}$$

The inverse equation is

$$\overset{\circ}{\underline{\boldsymbol{\tau}}} = \left[\underline{\mathcal{L}}_{(0)} - \frac{2\mu}{1 + h_{\mathrm{t}}^{\mathrm{p}}/\mu} \frac{(\boldsymbol{\sigma}' - \boldsymbol{\alpha}) \otimes (\boldsymbol{\sigma}' - \boldsymbol{\alpha})}{(\boldsymbol{\sigma}' - \boldsymbol{\alpha}) : (\boldsymbol{\sigma}' - \boldsymbol{\alpha})} \right] : \mathbf{D}, \tag{9.4.43}$$

provided that

$$(\boldsymbol{\sigma}' - \boldsymbol{\alpha}) : \mathbf{D} > 0. \tag{9.4.44}$$

In retrospect, the evolution equation for the back stress

$$\overset{\circ}{\underline{\boldsymbol{\alpha}}} = 2h_{\mathrm{t}}^{\mathrm{p}}\, \mathbf{D}^{\mathrm{p}} \tag{9.4.45}$$

can be expressed in terms of the stress rate or the rate of deformation as

$$\overset{\circ}{\underline{\boldsymbol{\alpha}}} = \frac{(\boldsymbol{\sigma}' - \boldsymbol{\alpha}) \otimes (\boldsymbol{\sigma}' - \boldsymbol{\alpha})}{(\boldsymbol{\sigma}' - \boldsymbol{\alpha}) : (\boldsymbol{\sigma}' - \boldsymbol{\alpha})} : \overset{\circ}{\underline{\boldsymbol{\tau}}}, \quad (\boldsymbol{\sigma}' - \boldsymbol{\alpha}) : \overset{\circ}{\underline{\boldsymbol{\tau}}} > 0, \tag{9.4.46}$$

$$\overset{\circ}{\underline{\boldsymbol{\alpha}}} = \frac{2h_{\mathrm{t}}^{\mathrm{p}}}{1 + h_{\mathrm{t}}^{\mathrm{p}}/\mu} \frac{(\boldsymbol{\sigma}' - \boldsymbol{\alpha}) \otimes (\boldsymbol{\sigma}' - \boldsymbol{\alpha})}{(\boldsymbol{\sigma}' - \boldsymbol{\alpha}) : (\boldsymbol{\sigma}' - \boldsymbol{\alpha})} : \mathbf{D}, \quad (\boldsymbol{\sigma}' - \boldsymbol{\alpha}) : \mathbf{D} > 0. \tag{9.4.47}$$

A nonlinear kinematic hardening model of Armstrong and Frederick (1966) is obtained if $\mathbf{C}$ in Eq. (9.4.33) is taken to be proportional to $\boldsymbol{\alpha}$,

$$\mathbf{C} = -c_0\, \boldsymbol{\alpha}, \tag{9.4.48}$$

where c_0 is a constant material parameter. In this case

$$\overset{\circ}{\underline{\boldsymbol{\alpha}}} = 2h\, \mathbf{D}^{\mathrm{p}} - c_0\, \boldsymbol{\alpha}\, (\mathbf{D}^{\mathrm{p}} : \mathbf{D}^{\mathrm{p}})^{1/2}, \tag{9.4.49}$$

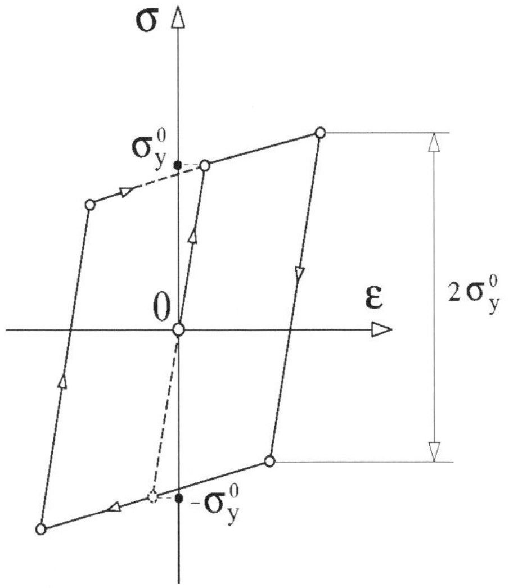

FIGURE 9.6. One-dimensional stress-strain response according to linear kinematic hardening model.

with h as another material parameter. The added nonlinear term in Eq. (9.4.49), referred to as a recall term, gives rise to hardening moduli for reversed plastic loading that are in better agreement with experimental data. It follows that

$$\mathbf{D}^{\mathrm{p}} = \frac{1}{2h(1-m)} \frac{(\boldsymbol{\sigma}' - \boldsymbol{\alpha}) \otimes (\boldsymbol{\sigma}' - \boldsymbol{\alpha})}{(\boldsymbol{\sigma}' - \boldsymbol{\alpha}) : (\boldsymbol{\sigma}' - \boldsymbol{\alpha})} : \overset{\circ}{\underline{\boldsymbol{\tau}}}, \tag{9.4.50}$$

where

$$m = \frac{c_0}{2\,h} \frac{(\boldsymbol{\sigma}' - \boldsymbol{\alpha}) : \boldsymbol{\alpha}}{[(\boldsymbol{\sigma}' - \boldsymbol{\alpha}) : (\boldsymbol{\sigma}' - \boldsymbol{\alpha})]^{1/2}}\,. \tag{9.4.51}$$

In modeling cyclic plasticity it may be convenient to additively decompose the back stress $\boldsymbol{\alpha}$ into two or more constituents, and construct separate evolution equation for each of these. For details, see Moosbrugger and McDowell (1989), Ohno and Wang (1993), and Jiang and Kurath (1996).

Ziegler (1959) used an evolution equation for back stress in the form

$$\overset{\circ}{\underline{\boldsymbol{\alpha}}} = \dot{\beta}\,(\boldsymbol{\sigma}' - \boldsymbol{\alpha}). \tag{9.4.52}$$

The proportionality factor $\dot{\beta}$ can be determined from the consistency condition in terms of $\boldsymbol{\sigma}$ and $\boldsymbol{\alpha}$ (Fig. 9.7). Detailed analysis is available in the book by Chakrabarty (1987). Duszek and Perzyna (1991) suggested an

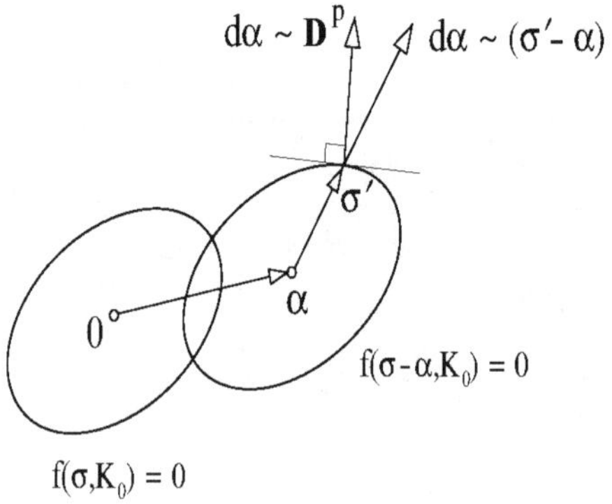

FIGURE 9.7. Translation of the yield surface according to kinematic hardening model. The center of the yield surface is the back stress $\boldsymbol{\alpha}$. Its evolution is governed by $\mathrm{d}\boldsymbol{\alpha} \sim \mathbf{D}^{\mathrm{p}}$ according to Prager's model, and by $\mathrm{d}\boldsymbol{\alpha} \sim (\boldsymbol{\sigma}' - \boldsymbol{\alpha})$ according to Ziegler's model.

evolution equation that is a linear combination of the Prager and Ziegler hardening rules. See also Ishlinsky (1954), Backhaus (1968, 1972), Eisenberg and Phillips (1968), and Lehmann (1972).

9.4.3. Combined Isotropic–Kinematic Hardening

In this hardening model the yield surface expands and translates during plastic deformation (Fig. 9.8), so that

$$f\,(\boldsymbol{\sigma}' - \boldsymbol{\alpha}, K_\alpha) = 0, \quad K_\alpha = K_\alpha(\vartheta). \tag{9.4.53}$$

The scalar function $K_\alpha(\vartheta)$, with ϑ defined by Eq. (9.4.3), specifies expansion of the yield surface, while (9.4.33) specifies its translation. The resulting constitutive equation for the plastic part of rate of deformation is

$$\mathbf{D}^{\mathrm{p}} = \dot{\gamma}\,\frac{\partial f}{\partial \boldsymbol{\sigma}}, \quad \dot{\gamma} = \frac{1}{H}\left(\frac{\partial f}{\partial \boldsymbol{\sigma}} : \overset{\circ}{\boldsymbol{\tau}}\right), \tag{9.4.54}$$

with

$$H = c\left(\frac{\partial f}{\partial \boldsymbol{\sigma}} : \frac{\partial f}{\partial \boldsymbol{\sigma}}\right) + \left(\frac{\partial f}{\partial \boldsymbol{\sigma}} : \frac{\partial f}{\partial \boldsymbol{\sigma}}\right)^{1/2}\left(\mathbf{C} : \frac{\partial f}{\partial \boldsymbol{\sigma}} - \sqrt{2}\,h^{\mathrm{p}}_{\alpha}\,\frac{\partial f}{\partial K_\alpha}\right). \tag{9.4.55}$$

The rate of the yield surface expansion is

$$h^{\mathrm{p}}_{\alpha} = \frac{\mathrm{d}K_\alpha}{\mathrm{d}\vartheta}. \tag{9.4.56}$$

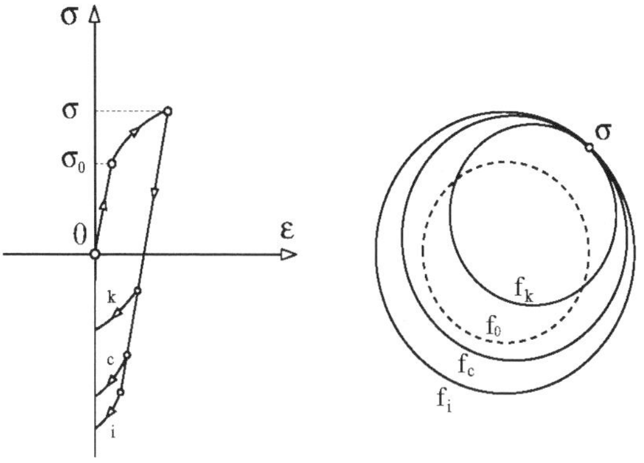

FIGURE 9.8. Geometric illustration of isotropic, kinematic and combined hardening. The initial yield surface (f_0) expands in the case of isotropic, translates in the case of kinematic (f_k), and expands and translates in the case of combined or mixed hardening (f_c).

If the yield surface is

$$\frac{1}{2} \left(\boldsymbol{\sigma}' - \boldsymbol{\alpha} \right) : \left(\boldsymbol{\sigma}' - \boldsymbol{\alpha} \right) = K_\alpha^2(\vartheta), \qquad (9.4.57)$$

where $\boldsymbol{\alpha}$ represents its current center, and $K_\alpha(\vartheta)$ its current radius, and if the evolution equation for back stress $\boldsymbol{\alpha}$ is given by Eq. (9.4.49), we obtain

$$\mathbf{D}^{\mathrm{p}} = \frac{1}{2h_\alpha^{\mathrm{p}} + 2h(1-m)} \frac{\left(\boldsymbol{\sigma}' - \boldsymbol{\alpha} \right) \otimes \left(\boldsymbol{\sigma}' - \boldsymbol{\alpha} \right)}{\left(\boldsymbol{\sigma}' - \boldsymbol{\alpha} \right) : \left(\boldsymbol{\sigma}' - \boldsymbol{\alpha} \right)} : \overset{\circ}{\boldsymbol{\tau}}. \qquad (9.4.58)$$

The parameter m is again specified by Eq. (9.4.51). This clearly encompasses the previously considered purely isotropic and kinematic hardening models. For purely kinematic hardening $h_\alpha^{\mathrm{p}} = 0$, and for purely isotropic hardening $h_\alpha^{\mathrm{p}} = h_{\mathrm{t}}^{\mathrm{p}}$ (plastic tangent modulus in simple shear).

For example, when $m = c_0 = 0$, and when the hardening is linear with the yield stress $K = K_0 + h_{\mathrm{t}}^{\mathrm{p}} \vartheta$, where $h_{\mathrm{t}}^{\mathrm{p}} = \text{const.}$, we can write

$$h = (1-r)h_{\mathrm{t}}^{\mathrm{p}}, \quad K_\alpha = K_0 + r h_{\mathrm{t}}^{\mathrm{p}} \vartheta, \qquad (9.4.59)$$

and $h_\alpha^{\mathrm{p}} = r h_{\mathrm{t}}^{\mathrm{p}}$. The parameter $0 \le r \le 1$ defines the amount of combined hardening. The value $r = 1$ corresponds to purely isotropic, and $r = 0$ to purely kinematic hardening. Equation (9.4.59) can be extended to the case of nonlinear hardening $K = K(\vartheta)$ by defining

$$h = (1-r)\frac{\mathrm{d}K}{\mathrm{d}\vartheta}, \quad K_\alpha = K_0 + r(K - K_0). \qquad (9.4.60)$$

Moreton, Moffat, and Parkinson (1981) observed large translations together with moderately small isotropic expansion and distortion of the yield surface in experiments with pressure vessel steels. Detailed description of the measured yield loci can be found in Naghdi, Essenburg, and Koff (1958), Bertsch and Findley (1962), Hecker (1976), Phillips and Lee (1979), Shiratori, Ikegami, and Yoshida (1979), Phillips and Das (1985), Stout, Martin, Helling, and Canova (1985), Wu, Lu, and Pan (1995), and Barlat *et al.* (1997).

9.4.4. Mróz Multisurface Model

More involved hardening models were suggested to better treat nonlinearities in stress-strain loops, cyclic hardening or softening, cyclic creep and stress relaxation. In order to describe nonlinear hardening and provide gradual transition from elastic to plastic deformation, Mróz (1967, 1976) introduced a multiyield surface model in which there is a field of hardening moduli, one for each yield surface. Initially the yield surfaces are assumed to be concentric (Fig. 9.9). When the stress point reaches the innermost surface $f_{<1>} = 0$, the plastic deformation develops according to linear hardening model with the plastic tangent modulus $h_{\mathrm{t}}^{\mathrm{p}}{}_{<1>}$, until the activated yield surface reaches the next surface $f_{<2>} = 0$. Subsequent plastic deformation develops according to linear hardening model with the plastic tangent modulus $h_{\mathrm{t}}^{\mathrm{p}}{}_{<2>}$, until the next surface is reached, etc. Suppose that pressure-independent yield surfaces are defined by

$$f_{<i>} = \frac{1}{2}(\boldsymbol{\sigma}' - \boldsymbol{\alpha}_{<i>}) : (\boldsymbol{\sigma}' - \boldsymbol{\alpha}_{<i>}) - K_{<i>}^2 = 0, \quad i = 1, 2, \cdots N. \tag{9.4.61}$$

The centers of the individual surfaces are $\boldsymbol{\alpha}_{<i>}$, and their sizes are specified by the constants $K_{<i>}$ (determined by fitting the nonlinear stress-strain curve in pure shear test). For simplicity, only translation of the yield surfaces is considered. To ascertain that two surfaces in contact have coincident outward normals, the active yield surface

$$f_{<i>} = 0 \tag{9.4.62}$$

translates in the direction of the stress difference $\boldsymbol{\sigma}'_{<i+1>} - \boldsymbol{\sigma}'$, where $\boldsymbol{\sigma}'$ is the current stress state on the yield surface $f_{<i>} = 0$, and $\boldsymbol{\sigma}'_{<i+1>}$ is the

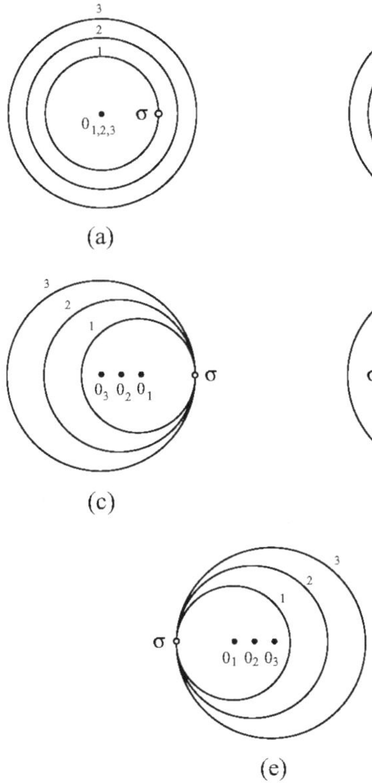

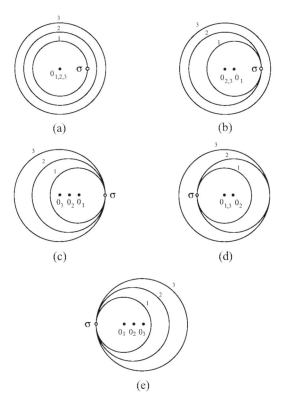

FIGURE 9.9. Illustration of the Mróz multisurface hardening model with the help of three initially concentric surfaces. Sequential translation of the surfaces are indicated corresponding to uniaxial monotonic loading in (b) and (c), and reversed loading in (d) and (e).

stress state on the subsequent yield surface

$$f_{<i+1>} = 0. \tag{9.4.63}$$

This stress state is defined by the requirement that the yield surface normals at $\boldsymbol{\sigma}'$ and $\boldsymbol{\sigma}'_{<i+1>}$ are parallel (Fig. 9.10). Thus, the evolution law for back stress is

$$\overset{\circ}{\underline{\boldsymbol{\alpha}}}_{<i>} = \dot{\beta}_{<i>} \left(\boldsymbol{\sigma}'_{<i+1>} - \boldsymbol{\sigma}'\right), \tag{9.4.64}$$

where

$$\frac{1}{K_{<i+1>}} \left(\boldsymbol{\sigma}'_{<i+1>} - \boldsymbol{\alpha}_{<i+1>}\right) = \frac{1}{K_{<i>}} \left(\boldsymbol{\sigma}' - \boldsymbol{\alpha}_{<i>}\right). \tag{9.4.65}$$

Inserting Eq. (9.4.65) into Eq. (9.4.64),

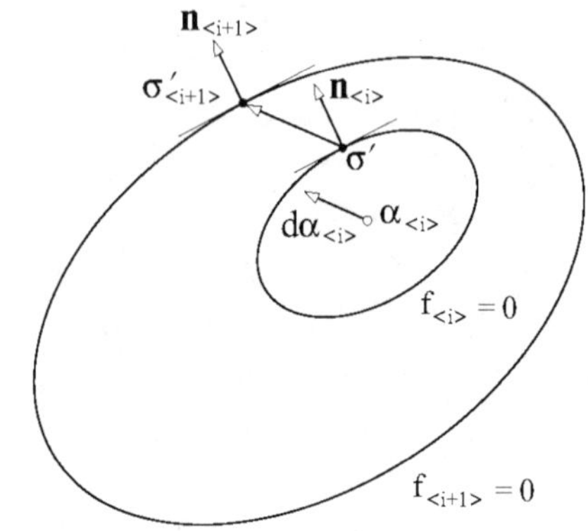

FIGURE 9.10. Translation of the surface $f_{<i>} = 0$ in Mróz's model is specified by $d\boldsymbol{\alpha}_{<i>} \sim (\boldsymbol{\sigma}'_{<i+1>} - \boldsymbol{\sigma}')$, where $\boldsymbol{\sigma}'_{<i+1>}$ is the stress state on the surface $f_{<i+1>} = 0$, with the normal $\mathbf{n}_{<i+1>}$ parallel to $\mathbf{n}_{<i>}$ at the state of stress $\boldsymbol{\sigma}'$ on the yield surface $f_{<i>} = 0$.

$$\overset{\circ}{\underline{\boldsymbol{\alpha}}}_{<i>} = \dot{\beta}_{<i>} \left[\frac{K_{<i+1>}}{K_{<i>}} (\boldsymbol{\sigma}' - \boldsymbol{\alpha}_{<i>}) - (\boldsymbol{\sigma}' - \boldsymbol{\alpha}_{<i+1>}) \right]. \tag{9.4.66}$$

The consistency condition

$$\dot{f}_{<i>} = 0 \tag{9.4.67}$$

gives

$$(\boldsymbol{\sigma}' - \boldsymbol{\alpha}_{<i>}) : \overset{\circ}{\underline{\boldsymbol{\tau}}} = (\boldsymbol{\sigma}' - \boldsymbol{\alpha}_{<i>}) : \overset{\circ}{\underline{\boldsymbol{\alpha}}}_i. \tag{9.4.68}$$

Combined with Eq. (9.4.66), this defines

$$\dot{\beta}_{<i>} = \frac{1}{2B_{<i>}} (\boldsymbol{\sigma}' - \boldsymbol{\alpha}_{<i>}) : \overset{\circ}{\underline{\boldsymbol{\tau}}}, \tag{9.4.69}$$

where

$$B_{<i>} = K_{<i>} K_{<i+1>} - \frac{1}{2} (\boldsymbol{\sigma}' - \boldsymbol{\alpha}_{<i>}) : (\boldsymbol{\sigma}' - \boldsymbol{\alpha}_{<i+1>}). \tag{9.4.70}$$

The plastic part of the rate of deformation tensor, during the loading between the active yield surface $f_{<i>} = 0$ and the nearby surface $f_{<i+1>} = 0$, is defined by the linear kinematic hardening law with the plastic tangent modulus $h^{\mathrm{p}}_{\mathrm{t}\,<i>}$. This gives, from Eq. (9.4.41),

$$\mathbf{D}^{\mathrm{p}} = \frac{1}{2h^{\mathrm{p}}_{\mathrm{t}\,<i>}} \left[\frac{(\boldsymbol{\sigma}' - \boldsymbol{\alpha}_{<i>}) \otimes (\boldsymbol{\sigma}' - \boldsymbol{\alpha}_{<i>})}{(\boldsymbol{\sigma}' - \boldsymbol{\alpha}_{<i>}) : (\boldsymbol{\sigma}' - \boldsymbol{\alpha}_{<i>})} \right] : \overset{\circ}{\underline{\boldsymbol{\tau}}}. \tag{9.4.71}$$

Further details, including the incorporation of isotropic component of hardening and determination of material parameters, can be found in cited Mróz's

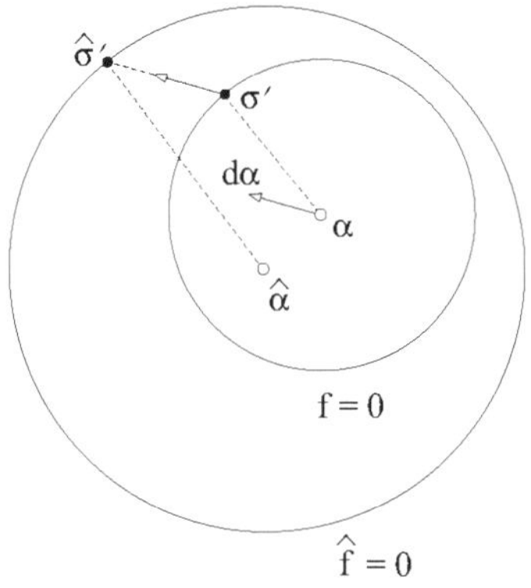

FIGURE 9.11. Schematic representation of the loading and bounding surface in the two-surface hardening model. The loading surface translates toward the bounding surface in the direction $\hat{\boldsymbol{\sigma}}' - \boldsymbol{\sigma}'$.

papers. See also Iwan (1967), Desai and Siriwardane (1984), Khan and Huang (1995), and Jiang and Sehitoglu (1996a,b).

9.4.5. Two-Surface Model

Dafalias and Popov (1975,1976), and Krieg (1975) suggested the hardening model which uses the yield (loading) surface and the limit (bounding) surface (Fig. 9.11). A smooth transition from elastic to plastic regions on loading is assured by introducing a continuous variation of the plastic tangent modulus between the two surfaces, i.e.,

$$h_{\mathrm{t}}^{\mathrm{p}} = h_{\mathrm{t}}^{\mathrm{p}}(\eta, \vartheta). \tag{9.4.72}$$

The scalar

$$\eta = \left[(\hat{\boldsymbol{\sigma}}' - \boldsymbol{\sigma}') : (\hat{\boldsymbol{\sigma}}' - \boldsymbol{\sigma}')\right]^{1/2} \tag{9.4.73}$$

is a measure of the distance between the current stress state $\boldsymbol{\sigma}'$ on the loading surface, and the corresponding, appropriately defined state of stress $\hat{\boldsymbol{\sigma}}'$ on the bounding surface. Only deviatoric parts of stress are used for pressure-independent plasticity. Suppose that the loading surface can translate and expand, such that

$$f = \frac{1}{2}(\boldsymbol{\sigma}' - \boldsymbol{\alpha}) : (\boldsymbol{\sigma}' - \boldsymbol{\alpha}) - K_{\alpha}^2(\vartheta) = 0, \quad \vartheta = \int_0^t (2\,\mathbf{D}^{\mathrm{p}} : \mathbf{D}^{\mathrm{p}})^{1/2}\,\mathrm{d}t. \tag{9.4.74}$$

The bounding surface is assumed to only translate, i.e.,

$$\hat{f} = \frac{1}{2}(\hat{\boldsymbol{\sigma}}' - \hat{\boldsymbol{\alpha}}) : (\hat{\boldsymbol{\sigma}}' - \hat{\boldsymbol{\alpha}}) - \hat{K}^2 = 0, \quad \hat{K} = \text{const.} \tag{9.4.75}$$

Translation of the loading surface is defined as in the Mróz's model, and it is in the direction of the stress difference $\hat{\boldsymbol{\sigma}}' - \boldsymbol{\sigma}'$. The current stress state on the loading surface $f = 0$ is $\boldsymbol{\sigma}'$, while $\hat{\boldsymbol{\sigma}}'$ is the stress state on the bounding surface $\hat{f} = 0$, where the surface normal is parallel to the loading surface normal at $\boldsymbol{\sigma}'$. Thus, the evolution law for back stress $\boldsymbol{\alpha}$ is

$$\overset{\circ}{\underline{\boldsymbol{\alpha}}} = \dot{\beta}\,(\hat{\boldsymbol{\sigma}}' - \boldsymbol{\sigma}') = \dot{\beta}\left[\frac{\hat{K}}{K_\alpha}(\boldsymbol{\sigma}' - \boldsymbol{\alpha}) - (\boldsymbol{\sigma}' - \hat{\boldsymbol{\alpha}})\right]. \tag{9.4.76}$$

The translation of the bounding surface is governed by a linear kinematic hardening rule

$$\overset{\circ}{\underline{\hat{\boldsymbol{\alpha}}}} = 2\hat{h}_{\mathrm{t}}^{\mathrm{p}}\,\mathbf{D}^{\mathrm{p}}, \tag{9.4.77}$$

where $\hat{h}_{\mathrm{t}}^{\mathrm{p}}$ is the corresponding, constant plastic tangent modulus. The plastic part of the rate of deformation tensor is taken to be

$$\mathbf{D}^{\mathrm{p}} = \frac{1}{2h_{\mathrm{t}}^{\mathrm{p}}}\left[\frac{(\boldsymbol{\sigma}' - \boldsymbol{\alpha}) \otimes (\boldsymbol{\sigma}' - \boldsymbol{\alpha})}{(\boldsymbol{\sigma}' - \boldsymbol{\alpha}) : (\boldsymbol{\sigma}' - \boldsymbol{\alpha})}\right] : \overset{\circ}{\underline{\boldsymbol{\tau}}}. \tag{9.4.78}$$

The consistency condition for the loading surface gives

$$(\boldsymbol{\sigma}' - \boldsymbol{\alpha}) : (\overset{\circ}{\underline{\boldsymbol{\tau}}} - \overset{\circ}{\underline{\boldsymbol{\alpha}}}) - 2\,h_\alpha^{\mathrm{p}}\,K_\alpha\,(2\,\mathbf{D}^{\mathrm{p}} : \mathbf{D}^{\mathrm{p}})^{1/2} = 0, \tag{9.4.79}$$

with $h_\alpha^{\mathrm{p}} = \mathrm{d}K_\alpha/\mathrm{d}\vartheta$. In view of Eq. (9.4.76) and (9.4.78), Eq. (9.4.79) defines

$$\dot{\beta} = \frac{1}{2B}\left(1 - \frac{h_\alpha^{\mathrm{p}}}{h_{\mathrm{t}}^{\mathrm{p}}}\right)(\boldsymbol{\sigma}' - \boldsymbol{\alpha}) : \overset{\circ}{\underline{\boldsymbol{\tau}}}, \tag{9.4.80}$$

where

$$B = K_\alpha \hat{K} - \frac{1}{2}(\boldsymbol{\sigma}' - \boldsymbol{\alpha}) : (\boldsymbol{\sigma}' - \hat{\boldsymbol{\alpha}}). \tag{9.4.81}$$

Finally, the consistency condition for the bounding surface,

$$(\hat{\boldsymbol{\sigma}}' - \hat{\boldsymbol{\alpha}}) : (\overset{\circ}{\underline{\hat{\boldsymbol{\tau}}}} - \overset{\circ}{\underline{\hat{\boldsymbol{\alpha}}}}) = 0, \tag{9.4.82}$$

specifies the stress rate $\overset{\circ}{\underline{\hat{\boldsymbol{\tau}}}}$ on the bounding surface that corresponds to a prescribed stress rate $\overset{\circ}{\underline{\boldsymbol{\tau}}}$ on the loading surface. Upon substitution of Eqs. (9.4.77) and (9.4.78) into Eq. (9.4.82), there follows

$$(\hat{\boldsymbol{\sigma}}' - \hat{\boldsymbol{\alpha}}) : \overset{\circ}{\underline{\hat{\boldsymbol{\tau}}}} = \frac{\hat{K}}{K_\alpha}\frac{\hat{h}_{\mathrm{t}}^{\mathrm{p}}}{h_{\mathrm{t}}^{\mathrm{p}}}(\boldsymbol{\sigma}' - \boldsymbol{\alpha}) : \overset{\circ}{\underline{\boldsymbol{\tau}}}. \tag{9.4.83}$$

The Mróz's assumption (9.4.65) was utilized, so that

$$\hat{\boldsymbol{\sigma}}' - \hat{\boldsymbol{\alpha}} = \frac{\hat{K}}{K_\alpha} (\boldsymbol{\sigma}' - \boldsymbol{\alpha}). \tag{9.4.84}$$

Further analysis, including the incorporation of isotropic component of hardening for the bounding surface, and the specification of material parameters, can be found in the cited papers. See also McDowell (1985,1987), Chaboche (1986), Hashiguchi (1981, 1988), and Ellyin (1989) for the generalization of the model and discussion of its performance. There has also been a study of cyclic hardening and softening using continuously evolving parameters and only one yield surface, presented by Haupt and Kamlah (1995), and Ristinmaa (1995). The papers by Caulk and Naghdi (1978), Drucker and Palgen (1981), and Naghdi and Nikkel (1986) address the modeling of saturation hardening under cyclic loading, and related problems.

9.5. Yield Surface with Vertex in Strain Space

Suppose that the yield surface in strain space (Fig. 9.12) has a pyramidal vertex, formed by k_0 intersecting segments (hyperplanes) such that, near the vertex,

$$\prod_{i=1}^{k_0} g_{(n)}^{<i>} \left(\mathbf{E}_{(n)}, \mathcal{H}\right) = 0, \quad k_0 \geq 2. \tag{9.5.1}$$

If the material obeys Ilyushin's postulate, from (8.5.10) it follows that $\mathrm{d}^{\mathrm{p}}\mathbf{T}_{(n)}$ lies within the cone of limiting inward normals to active segments of the yield vertex, i.e.,

$$\mathrm{d}^{\mathrm{p}}\mathbf{T}_{(n)} = -\sum_{i=1}^{k} \mathrm{d}\gamma_{(n)}^{<i>} \frac{\partial g_{(n)}^{<i>}}{\partial \mathbf{E}_{(n)}}, \quad \mathrm{d}\gamma_{(n)}^{<i>} > 0. \tag{9.5.2}$$

Thus,

$$\mathrm{d}\mathbf{T}_{(n)} = \boldsymbol{\Lambda}_{(n)} : \mathrm{d}\mathbf{E}_{(n)} - \sum_{i=1}^{k} \mathrm{d}\gamma_{(n)}^{<i>} \frac{\partial g_{(n)}^{<i>}}{\partial \mathbf{E}_{(n)}}, \tag{9.5.3}$$

where k is the number of active vertex segments ($\mathrm{d}\gamma_{(n)}^{<i>} = 0$ for $k < i \leq k_0$). If the strain rate is in a fully active range, so that plastic loading takes place with respect to all vertex segments, we have $k = k_0$. The scalars $\mathrm{d}\gamma_{(n)}^{<i>}$ depend on the current values of $\mathbf{E}_{(n)}$, $\mathcal{H}$, and their increments. The

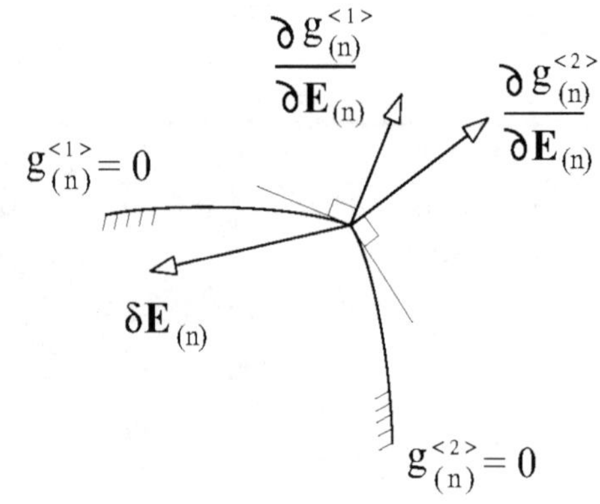

FIGURE 9.12. Yield surface vertex in strain space. Elastic strain increment $\delta\mathbf{E}_{(n)}$ is directed along or inside the vertex segments.

consistency condition for each active vertex segment is

$$\frac{\partial g_{(n)}^{<i>}}{\partial \mathbf{E}_{(n)}} : \mathrm{d}\mathbf{E}_{(n)} + \mathrm{d}^{\mathrm{p}} g_{(n)}^{<i>} = 0, \quad \mathrm{d}\gamma_{(n)}^{<i>} > 0, \tag{9.5.4}$$

where

$$\mathrm{d}^{\mathrm{p}} g_{(n)}^{<i>} = g_{(n)}^{<i>}\left(\mathbf{E}_{(n)}, \mathcal{H} + d\mathcal{H}\right) - g_{(n)}^{<i>}\left(\mathbf{E}_{(n)}, \mathcal{H}\right). \tag{9.5.5}$$

In the case when the vertex segment is not active,

$$\frac{\partial g_{(n)}^{<i>}}{\partial \mathbf{E}_{(n)}} : \mathrm{d}\mathbf{E}_{(n)} + \mathrm{d}^{\mathrm{p}} g_{(n)}^{<i>} \leq 0, \quad \mathrm{d}\gamma_{(n)}^{<i>} = 0. \tag{9.5.6}$$

It is assumed that the vertex segment can harden even if it is inactive, due to cross or latent hardening produced by the ongoing plastic deformation from the neighboring active vertex segments. Equality sign in (9.5.6) applies if the yield state remains on the intersection of active and inactive vertex segments.

Suppose that

$$\mathrm{d}^{\mathrm{p}} g_{(n)}^{<i>} = -\sum_{j=1}^{k} h_{(n)}^{<ij>}\, \mathrm{d}\gamma_{(n)}^{<j>} < 0, \tag{9.5.7}$$

where $h_{(n)}^{<ij>}$ are plastic moduli, in general nonsymmetric and dependent on the current plastic state. The quantity in (9.5.7) is negative because of (9.5.4), and because the scalar product of the increment of elastoplastic strain and the outer normal to any active yield segment at the vertex is positive,

$$\frac{\partial g_{(n)}^{<i>}}{\partial \mathbf{E}_{(n)}} : \mathrm{d}\mathbf{E}_{(n)} > 0, \quad i = 1, 2, \ldots, k. \tag{9.5.8}$$

Substitution into (9.5.4) and (9.5.6) gives

$$\frac{\partial g_{(n)}^{<i>}}{\partial \mathbf{E}_{(n)}} : \mathrm{d}\mathbf{E}_{(n)} = \sum_{j=1}^{k} h_{(n)}^{<ij>} \, \mathrm{d}\gamma_{(n)}^{<j>}, \quad \mathrm{d}\gamma_{(n)}^{<i>} > 0, \tag{9.5.9}$$

$$\frac{\partial g_{(n)}^{<i>}}{\partial \mathbf{E}_{(n)}} : \mathrm{d}\mathbf{E}_{(n)} \leq \sum_{j=1}^{k} h_{(n)}^{<ij>} \, \mathrm{d}\gamma_{(n)}^{<j>}, \quad \mathrm{d}\gamma_{(n)}^{<i>} = 0. \tag{9.5.10}$$

If the matrix of plastic moduli $h_{(n)}^{<ij>}$ is positive definite (thus, nonsingular), (9.5.9) gives a unique set of values

$$\mathrm{d}\gamma_{(n)}^{<i>} = \sum_{j=1}^{k} h_{(n)}^{<ij>-1} \, \frac{\partial g_{(n)}^{<j>}}{\partial \mathbf{E}_{(n)}} : \mathrm{d}\mathbf{E}_{(n)}, \tag{9.5.11}$$

for a prescribed strain increment $\mathrm{d}\mathbf{E}_{(n)}$. Elements of the matrix inverse to plastic moduli matrix $h_{(n)}^{<ij>}$ are denoted by $h_{(n)}^{<ij>-1}$. The substitution of Eq. (9.5.11) into Eq. (9.5.3) then gives

$$\dot{\mathbf{T}}_{(n)} = \left(\boldsymbol{\Lambda}_{(n)} - \sum_{i=1}^{k} \sum_{j=1}^{k} h_{(n)}^{<ij>-1} \, \frac{\partial g_{(n)}^{<i>}}{\partial \mathbf{E}_{(n)}} \otimes \frac{\partial g_{(n)}^{<j>}}{\partial \mathbf{E}_{(n)}} \right) : \dot{\mathbf{E}}_{(n)}. \tag{9.5.12}$$

This extends the constitutive structure (9.1.11) with a smooth yield surface in strain space to the case when the yield surface has a vertex.

The trace product of (9.5.2) with $\mathrm{d}\mathbf{E}_{(n)}$ yields, upon substitution of (9.5.9),

$$\dot{\mathbf{T}}_{(n)}^{\mathrm{p}} : \dot{\mathbf{E}}_{(n)} = - \sum_{i=1}^{k} \sum_{j=1}^{k} h_{(n)}^{<ij>} \, \dot{\gamma}_{(n)}^{<i>} \dot{\gamma}_{(n)}^{<j>}. \tag{9.5.13}$$

For positive definite matrix of plastic moduli this is clearly negative, in accord with (8.5.10) and Ilyushin's postulate. On the other hand, elastic increments from the yield state at the vertex are directed inside the yield surface and, thus, satisfy a set of k_0 inequalities

$$\frac{\partial g_{(n)}^{<i>}}{\partial \mathbf{E}_{(n)}} : \dot{\mathbf{E}}_{(n)} \leq 0, \quad i = 1, 2, \ldots, k_0. \tag{9.5.14}$$

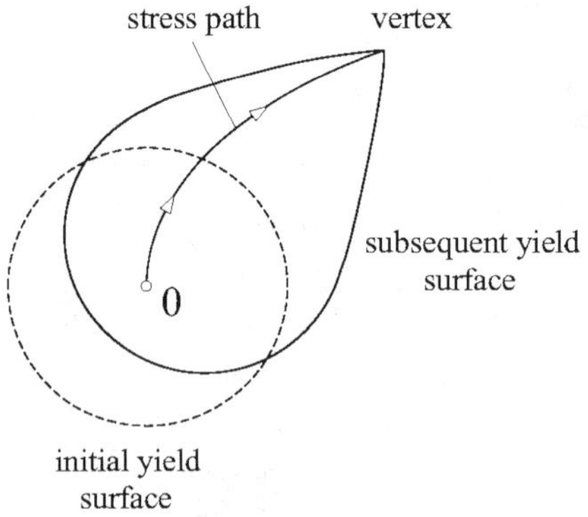

FIGURE 9.13. Development of the vertex at the loading point of the subsequent yield surface.

9.6. Yield Surface with Vertex in Stress Space

Physical theories of plasticity (Batdorf and Budiansky, 1949, 1954; Sanders, 1954; Hill, 1966, 1967b; Hutchinson, 1970) imply formation of the corner or vertex at the loading point of the yield surface (Fig. 9.13). Although sharp corners are seldom seen experimentally, the yield surfaces with relatively high curvature at the loading point are often observed (Hecker, 1972, 1976; Naghdi, 1990). Suppose then that the yield surface in stress space has a pyramidal vertex (Fig. 9.14), formed by k_0 intersecting segments such that, near the vertex,

$$\prod_{i=1}^{k_0} f_{(n)}^{<i>}\left(\mathbf{T}_{(n)}, \mathcal{H}\right) = 0, \quad k_0 \geq 2. \tag{9.6.1}$$

If the material obeys Ilyushin's postulate, from the analysis in Subsection 8.5.1 it follows that $\mathrm{d}^{\mathrm{p}}\mathbf{E}_{(n)}$ lies within the cone of limiting outward normals to active segments of the yield vertex, so that

$$\mathrm{d}^{\mathrm{p}}\mathbf{E}_{(n)} = \sum_{i=1}^{k} \mathrm{d}\gamma_{(n)}^{<i>} \frac{\partial f_{(n)}^{<i>}}{\partial \mathbf{T}_{(n)}}, \quad \mathrm{d}\gamma_{(n)}^{<i>} > 0, \tag{9.6.2}$$

and

$$\mathrm{d}\mathbf{E}_{(n)} = \mathbf{M}_{(n)} : \mathrm{d}\mathbf{T}_{(n)} + \sum_{i=1}^{k} \mathrm{d}\gamma_{(n)}^{<i>} \frac{\partial f_{(n)}^{<i>}}{\partial \mathbf{T}_{(n)}}. \tag{9.6.3}$$

It is assumed that plastic loading is taking place through k active vertex segments. If the stress rate is in a fully active range, so that plastic loading takes place with respect to all vertex segments, $k = k_0$. (Specification

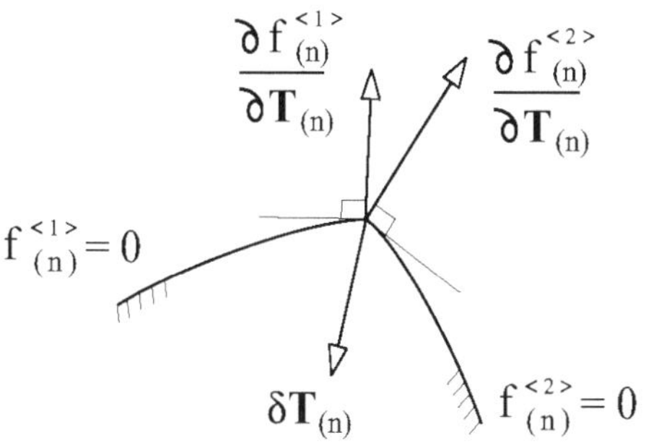

FIGURE 9.14. Yield surface vertex in stress space. Elastic stress increment $\delta\mathbf{T}_{(n)}$ is directed along or inside the vertex segments.

of fully active range and dissection of the stress rate space into pyramidal regions of partially active range is discussed, in the context of crystal plasticity, in Section 12.13). The scalars $\mathrm{d}\gamma^{<i>}_{(n)}$ depend on the current values of $\mathbf{T}_{(n)}$, $\mathcal{H}$, and their increments. The consistency condition for each active vertex segment is

$$\frac{\partial f^{<i>}_{(n)}}{\partial \mathbf{T}_{(n)}} : \mathrm{d}\mathbf{T}_{(n)} + \mathrm{d}^{\mathrm{p}} f^{<i>}_{(n)} = 0, \quad \mathrm{d}\gamma^{<i>}_{(n)} > 0, \tag{9.6.4}$$

where

$$\mathrm{d}^{\mathrm{p}} f^{<i>}_{(n)} = f^{<i>}_{(n)}\left(\mathbf{T}_{(n)}, \mathcal{H} + d\mathcal{H}\right) - f^{<i>}_{(n)}\left(\mathbf{T}_{(n)}, \mathcal{H}\right). \tag{9.6.5}$$

If the vertex segment is not active,

$$\frac{\partial f^{<i>}_{(n)}}{\partial \mathbf{T}_{(n)}} : \mathrm{d}\mathbf{T}_{(n)} + \mathrm{d}^{\mathrm{p}} f^{<i>}_{(n)} \le 0, \quad \mathrm{d}\gamma^{<i>}_{(n)} = 0. \tag{9.6.6}$$

Consistent with the analysis of the yield vertex in strain space, it is assumed that the vertex segment can harden even if it is inactive, due to cross or latent hardening produced by ongoing plastic deformation associated with the neighboring active vertex segments. Equality sign in (9.6.6) applies if the yield state remains on the intersection of active and inactive vertex segments.

Suppose that

$$\mathrm{d}^{\mathrm{p}} f^{<i>}_{(n)} = -\sum_{j=1}^{k} H^{<ij>}_{(n)}\, \mathrm{d}\gamma^{<j>}_{(n)} < 0, \tag{9.6.7}$$

where $H^{<ij>}_{(n)}$ are plastic moduli, in general nonsymmetric and dependent on the current plastic state. Substitution into (9.6.4) and (9.6.6) gives

$$\frac{\partial f^{<i>}_{(n)}}{\partial \mathbf{T}_{(n)}} : \mathrm{d}\mathbf{T}_{(n)} = \sum_{j=1}^{k} H^{<ij>}_{(n)}\, \mathrm{d}\gamma^{<j>}_{(n)}, \quad \mathrm{d}\gamma^{<i>}_{(n)} > 0, \tag{9.6.8}$$

$$\frac{\partial f^{<i>}_{(n)}}{\partial \mathbf{T}_{(n)}} : \mathrm{d}\mathbf{T}_{(n)} \leq \sum_{j=1}^{k} H^{<ij>}_{(n)}\, \mathrm{d}\gamma^{<j>}_{(n)}, \quad \mathrm{d}\gamma^{<i>}_{(n)} = 0. \tag{9.6.9}$$

The relationship between the moduli $H^{<ij>}_{(n)}$ and $h^{<ij>}_{(n)}$ can be derived by recalling that

$$f^{<i>}_{(n)}\left(\mathbf{T}_{(n)}, \mathcal{H}\right) = f^{<i>}_{(n)}\left[\mathbf{T}_{(n)}\left(\mathbf{E}_{(n)}, \mathcal{H}\right), \mathcal{H}\right] = g^{<i>}_{(n)}\left(\mathbf{E}_{(n)}, \mathcal{H}\right). \tag{9.6.10}$$

Thus,

$$\begin{aligned}
\mathrm{d}^{\mathrm{p}} g^{<i>}_{(n)} &= g^{<i>}_{(n)}\left(\mathbf{E}_{(n)}, \mathcal{H} + d\mathcal{H}\right) - g^{<i>}_{(n)}\left(\mathbf{E}_{(n)}, \mathcal{H}\right) \\
&= f^{<i>}_{(n)}\left[\mathbf{T}_{(n)}\left(\mathbf{E}_{(n)}, \mathcal{H} + d\mathcal{H}\right), \mathcal{H} + d\mathcal{H}\right] \\
&- f^{<i>}_{(n)}\left[\mathbf{T}_{(n)}\left(\mathbf{E}_{(n)}, \mathcal{H}\right), \mathcal{H}\right] \\
&= f^{<i>}_{(n)}\left[\mathbf{T}_{(n)}\left(\mathbf{E}_{(n)}, \mathcal{H}\right) + \mathrm{d}^{\mathrm{p}}\mathbf{T}_{(n)}, \mathcal{H} + d\mathcal{H}\right] \\
&- f^{<i>}_{(n)}\left[\mathbf{T}_{(n)}\left(\mathbf{E}_{(n)}, \mathcal{H}\right), \mathcal{H}\right],
\end{aligned} \tag{9.6.11}$$

which gives

$$\mathrm{d}^{\mathrm{p}} g^{<i>}_{(n)} = \mathrm{d}^{\mathrm{p}} f^{<i>}_{(n)} + \frac{\partial f^{<i>}_{(n)}}{\partial \mathbf{T}_{(n)}} : \mathrm{d}^{\mathrm{p}}\mathbf{T}_{(n)}. \tag{9.6.12}$$

Upon substitution of (9.5.2), (9.5.7), and (9.6.7) into Eq. (9.6.12), there follows

$$H^{<ij>}_{(n)} = h^{<ij>}_{(n)} - \frac{\partial f^{<i>}_{(n)}}{\partial \mathbf{T}_{(n)}} : \frac{\partial g^{<j>}_{(n)}}{\partial \mathbf{E}_{(n)}}. \tag{9.6.13}$$

Since

$$\frac{\partial f^{<i>}_{(n)}}{\partial \mathbf{T}_{(n)}} = \mathbf{M}_{(n)} : \frac{\partial g^{<j>}_{(n)}}{\partial \mathbf{E}_{(n)}}, \tag{9.6.14}$$

the differences of plastic moduli $H^{<ij>}_{(n)} - h^{<ij>}_{(n)}$ form a symmetric matrix, provided that the elastic moduli tensor $\mathbf{M}_{(n)}$ obeys the reciprocal symmetry.

If the matrix of plastic moduli $H^{<ij>}_{(n)}$ is nonsingular, inversion of (9.6.8) gives

$$\mathrm{d}\gamma^{<i>}_{(n)} = \sum_{j=1}^{k} H^{<ij>-1}_{(n)} \frac{\partial f^{<j>}_{(n)}}{\partial \mathbf{T}_{(n)}} : \mathrm{d}\mathbf{T}_{(n)}, \tag{9.6.15}$$

for a prescribed stress increment $d\mathbf{T}_{(n)}$. Elements of the matrix inverse to plastic moduli matrix $H_{(n)}^{<ij>}$ are denoted by $H_{(n)}^{<ij>-1}$. The substitution of Eq. (9.6.15) into Eq. (9.6.3) gives

$$\dot{\mathbf{E}}_{(n)} = \left(\mathbf{M}_{(n)} + \sum_{i=1}^{k} \sum_{j=1}^{k} H_{(n)}^{<ij>-1} \frac{\partial f_{(n)}^{<i>}}{\partial \mathbf{T}_{(n)}} \otimes \frac{\partial f_{(n)}^{<j>}}{\partial \mathbf{T}_{(n)}} \right) : \dot{\mathbf{T}}_{(n)}. \tag{9.6.16}$$

This extends the constitutive structure (9.2.4) with a smooth yield surface in stress space to the case when the yield surface has a vertex.

Upon substitution of (9.6.8), the trace product of (9.6.2) with $d\mathbf{T}_{(n)}$ yields

$$\dot{\mathbf{T}}_{(n)} : \dot{\mathbf{E}}_{(n)}^{\mathrm{p}} = \sum_{i=1}^{k} \sum_{j=1}^{k} H_{(n)}^{<ij>} \dot{\gamma}_{(n)}^{<i>} \dot{\gamma}_{(n)}^{<j>}. \tag{9.6.17}$$

In the hardening range the plastic moduli $H_{(n)}^{<ij>}$ form a positive definite matrix, so that the quantity in (9.6.17) is positive. In this case, for a prescribed rate of stress $\dot{\mathbf{T}}_{(n)}$, the plastic response is unique and given by (9.6.16). In the softening range the quantity in (9.6.17) is negative. For a prescribed rate of stress, either plastic response given by (9.6.16) applies, or elastic response $\dot{\mathbf{E}}_{(n)} = \mathbf{M}_{(n)} : \dot{\mathbf{T}}_{(n)}$ takes place. In the case of ideal plasticity (vanishing self and latent hardening rates), $\dot{\gamma}_{(n)}^{<i>}$ in Eq. (9.6.2) are indeterminate by the constitutive analysis.

Elastic increments from the yield state at the vertex are always directed inside the yield surface and thus satisfy a set of k_0 inequalities

$$\frac{\partial f_{(n)}^{<i>}}{\partial \mathbf{T}_{(n)}} : \dot{\mathbf{T}}_{(n)} \leq 0, \quad i = 1, 2, \ldots, k_0, \tag{9.6.18}$$

which are dual to (9.5.14).

The papers by Koiter (1953), Mandel (1965), Sewell (1974), Hill (1978), and Ottosen and Ristinmaa (1996) offer further analysis of the plasticity theory with yield corners or vertices.

9.7. Pressure-Dependent Plasticity

For porous metals, concrete and geomaterials like soils and rocks, plastic deformation has its origin in pressure dependent microscopic processes. The corresponding yield condition depends on both deviatoric and hydrostatic

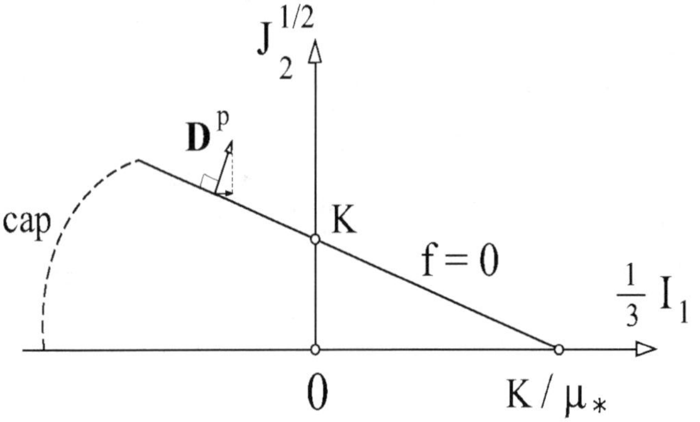

FIGURE 9.15. The Drucker–Prager yield condition shown in the coordinates of stress invariants I_1 and J_2. The yield stress in pure shear is K, and the frictional parameter is μ_*. The horizontal projection of the plastic rate of deformation indicates plastic dilatation according to normality and associative flow rule. At high pressure a cap is used to close the cone.

parts of the stress tensor. Constitutive modeling of such materials is the concern of this section.

9.7.1. Drucker–Prager Condition for Geomaterials

Drucker and Prager (1952) suggested that the yielding in soils occurs when the shear stress on octahedral planes overcomes cohesive and frictional resistance to sliding on these planes, i.e., when

$$\tau_{\text{oct}} = \tau_{\text{frict}} + \sqrt{\frac{2}{3}}\, K, \tag{9.7.1}$$

where

$$\tau_{\text{oct}} = \left(\frac{2}{3}\, J_2\right)^{1/2}, \quad \tau_{\text{frict}} = -\mu^* \sigma_{\text{oct}} = -\frac{1}{3}\, \mu^* I_1. \tag{9.7.2}$$

The coefficient of internal friction (material parameter) is μ^*. The first invariant of the Cauchy stress tensor is I_1, and J_2 is the second invariant of deviatoric part of the Cauchy stress,

$$I_1 = \operatorname{tr} \boldsymbol{\sigma}, \quad J_2 = \frac{1}{2}\, \boldsymbol{\sigma}' : \boldsymbol{\sigma}'. \tag{9.7.3}$$

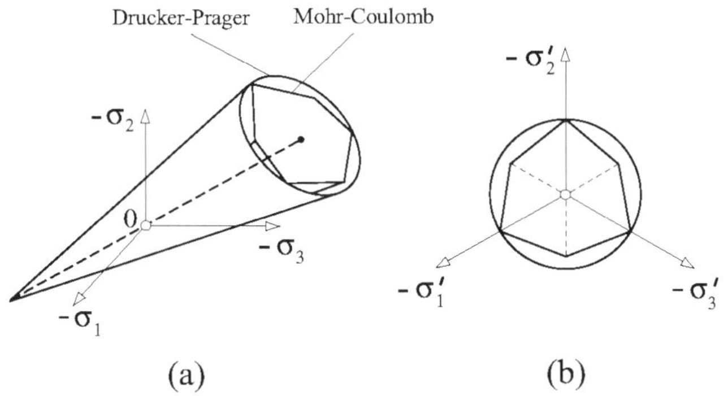

FIGURE 9.16. The Drucker–Prager cone and the Mohr–Coulomb pyramid matched along the compressive meridian, shown in (a) principal stress space, and (b) deviatoric plane.

The yield condition is consequently

$$f = J_2^{1/2} + \frac{1}{3}\,\mu_* I_1 - K = 0, \tag{9.7.4}$$

where the parameter

$$\mu_* = \sqrt{\frac{3}{2}}\,\mu^* \tag{9.7.5}$$

is conveniently introduced (Fig. 9.15). This geometrically represents a cone in the principal stress space with its axis parallel to the hydrostatic axis (Fig. 9.16). The radius of the circle in the deviatoric (π) plane is $\sqrt{2}\,K$, where K is the yield stress in simple shear. The angle of the cone is $\tan^{-1}(\sqrt{2}\,\mu_*/3)$. The yield stresses in uniaxial tension and compression are, according to Eq. (9.7.4),

$$Y^+ = \frac{\sqrt{3}\,K}{1+\mu_*/\sqrt{3}}, \quad Y^- = \frac{\sqrt{3}\,K}{1-\mu_*/\sqrt{3}}. \tag{9.7.6}$$

For the yield condition to be physically meaningful, the restriction must hold

$$\mu_* < \sqrt{3}. \tag{9.7.7}$$

If the compressive states of stress are considered positive (as commonly done in geomechanics, e.g., Jaeger and Cook, 1976; Salençon, 1977), a minus sign appears in front of the second term of f in Eq. (9.7.4). For the effects of the third stress invariant on plastic deformation of pressure sensitive materials, see Bardet (1990) and the references therein. The second and third deviatoric stress invariants define the Lode angle θ by (e.g., Chen and Han, 1988)

$$\cos(3\theta) = \left(\frac{27J_3^2}{4J_2^3}\right)^{1/2}. \tag{9.7.8}$$

When the Drucker–Prager cone is applied to porous rocks, it overestimates the yield stress at higher pressures, and inadequately predicts inelastic volume changes. To circumvent the former, DiMaggio and Sandler (1971) introduced an ellipsoidal cap to close the cone at certain level of pressure. Other shapes of the cap were also used. Details can be found in Chen and Han (1988), and Lubarda, Mastilovic, and Knap (1996).

Constitutive analysis of inelastic response of concrete has been studied extensively. Representative references include Ortiz and Popov (1982), Ortiz (1985), Pietruszczak, Jiang, and Mirza (1988), Faruque and Chang (1990), Voyiadjis and Abu-Lebdeh (1994), Lubarda, Krajcinovic, and Mastilovic (1994), and Lade and Kim (1995). Pressure-dependent response of granular materials was modeled by Mehrabadi and Cowin (1981), Christoffersen, Mehrabadi, and Nemat-Nasser (1981), Dorris and Nemat-Nasser (1982), Anand (1983), Chandler (1985), Harris (1992), and others.

9.7.2. Gurson Yield Condition for Porous Metals

Based on a rigid-perfectly plastic analysis of spherically symmetric deformation around a spherical cavity, Gurson (1977) suggested a yield condition for porous metals in the form

$$f = J_2 + \frac{2}{3}\,vY_0^2 \cosh\left(\frac{I_1}{2Y_0}\right) - (1+v^2)\frac{Y_0^2}{3} = 0, \tag{9.7.9}$$

where v is the porosity (void/volume fraction), and $Y_0 = \text{const.}$ is the tensile yield stress of the matrix material (Fig. 9.17). Generalization to include hardening matrix material is also possible. The change in porosity during plastic deformation is given by the evolution equation

$$\dot{v} = (1-v)\,\mathrm{tr}\,\mathbf{D}^{\mathrm{p}}. \tag{9.7.10}$$

Other evolution equations, which take into account nucleation and growth of voids, have been considered (e.g., Tvergaard and Needleman, 1984). To improve its predictions and agreement with experimental data, Tvergaard (1982) introduced two additional material parameters in the structure of the Gurson yield criterion. Mear and Hutchinson (1985) incorporated the effects of anisotropic (kinematic) hardening by replacing J_2 in Eq. (9.7.9)

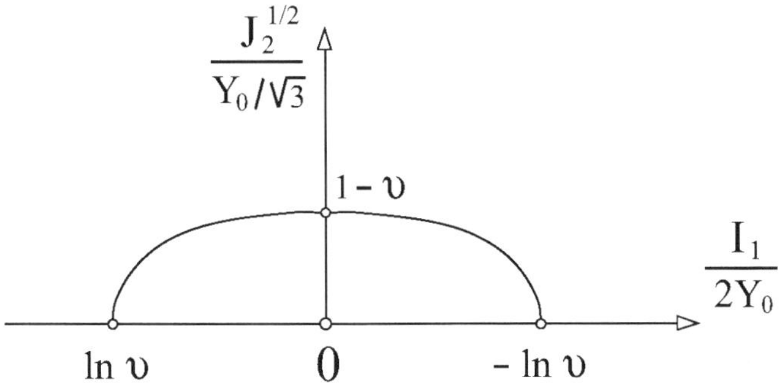

FIGURE 9.17. Gurson yield condition for porous metals with the void/volume fraction v. The tensile yield stress of the matrix material is Y_0.

with $(1/2)(\boldsymbol{\sigma}' - \boldsymbol{\alpha}) : (\boldsymbol{\sigma}' - \boldsymbol{\alpha})$, where $\boldsymbol{\alpha}$ defines the intersection of the current axis of the yield surface, parallel to hydrostatic axis, with the deviatoric plane. Yield functions and flow rules for porous pressure-dependent polymeric materials were analyzed by Lee and Oung (2000).

9.7.3. Constitutive Equations

The pressure-dependent yield conditions considered in two previous subsections are of the type

$$f(J_2, I_1, \mathcal{H}) = 0, \tag{9.7.11}$$

where $\mathcal{H}$ designates the appropriate history parameters. If it is assumed that the considered materials obey Ilyushin's postulate, the plastic part of the rate of deformation tensor is normal to the yield surface, and

$$\mathbf{D}^{\mathrm{p}} = \dot{\gamma}\,\frac{\partial f}{\partial \boldsymbol{\sigma}}, \quad \frac{\partial f}{\partial \boldsymbol{\sigma}} = \frac{\partial f}{\partial J_2}\,\boldsymbol{\sigma}' + \frac{\partial f}{\partial I_1}\,\mathbf{I}. \tag{9.7.12}$$

The loading index can be expressed as

$$\dot{\gamma} = \frac{1}{H}\left(\frac{\partial f}{\partial J_2}\,\boldsymbol{\sigma}' + \frac{\partial f}{\partial I_1}\,\mathbf{I}\right) : \overset{\circ}{\underline{\boldsymbol{\tau}}}, \tag{9.7.13}$$

where H is an appropriate hardening modulus. The plastic part of the rate of deformation, corresponding to $\overset{\circ}{\underline{\boldsymbol{\tau}}}$, is again denoted by $\mathbf{D}^{\mathrm{p}}$. Substitution of Eq. (9.7.13) into Eq. (9.7.12), therefore, gives

$$\mathbf{D}^{\mathrm{p}} = \frac{1}{H}\left[\left(\frac{\partial f}{\partial J_2}\,\boldsymbol{\sigma}' + \frac{\partial f}{\partial I_1}\,\mathbf{I}\right) \otimes \left(\frac{\partial f}{\partial J_2}\,\boldsymbol{\sigma}' + \frac{\partial f}{\partial I_1}\,\mathbf{I}\right)\right] : \overset{\circ}{\underline{\boldsymbol{\tau}}}. \tag{9.7.14}$$

The volumetric part of the plastic rate of deformation is

$$\operatorname{tr} \mathbf{D}^{\mathrm{p}} = \frac{3}{H} \frac{\partial f}{\partial I_1} \left(\frac{\partial f}{\partial J_2} \boldsymbol{\sigma}' + \frac{\partial f}{\partial I_1} \mathbf{I} \right) : \overset{\circ}{\underline{\boldsymbol{\tau}}}. \tag{9.7.15}$$

Geomaterials

For the Drucker–Prager yield condition,

$$\frac{\partial f}{\partial J_2} = \frac{1}{2} J_2^{-1/2}, \quad \frac{\partial f}{\partial I_1} = \frac{1}{3} \mu_*, \tag{9.7.16}$$

and

$$H = h_{\mathrm{t}}^{\mathrm{p}} = \frac{\mathrm{d}K}{\mathrm{d}\vartheta}, \quad \vartheta = \int_0^t \left(2\,\mathbf{D}^{\mathrm{p}\prime} : \mathbf{D}^{\mathrm{p}\prime}\right)^{1/2} \mathrm{d}t. \tag{9.7.17}$$

The relationship $K = K(\vartheta)$ between the shear yield stress K, under given superimposed pressure, and the generalized shear plastic strain ϑ is assumed to be known. Note that $\dot{\vartheta} = \dot{\gamma}$.

Alternatively, the hardening modulus can be expressed as

$$H = \frac{1}{3} \left(1 - \frac{\mu_*}{\sqrt{3}} \right)^2 \frac{\mathrm{d}Y^-}{\mathrm{d}\vartheta}, \tag{9.7.18}$$

where Y^- is the yield stress in uniaxial compression. The generalized plastic strain is in this case defined by

$$\vartheta = \frac{1 - \mu_*/\sqrt{3}}{(1 + 2\mu_*^2/3)^{1/2}} \int_0^t \left(\frac{2}{3} \mathbf{D}^{\mathrm{p}} : \mathbf{D}^{\mathrm{p}} \right)^{1/2} \mathrm{d}t, \tag{9.7.19}$$

which coincides with the longitudinal strain in uniaxial compression test. The relationship between ϑ and γ is

$$\frac{\mathrm{d}\vartheta}{\mathrm{d}\gamma} = \frac{1}{\sqrt{3}} \left(1 - \frac{\mu_*}{\sqrt{3}} \right). \tag{9.7.20}$$

Porous Metals

For the Gurson yield condition we have

$$\frac{\partial f}{\partial J_2} = 1, \quad \frac{\partial f}{\partial I_1} = \frac{1}{3} \upsilon\, Y_0 \sinh\left(\frac{I_1}{2Y_0} \right), \tag{9.7.21}$$

and

$$H = \frac{2}{3} \upsilon(1-\upsilon) Y_0^3 \sinh\left(\frac{I_1}{2Y_0} \right) \left[\upsilon - \cosh\left(\frac{I_1}{2Y_0} \right) \right]. \tag{9.7.22}$$

From Eqs. (9.7.10) and (9.7.12) it follows that the porosity evolves according to

$$\dot{\upsilon} = \dot{\gamma}\, \upsilon(1-\upsilon) Y_0 \sinh\left(\frac{I_1}{2Y_0} \right). \tag{9.7.23}$$

Further analysis of inelastic deformation of porous materials can be found in Lee (1988), Cocks (1989), Qiu and Weng (1993), and Sun (1995).

9.8. Nonassociative Plasticity

Constitutive equations in which plastic part of the rate of strain is normal to a locally smooth yield surface $f_{(n)} = 0$ in the conjugate stress space,

$$\dot{\mathbf{E}}^{\mathrm{p}}_{(n)} = \dot{\gamma}_{(n)} \frac{\partial f_{(n)}}{\partial \mathbf{T}_{(n)}}, \tag{9.8.1}$$

are referred to as the associative flow rules. As discussed in Section 8.5, a sufficient condition for this constitutive structure is that the material obeys Ilyushin's postulate. However, many pressure-dependent dilatant materials, with internal frictional effects, are not well described by associative flow rules. For example, associative flow rules largely overestimate inelastic volume changes in geomaterials like rocks and soils (Rudnicki and Rice, 1975; Rice, 1977), and in certain high-strength steels exhibiting the strength-differential effect by which the yield strength is higher in compression than in tension (Spitzig, Sober, and Richmond, 1975; Casey and Sullivan, 1985; Lee, 1988). For such materials, plastic part of the rate of strain is taken to be normal to the plastic potential surface

$$\pi_{(n)} = 0, \tag{9.8.2}$$

which is distinct from the yield surface

$$f_{(n)} = 0. \tag{9.8.3}$$

The resulting constitutive structure,

$$\dot{\mathbf{E}}^{\mathrm{p}}_{(n)} = \dot{\gamma}_{(n)} \frac{\partial \pi_{(n)}}{\partial \mathbf{T}_{(n)}}, \tag{9.8.4}$$

is known as the nonassociative flow rule (e.g., Mróz, 1963; Nemat-Nasser, 1983; Runesson and Mróz, 1989).

The consistency condition $\dot{f}_{(n)} = 0$ gives

$$\dot{\gamma}_{(n)} = \frac{1}{H_{(n)}} \frac{\partial f_{(n)}}{\partial \mathbf{T}_{(n)}} : \dot{\mathbf{T}}_{(n)}, \tag{9.8.5}$$

where $H_{(n)}$ is an appropriate hardening modulus. Thus,

$$\dot{\mathbf{E}}^{\mathrm{p}}_{(n)} = \frac{1}{H_{(n)}} \left(\frac{\partial \pi_{(n)}}{\partial \mathbf{T}_{(n)}} \otimes \frac{\partial f_{(n)}}{\partial \mathbf{T}_{(n)}} \right) : \dot{\mathbf{T}}_{(n)}. \tag{9.8.6}$$

The overall constitutive structure is

$$\dot{\mathbf{E}}_{(n)} = \left[\mathbf{M}_{(n)} + \frac{1}{H_{(n)}}\left(\frac{\partial \pi_{(n)}}{\partial \mathbf{T}_{(n)}} \otimes \frac{\partial f_{(n)}}{\partial \mathbf{T}_{(n)}}\right)\right] : \dot{\mathbf{T}}_{(n)}. \tag{9.8.7}$$

Since

$$\pi_{(n)} \neq f_{(n)}, \tag{9.8.8}$$

the elastoplastic compliance tensor in Eq. (9.8.7) does not possess a reciprocal symmetry. In an inverted form, the constitutive equation (9.8.7) becomes

$$\dot{\mathbf{T}}_{(n)} = \left[\mathbf{\Lambda}_{(n)} - \frac{1}{h_{(n)}}\left(\mathbf{\Lambda}_{(n)} : \frac{\partial \pi_{(n)}}{\partial \mathbf{T}_{(n)}}\right) \otimes \left(\frac{\partial f_{(n)}}{\partial \mathbf{T}_{(n)}} : \mathbf{\Lambda}_{(n)}\right)\right] : \dot{\mathbf{E}}_{(n)}, \tag{9.8.9}$$

where

$$h_{(n)} = H_{(n)} + \frac{\partial f_{(n)}}{\partial \mathbf{T}_{(n)}} : \mathbf{\Lambda}_{(n)} : \frac{\partial \pi_{(n)}}{\partial \mathbf{T}_{(n)}}. \tag{9.8.10}$$

9.8.1. Plastic Potential for Geomaterials

To better describe inelastic behavior of geomaterials whose yield is governed by the Drucker–Prager yield condition of Eq. (9.7.4), a nonassociative flow rule can be used with the plastic potential (Fig. 9.18)

$$\pi = J_2^{1/2} + \frac{1}{3}\beta I_1 - K = 0. \tag{9.8.11}$$

The material parameter β is in general different from the friction parameter μ_* of Eq. (9.7.4). Thus,

$$\mathbf{D}^{\mathrm{p}} = \dot{\gamma}\frac{\partial \pi}{\partial \boldsymbol{\sigma}} = \dot{\gamma}\left(\frac{1}{2}J_2^{-1/2}\boldsymbol{\sigma}' + \frac{1}{3}\beta\,\mathbf{I}\right). \tag{9.8.12}$$

The loading index $\dot{\gamma}$ is determined from the consistency condition. Assuming known the relationship

$$K = K(\vartheta) \tag{9.8.13}$$

between the shear yield stress and the generalized shear plastic strain ϑ, defined by Eq. (9.7.17), the condition $\dot{f} = 0$ gives

$$\dot{\gamma} = \frac{1}{H}\left(\frac{1}{2}J_2^{-1/2}\boldsymbol{\sigma}' + \frac{1}{3}\mu_*\mathbf{I}\right) : \overset{\circ}{\boldsymbol{\tau}}, \quad H = h_{\mathrm{t}}^{\mathrm{p}} = \frac{\mathrm{d}K}{\mathrm{d}\vartheta}. \tag{9.8.14}$$

Alternatively, assuming known the relationship

$$Y^- = Y^-(\vartheta) \tag{9.8.15}$$

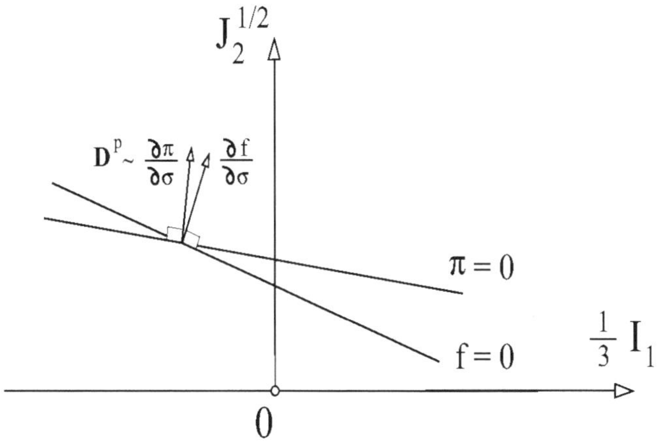

FIGURE 9.18. Illustration of a nonassociative flow rule. The plastic rate of deformation $\mathbf{D}^{\mathrm{p}}$ is normal to the flow potential $\pi = 0$, which is distinct from the yield surface $f = 0$.

between the yield stress in uniaxial compression and the generalized plastic strain

$$\vartheta = \frac{1 - \beta/\sqrt{3}}{(1 + 2\beta^2/3)^{1/2}} \int_0^t \left(\frac{2}{3}\mathbf{D}^{\mathrm{p}} : \mathbf{D}^{\mathrm{p}}\right)^{1/2} \mathrm{d}t, \tag{9.8.16}$$

the hardening modulus is

$$H = \frac{1}{3}\left(1 - \frac{\mu_*}{\sqrt{3}}\right)\left(1 - \frac{\beta}{\sqrt{3}}\right)\frac{\mathrm{d}Y^-}{\mathrm{d}\vartheta}\,. \tag{9.8.17}$$

The substitution of Eq. (9.8.14) into Eq. (9.8.12) gives

$$\mathbf{D}^{\mathrm{p}} = \frac{1}{H}\left[\left(\frac{1}{2}\,J_2^{-1/2}\boldsymbol{\sigma}' + \frac{1}{3}\,\beta\,\mathbf{I}\right) \otimes \left(\frac{1}{2}\,J_2^{-1/2}\boldsymbol{\sigma}' + \frac{1}{3}\,\mu_*\,\mathbf{I}\right)\right] : \overset{\circ}{\boldsymbol{\tau}}. \tag{9.8.18}$$

The deviatoric and spherical parts are

$$\mathbf{D}^{\mathrm{p}\prime} = \frac{1}{2H}\,\frac{\boldsymbol{\sigma}'}{J_2^{1/2}}\left(\frac{\boldsymbol{\sigma}' : \overset{\circ}{\boldsymbol{\tau}}}{2\,J_2^{1/2}} + \frac{1}{3}\,\mu_*\,\mathrm{tr}\,\overset{\circ}{\boldsymbol{\tau}}\right), \tag{9.8.19}$$

$$\mathrm{tr}\,\mathbf{D}^{\mathrm{p}} = \frac{\beta}{H}\left(\frac{\boldsymbol{\sigma}' : \overset{\circ}{\boldsymbol{\tau}}}{2\,J_2^{1/2}} + \frac{1}{3}\,\mu_*\,\mathrm{tr}\,\overset{\circ}{\boldsymbol{\tau}}\right). \tag{9.8.20}$$

To physically interpret the parameter β, we observe from Eq. (9.8.12) that

$$(2\,\mathbf{D}^{\mathrm{p}\prime} : \mathbf{D}^{\mathrm{p}\prime})^{1/2} = \dot{\gamma}, \quad \mathrm{tr}\,\mathbf{D}^{\mathrm{p}} = \beta\,\dot{\gamma}, \tag{9.8.21}$$

i.e.,

$$\beta = \frac{\operatorname{tr} \mathbf{D}^{\mathrm{p}}}{\left(2\, \mathbf{D}^{\mathrm{p}\prime} : \mathbf{D}^{\mathrm{p}\prime}\right)^{1/2}} . \tag{9.8.22}$$

Thus, β is the ratio of the volumetric and shear part of the plastic rate of deformation, which is often called the dilatancy factor (Rudnicki and Rice, 1975). Representative values of the friction coefficient and the dilatancy factor for fissured rocks, listed by Rudnicki and Rice (*op. cit.*), indicate that

$$\mu_* = 0.3 \div 1, \quad \beta = 0.1 \div 0.5. \tag{9.8.23}$$

The frictional parameter and inelastic dilatancy of the material actually change with the progression of inelastic deformation, but are here treated as constants. For a more elaborate analysis, which accounts for their variation, the paper by Nemat-Nasser and Shokooh (1980) can be consulted. Note also that

$$\dot{\gamma} = \frac{\boldsymbol{\sigma} : \mathbf{D}^{\mathrm{p}\prime}}{J_2^{1/2}} . \tag{9.8.24}$$

The deviatoric and spherical parts of the total rate of deformation are, respectively,

$$\mathbf{D}' = \frac{\overset{\circ}{\underline{\boldsymbol{\tau}}}{}'}{2\mu} + \frac{1}{2H} \frac{\boldsymbol{\sigma}'}{J_2^{1/2}} \left(\frac{\boldsymbol{\sigma}' : \overset{\circ}{\underline{\boldsymbol{\tau}}}}{2\, J_2^{1/2}} + \frac{1}{3} \mu_* \operatorname{tr} \overset{\circ}{\underline{\boldsymbol{\tau}}} \right), \tag{9.8.25}$$

$$\operatorname{tr} \mathbf{D} = \frac{1}{3\kappa} \operatorname{tr} \overset{\circ}{\underline{\boldsymbol{\tau}}} + \frac{\beta}{H} \left(\frac{\boldsymbol{\sigma}' : \overset{\circ}{\underline{\boldsymbol{\tau}}}}{2\, J_2^{1/2}} + \frac{1}{3} \mu_* \operatorname{tr} \overset{\circ}{\underline{\boldsymbol{\tau}}} \right). \tag{9.8.26}$$

These can be inverted to give the deviatoric and spherical parts of the stress rate as

$$\overset{\circ}{\underline{\boldsymbol{\tau}}}{}' = 2\mu \left[\mathbf{D}' - \frac{1}{c} \frac{\boldsymbol{\sigma}'}{J_2^{1/2}} \left(\frac{\boldsymbol{\sigma}' : \mathbf{D}}{2\, J_2^{1/2}} + \mu_* \frac{\kappa}{2\mu} \operatorname{tr} \mathbf{D} \right) \right], \tag{9.8.27}$$

$$\operatorname{tr} \overset{\circ}{\underline{\boldsymbol{\tau}}} = \frac{3\kappa}{c} \left[\left(1 + \frac{H}{\mu} \right) \operatorname{tr} \mathbf{D} - \beta \frac{\boldsymbol{\sigma}' : \mathbf{D}}{J_2^{1/2}} \right], \tag{9.8.28}$$

where

$$c = 1 + \frac{H}{\mu} + \mu_* \beta \frac{\kappa}{\mu} . \tag{9.8.29}$$

If the friction coefficient μ_* is equal to zero, Eqs. (9.8.27) and (9.8.28) reduce to

$$\overset{\circ}{\underline{\boldsymbol{\tau}}}{}' = 2\mu \left[\mathbf{D}' - \frac{1}{1 + H/\mu} \frac{(\boldsymbol{\sigma}' \otimes \boldsymbol{\sigma}') : \mathbf{D}}{2\, J_2} \right], \tag{9.8.30}$$

$$\operatorname{tr}\overset{\circ}{\underline{\tau}} = 3\kappa\left(\operatorname{tr}\mathbf{D} - \frac{\beta}{1+H/\mu}\,\frac{\boldsymbol{\sigma}':\mathbf{D}}{J_2^{1/2}}\right). \tag{9.8.31}$$

With a vanishing dilatancy factor ($\beta = 0$), Eqs. (9.8.30) and (9.8.31) coincide with the constitutive equations of isotropic hardening pressure-independent metal plasticity (Subsection 9.4.1). Other nonassociative models for geological materials are discussed by Desai and Hasmini (1989).

Constitutive Inequalities

Returning to Eq. (9.8.18), a trace product with $\overset{\circ}{\underline{\tau}}$ gives

$$\overset{\circ}{\underline{\tau}}:\mathbf{D}^{\mathrm{p}} = \frac{1}{H}\left[\left(\frac{1}{2}J_2^{-1/2}\boldsymbol{\sigma}' + \frac{1}{3}\beta\,\mathbf{I}\right):\overset{\circ}{\underline{\tau}}\right]\left[\left(\frac{1}{2}J_2^{-1/2}\boldsymbol{\sigma}' + \frac{1}{3}\mu_*\mathbf{I}\right):\overset{\circ}{\underline{\tau}}\right]. \tag{9.8.32}$$

In the hardening range ($H > 0$), from Eq. (9.8.14) it follows that

$$\left(\frac{1}{2}J_2^{-1/2}\boldsymbol{\sigma}' + \frac{1}{3}\mu_*\mathbf{I}\right):\overset{\circ}{\underline{\tau}} > 0, \tag{9.8.33}$$

since $\dot{\gamma} > 0$. Thus, from Eq. (9.8.32) the sign of $\overset{\circ}{\underline{\tau}}:\mathbf{D}^{\mathrm{p}}$ is determined by the sign of

$$\left(\frac{1}{2}J_2^{-1/2}\boldsymbol{\sigma}' + \frac{1}{3}\beta\,\mathbf{I}\right):\overset{\circ}{\underline{\tau}}. \tag{9.8.34}$$

Depending on the state of stress and the type of incipient loading, this can be either positive or negative. Therefore, in the framework of nonassociative plasticity, the quantity $\overset{\circ}{\underline{\tau}}:\mathbf{D}^{\mathrm{p}}$ can be negative even in the hardening range. This is in contrast to associative plasticity, where $\overset{\circ}{\underline{\tau}}:\mathbf{D}^{\mathrm{p}}$ is always positive in the hardening range, by Eq. (8.8.8). Similarly, $\overset{\circ}{\underline{\tau}}:\mathbf{D}^{\mathrm{p}}$ can be positive in the softening range. Illustrative examples can be found in the article by Lubarda, Mastilovic, and Knap (1996).

The fact that $\overset{\circ}{\underline{\tau}}:\mathbf{D}^{\mathrm{p}}$ can be negative in the hardening range does not necessarily imply that material becomes unstable. Whether an instability actually occurs at a given state of stress and material constitution is answered by a bifurcation-type analysis, such as used by Rudnicki and Rice (*op. cit.*). For example, they found that for certain states of stress, localization is possible even in the hardening range, for materials described by a

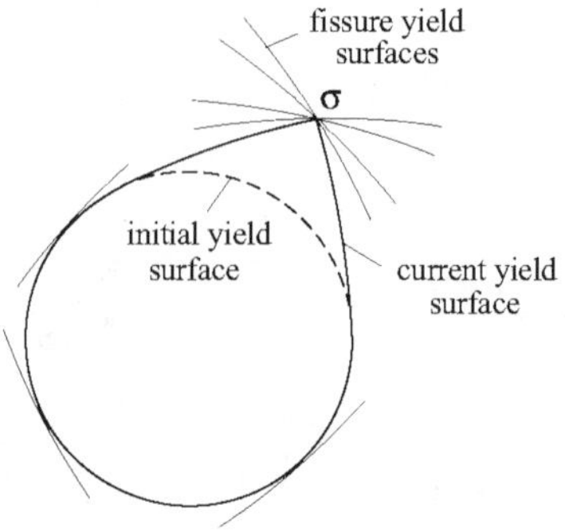

FIGURE 9.19. Macroscopic yield surface formed as an envelope of individual fissure yield surfaces. The yield vertex forms at the loading point due to sliding on favorably oriented fissure surfaces.

nonassociative flow rule. This is never the case for materials with an associative flow rule. Plastic instability and bifurcation analysis are considered in Chapter 10.

9.8.2. Yield Vertex Model for Fissured Rocks

In a brittle rock, modeled to contain a collection of randomly oriented fissures, inelastic deformation results from frictional sliding on the fissure surfaces. Inelastic dilatancy under overall compressive loads is a consequence of opening the fissures at asperities and local tensile fractures at some angle at the edges of fissures. Individual yield surface may be associated with each fissure. Expressed in terms of the resolved shear stress in the plane of fissure with normal $\mathbf{n}$, this is

$$\mathbf{n} \cdot \boldsymbol{\sigma} \cdot \mathbf{m} + \mu_* \, \mathbf{n} \cdot \boldsymbol{\sigma} \cdot \mathbf{n} = \text{const.}, \tag{9.8.35}$$

where μ_* is the friction coefficient between the surfaces of the fissure, and $\mathbf{m}$ is the sliding direction (direction of the maximum shear stress in the plane of fissure). The macroscopic yield surface is the envelope of individual yield surfaces (Fig. 9.19) for fissures of all orientations (Rudnicki and Rice, 1975). This is similar to slip model of metal plasticity (Batdorf and Budiansky, 1949,1954; Sanders, 1954; Hill, 1967b).

Continued stressing in the same direction will cause continuing sliding on favorably oriented (already activated) fissures, and will initiate sliding for a progressively greater number of orientations. After certain amount of inelastic deformation, the macroscopic yield envelope develops a vertex at the loading point. The stress increment normal to the original stress direction will initiate or continue sliding of fissure surfaces for some fissure orientations. In isotropic hardening idealization with a smooth yield surface, however, a stress increment tangential to the yield surface will cause only elastic deformation, overestimating the stiffness of the response. In order to take into account the effect of the yield vertex in an approximate way, Rudnicki and Rice (*op. cit.*) introduced a second plastic modulus H_1, which governs the response to part of the stress increment directed tangentially to what is taken to be the smooth yield surface through the same stress point (Fig. 9.20). Since no vertex formation is associated with hydrostatic stress increments, tangential stress increments are taken to be deviatoric, and Eq. (9.8.19) is replaced with

$$\mathbf{D}^{\mathrm{p}\prime} = \frac{1}{2H}\,\frac{\boldsymbol{\sigma}'}{J_2^{1/2}}\left(\frac{\boldsymbol{\sigma}' : \overset{\circ}{\underline{\boldsymbol{\tau}}}}{2\,J_2^{1/2}} + \frac{1}{3}\,\mu_*\,\mathrm{tr}\,\overset{\circ}{\underline{\boldsymbol{\tau}}}\right) + \frac{1}{2H_1}\left(\overset{\circ}{\underline{\boldsymbol{\tau}}}{}' - \frac{\boldsymbol{\sigma}' : \overset{\circ}{\underline{\boldsymbol{\tau}}}}{2\,J_2}\,\boldsymbol{\sigma}'\right). \tag{9.8.36}$$

The dilation induced by the small tangential stress increment is assumed to be negligible, so that Eq. (9.8.20) still applies for $\mathrm{tr}\,\mathbf{D}^{\mathrm{p}}$. The constitutive structure in Eq. (9.8.36) is intended to model the response at a yield surface vertex for small deviations from proportional ("straight ahead") loading $\overset{\circ}{\underline{\boldsymbol{\tau}}} \sim \boldsymbol{\sigma}'$.

The expressions for the rate of stress in terms of the rate of deformation are obtained by inversion of the expression for the rate of deformation corresponding to Eqs. (9.8.20) and (9.8.36). The results are

$$\overset{\circ}{\underline{\boldsymbol{\tau}}}{}' = 2\mu\left[\frac{1}{b}\,\mathbf{D}' - \frac{a}{bc}\,\frac{(\boldsymbol{\sigma}'\otimes\boldsymbol{\sigma}') : \mathbf{D}}{2\,J_2} - \frac{1}{c}\,\mu_*\,\frac{\kappa}{2\mu}\,\frac{\boldsymbol{\sigma}'}{J_2^{1/2}}\,\mathrm{tr}\,\mathbf{D}\right], \tag{9.8.37}$$

$$\mathrm{tr}\,\overset{\circ}{\underline{\boldsymbol{\tau}}} = \frac{3\kappa}{c}\left[\left(1 + \frac{H}{\mu}\right)\mathrm{tr}\,\mathbf{D} - \beta\,\frac{\boldsymbol{\sigma}' : \mathbf{D}}{J_2^{1/2}}\right]. \tag{9.8.38}$$

The parameters a and b are given by

$$a = 1 - \frac{H}{H_1} - \mu_*\beta\,\frac{\kappa}{H_1}\,, \quad b = 1 + \frac{\mu}{H_1}\,, \tag{9.8.39}$$

and c is defined by Eq. (9.8.29).

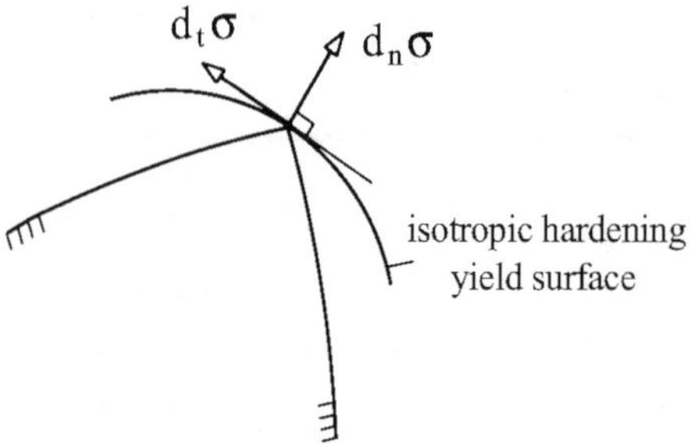

FIGURE 9.20. A stress increment from a yield vertex decomposed in the normal and tangential direction relative to an isotropic hardening smooth yield surface passing through the vertex. The tangential component $d_t\sigma$ does not cause plastic flow for smooth yield idealization, but it does for the yield vertex.

Another model in which the plastic rate of deformation depends on the component of stress rate tangential to smooth yield surface was proposed by Hashiguchi (1993).

9.9. Thermoplasticity

Nonisothermal plasticity is considered in this section assuming that the temperature is not too high, so that creep deformation can be neglected. The analysis may also be adequate for certain applications under high stresses of short duration, where the temperature increase is more pronounced but the viscous (creep) strains have no time to develop (Prager, 1958; Kachanov, 1971). Thus, infinitesimal changes of stress and temperature applied to the material at a given state produce a unique infinitesimal change of strain, independently of the speed with which these changes are made. Rate-dependent plasticity will be considered in Section 9.10.

The formulation of thermoplastic analysis under described conditions can proceed by introducing a nonisothermal yield condition in either stress or strain space. For example, the yield function in stress space is defined by

$$f_{(n)}\left(\mathbf{T}_{(n)}, \theta, \mathcal{H}\right) = 0, \tag{9.9.1}$$

where θ is the temperature, and $\mathcal{H}$ is the pattern of internal rearrangements. The response within the yield surface is thermoelastic. If the Gibbs energy per unit reference volume, relative to selected stress and strain measures, is

$$\Phi_{(n)} = \Phi_{(n)}\left(\mathbf{T}_{(n)}, \theta, \mathcal{H}\right), \tag{9.9.2}$$

the strain is

$$\mathbf{E}_{(n)} = \frac{\partial \Phi_{(n)}}{\partial \mathbf{T}_{(n)}}. \tag{9.9.3}$$

Consider the stress state $\mathbf{T}_{(n)}$ on the current yield surface. The rates of stress and temperature associated with thermoplastic loading satisfy the consistency condition $\dot{f}_{(n)} = 0$, which gives

$$\frac{\partial f_{(n)}}{\partial \mathbf{T}_{(n)}} : \dot{\mathbf{T}}_{(n)} + \frac{\partial f_{(n)}}{\partial \theta} : \dot{\theta} - H_{(n)}\,\dot{\gamma}_{(n)} = 0. \tag{9.9.4}$$

The hardening parameter is

$$H_{(n)} = H_{(n)}\left(\mathbf{T}_{(n)}, \theta, \mathcal{H}\right), \tag{9.9.5}$$

and the loading index

$$\dot{\gamma}_{(n)} > 0. \tag{9.9.6}$$

Three types of thermoplastic response are possible,

$$\begin{aligned}
&H_{(n)} > 0, \quad \frac{\partial f_{(n)}}{\partial \mathbf{T}_{(n)}} : \dot{\mathbf{T}}_{(n)} + \frac{\partial f_{(n)}}{\partial \theta} : \dot{\theta} > 0 \quad \text{thermoplastic hardening,}\\
&H_{(n)} < 0, \quad \frac{\partial f_{(n)}}{\partial \mathbf{T}_{(n)}} : \dot{\mathbf{T}}_{(n)} + \frac{\partial f_{(n)}}{\partial \theta} : \dot{\theta} < 0 \quad \text{thermoplastic softening,}\\
&H_{(n)} = 0, \quad \frac{\partial f_{(n)}}{\partial \mathbf{T}_{(n)}} : \dot{\mathbf{T}}_{(n)} + \frac{\partial f_{(n)}}{\partial \theta} : \dot{\theta} = 0 \quad \text{ideally thermoplastic.}
\end{aligned} \tag{9.9.7}$$

This parallels the isothermal classification of Eq. (9.2.8).

Since rate-independence is assumed, the constitutive relationship of thermoplasticity must be homogeneous of degree one in the rates of stress, strain and temperature. For thermoplastic part of the rate of strain this is satisfied by the normality structure

$$\dot{\mathbf{E}}^{\mathrm{p}}_{(n)} = \dot{\gamma}_{(n)}\,\frac{\partial f_{(n)}}{\partial \mathbf{T}_{(n)}}. \tag{9.9.8}$$

In view of Eq. (9.9.4), this becomes

$$\dot{\mathbf{E}}^{\mathrm{p}}_{(n)} = \frac{1}{H_{(n)}}\left(\frac{\partial f_{(n)}}{\partial \mathbf{T}_{(n)}} : \dot{\mathbf{T}}_{(n)} + \frac{\partial f_{(n)}}{\partial \theta} : \dot{\theta}\right)\frac{\partial f_{(n)}}{\partial \mathbf{T}_{(n)}}. \tag{9.9.9}$$

The strain rate is the sum of thermoelastic and thermoplastic parts,

$$\dot{\mathbf{E}}_{(n)} = \dot{\mathbf{E}}^{\mathrm{e}}_{(n)} + \dot{\mathbf{E}}^{\mathrm{p}}_{(n)}. \tag{9.9.10}$$

The thermoelastic part is governed by

$$\dot{\mathbf{E}}^{\mathrm{e}}_{(n)} = \frac{\partial^2 \Phi_{(n)}}{\partial \mathbf{T}_{(n)} \otimes \partial \mathbf{T}_{(n)}} : \dot{\mathbf{T}}_{(n)} + \frac{\partial^2 \Phi_{(n)}}{\partial \mathbf{T}_{(n)} \partial \theta}\, \dot{\theta}. \tag{9.9.11}$$

For example, if the Gibbs energy is taken to be

$$\begin{aligned}\Phi_{(n)} = {} & \frac{1}{4\mu_{(n)}} \left(\operatorname{tr} \mathbf{T}^2_{(n)} - \frac{\lambda_{(n)}}{3\lambda_{(n)} + 2\mu_{(n)}} \operatorname{tr}^2 \mathbf{T}_{(n)} \right) \\ & + \alpha_{(n)}(\theta) \operatorname{tr} \mathbf{T}_{(n)} + \beta_{(n)}(\theta, \mathcal{H}),\end{aligned} \tag{9.9.12}$$

we obtain

$$\dot{\mathbf{E}}^{\mathrm{e}}_{(n)} = \frac{1}{2\mu_{(n)}} \left(\boldsymbol{I} - \frac{\lambda_{(n)}}{2\mu_{(n)} + 3\lambda_{(n)}} \mathbf{I} \otimes \mathbf{I} \right) : \dot{\mathbf{T}}_{(n)} + \alpha'_{(n)}(\theta)\, \dot{\theta}\, \mathbf{I}. \tag{9.9.13}$$

The Lamé type elastic constants corresponding to selected stress and strain measures are $\lambda_{(n)}$ and $\mu_{(n)}$. The scalar function $\alpha_{(n)}$ is an appropriate function of the temperature. Its temperature gradient is $\alpha'_{(n)} = \mathrm{d}\alpha_{(n)}/\mathrm{d}\theta$.

9.9.1. Isotropic and Kinematic Hardening

Suppose that a nonisothermal yield condition in the Cauchy stress space is a temperature-dependent von Mises condition

$$f = \frac{1}{2} \boldsymbol{\sigma}' : \boldsymbol{\sigma}' - [\varphi(\theta)\, K(\vartheta)]^2 = 0. \tag{9.9.14}$$

The thermoplastic part of the deformation rate is then

$$\mathbf{D}^{\mathrm{p}} = \frac{1}{2\varphi h^{\mathrm{p}}_{\mathrm{t}}} \left(\frac{\boldsymbol{\sigma}' \otimes \boldsymbol{\sigma}'}{\boldsymbol{\sigma}' : \boldsymbol{\sigma}'} : \overset{\circ}{\underline{\boldsymbol{\tau}}} - \boldsymbol{\sigma}' \frac{\varphi'}{\varphi}\, \dot{\theta} \right), \tag{9.9.15}$$

where

$$h^{\mathrm{p}}_{\mathrm{t}} = \frac{\mathrm{d}K}{\mathrm{d}\vartheta}, \quad \varphi' = \frac{\mathrm{d}\varphi}{\mathrm{d}\theta}. \tag{9.9.16}$$

Combining Eqs. (9.9.13) and (9.9.15), the total rate of deformation becomes

$$\begin{aligned}\mathbf{D} = {} & \left[\frac{1}{2\mu} \left(\boldsymbol{I} - \frac{\lambda}{2\mu + 3\lambda} \mathbf{I} \otimes \mathbf{I} \right) + \frac{1}{2\varphi h^{\mathrm{p}}_{\mathrm{t}}} \frac{\boldsymbol{\sigma}' \otimes \boldsymbol{\sigma}'}{\boldsymbol{\sigma}' : \boldsymbol{\sigma}'} \right] : \overset{\circ}{\underline{\boldsymbol{\tau}}} \\ & + \left[\alpha'(\theta)\, \mathbf{I} - \frac{\varphi'}{2\varphi^2 h^{\mathrm{p}}_{\mathrm{t}}} \boldsymbol{\sigma}' \right] \dot{\theta}.\end{aligned} \tag{9.9.17}$$

The inverse constitutive equation for the stress rate is

$$\begin{aligned}\overset{\circ}{\underline{\boldsymbol{\tau}}} = {} & \left(\lambda \mathbf{I} \otimes \mathbf{I} + 2\mu \boldsymbol{I} - \frac{2\mu}{1 + \varphi h^{\mathrm{p}}_{\mathrm{t}}/\mu} \frac{\boldsymbol{\sigma}' \otimes \boldsymbol{\sigma}'}{\boldsymbol{\sigma}' : \boldsymbol{\sigma}'} \right) : \mathbf{D} \\ & - \left[(3\lambda + 2\mu)\, \alpha' \mathbf{I} - \frac{1}{1 + \varphi h^{\mathrm{p}}_{\mathrm{t}}/\mu} \frac{\varphi'}{\varphi} \boldsymbol{\sigma}' \right] \dot{\theta}.\end{aligned} \tag{9.9.18}$$

This can be viewed as a generalization of an infinitesimal strain formulation for a rigid-thermoplastic material, given by Prager (1958). See also Boley and Weiner (1960), Drucker (1960), Lee and Wierzbicki (1967), Lee (1969), Lubarda (1986,1989), and Naghdi (1960,1990).

In the case of thermoplasticity with linear kinematic hardening ($c = 2h_{\mathrm{t}}^{\mathrm{p}}$), and the temperature-dependent yield surface

$$f = \frac{1}{2}\,(\boldsymbol{\sigma}\,' - \boldsymbol{\alpha}) : (\boldsymbol{\sigma}\,' - \boldsymbol{\alpha}) - [\varphi(\theta)K]^2 = 0, \quad K = \text{const.}, \tag{9.9.19}$$

the thermoplastic rate of deformation is

$$\mathbf{D}^{\mathrm{p}} = \frac{1}{2h_{\mathrm{t}}^{\mathrm{p}}}\left[\frac{(\boldsymbol{\sigma}\,' - \boldsymbol{\alpha}) \otimes (\boldsymbol{\sigma}\,' - \boldsymbol{\alpha})}{(\boldsymbol{\sigma}\,' - \boldsymbol{\alpha}) : (\boldsymbol{\sigma}\,' - \boldsymbol{\alpha})} : \overset{\circ}{\underline{\boldsymbol{\tau}}} - \frac{\varphi'}{\varphi}\,(\boldsymbol{\sigma}\,' - \boldsymbol{\alpha})\,\dot{\theta}\right]. \tag{9.9.20}$$

Thermoelastic portion of the rate of deformation is as in Eq. (9.9.17), so that inversion of the expression for the total rate of deformation gives

$$\begin{aligned}\overset{\circ}{\underline{\boldsymbol{\tau}}} = {} & \left[\lambda\,\mathbf{I}\otimes\mathbf{I} + 2\mu\,\boldsymbol{I} - \frac{2\mu}{1 + h_{\mathrm{t}}^{\mathrm{p}}/\mu}\,\frac{(\boldsymbol{\sigma}\,' - \boldsymbol{\alpha}) \otimes (\boldsymbol{\sigma}\,' - \boldsymbol{\alpha})}{(\boldsymbol{\sigma}\,' - \boldsymbol{\alpha}) : (\boldsymbol{\sigma}\,' - \boldsymbol{\alpha})}\right] : \mathbf{D} \\ & - \left[(3\lambda + 2\mu)\,\alpha'\,\mathbf{I} - \frac{1}{1 + h_{\mathrm{t}}^{\mathrm{p}}/\mu}\,\frac{\varphi'}{\varphi}\,(\boldsymbol{\sigma}\,' - \boldsymbol{\alpha})\right]\dot{\theta}.\end{aligned} \tag{9.9.21}$$

Additional analysis of the rate-type constitutive equations of thermoplasticity was presented by Green and Naghdi (1965), De Boer (1977), Lehmann (1985), Zdebel and Lehmann (1987), Wang and Ohno (1991), McDowell (1992), Lucchesi and Šilhavý (1993), and Casey (1998). Experimental investigations of nonisothermal yield surfaces were reported by Phillips (1974, 1982), and others.

9.10. Rate-Dependent Plasticity

There are two types of constitutive equations used in modeling the rate-dependent plastic response of metals and alloys. In one approach, there is no yield surface in the model and plastic deformation commences from the onset of loading, although it may be exceedingly small below certain levels of applied stress. This type of modeling is particularly advocated by researchers in materials science, who view inelastic deformation process as inherently time-dependent. For example, this view is supported by the dislocation dynamics study of crystallographic slip in metals, as reported by Johnston and Gilman (1959). Since there is no separation of time-independent and creep effects, the modeling is often referred to as a unified creep–plasticity theory (Hart, 1970; Bodner and Partom, 1975; Miller, 1976,1987; Krieg, 1977; Estrin and Mecking, 1986). The second approach uses the notion of the static yield surface and dynamic loading surface, and is referred to as a viscoplastic modeling.

In his analysis of rate-dependent behavior of metals, Rice (1970,1971) showed that the plastic rate of strain can be derived from a scalar flow potential $\Omega_{(n)}$, as its gradient

$$\dot{\mathbf{E}}^{\mathrm{p}}_{(n)} = \frac{\partial \Omega_{(n)}\left(\mathbf{T}_{(n)}, \theta, \mathcal{H}\right)}{\partial \mathbf{T}_{(n)}}, \tag{9.10.1}$$

provided that the rate of shearing on any given slip system within a crystalline grain depends on local stresses only through the resolved shear stress. The history of deformation is represented by the pattern of internal rearrangements $\mathcal{H}$, and the absolute temperature is θ (Section 4.5). Geometrically, the plastic part of the strain rate is normal to surfaces of constant flow potential in stress space (see also Section 8.4). There is no yield surface in the model, and plastic deformation commences from the onset of loading. Time-independent behavior can be recovered, under certain idealizations – neglecting creep and rate effects, as an appropriate limit. In this limit, at each instant of deformation there is a range of stress space over which the flow potential is constant. The current yield surface is then a boundary of this range, a singular clustering of all surfaces of constant flow potential.

9.10.1. Power-Law and Johnson–Cook Models

The power-law representation of the flow potential in the Cauchy stress space is

$$\Omega = \frac{2\dot{\gamma}^0}{m+1}\left(\frac{J_2^{1/2}}{K}\right)^m J_2^{1/2}, \quad J_2 = \frac{1}{2}\boldsymbol{\sigma}' : \boldsymbol{\sigma}', \tag{9.10.2}$$

where $K = K(\theta, \mathcal{H})$ is the reference shear stress, $\dot{\gamma}^0$ is the reference shear strain rate to be selected for each material, and m is the material parameter (of the order of 100 for metals at room temperature and strain rates below $10^4\ \mathrm{s}^{-1}$; Nemat-Nasser, 1992). The corresponding plastic part of the rate of deformation is

$$\mathbf{D}^{\mathrm{p}} = \dot{\gamma}^0 \left(\frac{J_2^{1/2}}{K}\right)^m \frac{\boldsymbol{\sigma}'}{J_2^{1/2}}. \tag{9.10.3}$$

The equivalent plastic strain ϑ, defined by Eq. (9.4.3), is usually used as the only history parameter, and the reference shear stress depends on ϑ and θ according to

$$K = K^0 \left(1 + \frac{\vartheta}{\vartheta^0}\right)^{\alpha} \exp\left(-\beta\, \frac{\theta - \theta_0}{\theta_{\mathrm{m}} - \theta_0}\right). \tag{9.10.4}$$

Here, K^0 and ϑ^0 are the normalizing stress and strain, θ_0 and θ_{m} are the room and melting temperatures, and α and β are the material parameters. The total rate of deformation is

$$\mathbf{D} = \underline{\mathbf{M}} : \overset{\circ}{\underline{\boldsymbol{\tau}}} + \dot{\gamma}^0 \left(\frac{J_2^{1/2}}{K}\right)^m \frac{\boldsymbol{\sigma}'}{J_2^{1/2}}. \tag{9.10.5}$$

The instantaneous elastic compliance tensor $\underline{\mathbf{M}}$ is defined, for infinitesimal elasticity, by Eq. (9.4.16). From the onset of loading the deformation rate consists of elastic and plastic constituents, although for large m the plastic contribution may be small if J_2 is less than K. The inverted form of (9.10.5), expressing $\overset{\circ}{\underline{\boldsymbol{\tau}}}$ in terms of $\mathbf{D}$, is

$$\overset{\circ}{\underline{\boldsymbol{\tau}}} = \underline{\boldsymbol{\Lambda}} : \mathbf{D} - 2\mu\dot{\gamma}^0 \left(\frac{J_2^{1/2}}{K}\right)^m \frac{\boldsymbol{\sigma}'}{J_2^{1/2}}, \tag{9.10.6}$$

where $\underline{\boldsymbol{\Lambda}} = \underline{\mathbf{M}}^{-1}$. The elastic shear modulus is μ.

Another representation of the flow potential, constructed according to Johnson–Cook (1983) model, is

$$\Omega = \frac{2\dot{\gamma}^0}{a} K \exp\left[a\left(\frac{J_2^{1/2}}{K} - 1\right)\right]. \tag{9.10.7}$$

The reference shear stress is

$$K = K^0 \left[1 + b\left(\frac{\vartheta}{\vartheta^0}\right)^c\right] \left[1 - \left(\frac{\theta - \theta_0}{\theta_{\mathrm{m}} - \theta_0}\right)^d\right], \tag{9.10.8}$$

where a, b, c, d are the material parameters. The corresponding plastic part of the rate of deformation becomes

$$\mathbf{D}^{\mathrm{p}} = \dot{\gamma}^0 \exp\left[a\left(\frac{J_2^{1/2}}{K} - 1\right)\right] \frac{\boldsymbol{\sigma}'}{J_2^{1/2}}. \tag{9.10.9}$$

Similar expressions can be obtained for other models and the choices of the flow potential (e.g., Zerilli and Armstrong, 1987; see also a section on the physically based constitutive equations in the review by Meyers, 1999). Since there is no yield surface and loading/unloading criteria, some authors refer to these constitutive models as nonlinearly viscoelastic models (e.g.,

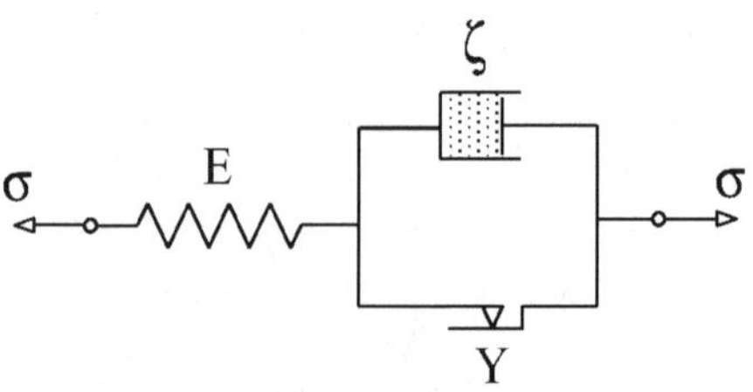

FIGURE 9.21. One-dimensional rheological model of elastic-viscoplastic response. The elastic modulus is E, the viscosity coefficient is ζ, and the yield stress of plastic element is Y.

Bardenhagen, Stout, and Gray, 1997). By selecting an appropriate large value of the parameter m, however, these rate-dependent models are able to reproduce almost rate-independent behavior. The function x^m in that sense can be considered to be a regularizing function (x stands for $J_2^{1/2}/K$). Other examples of regularizing functions are $\tanh(x/m)$, and $[\exp(x) - 1]^m$.

9.10.2. Viscoplasticity Models

For high strain rate applications in dynamic plasticity (e.g., Cristescu, 1967; Cristescu and Suliciu, 1982; Clifton, 1983, 1985) viscoplastic models are often used. One dimensional rheological model of viscoplastic response is shown in Fig. 9.21. There are two surfaces in viscoplastic modeling, a static yield surface and a dynamic loading surface. Consider a simple model of J_2 viscoplasticity. The flow potential can be taken as

$$\Omega = \frac{1}{\zeta}\langle\, J_2^{1/2} - K_s(\vartheta)\,\rangle^2, \tag{9.10.10}$$

where ζ is the viscosity coefficient, and $K_s(\vartheta)$ represents the shear stress – plastic strain relationship from the (quasi) static shear test. The Macauley brackets are used, such that

$$\langle\psi\rangle = \begin{cases} \psi, & \text{if } \psi \geq 0, \\ 0, & \text{if } \psi < 0, \end{cases} \tag{9.10.11}$$

i.e., $\langle\psi\rangle = (\psi + |\psi|)/2$. The positive difference

$$J_2^{1/2} - K_s(\vartheta) \tag{9.10.12}$$

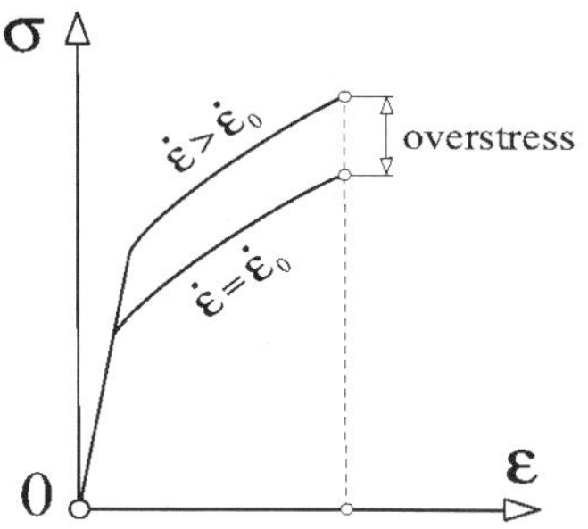

FIGURE 9.22. Stress-strain curves for quasi-static and dynamic loading conditions. The overstress measure is the difference between dynamic and static stress at a given amount of strain.

between the measures of the current dynamic stress and corresponding static stress (at a given level of equivalent plastic strain ϑ) is known as the overstress measure (Sokolovskii, 1948; Malvern, 1951). This is illustrated for uniaxial loading in Fig. 9.22. The plastic part of the rate of deformation is

$$\mathbf{D}^{\mathrm{p}} = \frac{1}{\zeta}\left[J_2^{1/2} - K_{\mathrm{s}}(\vartheta)\right]\frac{\boldsymbol{\sigma}'}{J_2^{1/2}}, \quad J_2^{1/2} - K_{\mathrm{s}}(\vartheta) > 0. \tag{9.10.13}$$

In the case of uniaxial loading σ with static yield stress $\sigma_Y > 0$, above gives

$$D^{\mathrm{p}} = \sqrt{\frac{2}{3}}\,\frac{1}{\zeta}\,(\sigma - \mathcal{P}\sigma), \quad \mathcal{P}\sigma = \sigma_Y\,\mathrm{sign}(\sigma). \tag{9.10.14}$$

This encompasses both tensile and compressive loading. When the operator $\mathcal{P}$ is applied to axial stress, it maps a tensile stress $\sigma > 0$ to σ_Y, and a compressive stress $\sigma < 0$ to $-\sigma_Y$ (Duvaut and Lions, 1976; Simo and Hughes, 1998).

The inverted form of Eq. (9.10.13) is

$$\boldsymbol{\sigma}' = \zeta\,\mathbf{D}^{\mathrm{p}} + 2K_{\mathrm{s}}(\vartheta)\,\frac{\mathbf{D}^{\mathrm{p}}}{\left(2\mathbf{D}^{\mathrm{p}} : \mathbf{D}^{\mathrm{p}}\right)^{1/2}}, \tag{9.10.15}$$

which shows that the rate-dependence in the model comes from the first term on the right-hand side only. In quasi-static tests, viscosity ζ is taken to be equal to zero, and Eq. (9.10.15) reduces to time-independent, von Mises isotropic hardening plasticity. In this case, the flow potential Ω is constant within the elastic range bounded by the yield surface $J_2^{1/2} = K_{\mathrm{s}}(\vartheta)$. The total rate of deformation is obtained by adding to (9.10.13) the elastic part of rate of deformation, such that

$$\mathbf{D} = \underline{\underline{\mathbf{M}}} : \overset{\circ}{\boldsymbol{\tau}} + \frac{1}{\zeta}\left[J_2^{1/2} - K_{\mathrm{s}}(\vartheta)\right]\frac{\boldsymbol{\sigma}'}{J_2^{1/2}}. \tag{9.10.16}$$

The inverted form of (9.10.16), in the case of infinitesimal elastic strain, is

$$\overset{\circ}{\boldsymbol{\tau}} = \underline{\underline{\boldsymbol{\Lambda}}} : \mathbf{D} - \frac{2\mu}{\zeta}\left[J_2^{1/2} - K_{\mathrm{s}}(\vartheta)\right]\frac{\boldsymbol{\sigma}'}{J_2^{1/2}}, \tag{9.10.17}$$

where μ is the elastic shear modulus.

Perzyna Model

More general representation for Ω is obtained by using the Perzyna (1963, 1966) viscoplastic model. For example, by taking

$$\Omega = \frac{C}{m+1}\langle\, f(\boldsymbol{\sigma}) - K_{\mathrm{s}}(\vartheta)\,\rangle^{m+1}, \tag{9.10.18}$$

we obtain

$$\mathbf{D}^{\mathrm{p}} = C\left[f(\boldsymbol{\sigma}) - K_{\mathrm{s}}(\vartheta)\right]^m \frac{\partial f}{\partial \boldsymbol{\sigma}}, \quad f(\boldsymbol{\sigma}) - K_{\mathrm{s}}(\vartheta) > 0. \tag{9.10.19}$$

If

$$f = J_2^{1/2}, \quad C = \frac{2}{\zeta}, \quad K_{\mathrm{s}}(\vartheta) = K^0 = \text{const.}, \tag{9.10.20}$$

Equation (9.10.19) gives

$$\mathbf{D}^{\mathrm{p}} = \frac{1}{\zeta}\left(J_2^{1/2} - K^0\right)^m \frac{\boldsymbol{\sigma}'}{J_2^{1/2}}. \tag{9.10.21}$$

This is is a generalization of the nonlinear Bingham model (e.g., Shames and Cozzarelli, 1992). In the case when

$$K_{\mathrm{s}}(\vartheta) = 0, \quad f = J_2^{1/2}, \quad C = \frac{2\dot{\gamma}^0}{K^m}, \tag{9.10.22}$$

Equation (9.10.19) reproduces the power-law J_2 creep of Eq. (9.10.3). See also Eisenberg and Yen (1981), and Bammann and Krieg (1987). The rate-dependent inelastic deformation of porous materials was studied by Duva and Hutchinson (1984), Haghi and Anand (1992), and Leblond, Perrin, and Suquet (1994).

Viscoplasticity with Isotropic–Kinematic Hardening

Other generalizations of Eq. (9.10.13) are possible. For example, suppose that the static yield condition is of a combined, isotropic–kinematic hardening type. The center of the yield surface is the back stress $\boldsymbol{\alpha}$ and the current radius of the yield surface is $K_\alpha(\vartheta)$. The dynamic loading condition is then

$$\hat{f} = \frac{1}{2}(\boldsymbol{\sigma}' - \boldsymbol{\alpha}) : (\boldsymbol{\sigma}' - \boldsymbol{\alpha}) - \hat{K}^2 = 0, \tag{9.10.23}$$

where $\hat{K}$ is the current radius of the loading surface. Consequently, the plastic rate of deformation becomes

$$\mathbf{D}^{\mathrm{p}} = \frac{1}{\zeta}\langle\, ||\boldsymbol{\sigma}' - \boldsymbol{\alpha}|| - \sqrt{2}\,K_\alpha\rangle\,\frac{\boldsymbol{\sigma}' - \boldsymbol{\alpha}}{||\boldsymbol{\sigma}' - \boldsymbol{\alpha}||}\,. \tag{9.10.24}$$

For convenience, we introduced the norm

$$||\boldsymbol{\sigma}' - \boldsymbol{\alpha}|| = \left[(\boldsymbol{\sigma}' - \boldsymbol{\alpha}) : (\boldsymbol{\sigma}' - \boldsymbol{\alpha})\right]^{1/2} = \sqrt{2}\,\hat{K}. \tag{9.10.25}$$

An accompanying evolution equation for the back stress $\boldsymbol{\alpha}$ is usually of the type given by Eq. (9.4.49). The viscosity parameter ζ can be a function of the introduced state variables. The potential function Ω, associated with Eq. (9.10.24), is

$$\Omega = \frac{1}{2\zeta}\langle\, ||\boldsymbol{\sigma}' - \boldsymbol{\alpha}|| - \sqrt{2}\,K_\alpha\rangle^2, \tag{9.10.26}$$

such that $\mathbf{D}^{\mathrm{p}} = \partial\Omega/\partial\boldsymbol{\sigma}$. Since

$$\mathbf{D}^{\mathrm{p}} = ||\mathbf{D}^{\mathrm{p}}||\,\frac{\boldsymbol{\sigma}' - \boldsymbol{\alpha}}{||\boldsymbol{\sigma}' - \boldsymbol{\alpha}||}\,, \tag{9.10.27}$$

comparison with (9.10.24) identifies

$$||\mathbf{D}^{\mathrm{p}}|| = \frac{1}{\zeta}\langle\, ||\boldsymbol{\sigma}' - \boldsymbol{\alpha}|| - \sqrt{2}\,K_\alpha\rangle. \tag{9.10.28}$$

Thus, the connection

$$\Omega = \frac{\zeta}{2}\,||\mathbf{D}^{\mathrm{p}}||^2\,. \tag{9.10.29}$$

The deviatoric symmetric tensor

$$\mathbf{d}^{\mathrm{p}} = \frac{\mathbf{D}^{\mathrm{p}}}{||\mathbf{D}^{\mathrm{p}}||} \tag{9.10.30}$$

has, in general, four independent components (since $||\mathbf{d}^{\mathrm{p}}|| = 1$). The representation $\mathbf{D}^{\mathrm{p}} = ||\mathbf{D}^{\mathrm{p}}||\,\mathbf{d}^{\mathrm{p}}$ is referred to as the polar representation of $\mathbf{D}^{\mathrm{p}}$ (Van Houtte, 1994).

More general expressions for the plastic rate of deformation have also been employed in the studies of viscoplastic response. Representative references include Chaboche (1989,1993,1996), Bammann (1990), McDowell (1992), and Freed and Walker (1991,1993). Nonassociative viscoplastic flow rules were considered by Marin and McDowell (1996), and for geomaterials by Cristescu (1994), who also gives the reference to other related work.

Generalized Duvaut–Lions Formulation

According to this model, the viscoplastic rate of deformation is postulated to be

$$\mathbf{D}^{\mathrm{p}} = \frac{1}{t_{\mathrm{d}}} \boldsymbol{\mathcal{M}} : (\boldsymbol{\sigma}' - \boldsymbol{\beta}), \quad f(\boldsymbol{\sigma}') \geq 0, \tag{9.10.31}$$

where t_{d} is the relaxation time, and $\boldsymbol{\mathcal{M}}$ is the elastic compliance tensor. For an isotropic material,

$$\boldsymbol{\mathcal{M}} = \frac{1}{2\mu} \boldsymbol{J} + \frac{1}{3\kappa} \boldsymbol{K}, \tag{9.10.32}$$

where μ and κ are the elastic shear and bulk moduli. The base tensors $\boldsymbol{J}$ and $\boldsymbol{K}$ sum to give the fourth-order unit tensor, $\boldsymbol{J} + \boldsymbol{K} = \boldsymbol{I}$, as discussed following Eq. (9.1.36). The deviatoric rest stress $\boldsymbol{\beta}$ in Eq. (9.10.31) is the stress corresponding to the inviscid solution, which satisfies the static yield condition $f(\boldsymbol{\beta}) = 0$. The rest stress is determined from the actual stress $\boldsymbol{\sigma}$ by the closest-point projection

$$\boldsymbol{\beta} = \boldsymbol{P} : \boldsymbol{\sigma}. \tag{9.10.33}$$

For example, if the operator $\boldsymbol{P}$ is defined by

$$\boldsymbol{P} = \sqrt{\frac{2}{3}}\, \sigma_Y \frac{\boldsymbol{J}}{\|\boldsymbol{\sigma}'\|}, \quad \|\boldsymbol{\sigma}'\| = (\boldsymbol{\sigma}' : \boldsymbol{\sigma}')^{1/2}, \tag{9.10.34}$$

there follows

$$\boldsymbol{\beta} = \sqrt{\frac{2}{3}}\, \sigma_Y \frac{\boldsymbol{\sigma}'}{\|\boldsymbol{\sigma}'\|}. \tag{9.10.35}$$

This corresponds to the static yield condition of the J_2 perfect plasticity, which is

$$f(\boldsymbol{\beta}) = \|\boldsymbol{\beta}\| - \sqrt{\frac{2}{3}}\, \sigma_Y = 0, \quad \sigma_Y = \text{const.} \tag{9.10.36}$$

The substitution of Eq. (9.10.32) into Eq. (9.10.31) gives the constitutive structure

$$\mathbf{D}^{\mathrm{p}} = \frac{1}{\zeta_{\mathrm{d}}} (\boldsymbol{\sigma}' - \boldsymbol{\beta}), \tag{9.10.37}$$

where

$$\zeta_{\mathrm{d}} = 2\mu t_{\mathrm{d}} > 0 \tag{9.10.38}$$

is the viscosity coefficient. Further analysis of the generalized Duvaut–Lions model and its numerical implementation can be found in the book by Simo and Hughes (1998). See also Krempl (1996), and Lubarda and Benson (2001).

Viscosity Tensor

The second-order viscosity tensor can be introduced as

$$\boldsymbol{Z} = \zeta_{\mathrm{d}}\,\boldsymbol{J} + \zeta_{\mathrm{v}}\,\boldsymbol{K}, \tag{9.10.39}$$

where ζ_{d} and ζ_{v} are the shear and bulk viscosities. The plastic rate of deformation of the generalized Duvaut–Lions model is then

$$\mathbf{D}^{\mathrm{p}} = \boldsymbol{Z}^{-1} : (\boldsymbol{\sigma}' - \boldsymbol{\beta}). \tag{9.10.40}$$

Introducing further the relaxation time tensor,

$$\boldsymbol{T} = t_{\mathrm{d}}\,\boldsymbol{J} + t_{\mathrm{v}}\,\boldsymbol{K}, \tag{9.10.41}$$

we have the connection

$$\boldsymbol{Z}^{-1} = \boldsymbol{T}^{-1} : \boldsymbol{\mathcal{M}}. \tag{9.10.42}$$

In particular, the relaxation time and viscosity coefficients are related by

$$\zeta_{\mathrm{d}} = 2\mu t_{\mathrm{d}}, \quad \zeta_{\mathrm{v}} = 3\kappa t_{\mathrm{v}}. \tag{9.10.43}$$

9.11. Deformation Theory of Plasticity

Simple plasticity theory has been suggested for proportional loading and small deformation by Hencky (1924) and Ilyushin (1947,1963). A large deformation version of this theory is here presented. It is convenient to cast the formulation by using the logarithmic strain

$$\mathbf{E}_{(0)} = \ln \mathbf{U}, \tag{9.11.1}$$

and its conjugate stress $\mathbf{T}_{(0)}$. The left stretch tensor is $\mathbf{U}$. Assume that the loading is such that all stress components increase proportionally, i.e.,

$$\mathbf{T}_{(0)} = c(t)\,\mathbf{T}^*_{(0)}, \tag{9.11.2}$$

where $\mathbf{T}^*_{(0)}$ is the stress tensor at an instant t^*, and $c(t)$ is a monotonically increasing function of t, with $c(t^*) = 1$. Evidently, Eq. (9.11.2) implies that the principal directions of $\mathbf{T}_{(0)}$ remain fixed during the deformation process, and parallel to those of $\mathbf{T}^*_{(0)}$.

Since the stress components proportionally increase, and no elastic unloading takes place, it is reasonable to assume that elastoplastic response can be described macroscopically by the constitutive structure of nonlinear elasticity, in which the total strain is a function of the total stress. Thus, we decompose the total strain tensor into elastic and plastic parts,

$$\mathbf{E}_{(0)} = \mathbf{E}^{\mathrm{e}}_{(0)} + \mathbf{E}^{\mathrm{p}}_{(0)}, \tag{9.11.3}$$

and assume that

$$\mathbf{E}^{\mathrm{e}}_{(0)} = \mathbf{M}_{(0)} : \mathbf{T}_{(0)}, \quad \mathbf{M}_{(0)} = \frac{1}{2\mu}\boldsymbol{J} + \frac{1}{3\kappa}\boldsymbol{K}, \tag{9.11.4}$$

$$\mathbf{E}^{\mathrm{p}}_{(0)} = \varphi\,\mathbf{T}'_{(0)}. \tag{9.11.5}$$

The shear and bulk moduli are μ and κ, the fourth-order tensors $\boldsymbol{J}$ and $\boldsymbol{K}$ are defined following Eq. (9.1.36), and φ is an appropriate scalar function to be determined in accord with experimental data. The prime designates a deviatoric part, so that plastic strain tensor is assumed to be traceless. More generally, a gradient of an isotropic function of $\mathbf{T}_{(0)}$ could be used in Eq. (9.11.5), in place of $\mathbf{T}'_{(0)}$ (Lubarda, 2000). This ensures that principal directions of plastic strain are parallel to those of $\mathbf{T}_{(0)}$. Since $\mathbf{M}_{(0)}$ in Eq. (9.11.4) corresponds to elastically isotropic material, principal directions of total strain $\mathbf{E}_{(0)}$ are also parallel to those of $\mathbf{T}_{(0)}$. Consequently, the stretch tensor $\mathbf{U}$ has its principal directions fixed during the deformation process, the matrix $\dot{\mathbf{U}}$ commutes with $\mathbf{U}$ and, by Eq. (3.6.18),

$$\dot{\mathbf{E}}_{(0)} = \dot{\mathbf{U}} \cdot \mathbf{U}^{-1}, \quad \mathbf{T}_{(0)} = \mathbf{R}^T \cdot \boldsymbol{\tau} \cdot \mathbf{R}. \tag{9.11.6}$$

The Kirchhoff stress is $\boldsymbol{\tau} = (\det \mathbf{F})\boldsymbol{\sigma}$, and $\mathbf{R}$ is the rotation tensor from the polar decomposition of deformation gradient $\mathbf{F} = \mathbf{R} \cdot \mathbf{U}$.

The requirement for the fixed principal directions of $\mathbf{U}$ severely restricts the class of admissible deformations. This is not surprising, because the premise of the deformation theory, the proportional stressing, imposes from outset the strong restrictions on the applicability of the analysis.

Introducing the spatial strain (see Subsection 2.3.2),

$$\boldsymbol{\mathcal{E}}_{(0)} = \mathbf{R}^T \cdot \mathbf{E}_{(0)} \cdot \mathbf{R}, \tag{9.11.7}$$

Equations (9.11.3)–(9.11.5) can be rewritten as

$$\boldsymbol{\mathcal{E}}_{(0)} = \boldsymbol{\mathcal{E}}^{\mathrm{e}}_{(0)} + \boldsymbol{\mathcal{E}}^{\mathrm{p}}_{(0)}, \tag{9.11.8}$$

$$\boldsymbol{\mathcal{E}}^{\mathrm{e}}_{(0)} = \mathbf{M}_{(0)} : \boldsymbol{\tau}, \tag{9.11.9}$$

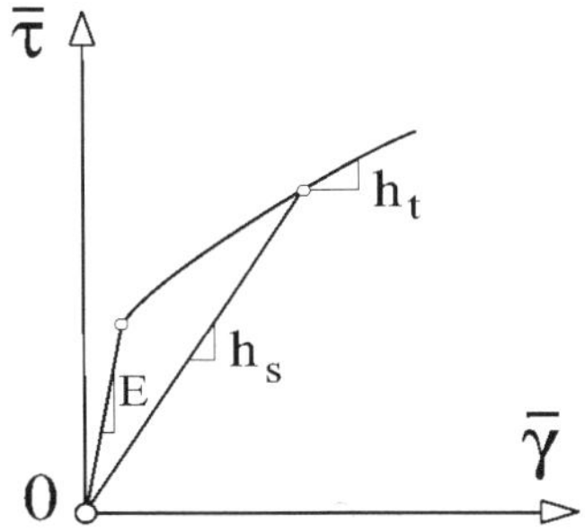

FIGURE 9.23. Nonlinear stress-strain response in pure shear. Indicated are the initial elastic modulus E, the secant modulus h_{s}, and the tangent modulus h_{t}.

$$\boldsymbol{\mathcal{E}}^{\mathrm{p}}_{(0)} = \varphi\, \boldsymbol{\tau}'. \tag{9.11.10}$$

It is noted that

$$\mathbf{T}'_{(0)} = \mathbf{R}^T \cdot \boldsymbol{\tau}' \cdot \mathbf{R}. \tag{9.11.11}$$

Suppose that a nonlinear relationship

$$\overline{\tau} = \overline{\tau}\,(\overline{\gamma}) \tag{9.11.12}$$

between the Kirchhoff stress and the logarithmic strain is available from the elastoplastic pure shear test ($E^{11}_{(0)} = \ln v$, $E^{22}_{(0)} = -\ln v$, all other $E^{ij}_{(0)}$ components being equal to zero; v is the amount of extension and compression in the two fixed principal directions 1 and 2). Let the secant and tangent moduli be defined by (Fig. 9.23)

$$h_{\mathrm{s}} = \frac{\overline{\tau}}{\overline{\gamma}}, \quad h_{\mathrm{t}} = \frac{\mathrm{d}\overline{\tau}}{\mathrm{d}\overline{\gamma}}, \tag{9.11.13}$$

and let

$$\overline{\tau} = \left(\frac{1}{2}\boldsymbol{\tau}' : \boldsymbol{\tau}'\right)^{1/2} = \left(\frac{1}{2}\mathbf{T}'_{(0)} : \mathbf{T}'_{(0)}\right)^{1/2}, \tag{9.11.14}$$

$$\overline{\gamma} = \left(2\,\boldsymbol{\mathcal{E}}'_{(0)} : \boldsymbol{\mathcal{E}}'_{(0)}\right)^{1/2} = \left(2\,\mathbf{E}'_{(0)} : \mathbf{E}'_{(0)}\right)^{1/2}. \tag{9.11.15}$$

Since, from Eqs. (9.11.9) and (9.11.10),

$$\boldsymbol{\mathcal{E}}'_{(0)} = \left(\frac{1}{2\mu} + \varphi\right)\boldsymbol{\tau}', \tag{9.11.16}$$

the substitution into Eq. (9.11.15) gives

$$\varphi = \frac{1}{2h_s} - \frac{1}{2\mu} . \tag{9.11.17}$$

Rate-Type Formulation of Deformation Theory

Although the deformation theory of plasticity is a total strain theory, the deformation theory can be cast in the rate-type form. This is important for later comparison with the flow theory of plasticity, and for extending the application of the resulting constitutive equations beyond the proportional loading. The rate-type formulation is also needed whenever the considered boundary value problem is being solved in an incremental manner.

Since $\dot{\mathbf{U}} \cdot \mathbf{U}^{-1}$ is symmetric, from the results in Section 2.6 we have

$$\mathbf{D} = \mathbf{R} \cdot \dot{\mathbf{E}}_{(0)} \cdot \mathbf{R}^T, \quad \mathbf{W} = \dot{\mathbf{R}} \cdot \mathbf{R}^{-1}. \tag{9.11.18}$$

Thus,

$$\dot{\mathbf{T}}_{(0)} = \mathbf{R}^T \cdot \overset{\circ}{\boldsymbol{\tau}} \cdot \mathbf{R}, \quad \overset{\circ}{\boldsymbol{\mathcal{E}}}_{(0)} = \mathbf{D}. \tag{9.11.19}$$

By differentiating Eqs. (9.11.3)–(9.11.5), or by applying the Jaumann derivative to Eqs. (9.11.8)–(9.11.10), there follows

$$\mathbf{D} = \mathbf{D}^{\mathrm{e}} + \mathbf{D}^{\mathrm{p}}, \tag{9.11.20}$$

$$\mathbf{D}^{\mathrm{e}} = \mathbf{M}_{(0)} : \overset{\circ}{\boldsymbol{\tau}}, \tag{9.11.21}$$

$$\mathbf{D}^{\mathrm{p}} = \dot{\varphi}\,\boldsymbol{\tau}' + \varphi\,\overset{\circ}{\boldsymbol{\tau}}{}'. \tag{9.11.22}$$

The deviatoric and spherical parts of the total rate of deformation tensor are accordingly

$$\mathbf{D}' = \dot{\varphi}\,\boldsymbol{\tau}' + \left(\frac{1}{2\mu} + \varphi\right) \overset{\circ}{\boldsymbol{\tau}}{}', \tag{9.11.23}$$

$$\operatorname{tr} \mathbf{D} = \frac{1}{3\kappa} \operatorname{tr} \overset{\circ}{\boldsymbol{\tau}}. \tag{9.11.24}$$

In order to derive an expression for the rate $\dot{\varphi}$, we differentiate Eqs. (9.11.14) and (9.11.15) to obtain

$$\bar{\tau}\dot{\bar{\tau}} = \frac{1}{2}\boldsymbol{\tau}' : \overset{\circ}{\boldsymbol{\tau}}, \quad \bar{\gamma}\dot{\bar{\gamma}} = 2\boldsymbol{\mathcal{E}}'_{(0)} : \mathbf{D}. \tag{9.11.25}$$

In view of Eqs. (9.11.13), (9.11.16), and (9.11.17), this gives

$$\frac{1}{2}\boldsymbol{\tau}' : \overset{\circ}{\boldsymbol{\tau}} = 2h_s h_t\, \boldsymbol{\mathcal{E}}'_{(0)} : \mathbf{D}' = h_t\, \boldsymbol{\tau}' : \mathbf{D}'. \tag{9.11.26}$$

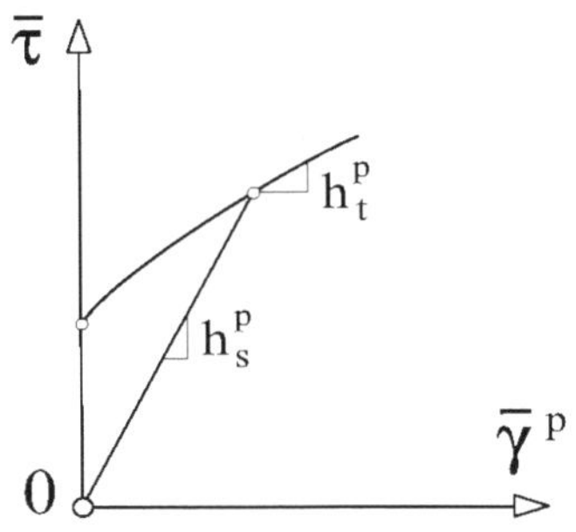

FIGURE 9.24. Shear stress vs. plastic shear strain. The plastic secant modulus is h_s^p, and the plastic tangent modulus is h_t^p.

When Eq. (9.11.23) is incorporated into Eq. (9.11.26), there follows

$$\dot{\varphi} = \frac{1}{2}\left(\frac{1}{h_\mathrm{t}} - \frac{1}{h_\mathrm{s}}\right)\frac{\boldsymbol{\tau}' : \overset{\circ}{\boldsymbol{\tau}}}{\boldsymbol{\tau}' : \boldsymbol{\tau}'}. \tag{9.11.27}$$

Substituting Eq. (9.11.27) into Eq. (9.11.23), the deviatoric part of the total rate of deformation becomes

$$\mathbf{D}' = \frac{1}{2h_\mathrm{s}}\left[\overset{\circ}{\boldsymbol{\tau}}{}' + \left(\frac{h_\mathrm{s}}{h_\mathrm{t}} - 1\right)\frac{(\boldsymbol{\tau}' \otimes \boldsymbol{\tau}') : \overset{\circ}{\boldsymbol{\tau}}}{\boldsymbol{\tau}' : \boldsymbol{\tau}'}\right]. \tag{9.11.28}$$

Equation (9.11.28) can be inverted to express the deviatoric part of $\overset{\circ}{\boldsymbol{\tau}}$ as

$$\overset{\circ}{\boldsymbol{\tau}}{}' = 2h_\mathrm{s}\left[\mathbf{D}' - \left(1 - \frac{h_\mathrm{t}}{h_\mathrm{s}}\right)\frac{(\boldsymbol{\tau}' \otimes \boldsymbol{\tau}') : \mathbf{D}}{\boldsymbol{\tau}' : \boldsymbol{\tau}'}\right]. \tag{9.11.29}$$

During initial, purely elastic stage of deformation, $h_\mathrm{t} = h_\mathrm{s} = \mu$. The onset of plasticity, beyond which Eqs. (9.11.28) and (9.11.29) apply, occurs when $\bar{\tau}$, defined by the second invariant of the deviatoric stress in Eq. (9.11.14), reaches the initial yield stress in shear. The resulting theory is often referred to as the J_2 deformation theory of plasticity.

If plastic secant and tangent moduli are used (Fig. 9.24), related to secant and tangent moduli with respect to total strain by

$$\frac{1}{h_\mathrm{t}} - \frac{1}{h_\mathrm{t}^\mathrm{p}} = \frac{1}{h_\mathrm{s}} - \frac{1}{h_\mathrm{s}^\mathrm{p}} = \frac{1}{\mu}, \tag{9.11.30}$$

the plastic part of the rate of deformation can be rewritten as

$$\mathbf{D}^\mathrm{p} = \frac{1}{2h_\mathrm{s}^\mathrm{p}}\overset{\circ}{\boldsymbol{\tau}}{}' + \left(\frac{1}{2h_\mathrm{t}^\mathrm{p}} - \frac{1}{2h_\mathrm{s}^\mathrm{p}}\right)\frac{(\boldsymbol{\tau}' \otimes \boldsymbol{\tau}') : \overset{\circ}{\boldsymbol{\tau}}}{\boldsymbol{\tau}' : \boldsymbol{\tau}'}. \tag{9.11.31}$$

9.11.1. Deformation vs. Flow Theory of Plasticity

For proportional loading, defined by Eq. (9.11.2), the stress rates are

$$\dot{\mathbf{T}}_{(0)} = \frac{\dot{c}}{c}\,\mathbf{T}_{(0)}, \quad \overset{\circ}{\boldsymbol{\tau}} = \frac{\dot{c}}{c}\,\boldsymbol{\tau}. \tag{9.11.32}$$

Consequently,

$$\frac{\boldsymbol{\tau}' : \overset{\circ}{\boldsymbol{\tau}}}{\boldsymbol{\tau}' : \boldsymbol{\tau}'} = \frac{\dot{c}}{c}, \tag{9.11.33}$$

and from Eq. (9.11.27) we have

$$\dot{\varphi} = \frac{1}{2}\left(\frac{1}{h_{\mathrm{t}}} - \frac{1}{h_{\mathrm{s}}}\right)\frac{\dot{c}}{c} = \frac{1}{2}\left(\frac{1}{h_{\mathrm{t}}^{\mathrm{p}}} - \frac{1}{h_{\mathrm{s}}^{\mathrm{p}}}\right)\frac{\dot{c}}{c}. \tag{9.11.34}$$

The plastic part of the rate of deformation reduces to

$$\mathbf{D}^{\mathrm{p}} = \frac{1}{2h_{\mathrm{t}}^{\mathrm{p}}}\,\frac{\dot{c}}{c}\,\boldsymbol{\tau}'. \tag{9.11.35}$$

On the other hand, in the case of the flow theory of plasticity,

$$\dot{\mathbf{E}}_{(0)} = \dot{\mathbf{E}}^{\mathrm{e}}_{(0)} + \dot{\mathbf{E}}^{\mathrm{p}}_{(0)}, \tag{9.11.36}$$

$$\dot{\mathbf{E}}^{\mathrm{e}}_{(0)} = \mathbf{M}_{(0)} : \dot{\mathbf{T}}_{(0)}, \quad \dot{\mathbf{E}}^{\mathrm{p}}_{(0)} = \dot{\gamma}\,\mathbf{T}'_{(0)}. \tag{9.11.37}$$

The yield surface is defined by

$$\frac{1}{2}\,\mathbf{T}'_{(0)} : \mathbf{T}'_{(0)} - \hat{k}^2(\vartheta) = 0, \quad \vartheta = \int_0^t \left(2\,\dot{\mathbf{E}}^{\mathrm{p}}_{(0)} : \dot{\mathbf{E}}^{\mathrm{p}}_{(0)}\right)^{1/2} \mathrm{d}t, \tag{9.11.38}$$

so that the consistency condition gives

$$\dot{\gamma} = \frac{1}{2h_{\mathrm{t}}^{\mathrm{p}}}\,\frac{\boldsymbol{\tau}' : \overset{\circ}{\boldsymbol{\tau}}}{\boldsymbol{\tau}' : \boldsymbol{\tau}'}. \tag{9.11.39}$$

The plastic tangent modulus is $h_{\mathrm{t}}^{\mathrm{p}} = \mathrm{d}\hat{k}/\mathrm{d}\vartheta$. The parameter $\hat{k}$ is related to k of Subsection 9.4.1 by $\hat{k} = (\det \mathbf{F})\,k$. Since

$$\mathbf{T}_{(0)} = \mathbf{R}^T \cdot \boldsymbol{\tau} \cdot \mathbf{R}, \quad \dot{\mathbf{E}}_{(0)} = \mathbf{R}^T \cdot \mathbf{D} \cdot \mathbf{R}, \tag{9.11.40}$$

the plastic part of the rate of deformation becomes

$$\mathbf{D}^{\mathrm{p}} = \frac{1}{2h_{\mathrm{t}}^{\mathrm{p}}}\,\frac{(\boldsymbol{\tau}' \otimes \boldsymbol{\tau}') : \overset{\circ}{\boldsymbol{\tau}}}{\boldsymbol{\tau}' : \boldsymbol{\tau}'}. \tag{9.11.41}$$

In the case of proportional loading, Eq. (9.11.41) reduces to Eq. (9.11.35). Illustrative examples can be found in Kachanov (1971), and Neale and Shrivastava (1990). Also, note the connection

$$\dot{\gamma} - \dot{\varphi} = \varphi\,\frac{\dot{c}}{c}. \tag{9.11.42}$$

A study of variational principles within the framework of deformation theory of plasticity is presented by Martin (1975), Temam (1985), Gao and Strang (1989), Ponte Castañeda (1992), and Han and Reddy (1999).

9.11.2. Application beyond Proportional Loading

Deformation theory agrees with flow theory of plasticity only under proportional loading, since then specification of the final state of stress also specifies the stress history. For general (nonproportional) loading, more accurate and physically appropriate is the flow theory of plasticity, particularly with an accurate modeling of the yield surface and the hardening characteristics. Budiansky (1959), however, indicated that deformation theory can be successfully used for certain nearly proportional loading paths, as well. The stress rate $\overset{\circ}{\boldsymbol{\tau}}{}'$ in Eq. (9.11.31) then does not have to be codirectional with $\boldsymbol{\tau}'$. The first and third term (both proportional to $1/2h_{\rm s}^{\rm p}$) in Eq. (9.11.31) do not cancel each other in this case (as they do for proportional loading), and the plastic part of the rate of deformation depends on both components of the stress rate $\overset{\circ}{\boldsymbol{\tau}}{}'$, one in the direction of $\boldsymbol{\tau}'$ and the other normal to it. In contrast, according to flow theory with the von Mises smooth yield surface, the component of the stress rate $\overset{\circ}{\boldsymbol{\tau}}{}'$ normal to $\boldsymbol{\tau}'$ (thus tangential to the yield surface) does not affect the plastic part of the rate of deformation. Physical theories of plasticity (Batdorf and Budiansky, 1954; Sanders, 1954; Hill, 1967b) indicate that the yield surface of a polycrystalline aggregate develops a vertex at its loading stress point, so that infinitesimal increments of stress in the direction normal to $\boldsymbol{\tau}'$ indeed cause further plastic flow ("vertex softening"). Since the structure of the deformation theory of plasticity under proportional loading does not use any notion of the yield surface, Budiansky (*op. cit.*) suggested that Eq. (9.11.31) can be adopted to describe the response when the yield surface develops a vertex. If Eq. (9.11.31) is rewritten in the form

$$\mathbf{D}^{\rm p} = \frac{1}{2h_{\rm s}^{\rm p}}\left[\overset{\circ}{\boldsymbol{\tau}}{}' - \frac{(\boldsymbol{\tau}'\otimes\boldsymbol{\tau}'):\overset{\circ}{\boldsymbol{\tau}}}{\boldsymbol{\tau}':\boldsymbol{\tau}'}\right] + \frac{1}{2h_{\rm t}^{\rm p}}\,\frac{(\boldsymbol{\tau}'\otimes\boldsymbol{\tau}'):\overset{\circ}{\boldsymbol{\tau}}}{\boldsymbol{\tau}':\boldsymbol{\tau}'}, \tag{9.11.43}$$

the first term on the right-hand side gives the response to component of the stress increment normal to $\boldsymbol{\tau}'$. The associated plastic modulus is $h_{\rm s}^{\rm p}$. The plastic modulus associated with the component of the stress increment in the direction of $\boldsymbol{\tau}'$ is h_t^p. Therefore, for continued plastic flow with small deviations from proportional loading (so that all yield segments which intersect at the vertex are active – fully active loading), Eq. (9.11.43) can be used as a model of a pointed vertex (Stören and Rice, 1975). The idea was used by Rudnicki and Rice (1975) in modeling the inelastic behavior of fissured rocks, as discussed in Subsection 9.8.2. See also Gotoh (1985), and Goya and Ito (1991).

For the full range of directions of the stress increment, the relationship between the rates of stress and plastic deformation is not necessarily linear, although it is homogeneous in these rates, in the absence of time-dependent (creep) effects. A corner theory that predicts continuous variation of the stiffness and allows increasingly nonproportional increments of stress was formulated by Christoffersen and Hutchinson (1979). This is discussed in the next subsection. When applied to the analysis of necking in thin sheets under biaxial stretching, the results were in better agreement with experiments than those obtained from the theory with a smooth yield characterization. Similar observations were long known in the field of elastoplastic buckling. Deformation theory predicts the buckling loads better than flow theory with a smooth yield surface (Hutchinson, 1974).

9.11.3. J_2 Corner Theory

In phenomenological J_2 corner theory of plasticity, proposed by Christoffersen and Hutchinson (1979), the instantaneous elastoplastic moduli for nearly proportional loading are chosen equal to the J_2 deformation theory moduli, while for increasing deviation from proportional loading the moduli increase smoothly until they coincide with elastic moduli for stress increments directed along or within the corner of the yield surface. The yield surface in the neighborhood of the loading point in deviatoric stress space (Fig. 9.25) is a cone around the axis

$$\boldsymbol{l} = \frac{\boldsymbol{\tau}'}{(\boldsymbol{\tau}' : \mathbf{M}_{\mathrm{def}}^{\mathrm{p}} : \boldsymbol{\tau}')^{1/2}}, \tag{9.11.44}$$

where $\mathbf{M}_{\mathrm{def}}^{\mathrm{p}}$ is the plastic compliance tensor of the deformation theory. The angular measure θ of the stress rate direction, relative to the cone axis, is defined by

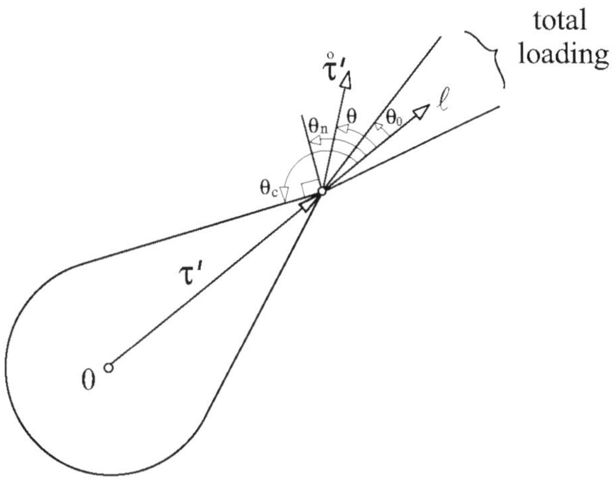

FIGURE 9.25. Near proportional or total loading range at the yield vertex of J_2 corner theory is a cone with the angle θ_0 around the axis $l \sim \boldsymbol{\tau}'$. The vertex cone is defined by the angle θ_c, and $\theta_n = \theta_c - \pi/2$.

$$\cos\theta = \frac{\boldsymbol{l} : \mathbf{M}^{p}_{def} : \overset{\circ}{\boldsymbol{\tau}}}{(\overset{\circ}{\boldsymbol{\tau}} : \mathbf{M}^{p}_{def} : \overset{\circ}{\boldsymbol{\tau}})^{1/2}} . \tag{9.11.45}$$

The conical surface separating elastic unloading and plastic loading is $\theta = \theta_c$, so that plastic rate of deformation falls within the range $0 \leq \theta \leq \theta_n$, where $\theta_n = \theta_c - \pi/2$. The range of near proportional loading is $0 \leq \theta \leq \theta_0$. The angle θ_0 is a suitable fraction of θ_n. The range of near proportional loading is the range of stress-rate directions for which no elastic unloading takes place on any of the yield vertex segments. This range is also called fully active or total loading range.

The stress-rate potential at the corner is defined by

$$\Pi = \Pi^{e} + \Pi^{p}, \quad \Pi^{p} = f(\theta)\Pi^{p}_{def}. \tag{9.11.46}$$

The elastic contribution to the stress-rate potential is

$$\Pi^{e} = \frac{1}{2}\overset{\circ}{\boldsymbol{\tau}} : \mathbf{M}^{e} : \overset{\circ}{\boldsymbol{\tau}}, \quad \mathbf{M}^{e} = \frac{1}{2\mu}\boldsymbol{J} + \frac{1}{3\kappa}\boldsymbol{K}. \tag{9.11.47}$$

The plastic stress-rate potential of the J_2 deformation theory can be written, from Eq. (9.11.28), as

$$\Pi^{p}_{def} = \frac{1}{2}\overset{\circ}{\boldsymbol{\tau}} : \mathbf{M}^{p}_{def} : \overset{\circ}{\boldsymbol{\tau}}, \quad \mathbf{M}^{p}_{def} = \frac{1}{2h_s}\left[\left(1 - \frac{h_s}{\mu}\right)\boldsymbol{J} + \left(\frac{h_s}{h_t} - 1\right)\frac{\boldsymbol{\tau}' \otimes \boldsymbol{\tau}'}{\boldsymbol{\tau}' : \boldsymbol{\tau}'}\right]. \tag{9.11.48}$$

The plastic stress-rate potential $\Pi^{\mathrm{p}}_{\mathrm{def}}$ is weighted by the cone transition function $f(\theta)$ to obtain the plastic stress-rate potential Π^{p} of the J_2 corner theory.

In the range of near proportional loading

$$0 \leq \theta \leq \theta_0, \quad f(\theta) = 1, \tag{9.11.49}$$

while in the elastic unloading range

$$\theta_{\mathrm{c}} \leq \theta \leq \pi, \quad f(\theta) = 0. \tag{9.11.50}$$

In the transition region $\theta_0 \leq \theta \leq \theta_{\mathrm{c}}$, the function $f(\theta)$ decreases monotonically and smoothly from one to zero in a way which ensures convexity of the plastic-rate potential,

$$\Pi^{\mathrm{p}}(\overset{\circ}{\boldsymbol{\tau}}_2) - \Pi^{\mathrm{p}}(\overset{\circ}{\boldsymbol{\tau}}_1) \geq \frac{\partial \Pi^{\mathrm{p}}}{\partial \overset{\circ}{\boldsymbol{\tau}}_1} : (\overset{\circ}{\boldsymbol{\tau}}_2 - \overset{\circ}{\boldsymbol{\tau}}_1). \tag{9.11.51}$$

A simple choice of $f(\theta)$ meeting these requirements is

$$f(\theta) = \cos^2\left(\frac{\pi}{2}\,\frac{\theta - \theta_0}{\theta_{\mathrm{c}} - \theta_0}\right), \quad \theta_0 \leq \theta \leq \theta_{\mathrm{c}}. \tag{9.11.52}$$

The specification of the angles θ_{c} and θ_0 in terms of the current stress measure is discussed by Christoffersen and Hutchinson (1979).

The rate-independence of the material response requires

$$\mathbf{D}^{\mathrm{p}} = \frac{\partial \Pi^{\mathrm{p}}}{\partial \overset{\circ}{\boldsymbol{\tau}}} = \frac{\partial^2 \Pi^{\mathrm{p}}}{\partial \overset{\circ}{\boldsymbol{\tau}} \otimes \partial \overset{\circ}{\boldsymbol{\tau}}} : \overset{\circ}{\boldsymbol{\tau}} = \mathbf{M}^{\mathrm{p}} : \overset{\circ}{\boldsymbol{\tau}} \tag{9.11.53}$$

to be a homogeneous function of degree one, and Π^{p} to be a homogeneous function of degree two in the stress rate $\overset{\circ}{\boldsymbol{\tau}}$. The function $\Pi^{\mathrm{p}}(\overset{\circ}{\boldsymbol{\tau}})$ is quadratic in the region of nearly proportional loading, but highly nonlinear in the transition region, due to nonlinearity associated with $f(\theta)$. The plastic rate of deformation is accordingly a linear function of $\overset{\circ}{\boldsymbol{\tau}}$ in the region of nearly proportional loading, but a nonlinear function in the transition region. Further details on the structure of J_2 corner theory, with its application to the study of sheet necking, are given in the Christoffersen and Hutchinson's paper. See also Needleman and Tvergaard (1982).

9.11.4. Pressure-Dependent Deformation Theory

To include pressure dependence and allow inelastic volume changes in deformation theory of plasticity, assume that, in place of Eq. (9.11.5), the plastic strain is related to stress by

$$\mathbf{E}^{\mathrm{p}}_{(0)} = \varphi \left[\mathbf{T}'_{(0)} + \frac{2}{3}\beta \left(\frac{1}{2}\mathbf{T}'_{(0)} : \mathbf{T}'_{(0)} \right)^{1/2} \mathbf{I}^0 \right], \tag{9.11.54}$$

where β is a material parameter, and $\mathbf{I}^0$ is the second-order unit tensor. It follows that the deviatoric and spherical parts of the plastic rate of deformation are

$$\mathbf{D}^{\mathrm{p}\prime} = \dot{\varphi}\,\boldsymbol{\tau}' + \varphi\,\overset{\circ}{\boldsymbol{\tau}}{}', \tag{9.11.55}$$

$$\operatorname{tr}\mathbf{D}^{\mathrm{p}} = 2\beta\, J_2^{1/2} \left(\dot{\varphi} + \varphi \frac{\boldsymbol{\tau}' : \overset{\circ}{\boldsymbol{\tau}}}{2\, J_2} \right). \tag{9.11.56}$$

The invariant

$$J_2 = \frac{1}{2}\boldsymbol{\tau}' : \boldsymbol{\tau}' \tag{9.11.57}$$

here represents the second invariant of the deviatoric part of the Kirchhoff stress.

Suppose that a nonlinear relationship $\bar{\tau} = \bar{\tau}(\bar{\gamma}^{\mathrm{p}})$ between the Kirchhoff stress and the plastic part of the logarithmic strain is available from the elastoplastic shear test. Let the plastic secant and tangent moduli be defined by

$$h^{\mathrm{p}}_{\mathrm{s}} = \frac{\bar{\tau}}{\bar{\gamma}^{\mathrm{p}}}, \quad h^{\mathrm{p}}_{\mathrm{t}} = \frac{\mathrm{d}\bar{\tau}}{\mathrm{d}\bar{\gamma}^{\mathrm{p}}}, \tag{9.11.58}$$

and let, in three-dimensional problems of overall compressive states of stress,

$$\bar{\tau} = J_2^{1/2} + \frac{1}{3}\mu_*\operatorname{tr}\boldsymbol{\tau}, \tag{9.11.59}$$

$$\bar{\gamma}^{\mathrm{p}} = \left(2\boldsymbol{\mathcal{E}}^{\mathrm{p}\ \prime}_{(0)} : \boldsymbol{\mathcal{E}}^{\mathrm{p}\ \prime}_{(0)} \right)^{1/2} = 2\,\varphi\, J_2^{1/2}. \tag{9.11.60}$$

Observe, from Eq. (9.11.54), that

$$\boldsymbol{\mathcal{E}}^{\mathrm{p}\ \prime}_{(0)} = \varphi\,\boldsymbol{\tau}'. \tag{9.11.61}$$

The friction-type coefficient in Eq. (9.11.59) is denoted by μ_*. By using the first of Eq. (9.11.58), therefore,

$$\varphi = \frac{1}{2h^{\mathrm{p}}_{\mathrm{s}}}\frac{\bar{\tau}}{J_2^{1/2}}. \tag{9.11.62}$$

In order to derive an expression for the rate $\dot{\varphi}$, differentiate Eqs. (9.11.59) and (9.11.60) to obtain

$$\dot{\bar{\tau}} = \frac{1}{2}J_2^{-1/2}(\boldsymbol{\tau}' : \overset{\circ}{\boldsymbol{\tau}}) + \frac{1}{3}\operatorname{tr}\overset{\circ}{\boldsymbol{\tau}}, \tag{9.11.63}$$

$$\dot{\bar{\gamma}}^{\mathrm{p}} = 2\left[\dot{\varphi}\, J_2^{1/2} + \frac{1}{2}\varphi\, J_2^{-1/2}(\boldsymbol{\tau}' : \overset{\circ}{\boldsymbol{\tau}})\right]. \tag{9.11.64}$$

Combining this with the second of Eq. (9.11.58) gives

$$\dot{\varphi} = \frac{1}{2}\left(\frac{1}{h_{\mathrm{t}}^{\mathrm{p}}} - \frac{1}{h_{\mathrm{s}}^{\mathrm{p}}}\frac{\bar{\tau}}{J_2^{1/2}}\right)\frac{\boldsymbol{\tau}' : \overset{\circ}{\boldsymbol{\tau}}}{2\,J_2} + \frac{1}{2h_{\mathrm{t}}^{\mathrm{p}}}\frac{1}{3}\mu_*\frac{\operatorname{tr}\overset{\circ}{\boldsymbol{\tau}}}{J_2^{1/2}}. \tag{9.11.65}$$

Consequently, by substituting Eqs. (9.11.62) and (9.11.65) into Eqs. (9.11.55) and (9.11.56), there follows

$$\begin{aligned}\mathbf{D}^{\mathrm{p}\prime} = {} & \frac{1}{2h_{\mathrm{s}}^{\mathrm{p}}}\frac{\bar{\tau}}{J_2^{1/2}}\overset{\circ}{\boldsymbol{\tau}}{}' + \frac{1}{2}\left(\frac{1}{h_{\mathrm{t}}^{\mathrm{p}}} - \frac{1}{h_{\mathrm{s}}^{\mathrm{p}}}\frac{\bar{\tau}}{J_2^{1/2}}\right)\frac{(\boldsymbol{\tau}'\otimes\boldsymbol{\tau}') : \overset{\circ}{\boldsymbol{\tau}}}{2\,J_2} \\ & + \frac{1}{2h_{\mathrm{t}}^{\mathrm{p}}}\frac{1}{3}\mu_*\frac{\operatorname{tr}\overset{\circ}{\boldsymbol{\tau}}}{J_2^{1/2}}\boldsymbol{\tau}',\end{aligned} \tag{9.11.66}$$

$$\operatorname{tr}\mathbf{D}^{\mathrm{p}} = \frac{\beta}{h_{\mathrm{t}}^{\mathrm{p}}}\left(\frac{\boldsymbol{\tau}' : \overset{\circ}{\boldsymbol{\tau}}}{2\,J_2^{1/2}} + \frac{1}{3}\mu_*\operatorname{tr}\overset{\circ}{\boldsymbol{\tau}}\right). \tag{9.11.67}$$

In the case when

$$\mu_* = 0, \quad \bar{\tau} = J_2^{1/2}, \tag{9.11.68}$$

Equation (9.11.66) simplifies and the deviatoric part of the plastic rate of deformation becomes

$$\mathbf{D}^{\mathrm{p}\prime} = \frac{1}{2h_{\mathrm{s}}^{\mathrm{p}}}\left[\overset{\circ}{\boldsymbol{\tau}}{}' + \left(\frac{h_{\mathrm{s}}^{\mathrm{p}}}{h_{\mathrm{t}}^{\mathrm{p}}} - 1\right)\frac{(\boldsymbol{\tau}'\otimes\boldsymbol{\tau}') : \overset{\circ}{\boldsymbol{\tau}}}{2\,J_2}\right], \tag{9.11.69}$$

while from Eq. (9.11.67) the volumetric part of the plastic rate of deformation is

$$\operatorname{tr}\mathbf{D}^{\mathrm{p}} = \frac{\beta}{2h_{\mathrm{t}}^{\mathrm{p}}}\frac{\boldsymbol{\tau}' : \overset{\circ}{\boldsymbol{\tau}}}{J_2^{1/2}}. \tag{9.11.70}$$

Noncoaxiality Factor

Equation (9.11.66) can be rewritten in an alternative form as

$$\mathbf{D}^{\mathrm{p}\prime} = \frac{1}{2h_{\mathrm{t}}^{\mathrm{p}}}\frac{\boldsymbol{\tau}'}{J_2^{1/2}}\left(\frac{\boldsymbol{\tau}' : \overset{\circ}{\boldsymbol{\tau}}}{2\,J_2^{1/2}} + \frac{1}{3}\mu_*\operatorname{tr}\overset{\circ}{\boldsymbol{\tau}}\right) + \frac{1}{2h_{\mathrm{s}}^{\mathrm{p}}}\frac{\bar{\tau}}{J_2^{1/2}}\left[\overset{\circ}{\boldsymbol{\tau}}{}' - \frac{(\boldsymbol{\tau}'\otimes\boldsymbol{\tau}') : \overset{\circ}{\boldsymbol{\tau}}}{2\,J_2}\right]. \tag{9.11.71}$$

The first part of $\mathbf{D}^{\mathrm{p}\prime}$ is coaxial with $\boldsymbol{\tau}'$. The second part is in the direction of the component of stress rate $\overset{\circ}{\boldsymbol{\tau}}{}'$ that is normal to $\boldsymbol{\tau}'$. There is no work associated with this part of the plastic rate of deformation, so that

$$\boldsymbol{\tau} : \mathbf{D}^{\mathrm{p}\,\prime} = \frac{1}{2h_{\mathrm{t}}^{\mathrm{p}}} \left(\boldsymbol{\tau}' : \overset{\circ}{\boldsymbol{\tau}} + \frac{2}{3}\,\mu_*\, J_2^{1/2}\, \mathrm{tr}\, \overset{\circ}{\boldsymbol{\tau}} \right). \tag{9.11.72}$$

Observe from Eqs. (9.11.67) and (9.11.72) that

$$\mathrm{tr}\, \mathbf{D}^{\mathrm{p}} = \beta\, \frac{\boldsymbol{\tau} : \mathbf{D}^{\mathrm{p}\,\prime}}{J_2^{1/2}}\,, \tag{9.11.73}$$

which offers a simple physical interpretation of the parameter β.

The coefficient

$$\varsigma = \frac{1}{2h_{\mathrm{s}}^{\mathrm{p}}}\, \frac{\overline{\tau}}{J_2^{1/2}} = \frac{1}{2h_{\mathrm{s}}^{\mathrm{p}}} \left(1 + \frac{1}{3}\,\mu_*\, \frac{\mathrm{tr}\, \boldsymbol{\tau}}{J_2^{1/2}} \right) \tag{9.11.74}$$

in Eq. (9.11.71) is a stress-dependent noncoaxiality factor. Other definitions of this factor have also been used in the literature (e.g., Nemat-Nasser, 1983).

Inverse Constitutive Equations

The deviatoric and volumetric part of the total rate of deformation are

$$\begin{aligned} \mathbf{D}' = {} & \left(\frac{1}{2\mu} + \frac{1}{2h_{\mathrm{s}}^{\mathrm{p}}}\, \frac{\overline{\tau}}{J_2^{1/2}} \right) \overset{\circ}{\boldsymbol{\tau}}{}' + \frac{1}{2} \left(\frac{1}{h_{\mathrm{t}}^{\mathrm{p}}} - \frac{1}{h_{\mathrm{s}}^{\mathrm{p}}}\, \frac{\overline{\tau}}{J_2^{1/2}} \right) \frac{(\boldsymbol{\tau}' \otimes \boldsymbol{\tau}') : \overset{\circ}{\boldsymbol{\tau}}}{2\, J_2} \\ & + \frac{1}{2h_{\mathrm{t}}^{\mathrm{p}}}\, \frac{1}{3}\, \mu_*\, \frac{\mathrm{tr}\, \overset{\circ}{\boldsymbol{\tau}}}{J_2^{1/2}}\, \boldsymbol{\tau}', \end{aligned} \tag{9.11.75}$$

$$\mathrm{tr}\, \mathbf{D} = \frac{1}{3} \left(\frac{1}{\kappa} + \frac{\mu_* \beta}{h_{\mathrm{t}}^{\mathrm{p}}} \right) \mathrm{tr}\, \overset{\circ}{\boldsymbol{\tau}} + \frac{\beta}{2h_{\mathrm{t}}^{\mathrm{p}}}\, \frac{\boldsymbol{\tau}' : \overset{\circ}{\boldsymbol{\tau}}}{J_2^{1/2}}\,. \tag{9.11.76}$$

The inverse relations are

$$\overset{\circ}{\underline{\boldsymbol{\tau}}}{}' = 2\mu \left[\frac{1}{b}\, \mathbf{D}' - \frac{a}{bc}\, \frac{(\boldsymbol{\tau}' \otimes \boldsymbol{\tau}') : \mathbf{D}}{2\, J_2} - \frac{1}{c}\, \mu_*\, \frac{\kappa}{2\mu}\, \frac{\boldsymbol{\tau}'}{J_2^{1/2}}\, \mathrm{tr}\, \mathbf{D} \right], \tag{9.11.77}$$

$$\mathrm{tr}\, \overset{\circ}{\underline{\boldsymbol{\tau}}} = \frac{3\kappa}{c} \left[\left(1 + \frac{h_{\mathrm{t}}^{\mathrm{p}}}{\mu} \right) \mathrm{tr}\, \mathbf{D} - \beta\, \frac{\boldsymbol{\tau}' : \mathbf{D}}{J_2^{1/2}} \right], \tag{9.11.78}$$

where

$$a = 1 - \frac{h_{\mathrm{t}}^{\mathrm{p}}}{h_{\mathrm{s}}^{\mathrm{p}}}\, \frac{\overline{\tau}}{J_2^{1/2}} \left(1 + \mu_* \beta\, \frac{\kappa}{h_{\mathrm{t}}^{\mathrm{p}}} \right), \quad b = 1 + \frac{\mu}{h_{\mathrm{s}}^{\mathrm{p}}}\, \frac{\overline{\tau}}{J_2^{1/2}}\,, \tag{9.11.79}$$

and

$$c = 1 + \frac{h_{\mathrm{t}}^{\mathrm{p}}}{\mu} + \mu_* \beta\, \frac{\kappa}{\mu}\,. \tag{9.11.80}$$

Comparing Eq. (9.8.36) of the modified flow theory with Eq. (9.11.71) of the pressure-dependent deformation theory of plasticity, it can be recognized that the two constitutive structures are equivalent, provided that the identification is made

$$H = h_t^p, \quad H_1 = h_s^p \frac{J_2^{1/2}}{\overline{\tau}} = \frac{1}{2\varsigma}. \tag{9.11.81}$$

With these connections, Eqs. (9.8.37) and (9.8.38) are also equivalent to Eqs. (9.11.77) and (9.11.78). The relationship between the two theories have been further discussed by Rudnicki (1982) and Nemat-Nasser (1982).

References

Anand, L. (1983), Plane deformations of ideal granular materials, *J. Mech. Phys. Solids*, Vol. 31, pp. 105–122.

Armstrong, P. J. and Frederick, C. O. (1966), A mathematical representation of the multiaxial Bauschinger effect, *G. E. G. B. Report RD/B/N*, 731.

Backhaus, G. (1968), Zur Fliessgrenze bei allgemeiner Verfestigung, *Z. angew. Math. Mech.*, Vol. 48, pp. 99–108.

Backhaus, G. (1972), Zur analytischen Erfassung des allgemeinen Bauschinger Effekts, *Acta Mech.*, Vol. 14, pp. 31–42.

Bammann, D. J. (1990), Modeling temperature and strain rate dependent large deformations of metals, *Appl. Mech. Rev.*, Vol. 43, No. 5, Part 2, pp. S312–S319.

Bammann, D. J. and Krieg, R. D. (1987), Summary and critique, in *Unified Constitutive Equations for Creep and Plasticity*, ed. A. K. Miller, pp. 303–336, Elsevier Applied Science, London.

Bardenhagen, S. G., Stout, M. G., and Gray, G. T. (1997), Three-dimensional, finite deformation, viscoplastic constitutive models for polymeric materials, *Mech. Mater.*, Vol. 25, pp. 235–253.

Bardet, J. P. (1990), Lode dependences for isotropic pressure-sensitive elastoplastic materials, *J. Appl. Mech.*, Vol. 57, pp. 498–506.

Barlat, F. *et al.* (1997), Yielding description for solution strengthened aluminum alloys, *Int. J. Plasticity*, Vol. 13, pp. 385–401.

Batdorf, S. S. and Budiansky, B. (1949), *A Mathematical Theory of Plasticity Based on the Concept of Slip*, N. A. C. A. TN 1871.

Batdorf, S. S. and Budiansky, B. (1954), Polyaxial stress-strain relations of a strain-hardening metal, *J. Appl. Mech.*, Vol. 21, pp. 323–326.

Bertsch, P. K. and Findley, W. N. (1962), An experimental study of subsequent yield surfaces – corners, normality, Bauschinger and allied effects, *Proc. 4th U.S. Natl. Congr. Appl. Mech.*, pp. 893–907, ASME, New York.

Bodner, S. R. and Partom, Y. (1975), Constitutive equations for elastic-viscoplastic strain-hardening materials, *J. Appl. Mech.*, Vol. 42, pp. 385–389.

Boley, B. A. and Weiner, J. H. (1960), *Theory of Thermal Stresses*, John Wiley, New York.

Budiansky, B. (1959), A reassessment of deformation theories of plasticity, *J. Appl. Mech.*, Vol. 26, pp. 259–264.

Casey, J. (1998), On elastic-thermo-plastic materials at finite deformations, *Int. J. Plasticity*, Vol. 14, pp. 173–191.

Casey, J. and Naghdi, P. M. (1981), On the characterization of strain-hardening in plasticity, *J. Appl. Mech.*, Vol. 48, pp. 285–296.

Casey, J. and Naghdi, P. M. (1983), On the nonequivalence of the stress and strain space formulations of plasticity theory, *J. Appl. Mech.*, Vol. 50, pp. 350–354.

Casey, J. and Sullivan, T. D. (1985), Pressure dependency, strength-differential effect, and plastic volume expansion in metals, *Int. J. Plasticity*, Vol. 1, pp. 39–61.

Caulk, D. A. and Naghdi, P. M. (1978), On the hardening response in small deformation of metals, *J. Appl. Mech.*, Vol. 45, pp. 755–764.

Chaboche, J.-L. (1986), Time independent constitutive theories for cyclic plasticity, *Int. J. Plasticity*, Vol. 2, pp. 149–188.

Chaboche, J.-L. (1989), Constitutive equations for cyclic plasticity and cyclic viscoplasticity, *Int. J. Plasticity*, Vol. 5, pp. 247–302.

Chaboche, J.-L. (1993), Cyclic viscoplastic constitutive equations, Part I: A thermodynamically consistent formulation, *J. Appl. Mech*, Vol. 60, pp. 813–821.

Chaboche, J.-L. (1996), Unified cyclic viscoplastic constitutive equations: Development, capabilities, and thermodynamic framework, in *Unified*

Constitutive Laws for Plastic Deformation, eds. A. S. Krausz and K. Krausz, pp. 1–68, Academic Press, San Diego.

Chakrabarty, J. (1987), *Theory of Plasticity*, McGraw-Hill, New York.

Chandler, H. W. (1985), A plasticity theory without Drucker's postulate, suitable for granular materials, *J. Mech. Phys. Solids*, Vol. 33, pp. 215–226.

Chen, W. F. and Han, D. J. (1988), *Plasticity for Structural Engineers*, Springer-Verlag, New York.

Christoffersen, J. and Hutchinson, J. W. (1979), A class of phenomenological corner theories of plasticity, *J. Mech. Phys. Solids*, Vol. 27, pp. 465–487.

Christoffersen, J., Mehrabadi, M. M., and Nemat-Nasser, S. (1981), A micromechanical description of granular material behavior, *J. Appl. Mech.*, Vol. 48, pp. 339–344.

Clifton, R. J. (1983), Dynamic plasticity, *J. Appl. Mech.*, Vol. 50, pp. 941–952.

Clifton, R. J. (1985), Experiments and the micromechanics of viscoplasticity, in *Plasticity Today – Modelling, Methods and Applications*, eds. A. Sawczuk and G. Bianchi, pp. 187–201, Elsevier Applied Science, London.

Cocks, A. C. F. (1989), Inelastic deformation of porous materials, *J. Mech. Phys. Solids*, Vol. 37, pp. 693–715.

Cristescu, N. (1967), *Dynamic Plasticity*, North-Holland, Amsterdam.

Cristescu, N. (1994), A procedure to determine nonassociated constitutive equations for geomaterials, *Int. J. Plasticity*, Vol. 10, pp. 103–131.

Cristescu, N. and Suliciu, I. (1982), *Viscoplasticity*, Martinus Nijhoff Publishers, The Hague.

Dafalias, Y. F. and Popov, E. P. (1975), A model of nonlinearly hardening materials for complex loading, *Acta Mech.*, Vol. 21, pp. 173–192.

Dafalias, Y. F. and Popov, E. P. (1976), Plastic internal variables formalism of cyclic plasticity, *J. Appl. Mech.*, Vol. 43, pp. 645–651.

De Boer, R. (1977), On non-isothermal elastic-plastic and elastic-viscoplastic deformations, *Int. J. Solids Struct.*, Vol. 13, pp. 1203–1217.

Desai, C. S. and Hasmini, Q. S. E. (1989), Analysis, evaluation, and implementation of a nonassociative model for geologic materials, *Int. J. Plasticity*, Vol. 5, pp. 397–420.

Desai, C. S. and Siriwardane, H. J. (1984), *Constitutive Laws for Engineering Materials – With Emphasis on Geological Materials*, Prentice-Hall, Inc., Englewood Cliffs, New Jersey.

DiMaggio, F. L. and Sandler, I. S. (1971), Material model for granular soils, *J. Engrg. Mech., ASCE*, Vol. 97, pp. 935–950.

Dorris, J. F. and Nemat-Nasser, S. (1982), A plasticity model for flow of granular materials under triaxial stress state, *Int. J. Solids Struct.*, Vol. 18, pp. 497–531.

Drucker, D. C. (1951), A more fundamental approach to plastic stress-strain relations, in *Proc. 1st U.S. Natl. Congr. Appl. Mech.*, ed. E. Sternberg, pp. 487–491, ASME, New York.

Drucker, D. C. (1959), A definition of stable inelastic material, *J. Appl. Mech.*, Vol. 26, pp. 101–106.

Drucker, D. C. (1960), Plasticity, in *Structural Mechanics – Proc. 1st Symp. Naval Struct. Mechanics*, eds. J. N. Goodier and N. J. Hoff, pp. 407–455, Pergamon Press, New York.

Drucker, D. C. (1960), Extension of the stability postulate with emphasis on temperature changes, in *Plasticity – Proc. 2nd Symp. Naval Struct. Mechanics*, eds. E. H. Lee and P. Symonds, pp. 170–184, Pergamon Press, New York.

Drucker, D. C. and Palgen, L. (1981), On stress-strain relations suitable for cyclic and other loading, *J. Appl. Mech.*, Vol. 48, pp. 479–485.

Drucker, D. C. and Prager, W. (1952), Soil mechanics and plastic analysis or limit design, *Q. Appl. Math.*, Vol. 10, pp. 157–165.

Duszek, M. K. and Perzyna, P. (1991), On combined isotropic and kinematic hardening effects in plastic flow processes, *Int. J. Plasticity*, Vol. 9, pp. 351–363.

Duva, J. M. and Hutchinson, J. W. (1984), Constitutive potentials for dilutely voided nonlinear materials, *Mech. Mater.*, Vol. 3, pp. 41–54.

Duvaut, G. and Lions, J.-L. (1976), *Inequalities in Mechanics and Physics*, Springer-Verlag, Berlin.

Eisenberg, M. A. and Phillips, A. (1968), On non-linear kinematic hardening, *Acta Mech.*, Vol. 5, pp. 1–13.

Eisenberg, M. A. and Yen, C.-F. (1981), A theory of multiaxial anisotropic viscoplasticity, *J. Appl. Mech.*, Vol. 48, pp. 276–284.

Ellyin, F. (1989), An anisotropic hardening rule for elastoplastic solids based on experimental observations, *J. Appl. Mech.*, Vol. 56, pp. 499–507.

Estrin, Y. and Mecking, H. (1986), An extension of the Bodner–Partom model of plastic deformation, *Int. J. Plasticity*, Vol. 2, pp. 73–85.

Faruque, M. O. and Chang, C. J. (1990), A constitutive model for pressure sensitive materials with particular reference to plain concrete, *Int. J. Plasticity*, Vol. 6, pp. 29–43.

Freed, A. D. and Walker, K. P. (1991), A viscoplastic theory with thermodynamic considerations, *Acta Mech.*, Vol. 90, pp. 155–174.

Freed, A. D. and Walker, K. P. (1993), Viscoplasticity with creep and plasticity bounds, *Int. J. Plasticity*, Vol. 9, pp. 213–242.

Gao, Y. and Strang, G. (1989), Dual extremum principles in finite deformation elastoplastic analysis, *Acta Applic. Math.*, Vol. 17, pp. 257–267.

Gotoh, M. (1985), A class of plastic constitutive equations with vertex effect – I. General theory, *Int. J. Solids Struct.*, Vol. 21, pp. 1101–1116.

Goya, M. and Ito, K. (1991), An expression of elastic-plastic constitutive law incorporating vertex formation and kinematic hardening, *J. Appl. Mech.*, Vol. 58, pp. 617–622.

Green, A. E. and Naghdi, P. M. (1965), A general theory of an elastic-plastic continuum, *Arch. Rat. Mech. Anal.*, Vol. 18, pp. 251–281.

Gupta, N. K. and Meyers, A. (1992), Considerations of translated stress deviators in describing yield surfaces, *Int. J. Plasticity*, Vol. 8, pp. 729–740.

Gupta, N. K. and Meyers, A. (1994), An alternative formulation for interpolating experimental yield surfaces, *Int. J. Plasticity*, Vol. 10, pp. 795–805.

Gurson, A. L. (1977), Continuum theory of ductile rapture by void nucleation and growth – Part I: Yield criteria and flow rules for porous ductile media, *J. Engng. Mater. Tech.*, Vol. 99, pp. 2–15.

Haghi, M. and Anand, L. (1992), A constitutive model for isotropic, porous, elastic-viscoplastic metals, *Mech. Mater.*, Vol. 13, pp. 37–53.

Han, W. and Reddy, B. D. (1999), *Plasticity – Mathematical Theory and Numerical Analysis*, Springer-Verlag, New York.

Harris, D. (1992), Plasticity models for soil, granular and jointed rock materials, *J. Mech. Phys. Solids*, Vol. 40, pp. 273–290.

Hart, E. W. (1970), A phenomenological theory for plastic deformation of polycrystalline metals, *Acta Metall.*, Vol. 18, pp. 599–610.

Hashiguchi, K. (1981), Constitutive equations of elastoplastic materials with anisotropic hardening and elastic-plastic transition, *J. Appl. Mech.*, Vol. 48, pp. 297–301.

Hashiguchi, K. (1988), A mathematical modification of two surface model formulation in plasticity, *Int. J. Solids Struct.*, Vol. 24, pp. 987–1001.

Hashiguchi, K. (1993), Fundamental requirements and formulation of elastoplastic constitutive equations with tangential plasticity, *Int. J. Plasticity*, Vol. 9, pp. 525–549.

Haupt, P. and Kamlah, M. (1995), Representation of cyclic hardening and softening using continuous variables, *Int. J. Plasticity*, Vol. 9, pp. 267–291.

Hecker, S. S. (1972), Experimental investigation of corners in the yield surface, *Acta Mech.*, Vol. 13, pp. 69–86.

Hecker, S. S. (1976), Experimental studies of yield phenomena in biaxially loaded metals, in *Constitutive Equations in Viscoplasticity: Computational and Engineering Aspects*, AMD Vol. 20, eds. J. A. Stricklin and K. J. Saczalski, pp. 1–33, ASME, New York.

Hencky, H. (1924), Zur Theorie plastischer Deformationen und der hierdurch im Material hervorgerufenen Nachspannungen, *Z. angew. Math. Mech.*, Vol. 4, pp. 323–334.

Hill, R. (1950), *The Mathematical Theory of Plasticity*, Oxford University Press, London.

Hill, R. (1965), Continuum micro-mechanics of elastoplastic polycrystals, *J. Mech. Phys. Solids*, Vol. 13, pp. 89–101.

Hill, R. (1966), Generalized constitutive relations for incremental deformation of metal crystals by multislip, *J. Mech. Phys. Solids*, Vol. 14, pp. 95–102.

Hill, R. (1967a), On the classical constitutive relations for elastic-plastic solids, in *Recent Progress in Applied Mechanics, Folke Odgvist Volume*, eds. B. Broberg, J. Hult and F. Niordson, pp. 241–249, Almqvist and Wiksel, Stockholm.

Hill, R. (1967b), The essential structure of constitutive laws for metal composites and polycrystals, *J. Mech. Phys. Solids*, Vol. 15, pp. 79–95.

Hill, R. (1978), Aspects of invariance in solid mechanics, *Adv. Appl. Mech.*, Vol. 18, pp. 1–75.

Hutchinson, J. W. (1970), Elastic-plastic behavior of polycrystalline metals and composites, *Proc. Roy. Soc. Lond. A*, Vol. 319, pp. 247–272.

Hutchinson, J. W. (1973), Finite strain analysis of elastic-plastic solids and structures, in *Numerical Solution of Nonlinear Structural Problems*, ed. R. F. Hartung, pp. 17–29, ASME, New York.

Hutchinson, J. W. (1974), Plastic buckling, *Adv. Appl. Mech.*, Vol. 14, pp. 67–144.

Ilyushin, A. A. (1947), Theory of plasticity at simple loading of the bodies exhibiting plastic hardening, *Prikl. Mat. Mekh.*, Vol. 11, pp. 291–296 (in Russian).

Ilyushin, A. A. (1963), *Plasticity. Foundations of the General Mathematical Theory*, Izd. Akad. Nauk SSSR, Moscow (in Russian).

Ishlinsky, A. Y. (1954), A general theory of plasticity with linear hardening, *Ukrain. Matem. Zhurnal*, Vol. 6, pp. 314–324 (in Russian).

Iwan, W. D. (1967), On a class of models for the yielding behavior of continuous and composite systems, *J. Appl. Mech.*, Vol. 34, pp. 612–617.

Jaeger, J. C. and Cook, N. G. W. (1976), *Fundamentals of Rock Mechanics*, Chapman and Hall, London.

Jiang, Y. and Kurath, P. (1996), Characteristics of the Armstrong–Frederick type plasticity models, *Int. J. Plasticity*, Vol. 12, pp. 387–415.

Jiang, Y. and Sehitoglu, H. (1996a), Comments on the Mróz multiple surface type plasticity models, *Int. J. Solids Struct.*, Vol. 33, pp. 1053–1068.

Jiang, Y. and Sehitoglu, H. (1996b), Modeling of cyclic ratchetting plasticity, Parts I and II, *J. Appl. Mech.*, Vol. 63, pp. 720–733.

Johnson, G. R. and Cook, W. H. (1983), A constitutive model and data for metals subjected to large strains, high strain rates, and high temperatures, in *Proceedings of the 7th International Symposium on Ballistics*, ADPA, pp. 1–7, The Hague.

Johnson, W. and Mellor, P. B. (1973), *Engineering Plasticity*, Van Nostrand Reinhold, London.

Johnston, W. G. and Gilman, J. J. (1959), Dislocation velocities, dislocation densities, and plastic flow in LiF crystals, *J. Appl. Phys.*, Vol. 30, pp. 129–144.

Kachanov, L. M. (1971), *Foundations of Theory of Plasticity*, North-Holland, Amsterdam.

Khan, A. S. and Huang, S. (1995), *Continuum Theory of Plasticity*, John Wiley & Sons, New York.

Klisinski, M., Mróz, Z., and Runesson, K. (1992), Structure of constitutive equations in plasticity for different choices of state and control variables, *Int. J. Plasticity*, Vol. 8, pp. 221–243.

Koiter, W. (1953), Stress-strain relations, uniqueness and variational theorems for elastic-plastic materials with a singular yield surface, *Q. Appl. Math.*, Vol. 11, pp. 350–354.

Krempl, E. (1996), A small-strain viscoplasticity theory based on overstress, in *Unified Constitutive Laws for Plastic Deformation*, eds. A. S. Krausz and K. Krausz, pp. 281–318, Academic Press, San Diego.

Krieg, R. D. (1975), A practical two surface plasticity theory, *J. Appl. Mech.*, Vol. 42, pp. 641–646.

Krieg, R. D. (1977), Numerical integration of some new unified plasticity-creep formulations, in *Proceedings of the 4th SMIRT Conference*, San Francisco, Paper M6/4.

Lade, P. V. and Kim, M. K. (1995), Single hardening constitutive model for soil, rock and concrete, *Int. J. Solids Struct.*, Vol. 32, pp. 1963–1978.

Leblond, J. B., Perrin, G., and Suquet, P. (1994), Exact results and approximate models for porous viscoplastic solids, *Int. J. Plasticity*, Vol. 10, pp. 213–235.

Lee, E. H. (1969), Elastic-plastic deformation at finite strains, *J. Appl. Mech.*, Vol. 36, pp. 1–6.

Lee, E. H. and Wierzbicki, T. (1967), Analysis of the propagation of plane elastic-plastic waves at finite strain, *J. Appl. Mech.*, Vol. 34, pp. 931–936.

Lee, J. H. (1988), Research note on a simple model for pressure-sensitive strain-hardening materials, *Int. J. Plasticity*, Vol. 4, pp. 265–278.

Lee, J. H. and Oung, J. (2000), Yield functions and flow rules for porous pressure-dependent strain-hardening polymeric materials, *J. Appl. Mech.*, Vol. 67, pp. 288–297.

Lee, Y. K. (1988), A finite elastoplastic flow theory for porous media, *Int. J. Plasticity*, Vol. 4, pp. 301–316.

Lehmann, Th. (1972), Einige Bemerkungen zu einer allgemeinen Klasse von Stoffgesetzen für grosse elasto-plastische Formänderungen, *Ing. Arch.*, Vol. 41, pp. 297–310.

Lehmann, Th. (1985), On a generalized constitutive law in thermo-plasticity taking into account different yield mechanisms, *Acta Mech.*, Vol. 57, pp. 1–23.

Lubarda, V. A. (1986), Finite compression of solids – Second order thermoelastic analysis, *Int. J. Solids Struct.*, Vol. 22, pp. 1517–1524.

Lubarda, V. A. (1989), Elasto-plastic constitutive behaviour of metals under high pressure, in *Advances in Plasticity 1989*, eds. A. S. Khan and M. Tokuda, pp. 633–636, Pergamon Press, Oxford.

Lubarda, V. A. (1994), Elastoplastic constitutive analysis with the yield surface in strain space, *J. Mech. Phys. Solids*, Vol. 42, pp. 931–952.

Lubarda, V. A. (2000), Deformation theory of plasticity revisited, *Proc. Montenegr. Acad. Sci. Arts*, Vol. 13, pp. 117–143.

Lubarda, V. A. and Benson, D. J. (2001), On the evolution equation for the rest stress in rate-dependent plasticity, *Int. J. Plasticity*, in press.

Lubarda, V. A., Krajcinovic, D., and Mastilovic, S. (1994), Damage model for brittle elastic solids with unequal tensile and compressive strengths, *Engng. Fracture Mech.*, Vol. 49, pp. 681–697.

Lubarda, V. A., Mastilovic, S., and Knap, J. (1996), Brittle-ductile transition in porous rocks by cap model, *ASCE J. Engr. Mech.*, Vol. 122, pp. 633–642.

Lubliner, J. (1990), *Plasticity Theory*, Macmillan Publishing Comp., New York.

Lucchesi, M. and Šilhavý, M. (1993), Thermoplastic materials with combined hardening, *Int. J. Plasticity*, Vol. 9, 291–315.

Malvern, L. E. (1951), The propagation of longitudinal waves of plastic deformation in a bar of material exhibiting a strain-rate effect, *J. Appl. Mech.*, Vol. 18, pp. 203–208.

Mandel, J. (1965), Generalisation de la theorie de plasticite de W. T. Koiter, *Int. J. Solids Struct.*, Vol. 1, 273–295.

Marin, E. B. and McDowell, D. L. (1996), Associative versus non-associative porous viscoplasticity based on internal state variable concepts, *Int. J. Plasticity*, Vol. 12, pp. 629–669.

Martin, J. B. (1975), *Plasticity: Fundamentals and General Results*, MIT Press, Cambridge, Massachusetts.

McDowell, D. L. (1985), A two surface model for transient nonproportional cyclic plasticity: Parts 1 and 2, *J. Appl. Mech.*, Vol. 52, pp. 298–308.

McDowell, D. L. (1987), An evaluation of recent developments in hardening and flow rules for rate-independent nonproportional cyclic plasticity, *J. Appl. Mech.*, Vol. 54, pp. 323–334.

McDowell, D. L. (1992), A nonlinear kinematic hardening theory for cyclic thermoplasticity and thermoviscoplasticity, *Int. J. Plasticity*, Vol. 8, pp. 695–728.

McMeeking, R. M. and Rice, J. R. (1975), Finite-element formulations for problems of large elastic-plastic deformation, *Int. J. Solids Struct.*, Vol. 11, pp. 601–616.

Mear, M. E. and Hutchinson, J. W. (1985), Influence of yield surface curvature on flow localization in dilatant plasticity, *Mech. Mater.*, Vol. 4, pp. 395–407.

Mehrabradi, M. M. and Cowin, S. C. (1981), On the double-sliding free-rotating model for the deformation of granular materials, *J. Mech. Phys. Solids*, Vol. 29, pp. 269–282.

Melan, E. (1938), Zur Plastizität des räumlichen Kontinuums, *Ing. Arch.*, Vol. 9, pp. 116–126.

Meyers, M. A. (1999), Dynamic deformation and failure, in *Mechanics and Materials – Fundamentals and Linkages*, eds. M. A. Meyers, R. W. Armstrong, and H. O. K. Kirchner, pp. 489–594, John Wiley & Sons, New York.

Miller, A. K. (1976), An inelastic constitutive model for monotonic, cyclic, and creep deformation, I and II, *J. Engng. Mater. Techn.*, Vol. 98, pp. 97–113.

Miller, A. K., ed. (1987), *Unified Constitutive Equations for Creep and Plasticity*, Elsevier Applied Science, London.

Moosbrugger, J. C. and McDowell, D. L. (1989), On a class of kinematic hardening rules for nonproportional cyclic plasticity, *J. Engng. Mater. Techn.*, Vol. 111, pp. 87–98.

Moreton, D. N., Moffat, D. G., and Parkinson, D. B. (1981), The yield surface behavior of pressure vessel steels, *J. Strain Anal.*, Vol. 16, pp. 127–136.

Mróz, Z. (1963), Non-associated flow laws in plasticity, *J. Mécanique*, Vol. 2, pp. 21–42.

Mróz, Z. (1967), On the description of anisotropic work-hardening, *J. Mech. Phys. Solids*, Vol. 15, pp. 163–175.

Mróz, Z. (1976), A non-linear hardening model and its application to cyclic plasticity, *Acta Mech.*, Vol. 25, pp. 51–61.

Naghdi, P. M. (1960), Stress-strain relations in plasticity and thermoplasticity, in *Plasticity – Proc. 2nd Symp. Naval Struct. Mechanics*, eds. E. H. Lee and P. Symonds, pp. 121–167, Pergamon Press, New York.

Naghdi, P. M. (1990), A critical review of the state of finite plasticity, *J. Appl. Math. Physics*, Vol. 41, pp. 315–394.

Naghdi, P. M., Essenburg, F., and Koff, W. (1958), An experimental study of initial and subsequent yield surfaces in plasticity, *J. Appl. Mech.*, Vol. 25, pp. 201–209.

Naghdi, P. M. and Nikkel, D. J. (1986), Two-dimensional strain cycling in plasticity, *J. Appl. Mech.*, Vol. 53, pp. 821–830.

Naghdi, P. M. and Trapp, J. A. (1975), The significance of formulating plasticity theory with reference to loading surfaces in strain space, *Int. J. Engng. Sci.*, Vol. 13, pp. 785–797.

Neale, K. W. (1981), Phenomenological constitutive laws in finite plasticity, *SM Archives*, Vol. 6, pp. 79–128.

Neale, K. W. and Shrivastava, S. C. (1990), Analytical solutions for circular bars subjected to large strain plastic torsion, *J. Appl. Mech.*, Vol. 57, pp. 298–306.

Needleman, A. (1982), Finite elements for finite strain plasticity problems, in *Plasticity of Metals at Finite Strain: Theory, Experiment and Computation*, eds. E. H. Lee and R. L. Mallett, pp. 387–443, Div. Appl. Mechanics, Stanford University, Stanford.

Needleman, A. and Tvergaard, V. (1982), Aspects of plastic postbuckling behavior, in *Mechanics of Solids – The Rodney Hill 60th Anniversary Volume*, eds. H. G. Hopkins and M. J. Sewell, pp. 453–498, Pergamon Press, Oxford.

Negahban, M. (1995), A study of theormodynamic restrictions, constraint conditions, and material symmetry in fully strain-space theories of plasticity, *Int. J. Plasticity*, Vol. 11, pp. 679–724.

Nemat-Nasser, S. (1982), Reply to Discussion of "On finite plastic flows of compressible materials with internal friction," *Int. J. Solids Struct.*, Vol. 18, pp. 361–366.

Nemat-Nasser, S. (1983), On finite plastic flow of crystalline solids and geomaterials, *J. Appl. Mech.*, Vol. 50, pp. 1114–1126.

Nemat-Nasser, S. (1992), Phenomenological theories of elastoplasticity and strain localization at high strain rates, *Appl. Mech. Rev.*, Vol. 45, No. 3, Part 2, pp. S19–S45.

Nemat-Nasser, S. and Shokooh, A. (1980), On finite plastic flows of compressible materials with internal friction, *Int. J. Solids Struct.*, Vol. 16, pp. 495–514.

Novozhilov, V. V. (1952), On the physical interpretation of the stress invariants used in the theory of plasticity, *Prikl. Math. Mech.*, Vol. 16, pp. 617–619 (in Russian).

Ohashi, Y., Tokuda, M., and Yamashita, H. (1975), Effect of third invariant of stress deviator on plastic deformation of mild steel, *J. Mech. Phys.*, Vol. 23, pp. 295–323.

Ohno, N. and Wang, J.-D. (1993), Kinematic hardening rules with critical state of dynamic recovery, I and II, *Int. J. Plasticity*, Vol. 9, pp. 375–403.

Ortiz, M. (1985), A constitutive theory for the inelastic behavior of concrete, *Mech. Mater.*, Vol. 4, pp. 67–93.

Ortiz, M. and Popov, E. P. (1982), A physical model for the inelasticity of concrete, *Proc. Roy. Soc. Lond. A*, Vol. 383, pp. 101–125.

Ottosen, N. S. and Ristinmaa, M. (1996), Corners in plasticity – Koiter's theory revisited, *Int. J. Solids Struct.*, Vol. 33, pp. 3697–3721.

Perzyna, P. (1963), The constitutive equations for rate sensitive plastic materials, *Q. Appl. Math.*, Vol. 20, pp. 321–332.

Perzyna, P. (1966), Fundamental problems in viscoplasticity, *Adv. Appl. Mech.*, Vol. 9, pp. 243–377.

Phillips, A. (1974), Experimental plasticity. Some thoughts on its present status and possible future trends, in *Foundations of Plasticity, Vol. II – Problems in Plasticity*, ed. A. Sawczuk, pp. 193–233, Noordhoff.

Phillips, A. (1982), Combined stress experiments in plasticity – The effects of temperature and time, in *Plasticity of Metals at Finite Strain: Theory, Experiment and Computation*, eds. E. H. Lee and R. L. Mallett, pp. 230–252, Div. Appl. Mechanics, Stanford University, Stanford.

Phillips, A. and Das, P. K. (1985), Yield surfaces and loading surfaces of aluminum and brass, *Int. J. Plasticity*, Vol. 1, pp. 89–109.

Phillips, A. and Lee, C.-W. (1979), Yield surfaces and loading surfaces. Experiments and recommendations, *Int. J. Solids Struct.*, Vol. 15, pp. 715–729.

Pietruszczak, S., Jiang, J., and Mirza, F. A. (1988), An elastoplastic constitutive model for concrete, *Int. J. Solids Struct.*, Vol. 24, pp. 705–722.

Ponte Castañeda, P. (1992), New variational principles in plasticity and their application to composite materials, *J. Mech. Phys. Solids*, Vol. 40, pp. 1757–1788.

Prager, W. (1955), The theory of plasticity: A survey of recent achievements (James Clayton Lecture), *Proc. Inst. Mech. Engrs.*, Vol. 169, pp. 41–57.

Prager, W. (1956), A new method of analyzing stresses and strains in work-hardening plastic solids, *J. Appl. Mech.*, Vol. 23, pp. 493–496.

Prager, W. (1958), Non-isothermal plastic deformation, *Konikl. Ned. Acad. Wetenschap. Proc.*, Vol. 61, pp. 176–182.

Qiu, Y. P. and Weng, G. J. (1993), Plastic potential and yield function of porous materials with aligned and randomly oriented spheroidal voids, *Int. J. Plasticity*, Vol. 9, pp. 271–290.

Rice, J. R. (1970), On the structure of stress-strain relations for time-dependent plastic deformation in metals, *J. Appl. Mech.*, Vol. 37, pp. 728–737.

Rice, J. R. (1971), Inelastic constitutive relations for solids: An internal variable theory and its application to metal plasticity, *J. Mech. Phys. Solids*, Vol. 19, pp. 433–455.

Rice, J. R. (1977), The localization of plastic deformation, in *Theoretical and Applied Mechanics*, ed. W. T. Koiter, pp. 207–220, North-Holland, Amsterdam.

Ristinmaa, M. (1995), Cyclic plasticity model using one yield surface only, *Int. J. Plasticity*, Vol. 11, pp. 163–181.

Rudnicki, J. W. (1982), Discussion of "On finite plastic flows of compressible materials with internal friction," *Int. J. Solids Struct.*, Vol. 18, pp. 357–360.

Rudnicki, J. W. and Rice, J. R. (1975), Conditions for the localization of deformation in pressure-sensitive dilatant materials, *J. Mech. Phys. Solids*, Vol. 23, pp. 371–394.

Runesson, K. and Mróz, Z. (1989), A note on nonassociated plastic flow rules, *Int. J. Plasticity*, Vol. 5, pp. 639–658.

Salençon, J. (1977), *Application of the Theory of Plasticity in Soil Mechanics*, John Wiley & Sons, Chichester, England.

Sanders, J. L. (1955), Plastic stress-strain relations based on infinitely many plane loading surfaces, in *Proc. 2nd U.S. Nat. Congr. Appl. Mech.*, ed. P. M. Naghdi, pp. 455–460, ASME, New York.

Sewell, M. J. (1974), A plastic flow rule at a yield vertex, *J. Mech. Phys. Solids*, Vol. 22, pp. 469–490.

Shames, I. H. and Cozzarelli, F. A. (1992), *Elastic and Inelastic Stress Analysis*, Prentice Hall, Englewood Cliffs, New Jersey.

Shiratori, E., Ikegami, K., and Yoshida, F. (1979), Analysis of stress-strain relations by use of an anisotropic hardening plastic potential, *J. Mech. Phys. Solids*, Vol. 27, pp. 213–229.

Simo, J. C. and Hughes, T. J. R. (1998), *Computational Inelasticity*, Springer-Verlag, New York.

Sokolovskii, V. V. (1948), Propagation of elastic-viscoplastic waves in bars, *Prikl. Mat. Mekh.*, Vol. 12, pp. 261–280 (in Russian).

Spitzig, W. A., Sober, R. J., and Richmond, O. (1975), Pressure dependence of yielding and associated volume expansion in tempered martensite, *Acta Metall.*, Vol. 23, pp. 885–893.

Stören, S. and Rice, J. R. (1975), Localized necking in thin sheets, *J. Mech. Phys. Solids*, Vol. 23, pp. 421–441.

Stout, M. G., Martin, P. L., Helling, D. E., and Canova, G. R. (1985), Multiaxial yield behavior of 1100 aluminum following various magnitudes of prestrain, *Int. J. Plasticity*, Vol. 1, pp. 163–174.

Sun, Y. (1995), Constitutive equations for ductile materials containing large and small voids, *Mech. Mater.*, Vol. 19, pp. 119–127.

Temam, R. (1985), *Mathematical Problems in Plasticity*, Gauthier-Villars, Paris (translation of French 1983 edition).

Tvergaard, V. (1982), On localization in ductile materials containing spherical voids, *Int. J. Fracture*, Vol. 18, pp. 237–252.

Tvergaard, V. and Needleman, A. (1984), Analysis of the cup-cone fracture in a round tensile bar, *Acta Metall.*, Vol. 32, pp. 157–169.

Van Houtte, P. (1994), Application of plastic potentials to strain rate sensitive and insensitive anisotropic materials, *Int. J. Plasticity*, Vol. 10, pp. 719–748.

Voyiadjis, G. Z. and Abu-Lebdeh, T. M. (1994), Plasticity model for concrete using the bounding surface concept, *Int. J. Plasticity*, Vol. 10, pp. 1–21.

Walpole, L. J. (1981), Elastic behavior of composite materials: Theoretical foundations, *Adv. Appl. Mech.*, Vol. 21, pp. 169–242.

Wang, J.-D. and Ohno, N. (1991), Two equivalent forms of nonlinear kinematic hardening: Application to nonisothermal plasticity, *Int. J. Plasticity*, Vol. 7, pp. 637–650.

Wu, H. C., Lu, J. K., and Pan, W. F. (1995), Some observations on yield surfaces for 304 stainless steel at large prestrain, *J. Appl. Mech.*, Vol. 62, pp. 626–632.

Yoder, P. J. and Iwan, W. D. (1981), On the formulation of strain space plasticity with multiple loading surfaces, *J. Appl. Mech.*, Vol. 48, pp. 773–778.

Zdebel, U. and Lehmann, Th. (1987), Some theoretical considerations and experimental investigations on a constitutive law in thermoplasticity, *Int. J. Plasticity*, Vol. 3, pp. 369–389.

Zerilli, F. J. and Armstrong, R. W. (1987), Dislocation-mechanics-based constitutive relations for material dynamics calculations, *J. Appl. Phys.*, Vol. 61, pp. 1816–1825.

Ziegler, H. (1959), A modification of Prager's hardening rule, *Q. Appl. Math.*, Vol. 17, pp. 55–65.

Zyczkowski, M. (1981), *Combined Loadings in the Theory of Plasticity*, PWN – Polish Scientific Publishers, Warszawa.

CHAPTER 10

PLASTIC STABILITY

Hill's theory of uniqueness and plastic stability is presented in this chapter. Exclusion functional and incrementally linear comparison material are first introduced. Eigenmodal deformations and acceleration waves in elastoplastic solids are then discussed. Fundamentals of Rice's localization analysis for various constitutive models are presented. Elastoplastic materials described by associative and nonassociative flow rules, as well as rigid-plastic materials are considered. The effects of yield vertices on localization predictions are examined.

10.1. Elastoplastic Rate-Potentials

The analysis is restricted to isothermal and rate-independent elastoplastic behavior. It was shown in Section 9.2 that the corresponding constitutive structure, for materials with a smooth yield surface, is bilinear and given by

$$\dot{\mathbf{T}}_{(n)} = \boldsymbol{\Lambda}^{\mathrm{ep}}_{(n)} : \dot{\mathbf{E}}_{(n)}. \tag{10.1.1}$$

One branch of the stiffness tensor $\boldsymbol{\Lambda}^{\mathrm{ep}}_{(n)}$ is associated with plastic loading, and the other with elastic unloading or neutral loading, such that

$$\boldsymbol{\Lambda}^{\mathrm{ep}}_{(n)} = \begin{cases} \boldsymbol{\Lambda}^{\mathrm{p}}_{(n)}, & \text{if } \ \frac{\partial f_{(n)}}{\partial \mathbf{T}_{(n)}} : \boldsymbol{\Lambda}_{(n)} : \dot{\mathbf{E}}_{(n)} > 0, \\ \\ \boldsymbol{\Lambda}_{(n)}, & \text{if } \ \frac{\partial f_{(n)}}{\partial \mathbf{T}_{(n)}} : \boldsymbol{\Lambda}_{(n)} : \dot{\mathbf{E}}_{(n)} \leq 0. \end{cases} \tag{10.1.2}$$

The stiffness tensor for plastic loading branch is defined by Eq. (9.2.10), i.e.,

$$\boldsymbol{\Lambda}^{\mathrm{p}}_{(n)} = \boldsymbol{\Lambda}_{(n)} - \frac{1}{h_{(n)}} \left(\boldsymbol{\Lambda}_{(n)} : \frac{\partial f_{(n)}}{\partial \mathbf{T}_{(n)}} \right) \otimes \left(\frac{\partial f_{(n)}}{\partial \mathbf{T}_{(n)}} : \boldsymbol{\Lambda}_{(n)} \right). \tag{10.1.3}$$

The elastic stiffness tensor is $\boldsymbol{\Lambda}_{(n)}$. More involved piecewise linear relations, with several or many branches, could be used to represent the behavior at the yield surface vertex (for example, for single crystals of metals deforming by multiple slip). Since $\mathbf{\Lambda}^{\mathrm{ep}}_{(n)}$ obeys the reciprocal symmetry, we can introduce the elastoplastic rate-potential function $\chi_{(n)}$, such that

$$\dot{\mathbf{T}}_{(n)} = \frac{\partial \chi_{(n)}}{\partial \dot{\mathbf{E}}_{(n)}}, \quad \chi_{(n)} = \frac{1}{2}\mathbf{\Lambda}^{\mathrm{ep}}_{(n)} :: \left(\dot{\mathbf{E}}_{(n)} \otimes \dot{\mathbf{E}}_{(n)}\right). \tag{10.1.4}$$

Alternatively, the elastoplastic constitutive structure can be expressed in terms of the rate of nominal stress and the rate of deformation tensor. By conveniently selecting $n = 1$ in Eq. (10.1.1), and by using the relationships

$$\dot{\mathbf{E}}_{(1)} = \frac{1}{2}\left(\mathbf{F}^T \cdot \dot{\mathbf{F}} + \dot{\mathbf{F}}^T \cdot \mathbf{F}\right), \quad \dot{\mathbf{T}}_{(1)} = \left(\dot{\mathbf{P}} - \mathbf{P} \cdot \mathbf{L}^T\right) \cdot \mathbf{F}^{-T}, \tag{10.1.5}$$

from Eqs. (3.8.8) and (3.8.14), it follows that

$$\dot{\mathbf{P}} = \mathbf{\Lambda}^{\mathrm{ep}} \cdot\cdot\, \dot{\mathbf{F}}. \tag{10.1.6}$$

The Cartesian components of elastoplastic moduli and pseudomoduli are related by

$$\Lambda^{\mathrm{ep}}_{JiLk} = \Lambda^{\mathrm{ep}\,(1)}_{JMLN} F_{iM} F_{kN} + T^{(1)}_{JL}\delta_{ik}, \tag{10.1.7}$$

as previously derived in Eq. (6.4.8). Since the pseudomoduli obey reciprocal symmetry ($\Lambda^{\mathrm{ep}}_{JiLk} = \Lambda^{\mathrm{ep}}_{LkJi}$), we can introduce the rate-potential function χ, such that

$$\dot{\mathbf{P}} = \frac{\partial \chi}{\partial \dot{\mathbf{F}}}, \quad \chi = \frac{1}{2}\mathbf{\Lambda}^{\mathrm{ep}} \cdots (\dot{\mathbf{F}} \otimes \dot{\mathbf{F}}). \tag{10.1.8}$$

The response over entire $\dot{\mathbf{F}}$ space is bilinear, since in the range of elastic unloading or neutral loading $\mathbf{\Lambda}^{\mathrm{ep}} = \mathbf{\Lambda}$ (tensor of elastic pseudomoduli), while in the range of plastic loading $\mathbf{\Lambda}^{\mathrm{ep}} = \mathbf{\Lambda}^{\mathrm{p}}$.

More generally, if inelastic rate response is thoroughly nonlinear (as in the description of actual behavior of polycrystals at yield vertices), we have

$$\dot{\mathbf{P}} = \frac{\partial \chi}{\partial \dot{\mathbf{F}}}, \quad \chi = \frac{1}{2}\dot{\mathbf{P}} \cdot\cdot\, \dot{\mathbf{F}}. \tag{10.1.9}$$

In the absence of time-dependent viscous effects, the rate-potential χ is necessarily homogeneous of degree two in $\dot{\mathbf{F}}$.

10.1.1. Current Configuration as Reference

When the current configuration is taken as the reference configuration, Eq. (10.1.8) becomes

$$\underline{\dot{\mathbf{P}}} = \frac{\partial \underline{\chi}}{\partial \mathbf{L}}, \quad \underline{\chi} = \frac{1}{2}\underline{\mathbf{\Lambda}}^{\mathrm{ep}} \cdots (\mathbf{L} \otimes \mathbf{L}), \tag{10.1.10}$$

since

$$\underline{\dot{\mathbf{F}}} = \mathbf{L}. \tag{10.1.11}$$

From Eqs. (6.3.4) and (6.4.16), or directly from Eq. (10.1.7), we have

$$\underline{\Lambda}^{\mathrm{ep}}_{jilk} = \underline{\mathcal{L}}^{\mathrm{ep}\,(1)}_{jilk} + \sigma_{jl}\delta_{ik}, \tag{10.1.12}$$

so that

$$\underline{\chi} = \frac{1}{2}\underline{\mathcal{L}}^{\mathrm{ep}}_{(1)} :: (\mathbf{D}\otimes\mathbf{D}) + \frac{1}{2}\boldsymbol{\sigma} : \left(\mathbf{L}^T\cdot\mathbf{L}\right). \tag{10.1.13}$$

Alternatively, in view of

$$\underline{\mathcal{L}}^{\mathrm{ep}}_{(0)} = \underline{\mathcal{L}}^{\mathrm{ep}}_{(1)} + 2\underline{\boldsymbol{\mathcal{S}}}, \tag{10.1.14}$$

where $\underline{\boldsymbol{\mathcal{S}}}$ is defined by Eq. (6.3.11), there follows

$$\underline{\chi} = \frac{1}{2}\underline{\mathcal{L}}^{\mathrm{ep}}_{(0)} :: (\mathbf{D}\otimes\mathbf{D}) + \frac{1}{2}\boldsymbol{\sigma} : \left(\mathbf{L}^T\cdot\mathbf{L} - 2\,\mathbf{D}^2\right). \tag{10.1.15}$$

The rate potentials $\underline{\chi}_{(n)}$ can be introduced such that

$$\underline{\dot{\mathbf{T}}}_{(n)} = \frac{\partial \underline{\chi}_{(n)}}{\partial \mathbf{D}}, \quad \underline{\chi}_{(n)} = \frac{1}{2}\underline{\mathcal{L}}^{\mathrm{ep}}_{(n)} :: (\mathbf{D}\otimes\mathbf{D}). \tag{10.1.16}$$

As in Section 7.6, the following relationships hold

$$\underline{\chi}_{(n)} = \underline{\chi}_{(0)} - n\boldsymbol{\sigma} : \mathbf{D}^2 = \underline{\chi}_{(1)} + (1-n)\boldsymbol{\sigma} : \mathbf{D}^2, \tag{10.1.17}$$

and

$$\underline{\chi} = \underline{\chi}_{(n)} + \frac{1}{2}\boldsymbol{\sigma} : \left[\mathbf{L}^T\cdot\mathbf{L} - 2(1-n)\mathbf{D}^2\right]. \tag{10.1.18}$$

In particular,

$$\overset{\circ}{\underline{\boldsymbol{\tau}}} = \underline{\mathcal{L}}^{\mathrm{ep}}_{(0)} : \mathbf{D} = \frac{\partial \underline{\chi}_{(0)}}{\partial \mathbf{D}}, \quad \underline{\chi}_{(0)} = \frac{1}{2}\underline{\mathcal{L}}^{\mathrm{ep}}_{(0)} :: (\mathbf{D}\otimes\mathbf{D}). \tag{10.1.19}$$

The tensor $\underline{\mathcal{L}}^{\mathrm{ep}}_{(0)}$ was explicitly given for various constitutive models in Chapter 9. In the range of elastic unloading or neutral loading it is equal to $\underline{\mathcal{L}}_{(0)}$, and in the range of plastic loading it is equal to $\underline{\mathcal{L}}^{\mathrm{p}}_{(0)}$. For example, in the case of isotropic hardening $\underline{\mathcal{L}}^{\mathrm{p}}_{(0)}$ is defined by Eq. (9.4.43), and in the case of linear kinematic hardening by Eq. (9.4.19).

If the response is thoroughly nonlinear,

$$\underline{\dot{\mathbf{P}}} = \frac{\partial \underline{\chi}}{\partial \mathbf{L}}, \quad \underline{\chi} = \frac{1}{2}\underline{\dot{\mathbf{P}}}\cdot\cdot\,\mathbf{L}, \tag{10.1.20}$$

where $\underline{\chi}$ is a homogeneous function of degree two in components of the velocity gradient $\mathbf{L}$.

10.2. Reciprocal Relations

For nonlinear incremental response (either thoroughly nonlinear or nonlinear on account of different behavior in loading and unloading), we can write

$$\dot{\mathbf{P}} \cdot\cdot \dot{\mathbf{F}} = 2\chi, \tag{10.2.1}$$

where χ is homogeneous of degree two in $\dot{\mathbf{F}}$. Taking the variation of Eq. (10.2.1), associated with an infinitesimal variation $\delta\dot{\mathbf{F}}$, gives

$$\delta\dot{\mathbf{P}} \cdot\cdot \dot{\mathbf{F}} + \dot{\mathbf{P}} \cdot\cdot \delta\dot{\mathbf{F}} = 2\delta\chi. \tag{10.2.2}$$

Since

$$\dot{\mathbf{P}} \cdot\cdot \delta\dot{\mathbf{F}} = \frac{\partial\chi}{\partial\dot{\mathbf{F}}} \cdot\cdot \delta\dot{\mathbf{F}} = \delta\chi, \tag{10.2.3}$$

we deduce from Eq. (10.2.2) the reciprocal relation

$$\delta\dot{\mathbf{P}} \cdot\cdot \dot{\mathbf{F}} = \dot{\mathbf{P}} \cdot\cdot \delta\dot{\mathbf{F}}. \tag{10.2.4}$$

This expression will be used in the derivation of the following reciprocal theorem. Consider a divergence expression

$$\boldsymbol{\nabla}^0 \cdot \left(\dot{\mathbf{P}} \cdot \delta\mathbf{v} - \delta\dot{\mathbf{P}} \cdot \mathbf{v}\right). \tag{10.2.5}$$

Since by Eq. (1.13.13),

$$\boldsymbol{\nabla}^0 \cdot \left(\dot{\mathbf{P}} \cdot \delta\mathbf{v}\right) = \left(\boldsymbol{\nabla}^0 \cdot \dot{\mathbf{P}}\right) \cdot \delta\mathbf{v} + \dot{\mathbf{P}} \cdot\cdot \delta\dot{\mathbf{F}}, \tag{10.2.6}$$

and similarly for the second term in (10.2.5), the divergence expression becomes

$$\boldsymbol{\nabla}^0 \cdot \left(\dot{\mathbf{P}} \cdot \delta\mathbf{v} - \delta\dot{\mathbf{P}} \cdot \mathbf{v}\right) = \left(\boldsymbol{\nabla}^0 \cdot \dot{\mathbf{P}}\right) \cdot \delta\mathbf{v} - \left(\boldsymbol{\nabla}^0 \cdot \delta\dot{\mathbf{P}}\right) \cdot \mathbf{v}. \tag{10.2.7}$$

The reciprocal relation (10.2.4) was utilized in the last step. Integrating Eq. (10.2.7) over the reference volume V^0, employing the equations of continuing equilibrium

$$\boldsymbol{\nabla}^0 \cdot \dot{\mathbf{P}} = -\rho^0\,\dot{\mathbf{b}}, \quad \boldsymbol{\nabla}^0 \cdot \delta\dot{\mathbf{P}} = -\rho^0\,\delta\dot{\mathbf{b}}, \tag{10.2.8}$$

and the Gauss theorem, gives

$$\begin{aligned} &\int_{V^0} \rho^0\,\dot{\mathbf{b}} \cdot \delta\mathbf{v}\,\mathrm{d}V^0 + \int_{S^0} \mathbf{n}^0 \cdot \dot{\mathbf{P}} \cdot \delta\mathbf{v}\,\mathrm{d}S^0 \\ &\qquad = \int_{V^0} \rho^0\,\delta\dot{\mathbf{b}} \cdot \mathbf{v}\,\mathrm{d}V^0 + \int_{S^0} \mathbf{n}^0 \cdot \delta\dot{\mathbf{P}} \cdot \mathbf{v}\,\mathrm{d}S^0. \end{aligned} \tag{10.2.9}$$

This is a reciprocal theorem for the considered incrementally nonlinear response (Hill, 1978).

For incrementally linear response, the variations $\delta\mathbf{v}$ and $\delta\dot{\mathbf{P}}$ can be replaced by (finite) differences $\mathbf{v}-\mathbf{v}^*$ and $\dot{\mathbf{P}}-\dot{\mathbf{P}}^*$ of any two (not necessarily nearby) equilibrium fields, and reciprocal relations of Eqs. (10.2.4) and (10.2.9) reduce to

$$\dot{\mathbf{P}}\cdot\cdot\,\dot{\mathbf{F}}^* = \dot{\mathbf{P}}^*\cdot\cdot\,\dot{\mathbf{F}}, \tag{10.2.10}$$

and

$$\begin{aligned}\int_{V^0}\rho^0\,\dot{\mathbf{b}}\cdot\mathbf{v}^*\,\mathrm{d}V^0 &+ \int_{S^0}\mathbf{n}^0\cdot\dot{\mathbf{P}}\cdot\mathbf{v}^*\,\mathrm{d}S^0\\ &= \int_{V^0}\rho^0\,\dot{\mathbf{b}}^*\cdot\mathbf{v}\,\mathrm{d}V^0 + \int_{S^0}\mathbf{n}^0\cdot\dot{\mathbf{P}}^*\cdot\mathbf{v}\,\mathrm{d}S^0.\end{aligned} \tag{10.2.11}$$

The latter is analogous to Betti's reciprocal theorem of classical elasticity, as discussed for incrementally linear elastic response in Subsection 7.5.1.

10.2.1. Clapeyron's Formula

Suppose that the stress rate field $\dot{\mathbf{P}}$ satisfies the equations of continuing equilibrium,

$$\boldsymbol{\nabla}^0\cdot\dot{\mathbf{P}}+\rho^0\,\dot{\mathbf{b}}=\mathbf{0}. \tag{10.2.12}$$

Then, for any analytically admissible velocity field $\mathbf{v}$, we have

$$\int_{V^0}\dot{\mathbf{P}}\cdot\cdot\,\dot{\mathbf{F}}\,\mathrm{d}V^0 = \int_{V^0}\rho^0\,\dot{\mathbf{b}}\cdot\mathbf{v}\,\mathrm{d}V^0 + \int_{S^0}\mathbf{n}^0\cdot\dot{\mathbf{P}}\cdot\mathbf{v}\,\mathrm{d}S^0, \tag{10.2.13}$$

by the Gauss theorem. For incrementally nonlinear response with $\dot{\mathbf{P}}$ defined by Eq. (10.1.9), χ being homogeneous of degree two in $\dot{\mathbf{F}}$, Eq. (10.2.13) becomes

$$2\int_{V^0}\chi\,\mathrm{d}V^0 = \int_{V^0}\rho^0\,\dot{\mathbf{b}}\cdot\mathbf{v}\,\mathrm{d}V^0 + \int_{S^0}\mathbf{n}^0\cdot\dot{\mathbf{P}}\cdot\mathbf{v}\,\mathrm{d}S^0. \tag{10.2.14}$$

The result is analogous to Clapeyron's formula of linear elasticity, and can be referred to as Clapeyron's formula of incrementally nonlinear response.

10.3. Variational Principle

If the stress rate field $\dot{\mathbf{P}}$ satisfies the equations of continuing equilibrium (10.2.12), then for any analytically admissible (not necessarily infinitesimal) velocity field $\delta\mathbf{v}$, it follows that

$$\int_{V^0}\dot{\mathbf{P}}\cdot\cdot\,\delta\dot{\mathbf{F}}\,\mathrm{d}V^0 = \int_{V^0}\rho^0\,\dot{\mathbf{b}}\cdot\delta\mathbf{v}\,\mathrm{d}V^0 + \int_{S^0}\mathbf{n}^0\cdot\dot{\mathbf{P}}\cdot\delta\mathbf{v}\,\mathrm{d}S^0, \tag{10.3.1}$$

again by the Gauss theorem. Recall that

$$\delta\dot{\mathbf{F}} = \delta\mathbf{v} \otimes \boldsymbol{\nabla}^0. \tag{10.3.2}$$

For incrementally nonlinear response with $\dot{\mathbf{P}}$ defined by Eq. (10.1.9), Eq. (10.3.1) becomes

$$\int_{V^0} \delta\chi \, \mathrm{d}V^0 = \int_{V^0} \rho^0 \,\dot{\mathbf{b}} \cdot \delta\mathbf{v} \, \mathrm{d}V^0 + \int_{S^0} \mathbf{n}^0 \cdot \dot{\mathbf{P}} \cdot \delta\mathbf{v} \, \mathrm{d}S^0. \tag{10.3.3}$$

Assuming that the rate of body forces is independent of the material response (deformation insensitive, dead body loading), Eq. (10.3.3) can be rewritten as

$$\delta\left(\int_{V^0} \chi \, \mathrm{d}V^0 - \int_{V^0} \rho^0 \,\dot{\mathbf{b}} \cdot \mathbf{v} \, \mathrm{d}V^0\right) = \int_{S_t^0} \dot{\mathbf{p}}_n \cdot \delta\mathbf{v} \, \mathrm{d}S_t^0, \tag{10.3.4}$$

provided that $\delta\mathbf{v}$ vanishes on $S_v^0 = S^0 - S_t^0$. If the current configuration is taken as the reference,

$$\delta\left(\int_V \underline{\chi} \, \mathrm{d}V - \int_V \rho \,\dot{\mathbf{b}} \cdot \mathbf{v} \, \mathrm{d}V\right) = \int_{S_t} \underline{\dot{\mathbf{p}}}_n \cdot \delta\mathbf{v} \, \mathrm{d}S_t, \tag{10.3.5}$$

since

$$\underline{\dot{\mathbf{p}}}_n \, \mathrm{d}S_t = \dot{\mathbf{p}}_n \, \mathrm{d}S_t^0. \tag{10.3.6}$$

The traction rate $\underline{\dot{\mathbf{p}}}_n$ is related to the rate of Cauchy traction $\dot{\mathbf{t}}_n$ by Eq. (3.9.18).

Suppose that the surface data over S_t consists of two parts,

$$\underline{\dot{\mathbf{p}}}_n = \underline{\dot{\mathbf{p}}}_n^{\mathrm{c}} + \underline{\dot{\mathbf{p}}}_n^{\mathrm{s}}, \tag{10.3.7}$$

where $\underline{\dot{\mathbf{p}}}_n^{\mathrm{c}}$ is the controllable part of the incremental loading (independent of material response), and $\underline{\dot{\mathbf{p}}}_n^{\mathrm{s}}$ is the deformation-sensitive part allowing for the deformability of both material and tool (linear homogeneous expression in $\mathbf{v}$ and $\mathbf{L}$), Hill (1978). For instance, in the case of fluid pressure, $\mathbf{t}_n = -p\,\mathbf{n}$, it follows that

$$\dot{\mathbf{t}}_n = -\dot{p}\,\mathbf{n} - p\,\dot{\mathbf{n}}, \tag{10.3.8}$$

where, from Eq. (2.4.18),

$$\dot{\mathbf{n}} = (\mathbf{n} \cdot \mathbf{D} \cdot \mathbf{n})\,\mathbf{n} - \mathbf{n} \cdot \mathbf{L}. \tag{10.3.9}$$

Thus, Eq. (3.9.18) gives

$$\underline{\dot{\mathbf{p}}}_n = -\dot{p}\,\mathbf{n} + p\,(\mathbf{n} \cdot \mathbf{L} - \mathbf{n}\,\mathrm{tr}\,\mathbf{D}). \tag{10.3.10}$$

The first term is deformation insensitive,

$$\dot{\underline{\mathbf{p}}}_n^{\mathrm{c}} = -\dot{p}\,\mathbf{n}, \tag{10.3.11}$$

while the remaining part is deformation sensitive,

$$\dot{\underline{\mathbf{p}}}_n^{\mathrm{s}} = p\,(\mathbf{n}\cdot\mathbf{L} - \mathbf{n}\,\mathrm{tr}\,\mathbf{D}). \tag{10.3.12}$$

A deformation-sensitive part of the incremental loading is self-adjoint if

$$\int_{S_t} \left(\dot{\underline{\mathbf{p}}}_n^{\mathrm{s}} \cdot \mathbf{v}^* - \dot{\underline{\mathbf{p}}}_n^{*\,\mathrm{s}} \cdot \mathbf{v} \right) \mathrm{d}S_t = 0, \tag{10.3.13}$$

for any two analytically admissible velocity fields $\mathbf{v}$ and $\mathbf{v}^*$ whose difference vanishes on S_v. Since $\dot{\underline{\mathbf{p}}}_n^{\mathrm{s}}$ is linear homogeneous, equivalent definitions are

$$\int_{S_t} \left(\dot{\underline{\mathbf{p}}}_n^{\mathrm{s}} \cdot \delta\mathbf{v} - \delta\dot{\underline{\mathbf{p}}}_n^{\mathrm{s}} \cdot \mathbf{v} \right) \mathrm{d}S_t = 0, \quad \text{i.e.,} \quad \int_{S_t} \dot{\underline{\mathbf{p}}}_n^{\mathrm{s}} \cdot \delta\mathbf{v}\,\mathrm{d}S_t = \frac{1}{2}\,\delta \int_{S_t} \dot{\underline{\mathbf{p}}}_n^{\mathrm{s}} \cdot \mathbf{v}\,\mathrm{d}S_t, \tag{10.3.14}$$

where $\delta\mathbf{v}$ is an analytically admissible infinitesimal variation of $\mathbf{v}$ that vanishes on S_v (Hill, *op. cit.*).

A true variational principle can be deduced from Eq. (10.3.5) when the surface data over S_t is self-adjoint in the sense of (10.3.14), since then

$$\delta\underline{\Xi} = 0, \tag{10.3.15}$$

with the variational integral

$$\underline{\Xi} = \int_V \underline{\chi}\,\mathrm{d}V - \int_V \rho\,\dot{\mathbf{b}}\cdot\mathbf{v}\,\mathrm{d}V - \int_{S_t} \left(\dot{\underline{\mathbf{p}}}_n^{\mathrm{c}} + \frac{1}{2}\,\dot{\underline{\mathbf{p}}}_n^{\mathrm{s}} \right) \cdot \mathbf{v}\,\mathrm{d}S_t. \tag{10.3.16}$$

Among all kinematically admissible velocity fields, the actual velocity field (whether unique or not) of the considered rate boundary-value problem renders stationary the functional $\underline{\Xi}(\mathbf{v})$. In Section 10.5 it will be shown that, under the uniqueness condition formulated in Section 10.4, the variational principle (10.3.15) with (10.3.16) can be strengthened to a minimum principle. Formulation of variational principles in the framework of infinitesimal strain is presented by Hill (1950), Drucker (1958,1960), and Koiter (1960). See also Ponter (1969), Neale (1972), and Sewell (1987).

10.3.1. Homogeneous Data

The incremental data is homogeneous at an instant of deformation process if

$$\dot{\mathbf{b}} = \mathbf{0} \ \text{ in } \ V, \qquad \mathbf{v} = \mathbf{0} \ \text{ on } \ S_v, \qquad \dot{\underline{\mathbf{p}}}_n^{\mathrm{c}} = \mathbf{0} \ \text{ on } \ S_t, \tag{10.3.17}$$

at that instant. The corresponding homogeneous boundary value problem is governed by the variational principle

$$\delta \underline{\Xi} = 0, \quad \underline{\Xi} = \int_V \underline{\chi} \, \mathrm{d}V - \frac{1}{2} \int_{S_t} \dot{\underline{\mathbf{p}}}^{\mathrm{s}}_n \cdot \mathbf{v} \, \mathrm{d}S_t. \tag{10.3.18}$$

In addition, the Clapeyron formula (10.2.14) reduces to

$$\int_V \underline{\chi} \, \mathrm{d}V = \frac{1}{2} \int_{S_t} \dot{\underline{\mathbf{p}}}^{\mathrm{s}}_{(n)} \cdot \mathbf{v} \, \mathrm{d}S_t. \tag{10.3.19}$$

A possible nontrivial solution is characterized by both

$$\delta \underline{\Xi} = 0 \quad \text{and} \quad \underline{\Xi} = 0. \tag{10.3.20}$$

For example, if $\underline{\chi}$ is given by Eq. (10.1.15), we have

$$\underline{\Xi} = \frac{1}{2} \int_V \left[\underline{\mathcal{L}}^{\mathrm{ep}}_{(0)} :: (\mathbf{D} \otimes \mathbf{D}) + \boldsymbol{\sigma} : \left(\mathbf{L}^T \cdot \mathbf{L} - 2\,\mathbf{D}^2 \right) \right] \mathrm{d}V - \frac{1}{2} \int_{S_t} \dot{\underline{\mathbf{p}}}^{\mathrm{s}}_n \cdot \mathbf{v} \, \mathrm{d}S_t. \tag{10.3.21}$$

Recall that the traction rate $\dot{\underline{\mathbf{p}}}^{\mathrm{s}}_n$ is related to the rate of Cauchy traction by an equation such as (3.9.18). When the geometry of the body is such that an admissible velocity field gives rise to large spins and small strain rates (as in slender beams), the terms proportional to stress within the volume integral in (10.3.21) can be of the same order as the terms proportional to elastoplastic moduli, even when the stress components are small compared to instantaneous moduli.

10.4. Uniqueness of Solution

In this section we consider the uniqueness of solution to incrementally nonlinear boundary-value problem, described by the equations of continuing equilibrium,

$$\boldsymbol{\nabla} \cdot \dot{\underline{\mathbf{P}}} + \rho \, \dot{\mathbf{b}} = \mathbf{0}, \tag{10.4.1}$$

and the boundary conditions

$$\mathbf{v} = \mathbf{v}_0 \quad \text{on} \quad S_v, \qquad \mathbf{n} \cdot \dot{\underline{\mathbf{P}}} = \dot{\underline{\mathbf{p}}}_n \quad \text{on} \quad S_t. \tag{10.4.2}$$

Material response is incrementally nonlinear and governed by Eq. (10.1.9). The incremental body loading is assumed to be deformation-insensitive, while deformation-sensitive part of incremental surface loading is self-adjoint in the spirit of Eq. (10.3.13).

Following Hill (1958,1961a,1978), suppose that there are two different solutions of Eqs. (10.4.1) and (10.4.2), $\mathbf{v}$ and $\mathbf{v}^*$. The corresponding velocity gradients are $\mathbf{L}$ and $\mathbf{L}^*$, and the rates of nominal stress are

$$\dot{\underline{\mathbf{P}}} = \frac{\partial \underline{\chi}}{\partial \mathbf{L}}, \quad \dot{\underline{\mathbf{P}}}^* = \frac{\partial \underline{\chi}}{\partial \mathbf{L}^*}. \tag{10.4.3}$$

Then, since

$$\nabla \cdot \left(\dot{\underline{\mathbf{P}}} - \dot{\underline{\mathbf{P}}}^*\right) = \mathbf{0}, \tag{10.4.4}$$

by the equations of equilibrium, the fields $\left(\dot{\underline{\mathbf{P}}}, \mathbf{L}\right)$ and $\left(\dot{\underline{\mathbf{P}}}^*, \mathbf{L}^*\right)$ necessarily satisfy the condition

$$\int_V \left(\dot{\underline{\mathbf{P}}}^* - \dot{\underline{\mathbf{P}}}\right) \cdot\cdot (\mathbf{L}^* - \mathbf{L})\, \mathrm{d}V = \int_{S_t} \left(\dot{\underline{\mathbf{p}}}_n^{*\,\mathrm{s}} - \dot{\underline{\mathbf{p}}}_n^{\mathrm{s}}\right) \cdot (\mathbf{v}^* - \mathbf{v})\, \mathrm{d}S_t. \tag{10.4.5}$$

This follows upon application of the Gauss divergence theorem. Consequently, from Eq. (10.4.5) the velocity field $\mathbf{v}$ is unique if

$$\int_V \left(\dot{\underline{\mathbf{P}}}^* - \dot{\underline{\mathbf{P}}}\right) \cdot\cdot (\mathbf{L}^* - \mathbf{L})\, \mathrm{d}V \neq \int_{S_t} \left(\dot{\underline{\mathbf{p}}}_n^{*\,\mathrm{s}} - \dot{\underline{\mathbf{p}}}_n^{\mathrm{s}}\right) \cdot (\mathbf{v}^* - \mathbf{v})\, \mathrm{d}S_t, \tag{10.4.6}$$

for all kinematically admissible $\mathbf{v}^*$ giving rise to

$$\mathbf{L}^* = \frac{\partial \mathbf{v}^*}{\partial \mathbf{x}}, \quad \dot{\underline{\mathbf{P}}}^* = \frac{\partial \underline{\chi}}{\partial \mathbf{L}^*}. \tag{10.4.7}$$

The stress rate $\dot{\underline{\mathbf{P}}}^*$ in (10.4.6) need not be statically admissible, so even if equality sign applies in (10.4.6) for some $\mathbf{v}^*$, the uniqueness is lost only if $\mathbf{v}^*$ gives rise to statically admissible stress-rate field $\dot{\underline{\mathbf{P}}}^*$. Therefore, a sufficient condition for uniqueness is

$$\int_V \left(\dot{\underline{\mathbf{P}}}^* - \dot{\underline{\mathbf{P}}}\right) \cdot\cdot (\mathbf{L}^* - \mathbf{L})\, \mathrm{d}V > \int_{S_t} \left(\dot{\underline{\mathbf{p}}}_n^{*\,\mathrm{s}} - \dot{\underline{\mathbf{p}}}_n^{\mathrm{s}}\right) \cdot (\mathbf{v}^* - \mathbf{v})\, \mathrm{d}S_t, \tag{10.4.8}$$

i.e.,

$$\int_V \left(\frac{\partial \underline{\chi}}{\partial \mathbf{L}^*} - \frac{\partial \underline{\chi}}{\partial \mathbf{L}}\right) \cdot\cdot (\mathbf{L}^* - \mathbf{L})\, \mathrm{d}V > \int_{S_t} \left(\dot{\underline{\mathbf{p}}}_n^{*\,\mathrm{s}} - \dot{\underline{\mathbf{p}}}_n^{\mathrm{s}}\right) \cdot (\mathbf{v}^* - \mathbf{v})\, \mathrm{d}S_t, \tag{10.4.9}$$

for the differences of all distinct kinematically admissible velocity fields $\mathbf{v}$ and $\mathbf{v}^*$.

For a piecewise linear response, the uniqueness condition (10.4.8) becomes

$$\int_V (\underline{\mathbf{\Lambda}}^{*\,\mathrm{ep}} \cdot\cdot \mathbf{L}^* - \underline{\mathbf{\Lambda}}^{\mathrm{ep}} \cdot\cdot \mathbf{L}) \cdot\cdot (\mathbf{L}^* - \mathbf{L})\, \mathrm{d}V > \int_{S_t} \left(\dot{\underline{\mathbf{p}}}_n^{*\,\mathrm{s}} - \dot{\underline{\mathbf{p}}}_n^{\mathrm{s}}\right) \cdot (\mathbf{v}^* - \mathbf{v})\, \mathrm{d}S_t. \tag{10.4.10}$$

The superimposed asterisk to one of the elastoplastic pseudomoduli tensors indicates that different loading branches (elastic or plastic) can correspond to different velocity fields $\mathbf{v}$ and $\mathbf{v}^*$ at each point of the continuum.

The condition (10.4.9), or (10.4.10), does not depend on prescribed $\dot{\underline{\mathbf{p}}}_n^{\mathrm{c}}$, nor does it depend on prescribed velocities on S_v, and is thus likely to be over-sufficient (i.e., not necessary).

The uniqueness condition (10.4.8) can be rewritten in terms of other stress measures. For example, it can be easily shown that

$$\dot{\underline{\mathbf{P}}} \cdot\cdot\, \mathbf{L}^* = \overset{\circ}{\boldsymbol{\tau}} : \mathbf{D}^* - \boldsymbol{\sigma} : \left(2\mathbf{D} \cdot \mathbf{D}^* - \mathbf{L}^T \cdot \mathbf{L}^*\right), \tag{10.4.11}$$

so that in (10.4.8) we have

$$\begin{aligned} \left(\dot{\underline{\mathbf{P}}}^* - \dot{\underline{\mathbf{P}}}\right) \cdot\cdot\, (\mathbf{L}^* - \mathbf{L}) = (\overset{\circ}{\boldsymbol{\tau}}{}^* - \overset{\circ}{\boldsymbol{\tau}}) : (\mathbf{D}^* - \mathbf{D}) \\ - \boldsymbol{\sigma} : \left[2(\mathbf{D}^* - \mathbf{D})^2 - \left(\mathbf{L}^{*T} - \mathbf{L}^T\right) \cdot (\mathbf{L}^* - \mathbf{L})\right]. \end{aligned} \tag{10.4.12}$$

10.4.1. Homogeneous Boundary Value Problem

A homogeneous boundary value problem for incrementally nonlinear material is described by

$$\boldsymbol{\nabla} \cdot \dot{\underline{\boldsymbol{P}}} = \mathbf{0}, \tag{10.4.13}$$

and the boundary conditions

$$\mathbf{w} = \mathbf{0} \quad \text{on} \quad S_v, \qquad \mathbf{n} \cdot \dot{\underline{\boldsymbol{P}}} = \dot{\underline{\boldsymbol{p}}}_n^{\mathrm{s}} \quad \text{on} \quad S_t, \tag{10.4.14}$$

where

$$\boldsymbol{L} = \frac{\partial \mathbf{w}}{\partial \mathbf{x}}, \quad \dot{\underline{\boldsymbol{P}}} = \frac{\partial \underline{\chi}}{\partial \boldsymbol{L}}. \tag{10.4.15}$$

This has always a null solution $\mathbf{w} = \mathbf{0}$. If the homogeneous problem also has a nontrivial solution $\mathbf{w} \neq \mathbf{0}$, then from (10.4.5)

$$\int_V \underline{\chi}\, \mathrm{d}V = \frac{1}{2} \int_{S_t} \dot{\underline{\boldsymbol{p}}}_n^{\mathrm{s}} \cdot \mathbf{w}\, \mathrm{d}S_t, \quad 2\,\underline{\chi} = \dot{\underline{\boldsymbol{P}}} \cdot\cdot\, \boldsymbol{L}. \tag{10.4.16}$$

Thus, if the exclusion functional is positive,

$$\underline{\mathcal{F}}(\mathbf{w}) = \int_V \underline{\chi}(\mathbf{w})\, \mathrm{d}V - \frac{1}{2} \int_{S_t} \dot{\underline{\boldsymbol{p}}}_n^{\mathrm{s}}(\mathbf{w}) \cdot \mathbf{w}\, \mathrm{d}S_t > 0, \tag{10.4.17}$$

for any kinematically admissible $\mathbf{w}$ giving rise to $\boldsymbol{L} = \partial \mathbf{w}/\partial \mathbf{x}$, the current state of material is incrementally unique (i.e., eigenstates under homogeneous data are excluded). In an eigenstate

$$\underline{\mathcal{F}}(\mathbf{w}) = 0, \tag{10.4.18}$$

for some kinematically admissible $\mathbf{w}$. Such an eigenmode $\mathbf{w}$ makes the exclusion functional stationary within the class of kinematically admissible variations $\delta\mathbf{w}$. Conversely, any kinematically admissible velocity field $\mathbf{w}$ that makes $\underline{\mathcal{F}}$ stationary is an eigenmode. This follows because for homogeneous problem the variational integral of Eq. (10.3.18) is equal to the exclusion functional,

$$\underline{\Xi} = \underline{\mathcal{F}}. \tag{10.4.19}$$

10.4.2. Incrementally Linear Comparison Material

In contrast to incrementally linear response, for incrementally nonlinear and piecewise linear response the difference $\underline{\dot{\mathbf{P}}} - \underline{\dot{\mathbf{P}}}^*$ is not a single-valued function of $\mathbf{v} - \mathbf{v}^*$, but of $\mathbf{v}$ and $\mathbf{v}^*$ individually. This makes direct application of the uniqueness criterion (10.4.8) and (10.4.10) for these materials more difficult. An indirect approach was introduced by Hill (1958, 1959, 1967). It is based on the notion of an incrementally linear comparison material, that is in a sense less stiff than the original material. Denote its rate potential by

$$\underline{\chi}^l = \frac{1}{2}\underline{\boldsymbol{\Lambda}}^l \cdots (\mathbf{L} \otimes \mathbf{L}). \tag{10.4.20}$$

If $\mathbf{v}$ and $\mathbf{v}^*$ are both solutions of the inhomogeneous boundary value problem corresponding to incrementally linear comparison material, then from (10.4.5)

$$\int_V \underline{\boldsymbol{\Lambda}}^l \cdots [(\mathbf{L}^* - \mathbf{L}) \otimes (\mathbf{L}^* - \mathbf{L})]\,\mathrm{d}V = \int_{S_t} \left(\underline{\dot{\mathbf{p}}}_n^{*\,\mathrm{s}} - \underline{\dot{\mathbf{p}}}_n^{\mathrm{s}}\right) \cdot (\mathbf{v}^* - \mathbf{v})\,\mathrm{d}S_t. \tag{10.4.21}$$

A sufficient condition for uniqueness is therefore

$$\int_V \underline{\boldsymbol{\Lambda}}^l \cdots [(\mathbf{L}^* - \mathbf{L}) \otimes (\mathbf{L}^* - \mathbf{L})]\,\mathrm{d}V > \int_{S_t} \left(\underline{\dot{\mathbf{p}}}_n^{*\,\mathrm{s}} - \underline{\dot{\mathbf{p}}}_n^{\mathrm{s}}\right) \cdot (\mathbf{v}^* - \mathbf{v})\,\mathrm{d}S_t, \tag{10.4.22}$$

for the difference of all distinct kinematically admissible velocity fields $\mathbf{v}$ and $\mathbf{v}^*$.

Following the development of Section 7.8 for incrementally linear elastic material, consider a homogeneous problem described by (10.4.13) and (10.4.14), where

$$\boldsymbol{L} = \frac{\partial \mathbf{w}}{\partial \mathbf{x}}, \quad \underline{\dot{\boldsymbol{P}}} = \underline{\boldsymbol{\Lambda}}^l \cdot\cdot\, \boldsymbol{L}. \tag{10.4.23}$$

There is always a null solution $\mathbf{w} = \mathbf{0}$ to this problem. If the homogeneous problem also has a nontrivial solution $\mathbf{w} \neq \mathbf{0}$, then from (10.4.21)

$$\frac{1}{2}\int_V \underline{\mathbf{\Lambda}}^l \cdot\cdot\cdot\cdot (\boldsymbol{L} \otimes \boldsymbol{L})\, \mathrm{d}V = \int_{S_t} \underline{\dot{\boldsymbol{p}}}_n^{\mathrm{s}} \cdot \mathbf{w}\, \mathrm{d}S_t. \tag{10.4.24}$$

The examination of the uniqueness of solution to inhomogeneous problem for incrementally linear comparison material is thus equivalent to examination of the uniqueness of solution to the associated homogeneous problem. Consequently, the uniqueness is assured, i.e., the inequality (10.4.22) is satisfied, if

$$\underline{\mathcal{F}} = \int_V \underline{\chi}^l(\mathbf{w})\, \mathrm{d}V - \frac{1}{2}\int_{S_t} \underline{\dot{\boldsymbol{p}}}_n^{\mathrm{s}}(\mathbf{w}) \cdot \mathbf{w}\, \mathrm{d}S_t > 0,\ \underline{\chi}^l(\mathbf{w}) = \frac{1}{2}\underline{\mathbf{\Lambda}}^l \cdot\cdot\cdot\cdot (\boldsymbol{L} \otimes \boldsymbol{L}), \tag{10.4.25}$$

for any kinematically admissible $\mathbf{w}$ giving rise to $\boldsymbol{L} = \partial \mathbf{w}/\partial \mathbf{x}$.

Suppose that, at the given state of deformation, the exclusion condition (10.4.25) is satisfied for incrementally linear material with the rate potential $\underline{\chi}^l$. Then, if

$$\underline{\chi}^l \leq \underline{\chi} \tag{10.4.26}$$

at each point (linear comparison material in this sense being less stiff), the exclusion functional (10.4.17) for incrementally nonlinear material with the rate potential $\underline{\chi}$ is also satisfied, precluding eigenstates under homogeneous data.

More strongly, if (10.4.25) is satisfied and the function $\underline{\chi} - \underline{\chi}^l$ is convex at each point, bifurcation is ruled out for any associated inhomogeneous data (Hill, 1978). Indeed, for convex function $\underline{\chi} - \underline{\chi}^l$, by definition of convexity we can write

$$\underline{\chi}(\mathbf{L}^*) - \underline{\chi}^l(\mathbf{L}^*) - [\underline{\chi}(\mathbf{L}) - \underline{\chi}^l(\mathbf{L})] \geq \frac{\partial(\underline{\chi} - \underline{\chi}^l)}{\partial \mathbf{L}} \cdot\cdot (\mathbf{L}^* - \mathbf{L}), \tag{10.4.27}$$

and likewise

$$\underline{\chi}(\mathbf{L}) - \underline{\chi}^l(\mathbf{L}) - [\underline{\chi}(\mathbf{L}^*) - \underline{\chi}^l(\mathbf{L}^*)] \geq \frac{\partial(\underline{\chi} - \underline{\chi}^l)}{\partial \mathbf{L}^*} \cdot\cdot (\mathbf{L} - \mathbf{L}^*). \tag{10.4.28}$$

The convexity condition (10.4.27) is schematically depicted in (Fig. 10.1). By summing up the above two inequalities, we obtain

$$\left[\frac{\partial(\underline{\chi} - \underline{\chi}^l)}{\partial \mathbf{L}^*} - \frac{\partial(\underline{\chi} - \underline{\chi}^l)}{\partial \mathbf{L}}\right] \cdot\cdot (\mathbf{L}^* - \mathbf{L}) \geq 0. \tag{10.4.29}$$

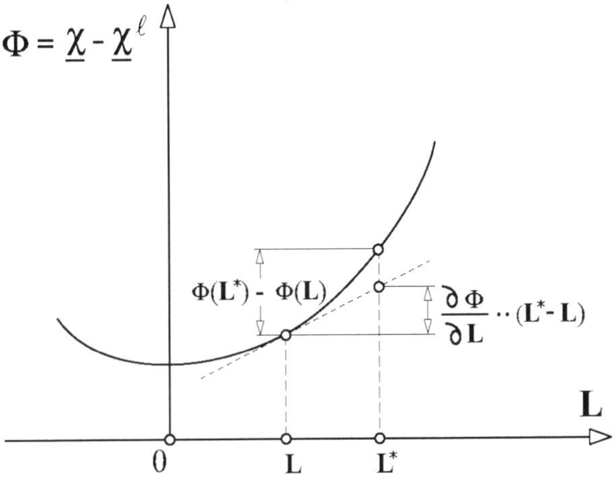

FIGURE 10.1. Schematic illustration of the convexity condition (10.4.27).

In view of Eq. (10.4.3) for the rates of nominal stress, and Eq. (10.4.20) for the rate potential, the inequality (10.4.29) can be recast in the following form

$$\begin{aligned}\left(\dot{\underline{\mathbf{P}}}^{*}-\dot{\underline{\mathbf{P}}}\right)\cdot\cdot(\mathbf{L}^{*}-\mathbf{L}) &\geq \left(\dot{\underline{\mathbf{P}}}^{*}-\dot{\underline{\mathbf{P}}}\right)^{l}\cdot\cdot(\mathbf{L}^{*}-\mathbf{L}) \\ &= \underline{\mathbf{\Lambda}}^{l}\cdot\cdot\cdot\cdot[(\mathbf{L}^{*}-\mathbf{L})\otimes(\mathbf{L}^{*}-\mathbf{L})].\end{aligned} \tag{10.4.30}$$

Therefore, if the inequality (10.4.25), implying (10.4.22), is satisfied for incrementally linear comparison material (i.e., if there is no bifurcation for incrementally linear comparison material), the convexity of the function $\underline{\chi}-\underline{\chi}^{l}$, leading to (10.4.30), assures that the inequality (10.4.8) is also satisfied, ruling out any bifurcation of incrementally nonlinear material at the considered state. On the other hand, if the current configuration is a primary eigenstate for $\underline{\chi}^{l}$ material, i.e., $\underline{\mathcal{F}}=0$ in (10.4.25), the bifurcation may still be excluded for $\underline{\chi}$ material, if $\underline{\chi}-\underline{\chi}^{l}$ is strictly convex in $\boldsymbol{L}$ (strict inequality applies in (10.4.29)).

For an analysis of uniqueness in the case of an incrementally nonlinear material model without a rate potential function, see the paper by Chambon and Caillerie (1999).

10.4.3. Comparison Material for Elastoplastic Response

For elastoplastic response with a piecewise linear relation defined by

$$\underline{\dot{\mathbf{P}}} = \underline{\mathbf{\Lambda}}^{\mathrm{ep}} \cdot\cdot\, \mathbf{L}, \quad \underline{\mathbf{\Lambda}}^{\mathrm{ep}} = \begin{cases} \underline{\mathbf{\Lambda}}^{\mathrm{p}}, & \text{for plastic loading,} \\ \underline{\mathbf{\Lambda}}, & \text{for elastic unloading or neutral loading,} \end{cases} \tag{10.4.31}$$

an incrementally linear comparison material can be taken to be the material whose stiffness is equal to $\underline{\mathbf{\Lambda}}^{\mathrm{p}}$ at plastically stressed points of the continuum. Elsewhere in the continuum, i.e., at elastically stressed points, the comparison material has the stiffness equal to $\underline{\mathbf{\Lambda}}$. The following is a proof of the required condition,

$$(\underline{\mathbf{\Lambda}}^{*\,\mathrm{ep}} \cdot\cdot\, \mathbf{L}^* - \underline{\mathbf{\Lambda}}^{\mathrm{ep}} \cdot\cdot\, \mathbf{L}) \cdot\cdot\, (\mathbf{L}^* - \mathbf{L}) \geq \underline{\mathbf{\Lambda}}^{\mathrm{p}} \cdot\cdot\cdot\cdot\, [(\mathbf{L}^* - \mathbf{L}) \otimes (\mathbf{L}^* - \mathbf{L})] \tag{10.4.32}$$

for the identification of selected incrementally linear comparison material, with the stiffness $\underline{\mathbf{\Lambda}}^{\mathrm{p}}$.

From Eqs. (9.1.1) and (9.1.4), a piecewise linear elastoplastic response is governed by

$$\dot{\mathbf{T}}_{(n)} = \mathbf{\Lambda}^{\mathrm{p}}_{(n)} : \dot{\mathbf{E}}_{(n)} - \left[\dot{\gamma}_{(n)} - \frac{1}{h_{(n)}} \left(\frac{\partial g_{(n)}}{\partial \mathbf{E}_{(n)}} : \dot{\mathbf{E}}_{(n)}\right)\right] \frac{\partial g_{(n)}}{\partial \mathbf{E}_{(n)}}, \tag{10.4.33}$$

where, by Eq. (9.1.13),

$$\mathbf{\Lambda}^{\mathrm{p}}_{(n)} = \mathbf{\Lambda}_{(n)} - \frac{1}{h_{(n)}} \left(\frac{\partial g_{(n)}}{\partial \mathbf{E}_{(n)}} \otimes \frac{\partial g_{(n)}}{\partial \mathbf{E}_{(n)}}\right). \tag{10.4.34}$$

The loading index is

$$\dot{\gamma}_{(n)} = \frac{1}{h_{(n)}} \left(\frac{\partial g_{(n)}}{\partial \mathbf{E}_{(n)}} : \dot{\mathbf{E}}_{(n)}\right) > 0 \tag{10.4.35}$$

for plastic loading, and $\dot{\gamma}_{(n)} = 0$ for elastic unloading or neutral loading. Consequently,

$$\begin{aligned} (\dot{\mathbf{T}}^*_{(n)} - \dot{\mathbf{T}}_{(n)}) : (\dot{\mathbf{E}}^*_{(n)} - \dot{\mathbf{E}}_{(n)}) &= \mathbf{\Lambda}^{\mathrm{p}}_{(n)} :: \left[\left(\dot{\mathbf{E}}^*_{(n)} - \dot{\mathbf{E}}_{(n)}\right) \otimes \left(\dot{\mathbf{E}}^*_{(n)} - \dot{\mathbf{E}}_{(n)}\right)\right] \\ - \left[\dot{\gamma}^*_{(n)} - \dot{\gamma}_{(n)} - \frac{1}{h_{(n)}} \frac{\partial g_{(n)}}{\partial \mathbf{E}_{(n)}} : \left(\dot{\mathbf{E}}^*_{(n)} - \dot{\mathbf{E}}_{(n)}\right)\right] &\frac{\partial g_{(n)}}{\partial \mathbf{E}_{(n)}} : \left(\dot{\mathbf{E}}^*_{(n)} - \dot{\mathbf{E}}_{(n)}\right). \end{aligned} \tag{10.4.36}$$

If both $\dot{\mathbf{E}}_{(n)}$ and $\dot{\mathbf{E}}^*_{(n)}$ correspond to plastic loading from the current state, the terms within square brackets in the second line of Eq. (10.4.36) cancel each other. If one strain rate corresponds to plastic loading and the other to elastic unloading, or if both strain rates correspond to elastic unloading, or if one strain rate corresponds to elastic unloading and the other to neutral loading, the whole expression

$$\left[\dot{\gamma}^*_{(n)} - \dot{\gamma}_{(n)} - \frac{1}{h_{(n)}} \frac{\partial g_{(n)}}{\partial \mathbf{E}_{(n)}} : \left(\dot{\mathbf{E}}^*_{(n)} - \dot{\mathbf{E}}_{(n)}\right)\right] \frac{\partial g_{(n)}}{\partial \mathbf{E}_{(n)}} : \left(\dot{\mathbf{E}}^*_{(n)} - \dot{\mathbf{E}}_{(n)}\right) \tag{10.4.37}$$

is negative. If both strain rates correspond to neutral loading, or one to neutral loading and the other to plastic loading, the above expression vanishes. Thus, from Eq. (10.4.36) it follows that

$$\left(\dot{\mathbf{T}}^*_{(n)} - \dot{\mathbf{T}}_{(n)}\right) : \left(\dot{\mathbf{E}}^*_{(n)} - \dot{\mathbf{E}}_{(n)}\right) \geq \mathbf{\Lambda}^{\mathrm{p}}_{(n)} :: \left[\left(\dot{\mathbf{E}}^*_{(n)} - \dot{\mathbf{E}}_{(n)}\right) \otimes \left(\dot{\mathbf{E}}^*_{(n)} - \dot{\mathbf{E}}_{(n)}\right)\right]. \tag{10.4.38}$$

This means that actual piecewise linear response is more convex than a hypothetical linear response with the stiffness moduli $\mathbf{\Lambda}^{\mathrm{p}}_{(n)}$ over the entire $\dot{\mathbf{E}}_{(n)}$ space.

If the current configuration is taken as the reference, (10.4.38) becomes

$$\left(\underline{\dot{\mathbf{T}}}^*_{(n)} - \underline{\dot{\mathbf{T}}}_{(n)}\right) : (\mathbf{D}^* - \mathbf{D}) \geq \underline{\mathbf{\Lambda}}^{\mathrm{p}}_{(n)} :: [(\mathbf{D}^* - \mathbf{D}) \otimes (\mathbf{D}^* - \mathbf{D})], \tag{10.4.39}$$

or, since $\underline{\dot{\mathbf{T}}}_{(n)}$ and $\underline{\mathbf{\Lambda}}_{(n)}$ are fully symmetric tensors,

$$\left(\underline{\dot{\mathbf{T}}}^*_{(n)} - \underline{\dot{\mathbf{T}}}_{(n)}\right) \cdot\cdot (\mathbf{L}^* - \mathbf{L}) \geq \underline{\mathbf{\Lambda}}^{\mathrm{p}}_{(n)} \cdot\cdot\cdot\cdot [(\mathbf{L}^* - \mathbf{L}) \otimes (\mathbf{L}^* - \mathbf{L})]. \tag{10.4.40}$$

To express this condition in terms of $\underline{\dot{\mathbf{P}}}$ and $\underline{\mathbf{\Lambda}}^{\mathrm{p}}$, the choice $n = 1$ is conveniently made in Eq. (10.4.40). Since

$$\underline{\dot{\mathbf{T}}}_{(1)} = \underline{\dot{\mathbf{P}}} - \boldsymbol{\sigma} \cdot \mathbf{L}^T, \tag{10.4.41}$$

from the second of Eq. (10.1.5), and recalling the relationship between the components of elastoplastic moduli and pseudomoduli given by Eq. (10.1.7), the substitution into Eq. (10.4.40) gives

$$\left(\underline{\dot{\mathbf{P}}}^* - \underline{\dot{\mathbf{P}}}\right) \cdot\cdot (\mathbf{L}^* - \mathbf{L}) \geq \underline{\mathbf{\Lambda}}^{\mathrm{p}} \cdot\cdot\cdot\cdot [(\mathbf{L}^* - \mathbf{L}) \otimes (\mathbf{L}^* - \mathbf{L})]. \tag{10.4.42}$$

This is precisely the condition (10.4.32).

In conclusion, the bifurcation problem for a piecewise linear elastoplastic material with the stiffness $\underline{\mathbf{\Lambda}}^{\mathrm{ep}}$ is reduced to determining primary eigenstate of incrementally linear comparison material with the stiffness $\underline{\mathbf{\Lambda}}^{\mathrm{p}}$. Among infinitely many deformation modes that are all solutions of given inhomogeneous problem for $\underline{\mathbf{\Lambda}}^{\mathrm{p}}$ material at that state (these being the sums of the increment of the fundamental solution and any multiple of the eigenmode solution), there may be those for which the strain rate at every plastically stressed point is in the plastic loading range of the $\underline{\mathbf{\Lambda}}^{\text{ep}}$ material itself. Such deformation modes are then also solutions of the given inhomogeneous rate problem of the $\underline{\mathbf{\Lambda}}^{\text{ep}}$ material, which means that a primary bifurcation for this material has been identified (Hill, 1978).

For incrementally linear comparison material bifurcation can occur in an eigenstate for any prescribed traction rates on S_t, and velocities on S_v. In an actual elastoplastic material bifurcation occurs only for those traction rates and prescribed velocities for which there is no elastic unloading in the current plastic region of the body. See also Nguyen (1987, 1994), Triantafyllidis (1983), and Petryk (1989).

For solids with corners on their yield surfaces, comparison material is defined as a hypothetical material whose every yield system is active. For example, with a pyramidal vertex formed by k_0 intersecting segments, from Section 9.5 it follows that

$$\mathbf{\Lambda}^{\text{p}} = \mathbf{\Lambda}_{(n)} - \sum_{i=1}^{k_0}\sum_{j=1}^{k_0} h_{(n)}^{<ij>-1} \left(\mathbf{\Lambda}_{(n)} : \frac{\partial f_{(n)}^{<i>}}{\partial \mathbf{T}_{(n)}}\right) \otimes \left(\frac{\partial f_{(n)}^{<j>}}{\partial \mathbf{T}_{(n)}} : \mathbf{\Lambda}_{(n)}\right). \tag{10.4.43}$$

The range of strain rate space in which no elastic unloading occurs on any yield segment is called fully active or total loading range (Sewell, 1972; Hutchinson, 1974). In the context of crystal plasticity, this is further discussed in Chapter 12.

10.5. Minimum Principle

If the uniqueness condition (10.4.8) applies, the variational principle (10.3.15) with (10.3.16) can be strengthened to a minimum principle. Let $\mathbf{v}$ be the actual unique solution of the considered problem, and $\mathbf{v}^*$ any kinematically admissible velocity field. First, it is observed that

$$\begin{aligned}\Xi(\mathbf{v}^*) - \Xi(\mathbf{v}) &= \int_V (\underline{\chi}^* - \underline{\chi})\,\mathrm{d}V - \int_V \rho\dot{\mathbf{b}}\cdot(\mathbf{v}^* - \mathbf{v})\,\mathrm{d}V \\ &- \int_{S_t} \underline{\dot{\mathbf{p}}}_n^{\text{c}} \cdot (\mathbf{v}^* - \mathbf{v})\,\mathrm{d}S_t - \frac{1}{2}\int_{S_t}\left(\underline{\dot{\mathbf{p}}}_n^{*\text{s}} + \underline{\dot{\mathbf{p}}}_n^{\text{s}}\right)\cdot(\mathbf{v}^* - \mathbf{v})\,\mathrm{d}S_t,\end{aligned} \tag{10.5.1}$$

and

$$\frac{1}{2}\left(\underline{\dot{\mathbf{p}}}_n^{*\text{s}} + \underline{\dot{\mathbf{p}}}_n^{\text{s}}\right) = \frac{1}{2}\left(\underline{\dot{\mathbf{p}}}_n^{*\text{s}} - \underline{\dot{\mathbf{p}}}_n^{\text{s}}\right) + \underline{\dot{\mathbf{p}}}_n^{\text{s}}. \tag{10.5.2}$$

Thus,

$$\Xi(\mathbf{v}^*) - \Xi(\mathbf{v}) = \int_V (\underline{\chi}^* - \underline{\chi})\,\mathrm{d}V - \frac{1}{2}\int_{S_t}\left(\dot{\underline{\mathbf{p}}}_n^{*\,\mathrm{s}} - \dot{\underline{\mathbf{p}}}_n^{\mathrm{s}}\right)\cdot(\mathbf{v}^* - \mathbf{v})\,\mathrm{d}S_t$$
$$- \int_V \rho\dot{\mathbf{b}}\cdot(\mathbf{v}^* - \mathbf{v})\,\mathrm{d}V - \int_{S_t}\dot{\underline{\mathbf{p}}}_n\cdot(\mathbf{v}^* - \mathbf{v})\,\mathrm{d}S_t. \tag{10.5.3}$$

The two surface integrals can be expressed by the Gauss theorem as the volume integrals, see (10.2.13), with the result

$$\Xi(\mathbf{v}^*) - \Xi(\mathbf{v}) = \int_V (\underline{\chi}^* - \underline{\chi})\,\mathrm{d}V - \int_V \dot{\underline{\mathbf{P}}}\cdot\cdot(\mathbf{L}^* - \mathbf{L})\,\mathrm{d}V$$
$$- \frac{1}{2}\int_{S_t}\left(\dot{\underline{\mathbf{p}}}_n^{*\,\mathrm{s}} - \dot{\underline{\mathbf{p}}}_n^{\mathrm{s}}\right)\cdot(\mathbf{v}^* - \mathbf{v})\,\mathrm{d}S_t. \tag{10.5.4}$$

For the minimum principle to hold, it is required to prove that the right hand side of Eq. (10.5.4) is positive. Following Hill (1978), introduce a continuous sequence of kinematically admissible fields

$$\mathbf{v}^+(\alpha) = \mathbf{v} + \alpha(\mathbf{v}^+ - \mathbf{v}), \quad 0 \le \alpha \le 1, \tag{10.5.5}$$

the parameter α being uniform throughout the body. Then, by Eq. (10.5.4),

$$\Xi(\mathbf{v}^+) - \Xi(\mathbf{v}) = \int_V (\underline{\chi}^+ - \underline{\chi})\,\mathrm{d}V - \int_V \dot{\underline{\mathbf{P}}}\cdot\cdot(\mathbf{L}^+ - \mathbf{L})\,\mathrm{d}V$$
$$- \frac{1}{2}\int_{S_t}\left(\dot{\underline{\mathbf{p}}}_n^{+\,\mathrm{s}} - \dot{\underline{\mathbf{p}}}_n^{\mathrm{s}}\right)\cdot(\mathbf{v}^+ - \mathbf{v})\,\mathrm{d}S_t. \tag{10.5.6}$$

Here,

$$\mathbf{L}^+ = \mathbf{L} + \alpha(\mathbf{L}^* - \mathbf{L}), \tag{10.5.7}$$

and, since $\dot{\underline{\mathbf{p}}}_n^{\mathrm{s}}$ is linear homogeneous in velocity gradient,

$$\dot{\underline{\mathbf{p}}}_n^{+\,\mathrm{s}} = \dot{\underline{\mathbf{p}}}_n^{\mathrm{s}} + \alpha\left(\dot{\underline{\mathbf{p}}}_n^{*\,\mathrm{s}} - \dot{\underline{\mathbf{p}}}_n^{\mathrm{s}}\right). \tag{10.5.8}$$

Consequently,

$$\alpha\frac{\mathrm{d}}{\mathrm{d}\alpha}\left[\Xi(\mathbf{v}^+) - \Xi(\mathbf{v})\right] = \int_V \left(\dot{\underline{\mathbf{P}}}^+ - \dot{\underline{\mathbf{P}}}\right)\cdot\cdot(\mathbf{L}^+ - \mathbf{L})\,\mathrm{d}V$$
$$- \int_{S_t}\left(\dot{\underline{\mathbf{p}}}_n^{+\,\mathrm{s}} - \dot{\underline{\mathbf{p}}}_n^{\mathrm{s}}\right)\cdot(\mathbf{v}^+ - \mathbf{v})\,\mathrm{d}S_t > 0, \tag{10.5.9}$$

which is positive by the uniqueness condition (10.4.8), applied to fields $\mathbf{v}$ and $\mathbf{v}^+$. In the derivation it is recalled that $\underline{\chi}$ is a homogeneous function of degree two, so that

$$\alpha\frac{\mathrm{d}\chi^+}{\mathrm{d}\alpha} = \frac{\partial\chi^+}{\partial\mathbf{L}^+}\cdot\cdot(\mathbf{L}^+ - \mathbf{L}) = 2\,\chi^+ - \dot{\underline{\mathbf{P}}}^+\cdot\cdot\mathbf{L} = \dot{\underline{\mathbf{P}}}^+\cdot\cdot(\mathbf{L}^+ - \mathbf{L}). \tag{10.5.10}$$

Therefore, in the range $0 < \alpha \leq 1$ the function $\Xi(\mathbf{v}^+) - \Xi(\mathbf{v})$ has a positive gradient,

$$\frac{\mathrm{d}}{\mathrm{d}\alpha}\left[\Xi(\mathbf{v}^+) - \Xi(\mathbf{v})\right] > 0. \tag{10.5.11}$$

Since $\Xi(\mathbf{v}^+) - \Xi(\mathbf{v})$ is equal to zero for $\alpha = 0$, it follows that

$$\Xi(\mathbf{v}^+) - \Xi(\mathbf{v}) > 0, \quad 0 < \alpha \leq 1, \tag{10.5.12}$$

which is a desired result. Thus,

$$\Xi(\mathbf{v}^*) > \Xi(\mathbf{v}) \tag{10.5.13}$$

for all kinematically admissible velocity fields $\mathbf{v}^*$, which implies a minimum principle.

10.6. Stability of Equilibrium

Consider an equilibrium state of the body whose response is incrementally nonlinear with the rate potential χ. Let the current equilibrium stress field be $\mathbf{P}$, associated with the body force $\mathbf{b}$ within V^0, the traction $\mathbf{p}_n$ over S_t^0, and prescribed displacement on the remaining part of the boundary $S^0 - S_t^0$. Assume that an infinitesimal virtual displacement field $\delta\mathbf{u}$ is imposed on the body ($\delta\mathbf{u} = \mathbf{0}$ on $S^0 - S_t^0$), under dead body force and unchanged controllable part of the surface loading. The work done by applied forces on this virtual displacement is

$$\int_{V^0} \rho^0\, \mathbf{b} \cdot \delta\mathbf{u}\, \mathrm{d}V^0 + \int_{S_t^0} \left(\mathbf{p}_n + \frac{1}{2}\,\delta\mathbf{p}_n^{\mathrm{s}}\right) \cdot \delta\mathbf{u}\, \mathrm{d}S_t^0, \tag{10.6.1}$$

since deformation-sensitive change $\delta\mathbf{p}_n^{\mathrm{s}}$, induced by $\delta\mathbf{u}$, is linear in $\delta\mathbf{u}$. The stress field $\mathbf{P}$ changes to $\mathbf{P} + \delta\mathbf{P}$, where $\delta\mathbf{P}$ is constitutively associated with the displacement increment $\delta\mathbf{u}$ through

$$\delta\mathbf{P} = \frac{\partial\chi}{\partial(\delta\mathbf{F})}, \quad \delta\mathbf{F} = \frac{\partial(\delta\mathbf{u})}{\partial\mathbf{X}}. \tag{10.6.2}$$

Kinematically admissible neighboring configurations need not be equilibrium configurations, i.e., the stress field $\mathbf{P} + \delta\mathbf{P}$ need not be an equilibrium field. The increment of internal energy associated with virtual change $\delta\mathbf{u}$ is, to second order,

$$\int_{V^0} \left(\mathbf{P} + \frac{1}{2}\,\delta\mathbf{P}\right) \cdot\cdot\, \delta\mathbf{F}\, \mathrm{d}V^0. \tag{10.6.3}$$

According to the energy criterion of stability, the underlying equilibrium configuration is stable if the increase of internal energy due to $\delta\mathbf{u}$ is greater than the work done by already applied forces in the virtual transition. Upon using (10.6.1), (10.6.3) and the formula (3.12.1), the stability condition becomes (Hill, 1958,1978)

$$\int_{V^0} \delta\mathbf{P} \cdot\cdot\, \delta\mathbf{F}\, \mathrm{d}V^0 > \int_{S_t^0} \delta\mathbf{p}_n^{\mathrm{s}} \cdot \delta\mathbf{u}\, \mathrm{d}S_t^0, \tag{10.6.4}$$

or

$$\int_{V^0} \dot{\mathbf{P}} \cdot\cdot\, \dot{\mathbf{F}}\, \mathrm{d}V^0 > \int_{S_t^0} \dot{\mathbf{p}}_n^{\mathrm{s}} \cdot \mathbf{v}\, \mathrm{d}S_t^0, \tag{10.6.5}$$

for all admissible velocity fields $\mathbf{v}$ vanishing on $S_v^0 = S^0 - S_t^0$. Since χ is a homogeneous function of $\dot{\mathbf{F}}$ of degree two, and $\dot{\mathbf{P}} = \partial\chi/\partial\dot{\mathbf{F}}$, (10.6.5) can be rewritten as

$$\int_{V^0} \chi\, \mathrm{d}V^0 > \frac{1}{2}\int_{S_t^0} \dot{\mathbf{p}}_n^{\mathrm{s}} \cdot \mathbf{v}\, \mathrm{d}S_t^0. \tag{10.6.6}$$

If the current configuration is taken as the reference, the stability criterion becomes

$$\int_V \underline{\chi}\, \mathrm{d}V = \frac{1}{2}\int_V \underline{\dot{\mathbf{P}}} \cdot\cdot\, \mathbf{L}\, \mathrm{d}V > \frac{1}{2}\int_{S_t} \underline{\dot{\mathbf{p}}}_n^{\mathrm{s}} \cdot \mathbf{v}\, \mathrm{d}S_t. \tag{10.6.7}$$

10.7. Relationship between Uniqueness and Stability Criteria

In this section we compare the uniqueness criterion from Section 10.4,

$$\int_V \left(\underline{\dot{\mathbf{P}}}^* - \underline{\dot{\mathbf{P}}}\right) \cdot\cdot\, (\mathbf{L}^* - \mathbf{L})\, \mathrm{d}V > \int_{S_t} \left(\underline{\dot{\mathbf{p}}}_n^{*\,\mathrm{s}} - \underline{\dot{\mathbf{p}}}_n^{\mathrm{s}}\right) \cdot (\mathbf{v}^* - \mathbf{v})\, \mathrm{d}S_t, \tag{10.7.1}$$

with the stability condition

$$\int_V \underline{\dot{\mathbf{P}}} \cdot\cdot\, \mathbf{L}\, \mathrm{d}V > \int_{S_t} \underline{\dot{\mathbf{p}}}_n^{\mathrm{s}} \cdot \mathbf{v}\, \mathrm{d}S_t. \tag{10.7.2}$$

For kinematically admissible fields $\mathbf{v}$ and $\mathbf{v}^*$ vanishing on $S_v = S - S_t$, the field $\mathbf{v} - \mathbf{v}^*$ also vanishes on S_v and can be used as an admissible field in (10.7.2). The condition (10.7.2) is then equivalent to (10.7.1) only if the response is incrementally linear, so that $\underline{\dot{\mathbf{P}}}^* - \underline{\dot{\mathbf{P}}}$ is a linear function of $\mathbf{v} - \mathbf{v}^*$. For nonlinear and piecewise linear response this is not the case and two conditions are not equivalent.

Suppose the uniqueness condition (10.7.1) is satisfied when S_v is rigidly constrained or absent. Since the field $\mathbf{v}^* = \mathbf{0}$ is an admissible field for this boundary condition, it can be combined in (10.7.1) with any other nonzero admissible field $\mathbf{v}$, reproducing (10.7.2). Thus, when the sufficient condition for uniqueness of the rate boundary value problem at given state is satisfied for rigidly constrained or absent S_v, the underlying equilibrium state is also stable. The converse is not necessarily true for incrementally nonlinear material. The boundary value problem need not have unique solution in a stable state, i.e., (10.7.2) may be satisfied but not (10.7.1). A stable bifurcation could occur, although not under dead loading ($\dot{\mathbf{b}} = \mathbf{0}$ and $\dot{\underline{\mathbf{p}}}_n^{\mathrm{c}} = \mathbf{0}$), since this would imply that

$$\int_V \dot{\underline{\mathbf{P}}} \cdot\cdot \mathbf{L}\,\mathrm{d}V = \int_{S_t} \dot{\underline{\mathbf{p}}}_n^{\mathrm{s}} \cdot \mathbf{v}\,\mathrm{d}S_t, \tag{10.7.3}$$

for the actual velocity field (by the divergence theorem). The loading would have to change such that

$$\begin{aligned} &\int_V \rho\,\dot{\mathbf{b}} \cdot \mathbf{v}\,\mathrm{d}V + \int_{S_t} \dot{\mathbf{p}}_n^{\mathrm{c}} \cdot \mathbf{v}\,\mathrm{d}S_t \\ &\qquad = \int_V \dot{\underline{\mathbf{P}}} \cdot\cdot \mathbf{L}\,\mathrm{d}V - \int_{S_t} \dot{\underline{\mathbf{p}}}_n^{\mathrm{s}} \cdot \mathbf{v}\,\mathrm{d}S_t > 0, \end{aligned} \tag{10.7.4}$$

for any actual field at the bifurcation.

Denote by V^{e} the elastically stressed part of the body, and by V^{p} the remaining plastically stressed part (i.e., the part that is at the state of incipient yield), and assume that $\dot{\underline{\mathbf{p}}}_n^{\mathrm{s}} = \mathbf{0}$ on S_t for any kinematically admissible velocity. For rigidly constrained or absent S_v, the uniqueness condition becomes

$$\int_V \dot{\underline{\mathbf{P}}} \cdot\cdot \mathbf{L}\,\mathrm{d}V = \int_{V^{\mathrm{e}}} \dot{\underline{\mathbf{P}}} \cdot\cdot \mathbf{L}\,\mathrm{d}V^{\mathrm{e}} + \int_{V^{\mathrm{p}}} \dot{\underline{\mathbf{P}}} \cdot\cdot \mathbf{L}\,\mathrm{d}V^{\mathrm{p}} > 0. \tag{10.7.5}$$

In the elastic region V^{e} the response is incrementally linear, $\dot{\underline{\mathbf{P}}} = \underline{\mathbf{\Lambda}} \cdot\cdot \mathbf{L}$. For incrementally linear comparison material in the plastic region V^{p}, we have $\dot{\underline{\mathbf{P}}} = \underline{\mathbf{\Lambda}}^{\mathrm{p}} \cdot \mathbf{L}$. Thus, (10.7.5) is replaced with

$$\int_{V^{\mathrm{e}}} \underline{\mathbf{\Lambda}} \cdot\cdot\cdot\cdot (\mathbf{L} \otimes \mathbf{L})\,\mathrm{d}V^{\mathrm{e}} + \int_{V^{\mathrm{p}}} \underline{\mathbf{\Lambda}}^{\mathrm{p}} \cdot\cdot\cdot\cdot (\mathbf{L} \otimes \mathbf{L})\,\mathrm{d}V^{\mathrm{p}} > 0, \tag{10.7.6}$$

or, in view of (10.1.10) and (10.1.15),

$$\begin{aligned} &\int_{V^{\mathrm{e}}} \underline{\boldsymbol{\mathcal{L}}}_{(0)} :: (\mathbf{D} \otimes \mathbf{D})\,\mathrm{d}V^{\mathrm{e}} + \int_{V^{\mathrm{p}}} \underline{\boldsymbol{\mathcal{L}}}_{(0)}^{\mathrm{p}} :: (\mathbf{D} \otimes \mathbf{D})\mathrm{d}V^{\mathrm{p}} \\ &\qquad > \int_V \boldsymbol{\sigma} : \left(2\,\mathbf{D}^2 - \mathbf{L}^T \cdot \mathbf{L}\right) \mathrm{d}V. \end{aligned} \tag{10.7.7}$$

For example, consider pressure-independent isotropic hardening plasticity for which, by Eq. (9.4.26),

$$\underline{\boldsymbol{\mathcal{L}}}_{(0)}^{\mathrm{p}} = \underline{\boldsymbol{\mathcal{L}}}_{(0)} - \frac{2\mu}{1 + h^{\mathrm{p}}/\mu} (\mathbb{M} \otimes \mathbb{M}), \tag{10.7.8}$$

where $\mathbb{M}$ is a deviatoric normalized tensor in the direction of the yield surface normal,

$$\mathbb{M} = \frac{\partial f/\partial \boldsymbol{\sigma}}{\|\partial f/\partial \boldsymbol{\sigma}\|}, \quad \left\|\frac{\partial f}{\partial \boldsymbol{\sigma}}\right\| = \left(\frac{\partial f}{\partial \boldsymbol{\sigma}} : \frac{\partial f}{\partial \boldsymbol{\sigma}}\right)^{1/2}. \tag{10.7.9}$$

The uniqueness condition (10.7.7) then becomes

$$H > \Sigma, \tag{10.7.10}$$

for all kinematically admissible velocity fields, where (Hill, 1958)

$$H = \int_V \underline{\boldsymbol{\mathcal{L}}}_{(0)} :: (\mathbf{D} \otimes \mathbf{D})\, \mathrm{d}V - \int_{V^{\mathrm{p}}} \frac{2\mu}{1 + h^{\mathrm{p}}/\mu} (\mathbb{M} : \mathbf{D})^2\, \mathrm{d}V^{\mathrm{p}}, \tag{10.7.11}$$

$$\Sigma = \int_V \boldsymbol{\sigma} : \left(2\mathbf{D}^2 - \mathbf{L}^T \cdot \mathbf{L}\right) \mathrm{d}V. \tag{10.7.12}$$

If the rate of hardening $h^{\mathrm{p}} \to \infty$, we have

$$H^\infty = \int_V \underline{\boldsymbol{\mathcal{L}}}_{(0)} :: (\mathbf{D} \otimes \mathbf{D})\, \mathrm{d}V = \int_V \left[\lambda\, (\mathrm{tr}\, \mathbf{D})^2 + 2\mu\, (\mathbf{D} : \mathbf{D})\right] \mathrm{d}V > 0. \tag{10.7.13}$$

In the ideally plastic limit, $h^{\mathrm{p}} \to 0$ and

$$H^0 = H^\infty - 2\mu \int_{V^{\mathrm{p}}} (\mathbb{M} : \mathbf{D})^2\, \mathrm{d}V^{\mathrm{p}} < H^\infty. \tag{10.7.14}$$

For any positive rate of hardening h^{p}, then,

$$H^0 \le H \le H^\infty. \tag{10.7.15}$$

When h^{p} is the same throughout the volume V^{p}, the uniqueness condition $H > \Sigma$ becomes

$$H^\infty - \frac{2\mu}{1 + h^{\mathrm{p}}/\mu} \int_{V^{\mathrm{p}}} (\mathbf{n} : \mathbf{D})^2\, \mathrm{d}V^{\mathrm{p}} > \Sigma. \tag{10.7.16}$$

Using (10.7.14) to eliminate the integral over V^{p}, this gives

$$H^\infty - \Sigma > \frac{1}{1 + h^{\mathrm{p}}/\mu} \left(H^\infty - H^0\right), \tag{10.7.17}$$

i.e.,

$$\frac{h^{\mathrm{p}}}{\mu} > \frac{\Sigma - H^0}{H^\infty - \Sigma}. \tag{10.7.18}$$

Thus, the solution is certainly unique if, for all kinematically admissible $\mathbf{v}$,

$$\frac{h^{\mathrm{p}}}{\mu} > \beta, \quad \beta = \max_{\mathbf{v}} \left(\frac{\Sigma - H^0}{H^\infty - \Sigma}\right). \tag{10.7.19}$$

Consider next stability of the underlying equilibrium configuration. The stability criterion is also given by (10.7.5). This further becomes

$$\int_{V^e} \underline{\boldsymbol{\Lambda}} \cdots (\mathbf{L} \otimes \mathbf{L})\, \mathrm{d}V^e + \int_{V^p_u} \underline{\boldsymbol{\Lambda}} \cdots (\mathbf{L} \otimes \mathbf{L})\, \mathrm{d}V^p_u + \int_{V^p_l} \underline{\boldsymbol{\Lambda}}^p \cdots (\mathbf{L} \otimes \mathbf{L})\, \mathrm{d}V^p_l > 0, \tag{10.7.20}$$

or, in view of (10.1.10) and (10.1.15),

$$\int_{V^e} \underline{\boldsymbol{\mathcal{L}}}_{(0)} :: (\mathbf{D} \otimes \mathbf{D})\, \mathrm{d}V^e + \int_{V^p_u} \underline{\boldsymbol{\mathcal{L}}}_{(0)} :: (\mathbf{D} \otimes \mathbf{D}) \mathrm{d}V^p_u + \int_{V^p_l} \underline{\boldsymbol{\mathcal{L}}}^p_{(0)} :: (\mathbf{D} \otimes \mathbf{D})\, \mathrm{d}V^p_l > \int_V \boldsymbol{\sigma} : \left(2\,\mathbf{D}^2 - \mathbf{L}^T \cdot \mathbf{L}\right) \mathrm{d}V. \tag{10.7.21}$$

Here, V^p_l is the part of V^p where plastic loading takes place, while V^p_u is the part of V^p where elastic unloading or neutral loading takes place, for the prescribed $\mathbf{v}$. When Eq. (10.7.8) is incorporated, this becomes

$$H_l > \Sigma, \tag{10.7.22}$$

for all kinematically admissible velocity fields, where

$$H_l = \int_V \underline{\boldsymbol{\mathcal{L}}}_{(0)} :: (\mathbf{D} \otimes \mathbf{D})\, \mathrm{d}V - \int_{V^p_l} \frac{2\mu}{1 + h^p/\mu} (\mathbb{M} : \mathbf{D})^2\, \mathrm{d}V^p_l. \tag{10.7.23}$$

The plastic loading condition in V^p_l is

$$\mathbb{M} : \mathbf{D} > 0. \tag{10.7.24}$$

If we define

$$H^0_l = H^\infty - 2\mu \int_{V^p_l} (\mathbb{M} : \mathbf{D})^2\, \mathrm{d}V^p_l, \tag{10.7.25}$$

the equilibrium is stable when

$$\frac{h^p}{\mu} > \beta_l, \quad \beta_l = \max_{\mathbf{v}} \left(\frac{\Sigma - H^0_l}{H^\infty - \Sigma} \right), \tag{10.7.26}$$

for all kinematically admissible $\mathbf{v}$.

Evidently, since $V^p_l \le V^p$, we have

$$H^0 \le H^0_l, \quad \beta_l \le \beta. \tag{10.7.27}$$

Thus, for certain problems and deformation paths, a state of bifurcation can be reached at an earlier stage than a failure of stability. This could occur at the hardening rate $h^p = \beta\,\mu$, when (10.7.1) fails and uniqueness is no longer certain. If such stable bifurcation occurs, the loading must change with further deformation according to (10.7.4). Assuming that the hardening rate gradually decreases as the deformation proceeds, the stability of equilibrium configuration would be lost at the lower hardening rate $h^{\mathrm{p}} = \beta_l\, \mu$.

10.8. Uniqueness and Stability for Rigid-Plastic Materials

If elastic moduli are assigned infinitely large values, only plastic strain can take place and the model of rigid-plastic behavior is obtained. For example, in the case of isotropic hardening, the rate of deformation is

$$\mathbf{D} = \frac{1}{2h}\,(\mathbb{M} : \overset{\circ}{\boldsymbol{\sigma}})\,\mathbb{M}, \tag{10.8.1}$$

provided that $\mathbb{M} : \overset{\circ}{\boldsymbol{\sigma}} > 0$. The hardening modulus is h (with the von Mises yield criterion, $h = h_{\mathrm{t}}$, the tangent modulus in shear test). The response is incompressible and bilinear, since in the hardening range

$$\mathbf{D} = \mathbf{0}, \quad \text{when} \quad \mathbb{M} : \overset{\circ}{\boldsymbol{\sigma}} \leq 0. \tag{10.8.2}$$

Also note that

$$\overset{\circ}{\underline{\boldsymbol{\tau}}} = \overset{\circ}{\boldsymbol{\sigma}}, \tag{10.8.3}$$

since $\operatorname{tr}\mathbf{D} = 0$. By taking the inner product of $\mathbf{D}$ with $\overset{\circ}{\boldsymbol{\sigma}}$ and with itself, it follows that

$$\overset{\circ}{\boldsymbol{\sigma}} : \mathbf{D} = \frac{1}{2h}\,(\mathbb{M} : \overset{\circ}{\boldsymbol{\sigma}})^2, \quad \mathbf{D} : \mathbf{D} = \frac{1}{4h^2}\,(\mathbb{M} : \overset{\circ}{\boldsymbol{\sigma}})^2, \tag{10.8.4}$$

so that

$$\overset{\circ}{\boldsymbol{\sigma}} : \mathbf{D} = 2h\,(\mathbf{D} : \mathbf{D}). \tag{10.8.5}$$

It can be readily shown, when both rates of deformation vanish, or when both rates are different from zero ($\mathbf{D} = \mathbf{0}$ and $\mathbf{D}^* = \mathbf{0}$, or $\mathbf{D} \neq \mathbf{0}$ and $\mathbf{D}^* \neq \mathbf{0}$),

$$(\overset{\circ}{\boldsymbol{\sigma}}{}^* - \overset{\circ}{\boldsymbol{\sigma}}) : (\mathbf{D}^* - \mathbf{D}) = 2h\,(\mathbf{D}^* - \mathbf{D}) : (\mathbf{D}^* : \mathbf{D}). \tag{10.8.6}$$

If one rate of deformation vanishes and the other does not (e.g., $\mathbf{D}^* = \mathbf{0}$ and $\mathbf{D} \neq \mathbf{0}$),

$$\begin{aligned}(\overset{\circ}{\boldsymbol{\sigma}}{}^* - \overset{\circ}{\boldsymbol{\sigma}}) : (\mathbf{D}^* - \mathbf{D}) &= \overset{\circ}{\boldsymbol{\sigma}} : \mathbf{D} - \overset{\circ}{\boldsymbol{\sigma}}{}^* : \mathbf{D} \\ &> 2h\,(\mathbf{D} : \mathbf{D}) = 2h\,(\mathbf{D}^* - \mathbf{D}) : (\mathbf{D}^* : \mathbf{D}),\end{aligned} \tag{10.8.7}$$

since $\overset{\circ}{\boldsymbol{\sigma}}{}^* : \mathbf{D} \leq 0$. Thus, for all pairs $\mathbf{D}$ and $\mathbf{D}^*$, we have

$$(\overset{\circ}{\boldsymbol{\sigma}}{}^* - \overset{\circ}{\boldsymbol{\sigma}}) : (\mathbf{D}^* - \mathbf{D}) \geq 2h\,(\mathbf{D}^* - \mathbf{D}) : (\mathbf{D}^* : \mathbf{D}). \tag{10.8.8}$$

The uniqueness condition (10.4.8) for an elastoplastic material can be written, in view of Eq. (10.4.12), as

$$\int_V \Big\{(\overset{\circ}{\underline{\tau}}{}^* - \overset{\circ}{\underline{\tau}}) : (\mathbf{D}^* - \mathbf{D}) - \boldsymbol{\sigma} : \big[2\,(\mathbf{D}^* - \mathbf{D})^2$$
$$- \left(\mathbf{L}^{*T} - \mathbf{L}^T\right) \cdot (\mathbf{L}^* - \mathbf{L})\,\big]\Big\}\,\mathrm{d}V > \int_{S_t} \left(\dot{\underline{\mathbf{p}}}{}_n^{*\,\mathrm{s}} - \dot{\underline{\mathbf{p}}}{}_n^{\mathrm{s}}\right) \cdot (\mathbf{v}^* - \mathbf{v})\,\mathrm{d}S_t. \tag{10.8.9}$$

Having regard to inequality (10.8.8), and $\overset{\circ}{\underline{\tau}} = \overset{\circ}{\boldsymbol{\sigma}}$, a sufficient condition for uniqueness of the boundary value problem for rigid-plastic material is

$$\int_V \{2h\,(\mathbf{D}^* - \mathbf{D}) : (\mathbf{D}^* - \mathbf{D}) - \boldsymbol{\sigma} : \big[2\,(\mathbf{D}^* - \mathbf{D})^2$$
$$- \left(\mathbf{L}^{*T} - \mathbf{L}^T\right) \cdot (\mathbf{L}^* - \mathbf{L})\,\big]\}\,\mathrm{d}V > \int_{S_t} \left(\dot{\underline{\mathbf{p}}}{}_n^{*\,\mathrm{s}} - \dot{\underline{\mathbf{p}}}{}_n^{\mathrm{s}}\right) \cdot (\mathbf{v}^* - \mathbf{v})\,\mathrm{d}S_t. \tag{10.8.10}$$

This also directly follows from the notion of an incrementally linear comparison material that reacts at every plastically stressed point according to plastic loading branch (10.8.1). Although $\boldsymbol{\sigma}$ is undetermined in rigid regions, the integrals in (10.8.10) can be taken over the whole volume, since there is no contribution from rigid regions ($\mathbf{v}$ there being equal to $\mathbf{v}^*$). Furthermore, since for isotropic behavior the principal directions of $\boldsymbol{\sigma}$ and $(\mathbf{D}^* - \mathbf{D})$ coincide, the tensor $\boldsymbol{\sigma} \cdot (\mathbf{D}^* - \mathbf{D})$ is symmetric, and

$$\boldsymbol{\sigma} : \left[2\,(\mathbf{D}^* - \mathbf{D})^2 - \left(\mathbf{L}^{*T} - \mathbf{L}^T\right) \cdot (\mathbf{L}^* - \mathbf{L})\right] = \boldsymbol{\sigma} : (\mathbf{L}^* - \mathbf{L})^2. \tag{10.8.11}$$

Consequently, the uniqueness is assured if the exclusion functional is positive

$$\underline{\mathcal{F}}(\mathbf{w}) = \int_V \left[2h\,(\boldsymbol{D} : \boldsymbol{D}) - \boldsymbol{\sigma} : \boldsymbol{L}^2\right]\,\mathrm{d}V - \int_{S_t} \dot{\underline{\boldsymbol{p}}}{}_n^{\mathrm{s}}(\mathbf{w}) \cdot \mathbf{w}\,\mathrm{d}S_t > 0, \tag{10.8.12}$$

for any incompressible kinematically admissible velocity field $\mathbf{w}$, which gives rise to rate of deformation $\boldsymbol{D}$ (symmetric part of $\boldsymbol{L} = \partial\mathbf{w}/\partial\mathbf{x}$) that is codirectional with $\mathbb{M}$ in the plastic region (though not necessarily in the same sense, since $\mathbf{D}^* - \mathbf{D}$ in (10.8.10) can be in either $\mathbb{M}$ or $-\mathbb{M}$ direction), and equal to zero in the rigid region.

If h is constant throughout plastically stressed region V^{p}, and if $\dot{\underline{\boldsymbol{p}}}{}_n^{\mathrm{s}} = \mathbf{0}$ on S_t, the uniqueness is certain when

$$2\,h > \max_{\mathbf{w}} \frac{\int_V (\boldsymbol{\sigma} : \boldsymbol{L}^2)\,\mathrm{d}V}{\int_V (\boldsymbol{D} : \boldsymbol{D})\,\mathrm{d}V}\,. \tag{10.8.13}$$

The underlying equilibrium configuration is stable if (10.8.13) holds, but the class of admissible velocity fields is further restricted by the requirement that $\mathbb{M} : \boldsymbol{D}$ is non-negative in the plastic region ($\mathbb{M} : \boldsymbol{D}$ can be either positive, negative or zero in the plastic region for admissible velocity fields in the uniqueness condition, so that this class is wider than the class of admissible velocity fields in the stability condition).

10.8.1. Uniaxial Tension

In the tension test of a specimen with uniform cross-section, the state of stress at an incipient bifurcation is uniform tension $\sigma_{11} = \sigma$, other stress components being equal to zero. An admissible velocity field for the uniqueness condition must be incompressible and give rise to the rate of deformation tensor $\boldsymbol{D}$ parallel to $\boldsymbol{\sigma}'$ (the yield surface normal). This is satisfied when

$$D_{22} = D_{33} = -\frac{1}{2} D_{11}, \quad D_{12} = D_{23} = D_{31} = 0. \tag{10.8.14}$$

Thus,

$$\boldsymbol{\sigma} : \boldsymbol{L}^2 = \sigma (D_{11}^2 - L_{12}^2 - L_{13}^2), \quad \boldsymbol{D} : \boldsymbol{D} = \frac{3}{2} D_{11}^2, \tag{10.8.15}$$

and the condition (10.8.13) gives

$$3\frac{h}{\sigma} > \max_{\mathbf{w}} \frac{\int_V (D_{11}^2 - L_{12}^2 - L_{13}^2)\, \mathrm{d}V}{\int_V D_{11}^2\, \mathrm{d}V}. \tag{10.8.16}$$

The right-hand side is always smaller than one (irrespective of the boundary conditions at the ends and specific representation of admissible functions $\mathbf{w}$), so that fundamental mode of deformation (uniform straining) is certainly unique (and underlying equilibrium configuration stable) for $h > \sigma/3$. With the von Mises yield criterion, $h = h_t$, and since $h_t = (1/3)\, \mathrm{d}\sigma/\mathrm{d}e$, where e denotes longitudinal logarithmic strain in uniaxial tension, the deformation mode is unique when the slope of the true stress-strain curve exceeds the current yield stress. As is well-known, at the critical value $\mathrm{d}\sigma/\mathrm{d}e = \sigma$, the applied load attains its maximum value and either further uniform straining or localized necking is possible in principle. Hutchinson and Miles (1974) have demonstrated that in the case of circular cylinder of incompressible elastic-plastic material, an axially symmetric bifurcation of a necking type exists when the true stress reaches a critical value slightly greater than the stress corresponding to the maximum load. The shear free ends of the cylinder with traction-free lateral surface were subject to uniform longitudinal relative displacement. A numerical study of necking in elastoplastic circular cylinders under uniaxial tension with different boundary conditions at the ends was performed by Needleman (1972). Hill and Hutchinson (1975) gave a comprehensive analysis of bifurcation modes from a state of homogeneous in-plane tension of an incompressible rectangular block under plane deformation. The sides of the block were traction-free and the shear-free ends were subject to uniform longitudinal relative displacement. See also Burke and Nix (1979), and Bardet (1991). For the effects of plastic non-normality on bifurcation prediction, see Needleman (1979) and Kleiber (1986). Bufurcation of an incompressible plate under pure bending in plane strain was studied by Triantafyllidis (1980).

10.8.2. Compression of Column

Consider a column of uniform cross-sectional area A, built at one end and loaded at the other by an increasing axial load N. The state of stress is uniaxial compression of amount $\sigma_{11} = -N/A$, except possibly near the ends. For sufficiently long, slender columns possible nonuniformities near the ends can be neglected and the uniqueness condition (10.8.13) gives (Hill, 1957)

$$3\,\frac{A\,h}{N} > \max_{\mathbf{w}} \frac{\int_V \left(L_{12}^2 + L_{13}^2 - D_{11}^2\right) \mathrm{d}V}{\int_V D_{11}^2\, \mathrm{d}V}\,. \tag{10.8.17}$$

The admissible velocity field $\mathbf{w}$ again satisfies the conditions $D_{22} = D_{33} = -D_{11}/2$, and $D_{12} = D_{23} = D_{31} = 0$, i.e.,

$$\begin{gathered}\frac{\partial w_2}{\partial x_2} = \frac{\partial w_3}{\partial x_3} = -\frac{1}{2}\frac{\partial w_1}{\partial x_1}\,,\\ \frac{\partial w_1}{\partial x_2} + \frac{\partial w_2}{\partial x_1} = 0, \quad \frac{\partial w_2}{\partial x_3} + \frac{\partial w_3}{\partial x_2} = 0, \quad \frac{\partial w_3}{\partial x_1} + \frac{\partial w_1}{\partial x_3} = 0.\end{gathered} \tag{10.8.18}$$

These have the general solution

$$\begin{aligned} w_1 &= a\,x_1x_2 + b\,x_1x_3 + c\,(2x_1^2 + x_2^2 + x_3^2) + d\,x_1,\\ w_2 &= -\frac{1}{2}\,b\,x_2x_3 - 2\,c\,x_1x_2 - \frac{1}{4}\,a\,(2x_1^2 + x_2^2 - x_3^2) - \frac{1}{2}\,d\,x_2,\\ w_3 &= -\frac{1}{2}\,a\,x_2x_3 - 2\,c\,x_1x_3 - \frac{1}{4}\,b\,(2x_1^2 - x_2^2 + x_3^2) - \frac{1}{2}\,d\,x_3. \end{aligned} \tag{10.8.19}$$

By taking the origin of the coordinate system at the centroid of the fixed end, above functions satisfy the end conditions

$$w_1 = w_2 = w_3 = 0, \quad \frac{\partial w_2}{\partial x_1} = \frac{\partial w_3}{\partial x_1} = 0 \tag{10.8.20}$$

at the origin. Selecting the axes x_2 and x_3 to be the principal centroidal axes of the cross section, the substitution of the expression for w_1 from (10.8.19) into (10.8.17) gives

$$3\,\frac{h}{Nl^2} > \max_{\mathbf{w}} \frac{a^2\left(\frac{1}{3} - \frac{I_3}{Al^2}\right) + b^2\left(\frac{1}{3} - \frac{I_2}{Al^2}\right) + 4c^2\,\frac{I_2+I_3}{Al^2} - \left[\frac{4}{3}\,c^2 + \left(2c + \frac{d}{l}\right)^2\right]}{a^2\,I_3 + b^2\,I_2 + \left[\frac{4}{3}\,c^2 + \left(2c + \frac{d}{l}\right)^2\right]Al^2}\,. \tag{10.8.21}$$

The second moments of the cross sectional area about the x_2 and x_3 axes are I_2 and I_3. The right-hand side in (10.8.21) has a maximum value when the square bracketed term vanishes, which occurs for $c = d = 0$ (for slender columns, $I_2 + I_3 << Al^2$). Thus,

$$3\,\frac{h}{Nl^2} > \frac{1}{3}\max_{\mathbf{w}}\left(\frac{a^2 + b^2}{a^2\,I_3 + b^2\,I_2}\right) - \frac{1}{Al^2}\,. \tag{10.8.22}$$

The term $(Al^2)^{-1}$ can be neglected for slender columns, and

$$\frac{h}{Nl^2} > \frac{1}{9\,I_{\min}}\,. \tag{10.8.23}$$

If $I_3 > I_2$, the maximum occurs for $a = 0$; if $I_2 > I_3$, the maximum occurs for $b = 0$; if $I_2 = I_3$, any ratio a/b can be used. In each case the $\mathbf{w}$ field reduces to pure bending. For example, for circular cross-section of radius R, we obtain

$$h > \frac{4}{9}\,\frac{N}{A}\left(\frac{l}{R}\right)^2. \tag{10.8.24}$$

In the consideration of stability the constants a, b, c, d are not entirely arbitrary in the expressions for admissible functions (10.8.19), but are subject to condition

$$\boldsymbol{\sigma}' : \boldsymbol{D} \geq 0, \quad \text{i.e.,} \quad D_{11} \leq 0. \tag{10.8.25}$$

This gives (Hill, 1957)

$$a\,x_2 + b\,x_3 + 4\,c\,x_1 + d \leq 0, \tag{10.8.26}$$

everywhere in the body. The expression (10.8.21) attains its maximum for $c = 0$, so that

$$3\,\frac{h}{Nl^2} > \max_{\mathbf{w}} \frac{a^2\left(\frac{1}{3} - \frac{I_3}{Al^2}\right) + b^2\left(\frac{1}{3} - \frac{I_2}{Al^2}\right) - \frac{d^2}{l^2}}{a^2\, I_3 + b^2\, I_2 + Ad^2}\,. \tag{10.8.27}$$

Suppose that the cross-section is a circle of radius R. The value of d which makes the right-hand side of (10.8.27) maximum and fulfills the condition (10.8.26) with $c = 0$ is readily found to be

$$d = -R(a^2 + b^2)^{1/2}. \tag{10.8.28}$$

The condition (10.8.27) consequently becomes

$$\frac{h}{Nl^2} > \frac{4}{45AR^2} - \frac{1}{3Al^2}\,. \tag{10.8.29}$$

Upon neglecting $(Al^2)^{-1}$ term,

$$h > \frac{4}{45}\,\frac{N}{A}\left(\frac{l}{R}\right)^2. \tag{10.8.30}$$

The obtained critical hardening rate for stability of column is 1/5 of that obtained from the condition of uniqueness, which is given by (10.8.24). More general elastoplastic analysis of column failure is presented by Hill and Sewell (1960,1962). A comprehensive treatment of plastic buckling and post-buckling behavior of columns and other structures is given by Hutchinson (1973,1974), and Bažant and Cedolin (1991). See also Storåkers (1971, 1977), Sewell (1973), Young (1976), Needleman and Tvergaard (1982), and Nguyen (1994).

10.9. Eigenmodal Deformations

From the analysis in preceding sections it is recognized that there may be particular configurations of the body where nominal tractions are momentarily constant as the body is incrementally deformed in certain ways. The corresponding instantaneous velocity fields are then nontrivial solutions of a homogeneous boundary-value problem. These velocity fields are referred to as eigenmodes. The underlying configurations are the eigenstates. An uniaxial tension specimen of a ductile metal at maximum load is an example of an eigenstate configuration. The presented theory is originally due to Hill (1967).

10.9.1. Eigenstates and Eigenmodes

Consider a solid body whose entire bounding surface is unconstrained ($S_t = S$). The exclusion functional of Eq. (10.4.17) is then

$$\underline{\mathcal{F}}(\mathbf{w}) = \int_V \underline{\chi}(\mathbf{w})\,\mathrm{d}V - \frac{1}{2}\int_S \underline{\dot{\boldsymbol{p}}}^{\mathrm{s}}_n(\mathbf{w})\cdot\mathbf{w}\,\mathrm{d}S. \tag{10.9.1}$$

If equilibrium configuration of an incrementally linear material is stable under all-around dead loads, the strain path cannot bifurcate from that state for any loading rates applied to the state. A sufficient condition for stability and uniqueness is that $\underline{\mathcal{F}}(\mathbf{w}) > 0$ for all admissible velocity fields $\mathbf{w}$. Bifurcation can occur only when a primary eigenstate is reached (first eigenstate reached on a given deformation path), where

$$\underline{\mathcal{F}}(\mathbf{w}) \geq 0, \tag{10.9.2}$$

with the equality sign for some velocity field (eigenmode velocity field).

For a piecewise linear or thoroughly nonlinear material response with the rate potential $\underline{\chi}$, a deformation path could bifurcate under varying load before the primary eigenstate is reached and stability lost. As discussed in Subsection 10.4.2, to prevent bifurcation before an eigenstate is reached, it is sufficient that configuration is stable for incrementally linear comparison material $\underline{\chi}^l$, and that $\underline{\chi} - \underline{\chi}^l$ is a convex function of $\boldsymbol{L}$. The bifurcation may be excluded for $\underline{\chi}$ material even if the configuration is an eigenstate for $\underline{\chi}^l$ material, but $\underline{\chi} - \underline{\chi}^l$ is strictly convex function in that configuration. If the current configuration is a primary eigenstate for $\underline{\chi}^l$ material, and $\underline{\chi} - \underline{\chi}^l$ is merely convex, the configuration may be a primary eigenstate for $\underline{\chi}$ material, provided there is an eigenmode of $\underline{\chi}^l$ material that is also an eigenmode of $\underline{\chi}$ material (giving rise to plastic loading throughout plastically stressed region of $\underline{\chi}$ material).

Suppose that for, either incrementally linear or incrementally nonlinear material, $\underline{\mathcal{F}}(\mathbf{w})$ is positive definite along a loading path from the undeformed state, until a primary eigenstate is reached where $\underline{\mathcal{F}}(\mathbf{w}) \geq 0$ (with equality sign for an eigenmodal field). Since $\underline{\mathcal{F}}(\mathbf{w})$ is non-negative in an eigenstate, vanishing only in an eigenmode, its first variation $\delta\underline{\mathcal{F}}$ must be zero in an eigenmode,

$$\delta\left[\int_V \underline{\chi}(\mathbf{w})\,\mathrm{d}V - \frac{1}{2}\int_S \underline{\dot{\boldsymbol{p}}}^{\mathrm{s}}_n(\mathbf{w})\cdot\mathbf{w}\,\mathrm{d}S\right] = 0. \tag{10.9.3}$$

Thus, in an eigenmode field $\mathbf{w}$,

$$\int_V \underline{\dot{\boldsymbol{P}}} \cdot\cdot\, \delta \boldsymbol{L}\, \mathrm{d}V - \int_S \underline{\dot{\boldsymbol{p}}}^{\mathrm{s}}_n \cdot \delta \mathbf{w}\, \mathrm{d}S = 0, \tag{10.9.4}$$

for all admissible variations $\delta\mathbf{w}$. In addition, the functional itself vanishes in an eigenmode,

$$\int_V \underline{\chi}(\mathbf{w})\, \mathrm{d}V - \frac{1}{2}\int_S \underline{\dot{\boldsymbol{p}}}^{\mathrm{s}}_n(\mathbf{w}) \cdot \mathbf{w}\, \mathrm{d}S = 0. \tag{10.9.5}$$

Under all-around deformation-insensitive dead loading, the above two conditions reduce to

$$\int_V \underline{\dot{\boldsymbol{P}}} \cdot\cdot\, \delta \boldsymbol{L}\, \mathrm{d}V = 0, \quad \int_V \underline{\chi}(\mathbf{w})\, \mathrm{d}V = 0. \tag{10.9.6}$$

An eigenmode is in this case a nontrivial solution of homogeneous boundary value problem described by

$$\boldsymbol{\nabla} \cdot \underline{\dot{\boldsymbol{P}}} = \mathbf{0} \quad \text{in} \quad V, \quad \text{and} \quad \mathbf{n} \cdot \underline{\dot{\boldsymbol{P}}} = \mathbf{0} \quad \text{on} \quad S. \tag{10.9.7}$$

10.9.2. Eigenmodal Spin

Suppose that a homogeneous body is uniformly strained from its undeformed configuration to a primary eigenstate configuration. The state of stress and material properties are then uniform at each instant of deformation, and $\underline{\chi}$ is the same function of velocity gradient at every point of the body in the considered configuration. By choosing velocity fields with arbitrary uniform gradient $\boldsymbol{L}$, it follows that $\underline{\mathcal{F}} > 0$ if and only if $\underline{\chi} > 0$ along stable segment of deformation path, and that $\underline{\chi} \geq 0$ in a primary eigenstate. Equality $\underline{\chi} = 0$ applies for an eigenmode velocity field, which also makes $\underline{\chi}$ stationary. Since

$$\delta\underline{\chi} = \underline{\dot{\boldsymbol{P}}} \cdot\cdot\, \delta\boldsymbol{L} = 0 \tag{10.9.8}$$

in an eigenmode for all $\delta\boldsymbol{L}$, we conclude that

$$\underline{\dot{\boldsymbol{P}}} = \frac{\partial\underline{\chi}}{\partial\boldsymbol{L}} = \mathbf{0}. \tag{10.9.9}$$

This means that the nominal stress is stationary in an eigenmode (momentarily constant as the body is incrementally deformed along an eigenmode field).

Since from Section 3.9,

$$\underline{\dot{\boldsymbol{T}}}_{(1)} = \underline{\dot{\boldsymbol{P}}} - \boldsymbol{\sigma} \cdot \boldsymbol{L}^T, \tag{10.9.10}$$

and since local rotational balance requires $\dot{\underline{\boldsymbol{T}}}_{(1)}$ to be symmetric, from (10.9.9) it follows that in an eigenmode

$$\boldsymbol{L} \cdot \boldsymbol{\sigma} = \boldsymbol{\sigma} \cdot \boldsymbol{L}^T, \tag{10.9.11}$$

so that

$$\boldsymbol{\sigma} \cdot \boldsymbol{W} + \boldsymbol{W} \cdot \boldsymbol{\sigma} = \boldsymbol{\sigma} \cdot \boldsymbol{D} - \boldsymbol{D} \cdot \boldsymbol{\sigma}. \tag{10.9.12}$$

This can be solved for $\boldsymbol{W}$ in terms of $\boldsymbol{\sigma}$ and $\boldsymbol{D}$ by using (1.12.12). The solution is an expression for the eigenmodal spin in terms of stress and eigenmodal rate of deformation,

$$\boldsymbol{W} = (\operatorname{tr} \mathbf{S})(\boldsymbol{\sigma} \cdot \boldsymbol{D} - \boldsymbol{D} \cdot \boldsymbol{\sigma}) - \mathbf{S} \cdot (\boldsymbol{\sigma} \cdot \boldsymbol{D} - \boldsymbol{D} \cdot \boldsymbol{\sigma}) - (\boldsymbol{\sigma} \cdot \boldsymbol{D} - \boldsymbol{D} \cdot \boldsymbol{\sigma}) \cdot \mathbf{S}, \tag{10.9.13}$$

where

$$\mathbf{S} = [(\operatorname{tr} \boldsymbol{\sigma})\, \mathbf{I} - \boldsymbol{\sigma}]^{-1}. \tag{10.9.14}$$

It is assumed that $\mathbf{S}$ exists. When written in terms of components on the principal axes of stress $\boldsymbol{\sigma}$, the required condition for the inverse in Eq. (10.9.14) to exist is

$$\det[(\operatorname{tr} \boldsymbol{\sigma})\, \mathbf{I} - \boldsymbol{\sigma}] = (\sigma_1 + \sigma_2)(\sigma_2 + \sigma_3)(\sigma_3 + \sigma_1) \neq 0. \tag{10.9.15}$$

The eigenmodal spin components on the principal stress axes are

$$W_{12} = \frac{\sigma_1 - \sigma_2}{\sigma_1 + \sigma_2}\, D_{12}, \quad W_{23} = \frac{\sigma_2 - \sigma_3}{\sigma_2 + \sigma_3}\, D_{23}, \quad W_{31} = \frac{\sigma_3 - \sigma_1}{\sigma_3 + \sigma_1}\, D_{31}. \tag{10.9.16}$$

Evidently, if the principal axes of $\boldsymbol{D}$ happen to coincide with those of $\boldsymbol{\sigma}$ (as in the case of rigid-plastic von Mises plasticity), the spin of an eigenmode field entirely vanishes. If the stress field has an axis of equilibrium, for example axis 1 in the case when $\sigma_2 + \sigma_3 = 0$, W_{23} is undetermined and D_{23} must vanish. On the other hand, when the stress state is uniaxial, $\sigma_2 = \sigma_3 = 0$, there is no restriction on D_{23} but W_{23} is still undetermined.

It can be readily verified that among all velocity gradients with the fixed strain rates, $\underline{\chi}$ attains its minimum when $\sigma_1 + \sigma_2 > 0$, $\sigma_2 + \sigma_3 > 0$, $\sigma_3 + \sigma_1 > 0$, and when the spin components are determined by (10.9.16).

Indeed, for an elastoplastic material, $\underline{\chi}$ can be written from (10.1.15) as

$$\underline{\chi} = \frac{1}{2}\boldsymbol{\mathcal{L}}^{\mathrm{p}}_{(0)} :: (\boldsymbol{D}\otimes\boldsymbol{D}) - \frac{1}{2}\boldsymbol{\sigma} : \mathbf{D}^2$$

$$+ \frac{1}{2}\left[(\sigma_1+\sigma_2)\, W_{12}^2 + (\sigma_2+\sigma_3)\, W_{23}^2 + (\sigma_3+\sigma_1)\, W_{31}^2\right] \qquad (10.9.17)$$

$$- (\sigma_1-\sigma_2)\, D_{12}\, W_{12} - (\sigma_2-\sigma_3)\, D_{23}\, W_{23} - (\sigma_3-\sigma_1)\, D_{31}\, W_{31}.$$

The stationary conditions

$$\frac{\partial\underline{\chi}}{\partial W_{ij}} = 0 \qquad (10.9.18)$$

clearly reproduce (10.9.16). The corresponding minimum of $\underline{\chi}$ is

$$\underline{\chi}^0 = \frac{1}{2}\boldsymbol{\mathcal{L}}^{\mathrm{p}}_{(0)} :: (\boldsymbol{D}\otimes\boldsymbol{D}) - \frac{1}{2}\boldsymbol{\sigma} : \mathbf{D}^2$$

$$- \frac{1}{2}\left[\frac{(\sigma_1-\sigma_2)^2}{\sigma_1+\sigma_2}\, D_{12}^2 + \frac{(\sigma_2-\sigma_3)^2}{\sigma_2+\sigma_3}\, D_{23}^2 + \frac{(\sigma_3-\sigma_1)^2}{\sigma_3+\sigma_1}\, D_{31}^2\right]. \qquad (10.9.19)$$

For isotropic hardening plasticity, from (9.8.14) we obtain

$$\frac{1}{2}\boldsymbol{\mathcal{L}}^{\mathrm{p}}_{(0)} :: (\boldsymbol{D}\otimes\boldsymbol{D}) = \frac{1}{2}\lambda\, (\mathrm{tr}\,\boldsymbol{D})^2 + \mu\,\boldsymbol{D} : \boldsymbol{D} - \frac{\mu}{1+h^{\mathrm{p}}/\mu}\, (\mathbb{M} : \boldsymbol{D})^2. \qquad (10.9.20)$$

Since, for isotropic smooth yield surface, $\mathbb{M}$ has the principal directions parallel to those of stress, $\mathbb{M}_{ij} = 0$ for $i \neq j$ on the coordinate axes parallel to the principal stress axes.

If $\boldsymbol{D}$ is the rate of deformation in an eigenmode, then

$$\underline{\chi}^0(\boldsymbol{D}) = 0. \qquad (10.9.21)$$

For all other rates of deformation in an eigenstate, $\underline{\chi}^0 > 0$. The uniqueness and stability are assured in any configuration before primary eigenstate is reached if $\underline{\chi}^0$, defined by (10.9.19), is positive definite in that configuration, since then $\underline{\chi}$ is also positive definite in that configuration.

In order that the configuration can qualify as stable by the criterion $\underline{\chi} > 0$ for all $\boldsymbol{L}$, the stress state has to be such that

$$\sigma_1+\sigma_2 > 0, \quad \sigma_2+\sigma_3 > 0, \quad \sigma_3+\sigma_1 > 0, \qquad (10.9.22)$$

which means that tension acts on the planes of maximum shear stress. This follows from (10.9.17) by choosing $\boldsymbol{L}$ to be an arbitrary antisymmetric (spin) tensor, so that

$$\underline{\chi} = \frac{1}{2}\left[(\sigma_1 + \sigma_2)\, W_{12}^2 + (\sigma_2 + \sigma_3)\, W_{23}^2 + (\sigma_3 + \sigma_1)\, W_{31}^2\right]. \tag{10.9.23}$$

Physically, (10.9.22) is imposed, because the opposite inequalities would allow dead loads to do positive work in certain virtual rotations of the body. Note, however, that pure spin cannot by itself be an eigenmode field under triaxial state of stress, since equations of continuing rotational equilibrium (10.9.12) would require that

$$(\sigma_1 + \sigma_2)\, W_{12} = 0, \quad (\sigma_2 + \sigma_3)\, W_{23} = 0, \quad (\sigma_3 + \sigma_1)\, W_{31} = 0. \tag{10.9.24}$$

Thus, unless the stress state has an axis of equilibrium, each spin component must vanish. This is also clear from (10.9.16); if the rate of deformation components are zero in an eigenmode, the eigenmode spin also vanishes. If $\sigma_1 + \sigma_2 = 0$, the spin W_{23} could be nonzero (but would be permissible as an actual mode only if it does not alter the applied tractions, keeping them dead in magnitude and direction, as in the case of uniaxial tension and a spin around the axis of loading).

10.9.3. Eigenmodal Rate of Deformation

The components of rate of deformation D_{ij} of an eigenmode velocity field are nontrivial solutions of the homogeneous system of equations resulting from (10.9.9). Since

$$\dot{\underline{\boldsymbol{P}}} = \overset{\circ}{\underline{\boldsymbol{\tau}}} - \boldsymbol{D}\cdot\boldsymbol{\sigma} - \boldsymbol{\sigma}\cdot\boldsymbol{W}, \tag{10.9.25}$$

the system of equations is

$$\underline{\boldsymbol{\mathcal{L}}}^{\mathrm{p}}_{(0)} : \boldsymbol{D} - \boldsymbol{D}\cdot\boldsymbol{\sigma} - \boldsymbol{\sigma}\cdot\boldsymbol{W} = \boldsymbol{0}, \tag{10.9.26}$$

where $\boldsymbol{W}$ is defined in terms of $\boldsymbol{\sigma}$ and $\boldsymbol{D}$ by (10.9.16). Specifically,

$$\boldsymbol{D}\cdot\boldsymbol{\sigma} + \boldsymbol{\sigma}\cdot\boldsymbol{W} = \begin{bmatrix} \sigma_1\, D_{11} & \frac{\sigma_1^2+\sigma_2^2}{\sigma_1+\sigma_2}\, D_{12} & \frac{\sigma_1^2+\sigma_3^2}{\sigma_1+\sigma_3}\, D_{13} \\ \frac{\sigma_1^2+\sigma_2^2}{\sigma_1+\sigma_2}\, D_{12} & \sigma_2\, D_{22} & \frac{\sigma_2^2+\sigma_3^2}{\sigma_2+\sigma_3}\, D_{23} \\ \frac{\sigma_1^2+\sigma_3^2}{\sigma_1+\sigma_3}\, D_{13} & \frac{\sigma_2^2+\sigma_3^2}{\sigma_2+\sigma_3}\, D_{23} & \sigma_3\, D_{33} \end{bmatrix}. \tag{10.9.27}$$

For a nontrivial solution of the system of six equations for six unknown components of the rate of deformation to exist, the determinant of the system

(10.9.26) must vanish. This provides a relationship between the instantaneous moduli and applied stress, which characterizes the primary eigenstate.

10.9.4. Uniaxial Tension of Elastic-Plastic Material

If the stress state has an axis of equilibrium, say corresponding to $\sigma_2 + \sigma_3 = 0$, there is only one term proportional to W_{23} that remains in (10.9.17), and for $\sigma_2 \neq \sigma_3$ this term can be made arbitrarily large and negative by appropriately adjusting the sign and magnitude of W_{23}. This means that $\underline{\chi}$ can be negative for some velocity gradients, implying that configuration under stress state with an axis of equilibrium could not qualify as stable. However, if $\sigma_2 = \sigma_3 = 0$, and $\sigma_1 > 0$, $\underline{\chi}$ in (10.9.17) does not depend on W_{23}, having a minimum

$$\underline{\chi}^0 = \frac{1}{2}\underline{\boldsymbol{\mathcal{L}}}^{\mathrm{p}}_{(0)} :: (\boldsymbol{D} \otimes \boldsymbol{D}) - \sigma_1 \left(\frac{1}{2}D_{11}^2 + D_{12}^2 + D_{13}^2\right) \tag{10.9.28}$$

in an eigenmode with the spin components

$$W_{12} = D_{12}, \quad W_{31} = -D_{13}. \tag{10.9.29}$$

The configuration under uniaxial tension is thus stable if

$$\begin{aligned}\underline{\chi}^0 &= \frac{1}{2}\lambda\,(D_{11} + D_{22} + D_{33})^2 + \mu(D_{11}^2 + D_{22}^2 + D_{33}^2 + 2D_{12}^2 + 2D_{23}^2 + 2D_{31}^2)\\ &- \frac{2\mu/3}{1 + h^{\mathrm{p}}/\mu}\left(D_{11} - \frac{1}{2}D_{22} - \frac{1}{2}D_{33}\right)^2 - \sigma_1\left(\frac{1}{2}D_{11}^2 + D_{12}^2 + D_{13}^2\right) > 0.\end{aligned} \tag{10.9.30}$$

Note that in uniaxial tension

$$\mathbb{M}_{22} = \mathbb{M}_{33} = -\frac{1}{2}\,\mathbb{M}_{11}, \tag{10.9.31}$$

since deviatoric components of uniaxial stress are so related. Thus, $\mathbb{M}_{11} = \sqrt{2/3}$. The function $\underline{\chi}^0$ can be split into two parts. The first part,

$$(2\mu - \sigma_1)(D_{12}^2 + D_{31}^2) + 2\mu D_{23}^2, \tag{10.9.32}$$

is positive for $\sigma_1 < 2\mu$. The function $\underline{\chi}^0$ will be certainly positive if the remaining term is also positive. We then require

$$\begin{aligned}&\frac{1}{2}\lambda\,(D_{11} + D_{22} + D_{33})^2 + \mu(D_{11}^2 + D_{22}^2 + D_{33}^2)\\ &\quad - \frac{2\mu/3}{1 + h^{\mathrm{p}}/\mu}\left(D_{11} - \frac{1}{2}D_{22} - \frac{1}{2}D_{33}\right)^2 - \frac{1}{2}\sigma_1\, D_{11}^2 > 0.\end{aligned} \tag{10.9.33}$$

This quadratic form in D_{11}, D_{22}, D_{33} is positive definite if the principal minors of associated matrix are positive definite. The first one is

$$\frac{1}{2}\lambda + \mu - \frac{2\mu/3}{1 + h^{\mathrm{p}}/\mu} - \frac{1}{2}\sigma_1 > 0, \tag{10.9.34}$$

which is fulfilled for realistic stress levels. The second one is fulfilled, as well. It remains to examine the determinant

$$\Delta = \begin{vmatrix} \frac{1}{2}\lambda + \mu - \frac{2}{3}\,\alpha\,\mu - \frac{1}{2}\sigma_1 & \frac{1}{2}\lambda + \frac{1}{3}\alpha\,\mu & \frac{1}{2}\lambda + \frac{1}{3}\alpha\,\mu \\ \frac{1}{2}\lambda + \frac{1}{3}\alpha\,\mu & \frac{1}{2}\lambda + \mu - \frac{1}{6}\alpha\,\mu & \frac{1}{2}\lambda - \frac{1}{6}\alpha\,\mu \\ \frac{1}{2}\lambda + \frac{1}{3}\alpha\,\mu & \frac{1}{2}\lambda - \frac{1}{6}\alpha\,\mu & \frac{1}{2}\lambda + \mu - \frac{1}{6}\alpha\,\mu \end{vmatrix}, \tag{10.9.35}$$

where

$$\alpha = \left(1 + \frac{h^{\mathrm{p}}}{\mu}\right)^{-1}. \tag{10.9.36}$$

Upon expansion,

$$\Delta = \frac{1}{2}\,\mu^2\left[(3\lambda + 2\mu)\,(1 - \alpha) - \sigma_1\left(1 + \frac{\lambda}{\mu} - \frac{1}{3}\,\alpha\right)\right], \tag{10.9.37}$$

which is positive when

$$h^{\mathrm{p}} > \frac{\sigma_1/3}{1 - \sigma_1/E}. \tag{10.9.38}$$

Here, E stands for the Young's modulus, related to Lamé constants by

$$E = \frac{3\lambda + 2\mu}{1 + \lambda/\mu}. \tag{10.9.39}$$

Since physically attainable values of stress are much smaller that the elastic modulus, stability and uniqueness are both practically assured for $\sigma_1 < 3h^{\mathrm{p}}$. The results for triaxial tension of compressible elastic-plastic materials were obtained by Miles (1975). In the next subsection we proceed with a less involved analysis for incompressible materials.

10.9.5. Triaxial Tension of Incompressible Material

For incompressible elastic-plastic material $\underline{\chi}^0$ is the sum of two parts,

$$\begin{aligned} &\mu\left(D_{11}^2 + D_{22}^2 + D_{33}^2\right) - \alpha\,\mu\,(\mathbb{M}_{11}\,D_{11} + \mathbb{M}_{22}\,D_{22} + \mathbb{M}_{33}\,D_{33})^2 \\ &\qquad - \frac{1}{2}\left(\sigma_1\,D_{11}^2 + \sigma_2\,D_2^2 + \sigma_3\,D_{33}^2\right), \end{aligned} \tag{10.9.40}$$

where $D_{33} = -(D_{11} + D_{22})$, and

$$\left(2\,\mu - \frac{\sigma_1^2 + \sigma_2^2}{\sigma_1 + \sigma_2}\right) D_{12}^2 + \left(2\,\mu - \frac{\sigma_2^2 + \sigma_3^2}{\sigma_2 + \sigma_3}\right) D_{23}^2 + \left(2\,\mu - \frac{\sigma_3^2 + \sigma_1^2}{\sigma_3 + \sigma_1}\right) D_{31}^2. \tag{10.9.41}$$

The second part is certainly positive for

$$\frac{\sigma_1^2 + \sigma_2^2}{\sigma_1 + \sigma_2} < 2\,\mu, \quad \frac{\sigma_2^2 + \sigma_3^2}{\sigma_2 + \sigma_3} < 2\,\mu, \quad \frac{\sigma_3^2 + \sigma_1^2}{\sigma_3 + \sigma_1} < 2\,\mu, \tag{10.9.42}$$

which is expected to be always the case within attainable range of applied stress. For positive definiteness of $\underset{\sim}{\chi}^0$ it is then sufficient to prove the positive definiteness of (10.9.40) for all volume preserving rate of deformation components. The elements of 2×2 determinant of the corresponding quadratic form are

$$\Delta_{11} = 2\,\mu - \alpha\,\mu\,(\mathbb{M}_{11} - \mathbb{M}_{33})^2 - \frac{1}{2}\,(\sigma_1 + \sigma_3), \tag{10.9.43}$$

$$\Delta_{22} = 2\,\mu - \alpha\,\mu\,(\mathbb{M}_{22} - \mathbb{M}_{33})^2 - \frac{1}{2}\,(\sigma_2 + \sigma_3), \tag{10.9.44}$$

$$\Delta_{12} = \Delta_{21} = \mu - \alpha\,\mu\,(\mathbb{M}_{11} - \mathbb{M}_{33})(\mathbb{M}_{22} - \mathbb{M}_{33}) - \frac{1}{2}\,\sigma_3. \tag{10.9.45}$$

The determinant Δ is accordingly

$$\begin{aligned}\frac{\Delta}{\mu^2} &= 3 - \frac{1}{\mu}\,(\sigma_1 + \sigma_2 + \sigma_3) + \frac{1}{4\,\mu^2}\,(\sigma_1\sigma_2 + \sigma_2\sigma_3 + \sigma_3\sigma_1) \\ &- \alpha\left\{3 - \frac{1}{2\mu}\left[(\mathbb{M}_{22} - \mathbb{M}_{33})^2\,\sigma_1 + (\mathbb{M}_{33} - \mathbb{M}_{11})^2\,\sigma_2 + (\mathbb{M}_{11} - \mathbb{M}_{22})^2\,\sigma_3\right]\right\}.\end{aligned} \tag{10.9.46}$$

This is positive when

$$h^{\mathrm{p}} > \frac{\frac{1}{2}\,(\mathbb{M}_{11}^2\sigma_1 + \mathbb{M}_{22}^2\sigma_2 + \mathbb{M}_{33}^2\sigma_3) - \frac{1}{12\mu}\,(\sigma_1\sigma_2 + \sigma_2\sigma_3 + \sigma_3\sigma_1)}{1 - \frac{1}{3\,\mu}\,(\sigma_1 + \sigma_2 + \sigma_3) + \frac{1}{12\,\mu^2}\,(\sigma_1\sigma_2 + \sigma_2\sigma_3 + \sigma_3\sigma_1)}. \tag{10.9.47}$$

It is recalled that $\mathbb{M}$ is deviatoric and normalized, so that

$$\mathbb{M}_{11} + \mathbb{M}_{22} + \mathbb{M}_{33} = 0, \quad \mathbb{M}_{11}^2 + \mathbb{M}_{22}^2 + \mathbb{M}_{33}^2 = 1. \tag{10.9.48}$$

The critical hardening rate therefore depends on the state of stress, elastic shear modulus μ, and the components of the tensor $\mathbb{M}$ which is normal to the yield surface.

For biaxial tension with $\sigma_3 = 0$, the uniqueness and stability are certain for

$$h^{\mathrm{p}} > \frac{\frac{1}{2}\left(\mathbb{M}_{11}^2\sigma_1 + \mathbb{M}_{22}^2\sigma_2\right) - \frac{1}{12\mu}\sigma_1\sigma_2}{1 - \frac{1}{3\mu}(\sigma_1+\sigma_2) + \frac{1}{12\mu^2}\sigma_1\sigma_2}. \tag{10.9.49}$$

For example, for the von Mises yield criterion,

$$\mathbb{M}_{11} = \frac{2\sigma_1 - \sigma_2}{[6(\sigma_1^2 - \sigma_1\sigma_2 + \sigma_2^2)]^{1/2}}, \quad \mathbb{M}_{22} = \frac{2\sigma_2 - \sigma_1}{[6(\sigma_1^2 - \sigma_1\sigma_2 + \sigma_2^2)]^{1/2}}, \tag{10.9.50}$$

and the condition (10.9.49) becomes

$$h^{\mathrm{p}} = h_{\mathrm{t}}^{\mathrm{p}} > \frac{4\sigma_1^3 - 3\sigma_1^2\sigma_2 - 3\sigma_1\sigma_2^2 + 4\sigma_2^3}{12\,(\sigma_1^2 - \sigma_1\sigma_2 + \sigma_2^2)}, \tag{10.9.51}$$

neglecting terms of the order σ/μ and smaller. For equal biaxial tension $\sigma_1 = \sigma_2 = \sigma$, we have by symmetry

$$\mathbb{M}_{11} = \mathbb{M}_{22} = \frac{1}{\sqrt{6}}, \tag{10.9.52}$$

for any isotropic smooth yield surface, and

$$h^{\mathrm{p}} > \frac{\sigma/6}{1 - \sigma/6\,\mu}. \tag{10.9.53}$$

For uniaxial tension with $\sigma_2 = \sigma_3 = 0$, $\mathbb{M}_{11} = \sqrt{2/3}$ and the condition (10.9.49) reduces to

$$h^{\mathrm{p}} > \frac{\sigma_1/3}{1 - \sigma_1/3\,\mu}. \tag{10.9.54}$$

Since for incompressible elasticity $E = 3\,\mu$, the condition (10.9.54) is in accord with the condition (10.9.38).

10.9.6. Triaxial Tension of Rigid-Plastic Material

For a rigid-plastic material model with isotropic smooth yield surface, the principal directions of the rate of deformation tensor are parallel to those of stress, and eigenmodal spin components are identically equal to zero. The bifurcation and instability are thus both excluded if

$$\underline{\chi} = h\,(\boldsymbol{D} : \boldsymbol{D}) - \frac{1}{2}\boldsymbol{\sigma} : \boldsymbol{D}^2 > 0. \tag{10.9.55}$$

Since constitutively admissible $\boldsymbol{D}$ (and thus any eigenmodal rate of deformation) must be codirectional with the stress, the condition (10.9.55) is met when the modulus h satisfies

$$h > \frac{1}{2}\boldsymbol{\sigma} : \mathbb{M}^2 = \frac{1}{2}\left(\mathbb{M}_{11}^2\sigma_1 + \mathbb{M}_{22}^2\sigma_2 + \mathbb{M}_{33}^2\sigma_3\right). \tag{10.9.56}$$

The tensor $\mathbb{M}$ is normal to the smooth yield surface $f = 0$, having principal directions parallel to those of stress. Equivalently, we can write

$$h > \frac{1}{2}\left(\sigma + \boldsymbol{\sigma}' : \mathbb{M}^2\right), \quad \sigma = \frac{1}{3}\operatorname{tr}\boldsymbol{\sigma}. \tag{10.9.57}$$

Expressed in terms of the principal stress components, and with the von Mises yield condition, this gives for biaxial tension

$$h > \frac{4\sigma_1^3 - 3\sigma_1^2\sigma_2 - 3\sigma_1\sigma_2^2 + 4\sigma_2^3}{12(\sigma_1^2 - \sigma_1\sigma_2 + \sigma_2^2)}, \tag{10.9.58}$$

as originally derived by Swift (1952)[1], and for triaxial tension

$$h > \frac{1}{2}\left[\sigma + \frac{(\sigma_1 - \sigma)^3 + (\sigma_2 - \sigma)^3 + (\sigma_3 - \sigma)^3}{(\sigma_1 - \sigma)^2 + (\sigma_2 - \sigma)^2 + (\sigma_3 - \sigma)^2}\right], \tag{10.9.59}$$

as derived by Hill (1967). For equal biaxial tension $\sigma_1 = \sigma_2 = \sigma$, $h > \sigma/6$, while for uniaxial tension with $\sigma_2 = \sigma_3$, $h > \sigma_1/3$, for any isotropic smooth yield surface, in accord with the results from previous subsections.

10.10. Acceleration Waves in Elastoplastic Solids

During wave propagation in a medium, certain field variables can be discontinuous across the wave front. If displacement discontinuity is precluded by assumption that the failure does not occur, the strongest possible discontinuity is in the velocity of the particle. This is called a shock wave. If the velocity is continuous, but acceleration is discontinuous across the wave front, the wave is called an acceleration wave. Weaker waves are characterized by discontinuities in higher time derivatives of the velocity field (e.g., Janssen, Datta, and Jahsman, 1972; Clifton, 1974; Ting, 1976).

Consider a portion of the deforming body momentarily bounded in part by the surface S, embedded in the material and deforming with it, and in part by the surface Σ which propagates relative to the material. If the enclosed volume at the considered instant is V, then, for any continuous differentiable field $\mathbf{T} = \mathbf{T}(\mathbf{x}, t)$,

$$\frac{\mathrm{d}}{\mathrm{d}t}\int_V \rho\,\mathbf{T}\,\mathrm{d}V = \int_V \frac{\partial}{\partial t}(\rho\,\mathbf{T})\,\mathrm{d}V + \int_S \rho\,\mathbf{T}\,\mathbf{v}\cdot\mathrm{d}\mathbf{S} + \int_\Sigma \rho\,\mathbf{T}\,c\,\mathrm{d}\Sigma. \tag{10.10.1}$$

The particle velocity is $\mathbf{v}$, and c is the propagation speed of the surface Σ in the direction of its outward normal, both relative to a fixed observer.

[1]Published as the first paper in the first volume of the *Journal of the Mechanics and Physics of Solids*.

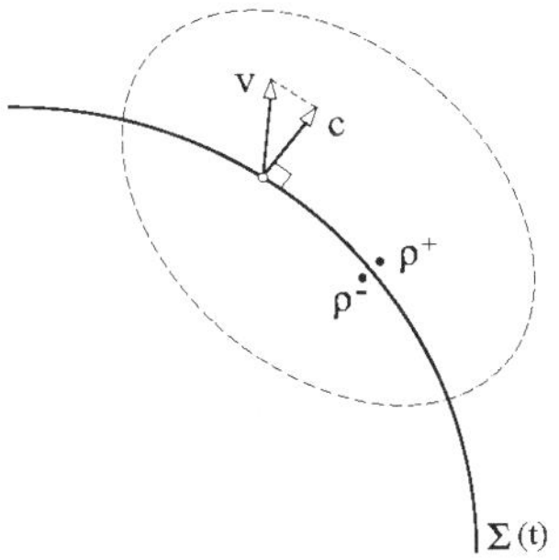

FIGURE 10.2. A surface of discontinuity $\Sigma(t)$ propagates relative to material with the speed c in the direction of its outward normal. The mass densities ahead and behind Σ are ρ^+ and ρ^-.

The above formula, which can be viewed as a modified Reynolds transport theorem of Eq. (3.2.6), will be used to derive the jump conditions across the wave front.

10.10.1. Jump Conditions for Shock Waves

Suppose that a mass density is discontinuous across Σ. Then, take

$$\mathbf{T} = 1, \tag{10.10.2}$$

and apply Eq. (10.10.1) to a thin slice of material immediately ahead and behind Σ. Summing up the resulting expressions, and implementing the conservation of mass condition, gives in the limit

$$c[\![\rho]\!] - [\![\rho\mathbf{v}]\!] \cdot \mathbf{n} = 0, \tag{10.10.3}$$

where $\mathbf{n}$ is the unit normal to Σ in the direction of propagation of Σ (Thomas, 1961). The brackets $[\![\ ]\!]$ designate the jump of the enclosed quantity across the surface Σ, e.g.,

$$[\![\rho]\!] = \rho^+ - \rho^-. \tag{10.10.4}$$

The superposed plus indicates the value at the point just ahead of Σ, and minus just behind the Σ (Fig. 10.2).

By taking

$$\mathbf{T} = \mathbf{v} \tag{10.10.5}$$

in Eq. (10.10.1), and by implementing Eq. (3.3.1), we similarly obtain

$$\mathbf{n}\cdot[\![\boldsymbol{\sigma}]\!]+\rho^{-}\left(c-\mathbf{v}^{-}\cdot\mathbf{n}\right)[\![\mathbf{v}]\!]=\mathbf{0}, \tag{10.10.6}$$

which relates the discontinuities in stress and velocity across the surface Σ. For further analysis of shock waves in elastic-plastic solids, see Wilkins (1964), Germain and Lee (1973), Ting (1976), and Drugan and Shen (1987).

10.10.2. Jump Conditions for Acceleration Waves

In an acceleration wave, the velocity and stress fields are continuous across Σ, but the acceleration $\dot{\mathbf{v}}=\mathrm{d}\mathbf{v}/\mathrm{d}t$ is not. To derive the corresponding jump condition across Σ, substitute

$$\mathbf{T}=\dot{\mathbf{v}} \tag{10.10.7}$$

in Eq. (10.10.1). In view of equations of motion and the relationship between the true and nominal tractions, we first have

$$\begin{aligned}\frac{\mathrm{d}}{\mathrm{d}t}\int_{V}\rho\,\dot{\mathbf{v}}\,\mathrm{d}V&=\frac{\mathrm{d}}{\mathrm{d}t}\int_{S}\mathbf{t}_n\,\mathrm{d}S+\int_{V}\rho\,\dot{\mathbf{b}}\,\mathrm{d}V\\&=\frac{\mathrm{d}}{\mathrm{d}t}\int_{S^0}\mathbf{p}_n\,\mathrm{d}S^0+\int_{V}\rho\,\dot{\mathbf{b}}\,\mathrm{d}V=\int_{S^0}\dot{\mathbf{P}}^T\cdot\mathbf{n}^0\,\mathrm{d}S^0+\int_{V}\rho\,\dot{\mathbf{b}}\,\mathrm{d}V.\end{aligned} \tag{10.10.8}$$

Further, the Nanson's relation (2.2.17) and Eq. (3.9.17) give

$$\int_{S^0}\dot{\mathbf{P}}^T\cdot\mathbf{n}^0\,\mathrm{d}S^0=\int_{S}\dot{\underline{\mathbf{P}}}^T\cdot\mathbf{n}\,\mathrm{d}S, \tag{10.10.9}$$

so that Eq. (10.10.1) becomes

$$\begin{aligned}&\int_{S}\mathbf{n}\cdot\dot{\underline{\mathbf{P}}}\,\mathrm{d}S+\int_{V}\rho\,\dot{\mathbf{b}}\,\mathrm{d}V\\&\quad=\int_{V}\frac{\partial}{\partial t}(\rho\,\dot{\mathbf{v}})\,\mathrm{d}V+\int_{S}\rho\,\dot{\mathbf{v}}\,\mathbf{v}\cdot\mathrm{d}\mathbf{S}+\int_{\Sigma}\rho\,\dot{\mathbf{v}}\,c\,\mathrm{d}\Sigma.\end{aligned} \tag{10.10.10}$$

Applying this to a thin slice of material just ahead and behind of Σ, the volume integrals vanish in the limit, and the summation yields

$$\mathbf{n}\cdot[\![\dot{\underline{\mathbf{P}}}]\!]+\rho\,c_r\,[\![\dot{\mathbf{v}}]\!]=\mathbf{0}. \tag{10.10.11}$$

Here,

$$c_r=c-\mathbf{v}\cdot\mathbf{n} \tag{10.10.12}$$

is the speed of Σ relative to the material. Equation (10.10.11) relates the jumps in the acceleration and stress rate across the surface Σ.

A characteristic segment of the wave is defined as the discontinuity in the gradient of the particle velocity across the wave front,

$$\boldsymbol{\eta} = \left[\!\left[\frac{\partial \mathbf{v}}{\partial n} \right]\!\right]. \tag{10.10.13}$$

The geometric and kinematic conditions of compatibility for the velocity field (Thomas, 1961; Hill, 1961b) give

$$[\![\mathbf{L}]\!] = \left[\!\left[\frac{\partial \mathbf{v}}{\partial \mathbf{x}} \right]\!\right] = \boldsymbol{\eta} \otimes \mathbf{n}, \quad \left[\!\left[\frac{\partial \mathbf{v}}{\partial t} \right]\!\right] = -c\,\boldsymbol{\eta}, \tag{10.10.14}$$

provided that $\mathbf{v}$ is continuous across Σ. Since

$$\dot{\mathbf{v}} = \frac{\partial \mathbf{v}}{\partial t} + \mathbf{L} \cdot \mathbf{v}, \tag{10.10.15}$$

a discontinuity in the acceleration is related to discontinuity in the velocity gradient by

$$[\![\dot{\mathbf{v}}]\!] = -c_r\,\boldsymbol{\eta}. \tag{10.10.16}$$

10.10.3. Propagation Condition

Substitution of Eq. (10.10.16) into Eq. (10.10.11) gives

$$\mathbf{n} \cdot [\![\underline{\dot{\mathbf{P}}}]\!] = \rho\, c_r^2\, \boldsymbol{\eta}. \tag{10.10.17}$$

Suppose that on both sides of Σ the plastic loading takes place. Since the stress and pseudomoduli are continuous across Σ in an acceleration wave, we have

$$[\![\underline{\dot{\mathbf{P}}}]\!] = [\![\underline{\mathbf{\Lambda}}^{\mathrm{p}} \cdot\cdot\, \mathbf{L}]\!] = \underline{\mathbf{\Lambda}}^{\mathrm{p}} \cdot\cdot\, [\![\mathbf{L}]\!] = \underline{\mathbf{\Lambda}}^{\mathrm{p}} \cdot\cdot\, (\boldsymbol{\eta} \otimes \mathbf{n}). \tag{10.10.18}$$

Combining Eqs. (10.10.17) and (10.10.18), therefore,

$$\mathbf{n} \cdot \underline{\mathbf{\Lambda}}^{\mathrm{p}} : (\mathbf{n} \otimes \boldsymbol{\eta}) = \rho\, c_r^2\, \boldsymbol{\eta}, \tag{10.10.19}$$

i.e.,

$$\underline{\mathbf{A}}^{\mathrm{p}} \cdot \boldsymbol{\eta} = \rho\, c_r^2\, \boldsymbol{\eta}. \tag{10.10.20}$$

The rectangular components of the real matrix $\underline{\mathbf{A}}^{\mathrm{p}}$ are

$$\underline{A}^{\mathrm{p}}_{ij} = \underline{\Lambda}^{\mathrm{p}}_{kilj} n_k n_l. \tag{10.10.21}$$

They depend on the current state of stress and material properties (embedded in $\underline{\mathbf{\Lambda}}^{\mathrm{p}}$), and the direction of propagation $\mathbf{n}$. In view of reciprocal symmetry ($\underline{\Lambda}^{\mathrm{p}}_{kilj} = \underline{\Lambda}^{\mathrm{p}}_{ljki}$), it follows that, in addition to be real, $\underline{\mathbf{A}}^{\mathrm{p}}$ is also symmetric ($\underline{A}^{\mathrm{p}}_{ij} = \underline{A}^{\mathrm{p}}_{ji}$). Thus, the eigenvalues $\rho\, c_r^2$ in Eq. (10.10.20) are all real. There is a wave propagating in the direction $\mathbf{n}$, carrying a discontinuity

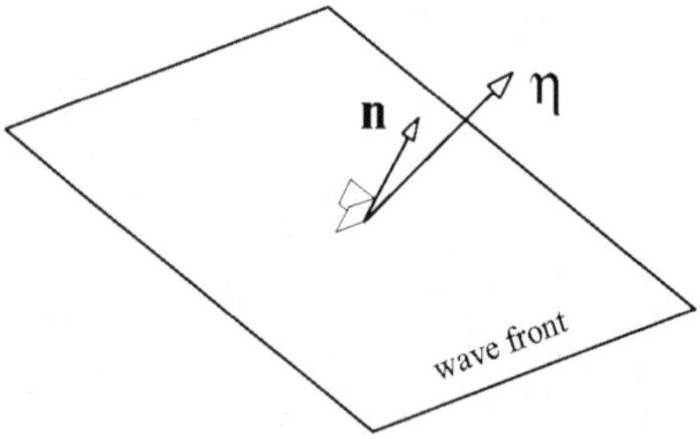

FIGURE 10.3. Plane wave propagating in the direction $\mathbf{n}$. The vector $\boldsymbol{\eta}$ is the polarization of the wave, which defines direction of the particle velocity.

$\boldsymbol{\eta}$, if the corresponding c_r^2 is positive. This is assured in the states where $\underline{\mathbf{A}}^{\mathrm{p}}$ is positive definite, since

$$\boldsymbol{\eta} \cdot \underline{\mathbf{A}}^{\mathrm{p}} \cdot \boldsymbol{\eta} = \rho\, c_r^2\, (\boldsymbol{\eta} \cdot \boldsymbol{\eta}). \tag{10.10.22}$$

The condition for nontrivial $\boldsymbol{\eta}$ to exist in the eigenvalue problem (10.10.20) is

$$\det(\underline{\mathbf{A}}^{\mathrm{p}} - \rho\, c_r^2\, \mathbf{I}) = 0. \tag{10.10.23}$$

A wave that carries a discontinuity in the velocity gradient also carries a discontinuity in the stress gradient. This is (Hill, 1961b)

$$\left[\!\left[\frac{\partial \sigma_{ij}}{\partial x_m} \right]\!\right] = \frac{1}{c_r} \left(\sigma_{ij}\delta_{kl} - \sigma_{jk}\delta_{il} - \underline{\Lambda}^{\mathrm{p}}_{ijkl} \right) n_k n_m\, \eta_l. \tag{10.10.24}$$

In view of the relationship between the moduli in Eq. (10.1.15), we also have

$$\left(\underline{\mathcal{L}}^{\mathrm{p}\,(1)}_{kilj} n_k n_l \right) \eta_j = (\rho\, c_r^2 - \sigma_n)\eta_i, \tag{10.10.25}$$

where $\sigma_n = \sigma_{ij} n_i n_j$ is the normal stress over Σ.

Propagation of Plane Waves

There is an analogy between governing equations for acceleration waves and plane waves. Indeed, consider the rate equations of motion,

$$\boldsymbol{\nabla} \cdot \underline{\dot{\mathbf{P}}} = \rho\, \frac{\mathrm{d}^2\mathbf{v}}{\mathrm{d}t^2}, \quad \underline{\dot{\mathbf{P}}} = \underline{\boldsymbol{\Lambda}} \cdot\cdot\, \mathbf{L}, \tag{10.10.26}$$

whose solutions are sought in the form of a plane wave propagating with a speed c in the direction $\mathbf{n}$,

$$\mathbf{v} = \boldsymbol{\eta} f(\mathbf{n} \cdot \mathbf{x} - ct). \tag{10.10.27}$$

The vector η is the polarization of the wave (Fig. 10.3). On substituting (10.10.27) into (10.10.26), we obtain the propagation condition

$$\underline{\mathbf{A}}^{\mathrm{p}} \cdot \eta = \rho c^2 \eta, \quad \underline{A}^{\mathrm{p}}_{ij} = \underline{\Lambda}^{\mathrm{p}}_{kilj} n_k n_l. \tag{10.10.28}$$

The second-order tensor $\underline{\mathbf{A}}^{\mathrm{p}}$ is referred to as the acoustic tensor. Thus, ρc^2 is an eigenvalue and η is an eigenvector of $\underline{\mathbf{A}}^{\mathrm{p}}$. Since $\underline{\mathbf{A}}^{\mathrm{p}}$ is real and symmetric, c^2 must be real. If $c^2 > 0$ (assured by positive definiteness of $\underline{\mathbf{A}}^{\mathrm{p}}$), there is a stability with respect to propagation of small disturbances, superposed to finitely deformed current state. Equation (10.10.28) then admits three linearly independent plane progressive waves for each direction of propagation $\mathbf{n}$. Small amplitude plane waves can propagate along a given direction in three distinct, mutually orthogonal modes. These modes are generally neither longitudinal nor transverse (i.e., η is neither parallel nor normal to $\mathbf{n}$). For $c^2 = 0$, there is a transition from stability to instability. The latter is associated with $c^2 < 0$, and a divergent growth of initial disturbance. These fundamental results were established by Hadamar (1903) in the context of elastic stability, and for inelasticity by Thomas (1961), Hill (1962), and Mandel (1966).

If $\underline{\boldsymbol{\Lambda}}^{\mathrm{p}}$ does not possess reciprocal symmetry (nonassociative plasticity), $\underline{\mathbf{A}}^{\mathrm{p}}$ is not symmetric, and it may happen that at some states of deformation and material parameters two eigenvalues in Eq. (10.10.28) are complex conjugates (one is always real), which means that a flutter type instability may occur (Rice, 1977; Bigoni, 1995). Uniqueness and stability criteria for elastoplastic materials with nonassociative flow rules were studied by Maier (1970), Raniecki (1979), Needleman (1979), Raniecki and Bruhns (1981), Bruhns (1984), Bigoni and Hueckel (1991), Ottosen and Runesson (1991), Bigoni and Zaccaria (1992,1993), Neilsen and Schreyer (1993), and others.

10.10.4. Stationary Discontinuity

When the matrix $\underline{\mathbf{A}}^{\mathrm{p}}$ has a zero eigenvalue ($c_r = 0$), there is a discontinuity surface that does not travel relative to the material (stationary discontinuity, in Hadamar's terminology). This happens if and only if a discontinuity surface normal $\mathbf{n}$ satisfies

$$\det\left(\underline{\mathbf{A}}^{\mathrm{p}}\right) = \det\left(\underline{\Lambda}^{\mathrm{p}}_{ijkl} n_i n_k\right) = 0. \tag{10.10.29}$$

The corresponding eigenvector $\boldsymbol{\eta}$ is a nontrivial solution of the homogeneous system of equations

$$\underline{\mathbf{A}}^{\mathrm{p}} \cdot \boldsymbol{\eta} = \mathbf{0}. \tag{10.10.30}$$

Equation (10.10.24) does not determine discontinuity in the stress gradient across stationary discontinuity (since $c_r = 0$), but it does impose a condition on the current moduli and stress components there. This is

$$(\boldsymbol{\eta} \cdot \mathbf{n}\, \delta_{ik} - \eta_i n_k)\, \sigma_{kj} = \underline{\Lambda}^{\mathrm{p}}_{ijkl}\, n_k \eta_l, \tag{10.10.31}$$

if discontinuity in the velocity gradient actually occurs. Since material particles remain on the surface of stationary discontinuity, there is no jump in acceleration or nominal traction rate across Σ, so that

$$[\![\, \dot{\mathbf{v}} \,]\!] = \mathbf{0}, \quad \mathbf{n} \cdot [\![\, \dot{\underline{\mathbf{P}}} \,]\!] = \mathbf{0}. \tag{10.10.32}$$

Note that

$$\mathbf{n} \cdot [\![\, \dot{\underline{\mathbf{P}}} \,]\!] = \underline{\mathbf{A}}^{\mathrm{p}} \cdot \boldsymbol{\eta}. \tag{10.10.33}$$

Furthermore, since

$$\dot{\underline{\mathbf{P}}} = \dot{\boldsymbol{\sigma}} + \boldsymbol{\sigma}\, \mathrm{tr}\, \mathbf{D} - \mathbf{L} \cdot \boldsymbol{\sigma}, \tag{10.10.34}$$

it follows that

$$\mathbf{n} \cdot [\![\, \dot{\underline{\mathbf{P}}} \,]\!] = \mathbf{n} \cdot [\![\, \dot{\boldsymbol{\sigma}} \,]\!]. \tag{10.10.35}$$

In proof, let

$$\boldsymbol{\eta} = \mathbf{b} + g\mathbf{n}, \tag{10.10.36}$$

where $\mathbf{b} \cdot \mathbf{n} = 0$ and g is a scalar function. Then, since $[\![\, \mathbf{L} \,]\!] = \boldsymbol{\eta} \otimes \mathbf{n}$, we have

$$\mathrm{tr}\, [\![\, \mathbf{D} \,]\!] = g, \quad \mathbf{n} \cdot [\![\, \mathbf{L} \,]\!] = g\mathbf{n}, \tag{10.10.37}$$

thus the result.

10.11. Analysis of Plastic Flow Localization

Consider an equilibrium configuration of uniformly strained homogeneous body. Suppose that increments of deformation (velocity) are prescribed on the boundary of the body, giving rise to uniform velocity gradient $\mathbf{L}^0$ throughout the body. The question is if there could be another statically and constitutively admissible velocity gradient field, associated with the same velocity boundary conditions. All-around displacement conditions are imposed to rule out geometric instabilities, such as buckling or necking, which could precede localization. We wish to examine if the bifurcation field can be characterized by localization of deformation within a planar band with normal $\mathbf{n}$, such that

$$\mathbf{L} = \mathbf{L}^0 + \boldsymbol{\eta} \otimes \mathbf{n}, \quad \text{i.e.,} \quad [\![\mathbf{L}]\!] = \boldsymbol{\eta} \otimes \mathbf{n}, \tag{10.11.1}$$

across the band. As discussed in the previous subsection, this could happen in the band whose normal $\mathbf{n}$ satisfies the condition (10.10.29), assuring that there is a nontrivial solution for $\boldsymbol{\eta}$ in equations

$$\underline{\Lambda}^{\mathrm{p}}_{ijkl} n_i n_k \eta_l = 0. \tag{10.11.2}$$

Here,

$$\underline{\Lambda}^{\mathrm{p}}_{ijkl} = \underline{\mathcal{L}}^{\mathrm{p}\,(1)}_{ijkl} + \sigma_{ik}\delta_{jl} = \underline{\mathcal{L}}^{\mathrm{p}\,(0)}_{ijkl} + \mathcal{R}_{ijkl}, \tag{10.11.3}$$

and

$$\mathcal{R}_{ijkl} = \frac{1}{2}\left(\sigma_{ik}\delta_{jl} - \sigma_{jk}\delta_{il} - \sigma_{il}\delta_{jk} - \sigma_{jl}\delta_{ik}\right). \tag{10.11.4}$$

It is noted that Eq. (10.11.2) can also be deduced through an eigenmodal analysis of the type used in Section 7.9.

10.11.1. Elastic-Plastic Materials

Following Rice (1977), suppose that elastoplastic response is described by a nonassociative flow rule, with the instantaneous elastoplastic stiffness

$$\underline{\mathcal{L}}^{\mathrm{p}}_{(0)} = \underline{\mathcal{L}}_{(0)} - \frac{1}{\hat{\mathbb{Q}} : \mathbb{P} + H}\, \hat{\mathbb{P}} \otimes \hat{\mathbb{Q}}, \tag{10.11.5}$$

where

$$\mathbb{P} = \frac{\partial \pi}{\partial \boldsymbol{\sigma}}, \quad \mathbb{Q} = \frac{\partial f}{\partial \boldsymbol{\sigma}}. \tag{10.11.6}$$

The potential function and the yield function are denoted by π and f, and

$$\hat{\mathbb{Q}} = \underline{\mathcal{L}}_{(0)} : \mathbb{Q}, \quad \hat{\mathbb{P}} = \underline{\mathcal{L}}_{(0)} : \mathbb{P}, \quad \hat{\mathbb{Q}} : \mathbb{P} = \mathbb{Q} : \underline{\mathcal{L}}_{(0)} : \mathbb{P}. \tag{10.11.7}$$

Equation (10.11.5) can be derived from the general expression (9.8.9), with the current state used as the reference, and with elastic and plastic parts of the rate of deformation defined with respect to stress rate $\overset{\circ}{\underline{\boldsymbol{\tau}}}$. Note that $\mathbb{P}$ and $\mathbb{Q}$ are not normalized. In particular, with isotropic elastic stiffness,

$$\underline{\mathcal{L}}_{(0)} = \lambda\, \mathbf{I} \otimes \mathbf{I} + 2\mu\, \boldsymbol{I}, \tag{10.11.8}$$

we have

$$\hat{\mathbb{Q}} \otimes \hat{\mathbb{P}} = (\lambda \operatorname{tr} \mathbb{Q}\, \mathbf{I} + 2\mu\, \mathbb{Q}) \otimes (\lambda \operatorname{tr} \mathbb{P}\, \mathbf{I} + 2\mu\, \mathbb{P}), \tag{10.11.9}$$

and

$$\hat{\mathbb{Q}} : \mathbb{P} = \lambda \, (\operatorname{tr} \mathbb{Q})(\operatorname{tr} \mathbb{P}) + 2\mu \, \mathbb{Q} : \mathbb{P}. \tag{10.11.10}$$

A nontrivial solution for $\boldsymbol{\eta}$ is sought in equations

$$\left(\underline{\mathcal{L}}^{(0)}_{ijkl} - \frac{1}{\hat{\mathbb{Q}} : \mathbb{P} + H} \, \hat{\mathbb{P}}_{ij} \hat{\mathbb{Q}}_{kl} + \mathcal{R}_{ijkl} \right) n_i n_k \eta_l = 0. \tag{10.11.11}$$

They can be rewritten in direct notation as

$$\mathbf{C} \cdot \boldsymbol{\eta} - \frac{1}{\hat{\mathbb{Q}} : \mathbb{P} + H} \, \hat{\mathbb{P}} \cdot \mathbf{n} \, (\mathbf{n} \cdot \hat{\mathbb{Q}} \cdot \boldsymbol{\eta}) + \mathbf{R} \cdot \boldsymbol{\eta} = \mathbf{0}. \tag{10.11.12}$$

The second-order tensors $\mathbf{C}$ and $\mathbf{R}$ are introduced by

$$C_{jl} = \underline{\mathcal{L}}^{(0)}_{ijkl} n_i n_k, \quad R_{jl} = \mathcal{R}_{ijkl} n_i n_k. \tag{10.11.13}$$

In view of the representation for $\underline{\mathcal{L}}_{(0)}$, the tensor $\mathbf{C}$ and its inverse are explicitly given by

$$\mathbf{C} = \mu \left(\mathbf{I} + \frac{1}{1 - 2\nu} \, \mathbf{n} \otimes \mathbf{n} \right), \quad \mathbf{C}^{-1} = \frac{1}{\mu} \left[\mathbf{I} - \frac{1}{2(1 - \nu)} \, \mathbf{n} \otimes \mathbf{n} \right], \tag{10.11.14}$$

where ν is the Poisson ratio. Multiplying (10.11.12) by $\mathbf{C}^{-1}$ gives

$$(\mathbf{I} + \mathbf{B}) \cdot \boldsymbol{\eta} = \frac{1}{\hat{\mathbb{Q}} : \mathbb{P} + H} \, \mathbf{C}^{-1} \cdot \hat{\mathbb{P}} \cdot \mathbf{n} \, (\mathbf{n} \cdot \hat{\mathbb{Q}} \cdot \boldsymbol{\eta}), \tag{10.11.15}$$

i.e.,

$$\boldsymbol{\eta} = \frac{1}{\hat{\mathbb{Q}} : \mathbb{P} + H} \, (\mathbf{I} + \mathbf{B})^{-1} \cdot \mathbf{C}^{-1} \cdot \hat{\mathbb{P}} \cdot \mathbf{n} \, (\mathbf{n} \cdot \hat{\mathbb{Q}} \cdot \boldsymbol{\eta}), \tag{10.11.16}$$

where

$$\mathbf{B} = \mathbf{C}^{-1} \cdot \mathbf{R}. \tag{10.11.17}$$

Since the components of the matrix $\mathbf{R}$ are of the order of stress, which is ordinarily much smaller than the elastic modulus, the components of matrix $\mathbf{B}$ are small comparing to one. Thus the inverse matrix $(\mathbf{I} + \mathbf{B})^{-1}$ can be determined accurately by retaining few leading terms in the expansion

$$(\mathbf{I} + \mathbf{B})^{-1} = \mathbf{I} - \mathbf{B} + \mathbf{B} \cdot \mathbf{B} - \cdots . \tag{10.11.18}$$

Equation (10.11.16) enables an easy identification of the critical hardening rate for the localization. Upon multiplication by $\mathbf{n} \cdot \hat{\mathbb{Q}}$ and the cancellation of $\mathbf{n} \cdot \hat{\mathbb{Q}} \cdot \boldsymbol{\eta}$, there follows

$$H = \mathbf{n} \cdot \hat{\mathbb{Q}} \cdot (\mathbf{I} + \mathbf{B})^{-1} \cdot \mathbf{C}^{-1} \cdot \hat{\mathbb{P}} \cdot \mathbf{n} - \hat{\mathbb{Q}} : \mathbb{P}. \tag{10.11.19}$$

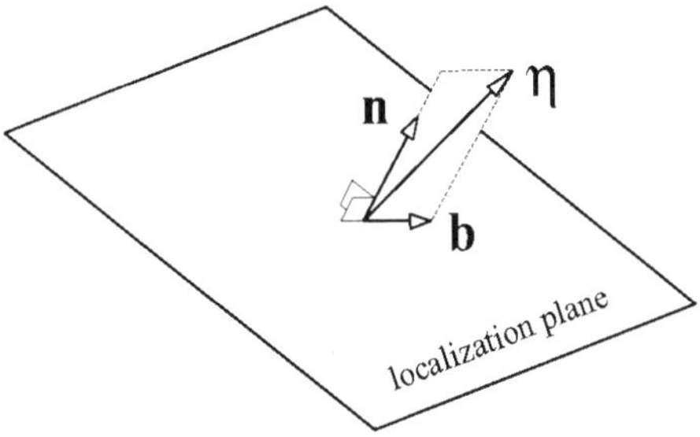

FIGURE 10.4. Localization plane (stationary discontinuity) with normal $\mathbf{n}$. The localization vector $\boldsymbol{\eta}$ defines velocity discontinuity across the plane. The component $\mathbf{b}$ in the plane of localization corresponds to shear band localization.

Furthermore, Eq. (10.11.16) by inspection gives the characteristic segment (localization vector)

$$\boldsymbol{\eta} \propto (\mathbf{I}+\mathbf{B})^{-1} \cdot \mathbf{C}^{-1} \cdot \hat{\mathbb{P}} \cdot \mathbf{n}, \qquad (10.11.20)$$

to within a scalar multiple.

If the $\mathbf{B}$ components are neglected (which is equivalent to approximating $\overset{\circ}{\underline{\boldsymbol{\tau}}}$ with $\dot{\boldsymbol{\sigma}}$ in the elastoplastic constitutive structure), Eq. (10.11.19) becomes

$$\begin{aligned}\frac{H}{\mu} &= 4\,\mathbf{n}\cdot\mathbb{Q}\cdot\mathbb{P}\cdot\mathbf{n} - \frac{2}{1-\nu}\,\mathbb{Q}_n\,\mathbb{P}_n - 2\,\mathbb{Q}:\mathbb{P} \\ &+ \frac{2\nu}{1-\nu}\left[(\operatorname{tr}\mathbb{Q})\mathbb{P}_n + (\operatorname{tr}\mathbb{P})\mathbb{Q}_n - (\operatorname{tr}\mathbb{Q})(\operatorname{tr}\mathbb{P})\right],\end{aligned} \qquad (10.11.21)$$

where

$$\mathbb{Q}_n = \mathbf{n}\cdot\mathbb{Q}\cdot\mathbf{n}, \quad \mathbb{P}_n = \mathbf{n}\cdot\mathbb{P}\cdot\mathbf{n}. \qquad (10.11.22)$$

The localization vector is

$$\boldsymbol{\eta} \propto \mathbb{P}\cdot\mathbf{n} - \frac{1}{2(1-\nu)}\left(\mathbb{P}_n - 2\nu\operatorname{tr}\mathbb{P}\right)\mathbf{n}. \qquad (10.11.23)$$

Observe that

$$\mathbf{n}\cdot\boldsymbol{\eta} \propto \frac{1}{2(1-\nu)}\left[(1-2\nu)\mathbb{P}_n + 2\nu\operatorname{tr}\mathbb{P}\right], \qquad (10.11.24)$$

so that the component of $\boldsymbol{\eta}$ in the plane of localization (Fig. 10.4) is

$$\mathbf{b} = \boldsymbol{\eta} - (\mathbf{n}\cdot\boldsymbol{\eta})\mathbf{n} \propto \mathbb{P}\cdot\mathbf{n} - \mathbb{P}_n\,\mathbf{n}. \qquad (10.11.25)$$

If

$$\mu\,\mathbb{P}_n + \lambda\operatorname{tr}\mathbb{P} = 0, \qquad (10.11.26)$$

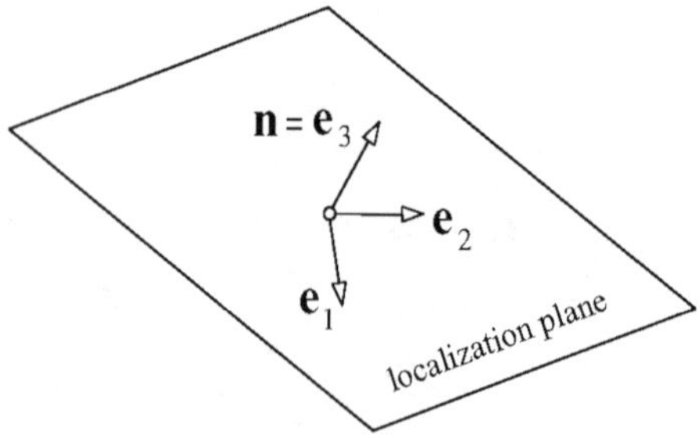

FIGURE 10.5. Localization plane with normal $\mathbf{n}$ in the coordinate direction $\mathbf{e}_3$. The other two coordinate directions $\mathbf{e}_1$ and $\mathbf{e}_2$ are in the plane of localization.

the shear band localization occurs ($\boldsymbol{\eta} = \mathbf{b}$).

Particularly simple representation of the expression for the critical hardening rate is obtained when the coordinate system is used with one axis in the direction $\mathbf{n}$ ($n_i = \delta_{i3}$). This is (Rice, 1977)

$$\frac{H}{\mu} = -2\,\mathbb{Q}_{\alpha\beta}\,\mathbb{P}_{\alpha\beta} - \frac{2\nu}{1-\nu}\,\mathbb{Q}_{\alpha\alpha}\,\mathbb{P}_{\beta\beta}, \tag{10.11.27}$$

where $\alpha, \beta = 1, 2$ denote the components on orthogonal axes in the plane of localization (Fig. 10.5). In the case of associative plasticity ($\mathbb{Q} = \mathbb{P}$), Eq. (10.11.27) shows that H at localization cannot be positive (i.e., softening is required for localization), at least when $\mathbf{B}$ terms are neglected, as assumed in (10.11.27).

A study of bifurcation in the form of shear bands from the nonhomogeneous stress state in the necked region of a tensile specimen is given by Iwakuma and Nemat-Nasser (1982). See also Ortiz, Leroy, and Needleman (1987), and Ramakrishnan and Atluri (1994). For the effects of elastic anisotropy on strain localization, the paper by Rizzi and Loret (1997) can be consulted.

10.11.2. Localization in Pressure-Sensitive Materials

For pressure-sensitive dilatant materials considered in Subsection 9.8.1, the yield and potential functions are such that

$$\mathbb{Q} = \frac{\boldsymbol{\sigma}'}{2J_2^{1/2}} + \frac{1}{3}\,\mu^*\,\mathbf{I}, \quad \mathbb{P} = \frac{\boldsymbol{\sigma}'}{2J_2^{1/2}} + \frac{1}{3}\,\beta\,\mathbf{I}, \tag{10.11.28}$$

where μ^* is the frictional parameter, and β the dilatancy factor. Thus,

$$\mathbb{Q}_n = \frac{\sigma'_n}{2J_2^{1/2}} + \frac{1}{3}\,\mu^*, \quad \mathbb{P}_n = \frac{\sigma'_n}{2J_2^{1/2}} + \frac{1}{3}\,\beta, \tag{10.11.29}$$

$$\mathbf{n}\cdot\mathbb{Q}\cdot\mathbb{P}\cdot\mathbf{n} = \frac{\mathbf{n}\cdot\boldsymbol{\sigma}'\cdot\boldsymbol{\sigma}'\cdot\mathbf{n}}{4J_2} + \frac{\sigma'_n}{6J_2^{1/2}}\,(\beta+\mu^*) + \frac{1}{9}\,\beta\mu^*, \tag{10.11.30}$$

$$\mathbb{Q}:\mathbb{P}_n = \frac{1}{2} + \frac{1}{3}\,\beta\mu^*, \quad \operatorname{tr}\mathbb{Q} = \mu^*, \quad \operatorname{tr}\mathbb{P} = \beta. \tag{10.11.31}$$

The deviatoric normal stress in the localization plane is $\sigma'_n = \mathbf{n}\cdot\boldsymbol{\sigma}'\cdot\mathbf{n}$. Substitution into Eq. (10.11.21), therefore, gives

$$\begin{aligned}\frac{H}{\mu} = \frac{\mathbf{n}\cdot\boldsymbol{\sigma}'\cdot\boldsymbol{\sigma}'\cdot\mathbf{n}}{J_2} &- \frac{1}{2(1-\nu)}\,\frac{\sigma_n'^{\,2}}{J_2} + \frac{1+\nu}{3(1-\nu)}\,\frac{\sigma'_n}{J_2^{1/2}}\,(\beta+\mu^*)\\ &- \frac{4(1+\nu)}{9(1-\nu)}\,\beta\mu^* - 1.\end{aligned} \tag{10.11.32}$$

If localization occurs, it will take place in the plane whose normal $\mathbf{n}$ maximizes the hardening rate H in Eq. (10.11.32) (H being a nonincreasing function of the amount of deformation imposed on the material). The problem was originally formulated and solved by Rudnicki and Rice (1975). To find the localization plane and the corresponding critical hardening rate, it is convenient to choose the coordinate axes parallel to principal stress axes. With respect to these axes,

$$\sigma'_n = (2\sigma'_2+\sigma'_3)n_2^2 + (2\sigma'_3+\sigma'_2)n_3^2 - (\sigma'_2+\sigma'_3), \tag{10.11.33}$$

and

$$\mathbf{n}\cdot\boldsymbol{\sigma}'\cdot\boldsymbol{\sigma}'\cdot\mathbf{n} = (\sigma'_2+\sigma'_3)^2 - \sigma'_3(2\sigma'_2+\sigma'_3)n_2^2 - \sigma'_2(2\sigma'_3+\sigma'_2)n_3^2, \tag{10.11.34}$$

since

$$n_1^2+n_2^2+n_3^2 = 1, \quad \sigma'_1+\sigma'_2+\sigma'_3 = 0. \tag{10.11.35}$$

Consequently, Eq. (10.11.32) becomes

$$\begin{aligned}\frac{H}{\mu} = \frac{1}{J_2}\Big\{&\sigma'_2\sigma'_3 - \sigma'_3(2\sigma'_2+\sigma'_3)n_2^2 - \sigma'_2(2\sigma'_3+\sigma'_2)n_3^2\\ &- \frac{1}{2(1-\nu)}\left[(2\sigma'_2+\sigma'_3)n_2^2 + (2\sigma'_3+\sigma'_2)n_3^2 - (\sigma'_2+\sigma'_3)\right]^2\\ &+ \frac{1+\nu}{3(1-\nu)}\,J_2^{1/2}\,(\beta+\mu^*)\left[(2\sigma'_2+\sigma'_3)n_2^2 + (2\sigma'_3+\sigma'_2)n_3^2 - (\sigma'_2+\sigma'_3)\right]\\ &- \frac{4(1+\nu)}{9(1-\nu)}\,J_2\beta\mu^*\Big\}.\end{aligned} \tag{10.11.36}$$

The stationary conditions

$$\frac{\partial H}{\partial n_2} = 0, \quad \frac{\partial H}{\partial n_3} = 0 \tag{10.11.37}$$

then yield

$$\begin{aligned}(2\sigma_2' + \sigma_3')n_2\,[\sigma_2' + \nu\sigma_3' + \frac{1+\nu}{3}\,J_2^{1/2}\,(\beta + \mu^*) \\ - (2\sigma_2' + \sigma_3')n_2^2 - (2\sigma_3' + \sigma_2')n_3^2] = 0,\end{aligned} \tag{10.11.38}$$

$$\begin{aligned}(2\sigma_3' + \sigma_2')n_3\,[\sigma_3' + \nu\sigma_2' + \frac{1+\nu}{3}\,J_2^{1/2}\,(\beta + \mu^*) \\ - (2\sigma_2' + \sigma_3')n_2^2 - (2\sigma_3' + \sigma_2')n_3^2] = 0.\end{aligned} \tag{10.11.39}$$

Note that

$$2\sigma_2' + \sigma_3' = \sigma_2 - \sigma_1 \le 0, \quad 2\sigma_3' + \sigma_2' = \sigma_3 - \sigma_1 \le 0. \tag{10.11.40}$$

If all principal stresses are distinct, there are three possibilities to satisfy Eqs. (10.11.38) and (10.11.39). These are

$$\begin{gathered} n_2 = 0, \quad n_3 \neq 0, \\ n_2 \neq 0, \quad n_3 = 0, \\ n_2 = n_3 = 0. \end{gathered} \tag{10.11.41}$$

If $n_2 = 0$, Eq. (10.11.39) gives

$$(2\sigma_3' + \sigma_2')n_3^2 - (\sigma_3' + \nu\sigma_2') = \frac{1+\nu}{3}\,J_2^{1/2}(\beta + \mu^*), \tag{10.11.42}$$

i.e.,

$$n_3^2 = \frac{\sigma_2 - \sigma_3}{\sigma_1 - \sigma_3} - (1+\nu)\,\frac{J_2^{1/2}}{\sigma_1 - \sigma_3}\left(\frac{\sigma_2'}{J_2^{1/2}} + \frac{\beta + \mu^*}{3}\right). \tag{10.11.43}$$

The value of n_3^2 must be between zero and one, $0 \le n_3^2 \le 1$. For positive β and μ^*, this is assured if

$$\beta + \mu^* \le \sqrt{3}. \tag{10.11.44}$$

In proof, one can use the connections

$$\frac{\sigma_1'}{J_2^{1/2}} = \left(1 - \frac{3}{4}\,\frac{\sigma_2'^{\,2}}{J_2}\right)^{1/2} - \frac{1}{2}\,\frac{\sigma_2'}{J_2^{1/2}}, \quad \frac{\sigma_3'}{J_2^{1/2}} = -\left(1 - \frac{3}{4}\,\frac{\sigma_2'^{\,2}}{J_2}\right)^{1/2} - \frac{1}{2}\,\frac{\sigma_2'}{J_2^{1/2}}, \tag{10.11.45}$$

which follow, for example, by solving

$$\sigma_2'^{\,2} + \sigma_3'^{\,2} + \sigma_2'\sigma_3' = J_2 \tag{10.11.46}$$

as a quadratic equation for σ'_3 in terms of σ'_2 and J_2. It is observed that

$$-\frac{1}{\sqrt{3}} \leq \frac{\sigma'_2}{J_2^{1/2}} \leq \frac{1}{\sqrt{3}}. \qquad (10.11.47)$$

The lower bound is associated with axially-symmetric tension ($\sigma_1 > \sigma_2 = \sigma_3$), and the upper bound with axially-symmetric compression ($\sigma_1 = \sigma_2 > \sigma_3$). Substituting $n_2 = 0$ and Eq. (10.11.42) into Eq. (10.11.36) gives the critical hardening rate associated with the choice $n_2 = 0$,

$$\frac{H_{(2)}}{\mu} = -\frac{{\sigma'_2}^2}{J_2} + \frac{1-\nu}{2}\left(\frac{1+\nu}{1-\nu}\,\frac{\beta+\mu^*}{3} - \frac{\sigma'_2}{J_2^{1/2}}\right)^2 - \frac{4(1+\nu)}{9(1-\nu)}\,\beta\mu^*. \qquad (10.11.48)$$

This can be rearranged as

$$\frac{H_{(2)}}{\mu} = \frac{1+\nu}{9(1-\nu)}(\beta-\mu^*)^2 - \frac{1+\nu}{2}\left(\frac{\sigma'_2}{J_2^{1/2}} + \frac{\beta+\mu^*}{3}\right)^2, \qquad (10.11.49)$$

which was originally derived by Rudnicki and Rice (1975). See also Perrin and Leblond (1993).

The second solution of Eqs. (10.11.38) and (10.11.39) is associated with $n_3 = 0$. In this case

$$n_2^2 = -\frac{\sigma_2-\sigma_3}{\sigma_1-\sigma_2} - (1+\nu)\,\frac{J_2^{1/2}}{\sigma_1-\sigma_2}\left(\frac{\sigma'_3}{J_2^{1/2}} + \frac{\beta+\mu^*}{3}\right), \qquad (10.11.50)$$

which must meet the condition $0 \leq n_2^2 \leq 1$. The critical hardening rate is consequently

$$\frac{H_{(3)}}{\mu} = \frac{1+\nu}{9(1-\nu)}(\beta-\mu^*)^2 - \frac{1+\nu}{2}\left(\frac{\sigma'_3}{J_2^{1/2}} + \frac{\beta+\mu^*}{3}\right)^2. \qquad (10.11.51)$$

The remaining solution of Eqs. (10.11.38) and (10.11.39) is associated with $n_2 = n_3 = 0$. The corresponding critical hardening rate $H_{(2,3)}$ can be calculated from Eq. (10.11.36).

Among the three values $H_{(2)}$, $H_{(3)}$ and $H_{(2,3)}$, the truly critical hardening rate is the largest of them. For realistic values of material properties β and μ^*, $H_{(2,3)}$ is always smaller than $H_{(2)}$ and $H_{(3)}$. This is expected on physical grounds because there is no shear stress in the localization plane associated with $H_{(2,3)}$ (localization plane being the principal stress plane), which greatly diminishes a tendency toward localization. We thus examine the inequality $H_{(2)} > H_{(3)}$. From (10.11.49) and (10.11.51), this is satisfied

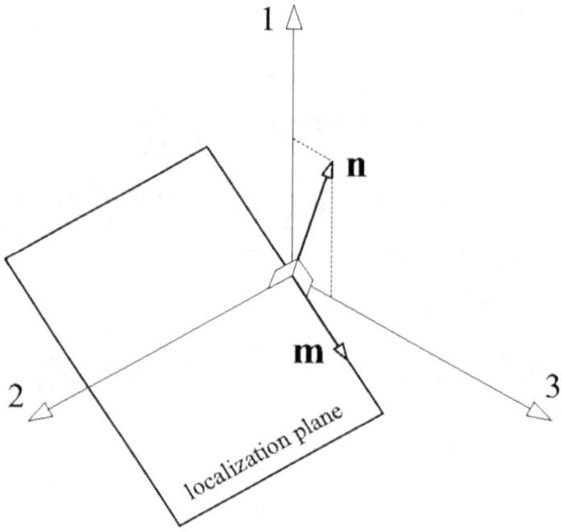

FIGURE 10.6. The localization plane according to considered pressure sensitive material model has its normal $\mathbf{n}$ perpendicular to the intermediate principal stress σ_2, so that in the coordinate system of principal stresses $\mathbf{n} = \{n_1, 0, (1 - n_1^2)^{1/2}\}$.

if

$$\left(\frac{\sigma_2'}{J_2^{1/2}} + \frac{\beta + \mu^*}{3}\right)^2 < \left(\frac{\sigma_3'}{J_2^{1/2}} + \frac{\beta + \mu^*}{3}\right)^2, \tag{10.11.52}$$

i.e.,

$$\frac{\sigma_1'}{J_2^{1/2}} > \frac{2}{3}(\beta + \mu^*). \tag{10.11.53}$$

The result can be expressed by using the first of expressions (10.11.45) as

$$\left(1 - \frac{3}{4}\frac{\sigma_2'^2}{J_2}\right)^{1/2} - \frac{1}{2}\frac{\sigma_2'}{J_2^{1/2}} > \frac{2}{3}(\beta + \mu^*). \tag{10.11.54}$$

In view of (10.11.47), a conservative bound assuring that $H_{(2)} > H_{(3)}$ is

$$\beta + \mu^* < \frac{\sqrt{3}}{2}, \tag{10.11.55}$$

whereas the condition

$$\beta + \mu^* > \sqrt{3} \tag{10.11.56}$$

assures that $H_{(3)} > H_{(2)}$. For the range of β and μ^* values used in constitutive modeling of fissured rocks, the latter case appears to be exceptional (Rudnicki and Rice, *op. cit.*). Thus, the localization will most likely occur in the plane whose normal is perpendicular to σ_2 direction ($n_2 = 0$), and the critical hardening rate is defined by Eq. (10.11.49); see Fig. 10.6.

It remains to examine a possibility for localization in the plane whose normal is perpendicular to σ_1 direction ($n_1 = 0$). The corresponding critical hardening rate would be

$$\frac{H_{(1)}}{\mu} = \frac{1+\nu}{9(1-\nu)}(\beta-\mu^*)^2 - \frac{1+\nu}{2}\left(\frac{\sigma_1'}{J_2^{1/2}} + \frac{\beta+\mu^*}{3}\right)^2. \tag{10.11.57}$$

This is greater than $H_{(2)}$ if

$$\frac{\sigma_3'}{J_2^{1/2}} > \frac{2}{3}(\beta+\mu^*). \tag{10.11.58}$$

However, from the second of expressions (10.11.45) it can be observed that $\sigma_3'/J_2^{1/2}$ is always negative in the range defined by (10.11.47). For frictional materials showing positive dilatancy, $\beta+\mu^* > 0$, the condition (10.11.58) is, therefore, never met. It could, however, be of interest in the study of loose granular materials which compact during shear, and thus exhibit negative dilatancy.

The expression for the critical hardening rate (10.11.49) reveals that localization in considered pressure-dependent dilatant materials is possible with positive hardening rate, depending on the value of $\sigma_3'/J_2^{1/2}$. The most critical (prompt to localization) is the state of stress

$$\frac{\sigma_2'}{J_2^{1/2}} = -\frac{\beta+\mu^*}{3}, \tag{10.11.59}$$

for which the critical hardening rate is

$$\frac{H_{(2)}}{\mu} = \frac{1+\nu}{9(1-\nu)}(\beta-\mu^*)^2. \tag{10.11.60}$$

The localization occurs in the plane whose normal is defined by

$$n_1^2 = \frac{\sigma_1-\sigma_2}{\sigma_1-\sigma_3}, \quad n_2 = 0, \quad n_3^2 = \frac{\sigma_2-\sigma_3}{\sigma_1-\sigma_3}. \tag{10.11.61}$$

Returning to Eqs. (10.11.38) and (10.11.39), if $\sigma_1 = \sigma_2 > \sigma_3$, n_2 remains unspecified by Eq. (10.11.38), which is satisfied by $2\sigma_2' + \sigma_3' = 0$, while Eq. (10.11.39) determines n_3. The critical hardening rate is still defined by Eq. (10.11.49), with $\sigma_2' = (\sigma_2 - \sigma_3)/3$.

The presented analysis in this subsection is based on the expression (10.11.21), which does not account for **B** terms, of the order of stress divided by elastic modulus. Inclusion of these terms and examination of their effects on the critical hardening rate and localization is given in the paper by

Rudnicki and Rice (1975). Further analysis of stability in the absence of plastic normality is available in Rice and Rudnicki (1980), Chau and Rudnicki (1990), and Li and Drucker (1994). Shear band formation in concrete was studied by Ortiz (1987). The book by Bažant and Cedolin (1991) provides additional references.

10.11.3. Rigid-Plastic Materials

For rigid-plastic materials the stress rate cannot be expressed in terms of the rate of deformation, so that localization condition cannot be put in the form (10.11.2). Instead, we impose conditions

$$[\![\mathbf{L}]\!] = \boldsymbol{\eta} \otimes \mathbf{n}, \quad \mathbf{n} \cdot [\![\dot{\boldsymbol{\sigma}}]\!] = \mathbf{0} \tag{10.11.62}$$

directly, following the procedure by Rice (1977). The constitutive structure for nonassociative rigid-plastic response is

$$\mathbf{D} = \frac{1}{H} \mathbb{P} \otimes \mathbb{Q} : \overset{\circ}{\boldsymbol{\sigma}}, \tag{10.11.63}$$

so that

$$[\![\mathbf{D}]\!] = \frac{1}{H} \mathbb{P} \otimes \mathbb{Q} : [\![\overset{\circ}{\boldsymbol{\sigma}}]\!], \quad [\![\overset{\circ}{\boldsymbol{\sigma}}]\!] = [\![\dot{\boldsymbol{\sigma}}]\!] - [\![\mathbf{W}]\!] \cdot \boldsymbol{\sigma} + \boldsymbol{\sigma} \cdot [\![\mathbf{W}]\!]. \tag{10.11.64}$$

Consequently,

$$\begin{aligned} \frac{1}{2} (\boldsymbol{\eta} \otimes \mathbf{n} + \mathbf{n} \otimes \boldsymbol{\eta}) = \frac{1}{H} \mathbb{P} \otimes \mathbb{Q} : \Big[[\![\dot{\boldsymbol{\sigma}}]\!] &- \frac{1}{2} (\boldsymbol{\eta} \otimes \mathbf{n} - \mathbf{n} \otimes \boldsymbol{\eta}) \cdot \boldsymbol{\sigma} \\ &+ \frac{1}{2} \boldsymbol{\sigma} \cdot (\boldsymbol{\eta} \otimes \mathbf{n} - \mathbf{n} \otimes \boldsymbol{\eta}) \Big]. \end{aligned} \tag{10.11.65}$$

This is evidently satisfied if $\mathbb{P}$ has the representation

$$\mathbb{P} = \frac{1}{2} (\boldsymbol{\nu} \otimes \mathbf{n} + \mathbf{n} \otimes \boldsymbol{\nu}), \tag{10.11.66}$$

for some vector $\boldsymbol{\nu}$, and if the localization vector is codirectional with $\boldsymbol{\nu}$,

$$\boldsymbol{\eta} = k\, \boldsymbol{\nu}. \tag{10.11.67}$$

Therefore, the localization can occur on the plane with normal $\mathbf{n}$ only if the state of stress is such that $\mathbb{P}$ has a special, rather restrictive representation given by (10.11.66). If the coordinate axes are selected with one axis parallel to $\mathbf{n}$ ($n_i = \delta_{i3}$), we have

$$\mathbb{P}_{\alpha\beta} = 0, \quad \alpha, \beta = 1, 2. \tag{10.11.68}$$

The intermediate principal value of such tensor is equal to zero ($\mathbb{P}_2 = 0$), so that $\mathbb{P}$ is a biaxial tensor with a spectral representation

$$\mathbb{P} = \mathbb{P}_1 \mathbf{e}_1 \otimes \mathbf{e}_1 + \mathbb{P}_3 \mathbf{e}_3 \otimes \mathbf{e}_3, \tag{10.11.69}$$

where $\mathbb{P}_1 \geq 0$, $\mathbb{P}_3 \leq 0$, and $\mathbf{e}_1$, $\mathbf{e}_2$, $\mathbf{e}_3$ are the principal directions of $\mathbb{P}$. It follows that

$$\mathbf{n} = \frac{1}{\sqrt{\mathbb{P}_1 - \mathbb{P}_3}} \left(\sqrt{\mathbb{P}_1}\, \mathbf{e}_1 + \sqrt{-\mathbb{P}_3}\, \mathbf{e}_3\right), \tag{10.11.70}$$

$$\boldsymbol{\nu} = \sqrt{\mathbb{P}_1 - \mathbb{P}_3} \left(\sqrt{\mathbb{P}_1}\, \mathbf{e}_1 - \sqrt{-\mathbb{P}_3}\, \mathbf{e}_3\right). \tag{10.11.71}$$

For example, it can be readily verified that this complies with

$$\mathbb{P} = \mathbb{P}_1 \mathbf{e}_1 \otimes \mathbf{e}_1 + \mathbb{P}_3 \mathbf{e}_3 \otimes \mathbf{e}_3 = \frac{1}{2} (\boldsymbol{\nu} \otimes \mathbf{n} + \mathbf{n} \otimes \boldsymbol{\nu}). \tag{10.11.72}$$

If neither $\mathbb{P}_1$ nor $\mathbb{P}_3$ vanishes, there are two possible localization planes, one with normal $\mathbf{n}$ defined by (10.11.70) and localization vector proportional to (10.11.71), and the other with

$$\mathbf{n} = \frac{1}{\sqrt{\mathbb{P}_1 - \mathbb{P}_3}} \left(\sqrt{\mathbb{P}_1}\, \mathbf{e}_1 - \sqrt{-\mathbb{P}_3}\, \mathbf{e}_3\right), \tag{10.11.73}$$

$$\boldsymbol{\nu} = \sqrt{\mathbb{P}_1 - \mathbb{P}_3} \left(\sqrt{\mathbb{P}_1}\, \mathbf{e}_1 + \sqrt{-\mathbb{P}_3}\, \mathbf{e}_3\right), \tag{10.11.74}$$

since $\boldsymbol{\eta}$ and $\mathbf{n}$ appear symmetrically in (10.11.66). If either $\mathbb{P}_1$ or $\mathbb{P}_3$ vanishes, there is one possible plane of localization. For instance, if $\mathbb{P}_3 = 0$, the localization plane has the normal $\mathbf{n} = \mathbf{e}_1$, and the corresponding $\boldsymbol{\nu} = \mathbb{P}_1 \mathbf{n}$. Observe that, in general,

$$\mathbf{n} \cdot \mathbb{P} \cdot \mathbf{n} = \mathbf{n} \cdot \boldsymbol{\nu} = \mathbb{P}_1 + \mathbb{P}_3, \quad \boldsymbol{\nu} \cdot \boldsymbol{\nu} = (\mathbb{P}_1 - \mathbb{P}_3)^2, \tag{10.11.75}$$

$$\boldsymbol{\nu} \cdot \mathbb{P} \cdot \boldsymbol{\nu} = (\boldsymbol{\nu} \cdot \boldsymbol{\nu}) \mathbf{n} \cdot \mathbb{P} \cdot \mathbf{n} = (\boldsymbol{\nu} \cdot \boldsymbol{\nu})(\mathbf{n} \cdot \boldsymbol{\nu}). \tag{10.11.76}$$

The component of the localization vector in the plane of localization is

$$\mathbf{b} = \boldsymbol{\eta} - (\mathbf{n} \cdot \boldsymbol{\eta})\mathbf{n} = 2k \sqrt{\frac{-\mathbb{P}_1 \mathbb{P}_3}{\mathbb{P}_1 - \mathbb{P}_3}} \left(\sqrt{-\mathbb{P}_3}\, \mathbf{e}_1 - \sqrt{\mathbb{P}_1}\, \mathbf{e}_3\right). \tag{10.11.77}$$

In the case of incompressible plastic flow, $\operatorname{tr} \mathbb{P} = \mathbb{P}_1 + \mathbb{P}_3 = 0$, and

$$\mathbf{n} \cdot \boldsymbol{\nu} = 0, \tag{10.11.78}$$

so that bifurcation vector $\boldsymbol{\eta}$ is in the plane of localization. The plane of localization is in this case the plane of maximum shear stress, since from Eq. (10.11.73),

$$\mathbf{n} = \frac{1}{\sqrt{2}}(\mathbf{e}_1 - \mathbf{e}_3). \tag{10.11.79}$$

Returning to Eq. (10.11.65), the substitution of (10.11.66) and (10.11.67) yields

$$k\left\{H + \frac{1}{2}\mathbb{Q} : [(\boldsymbol{\nu}\otimes\mathbf{n} - \mathbf{n}\otimes\boldsymbol{\nu})\cdot\boldsymbol{\sigma} - \boldsymbol{\sigma}\cdot(\boldsymbol{\nu}\otimes\mathbf{n} - \mathbf{n}\otimes\boldsymbol{\nu})]\right\} = \mathbb{Q} : [\![\dot{\boldsymbol{\sigma}}]\!]. \tag{10.11.80}$$

We impose now the remaining discontinuity condition $\mathbf{n}\cdot[\![\dot{\boldsymbol{\sigma}}]\!] = \mathbf{0}$. With the orthogonal axes 1, 2 in the plane of localization, and the axis 3 in the direction of $\mathbf{n}$, it follows that

$$[\![\dot{\sigma}_{3j}]\!] = 0, \quad (j = 1, 2, 3) \tag{10.11.81}$$

and

$$\mathbb{Q} : [\![\dot{\boldsymbol{\sigma}}]\!] = \mathbb{Q}_{\alpha\beta}[\![\dot{\sigma}_{\alpha\beta}]\!], \quad \alpha, \beta = 1, 2. \tag{10.11.82}$$

The condition (10.11.80) is accordingly

$$k\left\{H + \frac{1}{2}\mathbb{Q} : [(\boldsymbol{\nu}\otimes\mathbf{n} - \mathbf{n}\otimes\boldsymbol{\nu})\cdot\boldsymbol{\sigma} - \boldsymbol{\sigma}\cdot(\boldsymbol{\nu}\otimes\mathbf{n} - \mathbf{n}\otimes\boldsymbol{\nu})]\right\} = \mathbb{Q}_{\alpha\beta}[\![\dot{\sigma}_{\alpha\beta}]\!]. \tag{10.11.83}$$

Suppose that plastic normality is obeyed, so that $\mathbb{P} = \mathbb{Q}$ (associative plasticity). The right-hand side of (10.11.83) is then equal to zero, because $\mathbb{P}_{\alpha\beta} = 0$ by Eq. (10.11.68). Thus, if localization occurs ($k \neq 0$), the bracketed term on the left-hand side of (10.11.83) must vanish. This gives the critical hardening rate

$$H = \frac{1}{2}[\boldsymbol{\nu}\cdot\boldsymbol{\sigma}\cdot\boldsymbol{\nu} - (\boldsymbol{\nu}\cdot\boldsymbol{\nu})\,\mathbf{n}\cdot\boldsymbol{\sigma}\cdot\mathbf{n}]. \tag{10.11.84}$$

If the principal directions of stress tensor $\boldsymbol{\sigma}$ are parallel to those of $\mathbf{D}$ and thus $\mathbb{P}$, its spectral decomposition is

$$\boldsymbol{\sigma} = \sigma_1\mathbf{e}_1\otimes\mathbf{e}_1 + \sigma_2\mathbf{e}_2\otimes\mathbf{e}_2 + \sigma_3\mathbf{e}_3\otimes\mathbf{e}_3. \tag{10.11.85}$$

In view of (10.11.70) and (10.11.71), then,

$$\mathbf{n}\cdot\boldsymbol{\sigma}\cdot\mathbf{n} = \frac{1}{\mathbb{P}_1 - \mathbb{P}_3}(\mathbb{P}_1\sigma_1 - \mathbb{P}_3\sigma_3), \quad \boldsymbol{\nu}\cdot\boldsymbol{\sigma}\cdot\boldsymbol{\nu} = (\mathbb{P}_1 - \mathbb{P}_3)(\mathbb{P}_1\sigma_1 - \mathbb{P}_3\sigma_3). \tag{10.11.86}$$

Since

$$\boldsymbol{\nu} \cdot \boldsymbol{\nu} = (\mathbb{P}_1 - \mathbb{P}_3)^2, \tag{10.11.87}$$

Equation (10.11.86) shows that

$$\boldsymbol{\nu} \cdot \boldsymbol{\sigma} \cdot \boldsymbol{\nu} = (\boldsymbol{\nu} \cdot \boldsymbol{\nu})\, \mathbf{n} \cdot \boldsymbol{\sigma} \cdot \mathbf{n}, \tag{10.11.88}$$

and from Eq. (10.11.84) the critical hardening rate is

$$H = 0. \tag{10.11.89}$$

If principal directions of $\boldsymbol{\sigma}$ are not parallel to those of $\mathbf{D}$ (as in the case of anisotropic hardening rigid-plastic response), the critical hardening rate is not necessarily equal to zero. Furthermore, in the case of nonassociative plastic response (plastic non-normality) it is possible that some of the components $\mathbb{Q}_{\alpha\beta}$ are nonzero. In that case, since the components $[\![\dot{\sigma}_{\alpha\beta}]\!]$ are unrestricted, the condition (10.11.83) permits $k \neq 0$, and thus localization for any value of the hardening rate H. Rice (1977) indicates that the inclusion of elastic effects mitigates this strong tendency for localization in the absence of normality, but the tendency remains.

Since $\mathbb{P}$ and $\mathbf{D}$ are coaxial tensors by (10.11.63), from Eq. (10.11.68) it follows that

$$D_{\alpha\beta} = 0, \quad (\alpha, \beta = 1, 2) \tag{10.11.90}$$

in the plane of localization. Therefore, if the deformation field is such that a nondeforming plane does not exist, the localization cannot occur within the considered constitutive and localization framework. For example, it has been long known that rigid-plastic model with a smooth yield surface predicts an unlimited ductility in thin sheets under positive in-plane principal stretch rates (e.g., with von Mises yield condition and associative flow rule, $2\sigma_2 > \sigma_1$ for positive stretch rate D_2, contrary to the requirement $2\sigma_2 = \sigma_1$ for the existence of nondeforming plane of localization). Since localization actually occurs in these experiments, constitutive models simulating the yield-vertex have been employed to explain the experimental observations (Stören and Rice, 1975). Alternatively, imperfection studies were used in which, rather than being perfectly homogeneous, the sheet was assumed to contain an imperfection in the form of a long thin slice of material with slightly different properties from the material outside (Marciniak and Kuczynski, 1967;

Anand and Spitzig, 1980). Detailed summary and results for various material models can be found in the papers by Needleman (1976), and Needleman and Tvergaard (1983,1992). See also Petryk and Thermann (1996). We discuss below the yield vertex effects on localization in rigid-plastic, and incompressible elastic-plastic materials.

10.11.4. Yield Vertex Effects on Localization

A constitutive model simulating formation and effects of the vertex at the loading point of the yield surface was presented in Subsections 9.8.2 and 9.11.2. In the case of rigid-plasticity with pressure-independent associative flow rule, the rate of deformation is defined by

$$\mathbf{D} = \frac{1}{h}(\mathbb{M}\otimes\mathbb{M}) : \overset{\circ}{\boldsymbol{\sigma}} + \frac{1}{h_1}\left[\overset{\circ}{\boldsymbol{\sigma}}{}' - (\mathbb{M}\otimes\mathbb{M}) : \overset{\circ}{\boldsymbol{\sigma}}\right]. \tag{10.11.91}$$

The normalized tensor

$$\mathbb{M} = \frac{\dfrac{\partial f}{\partial \boldsymbol{\sigma}}}{\left(\dfrac{\partial f}{\partial \boldsymbol{\sigma}} : \dfrac{\partial f}{\partial \boldsymbol{\sigma}}\right)^{1/2}}, \tag{10.11.92}$$

is a deviatoric second-order tensor, f being a pressure-independent yield function. For example,

$$\mathbb{M} = \frac{\boldsymbol{\sigma}'}{(2J_2)^{1/2}}, \quad \text{if} \quad f = J_2^{1/2} = \left(\frac{1}{2}\boldsymbol{\sigma}' : \boldsymbol{\sigma}'\right)^{1/2}. \tag{10.11.93}$$

The hardening modulus of the vertex response

$$h_1 > h \tag{10.11.94}$$

governs the response to part of the stress increment directed tangentially to what is taken to be a smooth yield surface through the considered stress point. Since

$$\mathbb{M} : \mathbf{D} = \frac{1}{h}\,(\mathbb{M} : \overset{\circ}{\boldsymbol{\sigma}}), \tag{10.11.95}$$

the inverse constitutive expression is

$$\overset{\circ}{\boldsymbol{\sigma}}{}' = h_1\mathbf{D} - (h_1 - h)(\mathbb{M}\otimes\mathbb{M}) : \mathbf{D}, \tag{10.11.96}$$

i.e.,

$$\overset{\circ}{\boldsymbol{\sigma}} = \dot{\sigma}\,\mathbf{I} + h_1\mathbf{D} - (h_1 - h)(\mathbb{M}\otimes\mathbb{M}) : \mathbf{D}. \tag{10.11.97}$$

Here,

$$\sigma = \frac{1}{3}\operatorname{tr}\boldsymbol{\sigma}, \quad \operatorname{tr}\mathbf{D} = \operatorname{tr}\mathbb{M} = 0. \tag{10.11.98}$$

The jump condition $\mathbf{n}\cdot[\![\,\dot{\boldsymbol{\sigma}}\,]\!]=\mathbf{0}$ is consequently

$$\begin{aligned}[\![\,\dot{\sigma}\,]\!]\,\mathbf{n}+\mathbf{n}\cdot[\![\,\mathbf{W}\,]\!]\cdot\boldsymbol{\sigma}-\mathbf{n}\cdot\boldsymbol{\sigma}\cdot[\![\,\mathbf{W}\,]\!]+h_1\mathbf{n}\cdot[\![\,\mathbf{D}\,]\!]\\ -(h_1-h)(\mathbf{n}\cdot\mathbb{M})(\mathbb{M}:[\![\,\mathbf{D}\,]\!])=\mathbf{0}.\end{aligned} \tag{10.11.99}$$

Since $[\![\,\mathbf{L}\,]\!]=\boldsymbol{\eta}\otimes\mathbf{n}$, and $\operatorname{tr}\mathbf{L}=0$ for incompressible material, $\boldsymbol{\eta}$ must be perpendicular to $\mathbf{n}$. Hence,

$$\boldsymbol{\eta}=g\,\mathbf{m},\quad \mathbf{m}\cdot\mathbf{n}=0, \tag{10.11.100}$$

where g is a scalar function (bifurcation amplitude), and $\mathbf{m}$ is a unit vector in the plane of localization. Therefore,

$$[\![\,\mathbf{L}\,]\!]=g(\mathbf{m}\otimes\mathbf{n}), \tag{10.11.101}$$

and (10.11.99) becomes

$$[\![\,\dot{\sigma}\,]\!]\,\mathbf{n}-\frac{1}{2}\,g\big[\mathbf{m}\cdot\boldsymbol{\sigma}+\sigma_{mn}\,\mathbf{n}-(h_1+\sigma_n)\,\mathbf{m}+2(h_1-h)\mathbb{M}_{mn}\,(\mathbf{n}\cdot\mathbb{M})\big]=\mathbf{0}, \tag{10.11.102}$$

where

$$\sigma_{mn}=\mathbf{m}\cdot\boldsymbol{\sigma}\cdot\mathbf{n},\quad \sigma_n=\mathbf{n}\cdot\boldsymbol{\sigma}\cdot\mathbf{n},\quad \mathbb{M}_{mn}=\mathbf{m}\cdot\mathbb{M}\cdot\mathbf{n}. \tag{10.11.103}$$

Performing a scalar product of Eq. (10.11.102) with unit vectors $\mathbf{n}$, $\mathbf{m}$ and $\mathbf{p}=\mathbf{m}\times\mathbf{n}$ ($\mathbf{m}$ and $\mathbf{p}$ thus both being in the plane of localization), yields

$$g\big[\sigma_{mn}+(h_1-h)\mathbb{M}_{mn}\,\mathbb{M}_n\big]=[\![\,\dot{\sigma}\,]\!], \tag{10.11.104}$$

$$\sigma_m-\sigma_n-h_1+2(h_1-h)\mathbb{M}_{mn}^2=0, \tag{10.11.105}$$

$$\sigma_{mp}+2(h_1-h)\mathbb{M}_{mn}\,\mathbb{M}_{np}=0, \tag{10.11.106}$$

with no summation over repeated index n.

If h_1 is considered to be a constant vertex hardening modulus, localization will occur in the plane for which h is maximum. By taking a variation of (10.11.105) corresponding to $\boldsymbol{\delta}\mathbf{n}\propto\mathbf{p}$ (so that $\mathbf{m}$ remains perpendicular to $\mathbf{n}+\boldsymbol{\delta}\mathbf{n}$, i.e., $\boldsymbol{\delta}\mathbf{m}=\mathbf{0}$), and by setting $\delta h=0$, it follows that

$$\sigma_{np}-2(h_1-h)\mathbb{M}_{mn}\,\mathbb{M}_{mp}=0, \tag{10.11.107}$$

with no sum on m. Equations (10.11.106) and (10.11.107) are both satisfied if the axis $\mathbf{p}$ is along one of the principal stress axes, provided that $\mathbb{M}$ and $\boldsymbol{\sigma}$ are coaxial tensors (isotropic hardening), for then

$$\sigma_{mp}=\sigma_{np}=0,\quad \mathbb{M}_{mp}=\mathbb{M}_{np}=0. \tag{10.11.108}$$

In the case of von Mises yield condition, $\mathbb{M}$ is given by Eq. (10.11.93), and Eqs. (10.11.106) and (10.11.107) are satisfied only if

$$\sigma_{mp} = \sigma_{np} = 0, \tag{10.11.109}$$

so that the axis $\mathbf{p}$ must be codirectional with one of the principal stress axes (Rice, 1977). In the case of a plasticity model without a vertex, we have found in the previous subsection that the axis of intermediate principal stress is in the plane of localization. Since the vertex model reduces to a nonvertex model in the limit $h_1 \to \infty$, we conclude that $\mathbf{p} = \mathbf{e}_2$, and therefore

$$\mathbf{n} = n_1\, \mathbf{e}_1 + (1 - n_1^2)^{1/2}\, \mathbf{e}_3, \quad \mathbf{m} = -(1 - n_1^2)^{1/2}\, \mathbf{e}_1 + n_1\, \mathbf{e}_3. \tag{10.11.110}$$

Consequently,

$$\sigma_m - \sigma_n = (\sigma_1 - \sigma_3)(1 - 2n_1^2), \quad \mathbb{M}_{mn}^2 = (\mathbb{M}_1 - \mathbb{M}_3)^2 n_1^2 (1 - n_1^2), \tag{10.11.111}$$

so that Eq. (10.11.105) becomes

$$2(h_1 - h)(\mathbb{M}_1 - \mathbb{M}_3)^2 n_1^2 (1 - n_1^2) + (\sigma_1 - \sigma_3)(1 - 2n_1^2) - h_1 = 0. \tag{10.11.112}$$

Performing the variation corresponding to δn_1 and setting $\delta h = 0$ gives

$$n_1^2 = \frac{1}{2}\left[1 - \frac{\sigma_1 - \sigma_3}{(h_1 - h)(\mathbb{M}_1 - \mathbb{M}_3)^2}\right]. \tag{10.11.113}$$

For this to be acceptable, $0 \le n_1^2 \le 1$. The condition $n_1^2 \le 1$ is satisfied for $h_1 > h$, while $n_1^2 \ge 0$ gives

$$h_1 - h \ge \frac{\sigma_1 - \sigma_3}{(\mathbb{M}_1 - \mathbb{M}_3)^2}. \tag{10.11.114}$$

If vertex effects are neglected ($h_1 \to \infty$), Eq. (10.11.113) reproduces the result $n_1^2 = 1/2$ from the previous subsection. Substituting (10.11.113) back into (10.11.112) gives a quadratic equation for the critical hardening rate h,

$$(h_1 - h)^2 - \frac{2h_1}{(\mathbb{M}_1 - \mathbb{M}_3)^2}(h_1 - h) + \frac{(\sigma_1 - \sigma_3)^2}{(\mathbb{M}_1 - \mathbb{M}_3)^4} = 0. \tag{10.11.115}$$

With the von Mises yield condition, we have

$$\mathbb{M}_1 - \mathbb{M}_3 = \frac{\sigma_1' - \sigma_3'}{(2J_2)^{1/2}}, \tag{10.11.116}$$

and since, by Eqs. (10.11.45),

$$\frac{\sigma_1' - \sigma_3'}{J_2^{1/2}} = 2\left(1 - \frac{3}{4}\frac{\sigma_2'^{\,2}}{J_2}\right)^{1/2}, \tag{10.11.117}$$

we obtain

$$n_1^2 = \frac{1}{2}\left[1 - \frac{1}{h_1 - h}\,\frac{J_2^{1/2}}{\left(1 - \frac{3}{4}\frac{\sigma_2'^{\,2}}{J_2}\right)^{1/2}}\right], \tag{10.11.118}$$

and

$$\left(1 - \frac{3}{4}\frac{\sigma_2'^{\,2}}{J_2}\right)(h_1 - h)^2 - h_1(h_1 - h) + J_2 = 0. \tag{10.11.119}$$

Alternatively, Eq. (10.11.119) can be written as (Rice, 1977)

$$\frac{3}{4}\frac{\sigma_2'^{\,2}}{J_2}(h_1 - h)^2 + h(h_1 - h) - J_2 = 0. \tag{10.11.120}$$

In order that $n_1^2 \geq 0$, from Eq. (10.11.118) it follows that the hardening rate at localization must satisfy the condition

$$\frac{h}{h_1} \leq 1 - 2\frac{J_2}{h_1^2}\,. \tag{10.11.121}$$

Under this condition, the critical hardening rate is, from Eq. (10.11.119),

$$\frac{h}{h_1} = 1 - \frac{1 \pm \sqrt{1 - 4(1-u)J_2/h_1^2}}{2(1-u)}, \tag{10.11.122}$$

where

$$u = \frac{3\sigma_2'^{\,2}}{4J_2}\,. \tag{10.11.123}$$

Plus sign should be used if localization occurs at negative h, and minus sign if it occurs at positive h, provided that h meets the condition (10.11.121). If the ratio J_2/h_1^2 is sufficiently small, the condition (10.11.122) gives

$$\frac{h}{h_1} = -\frac{u}{1-u} + \frac{J_2}{h_1^2} + \cdots\,. \tag{10.11.124}$$

In this case, unless plane stain conditions prevail ($u \to 0$), strain softening is required for localization ($h < 0$).

An analysis of localization for elastic-plastic materials with yield vertex effects is more involved, but for an incompressible elastic-plastic material the results can be easily deduced from the rigid-plastic analysis. Addition of elastic part of the rate of deformation ($\mathbf{D}^{\mathrm{e}} = \overset{\circ}{\boldsymbol{\sigma}}{}'/2\mu$) to plastic part gives

$$\mathbf{D} = \left(\frac{1}{h} - \frac{1}{h_1}\right)(\mathbb{M} \otimes \mathbb{M}) : \overset{\circ}{\boldsymbol{\sigma}} + \left(\frac{1}{h_1} + \frac{1}{2\mu}\right)\overset{\circ}{\boldsymbol{\sigma}}{}'. \tag{10.11.125}$$

Evidently, the corresponding localization results can be directly obtained from previously derived results for rigid-plastic material, if the replacements

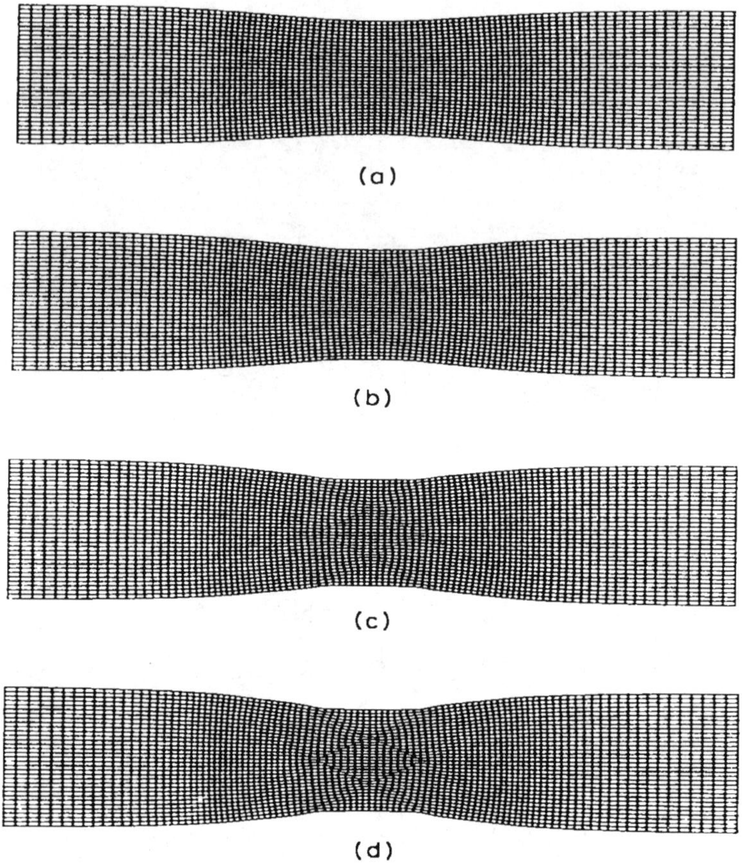

FIGURE 10.7. The neck development obtained by finite element calculations and J_2 corner theory. An initial thickness inhomogeneity grows into the necking mode. At high local strain levels the bands of intense shear deformation develop in the necked region (from Tvergaard, Needleman, and Lo, 1981; with permission from Elsevier Science).

are made

$$\frac{1}{h} \to \frac{1}{h} + \frac{1}{2\mu}, \quad \frac{1}{h_1} \to \frac{1}{h_1} + \frac{1}{2\mu}. \tag{10.11.126}$$

Numerical evaluations reveal that the critical h for localization at states other than plane strain is considerably less negative than the critical h predicted by an analysis without the yield vertex effects (Rudnicki and Rice, 1975; Rice, 1977).

There has been a number of localization studies based on the more involved corner theories of plasticity. The phenomenological J_2 corner theory of Christoffersen and Hutchinson (1979) has been frequently utilized (e.g., Hutchinson and Tvergaard, 1981; Tvergaard, Needleman, and Lo, 1981). Details of the localization predictions can be found in the original papers and reviews (Tvergaard, 1992; Needleman and Tvergaard, 1992). For example, Fig. 10.7 from Tvergaard, Needleman, and Lo (1981) shows the neck development obtained by finite element calculations and J_2 corner theory. An initial imperfection in the form of a long wave-length thickness inhomogeneity grows into the necking mode. Subsequently, at sufficiently high local strain levels, the bands of intense shear deformation develop in the necked region. The localization in rate-dependent solids and under dynamic loading conditions was studied by Anand, Kim, and Shawki (1987), Needleman (1988, 1989), Batra and Kim (1990), Xu and Needleman (1992), and others.

References

Anand, L., Kim, K. H., and Shawki, T. G. (1987), Onset of shear localization in viscoplastic solids, *J. Mech. Phys. Solids*, Vol. 35, pp. 407–429.

Anand, L. and Spitzig, W. A. (1980), Initiation of localized shear bands in plane strain, *J. Mech. Phys. Solids*, Vol. 28, pp. 113–128.

Bardet, J. P. (1991), Analytical solutions for the plane-strain bifurcation of compressible solids, *J. Appl. Mech.*, Vol. 58, pp. 651–657.

Batra, R. C. and Kim, C. H. (1990), Effect of viscoplastic flow rules on the initiation and growth of shear bands at high strain rates, *J. Mech. Phys. Solids*, Vol. 38, pp. 859–874.

Bažant, Z. P. and Cedolin, L. (1991), *Stability of Structures: Elastic, Inelastic, Fracture, and Damage Theories*, Oxford University Press, New York.

Burke, M. A. and Nix, W. D. (1979), A numerical study of necking in the plane tension test, *Int. J. Solids Struct.*, Vol. 15, pp. 379–393.

Bigoni, D. (1995), On flutter instability in elastoplastic constitutive models, *Int. J. Solids Struct.*, Vol. 32, pp. 3167–3189.

Bigoni, D. and Hueckel, T. (1991), Uniqueness and localization – associative and nonassociative elastoplasticity, *Int. J. Solids Struct.*, Vol. 28, pp. 197–213.

Bigoni, D. and Zaccaria, D. (1992), Loss of strong ellipticity in nonassociative elastoplasticity, *J. Mech. Phys. Solids*, Vol. 40, pp. 1313–1331.

Bigoni, D. and Zaccaria, D. (1993), On strain localization analysis of elastoplastic materials at finite strains, *Int. J. Plasticity*, Vol. 9, pp. 21–33.

Bruhns, O. T. (1984), Bifurcation problems in plasticity, in *The Constitutive Law in Thermoplasticity*, ed. Th. Lehmann, pp. 465–540, Springer-Verlag, Wien.

Chambon, R. and Caillerie, D. (1996), Existence and uniqueness theorems for boundary value problems involving incrementally nonlinear models, *Int. J. Solids Struct.*, Vol. 36, pp. 5089–5099.

Chau, K.-T. and Rudnicki, J. W. (1990), Bifurcations of compressible pressure-sensitive materials in plane strain tension and compression, *J. Mech. Phys. Solids*, Vol. 38, pp. 875–898.

Clifton, R. J. (1974), Plastic waves: Theory and experiment, in *Mechanics Today*, Vol. 1, ed. S. Nemat-Nasser, pp. 102–167, Pergamon Press, New York.

Christoffersen, J. and Hutchinson, J. W. (1979), A class of phenomenological corner theories of plasticity, *J. Mech. Phys. Solids*, Vol. 27, pp. 465–487.

Drucker, D. C. (1958), Variational principles in mathematical theory of plasticity, *Proc. Symp. Appl. Math.*, Vol. 8, pp. 7–22, McGraw-Hill, New York.

Drucker, D. C. (1960), Plasticity, in *Structural Mechanics – Proc. 1st Symp. Naval Struct. Mechanics*, eds. J. N. Goodier and N. J. Hoff, pp. 407–455, Pergamon Press, New York.

Drugan, W. J. and Shen, Y. (1987), Restrictions on dynamically propagating surfaces of strong discontinuity in elastic-plastic solids, *J. Mech. Phys. Solids*, Vol. 35, pp. 771–787.

Germain, P. and Lee, E. H. (1973), On shock waves in elastic-plastic solids, *J. Mech. Phys. Solids*, Vol. 21, pp. 359–382.

Hadamar, J. (1903), *Leçons sur la Propagation des Ondes et les Équations de L'Hydrodinamique*, Hermann, Paris.

Hill, R. (1950), *The Mathematical Theory of Plasticity*, Oxford University Press, London.

Hill, R. (1957), On the problem of uniqueness in the theory of a rigid-plastic solid – III, *J. Mech. Phys. Solids*, Vol. 5, pp. 153–161.

Hill, R. (1958), A general theory of uniqueness and stability in elastic-plastic solids, *J. Mech. Phys. Solids*, Vol. 6, pp. 236–249.

Hill, R. (1959), Some basic principles in the mechanics of solids without natural time, *J. Mech. Phys. Solids*, Vol. 7, pp. 209–225.

Hill, R. (1961a), Bifurcation and uniqueness in non-linear mechanics of continua, in *Problems of Continuum Mechanics – N. I. Muskhelishvili Anniversary Volume*, pp. 155-164, SIAM, Philadelphia.

Hill, R. (1961b), Discontinuity relations in mechanics of solids, in *Progress in Solid Mechanics*, eds. I. N. Sneddon and R. Hill, Vol. 2, pp. 244–276, North-Holland, Amsterdam.

Hill, R. (1962), Acceleration waves in solids, *J. Mech. Phys. Solids*, Vol. 10, pp. 1–16.

Hill, R. (1967), Eigenmodal deformations in elastic/plastic continua, *J. Mech. Phys. Solids*, Vol. 15, pp. 371–385.

Hill, R. (1978), Aspects of invariance in solid mechanics, *Adv. Appl. Mech.*, Vol. 18, pp. 1–75.

Hill, R. and Hutchinson, J. W. (1975), Bifurcation phenomena in the plane tension test, *J. Mech. Phys. Solids*, Vol. 23, pp. 239–264.

Hill, R. and Sewell, M. J. (1960), A general theory of inelastic column failure – I and II, *J. Mech. Phys. Solids*, Vol. 8, pp. 105–118.

Hill, R. and Sewell, M. J. (1962), A general theory of inelastic column failure – III, *J. Mech. Phys. Solids*, Vol. 10, pp. 285–300.

Hutchinson, J. W. (1973), Post-bifurcation behavior in the plastic range, *J. Mech. Phys. Solids*, Vol. 21, pp. 163–190.

Hutchinson, J. W. (1974), Plastic buckling, *Adv. Appl. Mech.*, Vol. 14, pp. 67–144.

Hutchinson, J. W. and Miles, J. P. (1974), Bifurcation analysis of the onset of necking in an elastic-plastic cylinder under uniaxial tension, *J. Mech. Phys. Solids*, Vol. 22, pp. 61–71.

Hutchinson, J. W. and Tvergaard, V. (1981), Shear band formation in plane strain, *Int. J. Solids Struct.*, Vol. 17, pp. 451–470.

Iwakuma, T. and Nemat-Nasser, S. (1982), An analytical estimate of shear band initiation in a necked bar, *Int. J. Solids Struct.*, Vol. 18, pp. 69–83.

Janssen, D. M., Datta, S. K., and Jahsman, W. E. (1972), Propagation of weak waves in elastic-plastic solids, *J. Mech. Phys. Solids*, Vol. 20, pp. 1–18.

Kleiber, M. (1986), On plastic localization and failure in plane strain and round void containing tensile bars, *Int. J. Plasticity*, Vol. 2, pp. 205–221.

Koiter, W. (1960), General theorems for elastic-plastic solids, in *Progress in Solid Mechanics*, eds. I. N. Sneddon and R. Hill, Vol. 1, pp. 165–221, North-Holland, Amsterdam.

Li, M. and Drucker, D. C. (1994), Instability and bifurcation of a nonassociated extended Mises model in the hardening regime, *J. Mech. Phys. Solids*, Vol. 42, pp. 1883–1904.

Maier, G. (1970), A minimum principle for incremental elastoplasticity with non-associated flow laws, *J. Mech. Phys. Solids*, Vol. 18, pp. 319–330.

Mandel, J. (1966), Conditions de stabilité et postulat de Drucker, in *Rheology and Soil Mechanics*, eds. J. Kravtchenko and P. M. Sirieys, pp. 58–68, Springer-Verlag, Berlin.

Marciniak, K. and Kuczynski, K. (1967), Limit strains in the process of stretch forming sheet metal, *Int. J. Mech. Sci.*, Vol. 9, pp. 609–620.

Miles, J. P. (1975), The initiation of necking in rectangular elastic-plastic specimens under uniaxial and biaxial tension, *J. Mech. Phys. Solids*, Vol. 23, pp. 197–213.

Neale, K. W. (1972), A general variational theorem for the rate problem in elasto-plasticity, *Int. J. Solids Struct.*, Vol. 8, pp. 865–876.

Needleman, A. (1972), A numerical study of necking in cylindrical bars, *J. Mech. Phys. Solids*, Vol. 20, pp. 111–127.

Needleman, A. (1976), Necking of pressurized spherical membranes, *J. Mech. Phys. Solids*, Vol. 24, pp. 339–359.

Needleman, A. (1979), Non-normality and bifurcation in plane strain tension and compression, *J. Mech. Phys. Solids*, Vol. 27, pp. 231–254.

Needleman, A. (1988), Material rate dependence and mesh sensitivity in localization problems, *Comput. Meth. Appl. Mech. Engrg.*, Vol. 67, pp. 69–85.

Needleman, A. (1989), Dynamic shear band development in plane strain, *J. Appl. Mech.*, Vol. 56, pp. 1–9.

Needleman, A. and Tvergaard, V. (1982), Aspects of plastic postbuckling behavior, in *Mechanics of Solids – The Rodney Hill 60th Anniversary Volume*, eds. H. G. Hopkins and M. J. Sewell, pp. 453–498, Pergamon Press, Oxford.

Needleman, A. and Tvergaard, V. (1983), Finite element analysis of localization in plasticity, in *Finite Elements Special Problems in Solid Mechanics, Vol. 5*, eds. J. T. Oden and G. F. Carey, pp. 94–157, Prentice-Hall, Englewood Cliffs, New Jersey.

Needleman, A. and Tvergaard, V. (1992), Analyses of plastic flow localization in metals, *Appl. Mech. Rev.*, Vol. 47, No. 3, Part 2, pp. S3–S18.

Neilsen, M. K. and Schreyer, H. L. (1993), Bifurcation in elastic-plastic materials, *Int. J. Solids Struct.*, Vol. 30, pp. 521–544.

Nguyen, Q. S. (1987), Bifurcation and post-bifurcation analysis in plasticity and brittle fracture, *J. Mech. Phys. Solids*, Vol. 35, pp. 303–324.

Nguyen, Q. S. (1994), Bifurcation and stability in dissipative media (plasticity, friction, fracture), *Appl. Mech. Rev.*, Vol. 47, No. 1, Part 1, pp. 1–31.

Ortiz, M. (1987), An analytical study of the localized failure modes of concrete, *Mech. Mater.*, Vol. 6, pp. 159–174.

Ortiz, M., Leroy, Y., and Needleman, A. (1987), A finite element method for localized failure analysis, *Comp. Meth. Appl. Mech. Engin.*, Vol. 61, pp. 189–214.

Ottosen, N. S. and Runesson, K. (1991), Discontinuous bifurcations in a nonassociated Mohr material, *Mech. Mater.*, Vol. 12, pp. 255–265.

Perrin, G. and Leblond, J. B. (1993), Rudnicki and Rice's analysis of strain localization revisited, *J. Appl. Mech.*, Vol. 60, pp. 842–846.

Petryk, H. (1989), On constitutive inequalities and bifurcation in elastic-plastic solids with a yield-surface vertex, *J. Mech. Phys. Solids*, Vol. 37, pp. 265–291.

Petryk, H. and Thermann, K. (1996), Post-critical plastic deformation of biaxially stretched sheets, *Int. J. Solids Struct.*, Vol. 33, pp. 689–705.

Ponter, A. R. S. (1969), Energy theorems and deformation bounds for constitutive relations associated with creep and plastic deformation of metals, *J. Mech. Phys. Solids*, Vol. 17, pp. 493–509.

Ramakrishnan, N. and Atluri, S. N. (1994), On shear band formation: I. Constitutive relationship for a dual yield model, *Int. J. Plasticity*, Vol. 10, pp. 499–520.

Raniecki, B. (1979), Uniqueness criteria in solids with non-associated plastic flow laws at finite deformations, *Bull. Acad. Polon. Sci. Techn.*, Vol. 27, pp. 391–399.

Raniecki, B. and Bruhns, O. T. (1981), Bounds to bifurcation stresses in solids with non-associated plastic flow law at finite strains, *J. Mech. Phys. Solids*, Vol. 29, pp. 153–172.

Rice, J. R. (1977), The localization of plastic deformation, in *Theoretical and Applied Mechanics – Proc. 14th Int. Congr. Theor. Appl. Mech.*, ed. W. T. Koiter, pp. 207–220, North-Holland, Amsterdam.

Rice, J. R. and Rudnicki, J. W. (1980), A note on some features of the theory of localization of deformation, *Int. J. Solids Struct.*, Vol. 16, pp. 597–605.

Rizzi, E. and Loret, B. (1997), Qualitative analysis of strain localization. Part I: Transversely isotropic elasticity and isotropic plasticity, *Int. J. Plasticity*, Vol. 13, pp. 461–499.

Rudnicki, J. W. and Rice, J. R. (1975), Conditions for the localization of deformation in pressure-sensitive dilatant materials, *J. Mech. Phys. Solids*, Vol. 23, pp. 371–394.

Sewell, M. J. (1972), A survey of plastic buckling, in *Stability*, ed. H. Leipholz, pp. 85–197, University of Waterloo Press, Ontario.

Sewell, M. J. (1973), A yield-surface corner lowers the buckling stress of an elastic-plastic plate under compression, *J. Mech. Phys. Solids*, Vol. 21, pp. 19–45.

Sewell, M. J. (1987), *Maximum and Minimum Principles – A Unified Approach with Applications*, Cambridge University Press, Cambridge.

Storåkers, B. (1971), Bifurcation and instability modes in thick-walled rigid-plastic cylinders under pressure, *J. Mech. Phys. Solids*, Vol. 19, pp. 339–351.

Storåkers, B. (1977), On uniqueness and stability under configuration-dependent loading of solids with or without a natural time, *J. Mech. Phys. Solids*, Vol. 25, pp. 269–287.

Stören, S. and Rice, J. R. (1975), Localized necking in thin sheets, *J. Mech. Phys. Solids*, Vol. 23, pp. 421–441.

Swift, H. W. (1952), Plastic instability under plane stress, *J. Mech. Phys. Solids*, Vol. 1, pp. 1–18.

Thomas, T. Y. (1961), *Plastic Flow and Fracture in Solids*, Academic Press, New York.

Ting, T. C. T. (1976), Shock waves and weak discontinuities in anisotropic elastic-plastic media, in *Propagation of Shock Waves in Solids*, ed. E. Varley, pp. 41–64, ASME, New York.

Triantafyllidis, N. (1980), Bifurcation phenomena in pure bending, *J. Mech. Phys. Solids*, Vol. 28, pp. 221–245.

Triantafyllidis, N. (1983), On the bifurcation and postbifurcation analysis of elastic-plastic solids under general prebifurcation conditions, *J. Mech. Phys. Solids*, Vol. 31, pp. 499–510.

Tvergaard, V. (1992), On the computational prediction of plastic strain localization, in *Mechanical Behavior of Materials – VI*, eds. M. Jono and T. Inoue, pp. 189–196, Pergamon Press, Oxford.

Tvergaard, V., Needleman, A., and Lo, K. K. (1981), Flow localization in the plane strain tensile test, *J. Mech. Phys. Solids*, Vol. 29, pp. 115–142.

Wilkins, M. L. (1964), Calculation of plastic flow, in *Methods in Computational Physics – Advances in Research and Applications*, eds. B. Alder, S. Fernbach, and M. Rotenberg, pp. 211–263, Academic Press, New York.

Xu, X.-P. and Needleman, A. (1992), The influence of nucleation criterion on shear localization in rate-sensitive porous plastic solids, *Int. J. Plasticity*, Vol. 8, pp. 315–330.

Young, N. J. B. (1976), Bifurcation phenomena in the plane compression test, *J. Mech. Phys. Solids*, Vol. 24, pp. 77–91.

CHAPTER 11

MULTIPLICATIVE DECOMPOSITION

This chapter deals with the formulation of the constitutive theory for large elastoplastic deformations within the framework of Lee's multiplicative decomposition of the deformation gradient. Kinematic and kinetic aspects of the theory are presented, with a particular accent given to the partition of the rate of deformation tensor into its elastic and plastic parts. The significance of plastic spin in the phenomenological theory is discussed. Isotropic and orthotropic materials are considered, and an introductory treatment of the damage-elastoplasticity is given.

11.1. Multiplicative Decomposition $\mathbf{F} = \mathbf{F}^{\mathrm{e}} \cdot \mathbf{F}^{\mathrm{p}}$

Consider the current elastoplastically deformed configuration of the material sample $\mathcal{B}$, whose initial undeformed configuration was $\mathcal{B}^0$. Let $\mathbf{F}$ be the deformation gradient that maps an infinitesimal material element $\mathrm{d}\mathbf{X}$ from $\mathcal{B}^0$ to $\mathrm{d}\mathbf{x}$ in $\mathcal{B}$, such that

$$\mathrm{d}\mathbf{x} = \mathbf{F} \cdot \mathrm{d}\mathbf{X}. \tag{11.1.1}$$

The initial and current location of the material particle are both referred to the same, fixed set of Cartesian coordinate axes. Introduce an intermediate configuration $\mathcal{B}^{\mathrm{p}}$ by elastically destressing the current configuration $\mathcal{B}$ to zero stress (Fig. 11.1). Such configuration differs from the initial configuration by residual (plastic) deformation, and from the current configuration by reversible (elastic) deformation. If $\mathrm{d}\mathbf{x}^{\mathrm{p}}$ is the material element in $\mathcal{B}^{\mathrm{p}}$, corresponding to $\mathrm{d}\mathbf{x}$ in $\mathcal{B}$, then

$$\mathrm{d}\mathbf{x} = \mathbf{F}^{\mathrm{e}} \cdot \mathrm{d}\mathbf{x}^{\mathrm{p}}, \tag{11.1.2}$$

where $\mathbf{F}^{\mathrm{e}}$ represents the deformation gradient associated with the elastic loading from $\mathcal{B}^{\mathrm{p}}$ to $\mathcal{B}$. If the deformation gradient of the transformation

$\mathcal{B}^0 \rightarrow \mathcal{B}^p$ is $\mathbf{F}^p$, such that

$$d\mathbf{x}^p = \mathbf{F}^p \cdot d\mathbf{X}, \tag{11.1.3}$$

the multiplicative decomposition of the total deformation gradient into its elastic and plastic parts follows

$$\mathbf{F} = \mathbf{F}^e \cdot \mathbf{F}^p. \tag{11.1.4}$$

The decomposition was introduced in the phenomenological theory of plasticity by Lee and Liu (1967), and Lee (1969). Early contributions also include Fox (1968), Willis (1969), Mandel (1971, 1973), and Kröner and Teodosiu (1973). For inhomogeneous deformations only $\mathbf{F}$ is a true deformation gradient, whose components are the partial derivatives $\partial\mathbf{x}/\partial\mathbf{X}$. The mappings $\mathcal{B}^p \rightarrow \mathcal{B}$ and $\mathcal{B}^0 \rightarrow \mathcal{B}^p$ are not, in general, continuous one-to-one mappings, so that $\mathbf{F}^e$ and $\mathbf{F}^p$ are not defined as the gradients of the respective mappings (which may not exist), but as the point functions (local deformation gradients). In the case when elastic destressing to zero stress ($\mathcal{B} \rightarrow \mathcal{B}^p$) is not physically achievable due to possible onset of reverse inelastic deformation before the state of zero stress is reached (which could occur at advanced stages of deformation due to anisotropic hardening and strong Bauschinger effect), the intermediate configuration can be conceptually introduced by virtual destressing to zero stress, locking all inelastic structural changes that would take place during the actual destressing.

There is a similar decomposition of the deformation gradient in thermoelasticity, where the total deformation gradient is expressed as the product of the elastic and thermal part. This has been studied by many, with a recent contribution given by Imam and Johnson (1998).

11.1.1. Nonuniqueness of Decomposition

The deformation gradients $\mathbf{F}^e$ and $\mathbf{F}^p$ are not uniquely defined because the intermediate unstressed configuration is not unique. Arbitrary local material rotations can be superposed to the intermediate configuration, preserving it unstressed. Thus, we can write

$$\mathbf{F} = \mathbf{F}^e \cdot \mathbf{F}^p = \hat{\mathbf{F}}^e \cdot \hat{\mathbf{F}}^p, \tag{11.1.5}$$

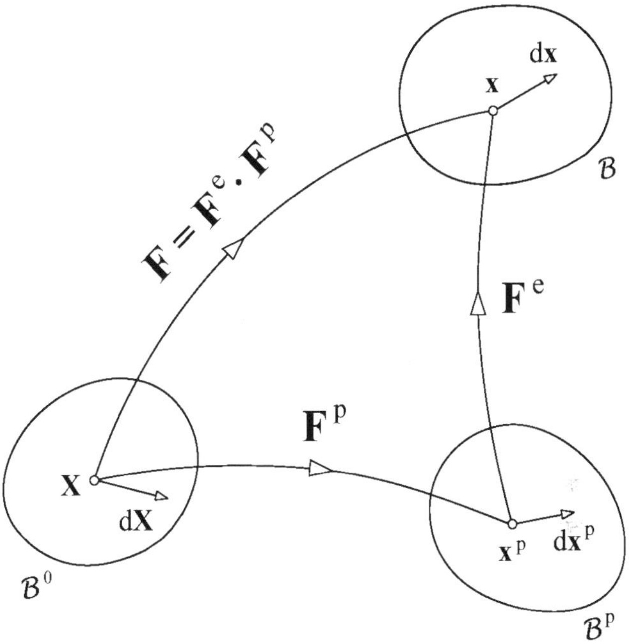

FIGURE 11.1. Schematic representation of the multiplicative decomposition of deformation gradient into its elastic and plastic parts. The intermediate configuration $\mathcal{B}^p$ is obtained from the current configuration $\mathcal{B}$ by elastic destressing to zero stress.

where

$$\hat{\mathbf{F}}^e = \mathbf{F}^e \cdot \hat{\mathbf{Q}}^T, \quad \hat{\mathbf{F}}^p = \hat{\mathbf{Q}} \cdot \mathbf{F}^p. \tag{11.1.6}$$

A local rotation is represented by an orthogonal tensor $\hat{\mathbf{Q}}$. If polar decompositions of deformation gradients are used,

$$\mathbf{F}^e = \mathbf{V}^e \cdot \mathbf{R}^e, \quad \mathbf{F}^p = \mathbf{R}^p \cdot \mathbf{U}^p, \tag{11.1.7}$$

it follows that only $\mathbf{R}^e$ and $\mathbf{R}^p$, not $\mathbf{V}^e$ and $\mathbf{U}^p$, are affected by the rotation of the intermediate state, i.e.,

$$\hat{\mathbf{R}}^e = \mathbf{R}^e \cdot \hat{\mathbf{Q}}^T, \quad \hat{\mathbf{R}}^p = \hat{\mathbf{Q}} \cdot \mathbf{R}^p, \tag{11.1.8}$$

while

$$\hat{\mathbf{V}}^e = \mathbf{V}^e, \quad \hat{\mathbf{U}}^p = \mathbf{U}^p. \tag{11.1.9}$$

Further discussion of the nonuniqueness of the decomposition can be found in the articles by Green and Naghdi (1971), Casey and Naghdi (1980), and

Naghdi (1990). Note that there is a unique decomposition

$$\mathbf{F} = \mathbf{V}^{\mathrm{e}} \cdot \mathbf{R}^{\mathrm{ep}} \cdot \mathbf{U}^{\mathrm{p}}, \tag{11.1.10}$$

since the rotation tensor

$$\mathbf{R}^{\mathrm{ep}} = \mathbf{R}^{\mathrm{e}} \cdot \mathbf{R}^{\mathrm{p}} = \hat{\mathbf{R}}^{\mathrm{e}} \cdot \hat{\mathbf{R}}^{\mathrm{p}} \tag{11.1.11}$$

is a unique tensor ($\hat{\mathbf{R}}^{\mathrm{ep}} = \mathbf{R}^{\mathrm{ep}}$).

In applications, the decomposition (11.1.4) can be made unique by additional requirements or specifications, dictated by the nature of the considered material model. For example, for elastically isotropic materials the stress response from $\mathcal{B}^{\mathrm{p}}$ to $\mathcal{B}$ depends only on the elastic stretch $\mathbf{V}^{\mathrm{e}}$, and not on the rotation $\mathbf{R}^{\mathrm{e}}$. Consequently, the intermediate configuration can be specified uniquely by requiring that the elastic unloading takes place without rotation,

$$\mathbf{F}^{\mathrm{e}} = \mathbf{V}^{\mathrm{e}}. \tag{11.1.12}$$

On the other hand, in single crystal plasticity (see Chapter 12), the orientation of the intermediate configuration is specified by a fixed orientation of the crystalline lattice, through which the material flows by crystallographic slip in the mapping from $\mathcal{B}^0$ to $\mathcal{B}^{\mathrm{p}}$. In Mandel's (1973, 1983) model, if the triad of orthogonal (director) vectors is attached to the initial configuration, and if this triad remains unaltered by plastic deformation, the intermediate configuration is referred to as isoclinic. Such configuration is unique at a given stage of elastoplastic deformation, because a superposed rotation $\hat{\mathbf{Q}} \neq \mathbf{I}$ would change the orientation of the director vectors, and the intermediate configuration would not remain isoclinic. For additional discussion, see the papers by Sidoroff (1975), and Kleiber and Raniecki (1985).

11.2. Decomposition of Strain Tensors

The left and right Cauchy–Green deformation tensors,

$$\mathbf{B} = \mathbf{F} \cdot \mathbf{F}^{T}, \quad \mathbf{C} = \mathbf{F}^{T} \cdot \mathbf{F}, \tag{11.2.1}$$

can be decomposed as

$$\mathbf{B} = \mathbf{F}^{\mathrm{e}} \cdot \mathbf{B}^{\mathrm{p}} \cdot \mathbf{F}^{\mathrm{e}T}, \quad \mathbf{C} = \mathbf{F}^{\mathrm{p}T} \cdot \mathbf{C}^{\mathrm{e}} \cdot \mathbf{F}^{\mathrm{p}}. \tag{11.2.2}$$

If the Lagrangian strains corresponding to deformation gradients $\mathbf{F}^{\mathrm{e}}$ and $\mathbf{F}^{\mathrm{p}}$ are defined by

$$\mathbf{E}^{\mathrm{e}} = \frac{1}{2}\left(\mathbf{C}^{\mathrm{e}} - \mathbf{I}\right), \quad \mathbf{E}^{\mathrm{p}} = \frac{1}{2}\left(\mathbf{C}^{\mathrm{p}} - \mathbf{I}\right), \tag{11.2.3}$$

where

$$\mathbf{C}^{\mathrm{e}} = \mathbf{F}^{\mathrm{e}T} \cdot \mathbf{F}^{\mathrm{e}}, \quad \mathbf{C}^{\mathrm{p}} = \mathbf{F}^{\mathrm{p}T} \cdot \mathbf{F}^{\mathrm{p}}, \tag{11.2.4}$$

the total Lagrangian strain (the strain measure from the family of material strain tensors (2.3.1) corresponding to $n = 1$) can be expressed as

$$\mathbf{E} = \frac{1}{2}\left(\mathbf{C} - \mathbf{I}\right) = \mathbf{E}^{\mathrm{p}} + \mathbf{F}^{\mathrm{p}T} \cdot \mathbf{E}^{\mathrm{e}} \cdot \mathbf{F}^{\mathrm{p}}. \tag{11.2.5}$$

The elastic and plastic strains $\mathbf{E}^{\mathrm{e}}$ and $\mathbf{E}^{\mathrm{p}}$ do not sum to give the total strain $\mathbf{E}$, because $\mathbf{E}$ and $\mathbf{E}^{\mathrm{p}}$ are defined relative to the initial configuration $\mathcal{B}^0$ as the reference configuration, while $\mathbf{E}^{\mathrm{e}}$ is defined relative to the intermediate configuration $\mathcal{B}^{\mathrm{p}}$ as the reference configuration. Consequently, it is the strain $\mathbf{F}^{\mathrm{p}T} \cdot \mathbf{E}^{\mathrm{e}} \cdot \mathbf{F}^{\mathrm{p}}$, induced from the elastic strain $\mathbf{E}^{\mathrm{e}}$ by plastic deformation $\mathbf{F}^{\mathrm{p}}$, that sums up with the plastic strain $\mathbf{E}^{\mathrm{p}}$ to give the total strain $\mathbf{E}$.

If Eulerian strains corresponding to deformation gradients $\mathbf{F}^{\mathrm{e}}$ and $\mathbf{F}^{\mathrm{p}}$ are introduced as

$$\mathcal{E}^{\mathrm{e}} = \frac{1}{2}\left(\mathbf{I} - \mathbf{B}^{\mathrm{e}-1}\right), \quad \mathcal{E}^{\mathrm{p}} = \frac{1}{2}\left(\mathbf{I} - \mathbf{B}^{\mathrm{p}-1}\right), \tag{11.2.6}$$

where

$$\mathbf{B}^{\mathrm{e}} = \mathbf{F}^{\mathrm{e}} \cdot \mathbf{F}^{\mathrm{e}T}, \quad \mathbf{B}^{\mathrm{p}} = \mathbf{F}^{\mathrm{p}} \cdot \mathbf{F}^{\mathrm{p}T}, \tag{11.2.7}$$

the total Eulerian strain (the strain measure from the family of spatial strain tensors (2.3.14) corresponding to $n = -1$) can be written as

$$\mathcal{E} = \frac{1}{2}\left(\mathbf{I} - \mathbf{B}^{-1}\right) = \mathcal{E}^{\mathrm{e}} + \mathbf{F}^{\mathrm{e}-T} \cdot \mathcal{E}^{\mathrm{p}} \cdot \mathbf{F}^{\mathrm{e}-1}. \tag{11.2.8}$$

The additive decomposition does not hold for the Eulerian strains either, because the elastic and total strain measures, $\mathcal{E}^{\mathrm{e}}$ and $\mathcal{E}$, are defined relative to the current configuration $\mathcal{B}$, while plastic strain $\mathcal{E}^{\mathrm{p}}$ is defined relative to the intermediate configuration $\mathcal{B}^{\mathrm{p}}$. Since

$$\mathbf{E} = \mathbf{F}^{T} \cdot \mathcal{E} \cdot \mathbf{F}, \tag{11.2.9}$$

there is a useful relationship

$$\mathbf{E} - \mathbf{E}^{\mathrm{p}} = \mathbf{F}^{T} \cdot \mathcal{E}^{\mathrm{e}} \cdot \mathbf{F}, \tag{11.2.10}$$

which shows that the difference between the total and plastic Lagrangian strain tensors is equal to the strain tensor induced from the Eulerian elastic strain $\boldsymbol{\mathcal{E}}^{\mathrm{e}}$ by the deformation $\mathbf{F}$. Dually, we have

$$\boldsymbol{\mathcal{E}} - \boldsymbol{\mathcal{E}}^{\mathrm{e}} = \mathbf{F}^{-T} \cdot \mathbf{E}^{\mathrm{p}} \cdot \mathbf{F}^{-1}. \tag{11.2.11}$$

11.3. Velocity Gradient and Strain Rates

Consider the velocity gradient in the current configuration at time t, defined by

$$\mathbf{L} = \dot{\mathbf{F}} \cdot \mathbf{F}^{-1}. \tag{11.3.1}$$

The superposed dot designates the material time derivative. By introducing the multiplicative decomposition of deformation gradient (11.1.4), the velocity gradient becomes

$$\mathbf{L} = \dot{\mathbf{F}}^{\mathrm{e}} \cdot \mathbf{F}^{\mathrm{e}-1} + \mathbf{F}^{\mathrm{e}} \cdot \left(\dot{\mathbf{F}}^{\mathrm{p}} \cdot \mathbf{F}^{\mathrm{p}-1}\right) \cdot \mathbf{F}^{\mathrm{e}-1}. \tag{11.3.2}$$

The rate of deformation $\mathbf{D}$ and the spin $\mathbf{W}$ are, respectively, the symmetric and antisymmetric part of $\mathbf{L}$,

$$\mathbf{D} = \left(\dot{\mathbf{F}}^{\mathrm{e}} \cdot \mathbf{F}^{\mathrm{e}-1}\right)_{\mathrm{s}} + \left[\mathbf{F}^{\mathrm{e}} \cdot \left(\dot{\mathbf{F}}^{\mathrm{p}} \cdot \mathbf{F}^{\mathrm{p}-1}\right) \cdot \mathbf{F}^{\mathrm{e}-1}\right]_{\mathrm{s}}, \tag{11.3.3}$$

$$\mathbf{W} = \left(\dot{\mathbf{F}}^{\mathrm{e}} \cdot \mathbf{F}^{\mathrm{e}-1}\right)_{\mathrm{a}} + \left[\mathbf{F}^{\mathrm{e}} \cdot \left(\dot{\mathbf{F}}^{\mathrm{p}} \cdot \mathbf{F}^{\mathrm{p}-1}\right) \cdot \mathbf{F}^{\mathrm{e}-1}\right]_{\mathrm{a}}. \tag{11.3.4}$$

For later purposes, it is convenient to identify the spin

$$\boldsymbol{\omega}^{\mathrm{p}} = \left[\mathbf{F}^{\mathrm{e}} \cdot \left(\dot{\mathbf{F}}^{\mathrm{p}} \cdot \mathbf{F}^{\mathrm{p}-1}\right) \cdot \mathbf{F}^{\mathrm{e}-1}\right]_{\mathrm{a}}. \tag{11.3.5}$$

Since

$$\dot{\mathbf{E}} = \mathbf{F}^{T} \cdot \mathbf{D} \cdot \mathbf{F}, \tag{11.3.6}$$

the following expressions hold for the rates of the introduced Lagrangian strains

$$\dot{\mathbf{E}}^{\mathrm{e}} = \mathbf{F}^{\mathrm{p}-T} \cdot \dot{\mathbf{E}} \cdot \mathbf{F}^{\mathrm{p}-1} - \left[\mathbf{C}^{\mathrm{e}} \cdot \left(\dot{\mathbf{F}}^{\mathrm{p}} \cdot \mathbf{F}^{\mathrm{p}-1}\right)\right]_{\mathrm{s}}, \tag{11.3.7}$$

$$\dot{\mathbf{E}}^{\mathrm{p}} = \mathbf{F}^{\mathrm{p}T} \cdot \left(\dot{\mathbf{F}}^{\mathrm{p}} \cdot \mathbf{F}^{\mathrm{p}-1}\right)_{\mathrm{s}} \cdot \mathbf{F}^{\mathrm{p}}. \tag{11.3.8}$$

These are here cast in terms of the strain rate $\dot{\mathbf{E}}$ and the velocity gradient $\dot{\mathbf{F}}^{\mathrm{p}} \cdot \mathbf{F}^{\mathrm{p}-1}$.

In Section 11.13 it will be shown that the elastic part of the rate of Lagrangian strain,

$$(\dot{\mathbf{E}})^{\mathrm{e}} = \mathbf{\Lambda}_{(1)}^{-1} : \dot{\mathbf{T}}, \tag{11.3.9}$$

where $\mathbf{T}$ is the symmetric Piola–Kirchhoff stress tensor, is in general different from the rate of strain $\dot{\mathbf{E}}^{\mathrm{e}}$. Similarly,

$$(\dot{\mathbf{E}})^{\mathrm{p}} = \dot{\mathbf{E}} - (\dot{\mathbf{E}})^{\mathrm{e}} \neq \dot{\mathbf{E}}^{\mathrm{p}}. \tag{11.3.10}$$

While $(\dot{\mathbf{E}})^{\mathrm{e}}$ and $(\dot{\mathbf{E}})^{\mathrm{p}}$ sum up to give $\dot{\mathbf{E}}$, in general

$$\dot{\mathbf{E}}^{\mathrm{e}} + \dot{\mathbf{E}}^{\mathrm{p}} \neq \dot{\mathbf{E}}. \tag{11.3.11}$$

For the rates of Eulerian strains we have

$$\dot{\boldsymbol{\mathcal{E}}}^{\mathrm{e}} = \left[\mathbf{B}^{\mathrm{e}-1} \cdot \left(\dot{\mathbf{F}}^{\mathrm{e}} \cdot \mathbf{F}^{\mathrm{e}-1}\right)\right]_{\mathrm{s}}, \tag{11.3.12}$$

$$\dot{\boldsymbol{\mathcal{E}}}^{\mathrm{p}} = \mathbf{F}^{\mathrm{e}T} \cdot \dot{\boldsymbol{\mathcal{E}}} \cdot \mathbf{F}^{\mathrm{e}} - \left[\mathbf{B}^{-1} \cdot \left(\dot{\mathbf{F}}^{\mathrm{e}} \cdot \mathbf{F}^{\mathrm{e}-1}\right)\right]_{\mathrm{s}}, \tag{11.3.13}$$

expressed in terms of the strain rate $\dot{\boldsymbol{\mathcal{E}}}$ and the velocity gradient $\dot{\mathbf{F}}^{\mathrm{e}} \cdot \mathbf{F}^{\mathrm{e}-1}$.

11.4. Objectivity Requirements

Upon superimposing a time-dependent rigid-body rotation $\mathbf{Q}$ to the current configuration $\mathcal{B}$, the deformation gradient $\mathbf{F}$ becomes

$$\mathbf{F}^{*} = \mathbf{Q} \cdot \mathbf{F}, \tag{11.4.1}$$

while the elastic and plastic parts $\mathbf{F}^{\mathrm{e}}$ and $\mathbf{F}^{\mathrm{p}}$ change to

$$\mathbf{F}^{\mathrm{e}*} = \mathbf{Q} \cdot \mathbf{F}^{\mathrm{e}} \cdot \hat{\mathbf{Q}}^{T}, \quad \mathbf{F}^{\mathrm{p}*} = \hat{\mathbf{Q}} \cdot \mathbf{F}^{\mathrm{p}}. \tag{11.4.2}$$

The rotation tensor $\hat{\mathbf{Q}}$ is imposed on the intermediate configuration $\mathcal{B}^{\mathrm{p}}$. This rotation depends on the rotation $\mathbf{Q}$ of the current configuration, and on the definition of the intermediate configuration used in the particular constitutive model (Lubarda, 1991 a). For example, if the intermediate configuration is defined to be isoclinic, then necessarily

$$\hat{\mathbf{Q}} = \mathbf{I}. \tag{11.4.3}$$

If the intermediate configuration is obtained from the current configuration by destressing without rotation $\mathbf{F}^{\mathrm{e}} = \mathbf{V}^{\mathrm{e}}$, then

$$\hat{\mathbf{Q}} = \mathbf{Q}, \tag{11.4.4}$$

in order that $\mathbf{F}^{\mathrm{e}}$ remains symmetric. Thus, with $\hat{\mathbf{Q}}$ appropriately specified in any particular case, the following transformation rules apply

$$\mathbf{V}^{\mathrm{e}*} = \mathbf{Q} \cdot \mathbf{V}^{\mathrm{e}} \cdot \mathbf{Q}^T, \quad \mathbf{V}^{\mathrm{p}*} = \hat{\mathbf{Q}} \cdot \mathbf{V}^{\mathrm{p}} \cdot \hat{\mathbf{Q}}^T, \tag{11.4.5}$$

$$\mathbf{R}^{\mathrm{e}*} = \mathbf{Q} \cdot \mathbf{R}^{\mathrm{e}} \cdot \hat{\mathbf{Q}}^T, \quad \mathbf{R}^{\mathrm{p}*} = \hat{\mathbf{Q}} \cdot \mathbf{R}^{\mathrm{p}}, \tag{11.4.6}$$

$$\mathbf{U}^{\mathrm{e}*} = \hat{\mathbf{Q}} \cdot \mathbf{U}^{\mathrm{e}} \cdot \hat{\mathbf{Q}}^T, \quad \mathbf{U}^{\mathrm{p}*} = \mathbf{U}^{\mathrm{p}}, \tag{11.4.7}$$

$$\mathbf{B}^{\mathrm{e}*} = \mathbf{Q} \cdot \mathbf{B}^{\mathrm{e}} \cdot \mathbf{Q}^T, \quad \mathbf{B}^{\mathrm{p}*} = \hat{\mathbf{Q}} \cdot \mathbf{B}^{\mathrm{p}} \cdot \hat{\mathbf{Q}}^T, \tag{11.4.8}$$

$$\mathbf{C}^{\mathrm{e}*} = \hat{\mathbf{Q}} \cdot \mathbf{C}^{\mathrm{e}} \cdot \hat{\mathbf{Q}}^T, \quad \mathbf{C}^{\mathrm{p}*} = \mathbf{C}^{\mathrm{p}}, \tag{11.4.9}$$

$$\mathbf{E}^{\mathrm{e}*} = \hat{\mathbf{Q}} \cdot \mathbf{E}^{\mathrm{e}} \cdot \hat{\mathbf{Q}}^T, \quad \mathbf{E}^{\mathrm{p}*} = \mathbf{E}^{\mathrm{p}}, \tag{11.4.10}$$

$$\boldsymbol{\mathcal{E}}^{\mathrm{e}*} = \mathbf{Q} \cdot \boldsymbol{\mathcal{E}}^{\mathrm{e}} \cdot \mathbf{Q}^T, \quad \boldsymbol{\mathcal{E}}^{\mathrm{p}*} = \hat{\mathbf{Q}} \cdot \boldsymbol{\mathcal{E}}^{\mathrm{p}} \cdot \hat{\mathbf{Q}}^T. \tag{11.4.11}$$

The transformation rules for the velocity gradients associated with $\mathbf{F}^{\mathrm{e}}$ and $\mathbf{F}^{\mathrm{p}}$ are

$$\dot{\mathbf{F}}^{\mathrm{e}*} \cdot \mathbf{F}^{\mathrm{e}*-1} = \dot{\mathbf{Q}} \cdot \mathbf{Q}^{-1} + \mathbf{Q} \cdot \left(\dot{\mathbf{F}}^{\mathrm{e}} \cdot \mathbf{F}^{\mathrm{e}-1}\right) \cdot \mathbf{Q}^T - \mathbf{F}^{\mathrm{e}*} \cdot \left(\dot{\hat{\mathbf{Q}}} \cdot \hat{\mathbf{Q}}^{-1}\right) \cdot \mathbf{F}^{\mathrm{e}*-1}, \tag{11.4.12}$$

$$\dot{\mathbf{F}}^{\mathrm{p}*} \cdot \mathbf{F}^{\mathrm{p}*-1} = \dot{\hat{\mathbf{Q}}} \cdot \hat{\mathbf{Q}}^{-1} + \hat{\mathbf{Q}} \cdot \left(\dot{\mathbf{F}}^{\mathrm{p}} \cdot \mathbf{F}^{\mathrm{p}-1}\right) \cdot \hat{\mathbf{Q}}^{\mathbf{T}}. \tag{11.4.13}$$

The corresponding Lagrangian and Eulerian strain rates transform according to

$$\dot{\mathbf{E}}^{\mathrm{e}*} = \hat{\mathbf{Q}} \cdot \dot{\mathbf{E}}^{\mathrm{e}} \cdot \hat{\mathbf{Q}}^T + \left(\dot{\hat{\mathbf{Q}}} \cdot \hat{\mathbf{Q}}^{-1}\right) \cdot \mathbf{E}^{\mathrm{e}*} - \mathbf{E}^{\mathrm{e}*} \cdot \left(\dot{\hat{\mathbf{Q}}} \cdot \hat{\mathbf{Q}}^{-1}\right), \tag{11.4.14}$$

$$\dot{\mathbf{E}}^{\mathrm{p}*} = \dot{\mathbf{E}}^{\mathrm{p}}, \tag{11.4.15}$$

$$\dot{\boldsymbol{\mathcal{E}}}^{\mathrm{e}*} = \mathbf{Q} \cdot \dot{\boldsymbol{\mathcal{E}}}^{\mathrm{e}} \cdot \mathbf{Q}^T + \left(\dot{\mathbf{Q}} \cdot \mathbf{Q}^{-1}\right) \cdot \boldsymbol{\mathcal{E}}^{\mathrm{e}*} - \boldsymbol{\mathcal{E}}^{\mathrm{e}*} \cdot \left(\dot{\mathbf{Q}} \cdot \mathbf{Q}^{-1}\right), \tag{11.4.16}$$

$$\dot{\boldsymbol{\mathcal{E}}}^{\mathrm{p}*} = \hat{\mathbf{Q}} \cdot \dot{\boldsymbol{\mathcal{E}}}^{\mathrm{p}} \cdot \hat{\mathbf{Q}}^T + \left(\dot{\hat{\mathbf{Q}}} \cdot \hat{\mathbf{Q}}^{-1}\right) \cdot \boldsymbol{\mathcal{E}}^{\mathrm{p}*} - \boldsymbol{\mathcal{E}}^{\mathrm{p}*} \cdot \left(\dot{\hat{\mathbf{Q}}} \cdot \hat{\mathbf{Q}}^{-1}\right). \tag{11.4.17}$$

Finally, the transformation rules for the total velocity gradient, and the total strain rates are

$$\mathbf{L}^* = \dot{\mathbf{Q}} \cdot \mathbf{Q}^{-1} + \mathbf{Q} \cdot \mathbf{L} \cdot \mathbf{Q}^T, \tag{11.4.18}$$

$$\dot{\mathbf{E}}^* = \dot{\mathbf{E}}, \tag{11.4.19}$$

$$\dot{\boldsymbol{\mathcal{E}}}^* = \mathbf{Q} \cdot \dot{\boldsymbol{\mathcal{E}}} \cdot \mathbf{Q}^T + \left(\dot{\mathbf{Q}} \cdot \mathbf{Q}^{-1}\right) \cdot \boldsymbol{\mathcal{E}}^* - \boldsymbol{\mathcal{E}}^* \cdot \left(\dot{\mathbf{Q}} \cdot \mathbf{Q}^{-1}\right), \tag{11.4.20}$$

as previously discussed in Section 2.9.

The objectivity requirements that need to be imposed in the theory of elastoplasticity based on the multiplicative decomposition of deformation gradient have been extensively discussed in the literature. Some of the representative references include Naghdi and Trapp (1974), Lubarda and Lee (1981), Casey and Naghdi (1981), Dashner (1986a,b), Casey (1987), Dafalias (1987,1988), Naghdi (1990), and Xiao, Bruhns, and Meyers (2000).

11.5. Jaumann Derivative of Elastic Deformation Gradient

In the context of the multiplicative decomposition of deformation gradient based on the intermediate configuration, it is convenient to introduce a particular type of the Jaumann derivative of elastic deformation gradient $\mathbf{F}^{\mathrm{e}}$. This is defined as the time derivative observed in two rotating coordinate systems, one rotating with the spin $\boldsymbol{\Omega}$ in the current configuration $\mathcal{B}$ and the other rotating with the spin $\boldsymbol{\Omega}^{\mathrm{p}}$ in the intermediate configuration $\mathcal{B}^{\mathrm{p}}$, such that (Lubarda, 1991a; Lubarda and Shih, 1994)

$$\overset{\bullet}{\mathbf{F}}{}^{\mathrm{e}} = \dot{\mathbf{F}}^{\mathrm{e}} - \boldsymbol{\Omega} \cdot \mathbf{F}^{\mathrm{e}} + \mathbf{F}^{\mathrm{e}} \cdot \boldsymbol{\Omega}^{\mathrm{p}}. \tag{11.5.1}$$

The spin tensors $\boldsymbol{\Omega}$ and $\boldsymbol{\Omega}^{\mathrm{p}}$ are at this point unspecified. They can be different or equal to each other, depending on the selected intermediate configuration and the intended application (see also Section 2.8). In any case, they transform under rigid-body rotations $\mathbf{Q}$ and $\hat{\mathbf{Q}}$ of the current and intermediate configurations according to

$$\boldsymbol{\Omega}^* = \dot{\mathbf{Q}} \cdot \mathbf{Q}^{-1} + \mathbf{Q} \cdot \boldsymbol{\Omega} \cdot \mathbf{Q}^T, \quad \boldsymbol{\Omega}^{\mathrm{p}*} = \dot{\hat{\mathbf{Q}}} \cdot \hat{\mathbf{Q}}^{-1} + \hat{\mathbf{Q}} \cdot \boldsymbol{\Omega}^{\mathrm{p}} \cdot \hat{\mathbf{Q}}^{\mathrm{T}}. \tag{11.5.2}$$

The Jaumann derivatives of $\mathbf{V}^{\mathrm{e}}$ and $\mathbf{R}^{\mathrm{e}}$, corresponding to Eq. (11.5.1), are

$$\overset{\bullet}{\mathbf{V}}{}^{\mathrm{e}} = \dot{\mathbf{V}}^{\mathrm{e}} - \boldsymbol{\Omega} \cdot \mathbf{V}^{\mathrm{e}} + \mathbf{V}^{\mathrm{e}} \cdot \boldsymbol{\Omega}, \quad \overset{\bullet}{\mathbf{R}}{}^{\mathrm{e}} = \dot{\mathbf{R}}^{\mathrm{e}} - \boldsymbol{\Omega} \cdot \mathbf{R}^{\mathrm{e}} + \mathbf{R}^{\mathrm{e}} \cdot \boldsymbol{\Omega}^{\mathrm{p}}, \tag{11.5.3}$$

while those of $\mathbf{B}^{\mathrm{e}}$ and $\mathbf{C}^{\mathrm{e}}$ are

$$\overset{\bullet}{\mathbf{B}}{}^{\mathrm{e}} = \dot{\mathbf{B}}^{\mathrm{e}} - \boldsymbol{\Omega} \cdot \mathbf{B}^{\mathrm{e}} + \mathbf{B}^{\mathrm{e}} \cdot \boldsymbol{\Omega}, \quad \overset{\bullet}{\mathbf{C}}{}^{\mathrm{e}} = \dot{\mathbf{C}}^{\mathrm{e}} - \boldsymbol{\Omega}^{\mathrm{p}} \cdot \mathbf{C}^{\mathrm{e}} + \mathbf{C}^{\mathrm{e}} \cdot \boldsymbol{\Omega}^{\mathrm{p}}. \tag{11.5.4}$$

It is easily verified that

$$\overset{\bullet}{\mathbf{F}}^{\mathrm{e}} \cdot \mathbf{F}^{\mathrm{e}-1} = \dot{\mathbf{F}}^{\mathrm{e}} \cdot \mathbf{F}^{\mathrm{e}-1} + \mathbf{F}^{\mathrm{e}} \cdot \boldsymbol{\Omega}^{\mathrm{p}} \cdot \mathbf{F}^{\mathrm{e}-1} - \boldsymbol{\Omega}, \tag{11.5.5}$$

$$\overset{\bullet}{\mathbf{V}}^{\mathrm{e}} \cdot \mathbf{V}^{\mathrm{e}-1} = \dot{\mathbf{V}}^{\mathrm{e}} \cdot \mathbf{V}^{\mathrm{e}-1} + \mathbf{V}^{\mathrm{e}} \cdot \boldsymbol{\Omega} \cdot \mathbf{V}^{\mathrm{e}-1} - \boldsymbol{\Omega}. \tag{11.5.6}$$

Under rigid-body rotations $\mathbf{Q}$ and $\hat{\mathbf{Q}}$, the introduced Jaumann derivatives transform as

$$\overset{\bullet}{\mathbf{F}}^{\mathrm{e}*} = \mathbf{Q} \cdot \overset{\bullet}{\mathbf{F}}^{\mathrm{e}} \cdot \hat{\mathbf{Q}}^T, \quad \overset{\bullet}{\mathbf{V}}^{\mathrm{e}*} = \mathbf{Q} \cdot \overset{\bullet}{\mathbf{V}}^{\mathrm{e}} \cdot \mathbf{Q}^T, \quad \overset{\bullet}{\mathbf{R}}^{\mathrm{e}*} = \mathbf{Q} \cdot \overset{\bullet}{\mathbf{R}}^{\mathrm{e}} \cdot \hat{\mathbf{Q}}^T, \tag{11.5.7}$$

$$\overset{\bullet}{\mathbf{B}}^{\mathrm{e}*} = \mathbf{Q} \cdot \overset{\bullet}{\mathbf{B}}^{\mathrm{e}} \cdot \mathbf{Q}^T, \quad \overset{\bullet}{\mathbf{C}}^{\mathrm{e}*} = \hat{\mathbf{Q}} \cdot \overset{\bullet}{\mathbf{C}}^{\mathrm{e}} \cdot \hat{\mathbf{Q}}^T. \tag{11.5.8}$$

Consequently,

$$\overset{\bullet}{\mathbf{F}}^{\mathrm{e}*} \cdot \mathbf{F}^{\mathrm{e}*-1} = \mathbf{Q} \cdot \left(\overset{\bullet}{\mathbf{F}}^{\mathrm{e}} \cdot \mathbf{F}^{\mathrm{e}-1} \right) \cdot \mathbf{Q}^T, \tag{11.5.9}$$

and likewise for the corresponding quantities associated with the Jaumann derivatives of $\mathbf{V}^{\mathrm{e}}$ and $\mathbf{R}^{\mathrm{e}}$.

11.6. Partition of Elastoplastic Rate of Deformation

In this section it is assumed that the material is elastically isotropic in its initial undeformed state, and that plastic deformation does not affect its elastic properties. The elastic response from $\mathcal{B}^{\mathrm{p}}$ to $\mathcal{B}$ is then independent of the rotation superposed to the intermediate configuration, and is given by

$$\boldsymbol{\tau} = \mathbf{F}^{\mathrm{e}} \cdot \frac{\partial \Psi^{\mathrm{e}}(\mathbf{E}^{\mathrm{e}})}{\partial \mathbf{E}^{\mathrm{e}}} \cdot \mathbf{F}^{\mathrm{e}T}. \tag{11.6.1}$$

The elastic strain energy per unit unstressed volume, Ψ^{e}, is an isotropic function of the Lagrangian strain $\mathbf{E}^{\mathrm{e}} = \left(\mathbf{F}^{\mathrm{e}T} \cdot \mathbf{F}^{\mathrm{e}} - \mathbf{I}\right)/2$. Plastic deformation is assumed to be incompressible ($\det \mathbf{F}^{\mathrm{e}} = \det \mathbf{F}$), so that $\boldsymbol{\tau} = (\det \mathbf{F})\boldsymbol{\sigma}$ is the Kirchhoff stress (the Cauchy stress $\boldsymbol{\sigma}$ weighted by $\det \mathbf{F}$).

By differentiating Eq. (11.6.1), we obtain

$$\dot{\boldsymbol{\tau}} - \left(\dot{\mathbf{F}}^{\mathrm{e}} \cdot \mathbf{F}^{\mathrm{e}-1}\right) \cdot \boldsymbol{\tau} - \boldsymbol{\tau} \cdot \left(\dot{\mathbf{F}}^{\mathrm{e}} \cdot \mathbf{F}^{\mathrm{e}-1}\right)^T = \boldsymbol{\mathcal{L}}_{(1)} : \left(\dot{\mathbf{F}}^{\mathrm{e}} \cdot \mathbf{F}^{\mathrm{e}-1}\right)_{\mathrm{s}}. \tag{11.6.2}$$

The subscript (1) is attached to the elastic moduli tensor $\boldsymbol{\mathcal{L}}_{(1)}$ to make the contact with the notation used in Section 6.2, e.g., Eq. (6.2.4). The rectangular components of $\boldsymbol{\mathcal{L}}_{(1)}$ are

$$\mathcal{L}_{ijkl}^{(1)} = F_{iM}^{\mathrm{e}} F_{jN}^{\mathrm{e}} \frac{\partial^2 \Psi^{\mathrm{e}}}{\partial E_{MN}^{\mathrm{e}} \partial E_{PQ}^{\mathrm{e}}} F_{kP}^{\mathrm{e}} F_{lQ}^{\mathrm{e}}. \tag{11.6.3}$$

Equation (11.6.2) can be equivalently written, in terms of the Jaumann derivative of $\boldsymbol{\tau}$ with respect to spin $\left(\dot{\mathbf{F}}^{\mathrm{e}}\cdot\mathbf{F}^{\mathrm{e}-1}\right)_{\mathrm{a}}$, as

$$\dot{\boldsymbol{\tau}}-\left(\dot{\mathbf{F}}^{\mathrm{e}}\cdot\mathbf{F}^{\mathrm{e}-1}\right)_{\mathrm{a}}\cdot\boldsymbol{\tau}+\boldsymbol{\tau}\cdot\left(\dot{\mathbf{F}}^{\mathrm{e}}\cdot\mathbf{F}^{\mathrm{e}-1}\right)_{\mathrm{a}}=\boldsymbol{\mathcal{L}}_{(0)}:\left(\dot{\mathbf{F}}^{\mathrm{e}}\cdot\mathbf{F}^{\mathrm{e}-1}\right)_{\mathrm{s}}. \tag{11.6.4}$$

The elastic moduli tensor $\boldsymbol{\mathcal{L}}_{(0)}$ has the components

$$\mathcal{L}^{(0)}_{ijkl}=\mathcal{L}^{(1)}_{ijkl}+\frac{1}{2}\left(\tau_{ik}\delta_{jl}+\tau_{jk}\delta_{il}+\tau_{il}\delta_{jk}+\tau_{jl}\delta_{ik}\right), \tag{11.6.5}$$

as in Eq. (6.2.15).

The elastic deformation gradient $\mathbf{F}^{\mathrm{e}}$ is defined relative to the intermediate configuration, which is changing during the ongoing elastoplastic deformation. This causes two difficulties in the identification of the elastic rate of deformation $\mathbf{D}^{\mathrm{e}}$. First, since $\mathbf{F}^{\mathrm{e}}$ and $\mathbf{F}^{\mathrm{p}}$ are specified only to within an arbitrary rotation $\hat{\mathbf{Q}}$, the velocity gradient $\dot{\mathbf{F}}^{\mathrm{e}}\cdot\mathbf{F}^{\mathrm{e}-1}$ and its symmetric and antisymmetric parts are not unique. Secondly, the deforming intermediate configuration also makes contribution to the elastic rate of deformation, so that this is not in general given only by $\left(\dot{\mathbf{F}}^{\mathrm{e}}\cdot\mathbf{F}^{\mathrm{e}-1}\right)_{\mathrm{s}}$. To overcome these difficulties, we resort to kinetic definition of the elastic strain increment $\mathbf{D}^{\mathrm{e}}\,\mathrm{d}t$, which is a reversible part of the total strain increment $\mathbf{D}\,\mathrm{d}t$ recovered upon loading–unloading cycle of the stress increment $\overset{\circ}{\boldsymbol{\tau}}\mathrm{d}t$. The Jaumann derivative of the Kirchhoff stress relative to material spin $\mathbf{W}$ is $\overset{\circ}{\boldsymbol{\tau}}$. Thus, we define

$$\mathbf{D}^{\mathrm{e}}=\boldsymbol{\mathcal{L}}_{(0)}^{-1}:\overset{\circ}{\boldsymbol{\tau}},\quad \overset{\circ}{\boldsymbol{\tau}}=\dot{\boldsymbol{\tau}}-\mathbf{W}\cdot\boldsymbol{\tau}+\boldsymbol{\tau}\cdot\mathbf{W}. \tag{11.6.6}$$

The remaining part of the total rate of deformation,

$$\mathbf{D}^{\mathrm{p}}=\mathbf{D}-\mathbf{D}^{\mathrm{e}}, \tag{11.6.7}$$

is the plastic part, which gives a residual strain increment left upon considered infinitesimal cycle of stress. When the material obeys Ilyushin's postulate, so defined plastic rate of deformation $\mathbf{D}^{\mathrm{p}}$ is codirectional with the outward normal to a locally smooth yield surface in the Cauchy stress space.

Therefore, to identify in Eq. (11.6.4) the elastic strain rate, according to the kinetic definition (11.6.6), we use Eq. (11.3.4) to eliminate $\left(\dot{\mathbf{F}}^{\mathrm{e}}\cdot\mathbf{F}^{\mathrm{e}-1}\right)_{\mathrm{a}}$ and obtain

$$\overset{\circ}{\boldsymbol{\tau}}=\boldsymbol{\mathcal{L}}_{(0)}:\left(\dot{\mathbf{F}}^{\mathrm{e}}\cdot\mathbf{F}^{\mathrm{e}-1}\right)_{\mathrm{s}}-\boldsymbol{\omega}^{\mathrm{p}}\cdot\boldsymbol{\tau}+\boldsymbol{\tau}\cdot\boldsymbol{\omega}^{\mathrm{p}}. \tag{11.6.8}$$

The spin $\boldsymbol{\omega}^{\mathrm{p}}$ is defined by Eq. (11.3.5). Consequently, the elastic rate of deformation is

$$\mathbf{D}^{\mathrm{e}} = \left(\dot{\mathbf{F}}^{\mathrm{e}} \cdot \mathbf{F}^{\mathrm{e}-1}\right)_{\mathrm{s}} - \boldsymbol{\mathcal{L}}_{(0)}^{-1} : (\boldsymbol{\omega}^{\mathrm{p}} \cdot \boldsymbol{\tau} - \boldsymbol{\tau} \cdot \boldsymbol{\omega}^{\mathrm{p}}) . \tag{11.6.9}$$

From Eq. (11.6.7), the corresponding plastic rate of deformation is given by

$$\mathbf{D}^{\mathrm{p}} = \left[\mathbf{F}^{\mathrm{e}} \cdot \left(\dot{\mathbf{F}}^{\mathrm{p}} \cdot \mathbf{F}^{\mathrm{p}-1}\right) \cdot \mathbf{F}^{\mathrm{e}-1}\right]_{\mathrm{s}} + \boldsymbol{\mathcal{L}}_{(0)}^{-1} : (\boldsymbol{\omega}^{\mathrm{p}} \cdot \boldsymbol{\tau} - \boldsymbol{\tau} \cdot \boldsymbol{\omega}^{\mathrm{p}}) . \tag{11.6.10}$$

Since $\boldsymbol{\mathcal{L}}_{(0)}^{-1}$ and $\overset{\circ}{\boldsymbol{\tau}}$ in (11.6.6) are independent of a superposed rotation to the intermediate configuration, Eq. (11.6.9) specifies $\mathbf{D}^{\mathrm{e}}$ uniquely. In contrast, its constituents, $\left(\dot{\mathbf{F}}^{\mathrm{e}} \cdot \mathbf{F}^{\mathrm{e}-1}\right)_{\mathrm{s}}$ and the term associated with the spin $\boldsymbol{\omega}^{\mathrm{p}}$, do depend on the choice of the intermediate configuration. Similar remarks apply to plastic rate of deformation $\mathbf{D}^{\mathrm{p}}$ in its representation (11.6.10).

As we have shown, the right hand side of (11.6.9) is the correct expression for the elastic rate of deformation, and not $\left(\dot{\mathbf{F}}^{\mathrm{e}} \cdot \mathbf{F}^{\mathrm{e}-1}\right)_{\mathrm{s}}$ alone. Only if the intermediate configuration (i.e., the rotation $\mathbf{R}^{\mathrm{e}}$ during destressing program) is chosen such that the spin $\boldsymbol{\omega}^{\mathrm{p}}$ vanishes,

$$\boldsymbol{\omega}^{\mathrm{p}} = \left[\mathbf{F}^{\mathrm{e}} \cdot \left(\dot{\mathbf{F}}^{\mathrm{p}} \cdot \mathbf{F}^{\mathrm{p}-1}\right) \cdot \mathbf{F}^{\mathrm{e}-1}\right]_{\mathrm{a}} = \mathbf{0}, \tag{11.6.11}$$

the rate of deformation $\left(\dot{\mathbf{F}}^{\mathrm{e}} \cdot \mathbf{F}^{\mathrm{e}-1}\right)_{\mathrm{s}}$ is exactly equal to $\mathbf{D}^{\mathrm{e}}$. Within the framework under discussion, this choice of the spin represents purely geometric (kinematic) specification of the intermediate configuration. It is not a constitutive assumption and has no consequences on Eq. (11.6.9). We could just as well define the intermediate configuration by requiring that the spin $\left(\dot{\mathbf{F}}^{\mathrm{e}} \cdot \mathbf{F}^{\mathrm{e}-1}\right)_{\mathrm{a}}$ vanishes identically. In this case,

$$\boldsymbol{\omega}^{\mathrm{p}} = \left[\mathbf{F}^{\mathrm{e}} \cdot \left(\dot{\mathbf{F}}^{\mathrm{p}} \cdot \mathbf{F}^{\mathrm{p}-1}\right) \cdot \mathbf{F}^{\mathrm{e}-1}\right]_{\mathrm{a}} = \mathbf{W}. \tag{11.6.12}$$

The end result is still Eq. (11.6.9), as can be checked by inspection.

The partition of $\mathbf{D}$ into its elastic and plastic parts within the framework of the multiplicative decomposition has been a topic of active research and discussion. Some of the representative references include Lee (1969), Freund (1970), Kratochvil (1973), Kleiber (1975), Nemat-Nasser (1979,1982), Lubarda and Lee (1981), Lee (1981,1985), Sidoroff (1982), Dafalias (1987), and Lubarda and Shih (1994).

We note that the second part of the rate of deformation $\mathbf{D}^{\mathrm{e}}$, in its representation (11.6.9), makes no contribution to elastic work. This follows by observing that, in view of elastic isotropy, the part of the rate of deformation

$$\mathcal{L}_{(0)}^{-1} : (\boldsymbol{\omega}^{\mathrm{p}} \cdot \boldsymbol{\tau} - \boldsymbol{\tau} \cdot \boldsymbol{\omega}^{\mathrm{p}}) \tag{11.6.13}$$

has its principal directions parallel to those of the associated stress rate $(\boldsymbol{\omega}^{\mathrm{p}} \cdot \boldsymbol{\tau} - \boldsymbol{\tau} \cdot \boldsymbol{\omega}^{\mathrm{p}})$. Since direction of this stress rate is normal to $\boldsymbol{\tau}$, their trace is zero, hence

$$\dot{\Psi}^{\mathrm{e}} = \boldsymbol{\tau} : \mathbf{D}^{\mathrm{e}} = \boldsymbol{\tau} : \left(\dot{\mathbf{F}}^{\mathrm{e}} \cdot \mathbf{F}^{\mathrm{e}-1}\right)_{\mathrm{s}}. \tag{11.6.14}$$

11.7. Analysis of Elastic Rate of Deformation

We present an alternative derivation of the expression for the elastic rate of deformation of elastically isotropic materials, which gives additional insight in the kinematics of elastoplastic deformation and the partitioning of the rate of deformation. We show that $\mathbf{D}^{\mathrm{e}}$ can be expressed as

$$\mathbf{D}^{\mathrm{e}} = \left(\overset{\bullet}{\mathbf{F}}{}^{\mathrm{e}} \cdot \mathbf{F}^{\mathrm{e}-1}\right)_{\mathrm{s}} = \left(\dot{\mathbf{F}}^{\mathrm{e}} \cdot \mathbf{F}^{\mathrm{e}-1}\right)_{\mathrm{s}} + \left(\mathbf{F}^{\mathrm{e}} \cdot \boldsymbol{\Omega}^{\mathrm{p}} \cdot \mathbf{F}^{\mathrm{e}-1}\right)_{\mathrm{s}}. \tag{11.7.1}$$

The Jaumann derivative $\overset{\bullet}{\mathbf{F}}{}^{\mathrm{e}}$ is defined by Eq. (11.5.1) with $\boldsymbol{\Omega} = \boldsymbol{\Omega}^{\mathrm{p}}$, i.e.,

$$\overset{\bullet}{\mathbf{F}}{}^{\mathrm{e}} = \dot{\mathbf{F}}^{\mathrm{e}} - \boldsymbol{\Omega}^{\mathrm{p}} \cdot \mathbf{F}^{\mathrm{e}} + \mathbf{F}^{\mathrm{e}} \cdot \boldsymbol{\Omega}^{\mathrm{p}}. \tag{11.7.2}$$

This represents the rate of $\mathbf{F}^{\mathrm{e}}$ observed in the coordinate systems that rotate with the spin $\boldsymbol{\Omega}^{\mathrm{p}}$ in both current and intermediate configurations. The spin $\boldsymbol{\Omega}^{\mathrm{p}}$ is defined as the solution of the matrix equation

$$\left(\dot{\mathbf{F}}^{\mathrm{e}} \cdot \mathbf{F}^{\mathrm{e}-1}\right)_{\mathrm{a}} + \left(\mathbf{F}^{\mathrm{e}} \cdot \boldsymbol{\Omega}^{\mathrm{p}} \cdot \mathbf{F}^{\mathrm{e}-1}\right)_{\mathrm{a}} = \mathbf{W}. \tag{11.7.3}$$

In proof, the application of the Jaumann derivative with respect to spin $\boldsymbol{\Omega}^{\mathrm{p}}$ to Eq. (11.6.1) gives

$$\overset{\bullet}{\boldsymbol{\tau}} = \left(\overset{\bullet}{\mathbf{F}}{}^{\mathrm{e}} \cdot \mathbf{F}^{\mathrm{e}-1}\right) \cdot \boldsymbol{\tau} + \boldsymbol{\tau} \cdot \left(\overset{\bullet}{\mathbf{F}}{}^{\mathrm{e}} \cdot \mathbf{F}^{\mathrm{e}-1}\right)^{T} + \mathbf{F}^{\mathrm{e}} \cdot \left(\frac{\partial^2 \Psi^{\mathrm{e}}}{\partial \mathbf{E}^{\mathrm{e}} \otimes \partial \mathbf{E}^{\mathrm{e}}} : \overset{\bullet}{\mathbf{E}}{}^{\mathrm{e}}\right) \cdot \mathbf{F}^{\mathrm{e}T}, \tag{11.7.4}$$

where

$$\overset{\bullet}{\boldsymbol{\tau}} = \dot{\boldsymbol{\tau}} - \boldsymbol{\Omega}^{\mathrm{p}} \cdot \boldsymbol{\tau} + \boldsymbol{\tau} \cdot \boldsymbol{\Omega}^{\mathrm{p}}, \quad \overset{\bullet}{\mathbf{E}}{}^{\mathrm{e}} = \mathbf{F}^{\mathrm{e}T} \cdot \left(\overset{\bullet}{\mathbf{F}}{}^{\mathrm{e}} \cdot \mathbf{F}^{\mathrm{e}-1}\right)_{\mathrm{s}} \cdot \mathbf{F}^{\mathrm{e}}. \tag{11.7.5}$$

Therefore, if Eqs. (11.7.1) and (11.7.3) hold, so that

$$\dot{\mathbf{F}}^{\mathrm{e}} \cdot \mathbf{F}^{\mathrm{e}-1} + \mathbf{F}^{\mathrm{e}} \cdot \boldsymbol{\Omega}^{\mathrm{p}} \cdot \mathbf{F}^{\mathrm{e}-1} = \mathbf{D}^{\mathrm{e}} + \mathbf{W}, \tag{11.7.6}$$

$$\overset{\bullet}{\mathbf{F}}{}^{\mathrm{e}} \cdot \mathbf{F}^{\mathrm{e}-1} = \mathbf{D}^{\mathrm{e}} + \mathbf{W} - \boldsymbol{\Omega}^{\mathrm{p}}, \tag{11.7.7}$$

the substitution into Eq. (11.7.4) yields

$$\overset{\circ}{\boldsymbol{\tau}} = \boldsymbol{\mathcal{L}}_{(0)} : \mathbf{D}^{\mathrm{e}}, \quad \mathbf{D}^{\mathrm{e}} = \left(\overset{\bullet}{\mathbf{F}^{\mathrm{e}}} \cdot \mathbf{F}^{\mathrm{e}-1}\right)_{\mathrm{s}}. \tag{11.7.8}$$

The two contributions to the elastic rate of deformation $\mathbf{D}^{\mathrm{e}}$ in Eq. (11.7.1) both depend on the choice of the intermediate configuration, i.e., the elastic rotation $\mathbf{R}^{\mathrm{e}}$ of the destressing program, but their sum giving $\mathbf{D}^{\mathrm{e}}$ does not. If elastic destressing is performed without rotation ($\mathbf{R}^{\mathrm{e}} = \mathbf{I}$), the spin $\boldsymbol{\Omega}^{\mathrm{p}} = \boldsymbol{\Omega}^{\mathrm{p}}_{I}$ is the solution of

$$\left(\dot{\mathbf{V}}^{\mathrm{e}} \cdot \mathbf{V}^{\mathrm{e}-1}\right)_{\mathrm{a}} + \left(\mathbf{V}^{\mathrm{e}} \cdot \boldsymbol{\Omega}^{\mathrm{p}}_{I} \cdot \mathbf{V}^{\mathrm{e}-1}\right)_{\mathrm{a}} = \mathbf{W}. \tag{11.7.9}$$

This defines the spin $\boldsymbol{\Omega}^{\mathrm{p}}_{I}$ uniquely in terms of $\mathbf{W}$, $\mathbf{V}^{\mathrm{e}}$ and $\dot{\mathbf{V}}^{\mathrm{e}}$. The expression for the elastic rate of deformation (11.7.1) is in this case

$$\mathbf{D}^{\mathrm{e}} = \left(\overset{\bullet}{\mathbf{V}^{\mathrm{e}}} \cdot \mathbf{V}^{\mathrm{e}-1}\right)_{\mathrm{s}} = \left(\dot{\mathbf{V}}^{\mathrm{e}} \cdot \mathbf{V}^{\mathrm{e}-1}\right)_{\mathrm{s}} + \left(\mathbf{V}^{\mathrm{e}} \cdot \boldsymbol{\Omega}^{\mathrm{p}}_{I} \cdot \mathbf{V}^{\mathrm{e}-1}\right)_{\mathrm{s}}. \tag{11.7.10}$$

The first term on the right-hand side represents the contribution to $\mathbf{D}^{\mathrm{e}}$ from the elastic stretching rate $\left(\dot{\mathbf{V}}^{\mathrm{e}} \cdot \mathbf{V}^{\mathrm{e}-1}\right)_{\mathrm{s}}$, while the second depends on the spin $\boldsymbol{\Omega}^{\mathrm{p}}_{I}$ and accounts for the effects of the deforming and rotating intermediate configuration.

Since

$$\dot{\mathbf{F}}^{\mathrm{e}} \cdot \mathbf{F}^{\mathrm{e}-1} + \mathbf{F}^{\mathrm{e}} \cdot \boldsymbol{\Omega}^{\mathrm{p}} \cdot \mathbf{F}^{\mathrm{e}-1} = \dot{\mathbf{V}}^{\mathrm{e}} \cdot \mathbf{V}^{\mathrm{e}-1} + \mathbf{V}^{\mathrm{e}} \cdot \boldsymbol{\Omega}^{\mathrm{p}}_{I} \cdot \mathbf{V}^{\mathrm{e}-1}, \tag{11.7.11}$$

and

$$\dot{\mathbf{F}}^{\mathrm{e}} \cdot \mathbf{F}^{\mathrm{e}-1} = \dot{\mathbf{V}}^{\mathrm{e}} \cdot \mathbf{V}^{\mathrm{e}-1} + \mathbf{V}^{\mathrm{e}} \cdot \left(\dot{\mathbf{R}}^{\mathrm{e}} \cdot \mathbf{R}^{\mathrm{e}-1}\right) \cdot \mathbf{V}^{\mathrm{e}-1}, \tag{11.7.12}$$

it follows that, for any other choice of the rotation $\mathbf{R}^{\mathrm{e}}$, the corresponding spin in the expression for the elastic rate of deformation (11.7.1) is

$$\boldsymbol{\Omega}^{\mathrm{p}} = \mathbf{R}^{\mathrm{e}T} \cdot \left(\boldsymbol{\Omega}^{\mathrm{p}}_{I} - \dot{\mathbf{R}}^{\mathrm{e}} \cdot \mathbf{R}^{\mathrm{e}-1}\right) \cdot \mathbf{R}^{\mathrm{e}}. \tag{11.7.13}$$

The expression for the elastic rate of deformation in Eq. (11.7.1) involves only kinematic quantities ($\mathbf{F}^{\mathrm{e}}$ and $\boldsymbol{\Omega}^{\mathrm{p}}$), while the previously derived expression (11.6.9) involves both kinematic and kinetic quantities. Clearly, there is a connection

$$\left(\mathbf{F}^{\mathrm{e}} \cdot \boldsymbol{\Omega}^{\mathrm{p}} \cdot \mathbf{F}^{\mathrm{e}-1}\right)_{\mathrm{s}} = -\boldsymbol{\mathcal{L}}^{-1}_{(0)} : \left(\boldsymbol{\omega}^{\mathrm{p}} \cdot \boldsymbol{\tau} - \boldsymbol{\tau} \cdot \boldsymbol{\omega}^{\mathrm{p}}\right). \tag{11.7.14}$$

Note also that Eq. (11.7.1) can be recast in the form

$$\mathbf{D}^{\mathrm{e}} = \frac{1}{2}\mathbf{F}^{\mathrm{e}-T} \cdot \overset{\bullet}{\mathbf{C}^{\mathrm{e}}} \cdot \mathbf{F}^{\mathrm{e}-1}, \quad \overset{\bullet}{\mathbf{C}^{\mathrm{e}}} = \dot{\mathbf{C}}^{\mathrm{e}} - \boldsymbol{\Omega}^{\mathrm{p}} \cdot \mathbf{C}^{\mathrm{e}} + \mathbf{C}^{\mathrm{e}} \cdot \boldsymbol{\Omega}^{\mathrm{p}}. \tag{11.7.15}$$

This expression, as well as (11.7.1), holds for elastoplastic deformation of elastically isotropic materials, regardless of whether the material hardens isotropically or anisotropically in the course of plastic deformation.

11.7.1. Analysis of Spin $\boldsymbol{\Omega}^{\mathrm{p}}$

The spin $\boldsymbol{\Omega}_I^{\mathrm{p}}$, obtained as the solution of Eq. (11.7.9), depends on $\mathbf{W}$, $\mathbf{V}^{\mathrm{e}}$, and $\dot{\mathbf{V}}^{\mathrm{e}}$. It is possible to derive an expression for this spin in terms of $\mathbf{W}$, $\mathbf{V}^{\mathrm{e}}$, and $\mathbf{D}^{\mathrm{e}}$. Proceeding as in Section 2.7, we first observe the identity

$$\mathbf{V}^{\mathrm{e}-1}\cdot\left(\dot{\mathbf{V}}^{\mathrm{e}}\cdot\mathbf{V}^{\mathrm{e}-1}\right)=\left(\dot{\mathbf{V}}^{\mathrm{e}}\cdot\mathbf{V}^{\mathrm{e}-1}\right)^{T}\cdot\mathbf{V}^{\mathrm{e}-1}, \tag{11.7.16}$$

which can be rewritten in the form

$$\mathbf{V}^{\mathrm{e}-1}\cdot\left(\dot{\mathbf{V}}^{\mathrm{e}}\cdot\mathbf{V}^{\mathrm{e}-1}\right)_{\mathrm{a}}+\left(\dot{\mathbf{V}}^{\mathrm{e}}\cdot\mathbf{V}^{\mathrm{e}-1}\right)_{\mathrm{a}}\cdot\mathbf{V}^{\mathrm{e}-1}=\mathbf{D}^{\mathrm{e}}\cdot\mathbf{V}^{\mathrm{e}-1}-\mathbf{V}^{\mathrm{e}-1}\cdot\mathbf{D}^{\mathrm{e}}. \tag{11.7.17}$$

This equation can be solved for $\left(\dot{\mathbf{V}}^{\mathrm{e}}\cdot\mathbf{V}^{\mathrm{e}-1}\right)_{\mathrm{a}}$ as

$$\begin{aligned}\left(\dot{\mathbf{V}}^{\mathrm{e}}\cdot\mathbf{V}^{\mathrm{e}-1}\right)_{\mathrm{a}} &= K_1\left(\mathbf{D}^{\mathrm{e}}\cdot\mathbf{V}^{\mathrm{e}-1}-\mathbf{V}^{\mathrm{e}-1}\cdot\mathbf{D}^{\mathrm{e}}\right)\\ &-\left(J_1\mathbf{I}-\mathbf{V}^{\mathrm{e}-1}\right)^{-1}\cdot\left(\mathbf{D}^{\mathrm{e}}\cdot\mathbf{V}^{\mathrm{e}-1}-\mathbf{V}^{\mathrm{e}-1}\cdot\mathbf{D}^{\mathrm{e}}\right)\\ &-\left(\mathbf{D}^{\mathrm{e}}\cdot\mathbf{V}^{\mathrm{e}-1}-\mathbf{V}^{\mathrm{e}-1}\cdot\mathbf{D}^{\mathrm{e}}\right)\cdot\left(J_1\mathbf{I}-\mathbf{V}^{\mathrm{e}-1}\right)^{-1},\end{aligned} \tag{11.7.18}$$

where

$$J_1=\operatorname{tr}\mathbf{V}^{\mathrm{e}-1},\quad K_1=\operatorname{tr}\left(J_1\mathbf{I}-\mathbf{V}^{\mathrm{e}-1}\right)^{-1}. \tag{11.7.19}$$

The left-hand side of Eq. (11.7.18) is also equal to $\mathbf{W}-\boldsymbol{\Omega}_I^{\mathrm{p}}$, by Eq. (11.7.7). Therefore,

$$\begin{aligned}\boldsymbol{\Omega}_I^{\mathrm{p}}=\mathbf{W}&-K_1\left(\mathbf{D}^{\mathrm{e}}\cdot\mathbf{V}^{\mathrm{e}-1}-\mathbf{V}^{\mathrm{e}-1}\cdot\mathbf{D}^{\mathrm{e}}\right)\\ &+\left(J_1\mathbf{I}-\mathbf{V}^{\mathrm{e}-1}\right)^{-1}\cdot\left(\mathbf{D}^{\mathrm{e}}\cdot\mathbf{V}^{\mathrm{e}-1}-\mathbf{V}^{\mathrm{e}-1}\cdot\mathbf{D}^{\mathrm{e}}\right)\\ &+\left(\mathbf{D}^{\mathrm{e}}\cdot\mathbf{V}^{\mathrm{e}-1}-\mathbf{V}^{\mathrm{e}-1}\cdot\mathbf{D}^{\mathrm{e}}\right)\cdot\left(J_1\mathbf{I}-\mathbf{V}^{\mathrm{e}-1}\right)^{-1}.\end{aligned} \tag{11.7.20}$$

The expression for $\boldsymbol{\Omega}^{\mathrm{p}}$ is obtained by substituting Eq. (11.7.20) into Eq. (11.7.13). The result is

$$\begin{aligned}\boldsymbol{\Omega}^{\mathrm{p}}=\mathbf{R}^{\mathrm{e}T}\cdot\left(\mathbf{W}-\dot{\mathbf{R}}^{\mathrm{e}}\cdot\mathbf{R}^{\mathrm{e}-1}\right)\cdot\mathbf{R}^{\mathrm{e}}&-K_1\left(\hat{\mathbf{D}}^{\mathrm{e}}\cdot\mathbf{U}^{\mathrm{e}-1}-\mathbf{U}^{\mathrm{e}-1}\cdot\hat{\mathbf{D}}^{\mathrm{e}}\right)\\ &+\left(J_1\mathbf{I}-\mathbf{U}^{\mathrm{e}-1}\right)^{-1}\cdot\left(\hat{\mathbf{D}}^{\mathrm{e}}\cdot\mathbf{U}^{\mathrm{e}-1}-\mathbf{U}^{\mathrm{e}-1}\cdot\hat{\mathbf{D}}^{\mathrm{e}}\right)\\ &+\left(\hat{\mathbf{D}}^{\mathrm{e}}\cdot\mathbf{U}^{\mathrm{e}-1}-\mathbf{U}^{\mathrm{e}-1}\cdot\hat{\mathbf{D}}^{\mathrm{e}}\right)\cdot\left(J_1\mathbf{I}-\mathbf{U}^{\mathrm{e}-1}\right)^{-1},\end{aligned} \tag{11.7.21}$$

where

$$\hat{\mathbf{D}}^{\mathrm{e}} = \mathbf{R}^{\mathrm{e}T} \cdot \mathbf{D}^{\mathrm{e}} \cdot \mathbf{R}^{\mathrm{e}}, \quad \mathbf{U}^{\mathrm{e}-1} = \mathbf{R}^{\mathrm{e}T} \cdot \mathbf{V}^{\mathrm{e}-1} \cdot \mathbf{R}^{\mathrm{e}}. \tag{11.7.22}$$

With a specified rotation $\mathbf{R}^{\mathrm{e}}$ of the destressing program, Eq. (11.7.21) determines the corresponding spin $\boldsymbol{\Omega}^{\mathrm{p}}$.

11.8. Analysis of Plastic Rate of Deformation

Having defined the elastic rate of deformation by Eq. (11.7.1), the remaining plastic rate of deformation is

$$\mathbf{D}^{\mathrm{p}} = \left[\mathbf{F}^{\mathrm{e}} \cdot \left(\dot{\mathbf{F}}^{\mathrm{p}} \cdot \mathbf{F}^{\mathrm{p}-1}\right) \cdot \mathbf{F}^{\mathrm{e}-1}\right]_{\mathrm{s}} - \left(\mathbf{F}^{\mathrm{e}} \cdot \boldsymbol{\Omega}^{\mathrm{p}} \cdot \mathbf{F}^{\mathrm{e}-1}\right)_{\mathrm{s}}. \tag{11.8.1}$$

In view of Eq. (11.7.7), we also have

$$\mathbf{D}^{\mathrm{p}} = \mathbf{F}^{\mathrm{e}} \cdot \left(\dot{\mathbf{F}}^{\mathrm{p}} \cdot \mathbf{F}^{\mathrm{p}-1}\right) \cdot \mathbf{F}^{\mathrm{e}-1} - \mathbf{F}^{\mathrm{e}} \cdot \boldsymbol{\Omega}^{\mathrm{p}} \cdot \mathbf{F}^{\mathrm{e}-1}, \tag{11.8.2}$$

since

$$\mathbf{D}^{\mathrm{e}} + \mathbf{D}^{\mathrm{p}} + \mathbf{W} = \mathbf{L}, \tag{11.8.3}$$

as given by Eq. (11.3.2). Alternatively, Eq. (11.8.2) can be written as

$$\mathbf{D}^{\mathrm{p}} = \mathbf{F}^{\mathrm{e}} \cdot \left(\overset{\bullet}{\mathbf{F}}{}^{\mathrm{p}} \cdot \mathbf{F}^{\mathrm{p}-1}\right) \cdot \mathbf{F}^{\mathrm{e}-1}, \quad \overset{\bullet}{\mathbf{F}}{}^{\mathrm{p}} = \dot{\mathbf{F}}^{\mathrm{p}} - \boldsymbol{\Omega}^{\mathrm{p}} \cdot \mathbf{F}^{\mathrm{p}}. \tag{11.8.4}$$

By taking the antisymmetric part of Eq. (11.8.2), therefore,

$$\left(\mathbf{F}^{\mathrm{e}} \cdot \boldsymbol{\Omega}^{\mathrm{p}} \cdot \mathbf{F}^{\mathrm{e}-1}\right)_{\mathrm{a}} = \left[\mathbf{F}^{\mathrm{e}} \cdot \left(\dot{\mathbf{F}}^{\mathrm{p}} \cdot \mathbf{F}^{\mathrm{p}-1}\right) \cdot \mathbf{F}^{\mathrm{e}-1}\right]_{\mathrm{a}}. \tag{11.8.5}$$

Furthermore, from Eq. (11.8.2) we have

$$\mathcal{D}^{\mathrm{p}} = \left(\mathbf{F}^{\mathrm{e}-1} \cdot \mathbf{D}^{\mathrm{p}} \cdot \mathbf{F}^{\mathrm{e}}\right)_{\mathrm{s}}, \tag{11.8.6}$$

$$\mathcal{W}^{\mathrm{p}} = \boldsymbol{\Omega}^{\mathrm{p}} + \left(\mathbf{F}^{\mathrm{e}-1} \cdot \mathbf{D}^{\mathrm{p}} \cdot \mathbf{F}^{\mathrm{e}}\right)_{\mathrm{a}}. \tag{11.8.7}$$

For convenience, the rate of deformation and the spin of the intermediate configuration are denoted by

$$\mathcal{D}^{\mathrm{p}} = \left(\dot{\mathbf{F}}^{\mathrm{p}} \cdot \mathbf{F}^{\mathrm{p}-1}\right)_{\mathrm{s}}, \quad \mathcal{W}^{\mathrm{p}} = \left(\dot{\mathbf{F}}^{\mathrm{p}} \cdot \mathbf{F}^{\mathrm{p}-1}\right)_{\mathrm{a}}. \tag{11.8.8}$$

These quantities, of course, depend on the choice of the intermediate configuration.

To elaborate, we start from the identity

$$\mathbf{C}^{\mathrm{e}} \cdot \left(\mathbf{F}^{\mathrm{e}-1} \cdot \mathbf{D}^{\mathrm{p}} \cdot \mathbf{F}^{\mathrm{e}}\right) = \left(\mathbf{F}^{\mathrm{e}-1} \cdot \mathbf{D}^{\mathrm{p}} \cdot \mathbf{F}^{\mathrm{e}}\right)^{T} \cdot \mathbf{C}^{\mathrm{e}}, \tag{11.8.9}$$

which can be recast as

$$\mathbf{C}^{\mathrm{e}}\cdot\left(\mathbf{F}^{\mathrm{e}-1}\cdot\mathbf{D}^{\mathrm{p}}\cdot\mathbf{F}^{\mathrm{e}}\right)_{\mathrm{a}}+\left(\mathbf{F}^{\mathrm{e}-1}\cdot\mathbf{D}^{\mathrm{p}}\cdot\mathbf{F}^{\mathrm{e}}\right)_{\mathrm{a}}\cdot\mathbf{C}^{\mathrm{e}}=\mathcal{D}^{\mathrm{p}}\cdot\mathbf{C}^{\mathrm{e}}-\mathbf{C}^{\mathrm{e}}\cdot\mathcal{D}^{\mathrm{p}}. \quad (11.8.10)$$

The last equation can be solved for $\left(\mathbf{F}^{\mathrm{e}-1}\cdot\mathbf{D}^{\mathrm{p}}\cdot\mathbf{F}^{\mathrm{e}}\right)_{\mathrm{a}}$ in terms of $\mathbf{C}^{\mathrm{e}}$ and $\mathcal{D}^{\mathrm{p}}$. The result is

$$\begin{aligned}\left(\mathbf{F}^{\mathrm{e}-1}\cdot\mathbf{D}^{\mathrm{p}}\cdot\mathbf{F}^{\mathrm{e}}\right)_{\mathrm{a}}&=k_1\left(\mathcal{D}^{\mathrm{p}}\cdot\mathbf{C}^{\mathrm{e}}-\mathbf{C}^{\mathrm{e}}\cdot\mathcal{D}^{\mathrm{p}}\right)\\&-\left(j_1\mathbf{I}-\mathbf{C}^{\mathrm{e}}\right)^{-1}\cdot\left(\mathcal{D}^{\mathrm{p}}\cdot\mathbf{C}^{\mathrm{e}}-\mathbf{C}^{\mathrm{e}}\cdot\mathcal{D}^{\mathrm{p}}\right)\\&-\left(\mathcal{D}^{\mathrm{p}}\cdot\mathbf{C}^{\mathrm{e}}-\mathbf{C}^{\mathrm{e}}\cdot\mathcal{D}^{\mathrm{p}}\right)\cdot\left(j_1\mathbf{I}-\mathbf{C}^{\mathrm{e}}\right)^{-1},\end{aligned} \quad (11.8.11)$$

where

$$j_1=\operatorname{tr}\mathbf{C}^{\mathrm{e}},\quad k_1=\operatorname{tr}\left(j_1\mathbf{I}-\mathbf{C}^{\mathrm{e}}\right)^{-1}. \quad (11.8.12)$$

The substitution of Eqs. (11.8.11) and (11.7.21) into Eq. (11.8.7) gives

$$\begin{aligned}\dot{\mathbf{R}}^{\mathrm{e}}\cdot\mathbf{R}^{\mathrm{e}-1}+\mathbf{R}^{\mathrm{e}}\cdot\mathcal{W}^{\mathrm{p}}\cdot\mathbf{R}^{\mathrm{e}T}&=\mathbf{W}-K_1\left(\mathbf{D}^{\mathrm{e}}\cdot\mathbf{V}^{\mathrm{e}-1}-\mathbf{V}^{\mathrm{e}-1}\cdot\mathbf{D}^{\mathrm{e}}\right)\\&+\left(J_1\mathbf{I}-\mathbf{V}^{\mathrm{e}-1}\right)^{-1}\cdot\left(\mathbf{D}^{\mathrm{e}}\cdot\mathbf{V}^{\mathrm{e}-1}-\mathbf{V}^{\mathrm{e}-1}\cdot\mathbf{D}^{\mathrm{e}}\right)\\&+\left(\mathbf{D}^{\mathrm{e}}\cdot\mathbf{V}^{\mathrm{e}-1}-\mathbf{V}^{\mathrm{e}-1}\cdot\mathbf{D}^{\mathrm{e}}\right)\cdot\left(J_1\mathbf{I}-\mathbf{V}^{\mathrm{e}-1}\right)^{-1}\\&+k_1\left(\bar{\mathcal{D}}^{\mathrm{p}}\cdot\mathbf{B}^{\mathrm{e}}-\mathbf{B}^{\mathrm{e}}\cdot\bar{\mathcal{D}}^{\mathrm{p}}\right)\\&-\left(j_1\mathbf{I}-\mathbf{B}^{\mathrm{e}}\right)^{-1}\cdot\left(\bar{\mathcal{D}}^{\mathrm{p}}\cdot\mathbf{B}^{\mathrm{e}}-\mathbf{B}^{\mathrm{e}}\cdot\bar{\mathcal{D}}^{\mathrm{p}}\right)\\&-\left(\bar{\mathcal{D}}^{\mathrm{p}}\cdot\mathbf{B}^{\mathrm{e}}-\mathbf{B}^{\mathrm{e}}\cdot\bar{\mathcal{D}}^{\mathrm{p}}\right)\cdot\left(j_1\mathbf{I}-\mathbf{B}^{\mathrm{e}}\right)^{-1}.\end{aligned} \quad (11.8.13)$$

The tensor

$$\bar{\mathcal{D}}^{\mathrm{p}}=\mathbf{R}^{\mathrm{e}}\cdot\mathcal{D}^{\mathrm{p}}\cdot\mathbf{R}^{\mathrm{e}T} \quad (11.8.14)$$

is actually independent of the rotation $\mathbf{R}^{\mathrm{e}}$, since it can be expressed from Eq. (11.8.2) as

$$\bar{\mathcal{D}}^{\mathrm{p}}=\left(\mathbf{V}^{\mathrm{e}-1}\cdot\mathbf{D}^{\mathrm{p}}\cdot\mathbf{V}^{\mathrm{e}}\right)_{\mathrm{s}}. \quad (11.8.15)$$

Note that

$$\operatorname{tr}\mathbf{C}^{\mathrm{e}}=\operatorname{tr}\mathbf{B}^{\mathrm{e}},\quad \operatorname{tr}\left(j_1\mathbf{I}-\mathbf{C}^{\mathrm{e}}\right)^{-1}=\operatorname{tr}\left(j_1\mathbf{I}-\mathbf{B}^{\mathrm{e}}\right)^{-1}. \quad (11.8.16)$$

Therefore, the spin

$$\dot{\mathbf{R}}^{\mathrm{e}}\cdot\mathbf{R}^{\mathrm{e}-1}+\mathbf{R}^{\mathrm{e}}\cdot\mathcal{W}^{\mathrm{p}}\cdot\mathbf{R}^{\mathrm{e}T} \quad (11.8.17)$$

in Eq. (11.8.13) is expressed in terms of $\mathbf{V}^{\mathrm{e}}$, $\mathbf{W}$, $\mathbf{D}^{\mathrm{e}}$, and $\mathbf{D}^{\mathrm{p}}$. For example, if destressing is without rotation ($\mathbf{R}^{\mathrm{e}}=\mathbf{I}$), Eq. (11.8.13) defines the corresponding spin $\mathcal{W}^{\mathrm{p}}$ of the intermediate configuration. On the other hand, if destressing program is defined such that the spin of intermediate configuration vanishes ($\mathcal{W}^{\mathrm{p}}=\mathbf{0}$), Eq. (11.8.13) defines the corresponding rotation $\mathbf{R}^{\mathrm{e}}$

of the destressing program. These (different) choices, however, do not affect the end result and the values of the components of the elastic and plastic rates of deformation $\mathbf{D}^{\mathrm{e}}$ and $\mathbf{D}^{\mathrm{p}}$.

11.8.1. Relationship between $\mathbf{D}^{\mathrm{p}}$ and $\boldsymbol{\mathcal{D}}^{\mathrm{p}}$

Equation (11.8.6), which expresses $\boldsymbol{\mathcal{D}}^{\mathrm{p}}$ in terms of $\mathbf{D}^{\mathrm{p}}$, can be rewritten as

$$\boldsymbol{\mathcal{D}}^{\mathrm{p}} = \frac{1}{2}\,\mathbf{F}^{\mathrm{e}-1}\cdot\left(\mathbf{D}^{\mathrm{p}}\cdot\mathbf{B}^{\mathrm{e}} + \mathbf{B}^{\mathrm{e}}\cdot\mathbf{D}^{\mathrm{p}}\right)\cdot\mathbf{F}^{\mathrm{e}-T}, \tag{11.8.18}$$

or,

$$\mathbf{D}^{\mathrm{p}}\cdot\mathbf{B}^{\mathrm{e}} + \mathbf{B}^{\mathrm{e}}\cdot\mathbf{D}^{\mathrm{p}} = 2\,\mathbf{F}^{\mathrm{e}}\cdot\boldsymbol{\mathcal{D}}^{\mathrm{p}}\cdot\mathbf{F}^{\mathrm{e}T}. \tag{11.8.19}$$

The solution for $\mathbf{D}^{\mathrm{p}}$ in terms of $\boldsymbol{\mathcal{D}}^{\mathrm{p}}$ is

$$\begin{aligned}\mathbf{D}^{\mathrm{p}} = 2k_1\left(\mathbf{F}^{\mathrm{e}}\cdot\boldsymbol{\mathcal{D}}^{\mathrm{p}}\cdot\mathbf{F}^{\mathrm{e}T}\right) &- 2\left(j_1\mathbf{I}-\mathbf{B}^{\mathrm{e}}\right)^{-1}\cdot\left(\mathbf{F}^{\mathrm{e}}\cdot\boldsymbol{\mathcal{D}}^{\mathrm{p}}\cdot\mathbf{F}^{\mathrm{e}T}\right)\\ &- 2\left(\mathbf{F}^{\mathrm{e}}\cdot\boldsymbol{\mathcal{D}}^{\mathrm{p}}\cdot\mathbf{F}^{\mathrm{e}T}\right)\cdot\left(j_1\mathbf{I}-\mathbf{B}^{\mathrm{e}}\right)^{-1}.\end{aligned} \tag{11.8.20}$$

Alternatively, we can start from Eqs. (11.8.2) and (11.8.7), i.e.,

$$\mathbf{D}^{\mathrm{p}} = \mathbf{F}^{\mathrm{e}}\cdot\left(\boldsymbol{\mathcal{D}}^{\mathrm{p}} + \boldsymbol{\mathcal{W}}^{\mathrm{p}} - \boldsymbol{\Omega}^{\mathrm{p}}\right)\cdot\mathbf{F}^{\mathrm{e}-1} = \mathbf{F}^{\mathrm{e}}\cdot\left[\boldsymbol{\mathcal{D}}^{\mathrm{p}} + \left(\mathbf{F}^{\mathrm{e}-1}\cdot\mathbf{D}^{\mathrm{p}}\cdot\mathbf{F}^{\mathrm{e}}\right)_{\mathrm{a}}\right]\cdot\mathbf{F}^{\mathrm{e}-1}. \tag{11.8.21}$$

The substitution of Eq. (11.8.11) gives

$$\begin{aligned}\mathbf{D}^{\mathrm{p}} = \mathbf{F}^{\mathrm{e}}\cdot\big[&\boldsymbol{\mathcal{D}}^{\mathrm{p}} + (\operatorname{tr}\mathbf{A}^{\mathrm{e}})\left(\boldsymbol{\mathcal{D}}^{\mathrm{p}}\cdot\mathbf{C}^{\mathrm{e}} - \mathbf{C}^{\mathrm{e}}\cdot\boldsymbol{\mathcal{D}}^{\mathrm{p}}\right)\\ &- \mathbf{A}^{\mathrm{e}}\cdot\left(\boldsymbol{\mathcal{D}}^{\mathrm{p}}\cdot\mathbf{C}^{\mathrm{e}} - \mathbf{C}^{\mathrm{e}}\cdot\boldsymbol{\mathcal{D}}^{\mathrm{p}}\right) - \left(\boldsymbol{\mathcal{D}}^{\mathrm{p}}\cdot\mathbf{C}^{\mathrm{e}} - \mathbf{C}^{\mathrm{e}}\cdot\boldsymbol{\mathcal{D}}^{\mathrm{p}}\right)\cdot\mathbf{A}^{\mathrm{e}}\big]\cdot\mathbf{F}^{\mathrm{e}-1},\end{aligned} \tag{11.8.22}$$

where

$$\mathbf{A}^{\mathrm{e}} = \left(j_1\mathbf{I}-\mathbf{C}^{\mathrm{e}}\right)^{-1}. \tag{11.8.23}$$

The antisymmetric part of Eq. (11.8.22) vanishes identically.

11.9. Expression for $\mathbf{D}^{\mathrm{e}}$ in Terms of $\mathbf{F}^{\mathrm{e}}$, $\mathbf{F}^{\mathrm{p}}$, and Their Rates

In Eq. (11.7.1) the elastic rate of deformation $\mathbf{D}^{\mathrm{e}}$ was the sum of two terms, the second term being dependent on the spin $\boldsymbol{\Omega}^{\mathrm{p}}$. It is possible to express this term as an explicit function of $\mathbf{F}^{\mathrm{e}}$ and $\mathbf{F}^{\mathrm{p}}$, and their rates. To that goal, consider the identity

$$\mathbf{B}^{\mathrm{e}-1}\cdot\left(\mathbf{F}^{\mathrm{e}}\cdot\boldsymbol{\Omega}^{\mathrm{p}}\cdot\mathbf{F}^{\mathrm{e}-1}\right) = -\left(\mathbf{F}^{\mathrm{e}}\cdot\boldsymbol{\Omega}^{\mathrm{p}}\cdot\mathbf{F}^{\mathrm{e}-1}\right)^{T}\cdot\mathbf{B}^{\mathrm{e}-1}, \tag{11.9.1}$$

which can be rewritten as

$$\begin{aligned}
&\mathbf{B}^{e-1}\cdot\left(\mathbf{F}^e\cdot\boldsymbol{\Omega}^p\cdot\mathbf{F}^{e-1}\right)_s+\left(\mathbf{F}^e\cdot\boldsymbol{\Omega}^p\cdot\mathbf{F}^{e-1}\right)_s\cdot\mathbf{B}^{e-1}\\
&=\left(\mathbf{F}^e\cdot\boldsymbol{\Omega}^p\cdot\mathbf{F}^{e-1}\right)_a\cdot\mathbf{B}^{e-1}-\mathbf{B}^{e-1}\cdot\left(\mathbf{F}^e\cdot\boldsymbol{\Omega}^p\cdot\mathbf{F}^{e-1}\right)_a\\
&=\boldsymbol{\omega}^p\cdot\mathbf{B}^{e-1}-\mathbf{B}^{e-1}\cdot\boldsymbol{\omega}^p.
\end{aligned}\tag{11.9.2}$$

Expression (11.8.5) was used in the last step. Equation (11.9.2) can be solved for $\left(\mathbf{F}^e\cdot\boldsymbol{\Omega}^p\cdot\mathbf{F}^{e-1}\right)_s$ in terms of $\mathbf{B}^{e-1}$ and the spin $\boldsymbol{\omega}^p$, with the result

$$\begin{aligned}
\left(\mathbf{F}^e\cdot\boldsymbol{\Omega}^p\cdot\mathbf{F}^{e-1}\right)_s&=k_1'\left(\boldsymbol{\omega}^p\cdot\mathbf{B}^{e-1}-\mathbf{B}^{e-1}\cdot\boldsymbol{\omega}^p\right)\\
&-\left(j_1'\mathbf{I}-\mathbf{B}^{e-1}\right)^{-1}\cdot\left(\boldsymbol{\omega}^p\cdot\mathbf{B}^{e-1}-\mathbf{B}^{e-1}\cdot\boldsymbol{\omega}^p\right)\\
&-\left(\boldsymbol{\omega}^p\cdot\mathbf{B}^{e-1}-\mathbf{B}^{e-1}\cdot\boldsymbol{\omega}^p\right)\cdot\left(j_1'\mathbf{I}-\mathbf{B}^{e-1}\right)^{-1},
\end{aligned}\tag{11.9.3}$$

where

$$j_1'=\operatorname{tr}\mathbf{B}^{e-1},\quad k_1'=\operatorname{tr}\left(j_1\mathbf{I}-\mathbf{B}^{e-1}\right)^{-1}.\tag{11.9.4}$$

Consequently, incorporating Eq. (11.9.3) into Eq. (11.7.1) gives an expression for the elastic rate of deformation, solely in terms of $\mathbf{F}^e$ and $\mathbf{F}^p$, and their rates. This is

$$\begin{aligned}
\mathbf{D}^e=\left(\dot{\mathbf{F}}^e\cdot\mathbf{F}^{e-1}\right)_s&+k_1'\left(\boldsymbol{\omega}^p\cdot\mathbf{B}^{e-1}-\mathbf{B}^{e-1}\cdot\boldsymbol{\omega}^p\right)\\
&-\left(j_1'\mathbf{I}-\mathbf{B}^{e-1}\right)^{-1}\cdot\left(\boldsymbol{\omega}^p\cdot\mathbf{B}^{e-1}-\mathbf{B}^{e-1}\cdot\boldsymbol{\omega}^p\right)\\
&-\left(\boldsymbol{\omega}^p\cdot\mathbf{B}^{e-1}-\mathbf{B}^{e-1}\cdot\boldsymbol{\omega}^p\right)\cdot\left(j_1'\mathbf{I}-\mathbf{B}^{e-1}\right)^{-1}.
\end{aligned}\tag{11.9.5}$$

11.9.1. Intermediate Configuration with $\boldsymbol{\omega}^p = 0$

The three most appealing choices of the intermediate configuration correspond to

$$\begin{aligned}
\mathbf{R}^e&=\mathbf{I},\\
\boldsymbol{\mathcal{W}}^p&=\mathbf{0},\\
\boldsymbol{\omega}^p&=\mathbf{0}.
\end{aligned}\tag{11.9.6}$$

We discuss here the last choice, i.e., we consider the intermediate configuration obtained by the destressing program such that

$$\boldsymbol{\omega}^p=\left[\mathbf{F}^e\cdot\left(\dot{\mathbf{F}}^p\cdot\mathbf{F}^{p-1}\right)\cdot\mathbf{F}^{e-1}\right]_a=\mathbf{0}.\tag{11.9.7}$$

From Eqs. (11.8.5) and (11.9.3) it follows that

$$\mathbf{F}^e\cdot\boldsymbol{\Omega}^p\cdot\mathbf{F}^{e-1}=\mathbf{0},\quad\text{i.e.,}\quad\boldsymbol{\Omega}^p=\mathbf{0}.\tag{11.9.8}$$

The corresponding rotation is, from Eq. (11.7.13),

$$\dot{\mathbf{R}}^e\cdot\mathbf{R}^{e-1}=\boldsymbol{\Omega}^p_I,\tag{11.9.9}$$

where $\boldsymbol{\Omega}_J^{\mathrm{p}}$ is defined by Eq. (11.7.20). Furthermore, from Eqs. (11.8.7) and (11.8.11), the spin of the intermediate configuration is

$$\begin{aligned}\mathcal{W}^{\mathrm{p}} = \left(\mathbf{F}^{\mathrm{e}-1}\cdot\mathbf{D}^{\mathrm{p}}\cdot\mathbf{F}^{\mathrm{e}}\right)_{\mathrm{a}} &= k_1\left(\mathcal{D}^{\mathrm{p}}\cdot\mathbf{C}^{\mathrm{e}} - \mathbf{C}^{\mathrm{e}}\cdot\mathcal{D}^{\mathrm{p}}\right)\\ &\quad - \left(j_1\mathbf{I}-\mathbf{C}^{\mathrm{e}}\right)^{-1}\cdot\left(\mathcal{D}^{\mathrm{p}}\cdot\mathbf{C}^{\mathrm{e}} - \mathbf{C}^{\mathrm{e}}\cdot\mathcal{D}^{\mathrm{p}}\right)\\ &\quad - \left(\mathcal{D}^{\mathrm{p}}\cdot\mathbf{C}^{\mathrm{e}} - \mathbf{C}^{\mathrm{e}}\cdot\mathcal{D}^{\mathrm{p}}\right)\cdot\left(j_1\mathbf{I}-\mathbf{C}^{\mathrm{e}}\right)^{-1}.\end{aligned} \tag{11.9.10}$$

The elastic and plastic rates of deformation are

$$\mathbf{D}^{\mathrm{e}} = \left(\dot{\mathbf{F}}^{\mathrm{e}}\cdot\mathbf{F}^{\mathrm{e}-1}\right)_{\mathrm{s}}, \tag{11.9.11}$$

$$\mathbf{D}^{\mathrm{p}} = \mathbf{F}^{\mathrm{e}}\cdot\left(\dot{\mathbf{F}}^{\mathrm{p}}\cdot\mathbf{F}^{\mathrm{p}-1}\right)\cdot\mathbf{F}^{\mathrm{e}-1}. \tag{11.9.12}$$

For any other choice of the intermediate configuration, not associated with the choice (11.9.7), the symmetric part of $\dot{\mathbf{F}}^{\mathrm{e}}\cdot\mathbf{F}^{\mathrm{e}-1}$ is not all, but only a portion of the elastic rate of deformation $\mathbf{D}^{\mathrm{e}}$.

11.10. Isotropic Hardening

In the case of isotropic hardening the yield function is an isotropic function of the Cauchy stress $\boldsymbol{\sigma}$. Thus, if the normality rule applies, the plastic rate of deformation $\mathbf{D}^{\mathrm{p}}$ is codirectional with the outward normal to a locally smooth yield surface in stress space, and its principal directions are parallel to those of the stress $\boldsymbol{\sigma}$. Since for elastically isotropic material $\mathbf{V}^{\mathrm{e}}$ and $\mathbf{B}^{\mathrm{e}}$ are also coaxial with $\boldsymbol{\sigma}$, their matrix products commute, and Eqs. (11.8.6) and (11.8.7) become

$$\mathcal{D}^{\mathrm{p}} = \mathbf{R}^{\mathrm{e}T}\cdot\mathbf{D}^{\mathrm{p}}\cdot\mathbf{R}^{\mathrm{e}}, \quad \mathcal{W}^{\mathrm{p}} = \boldsymbol{\Omega}^{\mathrm{p}}, \tag{11.10.1}$$

because

$$\left(\mathbf{F}^{\mathrm{e}-1}\cdot\mathbf{D}^{\mathrm{p}}\cdot\mathbf{F}^{\mathrm{e}}\right)_{\mathrm{a}} = \left(\mathbf{R}^{\mathrm{e}T}\cdot\mathbf{D}^{\mathrm{p}}\cdot\mathbf{R}^{\mathrm{e}}\right)_{\mathrm{a}} = \mathbf{0}. \tag{11.10.2}$$

Furthermore, since from Eq. (11.8.15) in the case of isotropic hardening,

$$\bar{\mathcal{D}}^{\mathrm{p}} = \mathbf{D}^{\mathrm{p}}, \tag{11.10.3}$$

Equation (11.8.13) reduces to

$$\begin{aligned}\dot{\mathbf{R}}^{\mathrm{e}}\cdot\mathbf{R}^{\mathrm{e}-1} + \mathbf{R}^{\mathrm{e}}\cdot\mathcal{W}^{\mathrm{p}}\cdot\mathbf{R}^{\mathrm{e}T} &= \mathbf{W} - K_1\left(\mathbf{D}^{\mathrm{e}}\cdot\mathbf{V}^{\mathrm{e}-1} - \mathbf{V}^{\mathrm{e}-1}\cdot\mathbf{D}^{\mathrm{e}}\right)\\ &\quad + \left(J_1\mathbf{I}-\mathbf{V}^{\mathrm{e}-1}\right)^{-1}\cdot\left(\mathbf{D}^{\mathrm{e}}\cdot\mathbf{V}^{\mathrm{e}-1} - \mathbf{V}^{\mathrm{e}-1}\cdot\mathbf{D}^{\mathrm{e}}\right)\\ &\quad + \left(\mathbf{D}^{\mathrm{e}}\cdot\mathbf{V}^{\mathrm{e}-1} - \mathbf{V}^{\mathrm{e}-1}\cdot\mathbf{D}^{\mathrm{e}}\right)\cdot\left(J_1\mathbf{I}-\mathbf{V}^{\mathrm{e}-1}\right)^{-1}.\end{aligned} \tag{11.10.4}$$

This is precisely the spin $\boldsymbol{\Omega}_I^{\rm p}$ of Eq. (11.7.20). In addition, we have

$$\boldsymbol{\omega}^{\rm p} = \left[\mathbf{F}^{\rm e} \cdot \left(\dot{\mathbf{F}}^{\rm p} \cdot \mathbf{F}^{\rm p-1}\right) \cdot \mathbf{F}^{\rm e-1}\right]_{\rm a} = \left(\mathbf{F}^{\rm e} \cdot \boldsymbol{\mathcal{W}}^{\rm p} \cdot \mathbf{F}^{\rm e-1}\right)_{\rm a}. \tag{11.10.5}$$

If the intermediate configuration is selected so that $\mathbf{R}^{\rm e} = \mathbf{I}$, Eq. (11.10.4) specifies the corresponding spin (Lubarda and Lee, 1981), as

$$\boldsymbol{\mathcal{W}}^{\rm p} = \boldsymbol{\Omega}_I^{\rm p}. \tag{11.10.6}$$

If the intermediate configuration is selected so that $\boldsymbol{\omega}^{\rm p} = \mathbf{0}$, then

$$\boldsymbol{\mathcal{W}}^{\rm p} = \mathbf{0} \tag{11.10.7}$$

(and *vice versa*, for isotropic hardening). The right-hand side of Eq. (11.10.4) defines the spin due to $\mathbf{R}^{\rm e}$, i.e.,

$$\dot{\mathbf{R}}^{\rm e} \cdot \mathbf{R}^{\rm e-1} = \boldsymbol{\Omega}_I^{\rm p}. \tag{11.10.8}$$

11.11. Kinematic Hardening

To approximately account for the Bauschinger effect and anisotropic hardening, the kinematic hardening model was introduced in Subsection 9.4.2. Translation of the yield surface in stress space is prescribed by the evolution equation for the back stress $\boldsymbol{\alpha}$ (center of the yield surface). A fairly general objective equation for this evolution is

$$\overset{\circ}{\boldsymbol{\alpha}} = \dot{\boldsymbol{\alpha}} - \mathbf{W} \cdot \boldsymbol{\alpha} + \boldsymbol{\alpha} \cdot \mathbf{W} = \mathbf{A}(\boldsymbol{\alpha}, \mathbf{D}^{\rm p}), \tag{11.11.1}$$

where $\mathbf{A}$ is an isotropic function of both $\boldsymbol{\alpha}$ and $\mathbf{D}^{\rm p}$. Its polynomial representation is given by Eq. (1.11.10). Assuming that $\boldsymbol{\alpha}$ is deviatoric and that the material response is rate-independent, the function $\mathbf{A}$ can be written as

$$\mathbf{A}(\boldsymbol{\alpha}, \mathbf{D}^{\rm p}) = \mathbf{G}(\boldsymbol{\alpha}, \mathbf{D}^{\rm p}) + \boldsymbol{\alpha} \cdot \hat{\mathbf{W}} - \hat{\mathbf{W}} \cdot \boldsymbol{\alpha}. \tag{11.11.2}$$

The tensor function $\mathbf{G}$ is

$$\begin{aligned} \mathbf{G}(\boldsymbol{\alpha}, \mathbf{D}^{\rm p}) = {} & \eta_1\, \mathbf{D}^{\rm p} + \eta_2\, D^{\rm p}\, \boldsymbol{\alpha} + \eta_3\, D^{\rm p} \left[\boldsymbol{\alpha}^2 - \frac{1}{3} \operatorname{tr}\left(\boldsymbol{\alpha}^2\right) \mathbf{I}\right] \\ & + \eta_4 \left[\boldsymbol{\alpha} \cdot \mathbf{D}^{\rm p} + \mathbf{D}^{\rm p} \cdot \boldsymbol{\alpha} - \frac{2}{3} \operatorname{tr}\left(\boldsymbol{\alpha} \cdot \mathbf{D}^{\rm p}\right) \mathbf{I}\right] \\ & + \eta_5 \left[\boldsymbol{\alpha}^2 \cdot \mathbf{D}^{\rm p} + \mathbf{D}^{\rm p} \cdot \boldsymbol{\alpha}^2 - \frac{2}{3} \operatorname{tr}\left(\boldsymbol{\alpha}^2 \cdot \mathbf{D}^{\rm p}\right) \mathbf{I}\right], \end{aligned} \tag{11.11.3}$$

and the spin

$$\begin{aligned} \hat{\mathbf{W}} = {} & \vartheta_1 \left(\boldsymbol{\alpha} \cdot \mathbf{D}^{\rm p} - \mathbf{D}^{\rm p} \cdot \boldsymbol{\alpha}\right) + \vartheta_2 \left(\boldsymbol{\alpha}^2 \cdot \mathbf{D}^{\rm p} - \mathbf{D}^{\rm p} \cdot \boldsymbol{\alpha}^2\right) + \\ & + \vartheta_3 \left(\boldsymbol{\alpha}^2 \cdot \mathbf{D}^{\rm p} \cdot \boldsymbol{\alpha} - \boldsymbol{\alpha} \cdot \mathbf{D}^{\rm p} \cdot \boldsymbol{\alpha}^2\right). \end{aligned} \tag{11.11.4}$$

The scalar

$$D^{\mathrm{p}} = (2\,\mathbf{D}^{\mathrm{p}} : \mathbf{D}^{\mathrm{p}})^{1/2} \tag{11.11.5}$$

is a homogeneous function of degree one in the components of plastic rate of deformation, while $\eta_i\,(i = 1, 2, \ldots, 5)$ and $\vartheta_i\,(i = 1, 2, 3)$ are scalar functions of the invariants of $\boldsymbol{\alpha}$. The representation of antisymmetric function $\hat{\mathbf{W}}$ in terms of $\boldsymbol{\alpha}$ and $\mathbf{D}^{\mathrm{p}}$ is constructed according to Eq. (1.11.11). The combination of terms $(\boldsymbol{\alpha}\cdot\hat{\mathbf{W}} - \hat{\mathbf{W}}\cdot\boldsymbol{\alpha})$, which is an isotropic symmetric function of $\boldsymbol{\alpha}$ and $\mathbf{D}^{\mathrm{p}}$, is given separately in the representation (11.11.2), so that the function $\mathbf{G}$ incorporates direct influence of the rate of deformation on the evolution of $\boldsymbol{\alpha}$, while $(\boldsymbol{\alpha}\cdot\hat{\mathbf{W}} - \hat{\mathbf{W}}\cdot\boldsymbol{\alpha})$ incorporates the influence of deformation imposed rotation of the lines of material elements considered to carry the embedded back stress (Agah-Tehrani, Lee, Mallett, and Onat, 1987). Such rotation can have a significant effect on the evolution, quite independently of the overall material spin $\mathbf{W}$. An example in the case of straining in simple shear is given by Lee, Mallett, and Wertheimer (1983). See also Dafalias (1983), Atluri (1984), Johnson and Bammann (1984), and Van der Giessen (1989).

The following relationships are further observed

$$\boldsymbol{\alpha}^2\cdot\mathbf{D}^{\mathrm{p}} - \mathbf{D}^{\mathrm{p}}\cdot\boldsymbol{\alpha}^2 = \boldsymbol{\alpha}\cdot(\boldsymbol{\alpha}\cdot\mathbf{D}^{\mathrm{p}} - \mathbf{D}^{\mathrm{p}}\cdot\boldsymbol{\alpha}) + (\boldsymbol{\alpha}\cdot\mathbf{D}^{\mathrm{p}} - \mathbf{D}^{\mathrm{p}}\cdot\boldsymbol{\alpha})\cdot\boldsymbol{\alpha}, \tag{11.11.6}$$

and

$$\begin{aligned}\boldsymbol{\alpha}^2\cdot\mathbf{D}^{\mathrm{p}}\cdot\boldsymbol{\alpha} - \boldsymbol{\alpha}\cdot\mathbf{D}^{\mathrm{p}}\cdot\boldsymbol{\alpha}^2 &= \boldsymbol{\alpha}\cdot\left(\boldsymbol{\alpha}^2\cdot\mathbf{D}^{\mathrm{p}} - \mathbf{D}^{\mathrm{p}}\cdot\boldsymbol{\alpha}^2\right)\\ &+ \left(\boldsymbol{\alpha}^2\cdot\mathbf{D}^{\mathrm{p}} - \mathbf{D}^{\mathrm{p}}\cdot\boldsymbol{\alpha}^2\right)\cdot\boldsymbol{\alpha} - \frac{1}{2}\,\mathrm{tr}\left(\boldsymbol{\alpha}^2\right)\left(\boldsymbol{\alpha}\cdot\mathbf{D}^{\mathrm{p}} - \mathbf{D}^{\mathrm{p}}\cdot\boldsymbol{\alpha}\right).\end{aligned} \tag{11.11.7}$$

The second of these can be expressed as

$$\begin{aligned}\boldsymbol{\alpha}^2\cdot\mathbf{D}^{\mathrm{p}}\cdot\boldsymbol{\alpha} - \boldsymbol{\alpha}\cdot\mathbf{D}^{\mathrm{p}}\cdot\boldsymbol{\alpha}^2 &= -\boldsymbol{\alpha}^2\cdot(\boldsymbol{\alpha}\cdot\mathbf{D}^{\mathrm{p}} - \mathbf{D}^{\mathrm{p}}\cdot\boldsymbol{\alpha})\\ &- (\boldsymbol{\alpha}\cdot\mathbf{D}^{\mathrm{p}} - \mathbf{D}^{\mathrm{p}}\cdot\boldsymbol{\alpha})\cdot\boldsymbol{\alpha}^2 + \frac{1}{2}\,\mathrm{tr}\left(\boldsymbol{\alpha}^2\right)\left(\boldsymbol{\alpha}\cdot\mathbf{D}^{\mathrm{p}} - \mathbf{D}^{\mathrm{p}}\cdot\boldsymbol{\alpha}\right).\end{aligned} \tag{11.11.8}$$

This is easily verified by recalling that $\boldsymbol{\alpha}$ is deviatoric ($\mathrm{tr}\,\boldsymbol{\alpha} = 0$) and that, from the Cayley–Hamilton theorem (1.4.1),

$$\boldsymbol{\alpha}^3 = \frac{1}{2}\,\mathrm{tr}\left(\boldsymbol{\alpha}^2\right)\boldsymbol{\alpha} + (\det\boldsymbol{\alpha})\,\mathbf{I}. \tag{11.11.9}$$

Substitution of Eqs. (11.11.6) and (11.11.8) into Eq. (11.11.4) thus yields

$$\hat{\mathbf{W}} = -\mathbf{H}\cdot(\boldsymbol{\alpha}\cdot\mathbf{D}^{\mathrm{p}} - \mathbf{D}^{\mathrm{p}}\cdot\boldsymbol{\alpha}) - (\boldsymbol{\alpha}\cdot\mathbf{D}^{\mathrm{p}} - \mathbf{D}^{\mathrm{p}}\cdot\boldsymbol{\alpha})\cdot\mathbf{H} + (\operatorname{tr}\mathbf{H})\,(\boldsymbol{\alpha}\cdot\mathbf{D}^{\mathrm{p}} - \mathbf{D}^{\mathrm{p}}\cdot\boldsymbol{\alpha})\,. \tag{11.11.10}$$

This expresses $\hat{\mathbf{W}}$ in terms of a basic antisymmetric tensor $(\boldsymbol{\alpha}\cdot\mathbf{D}^{\mathrm{p}} - \mathbf{D}^{\mathrm{p}}\cdot\boldsymbol{\alpha})$ and an isotropic tensor function $\mathbf{H}(\boldsymbol{\alpha})$, defined by

$$\mathbf{H} = \vartheta_1\,\mathbf{I} - \vartheta_2\,\boldsymbol{\alpha} + \vartheta_3\left[\boldsymbol{\alpha}^2 - \frac{1}{2}\operatorname{tr}\left(\boldsymbol{\alpha}^2\right)\mathbf{I}\right]. \tag{11.11.11}$$

The evolution equation for the back stress (11.11.1) consequently becomes

$$\overset{\bullet}{\boldsymbol{\alpha}} = \mathbf{G}(\boldsymbol{\alpha}, \mathbf{D}^{\mathrm{p}}), \tag{11.11.12}$$

where, in view of Eq. (11.11.2),

$$\overset{\bullet}{\boldsymbol{\alpha}} = \overset{\circ}{\boldsymbol{\alpha}} + \hat{\mathbf{W}}\cdot\boldsymbol{\alpha} - \boldsymbol{\alpha}\cdot\hat{\mathbf{W}} = \dot{\boldsymbol{\alpha}} - \hat{\boldsymbol{\omega}}\cdot\boldsymbol{\alpha} + \boldsymbol{\alpha}\cdot\hat{\boldsymbol{\omega}}. \tag{11.11.13}$$

The spin used to define the Jaumann derivative $\overset{\bullet}{\boldsymbol{\alpha}}$ is

$$\hat{\boldsymbol{\omega}} = \mathbf{W} - \hat{\mathbf{W}}. \tag{11.11.14}$$

Either the spin $\hat{\mathbf{W}}$, associated with the angular velocity of the embedded back stress, or the relative spin $\hat{\boldsymbol{\omega}}$ (relative to the deforming material), can be referred to as the plastic spin. The constitutive equation for $\hat{\mathbf{W}}$ is given by Eqs. (11.11.10) and (11.11.11), with the appropriately specified parameters $\vartheta_i\,(i = 1, 2, 3)$.

The introduction of the plastic spin as an ingredient of the phenomenological theory of plasticity was motivated by the attempts to eliminate spurious oscillations of shear stress, obtained under monotonically increasing straining in simple shear, within the model of kinematic hardening and simple evolution equation for the back stress $\overset{\circ}{\boldsymbol{\alpha}} \propto \mathbf{D}^{\mathrm{p}}$ (Nagtegaal and de Jong, 1982; Lee, Mallett, and Wertheimer, 1983). Further research on plastic spin was subsequently stimulated by the work of Loret (1983) and Dafalias (1983, 1985). Various aspects of this work have been discussed or reviewed by Aifantis (1987), Zbib and Aifantis (1988), Van der Giessen (1991), Nemat-Nasser (1992), Lubarda and Shih (1994), Besseling and Van der Giessen (1994), and Dafalias (1999). The survey paper by Dafalias (1999) contains additional references. Research on the plastic spin in crystal plasticity is discussed in Chapter 12. An analysis of plastic spin in the corner theory of plasticity was presented by Kuroda (1995).

The elastoplastic behavior of amorphous polymers was studied within the framework of multiplicative decomposition by Boyce, Parks, and Argon (1988), and Boyce, Weber, and Parks (1990). Other viscoplastic solids were considered by Weber and Anand (1990). See also Anand (1980) for an application to pressure sensitive dilatant materials. Computational aspects of finite deformation elastoplasticity based on the multiplicative decomposition were examined by Needleman (1985), Simo and Ortiz (1985), Moran, Ortiz, and Shih (1990), Simo (1998), Simo and Hughes (1998), and Belytschko, Liu, and Moran (2000).

11.12. Rates of Deformation Due to Convected Stress Rate

The rate of deformation tensor was partitioned in Section 11.6 into its elastic and plastic parts by using the Jaumann rate of the Kirchhoff stress, such that

$$\mathbf{D}^{\mathrm{e}}_{(0)} = \boldsymbol{\mathcal{L}}^{-1}_{(0)} : \overset{\circ}{\boldsymbol{\tau}}, \quad \mathbf{D}^{\mathrm{p}}_{(0)} = \mathbf{D} - \mathbf{D}^{\mathrm{e}}_{(0)}. \tag{11.12.1}$$

The subscript (0) is added to indicate that the partition was with respect to the stress rate $\overset{\circ}{\boldsymbol{\tau}}$. In terms of $\mathbf{F}^{\mathrm{e}}$ and $\mathbf{F}^{\mathrm{p}}$, and their rates, it was found that

$$\begin{aligned}\mathbf{D}^{\mathrm{e}}_{(0)} = \left(\dot{\mathbf{F}}^{\mathrm{e}} \cdot \mathbf{F}^{\mathrm{e}-1}\right)_{\mathrm{s}} - \boldsymbol{\mathcal{L}}^{-1}_{(0)} : \Big\{&\left[\mathbf{F}^{\mathrm{e}} \cdot \left(\dot{\mathbf{F}}^{\mathrm{p}} \cdot \mathbf{F}^{\mathrm{p}-1}\right) \cdot \mathbf{F}^{\mathrm{e}-1}\right]_{\mathrm{a}} \cdot \boldsymbol{\tau} \\ &- \boldsymbol{\tau} \cdot \left[\mathbf{F}^{\mathrm{e}} \cdot \left(\dot{\mathbf{F}}^{\mathrm{p}} \cdot \mathbf{F}^{\mathrm{p}-1}\right) \cdot \mathbf{F}^{\mathrm{e}-1}\right]_{\mathrm{a}}\Big\},\end{aligned} \tag{11.12.2}$$

$$\begin{aligned}\mathbf{D}^{\mathrm{p}}_{(0)} = \left[\mathbf{F}^{\mathrm{e}} \cdot \left(\dot{\mathbf{F}}^{\mathrm{p}} \cdot \mathbf{F}^{\mathrm{p}-1}\right) \cdot \mathbf{F}^{\mathrm{e}-1}\right]_{\mathrm{s}} + \boldsymbol{\mathcal{L}}^{-1}_{(0)} : \Big\{&\left[\mathbf{F}^{\mathrm{e}} \cdot \left(\dot{\mathbf{F}}^{\mathrm{p}} \cdot \mathbf{F}^{\mathrm{p}-1}\right) \cdot \mathbf{F}^{\mathrm{e}-1}\right]_{\mathrm{a}} \cdot \boldsymbol{\tau} \\ &- \boldsymbol{\tau} \cdot \left[\mathbf{F}^{\mathrm{e}} \cdot \left(\dot{\mathbf{F}}^{\mathrm{p}} \cdot \mathbf{F}^{\mathrm{p}-1}\right) \cdot \mathbf{F}^{\mathrm{e}-1}\right]_{\mathrm{a}}\Big\}.\end{aligned} \tag{11.12.3}$$

The corresponding elastic and plastic parts of the stress rate $\overset{\circ}{\boldsymbol{\tau}}$ are

$$\overset{\circ}{\boldsymbol{\tau}}{}^{\mathrm{e}} = \boldsymbol{\mathcal{L}}_{(0)} : \mathbf{D}, \quad \overset{\circ}{\boldsymbol{\tau}}{}^{\mathrm{p}} = -\boldsymbol{\mathcal{L}}_{(0)} : \mathbf{D}^{\mathrm{p}}_{(0)}. \tag{11.12.4}$$

An alternative partition of the rate of deformation tensor can be obtained by using the convected rate of the Kirchhoff stress $\overset{\triangle}{\boldsymbol{\tau}}$, such that

$$\mathbf{D}^{\mathrm{e}}_{(1)} = \boldsymbol{\mathcal{L}}^{-1}_{(1)} : \overset{\triangle}{\boldsymbol{\tau}}, \quad \mathbf{D}^{\mathrm{p}}_{(1)} = \mathbf{D} - \mathbf{D}^{\mathrm{e}}_{(1)}. \tag{11.12.5}$$

Indeed, from Eq. (11.6.2) it follows that

$$\overset{\triangle}{\boldsymbol{\tau}} = \boldsymbol{\mathcal{L}}_{(1)} : \left(\dot{\mathbf{F}}^{\mathrm{e}} \cdot \mathbf{F}^{\mathrm{e}-1}\right)_{\mathrm{s}} - \left\{\left[\mathbf{F}^{\mathrm{e}} \cdot \left(\dot{\mathbf{F}}^{\mathrm{p}} \cdot \mathbf{F}^{\mathrm{p}-1}\right) \cdot \mathbf{F}^{\mathrm{e}-1}\right] \cdot \boldsymbol{\tau} + \boldsymbol{\tau} \cdot \left[\mathbf{F}^{\mathrm{e}} \cdot \left(\dot{\mathbf{F}}^{\mathrm{p}} \cdot \mathbf{F}^{\mathrm{p}-1}\right) \cdot \mathbf{F}^{\mathrm{e}-1}\right]^{T}\right\}. \tag{11.12.6}$$

This defines the elastic part of the rate of deformation corresponding to the stress rate $\overset{\triangle}{\boldsymbol{\tau}}$, which is

$$\mathbf{D}^{\mathrm{e}}_{(1)} = \left(\dot{\mathbf{F}}^{\mathrm{e}} \cdot \mathbf{F}^{\mathrm{e}-1}\right)_{\mathrm{s}} - \boldsymbol{\mathcal{L}}^{-1}_{(1)} : \left\{\left[\mathbf{F}^{\mathrm{e}} \cdot \left(\dot{\mathbf{F}}^{\mathrm{p}} \cdot \mathbf{F}^{\mathrm{p}-1}\right) \cdot \mathbf{F}^{\mathrm{e}-1}\right] \cdot \boldsymbol{\tau} + \boldsymbol{\tau} \cdot \left[\mathbf{F}^{\mathrm{e}} \cdot \left(\dot{\mathbf{F}}^{\mathrm{p}} \cdot \mathbf{F}^{\mathrm{p}-1}\right) \cdot \mathbf{F}^{\mathrm{e}-1}\right]^{T}\right\}. \tag{11.12.7}$$

The remaining part of the rate of deformation is the plastic part,

$$\mathbf{D}^{\mathrm{p}}_{(1)} = \left[\mathbf{F}^{\mathrm{e}} \cdot \left(\dot{\mathbf{F}}^{\mathrm{p}} \cdot \mathbf{F}^{\mathrm{p}-1}\right) \cdot \mathbf{F}^{\mathrm{e}-1}\right]_{\mathrm{s}} + \boldsymbol{\mathcal{L}}^{-1}_{(1)} : \left\{\left[\mathbf{F}^{\mathrm{e}} \cdot \left(\dot{\mathbf{F}}^{\mathrm{p}} \cdot \mathbf{F}^{\mathrm{p}-1}\right) \cdot \mathbf{F}^{\mathrm{e}-1}\right] \cdot \boldsymbol{\tau} + \boldsymbol{\tau} \cdot \left[\mathbf{F}^{\mathrm{e}} \cdot \left(\dot{\mathbf{F}}^{\mathrm{p}} \cdot \mathbf{F}^{\mathrm{p}-1}\right) \cdot \mathbf{F}^{\mathrm{e}-1}\right]^{T}\right\}. \tag{11.12.8}$$

The corresponding elastic and plastic parts of the stress rate $\overset{\triangle}{\boldsymbol{\tau}}$ are

$$\overset{\triangle}{\boldsymbol{\tau}}{}^{\mathrm{e}} = \boldsymbol{\mathcal{L}}_{(1)} : \mathbf{D}, \quad \overset{\triangle}{\boldsymbol{\tau}}{}^{\mathrm{p}} = -\boldsymbol{\mathcal{L}}_{(1)} : \mathbf{D}^{\mathrm{p}}_{(1)}. \tag{11.12.9}$$

It is readily verified that

$$\overset{\circ}{\boldsymbol{\tau}}{}^{\mathrm{p}} = \overset{\triangle}{\boldsymbol{\tau}}{}^{\mathrm{p}}. \tag{11.12.10}$$

The partition of the rate of deformation based on the convected rate of the Kirchhoff stress, which involves in its definition both the spin and the rate of deformation, may appear less appealing than the partition based on the Jaumann rate, which involves only the spin part of the velocity gradient. However, the partition based on the convected rate is inherent in the constitutive formulation based on the Lagrangian strain and its conjugate, symmetric Piola–Kirchhoff stress. Since $\dot{\mathbf{E}} = \mathbf{F}^T \cdot \mathbf{D} \cdot \mathbf{F}$, the elastic and plastic parts of the rate of Lagrangian strain are (Lubarda, 1994a)

$$(\dot{\mathbf{E}})^{\mathrm{e}} = \mathbf{F}^T \cdot \mathbf{D}^{\mathrm{e}}_{(1)} \cdot \mathbf{F}, \quad (\dot{\mathbf{E}})^{\mathrm{p}} = \mathbf{F}^T \cdot \mathbf{D}^{\mathrm{p}}_{(1)} \cdot \mathbf{F}. \tag{11.12.11}$$

These are defined such that

$$(\dot{\mathbf{E}})^{\mathrm{e}} = \boldsymbol{\Lambda}^{-1}_{(1)} : \dot{\mathbf{T}}, \quad (\dot{\mathbf{E}})^{\mathrm{p}} = \dot{\mathbf{E}} - (\dot{\mathbf{E}})^{\mathrm{e}}, \tag{11.12.12}$$

where $\mathbf{T}$ is the symmetric Piola–Kirchhoff stress tensor, conjugate to the Lagrangian strain $\mathbf{E}$ (the conjugate measures from Chapters 2 and 3 corresponding to $n = 1$; for simplicity we omit here the subscript (1) in the notation for $\mathbf{T}_{(1)}$ and $\mathbf{E}_{(1)}$). The plastic part of the rate of Lagrangian strain is normal to a locally smooth yield surface in the Piola–Kirchhoff stress space, and is within the cone of outward normals at the vertex of the yield surface. An independent derivation of the partition of the rate of Lagrangian strain into its elastic and plastic parts is presented in the following section.

11.13. Partition of the Rate of Lagrangian Strain

If elastic strain energy per unit unstressed volume is an isotropic function of the Lagrangian strain $\mathbf{E}^{\mathrm{e}}$, it can be expressed, with the help of Eq. (11.2.5), as

$$\Psi^{\mathrm{e}} = \Psi^{\mathrm{e}}(\mathbf{E}^{\mathrm{e}}) = \Psi^{\mathrm{e}}\left[\mathbf{F}^{\mathrm{p}-T} \cdot (\mathbf{E} - \mathbf{E}^{\mathrm{p}}) \cdot \mathbf{F}^{\mathrm{p}-1}\right]. \tag{11.13.1}$$

From this we deduce that

$$\mathbf{T}^{\mathrm{e}} = \frac{\partial \Psi^{\mathrm{e}}}{\partial \mathbf{E}^{\mathrm{e}}}, \quad \mathbf{T} = \frac{\partial \Psi^{\mathrm{e}}}{\partial \mathbf{E}}, \tag{11.13.2}$$

with a connection between the two stress tensors

$$\mathbf{T}^{\mathrm{e}} = \mathbf{F}^{\mathrm{p}} \cdot \mathbf{T} \cdot \mathbf{F}^{\mathrm{p}T}. \tag{11.13.3}$$

The stress tensors $\mathbf{T}^{\mathrm{e}}$ and $\mathbf{T}$ are related to the Kirchhoff stress $\boldsymbol{\tau}$ by

$$\mathbf{T}^{\mathrm{e}} = \mathbf{F}^{\mathrm{e}-1} \cdot \boldsymbol{\tau} \cdot \mathbf{F}^{\mathrm{e}-T}, \quad \mathbf{T} = \mathbf{F}^{-1} \cdot \boldsymbol{\tau} \cdot \mathbf{F}^{-T}. \tag{11.13.4}$$

The plastic incompressibility is assumed, so that

$$\det \mathbf{F}^{\mathrm{e}} = \det \mathbf{F}. \tag{11.13.5}$$

The two moduli tensors are defined by

$$\boldsymbol{\Lambda}^{\mathrm{e}}_{(1)} = \frac{\partial^2 \Psi^{\mathrm{e}}}{\partial \mathbf{E}^{\mathrm{e}} \otimes \partial \mathbf{E}^{\mathrm{e}}}, \quad \boldsymbol{\Lambda}_{(1)} = \frac{\partial^2 \Psi^{\mathrm{e}}}{\partial \mathbf{E} \otimes \partial \mathbf{E}}, \tag{11.13.6}$$

such that

$$\boldsymbol{\Lambda}_{(1)} = \mathbf{F}^{\mathrm{p}-1}\,\mathbf{F}^{\mathrm{p}-1}\,\boldsymbol{\Lambda}^{\mathrm{e}}_{(1)}\,\mathbf{F}^{\mathrm{p}-T}\,\mathbf{F}^{\mathrm{p}-T}. \tag{11.13.7}$$

In addition, the moduli tensor $\boldsymbol{\mathcal{L}}_{(1)}$ is

$$\boldsymbol{\mathcal{L}}_{(1)} = \mathbf{F}^{\mathrm{e}}\,\mathbf{F}^{\mathrm{e}}\,\boldsymbol{\Lambda}^{\mathrm{e}}_{(1)}\,\mathbf{F}^{\mathrm{e}T}\,\mathbf{F}^{\mathrm{e}T} = \mathbf{F}\,\mathbf{F}\,\boldsymbol{\Lambda}_{(1)}\,\mathbf{F}^{T}\,\mathbf{F}^{T}. \tag{11.13.8}$$

The tensor products are here defined as in Eq. (11.6.3).

By differentiating the first expression in (11.13.2), there follows

$$\dot{\mathbf{T}}^{\mathrm{e}} = \boldsymbol{\Lambda}^{\mathrm{e}}_{(1)} : \dot{\mathbf{E}}^{\mathrm{e}}, \tag{11.13.9}$$

while differentiation of Eq. (11.13.3) gives

$$\dot{\mathbf{T}}^{\mathrm{e}} = \mathbf{F}^{\mathrm{p}} \cdot \left(\dot{\mathbf{T}} + \mathbf{Z}^{\mathrm{p}} \cdot \mathbf{T} + \mathbf{T} \cdot \mathbf{Z}^{\mathrm{p}T}\right) \cdot \mathbf{F}^{\mathrm{p}T}. \tag{11.13.10}$$

The second-order tensor $\mathbf{Z}^{\mathrm{p}}$ is

$$\mathbf{Z}^{\mathrm{p}} = \mathbf{F}^{\mathrm{p}-1} \cdot \left(\dot{\mathbf{F}}^{\mathrm{p}} \cdot \mathbf{F}^{\mathrm{p}-1}\right) \mathbf{F}^{\mathrm{p}}. \tag{11.13.11}$$

Since, from Eq. (11.3.7),

$$\dot{\mathbf{E}}^{\mathrm{e}} = \mathbf{F}^{\mathrm{p}-T} \cdot \left\{\dot{\mathbf{E}} - \mathbf{F}^{\mathrm{p}T} \cdot \left[\mathbf{C}^{\mathrm{e}} \cdot \left(\dot{\mathbf{F}}^{\mathrm{p}} \cdot \mathbf{F}^{\mathrm{p}-1}\right)\right]_{\mathrm{s}} \cdot \mathbf{F}^{\mathrm{p}}\right\} \cdot \mathbf{F}^{\mathrm{p}-1}, \tag{11.13.12}$$

the substitution of Eqs. (11.13.10) and (11.13.12) into Eq. (11.13.9) yields

$$\dot{\mathbf{T}} = \boldsymbol{\Lambda}_{(1)} : \left\{\dot{\mathbf{E}} - \mathbf{F}^{\mathrm{p}T} \cdot \left[\mathbf{C}^{\mathrm{e}} \cdot \left(\dot{\mathbf{F}}^{\mathrm{p}} \cdot \mathbf{F}^{\mathrm{p}-1}\right)\right]_{\mathrm{s}} \cdot \mathbf{F}^{\mathrm{p}}\right\} - \left(\mathbf{Z}^{\mathrm{p}} \cdot \mathbf{T} + \mathbf{T} \cdot \mathbf{Z}^{\mathrm{p}T}\right). \tag{11.13.13}$$

The elastic part of the rate of Lagrangian strain is defined by

$$(\dot{\mathbf{E}})^{\mathrm{e}} = \boldsymbol{\Lambda}^{-1}_{(1)} : \dot{\mathbf{T}}. \tag{11.13.14}$$

Consequently, upon partitioning the total rate of strain as (Fig. 11.2)

$$\dot{\mathbf{E}} = (\dot{\mathbf{E}})^{\mathrm{e}} + (\dot{\mathbf{E}})^{\mathrm{p}}, \tag{11.13.15}$$

we identify from Eq. (11.13.13) the plastic part of the rate of Lagrangian strain as

$$(\dot{\mathbf{E}})^{\mathrm{p}} = \mathbf{F}^{\mathrm{p}T} \cdot \left[\mathbf{C}^{\mathrm{e}} \cdot \left(\dot{\mathbf{F}}^{\mathrm{p}} \cdot \mathbf{F}^{\mathrm{p}-1}\right)\right]_{\mathrm{s}} \cdot \mathbf{F}^{\mathrm{p}} + \boldsymbol{\Lambda}^{-1}_{(1)} : \left(\mathbf{Z}^{\mathrm{p}} \cdot \mathbf{T} + \mathbf{T} \cdot \mathbf{Z}^{\mathrm{p}T}\right). \tag{11.13.16}$$

The elastic part is then

$$(\dot{\mathbf{E}})^{\mathrm{e}} = \mathbf{F}^{\mathrm{p}T} \cdot \dot{\mathbf{E}}^{\mathrm{e}} \cdot \mathbf{F}^{\mathrm{p}} - \boldsymbol{\Lambda}^{-1}_{(1)} : \left(\mathbf{Z}^{\mathrm{p}} \cdot \mathbf{T} + \mathbf{T} \cdot \mathbf{Z}^{\mathrm{p}T}\right), \tag{11.13.17}$$

where

$$\dot{\mathbf{E}}^{\mathrm{e}} = \mathbf{F}^{\mathrm{e}T} \cdot \left(\dot{\mathbf{F}}^{\mathrm{e}} \cdot \mathbf{F}^{\mathrm{e}-1}\right)_{\mathrm{s}} \cdot \mathbf{F}^{\mathrm{e}}. \tag{11.13.18}$$

It can be easily verified that the expressions (11.13.16) and(11.13.17) agree with the expressions (11.12.11), provided that $\mathbf{D}^{\mathrm{e}}_{(1)}$ and $\mathbf{D}^{\mathrm{p}}_{(1)}$ are defined by Eqs. (11.12.7) and (11.12.8).

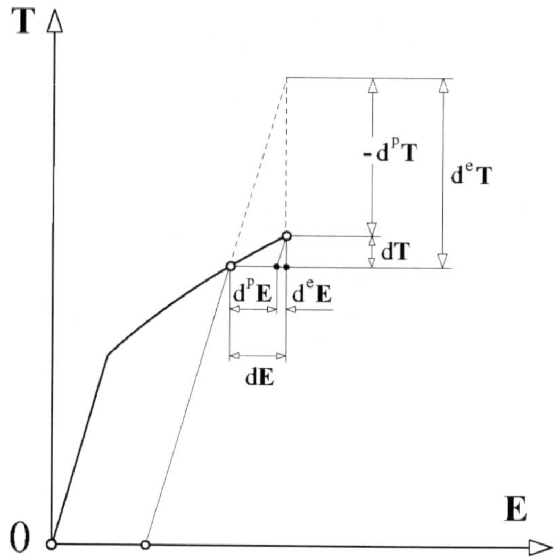

FIGURE 11.2. Geometric interpretation of the partition of the stress and strain increments into their elastic and plastic parts.

11.14. Partition of the Rate of Deformation Gradient

The rate of deformation gradient $\dot{\mathbf{F}}$ can also be partitioned into its elastic and plastic parts,

$$\dot{\mathbf{F}} = (\dot{\mathbf{F}})^{\mathrm{e}} + (\dot{\mathbf{F}})^{\mathrm{p}}. \tag{11.14.1}$$

The elastic part is defined by

$$(\dot{\mathbf{F}})^{\mathrm{e}} = \boldsymbol{\Lambda}^{-1} \cdot\cdot\, \dot{\mathbf{P}}. \tag{11.14.2}$$

It is assumed that the elastic pseudomoduli tensor $\boldsymbol{\Lambda}$ has its inverse, the elastic pseudocompliances tensor $\boldsymbol{\Lambda}^{-1}$, such that

$$\boldsymbol{\Lambda} \cdot\cdot\, \boldsymbol{\Lambda}^{-1} = \boldsymbol{\Lambda}^{-1} \cdot\cdot\, \boldsymbol{\Lambda} = \boldsymbol{I}, \tag{11.14.3}$$

where $I_{ijkl} = \delta_{il}\delta_{jk}$ (with the components of $\boldsymbol{\Lambda}$ and $\boldsymbol{\Lambda}^{-1}$ expressed in the same rectangular coordinate system).

In the derivation, first note that the elastic nominal stress and the overall nominal stress,

$$\mathbf{P}^{\mathrm{e}} = \mathbf{T}^{\mathrm{e}} \cdot \mathbf{F}^{\mathrm{e}T}, \quad \mathbf{P} = \mathbf{T} \cdot \mathbf{F}^{T}, \tag{11.14.4}$$

are derived from the elastic strain energy Ψ^{e} as (Lubarda and Benson, 2001)

$$\mathbf{P}^{\mathrm{e}} = \frac{\partial \Psi^{\mathrm{e}}}{\partial \mathbf{F}^{\mathrm{e}}}, \quad \mathbf{P} = \frac{\partial \Psi^{\mathrm{e}}}{\partial \mathbf{F}}. \tag{11.14.5}$$

The connection between the two tensors is

$$\mathbf{P}^{\mathrm{e}} = \mathbf{F}^{\mathrm{p}} \cdot \mathbf{P}. \tag{11.14.6}$$

The corresponding pseudomoduli tensors are

$$\boldsymbol{\Lambda}^{\mathrm{e}} = \frac{\partial^2 \Psi^{\mathrm{e}}}{\partial \mathbf{F}^{\mathrm{e}} \otimes \partial \mathbf{F}^{\mathrm{e}}}, \quad \boldsymbol{\Lambda} = \frac{\partial^2 \Psi^{\mathrm{e}}}{\partial \mathbf{F} \otimes \partial \mathbf{F}}. \tag{11.14.7}$$

It can be readily verified by partial differentiation that the components of the two pseudomoduli tensors (in the same rectangular coordinate system) are related by

$$\Lambda^{\mathrm{e}}_{ijkl} = F^{\mathrm{p}}_{im}\, \Lambda_{mjnl}\, F^{\mathrm{p}}_{kn}. \tag{11.14.8}$$

The pseudomoduli tensor $\boldsymbol{\Lambda}^{\mathrm{e}}$ appears in the expression

$$\dot{\mathbf{P}}^{\mathrm{e}} = \boldsymbol{\Lambda}^{\mathrm{e}} \cdot\cdot\, \dot{\mathbf{F}}^{\mathrm{e}}. \tag{11.14.9}$$

By differentiating Eq. (11.14.6), there follows

$$\dot{\mathbf{P}}^{\mathrm{e}} = \mathbf{F}^{\mathrm{p}} \cdot \dot{\mathbf{P}} + \dot{\mathbf{F}}^{\mathrm{p}} \cdot \mathbf{P}. \tag{11.14.10}$$

Substitution of Eqs. (11.14.10) and (11.14.8) into Eq. (11.14.9) gives

$$\dot{\mathbf{P}} = \boldsymbol{\Lambda} \cdot\cdot \left(\dot{\mathbf{F}}^{\mathrm{e}} \cdot \mathbf{F}^{\mathrm{p}}\right) - \mathbf{F}^{\mathrm{p}-1} \cdot \dot{\mathbf{F}}^{\mathrm{p}} \cdot \mathbf{P}. \tag{11.14.11}$$

On the other hand, by differentiating the multiplicative decomposition $\mathbf{F} = \mathbf{F}^{\mathrm{e}} \cdot \mathbf{F}^{\mathrm{p}}$, the rate of deformation gradient is

$$\dot{\mathbf{F}} = \dot{\mathbf{F}}^{\mathrm{e}} \cdot \mathbf{F}^{\mathrm{p}} + \mathbf{F}^{\mathrm{e}} \cdot \dot{\mathbf{F}}^{\mathrm{p}}. \tag{11.14.12}$$

Using this, Eq. (11.14.11) can be rewritten as

$$\dot{\mathbf{P}} = \boldsymbol{\Lambda} \cdot\cdot \left(\dot{\mathbf{F}} - \mathbf{F}^{\mathrm{e}} \cdot \dot{\mathbf{F}}^{\mathrm{p}}\right) - \mathbf{F}^{\mathrm{p}-1} \cdot \dot{\mathbf{F}}^{\mathrm{p}} \cdot \mathbf{P}, \tag{11.14.13}$$

i.e.,

$$\dot{\mathbf{P}} = \boldsymbol{\Lambda} \cdot\cdot \left[\dot{\mathbf{F}} - \mathbf{F}^{\mathrm{e}} \cdot \dot{\mathbf{F}}^{\mathrm{p}} - \boldsymbol{\Lambda}^{-1} \cdot\cdot \left(\mathbf{F}^{\mathrm{p}-1} \cdot \dot{\mathbf{F}}^{\mathrm{p}} \cdot \mathbf{P}\right)\right]. \tag{11.14.14}$$

From Eq. (11.14.14) we now identify the plastic part of the rate of deformation gradient as

$$(\dot{\mathbf{F}})^{\mathrm{p}} = \mathbf{F}^{\mathrm{e}} \cdot \dot{\mathbf{F}}^{\mathrm{p}} + \boldsymbol{\Lambda}^{-1} \cdot\cdot \left(\mathbf{F}^{\mathrm{p}-1} \cdot \dot{\mathbf{F}}^{\mathrm{p}} \cdot \mathbf{P}\right). \tag{11.14.15}$$

The remaining part of $\dot{\mathbf{F}}$ is the elastic part,

$$(\dot{\mathbf{F}})^{\mathrm{e}} = \dot{\mathbf{F}}^{\mathrm{e}} \cdot \mathbf{F}^{\mathrm{p}} - \mathbf{\Lambda}^{-1} \cdot\cdot \left(\mathbf{F}^{\mathrm{p}-1} \cdot \dot{\mathbf{F}}^{\mathrm{p}} \cdot \mathbf{P} \right), \tag{11.14.16}$$

complying with Eq. (11.14.2).

Equation (11.14.14) also serves to identify the elastic and plastic parts of the rate of nominal stress. These are

$$(\dot{\mathbf{P}})^{\mathrm{e}} = \mathbf{\Lambda} \cdot\cdot \dot{\mathbf{F}}, \tag{11.14.17}$$

$$(\dot{\mathbf{P}})^{\mathrm{p}} = -\left[\mathbf{F}^{\mathrm{p}-1} \cdot \dot{\mathbf{F}}^{\mathrm{p}} \cdot \mathbf{P} + \mathbf{\Lambda} \cdot\cdot \left(\mathbf{F}^{\mathrm{e}} \cdot \dot{\mathbf{F}}^{\mathrm{p}} \right) \right], \tag{11.14.18}$$

such that

$$\dot{\mathbf{P}} = (\dot{\mathbf{P}})^{\mathrm{e}} + (\dot{\mathbf{P}})^{\mathrm{p}}. \tag{11.14.19}$$

Evidently, by comparing Eqs. (11.14.15) and (11.14.18), there is a relationship between plastic parts of the rate of nominal stress and deformation gradient,

$$(\dot{\mathbf{P}})^{\mathrm{p}} = -\mathbf{\Lambda} \cdot\cdot (\dot{\mathbf{F}})^{\mathrm{p}}. \tag{11.14.20}$$

11.15. Relationship between $(\dot{\mathbf{P}})^{\mathrm{p}}$ and $(\dot{\mathbf{T}})^{\mathrm{p}}$

To derive the relationship between plastic parts of the rates of nominal and symmetric Piola–Kirchhoff stress,

$$(\dot{\mathbf{P}})^{\mathrm{p}} = \dot{\mathbf{P}} - \mathbf{\Lambda} \cdot\cdot \dot{\mathbf{F}}, \quad (\dot{\mathbf{T}})^{\mathrm{p}} = \dot{\mathbf{T}} - \mathbf{\Lambda}_{(1)} : \dot{\mathbf{E}}, \tag{11.15.1}$$

we first recall the relationships between $\dot{\mathbf{P}}$ and $\dot{\mathbf{T}}$, and $\mathbf{\Lambda}$ and $\mathbf{\Lambda}_{(1)}$, which were derived in Section 6.4. Following Hill (1984), these can be conveniently cast as

$$\mathbf{\Lambda} = \boldsymbol{\mathcal{K}}^T : \mathbf{\Lambda}_{(1)} : \boldsymbol{\mathcal{K}} + \boldsymbol{\mathcal{T}}, \quad \dot{\mathbf{P}} = \boldsymbol{\mathcal{K}}^T : \dot{\mathbf{T}} + \boldsymbol{\mathcal{T}} \cdot\cdot \dot{\mathbf{F}}. \tag{11.15.2}$$

The tensor $\mathbf{\Lambda}_{(1)}$ possesses the reciprocal symmetry $ij \leftrightarrow kl$. The rectangular components of the fourth-order tensors $\boldsymbol{\mathcal{K}}$ and $\boldsymbol{\mathcal{T}}$ are

$$\mathcal{K}_{ijkl} = \frac{1}{2}\left(\delta_{ik} F_{lj} + \delta_{jk} F_{li}\right), \quad \mathcal{T}_{ijkl} = T_{ik}\delta_{jl}. \tag{11.15.3}$$

They obey the symmetry

$$\mathcal{K}_{ijkl} = \mathcal{K}_{jikl}, \quad \mathcal{T}_{ijkl} = \mathcal{T}_{klij}. \tag{11.15.4}$$

The tensor $\boldsymbol{\mathcal{K}}$ is particularly convenient, because in the trace operation with a second-order tensor $\mathbf{A}$ it behaves such that

$$\boldsymbol{\mathcal{K}} \cdot\cdot \mathbf{A} = \mathbf{A} \cdot\cdot \boldsymbol{\mathcal{K}}^T = \frac{1}{2}\left(\mathbf{F}^T \cdot \mathbf{A} + \mathbf{A}^T \cdot \mathbf{F}\right), \tag{11.15.5}$$

$$\boldsymbol{\mathcal{K}}^T \cdot\cdot \mathbf{A} = \mathbf{A} \cdot\cdot \boldsymbol{\mathcal{K}} = \frac{1}{2}\left(\mathbf{A} + \mathbf{A}^T\right) \cdot \mathbf{F}^T. \tag{11.15.6}$$

In particular,

$$\boldsymbol{\mathcal{K}} \cdot\cdot \dot{\mathbf{F}} = \dot{\mathbf{F}} \cdot\cdot \boldsymbol{\mathcal{K}}^T = \dot{\mathbf{E}}, \tag{11.15.7}$$

$$\boldsymbol{\mathcal{K}}^T : \mathbf{T} = \mathbf{T} : \boldsymbol{\mathcal{K}} = \mathbf{T} \cdot \mathbf{F}^T = \mathbf{P}. \tag{11.15.8}$$

If $\mathbf{A}$ is symmetric, the trace product $\cdot\cdot$ can be replaced by $:$ product in Eqs. (11.15.5) and (11.15.6).

The relationship between $(\dot{\mathbf{P}})^{\mathrm{p}}$ and $(\dot{\mathbf{T}})^{\mathrm{p}}$ now follows by taking the trace product of the second equation in (11.15.1) with $\boldsymbol{\mathcal{K}}^T$ from the left. Upon using Eq. (11.15.2), this gives

$$(\dot{\mathbf{P}})^{\mathrm{p}} = \boldsymbol{\mathcal{K}}^T : (\dot{\mathbf{T}})^{\mathrm{p}}. \tag{11.15.9}$$

Since

$$(\dot{\mathbf{P}})^{\mathrm{p}} = -\boldsymbol{\Lambda} \cdot\cdot (\dot{\mathbf{F}})^{\mathrm{p}}, \quad (\dot{\mathbf{T}})^{\mathrm{p}} = -\boldsymbol{\Lambda}_{(1)} : (\dot{\mathbf{E}})^{\mathrm{p}}, \tag{11.15.10}$$

we, in addition, have

$$(\dot{\mathbf{F}})^{\mathrm{p}} = \boldsymbol{\Lambda}^{-1} \cdot\cdot \boldsymbol{\mathcal{K}}^T : \boldsymbol{\Lambda}_{(1)} : (\dot{\mathbf{E}})^{\mathrm{p}}. \tag{11.15.11}$$

Note that

$$\dot{\mathbf{F}} \cdot\cdot (\dot{\mathbf{P}})^{\mathrm{p}} = \dot{\mathbf{E}} : (\dot{\mathbf{T}})^{\mathrm{p}}, \tag{11.15.12}$$

which directly follows by taking the trace product of (11.15.9) with $\dot{\mathbf{F}}$ from the left, and by using Eq. (11.15.7).

11.16. Normality Properties

If increments rather than rates are used, we can write Eq. (11.15.12) as

$$\mathrm{d}\mathbf{F} \cdot\cdot \mathrm{d}^{\mathrm{p}}\mathbf{P} = \mathrm{d}\mathbf{E} : \mathrm{d}^{\mathrm{p}}\mathbf{T}. \tag{11.16.1}$$

An analogous expression holds when the increments of $\mathbf{F}$ and $\mathbf{E}$ are used along an unloading elastic branch of the response, i.e.,

$$\delta\mathbf{F} \cdot\cdot \mathrm{d}^{\mathrm{p}}\mathbf{P} = \delta\mathbf{E} : \mathrm{d}^{\mathrm{p}}\mathbf{T}. \tag{11.16.2}$$

If this is positive, we say that the material complies with the normality rule in strain space. Since

$$\mathrm{d}^{\mathrm{p}}\mathbf{P} = -\mathbf{\Lambda} \cdot\cdot\, \mathrm{d}^{\mathrm{p}}\mathbf{F}, \quad \mathrm{d}^{\mathrm{p}}\mathbf{T} = -\mathbf{\Lambda}_{(1)} : \mathrm{d}^{\mathrm{p}}\mathbf{E}, \tag{11.16.3}$$

and

$$\delta\mathbf{P} = \mathbf{\Lambda} \cdot\cdot\, \delta\mathbf{F}, \quad \delta\mathbf{T} = \mathbf{\Lambda}_{(1)} : \delta\mathbf{E}, \tag{11.16.4}$$

the substitution into Eq. (11.16.2) yields a dual relationship

$$\delta\mathbf{P} \cdot\cdot\, \mathrm{d}^{\mathrm{p}}\mathbf{F} = \delta\mathbf{T} : \mathrm{d}^{\mathrm{p}}\mathbf{E}. \tag{11.16.5}$$

When this is negative, the material complies with the normality rule in stress space. We recall from Section 8.5, if the material complies with Ilyushin's postulate of positive net work in an isothermal cycle of strain that involves plastic deformation, the quantity in (11.16.1) must be negative, i.e.,

$$\mathrm{d}\mathbf{F} \cdot\cdot\, \mathrm{d}^{\mathrm{p}}\mathbf{P} = \mathrm{d}\mathbf{E} : \mathrm{d}^{\mathrm{p}}\mathbf{T} < 0. \tag{11.16.6}$$

Equation (11.16.1) does not have a dual relationship, since

$$\mathrm{d}\mathbf{P} \cdot\cdot\, \mathrm{d}^{\mathrm{p}}\mathbf{F} \neq \mathrm{d}\mathbf{T} : \mathrm{d}^{\mathrm{p}}\mathbf{E}. \tag{11.16.7}$$

Instead, we can only write

$$\mathrm{d}\mathbf{F} \cdot\cdot\, \mathbf{\Lambda} \cdot\cdot\, \mathrm{d}^{\mathrm{p}}\mathbf{F} = \mathrm{d}\mathbf{E} : \mathbf{\Lambda}_{(1)} : \mathrm{d}^{\mathrm{p}}\mathbf{E}, \tag{11.16.8}$$

or

$$\mathrm{d}\mathbf{P} \cdot\cdot\, \mathrm{d}^{\mathrm{p}}\mathbf{F} + \mathrm{d}^{\mathrm{p}}\mathbf{F} \cdot\cdot\, \mathbf{\Lambda} \cdot\cdot\, \mathrm{d}^{\mathrm{p}}\mathbf{F} = \mathrm{d}\mathbf{T} : \mathrm{d}^{\mathrm{p}}\mathbf{E} + \mathrm{d}^{\mathrm{p}}\mathbf{E} : \mathbf{\Lambda}_{(1)} : \mathrm{d}^{\mathrm{p}}\mathbf{E}. \tag{11.16.9}$$

If the material is in the hardening range relative to the conjugate measures $\mathbf{E}$ and $\mathbf{T}$, the stress increment $\mathrm{d}\mathbf{T}$, producing plastic deformation $\mathrm{d}^{\mathrm{p}}\mathbf{E}$, is directed outside the yield surface, satisfying $\mathrm{d}\mathbf{T} : \mathrm{d}^{\mathrm{p}}\mathbf{E} > 0$. If the material is in the softening range, the stress increment producing plastic deformation is directed inside the yield surface, satisfying the reversed inequality. The quantity $\mathrm{d}\mathbf{T} : \mathrm{d}^{\mathrm{p}}\mathbf{E}$, however, is not invariant under the change of strain measure, and the material judged to be in the hardening range relative to one pair of the conjugate stress and strain measures, may be in the softening range relative to another pair.

As an illustration, consider a uniaxial tension of an incompressible rigid-plastic material whose response in the Cauchy stress vs. logarithmic strain

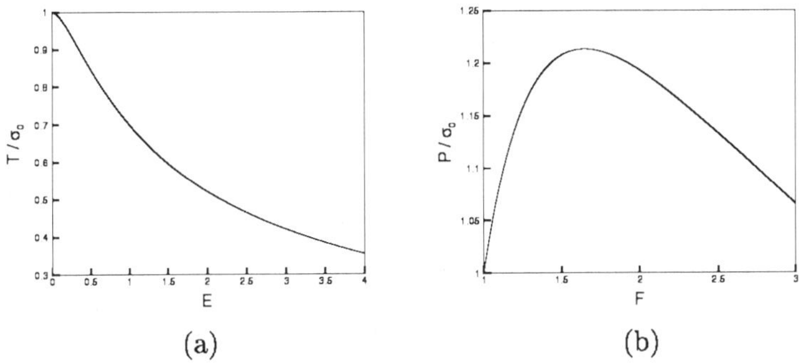

FIGURE 11.3. (a) Piola–Kirchhoff vs. Lagrangian strain, and (b) nominal stress vs. deformation gradient in uniaxial tension of rigid-plastic material with the constant hardening modulus $k = 2\sigma_0$ relative to the Cauchy stress and logarithmic strain measures.

space is described by the linear hardening

$$\sigma = k \ln \lambda + \sigma_0. \tag{11.16.10}$$

The constant rate of hardening is k, the initial yield stress is σ_0, and λ is the longitudinal stretch ratio. The corresponding response observed relative to $T = \sigma/\lambda^2$ vs. $E = (\lambda^2 - 1)/2$ measures is

$$T = \frac{1}{1+2E}\left[\sigma_0 + \frac{1}{2}\, k \ln(1+2E)\right]. \tag{11.16.11}$$

A transition from the hardening to softening occurs at

$$E_0 = \frac{1}{2}\left[\exp\left(1 - \frac{2\sigma_0}{k}\right) - 1\right]. \tag{11.16.12}$$

The response observed relative to $P = \sigma/\lambda$ vs. $F = \lambda$ measures is

$$P = \frac{1}{F}\,(\sigma_0 + k \ln F). \tag{11.16.13}$$

A transition from the hardening to softening occurs at

$$F_0 = \exp\left(1 - \frac{\sigma_0}{k}\right). \tag{11.16.14}$$

For example, for $k = 2\sigma_0$ the softening commences at $F_0 = \sqrt{e}$ in the nominal stress vs. deformation gradient, and from the onset of deformation in the Piola–Kirchhoff stress vs. Lagrangian strain measures ($E_0 = 0$). However, a necking of the specimen begins when $\mathrm{d}\sigma/\mathrm{d}(\ln\lambda) = \sigma$, i.e., when

$\lambda = \exp(1 - \sigma_0/k)$. Thus, the necking takes place when

$$\ln \lambda_* = 1 - \frac{\sigma_0}{k}, \quad 2E_* = \exp\left[2\left(1 - \frac{\sigma_0}{k}\right)\right] - 1, \quad F_* = \exp\left(1 - \frac{\sigma_0}{k}\right). \tag{11.16.15}$$

For $k = 2\sigma_0$, this gives

$$\ln \lambda_* = \frac{1}{2} \quad E_* = \frac{1}{2}(e - 1), \quad F_* = \sqrt{e}. \tag{11.16.16}$$

While the onset of softening coincides with the onset of necking when (P, F) measures are used, with $k = 2\sigma_0$ the necking occurs in the softening range relative to (T, E) measures, and in the hardening range relative to $(\sigma, \ln \lambda)$ measures (Fig. 11.3).

11.17. Elastoplastic Deformation of Orthotropic Materials

11.17.1. Principal Axes of Orthotropy

Consider an elastically orthotropic material in its undeformed configuration $\mathcal{B}^0$. Let the unit vectors $\mathbf{a}_i^0$ $(i = 1, 2, 3)$ define the corresponding principal axes of orthotropy (Fig. 11.4). The elastic strain energy function can be most conveniently expressed in the coordinate system with the axes parallel to $\mathbf{a}_i^0$. Denote this representation by

$$\Psi^{\mathrm{e}} = \Psi^{\mathrm{e}}(\mathbf{E}^{\mathrm{e}}), \tag{11.17.1}$$

where $\mathbf{E}^{\mathrm{e}}$ is the Lagrangian strain of purely elastic deformation from $\mathcal{B}^0$. If it is assumed that the material remains orthotropic during elastoplastic deformation, the principal axes of orthotropy in the intermediate configuration $\mathcal{B}^{\mathrm{p}}$ are defined by the unit vectors

$$\hat{\mathbf{a}}_i = \boldsymbol{\mathcal{R}} \cdot \mathbf{a}_i^0, \tag{11.17.2}$$

where $\boldsymbol{\mathcal{R}}$ is an orthogonal rotation tensor. The elastic strain energy relative to the unstressed intermediate configuration, expressed in the original coordinate system with the axes parallel to $\mathbf{a}_i^0$, is (Lubarda, 1991b)

$$\Psi^{\mathrm{e}} = \Psi^{\mathrm{e}}(\boldsymbol{\mathcal{R}}^T \cdot \mathbf{E}^{\mathrm{e}} \cdot \boldsymbol{\mathcal{R}}). \tag{11.17.3}$$

The function Ψ^{e} here is the same function as that used in Eq. (11.17.1) to describe elastic response from the initial undeformed configuration, but its arguments are the components of the rotated strain tensor

$$\hat{\mathbf{E}}^{\mathrm{e}} = \boldsymbol{\mathcal{R}}^T \cdot \mathbf{E}^{\mathrm{e}} \cdot \boldsymbol{\mathcal{R}}, \quad \mathbf{E}^{\mathrm{e}} = \frac{1}{2}\left(\mathbf{F}^{\mathrm{e}T} \cdot \mathbf{F}^{\mathrm{e}} - \mathbf{I}\right). \tag{11.17.4}$$

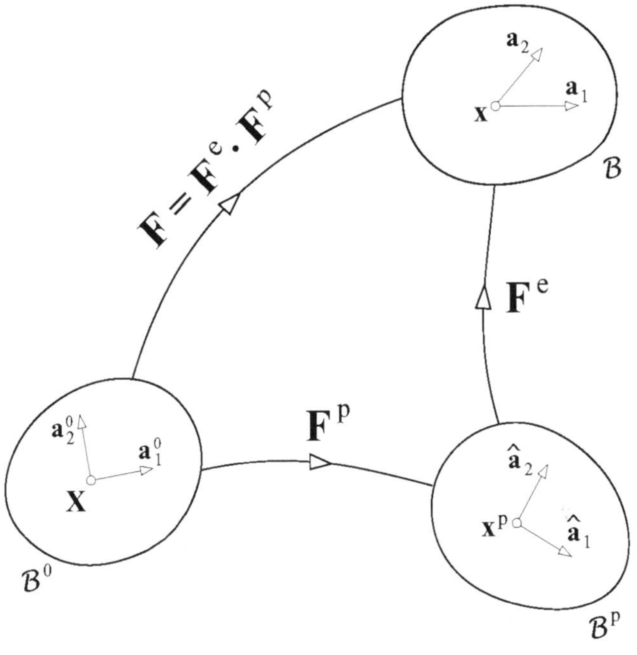

FIGURE 11.4. Multiplicative decomposition of deformation gradient for an orthotropic material. Principal directions of orthotropy $\mathbf{a}_i^0$ in the initial configuration $\mathcal{B}^0$ are rotated to $\hat{\mathbf{a}}_i = \boldsymbol{\mathcal{R}} \cdot \mathbf{a}_i^0$ in the intermediate configuration $\mathcal{B}^{\mathrm{p}}$. They are then convected to the current configuration $\mathcal{B}$ by elastic deformation such that $\mathbf{a}_i = \mathbf{F}^{\mathrm{e}} \cdot \hat{\mathbf{a}}_i$.

The components of the strain tensor $\mathbf{E}^{\mathrm{e}}$, observed in the coordinate system with the axes parallel to $\mathbf{a}_i^0$, are numerically equal to the components of the strain tensor $\hat{\mathbf{E}}^{\mathrm{e}}$, observed in the coordinate system with the axes parallel to $\hat{\mathbf{a}}_i$.

Due to possible discontinuities in displacements and rotations of the material elements at the microscale, caused by plastic deformation, the rotation tensor $\boldsymbol{\mathcal{R}}$ is in general not specified by the overall plastic deformation gradient $\mathbf{F}^{\mathrm{p}}$. In particular, the vectors $\mathbf{a}_i^0$ are not simply convected with the material in the transformation from $\mathcal{B}^0$ to $\mathcal{B}^{\mathrm{p}}$. In contrast, the unit vectors $\hat{\mathbf{a}}_i$ can be considered as embedded in the material during the elastic deformation from $\mathcal{B}^{\mathrm{p}}$ to $\mathcal{B}$. Thus, they become

$$\mathbf{a}_i = \mathbf{F}^{\mathrm{e}} \cdot \hat{\mathbf{a}}_i = \mathbf{F}^{\mathrm{e}} \cdot \boldsymbol{\mathcal{R}} \cdot \mathbf{a}_i^0 \tag{11.17.5}$$

in the elastoplastically deformed configuration $\mathcal{B}$. By differentiation, their rate of change is

$$\dot{\mathbf{a}}_i = \left[\dot{\mathbf{F}}^{\mathrm{e}} \cdot \mathbf{F}^{\mathrm{e}-1} + \mathbf{F}^{\mathrm{e}} \cdot \left(\dot{\boldsymbol{\mathcal{R}}} \cdot \boldsymbol{\mathcal{R}}^{-1}\right) \cdot \mathbf{F}^{\mathrm{e}-1}\right] \cdot \mathbf{a}_i. \tag{11.17.6}$$

If $\mathbf{R}^{\mathrm{e}}$ is the rotation tensor from the polar decomposition

$$\mathbf{F}^{\mathrm{e}} = \mathbf{V}^{\mathrm{e}} \cdot \mathbf{R}^{\mathrm{e}} = \mathbf{R}^{\mathrm{e}} \cdot \mathbf{U}^{\mathrm{e}}, \tag{11.17.7}$$

we have

$$\mathbf{a}_i = \mathbf{V}^{\mathrm{e}} \cdot \boldsymbol{\mathcal{R}}^{\mathrm{e}} \cdot \mathbf{a}_i^0, \quad \boldsymbol{\mathcal{R}}^{\mathrm{e}} = \mathbf{R}^{\mathrm{e}} \cdot \boldsymbol{\mathcal{R}}. \tag{11.17.8}$$

While both $\boldsymbol{\mathcal{R}}$ and $\mathbf{R}^{\mathrm{e}}$ depend on the choice of the intermediate configuration (superposed rotation $\hat{\mathbf{Q}}$), so that

$$\boldsymbol{\mathcal{R}}^* = \hat{\mathbf{Q}} \cdot \boldsymbol{\mathcal{R}}, \quad \mathbf{R}^{\mathrm{e}*} = \mathbf{R}^{\mathrm{e}} \cdot \hat{\mathbf{Q}}^T, \tag{11.17.9}$$

the rotation $\boldsymbol{\mathcal{R}}^{\mathrm{e}}$ is a unique quantity, independent of $\hat{\mathbf{Q}}$, i.e.,

$$\boldsymbol{\mathcal{R}}^{\mathrm{e}*} = \boldsymbol{\mathcal{R}}^{\mathrm{e}}. \tag{11.17.10}$$

Apart from rotation of the material elements caused by elastic stretching $\mathbf{V}^{\mathrm{e}}$, the directions of the principal axes of orthotropy in the deformed configuration $\mathcal{B}$ are completely specified by the rotation tensor $\boldsymbol{\mathcal{R}}^{\mathrm{e}}$.

11.17.2. Partition of the Rate of Deformation

The stress response from $\mathcal{B}^{\mathrm{p}}$ to $\mathcal{B}$ is given by

$$\boldsymbol{\tau} = \mathbf{F}^{\mathrm{e}} \cdot \frac{\partial \Psi^{\mathrm{e}}(\hat{\mathbf{E}}^{\mathrm{e}})}{\partial \mathbf{E}^{\mathrm{e}}} \cdot \mathbf{F}^{\mathrm{e}T}, \tag{11.17.11}$$

where $\boldsymbol{\tau} = (\det \mathbf{F})\boldsymbol{\sigma}$ is the Kirchhoff stress. The elastic strain energy function is reckoned per unit unstressed volume, and plastic deformation is assumed to be incompressible. Upon differentiation, we obtain

$$\begin{aligned} \dot{\boldsymbol{\tau}} - \left(\dot{\mathbf{F}}^{\mathrm{e}} \cdot \mathbf{F}^{\mathrm{e}-1}\right) \cdot \boldsymbol{\tau} - \boldsymbol{\tau} \cdot \left(\dot{\mathbf{F}}^{\mathrm{e}} \cdot \mathbf{F}^{\mathrm{e}-1}\right)^T &= \boldsymbol{\mathcal{L}}_{(1)} : \left(\dot{\mathbf{F}}^{\mathrm{e}} \cdot \mathbf{F}^{\mathrm{e}-1}\right)_{\mathrm{s}} \\ &+ \mathbf{F}^{\mathrm{e}} \cdot \left(\frac{\partial^2 \Psi^{\mathrm{e}}}{\partial \mathbf{E}^{\mathrm{e}} \otimes \partial \boldsymbol{\mathcal{R}}} \cdot\cdot\, \dot{\boldsymbol{\mathcal{R}}}\right) \cdot \mathbf{F}^{\mathrm{e}T}. \end{aligned} \tag{11.17.12}$$

The rectangular components of $\boldsymbol{\mathcal{L}}_{(1)}$ are

$$\mathcal{L}_{ijkl}^{(1)} = F_{iM}^{\mathrm{e}} F_{jN}^{\mathrm{e}} \frac{\partial^2 \Psi^{\mathrm{e}}(\hat{\mathbf{E}}^{\mathrm{e}})}{\partial E_{MN}^{\mathrm{e}}\, \partial E_{PQ}^{\mathrm{e}}} F_{kP}^{\mathrm{e}} F_{lQ}^{\mathrm{e}}. \tag{11.17.13}$$

The last term on the right-hand side of Eq. (11.17.12) can be conveniently rewritten as

$$\mathbf{F}^{\mathrm{e}} \cdot \left(\frac{\partial^2 \Psi^{\mathrm{e}}}{\partial \mathbf{E}^{\mathrm{e}} \otimes \partial \boldsymbol{\mathcal{R}}} \cdot \cdot \dot{\boldsymbol{\mathcal{R}}} \right) \cdot \mathbf{F}^{\mathrm{e}T} = -\boldsymbol{\mathcal{L}}_{(1)} : \mathbf{Z}_{\mathrm{s}} - \boldsymbol{\tau} \cdot \mathbf{Z} - \mathbf{Z}^T \cdot \boldsymbol{\tau}, \tag{11.17.14}$$

where

$$\mathbf{Z} = \mathbf{F}^{\mathrm{e}-T} \cdot \left(\dot{\boldsymbol{\mathcal{R}}} \cdot \boldsymbol{\mathcal{R}}^{-1} \right) \cdot \mathbf{F}^{\mathrm{e}T}. \tag{11.17.15}$$

In the transition, the following expressions were utilized

$$\frac{\partial \Psi^{\mathrm{e}}}{\partial \boldsymbol{\mathcal{R}}} = 2 \boldsymbol{\mathcal{R}}^T \cdot \frac{\partial \Psi^{\mathrm{e}}}{\partial \mathbf{E}^{\mathrm{e}}} \cdot \mathbf{E}^{\mathrm{e}}, \tag{11.17.16}$$

$$\frac{\partial^2 \Psi^{\mathrm{e}}}{\partial E^{\mathrm{e}}_{IJ}\, \partial \mathcal{R}_{KL}} = \mathcal{R}_{ML} \left(2 \frac{\partial^2 \Psi^{\mathrm{e}}}{\partial E^{\mathrm{e}}_{IJ}\, \partial E^{\mathrm{e}}_{MN}} E^{\mathrm{e}}_{NK} + \frac{\partial \Psi^{\mathrm{e}}}{\partial E^{\mathrm{e}}_{MI}} \delta_{JK} + \frac{\partial \Psi^{\mathrm{e}}}{\partial E^{\mathrm{e}}_{MJ}} \delta_{IK} \right). \tag{11.17.17}$$

The right-hand side of Eq. (11.17.14) is also equal to

$$-\boldsymbol{\mathcal{L}}_{(1)} : \mathbf{Z}_{\mathrm{s}} - \boldsymbol{\tau} \cdot \mathbf{Z} - \mathbf{Z}^T \cdot \boldsymbol{\tau} = -\boldsymbol{\mathcal{L}}_{(0)} : \mathbf{Z}_{\mathrm{s}} - \boldsymbol{\tau} \cdot \mathbf{Z}_{\mathrm{a}} + \mathbf{Z}_{\mathrm{a}} \cdot \boldsymbol{\tau}. \tag{11.17.18}$$

The components of the elastic moduli tensor $\boldsymbol{\mathcal{L}}_{(0)}$ are

$$\mathcal{L}^{(0)}_{ijkl} = \mathcal{L}^{(1)}_{ijkl} + \frac{1}{2} \left(\tau_{ik} \delta_{jl} + \tau_{jk} \delta_{il} + \tau_{il} \delta_{jk} + \tau_{jl} \delta_{ik} \right). \tag{11.17.19}$$

Thus, Eq. (11.17.12) becomes

$$\begin{aligned} \dot{\boldsymbol{\tau}} - \left(\dot{\mathbf{F}}^{\mathrm{e}} \cdot \mathbf{F}^{\mathrm{e}-1} \right)_{\mathrm{a}} \cdot \boldsymbol{\tau} + \boldsymbol{\tau} \cdot \left(\dot{\mathbf{F}}^{\mathrm{e}} \cdot \mathbf{F}^{\mathrm{e}-1} \right)_{\mathrm{a}} &= \boldsymbol{\mathcal{L}}_{(0)} : \left[\left(\dot{\mathbf{F}}^{\mathrm{e}} \cdot \mathbf{F}^{\mathrm{e}-1} \right)_{\mathrm{s}} - \mathbf{Z}_{\mathrm{s}} \right] \\ &+ \mathbf{Z}_{\mathrm{a}} \cdot \boldsymbol{\tau} - \boldsymbol{\tau} \cdot \mathbf{Z}_{\mathrm{a}}. \end{aligned} \tag{11.17.20}$$

To proceed with the analysis, we recall from Eq. (11.3.4) that

$$\left(\dot{\mathbf{F}}^{\mathrm{e}} \cdot \mathbf{F}^{\mathrm{e}-1} \right)_{\mathrm{a}} = \mathbf{W} - \boldsymbol{\omega}^{\mathrm{p}}, \quad \boldsymbol{\omega}^{\mathrm{p}} = \left[\mathbf{F}^{\mathrm{e}} \cdot \left(\dot{\mathbf{F}}^{\mathrm{p}} \cdot \mathbf{F}^{\mathrm{p}-1} \right) \cdot \mathbf{F}^{\mathrm{e}-1} \right]_{\mathrm{a}}. \tag{11.17.21}$$

When this is substituted into Eq. (11.17.20), there follows

$$\overset{\circ}{\boldsymbol{\tau}} = \boldsymbol{\mathcal{L}}_{(0)} : \left[\left(\dot{\mathbf{F}}^{\mathrm{e}} \cdot \mathbf{F}^{\mathrm{e}-1} \right)_{\mathrm{s}} - \mathbf{Z}_{\mathrm{s}} \right] - \left(\boldsymbol{\omega}^{\mathrm{p}} - \mathbf{Z}_{\mathrm{a}} \right) \cdot \boldsymbol{\tau} + \boldsymbol{\tau} \cdot \left(\boldsymbol{\omega}^{\mathrm{p}} - \mathbf{Z}_{\mathrm{a}} \right). \tag{11.17.22}$$

Consequently, the elastic rate of deformation is given by

$$\begin{aligned} \mathbf{D}^{\mathrm{e}} &= \left(\dot{\mathbf{F}}^{\mathrm{e}} \cdot \mathbf{F}^{\mathrm{e}-1} \right)_{\mathrm{s}} - \mathbf{Z}_{\mathrm{s}} \\ &- \boldsymbol{\mathcal{L}}^{-1}_{(0)} : \left[\left(\boldsymbol{\omega}^{\mathrm{p}} - \mathbf{Z}_{\mathrm{a}} \right) \cdot \boldsymbol{\tau} - \boldsymbol{\tau} \cdot \left(\boldsymbol{\omega}^{\mathrm{p}} - \mathbf{Z}_{\mathrm{a}} \right) \right]. \end{aligned} \tag{11.17.23}$$

The remaining, plastic part of the rate of deformation is

$$\begin{aligned} \mathbf{D}^{\mathrm{p}} &= \left[\mathbf{F}^{\mathrm{e}} \cdot \left(\dot{\mathbf{F}}^{\mathrm{p}} \cdot \mathbf{F}^{\mathrm{p}-1} \right) \cdot \mathbf{F}^{\mathrm{e}-1} \right]_{\mathrm{s}} + \mathbf{Z}_{\mathrm{s}} \\ &+ \boldsymbol{\mathcal{L}}^{-1}_{(0)} : \left[\left(\boldsymbol{\omega}^{\mathrm{p}} - \mathbf{Z}_{\mathrm{a}} \right) \cdot \boldsymbol{\tau} - \boldsymbol{\tau} \cdot \left(\boldsymbol{\omega}^{\mathrm{p}} - \mathbf{Z}_{\mathrm{a}} \right) \right]. \end{aligned} \tag{11.17.24}$$

The spin $(\boldsymbol{\omega}^{\mathrm{p}} - \mathbf{Z}_{\mathrm{a}})$, appearing in the previous equations, can be expressed from Eqs. (11.17.15) and (11.17.21) as

$$\boldsymbol{\omega}^{\mathrm{p}} - \mathbf{Z}_{\mathrm{a}} = \left[\mathbf{F}^{\mathrm{e}} \cdot \left(\dot{\mathbf{F}}^{\mathrm{p}} \cdot \mathbf{F}^{\mathrm{p}-1} - \dot{\boldsymbol{\mathcal{R}}} \cdot \boldsymbol{\mathcal{R}}^{-1}\right) \cdot \mathbf{F}^{\mathrm{e}-1}\right]_{\mathrm{a}}. \tag{11.17.25}$$

11.17.3. Isoclinic Intermediate Configuration

If intermediate configuration is specified by

$$\boldsymbol{\mathcal{R}} = \mathbf{I}, \quad \text{i.e.,} \quad \hat{\mathbf{a}}_i = \mathbf{a}_i^0, \tag{11.17.26}$$

it is referred to as an isoclinic intermediate configuration. The terminology is originally due to Mandel (1973). For an isoclinic intermediate configuration, therefore,

$$\mathbf{a}_i = \mathbf{F}^{\mathrm{e}} \cdot \mathbf{a}_i^0 = \mathbf{V}^{\mathrm{e}} \cdot \mathbf{R}^{\mathrm{e}} \cdot \mathbf{a}_i^0. \tag{11.17.27}$$

If the rotation $\mathbf{R}^{\mathrm{e}}$ is determined by the integration from an appropriately constructed constitutive expression for the spin

$$\boldsymbol{\Omega}^{\mathrm{e}} = \dot{\mathbf{R}}^{\mathrm{e}} \cdot \mathbf{R}^{\mathrm{e}-1}, \tag{11.17.28}$$

the stress response and the elastic moduli of an orthotropic material are derived from

$$\boldsymbol{\tau} = 2\,\mathbf{V}^{\mathrm{e}} \cdot \frac{\partial \Psi^{\mathrm{e}}(\mathbf{R}^{\mathrm{e}T} \cdot \mathbf{B}^{\mathrm{e}} \cdot \mathbf{R}^{\mathrm{e}})}{\partial \mathbf{B}^{\mathrm{e}}} \cdot \mathbf{V}^{\mathrm{e}}, \tag{11.17.29}$$

$$\mathcal{L}_{ijkl}^{(1)} = 4V_{im}^{\mathrm{e}} V_{jn}^{\mathrm{e}} \frac{\partial^2 \Psi^{\mathrm{e}}(\mathbf{R}^{\mathrm{e}T} \cdot \mathbf{B}^{\mathrm{e}} \cdot \mathbf{R}^{\mathrm{e}})}{\partial B_{mn}^{\mathrm{e}}\, \partial B_{pq}^{\mathrm{e}}} V_{kp}^{\mathrm{e}} V_{lq}^{\mathrm{e}}, \tag{11.17.30}$$

in terms of $\mathbf{V}^{\mathrm{e}}$ and $\mathbf{R}^{\mathrm{e}}$.

Since $\dot{\boldsymbol{\mathcal{R}}} = \mathbf{0}$ for an isoclinic intermediate configuration, we have $\mathbf{Z} = \mathbf{0}$ in Eq. (11.17.15). Consequently, from Eqs. (11.17.23) and (11.17.24), the elastic and plastic parts of the rate of deformation become

$$\mathbf{D}^{\mathrm{e}} = \left(\dot{\mathbf{F}}^{\mathrm{e}} \cdot \mathbf{F}^{\mathrm{e}-1}\right)_{\mathrm{s}} - \boldsymbol{\mathcal{L}}_{(0)}^{-1} : (\boldsymbol{\omega}^{\mathrm{p}} \cdot \boldsymbol{\tau} - \boldsymbol{\tau} \cdot \boldsymbol{\omega}^{\mathrm{p}}), \tag{11.17.31}$$

$$\mathbf{D}^{\mathrm{p}} = \left[\mathbf{F}^{\mathrm{e}} \cdot \left(\dot{\mathbf{F}}^{\mathrm{p}} \cdot \mathbf{F}^{\mathrm{p}-1}\right) \cdot \mathbf{F}^{\mathrm{e}-1}\right]_{\mathrm{s}} + \boldsymbol{\mathcal{L}}_{(0)}^{-1} : (\boldsymbol{\omega}^{\mathrm{p}} \cdot \boldsymbol{\tau} - \boldsymbol{\tau} \cdot \boldsymbol{\omega}^{\mathrm{p}}). \tag{11.17.32}$$

In particular, if the principal directions of stress remain parallel to $\mathbf{a}_i^0$ during the deformation, the orientation of the principal directions of orthotropy are fixed, and $\mathbf{R}^{\mathrm{e}} = \mathbf{I}$.

11.17.4. Orthotropic Yield Criterion

The yield criterion of an orthotropic material can be constructed by using an orthotropic function of the rotated-axes components of the Cauchy stress, i.e.,

$$f(\hat{\boldsymbol{\sigma}}, k) = 0, \quad \hat{\boldsymbol{\sigma}} = \boldsymbol{\mathcal{R}}^{\mathrm{e}T} \cdot \boldsymbol{\sigma} \cdot \boldsymbol{\mathcal{R}}^{\mathrm{e}}. \tag{11.17.33}$$

The scalar k specifies the current size of the yield surface. For isotropic hardening, this is a function of an equivalent or generalized plastic strain. Using Hill's (1948) orthotropic criterion, the function f can be expressed as

$$\begin{aligned} f = \big[f_0(\hat{\sigma}_{22} - \hat{\sigma}_{33})^2 + g_0(\hat{\sigma}_{33} - \hat{\sigma}_{11})^2 + h_0(\hat{\sigma}_{11} - \hat{\sigma}_{22})^2 \\ +2l_0\,\hat{\sigma}_{23}^2 + 2m_0\,\hat{\sigma}_{31}^2 + 2n_0\,\hat{\sigma}_{12}^2\big]^{1/2} - k. \end{aligned} \tag{11.17.34}$$

The plastic part of the rate of deformation is assumed to be normal to the yield surface, and given by

$$\mathbf{D}^{\mathrm{p}} = \frac{1}{H}\left(\frac{\partial f}{\partial \boldsymbol{\sigma}} \otimes \frac{\partial f}{\partial \boldsymbol{\sigma}}\right) : \overset{\circ}{\boldsymbol{\tau}}. \tag{11.17.35}$$

The scalar H is determined from the consistency condition $\dot{f} = 0$. For an isoclinic intermediate configuration,

$$\boldsymbol{\mathcal{R}}^{\mathrm{e}} = \mathbf{R}^{\mathrm{e}}, \quad \hat{\boldsymbol{\sigma}} = \mathbf{R}^{\mathrm{e}T} \cdot \boldsymbol{\sigma} \cdot \mathbf{R}^{\mathrm{e}}. \tag{11.17.36}$$

If the constitutive expression for the spin $\boldsymbol{\Omega}^{\mathrm{e}}$ is available, the rotation $\mathbf{R}^{\mathrm{e}}$ is determined by the integration from Eq. (11.17.28). Equation (11.17.35) then defines the plastic part of the rate of deformation for an orthotropic material.

Additional analysis of the yield criteria and constitutive theory for orthotropic materials is available in Hill (1979,1990,1993), Boehler (1982, 1987), Betten (1988), Ferron, Makkouk, and Morreale (1994), Steinmann, Miehe, and Stein (1996), and Vial-Edwards (1997). For an elastoplastic analysis of the transversely isotropic materials, see Aravas (1992).

11.18. Damage-Elastoplasticity

11.18.1. Damage Variables

If plastic deformation affects the elastic properties, which, for example, can happen due to grain (lattice) rotations in a polycrystalline metal sample and resulting crystallographic texture, additional variables need to be introduced in the constitutive framework to describe these changes. They are referred to as the damage variables. They describe a degradation of the elastic properties and their directional changes produced by plastic deformation. Damage variables may be scalars, vectors, second- or higher-order tensors. Derivation in this section will be restricted to damage variables that are either scalars, second- or fourth-order symmetric tensors, collectively denoted by $\mathbf{d}$.

Damage variables change only during plastic deformation, remaining unaltered by elastic unloading or reverse elastic loading, except for the elastic embedding which convects them with the material (Lubarda, 1994b). Thus, if a damage variable in the configuration $\mathcal{B}$ is $\mathbf{d}$, it becomes $\hat{\mathbf{d}}$ in the intermediate configuration $\mathcal{B}^{\mathrm{p}}$, where $\hat{\mathbf{d}}$ is induced from $\mathbf{d}$ by the elastic deformation $\mathbf{F}^{\mathrm{e}}$. For example, the induced damage variable can be defined by the transformation of a weighted contravariant or covariant type. For the second-order tensor these are

$$\hat{\mathbf{d}} = (\det \mathbf{F}^{\mathrm{e}})^{w}\, \mathbf{F}^{\mathrm{e}-1} \cdot \mathbf{d} \cdot \mathbf{F}^{\mathrm{e}-T}, \quad \hat{\mathbf{d}} = (\det \mathbf{F}^{\mathrm{e}})^{-w}\, \mathbf{F}^{\mathrm{e}T} \cdot \mathbf{d} \cdot \mathbf{F}^{\mathrm{e}}, \tag{11.18.1}$$

where w is the weight. Transformations of mixed type could also be considered. For the fourth-order tensors the weighted contravariant and covariant transformations are

$$\hat{\mathbf{d}} = (\det \mathbf{F}^{\mathrm{e}})^{w}\, \mathbf{F}^{\mathrm{e}-1}\, \mathbf{F}^{\mathrm{e}-1}\, \mathbf{d}\, \mathbf{F}^{\mathrm{e}-T}\, \mathbf{F}^{\mathrm{e}-T}, \quad \hat{\mathbf{d}} = (\det \mathbf{F}^{\mathrm{e}})^{-w}\, \mathbf{F}^{\mathrm{e}T}\, \mathbf{F}^{\mathrm{e}T}\, \mathbf{d}\, \mathbf{F}^{\mathrm{e}}\, \mathbf{F}^{\mathrm{e}}. \tag{11.18.2}$$

The products in (11.18.2) are defined such that, for example, the components of the covariant transformation are

$$\hat{d}_{IJKL} = (\det \mathbf{F}^{\mathrm{e}})^{-w}\, F^{\mathrm{e}}_{mI}\, F^{\mathrm{e}}_{nJ}\, d_{mnpq}\, F^{\mathrm{e}}_{pK}\, F^{\mathrm{e}}_{qL}. \tag{11.18.3}$$

The elastic strain energy per unit unstressed volume in the configuration $\mathcal{B}^{\mathrm{p}}$ is

$$\Psi^{\mathrm{e}} = \Psi^{\mathrm{e}}(\mathbf{E}^{\mathrm{e}}, \hat{\mathbf{d}}). \tag{11.18.4}$$

The elastic strain energy per unit initial volume in the configuration $\mathcal{B}^0$ is then

$$\Psi = (\det \mathbf{F}^{\mathrm{p}})\, \Psi^{\mathrm{e}} = \Psi(\mathbf{E}^{\mathrm{e}}, \hat{\mathbf{d}}), \tag{11.18.5}$$

which is equal to Ψ^{e} only when the plastic deformation is incompressible. The function Ψ is an isotropic function of both $\mathbf{E}^{\mathrm{e}}$ and $\hat{\mathbf{d}}$. This means that, under a rigid-body rotation $\hat{\mathbf{Q}}$, superposed to the intermediate configuration,

$$\Psi(\hat{\mathbf{Q}} \cdot \mathbf{E}^{\mathrm{e}} \cdot \hat{\mathbf{Q}}^T, \hat{\mathbf{Q}} \cdot \hat{\mathbf{d}} \cdot \hat{\mathbf{Q}}^T) = \Psi(\mathbf{E}^{\mathrm{e}}, \hat{\mathbf{d}}). \tag{11.18.6}$$

The damage variable in this expression is assumed to be a second-order symmetric tensor. The elastic stress response from $\mathcal{B}^{\mathrm{p}}$ to $\mathcal{B}$ is consequently

$$(\det \mathbf{F}^{\mathrm{e}})\, \boldsymbol{\sigma} = \mathbf{F}^{\mathrm{e}} \cdot \frac{\partial \Psi^{\mathrm{e}}(\mathbf{E}^{\mathrm{e}}, \hat{\mathbf{d}})}{\partial \mathbf{E}^{\mathrm{e}}} \cdot \mathbf{F}^{\mathrm{e}T}, \tag{11.18.7}$$

or

$$\boldsymbol{\tau} = \mathbf{F}^{\mathrm{e}} \cdot \frac{\partial \Psi(\mathbf{E}^{\mathrm{e}}, \hat{\mathbf{d}})}{\partial \mathbf{E}^{\mathrm{e}}} \cdot \mathbf{F}^{\mathrm{e}T}, \tag{11.18.8}$$

where $\boldsymbol{\tau} = (\det \mathbf{F})\boldsymbol{\sigma}$ is the Kirchhoff stress.

11.18.2. Inelastic and Damage Rates of Deformation

Upon differentiation of Eq. (11.18.8), we obtain

$$\begin{aligned} \dot{\boldsymbol{\tau}} - \left(\dot{\mathbf{F}}^{\mathrm{e}} \cdot \mathbf{F}^{\mathrm{e}-1}\right) \cdot \boldsymbol{\tau} - \boldsymbol{\tau} \cdot \left(\dot{\mathbf{F}}^{\mathrm{e}} \cdot \mathbf{F}^{\mathrm{e}-1}\right)^T &= \boldsymbol{\mathcal{L}}_{(1)} : \left(\dot{\mathbf{F}}^{\mathrm{e}} \cdot \mathbf{F}^{\mathrm{e}-1}\right)_{\mathrm{s}} \\ &+ \mathbf{F}^{\mathrm{e}} \cdot \left(\frac{\partial^2 \Psi}{\partial \mathbf{E}^{\mathrm{e}} \otimes \partial \hat{\mathbf{d}}} : \dot{\hat{\mathbf{d}}}\right) \cdot \mathbf{F}^{\mathrm{e}T}. \end{aligned} \tag{11.18.9}$$

The rectangular components of $\boldsymbol{\mathcal{L}}_{(1)}$ are

$$\mathcal{L}^{(1)}_{ijkl} = F^{\mathrm{e}}_{iM} F^{\mathrm{e}}_{jN} \frac{\partial^2 \Psi(\mathbf{E}^{\mathrm{e}}, \hat{\mathbf{d}})}{\partial E^{\mathrm{e}}_{MN}\, \partial E^{\mathrm{e}}_{PQ}} F^{\mathrm{e}}_{kP} F^{\mathrm{e}}_{lQ}. \tag{11.18.10}$$

The last term on the right-hand side of Eq. (11.18.9) can be conveniently rewritten as

$$\mathbf{F}^{\mathrm{e}} \cdot \left(\frac{\partial^2 \Psi}{\partial \mathbf{E}^{\mathrm{e}} \otimes \partial \hat{\mathbf{d}}} : \dot{\hat{\mathbf{d}}}\right) \cdot \mathbf{F}^{\mathrm{e}T} = \frac{\partial \boldsymbol{\tau}}{\partial \hat{\mathbf{d}}} : \dot{\hat{\mathbf{d}}}. \tag{11.18.11}$$

Substitution of Eqs. (11.18.11) and (11.17.21) into Eq. (11.18.9) then gives

$$\overset{\circ}{\boldsymbol{\tau}} = \boldsymbol{\mathcal{L}}_{(0)} : \left(\dot{\mathbf{F}}^{\mathrm{e}} \cdot \mathbf{F}^{\mathrm{e}-1}\right)_{\mathrm{s}} - \boldsymbol{\omega}^{\mathrm{p}} \cdot \boldsymbol{\tau} + \boldsymbol{\tau} \cdot \boldsymbol{\omega}^{\mathrm{p}} + \frac{\partial \boldsymbol{\tau}}{\partial \hat{\mathbf{d}}} : \dot{\hat{\mathbf{d}}}. \tag{11.18.12}$$

The elastic part of the rate of deformation,

$$\mathbf{D}^{\mathrm{e}} = \boldsymbol{\mathcal{L}}_{(0)}^{-1} : \overset{\circ}{\boldsymbol{\tau}}, \tag{11.18.13}$$

is identified from Eq. (11.18.12) as

$$\mathbf{D}^{\mathrm{e}} = \left(\dot{\mathbf{F}}^{\mathrm{e}} \cdot \mathbf{F}^{\mathrm{e}-1}\right)_{\mathrm{s}} - \boldsymbol{\mathcal{L}}_{(0)}^{-1} : (\boldsymbol{\omega}^{\mathrm{p}} \cdot \boldsymbol{\tau} - \boldsymbol{\tau} \cdot \boldsymbol{\omega}^{\mathrm{p}}) + \boldsymbol{\mathcal{L}}_{(0)}^{-1} : \left(\frac{\partial \boldsymbol{\tau}}{\partial \hat{\mathbf{d}}} : \dot{\hat{\mathbf{d}}}\right). \tag{11.18.14}$$

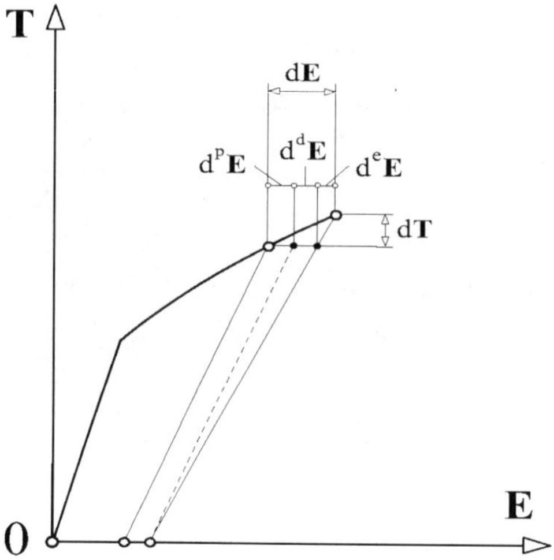

FIGURE 11.5. Geometric interpretation of the partition of the strain increment into its elastic, damage, and plastic parts.

The remaining part of the total rate of deformation is the inelastic part

$$\mathbf{D}^{\mathrm{i}} = \mathbf{D} - \mathbf{D}^{\mathrm{e}} = \left[\mathbf{F}^{\mathrm{e}} \cdot \left(\dot{\mathbf{F}}^{\mathrm{p}} \cdot \mathbf{F}^{\mathrm{p}-1}\right) \cdot \mathbf{F}^{\mathrm{e}-1}\right]_{\mathrm{s}} + \boldsymbol{\mathcal{L}}_{(0)}^{-1} : (\boldsymbol{\omega}^{\mathrm{p}} \cdot \boldsymbol{\tau} - \boldsymbol{\tau} \cdot \boldsymbol{\omega}^{\mathrm{p}}) - \boldsymbol{\mathcal{L}}_{(0)}^{-1} : \left(\frac{\partial \boldsymbol{\tau}}{\partial \hat{\mathrm{d}}} : \dot{\hat{\mathrm{d}}}\right). \tag{11.18.15}$$

The first two terms on the right-hand side of Eq. (11.18.15) represent the plastic part

$$\mathbf{D}^{\mathrm{p}} = \left[\mathbf{F}^{\mathrm{e}} \cdot \left(\dot{\mathbf{F}}^{\mathrm{p}} \cdot \mathbf{F}^{\mathrm{p}-1}\right) \cdot \mathbf{F}^{\mathrm{e}-1}\right]_{\mathrm{s}} + \boldsymbol{\mathcal{L}}_{(0)}^{-1} : (\boldsymbol{\omega}^{\mathrm{p}} \cdot \boldsymbol{\tau} - \boldsymbol{\tau} \cdot \boldsymbol{\omega}^{\mathrm{p}}), \tag{11.18.16}$$

while

$$\mathbf{D}^{\mathrm{d}} = -\boldsymbol{\mathcal{L}}_{(0)}^{-1} : \left(\frac{\partial \boldsymbol{\tau}}{\partial \hat{\mathrm{d}}} : \dot{\hat{\mathrm{d}}}\right) \tag{11.18.17}$$

is the damage part of the rate of deformation tensor (Fig. 11.5). These are such that

$$\mathbf{D}^{\mathrm{i}} = \mathbf{D}^{\mathrm{p}} + \mathbf{D}^{\mathrm{d}}, \tag{11.18.18}$$

and

$$\mathbf{D} = \mathbf{D}^{\mathrm{e}} + \mathbf{D}^{\mathrm{i}} = \mathbf{D}^{\mathrm{e}} + \mathbf{D}^{\mathrm{p}} + \mathbf{D}^{\mathrm{d}}. \tag{11.18.19}$$

If the material behavior complies with Ilyushin's postulate, the inelastic part $\mathbf{D}^{\mathrm{i}}$ of the rate of deformation tensor is normal to a locally smooth yield surface in the Cauchy stress space.

11.18.3. Rates of Damage Tensors

For a scalar damage variable, which transforms during the elastic deformation according to

$$\hat{d} = (\det \mathbf{F}^{\mathrm{e}})^{w}\, d, \tag{11.18.20}$$

the rates of d and $\hat{d}$ are related by

$$\dot{\hat{d}} = (\det \mathbf{F}^{\mathrm{e}}) \left[\dot{d} + w\, d \operatorname{tr}\left(\dot{\mathbf{F}}^{\mathrm{e}} \cdot \mathbf{F}^{\mathrm{e}-1}\right)\right]. \tag{11.18.21}$$

For a second-order damage tensor $\mathbf{d}$ and a covariant type transformation, we have

$$\hat{\mathbf{d}} = (\det \mathbf{F}^{\mathrm{e}})^{-w}\, \mathbf{F}^{\mathrm{e}T} \cdot \mathbf{d} \cdot \mathbf{F}^{\mathrm{e}}, \quad \dot{\hat{\mathbf{d}}} = (\det \mathbf{F}^{\mathrm{e}})^{-w}\, \mathbf{F}^{\mathrm{e}T} \cdot \overset{\blacktriangle}{\mathbf{d}} \cdot \mathbf{F}^{\mathrm{e}}. \tag{11.18.22}$$

Here,

$$\overset{\blacktriangle}{\mathbf{d}} = \dot{\mathbf{d}} + \mathbf{d} \cdot \left(\dot{\mathbf{F}}^{\mathrm{e}} \cdot \mathbf{F}^{\mathrm{e}-1}\right) + \left(\dot{\mathbf{F}}^{\mathrm{e}} \cdot \mathbf{F}^{\mathrm{e}-1}\right)^{T} \cdot \mathbf{d} - w\, \mathbf{d} \operatorname{tr}\left(\dot{\mathbf{F}}^{\mathrm{e}} \cdot \mathbf{F}^{\mathrm{e}-1}\right) \tag{11.18.23}$$

represents the convected rate associated with a weighted covariant transformation. If the induced tensor $\hat{\mathbf{d}}$ is obtained from $\mathbf{d}$ by a contravariant transformation, then

$$\hat{\mathbf{d}} = (\det \mathbf{F}^{\mathrm{e}})^{w}\, \mathbf{F}^{\mathrm{e}-1} \cdot \mathbf{d} \cdot \mathbf{F}^{\mathrm{e}-T}, \quad \dot{\hat{\mathbf{d}}} = (\det \mathbf{F}^{\mathrm{e}})^{w}\, \mathbf{F}^{\mathrm{e}-1} \cdot \overset{\blacktriangledown}{\mathbf{d}} \cdot \mathbf{F}^{\mathrm{e}-T}. \tag{11.18.24}$$

The convected rate associated with a weighted contravariant transformation is

$$\overset{\blacktriangledown}{\mathbf{d}} = \dot{\mathbf{d}} - \left(\dot{\mathbf{F}}^{\mathrm{e}} \cdot \mathbf{F}^{\mathrm{e}-1}\right) \cdot \mathbf{d} - \mathbf{d} \cdot \left(\dot{\mathbf{F}}^{\mathrm{e}} \cdot \mathbf{F}^{\mathrm{e}-1}\right)^{T} + w\, \mathbf{d} \operatorname{tr}\left(\dot{\mathbf{F}}^{\mathrm{e}} \cdot \mathbf{F}^{\mathrm{e}-1}\right). \tag{11.18.25}$$

For the fourth-order damage tensor with a covariant transformation, we similarly have

$$\hat{\mathbf{d}} = (\det \mathbf{F}^{\mathrm{e}})^{-w}\, \mathbf{F}^{\mathrm{e}T}\, \mathbf{F}^{\mathrm{e}T} \mathbf{d}\, \mathbf{F}^{\mathrm{e}}\, \mathbf{F}^{\mathrm{e}}, \tag{11.18.26}$$

$$\dot{\hat{\mathbf{d}}} = (\det \mathbf{F}^{\mathrm{e}})^{-w}\, \mathbf{F}^{\mathrm{e}T}\, \mathbf{F}^{\mathrm{e}T}\, \overset{\blacktriangle}{\mathbf{d}}\, \mathbf{F}^{\mathrm{e}}\, \mathbf{F}^{\mathrm{e}}. \tag{11.18.27}$$

The rectangular components of $\overset{\blacktriangle}{\mathbf{d}}$ are

$$\begin{aligned} \overset{\blacktriangle}{d}_{ijkl} = \dot{d}_{ijkl} &+ L^{\mathrm{e}}_{mi}\, d_{mjkl} + L^{\mathrm{e}}_{mj}\, d_{imkl} + L^{\mathrm{e}}_{mk}\, d_{ijml} + L^{\mathrm{e}}_{ml}\, d_{ijkm} \\ &- w\, L^{\mathrm{e}}_{mm}\, d_{ijkl}. \end{aligned} \tag{11.18.28}$$

The notation $\mathbf{L}^{\mathrm{e}} = \dot{\mathbf{F}}^{\mathrm{e}} \cdot \mathbf{F}^{\mathrm{e}-1}$ is used in Eq. (11.18.28). If the induced tensor $\hat{\mathbf{d}}$ is obtained from the fourth-order damage tensor $\mathbf{d}$ by a contravariant transformation, there follows

$$\hat{\mathbf{d}} = (\det \mathbf{F}^e)^w \, \mathbf{F}^{e-1} \, \mathbf{F}^{e-1} \, \mathbf{d} \, \mathbf{F}^{e-T} \, \mathbf{F}^{e-T}, \qquad (11.18.29)$$

$$\overset{\mathbf{\cdot}}{\hat{\mathbf{d}}} = (\det \mathbf{F}^e)^w \, \mathbf{F}^{e-1} \, \mathbf{F}^{e-1} \, \overset{\blacktriangledown}{\mathbf{d}} \, \mathbf{F}^{e-T} \, \mathbf{F}^{e-T}, \qquad (11.18.30)$$

where

$$\overset{\blacktriangledown}{d}_{ijkl} = \dot{d}_{ijkl} - L^e_{im} \, d_{mjkl} - L^e_{jm} \, d_{imkl} - L^e_{km} \, d_{ijml} - L^e_{lm} \, d_{ijkm}$$
$$+ \, w \, L^e_{mm} \, d_{ijkl}. \qquad (11.18.31)$$

Substituting the expression for $\dot{\hat{\mathbf{d}}}$, corresponding to the tensorial order of the introduced damage variable $\mathbf{d}$ and the transformation rule between $\hat{\mathbf{d}}$ and $\mathbf{d}$, into the expression for the damage part of the rate of deformation, gives

$$\frac{\partial \boldsymbol{\tau}}{\partial \hat{\mathbf{d}}} : \dot{\hat{\mathbf{d}}} = \frac{\partial \boldsymbol{\tau}}{\partial \mathbf{d}} : \overset{\blacktriangle}{\mathbf{d}}, \quad \text{or} \quad \frac{\partial \boldsymbol{\tau}}{\partial \hat{\mathbf{d}}} : \dot{\hat{\mathbf{d}}} = \frac{\partial \boldsymbol{\tau}}{\partial \mathbf{d}} : \overset{\blacktriangledown}{\mathbf{d}}, \qquad (11.18.32)$$

and

$$\mathbf{D}^d = -\boldsymbol{\mathcal{L}}_{(0)}^{-1} : \left(\frac{\partial \boldsymbol{\tau}}{\partial \mathbf{d}} : \overset{\blacktriangle}{\mathbf{d}} \right), \quad \text{or} \quad \mathbf{D}^d = -\boldsymbol{\mathcal{L}}_{(0)}^{-1} : \left(\frac{\partial \boldsymbol{\tau}}{\partial \mathbf{d}} : \overset{\blacktriangledown}{\mathbf{d}} \right). \qquad (11.18.33)$$

With the specified evolution equation for $\overset{\blacktriangle}{\mathbf{d}}$ or $\overset{\blacktriangledown}{\mathbf{d}}$, this determines the damage part of the rate of deformation.

Further elaboration on the constitutive theory of damage-elastoplasticity can be found in the papers by Simo and Ju (1987), Lehmann (1991), Hansen and Schreyer (1994), Lubarda (1994b), and Lubarda and Krajcinovic (1995). See also the books by Lemaitre and Chaboche (1990), Maugin (1992), Krajcinovic (1996), and Voyiadjis and Kattan (1999).

11.19. Reversed Decomposition $\mathbf{F} = \mathbf{F}_p \cdot \mathbf{F}_e$

In the wake of Lee's decomposition $\mathbf{F} = \mathbf{F}^e \cdot \mathbf{F}^p$, the suggestions were made for an alternative, reversed decomposition $\mathbf{F} = \mathbf{F}_p \cdot \mathbf{F}_e$ (e.g., Clifton, 1972; Nemat-Nasser, 1979). This decomposition, however, remained far less employed than the original Lee's decomposition. Lubarda (1999) recently demonstrated that the constitutive analysis of elastoplastic behavior can be developed by using the reversed decomposition quite analogously as using Lee's decomposition. The two formulations can be viewed in many respects as dual to each other, both leading to the same final structure of the constitutive equations, although some of the derivation and interpretations are simpler in the case of Lee's decomposition.

The reversed decomposition is introduced as follows. An arbitrary state of elastoplastic deformation, corresponding to the deformation gradient $\mathbf{F}$, is imagined to be reached in two stages. First, it is assumed that all internal mechanisms responsible for plastic deformation are frozen, so that, for example, the critical forces needed to drive dislocations, or the critical resolved shear stresses of the crystalline slip systems, are assigned infinitely large values. The application of the total stress to such material, incapable of plastic deformation, results in the pure elastic deformation $\mathbf{F}_{\mathrm{e}}$. This carries the material from its initial configuration $\mathcal{B}^0$ to the intermediate configuration $\mathcal{B}_{\mathrm{e}}$. Subsequently, the material is plastically unlocked, by defreezing the mechanisms of plastic deformation, which enables the material to flow at the constant stress. The corresponding part of the deformation gradient, associated with the transition from the intermediate $\mathcal{B}_{\mathrm{e}}$ to the final configuration $\mathcal{B}$, is the plastic part of deformation gradient $\mathbf{F}_{\mathrm{p}}$ (Fig. 11.6). Thus, the reversed decomposition

$$\mathbf{F} = \mathbf{F}_{\mathrm{p}} \cdot \mathbf{F}_{\mathrm{e}}. \tag{11.19.1}$$

The intermediate elastically deformed configuration $\mathcal{B}_{\mathrm{e}}$ is unique, since a superposed rotation to $\mathcal{B}_{\mathrm{e}}$ would rotate the stress state, and the plastic flow from $\mathcal{B}_{\mathrm{e}}$ to $\mathcal{B}$ would not take place at the constant state of stress. In the subsequent analysis it will be assumed that plastic flow is incompressible and that elastic properties of the material are not affected by plastic deformation. Relative to a given orientation of the principal directions of elastic anisotropy, there is in this case a unique

$$\mathbf{F}_{\mathrm{e}} = \mathbf{F}^{\mathrm{e}} \tag{11.19.2}$$

that gives rise to total stress in $\mathcal{B}_{\mathrm{e}}$ and $\mathcal{B}$. This stress is

$$\boldsymbol{\tau} = \mathbf{F}^{\mathrm{e}} \cdot \frac{\partial \Psi^{\mathrm{e}}}{\partial \mathbf{E}^{\mathrm{e}}} \cdot \mathbf{F}^{\mathrm{e}T}. \tag{11.19.3}$$

The elastic strain energy per unit initial volume is Ψ^{e}, while $\mathbf{E}^{\mathrm{e}}$ is the Lagrangian elastic strain relative to its ground state ($\mathcal{B}^{\mathrm{p}}$ in the case of Lee's

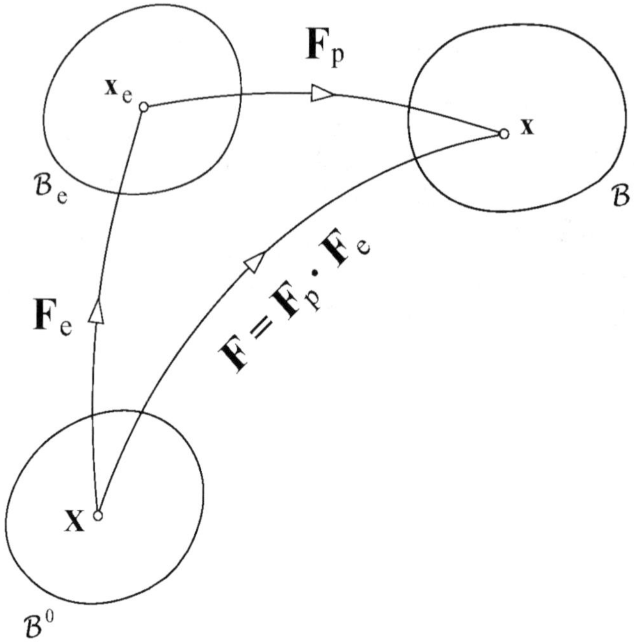

FIGURE 11.6. Schematic representation of the reversed multiplicative decomposition of deformation gradient into its elastic and plastic parts. The intermediate configuration $\mathcal{B}_e$ is obtained from the initial configuration $\mathcal{B}^0$ by elastic loading to the current stress level, assuming that all inelastic mechanisms of the deformation are momentarily frozen.

decomposition and $\mathcal{B}^0$ in the case of the reversed decomposition, both having the same orientation of the principal axes of anisotropy, relative to the fixed frame of reference). The Kirchhoff stress is $\boldsymbol{\tau} = (\det \mathbf{F})\boldsymbol{\sigma}$, where $\boldsymbol{\sigma}$ designates the Cauchy stress. Therefore, we can write

$$\mathbf{F} = \mathbf{F}^{e} \cdot \mathbf{F}^{p} = \mathbf{F}_{p} \cdot \mathbf{F}^{e}. \tag{11.19.4}$$

The same elastic deformation gradient $\mathbf{F}^e$ appears in both decompositions. The relationship between plastic parts of the deformation gradient is consequently

$$\mathbf{F}^{p} = \mathbf{F}^{e-1} \cdot \mathbf{F}_{p} \cdot \mathbf{F}^{e}. \tag{11.19.5}$$

If the material is elastically isotropic, an initial rotation $\mathbf{R}^e$ of $\mathcal{B}^0$ does not affect the stress response, and the relevant part of the total deformation gradient for the constitutive analysis is

$$\mathbf{F} = \mathbf{F}_{\rm p} \cdot \mathbf{V}^{\rm e}. \tag{11.19.6}$$

In this case, therefore, we can write

$$\mathbf{F} = \mathbf{V}^{\rm e} \cdot \mathbf{F}^{\rm p} = \mathbf{F}_{\rm p} \cdot \mathbf{V}^{\rm e}. \tag{11.19.7}$$

11.19.1. Elastic Unloading

During elastic loading from $\mathcal{B}^{\rm p}$ to $\mathcal{B}$, or elastic unloading from $\mathcal{B}$ to $\mathcal{B}^{\rm p}$, the plastic deformation gradient $\mathbf{F}^{\rm p}$ of the decomposition $\mathbf{F} = \mathbf{F}^{\rm e} \cdot \mathbf{F}^{\rm p}$ remains constant. This greatly simplifies the derivation of the corresponding constitutive equations. As shown in Section 11.3, the velocity gradient in $\mathcal{B}$ is

$$\mathbf{L} = \dot{\mathbf{F}}^{\rm e} \cdot \mathbf{F}^{\rm e-1} + \mathbf{F}^{\rm e} \cdot \left(\dot{\mathbf{F}}^{\rm p} \cdot \mathbf{F}^{\rm p-1} \right) \cdot \mathbf{F}^{\rm e-1}, \tag{11.19.8}$$

so that during elastic unloading

$$\dot{\mathbf{F}}^{\rm p} = \mathbf{0}, \quad \mathbf{L} = \dot{\mathbf{F}}^{\rm e} \cdot \mathbf{F}^{\rm e-1}. \tag{11.19.9}$$

In the framework of the reversed decomposition $\mathbf{F} = \mathbf{F}_{\rm p} \cdot \mathbf{F}^{\rm e}$, however, the plastic part of deformation gradient $\mathbf{F}_{\rm p}$ does not remain constant during elastic unloading. In fact, upon complete unloading from an elastoplastic state of deformation to zero stress, the configuration $\mathcal{B}^{\rm p}$ is reached, and $\mathbf{F}_{\rm p} = \mathbf{F}^{\rm p}$ at that instant (Fig. 11.7). Therefore, $\dot{\mathbf{F}}_{\rm p} \neq \mathbf{0}$ during elastic unloading. This can also be recognized from the general relationship between $\mathbf{F}^{\rm p}$ and $\mathbf{F}_{\rm p}$. By differentiating Eq. (11.19.5), we obtain

$$\dot{\mathbf{F}}^{\rm p} = \mathbf{F}^{\rm e-1} \cdot \overset{*}{\mathbf{F}}_{\rm p} \cdot \mathbf{F}^{\rm e}, \tag{11.19.10}$$

where

$$\overset{*}{\mathbf{F}}_{\rm p} = \dot{\mathbf{F}}_{\rm p} - \left(\dot{\mathbf{F}}^{\rm e} \cdot \mathbf{F}^{\rm e-1} \right) \cdot \mathbf{F}_{\rm p} + \mathbf{F}_{\rm p} \cdot \left(\dot{\mathbf{F}}^{\rm e} \cdot \mathbf{F}^{\rm e-1} \right) \tag{11.19.11}$$

is a convected-type derivative of $\mathbf{F}_{\rm p}$ relative to elastic deformation. Consequently,

$$\overset{*}{\mathbf{F}}_{\rm p} = \mathbf{0}, \quad \text{if} \quad \dot{\mathbf{F}}^{\rm p} = \mathbf{0}, \tag{11.19.12}$$

and in this case

$$\dot{\mathbf{F}}_{\rm p} = \left(\dot{\mathbf{F}}^{\rm e} \cdot \mathbf{F}^{\rm e-1} \right) \cdot \mathbf{F}_{\rm p} - \mathbf{F}_{\rm p} \cdot \left(\dot{\mathbf{F}}^{\rm e} \cdot \mathbf{F}^{\rm e-1} \right). \tag{11.19.13}$$

The last expression defines the change of $\mathbf{F}_{\rm p}$ during elastic unloading.

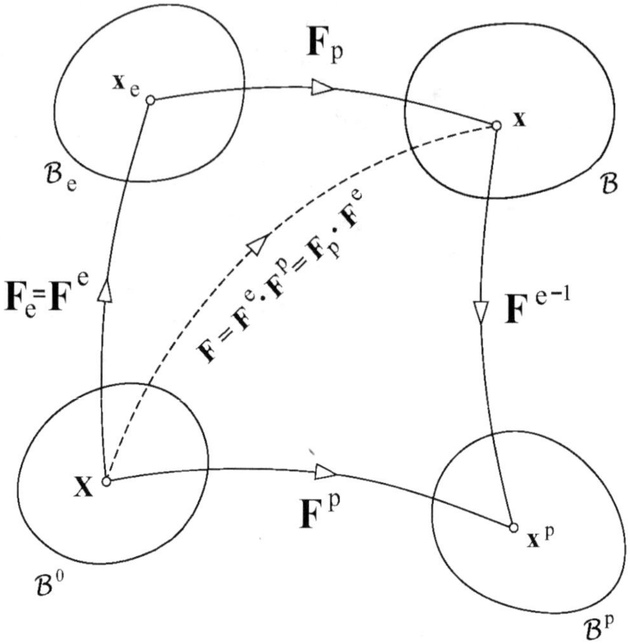

FIGURE 11.7. Plastic part of deformation gradient $\mathbf{F}_p$ does not remain constant during elastic unloading. Upon complete unloading to zero stress, the configuration $\mathcal{B}^p$ is reached, and $\mathbf{F}_p = \mathbf{F}^p$ at that instant.

Furthermore, from Eq. (11.19.10),

$$\dot{\mathbf{F}}^p \cdot \mathbf{F}^{p-1} = \mathbf{F}^{e-1} \cdot \left(\overset{*}{\mathbf{F}}_p \cdot \mathbf{F}_p^{-1} \right) \cdot \mathbf{F}^e, \tag{11.19.14}$$

and the substitution into Eq. (11.19.8) gives

$$\mathbf{L} = \dot{\mathbf{F}}^e \cdot \mathbf{F}^{e-1} + \overset{*}{\mathbf{F}}_p \cdot \mathbf{F}_p^{-1}. \tag{11.19.15}$$

11.19.2. Elastic and Plastic Rates of Deformation

If the elastic part of the rate of deformation tensor is defined by a kinetic relation

$$\mathbf{D}^e = \boldsymbol{\mathcal{L}}_{(0)}^{-1} : \overset{\circ}{\boldsymbol{\tau}}, \quad \overset{\circ}{\boldsymbol{\tau}} = \dot{\boldsymbol{\tau}} - \mathbf{W} \cdot \boldsymbol{\tau} + \boldsymbol{\tau} \cdot \mathbf{W}, \tag{11.19.16}$$

it follows that

$$\mathbf{D}^e = \left(\dot{\mathbf{F}}^e \cdot \mathbf{F}^{e-1} \right)_s - \boldsymbol{\mathcal{L}}_{(0)}^{-1} : \left(\boldsymbol{\omega}^p \cdot \boldsymbol{\tau} - \boldsymbol{\tau} \cdot \boldsymbol{\omega}^p \right), \tag{11.19.17}$$

$$\mathbf{D}^\mathrm{p} = \left(\overset{*}{\mathbf{F}}_\mathrm{p} \cdot \mathbf{F}_\mathrm{p}^{-1}\right)_\mathrm{s} + \boldsymbol{\mathcal{L}}_{(0)}^{-1} : (\boldsymbol{\omega}^\mathrm{p} \cdot \boldsymbol{\tau} - \boldsymbol{\tau} \cdot \boldsymbol{\omega}^\mathrm{p}) . \tag{11.19.18}$$

The spin $\boldsymbol{\omega}^\mathrm{p}$ is

$$\boldsymbol{\omega}^\mathrm{p} = \left(\overset{*}{\mathbf{F}}_\mathrm{p} \cdot \mathbf{F}_\mathrm{p}^{-1}\right)_\mathrm{a} = \left[\mathbf{F}^\mathrm{e} \cdot \left(\dot{\mathbf{F}}^\mathrm{p} \cdot \mathbf{F}^{\mathrm{p}-1}\right) \cdot \mathbf{F}^{\mathrm{e}-1}\right]_\mathrm{a} . \tag{11.19.19}$$

The elastic part of the rate of deformation can also be expressed as

$$\mathbf{D}^\mathrm{e} = \left(\overset{\bullet}{\mathbf{F}}{}^\mathrm{e} \cdot \mathbf{F}^{\mathrm{e}-1}\right)_\mathrm{s} , \tag{11.19.20}$$

where

$$\overset{\bullet}{\mathbf{F}}{}^\mathrm{e} = \dot{\mathbf{F}}^\mathrm{e} - \boldsymbol{\Omega}^\mathrm{p} \cdot \mathbf{F}^\mathrm{e} + \mathbf{F}^\mathrm{e} \cdot \boldsymbol{\Omega}^\mathrm{p} . \tag{11.19.21}$$

The spin $\boldsymbol{\Omega}^\mathrm{p}$ is the solution of the matrix equation

$$\mathbf{W} = \left(\dot{\mathbf{F}}^\mathrm{e} \cdot \mathbf{F}^{\mathrm{e}-1}\right)_\mathrm{a} + \left(\mathbf{F}^\mathrm{e} \cdot \boldsymbol{\Omega}^\mathrm{p} \cdot \mathbf{F}^{\mathrm{e}-1}\right)_\mathrm{a} . \tag{11.19.22}$$

This is analogous to the derivation from Section 11.7, based on Lee's decomposition. Therefore,

$$\mathbf{D}^\mathrm{e} = \left(\dot{\mathbf{F}}^\mathrm{e} \cdot \mathbf{F}^{\mathrm{e}-1}\right)_\mathrm{s} + \left(\mathbf{F}^\mathrm{e} \cdot \boldsymbol{\Omega}^\mathrm{p} \cdot \mathbf{F}^{\mathrm{e}-1}\right)_\mathrm{s} , \tag{11.19.23}$$

$$\mathbf{D}^\mathrm{p} = \left(\overset{*}{\mathbf{F}}_\mathrm{p} \cdot \mathbf{F}_\mathrm{p}^{-1}\right)_\mathrm{s} - \left(\mathbf{F}^\mathrm{e} \cdot \boldsymbol{\Omega}^\mathrm{p} \cdot \mathbf{F}^{\mathrm{e}-1}\right)_\mathrm{s} . \tag{11.19.24}$$

Summing up Eqs. (11.19.22)–(11.19.24), we obtain an expression for the velocity gradient $\mathbf{L}$. The comparison with Eq. (11.19.15) then gives

$$\mathbf{0} = \left(\overset{*}{\mathbf{F}}_\mathrm{p} \cdot \mathbf{F}_\mathrm{p}^{-1}\right)_\mathrm{a} - \left(\mathbf{F}^\mathrm{e} \cdot \boldsymbol{\Omega}^\mathrm{p} \cdot \mathbf{F}^{\mathrm{e}-1}\right)_\mathrm{a} . \tag{11.19.25}$$

Thus, the plastic part of the rate of deformation can be alternatively written as

$$\mathbf{D}^\mathrm{p} = \overset{*}{\mathbf{F}}_\mathrm{p} \cdot \mathbf{F}_\mathrm{p}^{-1} - \mathbf{F}^\mathrm{e} \cdot \boldsymbol{\Omega}^\mathrm{p} \cdot \mathbf{F}^{\mathrm{e}-1} . \tag{11.19.26}$$

The result is in accord with Eq. (11.8.2), as can be verified by using the definitions of $\overset{\bullet}{\mathbf{F}}{}^\mathrm{p}$ and $\overset{*}{\mathbf{F}}_\mathrm{p}$, and the relationship

$$\mathbf{F}_\mathrm{p}^{-1} = \mathbf{F}^\mathrm{e} \cdot \mathbf{F}^{\mathrm{p}-1} \cdot \mathbf{F}^{\mathrm{e}-1} . \tag{11.19.27}$$

The presented derivation demonstrates a duality in the constitutive formulation of large-deformation elastoplasticity based on Lee's decomposition $\mathbf{F} = \mathbf{F}^\mathrm{e} \cdot \mathbf{F}^\mathrm{p}$ and the reversed decomposition $\mathbf{F} = \mathbf{F}_\mathrm{p} \cdot \mathbf{F}_\mathrm{e}$, at least for the considered material models. The structure of the kinematic expressions is more involved in the case of the reversed decomposition, partly because during elastic unloading the plastic deformation gradient $\mathbf{F}^{\mathrm{p}}$ of Lee's decomposition remains constant, while $\mathbf{F}_{\mathrm{p}}$ of the reversed decomposition changes, albeit in a definite manner specified by Eq. (11.19.13). It is possible, however, that in some applications the reversed decomposition may have certain advantages. For example, Clifton (1972) found that it is slightly more convenient in the analysis of one-dimensional wave propagation in elastic-viscoplastic solids.

Lee's decomposition has definite advantages in modeling the plasticity with evolving elastic properties. In this case, a set of damage or structural tensors can be attached to the intermediate configuration $\mathcal{B}^{\mathrm{p}}$ to represent its current state of elastic anisotropy. The structural tensors evolve during plastic deformation, depending on the nature of microscopic inelastic processes, as represented by the appropriate evolution equations. The stress response at each instant of deformation is given in terms of the gradient of elastic strain energy with respect to elastic strain, at the current values of the structural tensors. This has been discussed in Section 11.18. In the case of the reversed decomposition, however, the elastic response is defined relative to the initial configuration $\mathcal{B}^0$, which does not contain any information about the evolving elastic properties or subsequently developed elastic anisotropy. Additional remedy has to be introduced to deal with these features of the material response, which is likely to make the reversed decomposition less attractive than the original Lee's decomposition.

References

Agah-Tehrani, A., Lee, E. H., Mallett, R. L., and Onat, E. T. (1987), The theory of elastic-plastic deformation at finite strain with induced anisotropy modeled as combined isotropic–kinemetic hardening, *J. Mech. Phys. Solids*, Vol. 35, pp. 519–539.

Aifantis, E. C. (1987), The physics of plastic deformation, *Int. J. Plasticity*, Vol. 3, pp. 211–247.

Anand, L. (1980), Constitutive equations for rate-independent, isotropic, elastic-plastic solids exhibiting pressure-sensitive yielding and plastic dilatancy, *J. Appl. Mech.*, Vol. 47, pp. 439–441.

Aravas, N. (1992), Finite elastoplastic transformations of transversely isotropic metals, *Int. J. Solids Struct.*, Vol. 29, pp. 2137–2157.

Atluri, S. N. (1984), On constitutive relations at finite strain: Hypoelasticity and elastoplasticity with isotropic or kinematic hardening, *Comput. Meth. Appl. Mech. Engrg.*, Vol. 43, pp. 137–171.

Belytschko, T., Liu, W. K., and Moran, B. (2000), *Nonlinear Finite Elements for Continua and Structures*, John Wiley & Sons, Inc., New York.

Besseling, J. F. and Van der Giessen, E. (1994), *Mathematical Modelling of Inelastic Deformation*, Chapman & Hall, London.

Betten, J. (1988), Applications of tensor functions to the formulation of yield criteria for anisotropic materials, *Int. J. Plasticity*, Vol. 4, pp. 29–46.

Boehler, J. P., ed. (1982), *Mechanical Behavior of Anisotropic Solids*, Martinus Nijhoff Publishers, The Hague.

Boehler, J. P., ed. (1987), *Application of Tensor Functions in Solid Mechanics*, Springer, Wien.

Boyce, M. C., Parks, D. M., and Argon, A. S. (1988), Large inelastic deformation of glassy polymers. Part I: Rate-dependent constitutive model, *Mech. Mater.*, Vol. 7, pp. 15–33.

Boyce, M. C., Weber, G. G., and Parks, D. M. (1989), On the kinematics of finite strain plasticity, *J. Mech. Phys. Solids*, Vol. 37, pp. 647–665.

Casey, J. (1987), Discussion of "Invariance considerations in large strain elasto-plasticity," *J. Appl. Mech.*, Vol. 54, pp. 247–248.

Casey, J. and Naghdi, P. M. (1980), Remarks on the use of the decomposition $F = F_e F_p$ in plasticity, *J. Appl. Mech.*, Vol. 47, pp. 672–675.

Casey, J. and Naghdi, P. M. (1981), Discussion of "A correct definition of elastic and plastic deformation and its computational significance," *J. Appl. Mech.*, Vol. 48, pp. 983–984.

Clifton, R. J. (1972), On the equivalence of $\mathbf{F}^e\mathbf{F}^p$ and $\bar{\mathbf{F}}^p\bar{\mathbf{F}}^e$, *J. Appl. Mech.*, Vol. 39, pp. 287–289.

Dafalias, Y. F. (1983), Corotational rates for kinematic hardening at large plastic deformations, *J. Appl. Mech.*, Vol. 50, pp. 561–565.

Dafalias, Y. F. (1985), The plastic spin, *J. Appl. Mech.*, Vol. 52, pp. 865–871.

Dafalias, Y. F. (1987), Issues in constitutive formulation at large elastoplastic deformation. Part I: Kinematics, *Acta Mech.*, Vol. 69, pp. 119–138.

Dafalias, Y. F. (1988), Issues in constitutive formulation at large elastoplastic deformation. Part I: Kinetics, *Acta Mech.*, Vol. 73, pp. 121–146.

Dafalias, Y. F. (1998), Plastic spin: Necessity or redundancy, *Int. J. Plasticity*, Vol. 14, pp. 909–931.

Dashner, P. A. (1986a), Invariance considerations in large strain elastoplasticity, *J. Appl. Mech.*, Vol. 53, pp. 55–60.

Dashner, P. A. (1986b), Plastic potential theory in large strain elastoplasticity, *Int. J. Solids Struct.*, Vol. 22, pp. 593–623.

Ferron, G., Makkouk, R., and Morreale, J. (1994), A parametric description of orthotropic plasticity in metal sheets, *Int. J. Plasticity*, Vol. 10, pp. 431–449.

Fox, N. (1968), On the continuum theories of dislocations and plasticity, *Q. J. Mech. Appl. Math.*, Vol. 21, pp. 67–75.

Freund, L. B. (1970), Constitutive equations for elastic-plastic materials at finite strain, *Int. J. Solids Struct.*, Vol. 6, pp. 1193–1209.

Green, A. E. and Naghdi P. M. (1971), Some remarks on elastic-plastic deformation at finite strain, *Int. J. Engng. Sci.*, Vol. 9, pp. 1219–1229.

Hansen, N. R. and Schreyer, H. L. (1994), A thermodynamically consistent framework for theories of elastoplasticity coupled with damage, *Int. J. Solids Struct.*, Vol. 31, pp. 359–389.

Hill, R. (1948), A theory of the yielding and plastic flow of anisotropic metals, *Proc. Roy. Soc. Lond. A*, Vol. 193, pp. 281–297.

Hill, R. (1979), Theoretical plasticity of textured aggregates, *Math. Proc. Camb. Phil. Soc.*, Vol. 85, pp. 179–191.

Hill, R. (1984), On macroscopic effects of heterogeneity in elastoplastic media at finite strain, *Math. Proc. Camb. Phil. Soc.*, Vol. 95, pp. 481–494.

Hill, R. (1990), Constitutive modelling of orthotropic plasticity in sheet metals, *J. Mech. Phys. Solids*, Vol. 38, pp. 405–417.

Hill, R. (1993), A user-friendly theory of orthotropic plasticity in sheet metals, *Int. J. Mech. Sci.*, Vol. 35, pp. 19–25.

Imam, A. and Johnson, G. C. (1998), Decomposition of the deformation gradient in thermoelasticity, *J. Appl. Mech.*, Vol. 65, pp. 362–366.

Johnson, G. C. and Bammann, D. J. (1984), A discussion of stress rates in finite deformation problems, *Int. J. Solids Struct.*, Vol. 20, pp. 725–737.

Kleiber, M. (1975), Kinematics of deformation processes in materials subjected to finite elastic-plastic strains, *Int. J. Engng. Sci.*, Vol. 13, pp. 513–525.

Kleiber, M. and Raniecki, B. (1985), Elastic-plastic materials at finite strains, in *Plasticity Today – Modelling, Methods and Applications*, eds. A. Sawczuk and G. Bianchi, pp. 3–46, Elsevier Applied Science, London.

Krajcinovic, D. (1996), *Damage Mechanics*, Elsevier, New York.

Kröner, E. and Teodosiu, C. (1973), Lattice defect approach to plasticity and viscoplasticity, in *Problems of Plasticity*, ed. A. Sawczuk, pp. 45–88, Noordhoff, Leyden.

Kratochvil, J. (1973), On a finite strain theory of elastic-inelastic materials, *Acta Mech.*, Vol. 16, pp. 127–142.

Kuroda, M. (1995), Plastic spin associated with a corner theory of plasticity, *Int. J. Plasticity*, Vol. 11, pp. 547–570.

Lee, E. H. (1969), Elastic-plastic deformation at finite strains, *J. Appl. Mech.*, Vol. 36, pp. 1–6.

Lee, E. H. (1981), Some comments on elastic-plastic analysis, *Int. J. Solids Struct.*, Vol. 17, pp. 859–872.

Lee, E. H. (1985), Finite deformation effects in plasticity analysis, in *Plasticity Today: Modelling, Methods and Applications*, eds. A. Sawczuk and G. Bianchi, pp. 61–77, Elsevier Applied Science, London.

Lee, E. H. and Liu, D. T. (1967), Finite-strain elastic-plastic theory particularly for plane wave analysis, *J. Appl. Phys.*, Vol. 38, pp. 19–27.

Lee, E. H., Mallett, R. L., and Wertheimer, T. B. (1983), Stress analysis for anisotropic hardening in finite-deformation plasticity, *J. Appl. Mech.*, Vol. 50, pp. 554–560.

Lehmann, Th. (1991), Thermodynamical foundations of large inelastic deformations of solid bodies including damage, *Int. J. Plasticity*, Vol. 7, pp. 79–98.

Lemaitre, J. and Chaboche, J.-L. (1990), *Mechanics of Solid Materials*, Cambridge University Press, Cambridge.

Loret, B. (1983), On the effects of plastic rotation in the finite deformation of anisotropic elastoplastic materials, *Mech. Mater.*, Vol. 2, pp. 287–304.

Lubarda, V. A. (1991 a), Constitutive analysis of large elasto-plastic deformation based on the multiplicative decomposition of deformation gradient, *Int. J. Solids Struct.*, Vol. 27, pp. 885–895.

Lubarda, V. A. (1991 b), Some aspects of elasto-plastic constitutive analysis of elastically anisotropic materials, *Int. J. Plasticity*, Vol. 7, pp. 625–636.

Lubarda, V. A. (1994 a), Elastoplastic constitutive analysis with the yield surface in strain space, *J. Mech. Phys. Solids*, Vol. 42, pp. 931–952.

Lubarda, V. A. (1994 b), An analysis of large-strain damage elastoplasticity, *Int. J. Solids Struct.*, Vol. 31, pp. 2951–2964.

Lubarda, V. A. (1999), Duality in constitutive formulation of finite-strain elastoplasticity based on $\mathbf{F} = \mathbf{F}_{\mathrm{e}}\mathbf{F}_{\mathrm{p}}$ and $\mathbf{F} = \mathbf{F}^{\mathrm{p}}\mathbf{F}^{\mathrm{e}}$ decompositions, *Int. J. Plasticity*, Vol. 15, pp. 1277–1290.

Lubarda, V. A. and Benson, D. J. (2001), On the partitioning of the rate of deformation gradient in phenomenological plasticity, *Int. J. Solids Struct.*, in press.

Lubarda, V. A. and Krajcinovic, D. (1995), Some fundamental issues in rate theory of damage-elastoplasticity, *Int. J. Plasticity*, Vol. 11, pp. 763–797.

Lubarda, V. A. and Lee, E. H. (1981), A correct definition of elastic and plastic deformation and its computational significance, *J. Appl. Mech.*, Vol. 48, pp. 35–40.

Lubarda, V. A. and Shih, C. F. (1994), Plastic spin and related issues in phenomenological plasticity, *J. Appl. Mech.*, Vol. 61, pp. 524–529.

Mandel, J. (1971), Plasticité classique et viscoplasticité, *Courses and Lectures, No. 97*, International Center for Mechanical Sciences, Udine, Springer, New York.

Mandel, J. (1973), Equations constitutives et directeurs dans les milieux plastiques et viscoplastiques, *Int. J. Solids Struct.*, Vol. 9, pp. 725–740.

Mandel, J. (1983), Sur la definition de la vittese de deformation elastique en grande transformation elastoplastique, *Int. J. Solids Struct.*, Vol. 19, pp. 573–578.

Maugin, G. A. (1992), *The Thermomechanics of Plasticity and Fracture*, Cambridge University Press, Cambridge.

Moran, B., Ortiz, M., and Shih, C. F. (1990), Formulation of implicit finite element methods for multiplicative finite deformation plasticity, *Int. J. Numer. Methods Eng.*, Vol. 29, pp. 483–514.

Naghdi, P. M. and Trapp, J. A. (1974), On finite elastic-plastic deformation of metals, *J. Appl. Mech.*, Vol. 41, pp. 254–260.

Naghdi, P. M. (1990), A critical review of the state of finite plasticity, *Z. angew. Math. Phys.*, Vol. 41, pp. 315–394.

Nagtegaal, J. C. and de Jong, J. E. (1982), Some aspects of nonisotropic work hardening in finite strain plasticity, in *Plasticity of Metals at Finite Strain: Theory, Experiment and Computation*, eds. E. H. Lee and R. L. Mallett, pp. 65–102, Stanford University, Stanford.

Needleman, A. (1985), On finite element formulations for large elastic–plastic deformations, *Comp. Struct.*, Vol. 20, pp. 247–257.

Nemat-Nasser, S. (1979), Decomposition of strain measures and their rates in finite deformation elastoplasticity, *Int. J. Solids Struct.*, Vol. 15, pp. 155–166.

Nemat-Nasser, S. (1982), On finite deformation elasto-plasticity, *Int. J. Solids Struct.*, Vol. 18, pp. 857–872.

Nemat-Nasser, S. (1992), Phenomenological theories of elastoplasticity and strain localization at high strain rates, *Appl. Mech. Rev.*, Vol. 45, No. 3, Part 2, pp. S19–S45.

Sidoroff, F. (1975), On the formulation of plasticity and viscoplasticity with internal variables, *Arch. Mech.*, Vol. 27, pp. 807–819.

Sidoroff, F. (1982), Incremental constitutive equations for large strain elasto-plasticity, *Int. J. Engng. Sci.*, Vol. 20, pp. 19–28.

Simo, J. C. (1998), Topics on numerical analysis and simulation of plasticity, in *Handbook of Numerical Analysis*, Vol. VI, eds. P. G. Ciarlet and J.-L. Lions, pp. 183–499, North-Holland, Amsterdam.

Simo, J. C. and Ju, J. W. (1987), Strain- and stress-based continuum damage models – I. Formulation, *Int. J. Solids Struct.*, Vol. 23, pp. 821–840.

Simo, J. C. and Hughes, T. J. R. (1998), *Computational Inelasticity*, Springer-Verlag, New York.

Simo, J. C. and Ortiz, M. (1985), A unified approach to finite deformation elastoplastic analysis based on the use of hyperelastic constitutive equations, *Comput. Meth. Appl. Mech. Engrg.*, Vol. 49, pp. 221–245.

Steinmann, P., Miehe, C., and Stein, E. (1996), Fast transient dynamic plane stress analysis of orthotropic Hill-type solids at finite elastoplastic strain, *Int. J. Solids Struct.*, Vol. 33, pp. 1543–1562.

Van der Giessen, E. (1989), Continuum models of large deformation plasticity, Parts I and II, *Eur. J. Mech., A/Solids*, Vol. 8, pp. 15–34 and 89–108.

Van der Giessen, E. (1991), Micromechanical and thermodynamic aspects of the plastic spin, *Int. J. Plasticity*, Vol. 7, pp. 365–386.

Vial-Edwards, C. (1997), Yield loci of FCC and BCC sheet metals, *Int. J. Plasticity*, Vol. 13, pp. 521–531.

Voyiadjis, G. Z. and Kattan, P. I. (1999), *Advances in Damage Mechanics: Metals and Metal Matrix Composites*, Elsevier, Amsterdam.

Weber, G. and Anand, L. (1990), Finite deformation constitutive equations and a time integration procedure for isotropic, hyperelastic-viscoplastic solids, *Comput. Meth. Appl. Mech. Engrg.*, Vol. 79, pp. 173–202.

Willis, J. R. (1969), Some constitutive equations applicable to problems of large dynamic plastic deformation, *J. Mech. Phys. Solids*, Vol. 17, pp. 359–369.

Xiao, H., Bruhns, O. T., and Meyers, A. (2000), A consistent finite elastoplasticity theory combining additive and multiplicative decomposition of the stretching and the deformation gradient, *Int. J. Plasticity*, Vol. 16, pp. 143–177.

Zbib, H. M. and Aifantis, E. C. (1988), On the concept of relative and plastic spins and its implications to large deformation theories. Part II: Anisotropic hardening, *Acta Mech.*, Vol. 75, pp. 35–56.

CHAPTER 12

CRYSTAL PLASTICITY

Previous chapters were devoted to phenomenological theory of plasticity, in which microscopic structure and mechanisms causing plastic flow were not included explicitly, but only implicitly through macroscopic variables, such as the generalized plastic strain, the radius of the yield surface, or the back stress. This chapter deals with plastic deformation of single crystals. The discrete dislocation substructure is still ignored, but plastic deformation is considered to occur in the form of smooth shearing on the slip planes and in the slip directions. Such continuum model of slip has its origin in the pioneering work of Taylor (1938). The model was further developed by Hill (1966) in the case of elastoplastic deformation with small elastic component of deformation, and by Rice (1971), Kratochvil (1971), Hill and Rice (1972), Havner (1973), Mandel (1974), Asaro and Rice (1977), and Hill and Havner (1982) in the case of finite elastic and plastic deformations. Since the theory explicitly accounts for the specific microscopic process (crystallographic slip), it is also referred to as the physical theory of plasticity. Optical micrographs of crystallographic slip are shown in Fig. 12.1. Other mechanisms of plastic deformation, such as twinning, displacive (martensitic) transformations, and diffusional processes are not considered in this chapter.

12.1. Kinematics of Crystal Deformation

The kinematic representation of elastoplastic deformation of single crystals (monocrystals), in which crystallographic slip is assumed to be the only mechanism of plastic deformation, is shown in Fig. 12.2. The material flows through the crystalline lattice via dislocation motion, while the lattice itself, with the material embedded to it, undergoes elastic deformation and rotation. The plastic deformation is considered to occur in the form of smooth

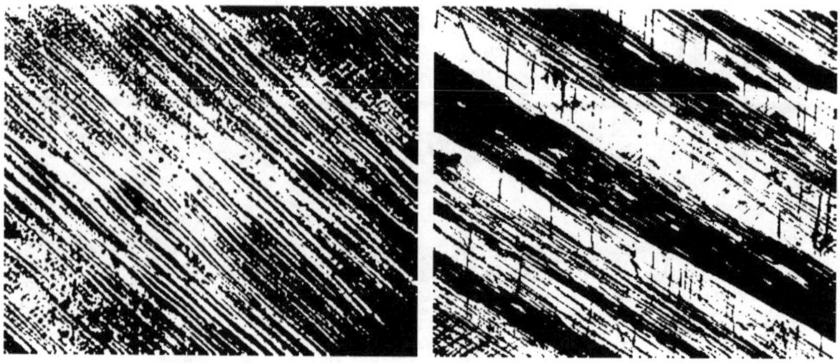

(a) (b)

FIGURE 12.1. (a) Optical micrograph of crystallographic slip in an Au crystal, and (b) bands of primary and secondary slip in an aluminum crystal (from Sawkill and Honeycombe, 1954; with permission from Elsevier Science).

shearing on the slip planes and in the slip directions. The deformation gradient $\mathbf{F}$ is decomposed as

$$\mathbf{F} = \mathbf{F}^* \cdot \mathbf{F}^{\mathrm{p}}, \tag{12.1.1}$$

where $\mathbf{F}^{\mathrm{p}}$ is the part of $\mathbf{F}$ due to slip only, while $\mathbf{F}^*$ is the part due to lattice stretching and rotation. This decomposition is formally analogous to Lee's (1969) multiplicative decomposition, discussed in the previous chapter. The deformation gradient remaining after elastic destressing and upon returning the lattice to its original orientation is $\mathbf{F}^{\mathrm{p}} = \mathbf{F} \cdot \mathbf{F}^{*-1}$. Denote the unit vector in the slip direction and the unit vector normal to the corresponding slip plane in the undeformed configuration by $\mathbf{s}_0^\alpha$ and $\mathbf{m}_0^\alpha$, where α designates the slip system. The same vectors are attached to the lattice in the intermediate configuration, because the lattice does not deform or rotate during the slip induced transformation $\mathbf{F}^{\mathrm{p}}$. The vector $\mathbf{s}_0^\alpha$ is embedded in the lattice, so that it becomes $\mathbf{s}^\alpha = \mathbf{F}^* \cdot \mathbf{s}_0^\alpha$ in the deformed configuration. The normal to the slip plane in the deformed configuration is defined by the reciprocal vector $\mathbf{m}^\alpha = \mathbf{m}_0^\alpha \cdot \mathbf{F}^{*-1}$. Thus,

$$\mathbf{s}^\alpha = \mathbf{F}^* \cdot \mathbf{s}_0^\alpha, \quad \mathbf{m}^\alpha = \mathbf{m}_0^\alpha \cdot \mathbf{F}^{*-1}. \tag{12.1.2}$$

In general, $\mathbf{s}^\alpha$ and $\mathbf{m}^\alpha$ are not unit vectors, but are orthogonal to each other, $\mathbf{s}^\alpha \cdot \mathbf{m}^\alpha = 0$.

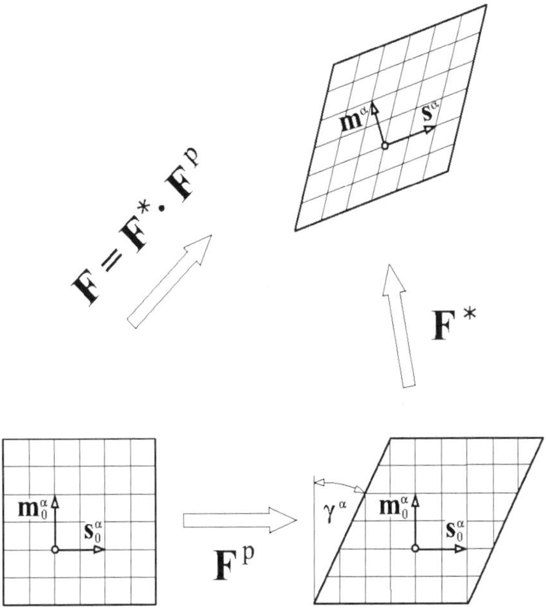

FIGURE 12.2. Kinematic model of elastoplastic deformation of single crystal. The material flows through the crystalline lattice by crystallographic slip, which gives rise to deformation gradient $\mathbf{F}^{\mathrm{p}}$. Subsequently, the material with the embedded lattice is deformed elastically from the intermediate to the current configuration. The lattice vectors in the two configurations are related by $\mathbf{s}^\alpha = \mathbf{F}^* \cdot \mathbf{s}_0^\alpha$ and $\mathbf{m}^\alpha = \mathbf{s}_0^\alpha \cdot \mathbf{F}^{*-1}$.

Velocity Gradient

In view of the decomposition (12.1.1), the velocity gradient $\mathbf{L} = \dot{\mathbf{F}} \cdot \mathbf{F}^{-1}$ can be expressed as

$$\mathbf{L} = \mathbf{L}^* + \mathbf{F}^* \cdot \left(\dot{\mathbf{F}}^{\mathrm{p}} \cdot \mathbf{F}^{\mathrm{p}-1}\right) \cdot \mathbf{F}^{*-1}, \tag{12.1.3}$$

where $\mathbf{L}^*$ is the lattice velocity gradient,

$$\mathbf{L}^* = \dot{\mathbf{F}}^* \cdot \mathbf{F}^{*-1}. \tag{12.1.4}$$

The velocity gradient in the intermediate configuration is produced by the slip rates $\dot{\gamma}^\alpha$ on n active slip systems, such that

$$\dot{\mathbf{F}}^{\mathrm{p}} \cdot \mathbf{F}^{\mathrm{p}-1} = \sum_{\alpha=1}^{n} \dot{\gamma}^\alpha \, \mathbf{s}_0^\alpha \otimes \mathbf{m}_0^\alpha. \tag{12.1.5}$$

The slip systems $(\mathbf{s}^0, \mathbf{m}^0)$ and $(-\mathbf{s}^0, \mathbf{m}^0)$ are considered as separate slip systems, on each of which only the positive slip rate is allowed. For example, with this convention, the total number of available slip systems in f.c.c. crystals is 24. Using Eq. (12.1.2), the corresponding tensor in the deformed configuration is

$$\mathbf{F}^* \cdot \left(\dot{\mathbf{F}}^{\mathrm{p}} \cdot \mathbf{F}^{\mathrm{p}-1}\right) \cdot \mathbf{F}^{*-1} = \sum_{\alpha=1}^{n} \dot{\gamma}^\alpha \, \mathbf{s}^\alpha \otimes \mathbf{m}^\alpha. \tag{12.1.6}$$

The right-hand side of Eq. (12.1.6) can be decomposed into its symmetric and anti-symmetric parts as

$$\sum_{\alpha=1}^{n} \dot{\gamma}^\alpha \, \mathbf{s}^\alpha \otimes \mathbf{m}^\alpha = \sum_{\alpha=1}^{n} \left(\mathbf{P}^\alpha + \mathbf{Q}^\alpha\right) \dot{\gamma}^\alpha. \tag{12.1.7}$$

The second-order (slip orientation) tensors $\mathbf{P}^\alpha$ and $\mathbf{Q}^\alpha$ are defined by (e.g., Asaro, 1983a)

$$\mathbf{P}^\alpha = \frac{1}{2} \left(\mathbf{s}^\alpha \otimes \mathbf{m}^\alpha + \mathbf{m}^\alpha \otimes \mathbf{s}^\alpha\right), \tag{12.1.8}$$

$$\mathbf{Q}^\alpha = \frac{1}{2} \left(\mathbf{s}^\alpha \otimes \mathbf{m}^\alpha - \mathbf{m}^\alpha \otimes \mathbf{s}^\alpha\right). \tag{12.1.9}$$

Thus, the velocity gradient can be expressed as

$$\mathbf{L} = \mathbf{L}^* + \sum_{\alpha=1}^{n} \left(\mathbf{P}^\alpha + \mathbf{Q}^\alpha\right) \dot{\gamma}^\alpha. \tag{12.1.10}$$

Upon using the decomposition of the lattice velocity gradient $\mathbf{L}^*$ into its symmetric and anti-symmetric parts, the lattice rate of deformation $\mathbf{D}^*$ and the lattice spin $\mathbf{W}^*$, i.e.,

$$\mathbf{L}^* = \mathbf{D}^* + \mathbf{W}^*, \tag{12.1.11}$$

we can split Eq. (12.1.10) into

$$\mathbf{D} = \mathbf{D}^* + \sum_{\alpha=1}^{n} \mathbf{P}^\alpha \, \dot{\gamma}^\alpha, \tag{12.1.12}$$

$$\mathbf{W} = \mathbf{W}^* + \sum_{\alpha=1}^{n} \mathbf{Q}^\alpha \, \dot{\gamma}^\alpha. \tag{12.1.13}$$

The time-rate of the Schmid orientation tensor $\mathbf{P}^\alpha$ can be found by differentiating Eq. (12.1.8). Since

$$\dot{\mathbf{s}}^\alpha = \mathbf{L}^* \cdot \mathbf{s}^\alpha, \quad \dot{\mathbf{m}}^\alpha = -\mathbf{m}^\alpha \cdot \mathbf{L}^*, \tag{12.1.14}$$

there follows

$$\begin{aligned}\dot{\mathbf{P}}^\alpha =& \frac{1}{2}\left[\mathbf{L}^* \cdot (\mathbf{s}^\alpha \otimes \mathbf{m}^\alpha) - (\mathbf{s}^\alpha \otimes \mathbf{m}^\alpha) \cdot \mathbf{L}^*\right. \\ & \left. - \mathbf{L}^{*T} \cdot (\mathbf{m}^\alpha \otimes \mathbf{s}^\alpha) + (\mathbf{m}^\alpha \otimes \mathbf{s}^\alpha) \cdot \mathbf{L}^{*T}\right].\end{aligned} \tag{12.1.15}$$

This can be rewritten in terms of the Jaumann derivative of $\mathbf{P}^\alpha$ with respect to the lattice spin as

$$\overset{\bullet}{\mathbf{P}}{}^\alpha = \dot{\mathbf{P}}^\alpha - \mathbf{W}^* \cdot \mathbf{P}^\alpha + \mathbf{P}^\alpha \cdot \mathbf{W}^* = \mathbf{D}^* \cdot \mathbf{Q}^\alpha - \mathbf{Q}^\alpha \cdot \mathbf{D}^*. \tag{12.1.16}$$

Rate of Lagrangian Strain

The following Lagrangian strain measures, relative to the initial reference configuration, can be introduced

$$\mathbf{E} = \frac{1}{2}\left(\mathbf{F}^T \cdot \mathbf{F} - \mathbf{I}\right), \quad \mathbf{E}^{\mathrm{p}} = \frac{1}{2}\left(\mathbf{F}^{\mathrm{p}T} \cdot \mathbf{F}^{\mathrm{p}} - \mathbf{I}\right), \tag{12.1.17}$$

where $\mathbf{I}$ is the second-order unit tensor. The Lagrangian lattice strain, with respect to the intermediate configuration, is

$$\mathbf{E}^* = \frac{1}{2}\left(\mathbf{F}^{*T} \cdot \mathbf{F}^* - \mathbf{I}\right). \tag{12.1.18}$$

The introduced strain measures are related by

$$\mathbf{E} = \mathbf{F}^{\mathrm{p}T} \cdot \mathbf{E}^* \cdot \mathbf{F}^{\mathrm{p}} + \mathbf{E}^{\mathrm{p}}. \tag{12.1.19}$$

By differentiating Eq. (12.1.19), the rate of total Lagrangian strain is

$$\dot{\mathbf{E}} = \mathbf{F}^{\mathrm{p}T} \cdot \left\{\dot{\mathbf{E}}^* + \frac{1}{2}\left[\mathbf{C}^* \cdot \left(\dot{\mathbf{F}}^{\mathrm{p}} \cdot \mathbf{F}^{\mathrm{p}-1}\right) + \left(\dot{\mathbf{F}}^{\mathrm{p}} \cdot \mathbf{F}^{\mathrm{p}-1}\right)^T \cdot \mathbf{C}^*\right]\right\} \cdot \mathbf{F}^{\mathrm{p}}, \tag{12.1.20}$$

where

$$\mathbf{C}^* = \mathbf{F}^{*T} \cdot \mathbf{F}^* \tag{12.1.21}$$

is the lattice deformation tensor. After Eq. (12.1.5) is substituted into Eq. (12.1.20), the rate of the Lagrangian strain becomes

$$\dot{\mathbf{E}} = \mathbf{F}^{\mathrm{p}T} \cdot \dot{\mathbf{E}}^* \cdot \mathbf{F}^{\mathrm{p}} + \sum_{\alpha=1}^{n} \mathbf{P}_0^\alpha \, \dot{\gamma}^\alpha. \tag{12.1.22}$$

The symmetric second-order tensor $\mathbf{P}_0^\alpha$ is defined by

$$\mathbf{P}_0^\alpha = \frac{1}{2}\mathbf{F}^{\mathrm{p}T} \cdot \left[\mathbf{C}^* \cdot (\mathbf{s}_0^\alpha \otimes \mathbf{m}_0^\alpha) + (\mathbf{m}_0^\alpha \otimes \mathbf{s}_0^\alpha) \cdot \mathbf{C}^*\right] \cdot \mathbf{F}^{\mathrm{p}}. \tag{12.1.23}$$

It can be easily verified that $\mathbf{P}_0^\alpha$ is induced from $\mathbf{P}^\alpha$ by the deformation $\mathbf{F}$, so that

$$\mathbf{P}_0^\alpha = \mathbf{F}^T \cdot \mathbf{P}^\alpha \cdot \mathbf{F}. \tag{12.1.24}$$

An equivalent representation of the tensor $\mathbf{P}_0^\alpha$ is

$$\mathbf{P}_0^\alpha = \frac{1}{2}\left(\mathbf{C}\cdot\mathbf{Z}_0^\alpha + \mathbf{Z}_0^{\alpha\,T}\cdot\mathbf{C}\right), \tag{12.1.25}$$

where $\mathbf{C} = \mathbf{F}^T\cdot\mathbf{F}$, and

$$\mathbf{Z}_0^\alpha = \mathbf{F}^{\mathrm{p}-1}\cdot(\mathbf{s}_0^\alpha\otimes\mathbf{m}_0^\alpha)\cdot\mathbf{F}^{\mathrm{p}}. \tag{12.1.26}$$

Since

$$\mathbf{P}^\alpha + \mathbf{Q}^\alpha = \mathbf{F}^*\cdot(\mathbf{s}_0^\alpha\otimes\mathbf{m}_0^\alpha)\cdot\mathbf{F}^{*-1}, \tag{12.1.27}$$

there is a connection

$$\mathbf{Z}_0^\alpha = \mathbf{F}^{-1}\cdot(\mathbf{P}^\alpha + \mathbf{Q}^\alpha)\cdot\mathbf{F}. \tag{12.1.28}$$

This shows that the tensor $\mathbf{Z}_0^\alpha$ is induced from the tensor $(\mathbf{P}^\alpha + \mathbf{Q}^\alpha)$ by the deformation $\mathbf{F}$. Its rate is

$$\dot{\mathbf{Z}}_0^\alpha = \sum_{\beta=1}^{n}\left(\mathbf{Z}_0^\alpha\cdot\mathbf{Z}_0^\beta - \mathbf{Z}_0^\beta\cdot\mathbf{Z}_0^\alpha\right)\dot{\gamma}^\beta. \tag{12.1.29}$$

The rate of $\mathbf{P}_0^\alpha$ is obtained by differentiating Eq. (12.1.23). The result is

$$\begin{aligned}\dot{\mathbf{P}}_0^\alpha &= \left(\mathbf{F}^{\mathrm{p}T}\cdot\dot{\mathbf{E}}^*\cdot\mathbf{F}^{\mathrm{p}}\right)\cdot\mathbf{Z}_0^\alpha + \mathbf{Z}_0^{\alpha\,T}\cdot\left(\mathbf{F}^{\mathrm{p}T}\cdot\dot{\mathbf{E}}^*\cdot\mathbf{F}^{\mathrm{p}}\right)\\ &+\sum_{\beta=1}^{n}\left(\mathbf{P}_0^\alpha\cdot\mathbf{Z}_0^\beta + \mathbf{Z}_0^{\beta\,T}\cdot\mathbf{P}_0^\alpha\right)\dot{\gamma}^\beta.\end{aligned} \tag{12.1.30}$$

The second-order tensors $\mathbf{P}_0^\alpha$ and $\mathbf{Z}_0^\alpha$ were originally introduced by Hill and Havner (1982) ($\tilde{\nu}$ and $\tilde{\mathbf{C}}$ in their notation).

12.2. Kinetic Preliminaries

In the following derivation it will be assumed that elastic properties of the crystal are not affected by crystallographic slip. Since slip is an isochoric deformation process, the elastic strain energy per unit initial volume can be written as

$$\Psi^{\mathrm{e}} = \Psi^{\mathrm{e}}(\mathbf{E}^*) = \Psi^{\mathrm{e}}\left[\mathbf{F}^{\mathrm{p}-T}\cdot(\mathbf{E}-\mathbf{E}^{\mathrm{p}})\cdot\mathbf{F}^{\mathrm{p}-1}\right], \tag{12.2.1}$$

in view of Eq. (12.1.19). The function Ψ^{e} is expressed in the coordinate system that has a fixed orientation relative to the lattice orientation in $\mathcal{B}^0$ and $\mathcal{B}^{\mathrm{p}}$. The symmetric Piola–Kirchhoff stress tensors, relative to the lattice and total deformation, are derived from Ψ^{e} by the gradient operations

$$\mathbf{T}^* = \frac{\partial\Psi^{\mathrm{e}}}{\partial\mathbf{E}^*}, \quad \mathbf{T} = \frac{\partial\Psi^{\mathrm{e}}}{\partial\mathbf{E}}. \tag{12.2.2}$$

They are related by

$$\mathbf{T}^* = \mathbf{F}^{\mathrm{p}} \cdot \mathbf{T} \cdot \mathbf{F}^{\mathrm{p}T}. \tag{12.2.3}$$

The stress tensors $\mathbf{T}^*$ and $\mathbf{T}$, expressed in terms of the Kirchhoff stress $\boldsymbol{\tau} = (\det \mathbf{F})\boldsymbol{\sigma}$ ($\boldsymbol{\sigma}$ denotes the Cauchy stress), are

$$\mathbf{T}^* = \mathbf{F}^{*-1} \cdot \boldsymbol{\tau} \cdot \mathbf{F}^{*-T}, \quad \mathbf{T} = \mathbf{F}^{-1} \cdot \boldsymbol{\tau} \cdot \mathbf{F}^{-T}. \tag{12.2.4}$$

Since plastic incompressibility is assumed, we have

$$\det \mathbf{F}^* = \det \mathbf{F}. \tag{12.2.5}$$

The rates of the Piola–Kirchhoff stresses $\dot{\mathbf{T}}^*$ and $\dot{\mathbf{T}}$ can be cast in terms of the convected rates of the Kirchhoff stress as

$$\dot{\mathbf{T}}^* = \mathbf{F}^{*-1} \cdot \overset{\blacktriangle}{\boldsymbol{\tau}} \cdot \mathbf{F}^{*-T}, \quad \dot{\mathbf{T}} = \mathbf{F}^{-1} \cdot \overset{\triangle}{\boldsymbol{\tau}} \cdot \mathbf{F}^{-T}. \tag{12.2.6}$$

The convected rates of the Kirchhoff stress, with respect to the lattice and total deformation, are

$$\overset{\blacktriangle}{\boldsymbol{\tau}} = \dot{\boldsymbol{\tau}} - \mathbf{L}^* \cdot \boldsymbol{\tau} - \boldsymbol{\tau} \cdot \mathbf{L}^{*T}, \quad \overset{\triangle}{\boldsymbol{\tau}} = \dot{\boldsymbol{\tau}} - \mathbf{L} \cdot \boldsymbol{\tau} - \boldsymbol{\tau} \cdot \mathbf{L}^{T}, \tag{12.2.7}$$

so that

$$\overset{\blacktriangle}{\boldsymbol{\tau}} = \overset{\triangle}{\boldsymbol{\tau}} + (\mathbf{L} - \mathbf{L}^*) \cdot \boldsymbol{\tau} + \boldsymbol{\tau} \cdot (\mathbf{L} - \mathbf{L}^*)^T. \tag{12.2.8}$$

The difference between the total and lattice velocity gradients is obtained from Eq. (12.1.10),

$$\mathbf{L} - \mathbf{L}^* = \sum_{\alpha=1}^{n} (\mathbf{P}^\alpha + \mathbf{Q}^\alpha)\, \dot{\gamma}^\alpha. \tag{12.2.9}$$

When this is substituted into Eq. (12.2.8), we obtain the relationship between the two convected stress rates,

$$\overset{\blacktriangle}{\boldsymbol{\tau}} = \overset{\triangle}{\boldsymbol{\tau}} + \sum_{\alpha=1}^{n} (\mathbf{P}^\alpha \cdot \boldsymbol{\tau} + \boldsymbol{\tau} \cdot \mathbf{P}^\alpha)\, \dot{\gamma}^\alpha + \sum_{\alpha=1}^{n} (\mathbf{Q}^\alpha \cdot \boldsymbol{\tau} - \boldsymbol{\tau} \cdot \mathbf{Q}^\alpha)\, \dot{\gamma}^\alpha. \tag{12.2.10}$$

Similarly, the Jaumann rates

$$\overset{\bullet}{\boldsymbol{\tau}} = \dot{\boldsymbol{\tau}} - \mathbf{W}^* \cdot \boldsymbol{\tau} + \boldsymbol{\tau} \cdot \mathbf{W}^*, \quad \overset{\circ}{\boldsymbol{\tau}} = \dot{\boldsymbol{\tau}} - \mathbf{W} \cdot \boldsymbol{\tau} + \boldsymbol{\tau} \cdot \mathbf{W} \tag{12.2.11}$$

are related by

$$\overset{\bullet}{\boldsymbol{\tau}} = \overset{\circ}{\boldsymbol{\tau}} + (\mathbf{W} - \mathbf{W}^*) \cdot \boldsymbol{\tau} - \boldsymbol{\tau} \cdot (\mathbf{W} - \mathbf{W}^*). \tag{12.2.12}$$

Since the difference between the total and lattice spin is, from Eq. (12.1.13),

$$\mathbf{W} - \mathbf{W}^* = \sum_{\alpha=1}^{n} \mathbf{Q}^\alpha\, \dot{\gamma}^\alpha, \tag{12.2.13}$$

the substitution into Eq. (12.2.12) gives

$$\overset{\bullet}{\boldsymbol{\tau}} = \overset{\circ}{\boldsymbol{\tau}} + \sum_{\alpha=1}^{n} (\mathbf{Q}^{\alpha} \cdot \boldsymbol{\tau} - \boldsymbol{\tau} \cdot \mathbf{Q}^{\alpha})\, \dot{\gamma}^{\alpha}. \tag{12.2.14}$$

The relationship between the rates of stress tensors $\mathbf{T}^*$ and $\mathbf{T}$ is obtained by differentiating Eq. (12.2.3), i.e.,

$$\dot{\mathbf{T}}^* = \mathbf{F}^{\mathrm{p}} \cdot \dot{\mathbf{T}} \cdot \mathbf{F}^{\mathrm{p}T} + \left(\dot{\mathbf{F}}^{\mathrm{p}} \cdot \mathbf{F}^{\mathrm{p}-1}\right) \cdot \mathbf{T}^* + \mathbf{T}^* \cdot \left(\dot{\mathbf{F}}^{\mathrm{p}} \cdot \mathbf{F}^{\mathrm{p}-1}\right)^T. \tag{12.2.15}$$

This can be rewritten as

$$\begin{aligned} \dot{\mathbf{T}} = \mathbf{F}^{\mathrm{p}-1} \cdot \dot{\mathbf{T}}^* \cdot \mathbf{F}^{\mathrm{p}-T} &- \left[\mathbf{F}^{\mathrm{p}-1} \cdot \left(\dot{\mathbf{F}}^{\mathrm{p}} \cdot \mathbf{F}^{\mathrm{p}-1}\right) \cdot \mathbf{F}^{\mathrm{p}}\right] \cdot \mathbf{T} \\ &- \mathbf{T} \cdot \left[\mathbf{F}^{\mathrm{p}-1} \cdot \left(\dot{\mathbf{F}}^{\mathrm{p}} \cdot \mathbf{F}^{\mathrm{p}-1}\right) \cdot \mathbf{F}^{\mathrm{p}}\right]^T. \end{aligned} \tag{12.2.16}$$

Upon using Eq. (12.1.5), we obtain

$$\dot{\mathbf{T}} = \mathbf{F}^{\mathrm{p}-1} \cdot \dot{\mathbf{T}}^* \cdot \mathbf{F}^{\mathrm{p}-T} - \sum_{\alpha=1}^{n} \left(\mathbf{Z}_0^{\alpha} \cdot \mathbf{T} + \mathbf{T} \cdot \mathbf{Z}_0^{\alpha\,T}\right) \dot{\gamma}^{\alpha}. \tag{12.2.17}$$

Additional kinematic and kinetic analysis can be found in Gurtin (2000).

Along purely elastic branch of the response (e.g., during elastic unloading), we have

$$\dot{\mathbf{T}} = \mathbf{F}^{\mathrm{p}-1} \cdot \dot{\mathbf{T}}^* \cdot \mathbf{F}^{\mathrm{p}-T}, \quad \dot{\mathbf{E}} = \mathbf{F}^{\mathrm{p}T} \cdot \dot{\mathbf{E}}^* \cdot \mathbf{F}^{\mathrm{p}}, \tag{12.2.18}$$

since then

$$\dot{\gamma}^{\alpha} = 0 \quad \text{and} \quad \dot{\mathbf{F}}^{\mathrm{p}} = \mathbf{0}. \tag{12.2.19}$$

12.3. Lattice Response

The tensors of elastic moduli corresponding to strain measures $\mathbf{E}^*$ and $\mathbf{E}$ are

$$\boldsymbol{\Lambda}^*_{(1)} = \frac{\partial^2 \Psi^{\mathrm{e}}}{\partial \mathbf{E}^* \otimes \partial \mathbf{E}^*}, \quad \boldsymbol{\Lambda}_{(1)} = \frac{\partial^2 \Psi^{\mathrm{e}}}{\partial \mathbf{E} \otimes \partial \mathbf{E}}, \tag{12.3.1}$$

with the connection

$$\boldsymbol{\Lambda}_{(1)} = \mathbf{F}^{\mathrm{p}-1}\, \mathbf{F}^{\mathrm{p}-1}\, \boldsymbol{\Lambda}^*_{(1)}\, \mathbf{F}^{\mathrm{p}-T}\, \mathbf{F}^{\mathrm{p}-T}. \tag{12.3.2}$$

Taking the time derivative in Eq. (12.2.2), there follows

$$\dot{\mathbf{T}}^* = \boldsymbol{\Lambda}^*_{(1)} : \dot{\mathbf{E}}^*. \tag{12.3.3}$$

Substituting the first of (12.2.6), and

$$\dot{\mathbf{E}}^* = \mathbf{F}^{*T} \cdot \mathbf{D}^* \cdot \mathbf{F}^*, \tag{12.3.4}$$

into Eq. (12.3.3), yields

$$\overset{\blacktriangle}{\boldsymbol{\tau}} = \boldsymbol{\mathcal{L}}_{(1)} : \mathbf{D}^{*}. \tag{12.3.5}$$

The relationship between the introduced elastic moduli is

$$\boldsymbol{\mathcal{L}}_{(1)} = \mathbf{F}^{*}\,\mathbf{F}^{*}\,\boldsymbol{\Lambda}^{*}_{(1)}\,\mathbf{F}^{*T}\,\mathbf{F}^{*T} = \mathbf{F}\,\mathbf{F}\,\boldsymbol{\Lambda}_{(1)}\,\mathbf{F}^{T}\,\mathbf{F}^{T}. \tag{12.3.6}$$

The tensor products are such that, in the component form,

$$\mathcal{L}^{(1)}_{ijkl} = F^{*}_{im}\,F^{*}_{jn}\,\Lambda^{*\,(1)}_{mnpq}\,F^{*\,T}_{pk}\,F^{*\,T}_{ql}. \tag{12.3.7}$$

If the Jaumann rate corotational with the lattice spin $\mathbf{W}^{*}$ is used, Eq. (12.3.5) can be recast in the form

$$\overset{\bullet}{\boldsymbol{\tau}} = \boldsymbol{\mathcal{L}}_{(0)} : \mathbf{D}^{*}. \tag{12.3.8}$$

The relationship between the corresponding elastic moduli tensors is

$$\boldsymbol{\mathcal{L}}_{(0)} = \boldsymbol{\mathcal{L}}_{(1)} + 2\,\boldsymbol{\mathcal{S}}, \tag{12.3.9}$$

which follows by recalling that

$$\overset{\blacktriangle}{\boldsymbol{\tau}} = \overset{\bullet}{\boldsymbol{\tau}} - \mathbf{D}^{*}\cdot\boldsymbol{\tau} - \boldsymbol{\tau}\cdot\mathbf{D}^{*}. \tag{12.3.10}$$

The rectangular components of the fourth-order tensor $\boldsymbol{\mathcal{S}}$ are

$$\mathcal{S}_{ijkl} = \frac{1}{4}\left(\tau_{ik}\delta_{jl} + \tau_{jk}\delta_{il} + \tau_{il}\delta_{jk} + \tau_{jl}\delta_{ik}\right), \tag{12.3.11}$$

as previously discussed in Section 6.2. Along an elastic branch of the response (elastic unloading from elastoplastic state), the total and lattice velocity gradients coincide, so that

$$\mathbf{L}^{*} = \mathbf{L}, \quad \overset{\blacktriangle}{\boldsymbol{\tau}} = \overset{\triangle}{\boldsymbol{\tau}}, \quad \overset{\bullet}{\boldsymbol{\tau}} = \overset{\circ}{\boldsymbol{\tau}}. \tag{12.3.12}$$

12.4. Elastoplastic Constitutive Framework

The rate-type constitutive framework for the elastoplastic loading of a single crystal is obtained by substituting Eq. (12.2.10), and

$$\mathbf{D}^{*} = \mathbf{D} - \sum_{\alpha=1}^{n} \mathbf{P}^{\alpha}\,\dot{\gamma}^{\alpha}, \tag{12.4.1}$$

into Eq. (12.3.5). The result is

$$\overset{\triangle}{\boldsymbol{\tau}} = \boldsymbol{\mathcal{L}}_{(1)} : \mathbf{D} - \sum_{\alpha=1}^{n} \mathbf{C}^{\alpha}\,\dot{\gamma}^{\alpha}, \tag{12.4.2}$$

where

$$\mathbf{C}^{\alpha} = \boldsymbol{\mathcal{L}}_{(1)} : \mathbf{P}^{\alpha} + \left(\mathbf{P}^{\alpha}\cdot\boldsymbol{\tau} + \boldsymbol{\tau}\cdot\mathbf{P}^{\alpha}\right) + \left(\mathbf{Q}^{\alpha}\cdot\boldsymbol{\tau} - \boldsymbol{\tau}\cdot\mathbf{Q}^{\alpha}\right). \tag{12.4.3}$$

Alternatively, if Eqs. (12.2.12) and (12.4.1) are substituted into Eq. (12.3.8), there follows

$$\overset{\circ}{\boldsymbol{\tau}} = \boldsymbol{\mathcal{L}}_{(0)} : \mathbf{D} - \sum_{\alpha=1}^{n} \mathbf{C}^{\alpha}\, \dot{\gamma}^{\alpha}, \tag{12.4.4}$$

where

$$\mathbf{C}^{\alpha} = \boldsymbol{\mathcal{L}}_{(0)} : \mathbf{P}^{\alpha} + (\mathbf{Q}^{\alpha} \cdot \boldsymbol{\tau} - \boldsymbol{\tau} \cdot \mathbf{Q}^{\alpha})\,. \tag{12.4.5}$$

Having in mind the connection (12.3.9) between the elastic moduli tensors $\boldsymbol{\mathcal{L}}_{(0)}$ and $\boldsymbol{\mathcal{L}}_{(1)}$, it is readily verified that the right-hand sides of Eqs. (12.4.3) and (12.4.5) are equal to each other.

An equivalent constitutive structure can be obtained relative to the Lagrangian strain and its conjugate symmetric Piola–Kirchhoff stress. The substitution of

$$\dot{\mathbf{T}}^{*} = \mathbf{F}^{\mathrm{p}} \cdot \left[\dot{\mathbf{T}} + \sum_{\alpha=1}^{n} \left(\mathbf{Z}_0^{\alpha} \cdot \mathbf{T} + \mathbf{T} \cdot \mathbf{Z}_0^{\alpha\, T} \right) \dot{\gamma}^{\alpha} \right] \cdot \mathbf{F}^{\mathrm{p}T} \tag{12.4.6}$$

and

$$\dot{\mathbf{E}}^{*} = \mathbf{F}^{\mathrm{p}-T} \cdot \left(\dot{\mathbf{E}} - \sum_{\alpha=1}^{n} \mathbf{P}_0^{\alpha}\, \dot{\gamma}^{\alpha} \right) \cdot \mathbf{F}^{\mathrm{p}-1}, \tag{12.4.7}$$

which follow from Eqs. (12.2.17) and (12.1.22), into Eq. (12.3.3) gives

$$\dot{\mathbf{T}} + \sum_{\alpha=1}^{n} \left(\mathbf{Z}_0^{\alpha} \cdot \mathbf{T} + \mathbf{T} \cdot \mathbf{Z}_0^{\alpha\, T} \right) \dot{\gamma}^{\alpha} = \boldsymbol{\Lambda}_{(1)} : \left(\dot{\mathbf{E}} - \sum_{\alpha=1}^{n} \mathbf{P}_0^{\alpha}\, \dot{\gamma}^{\alpha} \right). \tag{12.4.8}$$

The relationship (12.3.2) between the moduli $\boldsymbol{\Lambda}_{(1)}$ and $\boldsymbol{\Lambda}^{*}_{(1)}$ was also utilized. Consequently,

$$\dot{\mathbf{T}} = \boldsymbol{\Lambda}_{(1)} : \dot{\mathbf{E}} - \sum_{\alpha=1}^{n} \mathbf{C}_0^{\alpha}\, \dot{\gamma}^{\alpha}, \tag{12.4.9}$$

where

$$\mathbf{C}_0^{\alpha} = \boldsymbol{\Lambda}_{(1)} : \mathbf{P}_0^{\alpha} + \mathbf{Z}_0^{\alpha} \cdot \mathbf{T} + \mathbf{T} \cdot \mathbf{Z}_0^{\alpha\, T}. \tag{12.4.10}$$

Recalling the expressions (12.1.24) and (12.1.28), and

$$\boldsymbol{\Lambda}_{(1)} = \mathbf{F}^{-1}\, \mathbf{F}^{-1}\, \boldsymbol{\mathcal{L}}_{(1)}\, \mathbf{F}^{-T}\, \mathbf{F}^{-T}, \tag{12.4.11}$$

we deduce the relationship between the tensors $\mathbf{C}_0^{\alpha}$ and $\mathbf{C}^{\alpha}$. This is

$$\mathbf{C}_0^{\alpha} = \mathbf{F}^{-1} \cdot \mathbf{C}^{\alpha} \cdot \mathbf{F}^{-T}. \tag{12.4.12}$$

In view of Eq. (12.1.24), there is also an identity

$$\mathbf{C}_0^{\alpha} : \mathbf{P}_0^{\alpha} = \mathbf{C}^{\alpha} : \mathbf{P}^{\alpha}. \tag{12.4.13}$$

12.5. Partition of Stress and Strain Rates

The elastic parts of the stress rates $\overset{\circ}{\boldsymbol{\tau}}$, $\overset{\triangle}{\boldsymbol{\tau}}$ and $\dot{\mathbf{T}}$ are defined by

$$\overset{\circ}{\boldsymbol{\tau}}{}^{\mathrm{e}} = \boldsymbol{\mathcal{L}}_{(0)} : \mathbf{D}, \quad \overset{\triangle}{\boldsymbol{\tau}}{}^{\mathrm{e}} = \boldsymbol{\mathcal{L}}_{(1)} : \mathbf{D}, \quad (\dot{\mathbf{T}})^{\mathrm{e}} = \boldsymbol{\Lambda}_{(1)} : \dot{\mathbf{E}}, \tag{12.5.1}$$

since, from Eqs. (12.4.2), (12.4.4), and (12.4.9), only the remaining parts of stress rates depend on the slip rates $\dot{\gamma}^{\alpha}$. These are the plastic parts

$$\overset{\circ}{\boldsymbol{\tau}}{}^{\mathrm{p}} = \overset{\triangle}{\boldsymbol{\tau}}{}^{\mathrm{p}} = -\sum_{\alpha=1}^{n} \mathbf{C}^{\alpha}\,\dot{\gamma}^{\alpha}, \quad (\dot{\mathbf{T}})^{\mathrm{p}} = -\sum_{\alpha=1}^{n} \mathbf{C}_0^{\alpha}\,\dot{\gamma}^{\alpha}. \tag{12.5.2}$$

In view of the connection (12.4.12), we have

$$(\dot{\mathbf{T}})^{\mathrm{p}} = \mathbf{F}^{-1} \cdot \overset{\triangle}{\boldsymbol{\tau}}{}^{\mathrm{p}} \cdot \mathbf{F}^{-T}. \tag{12.5.3}$$

This relationship was anticipated from the previously established relationship given by the second expression in Eq. (12.2.6). Physically, the plastic stress rate $(\dot{\mathbf{T}})^{\mathrm{p}}$ gives a residual stress decrement $(\dot{\mathbf{T}})^{\mathrm{p}}\,\mathrm{d}t$ in an infinitesimal strain cycle, associated with application and removal of the strain increment $\dot{\mathbf{E}}\,\mathrm{d}t$.

The rate of deformation tensor and the rate of Lagrangian strain can be expressed from Eqs. (12.4.2), (12.4.4) and (12.4.9) as

$$\mathbf{D} = \boldsymbol{\mathcal{M}}_{(0)} : \overset{\circ}{\boldsymbol{\tau}} + \sum_{\alpha=1}^{n} \boldsymbol{\mathcal{M}}_{(0)} : \mathbf{C}^{\alpha}\,\dot{\gamma}^{\alpha}, \tag{12.5.4}$$

$$\mathbf{D} = \boldsymbol{\mathcal{M}}_{(1)} : \overset{\triangle}{\boldsymbol{\tau}} + \sum_{\alpha=1}^{n} \boldsymbol{\mathcal{M}}_{(1)} : \mathbf{C}^{\alpha}\,\dot{\gamma}^{\alpha}, \tag{12.5.5}$$

$$\dot{\mathbf{E}} = \mathbf{M}_{(1)} : \dot{\mathbf{T}} + \sum_{\alpha=1}^{n} \mathbf{M}_{(1)} : \mathbf{C}_0^{\alpha}\,\dot{\gamma}^{\alpha}. \tag{12.5.6}$$

The introduced elastic compliances tensors are

$$\boldsymbol{\mathcal{M}}_{(0)} = \boldsymbol{\mathcal{L}}_{(0)}^{-1}, \quad \boldsymbol{\mathcal{M}}_{(1)} = \boldsymbol{\mathcal{L}}_{(1)}^{-1}, \quad \mathbf{M}_{(1)} = \boldsymbol{\Lambda}_{(1)}^{-1}. \tag{12.5.7}$$

The elastic parts of the rate of deformation tensor $\mathbf{D}$, corresponding to the Jaumann and convected rates of the Kirchhoff stress, and the elastic part of the rate of Lagrangian strain $\dot{\mathbf{E}}$, are defined by

$$\mathbf{D}_{(0)}^{\mathrm{e}} = \boldsymbol{\mathcal{M}}_{(0)} : \overset{\circ}{\boldsymbol{\tau}}, \quad \mathbf{D}_{(1)}^{\mathrm{e}} = \boldsymbol{\mathcal{M}}_{(1)} : \overset{\triangle}{\boldsymbol{\tau}}, \quad (\dot{\mathbf{E}})^{\mathrm{e}} = \mathbf{M}_{(1)} : \dot{\mathbf{T}}. \tag{12.5.8}$$

The remaining parts of $\mathbf{D}$ and $\dot{\mathbf{E}}$ depend on the slip rates $\dot{\gamma}^\alpha$. They are the plastic parts

$$\mathbf{D}^{\mathrm{p}}_{(0)} = \sum_{\alpha=1}^{n} \mathbf{H}^\alpha \dot{\gamma}^\alpha, \quad \mathbf{D}^{\mathrm{p}}_{(1)} = \sum_{\alpha=1}^{n} \mathbf{G}^\alpha \dot{\gamma}^\alpha, \quad (\dot{\mathbf{E}})^{\mathrm{p}} = \sum_{\alpha=1}^{n} \mathbf{G}^\alpha_0 \dot{\gamma}^\alpha, \tag{12.5.9}$$

where

$$\mathbf{H}^\alpha = \boldsymbol{\mathcal{M}}_{(0)} : \mathbf{C}^\alpha, \quad \mathbf{G}^\alpha = \boldsymbol{\mathcal{M}}_{(1)} : \mathbf{C}^\alpha, \quad \mathbf{G}^\alpha_0 = \mathbf{M}_{(1)} : \mathbf{C}^\alpha_0. \tag{12.5.10}$$

By comparing Eqs. (12.5.2) and (12.5.9), the plastic parts of the stress and strain rates are related by

$$\overset{\circ}{\boldsymbol{\tau}}{}^{\mathrm{p}} = -\boldsymbol{\mathcal{L}}_{(0)} : \mathbf{D}^{\mathrm{p}}_{(0)}, \quad \overset{\triangle}{\boldsymbol{\tau}}{}^{\mathrm{p}} = -\boldsymbol{\mathcal{L}}_{(1)} : \mathbf{D}^{\mathrm{p}}_{(1)}, \quad (\dot{\mathbf{T}})^{\mathrm{p}} = -\boldsymbol{\Lambda}_{(1)} : (\dot{\mathbf{E}})^{\mathrm{p}}. \tag{12.5.11}$$

Since

$$\mathbf{M}_{(1)} = \mathbf{F}^T \mathbf{F}^T \boldsymbol{\mathcal{M}}_{(1)} \mathbf{F} \mathbf{F}, \tag{12.5.12}$$

and recalling the relationship (12.4.12) between the tensors $\mathbf{C}^\alpha$ and $\mathbf{C}^\alpha_0$, there is a connection

$$\mathbf{G}^\alpha_0 = \mathbf{F}^T \cdot \mathbf{G}^\alpha \cdot \mathbf{F}. \tag{12.5.13}$$

Thus

$$(\dot{\mathbf{E}})^{\mathrm{p}} = \mathbf{F}^T \cdot \mathbf{D}^{\mathrm{p}}_{(1)} \cdot \mathbf{F}, \tag{12.5.14}$$

as anticipated from the general expression $\dot{\mathbf{E}} = \mathbf{F}^T \cdot \mathbf{D} \cdot \mathbf{F}$. The strain increment $(\dot{\mathbf{E}})^{\mathrm{p}}\, dt$ represents a residual strain increment left in the crystal upon an infinitesimal loading/unloading cycle associated with the stress increment $\dot{\mathbf{T}}\, dt$. The strain increment $\mathbf{D}^{\mathrm{p}}_{(0)}\, dt$ is a residual strain increment left in the crystal upon an infinitesimal loading/unloading cycle associated with the stress increment $\overset{\circ}{\boldsymbol{\tau}} dt$. Here,

$$\overset{\circ}{\boldsymbol{\tau}} dt = (\det \mathbf{F})\, \overset{\circ}{\underline{\boldsymbol{\tau}}} dt, \tag{12.5.15}$$

where $\overset{\circ}{\underline{\boldsymbol{\tau}}} dt$ is the increment of stress conjugate to the logarithmic strain, when the reference configuration is taken to momentarily coincide with the current configuration. This has been discussed in more details in Section 3.9. Finally, $\overset{\triangle}{\underline{\boldsymbol{\tau}}}\, dt$ is the increment of the symmetric Piola–Kirchhoff stress, conjugate to the Lagrangian strain, when the reference configuration is taken to be the current configuration.

The relationship between the plastic parts of the rate of deformation $\mathbf{D}^{\mathrm{p}}_{(0)}$ and $\mathbf{D}^{\mathrm{p}}_{(1)}$ can be obtained by substituting the first two expressions from (12.5.11) into the identity $\overset{\circ}{\boldsymbol{\tau}}^{\mathrm{p}} = \overset{\triangle}{\boldsymbol{\tau}}^{\mathrm{p}}$. This gives

$$\boldsymbol{\mathcal{L}}_{(0)} : \mathbf{D}^{\mathrm{p}}_{(0)} = \boldsymbol{\mathcal{L}}_{(1)} : \mathbf{D}^{\mathrm{p}}_{(1)}. \tag{12.5.16}$$

Since $\boldsymbol{\mathcal{L}}_{(0)} = \boldsymbol{\mathcal{L}}_{(1)} + 2\boldsymbol{\mathcal{S}}$, we obtain

$$\mathbf{D}^{\mathrm{p}}_{(0)} = \mathbf{D}^{\mathrm{p}}_{(1)} - 2\boldsymbol{\mathcal{M}}_{(0)} : \boldsymbol{\mathcal{S}} : \mathbf{D}^{\mathrm{p}}_{(1)}, \tag{12.5.17}$$

$$\mathbf{D}^{\mathrm{p}}_{(1)} = \mathbf{D}^{\mathrm{p}}_{(0)} + 2\boldsymbol{\mathcal{M}}_{(1)} : \boldsymbol{\mathcal{S}} : \mathbf{D}^{\mathrm{p}}_{(0)}. \tag{12.5.18}$$

Thus, the relative difference between the components of $\mathbf{D}^{\mathrm{p}}_{(0)}$ and $\mathbf{D}^{\mathrm{p}}_{(1)}$ is of the order of stress over elastic modulus (Lubarda, 1999).

The plastic strain rates can be expressed in terms of the previously introduced tensors $\mathbf{P}^{\alpha}$, $\mathbf{Q}^{\alpha}$ and $\mathbf{Z}^{\alpha}_{0}$ by using Eqs. (12.4.3), (12.4.5), and (12.4.10). The results are

$$\mathbf{D}^{\mathrm{p}}_{(0)} = \sum_{\alpha=1}^{n} \left[\mathbf{P}^{\alpha} + \boldsymbol{\mathcal{M}}_{(0)} : (\mathbf{Q}^{\alpha} \cdot \boldsymbol{\tau} - \boldsymbol{\tau} \cdot \mathbf{Q}^{\alpha})\right] \dot{\gamma}^{\alpha}, \tag{12.5.19}$$

$$\mathbf{D}^{\mathrm{p}}_{(1)} = \sum_{\alpha=1}^{n} \left\{\mathbf{P}^{\alpha} + \boldsymbol{\mathcal{M}}_{(1)} : \left[(\mathbf{P}^{\alpha} \cdot \boldsymbol{\tau} + \boldsymbol{\tau} \cdot \mathbf{P}^{\alpha}) + (\mathbf{Q}^{\alpha} \cdot \boldsymbol{\tau} - \boldsymbol{\tau} \cdot \mathbf{Q}^{\alpha})\right]\right\} \dot{\gamma}^{\alpha}, \tag{12.5.20}$$

$$(\dot{\mathbf{E}})^{\mathrm{p}} = \sum_{\alpha=1}^{n} \left[\mathbf{P}^{\alpha}_{0} + \mathbf{M}_{(1)} : \left(\mathbf{Z}^{\alpha}_{0} \cdot \mathbf{T} + \mathbf{T} \cdot \mathbf{Z}^{\alpha\,T}_{0}\right)\right] \dot{\gamma}^{\alpha}. \tag{12.5.21}$$

As discussed by Hill and Rice (1972), and Hill and Havner (1982), although $\mathbf{D}^{\mathrm{p}}_{(0)} = \mathbf{D} - \boldsymbol{\mathcal{M}}_{(0)} : \overset{\circ}{\boldsymbol{\tau}}$ in Eq. (12.5.19) is commonly called the plastic rate of deformation, it does not come from the slip deformation only. There is a further net elastic contribution from the lattice,

$$\boldsymbol{\mathcal{M}}_{(0)} : (\overset{\bullet}{\boldsymbol{\tau}} - \overset{\circ}{\boldsymbol{\tau}}) = \boldsymbol{\mathcal{M}}_{(0)} : \sum_{\alpha=1}^{n} (\mathbf{Q}^{\alpha} \cdot \boldsymbol{\tau} - \boldsymbol{\tau} \cdot \mathbf{Q}^{\alpha})\, \dot{\gamma}^{\alpha}, \tag{12.5.22}$$

caused by the slip-induced rotation of the lattice relative to the stress, as embodied in (12.2.14). Similar comments apply to $\mathbf{D}^{\mathrm{p}}_{(1)}$ and $(\dot{\mathbf{E}})^{\mathrm{p}}$ in Eqs. (12.5.20) and (12.5.21).

12.6. Partition of Rate of Deformation Gradient

In this section we partition the rate of deformation gradient into its elastic and plastic parts, such that

$$\dot{\mathbf{F}} = (\dot{\mathbf{F}})^{\mathrm{e}} + (\dot{\mathbf{F}})^{\mathrm{p}}. \tag{12.6.1}$$

The derivation proceeds as in Section 11.14. The elastic part is defined by

$$(\dot{\mathbf{F}})^{\mathrm{e}} = \mathbf{M} \cdot\cdot\, \dot{\mathbf{P}}, \quad \mathbf{M} = \boldsymbol{\Lambda}^{-1}. \tag{12.6.2}$$

The lattice nominal stress and the overall nominal stress

$$\mathbf{P}^* = \mathbf{T}^* \cdot \mathbf{F}^{*T}, \quad \mathbf{P} = \mathbf{T} \cdot \mathbf{F}^T \tag{12.6.3}$$

are derived from the elastic strain energy as

$$\mathbf{P}^* = \frac{\partial \Psi^{\mathrm{e}}}{\partial \mathbf{F}^*}, \quad \mathbf{P} = \frac{\partial \Psi^{\mathrm{e}}}{\partial \mathbf{F}}, \tag{12.6.4}$$

with the connection

$$\mathbf{P}^* = \mathbf{F}^{\mathrm{p}} \cdot \mathbf{P}. \tag{12.6.5}$$

The corresponding pseudomoduli tensors are

$$\boldsymbol{\Lambda}^* = \frac{\partial^2 \Psi^{\mathrm{e}}}{\partial \mathbf{F}^* \otimes \partial \mathbf{F}^*}, \quad \boldsymbol{\Lambda} = \frac{\partial^2 \Psi^{\mathrm{e}}}{\partial \mathbf{F} \otimes \partial \mathbf{F}}. \tag{12.6.6}$$

Their components (in the same rectangular coordinate system) are related by

$$\Lambda^*_{ijkl} = F^{\mathrm{p}}_{im}\, \Lambda_{mjnl}\, F^{\mathrm{p}}_{kn}\,. \tag{12.6.7}$$

The lattice elasticity is governed by the rate-type constitutive equation

$$\dot{\mathbf{P}}^* = \boldsymbol{\Lambda}^* \cdot\cdot\, \dot{\mathbf{F}}^*. \tag{12.6.8}$$

By differentiating Eq. (12.6.5), there follows

$$\dot{\mathbf{P}}^* = \mathbf{F}^{\mathrm{p}} \cdot \dot{\mathbf{P}} + \dot{\mathbf{F}}^{\mathrm{p}} \cdot \mathbf{P}. \tag{12.6.9}$$

The substitution of Eqs. (12.6.9) and (12.6.7) into Eq. (12.6.8) gives

$$\dot{\mathbf{P}} = \boldsymbol{\Lambda} \cdot\cdot \left(\dot{\mathbf{F}}^* \cdot \mathbf{F}^{\mathrm{p}}\right) - \mathbf{F}^{\mathrm{p}-1} \cdot \dot{\mathbf{F}}^{\mathrm{p}} \cdot \mathbf{P}. \tag{12.6.10}$$

On the other hand, by differentiating the multiplicative decomposition $\mathbf{F} = \mathbf{F}^* \cdot \mathbf{F}^{\mathrm{p}}$, the rate of deformation gradient is

$$\dot{\mathbf{F}} = \dot{\mathbf{F}}^* \cdot \mathbf{F}^{\mathrm{p}} + \mathbf{F}^* \cdot \dot{\mathbf{F}}^{\mathrm{p}}. \tag{12.6.11}$$

Using this, Eq. (12.6.10) can be rewritten as

$$\dot{\mathbf{P}} = \boldsymbol{\Lambda} \cdot\cdot \left(\dot{\mathbf{F}} - \mathbf{F}^* \cdot \dot{\mathbf{F}}^{\mathrm{p}}\right) - \mathbf{F}^{\mathrm{p}-1} \cdot \dot{\mathbf{F}}^{\mathrm{p}} \cdot \mathbf{P}, \tag{12.6.12}$$

i.e.,

$$\dot{\mathbf{P}} = \boldsymbol{\Lambda} \cdot\cdot \left[\dot{\mathbf{F}} - \mathbf{F}^* \cdot \dot{\mathbf{F}}^{\mathrm{p}} - \mathbf{M} \cdot\cdot \left(\mathbf{F}^{\mathrm{p}-1} \cdot \dot{\mathbf{F}}^{\mathrm{p}} \cdot \mathbf{P}\right)\right]. \tag{12.6.13}$$

From Eq. (12.6.13) we identify the plastic part of the rate of deformation gradient as

$$(\dot{\mathbf{F}})^{\mathrm{p}} = \mathbf{F}^* \cdot \dot{\mathbf{F}}^{\mathrm{p}} + \mathbf{M} \cdot\cdot \left(\mathbf{F}^{\mathrm{p}-1} \cdot \dot{\mathbf{F}}^{\mathrm{p}} \cdot \mathbf{P} \right). \tag{12.6.14}$$

The remaining part of the rate of deformation gradient $\dot{\mathbf{F}}$ is the elastic part,

$$(\dot{\mathbf{F}})^{\mathrm{e}} = \dot{\mathbf{F}}^* \cdot \mathbf{F}^{\mathrm{p}} - \mathbf{M} \cdot\cdot \left(\mathbf{F}^{\mathrm{p}-1} \cdot \dot{\mathbf{F}}^{\mathrm{p}} \cdot \mathbf{P} \right), \tag{12.6.15}$$

complying with the definition (12.6.2).

Equation (12.6.13) also serves to identify the elastic and plastic parts of the rate of nominal stress. These are

$$(\dot{\mathbf{P}})^{\mathrm{e}} = \mathbf{\Lambda} \cdot\cdot \dot{\mathbf{F}}, \tag{12.6.16}$$

$$(\dot{\mathbf{P}})^{\mathrm{p}} = -\left[\mathbf{F}^{\mathrm{p}-1} \cdot \dot{\mathbf{F}}^{\mathrm{p}} \cdot \mathbf{P} + \mathbf{\Lambda} \cdot\cdot \left(\mathbf{F}^* \cdot \dot{\mathbf{F}}^{\mathrm{p}} \right) \right], \tag{12.6.17}$$

such that

$$\dot{\mathbf{P}} = (\dot{\mathbf{P}})^{\mathrm{e}} + (\dot{\mathbf{P}})^{\mathrm{p}}. \tag{12.6.18}$$

Evidently, by comparing Eqs. (12.6.14) and (12.6.17), there is a relationship between the plastic parts

$$(\dot{\mathbf{P}})^{\mathrm{p}} = -\mathbf{\Lambda} \cdot\cdot (\dot{\mathbf{F}})^{\mathrm{p}}. \tag{12.6.19}$$

To express the plastic parts of the rate of nominal stress and deformation gradient in terms of the slip rates $\dot{\gamma}^{\alpha}$, Eq. (12.1.5) is first rewritten as

$$\dot{\mathbf{F}}^{\mathrm{p}} = \sum_{\alpha=1}^{n} \dot{\gamma}^{\alpha} \left(\mathbf{s}_0^{\alpha} \otimes \mathbf{m}_0^{\alpha} \right) \cdot \mathbf{F}^{\mathrm{p}}. \tag{12.6.20}$$

Upon substitution into Eq. (12.6.14), the plastic part of the rate of deformation gradient becomes

$$(\dot{\mathbf{F}})^{\mathrm{p}} = \sum_{\alpha=1}^{n} \mathbf{A}^{\alpha} \dot{\gamma}^{\alpha}, \tag{12.6.21}$$

where

$$\mathbf{A}^{\alpha} = (\mathbf{s}^{\alpha} \otimes \mathbf{m}^{\alpha}) \cdot \mathbf{F} + \mathbf{M} \cdot\cdot \mathbf{F}^{-1} \cdot (\mathbf{s}^{\alpha} \otimes \mathbf{m}^{\alpha}) \cdot \mathbf{F} \cdot \mathbf{P}. \tag{12.6.22}$$

The plastic part of the rate of nominal stress is then

$$(\dot{\mathbf{P}})^{\mathrm{p}} = -\sum_{\alpha=1}^{n} \mathbf{B}^{\alpha} \dot{\gamma}^{\alpha}, \tag{12.6.23}$$

where

$$\mathbf{B}^{\alpha} = \mathbf{\Lambda} \cdot\cdot \mathbf{A}^{\alpha} = \mathbf{F}^{-1} \cdot (\mathbf{s}^{\alpha} \otimes \mathbf{m}^{\alpha}) \cdot \mathbf{F} \cdot \mathbf{P} + \mathbf{\Lambda} \cdot\cdot (\mathbf{s}^{\alpha} \otimes \mathbf{m}^{\alpha}) \cdot \mathbf{F}. \tag{12.6.24}$$

Relationship between $(\dot{\mathbf{P}})^{\mathrm{p}}$ *and* $(\dot{\mathbf{T}})^{\mathrm{p}}$

The relationship between the plastic parts of the rate of nominal and symmetric Piola–Kirchhoff stress,

$$(\dot{\mathbf{P}})^{\mathrm{p}} = \dot{\mathbf{P}} - \mathbf{\Lambda} \cdot\cdot\, \dot{\mathbf{F}}, \tag{12.6.25}$$

$$(\dot{\mathbf{T}})^{\mathrm{p}} = \dot{\mathbf{T}} - \mathbf{\Lambda}_{(1)} : \dot{\mathbf{E}}, \tag{12.6.26}$$

can be derived as follows. First, we recall that

$$\boldsymbol{\mathcal{K}} \cdot\cdot\, \dot{\mathbf{F}} = \dot{\mathbf{E}}, \quad \boldsymbol{\mathcal{K}}^T : \mathbf{T} = \mathbf{T} \cdot \mathbf{F}^T = \mathbf{P}, \tag{12.6.27}$$

and

$$\mathbf{\Lambda} = \boldsymbol{\mathcal{K}}^T : \mathbf{\Lambda}_{(1)} : \boldsymbol{\mathcal{K}} + \boldsymbol{\mathcal{T}}, \tag{12.6.28}$$

$$\dot{\mathbf{P}} = \boldsymbol{\mathcal{K}}^T : \dot{\mathbf{T}} + \boldsymbol{\mathcal{T}} \cdot\cdot\, \dot{\mathbf{F}}. \tag{12.6.29}$$

The rectangular components of the fourth-order tensors $\boldsymbol{\mathcal{K}}$ and $\boldsymbol{\mathcal{T}}$ are

$$\mathcal{K}_{ijkl} = \frac{1}{2}\left(\delta_{ik}F_{lj} + \delta_{jk}F_{li}\right), \quad \mathcal{T}_{ijkl} = T_{ik}\delta_{jl}. \tag{12.6.30}$$

Taking a trace product of Eq. (12.6.26) with $\boldsymbol{\mathcal{K}}^T$ from the left gives

$$(\dot{\mathbf{P}})^{\mathrm{p}} = \boldsymbol{\mathcal{K}}^T : (\dot{\mathbf{T}})^{\mathrm{p}}. \tag{12.6.31}$$

Furthermore, since

$$(\dot{\mathbf{P}})^{\mathrm{p}} = -\mathbf{\Lambda} \cdot\cdot\, (\dot{\mathbf{F}})^{\mathrm{p}}, \quad (\dot{\mathbf{T}})^{\mathrm{p}} = -\mathbf{\Lambda}_{(1)} : (\dot{\mathbf{E}})^{\mathrm{p}}, \tag{12.6.32}$$

we obtain

$$(\dot{\mathbf{F}})^{\mathrm{p}} = \mathbf{M} \cdot\cdot\, \boldsymbol{\mathcal{K}}^T : \mathbf{\Lambda}_{(1)} : (\dot{\mathbf{E}})^{\mathrm{p}}. \tag{12.6.33}$$

It is noted that

$$\dot{\mathbf{F}} \cdot\cdot\, (\dot{\mathbf{P}})^{\mathrm{p}} = \dot{\mathbf{E}} : (\dot{\mathbf{T}})^{\mathrm{p}}. \tag{12.6.34}$$

This follows by taking a trace product of Eq. (12.6.25) with $\dot{\mathbf{F}}$ from the left, and by using Eqs. (12.6.27)–(12.6.29). If crystalline behavior is in accord with Ilyushin's postulate of the positive net work in a cycle of strain that involves plastic slip, the quantity in (12.6.34) must be negative. On the other hand,

$$\dot{\mathbf{P}} \cdot\cdot\, (\dot{\mathbf{F}})^{\mathrm{p}} \neq \dot{\mathbf{T}} : (\dot{\mathbf{E}})^{\mathrm{p}}. \tag{12.6.35}$$

Finally, having in mind that

$$(\dot{\mathbf{P}})^{\mathrm{p}} = -\sum_{\alpha=1}^{n} \mathbf{B}^\alpha \dot{\gamma}^\alpha, \quad (\dot{\mathbf{T}})^{\mathrm{p}} = -\sum_{\alpha=1}^{n} \mathbf{C}_0^\alpha \dot{\gamma}^\alpha, \tag{12.6.36}$$

we obtain from Eq. (12.6.31)

$$\mathbf{B}^{\alpha} = \boldsymbol{\mathcal{K}}^{T} : \mathbf{C}_0^{\alpha}. \tag{12.6.37}$$

This relationship can be verified by using Eqs. (12.4.10) and (12.6.24), which explicitly specify the tensors $\mathbf{C}_0^{\alpha}$ and $\mathbf{B}^{\alpha}$, and by performing a trace product of $\boldsymbol{\mathcal{K}}^{T}$ with $\mathbf{C}_0^{\alpha}$. In the derivation, it is helpful to use the property of $\boldsymbol{\mathcal{K}}$ in the trace operation with a second-order tensor $\mathbf{A}$, i.e.,

$$\boldsymbol{\mathcal{K}}^{T} \cdot\cdot\, \mathbf{A} = \frac{1}{2}\,(\mathbf{A} + \mathbf{A}^{T}) \cdot \mathbf{F}^{T}. \tag{12.6.38}$$

In addition, we note that

$$\mathbf{F}^{-1} \cdot \mathbf{P}^{T} \cdot \mathbf{F}^{T} = \mathbf{P}, \tag{12.6.39}$$

$$\mathbf{P}_0^{\alpha} = \boldsymbol{\mathcal{K}} \cdot\cdot\, (\mathbf{s}^{\alpha} \otimes \mathbf{m}^{\alpha}) \cdot \mathbf{F}, \tag{12.6.40}$$

$$\boldsymbol{\mathcal{T}} \cdot\cdot\, (\mathbf{s}^{\alpha} \otimes \mathbf{m}^{\alpha}) \cdot \mathbf{F} = \mathbf{P} \cdot (\mathbf{m}^{\alpha} \otimes \mathbf{s}^{\alpha})\,. \tag{12.6.41}$$

12.7. Generalized Schmid Stress and Normality

For the rate-independent materials it is commonly assumed that plastic flow occurs on a slip system when the resolved shear stress (Schmid stress) on that system reaches the critical value (e.g., Schmid and Boas, 1968)

$$\tau^{\alpha} = \tau_{\mathrm{cr}}^{\alpha}. \tag{12.7.1}$$

In the finite strain context, τ^{α} can be defined as the work conjugate to slip rate $\dot{\gamma}^{\alpha}$, such that

$$\sum_{\alpha=1}^{n} \tau^{\alpha}\, \dot{\gamma}^{\alpha} = \mathbf{T} : \sum_{\alpha=1}^{n} \mathbf{P}_0^{\alpha}\, \dot{\gamma}^{\alpha} = \boldsymbol{\tau} : \sum_{\alpha=1}^{n} \mathbf{P}^{\alpha}\, \dot{\gamma}^{\alpha}. \tag{12.7.2}$$

Therefore,

$$\tau^{\alpha} = \mathbf{P}_0^{\alpha} : \mathbf{T} = \mathbf{P}^{\alpha} : \boldsymbol{\tau}. \tag{12.7.3}$$

This definition of τ^{α} will be referred to as the generalized Schmidt stress,

$$\tau^{\alpha} = \mathbf{s} \cdot \boldsymbol{\tau} \cdot \mathbf{m}. \tag{12.7.4}$$

With so defined τ^{α}, we prove that the plastic part of the strain rate $(\dot{\mathbf{E}})^{\mathrm{p}}$ lies within a pyramid of outward normals to the yield surface at $\mathbf{T}$, each normal being associated with an active slip system (Rice, 1971; Hill and Rice, 1972; Havner, 1982,1992). For example, for f.c.c. crystals the yield surface consists of 24 hyperplanes, forming a polyhedron within which the response is purely elastic. The direction of the normal to the yield plane $\tau^\alpha = \tau_{\rm cr}^\alpha$ at $\mathbf{T}$ is determined from

$$\frac{\partial \tau^\alpha}{\partial \mathbf{T}} = \mathbf{P}_0^\alpha + \frac{\partial \mathbf{P}_0^\alpha}{\partial \mathbf{T}} : \mathbf{T} = \mathbf{P}_0^\alpha + \frac{\partial \mathbf{P}_0^\alpha}{\partial \mathbf{E}} : \mathbf{M}_{(1)} : \mathbf{T}. \tag{12.7.5}$$

From Eq. (12.1.25) it follows that, at fixed slips (fixed $\mathbf{F}^{\rm p}$),

$$\mathbf{T} : \frac{\partial \mathbf{P}_0^\alpha}{\partial \mathbf{E}} = \mathbf{Z}_0^\alpha \cdot \mathbf{T} + \mathbf{T} \cdot \mathbf{Z}_0^{\alpha\,T}. \tag{12.7.6}$$

The substitution into Eq. (12.7.5) gives

$$\frac{\partial \tau^\alpha}{\partial \mathbf{T}} = \mathbf{P}_0^\alpha + \mathbf{M}_{(1)} : \left(\mathbf{Z}_0^\alpha \cdot \mathbf{T} + \mathbf{T} \cdot \mathbf{Z}_0^{\alpha\,T}\right). \tag{12.7.7}$$

Comparison with Eq. (12.5.21) confirms the normality property

$$(\dot{\mathbf{E}})^{\rm p} = \sum_{\alpha=1}^{n} \frac{\partial \tau^\alpha}{\partial \mathbf{T}}\, \dot\gamma^\alpha. \tag{12.7.8}$$

This also shows that the contribution to $(\dot{\mathbf{E}})^{\rm p}$ due to individual slip rate $\dot\gamma^\alpha$ is governed by the gradient $\partial\tau^\alpha/\partial\mathbf{T}$ of the corresponding resolved shear stress τ^α. Equation (12.7.8) can be rewritten as

$$(\dot{\mathbf{E}})^{\rm p} = \frac{\partial}{\partial \mathbf{T}} \sum_{\alpha=1}^{n} (\tau^\alpha\, \dot\gamma^\alpha), \tag{12.7.9}$$

with understanding that the partial differentiation is performed at fixed $\mathbf{F}^{\rm p}$ and $\dot\gamma^\alpha$. Relation (12.7.9) states that $\sum(\tau^\alpha\,\dot\gamma^\alpha)$ acts as the plastic potential for $(\dot{\mathbf{E}})^{\rm p}$ over an elastic domain in the stress $\mathbf{T}$ space (Havner, 1992).

Dually, in strain space we have

$$\frac{\partial \tau^\alpha}{\partial \mathbf{E}} = \mathbf{\Lambda}_{(1)} : \mathbf{P}_0^\alpha + \frac{\partial \mathbf{P}_0^\alpha}{\partial \mathbf{E}} : \mathbf{T}, \tag{12.7.10}$$

i.e.,

$$\frac{\partial \tau^\alpha}{\partial \mathbf{E}} = \mathbf{\Lambda}_{(1)} : \mathbf{P}_0^\alpha + \mathbf{Z}_0^\alpha \cdot \mathbf{T} + \mathbf{T} \cdot \mathbf{Z}_0^{\alpha\,T}. \tag{12.7.11}$$

The right-hand side is equal to $\mathbf{C}_0^\alpha$ of Eq. (12.4.10). Thus, in view of (12.5.2), we establish the normality property

$$(\dot{\mathbf{T}})^{\rm p} = -\sum_{\alpha=1}^{n} \frac{\partial \tau^\alpha}{\partial \mathbf{E}}\, \dot\gamma^\alpha. \tag{12.7.12}$$

The contribution to $(\dot{\mathbf{T}})^{\rm p}$ due to individual slip rate $\dot\gamma^\alpha$ is governed by the gradient $\partial\tau^\alpha/\partial\mathbf{E}$ of the corresponding resolved shear stress τ^α. Equation (12.7.12) can be rewritten as

$$(\dot{\mathbf{T}})^{\rm p} = -\frac{\partial}{\partial \mathbf{E}} \sum_{\alpha=1}^{n} (\tau^\alpha\, \dot\gamma^\alpha), \tag{12.7.13}$$

again with understanding that the partial differentiation is performed at fixed $\mathbf{F}^{\mathrm{p}}$ and $\dot{\gamma}^{\alpha}$. Relation (12.7.13) states that $-\sum(\tau^{\alpha}\,\dot{\gamma}^{\alpha})$ acts as the plastic potential for $(\dot{\mathbf{T}})^{\mathrm{p}}$ over an elastic domain in the strain $\mathbf{E}$ space.

The normality, here proved relative to the conjugate measures $\mathbf{E}$ and $\mathbf{T}$, holds with respect to any other conjugate measures of stress and strain (e.g., Hill and Havner, 1982). Deviations from the normality arise when τ^{α} in Eq. (12.7.1) is defined to be other than the generalized Schmid stress of Eq. (12.7.3). The resulting non-normality enhances a tendency toward localization of deformation, as discussed in a general context in Chapter 10. Indeed, in their study of strain localization in ductile crystals deforming by single slip, Asaro and Rice (1977) showed that the critical hardening rate for the onset of localization may be positive when the non-Schmid effects are present, i.e., when the stress components other than the resolved shear stress affect the slip. In contrast, when the slip is governed by the resolved shear stress only, the critical hardening rate for the onset of localization must be either negative or zero (i.e., ideally-plastic or strain softening state must be reached for the localization). The non-Schmid effects will not be further considered in this chapter. The reviews by Asaro (1983b) and Bassani (1993), and the book by Havner (1992) can be consulted. See also the papers by Qin and Bassani (1992a,b), Dao and Asaro (1996), and Brünig and Obrecht (1998).

Normality Rules for $(\dot{\mathbf{F}})^{\mathrm{p}}$ *and* $(\dot{\mathbf{P}})^{\mathrm{p}}$

If the nominal stress is used to express the resolved shear stress τ^{α}, the rate of work can be written as

$$\mathbf{P}\cdot\cdot\,\dot{\mathbf{F}} = \mathbf{P}\cdot\cdot\left(\dot{\mathbf{F}}^{*}\cdot\mathbf{F}^{\mathrm{p}} + \mathbf{F}^{*}\cdot\dot{\mathbf{F}}^{\mathrm{p}}\right). \tag{12.7.14}$$

The part associated with $\dot{\mathbf{F}}^{\mathrm{p}}$ is the rate of slip work, i.e.,

$$\sum_{\alpha=1}^{n}\tau^{\alpha}\,\dot{\gamma}^{\alpha} = \mathbf{P}\cdot\cdot\left(\mathbf{F}^{*}\cdot\dot{\mathbf{F}}^{\mathrm{p}}\right). \tag{12.7.15}$$

Substituting Eq. (12.6.20) for $\dot{\mathbf{F}}^{\mathrm{p}}$ gives

$$\sum_{\alpha=1}^{n}\tau^{\alpha}\,\dot{\gamma}^{\alpha} = \mathbf{P}\cdot\cdot\,\sum_{\alpha=1}^{n}\mathbf{F}^{*}\cdot(\mathbf{s}_0^{\alpha}\otimes\mathbf{m}_0^{\alpha})\cdot\mathbf{F}^{\mathrm{p}}\,\dot{\gamma}^{\alpha}. \tag{12.7.16}$$

From this we identify the generalized resolved shear stress in terms of the nominal stress,

$$\tau^\alpha = \mathbf{P} \cdot\cdot \left[\mathbf{F} \cdot \mathbf{F}^{\mathrm{p}-1} \cdot (\mathbf{s}_0^\alpha \otimes \mathbf{m}_0^\alpha) \cdot \mathbf{F}^{\mathrm{p}}\right]. \tag{12.7.17}$$

It is easily verified that τ^α given by Eq. (12.7.17) is equal to τ^α of Eq. (12.7.3).

The direction of the normal to the yield plane $\tau^\alpha = \tau_{\mathrm{cr}}^\alpha$ at $\mathbf{P}$ is determined from the gradient $\partial\tau^\alpha/\partial\mathbf{P}$. This is, by Eq. (12.7.17),

$$\frac{\partial\tau^\alpha}{\partial\mathbf{P}} = \mathbf{F}\cdot\mathbf{F}^{\mathrm{p}-1}\cdot(\mathbf{s}_0^\alpha \otimes \mathbf{m}_0^\alpha)\cdot\mathbf{F}^{\mathrm{p}} + \mathbf{M}\cdot\cdot \left[\mathbf{F}^{\mathrm{p}-1} \cdot (\mathbf{s}_0^\alpha \otimes \mathbf{m}_0^\alpha) \cdot \mathbf{F}^{\mathrm{p}} \cdot \mathbf{P}\right], \tag{12.7.18}$$

i.e.,

$$\frac{\partial\tau^\alpha}{\partial\mathbf{P}} = (\mathbf{s}^\alpha \otimes \mathbf{m}^\alpha) \cdot \mathbf{F} + \mathbf{M} \cdot\cdot \mathbf{F}^{-1} \cdot (\mathbf{s}^\alpha \otimes \mathbf{m}^\alpha) \cdot \mathbf{F} \cdot \mathbf{P}. \tag{12.7.19}$$

The right-hand side is equal to $\mathbf{A}^\alpha$ of Eq. (12.6.22), so that

$$\frac{\partial\tau^\alpha}{\partial\mathbf{P}} = \mathbf{A}^\alpha. \tag{12.7.20}$$

Thus, in view of (12.6.21), we establish the normality property for the plastic part of the rate of deformation gradient,

$$(\dot{\mathbf{F}})^{\mathrm{p}} = \sum_{\alpha=1}^{n} \frac{\partial\tau^\alpha}{\partial\mathbf{P}}\, \dot{\gamma}^\alpha. \tag{12.7.21}$$

Equation (12.7.21) can be rewritten as

$$(\dot{\mathbf{F}})^{\mathrm{p}} = \frac{\partial}{\partial\mathbf{P}} \sum_{\alpha=1}^{n} (\tau^\alpha\, \dot{\gamma}^\alpha), \tag{12.7.22}$$

with the partial differentiation performed at fixed $\mathbf{F}^{\mathrm{p}}$ and $\dot{\gamma}^\alpha$. This states that $\sum(\tau^\alpha\, \dot{\gamma}^\alpha)$ acts as the plastic potential for $(\dot{\mathbf{F}})^{\mathrm{p}}$ over an elastic domain in $\mathbf{P}$ space.

Dually, by taking the gradient of (12.7.17) with respect to $\mathbf{F}$, we obtain

$$\frac{\partial\tau^\alpha}{\partial\mathbf{F}} = \mathbf{\Lambda}\cdot\cdot \left[\mathbf{F} \cdot \mathbf{F}^{\mathrm{p}-1} \cdot (\mathbf{s}_0^\alpha \otimes \mathbf{m}_0^\alpha) \cdot \mathbf{F}^{\mathrm{p}}\right] + \mathbf{F}^{\mathrm{p}-1}\cdot(\mathbf{s}_0^\alpha \otimes \mathbf{m}_0^\alpha)\cdot\mathbf{F}^{\mathrm{p}}\cdot\mathbf{P}. \tag{12.7.23}$$

The right-hand side is equal to $\mathbf{B}^\alpha$ of Eq. (12.6.24). Thus, in view of (12.6.23), we establish the normality property for the plastic part of the rate of nominal stress,

$$(\dot{\mathbf{P}})^{\mathrm{p}} = -\sum_{\alpha=1}^{n} \frac{\partial\tau^\alpha}{\partial\mathbf{F}}\, \dot{\gamma}^\alpha. \tag{12.7.24}$$

Alternatively,

$$(\dot{\mathbf{P}})^{\mathrm{p}} = -\frac{\partial}{\partial \mathbf{F}} \sum_{\alpha=1}^{n} (\tau^{\alpha}\, \dot{\gamma}^{\alpha})\,, \tag{12.7.25}$$

with understanding that the partial differentiation is performed at fixed $\mathbf{F}^{\mathrm{p}}$ and $\dot{\gamma}^{\alpha}$. Relation (12.7.25) states that $-\sum(\tau^{\alpha}\, \dot{\gamma}^{\alpha})$ acts as the plastic potential for $(\dot{\mathbf{P}})^{\mathrm{p}}$ over an elastic domain in $\mathbf{F}$ space.

12.8. Rate of Plastic Work

In the previous section we defined the rate of slip work by

$$\dot{w}^{\mathrm{slip}} = \sum_{\alpha=1}^{n} \tau^{\alpha}\, \dot{\gamma}^{\alpha}\,. \tag{12.8.1}$$

This invariant quantity is not equal to $\mathbf{T} \cdot\cdot (\dot{\mathbf{E}})^{\mathrm{p}}$, nor $\mathbf{P} \cdot\cdot (\dot{\mathbf{F}})^{\mathrm{p}}$. It is of interest to elaborate on the relationships between $\dot{w}^{\mathrm{slip}}$ and these latter work quantities. First, from Eqs. (12.4.10) and (12.5.9) we express the rate of plastic work, associated with the plastic part of strain rate $(\dot{\mathbf{E}})^{\mathrm{p}}$, as

$$\begin{aligned} \mathbf{T} : (\dot{\mathbf{E}})^{\mathrm{p}} &= \mathbf{T} : \sum_{\alpha=1}^{n} \mathbf{G}_0^{\alpha}\, \dot{\gamma}^{\alpha} \\ &= \mathbf{T} : \mathbf{M}_{(1)} : \sum_{\alpha=1}^{n} \left(\boldsymbol{\Lambda}_{(1)} : \mathbf{P}_0^{\alpha} + \mathbf{Z}_0^{\alpha} \cdot \mathbf{T} + \mathbf{T} \cdot \mathbf{Z}_0^{\alpha\, T} \right) \dot{\gamma}^{\alpha}\,. \end{aligned} \tag{12.8.2}$$

Comparing with Eq. (12.7.2), i.e.,

$$\sum_{\alpha=1}^{n} \tau^{\alpha}\, \dot{\gamma}^{\alpha} = \mathbf{T} : \sum_{\alpha=1}^{n} \mathbf{P}_0^{\alpha}\, \dot{\gamma}^{\alpha}\,, \tag{12.8.3}$$

we establish the relationship

$$\mathbf{T} : (\dot{\mathbf{E}})^{\mathrm{p}} = \sum_{\alpha=1}^{n} \tau^{\alpha}\, \dot{\gamma}^{\alpha} + \mathbf{T} : \mathbf{M}_{(1)} : \sum_{\alpha=1}^{n} \left(\mathbf{Z}_0^{\alpha} \cdot \mathbf{T} + \mathbf{T} \cdot \mathbf{Z}_0^{\alpha\, T} \right) \dot{\gamma}^{\alpha}\,. \tag{12.8.4}$$

Similarly, from Eqs. (12.6.21) and (12.6.22), we can express the rate of plastic work, associated with the plastic part of rate of deformation tensor $(\dot{\mathbf{F}})^{\mathrm{p}}$, as

$$\begin{aligned} \mathbf{P} \cdot\cdot (\dot{\mathbf{F}})^{\mathrm{p}} &= \mathbf{P} \cdot\cdot \sum_{\alpha=1}^{n} \mathbf{A}^{\alpha}\, \dot{\gamma}^{\alpha} \\ &= \mathbf{P} \cdot\cdot \sum_{\alpha=1}^{n} \left[\mathbf{M} \cdot\cdot \mathbf{F}^{-1} \cdot (\mathbf{P}^{\alpha} + \mathbf{Q}^{\alpha}) \cdot \mathbf{F} \cdot \mathbf{P} + (\mathbf{P}^{\alpha} + \mathbf{Q}^{\alpha}) \cdot \mathbf{F} \right] \dot{\gamma}^{\alpha}\,. \end{aligned} \tag{12.8.5}$$

Since, from Eq. (12.7.16),

$$\sum_{\alpha=1}^{n} \tau^\alpha \dot{\gamma}^\alpha = \mathbf{P} \cdot\cdot \sum_{\alpha=1}^{n} (\mathbf{P}^\alpha + \mathbf{Q}^\alpha) \cdot \mathbf{F} \, \dot{\gamma}^\alpha \,, \tag{12.8.6}$$

we obtain

$$\mathbf{P} \cdot\cdot (\dot{\mathbf{F}})^{\mathrm{p}} = \sum_{\alpha=1}^{n} \tau^\alpha \dot{\gamma}^\alpha + \mathbf{P} \cdot\cdot \mathbf{M} \cdot\cdot \sum_{\alpha=1}^{n} \mathbf{F}^{-1} \cdot (\mathbf{P}^\alpha + \mathbf{Q}^\alpha) \cdot \mathbf{F} \cdot \mathbf{P} \, \dot{\gamma}^\alpha \,. \tag{12.8.7}$$

The plastic work quantities $\mathbf{P} \cdot\cdot (\dot{\mathbf{F}})^{\mathrm{p}}$ and $\mathbf{T} : (\dot{\mathbf{F}})^{\mathrm{p}}$ are not equal to each other. Recalling that $\mathbf{P} = \mathbf{T} : \boldsymbol{\mathcal{K}}$, and by using Eq. (12.6.33), we have the connection

$$\mathbf{P} \cdot\cdot (\dot{\mathbf{F}})^{\mathrm{p}} = \mathbf{T} : \left[\boldsymbol{\mathcal{K}} \cdot\cdot \mathbf{M} \cdot\cdot \boldsymbol{\mathcal{K}}^T : \boldsymbol{\Lambda}_{(1)} \right] : (\dot{\mathbf{E}})^{\mathrm{p}}. \tag{12.8.8}$$

The inequality

$$\mathbf{P} \cdot\cdot (\dot{\mathbf{F}})^{\mathrm{p}} \neq \mathbf{T} : (\dot{\mathbf{E}})^{\mathrm{p}} \tag{12.8.9}$$

is physically clear, because $\mathbf{P}$ and $\mathbf{T}$ do not cycle simultaneously in the deformation cycle involving plastic slip, since cycling $\mathbf{P}$ does not cycle $\mathbf{T}$, and *vice versa*.

Expressed in terms of the increments, we can write

$$\mathbf{P} \cdot\cdot \, (\mathrm{d}\mathbf{F} - \mathbf{M} \cdot\cdot \mathrm{d}\mathbf{P}) \neq \mathbf{T} : (\mathrm{d}\mathbf{E} - \mathbf{M}_{(1)} : \mathrm{d}\mathbf{T}). \tag{12.8.10}$$

We also recall that the increment of plastic work $\mathbf{T} : \mathrm{d}^{\mathrm{p}}\mathbf{E}$ is not invariant under the change of strain and conjugate stress measure (again because different stress measures do not cycle simultaneously).

Second-Order Work Quantities

The analysis of the relationship between the first- and second-order plastic work quantities, defined by $\mathbf{P} \cdot\cdot \mathrm{d}^{\mathrm{p}}\mathbf{F}$ and $\mathrm{d}\mathbf{P} \cdot\cdot \mathrm{d}^{\mathrm{p}}\mathbf{F}$, or by $\mathbf{T} : \mathrm{d}^{\mathrm{p}}\mathbf{E}$ and $\mathrm{d}\mathbf{T} : \mathrm{d}^{\mathrm{p}}\mathbf{E}$, can be pursued further. From the basic work identity

$$\mathbf{P} \cdot\cdot \mathrm{d}\mathbf{F} = \mathbf{T} : \mathrm{d}\mathbf{E}, \tag{12.8.11}$$

and from the partition of the increments of deformation gradient and strain tensor into their elastic and plastic parts, we have

$$\mathbf{P} \cdot\cdot \mathrm{d}^{\mathrm{p}}\mathbf{F} + \mathbf{P} \cdot\cdot \mathbf{M} \cdot\cdot \mathrm{d}\mathbf{P} = \mathbf{T} \cdot\cdot \mathrm{d}^{\mathrm{p}}\mathbf{E} + \mathbf{T} \cdot\cdot \mathbf{M}_{(1)} \cdot\cdot \mathrm{d}\mathbf{T}, \tag{12.8.12}$$

i.e.,

$$\mathbf{P} \cdot\cdot \mathrm{d}^{\mathrm{p}}\mathbf{F} = \mathbf{T} : \mathrm{d}^{\mathrm{p}}\mathbf{E} + \mathbf{T} : \mathbf{M}_{(1)} : \mathrm{d}\mathbf{T} - \mathbf{P} \cdot\cdot \mathbf{M} \cdot\cdot \mathrm{d}\mathbf{P}. \tag{12.8.13}$$

By eliminating $\mathbf{P}$ in terms of $\mathbf{T}$, this can be rewritten as

$$\mathbf{P}\cdot\cdot\,\mathrm{d}^{\mathrm{p}}\mathbf{F} = \mathbf{T} : \mathrm{d}^{\mathrm{p}}\mathbf{E} + \mathbf{T} : \left(\mathbf{M}_{(1)} - \boldsymbol{\mathcal{K}}\cdot\cdot\,\mathbf{M}\cdot\cdot\,\boldsymbol{\mathcal{K}}^{T}\right) : \mathrm{d}\mathbf{T}. \tag{12.8.14}$$

This is an explicit relationship between the first-order quantities $\mathbf{P}\cdot\cdot\,\mathrm{d}^{\mathrm{p}}\mathbf{F}$ and $\mathbf{T} : \mathrm{d}^{\mathrm{p}}\mathbf{E}$.

Regarding the second-order work contribution, we proceed from

$$\mathrm{d}\mathbf{P}\cdot\cdot\,\mathrm{d}^{\mathrm{p}}\mathbf{F} = \mathrm{d}\mathbf{P}\cdot\cdot\,\mathrm{d}\mathbf{F} - \mathrm{d}\mathbf{P}\cdot\cdot\,\mathbf{M}\cdot\cdot\,\mathrm{d}\mathbf{P}. \tag{12.8.15}$$

By substituting

$$\mathrm{d}\mathbf{P} = \mathrm{d}\mathbf{T} : \boldsymbol{\mathcal{K}} + \boldsymbol{\mathcal{T}}\cdot\cdot\,\mathrm{d}\mathbf{F}, \quad \mathrm{d}\mathbf{E} = \boldsymbol{\mathcal{K}}\cdot\cdot\,\mathrm{d}\mathbf{F}, \tag{12.8.16}$$

and by using the decomposition of $\mathrm{d}\mathbf{E}$ into its elastic and plastic parts, there follows

$$\mathrm{d}\mathbf{P}\cdot\cdot\,\mathrm{d}^{\mathrm{p}}\mathbf{F} = \mathrm{d}\mathbf{T} : \mathrm{d}^{\mathrm{p}}\mathbf{E} + \mathrm{d}\mathbf{T} : \mathbf{M}_{(1)} : \mathrm{d}\mathbf{T} - \mathrm{d}\mathbf{P}\cdot\cdot\,\mathbf{M}\cdot\cdot\,\mathrm{d}\mathbf{P} + \mathrm{d}\mathbf{F}\cdot\cdot\,\boldsymbol{\mathcal{T}}\cdot\cdot\,\mathrm{d}\mathbf{F}. \tag{12.8.17}$$

This relates the second-order work quantities $\mathrm{d}\mathbf{P}\cdot\cdot\,\mathrm{d}^{\mathrm{p}}\mathbf{F}$ and $\mathrm{d}\mathbf{T} : \mathrm{d}^{\mathrm{p}}\mathbf{E}$.

For completeness of the analysis, we record two more formulas. The first one is

$$\mathbf{F}\cdot\cdot\,\mathrm{d}^{\mathrm{p}}\mathbf{P} = \mathbf{F}\cdot\cdot\,\left(\boldsymbol{\mathcal{K}}^{T} : \mathrm{d}^{\mathrm{p}}\mathbf{T}\right) = \mathbf{C} : \mathrm{d}^{\mathrm{p}}\mathbf{T}, \tag{12.8.18}$$

where

$$\mathbf{C} = \mathbf{F}^{T}\cdot\mathbf{F} = \boldsymbol{\mathcal{K}}\cdot\cdot\,\mathbf{F} = \mathbf{F}\cdot\boldsymbol{\mathcal{K}}^{T}. \tag{12.8.19}$$

The second formula is

$$\mathrm{d}\mathbf{F}\cdot\cdot\,\mathrm{d}^{\mathrm{p}}\mathbf{P} = \mathbf{F}\cdot\cdot\,\left(\boldsymbol{\mathcal{K}}^{T} : \mathrm{d}^{\mathrm{p}}\mathbf{T}\right) = \mathrm{d}\mathbf{E} : \mathrm{d}^{\mathrm{p}}\mathbf{T}, \tag{12.8.20}$$

where

$$\mathrm{d}\mathbf{E} = \boldsymbol{\mathcal{K}}\cdot\cdot\,\mathrm{d}\mathbf{F} = \mathrm{d}\mathbf{F}\cdot\boldsymbol{\mathcal{K}}^{T}. \tag{12.8.21}$$

These formulas demonstrate the invariance of $\mathbf{C} : \mathrm{d}^{\mathrm{p}}\mathbf{T}$ and $\mathrm{d}\mathbf{E} : \mathrm{d}^{\mathrm{p}}\mathbf{T}$ under the change of the strain measure $\mathbf{E}$ and its conjugate stress $\mathbf{T}$ (because $\mathbf{F}\cdot\cdot\,\mathrm{d}^{\mathrm{p}}\mathbf{P}$ and $\mathrm{d}\mathbf{F} : \mathrm{d}^{\mathrm{p}}\mathbf{P}$ are independent of these measures).

The second-order quantity in Eq. (12.8.20) is proportional to the net expenditure of work in a cycle (application and removal) of $\mathrm{d}\mathbf{F}$, which is by the trapezoidal rule of quadrature

$$-\frac{1}{2}\,\mathrm{d}\mathbf{F}\cdot\cdot\,\mathrm{d}^{\mathrm{p}}\mathbf{P} = -\frac{1}{2}\,\mathrm{d}\mathbf{E} : \mathrm{d}^{\mathrm{p}}\mathbf{T}. \tag{12.8.22}$$

12.9. Hardening Rules and Slip Rates

The rate of change of the critical value of the resolved shear stress on a given slip system is defined by the hardening law

$$\dot{\tau}_{\mathrm{cr}}^{\alpha} = \sum_{\beta=1}^{n_0} h_{\alpha\beta}\,\dot{\gamma}^{\beta}, \quad \alpha = 1, 2, \ldots, N, \tag{12.9.1}$$

where N is the total number of all available slip systems, and n_0 is the number of critical (potentially active) slip systems, for which

$$\tau^{\alpha} = \tau_{\mathrm{cr}}^{\alpha}. \tag{12.9.2}$$

The coefficients $h_{\alpha\beta}$ are the slip-plane hardening rates (moduli). The moduli corresponding to $\alpha = \beta$ represent the self-hardening on a given slip system, while $\alpha \neq \beta$ moduli represent the latent hardening. When $\alpha > n_0$, $\beta \leq n_0$, the moduli represent latent hardening of the noncritical systems. The hardening moduli $h_{\alpha\beta}$ can be formally defined for $n_0 < \beta \leq N$, but their values are irrelevant since the corresponding $\dot{\gamma}^{\beta}$ are always zero.

The consistency condition for the slip on the critical system α is

$$\dot{\tau}^{\alpha} = \sum_{\beta=1}^{n} h_{\alpha\beta}\,\dot{\gamma}^{\beta}, \quad \dot{\gamma}^{\alpha} > 0. \tag{12.9.3}$$

The number of active slip systems is n, and the corresponding slips are labeled by $\dot{\gamma}^1, \dot{\gamma}^2, \ldots, \dot{\gamma}^n$. If the critical system becomes inactive,

$$\dot{\tau}^{\alpha} \leq \sum_{\beta=1}^{n} h_{\alpha\beta}\,\dot{\gamma}^{\beta}, \quad \dot{\gamma}^{\alpha} = 0. \tag{12.9.4}$$

Equality sign applies only if the system remains critical ($\dot{\tau}^{\alpha} = \dot{\tau}_{\mathrm{cr}}^{\alpha}$). For a noncritical system,

$$\tau^{\alpha} < \tau_{\mathrm{cr}}^{\alpha}, \quad \dot{\gamma}^{\alpha} = 0. \tag{12.9.5}$$

The rate of the generalized Schmid stress is obtained by differentiation from Eq. (12.7.3), i.e., either from

$$\dot{\tau}^{\alpha} = \dot{\mathbf{P}}_0^{\alpha} : \mathbf{T} + \mathbf{P}_0^{\alpha} : \dot{\mathbf{T}}, \tag{12.9.6}$$

or

$$\dot{\tau}^{\alpha} = \dot{\mathbf{P}}^{\alpha} : \boldsymbol{\tau} + \mathbf{P}^{\alpha} : \dot{\boldsymbol{\tau}}. \tag{12.9.7}$$

If Eq. (12.9.6) is used, from Eq. (12.1.30) we find

$$\dot{\mathbf{P}}_0^\alpha : \mathbf{T} = \left(\mathbf{Z}_0^\alpha \cdot \mathbf{T} + \mathbf{T} \cdot \mathbf{Z}_0^{\alpha\,T}\right) : \left(\mathbf{F}^{\mathrm{p}T} \cdot \dot{\mathbf{E}}^* \cdot \mathbf{F}^{\mathrm{p}}\right) + \sum_{\beta=1}^{n} \mathbf{P}_0^\alpha : \left(\mathbf{Z}_0^\beta \cdot \mathbf{T} + \mathbf{T} \cdot \mathbf{Z}_0^{\beta\,T}\right) \dot{\gamma}^\beta. \tag{12.9.8}$$

Since, from Eq. (12.2.17),

$$\mathbf{P}_0^\alpha : \dot{\mathbf{T}} = \mathbf{P}_0^\alpha : \left(\mathbf{F}^{\mathrm{p}-1} \cdot \dot{\mathbf{T}}^* \cdot \mathbf{F}^{\mathrm{p}-T}\right) - \sum_{\beta=1}^{n} \mathbf{P}_0^\alpha : \left(\mathbf{Z}_0^\beta \cdot \mathbf{T} + \mathbf{T} \cdot \mathbf{Z}_0^{\beta\,T}\right) \dot{\gamma}^\beta, \tag{12.9.9}$$

the substitution into Eq. (12.9.6) gives

$$\dot{\tau}^\alpha = \mathbf{P}_0^\alpha : \left(\mathbf{F}^{\mathrm{p}-1} \cdot \dot{\mathbf{T}}^* \cdot \mathbf{F}^{\mathrm{p}-T}\right) + \left(\mathbf{Z}_0^\alpha \cdot \mathbf{T} + \mathbf{T} \cdot \mathbf{Z}_0^{\alpha\,T}\right) : \left(\mathbf{F}^{\mathrm{p}T} \cdot \dot{\mathbf{E}}^* \cdot \mathbf{F}^{\mathrm{p}}\right). \tag{12.9.10}$$

Recalling that

$$\dot{\mathbf{T}}^* = \boldsymbol{\Lambda}_{(1)}^* : \dot{\mathbf{E}}^* = \left(\mathbf{F}^{\mathrm{p}}\, \mathbf{F}^{\mathrm{p}}\, \boldsymbol{\Lambda}_{(1)}\, \mathbf{F}^{\mathrm{p}T}\, \mathbf{F}^{\mathrm{p}T}\right) : \dot{\mathbf{E}}^*, \tag{12.9.11}$$

there follows

$$\dot{\tau}^\alpha = \left(\boldsymbol{\Lambda}_{(1)} : \mathbf{P}_0^\alpha + \mathbf{Z}_0^\alpha \cdot \mathbf{T} + \mathbf{T} \cdot \mathbf{Z}_0^{\alpha\,T}\right) : \left(\mathbf{F}^{\mathrm{p}T} \cdot \dot{\mathbf{E}}^* \cdot \mathbf{F}^{\mathrm{p}}\right). \tag{12.9.12}$$

Thus, in view of Eq. (12.4.10), we have

$$\dot{\tau}^\alpha = \mathbf{C}_0^\alpha : \left(\mathbf{F}^{\mathrm{p}T} \cdot \dot{\mathbf{E}}^* \cdot \mathbf{F}^{\mathrm{p}}\right), \tag{12.9.13}$$

which is a desired expression for the rate of the generalized Schmid stress.

The expression for $\dot{\tau}^\alpha$ can also be obtained by starting from Eq. (12.9.7). First, Eq. (12.1.16) gives

$$\dot{\mathbf{P}}^\alpha : \boldsymbol{\tau} = \left(\mathbf{D}^* \cdot \mathbf{Q}^\alpha - \mathbf{Q}^\alpha \cdot \mathbf{D}^*\right) : \boldsymbol{\tau} + \left(\mathbf{W}^* \cdot \mathbf{P}^\alpha - \mathbf{P}^\alpha \cdot \mathbf{W}^*\right) : \boldsymbol{\tau}. \tag{12.9.14}$$

By using Eq. (12.2.11), we obtain

$$\mathbf{P}^\alpha : \dot{\boldsymbol{\tau}} = \mathbf{P}^\alpha : \overset{\bullet}{\boldsymbol{\tau}} - \left(\mathbf{W}^* \cdot \mathbf{P}^\alpha - \mathbf{P}^\alpha \cdot \mathbf{W}^*\right) : \boldsymbol{\tau}. \tag{12.9.15}$$

The substitution into Eq. (12.9.7) then gives

$$\dot{\tau}^\alpha = \mathbf{P}^\alpha : \overset{\bullet}{\boldsymbol{\tau}} + \left(\mathbf{D}^* \cdot \mathbf{Q}^\alpha - \mathbf{Q}^\alpha \cdot \mathbf{D}^*\right) : \boldsymbol{\tau}. \tag{12.9.16}$$

Since $\overset{\bullet}{\boldsymbol{\tau}} = \boldsymbol{\mathcal{L}}_{(0)} : \mathbf{D}^*$ by Eq. (12.3.8), there follows

$$\dot{\tau}^\alpha = \left(\boldsymbol{\mathcal{L}}_{(0)} : \mathbf{P}^\alpha + \mathbf{Q}^\alpha \cdot \boldsymbol{\tau} - \boldsymbol{\tau} \cdot \mathbf{Q}^\alpha\right) : \mathbf{D}^*. \tag{12.9.17}$$

Consequently, in view of Eq. (12.4.5), we have

$$\dot{\tau}^\alpha = \mathbf{C}^\alpha : \mathbf{D}^*. \tag{12.9.18}$$

This parallels the previously derived expression (12.9.13). Recalling the relationship between $\mathbf{C}^\alpha$ and $\mathbf{C}_0^\alpha$, and between $\mathbf{D}^*$ and $\dot{\mathbf{E}}^*$, it is readily verified that the two expressions are equivalent.

When Eq. (12.1.22) is substituted into Eq. (12.9.13) to eliminate the term $\mathbf{F}^{\mathrm{p}T} \cdot \dot{\mathbf{E}}^* \cdot \mathbf{F}^{\mathrm{p}}$, or when Eq. (12.1.12) is substituted into Eq. (12.9.18) to eliminate $\mathbf{D}^*$, we obtain

$$\dot{\tau}^\alpha = \mathbf{C}_0^\alpha : \dot{\mathbf{E}} - \sum_{\beta=1}^{n} \mathbf{C}_0^\alpha : \mathbf{P}_0^\beta \, \dot{\gamma}^\beta, \tag{12.9.19}$$

$$\dot{\tau}^\alpha = \mathbf{C}^\alpha : \mathbf{D} - \sum_{\beta=1}^{n} \mathbf{C}^\alpha : \mathbf{P}^\beta \, \dot{\gamma}^\beta. \tag{12.9.20}$$

Combining with Eq. (12.9.3) yields

$$\mathbf{C}_0^\alpha : \dot{\mathbf{E}} = \sum_{\beta=1}^{n} \left(h_{\alpha\beta} + \mathbf{C}_0^\alpha : \mathbf{P}_0^\beta \right) \dot{\gamma}^\beta, \tag{12.9.21}$$

$$\mathbf{C}^\alpha : \mathbf{D} = \sum_{\beta=1}^{n} \left(h_{\alpha\beta} + \mathbf{C}^\alpha : \mathbf{P}^\beta \right) \dot{\gamma}^\beta. \tag{12.9.22}$$

Since

$$\mathbf{C}^\alpha = \mathbf{F} \cdot \mathbf{C}_0^\alpha \cdot \mathbf{F}^T, \quad \mathbf{D} = \mathbf{F}^{-T} \cdot \dot{\mathbf{E}} \cdot \mathbf{F}^{-1}, \tag{12.9.23}$$

there is a connection

$$\mathbf{C}_0^\alpha : \dot{\mathbf{E}} = \mathbf{C}^\alpha : \mathbf{D}, \tag{12.9.24}$$

and from Eqs. (12.9.21) and (12.9.22) we deduce the identity

$$\mathbf{C}_0^\alpha : \mathbf{P}_0^\beta = \mathbf{C}^\alpha : \mathbf{P}^\beta. \tag{12.9.25}$$

This also follows directly from

$$\mathbf{C}_0^\alpha = \mathbf{F}^{-1} \cdot \mathbf{C}^\alpha \cdot \mathbf{F}^{-T}, \quad \mathbf{P}_0^\beta = \mathbf{F}^T \cdot \mathbf{P}^\beta \cdot \mathbf{F}. \tag{12.9.26}$$

Therefore, by introducing the matrix with components

$$g_{\alpha\beta} = h_{\alpha\beta} + \mathbf{C}_0^\alpha : \mathbf{P}_0^\beta = h_{\alpha\beta} + \mathbf{C}^\alpha : \mathbf{P}^\beta, \tag{12.9.27}$$

equations (12.9.21) and (12.9.22) reduce to

$$\mathbf{C}_0^\alpha : \dot{\mathbf{E}} = \mathbf{C}^\alpha : \mathbf{D} = \sum_{\beta=1}^{n} g_{\alpha\beta} \, \dot{\gamma}^\beta, \quad \dot{\gamma}^\alpha > 0. \tag{12.9.28}$$

If the α system is inactive ($\dot{\tau}^\alpha \leq \dot{\tau}_{\mathrm{cr}}^\alpha$), we have

$$\mathbf{C}_0^\alpha : \dot{\mathbf{E}} = \mathbf{C}^\alpha : \mathbf{D} \leq \sum_{\beta=1}^{n} g_{\alpha\beta} \, \dot{\gamma}^\beta, \quad \dot{\gamma}^\alpha = 0. \tag{12.9.29}$$

Suppose that the matrix with components $g_{\alpha\beta}$ is nonsingular, so that the inverse matrix whose components are designated by $g_{\alpha\beta}^{-1}$ exists. Equation (12.9.28) can then be solved for the slip rates to give

$$\dot{\gamma}^\alpha = \sum_{\beta=1}^{n} g_{\alpha\beta}^{-1} \mathbf{C}_0^\beta : \dot{\mathbf{E}} = \sum_{\beta=1}^{n} g_{\alpha\beta}^{-1} \mathbf{C}^\beta : \mathbf{D}. \tag{12.9.30}$$

After substitution into Eq. (12.5.2), the plastic parts of the corresponding stress rates become

$$\overset{\circ}{\boldsymbol{\tau}}{}^{\mathrm{p}} = \overset{\triangle}{\boldsymbol{\tau}}{}^{\mathrm{p}} = -\sum_{\alpha=1}^{n}\sum_{\beta=1}^{n} g_{\alpha\beta}^{-1} \left(\mathbf{C}^\alpha \otimes \mathbf{C}^\beta\right) : \mathbf{D}, \tag{12.9.31}$$

$$(\dot{\mathbf{T}})^{\mathrm{p}} = -\sum_{\alpha=1}^{n}\sum_{\beta=1}^{n} g_{\alpha\beta}^{-1} \left(\mathbf{C}_0^\alpha \otimes \mathbf{C}_0^\beta\right) : \dot{\mathbf{E}}. \tag{12.9.32}$$

Combining with the elastic parts, defined by Eq. (12.5.1), finally yields

$$\overset{\circ}{\boldsymbol{\tau}} = \left(\boldsymbol{\mathcal{L}}_{(0)} - \sum_{\alpha=1}^{n}\sum_{\beta=1}^{n} g_{\alpha\beta}^{-1} \mathbf{C}^\alpha \otimes \mathbf{C}^\beta\right) : \mathbf{D}, \tag{12.9.33}$$

$$\overset{\triangle}{\boldsymbol{\tau}} = \left(\boldsymbol{\mathcal{L}}_{(1)} - \sum_{\alpha=1}^{n}\sum_{\beta=1}^{n} g_{\alpha\beta}^{-1} \mathbf{C}^\alpha \otimes \mathbf{C}^\beta\right) : \mathbf{D}, \tag{12.9.34}$$

$$\dot{\mathbf{T}} = \left(\boldsymbol{\Lambda}_{(1)} - \sum_{\alpha=1}^{n}\sum_{\beta=1}^{n} g_{\alpha\beta}^{-1} \mathbf{C}_0^\alpha \otimes \mathbf{C}_0^\beta\right) : \dot{\mathbf{E}}. \tag{12.9.35}$$

These are alternative representations of the constitutive structure for elastoplastic deformation of single crystals. The fourth-order tensors within the brackets are the crystalline elastoplastic moduli tensors.

12.10. Uniqueness of Slip Rates for Prescribed Strain Rate

Hill and Rice (1972) have shown that, for a prescribed rate of deformation, sufficient condition for the unique set of slip rates $\dot{\gamma}^\alpha$ is that the matrix with components $g_{\alpha\beta}$, over all n_0 critical systems, is positive definite. In proof, denote by

$$\Delta\dot{\gamma}^\alpha = \dot{\gamma}^\alpha - \dot{\bar{\gamma}}^\alpha \quad (\alpha = 1, 2, \ldots, n_0) \tag{12.10.1}$$

the difference between the slip rates in two different slip modes, both at the same stress and hardening state, one being associated with the rate of deformation $\mathbf{D}$ and the other with $\bar{\mathbf{D}}$. From Eq. (12.5.2), then,

$$-\Delta(\overset{\circ}{\boldsymbol{\tau}}^{\mathrm{p}}) = \sum_{\alpha=1}^{n_0} \mathbf{C}^\alpha \Delta\dot{\gamma}^\alpha, \tag{12.10.2}$$

and

$$-\Delta(\overset{\circ}{\boldsymbol{\tau}}^{\mathrm{p}}) : \Delta\mathbf{D} = \sum_{\alpha=1}^{n_0} (\mathbf{C}^\alpha : \Delta\mathbf{D})\, \Delta\dot{\gamma}^\alpha, \tag{12.10.3}$$

where

$$\Delta\mathbf{D} = \mathbf{D} - \bar{\mathbf{D}}. \tag{12.10.4}$$

If the slip system α is active in both modes,

$$\mathbf{C}^\alpha : \mathbf{D} - \sum_{\beta=1}^{n_0} g_{\alpha\beta}\, \dot{\gamma}^\beta = 0, \quad \dot{\gamma}^\alpha > 0, \tag{12.10.5}$$

$$\mathbf{C}^\alpha : \bar{\mathbf{D}} - \sum_{\beta=1}^{n_0} g_{\alpha\beta}\, \dot{\bar{\gamma}}^\beta = 0, \quad \dot{\bar{\gamma}}^\alpha > 0. \tag{12.10.6}$$

Consequently, in this case

$$\mathbf{C}^\alpha : \Delta\mathbf{D} - \sum_{\beta=1}^{n_0} g_{\alpha\beta}\, \Delta\dot{\gamma}^\beta = 0, \tag{12.10.7}$$

and, upon multiplication with $\Delta\dot{\gamma}^\alpha$,

$$(\mathbf{C}^\alpha : \Delta\mathbf{D})\, \Delta\dot{\gamma}^\alpha = \sum_{\beta=1}^{n_0} g_{\alpha\beta}\, \Delta\dot{\gamma}^\alpha\, \Delta\dot{\gamma}^\beta. \tag{12.10.8}$$

If the slip system α is active in the first mode, but inactive in the second mode, i.e.,

$$\mathbf{C}^\alpha : \mathbf{D} - \sum_{\beta=1}^{n_0} g_{\alpha\beta}\, \dot{\gamma}^\beta = 0, \quad \dot{\gamma}^\alpha > 0, \tag{12.10.9}$$

$$\mathbf{C}^\alpha : \bar{\mathbf{D}} - \sum_{\beta=1}^{n_0} g_{\alpha\beta}\, \dot{\bar{\gamma}}^\beta \leq 0, \quad \dot{\bar{\gamma}}^\alpha = 0, \tag{12.10.10}$$

then

$$\mathbf{C}^\alpha : \Delta\mathbf{D} - \sum_{\beta=1}^{n_0} g_{\alpha\beta}\, \Delta\dot{\gamma}^\beta \geq 0, \quad \Delta\dot{\gamma}^\alpha > 0. \tag{12.10.11}$$

Thus, upon multiplication with $\Delta\dot{\gamma}^\alpha$,

$$(\mathbf{C}^\alpha : \Delta\mathbf{D})\, \Delta\dot{\gamma}^\alpha \geq \sum_{\beta=1}^{n_0} g_{\alpha\beta}\, \Delta\dot{\gamma}^\alpha\, \Delta\dot{\gamma}^\beta. \tag{12.10.12}$$

Inequality (12.10.12) also holds in the case when α system is active in the second and inactive in the first mode, since then, in place of (12.10.11),

$$\mathbf{C}^\alpha : \Delta\mathbf{D} - \sum_{\beta=1}^{n_0} g_{\alpha\beta}\, \Delta\dot{\gamma}^\beta \leq 0, \quad \Delta\dot{\gamma}^\alpha < 0. \tag{12.10.13}$$

Finally, if the slip system α is inactive in both modes,

$$(\mathbf{C}^\alpha : \Delta\mathbf{D})\, \Delta\dot{\gamma}^\alpha = \sum_{\beta=1}^{n_0} g_{\alpha\beta}\, \Delta\dot{\gamma}^\alpha\, \Delta\dot{\gamma}^\beta, \tag{12.10.14}$$

because $\Delta\dot{\gamma}^\alpha = 0$. Therefore, (12.10.12) covers all cases, since either $=$ or $>$ sign applies. Summing over all critical systems gives

$$\sum_{\alpha=1}^{n_0} (\mathbf{C}^\alpha : \Delta\mathbf{D})\, \Delta\dot{\gamma}^\alpha \geq \sum_{\alpha=1}^{n_0} \sum_{\beta=1}^{n_0} g_{\alpha\beta}\, \Delta\dot{\gamma}^\alpha\, \Delta\dot{\gamma}^\beta. \tag{12.10.15}$$

From (12.10.15) we deduce that the positive definiteness of the matrix $g_{\alpha\beta}$ is a sufficient condition for the unique slip rates $\dot{\gamma}^\alpha$ under prescribed $\mathbf{D}$. Indeed, for a prescribed rate of deformation, the difference $\Delta\mathbf{D} = \mathbf{0}$, and if $g_{\alpha\beta}$ is positive definite, (12.10.15) can be satisfied only when $\Delta\dot{\gamma}^\alpha = 0$, for all α.

The positive definiteness of the matrix $g_{\alpha\beta}$ depends sensitively on the hardening moduli, stress state and the number and orientation of critical slip systems. The uniqueness is generally not guaranteed, particularly with higher rates of latent hardening (Hill, 1966; Hill and Rice, 1972; Havner, 1982; Asaro, 1983b; Franciosi and Zaoui, 1991).

12.11. Further Analysis of Constitutive Equations

Another route toward elastoplastic constitutive equations of single crystals is to proceed from

$$\dot{\tau}^\alpha = \mathbf{C}^\alpha : \mathbf{D}^* = \mathbf{C}^\alpha : \boldsymbol{\mathcal{M}}_{(0)} : \overset{\bullet}{\boldsymbol{\tau}}, \tag{12.11.1}$$

i.e.,

$$\dot{\tau}^\alpha = \mathbf{H}^\alpha : \overset{\bullet}{\boldsymbol{\tau}}, \quad \mathbf{H}^\alpha = \mathbf{C}^\alpha : \boldsymbol{\mathcal{M}}_{(0)} = \boldsymbol{\mathcal{M}}_{(0)} : \mathbf{C}^\alpha. \tag{12.11.2}$$

Since from Eqs. (12.2.14) and (12.4.5),

$$\overset{\bullet}{\boldsymbol{\tau}} = \overset{\circ}{\boldsymbol{\tau}} + \sum_{\beta=1}^{n} \left(\mathbf{C}^\beta - \boldsymbol{\mathcal{L}}_{(0)} : \mathbf{P}^\beta\right) \dot{\gamma}^\beta, \tag{12.11.3}$$

Equation (12.11.2) becomes

$$\dot{\tau}^{\alpha} = \mathbf{H}^{\alpha} : \overset{\circ}{\boldsymbol{\tau}} + \sum_{\beta=1}^{n} \left(\mathbf{H}^{\beta} - \mathbf{P}^{\beta}\right) \dot{\gamma}^{\beta}. \tag{12.11.4}$$

On an active slip system this must be equal to

$$\dot{\tau}^{\alpha} = \sum_{\beta=1}^{n} h_{\alpha\beta}\, \dot{\gamma}^{\beta}, \tag{12.11.5}$$

which gives

$$\mathbf{H}^{\alpha} : \overset{\circ}{\boldsymbol{\tau}} = \sum_{\beta=1}^{n} a_{\alpha\beta}\, \dot{\gamma}^{\beta}, \quad \dot{\gamma}^{\alpha} > 0, \tag{12.11.6}$$

where

$$a_{\alpha\beta} = h_{\alpha\beta} + \mathbf{C}^{\alpha} : \left(\mathbf{P}^{\beta} - \mathbf{H}^{\beta}\right) = g_{\alpha\beta} - \mathbf{C}^{\alpha} : \mathbf{H}^{\beta}. \tag{12.11.7}$$

When a slip system is inactive,

$$\mathbf{H}^{\alpha} : \overset{\circ}{\boldsymbol{\tau}} \leq \sum_{\beta=1}^{n} a_{\alpha\beta}\, \dot{\gamma}^{\beta}, \quad \dot{\gamma}^{\alpha} = 0. \tag{12.11.8}$$

If the inverse matrix, whose components are designated by $a_{\alpha\beta}^{-1}$, exists, Eq. (12.11.6) can be solved for the slip rates in terms of the stress rate as

$$\dot{\gamma}^{\alpha} = \sum_{\beta=1}^{n} a_{\alpha\beta}^{-1}\, \mathbf{H}^{\beta} : \overset{\circ}{\boldsymbol{\tau}}. \tag{12.11.9}$$

Substituting this into the first of equations (12.5.9) gives

$$\mathbf{D}_{(0)}^{\mathrm{p}} = \sum_{\alpha=1}^{n} \sum_{\beta=1}^{n} a_{\alpha\beta}^{-1} \left(\mathbf{H}^{\alpha} \otimes \mathbf{H}^{\beta}\right) : \overset{\circ}{\boldsymbol{\tau}}. \tag{12.11.10}$$

Combining with the elastic part, defined by Eq. (12.5.8), yields the constitutive equation for the elastoplastic loading of a single crystal,

$$\mathbf{D} = \left(\boldsymbol{\mathcal{M}}_{(0)} + \sum_{\alpha=1}^{n} \sum_{\beta=1}^{n} a_{\alpha\beta}^{-1}\, \mathbf{H}^{\alpha} \otimes \mathbf{H}^{\beta}\right) : \overset{\circ}{\boldsymbol{\tau}}. \tag{12.11.11}$$

The fourth-order tensor within the brackets is the crystalline elastoplastic compliances tensor.

If the convected rate of stress is used, we have

$$\mathbf{G}^{\alpha} : \overset{\triangle}{\boldsymbol{\tau}} = \sum_{\beta=1}^{n} b_{\alpha\beta}\, \dot{\gamma}^{\beta}, \tag{12.11.12}$$

where

$$b_{\alpha\beta} = h_{\alpha\beta} + \mathbf{C}^{\alpha} : \left(\mathbf{P}^{\beta} - \mathbf{G}^{\beta}\right) = g_{\alpha\beta} - \mathbf{C}^{\alpha} : \mathbf{G}^{\beta}, \tag{12.11.13}$$

and

$$\mathbf{G}_0^\alpha : \dot{\mathbf{T}} = \sum_{\beta=1}^{n} b_{\alpha\beta}^0 \, \dot{\gamma}^\beta, \quad \dot{\gamma}^\alpha > 0. \tag{12.11.14}$$

Here,

$$b_{\alpha\beta}^0 = h_{\alpha\beta} + \mathbf{C}_0^\alpha : \left(\mathbf{P}_0^\beta - \mathbf{G}_0^\beta\right) = g_{\alpha\beta} - \mathbf{C}_0^\alpha : \mathbf{G}_0^\beta. \tag{12.11.15}$$

However, the identity holds

$$\mathbf{C}_0^\alpha : \mathbf{G}_0^\beta = \mathbf{C}^\alpha : \mathbf{G}^\beta, \tag{12.11.16}$$

because

$$\mathbf{C}_0^\alpha = \mathbf{F}^{-1} \cdot \mathbf{C}^\alpha \cdot \mathbf{F}^{-T}, \quad \mathbf{G}_0^\beta = \mathbf{F}^T \cdot \mathbf{G}^\beta \cdot \mathbf{F}, \tag{12.11.17}$$

and, consequently,

$$b_{\alpha\beta}^0 = b_{\alpha\beta}. \tag{12.11.18}$$

This is also clear from Eqs. (12.11.12) and (12.11.14), and the identity

$$\mathbf{G}_0^\alpha : \dot{\mathbf{T}} = \mathbf{G}^\alpha : \overset{\triangle}{\boldsymbol{\tau}}. \tag{12.11.19}$$

If $b_{\alpha\beta}$ has an inverse matrix whose components are denoted by $b_{\alpha\beta}^{-1}$, the slip rates can be determined from

$$\dot{\gamma}^\alpha = \sum_{\beta=1}^{n} b_{\alpha\beta}^{-1} \, \mathbf{G}^\beta : \overset{\triangle}{\boldsymbol{\tau}} = \sum_{\beta=1}^{n} b_{\alpha\beta}^{-1} \, \mathbf{G}_0^\beta : \dot{\mathbf{T}}. \tag{12.11.20}$$

When Eq. (12.11.20) is substituted into (12.5.9), there follows

$$\mathbf{D}_{(1)}^{\mathrm{p}} = \sum_{\alpha=1}^{n} \sum_{\beta=1}^{n} b_{\alpha\beta}^{-1} \left(\mathbf{G}^\alpha \otimes \mathbf{G}^\beta\right) : \overset{\triangle}{\boldsymbol{\tau}}, \tag{12.11.21}$$

$$(\dot{\mathbf{E}})^{\mathrm{p}} = \sum_{\alpha=1}^{n} \sum_{\beta=1}^{n} b_{\alpha\beta}^{-1} \left(\mathbf{G}_0^\alpha \otimes \mathbf{G}_0^\beta\right) : \dot{\mathbf{T}}. \tag{12.11.22}$$

Combining with the elastic parts of Eq. (12.5.8) finally gives

$$\mathbf{D} = \left(\boldsymbol{\mathcal{M}}_{(1)} + \sum_{\alpha=1}^{n} \sum_{\beta=1}^{n} b_{\alpha\beta}^{-1} \, \mathbf{G}^\alpha \otimes \mathbf{G}^\beta\right) : \overset{\triangle}{\boldsymbol{\tau}}, \tag{12.11.23}$$

$$\dot{\mathbf{E}} = \left(\mathbf{M}_{(1)} + \sum_{\alpha=1}^{n} \sum_{\beta=1}^{n} b_{\alpha\beta}^{-1} \, \mathbf{G}_0^\alpha \otimes \mathbf{G}_0^\beta\right) : \dot{\mathbf{T}}. \tag{12.11.24}$$

These constitutive equations complement the previously derived constitutive equation (12.11.11), which was expressed in terms of the Jaumann rate of the Kirchhoff stress.

12.12. Uniqueness of Slip Rates for Prescribed Stress Rate

The uniqueness of the set of slip rates for the prescribed stress rate has to be examined separately for each selection of the strain and its conjugate stress measure. This is because the moduli $a_{\alpha\beta}$ and $b_{\alpha\beta}$ are different, while the moduli $g_{\alpha\beta}$ used in the proof given in Section 12.10 were measure invariant. Consequently, let us examine the uniqueness of $\dot{\gamma}^\alpha$ when $\dot{\mathbf{T}}$ is prescribed. Denote again by $\Delta\dot{\gamma}^\alpha = \dot{\gamma}^\alpha - \overset{*}{\dot{\gamma}}{}^\alpha$ $(\alpha = 1, 2, \ldots, n_0)$ the difference between the slip rates in two different slip modes, both at the same stress and hardening state. One mode is associated with the rate of stress $\dot{\mathbf{T}}$ and the other with $\overset{*}{\dot{\mathbf{T}}}$. From Eq. (12.5.9) we have

$$\Delta(\dot{\mathbf{E}})^{\mathrm{p}} = \sum_{\alpha=1}^{n_0} \mathbf{G}_0^\alpha \Delta\dot{\gamma}^\alpha, \tag{12.12.1}$$

and

$$\Delta(\dot{\mathbf{E}})^{\mathrm{p}} : \Delta\dot{\mathbf{T}} = \sum_{\alpha=1}^{n_0} \left(\mathbf{G}_0^\alpha : \Delta\dot{\mathbf{T}}\right) \Delta\dot{\gamma}^\alpha, \tag{12.12.2}$$

where

$$\Delta\dot{\mathbf{T}} = \dot{\mathbf{T}} - \overset{*}{\dot{\mathbf{T}}}. \tag{12.12.3}$$

If the slip system α is active in both modes,

$$\mathbf{G}_0^\alpha : \dot{\mathbf{T}} - \sum_{\beta=1}^{n_0} b_{\alpha\beta}\, \dot{\gamma}^\beta = 0, \quad \dot{\gamma}^\alpha > 0, \tag{12.12.4}$$

$$\mathbf{G}_0^\alpha : \overset{*}{\dot{\mathbf{T}}} - \sum_{\beta=1}^{n_0} b_{\alpha\beta}\, \overset{*}{\dot{\gamma}}{}^\beta = 0, \quad \overset{*}{\dot{\gamma}}{}^\alpha > 0. \tag{12.12.5}$$

In this case,

$$\mathbf{G}_0^\alpha : \Delta\dot{\mathbf{T}} - \sum_{\beta=1}^{n_0} b_{\alpha\beta}\, \Delta\dot{\gamma}^\beta = 0, \tag{12.12.6}$$

and, upon multiplication with $\Delta\dot{\gamma}^\alpha$,

$$\left(\mathbf{G}_0^\alpha : \Delta\dot{\mathbf{T}}\right) \Delta\dot{\gamma}^\alpha = \sum_{\beta=1}^{n_0} b_{\alpha\beta}\, \Delta\dot{\gamma}^\alpha\, \Delta\dot{\gamma}^\beta. \tag{12.12.7}$$

If the slip system α is active in the first mode, but inactive in the second mode, i.e.,

$$\mathbf{G}_0^\alpha : \dot{\mathbf{T}} - \sum_{\beta=1}^{n_0} b_{\alpha\beta}\, \dot{\gamma}^\beta = 0, \quad \dot{\gamma}^\alpha > 0, \tag{12.12.8}$$

$$\mathbf{G}_0^\alpha : \dot{\mathbf{T}} - \sum_{\beta=1}^{n_0} b_{\alpha\beta}\, \dot{\gamma}^\beta \leq 0, \quad \dot{\gamma}^\alpha = 0, \tag{12.12.9}$$

then

$$\mathbf{G}_0^\alpha : \Delta\dot{\mathbf{T}} - \sum_{\beta=1}^{n_0} b_{\alpha\beta}\, \Delta\dot{\gamma}^\beta \geq 0, \quad \Delta\dot{\gamma}^\alpha > 0. \tag{12.12.10}$$

Thus, upon multiplication with $\Delta\dot{\gamma}^\alpha$,

$$\left(\mathbf{G}_0^\alpha : \Delta\dot{\mathbf{T}}\right) \Delta\dot{\gamma}^\alpha \geq \sum_{\beta=1}^{n_0} b_{\alpha\beta}\, \Delta\dot{\gamma}^\alpha\, \Delta\dot{\gamma}^\beta. \tag{12.12.11}$$

Inequality (12.12.11) also holds in the case when α system is active in the second and inactive in the first mode, since then, in place of (12.12.10),

$$\mathbf{G}_0^\alpha : \Delta\dot{\mathbf{T}} - \sum_{\beta=1}^{n_0} b_{\alpha\beta}\, \Delta\dot{\gamma}^\beta \leq 0, \quad \Delta\dot{\gamma}^\alpha < 0. \tag{12.12.12}$$

Finally, if the slip system α is inactive in both modes,

$$\left(\mathbf{G}_0^\alpha : \Delta\dot{\mathbf{T}}\right) \Delta\dot{\gamma}^\alpha = \sum_{\beta=1}^{n_0} b_{\alpha\beta}\, \Delta\dot{\gamma}^\alpha\, \Delta\dot{\gamma}^\beta, \tag{12.12.13}$$

because $\Delta\dot{\gamma}^\alpha = 0$. Therefore, (12.12.11) encompasses all cases, since either $=$ or $>$ sign applies. Summing over all critical systems, therefore, gives

$$\sum_{\alpha=1}^{n_0} \left(\mathbf{G}_0^\alpha : \Delta\dot{\mathbf{T}}\right) \Delta\dot{\gamma}^\alpha \geq \sum_{\alpha=1}^{n_0}\sum_{\beta=1}^{n_0} b_{\alpha\beta}\, \Delta\dot{\gamma}^\alpha\, \Delta\dot{\gamma}^\beta. \tag{12.12.14}$$

From (12.12.14) we deduce that the positive definiteness of the matrix $b_{\alpha\beta}$ is a sufficient condition for the unique slip rates $\dot{\gamma}^\alpha$ under prescribed $\dot{\mathbf{T}}$. Indeed, for a prescribed stress rate, the difference $\Delta\dot{\mathbf{T}} = \mathbf{0}$, and if $b_{\alpha\beta}$ is positive definite, the inequality (12.12.14) can be satisfied only when $\Delta\dot{\gamma}^\alpha = 0$, for all α. The same applies if the stress rate $\overset{\triangle}{\boldsymbol{\tau}}$ is used ($\overset{\triangle}{\boldsymbol{\tau}}$ is proportional to $\dot{\mathbf{T}}$, if current configuration is taken for the reference). By an analogous prove, when the stress rate $\overset{\circ}{\boldsymbol{\tau}}$ is prescribed ($\overset{\circ}{\boldsymbol{\tau}}$ is proportional to the rate of stress conjugate to logarithmic strain, when the current configuration is taken as the reference), the slip rates $\dot{\gamma}^\alpha$ are guaranteed to be unique if the matrix with component $a_{\alpha\beta}$ is positive definite.

Finally, we note that, from Eqs. (12.5.10), (12.11.7) and (12.11.13),

$$\begin{aligned} g_{\alpha\beta} &= a_{\alpha\beta} + \mathbf{C}^\alpha : \boldsymbol{\mathcal{M}}_{(0)} : \mathbf{C}^\beta \\ &= b_{\alpha\beta} + \mathbf{C}^\alpha : \boldsymbol{\mathcal{M}}_{(1)} : \mathbf{C}^\beta = b_{\alpha\beta} + \mathbf{C}_0^\alpha : \mathbf{M}_{(1)} : \mathbf{C}_0^\beta. \end{aligned} \tag{12.12.15}$$

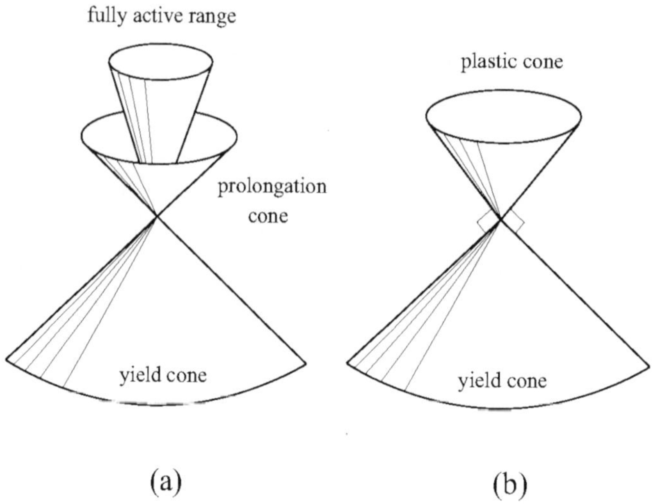

FIGURE 12.3. (a) The yield cone in stress space. Indicated also are prolongation of the yield cone and the cone of fully active range associated with the directions of stress rate for which all segments of the yield cone are active. (b) The plastic cone defining the range of possible directions of the plastic rate of strain.

Thus, if $a_{\alpha\beta}$ is positive definite, positive definiteness of $g_{\alpha\beta}$ is ensured if $\boldsymbol{\mathcal{M}}_{(0)}$ is positive definite. Likewise, if $b_{\alpha\beta}$ is positive definite, the positive definiteness of $g_{\alpha\beta}$ is ensured if $\boldsymbol{\mathcal{M}}_{(1)}$ or, equivalently, $\mathbf{M}_{(1)}$ is positive definite.

12.13. Fully Active or Total Loading Range

Suppose that the yield vertex in stress space $\mathbf{T}$ is a pyramid formed by n_0 intersecting hyperplanes corresponding to n_0 potentially active slip systems. The range of directions of the stress rate $\dot{\mathbf{T}}$ for which all n_0 vertex segments are active (slip takes place on all n_0 slip systems) is defined by n_0 inequalities

$$\sum_{\beta=1}^{n_0} b_{\alpha\beta}^{-1}\,\mathbf{G}_0^{\beta} : \dot{\mathbf{T}} > 0, \quad \alpha = 1, 2, \ldots, n_0. \tag{12.13.1}$$

These follow from (12.11.14) and the requirement that all slip rates are positive (the matrix with components $b_{\alpha\beta}$ is assumed to be positive definite). The corresponding range of the stress rate space is referred to as the fully active or total loading range (Fig. 12.3a). The terminology is borrowed from Hill (1966) for fully active, and from Sanders (1955) for total loading range. The elastic unloading takes place on all slip systems if $\dot{\mathbf{T}}$ falls within the range

$$\mathbf{G}_0^\alpha : \dot{\mathbf{T}} \le 0, \quad \alpha = 1, 2, \ldots, n_0, \tag{12.13.2}$$

which is the boundary or the interior of the pyramidal yield vertex. The outward normal to α segment of the vertex is codirectional with $\mathbf{G}_0^\alpha$. The remainder of the stress rate space is dissected into

$$\binom{n_0}{1} + \binom{n_0}{2} + \cdots + \binom{n_0}{n_0 - 1} = 2^{n_0} - 2 \tag{12.13.3}$$

pyramidal regions of partial loading ($n_0 \ge 2$). For example, there are n_0 pyramidal regions of single slip, and $n_0(n_0-1)/2$ pyramidal regions of double slip. There are also n_0 pyramidal regions of multislip over different sets of $(n_0 - 1)$ slip systems.

As an illustration, consider a pyramidal region of double slip on the first and second slip system ($\alpha = 1, 2$). From Eq. (12.11.14) there follows

$$\mathbf{G}_0^1 : \dot{\mathbf{T}} = b_{11}\dot{\gamma}^1 + b_{12}\dot{\gamma}^2, \quad \mathbf{G}_0^2 : \dot{\mathbf{T}} = b_{21}\dot{\gamma}^1 + b_{22}\dot{\gamma}^2, \tag{12.13.4}$$

$$\mathbf{G}_0^\alpha : \dot{\mathbf{T}} \le b_{\alpha 1}\dot{\gamma}^1 + b_{\alpha 2}\dot{\gamma}^2, \quad 3 \le \alpha \le n_0. \tag{12.13.5}$$

Since double slip is assumed to take place under prescribed $\dot{\mathbf{T}}$, the two equations in (12.13.4) can be solved for the slip rates to give

$$\dot{\gamma}^1 = \frac{1}{\Delta}(b_{22}\mathbf{G}_0^1 - b_{12}\mathbf{G}_0^2) : \dot{\mathbf{T}}, \quad \dot{\gamma}^2 = \frac{1}{\Delta}(b_{11}\mathbf{G}_0^2 - b_{21}\mathbf{G}_0^1) : \dot{\mathbf{T}}, \tag{12.13.6}$$

where

$$\Delta = b_{11}b_{22} - b_{12}b_{21} > 0. \tag{12.13.7}$$

Thus, since $\dot{\gamma}^1 > 0$ and $\dot{\gamma}^2 > 0$, we have

$$(b_{22}\mathbf{G}_0^1 - b_{12}\mathbf{G}_0^2) : \dot{\mathbf{T}} > 0, \quad (b_{11}\mathbf{G}_0^2 - b_{21}\mathbf{G}_0^1) : \dot{\mathbf{T}} > 0. \tag{12.13.8}$$

Furthermore, if (12.13.6) is substituted into (12.13.5), there follows

$$\begin{aligned} &[\,(b_{11}b_{22} - b_{12}b_{21})\mathbf{G}_0^\alpha + (b_{\alpha 2}b_{21} - b_{\alpha 1}b_{22})\mathbf{G}_0^1 \\ &\quad + (b_{\alpha 1}b_{12} - b_{\alpha 2}b_{11})\mathbf{G}_0^2] : \dot{\mathbf{T}} \le 0, \quad 3 \le \alpha \le n_0. \end{aligned} \tag{12.13.9}$$

The inequalities (12.13.8) and (12.13.9) define the pyramidal region of double slip over slip systems 1 and 2 at the vertex formed by $n_0 \ge 3$ yield segments.

Similarly, the pyramidal region of single slip over the slip system 1 is defined by the inequalities

$$\mathbf{G}_0^1 : \dot{\mathbf{T}} > 0, \quad (b_{11}\mathbf{G}_0^\alpha - b_{\alpha 1}\mathbf{G}_0^1) : \dot{\mathbf{T}} \leq 0, \quad 2 \leq \alpha \leq n_0. \tag{12.13.10}$$

Fully active range and the two regions of single slip for the case $n_0 = 2$ are schematically shown in Fig. 12.4a.

If there is no latent hardening ($h_{\alpha\beta} = 0$ for $\alpha \neq \beta$), the fully active range is just the prolongation of the yield vertex (prolongation cone in Fig. 12.3b). Thus, a pyramidal region of double slip on the first and second slip system ($\alpha = 1, 2$) is defined by

$$\mathbf{G}_0^1 : \dot{\mathbf{T}} > 0, \quad \mathbf{G}_0^2 : \dot{\mathbf{T}} > 0, \quad \mathbf{G}_0^\alpha : \dot{\mathbf{T}} \leq 0, \quad 3 \leq \alpha \leq n_0. \tag{12.13.11}$$

The pyramidal region of single slip over the slip system 1 is similarly

$$\mathbf{G}_0^1 : \dot{\mathbf{T}} > 0, \quad \mathbf{G}_0^\alpha : \dot{\mathbf{T}} \leq 0, \quad 2 \leq \alpha \leq n_0. \tag{12.13.12}$$

Fully active range and the two regions of single slip are in this case sketched in Fig. 12.4b. With no latent hardening, the range of possible directions for the plastic rate of deformation coincides with the fully active range. For an analysis of elastic-plastic crystals characterized by a smooth yield surface with rounded corners, see Gambin (1992).

12.14. Constitutive Inequalities

We first recall from Sections 12.5 and 12.9 that

$$\mathbf{C}^\alpha : \mathbf{D} = \mathbf{C}_0^\alpha : \dot{\mathbf{E}} = \sum_{\beta=1}^{n} g_{\alpha\beta}\, \dot{\gamma}^\beta, \tag{12.14.1}$$

and

$$\overset{\circ}{\boldsymbol{\tau}}{}^{\mathrm{p}} = \overset{\triangle}{\boldsymbol{\tau}}{}^{\mathrm{p}} = -\sum_{\alpha=1}^{n} \mathbf{C}^\alpha\, \dot{\gamma}^\alpha, \quad (\dot{\mathbf{T}})^{\mathrm{p}} = -\sum_{\alpha=1}^{n} \mathbf{C}_0^\alpha\, \dot{\gamma}^\alpha. \tag{12.14.2}$$

Thus,

$$\overset{\circ}{\boldsymbol{\tau}}{}^{\mathrm{p}} : \mathbf{D} = \overset{\triangle}{\boldsymbol{\tau}}{}^{\mathrm{p}} : \mathbf{D} = (\dot{\mathbf{T}})^{\mathrm{p}} : \dot{\mathbf{E}} = -\sum_{\alpha=1}^{n}\sum_{\beta=1}^{n} g_{\alpha\beta}\, \dot{\gamma}^\alpha\, \dot{\gamma}^\beta. \tag{12.14.3}$$

In this expression we can replace the number of active slip systems n with the number of critical slip systems n_0, because $\dot{\gamma}^\alpha = 0$ for inactive critical

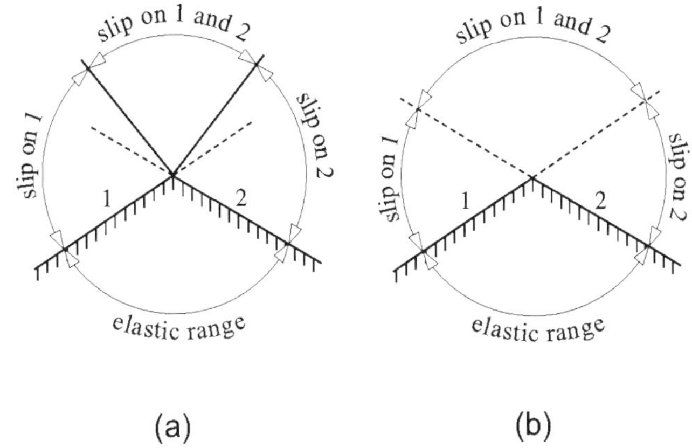

FIGURE 12.4. (a) The yield vertex formed by two segments 1 and 2. Indicated are the fully active range of slip on both slip systems, the two ranges of single slip, and the range of elastic unloading. (b) The same as in (a), but without latent hardening. The fully active range coincides with the prolongation of the yield vertex.

systems. Thus, if the matrix with components $g_{\alpha\beta}$ over all critical systems is positive definite, Eq. (12.14.3) yields

$$\mathring{\boldsymbol{\tau}}^{\mathrm{p}} : \mathbf{D} = \overset{\triangle}{\boldsymbol{\tau}}{}^{\mathrm{p}} : \mathbf{D} = (\dot{\mathbf{T}})^{\mathrm{p}} : \dot{\mathbf{E}} < 0. \tag{12.14.4}$$

The inequality holds regardless of whether the crystal is in the state of overall hardening or softening (Fig. 12.5). Recall that, in the context of general strain measures, the quantity $\mathrm{d}\mathbf{E} : \mathrm{d}^{\mathrm{p}}\mathbf{T}$ is measure invariant, i.e., it does not change its value with the change of strain $\mathbf{E}$ and its conjugate stress measure $\mathbf{T}$.

On the other hand,

$$\mathring{\boldsymbol{\tau}} : \mathbf{D}^{\mathrm{p}}_{(0)} \neq \overset{\triangle}{\boldsymbol{\tau}} : \mathbf{D}^{\mathrm{p}}_{(1)} = \dot{\mathbf{T}} : (\dot{\mathbf{E}})^{\mathrm{p}}. \tag{12.14.5}$$

This can be deduced from the derived equations in Sections 12.5 and 12.9, i.e., from

$$\mathbf{H}^{\alpha} : \mathring{\boldsymbol{\tau}} = \sum_{\beta=1}^{n} a_{\alpha\beta}\,\dot{\gamma}^{\beta}, \quad \mathbf{G}^{\alpha} : \overset{\triangle}{\boldsymbol{\tau}} = \mathbf{G}^{\alpha}_{0} : \dot{\mathbf{T}} = \sum_{\beta=1}^{n} b_{\alpha\beta}\,\dot{\gamma}^{\beta}, \tag{12.14.6}$$

$$\mathbf{D}^{\mathrm{p}}_{(0)} = \sum_{\alpha=1}^{n} \mathbf{H}^{\alpha}\,\dot{\gamma}^{\alpha}, \quad \mathbf{D}^{\mathrm{p}}_{(1)} = \sum_{\alpha=1}^{n} \mathbf{G}^{\alpha}\,\dot{\gamma}^{\alpha}, \quad (\dot{\mathbf{E}})^{\mathrm{p}} = -\sum_{\alpha=1}^{n} \mathbf{G}^{\alpha}_{0}\,\dot{\gamma}^{\alpha}. \tag{12.14.7}$$

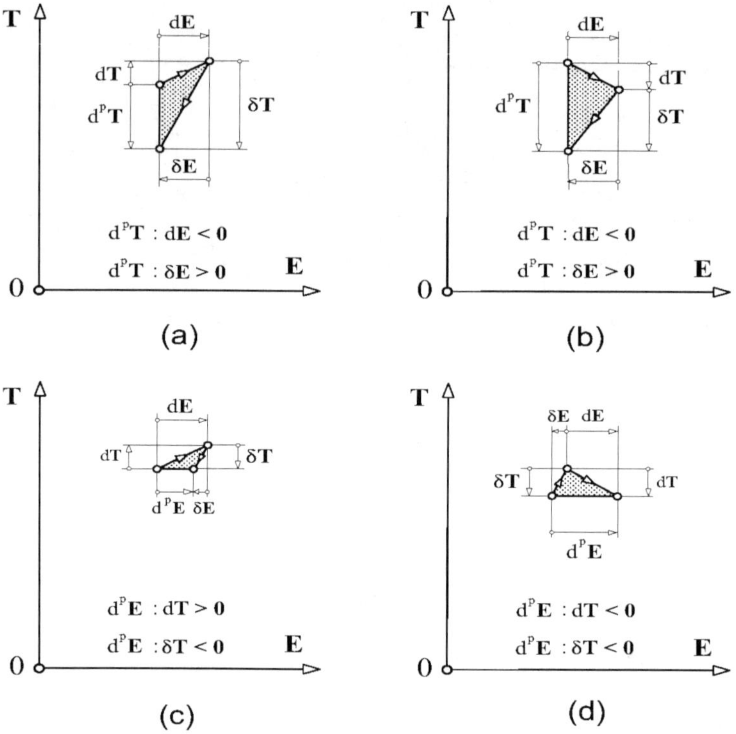

FIGURE 12.5. One-dimensional illustration of elastoplastic inequalities for the hardening and softening material response. Infinitesimal cycles of stress are shown in parts (a) and (b), and of strain in parts (c) and (d). Indicated stress and strain increments are positive when their arrows are directed in the positive coordinate directions.

These yield

$$\overset{\circ}{\boldsymbol{\tau}} : \mathbf{D}^{\mathrm{p}}_{(0)} = \sum_{\alpha=1}^{n}\sum_{\beta=1}^{n} a_{\alpha\beta}\,\dot{\gamma}^{\alpha}\,\dot{\gamma}^{\beta}, \quad \overset{\triangle}{\boldsymbol{\tau}} : \mathbf{D}^{\mathrm{p}}_{(1)} = \dot{\mathbf{T}} : (\dot{\mathbf{E}})^{\mathrm{p}} = \sum_{\alpha=1}^{n}\sum_{\beta=1}^{n} b_{\alpha\beta}\,\dot{\gamma}^{\alpha}\,\dot{\gamma}^{\beta}. \tag{12.14.8}$$

In particular, it may happen that $\overset{\circ}{\boldsymbol{\tau}} : \mathbf{D}^{\mathrm{p}}_{(0)} > 0$, implying the hardening relative to utilized measures of conjugate stress and strain, while $\overset{\triangle}{\boldsymbol{\tau}} : \mathbf{D}^{\mathrm{p}}_{(1)} < 0$, implying the softening relative to these measures.

In fact, by multiplying Eq. (12.11.4) with $\dot{\gamma}^\alpha$, summing over α, and by using the first of (12.5.9) gives

$$\overset{\circ}{\boldsymbol{\tau}} : \mathbf{D}^{\mathrm{p}}_{(0)} = \sum_{\alpha=1}^{n} \dot{\tau}^\alpha \dot{\gamma}^\alpha + \sum_{\alpha=1}^{n} \sum_{\beta=1}^{n} \mathbf{C}^\alpha : \left(\mathbf{P}^\beta - \mathbf{H}^\beta\right) \dot{\gamma}^\alpha \dot{\gamma}^\beta . \tag{12.14.9}$$

Similarly,

$$\overset{\triangle}{\boldsymbol{\tau}} : \mathbf{D}^{\mathrm{p}}_{(1)} = \sum_{\alpha=1}^{n} \dot{\tau}^\alpha \dot{\gamma}^\alpha + \sum_{\alpha=1}^{n} \sum_{\beta=1}^{n} \mathbf{C}^\alpha : \left(\mathbf{P}^\beta - \mathbf{G}^\beta\right) \dot{\gamma}^\alpha \dot{\gamma}^\beta . \tag{12.14.10}$$

Their difference is, thus,

$$\overset{\circ}{\boldsymbol{\tau}} : \mathbf{D}^{\mathrm{p}}_{(0)} - \overset{\triangle}{\boldsymbol{\tau}} : \mathbf{D}^{\mathrm{p}}_{(1)} = \sum_{\alpha=1}^{n} \sum_{\beta=1}^{n} \mathbf{C}^\alpha : \left(\mathbf{G}^\beta - \mathbf{H}^\beta\right) \dot{\gamma}^\alpha \dot{\gamma}^\beta , \tag{12.14.11}$$

or

$$\overset{\circ}{\boldsymbol{\tau}} : \mathbf{D}^{\mathrm{p}}_{(0)} - \overset{\triangle}{\boldsymbol{\tau}} : \mathbf{D}^{\mathrm{p}}_{(1)} = \sum_{\alpha=1}^{n} \sum_{\beta=1}^{n} \mathbf{C}^\alpha : \left(\boldsymbol{\mathcal{M}}_{(1)} - \boldsymbol{\mathcal{M}}_{(0)}\right) : \mathbf{C}^\beta \dot{\gamma}^\alpha \dot{\gamma}^\beta , \tag{12.14.12}$$

which can be either positive or negative.

In retrospect, the inequality in (12.14.5) was anticipated in the context of general strain measures, because the second-order work quantity $\mathrm{d}\mathbf{T} : \mathrm{d}^{\mathrm{p}}\mathbf{E}$ is not measure invariant, and changes its value with the change of strain and its conjugate stress measure.

In contrast to (12.14.5), there is an equality

$$\overset{\bullet}{\boldsymbol{\tau}} : \mathbf{D}^{\mathrm{p}}_{(0)} = \overset{\blacktriangle}{\boldsymbol{\tau}} : \mathbf{D}^{\mathrm{p}}_{(1)} = \left(\mathbf{F}^{\mathrm{p}-1} \cdot \dot{\mathbf{T}}^{*} \cdot \mathbf{F}^{\mathrm{p}-T}\right) : (\dot{\mathbf{E}})^{\mathrm{p}} = \sum_{\alpha=1}^{n} \dot{\tau}^\alpha \dot{\gamma}^\alpha . \tag{12.14.13}$$

Further Inequalities

If $\mathrm{d}^{\mathrm{p}}\mathbf{E}$ is the plastic part of the strain increment along plastic loading branch, while $\delta\mathbf{T}$ is the stress increment along elastic unloading branch, from Eq. (12.14.13) it follows that

$$\delta\mathbf{T} : \mathrm{d}^{\mathrm{p}}\mathbf{E} = \sum_{\alpha=1}^{n} \delta\tau^\alpha \, \mathrm{d}\gamma^\alpha < 0, \tag{12.14.14}$$

provided that elastic unloading is such that it reduces τ^α on each critical system ($\delta\tau^\alpha < 0$). The slip increments $\mathrm{d}\gamma^\alpha$ are assumed to be always positive during plastic loading, so that opposite directions of slip in the same glide plane are represented by distinct α's. The inequality (12.14.14) is measure invariant. The measure invariance is clear since

$$-\delta\mathbf{T} : \mathrm{d}^{\mathrm{p}}\mathbf{E} = \delta\mathbf{E} : \mathrm{d}^{\mathrm{p}}\mathbf{T} = \delta\mathbf{F} \cdot\cdot\, \mathrm{d}^{\mathrm{p}}\mathbf{P}. \tag{12.14.15}$$

This follows by recalling that

$$\mathrm{d}^{\mathrm{p}}\mathbf{P} = \boldsymbol{\mathcal{K}}^T : \mathrm{d}^{\mathrm{p}}\mathbf{T}, \quad \delta\mathbf{E} = \delta\mathbf{F} \cdot\cdot\, \boldsymbol{\mathcal{K}}^T, \tag{12.14.16}$$

and

$$\delta\mathbf{T} : \mathrm{d}^{\mathrm{p}}\mathbf{E} = \delta\mathbf{E} : \boldsymbol{\Lambda}_{(1)} : \mathrm{d}^{\mathrm{p}}\mathbf{E} = -\delta\mathbf{E} : \mathrm{d}^{\mathrm{p}}\mathbf{T}. \tag{12.14.17}$$

Thus

$$\delta\mathbf{E} : \mathrm{d}^{\mathrm{p}}\mathbf{T} = \delta\mathbf{F} \cdot\cdot\, \mathrm{d}^{\mathrm{p}}\mathbf{P} > 0. \tag{12.14.18}$$

The transition between the inequalities (12.14.14) and (12.14.18) can also be conveniently deduced from an invariant bilinear form, introduced in a more general context by Hill (1972). This is

$$\delta\mathbf{T} : \mathrm{d}^{\mathrm{p}}\mathbf{E} - \mathrm{d}^{\mathrm{p}}\mathbf{T} : \delta\mathbf{E} = \delta\mathbf{P} \cdot\cdot\, \mathrm{d}^{\mathrm{p}}\mathbf{F} - \mathrm{d}^{\mathrm{p}}\mathbf{P} \cdot\cdot\, \delta\mathbf{F}. \tag{12.14.19}$$

It is easily verified that

$$\delta\mathbf{T} : \mathrm{d}^{\mathrm{p}}\mathbf{E} - \mathrm{d}^{\mathrm{p}}\mathbf{T} : \delta\mathbf{E} = 2\delta\mathbf{T} : \mathrm{d}^{\mathrm{p}}\mathbf{E} = -2\delta\mathbf{E} : \mathrm{d}^{\mathrm{p}}\mathbf{T}. \tag{12.14.20}$$

Thus, if $\delta\mathbf{T} : \mathrm{d}^{\mathrm{p}}\mathbf{E} < 0$, then $\delta\mathbf{E} : \mathrm{d}^{\mathrm{p}}\mathbf{T} > 0$, and *vice versa*.

It is noted that

$$\delta\mathbf{E} : \mathbf{C}_0^{\alpha} \leq 0, \quad \alpha = 1, 2, \ldots, n_0. \tag{12.14.21}$$

These inequalities hold because the elastic strain increment is directed inside of the yield vertex in strain space formed by n_0 hyperplane segments (or along some of the vertex segments), while $\mathbf{C}_0^{\alpha}$ are in the directions of their outer normals (Fig. 12.6). From the inequalities (12.14.21) we can deduce the normality rule. Indeed, by multiplying (12.14.21) with $\mathrm{d}\gamma^{\alpha} \geq 0$ ($\mathrm{d}\gamma^{\alpha} = 0$ for $n_0 - n$ inactive critical systems at the vertex), and by summing over α, there follows

$$\delta\mathbf{E} : \sum_{\alpha=1}^{n_0} \mathbf{C}_0^{\alpha}\, \mathrm{d}\gamma^{\alpha} < 0. \tag{12.14.22}$$

In view of Eq. (12.5.2), this implies the normality

$$\delta\mathbf{E} : \mathrm{d}^{\mathrm{p}}\mathbf{T} > 0. \tag{12.14.23}$$

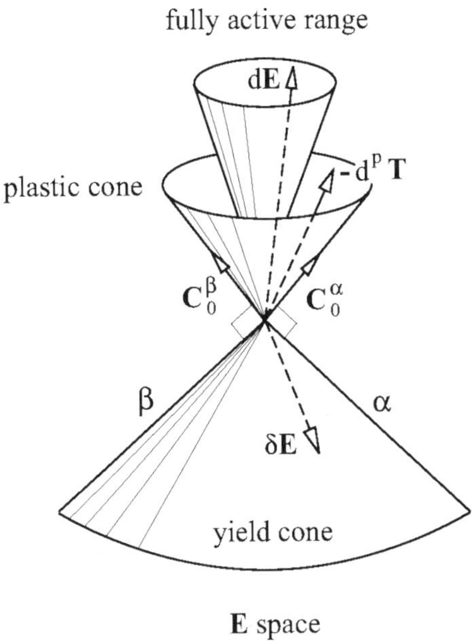

FIGURE 12.6. The yield cone in strain space. The plastic part of the rate of stress $-\mathrm{d}^{\mathrm{p}}\mathbf{T}$ falls within the plastic cone defined by the normals to individual yield segments, such as $\mathbf{C}_0^{\alpha}$ and $\mathbf{C}_0^{\beta}$. If the strain increment $\mathrm{d}\mathbf{E}$ is within fully active range, all yield segments are active and participate in plastic flow. The elastic unloading increment of strain $\delta\mathbf{E}$ is directed within the yield cone.

12.15. Implications of Ilyushin's Postulate

We demonstrate in this section that the inequality (12.14.14) is in accord with Ilyushin's postulate of positive net work in an isothermal cycle of strain that involves plastic slip,

$$\oint_E \mathbf{T} : \mathrm{d}\mathbf{E} > 0. \tag{12.15.1}$$

As discussed in Section 8.5, when Ilyushin's postulate is applied to an infinitesimal strain cycle emanating from the yield surface, the net expenditure of work must be positive. By the trapezoidal rule of quadrature this work is

$$-\frac{1}{2}\,\mathrm{d}^{\mathrm{p}}\mathbf{T} : \mathrm{d}\mathbf{E} > 0, \tag{12.15.2}$$

so that

$$\mathrm{d}^{\mathrm{p}}\mathbf{T} : \mathrm{d}\mathbf{E} < 0. \tag{12.15.3}$$

This inequality is often considered as a basic or fundamental inequality of crystal plasticity (Havner, 1992). Comparing with Eq. (12.14.3), we see that the positive definiteness of $g_{\alpha\beta}$ ensures that the crystal behavior is in accord with the inequality (12.15.3).

By considering the strain cycle with a sufficiently small segment along which the slip takes place, it was shown in Section 8.5 that Ilyushin's postulate implies, to first order,

$$\oint_E \mathbf{T} : \mathrm{d}\mathbf{E} = (\mathrm{d}^{\mathrm{p}}\Psi)^0 - \mathrm{d}^{\mathrm{p}}\Psi > 0. \qquad (12.15.4)$$

The plastic parts of the free energy at the strain levels $\mathbf{E}$ and $\mathbf{E}^0$, due to change in slip alone, are defined by

$$\mathrm{d}^{\mathrm{p}}\Psi = \Psi\left(\mathbf{E}, \mathcal{H} + d\mathcal{H}\right) - \Psi\left(\mathbf{E}, \mathcal{H}\right), \qquad (12.15.5)$$

$$(\mathrm{d}^{\mathrm{p}}\Psi)^0 = \Psi\left(\mathbf{E}^0, \mathcal{H} + d\mathcal{H}\right) - \Psi\left(\mathbf{E}^0, \mathcal{H}\right). \qquad (12.15.6)$$

Infinitesimal change of the pattern of internal rearrangements $d\mathcal{H}$ is fully described by the slip increments $\mathrm{d}\gamma^\alpha$. The state $(\mathbf{E}, \mathcal{H})$ is on the yield surface, while the other three states are inside the yield surface (Fig. 12.6). The plastic change of the free energy in the loading/unloading transition from $(\mathbf{E}, \mathcal{H})$ to $(\mathbf{E}, \mathcal{H} + d\mathcal{H})$ is equal to the negative of the work done on the increment of strain caused by the slip $\mathrm{d}\gamma^\alpha$. This is, to first order,

$$\mathrm{d}^{\mathrm{p}}\Psi = -\mathbf{T} : \sum_{\alpha=1}^{n} \mathbf{P}_0^\alpha \,\mathrm{d}\gamma^\alpha = -\sum_{\alpha=1}^{n} \tau^\alpha \,\mathrm{d}\gamma^\alpha. \qquad (12.15.7)$$

The resolved shear stress at the stress state $\mathbf{T}$ is $\tau^\alpha = \mathbf{T} : \mathbf{P}_0^\alpha$, by Eq. (12.7.3). The plastic change of the free energy in the loading/unloading transition from $(\mathbf{E}^0, \mathcal{H})$ to $(\mathbf{E}^0, \mathcal{H} + d\mathcal{H})$ is equal to the negative of the work done on slip increments $\mathrm{d}\gamma^\alpha$ by the resolved shear stress τ_0^α, corresponding to stress $\mathbf{T}^0$ at the state $(\mathbf{E}^0, \mathcal{H})$. Thus,

$$(\mathrm{d}^{\mathrm{p}}\Psi)^0 = -\mathbf{T}^0 : \sum_{\alpha=1}^{n} (\mathbf{P}_0^\alpha)^0 \,\mathrm{d}\gamma^\alpha = -\sum_{\alpha=1}^{n} \tau_0^\alpha \,\mathrm{d}\gamma^\alpha, \qquad (12.15.8)$$

where $\tau_0^\alpha = \mathbf{T}^0 : (\mathbf{P}_0^\alpha)^0$. Substitution into (12.15.4) gives

$$\sum_{\alpha=1}^{n} (\tau^\alpha - \tau_0^\alpha) \,\mathrm{d}\gamma^\alpha > 0. \qquad (12.15.9)$$

The inequality may be referred to as the maximum slip work inequality (analogous to maximum plastic work inequality discussed in Section 8.6). Introducing the elastic unloading increments of the resolved shear stress

$$\delta\tau^\alpha = \tau^\alpha - \tau_0^\alpha, \tag{12.15.10}$$

the inequality (12.15.9) becomes

$$\sum_{\alpha=1}^{n} \delta\tau^\alpha \, \mathrm{d}\gamma^\alpha < 0. \tag{12.15.11}$$

Since

$$\delta\mathbf{T} : \mathrm{d}^{\mathrm{p}}\mathbf{E} = -\delta\mathbf{E} : \mathrm{d}^{\mathrm{p}}\mathbf{T} = \sum_{\alpha=1}^{n} \delta\tau^\alpha \, \mathrm{d}\gamma^\alpha, \tag{12.15.12}$$

we conclude that Ilyushin's postulate (12.15.4), and the resulting inequality (12.15.11), ensure (12.14.14) and (12.14.18), and the normality properties for $\mathrm{d}^{\mathrm{p}}\mathbf{E}$ and $\mathrm{d}^{\mathrm{p}}\mathbf{T}$.

12.16. Lower Bound on Second-Order Work

In this section we prove that the symmetric positive definite matrix of moduli $g_{\alpha\beta}$, over all n_0 critical systems, guarantees that the second-order work $\mathrm{d}\mathbf{T} : \mathrm{d}\mathbf{E}$ in an actual crystal response, with $n < n_0$ active slip systems, is not less than it would be with all critical systems active (Sewell, 1972; Havner, 1992). To that goal, introduce the net resistance force on a critical system α by

$$\dot{f}^\alpha = \dot{\tau}_{\mathrm{cr}}^\alpha - \dot{\tau}^\alpha \begin{cases} = 0, & \dot{\gamma}^\alpha > 0, \\ \geq 0, & \dot{\gamma}^\alpha = 0. \end{cases} \tag{12.16.1}$$

The rates of the critical resolved shear stress and the resolved shear stress are defined by Eqs. (12.9.1) and (12.9.19), i.e.,

$$\dot{\tau}_{\mathrm{cr}}^\alpha = \sum_{\beta=1}^{n_0} h_{\alpha\beta} \, \dot{\gamma}^\beta, \tag{12.16.2}$$

$$\dot{\tau}^\alpha = \mathbf{C}_0^\alpha : \dot{\mathbf{E}} - \sum_{\beta=1}^{n_0} \mathbf{C}_0^\alpha : \mathbf{P}_0^\beta \, \dot{\gamma}^\beta. \tag{12.16.3}$$

Since, from Eq. (12.9.27), $h_{\alpha\beta} = g_{\alpha\beta} - \mathbf{C}_0^\alpha : \mathbf{P}_0^\beta$, the substitution of Eqs. (12.16.2) and (12.16.3) into (12.16.1) yields

$$\dot{f}^\alpha = \sum_{\beta=1}^{n_0} g_{\alpha\beta} \, \dot{\gamma}^\beta - \mathbf{C}_0^\alpha : \dot{\mathbf{E}}. \tag{12.16.4}$$

If the matrix $g_{\alpha\beta}$ is positive definite, it has an inverse, and Eq. (12.16.4) can be solved for $\dot{\gamma}^\alpha$ to give

$$\dot{\gamma}^\alpha = \sum_{\beta=1}^{n_0} g_{\alpha\beta}^{-1} \left(\dot{f}^\beta + \mathbf{C}_0^\beta : \dot{\mathbf{E}} \right). \tag{12.16.5}$$

The plastic part of the stress rate can then be expressed from Eq. (12.5.2) as

$$(\dot{\mathbf{T}})^{\mathrm{p}} = -\sum_{\alpha=1}^{n_0} \mathbf{C}_0^\alpha \, \dot{\gamma}^\alpha = -\sum_{\alpha=1}^{n_0} \sum_{\beta=1}^{n_0} \mathbf{C}_0^\alpha \, g_{\alpha\beta}^{-1} \left(\dot{f}^\beta + \mathbf{C}_0^\beta : \dot{\mathbf{E}} \right). \tag{12.16.6}$$

The substitution into

$$(\dot{\mathbf{T}})^{\mathrm{p}} = \mathbf{\Lambda}_{(1)} : \dot{\mathbf{E}} + (\dot{\mathbf{T}})^{\mathrm{p}} \tag{12.16.7}$$

gives

$$\dot{\mathbf{T}} = \mathbf{\Lambda}_{(1)}^{\mathrm{p}} : \dot{\mathbf{E}} - \sum_{\alpha=1}^{n_0} \sum_{\beta=1}^{n_0} \mathbf{C}_0^\alpha \, g_{\alpha\beta}^{-1} \, \dot{f}^\beta. \tag{12.16.8}$$

The tensor

$$\mathbf{\Lambda}_{(1)}^{\mathrm{p}} = \mathbf{\Lambda}_{(1)} - \sum_{\alpha=1}^{n_0} \sum_{\beta=1}^{n_0} g_{\alpha\beta}^{-1} \left(\mathbf{C}_0^\alpha \otimes \mathbf{C}_0^\beta \right) \tag{12.16.9}$$

is the stiffness tensor of fully plastic response, in which all critical systems are supposed to be active ($\dot{f}^\alpha = 0$ for $\alpha = 1, 2, \ldots, n_0$).

By taking a trace product of (12.16.8) with $\dot{\mathbf{E}}$, we obtain

$$\dot{\mathbf{T}} : \dot{\mathbf{E}} = \dot{\mathbf{E}} : \mathbf{\Lambda}_{(1)}^{\mathrm{p}} : \dot{\mathbf{E}} - \sum_{\alpha=1}^{n_0} \sum_{\beta=1}^{n_0} \left(\mathbf{C}_0^\alpha : \dot{\mathbf{E}} \right) g_{\alpha\beta}^{-1} \, \dot{f}^\beta. \tag{12.16.10}$$

The term involving a double sum on the right-hand side can be expressed, by substituting Eq. (12.16.4) to eliminate $\mathbf{C}_0^\alpha : \dot{\mathbf{E}}$, as

$$\sum_{\alpha=1}^{n_0} \sum_{\beta=1}^{n_0} \left(\mathbf{C}_0^\alpha : \dot{\mathbf{E}} \right) g_{\alpha\beta}^{-1} \, \dot{f}^\beta = -\sum_{\alpha=1}^{n_0} \sum_{\beta=1}^{n_0} g_{\alpha\beta}^{-1} \, \dot{f}^\alpha \dot{f}^\beta + \sum_{\alpha=1}^{n_0} \sum_{\beta=1}^{n_0} \sum_{\nu=1}^{n_0} g_{\alpha\nu} \, g_{\alpha\beta}^{-1} \, \dot{\gamma}^\nu \dot{f}^\beta. \tag{12.16.11}$$

If $g_{\alpha\beta}$ is a symmetric matrix, the sum over α of $g_{\alpha\nu} \, g_{\alpha\beta}^{-1}$ is equal to $\delta_{\nu\beta}$, and the triple sum on the right-hand side of (12.16.11) vanishes, because $\dot{\gamma}^\beta \, \dot{f}^\beta = 0$ for all β ($\dot{f}^\beta$ vanishing on active and $\dot{\gamma}^\beta$ on inactive slip systems). Therefore, Eq. (12.16.10) reduces to

$$\dot{\mathbf{T}} : \dot{\mathbf{E}} = \dot{\mathbf{E}} : \mathbf{\Lambda}_{(1)}^{\mathrm{p}} : \dot{\mathbf{E}} + \sum_{\alpha=1}^{n_0} \sum_{\beta=1}^{n_0} g_{\alpha\beta}^{-1} \, \dot{f}^\alpha \dot{f}^\beta. \tag{12.16.12}$$

Since $g_{\alpha\beta}$ is positive definite, we infer that

$$\dot{\mathbf{T}} : \dot{\mathbf{E}} \geq \dot{\mathbf{E}} : \boldsymbol{\Lambda}^{\mathrm{p}}_{(1)} : \dot{\mathbf{E}}. \tag{12.16.13}$$

The equality holds only if the actual response momentarily takes place with all critical systems active. Alternatively, expressed in terms of increments,

$$\left(\mathrm{d}\mathbf{T} - \boldsymbol{\Lambda}^{\mathrm{p}}_{(1)} : \mathrm{d}\mathbf{E}\right) : \mathrm{d}\mathbf{E} \geq 0, \tag{12.16.14}$$

which establishes a lower bound on the second-order work quantity $\mathrm{d}\mathbf{T} : \mathrm{d}\mathbf{E}$.

12.17. Rigid-Plastic Behavior

In the rigid-plastic idealization,

$$\mathbf{F} = \mathbf{R}^* \cdot \mathbf{F}^{\mathrm{p}}, \tag{12.17.1}$$

where $\mathbf{R}^*$ is the lattice rotation, which carries the lattice vector $\mathbf{s}_0^\alpha$ into $\mathbf{s}^\alpha = \mathbf{R}^* \cdot \mathbf{s}_0^\alpha$. The lattice rate of deformation vanishes ($\mathbf{D}^* = \mathbf{0}$), and the total rate of deformation is solely due to slip,

$$\mathbf{D} = \sum_{\alpha=1}^{n} \mathbf{P}^\alpha \, \dot{\gamma}^\alpha. \tag{12.17.2}$$

The spin tensor can be expressed as

$$\mathbf{W} = \mathbf{W}^* + \sum_{\alpha=1}^{n} \mathbf{Q}^\alpha \, \dot{\gamma}^\alpha. \tag{12.17.3}$$

The lattice spin is

$$\mathbf{W}^* = \dot{\mathbf{R}}^* \cdot \mathbf{R}^{*-1}, \tag{12.17.4}$$

while

$$\mathbf{P}^\alpha + \mathbf{Q}^\alpha = \mathbf{s}^\alpha \otimes \mathbf{m}^\alpha = \mathbf{R}^* \cdot (\mathbf{s}_0^\alpha \otimes \mathbf{m}_0^\alpha) \cdot \mathbf{R}^{*T}. \tag{12.17.5}$$

The rate of the generalized Schmid stress on an active slip system meets the consistency condition

$$\dot{\tau}^\alpha = \mathbf{P}^\alpha : \overset{\bullet}{\boldsymbol{\sigma}} = \sum_{\beta=1}^{n} h_{\alpha\beta} \, \dot{\gamma}^\beta. \tag{12.17.6}$$

It is noted that for the rigid-plastic model of crystal plasticity, the deformation is isochoric ($\det \mathbf{F} = 1$), so that the Kirchhoff and Cauchy stress coincide ($\boldsymbol{\tau} = \boldsymbol{\sigma}$). By substituting Eq. (12.2.14) for $\overset{\bullet}{\boldsymbol{\sigma}}$, there follows

$$\mathbf{P}^\alpha : \overset{\circ}{\boldsymbol{\sigma}} = \sum_{\beta=1}^{n} a_{\alpha\beta} \, \dot{\gamma}^\beta, \tag{12.17.7}$$

where

$$a_{\alpha\beta} = h_{\alpha\beta} - \mathbf{P}^{\alpha} : \left(\mathbf{Q}^{\beta} \cdot \boldsymbol{\sigma} - \boldsymbol{\sigma} \cdot \mathbf{Q}^{\beta}\right). \tag{12.17.8}$$

The slip rates are thus

$$\dot{\gamma}^{\alpha} = \sum_{\beta=1}^{n} a_{\alpha\beta}^{-1} \mathbf{P}^{\beta} : \overset{\circ}{\boldsymbol{\sigma}}, \tag{12.17.9}$$

provided that the inverse matrix $a_{\alpha\beta}^{-1}$ exists (see, also, Khan and Huang, 1995).

Alternative derivation proceeds from

$$\dot{\tau}^{\alpha} = \mathbf{P}_0^{\alpha} : \left(\mathbf{F}^{\mathrm{p}-1} \cdot \dot{\mathbf{T}}^{*} \cdot \mathbf{F}^{\mathrm{p}-T}\right) = \sum_{\beta=1}^{n} h_{\alpha\beta}\, \dot{\gamma}^{\beta}. \tag{12.17.10}$$

By substituting Eq. (12.2.17), we have

$$\mathbf{P}_0^{\alpha} : \dot{\mathbf{T}} = \sum_{\beta=1}^{n} b_{\alpha\beta}\, \dot{\gamma}^{\beta}, \tag{12.17.11}$$

where

$$b_{\alpha\beta} = h_{\alpha\beta} - \mathbf{P}_0^{\alpha} : \left(\mathbf{Z}_0^{\beta} \cdot \mathbf{T} + \mathbf{T} \cdot \mathbf{Z}_0^{\beta\,T}\right). \tag{12.17.12}$$

If this matrix is invertible, the slip rates are

$$\dot{\gamma}^{\alpha} = \sum_{\beta=1}^{n} b_{\alpha\beta}^{-1} \mathbf{P}_0^{\beta} : \dot{\mathbf{T}}. \tag{12.17.13}$$

When the convected derivative of the Kirchhoff stress is used, the slip rates can be expressed as

$$\dot{\gamma}^{\alpha} = \sum_{\beta=1}^{n} b_{\alpha\beta}^{-1} \mathbf{P}^{\beta} : \overset{\triangle}{\boldsymbol{\sigma}}, \tag{12.17.14}$$

with

$$b_{\alpha\beta} = h_{\alpha\beta} - \mathbf{P}^{\alpha} : \left[\left(\mathbf{P}^{\beta} + \mathbf{Q}^{\beta}\right) \cdot \boldsymbol{\sigma} + \boldsymbol{\sigma} \cdot \left(\mathbf{P}^{\beta} - \mathbf{Q}^{\beta}\right)\right]. \tag{12.17.15}$$

It is easily verified that

$$\mathbf{P}_0^{\alpha} : \dot{\mathbf{T}} = \mathbf{P}^{\alpha} : \overset{\triangle}{\boldsymbol{\sigma}}, \tag{12.17.16}$$

and

$$\mathbf{P}_0^{\alpha} : \left(\mathbf{Z}_0^{\beta} \cdot \mathbf{T} + \mathbf{T} \cdot \mathbf{Z}_0^{\beta\,T}\right) = \mathbf{P}^{\alpha} : \left[\left(\mathbf{P}^{\beta} + \mathbf{Q}^{\beta}\right) \cdot \boldsymbol{\sigma} + \boldsymbol{\sigma} \cdot \left(\mathbf{P}^{\beta} - \mathbf{Q}^{\beta}\right)\right]. \tag{12.17.17}$$

Evidently,

$$\overset{\circ}{\boldsymbol{\sigma}} : \mathbf{D} = \sum_{\alpha=1}^{n} \sum_{\beta=1}^{n} a_{\alpha\beta}\, \dot{\gamma}^{\alpha}\, \dot{\gamma}^{\beta}, \tag{12.17.18}$$

$$\overset{\triangle}{\boldsymbol{\sigma}} : \mathbf{D} = \dot{\mathbf{T}} : \dot{\mathbf{E}} = \sum_{\alpha=1}^{n} \sum_{\beta=1}^{n} b_{\alpha\beta}\, \dot{\gamma}^{\alpha}\, \dot{\gamma}^{\beta}. \tag{12.17.19}$$

The sign of these clearly depends on the positive definiteness of the matrices $a_{\alpha\beta}$ and $b_{\alpha\beta}$, respectively. In particular, one can be positive, the other can be negative.

12.18. Geometric Softening

A rigid-plastic model can be conveniently used to illustrate that the lattice rotation can cause an apparent softening of the crystal, even when the slip directions are still hardening. Consider a specimen under uniaxial tension oriented for single slip along the direction $\mathbf{s}_0$, on the slip plane with the normal $\mathbf{m}_0$ (Fig. 12.7). The corresponding rate of deformation and the spin tensors can be expressed from Eqs. (12.17.2), (12.17.3), and (12.17.9) as

$$\mathbf{D} = \frac{1}{a}\,(\mathbf{P} \otimes \mathbf{P}) : \overset{\circ}{\boldsymbol{\sigma}}, \tag{12.18.1}$$

$$\mathbf{W} = \mathbf{W}^* + \frac{1}{a}\,(\mathbf{Q} \otimes \mathbf{P}) : \overset{\circ}{\boldsymbol{\sigma}}, \tag{12.18.2}$$

where

$$a = h - \mathbf{P} : (\mathbf{Q} \cdot \boldsymbol{\sigma} - \boldsymbol{\sigma} \cdot \mathbf{Q}), \tag{12.18.3}$$

and

$$\mathbf{P} = \frac{1}{2}\,(\mathbf{s} \otimes \mathbf{m} + \mathbf{m} \otimes \mathbf{s}), \quad \mathbf{Q} = \frac{1}{2}\,(\mathbf{s} \otimes \mathbf{m} - \mathbf{m} \otimes \mathbf{s}). \tag{12.18.4}$$

Suppose that the specimen is under uniaxial tension in the direction $\mathbf{n}$, which is fixed by the grips of the loading machine. The Cauchy stress tensor is then

$$\boldsymbol{\sigma} = \sigma\, \mathbf{n} \otimes \mathbf{n}, \tag{12.18.5}$$

the material spin is $\mathbf{W} = \mathbf{0}$, and

$$\overset{\circ}{\boldsymbol{\sigma}} = \dot{\sigma}\, \mathbf{n} \otimes \mathbf{n}. \tag{12.18.6}$$

It follows that

$$\mathbf{P} : \overset{\circ}{\boldsymbol{\sigma}} = \dot{\sigma}\,(\mathbf{m} \cdot \mathbf{n})(\mathbf{s} \cdot \mathbf{n}) = \dot{\sigma}\, \cos\phi\, \cos\psi, \tag{12.18.7}$$

where ϕ is the angle between the current slip plane normal $\mathbf{m}$ and the loading direction $\mathbf{n}$, while ψ is the angle between the current slip direction $\mathbf{s}$ and the

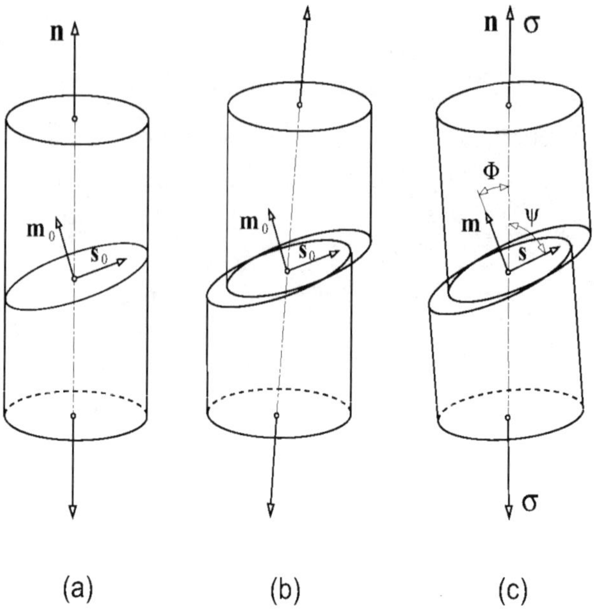

FIGURE 12.7. Single crystal under uniaxial tension oriented for single slip along the slip direction $\mathbf{s}_0$ in the slip plane with the normal $\mathbf{m}_0$; parts (a) and (b). The lattice rotates during deformation so that the slip direction $\mathbf{s}$ in the deformed configuration makes an angle ψ with the longitudinal direction $\mathbf{n}$; part (c). The angle between the slip plane normal $\mathbf{m}$ and the longitudinal direction is ϕ.

loading direction $\mathbf{n}$ (Fig. 12.7). It is easily found that

$$\mathbf{P} : (\mathbf{Q} \cdot \boldsymbol{\sigma} - \boldsymbol{\sigma} \cdot \mathbf{Q}) = \frac{1}{2}\sigma\left[(\mathbf{m}\cdot\mathbf{n})^2 - (\mathbf{s}\cdot\mathbf{n})^2\right] = \frac{1}{2}\sigma\left(\cos^2\phi - \cos^2\psi\right). \tag{12.18.8}$$

Therefore, upon substitution into Eq. (12.18.1),

$$\mathbf{D} = \frac{\dot{\sigma}\cos\phi\cos\psi}{h - \frac{1}{2}\left(\cos^2\phi - \cos^2\psi\right)}\,\mathbf{P}. \tag{12.18.9}$$

Denoting by e the longitudinal strain in the direction of the specimen axis $\mathbf{n}$, we can write

$$\dot{e} = \mathbf{n}\cdot\mathbf{D}\cdot\mathbf{n}, \tag{12.18.10}$$

and Eq. (12.18.9) yields

$$\dot{e} = \frac{\dot{\sigma}\cos^2\phi\cos^2\psi}{h - \frac{1}{2}\left(\cos^2\phi - \cos^2\psi\right)}, \tag{12.18.11}$$

i.e.,

$$\dot{\sigma} = \left[\frac{h}{\cos^2\phi \, \cos^2\psi} - \frac{\sigma\,(\cos^2\phi - \cos^2\psi)}{2\,\cos^2\phi\,\cos^2\psi} \right] \dot{e}. \tag{12.18.12}$$

Depending on the current orientation of the active slip system, the modulus in Eq. (12.18.12) can be positive, zero or negative. If the lattice has rotated such that

$$\cos^2\phi - \cos^2\psi > \frac{2h}{\sigma}, \tag{12.18.13}$$

the current modulus is negative, although the slip direction may still be hardening ($h > 0$). The resulting apparent softening is purely geometrical effect, due to rotation of the lattice caused by crystallographic slip, and is referred to as geometric softening. In the derivation it was assumed that the lattice rotation does not activate the slip on another slip system.

The product

$$M = \cos\phi \, \cos\psi \tag{12.18.14}$$

is known as the Schmid factor. The resolved shear stress in the slip direction, due to applied tension σ, is $\tau = M\,\sigma$. Since the rate of work can be expressed as $\dot{w} = \sigma\,\dot{e} = \tau\,\dot{\gamma}$, it follows that the slip rate $\dot{\gamma}$ can be expressed in terms of the longitudinal strain rate $\dot{e}$ as $\dot{\gamma} = \dot{e}/M$. Therefore, larger the Schmid factor M, larger the resolved shear stress on the slip system and smaller the corresponding slip rate.

Since the material spin vanishes in uniaxial tension ($\mathbf{W} = \mathbf{0}$), from Eq. (12.18.2) we obtain an expression for the lattice spin

$$\mathbf{W}^* = -\frac{\dot{\sigma}\,\cos\phi\,\cos\psi}{h - \frac{1}{2}\,(\cos^2\phi - \cos^2\psi)}\,\mathbf{Q}. \tag{12.18.15}$$

An analysis of lattice spin in an elastoplastic crystal under uniaxial tension is presented in the paper by Aravas and Aifantis (1991).

12.19. Minimum Shear and Maximum Work Principle

The only mechanism of deformation in rigid-plastic crystal, within the framework of this chapter, is the simple shearing on active slip systems. Therefore, if the slip rates $\dot{\gamma}^\alpha$ ($\alpha = 1, 2, \ldots, n$) are prescribed, the corresponding rate of deformation is uniquely determined from

$$\mathbf{D} = \sum_{\alpha=1}^{n} \mathbf{P}^\alpha \, \dot{\gamma}^\alpha. \tag{12.19.1}$$

On the other hand, when the components of $\mathbf{D}$ are prescribed, there are n_0 unknown slip rates on n_0 critical systems, and 5 independent equations between them ($\operatorname{tr}\mathbf{D}$ being equal to zero, since slip is an isochoric deformation process). If there are less than five available slip systems (as in hexagonal crystals), a combination of shears cannot be found that produces an arbitrary $\mathbf{D}$. If $n_0 = 5$, there is a unique set of slip rates provided that the determinant of the coefficients is not equal to zero (independent slip systems; e.g., if three slip systems are in the same plane, only two are independent). If $n_0 > 5$, a set of five slip systems can be selected in any one of $C_5^{n_0}$ ways; the corresponding slip rates can be found for those sets that consist of five independent slip systems (see Section 14.2). Of course, it may also be possible to find combinations of six or more slip rates that give rise to a prescribed $\mathbf{D}$.

Selection of the physically operative combination is greatly facilitated by the following Taylor's minimum shear principle: among all geometrically possible combinations of shears that can produce a prescribed strain, physically possible (operative) combination renders the sum of the absolute values of shears the least. If more than one combination is physically possible, the sums of the corresponding absolute values of shears are equal. The principle was proposed by Taylor (1938), and was proved by Bishop and Hill (1951). Indeed, let n slip rates $\dot{\gamma}^\alpha$ be actually operating set producing a prescribed $\mathbf{D}$, at the given state of stress $\boldsymbol{\sigma}$, i.e.,

$$\sum_{\alpha=1}^{n} \mathbf{P}^\alpha \dot{\gamma}^\alpha = \mathbf{D}, \quad |\, \tau^\alpha \,| = |\, \mathbf{P}^\alpha : \boldsymbol{\sigma} \,| = \tau_{\mathrm{cr}}^\alpha \quad (\alpha = 1, 2, \ldots, n). \qquad (12.19.2)$$

Here, for convenience, the slip in the opposite sense along the same slip direction is not considered as an independent slip system, so that $\dot{\gamma}^\alpha < 0$ when $\tau^\alpha < 0$. The Bauschinger effect along the slip direction is assumed to be absent in Eq. (12.19.2). Further, let $\bar{n}$ slip rates $\dot{\bar{\gamma}}^\alpha$ be geometrically possible, but not physically operating, set of shears associated with a prescribed $\mathbf{D}$, i.e.,

$$\sum_{\alpha=1}^{\bar{n}} \bar{\mathbf{P}}^\alpha \dot{\bar{\gamma}}^\alpha = \mathbf{D}, \quad |\, \bar{\tau}^\alpha \,| = |\, \bar{\mathbf{P}}^\alpha : \boldsymbol{\sigma} \,| \leq \bar{\tau}_{\mathrm{cr}}^\alpha \quad (\alpha = 1, 2, \ldots, \bar{n}). \qquad (12.19.3)$$

Then, we can write

$$\boldsymbol{\sigma} : \mathbf{D} = \sum_{\alpha=1}^{n} \mathbf{P}^\alpha : \boldsymbol{\sigma}\, \dot{\gamma}^\alpha = \sum_{\alpha=1}^{\bar{n}} \bar{\mathbf{P}}^\alpha : \boldsymbol{\sigma}\, \dot{\bar{\gamma}}^\alpha, \tag{12.19.4}$$

or,

$$\sum_{\alpha=1}^{n} \tau^\alpha\, \dot{\gamma}^\alpha = \sum_{\alpha=1}^{\bar{n}} \bar{\tau}^\alpha\, \dot{\bar{\gamma}}^\alpha. \tag{12.19.5}$$

Furthermore,

$$\sum_{\alpha=1}^{n} \tau^\alpha\, \dot{\gamma}^\alpha = \sum_{\alpha=1}^{n} |\, \tau^\alpha \,|\,|\, \dot{\gamma}^\alpha \,| = \sum_{\alpha=1}^{n} \tau_{\mathrm{cr}}^\alpha \,|\, \dot{\gamma}^\alpha \,|, \tag{12.19.6}$$

$$\sum_{\alpha=1}^{\bar{n}} \bar{\tau}^\alpha\, \dot{\bar{\gamma}}^\alpha = \sum_{\alpha=1}^{\bar{n}} |\, \bar{\tau}^\alpha \,|\,|\, \dot{\bar{\gamma}}^\alpha \,| \leq \sum_{\alpha=1}^{\bar{n}} \bar{\tau}_{\mathrm{cr}}^\alpha \,|\, \dot{\bar{\gamma}}^\alpha \,|\,. \tag{12.19.7}$$

Consequently, upon combination with Eq. (12.19.5),

$$\sum_{\alpha=1}^{n} \tau_{\mathrm{cr}}^\alpha \,|\, \dot{\gamma}^\alpha \,| \leq \sum_{\alpha=1}^{\bar{n}} \bar{\tau}_{\mathrm{cr}}^\alpha \,|\, \dot{\bar{\gamma}}^\alpha \,|\,. \tag{12.19.8}$$

This means that the work on physically operating slip rates is not greater than the work on the slip rates that are only geometrically possible. If the hardening on all slip systems is the same (isotropic hardening), the critical resolved shear stresses at a given stage of deformation are equal on all slip systems (regardless of how much slip actually occurred on individual slip systems), and (12.19.8) reduces to

$$\sum_{\alpha=1}^{n} |\, \dot{\gamma}^\alpha \,| \leq \sum_{\alpha=1}^{\bar{n}} |\, \dot{\bar{\gamma}}^\alpha \,|\,. \tag{12.19.9}$$

This is the minimum shear principle. Among all geometrically admissible sets of slip rates, the sum of absolute values of the slip rates is least for the physically operative set of slip rates.

Bishop and Hill (1951) also formulated and proved the maximum work principle for a rigid-plastic single crystal. If $\mathbf{D}$ is the rate of deformation that takes place at the state of stress $\boldsymbol{\sigma}$, then for any other state of stress $\boldsymbol{\sigma}_*$, which does not violate the yield condition on any slip system, the difference of the corresponding rates of work per unit volume is

$$(\boldsymbol{\sigma} - \boldsymbol{\sigma}_*) : \mathbf{D} = \sum_{\alpha=1}^{n} (\tau^\alpha - \tau_*^\alpha)\, \dot{\gamma}^\alpha. \tag{12.19.10}$$

The summation extends over all slip rates of a set giving rise to the rate of deformation tensor $\mathbf{D}$ at the state of Cauchy stress $\boldsymbol{\sigma}$. If $\dot{\gamma}^\alpha > 0$ in the direction α, then

$$\tau^\alpha - \tau_*^\alpha = \tau_{\text{cr+}}^\alpha - \tau_*^\alpha \geq 0. \tag{12.19.11}$$

If $\dot{\gamma}^\alpha < 0$ in the direction α, then

$$\tau^\alpha - \tau_*^\alpha = -\tau_{\text{cr-}}^\alpha - \tau_*^\alpha \leq 0, \tag{12.19.12}$$

since by hypothesis

$$-\tau_{\text{cr-}}^\alpha \leq \tau_*^\alpha \leq \tau_{\text{cr+}}^\alpha\,. \tag{12.19.13}$$

The microscopic Bauschinger effect is here allowed, so that the critical shear stresses in opposite directions may be different ($\tau_{\text{cr-}}^\alpha \neq \tau_{\text{cr+}}^\alpha$). All products in the sum on the right-hand side of Eq. (12.19.10) are thus positive or zero, and so

$$(\boldsymbol{\sigma} - \boldsymbol{\sigma}_*) : \mathbf{D} \geq 0, \tag{12.19.14}$$

which is the principle of maximum work. The equality in (12.19.14) holds only when $\tau^\alpha = \tau_*^\alpha$ for all active slip systems. If there are at least 5 of these, the stress states $\boldsymbol{\sigma}$ and $\boldsymbol{\sigma}_*$ can only differ by a hydrostatic stress.

12.20. Modeling of Latent Hardening

A diagonal term $h_{\alpha\alpha}$ of the hardening matrix represents the rate of self-hardening, i.e., the rate of hardening on the slip system α due to slip on that system itself. An off-diagonal term $h_{\alpha\beta}$ represents the rate of latent or cross hardening, i.e., the rate of hardening on the slip system α due to slip on the system β. It has been observed that the ratio of latent hardening to self-hardening is frequently in the range between 1 and 1.4 (Kocks, 1970; Asaro, 1983a; Bassani, 1990; Bassani and Wu, 1991). For slip systems within the same plane (coplanar systems), the ratio is closer to 1. Larger values are observed for systems on intersecting slip planes. Estimates of latent hardening are most commonly done by the measurements of the lattice rotation "overshoot". When the single crystal is deformed by tension in a single slip mode, the lattice rotates relative to the loading axis, so that the slip direction rotates toward the loading axis. After a finite amount of slip on the primary system, a second (conjugate) slip system becomes critical. If the latent hardening on the conjugate slip system is larger than the self-hardening

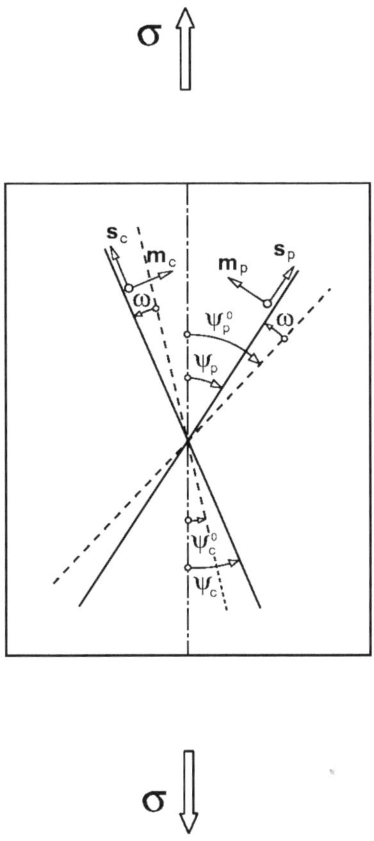

FIGURE 12.8. Plane model of a single crystal. Initially, the crystal deforms by single slip on the primary slip system $(\mathbf{s}_p, \mathbf{m}_p)$. As the lattice rotates through the angle ω, the conjugate slip system $(\mathbf{s}_c, \mathbf{m}_c)$ becomes critical, which results in double slip of the crystal.

on the primary system, the lattice rotation overshoots the symmetry position, at which the two slip directions are symmetric about the tensile axis, until the resolved shear stress on the conjugate system exceeds that on the primary system, and the conjugate slip begins. This is schematically illustrated in Fig. 12.8. Other methods for estimating latent hardening are also available. An optical micrograph showing the primary and conjugate slip is shown in Fig. 12.9. The primary slip system is designated by $(\mathbf{s}_p, \mathbf{m}_p)$, and the conjugate slip system by $(\mathbf{s}_c, \mathbf{m}_c)$.

The simplest model of latent hardening is associated with a symmetric matrix of the hardening rates

$$h_{\alpha\beta} = h_1 + (h - h_1)\,\delta_{\alpha\beta}, \tag{12.20.1}$$

where h is the rate of self-hardening, and h_1 is the rate of latent hardening ($1 \leq h_1/h \leq 1.4$). For $h_1 = h$, Taylor's (1938) isotropic hardening is obtained, i.e., $h_{\alpha\beta} = h$ for all slip systems, momentarily active or not. However, a symmetric form of the hardening matrix $h_{\alpha\beta}$ in Eq. (12.20.1) implies, from Eq. (12.9.27), a nonsymmetric matrix of the moduli $g_{\alpha\beta}$, and, thus, a nonsymmetric elastoplastic stiffness tensors in the constitutive equations (12.9.33)–(12.9.35). The lack of reciprocal symmetry of these tensors prevents the variational formulation of the boundary value problem, and makes analytical study of elastoplastic uniqueness and bifurcation problems more difficult. For localization in single crystals, see Asaro and Rice (1979), Pierce, Asaro, and Needleman (1982), Pierce (1983), and Perzyna and Korbel (1996).

To achieve the symmetry of $g_{\alpha\beta}$, Havner and Shalaby (1977,1978) proposed that

$$h_{\alpha\beta} = h + \mathbf{P}^\alpha : \left(\mathbf{Q}^\beta \cdot \boldsymbol{\tau} - \boldsymbol{\tau} \cdot \mathbf{Q}^\beta\right) = h + 2\left(\mathbf{P}^\alpha \cdot \mathbf{Q}^\beta\right) : \boldsymbol{\tau}, \tag{12.20.2}$$

since then

$$g_{\alpha\beta} = h + \mathbf{P}^\alpha : \boldsymbol{\mathcal{L}}_{(0)} : \mathbf{P}^\beta + 2\left(\mathbf{P}^\alpha \cdot \mathbf{Q}^\beta + \mathbf{P}^\beta \cdot \mathbf{Q}^\alpha\right) : \boldsymbol{\tau} \tag{12.20.3}$$

becomes symmetric ($g_{\alpha\beta} = g_{\beta\alpha}$). The hardening law of Eq. (12.9.1) can in this case be expressed, with the help of (12.1.13), as

$$\dot{\tau}^\alpha_{\rm cr} = h \sum_{\beta=1}^{n} \dot{\gamma}^\beta + 2\,\mathbf{P}^\alpha : \left[(\mathbf{W} - \mathbf{W}^*) \cdot \boldsymbol{\tau}\right]. \tag{12.20.4}$$

If the loading and orientation of the crystal are such that the lattice spin is equal to the material spin, the above reduces to Taylor's hardening model (Havner, 1992). Pierce, Asaro, and Needleman (1982) observed that the latent hardening rates from Eq. (12.20.2) are too high, and proposed instead

$$h_{\alpha\beta} = h_1 + (h - h_1)\,\delta_{\alpha\beta} + \left(\mathbf{P}^\alpha \cdot \mathbf{Q}^\beta - \mathbf{P}^\beta \cdot \mathbf{Q}^\alpha\right) : \boldsymbol{\tau}, \tag{12.20.5}$$

which gives

$$g_{\alpha\beta} = h_1 + (h - h_1)\,\delta_{\alpha\beta} + \mathbf{P}^\alpha : \boldsymbol{\mathcal{L}}_{(0)} : \mathbf{P}^\beta + \left(\mathbf{P}^\alpha \cdot \mathbf{Q}^\beta + \mathbf{P}^\beta \cdot \mathbf{Q}^\alpha\right) : \boldsymbol{\tau}. \tag{12.20.6}$$

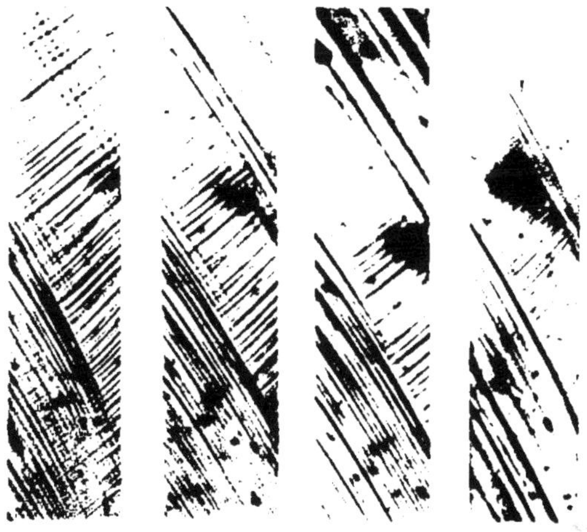

FIGURE 12.9. Optical micrographs of α-brass crystals deformed in tension, showing the primary and conjugate slips (from Asaro, 1983b; with permission from Academic Press).

Still, the predicted rates of latent hardening were above experimentally observed values.

Other models of latent hardening were also suggested in the literature. A two-parameter modification of Taylor's model was proposed by Nakada and Keh (1966). According to this model,

$$\dot{\tau}_{\rm cr}^{\alpha} = h_1 \sum_{i=1}^{n_1} \dot{\gamma}^i + h_2 \sum_{j=1}^{n_2} \dot{\gamma}^j, \tag{12.20.7}$$

where

$$\mathbf{m}^i = \mathbf{m}^\alpha, \quad \mathbf{m}^j \neq \mathbf{m}^\alpha, \quad h_2 > h_1 > 0. \tag{12.20.8}$$

The rate of hardening on the slip system (α) and all coplanar systems is h_1, while h_2 is the rate of hardening on other slip systems. The sum $n_1 + n_2 = n$ is the number of all active slip systems. Further analysis and the study of the response of f.c.c. and b.c.c. crystals based on the considered hardening models can be found in Havner (1985, 1992). For example, Havner (1992) demonstrated that, under infinitesimal lattice strain, all hardening models here considered are in accord with the basic inequality $\mathrm{d^pT} : \mathrm{dE} < 0$, and they all give rise to positive definite matrix $g_{\alpha\beta}$. See also Weng (1987).

12.21. Rate-Dependent Models

One of the difficulties with the rate-independent crystal plasticity is that the slip rates $\dot{\gamma}^\alpha$ may not be uniquely determined in terms of the prescribed deformation or stress rates. When the deformation rates are prescribed, uniqueness is not guaranteed when more than five linearly independent slip systems are potentially active. When the stress rates are prescribed, uniqueness is not guaranteed even with fewer than five active systems, particularly when a full range of realistic experimental data for strain hardening behavior is used (Pierce, Asaro, and Needleman, 1983). This has stimulated introduction of the rate-dependent models of crystal plasticity. The slip rates in the constitutive equations from Section 12.4, such as

$$\frac{\mathrm{d}\mathbf{T}}{dt} = \mathbf{\Lambda}_{(1)} : \frac{\mathrm{d}\mathbf{E}}{dt} - \sum_{\alpha=1}^{N} \mathbf{C}_0^\alpha \frac{\mathrm{d}\gamma^\alpha}{dt}, \tag{12.21.1}$$

are prescribed directly and uniquely in terms of the current stress state and the internal structure of the material. The derivatives in Eq. (12.21.1) are with respect to physical time t. In this formulation, there is no explicit yielding, or division of slip systems into active and inactive. All slip systems are active: if the resolved shear stress on a slip system is nonzero, the plastic shearing occurs.

An often utilized expression for the slip rates is the power-law of the type used by Hutchinson (1976) for polycrystalline creep, and by Pan and Rice (1983) to describe the influence of the rate sensitivity on the yield vertex behavior in single crystals. This is

$$\dot{\gamma}^\alpha = \dot{\gamma}_0^\alpha \operatorname{sgn}(\tau^\alpha) \left| \frac{\tau^\alpha}{\tau_{\mathrm{r}}^\alpha} \right|^{1/m}. \tag{12.21.2}$$

The resolved shear stress is $\tau^\alpha = \mathbf{s}^\alpha \cdot \boldsymbol{\tau} \cdot \mathbf{m}^\alpha$. The current strain-hardened state of slip systems is represented by the hardness parameters τ_{r}^α, $\dot{\gamma}_0^\alpha$ is the reference rate of shearing (which can be same for all slip systems), m characterizes the material rate sensitivity, and sgn is the sign function. The rate-independent response is achieved in the limit $m \to 0$. For sufficiently small values of m (say, $m \leq 0.02$), the slip rates $\dot{\gamma}^\alpha$ are exceedingly small when $\tau^\alpha < \tau_{\mathrm{r}}^\alpha$, so that "yielding" would appear to occur abruptly as τ^α

approaches the current value of $\tau_{\mathrm{r}}^{\alpha}$. The hardening parameters $\tau_{\mathrm{r}}^{\alpha}$ are positive. Their initial values are $\tau_{\mathrm{r}\,0}^{\alpha}$, and they change according to evolution equations

$$\dot{\tau}_{\mathrm{r}}^{\alpha} = \sum_{\beta=1}^{N} h_{\alpha\beta}\,|\dot{\gamma}^{\beta}|. \tag{12.21.3}$$

The slip hardening moduli, including self and latent hardening, are $h_{\alpha\beta}$. Since all slip systems are potentially active in the rate-dependent formulation, it is more convenient to consider $(\mathbf{s}^{\alpha}, \mathbf{m}^{\beta})$ and $(-\mathbf{s}^{\alpha}, \mathbf{m}^{\beta})$ as the same slip system, i.e., to permit $\dot{\gamma}^{\alpha}$ to be negative if the corresponding τ^{α} is negative. This sign convention is embodied in Eqs. (12.21.2) and (12.21.3). For example, the total number of slip systems in f.c.c. crystals is then $N = 12$. In practice, the functions $\tau_{\mathrm{r}}^{\alpha}$ would be fit to τ vs. γ curves, obtained from the crystal deformed in the single slip modes, and with latent hardening estimated from the measurements of the lattice rotation overshoots (Asaro, 1983 a). If all self-hardening moduli are equal to h and all latent hardening moduli are equal to h_1, we can write

$$h_{\alpha\beta} = h_1 + (h - h_1)\,\delta_{\alpha\beta}. \tag{12.21.4}$$

In their analysis of localization of deformation in rate-dependent single crystals subject to tensile loading, Pierce, Asaro, and Needleman (1983) used the following expression for the change of the self-hardening modulus during the slip,

$$h = h(\gamma) = h_0 \operatorname{sech}^2 \left| \frac{h_0 \gamma}{\tau_{\mathrm{s}} - \tau_0} \right|. \tag{12.21.5}$$

The initial hardening rate is h_0, the initial yield stress is τ_0, and γ is the cumulative shear strain on all slip systems,

$$\gamma = \sum_{\alpha=1}^{N} |\gamma^{\alpha}|. \tag{12.21.6}$$

The hardening rule (12.21.5) describes the material that saturates at large strains, as the flow stress approaches τ_{s}. The latent hardening modulus is taken to be $h_1 = q\,h$, where q is in the range $1 \leq q \leq 1.4$.

A described rate-dependent model of crystal plasticity allows an extension of the rate-independent calculations for various problems to much broader range of the material strain hardening properties and crystal geometry. For example, Pierce, Asaro, and Needleman (1983) found that even a

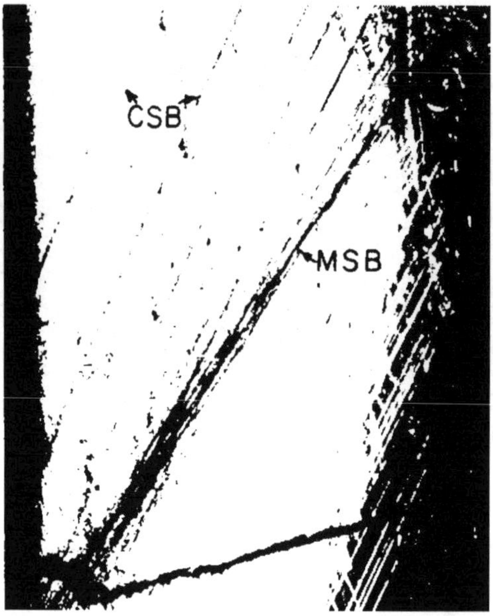

FIGURE 12.10. Formation of the macroscopic shear band (MSB) within clusters of coarse slip bands (CSB) in an aluminum-copper alloy crystal (from Chang and Asaro, 1981; with permission from Elsevier Science).

very moderate rate sensitivity had a noticeable influence on the development of localized deformation modes. Additional analysis is given by Zarka (1973), Canova, Molinari, Fressengeas, and Kocks (1988), and Teodosiu (1997). A micrograph of the coarse slip band and macroscopic shear band from experimental study of localized flow in single crystals by Chang and Asaro (1981) is shown in Fig. 12.10.

12.22. Flow Potential and Normality Rule

To make a contact with the rate-dependent analysis presented in Section 8.4, we derive the flow potential for the plastic part of the strain rate, corresponding to the slip rates of Eq. (12.21.2). To that goal, we first rewrite Eq. (12.21.1) as

$$\frac{\mathrm{d}\mathbf{E}}{\mathrm{d}t} = \mathbf{M}_{(1)} : \frac{\mathrm{d}\mathbf{T}}{\mathrm{d}t} + \sum_{\alpha=1}^{N} \mathbf{G}_0^\alpha \frac{\mathrm{d}\gamma^\alpha}{\mathrm{d}t}, \tag{12.22.1}$$

where $\mathbf{G}_0^\alpha = \mathbf{M}_{(1)} : \mathbf{C}_0^\alpha$ and N is the number of all available slip systems. The plastic contribution to the strain rate is

$$\frac{\mathrm{d}^{\mathrm{p}}\mathbf{E}}{\mathrm{d}t} = \sum_{\alpha=1}^{N} \mathbf{G}_0^\alpha \, \frac{\mathrm{d}\gamma^\alpha}{\mathrm{d}t} \,. \tag{12.22.2}$$

The multiplication with an instantaneously applied stress increment $\delta\mathbf{T}$, which would give rise to purely elastic strain increment $\delta\mathbf{E}$, yields

$$\delta\mathbf{T} : \frac{\mathrm{d}^{\mathrm{p}}\mathbf{E}}{\mathrm{d}t} = \sum_{\alpha=1}^{N} \left(\mathbf{G}_0^\alpha : \delta\mathbf{T}\right) \frac{\mathrm{d}\gamma^\alpha}{\mathrm{d}t} \,. \tag{12.22.3}$$

On the other hand, from Eq. (12.9.13),

$$\delta\tau^\alpha = \mathbf{C}_0^\alpha : \left(\mathbf{F}^{\mathrm{p}T} \cdot \mathrm{d}\mathbf{E}^* \cdot \mathbf{F}^{\mathrm{p}}\right) = \mathbf{C}_0^\alpha : \delta\mathbf{E} = \mathbf{G}_0^\alpha : \delta\mathbf{T}, \tag{12.22.4}$$

since $\delta\mathbf{E} = \mathbf{M}_{(1)} : \delta\mathbf{T}$. The substitution into Eq. (12.22.3) gives

$$\delta\mathbf{T} : \frac{\mathrm{d}^{\mathrm{p}}\mathbf{E}}{\mathrm{d}t} = \sum_{\alpha=1}^{N} \delta\tau^\alpha \, \frac{\mathrm{d}\gamma^\alpha}{\mathrm{d}t} \,. \tag{12.22.5}$$

The slip rates in Eq. (12.21.2) are prescribed as functions of the resolved shear stress τ^α and the hardness parameter τ_{r}^α. This implies that

$$\delta\tau^\alpha \, \frac{\mathrm{d}\gamma^\alpha}{\mathrm{d}t} = \delta\omega^\alpha \left(\tau^\alpha, \tau_{\mathrm{r}}^\alpha\right), \tag{12.22.6}$$

and

$$\frac{\mathrm{d}\gamma^\alpha}{\mathrm{d}t} = \frac{\partial\omega^\alpha}{\partial\tau^\alpha} \,. \tag{12.22.7}$$

Here,

$$\omega^\alpha = \frac{m}{m+1} \, \dot{\gamma}_0^\alpha \, \tau_{\mathrm{r}}^\alpha \left| \frac{\tau^\alpha}{\tau_{\mathrm{r}}^\alpha} \right|^{\frac{m+1}{m}} \tag{12.22.8}$$

is a scalar flow potential for the slip system α. Consequently, Eq. (12.22.5) becomes

$$\delta\mathbf{T} : \frac{\mathrm{d}^{\mathrm{p}}\mathbf{E}}{\mathrm{d}t} = \sum_{\alpha=1}^{N} \delta\omega^\alpha \left(\tau^\alpha, \tau_{\mathrm{r}}^\alpha\right) = \delta\Omega \left(\mathbf{T}, \tau_{\mathrm{r}}^1, \tau_{\mathrm{r}}^2, \ldots, \tau_{\mathrm{r}}^N\right). \tag{12.22.9}$$

This establishes the normality rule

$$\frac{\mathrm{d}^{\mathrm{p}}\mathbf{E}}{\mathrm{d}t} = \frac{\partial\Omega}{\partial\mathbf{T}} \,. \tag{12.22.10}$$

The overall (macroscopic) flow potential for the plastic part of strain rate is

$$\Omega = \sum_{\alpha=1}^{N} \omega^\alpha = \sum_{\alpha=1}^{N} \frac{m}{m+1} \, \dot{\gamma}_0^\alpha \, \tau_{\mathrm{r}}^\alpha \left| \frac{\tau^\alpha}{\tau_{\mathrm{r}}^\alpha} \right|^{\frac{m+1}{m}}, \quad \tau^\alpha = \mathbf{P}_0^\alpha : \mathbf{T}. \tag{12.22.11}$$

References

Aravas, N. and Aifantis, E. C. (1991), On the geometry of slip and spin in finite plasticity deformation, *Int. J. Plasticity*, Vol. 7, pp. 141–160.

Asaro, R. J. (1983a), Crystal plasticity, *J. Appl. Mech.*, Vol. 50, pp. 921–934.

Asaro, R. J. (1983b), Micromechanics of crystals and polycrystals, *Adv. Appl. Mech.*, Vol. 23, pp. 1–115.

Asaro, R. J. and Rice, J. R. (1977), Strain localization in ductile single crystals, *J. Mech. Phys. Solids*, Vol. 25, pp. 309–338.

Bassani, J. L. (1990), Single crystal hardening, *Appl. Mech. Rev.*, Vol. 43, No. 5, Part 2, pp. S320–S327.

Bassani, J. L. (1993), Plastic flow of crystals, *Adv. Appl. Mech.*, Vol. 30, pp. 191–258.

Bassani, J. L. and Wu, T. Y. (1991), Latent hardening in single crystals II. Analytical characterization and predictions, *Proc. Roy. Soc. Lond. A*, Vol. 435, pp. 21–41.

Bishop, J. F. W. and Hill, R. (1951), A theory of plastic distortion of a polycrystalline aggregate under combined stresses, *Phil. Mag.*, Vol. 42, pp. 414–427.

Brünig, M. and Obrecht, H. (1998), Finite elastic-plastic deformation behaviour of crystalline solids based on a non-associated macroscopic flow rule, *Int. J. Plasticity*, Vol. 14, pp. 1189–1208.

Canova, G. R., Molinari, A., Fressengeas, C., and Kocks, U. F. (1988), The effects of rate sensitivity on slip system activity and lattice rotation, *Acta Metall.*, Vol. 36, pp. 1961–1970.

Chang, Y. W. and Asaro, R. J. (1981), An experimental study of shear localization in aluminum-copper single crystals, *Acta Metall.*, Vol. 29, pp. 241–257.

Dao, M. and Asaro, R. J. (1996), Localized deformation modes and non-Schmid effects in crystalline solids. Part I. Critical conditions for localization, *Mech. Mater.*, Vol. 23, pp. 71–102.

Franciosi, P. and Zaoui, A. (1991), Crystal hardening and the issues of uniqueness, *Int. J. Plasticity*, Vol. 7, pp. 295–311.

Gambin, W. (1992), Refined analysis of elastic-plastic crystals, *Int. J. Solids Struct.*, Vol. 29, pp. 2013–2021.

Gurtin, M. E. (2000), On the plasticity of single crystals: free energy, microforces, plastic-strain gradients, *J. Mech. Phys. Solids*, Vol. 48, pp. 989–1036.

Havner, K. S. (1973), On the mechanics of crystalline solids, *J. Mech. Phys. Solids*, Vol. 21, pp. 383–394.

Havner, K. S. (1982), The theory of finite plastic deformation of crystalline solids, in *Mechanics of Solids – The Rodney Hill 60th Anniversary Volume*, eds. H. G. Hopkins and M. J. Sewell, pp. 265–302, Pergamon Press, Oxford.

Havner, K. S. (1985), Comparisons of crystal hardening laws in multiple slip, *Int. J. Plasticity*, Vol. 1, 111–124.

Havner, K. S. (1992), *Finite Plastic Deformation of Crystalline Solids*, Cambridge University Press, Cambridge.

Havner, K. S. and Shalaby, A. H. (1977), A simple mathematical theory of finite distortional latent hardening in single crystals, *Proc. Roy. Soc. Lond. A*, Vol. 358, pp. 47–70.

Havner, K. S. and Shalaby, A. H. (1978), Further investigation of a new hardening law in crystal plasticity, *J. Appl. Mech.*, Vol. 45, pp. 500–506.

Hill, R. (1966), Generalized constitutive relations for incremental deformation of metal crystals by multislip, *J. Mech. Phys. Solids*, Vol. 14, pp. 95–102.

Hill, R. (1972), On constitutive macro-variables for heterogeneous solids at finite strain, *Proc. Roy. Soc. Lond. A*, Vol. 326, pp. 131–147.

Hill, R. (1978), Aspects of invariance in solid mechanics, *Adv. Appl. Mech.*, Vol. 18, pp. 1–75.

Hill, R. (1984), On macroscopic effects of heterogeneity in elastoplastic media at finite strain, *Math. Proc. Camb. Phil. Soc.*, Vol. 95, pp. 481–494.

Hill, R. and Havner, K. S. (1982), Perspectives in the mechanics of elastoplastic crystals, *J. Mech. Phys. Solids*, Vol. 30, pp. 5–22.

Hill, R. and Rice, J. R. (1972), Constitutive analysis of elastic-plastic crystals at arbitrary strain, *J. Mech. Phys. Solids*, Vol. 20, pp. 401–413.

Hutchinson, J. W. (1976), Bounds and self-consistent estimates for creep of polycrystalline materials, *Proc. Roy. Soc. Lond. A*, Vol. 348, pp. 101–127.

Khan, A. S. and Huang, S. (1995), *Continuum Theory of Plasticity*, John Wiley & Sons, Inc., New York.

Kocks, U. F. (1970), The relation between polycrystalline deformation and single-crystal deformation, *Metall. Trans.*, Vol. 1, pp. 1121–1142.

Kratochvil, J. (1971), Finite-strain theory of crystalline elastic-inelastic materials, *J. Appl. Phys.*, Vol. 42, pp. 1104–1108.

Lubarda, V. A. (1999), On the partition of the rate of deformation in crystal plasticity, *Int. J. Plasticity*, Vol. 15, pp. 721–736.

Mandel, J. (1974), Thermodynamics and plasticity, in *Foundations of Continuum Thermodynamics*, eds. J. J. D. Domingos, M. N. R. Nina, and J. H. Whitelaw, pp. 283–311, McMillan Publ., London.

Nakada, Y. and Keh, A. S. (1966), Latent hardening in iron single crystals, *Acta Metall.*, Vol. 14, pp. 961–973.

Pan, J. and Rice, J. R. (1983), Rate sensitivity of plastic flow and implications for yield-surface vertices, *Int. J. Solids Struct.*, Vol. 19, pp. 973–987.

Perzyna, P. and Korbel, K. (1996), Analysis of the influence of the substructure of a crystal on shear band localization phenomena of plastic deformation, *Mech. Mater.*, Vol. 24, pp. 141–158.

Pierce, D. (1983), Shear band bifurcations in ductile single crystals, *J. Mech. Phys. Solids*, Vol. 31, pp. 133–153.

Pierce, D., Asaro, R. J., and Needleman, A. (1982), An analysis of nonuniform and localized deformation in ductile single crystals, *Acta Metall.*, Vol. 30, pp. 1087–1119.

Pierce, D., Asaro, R. J., and Needleman, A. (1983), Material rate dependence and localized deformation in crystalline solids, *Acta Metall.*, Vol. 31, pp. 1951–1976.

Qin, Q. and Bassani, J. L. (1992a), Non-Schmid yield behavior in single crystals, *J. Mech. Phys. Solids*, Vol. 40, pp. 813–833.

Qin, Q. and Bassani, J. L. (1992b), Non-associated plastic flow in single crystals, *J. Mech. Phys. Solids*, Vol. 40, pp. 835–862.

Rice, J. R. (1971), Inelastic constitutive relations for solids: An internal variable theory and its application to metal plasticity, *J. Mech. Phys. Solids*, Vol. 19, pp. 433–455.

Sanders, J. L. (1955), Plastic stress-strain relations based on infinitely many plane loading surfaces, in *Proc. 2nd U.S. Nat. Congr. of Appl. Mech.*, ed. P. M. Naghdi, pp. 455–460, ASME, New York.

Sawkill, J. and Honeycombe, R. W. K. (1954), Strain hardening in face-centred cubic metal crystals, *Acta Metall.*, Vol. 2, pp. 854–864.

Schmid, E. and Boas, W. (1968), *Plasticity of Crystals*, Chapman and Hall, London.

Sewell, M. J. (1972), A survey of plastic buckling, in *Stability*, ed. H. Leipholz, pp. 85–197, University of Waterloo Press, Ontario.

Taylor, G. I. (1938), Plastic strain in metals, *J. Inst. Metals*, Vol. 62, pp. 307–324.

Teodosiu, C. (1997), Dislocation modelling of crystalline plasticity, in *Large Plastic Deformation of Crystalline Aggregates*, ed. C. Teodosiu, pp. 21–80, Springer-Verlag, Wien.

Weng, G. J. (1987), Anisotropic hardening in single crystals and the plasticity of polycrystals, *Int. J. Plasticity*, Vol. 3, pp. 315–339.

Zarka, J. (1973), Etude du comportement des monocristaux métalliques. Application à la traction du monocristal c.f.c., *J. Mécanique*, Vol. 12, pp. 275–318.

CHAPTER 13

MICRO-TO-MACRO TRANSITION

Some fundamental aspects of the transition in the constitutive description of the material response from microlevel to macrolevel are discussed in this chapter. The analysis is aimed toward the derivation of the constitutive equations for polycrystalline aggregates based on the known constitutive equations for elastoplastic single crystals. The theoretical framework for this study was developed by Bishop and Hill (1951a,b), Hill (1963,1967,1972), Mandel (1966), Bui (1970), Rice (1970,1971,1975), Hill and Rice (1973), Havner (1973,1974), and others. The presentation in this chapter follows the large deformation formulation of Hill (1984,1985). The representative macroelement is defined, and the macroscopic measures of stress and strain, and their rates, are introduced. The corresponding elastoplastic moduli and pseudomoduli tensors, the macroscopic normality and the macroscopic plastic potentials are then discussed.

13.1. Representative Macroelement

A polycrystalline aggregate is considered to be macroscopically homogeneous by assuming that local microscopic heterogeneities (due to different orientation and state of hardening of individual crystal grains) are distributed in such a way that the material elements beyond some minimum scale have essentially the same overall macroscopic properties. This minimum scale defines the size of the representative macroelement or representative cell (Fig. 13.1). The representative macroelement can be viewed as a material point in the continuum mechanics of macroscopic aggregate behavior. To be statistically representative of the local properties of its microconstituents, the representative macroelement must include a sufficiently large number of microelements (Kröner, 1971; Sanchez-Palencia, 1980; Kunin, 1982). For

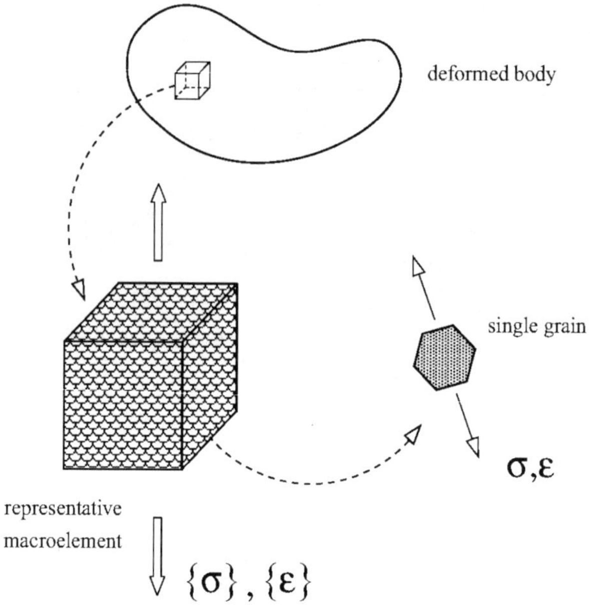

FIGURE 13.1. Representative macroelement of a deformed body consists of a large number of constituting microelements – single grains in the case of a polycrystalline aggregate (schematics adopted from Yang and Lee, 1993).

example, for relatively fine-grained metals, a representative macroelement of volume 1 mm^3 contains a minimum of 1000 crystal grains (Havner, 1992). The concept of the representative macroelement is used in various branches of the mechanics of heterogeneous materials, and is also referred to as the representative volume element (e.g., Mura, 1987; Suquet, 1987; Torquato, 1991; Maugin, 1992; Nemat-Nasser and Hori, 1993; Hori and Nemat-Nasser, 1999). See also Hashin (1964), Willis (1981), Sawicki (1983), Ortiz (1987), and Drugan and Willis (1996). For the linkage of atomistic and continuum models of the material response, the review by Ortiz and Phillips (1999) can be consulted.

13.2. Averages over a Macroelement

Experimental determination of the mechanical behavior of an aggregate is commonly based on the measured loads and displacements over its external surface. Consequently, the macrovariables introduced in the constitutive analysis should be expressible in terms of this surface data alone (Hill, 1972). Let

$$\mathbf{F}(\mathbf{X},t)=\frac{\partial\mathbf{x}}{\partial\mathbf{X}},\quad \det\mathbf{F}>0, \tag{13.2.1}$$

be the deformation gradient at the microlevel of description, associated with a (continuous and piecewise continuously differentiable) microdeformation within a crystalline grain, $\mathbf{x}=\mathbf{x}(\mathbf{X},t)$. The reference position of the particle is $\mathbf{X}$, and its current position at time t (on some quasi-static scale, for rate-independent response) is $\mathbf{x}$. The volume average of the deformation gradient over the reference volume V^0 of the macroelement is

$$\langle\mathbf{F}\rangle=\frac{1}{V^0}\int_{V^0}\mathbf{F}\,\mathrm{d}V^0=\frac{1}{V^0}\int_{S^0}\mathbf{x}\otimes\mathbf{n}^0\,\mathrm{d}S^0, \tag{13.2.2}$$

by the Gauss divergence theorem. The unit outward normal to the bounding surface S^0 of the macroelement volume is $\mathbf{n}^0$. In particular, with $\mathbf{F}=\mathbf{I}$ (unit tensor), Eq. (13.2.2) gives an identity

$$\frac{1}{V^0}\int_{S^0}\mathbf{X}\otimes\mathbf{n}^0\,\mathrm{d}S^0=\mathbf{I}. \tag{13.2.3}$$

The volume average of the rate of deformation gradient,

$$\dot{\mathbf{F}}(\mathbf{X},t)=\frac{\partial\mathbf{v}}{\partial\mathbf{X}},\quad \mathbf{v}=\dot{\mathbf{x}}(\mathbf{X},t), \tag{13.2.4}$$

where $\mathbf{v}$ is the velocity field, is

$$\langle\dot{\mathbf{F}}\rangle=\frac{1}{V^0}\int_{V^0}\dot{\mathbf{F}}\,\mathrm{d}V^0=\frac{1}{V^0}\int_{S^0}\mathbf{v}\otimes\mathbf{n}^0\,\mathrm{d}S^0. \tag{13.2.5}$$

If the current configuration is taken as the reference configuration ($\mathbf{x}=\mathbf{X}$, $\mathbf{F}=\mathbf{I}$, $\dot{\mathbf{F}}=\mathbf{L}=\partial\mathbf{v}/\partial\mathbf{x}$), Eq. (13.2.2) gives

$$\frac{1}{V}\int_{S}\mathbf{x}\otimes\mathbf{n}\,\mathrm{d}S=\mathbf{I}. \tag{13.2.6}$$

The current volume of the deformed macroelement is V, and S is its bounding surface with the unit outward normal $\mathbf{n}$. With this choice of the reference configuration, the volume average of the velocity gradient $\mathbf{L}$ is, from Eq. (13.2.5),

$$\{\mathbf{L}\}=\frac{1}{V}\int_{V}\mathbf{L}\,\mathrm{d}V=\frac{1}{V}\int_{S}\mathbf{v}\otimes\mathbf{n}\,\mathrm{d}S. \tag{13.2.7}$$

Enclosure within $\{\,\}$ brackets is used to indicate that the average is taken over the deformed volume of the macroelement.

Let $\mathbf{P} = \mathbf{P}(\mathbf{X}, t)$ be a nonsymmetric nominal stress field within the macroelement. In the absence of body forces, equations of translational balance are

$$\boldsymbol{\nabla}^0 \cdot \mathbf{P} = \mathbf{0} \quad \text{in} \quad V^0, \qquad \mathbf{n}^0 \cdot \mathbf{P} = \mathbf{p}_n \quad \text{on} \quad S^0. \tag{13.2.8}$$

Here, $\boldsymbol{\nabla}^0 = \partial/\partial\mathbf{X}$ is the gradient operator with respect to reference coordinates, and $\mathbf{p}_n$ is the nominal traction (related to the true traction $\mathbf{t}_n$ by $\mathbf{p}_n\,\mathrm{d}S^0 = \mathbf{t}_n\,\mathrm{d}S$). The rotational balance requires $\mathbf{F}\cdot\mathbf{P} = \boldsymbol{\tau}$ to be a symmetric tensor, where $\boldsymbol{\tau} = (\det\mathbf{F})\boldsymbol{\sigma}$ is the Kirchhoff stress, and $\boldsymbol{\sigma}$ is the true or Cauchy stress.

Equations of the continuing translational balance are

$$\boldsymbol{\nabla}^0 \cdot \dot{\mathbf{P}} = \mathbf{0} \quad \text{in} \quad V^0, \qquad \mathbf{n}^0 \cdot \dot{\mathbf{P}} = \dot{\mathbf{p}}_n \quad \text{on} \quad S^0. \tag{13.2.9}$$

The rates of nominal and true traction are related by

$$\dot{\mathbf{p}}_n\,\mathrm{d}S^0 = \left[\dot{\mathbf{t}}_n + (\operatorname{tr}\mathbf{D} - \mathbf{n}\cdot\mathbf{D}\cdot\mathbf{n})\,\mathbf{t}_n\right]\mathrm{d}S, \tag{13.2.10}$$

as in Eq. (3.8.16). The rate of deformation tensor is $\mathbf{D}$. By differentiating $\mathbf{F}\cdot\mathbf{P} = \mathbf{P}^T\cdot\mathbf{F}^T$ (expressing the symmetry of $\boldsymbol{\tau}$), we obtain the condition for the continuing rotational balance

$$\dot{\mathbf{F}}\cdot\mathbf{P} + \mathbf{F}\cdot\dot{\mathbf{P}} = \dot{\mathbf{P}}^T\cdot\mathbf{F}^T + \mathbf{P}^T\cdot\dot{\mathbf{F}}^T. \tag{13.2.11}$$

The volume averages of the nominal stress and its rate are (Hill, 1972)

$$\langle\mathbf{P}\rangle = \frac{1}{V^0}\int_{V^0}\mathbf{P}\,\mathrm{d}V^0 = \frac{1}{V^0}\int_{S^0}\mathbf{X}\otimes\mathbf{p}_n\,\mathrm{d}S^0, \tag{13.2.12}$$

$$\langle\dot{\mathbf{P}}\rangle = \frac{1}{V^0}\int_{V^0}\dot{\mathbf{P}}\,\mathrm{d}V^0 = \frac{1}{V^0}\int_{S^0}\mathbf{X}\otimes\dot{\mathbf{p}}_n\,\mathrm{d}S^0. \tag{13.2.13}$$

Both of these are expressed on the far right-hand sides solely in terms of the surface data $\mathbf{p}_n$ and $\dot{\mathbf{p}}_n$ over S^0. This follows from the divergence theorem and equilibrium equations (13.2.8) and (13.2.9). If current configuration is chosen as the reference ($\mathbf{P} = \boldsymbol{\sigma}$, $\mathbf{p}_n = \mathbf{t}_n$), Eq. (13.2.12) gives

$$\{\boldsymbol{\sigma}\} = \frac{1}{V}\int_V \boldsymbol{\sigma}\,\mathrm{d}V = \frac{1}{V}\int_S \mathbf{x}\otimes\mathbf{t}_n\,\mathrm{d}S. \tag{13.2.14}$$

With this choice of the reference configuration, the rate of nominal stress is from Eq. (3.9.10) equal to

$$\underline{\dot{\mathbf{P}}} = \dot{\boldsymbol{\sigma}} + \boldsymbol{\sigma}\operatorname{tr}\mathbf{D} - \mathbf{L}\cdot\boldsymbol{\sigma}. \tag{13.2.15}$$

Thus, in view of Eq. (13.2.10), the average in Eq. (13.2.13) becomes

$$\{\dot{\boldsymbol{\sigma}} + \boldsymbol{\sigma} \operatorname{tr} \mathbf{D} - \mathbf{L} \cdot \boldsymbol{\sigma}\} = \frac{1}{V} \int_S \mathbf{x} \otimes \left[\dot{\mathbf{t}}_n + (\operatorname{tr} \mathbf{D} - \mathbf{n} \cdot \mathbf{D} \cdot \mathbf{n})\, \mathbf{t}_n\right] \mathrm{d}S. \tag{13.2.16}$$

Note that, from Eq. (13.2.14),

$$\int_{V^0} \boldsymbol{\tau} \mathrm{d}V^0 = \int_V \boldsymbol{\sigma} \, \mathrm{d}V = \int_S \mathbf{x} \otimes \mathbf{t}_n \, \mathrm{d}S = \int_{S^0} \mathbf{x} \otimes \mathbf{p}_n \, \mathrm{d}S^0, \tag{13.2.17}$$

so that

$$\langle \boldsymbol{\tau} \rangle = \frac{1}{V^0} \int_{V^0} \boldsymbol{\tau} \mathrm{d}V^0 = \frac{1}{V^0} \int_{S^0} \mathbf{x} \otimes \mathbf{p}_n \, \mathrm{d}S^0. \tag{13.2.18}$$

Since $\boldsymbol{\tau} = \mathbf{F} \cdot \mathbf{P}$, from Eq. (13.2.18) we have

$$\langle \mathbf{F} \cdot \mathbf{P} \rangle = \frac{1}{V^0} \int_{V^0} \mathbf{F} \cdot \mathbf{P} \, \mathrm{d}V^0 = \frac{1}{V^0} \int_{S^0} \mathbf{x} \otimes \mathbf{p}_n \, \mathrm{d}S^0. \tag{13.2.19}$$

This also follows directly by integration and application of the divergence theorem and equilibrium equations. Similarly,

$$\langle \mathbf{F} \cdot \dot{\mathbf{P}} \rangle = \frac{1}{V^0} \int_{V^0} \mathbf{F} \cdot \dot{\mathbf{P}} \, \mathrm{d}V^0 = \frac{1}{V^0} \int_{S^0} \mathbf{x} \otimes \dot{\mathbf{p}}_n \, \mathrm{d}S^0, \tag{13.2.20}$$

$$\langle \dot{\mathbf{F}} \cdot \mathbf{P} \rangle = \frac{1}{V^0} \int_{V^0} \dot{\mathbf{F}} \cdot \mathbf{P} \, \mathrm{d}V^0 = \frac{1}{V^0} \int_{S^0} \mathbf{v} \otimes \mathbf{p}_n \, \mathrm{d}S^0, \tag{13.2.21}$$

$$\langle \dot{\mathbf{F}} \cdot \dot{\mathbf{P}} \rangle = \frac{1}{V^0} \int_{V^0} \dot{\mathbf{F}} \cdot \dot{\mathbf{P}} \, \mathrm{d}V^0 = \frac{1}{V^0} \int_{S^0} \mathbf{v} \otimes \dot{\mathbf{p}}_n \, \mathrm{d}S^0. \tag{13.2.22}$$

In the last four expressions, the $\mathbf{F}$ and $\mathbf{P}$ fields, and their rates, need not be constitutively related to each other.

13.3. Theorem on Product Averages

In the mechanics of macroscopic aggregate behavior it is of fundamental importance to express the volume averages of various kinematic and kinetic quantities in terms of the basic macroscopic variables $\langle \mathbf{F} \rangle$ and $\langle \mathbf{P} \rangle$, and their rates. We begin with the evaluation of the product average $\langle \mathbf{F} \cdot \mathbf{P} \rangle$ in terms of $\langle \mathbf{F} \rangle$ and $\langle \mathbf{P} \rangle$. Following Hill (1984), consider the identity

$$\langle \mathbf{F} \cdot \mathbf{P} \rangle - \langle \mathbf{F} \rangle \cdot \langle \mathbf{P} \rangle = \langle (\mathbf{F} - \langle \mathbf{F} \rangle) \cdot (\mathbf{P} - \langle \mathbf{P} \rangle) \rangle. \tag{13.3.1}$$

This identity holds because, for example,

$$\langle \mathbf{F} \cdot \langle \mathbf{P} \rangle \rangle = \langle \langle \mathbf{F} \rangle \cdot \mathbf{P} \rangle = \langle \mathbf{F} \rangle \cdot \langle \mathbf{P} \rangle. \tag{13.3.2}$$

The right-hand side of Eq. (13.3.1) can be expressed as

$$\langle(\mathbf{F}-\langle\mathbf{F}\rangle)\cdot(\mathbf{P}-\langle\mathbf{P}\rangle)\rangle = \frac{1}{V^0}\int_{S^0}(\mathbf{x}-\langle\mathbf{F}\rangle\cdot\mathbf{X})\otimes(\mathbf{P}-\langle\mathbf{P}\rangle)^T\cdot\mathbf{n}^0\,\mathrm{d}S^0, \tag{13.3.3}$$

which can be verified by the Gauss divergence theorem. This leads to Hill's (1972,1984) theorem on product averages: The product average decomposes into the product of averages,

$$\langle\mathbf{F}\cdot\mathbf{P}\rangle = \langle\mathbf{F}\rangle\cdot\langle\mathbf{P}\rangle, \tag{13.3.4}$$

provided that

$$\int_{S^0}(\mathbf{x}-\langle\mathbf{F}\rangle\cdot\mathbf{X})\otimes(\mathbf{P}-\langle\mathbf{P}\rangle)^T\cdot\mathbf{n}^0\,\mathrm{d}S^0 = 0. \tag{13.3.5}$$

The condition (13.3.5) is met, in particular, when the surface S^0 is deformed or loaded uniformly, i.e., when

$$\mathbf{x} = \overline{\mathbf{F}}(t)\cdot\mathbf{X} \quad \text{or} \quad \mathbf{p}_n = \mathbf{n}^0\cdot\overline{\mathbf{P}}(t) \quad \text{on} \quad S^0, \tag{13.3.6}$$

since then

$$\langle\mathbf{F}\rangle = \overline{\mathbf{F}}(t) \quad \text{or} \quad \langle\mathbf{P}\rangle = \overline{\mathbf{P}}(t), \tag{13.3.7}$$

which makes the integral in (13.3.5) identically equal to zero.

An analog of Eqs. (13.3.4) and (13.3.5), involving the rate of $\mathbf{P}$, is

$$\langle\mathbf{F}\cdot\dot{\mathbf{P}}\rangle = \langle\mathbf{F}\rangle\cdot\langle\dot{\mathbf{P}}\rangle, \tag{13.3.8}$$

provided that

$$\int_{S^0}(\mathbf{x}-\langle\mathbf{F}\rangle\cdot\mathbf{X})\otimes\left(\dot{\mathbf{P}}-\langle\dot{\mathbf{P}}\rangle\right)^T\cdot\mathbf{n}^0\,\mathrm{d}S^0 = 0. \tag{13.3.9}$$

The condition (13.3.9) is, for example, met when

$$\mathbf{x} = \overline{\mathbf{F}}(t)\cdot\mathbf{X} \quad \text{or} \quad \dot{\mathbf{p}}_n = \mathbf{n}^0\cdot\overline{\dot{\mathbf{P}}}(t) \quad \text{on} \quad S^0. \tag{13.3.10}$$

The other analogs are, evidently,

$$\langle\dot{\mathbf{F}}\cdot\mathbf{P}\rangle = \langle\dot{\mathbf{F}}\rangle\cdot\langle\mathbf{P}\rangle, \tag{13.3.11}$$

provided that

$$\int_{S^0}\left(\mathbf{v}-\langle\dot{\mathbf{F}}\rangle\cdot\mathbf{X}\right)\otimes(\mathbf{P}-\langle\mathbf{P}\rangle)^T\cdot\mathbf{n}^0\,\mathrm{d}S^0 = 0, \tag{13.3.12}$$

and

$$\langle\dot{\mathbf{F}}\cdot\dot{\mathbf{P}}\rangle = \langle\dot{\mathbf{F}}\rangle\cdot\langle\dot{\mathbf{P}}\rangle, \tag{13.3.13}$$

provided that

$$\int_{S^0} \left(\mathbf{v} - \langle\dot{\mathbf{F}}\rangle \cdot \mathbf{X}\right) \otimes \left(\dot{\mathbf{P}} - \langle\dot{\mathbf{P}}\rangle\right)^T \cdot \mathbf{n}^0 \, \mathrm{d}S^0 = 0. \tag{13.3.14}$$

For instance, the requirement (13.3.14) is met when

$$\mathbf{v} = \bar{\dot{\mathbf{F}}}(t) \cdot \mathbf{X} \quad \text{or} \quad \dot{\mathbf{p}}_n = \mathbf{n}^0 \cdot \bar{\dot{\mathbf{P}}}(t) \quad \text{on} \quad S^0. \tag{13.3.15}$$

It is noted that, with the current configuration as the reference, Eq. (13.3.11) gives

$$\{\mathbf{L} \cdot \boldsymbol{\sigma}\} = \{\mathbf{L}\} \cdot \{\boldsymbol{\sigma}\}. \tag{13.3.16}$$

Under the prescribed uniform boundary conditions (13.3.6), the overall rotational balance, expressed in terms of the macrovariables, is

$$\langle\mathbf{F}\rangle \cdot \langle\mathbf{P}\rangle = \langle\mathbf{P}\rangle^T \cdot \langle\mathbf{F}\rangle^T. \tag{13.3.17}$$

This follows from Eq. (13.3.4) by applying the transpose operation to both sides, and by using the symmetry condition at microlevel $\mathbf{F} \cdot \mathbf{P} = \mathbf{P}^T \cdot \mathbf{F}^T$. Similarly, by differentiating Eq. (13.3.4), we have

$$\langle\dot{\mathbf{F}} \cdot \mathbf{P} + \mathbf{F} \cdot \dot{\mathbf{P}}\rangle = \langle\dot{\mathbf{F}}\rangle \cdot \langle\mathbf{P}\rangle + \langle\mathbf{F}\rangle \cdot \langle\dot{\mathbf{P}}\rangle. \tag{13.3.18}$$

By applying the transpose operation to both sides of this equation and by imposing (13.2.11), we establish the condition for the overall continuing rotational balance, in terms of the macrovariables, and under prescribed uniform boundary conditions. This is

$$\langle\dot{\mathbf{F}}\rangle \cdot \langle\mathbf{P}\rangle + \langle\mathbf{F}\rangle \cdot \langle\dot{\mathbf{P}}\rangle = \langle\mathbf{P}\rangle^T \cdot \langle\dot{\mathbf{F}}\rangle^T + \langle\dot{\mathbf{P}}\rangle^T \cdot \langle\mathbf{F}\rangle^T. \tag{13.3.19}$$

Upon contraction operation in Eq. (13.3.4), we obtain

$$\langle\mathbf{F} \cdot\cdot\, \mathbf{P}\rangle = \langle\mathbf{F}\rangle \cdot\cdot \langle\mathbf{P}\rangle. \tag{13.3.20}$$

Since the trace product is commutative, we also have

$$\langle\mathbf{P} \cdot\cdot\, \mathbf{F}\rangle = \langle\mathbf{P}\rangle \cdot\cdot \langle\mathbf{F}\rangle. \tag{13.3.21}$$

Likewise,

$$\langle\mathbf{P} \cdot\cdot\, \dot{\mathbf{F}}\rangle = \langle\mathbf{P}\rangle \cdot\cdot \langle\dot{\mathbf{F}}\rangle, \tag{13.3.22}$$

$$\langle\dot{\mathbf{P}} \cdot\cdot\, \mathbf{F}\rangle = \langle\dot{\mathbf{P}}\rangle \cdot\cdot \langle\mathbf{F}\rangle, \tag{13.3.23}$$

$$\langle\dot{\mathbf{P}} \cdot\cdot\, \dot{\mathbf{F}}\rangle = \langle\dot{\mathbf{P}}\rangle \cdot\cdot \langle\dot{\mathbf{F}}\rangle. \tag{13.3.24}$$

In these expressions, $\mathbf{P}$ and $\dot{\mathbf{P}}$ are statically admissible, while $\mathbf{F}$ and $\dot{\mathbf{F}}$ are kinematically admissible fields, but they are not necessarily constitutively related to each other. For example, if $\mathrm{d}\mathbf{P}$ and $\delta\mathbf{F}$ are two unrelated increments of $\mathbf{P}$ and $\mathbf{F}$, we can write

$$\langle \mathrm{d}\mathbf{P} \cdot\cdot\, \delta\mathbf{F} \rangle = \langle \mathrm{d}\mathbf{P} \rangle \cdot\cdot\, \langle \delta\mathbf{F} \rangle. \tag{13.3.25}$$

When the current configuration is the reference, Eq. (13.3.22) becomes

$$\{\boldsymbol{\sigma} : \mathbf{L}\} = \{\boldsymbol{\sigma}\} : \{\mathbf{L}\}, \quad \text{i.e.,} \quad \{\boldsymbol{\sigma} : \mathbf{D}\} = \{\boldsymbol{\sigma}\} : \{\mathbf{D}\}, \tag{13.3.26}$$

while Eq. (13.3.24) gives

$$\{(\dot{\boldsymbol{\sigma}} + \boldsymbol{\sigma}\,\mathrm{tr}\,\mathbf{D} - \mathbf{L} \cdot \boldsymbol{\sigma}) \cdot\cdot\, \mathbf{L}\} = \{\,\dot{\boldsymbol{\sigma}} + \boldsymbol{\sigma}\,\mathrm{tr}\,\mathbf{D} - \mathbf{L} \cdot \boldsymbol{\sigma}\,\} \cdot\cdot \{\mathbf{L}\}. \tag{13.3.27}$$

Additional analysis of the averaging theorems can be found in the paper by Nemat-Nasser (1999).

13.4. Macroscopic Measures of Stress and Strain

The macroscopic or aggregate measure of the symmetric Piola–Kirchhoff stress, denoted by $[\mathbf{T}]$, is defined such that

$$\langle \mathbf{P} \rangle = \langle \mathbf{T} \cdot \mathbf{F}^T \rangle = [\mathbf{T}] \cdot \langle \mathbf{F} \rangle^T. \tag{13.4.1}$$

Enclosure within square [ ] rather than $\langle\,\rangle$ brackets is used to indicate that the macroscopic measure of the Piola–Kirchhoff stress in Eq. (13.4.1) is not equal to the volume average of the microscopic Piola–Kirchhoff stress, i.e.,

$$[\mathbf{T}] \neq \frac{1}{V^0} \int_{V^0} \mathbf{T}\,\mathrm{d}V^0. \tag{13.4.2}$$

However, $[\mathbf{T}]$ is a symmetric tensor, because the tensor $\langle \mathbf{F} \rangle \cdot \langle \mathbf{P} \rangle$ is symmetric, by Eq. (13.3.17).

Although $[\mathbf{T}]$ is not a direct volume average of $\mathbf{T}$, it is defined in Eq. (13.4.1) in terms of the volume averages of $\langle \mathbf{F} \rangle$ and $\langle \mathbf{P} \rangle$, both of which are expressible in terms of the surface data alone. Thus, $[\mathbf{T}]$ is a suitable macroscopic variable for the constitutive analysis. (Since there is no explicit connection between $[\mathbf{T}]$ and $\langle \mathbf{T} \rangle$, the latter average is actually not suitable as a macrovariable at all). When the current configuration is taken for the reference ($\mathbf{P} = \mathbf{T} = \boldsymbol{\sigma}$), Eq. (13.4.1) gives

$$\{\boldsymbol{\sigma}\} = [\boldsymbol{\sigma}]. \tag{13.4.3}$$

This shows that the macroscopic measure of the Cauchy stress is the volume average of the microscopic Cauchy stress.

The macroscopic measure of the Lagrangian strain is defined by

$$[\mathbf{E}] = \frac{1}{2}\left(\langle \mathbf{F}\rangle^T \cdot \langle \mathbf{F}\rangle - \mathbf{I}\right), \tag{13.4.4}$$

for then $[\mathbf{T}]$ is generated from $[\mathbf{E}]$ by the work conjugency

$$\langle \dot{w}\rangle = \langle \mathbf{P}\cdot\cdot\dot{\mathbf{F}}\rangle = \langle \mathbf{T} : \dot{\mathbf{E}}\rangle = [\mathbf{T}] : [\dot{\mathbf{E}}]. \tag{13.4.5}$$

Indeed,

$$\langle \mathbf{P}\cdot\cdot\dot{\mathbf{F}}\rangle = \langle \mathbf{P}\rangle\cdot\cdot\langle\dot{\mathbf{F}}\rangle = [\mathbf{T}]\cdot\langle \mathbf{F}\rangle^T\cdot\cdot\langle\dot{\mathbf{F}}\rangle = [\mathbf{T}] : [\dot{\mathbf{E}}], \tag{13.4.6}$$

where

$$[\dot{\mathbf{E}}] = \frac{1}{2}\left(\langle\dot{\mathbf{F}}\rangle^T\cdot\langle\mathbf{F}\rangle + \langle\mathbf{F}\rangle^T\cdot\langle\dot{\mathbf{F}}\rangle\right). \tag{13.4.7}$$

The trace property $\mathbf{A}\cdot\mathbf{B}\cdot\cdot\,\mathbf{C} = \mathbf{A}\cdot\cdot\,\mathbf{B}\cdot\mathbf{C}$ was used for the second-order tensors, such as $\mathbf{A}$, $\mathbf{B}$ and $\mathbf{C}$.

The macroscopic measure of the Lagrangian strain $[\mathbf{E}]$ is not a direct volume average of the microscopic Lagrangian strain, i.e.,

$$[\mathbf{E}] \neq \frac{1}{V^0}\int_{V^0} \mathbf{E}\,\mathrm{d}V^0, \tag{13.4.8}$$

because

$$\langle \mathbf{F}^T\cdot\mathbf{F}\rangle \neq \langle\mathbf{F}\rangle^T\cdot\langle\mathbf{F}\rangle. \tag{13.4.9}$$

The rates of the macroscopic nominal and symmetric Piola–Kirchhoff stress tensors are related by

$$\langle\dot{\mathbf{P}}\rangle = [\dot{\mathbf{T}}]\cdot\langle\mathbf{F}\rangle^T + [\mathbf{T}]\cdot\langle\dot{\mathbf{F}}\rangle^T, \tag{13.4.10}$$

which follows from Eq. (13.4.1) by differentiation. When this is subjected to the trace product with $\langle\dot{\mathbf{F}}\rangle$, we obtain

$$\langle\dot{\mathbf{P}}\rangle\cdot\cdot\langle\dot{\mathbf{F}}\rangle = [\dot{\mathbf{T}}] : [\dot{\mathbf{E}}] + \mathbf{T} : \left(\langle\dot{\mathbf{F}}\rangle^T\cdot\langle\dot{\mathbf{F}}\rangle\right). \tag{13.4.11}$$

If the current configuration is selected for the reference, the stress rate $\dot{\mathbf{T}}$ is equal to (see Section 3.8)

$$\overset{\triangle}{\underline{\boldsymbol{\tau}}} = \overset{\triangle}{\boldsymbol{\sigma}} + \boldsymbol{\sigma}\,\mathrm{tr}\,\mathbf{D}, \tag{13.4.12}$$

and Eq. (13.4.10) becomes

$$\{\,\dot{\boldsymbol{\sigma}} + \boldsymbol{\sigma}\,\mathrm{tr}\,\mathbf{D} - \mathbf{L}\cdot\boldsymbol{\sigma}\,\} = [\,\overset{\triangle}{\boldsymbol{\sigma}} + \boldsymbol{\sigma}\,\mathrm{tr}\,\mathbf{D}\,] + [\boldsymbol{\sigma}]\cdot\langle\mathbf{L}\rangle^T. \tag{13.4.13}$$

Since $[\boldsymbol{\sigma}] = \{\boldsymbol{\sigma}\}$, and since by direct integration

$$\{\boldsymbol{\sigma} \cdot \mathbf{L}^T\} = \{\boldsymbol{\sigma}\} \cdot \{\mathbf{L}\}^T, \tag{13.4.14}$$

we deduce from Eq. (13.4.13) that

$$\{\overset{\triangle}{\boldsymbol{\sigma}} + \boldsymbol{\sigma} \operatorname{tr} \mathbf{D}\} = [\overset{\triangle}{\boldsymbol{\sigma}} + \boldsymbol{\sigma} \operatorname{tr} \mathbf{D}], \tag{13.4.15}$$

i.e.,

$$[\overset{\triangle}{\underline{\boldsymbol{\tau}}}] = \{\overset{\triangle}{\underline{\boldsymbol{\tau}}}\}. \tag{13.4.16}$$

Furthermore, with the current configuration as the reference, Eq. (13.4.7) gives

$$[\mathbf{D}] = \{\mathbf{D}\}. \tag{13.4.17}$$

Thus, the macroscopic measure of the rate of deformation is the volume average of the microscopic rate of deformation.

The macroscopic infinitesimal deformation gradient and, thus, the macroscopic infinitesimal strain and rotation are also direct volume averages of the corresponding microscopic quantities. For the definition of the macroscopic measures of the rate of stress and deformation in the solids undergoing phase transformation, see Petryk (1998).

13.5. Influence Tensors of Elastic Heterogeneity

We consider materials for which the interior elastic fields depend uniquely and continuously on the surface data. Then, under uniform data on S^0, specified by (13.3.15), the fields $\dot{\mathbf{F}}$ and $\dot{\mathbf{P}}$ within V^0 depend uniquely on $\langle\dot{\mathbf{F}}\rangle$. For incrementally linear material response, this dependence is also linear. Thus, following Hill (1984), we introduce the influence tensors (functions) of elastic heterogeneity, denoted by $\boldsymbol{\mathcal{F}}$ and $\boldsymbol{\mathcal{P}}$, such that

$$\dot{\mathbf{F}} = \boldsymbol{\mathcal{F}} \cdot\cdot \langle\dot{\mathbf{F}}\rangle = \langle\dot{\mathbf{F}}\rangle \cdot\cdot \boldsymbol{\mathcal{F}}^T, \tag{13.5.1}$$

$$\dot{\mathbf{P}} = \boldsymbol{\mathcal{P}} \cdot\cdot \langle\dot{\mathbf{P}}\rangle = \langle\dot{\mathbf{P}}\rangle \cdot\cdot \boldsymbol{\mathcal{P}}^T, \tag{13.5.2}$$

where

$$\langle\boldsymbol{\mathcal{F}}\rangle = \boldsymbol{I}, \quad \langle\boldsymbol{\mathcal{P}}\rangle = \boldsymbol{I}. \tag{13.5.3}$$

The rectangular components of the fourth-order unit tensor $\boldsymbol{I}$ are

$$I_{ijkl} = \delta_{il}\delta_{jk}, \quad I_{ijkl} = I_{klij}. \tag{13.5.4}$$

The influence tensors $\boldsymbol{\mathcal{F}}$ and $\boldsymbol{\mathcal{P}}$ are functions of the current heterogeneities of stress and material properties within a macroelement. As pointed out by Hill (1984), kinematic data is never micro-uniform, since equivalent macroelements in a test specimen are constrained by one another, not by the apparatus. This results in fluctuations of $\dot{\mathbf{F}} \cdot \mathbf{X}$ on S^0 around $\langle \dot{\mathbf{F}} \rangle \cdot \mathbf{X}$, but the effect of these fluctuations decay rapidly with depth toward interior of the macroelement. Equations (13.5.1) and (13.5.2) can then be adopted for this macro-uniform surface data, as well, except within a negligible layer near the bounding surface of the macroelement. See also Mandel (1964) and Stolz (1997).

13.6. Macroscopic Free and Complementary Energy

The local free energy, per unit reference volume, is a potential for the local nominal stress, such that

$$\mathbf{P} = \frac{\partial \Psi}{\partial \mathbf{F}}, \quad \Psi = \Psi(\mathbf{F}, \mathcal{H}). \tag{13.6.1}$$

The pattern of internal rearrangement due to plastic deformation is designated by $\mathcal{H}$. The macroscopic free energy, per unit volume of the aggregate macroelement, is the volume average of Ψ,

$$\hat{\Psi} = \langle \Psi \rangle = \frac{1}{V^0} \int_{V^0} \Psi(\mathbf{F}, \mathcal{H})\, \mathrm{d}V^0. \tag{13.6.2}$$

This acts as a potential for the macroscopic nominal stress, such that

$$\langle \mathbf{P} \rangle = \frac{\partial \hat{\Psi}}{\partial \langle \mathbf{F} \rangle}, \quad \hat{\Psi} = \hat{\Psi}(\langle \mathbf{F} \rangle, \mathcal{H}). \tag{13.6.3}$$

Indeed,

$$\frac{\partial \hat{\Psi}}{\partial \langle \mathbf{F} \rangle} = \frac{\partial}{\partial \langle \mathbf{F} \rangle} \langle \Psi \rangle = \langle \frac{\partial \Psi}{\partial \langle \mathbf{F} \rangle} \rangle = \langle \frac{\partial \Psi}{\partial \mathbf{F}} \cdot\cdot \frac{\partial \mathbf{F}}{\partial \langle \mathbf{F} \rangle} \rangle = \langle \mathbf{P} \cdot\cdot \boldsymbol{\mathcal{F}} \rangle = \langle \mathbf{P} \rangle. \tag{13.6.4}$$

It is noted that, at fixed $\mathcal{H}$, from Eq. (13.5.1) we have

$$\delta \langle \mathbf{F} \rangle = \boldsymbol{\mathcal{F}} \cdot\cdot \delta \langle \mathbf{F} \rangle, \quad \text{i.e.,} \quad \frac{\partial \mathbf{F}}{\partial \langle \mathbf{F} \rangle} = \boldsymbol{\mathcal{F}}, \tag{13.6.5}$$

which was used after partial differentiation in Eq. (13.6.4). Also, under uniform boundary data,

$$\langle \mathbf{P} \cdot\cdot \boldsymbol{\mathcal{F}} \rangle = \langle \mathbf{P} \rangle, \tag{13.6.6}$$

because

$$\langle \mathbf{P} \rangle \cdot\cdot \delta \langle \mathbf{F} \rangle = \langle \mathbf{P} \cdot\cdot \delta \mathbf{F} \rangle = \langle \mathbf{P} \cdot\cdot \boldsymbol{\mathcal{F}} \cdot\cdot \delta \langle \mathbf{F} \rangle \rangle = \langle \mathbf{P} \cdot\cdot \boldsymbol{\mathcal{F}} \rangle \cdot\cdot \delta \langle \mathbf{F} \rangle. \tag{13.6.7}$$

The local complementary energy Φ, per unit reference volume, is a potential for the local deformation gradient. This is a Legendre transform of Ψ, such that

$$\mathbf{F} = \frac{\partial \Phi}{\partial \mathbf{P}}, \quad \Phi(\mathbf{P}, \mathcal{H}) = \mathbf{P} \cdot\cdot \mathbf{F} - \Psi(\mathbf{F}, \mathcal{H}). \tag{13.6.8}$$

The macroscopic free energy, per unit volume of the aggregate macroelement, is a potential for the macroscopic deformation gradient,

$$\langle \mathbf{F} \rangle = \frac{\partial \hat{\Phi}}{\partial \langle \mathbf{P} \rangle}, \quad \hat{\Phi}(\langle \mathbf{P} \rangle, \mathcal{H}) = \langle \mathbf{P} \rangle \cdot\cdot \langle \mathbf{F} \rangle - \hat{\Psi}(\langle \mathbf{F} \rangle, \mathcal{H}). \tag{13.6.9}$$

Under conditions allowing the product theorem $\langle \mathbf{P} \cdot\cdot \delta\mathbf{F} \rangle = \langle \mathbf{P} \rangle \cdot\cdot \delta\langle \mathbf{F} \rangle$ to be used, $\hat{\Phi}$ is the volume average of Φ, i.e.,

$$\hat{\Phi} = \langle \Phi \rangle. \tag{13.6.10}$$

In this case, the potential property of $\hat{\Phi}$ can be demonstrated through

$$\frac{\partial \hat{\Phi}}{\partial \langle \mathbf{P} \rangle} = \frac{\partial}{\partial \langle \mathbf{P} \rangle} \langle \Phi \rangle = \langle \frac{\partial \Phi}{\partial \langle \mathbf{P} \rangle} \rangle = \langle \frac{\partial \Phi}{\partial \mathbf{P}} \cdot\cdot \frac{\partial \mathbf{P}}{\partial \langle \mathbf{P} \rangle} \rangle = \langle \mathbf{F} \cdot\cdot \boldsymbol{\mathcal{P}} \rangle = \langle \mathbf{F} \rangle. \tag{13.6.11}$$

Again, at fixed $\mathcal{H}$, from Eq. (13.5.2) we have

$$\delta \langle \mathbf{P} \rangle = \boldsymbol{\mathcal{P}} \cdot\cdot \delta \langle \mathbf{P} \rangle, \quad \text{i.e.,} \quad \frac{\partial \mathbf{P}}{\partial \langle \mathbf{P} \rangle} = \boldsymbol{\mathcal{P}}, \tag{13.6.12}$$

which was used after partial differentiation in Eq. (13.6.11). In addition, under uniform boundary data,

$$\langle \mathbf{F} \cdot\cdot \boldsymbol{\mathcal{P}} \rangle = \langle \mathbf{F} \rangle, \tag{13.6.13}$$

because

$$\langle \mathbf{F} \rangle \cdot\cdot \delta \langle \mathbf{P} \rangle = \langle \mathbf{F} \cdot\cdot \delta \mathbf{P} \rangle = \langle \mathbf{F} \cdot\cdot \boldsymbol{\mathcal{P}} \cdot\cdot \delta \langle \mathbf{P} \rangle \rangle = \langle \mathbf{F} \cdot\cdot \boldsymbol{\mathcal{P}} \rangle \cdot\cdot \delta \langle \mathbf{P} \rangle. \tag{13.6.14}$$

13.7. Macroscopic Elastic Pseudomoduli

The tensor of macroscopic elastic pseudomoduli is defined by

$$[\boldsymbol{\Lambda}] = \frac{\partial^2 \hat{\Psi}}{\partial \langle \mathbf{F} \rangle \otimes \partial \langle \mathbf{F} \rangle} = \frac{\partial \langle \mathbf{P} \rangle}{\partial \langle \mathbf{F} \rangle} = \langle \frac{\partial \mathbf{P}}{\partial \langle \mathbf{F} \rangle} \rangle = \langle \frac{\partial \mathbf{P}}{\partial \mathbf{F}} \cdot\cdot \frac{\partial \mathbf{F}}{\partial \langle \mathbf{F} \rangle} \rangle = \langle \boldsymbol{\Lambda} \cdot\cdot \boldsymbol{\mathcal{F}} \rangle. \tag{13.7.1}$$

The tensor of local elastic pseudomoduli is $\boldsymbol{\Lambda}$. Along an elastic branch of the material response at microlevel, the rates of $\mathbf{P}$ and $\mathbf{F}$ are related by

$$\dot{\mathbf{P}} = \boldsymbol{\Lambda} \cdot\cdot \dot{\mathbf{F}}, \quad \boldsymbol{\Lambda} = \frac{\partial \mathbf{P}}{\partial \mathbf{F}}. \tag{13.7.2}$$

The macroscopic tensor of elastic pseudomoduli $[\mathbf{\Lambda}]$ relates $\langle\dot{\mathbf{P}}\rangle$ and $\langle\dot{\mathbf{F}}\rangle$, such that

$$\langle\dot{\mathbf{P}}\rangle = \langle\mathbf{\Lambda}\cdot\cdot\dot{\mathbf{F}}\rangle = [\mathbf{\Lambda}]\cdot\cdot\langle\dot{\mathbf{F}}\rangle. \tag{13.7.3}$$

An alternative derivation of the relationship between the local and macroscopic pseudomoduli, given in Eq. (13.7.1) is as follows. First, by substituting Eq. (13.7.3) into Eq. (13.5.2), we have

$$\dot{\mathbf{P}} = \boldsymbol{\mathcal{P}}\cdot\cdot\langle\dot{\mathbf{P}}\rangle = \boldsymbol{\mathcal{P}}\cdot\cdot[\mathbf{\Lambda}]\cdot\cdot\langle\dot{\mathbf{F}}\rangle. \tag{13.7.4}$$

On the other hand, introducing (13.7.2), and then (13.5.1), into Eq. (13.5.2) gives

$$\dot{\mathbf{P}} = \boldsymbol{\mathcal{P}}\cdot\cdot\langle\dot{\mathbf{P}}\rangle = \boldsymbol{\mathcal{P}}\cdot\cdot\langle\mathbf{\Lambda}\cdot\cdot\dot{\mathbf{F}}\rangle = \boldsymbol{\mathcal{P}}\cdot\cdot\langle\mathbf{\Lambda}\cdot\cdot\boldsymbol{\mathcal{F}}\rangle\cdot\cdot\langle\dot{\mathbf{F}}\rangle. \tag{13.7.5}$$

Comparing Eqs. (13.7.4) and (13.7.5), we obtain

$$[\mathbf{\Lambda}] = \langle\mathbf{\Lambda}\cdot\cdot\boldsymbol{\mathcal{F}}\rangle. \tag{13.7.6}$$

This shows that the tensor of macroscopic elastic pseudomoduli is a weighted volume average of the tensor of local elastic pseudomoduli $\mathbf{\Lambda}$, the weight being the influence tensor $\boldsymbol{\mathcal{F}}$ of elastic heterogeneity within a representative macroelement. In addition, since

$$\dot{\mathbf{P}} = \mathbf{\Lambda}\cdot\cdot\dot{\mathbf{F}} = \mathbf{\Lambda}\cdot\cdot\boldsymbol{\mathcal{F}}\cdot\cdot\langle\dot{\mathbf{F}}\rangle, \tag{13.7.7}$$

by comparing with (13.7.4) we observe that

$$\boldsymbol{\mathcal{P}}\cdot\cdot[\mathbf{\Lambda}] = \mathbf{\Lambda}:\boldsymbol{\mathcal{F}}. \tag{13.7.8}$$

The symmetry of elastic response at the microlevel is transmitted to the macrolevel, i.e.,

$$\text{if}\quad \mathbf{\Lambda}^T = \mathbf{\Lambda},\quad \text{then}\quad [\mathbf{\Lambda}]^T = [\mathbf{\Lambda}]. \tag{13.7.9}$$

This does not appear to be evident at first from Eq. (13.7.6) or Eq. (13.7.8). However, since

$$\langle\dot{\mathbf{F}}\cdot\cdot\dot{\mathbf{P}}\rangle = \langle\dot{\mathbf{F}}\rangle\cdot\cdot\langle\dot{\mathbf{P}}\rangle, \tag{13.7.10}$$

and in view of Eqs. (13.5.1) and (13.5.2) giving

$$\langle\dot{\mathbf{F}}\cdot\cdot\dot{\mathbf{P}}\rangle = \langle\dot{\mathbf{F}}\rangle\cdot\cdot\langle\boldsymbol{\mathcal{F}}^T\cdot\cdot\boldsymbol{\mathcal{P}}\rangle\cdot\cdot\langle\dot{\mathbf{P}}\rangle, \tag{13.7.11}$$

the comparison with Eq. (13.7.10) establishes

$$\langle\boldsymbol{\mathcal{F}}^T\cdot\cdot\boldsymbol{\mathcal{P}}\rangle = \boldsymbol{I}. \tag{13.7.12}$$

Therefore, upon taking a trace product of Eq. (13.7.8) with $\boldsymbol{\mathcal{F}}^T$ from the left, and upon the volume averaging over V^0, there follows

$$[\boldsymbol{\Lambda}] = \langle \boldsymbol{\mathcal{F}}^T \cdot\cdot \boldsymbol{\Lambda} \cdot\cdot \boldsymbol{\mathcal{F}} \rangle. \tag{13.7.13}$$

This demonstrates that $[\boldsymbol{\Lambda}]$ is indeed symmetric whenever $\boldsymbol{\Lambda}$ is.

When the current configuration is the reference, the previous formulas reduce to

$$\{\dot{\underline{\mathbf{P}}}\} = [\underline{\boldsymbol{\Lambda}}] \cdot\cdot \{\mathbf{L}\}, \tag{13.7.14}$$

$$\mathbf{L} = \underline{\boldsymbol{\mathcal{F}}} \cdot\cdot \{\mathbf{L}\}, \quad \dot{\underline{\mathbf{P}}} = \underline{\boldsymbol{\mathcal{P}}} \cdot\cdot \{\dot{\underline{\mathbf{P}}}\}, \tag{13.7.15}$$

and

$$[\underline{\boldsymbol{\Lambda}}] = \{\underline{\boldsymbol{\mathcal{F}}}^T \cdot\cdot \underline{\boldsymbol{\Lambda}} \cdot\cdot \underline{\boldsymbol{\mathcal{F}}}\}. \tag{13.7.16}$$

The underlined symbol indicates that the current configuration is taken for the reference.

13.8. Macroscopic Elastic Pseudocompliances

Suppose that the local elastic pseudomoduli tensor $\boldsymbol{\Lambda}$ has its inverse, the local elastic pseudocompliances tensor $\mathbf{M} = \boldsymbol{\Lambda}^{-1}$ (except possibly at isolated singular points within each crystal grain, whose contribution to volume integrals over the macroelement can be ignored in the micro-to-macro transition; Hill, 1984). We then write

$$\dot{\mathbf{F}} = \mathbf{M} \cdot\cdot \dot{\mathbf{P}}, \tag{13.8.1}$$

where

$$\boldsymbol{\Lambda} \cdot\cdot \mathbf{M} = \mathbf{M} \cdot\cdot \boldsymbol{\Lambda}^{-1} = \boldsymbol{I}. \tag{13.8.2}$$

The macroscopic tensor of elastic pseudocompliances $[\mathbf{M}]$ is introduced by requiring that

$$\langle \dot{\mathbf{F}} \rangle = \langle \mathbf{M} \cdot\cdot \dot{\mathbf{P}} \rangle = [\mathbf{M}] \cdot\cdot \langle \dot{\mathbf{P}} \rangle. \tag{13.8.3}$$

By substituting Eq. (13.8.3) into (13.5.1), we obtain

$$\dot{\mathbf{F}} = \boldsymbol{\mathcal{F}} \cdot\cdot \langle \dot{\mathbf{F}} \rangle = \boldsymbol{\mathcal{F}} \cdot\cdot [\mathbf{M}] \cdot\cdot \langle \dot{\mathbf{P}} \rangle. \tag{13.8.4}$$

On the other hand, introducing (13.7.2), and then (13.5.2), into Eq. (13.5.1) gives

$$\dot{\mathbf{F}} = \boldsymbol{\mathcal{F}} \cdot\cdot \langle \dot{\mathbf{F}} \rangle = \boldsymbol{\mathcal{F}} \cdot\cdot \langle \mathbf{M} \cdot\cdot \dot{\mathbf{P}} \rangle = \boldsymbol{\mathcal{F}} \cdot\cdot \langle \mathbf{M} \cdot\cdot \boldsymbol{\mathcal{P}} \rangle \cdot\cdot \langle \dot{\mathbf{P}} \rangle. \tag{13.8.5}$$

Comparing Eqs. (13.8.4) and (13.8.5) yields

$$[\mathbf{M}] = \langle \mathbf{M} \cdot\cdot \boldsymbol{\mathcal{P}} \rangle. \tag{13.8.6}$$

This shows that the tensor of macroscopic elastic pseudocompliances is a weighted volume average of the tensor of local elastic pseudocompliances $\mathbf{M}$, the weight being the influence tensor $\boldsymbol{\mathcal{P}}$ of elastic heterogeneity within a representative macroelement. In addition, since

$$\dot{\mathbf{F}} = \mathbf{M} \cdot\cdot \dot{\mathbf{P}} = \mathbf{M} \cdot\cdot \boldsymbol{\mathcal{P}} \cdot\cdot \langle \dot{\mathbf{P}} \rangle, \tag{13.8.7}$$

by comparing with (13.8.4) there follows

$$\boldsymbol{\mathcal{F}} \cdot\cdot [\mathbf{M}] = \mathbf{M} : \boldsymbol{\mathcal{P}}. \tag{13.8.8}$$

We now demonstrate, independently of the proof from the previous section, that the symmetry of elastic response at the microlevel is transmitted to the macrolevel. First, we note that

$$\langle \dot{\mathbf{P}} \cdot\cdot \dot{\mathbf{F}} \rangle = \langle \dot{\mathbf{P}} \rangle \cdot\cdot \langle \dot{\mathbf{F}} \rangle. \tag{13.8.9}$$

Since, by (13.5.1) and (13.5.2), we have

$$\langle \dot{\mathbf{P}} \cdot\cdot \dot{\mathbf{F}} \rangle = \langle \dot{\mathbf{P}} \rangle \cdot\cdot \langle \boldsymbol{\mathcal{P}}^T \cdot\cdot \boldsymbol{\mathcal{F}} \rangle \cdot\cdot \langle \dot{\mathbf{F}} \rangle, \tag{13.8.10}$$

the comparison with Eq. (13.8.9) gives

$$\langle \boldsymbol{\mathcal{P}}^T \cdot\cdot \boldsymbol{\mathcal{F}} \rangle = \boldsymbol{I}. \tag{13.8.11}$$

Therefore, upon taking a trace product of Eq. (13.8.8) with $\boldsymbol{\mathcal{P}}^T$ from the left, and upon the volume averaging, we obtain

$$[\mathbf{M}] = \langle \boldsymbol{\mathcal{P}}^T \cdot\cdot \mathbf{M} \cdot\cdot \boldsymbol{\mathcal{P}} \rangle. \tag{13.8.12}$$

Consequently, if there is a symmetry of elastic response at the microlevel, it is transmitted to the macrolevel, i.e.,

$$\text{if} \quad \mathbf{M}^T = \mathbf{M}, \quad \text{then} \quad [\mathbf{M}]^T = [\mathbf{M}]. \tag{13.8.13}$$

When the macroscopic complementary energy is used to define the elastic pseudocompliances tensor, we can write

$$[\mathbf{M}] = \frac{\partial^2 \hat{\Phi}}{\partial \langle \mathbf{P} \rangle \otimes \partial \langle \mathbf{P} \rangle} = \frac{\partial \langle \mathbf{F} \rangle}{\partial \langle \mathbf{P} \rangle} = \langle \frac{\partial \mathbf{F}}{\partial \langle \mathbf{P} \rangle} \rangle = \langle \frac{\partial \mathbf{F}}{\partial \mathbf{P}} \cdot\cdot \frac{\partial \mathbf{P}}{\partial \langle \mathbf{P} \rangle} \rangle = \langle \mathbf{M} \cdot\cdot \boldsymbol{\mathcal{P}} \rangle. \tag{13.8.14}$$

13.9. Macroscopic Elastic Moduli

The macroscopic elastic moduli tensor $[\mathbf{\Lambda}_{(1)}]$, corresponding to the macroscopic Lagrangian strain and its conjugate stress, is defined by requiring that

$$[\dot{\mathbf{T}}] = [\mathbf{\Lambda}_{(1)}] : [\dot{\mathbf{E}}]. \tag{13.9.1}$$

To obtain the relationship between $[\mathbf{\Lambda}_{(1)}]$ and $[\mathbf{\Lambda}]$, we use Eq. (13.4.10), which is here conveniently rewritten as

$$\langle \dot{\mathbf{P}} \rangle = \langle \boldsymbol{\mathcal{K}} \rangle^T : [\dot{\mathbf{T}}] + [\boldsymbol{\mathcal{T}}] \cdot\cdot \langle \dot{\mathbf{F}} \rangle. \tag{13.9.2}$$

The rectangular components of the fourth-order tensors $\langle \boldsymbol{\mathcal{K}} \rangle$ and $[\boldsymbol{\mathcal{T}}]$ are

$$\langle \mathcal{K} \rangle_{ijkl} = \frac{1}{2} \left(\delta_{ik} \langle F \rangle_{lj} + \delta_{jk} \langle F \rangle_{li} \right), \quad [\mathcal{T}]_{ijkl} = [T]_{ik}\, \delta_{jl}. \tag{13.9.3}$$

Substitution of Eq. (13.7.3) into Eq. (13.9.2) gives

$$[\mathbf{\Lambda}] = \langle \boldsymbol{\mathcal{K}} \rangle^T : [\mathbf{\Lambda}_{(1)}] : \langle \boldsymbol{\mathcal{K}} \rangle + [\boldsymbol{\mathcal{T}}]. \tag{13.9.4}$$

Expressed in rectangular components, this is

$$[\Lambda]_{ijkl} = [\Lambda_{(1)}]_{ipkq} \langle F \rangle_{jp} \langle F \rangle_{lq} + [T]_{ik}\, \delta_{jl}. \tag{13.9.5}$$

Clearly, the symmetry $ij \leftrightarrow kl$ of the macroscopic pseudomoduli imposes the same symmetry for the macroscopic moduli, and *vice versa*. Also, recall the symmetry $\boldsymbol{\mathcal{T}}^T = \boldsymbol{\mathcal{T}}$.

When the current configuration is the reference, Eq. (13.9.4) reduces to

$$[\underline{\mathbf{\Lambda}}] = [\underline{\mathbf{\Lambda}}_{(1)}] + [\underline{\boldsymbol{\mathcal{T}}}], \tag{13.9.6}$$

with the component form

$$[\underline{\Lambda}]_{ijkl} = [\underline{\Lambda}_{(1)}]_{ijkl} + \{\sigma\}_{ik}\, \delta_{jl}. \tag{13.9.7}$$

In addition, Eq. (13.9.1) becomes

$$\{\overset{\triangle}{\underline{\mathcal{T}}}\} = [\underline{\mathbf{\Lambda}}_{(1)}] : \{\mathbf{D}\}. \tag{13.9.8}$$

13.10. Plastic Increment of Macroscopic Nominal Stress

The increment of macroscopic nominal stress can be partitioned into elastic and plastic parts as

$$\mathrm{d}\langle \mathbf{P} \rangle = \mathrm{d}^{\mathrm{e}} \langle \mathbf{P} \rangle + \mathrm{d}^{\mathrm{p}} \langle \mathbf{P} \rangle. \tag{13.10.1}$$

The elastic part is defined by

$$\mathrm{d}^{\mathrm{e}} \langle \mathbf{P} \rangle = [\mathbf{\Lambda}] \cdot\cdot \mathrm{d}\langle \mathbf{F} \rangle. \tag{13.10.2}$$

The remaining part,

$$\mathrm{d}^{\mathrm{p}}\langle \mathbf{P} \rangle = \mathrm{d}\langle \mathbf{P} \rangle - [\mathbf{\Lambda}] \cdot\cdot\, \mathrm{d}\langle \mathbf{F} \rangle, \tag{13.10.3}$$

is the plastic part of the increment $\mathrm{d}\langle \mathbf{P} \rangle$. The macroscopic elastoplastic increment of the deformation gradient is $\mathrm{d}\langle \mathbf{F} \rangle$.

It is of interest to establish the relationship between the plastic increments of macroscopic and microscopic (local) nominal stress, $\mathrm{d}^{\mathrm{p}}\langle \mathbf{P} \rangle$ and $\mathrm{d}^{\mathrm{p}}\mathbf{P}$. To that goal, consider the volume average of the trace product between an elastic unloading increment of the local deformation gradient $\delta\mathbf{F}$ and the plastic increment of the local nominal stress $\mathrm{d}^{\mathrm{p}}\mathbf{P}$, i.e.,

$$\langle \delta\mathbf{F} \cdot\cdot\, \mathrm{d}^{\mathrm{p}}\mathbf{P} \rangle = \langle \delta\mathbf{F} \cdot\cdot\, (\mathrm{d}\mathbf{P} - \mathbf{\Lambda} \cdot\cdot\, \mathrm{d}\mathbf{F}) \rangle = \langle \delta\mathbf{F} \cdot\cdot\, \mathrm{d}\mathbf{P} \rangle - \langle \delta\mathbf{F} \cdot\cdot\, \mathbf{\Lambda} \cdot\cdot\, \mathrm{d}\mathbf{F} \rangle. \tag{13.10.4}$$

Since $\mathrm{d}\mathbf{F}$ and $\delta\mathbf{F}$ are kinematically admissible, and $\mathrm{d}\mathbf{P}$ and $\delta\mathbf{F} \cdot\cdot\, \mathbf{\Lambda}$ are statically admissible fields, we can use the product theorem of Section 13.3 to write

$$\langle \delta\mathbf{F} \cdot\cdot\, \mathrm{d}\mathbf{P} \rangle = \langle \delta\mathbf{F} \rangle \cdot\cdot\, \langle \mathrm{d}\mathbf{P} \rangle = \delta\langle \mathbf{F} \rangle \cdot\cdot\, \mathrm{d}\langle \mathbf{P} \rangle, \tag{13.10.5}$$

$$\langle \delta\mathbf{F} \cdot\cdot\, \mathbf{\Lambda} \cdot\cdot\, \mathrm{d}\mathbf{F} \rangle = \langle \delta\mathbf{F} \cdot\cdot\, \mathbf{\Lambda} \rangle \cdot\cdot\, \mathrm{d}\langle \mathbf{F} \rangle = \delta\langle \mathbf{F} \rangle \cdot\cdot\, \langle \boldsymbol{\mathcal{F}}^{T} \cdot\cdot\, \mathbf{\Lambda} \rangle \cdot\cdot\, \mathrm{d}\langle \mathbf{F} \rangle. \tag{13.10.6}$$

Upon substitution into Eq. (13.10.4), there follows

$$\langle \delta\mathbf{F} \cdot\cdot\, \mathrm{d}^{\mathrm{p}}\mathbf{P} \rangle = \delta\langle \mathbf{F} \rangle \cdot\cdot\, (\mathrm{d}\langle \mathbf{P} \rangle - [\mathbf{\Lambda}] \cdot\cdot\, \mathrm{d}\langle \mathbf{F} \rangle)\,. \tag{13.10.7}$$

Recall that $[\mathbf{\Lambda}]$ is symmetric, and

$$\delta\mathbf{F} = \boldsymbol{\mathcal{F}} \cdot\cdot\, \delta\langle \mathbf{F} \rangle = \delta\langle \mathbf{F} \rangle \cdot\cdot\, \boldsymbol{\mathcal{F}}^{T}, \tag{13.10.8}$$

so that

$$[\mathbf{\Lambda}] = \langle \mathbf{\Lambda} \cdot\cdot\, \boldsymbol{\mathcal{F}} \rangle = \langle \boldsymbol{\mathcal{F}}^{T} \cdot\cdot\, \mathbf{\Lambda} \rangle. \tag{13.10.9}$$

Also note that

$$\langle \mathrm{d}\mathbf{P} \rangle = \mathrm{d}\langle \mathbf{P} \rangle, \quad \langle \mathrm{d}\mathbf{F} \rangle = \mathrm{d}\langle \mathbf{F} \rangle, \tag{13.10.10}$$

and likewise for δ increments. Consequently,

$$\langle \delta\mathbf{F} \cdot\cdot\, \mathrm{d}^{\mathrm{p}}\mathbf{P} \rangle = \delta\langle \mathbf{F} \rangle \cdot\cdot\, \mathrm{d}^{\mathrm{p}}\langle \mathbf{P} \rangle. \tag{13.10.11}$$

Furthermore,

$$\langle \delta\mathbf{F} \cdot\cdot\, \mathrm{d}^{\mathrm{p}}\mathbf{P} \rangle = \delta\langle \mathbf{F} \rangle \cdot\cdot\, \mathrm{d}\langle \mathbf{P} \rangle - \delta\langle \mathbf{P} \rangle \cdot\cdot\, \mathrm{d}\langle \mathbf{F} \rangle, \tag{13.10.12}$$

which can be easily verified by substituting $\delta\langle \mathbf{P} \rangle = \delta\langle \mathbf{F} \rangle \cdot\cdot\, [\mathbf{\Lambda}]$, and by using Eq. (13.10.3).

On the other hand, from Eq. (13.5.1) we directly obtain

$$\langle \delta\mathbf{F} \cdot\cdot\, \mathrm{d}^{\mathrm{p}}\mathbf{P} \rangle = \delta\langle \mathbf{F} \rangle \cdot\cdot \langle \boldsymbol{\mathcal{F}}^T \cdot\cdot\, \mathrm{d}^{\mathrm{p}}\mathbf{P} \rangle. \tag{13.10.13}$$

The comparison of Eqs. (13.10.11) and (13.10.13) establishes

$$\mathrm{d}^{\mathrm{p}}\langle \mathbf{P} \rangle = \langle \boldsymbol{\mathcal{F}}^T \cdot\cdot\, \mathrm{d}^{\mathrm{p}}\mathbf{P} \rangle. \tag{13.10.14}$$

Therefore, the plastic part of the increment of macroscopic nominal stress is a weighted volume average of the plastic part of the increment of local nominal stress (Hill, 1984; Havner, 1992).

13.10.1. Plastic Potential and Normality Rule

From Eq. (13.10.11) it follows, if the normality rule applies at the microlevel, it is transmitted to the macrolevel, i.e.,

$$\delta\mathbf{F} \cdot\cdot\, \mathrm{d}^{\mathrm{p}}\mathbf{P} > 0 \quad \text{implies} \quad \delta\langle \mathbf{F} \rangle \cdot\cdot\, \mathrm{d}^{\mathrm{p}}\langle \mathbf{P} \rangle > 0. \tag{13.10.15}$$

We recall from Section 12.7 that $-\sum(\tau^\alpha\, \mathrm{d}\gamma^\alpha)$ acts as the plastic potential for $\mathrm{d}^{\mathrm{p}}\mathbf{P}$ over an elastic domain in $\mathbf{F}$ space, such that

$$\mathrm{d}^{\mathrm{p}}\mathbf{P} = -\frac{\partial}{\partial \mathbf{F}} \sum_{\alpha=1}^{n} (\tau^\alpha\, \mathrm{d}\gamma^\alpha). \tag{13.10.16}$$

The partial differentiation is performed at the fixed slip and slip increments $\mathrm{d}\gamma^\alpha$. The local resolved shear stress on the α slip system is τ^α, and n is the number of active slip systems. Substitution into Eq. (13.10.14) gives

$$\mathrm{d}^{\mathrm{p}}\langle \mathbf{P} \rangle = -\langle \boldsymbol{\mathcal{F}}^T \cdot\cdot\, \frac{\partial}{\partial \mathbf{F}} \sum_{\alpha=1}^{n} (\tau^\alpha\, \mathrm{d}\gamma^\alpha) \rangle. \tag{13.10.17}$$

Since, at the fixed slip,

$$\frac{\partial}{\partial \langle \mathbf{F} \rangle} = \frac{\partial}{\partial \mathbf{F}} \cdot\cdot\, \frac{\partial \mathbf{F}}{\partial \langle \mathbf{F} \rangle} = \frac{\partial}{\partial \mathbf{F}} \cdot\cdot\, \boldsymbol{\mathcal{F}} = \boldsymbol{\mathcal{F}}^T \cdot\cdot\, \frac{\partial}{\partial \mathbf{F}}, \tag{13.10.18}$$

Equation (13.10.17) becomes

$$\mathrm{d}^{\mathrm{p}}\langle \mathbf{P} \rangle = -\frac{\partial}{\partial \langle \mathbf{F} \rangle} \langle \sum_{\alpha=1}^{n} \tau^\alpha\, \mathrm{d}\gamma^\alpha \rangle. \tag{13.10.19}$$

This shows that $-\langle \sum \tau^\alpha\, \mathrm{d}\gamma^\alpha \rangle$ is a plastic potential for $\mathrm{d}^{\mathrm{p}}\langle \mathbf{P} \rangle$ over an elastic domain in $\langle \mathbf{F} \rangle$ space (Hill and Rice, 1973; Havner, 1986). Since the number n of active slip systems changes from grain to grain, depending on its orientation and the state of hardening, the sum in Eq. (13.10.19) is kept within the $\langle\ \rangle$ brackets, i.e., within the volume integral appearing in the definition of the $\langle\ \rangle$ average.

13.10.2. Local Residual Increment of Nominal Stress

The plastic part of the increment of macroscopic nominal stress $\mathrm{d}^{\mathrm{p}}\langle\mathbf{P}\rangle$ in Eq. (13.10.3) gives the macroscopic stress decrement after a cycle (application and removal) of the increment of macroscopic deformation gradient $\mathrm{d}\langle\mathbf{F}\rangle$. At the microlevel, however, the local decrement of stress $\mathrm{d}^{\mathrm{s}}\mathbf{P}$, after a cycle of the increment of macroscopic deformation gradient $\mathrm{d}\langle\mathbf{F}\rangle$, is obtained by subtracting from $\mathrm{d}\mathbf{P}$ the local stress increment associated with an imagined (conceptual) elastic removal of $\mathrm{d}\langle\mathbf{F}\rangle$. This is $\boldsymbol{\mathcal{P}}\cdot\cdot[\boldsymbol{\Lambda}]\cdot\cdot\mathrm{d}\langle\mathbf{F}\rangle$, so that (Hill, 1984; Havner, 1992)

$$\mathrm{d}^{\mathrm{s}}\mathbf{P} = \mathrm{d}\mathbf{P} - \boldsymbol{\mathcal{P}}\cdot\cdot[\boldsymbol{\Lambda}]\cdot\cdot\mathrm{d}\langle\mathbf{F}\rangle. \tag{13.10.20}$$

Upon a conceptual elastic removal of macroscopic $\mathrm{d}\langle\mathbf{F}\rangle$, the residual increment of the deformation gradient at microscopic level would be

$$\mathrm{d}^{\mathrm{s}}\mathbf{F} = \mathrm{d}\mathbf{F} - \boldsymbol{\mathcal{F}}\cdot\cdot\mathrm{d}\langle\mathbf{F}\rangle. \tag{13.10.21}$$

Recall from Eq. (13.7.8) that $\boldsymbol{\mathcal{P}}\cdot\cdot[\boldsymbol{\Lambda}] = \boldsymbol{\Lambda} : \boldsymbol{\mathcal{F}}$, so that

$$\mathrm{d}\mathbf{P} - \mathrm{d}^{\mathrm{s}}\mathbf{P} = \boldsymbol{\Lambda}\cdot\cdot\left(\mathrm{d}\mathbf{F} - \mathrm{d}^{\mathrm{s}}\mathbf{F}\right). \tag{13.10.22}$$

Note that $\mathrm{d}^{\mathrm{s}}\mathbf{F}$ is kinematically admissible field (because $\mathrm{d}\mathbf{F}$ and $\boldsymbol{\mathcal{F}}\cdot\cdot\mathrm{d}\langle\mathbf{F}\rangle$ are), while $\mathrm{d}^{\mathrm{s}}\mathbf{P}$ is statically admissible field (because $\mathrm{d}\mathbf{P}$ and $\boldsymbol{\Lambda}\cdot\cdot\boldsymbol{\mathcal{F}}\cdot\cdot\mathrm{d}\langle\mathbf{F}\rangle$ are).

The local increment of stress $\mathrm{d}^{\mathrm{s}}\mathbf{P}$ is different from the local plastic increment

$$\mathrm{d}^{\mathrm{p}}\mathbf{P} = \mathrm{d}\mathbf{P} - \boldsymbol{\Lambda}\cdot\cdot\mathrm{d}\mathbf{F}, \tag{13.10.23}$$

associated with a cycle of the increment of local deformation gradient $\mathrm{d}\mathbf{F}$. They are related by

$$\mathrm{d}^{\mathrm{s}}\mathbf{P} - \mathrm{d}^{\mathrm{p}}\mathbf{P} = \boldsymbol{\Lambda}\cdot\cdot\mathrm{d}^{\mathrm{s}}\mathbf{F}. \tag{13.10.24}$$

Also, it can be easily verified that

$$\mathrm{d}^{\mathrm{s}}\mathbf{F} - \mathrm{d}^{\mathrm{p}}\mathbf{F} = \mathbf{M}\cdot\cdot\mathrm{d}^{\mathrm{s}}\mathbf{P}. \tag{13.10.25}$$

On the other hand,

$$\langle\mathrm{d}^{\mathrm{s}}\mathbf{P}\rangle = \mathrm{d}^{\mathrm{p}}\langle\mathbf{P}\rangle, \quad \langle\mathrm{d}^{\mathrm{s}}\mathbf{F}\rangle = \mathbf{0}, \tag{13.10.26}$$

which follow from Eqs. (13.10.20) and (13.10.21), and $\langle \boldsymbol{\mathcal{F}} \rangle = \langle \boldsymbol{\mathcal{P}} \rangle = \boldsymbol{I}$.

Since $\mathrm{d}^{\mathrm{s}}\mathbf{F}$ is kinematically and $\mathrm{d}^{\mathrm{s}}\mathbf{P}$ is statically admissible field, by the theorem on product averages we obtain

$$\langle \mathrm{d}^{\mathrm{s}}\mathbf{P} \cdot\cdot\, \mathrm{d}^{\mathrm{s}}\mathbf{F} \rangle = \langle \mathrm{d}^{\mathrm{s}}\mathbf{P} \rangle \cdot\cdot \langle \mathrm{d}^{\mathrm{s}}\mathbf{F} \rangle = 0. \tag{13.10.27}$$

There is also an identity for the volume averages of the trace products

$$\langle \delta\mathbf{F} \cdot\cdot\, \mathrm{d}^{\mathrm{s}}\mathbf{P} \rangle = \langle \delta\mathbf{F} \cdot\cdot\, \mathrm{d}^{\mathrm{p}}\mathbf{P} \rangle, \tag{13.10.28}$$

where $\delta\mathbf{F}$ is an increment of the local deformation gradient along purely elastic branch of the response. Indeed,

$$\begin{aligned} \langle \delta\mathbf{F} \cdot\cdot\, \mathrm{d}^{\mathrm{s}}\mathbf{P} \rangle &= \langle \delta\mathbf{F} \cdot\cdot (\mathrm{d}\mathbf{P} - \boldsymbol{\mathcal{P}} \cdot\cdot [\boldsymbol{\Lambda}] \cdot\cdot\, \mathrm{d}\langle \mathbf{F} \rangle) \rangle \\ &= \delta\langle \mathbf{F} \rangle \cdot\cdot\, \mathrm{d}\langle \mathbf{P} \rangle - \langle \delta\mathbf{F} \cdot\cdot \boldsymbol{\mathcal{P}} \rangle \cdot\cdot [\boldsymbol{\Lambda}] \cdot\cdot\, \mathrm{d}\langle \mathbf{F} \rangle. \end{aligned} \tag{13.10.29}$$

It is observed that

$$\langle \delta\mathbf{F} \cdot\cdot \boldsymbol{\mathcal{P}} \rangle = \langle \delta\langle \mathbf{F} \rangle \cdot\cdot \boldsymbol{\mathcal{F}}^T \cdot\cdot \boldsymbol{\mathcal{P}} \rangle = \delta\langle \mathbf{F} \rangle \cdot\cdot \langle \boldsymbol{\mathcal{F}}^T \cdot\cdot \boldsymbol{\mathcal{P}} \rangle = \delta\langle \mathbf{F} \rangle, \tag{13.10.30}$$

because $\langle \boldsymbol{\mathcal{F}}^T \cdot\cdot \boldsymbol{\mathcal{P}} \rangle = \boldsymbol{I}$, by (13.7.12). Thus, Eq. (13.10.29) becomes

$$\langle \delta\mathbf{F} \cdot\cdot\, \mathrm{d}^{\mathrm{s}}\mathbf{P} \rangle = \delta\langle \mathbf{F} \rangle \cdot\cdot\, \mathrm{d}^{\mathrm{p}}\langle \mathbf{P} \rangle. \tag{13.10.31}$$

In view of Eq. (13.10.11), this reduces to Eq. (13.10.28). Furthermore, since $\langle \mathrm{d}^{\mathrm{s}}\mathbf{P} \rangle = \mathrm{d}^{\mathrm{p}}\langle \mathbf{P} \rangle$, Eq. (13.10.31) gives

$$\langle \delta\mathbf{F} \cdot\cdot\, \mathrm{d}^{\mathrm{s}}\mathbf{P} \rangle = \delta\langle \mathbf{F} \rangle \cdot\cdot \langle \mathrm{d}^{\mathrm{s}}\mathbf{P} \rangle. \tag{13.10.32}$$

This was anticipated from the theorem on product averages, because $\delta\mathbf{F}$ is kinematically admissible and $\mathrm{d}^{\mathrm{s}}\mathbf{P}$ is statically admissible field.

The following two identities are noted

$$\langle \mathrm{d}^{\mathrm{s}}\mathbf{F} \cdot\cdot \boldsymbol{\Lambda} \cdot\cdot\, \mathrm{d}^{\mathrm{p}}\mathbf{F} \rangle = \langle \mathrm{d}^{\mathrm{s}}\mathbf{F} \cdot\cdot \boldsymbol{\Lambda} \cdot\cdot\, \mathrm{d}^{\mathrm{s}}\mathbf{F} \rangle, \tag{13.10.33}$$

$$\langle \mathrm{d}^{\mathrm{s}}\mathbf{P} \cdot\cdot \mathbf{M} \cdot\cdot\, \mathrm{d}^{\mathrm{p}}\mathbf{P} \rangle = \langle \mathrm{d}^{\mathrm{s}}\mathbf{P} \cdot\cdot \mathbf{M} \cdot\cdot\, \mathrm{d}^{\mathrm{s}}\mathbf{P} \rangle. \tag{13.10.34}$$

They follow from Eqs. (13.10.24), (13.10.25), and (13.11.26).

13.11. Plastic Increment of Macroscopic Deformation Gradient

Dually to the analysis from the previous section, the increment of macroscopic deformation gradient can be partitioned into its elastic and plastic parts as

$$\mathrm{d}\langle \mathbf{F} \rangle = \mathrm{d}^{\mathrm{e}}\langle \mathbf{F} \rangle + \mathrm{d}^{\mathrm{p}}\langle \mathbf{F} \rangle. \tag{13.11.1}$$

The elastic part is defined by

$$\mathrm{d}^{\mathrm{e}}\langle\mathbf{F}\rangle = [\mathbf{M}]\cdot\cdot\,\mathrm{d}\langle\mathbf{P}\rangle, \tag{13.11.2}$$

while

$$\mathrm{d}^{\mathrm{p}}\langle\mathbf{F}\rangle = \mathrm{d}\langle\mathbf{F}\rangle - [\mathbf{M}]\cdot\cdot\,\mathrm{d}\langle\mathbf{P}\rangle \tag{13.11.3}$$

is the plastic part of the increment $\mathrm{d}\langle\mathbf{F}\rangle$.

To establish the relationship between the plastic increments of macroscopic and microscopic deformation gradients, $\mathrm{d}^{\mathrm{p}}\langle\mathbf{F}\rangle$ and $\mathrm{d}^{\mathrm{p}}\mathbf{F}$, consider the volume average of the trace product between an elastic unloading increment of the local nominal stress $\delta\mathbf{P}$ and the plastic increment of the local deformation gradient $\mathrm{d}^{\mathrm{p}}\mathbf{F}$, i.e.,

$$\langle\delta\mathbf{P}\cdot\cdot\,\mathrm{d}^{\mathrm{p}}\mathbf{F}\rangle = \langle\delta\mathbf{P}\cdot\cdot\,(\mathrm{d}\mathbf{F}-\mathbf{M}\cdot\cdot\,\mathrm{d}\mathbf{P})\rangle = \langle\delta\mathbf{P}\cdot\cdot\,\mathrm{d}\mathbf{F}\rangle - \langle\delta\mathbf{P}\cdot\cdot\,\mathbf{M}\cdot\cdot\,\mathrm{d}\mathbf{P}\rangle. \tag{13.11.4}$$

Since $\mathrm{d}\mathbf{P}$ and $\delta\mathbf{P}$ are statically admissible, and $\mathrm{d}\mathbf{F}$ and $\delta\mathbf{P}\cdot\cdot\,\mathbf{M}$ are kinematically admissible fields, we can use the product theorem of Section 13.3 to write

$$\langle\delta\mathbf{P}\cdot\cdot\,\mathrm{d}\mathbf{F}\rangle = \delta\langle\mathbf{P}\rangle\cdot\cdot\,\mathrm{d}\langle\mathbf{F}\rangle, \tag{13.11.5}$$

$$\langle\delta\mathbf{P}\cdot\cdot\,\mathbf{M}\cdot\cdot\,\mathrm{d}\mathbf{P}\rangle = \langle\delta\mathbf{P}\cdot\cdot\,\mathbf{M}\rangle\cdot\cdot\,\mathrm{d}\langle\mathbf{P}\rangle = \delta\langle\mathbf{P}\rangle\cdot\cdot\,\langle\boldsymbol{\mathcal{P}}^T\cdot\cdot\,\mathbf{M}\rangle\cdot\cdot\,\mathrm{d}\langle\mathbf{P}\rangle. \tag{13.11.6}$$

Upon substitution into Eq. (13.11.4), we obtain

$$\langle\delta\mathbf{P}\cdot\cdot\,\mathrm{d}^{\mathrm{p}}\mathbf{F}\rangle = \delta\langle\mathbf{P}\rangle\cdot\cdot\,(\mathrm{d}\langle\mathbf{F}\rangle - [\mathbf{M}]\cdot\cdot\,\mathrm{d}\langle\mathbf{P}\rangle)\,. \tag{13.11.7}$$

Recall that $[\mathbf{M}]$ is symmetric, and

$$\delta\mathbf{P} = \boldsymbol{\mathcal{P}}\cdot\cdot\,\delta\langle\mathbf{P}\rangle = \delta\langle\mathbf{P}\rangle\cdot\cdot\,\boldsymbol{\mathcal{P}}^T, \tag{13.11.8}$$

so that

$$[\mathbf{M}] = \langle\mathbf{M}\cdot\cdot\,\boldsymbol{\mathcal{P}}\rangle = \langle\boldsymbol{\mathcal{P}}^T\cdot\cdot\,\mathbf{M}\rangle. \tag{13.11.9}$$

Consequently,

$$\langle\delta\mathbf{P}\cdot\cdot\,\mathrm{d}^{\mathrm{p}}\mathbf{F}\rangle = \delta\langle\mathbf{P}\rangle\cdot\cdot\,\mathrm{d}^{\mathrm{p}}\langle\mathbf{F}\rangle. \tag{13.11.10}$$

Note that

$$\langle\delta\mathbf{P}\cdot\cdot\,\mathrm{d}^{\mathrm{p}}\mathbf{F}\rangle = \delta\langle\mathbf{P}\rangle\cdot\cdot\,\mathrm{d}\langle\mathbf{F}\rangle - \delta\langle\mathbf{F}\rangle\cdot\cdot\,\mathrm{d}\langle\mathbf{P}\rangle, \tag{13.11.11}$$

which can be easily verified by substituting $\delta\langle\mathbf{F}\rangle = \delta\langle\mathbf{P}\rangle\cdot\cdot\,[\mathbf{M}]$, and by using Eq. (13.11.3).

On the other hand, from (13.5.2) we have

$$\langle \delta \mathbf{P} \cdot\cdot \mathrm{d}^{\mathrm{p}}\mathbf{F} \rangle = \delta \langle \mathbf{P} \rangle \cdot\cdot \langle \boldsymbol{\mathcal{P}}^{T} \cdot\cdot \mathrm{d}^{\mathrm{p}}\mathbf{F} \rangle. \tag{13.11.12}$$

Comparison of Eqs. (13.11.10) and (13.11.12) yields

$$\mathrm{d}^{\mathrm{p}} \langle \mathbf{F} \rangle = \langle \boldsymbol{\mathcal{P}}^{T} \cdot\cdot \mathrm{d}^{\mathrm{p}}\mathbf{F} \rangle. \tag{13.11.13}$$

Therefore, the plastic part of the increment of macroscopic deformation gradient is a weighted volume average of the plastic part of the increment of local deformation gradient.

13.11.1. Plastic Potential and Normality Rule

From Eq. (13.11.10) it follows, if the normality rule applies at the microlevel, it is transmitted to the macrolevel, i.e.,

$$\delta \mathbf{P} \cdot\cdot \mathrm{d}^{\mathrm{p}}\mathbf{F} < 0 \quad \text{implies} \quad \delta \langle \mathbf{P} \rangle \cdot\cdot \mathrm{d}^{\mathrm{p}} \langle \mathbf{F} \rangle < 0. \tag{13.11.14}$$

From Section 12.7 we recall that $\sum(\tau^{\alpha}\, \mathrm{d}\gamma^{\alpha})$ acts as a plastic potential for $\mathrm{d}^{\mathrm{p}}\mathbf{F}$ over an elastic domain in $\mathbf{P}$ space, such that

$$\mathrm{d}^{\mathrm{p}}\mathbf{F} = \frac{\partial}{\partial \mathbf{P}} \sum_{\alpha=1}^{n} (\tau^{\alpha}\, \mathrm{d}\gamma^{\alpha}). \tag{13.11.15}$$

The partial differentiation is performed at the fixed slip and slip increments $\mathrm{d}\gamma^{\alpha}$. Substitution into Eq. (13.11.13) gives

$$\mathrm{d}^{\mathrm{p}} \langle \mathbf{F} \rangle = \langle \boldsymbol{\mathcal{P}}^{T} \cdot\cdot \frac{\partial}{\partial \mathbf{P}} \sum_{\alpha=1}^{n} (\tau^{\alpha}\, \mathrm{d}\gamma^{\alpha}) \rangle. \tag{13.11.16}$$

Since, at the fixed slip,

$$\frac{\partial}{\partial \langle \mathbf{P} \rangle} = \frac{\partial}{\partial \mathbf{P}} \cdot\cdot \frac{\partial \mathbf{P}}{\partial \langle \mathbf{P} \rangle} = \frac{\partial}{\partial \mathbf{P}} \cdot\cdot \boldsymbol{\mathcal{P}} = \boldsymbol{\mathcal{P}}^{T} \cdot\cdot \frac{\partial}{\partial \mathbf{P}}, \tag{13.11.17}$$

Equation (13.11.16) becomes

$$\mathrm{d}^{\mathrm{p}} \langle \mathbf{F} \rangle = \frac{\partial}{\partial \langle \mathbf{P} \rangle} \langle \sum_{\alpha=1}^{n} \tau^{\alpha}\, \mathrm{d}\gamma^{\alpha} \rangle. \tag{13.11.18}$$

This shows that $\langle \sum \tau^{\alpha}\, \mathrm{d}\gamma^{\alpha} \rangle$ is a plastic potential for $\mathrm{d}^{\mathrm{p}} \langle \mathbf{F} \rangle$ over an elastic domain in $\langle \mathbf{P} \rangle$ space.

13.11.2. Local Residual Increment of Deformation Gradient

The plastic part of the increment of macroscopic deformation gradient $\mathrm{d}^{\mathrm{p}}\langle\mathbf{F}\rangle$ in Eq. (13.11.3) represents a residual increment of macroscopic deformation gradient after a cycle of the increment of macroscopic nominal stress $\mathrm{d}\langle\mathbf{P}\rangle$. At the microlevel, however, the local residual increment of deformation gradient $\mathrm{d}^{\mathrm{s}}\mathbf{F}$, left upon a cycle of $\mathrm{d}\langle\mathbf{P}\rangle$, is obtained by subtracting from $\mathrm{d}\mathbf{F}$ the local deformation gradient increment associated with an imagined elastic removal of $\mathrm{d}\langle\mathbf{P}\rangle$. This is $\boldsymbol{\mathcal{F}}\cdot\cdot[\mathbf{M}]\cdot\cdot\mathrm{d}\langle\mathbf{P}\rangle$, so that

$$\mathrm{d}^{\mathrm{r}}\mathbf{F} = \mathrm{d}\mathbf{F} - \boldsymbol{\mathcal{F}}\cdot\cdot[\mathbf{M}]\cdot\cdot\mathrm{d}\langle\mathbf{P}\rangle. \tag{13.11.19}$$

Upon a conceptual elastic removal of macroscopic $\mathrm{d}\langle\mathbf{P}\rangle$, the residual change of the local nominal stress would be

$$\mathrm{d}^{\mathrm{r}}\mathbf{P} = \mathrm{d}\mathbf{P} - \boldsymbol{\mathcal{P}}\cdot\cdot\mathrm{d}\langle\mathbf{P}\rangle, \tag{13.11.20}$$

since $\boldsymbol{\mathcal{P}}\cdot\cdot\mathrm{d}\langle\mathbf{P}\rangle$ is the local stress due to $\mathrm{d}\langle\mathbf{P}\rangle$ in an imagined elastic response. Recall from Eq. (13.8.8) that $\boldsymbol{\mathcal{F}}\cdot\cdot[\mathbf{M}] = \mathbf{M}:\boldsymbol{\mathcal{P}}$, so that

$$\mathrm{d}\mathbf{F} - \mathrm{d}^{\mathrm{r}}\mathbf{F} = \mathbf{M}\cdot\cdot(\mathrm{d}\mathbf{P} - \mathrm{d}^{\mathrm{r}}\mathbf{P})\,. \tag{13.11.21}$$

Note that $\mathrm{d}^{\mathrm{r}}\mathbf{P}$ is statically admissible field (because $\mathrm{d}\mathbf{P}$ and $\boldsymbol{\mathcal{P}}\cdot\cdot\mathrm{d}\langle\mathbf{P}\rangle$ are), while $\mathrm{d}^{\mathrm{r}}\mathbf{F}$ is kinematically admissible field (because $\mathrm{d}\mathbf{F}$ and $\mathbf{M}\cdot\cdot\boldsymbol{\mathcal{P}}\cdot\cdot\mathrm{d}\langle\mathbf{P}\rangle$ are).

The local increment of deformation gradient $\mathrm{d}^{\mathrm{r}}\mathbf{F}$ is different from the local plastic increment

$$\mathrm{d}^{\mathrm{p}}\mathbf{F} = \mathrm{d}\mathbf{F} - \mathbf{M}\cdot\cdot\mathrm{d}\mathbf{P}, \tag{13.11.22}$$

associated with a cycle of the increment of local nominal stress $\mathrm{d}\mathbf{P}$. They are related by

$$\mathrm{d}^{\mathrm{r}}\mathbf{F} - \mathrm{d}^{\mathrm{p}}\mathbf{F} = \mathbf{M}\cdot\cdot\mathrm{d}^{\mathrm{r}}\mathbf{P}. \tag{13.11.23}$$

In addition, we have

$$\mathrm{d}^{\mathrm{r}}\mathbf{P} - \mathrm{d}^{\mathrm{p}}\mathbf{P} = \boldsymbol{\Lambda}\cdot\cdot\mathrm{d}^{\mathrm{r}}\mathbf{F}. \tag{13.11.24}$$

In general, neither $\mathrm{d}^{\mathrm{p}}\mathbf{F}$ is kinematically admissible, nor $\mathrm{d}^{\mathrm{p}}\mathbf{P}$ is statically admissible field. On the other hand,

$$\langle\mathrm{d}^{\mathrm{r}}\mathbf{F}\rangle = \mathrm{d}^{\mathrm{p}}\langle\mathbf{F}\rangle, \quad \langle\mathrm{d}^{\mathrm{r}}\mathbf{P}\rangle = \mathbf{0}, \tag{13.11.25}$$

which follow from Eqs. (13.11.19) and (13.11.20), and $\langle\boldsymbol{\mathcal{P}}\rangle = \langle\boldsymbol{\mathcal{F}}\rangle = \boldsymbol{I}$.

Since $\mathrm{d}^{\mathrm{r}}\mathbf{F}$ is kinematically and $\mathrm{d}^{\mathrm{r}}\mathbf{P}$ is statically admissible field, by the theorem on product averages we can write

$$\langle \mathrm{d}^{\mathrm{r}}\mathbf{P} \cdot\cdot\, \mathrm{d}^{\mathrm{r}}\mathbf{F} \rangle = \langle \mathrm{d}^{\mathrm{r}}\mathbf{P} \rangle \cdot\cdot \langle \mathrm{d}^{\mathrm{r}}\mathbf{F} \rangle = 0. \tag{13.11.26}$$

There is also an identity for the volume averages of the trace products

$$\langle \delta\mathbf{P} \cdot\cdot\, \mathrm{d}^{\mathrm{r}}\mathbf{F} \rangle = \langle \delta\mathbf{P} \cdot\cdot\, \mathrm{d}^{\mathrm{p}}\mathbf{F} \rangle, \tag{13.11.27}$$

where $\delta\mathbf{P}$ is an increment of the local nominal stress along purely elastic branch of the response. Indeed, by an analogous derivation as in Subsection 13.10.2, there follows

$$\begin{aligned} \langle \delta\mathbf{P} \cdot\cdot\, \mathrm{d}^{\mathrm{r}}\mathbf{F} \rangle &= \langle \delta\mathbf{P} \cdot\cdot (\mathrm{d}\mathrm{F} - \boldsymbol{\mathcal{F}} \cdot\cdot [\mathbf{M}] \cdot\cdot\, \mathrm{d}\langle \mathbf{P} \rangle) \rangle \\ &= \delta\langle \mathbf{P} \rangle \cdot\cdot\, \mathrm{d}\langle \mathbf{F} \rangle - \langle \delta\mathbf{P} \cdot\cdot \boldsymbol{\mathcal{F}} \rangle \cdot\cdot [\mathbf{M}] \cdot\cdot\, \mathrm{d}\langle \mathbf{P} \rangle. \end{aligned} \tag{13.11.28}$$

Furthermore,

$$\langle \delta\mathbf{P} \cdot\cdot \boldsymbol{\mathcal{F}} \rangle = \langle \delta\langle \mathbf{P} \rangle \cdot\cdot \boldsymbol{\mathcal{P}}^T \cdot\cdot \boldsymbol{\mathcal{F}} \rangle = \delta\langle \mathbf{P} \rangle \cdot\cdot \langle \boldsymbol{\mathcal{P}}^T \cdot\cdot \boldsymbol{\mathcal{F}} \rangle = \delta\langle \mathbf{P} \rangle, \tag{13.11.29}$$

because $\langle \boldsymbol{\mathcal{P}}^T \cdot\cdot \boldsymbol{\mathcal{F}} \rangle = \boldsymbol{I}$, by Eq. (13.8.11). Thus, Eq. (13.11.28) becomes

$$\langle \delta\mathbf{P} \cdot\cdot\, \mathrm{d}^{\mathrm{r}}\mathbf{F} \rangle = \delta\langle \mathbf{P} \rangle \cdot\cdot\, \mathrm{d}^{\mathrm{p}}\langle \mathbf{F} \rangle. \tag{13.11.30}$$

In view of Eq. (13.11.10) this reduces to Eq. (13.11.27). Also, since $\langle \mathrm{d}^{\mathrm{r}}\mathbf{F} \rangle = \mathrm{d}^{\mathrm{p}}\langle \mathbf{F} \rangle$, Eq. (13.11.30) gives

$$\langle \delta\mathbf{P} \cdot\cdot\, \mathrm{d}^{\mathrm{r}}\mathbf{F} \rangle = \delta\langle \mathbf{P} \rangle \cdot\cdot \langle \mathrm{d}^{\mathrm{r}}\mathbf{F} \rangle. \tag{13.11.31}$$

This was anticipated from the theorem on product averages, because $\delta\mathbf{P}$ is statically admissible and $\mathrm{d}^{\mathrm{r}}\mathbf{F}$ is kinematically admissible field.

The following two identities, which follow from Eqs. (13.11.23), (13.11.24), and (13.11.26), are noted

$$\langle \mathrm{d}^{\mathrm{r}}\mathbf{F} \cdot\cdot \boldsymbol{\Lambda} \cdot\cdot\, \mathrm{d}^{\mathrm{p}}\mathbf{F} \rangle = \langle \mathrm{d}^{\mathrm{r}}\mathbf{F} \cdot\cdot \boldsymbol{\Lambda} \cdot\cdot\, \mathrm{d}^{\mathrm{r}}\mathbf{F} \rangle, \tag{13.11.32}$$

$$\langle \mathrm{d}^{\mathrm{r}}\mathbf{P} \cdot\cdot \mathbf{M} \cdot\cdot\, \mathrm{d}^{\mathrm{p}}\mathbf{P} \rangle = \langle \mathrm{d}^{\mathrm{r}}\mathbf{P} \cdot\cdot \mathbf{M} \cdot\cdot\, \mathrm{d}^{\mathrm{r}}\mathbf{P} \rangle. \tag{13.11.33}$$

By comparing the results of this subsection with those from the Subsection 13.10.2, it can be easily verified that

$$\mathrm{d}^{\mathrm{r}}\mathbf{P} - \mathrm{d}^{\mathrm{s}}\mathbf{P} = \boldsymbol{\Lambda} \cdot\cdot (\mathrm{d}^{\mathrm{r}}\mathbf{F} - \mathrm{d}^{\mathrm{s}}\mathbf{F})\,. \tag{13.11.34}$$

The local residual quantities here discussed are of interest in the analysis of the work and energy-related macroscopic quantities considered in Section 13.14.

13.12. Plastic Increment of Macroscopic Piola–Kirchhoff Stress

The increment of the macroscopic symmetric Piola–Kirchhoff stress can be partitioned into its elastic and plastic parts, such that

$$\mathrm{d}[\mathbf{T}] = \mathrm{d}^{\mathrm{e}}[\mathbf{T}] + \mathrm{d}^{\mathrm{p}}[\mathbf{T}]. \tag{13.12.1}$$

The elastic part is defined by

$$\mathrm{d}^{\mathrm{e}}[\mathbf{T}] = [\mathbf{\Lambda}_{(1)}] : \mathrm{d}[\mathbf{E}]. \tag{13.12.2}$$

The remaining part,

$$\mathrm{d}^{\mathrm{p}}[\mathbf{T}] = \mathrm{d}[\mathbf{T}] - [\mathbf{\Lambda}_{(1)}] : \mathrm{d}[\mathbf{E}], \tag{13.12.3}$$

is the plastic part of the increment $\mathrm{d}[\mathbf{T}]$. The macroscopic elastoplastic increment of the Lagrangian strain is $\mathrm{d}[\mathbf{E}]$.

The plastic part $\mathrm{d}^{\mathrm{p}}[\mathbf{T}]$ can be related to $\mathrm{d}^{\mathrm{p}}\langle\mathbf{P}\rangle$ by substituting Eq. (13.9.4), and

$$\mathrm{d}\langle\mathbf{P}\rangle = \langle\boldsymbol{\mathcal{K}}\rangle^T : \mathrm{d}[\mathbf{T}] + [\boldsymbol{\mathcal{T}}]\cdot\cdot\,\mathrm{d}\langle\mathbf{F}\rangle, \tag{13.12.4}$$

$$\mathrm{d}[\mathbf{E}] = \langle\boldsymbol{\mathcal{K}}\rangle\cdot\cdot\,\mathrm{d}\langle\mathbf{F}\rangle, \tag{13.12.5}$$

into Eq. (13.10.3). The result is

$$\mathrm{d}^{\mathrm{p}}\langle\mathbf{P}\rangle = \langle\boldsymbol{\mathcal{K}}\rangle^T : \mathrm{d}^{\mathrm{p}}[\mathbf{T}]. \tag{13.12.6}$$

Normality Rules

To discuss the normality rules, we first observe that

$$\delta\langle\mathbf{F}\rangle\cdot\cdot\,\mathrm{d}^{\mathrm{p}}\langle\mathbf{P}\rangle = \delta\langle\mathbf{F}\rangle\cdot\cdot\,\langle\boldsymbol{\mathcal{K}}\rangle^T : \mathrm{d}^{\mathrm{p}}[\mathbf{T}] = \delta[\mathbf{E}] : \mathrm{d}^{\mathrm{p}}[\mathbf{T}]. \tag{13.12.7}$$

This shows, if the normality holds for the plastic part of the increment of macroscopic nominal stress, it also holds for the plastic part of the increment of macroscopic Piola–Kirchhoff stress, and *vice versa*, i.e.,

$$\delta\langle\mathbf{F}\rangle\cdot\cdot\,\mathrm{d}^{\mathrm{p}}\langle\mathbf{P}\rangle > 0 \quad \Longleftrightarrow \quad \delta[\mathbf{E}] : \mathrm{d}^{\mathrm{p}}[\mathbf{T}] > 0. \tag{13.12.8}$$

Furthermore, we have

$$\langle\delta\mathbf{F}\cdot\cdot\,\mathrm{d}^{\mathrm{p}}\mathbf{P}\rangle = \langle\delta\mathbf{E} : \mathrm{d}^{\mathrm{p}}\mathbf{T}\rangle, \tag{13.12.9}$$

because locally $\delta\mathbf{F}\cdot\cdot\,\mathrm{d}^{\mathrm{p}}\mathbf{P} = \delta\mathbf{E} : \mathrm{d}^{\mathrm{p}}\mathbf{T}$, as shown in Section 12.14. Thus, by comparing Eqs. (13.12.7) and (13.12.9), and having in mind Eq. (13.10.11), it follows that

$$\langle \delta\mathbf{E} : \mathrm{d}^{\mathrm{p}}\mathbf{T} \rangle = \delta[\mathbf{E}] : \mathrm{d}^{\mathrm{p}}[\mathbf{T}]. \tag{13.12.10}$$

Consequently, if the normality rule applies at the microlevel, it is transmitted to the macrolevel,

$$\delta\mathbf{E} : \mathrm{d}^{\mathrm{p}}\mathbf{T} > 0 \quad \Longrightarrow \quad \delta[\mathbf{E}] : \mathrm{d}^{\mathrm{p}}[\mathbf{T}] > 0. \tag{13.12.11}$$

We can derive an expression for $\mathrm{d}^{\mathrm{p}}\mathbf{T}$ in terms of the macroscopic plastic potential. To that goal, note that

$$\frac{\partial}{\partial\langle\mathbf{F}\rangle} = \langle\boldsymbol{\mathcal{K}}\rangle^{T} : \frac{\partial}{\partial[\mathbf{E}]}\,. \tag{13.12.12}$$

When this is substituted into Eq. (13.10.19), there follows

$$\mathrm{d}^{\mathrm{p}}\langle\mathbf{P}\rangle = -\frac{\partial}{\partial\langle\mathbf{F}\rangle}\langle\sum_{\alpha=1}^{n}\tau^{\alpha}\,\mathrm{d}\gamma^{\alpha}\rangle = -\langle\boldsymbol{\mathcal{K}}\rangle^{T} : \frac{\partial}{\partial[\mathbf{E}]}\langle\sum_{\alpha=1}^{n}\tau^{\alpha}\,\mathrm{d}\gamma^{\alpha}\rangle, \tag{13.12.13}$$

and the comparison with Eq. (13.12.6) establishes

$$\mathrm{d}^{\mathrm{p}}[\mathbf{T}] = -\frac{\partial}{\partial[\mathbf{E}]}\langle\sum_{\alpha=1}^{n}\tau^{\alpha}\,\mathrm{d}\gamma^{\alpha}\rangle. \tag{13.12.14}$$

This demonstrates that $-\langle\sum\tau^{\alpha}\,\mathrm{d}\gamma^{\alpha}\rangle$ is the plastic potential for $\mathrm{d}^{\mathrm{p}}[\mathbf{T}]$ over an elastic domain in $[\mathbf{E}]$ space . This result is originally due to Hill and Rice (1973).

13.13. Plastic Increment of Macroscopic Lagrangian Strain

The increment of the macroscopic Lagrangian strain is partitioned into its elastic and plastic parts as

$$\mathrm{d}[\mathbf{E}] = \mathrm{d}^{\mathrm{e}}[\mathbf{E}] + \mathrm{d}^{\mathrm{p}}[\mathbf{E}]. \tag{13.13.1}$$

The elastic part is

$$\mathrm{d}^{\mathrm{e}}[\mathbf{E}] = [\mathbf{M}_{(1)}] : \mathrm{d}[\mathbf{T}], \tag{13.13.2}$$

while

$$\mathrm{d}^{\mathrm{p}}[\mathbf{E}] = \mathrm{d}[\mathbf{E}] - [\mathbf{M}_{(1)}] : \mathrm{d}[\mathbf{T}] \tag{13.13.3}$$

represents the plastic part of the increment $\mathrm{d}[\mathbf{E}]$. The tensor of macroscopic elastic compliances is

$$[\mathbf{M}_{(1)}] = [\boldsymbol{\Lambda}_{(1)}]^{-1}. \tag{13.13.4}$$

From Eqs. (13.12.3) and (13.13.3), we observe the connections

$$\mathrm{d}^{\mathrm{p}}[\mathbf{T}] = -[\mathbf{\Lambda}_{(1)}] : \mathrm{d}[\mathbf{E}], \quad \mathrm{d}^{\mathrm{p}}[\mathbf{E}] = -[\mathbf{M}_{(1)}] : \mathrm{d}[\mathbf{T}]. \tag{13.13.5}$$

The plastic part $\mathrm{d}^{\mathrm{p}}[\mathbf{E}]$ can be related to $\mathrm{d}^{\mathrm{p}}\langle\mathbf{F}\rangle$ by substituting

$$\mathrm{d}^{\mathrm{p}}\langle\mathbf{P}\rangle = -[\mathbf{\Lambda}] : \mathrm{d}^{\mathrm{p}}\langle\mathbf{F}\rangle, \quad \mathrm{d}^{\mathrm{p}}[\mathbf{T}] = -[\mathbf{\Lambda}_{(1)}] : \mathrm{d}^{\mathrm{p}}[\mathbf{E}] \tag{13.13.6}$$

into Eq. (13.12.6). The result is

$$[\mathbf{\Lambda}] \cdot\cdot\, \mathrm{d}^{\mathrm{p}}\langle\mathbf{F}\rangle = \langle\boldsymbol{\mathcal{K}}\rangle^T : [\mathbf{\Lambda}_{(1)}] : \mathrm{d}^{\mathrm{p}}[\mathbf{E}], \tag{13.13.7}$$

i.e.,

$$\mathrm{d}^{\mathrm{p}}\langle\mathbf{F}\rangle = [\mathbf{M}] \cdot\cdot\, \langle\boldsymbol{\mathcal{K}}\rangle^T : [\mathbf{\Lambda}_{(1)}] : \mathrm{d}^{\mathrm{p}}[\mathbf{E}]. \tag{13.13.8}$$

Normality Rules

First, it is noted that

$$\delta\langle\mathbf{P}\rangle \cdot\cdot\, \mathrm{d}^{\mathrm{p}}\langle\mathbf{F}\rangle = \delta\langle\mathbf{P}\rangle \cdot\cdot\, [\mathbf{M}] \cdot\cdot\, \langle\boldsymbol{\mathcal{K}}\rangle^T : [\mathbf{\Lambda}_{(1)}] : \mathrm{d}^{\mathrm{p}}[\mathbf{E}]. \tag{13.13.9}$$

Since

$$\delta\langle\mathbf{P}\rangle \cdot\cdot\, [\mathbf{M}] \cdot\cdot\, \langle\boldsymbol{\mathcal{K}}\rangle^T = \delta\langle\mathbf{F}\rangle \cdot\cdot\, \langle\boldsymbol{\mathcal{K}}\rangle^T = \delta[\mathbf{E}], \tag{13.13.10}$$

and

$$\delta[\mathbf{E}] : [\mathbf{\Lambda}_{(1)}] = \delta[\mathbf{T}], \tag{13.13.11}$$

Equation (13.13.9) becomes

$$\delta\langle\mathbf{P}\rangle \cdot\cdot\, \mathrm{d}^{\mathrm{p}}\langle\mathbf{F}\rangle = \delta[\mathbf{T}] : \mathrm{d}^{\mathrm{p}}[\mathbf{E}]. \tag{13.13.12}$$

Therefore, if the normality holds for the plastic part of the increment of macroscopic deformation gradient, it also holds for the plastic part of the increment of macroscopic Lagrangian strain, and *vice versa*, i.e.,

$$\delta\langle\mathbf{P}\rangle \cdot\cdot\, \mathrm{d}^{\mathrm{p}}\langle\mathbf{F}\rangle < 0 \quad \Longleftrightarrow \quad \delta[\mathbf{T}] : \mathrm{d}^{\mathrm{p}}[\mathbf{E}] < 0. \tag{13.13.13}$$

Next, there is an identity

$$\langle\delta\mathbf{P} \cdot\cdot\, \mathrm{d}^{\mathrm{p}}\mathbf{F}\rangle = \langle\delta\mathbf{T} : \mathrm{d}^{\mathrm{p}}\mathbf{E}\rangle, \tag{13.13.14}$$

because locally $\delta\mathbf{P} \cdot\cdot\, \mathrm{d}^{\mathrm{p}}\mathbf{F} = \delta\mathbf{T} : \mathrm{d}^{\mathrm{p}}\mathbf{E}$, as can be inferred from the analysis in Section 12.14. Thus, by comparing Eqs. (13.13.12) and (13.13.14), and by recalling Eq. (13.11.10), it follows that

$$\langle\delta\mathbf{T} : \mathrm{d}^{\mathrm{p}}\mathbf{E}\rangle = \delta[\mathbf{T}] : \mathrm{d}^{\mathrm{p}}[\mathbf{E}]. \tag{13.13.15}$$

Consequently, if the normality rule applies at the microlevel, it is transmitted to the macrolevel (Hill, 1972), i.e.,

$$\delta\mathbf{T} : \mathrm{d}^{\mathrm{p}}\mathbf{E} < 0 \quad \Longrightarrow \quad \delta[\mathbf{T}] : \mathrm{d}^{\mathrm{p}}[\mathbf{E}] < 0. \tag{13.13.16}$$

In the context of small deformation the result was originally obtained by Mandel (1966) and Hill (1967).

An expression for $\mathrm{d}^{\mathrm{p}}\mathbf{E}$ can be derived in terms of the macroscopic plastic potential by using the chain rule,

$$\frac{\partial}{\partial[\mathbf{E}]} = \frac{\partial}{\partial[\mathbf{T}]} : [\mathbf{\Lambda}_{(1)}], \tag{13.13.17}$$

in Eq. (13.12.14). This gives

$$\mathrm{d}^{\mathrm{p}}[\mathbf{T}] = -\frac{\partial}{\partial[\mathbf{T}]} : [\mathbf{\Lambda}_{(1)}]\,\langle\sum_{\alpha=1}^{n} \tau^{\alpha}\,\mathrm{d}\gamma^{\alpha}\rangle. \tag{13.13.18}$$

Upon the trace product with $[\mathbf{M}_{(1)}]$, we obtain

$$\mathrm{d}^{\mathrm{p}}[\mathbf{E}] = \frac{\partial}{\partial[\mathbf{T}]}\,\langle\sum_{\alpha=1}^{n} \tau^{\alpha}\,\mathrm{d}\gamma^{\alpha}\rangle, \tag{13.13.19}$$

having regard to (13.13.5). This shows that $\langle\sum \tau^{\alpha}\,\mathrm{d}\gamma^{\alpha}\rangle$ is a plastic potential for $\mathrm{d}^{\mathrm{p}}[\mathbf{E}]$ over an elastic domain in $[\mathbf{T}]$ space.

13.14. Macroscopic Increment of Plastic Work

The macroscopic increment of slip work, per unit volume of the macroelement, is the volume average

$$\langle\mathrm{d}w^{\mathrm{slip}}\rangle = \langle\sum_{\alpha=1}^{n} \tau^{\alpha}\,\mathrm{d}\gamma^{\alpha}\rangle = \frac{1}{V^0}\int_{V^0}\left(\sum_{\alpha=1}^{n} \tau^{\alpha}\,\mathrm{d}\gamma^{\alpha}\right)\mathrm{d}V^0. \tag{13.14.1}$$

The number n of active slip systems changes from grain to grain within the macroelement, depending on the grain orientation and the state of hardening.

Another quantity, which will be referred to as the macroscopic increment of plastic work, can be introduced as follows. Consider a cycle of the application and removal of the macroscopic increment of nominal stress $\mathrm{d}\langle\mathbf{P}\rangle$. The corresponding macroscopic work can be determined by considering the volume average of the first-order work quantity

$$\mathbf{P}\cdot\cdot\,\mathrm{d}^{\mathrm{p}}\mathbf{F} = \mathbf{P}\cdot\cdot\,(\mathrm{d}^{\mathrm{r}}\mathbf{F} - \mathbf{M}\cdot\cdot\,\mathrm{d}^{\mathrm{r}}\mathbf{P})\,, \tag{13.14.2}$$

which is

$$\langle \mathbf{P} \cdot\cdot \mathrm{d}^{\mathrm{p}}\mathbf{F} \rangle = \langle \mathbf{P} \rangle \cdot\cdot \mathrm{d}^{\mathrm{p}}\langle \mathbf{F} \rangle - \langle \mathbf{P} \cdot\cdot \mathbf{M} \cdot\cdot \mathrm{d}^{\mathrm{r}}\mathbf{P} \rangle. \tag{13.14.3}$$

This follows because $\mathbf{P}$ is statically admissible and $\mathrm{d}^{\mathrm{r}}\mathbf{F}$ is kinematically admissible, so that

$$\langle \mathbf{P} \cdot\cdot \mathrm{d}^{\mathrm{r}}\mathbf{F} \rangle = \langle \mathbf{P} \rangle \cdot\cdot \langle \mathrm{d}^{\mathrm{r}}\mathbf{F} \rangle = \langle \mathbf{P} \rangle \cdot\cdot \mathrm{d}^{\mathrm{p}}\langle \mathbf{F} \rangle. \tag{13.14.4}$$

Thus,

$$\langle \mathbf{P} \rangle \cdot\cdot \mathrm{d}^{\mathrm{p}}\langle \mathbf{F} \rangle = \langle \mathbf{P} \cdot\cdot \mathrm{d}^{\mathrm{p}}\mathbf{F} \rangle + \langle \mathbf{P} \cdot\cdot \mathbf{M} \cdot\cdot \mathrm{d}^{\mathrm{r}}\mathbf{P} \rangle. \tag{13.14.5}$$

The result shows that the macroscopic first-order work quantity in the cycle of $\mathrm{d}\langle \mathbf{P} \rangle$ is not equal to the volume average of the local work quantity $\mathbf{P} \cdot\cdot \mathrm{d}^{\mathrm{p}}\mathbf{F}$. This was expected on physical grounds, because cycling $\mathrm{d}\langle \mathbf{P} \rangle$ macroscopically does not simultaneously cycle every $\mathrm{d}\mathbf{P}$ locally. In fact, the residual increment of stress left locally upon the cycle of $\mathrm{d}\langle \mathbf{P} \rangle$ is $\mathrm{d}^{\mathrm{r}}\langle \mathbf{P} \rangle$ of Eq. (13.11.20).

To analyze the increment of macroscopic plastic work with an accuracy to the second order, consider

$$\langle (\mathbf{P} + \frac{1}{2}\mathrm{d}\mathbf{P}) \cdot\cdot \mathrm{d}^{\mathrm{p}}\mathbf{F} \rangle = \langle \mathbf{P} \cdot\cdot \mathrm{d}^{\mathrm{p}}\mathbf{F} \rangle + \frac{1}{2}\langle \mathrm{d}\mathbf{P} \cdot\cdot \mathrm{d}^{\mathrm{p}}\mathbf{F} \rangle. \tag{13.14.6}$$

The second-order contribution can be expressed by using the identity

$$\mathrm{d}\mathbf{P} \cdot\cdot \mathrm{d}^{\mathrm{p}}\mathbf{F} = \mathrm{d}\mathbf{P} \cdot\cdot (\mathrm{d}^{\mathrm{r}}\mathbf{F} - \mathbf{M} \cdot\cdot \mathrm{d}^{\mathrm{r}}\mathbf{P}). \tag{13.14.7}$$

In view of (13.11.20), this can be rewritten as

$$\mathrm{d}\mathbf{P} \cdot\cdot \mathrm{d}^{\mathrm{p}}\mathbf{F} = \mathrm{d}\mathbf{P} \cdot\cdot \mathrm{d}^{\mathrm{r}}\mathbf{F} - \left(\mathrm{d}^{\mathrm{r}}\mathbf{P} + \mathrm{d}\langle \mathbf{P} \rangle \cdot\cdot \boldsymbol{\mathcal{P}}^T\right) \cdot\cdot \mathbf{M} \cdot\cdot \mathrm{d}^{\mathrm{r}}\mathbf{P}. \tag{13.14.8}$$

Since $\mathrm{d}^{\mathrm{r}}\mathbf{F}$ and $\mathrm{d}\langle \mathbf{P} \rangle \cdot\cdot \boldsymbol{\mathcal{P}}^T \cdot\cdot \mathbf{M} = \mathbf{M} \cdot\cdot \boldsymbol{\mathcal{P}} \cdot\cdot \mathrm{d}\langle \mathbf{P} \rangle$ are kinematically admissible fields, and since $\langle \mathrm{d}^{\mathrm{r}}\mathbf{F} \rangle = \mathrm{d}^{\mathrm{p}}\langle \mathbf{F} \rangle$ and $\langle \mathrm{d}^{\mathrm{r}}\mathbf{P} \rangle = \mathbf{0}$, upon the averaging of Eq. (13.14.8) we obtain

$$\langle \mathrm{d}\mathbf{P} \cdot\cdot \mathrm{d}^{\mathrm{p}}\mathbf{F} \rangle = \mathrm{d}\langle \mathbf{P} \rangle \cdot\cdot \mathrm{d}^{\mathrm{p}}\langle \mathbf{F} \rangle - \langle \mathrm{d}^{\mathrm{r}}\mathbf{P} \cdot\cdot \mathbf{M} \cdot\cdot \mathrm{d}^{\mathrm{r}}\mathbf{P} \rangle, \tag{13.14.9}$$

i.e.,

$$\mathrm{d}\langle \mathbf{P} \rangle \cdot\cdot \mathrm{d}^{\mathrm{p}}\langle \mathbf{F} \rangle = \langle \mathrm{d}\mathbf{P} \cdot\cdot \mathrm{d}^{\mathrm{p}}\mathbf{F} \rangle + \langle \mathrm{d}^{\mathrm{r}}\mathbf{P} \cdot\cdot \mathbf{M} \cdot\cdot \mathrm{d}^{\mathrm{r}}\mathbf{P} \rangle. \tag{13.14.10}$$

Combining Eqs. (13.14.4), (13.14.6), and (13.14.9), the increment of macroscopic plastic work, to second order, can be expressed as

$$\begin{aligned}(\langle \mathbf{P}\rangle + \frac{1}{2}\,\mathrm{d}\langle \mathbf{P}\rangle)\cdot\cdot\,\mathrm{d}^{\mathrm{p}}\langle \mathbf{F}\rangle &= \langle (\mathbf{P} + \frac{1}{2}\,\mathrm{d}\mathbf{P})\cdot\cdot\,\mathrm{d}^{\mathrm{p}}\mathbf{F}\rangle \\ &+ \langle (\mathbf{P} + \frac{1}{2}\,\mathrm{d}^{\mathrm{r}}\mathbf{P})\cdot\cdot\,\mathbf{M}\cdot\cdot\,\mathrm{d}^{\mathrm{r}}\mathbf{P}\rangle .\end{aligned} \tag{13.14.11}$$

The first- and second-order plastic work quantities, defined by $\mathbf{P}\cdot\cdot\,\mathrm{d}^{\mathrm{p}}\mathbf{F}$ and $\mathrm{d}\mathbf{P}\cdot\cdot\,\mathrm{d}^{\mathrm{p}}\mathbf{F}$, are not equal to $\mathbf{T} : \mathrm{d}^{\mathrm{p}}\mathbf{E}$ and $\mathrm{d}\mathbf{T} : \mathrm{d}^{\mathrm{p}}\mathbf{E}$, as discussed in Section 12.8. The latter quantities are actually not measure invariant, but change their values with the change of the strain and conjugate stress measure.

Related Work Expressions

When the Lagrangian strain and Piola–Kirchhoff stress are used, we have from Eqs. (12.8.13) and (12.8.17),

$$\mathbf{P}\cdot\cdot\,\mathrm{d}^{\mathrm{p}}\mathbf{F} = \mathbf{T} : \mathrm{d}^{\mathrm{p}}\mathbf{E} + \mathbf{T} : \mathbf{M}_{(1)} : \mathrm{d}\mathbf{T} - \mathbf{P}\cdot\cdot\,\mathbf{M}\cdot\cdot\,\mathrm{d}\mathbf{P}, \tag{13.14.12}$$

$$\mathrm{d}\mathbf{P}\cdot\cdot\,\mathrm{d}^{\mathrm{p}}\mathbf{F} = \mathrm{d}\mathbf{T} : \mathrm{d}^{\mathrm{p}}\mathbf{E} + \mathrm{d}\mathbf{T} : \mathbf{M}_{(1)} : \mathrm{d}\mathbf{T} - \mathrm{d}\mathbf{P}\cdot\cdot\,\mathbf{M}\cdot\cdot\,\mathrm{d}\mathbf{P} + \mathrm{d}\mathbf{F}\cdot\cdot\,\boldsymbol{\mathcal{T}}\cdot\cdot\,\mathrm{d}\mathbf{F}. \tag{13.14.13}$$

The corresponding expressions for the macroscopic quantities are readily obtained. The first one is

$$\begin{aligned}\langle \mathbf{P}\rangle\cdot\cdot\,\mathrm{d}^{\mathrm{p}}\langle \mathbf{F}\rangle &= \langle \mathbf{P}\rangle\cdot\cdot\,(\mathrm{d}\langle \mathbf{F}\rangle - [\mathbf{M}]\cdot\cdot\,\mathrm{d}\langle \mathbf{P}\rangle) \\ &= [\mathbf{T}] : \mathrm{d}[\mathbf{E}] - \langle \mathbf{P}\rangle\cdot\cdot\,[\mathbf{M}]\cdot\cdot\,\mathrm{d}\langle \mathbf{P}\rangle ,\end{aligned} \tag{13.14.14}$$

i.e.,

$$\langle \mathbf{P}\rangle\cdot\cdot\,\mathrm{d}^{\mathrm{p}}\langle \mathbf{F}\rangle = [\mathbf{T}] : \mathrm{d}^{\mathrm{p}}[\mathbf{E}] + [\mathbf{T}] : [\mathbf{M}_{(1)}] : \mathrm{d}[\mathbf{T}] - \langle \mathbf{P}\rangle\cdot\cdot\,[\mathbf{M}]\cdot\cdot\,\mathrm{d}\langle \mathbf{P}\rangle . \tag{13.14.15}$$

Similarly,

$$\mathrm{d}\langle \mathbf{P}\rangle\cdot\cdot\,\mathrm{d}^{\mathrm{p}}\langle \mathbf{F}\rangle = \mathrm{d}[\mathbf{T}] : \mathrm{d}[\mathbf{E}] - \mathrm{d}\langle \mathbf{P}\rangle\cdot\cdot\,[\mathbf{M}]\cdot\cdot\,\mathrm{d}\langle \mathbf{P}\rangle + \mathrm{d}\langle \mathbf{F}\rangle\cdot\cdot\,[\boldsymbol{\mathcal{T}}]\cdot\cdot\,\mathrm{d}\langle \mathbf{F}\rangle , \tag{13.14.16}$$

and

$$\begin{aligned}\mathrm{d}\langle \mathbf{P}\rangle\cdot\cdot\,\mathrm{d}^{\mathrm{p}}\langle \mathbf{F}\rangle &= \mathrm{d}[\mathbf{T}] : \mathrm{d}^{\mathrm{p}}[\mathbf{E}] + \mathrm{d}[\mathbf{T}] : [\mathbf{M}_{(1)}] : \mathrm{d}[\mathbf{T}] \\ &- \mathrm{d}\langle \mathbf{P}\rangle\cdot\cdot\,[\mathbf{M}]\cdot\cdot\,\mathrm{d}\langle \mathbf{P}\rangle + \mathrm{d}\langle \mathbf{F}\rangle\cdot\cdot\,[\boldsymbol{\mathcal{T}}]\cdot\cdot\,\mathrm{d}\langle \mathbf{F}\rangle .\end{aligned} \tag{13.14.17}$$

We now proceed to establish the relationships between the macroscopic quantities $[\mathbf{T}] : \mathrm{d}^{\mathrm{p}}[\mathbf{E}]$ and $\mathrm{d}[\mathbf{T}] : \mathrm{d}^{\mathrm{p}}[\mathbf{E}]$, and the volume averages $\langle \mathbf{T} : \mathrm{d}^{\mathrm{p}}\mathbf{E}\rangle$ and $\langle \mathrm{d}\mathbf{T} : \mathrm{d}^{\mathrm{p}}\mathbf{E}\rangle$. First, since from Eq. (13.4.5),

$$[\mathbf{T}] : \mathrm{d}[\mathbf{E}] = \langle \mathbf{T} : \mathrm{d}\mathbf{E}\rangle , \tag{13.14.18}$$

we obtain

$$[\mathbf{T}] : \left(\mathrm{d}^{\mathrm{p}}[\mathbf{E}] + [\mathbf{M}_{(1)}] : \mathrm{d}[\mathbf{T}]\right) = \langle \mathbf{T} : \left(\mathrm{d}^{\mathrm{p}}\mathbf{E} + \mathbf{M}_{(1)} : \mathrm{d}\mathbf{T}\right)\rangle. \tag{13.14.19}$$

Therefore,

$$[\mathbf{T}] : \mathrm{d}^{\mathrm{p}}[\mathbf{E}] = \langle \mathbf{T} : \mathrm{d}^{\mathrm{p}}\mathbf{E}\rangle + \langle \mathbf{T} : \mathbf{M}_{(1)} : \mathrm{d}\mathbf{T}\rangle - [\mathbf{T}] : [\mathbf{M}_{(1)}] : \mathrm{d}[\mathbf{T}]. \tag{13.14.20}$$

To derive the formula for the second-order work quantity, we begin by volume averaging of (13.14.13), i.e.,

$$\begin{aligned} \langle \mathrm{d}\mathbf{P} \cdot\cdot\, \mathrm{d}^{\mathrm{p}}\mathbf{F}\rangle = {} & \langle \mathrm{d}\mathbf{T} : \mathrm{d}^{\mathrm{p}}\mathbf{E}\rangle + \langle \mathrm{d}\mathbf{T} : \mathbf{M}_{(1)} : \mathrm{d}\mathbf{T}\rangle \\ & - \langle \mathrm{d}\mathbf{P} \cdot\cdot\, \mathbf{M} \cdot\cdot\, \mathrm{d}\mathbf{P}\rangle + \langle \mathrm{d}\mathbf{F} \cdot\cdot\, \boldsymbol{\mathcal{T}} \cdot\cdot\, \mathrm{d}\mathbf{F}\rangle. \end{aligned} \tag{13.14.21}$$

On the other hand, there is a relationship

$$\langle \mathrm{d}\mathbf{P} \cdot\cdot\, \mathbf{M} \cdot\cdot\, \mathrm{d}\mathbf{P}\rangle - \langle \mathrm{d}^{\mathrm{r}}\mathbf{P} \cdot\cdot\, \mathbf{M} \cdot\cdot\, \mathrm{d}^{\mathrm{r}}\mathbf{P}\rangle = \mathrm{d}\langle \mathbf{P}\rangle \cdot\cdot\, [\mathbf{M}] \cdot\cdot\, \mathrm{d}\langle \mathbf{P}\rangle. \tag{13.14.22}$$

The latter can be verified by subtracting

$$\langle \mathrm{d}^{\mathrm{r}}\mathbf{P} \cdot\cdot\, \mathbf{M} \cdot\cdot\, \mathrm{d}^{\mathrm{r}}\mathbf{P}\rangle = \langle \mathrm{d}^{\mathrm{r}}\mathbf{P} \cdot\cdot\, \mathbf{M} \cdot\cdot\, (\mathrm{d}\mathbf{P} - \boldsymbol{\mathcal{P}} \cdot\cdot\, \mathrm{d}\langle \mathbf{P}\rangle)\rangle \tag{13.14.23}$$

from

$$\langle \mathrm{d}\mathbf{P} \cdot\cdot\, \mathbf{M} \cdot\cdot\, \mathrm{d}\mathbf{P}\rangle = \langle \left(\mathrm{d}^{\mathrm{r}}\mathbf{P} + \mathrm{d}\langle \mathbf{P}\rangle \cdot\cdot\, \boldsymbol{\mathcal{P}}^T\right) \cdot\cdot\, \mathbf{M} \cdot\cdot\, \mathrm{d}\mathbf{P}\rangle, \tag{13.14.24}$$

and by using the theorem on product averages for the appropriate admissible fields. The results $\langle \boldsymbol{\mathcal{P}}^T \cdot\cdot\, \mathbf{M}\rangle = [\mathbf{M}]$ and $\langle \mathrm{d}^{\mathrm{r}}\mathbf{P}\rangle = \mathbf{0}$, from Eqs. (13.8.6) and (13.11.25), were also used. Substitution of Eq. (13.14.22) into (13.14.21) then gives

$$\begin{aligned} & \mathrm{d}\langle \mathbf{P}\rangle \cdot\cdot\, \mathrm{d}^{\mathrm{p}}\langle \mathbf{F}\rangle + \mathrm{d}\langle \mathbf{P}\rangle \cdot\cdot\, [\mathbf{M}] \cdot\cdot\, \mathrm{d}\langle \mathbf{P}\rangle \\ & \quad = \langle \mathrm{d}\mathbf{T} : \mathrm{d}^{\mathrm{p}}\mathbf{E}\rangle + \langle \mathrm{d}\mathbf{T} : \mathbf{M}_{(1)} : \mathrm{d}\mathbf{T}\rangle + \langle \mathrm{d}\mathbf{F} \cdot\cdot\, \boldsymbol{\mathcal{T}} \cdot\cdot\, \mathrm{d}\mathbf{F}\rangle. \end{aligned} \tag{13.14.25}$$

Equation (13.2.9) was used to eliminate $\langle \mathrm{d}\mathbf{P} \cdot\cdot\, \mathrm{d}^{\mathrm{p}}\mathbf{F}\rangle$ in terms of $\mathrm{d}\langle \mathbf{P}\rangle \cdot\cdot\, \mathrm{d}^{\mathrm{p}}\langle \mathbf{F}\rangle$. By combining Eq. (13.14.25) with Eq. (13.14.17), we finally obtain

$$\begin{aligned} \mathrm{d}[\mathbf{T}] : \mathrm{d}^{\mathrm{p}}[\mathbf{E}] = {} & \langle \mathrm{d}\mathbf{T} : \mathrm{d}^{\mathrm{p}}\mathbf{E}\rangle + \langle \mathrm{d}\mathbf{T} : \mathbf{M}_{(1)} : \mathrm{d}\mathbf{T}\rangle + \langle \mathrm{d}\mathbf{F} \cdot\cdot\, \boldsymbol{\mathcal{T}} \cdot\cdot\, \mathrm{d}\mathbf{F}\rangle \\ & - \mathrm{d}[\mathbf{T}] : [\mathbf{M}_{(1)}] : \mathrm{d}[\mathbf{T}] - \mathrm{d}\langle \mathbf{F}\rangle \cdot\cdot\, [\boldsymbol{\mathcal{T}}] \cdot\cdot\, \mathrm{d}\langle \mathbf{F}\rangle, \end{aligned} \tag{13.14.26}$$

which was originally derived by Hill (1985).

In the infinitesimal (ε) strain theory, there is no distinction between various stress and strain measures, and both (13.14.10) and (13.14.26) reduce to

$$\mathrm{d}\langle \boldsymbol{\sigma}\rangle : \mathrm{d}^{\mathrm{p}}\langle \boldsymbol{\varepsilon}\rangle = \langle \mathrm{d}\boldsymbol{\sigma} : \mathrm{d}^{\mathrm{p}}\boldsymbol{\varepsilon}\rangle + \langle \mathrm{d}^{\mathrm{r}}\boldsymbol{\sigma} : \mathbf{M} : \mathrm{d}^{\mathrm{r}}\boldsymbol{\sigma}\rangle. \tag{13.14.27}$$

The rotational effects on the stress rate are neglected if Eq. (13.14.27) is deduced from Eq. (13.14.26), and the Cauchy stress $\boldsymbol{\sigma}$ is used in place of $\mathbf{P}$ in Eq. (13.14.22). All elastic compliances are given by the tensor $\mathbf{M}$. Equation (13.14.27) was originally derived by Mandel (1966). With the positive definite $\mathbf{M}$, it follows that

$$\mathrm{d}\langle\boldsymbol{\sigma}\rangle : \mathrm{d}^{\mathrm{p}}\langle\boldsymbol{\varepsilon}\rangle > \langle \mathrm{d}\boldsymbol{\sigma} : \mathrm{d}^{\mathrm{p}}\boldsymbol{\varepsilon}\rangle. \tag{13.14.28}$$

Thus, within infinitesimal range, the stability at microlevel, $\mathrm{d}\boldsymbol{\sigma} : \mathrm{d}^{\mathrm{p}}\boldsymbol{\varepsilon} > 0$, ensures the stability at macrolevel, $\mathrm{d}\langle\boldsymbol{\sigma}\rangle : \mathrm{d}^{\mathrm{p}}\langle\boldsymbol{\varepsilon}\rangle > 0$.

13.15. Nontransmissibility of Basic Crystal Inequality

Consider a cycle of the application and removal of the macroscopic increment of deformation gradient $\mathrm{d}\langle\mathbf{F}\rangle$. Since

$$\mathbf{F} \cdot\cdot\, \mathrm{d}^{\mathrm{p}}\mathbf{P} = \mathbf{F} \cdot\cdot\, (\mathrm{d}^{\mathrm{s}}\mathbf{P} - \boldsymbol{\Lambda} \cdot\cdot\, \mathrm{d}^{\mathrm{s}}\mathbf{F})\,, \tag{13.15.1}$$

the volume average is

$$\langle \mathbf{F} \cdot\cdot\, \mathrm{d}^{\mathrm{p}}\mathbf{P}\rangle = \langle\mathbf{F}\rangle \cdot\cdot\, \mathrm{d}^{\mathrm{p}}\langle\mathbf{P}\rangle - \langle \mathbf{F} \cdot\cdot\, \boldsymbol{\Lambda} \cdot\cdot\, \mathrm{d}^{\mathrm{s}}\mathbf{F}\rangle. \tag{13.15.2}$$

This follows because $\mathbf{F}$ is kinematically admissible and $\mathrm{d}^{\mathrm{s}}\mathbf{P}$ is statically admissible, so that

$$\langle \mathbf{F} \cdot\cdot\, \mathrm{d}^{\mathrm{s}}\mathbf{P}\rangle = \langle\mathbf{F}\rangle \cdot\cdot\, \langle \mathrm{d}^{\mathrm{s}}\mathbf{P}\rangle = \langle\mathbf{F}\rangle \cdot\cdot\, \mathrm{d}^{\mathrm{p}}\langle\mathbf{P}\rangle. \tag{13.15.3}$$

Thus, dually to Eq. (13.14.5), we have

$$\langle\mathbf{F}\rangle \cdot\cdot\, \mathrm{d}^{\mathrm{p}}\langle\mathbf{P}\rangle = \langle \mathbf{F} \cdot\cdot\, \mathrm{d}^{\mathrm{p}}\mathbf{P}\rangle + \langle \mathbf{F} \cdot\cdot\, \boldsymbol{\Lambda} \cdot\cdot\, \mathrm{d}^{\mathrm{s}}\mathbf{F}\rangle. \tag{13.15.4}$$

This was expected on physical grounds, because cycling $\mathrm{d}\langle\mathbf{F}\rangle$ macroscopically does not simultaneously cycle every $\mathrm{d}\mathbf{F}$ locally. In fact, the residual increment of deformation left locally upon the cycle of $\mathrm{d}\langle\mathbf{F}\rangle$ is $\mathrm{d}^{\mathrm{s}}\langle\mathbf{F}\rangle$, given by Eq. (13.10.21).

Consider next the net expenditure of work in a cycle of $\mathrm{d}\langle\mathbf{F}\rangle$. By the trapezoidal rule of quadrature, the net work expended locally is

$$-\frac{1}{2}\,\mathrm{d}\mathbf{F} \cdot\cdot\, \mathrm{d}^{\mathrm{p}}\mathbf{P}, \tag{13.15.5}$$

to second-order. The quantity

$$\mathrm{d}\mathbf{F} \cdot\cdot\, \mathrm{d}^{\mathrm{p}}\mathbf{P} = \mathrm{d}\mathbf{F} \cdot\cdot\, (\mathrm{d}^{\mathrm{s}}\mathbf{P} - \boldsymbol{\Lambda} \cdot\cdot\, \mathrm{d}^{\mathrm{s}}\mathbf{F}) \tag{13.15.6}$$

can be rewritten, by using Eq. (13.10.21), as

$$\mathrm{d}\mathbf{F}\cdot\cdot\,\mathrm{d}^{\mathrm{p}}\mathbf{P} = \mathrm{d}\mathbf{F}\cdot\cdot\,\mathrm{d}^{\mathrm{s}}\mathbf{P} - \left(\mathrm{d}^{\mathrm{s}}\mathbf{F} + \mathrm{d}\langle\mathbf{F}\rangle\cdot\cdot\,\boldsymbol{\mathcal{F}}^{T}\right)\cdot\cdot\,\boldsymbol{\Lambda}\cdot\cdot\,\mathrm{d}^{\mathrm{s}}\mathbf{F}. \tag{13.15.7}$$

Since $\mathrm{d}^{\mathrm{s}}\mathbf{P}$ and $\mathrm{d}\langle\mathbf{F}\rangle\cdot\cdot\,\boldsymbol{\mathcal{F}}^{T}\cdot\cdot\,\boldsymbol{\Lambda} = \boldsymbol{\Lambda}\cdot\cdot\,\boldsymbol{\mathcal{F}}\cdot\cdot\,\mathrm{d}\langle\mathbf{F}\rangle$ are statically admissible fields, and since $\langle\mathrm{d}^{\mathrm{s}}\mathbf{P}\rangle = \mathrm{d}^{\mathrm{p}}\langle\mathbf{P}\rangle$ and $\langle\mathrm{d}^{\mathrm{s}}\mathbf{F}\rangle = \mathbf{0}$, upon the averaging of Eq. (13.15.7) we obtain

$$\langle\mathrm{d}\mathbf{F}\cdot\cdot\,\mathrm{d}^{\mathrm{p}}\mathbf{P}\rangle = \mathrm{d}\langle\mathbf{F}\rangle\cdot\cdot\,\mathrm{d}^{\mathrm{p}}\langle\mathbf{P}\rangle - \langle\mathrm{d}^{\mathrm{s}}\mathbf{F}\cdot\cdot\,\boldsymbol{\Lambda}\cdot\cdot\,\mathrm{d}^{\mathrm{s}}\mathbf{F}\rangle, \tag{13.15.8}$$

i.e.,

$$\mathrm{d}\langle\mathbf{F}\rangle\cdot\cdot\,\mathrm{d}^{\mathrm{p}}\langle\mathbf{P}\rangle = \langle\mathrm{d}\mathbf{F}\cdot\cdot\,\mathrm{d}^{\mathrm{p}}\mathbf{P}\rangle + \langle\mathrm{d}^{\mathrm{s}}\mathbf{F}\cdot\cdot\,\boldsymbol{\Lambda}\cdot\cdot\,\mathrm{d}^{\mathrm{s}}\mathbf{F}\rangle. \tag{13.15.9}$$

This shows that $\mathrm{d}\langle\mathbf{F}\rangle\cdot\cdot\,\mathrm{d}^{\mathrm{p}}\langle\mathbf{P}\rangle$ is not equal to the volume average of the local quantity $\mathrm{d}\mathbf{F}\cdot\cdot\,\mathrm{d}^{\mathrm{p}}\mathbf{P}$, because cycling $\mathrm{d}\langle\mathbf{F}\rangle$ macroscopically does not simultaneously cycle every $\mathrm{d}\mathbf{F}$ locally.

The second-order work quantity $\mathrm{d}\mathbf{F}\cdot\cdot\,\mathrm{d}^{\mathrm{p}}\mathbf{P}$ is equal to the measure invariant quantity $\mathrm{d}\mathbf{E} : \mathrm{d}^{\mathrm{p}}\mathbf{T}$, as discussed in Section 12.8. Thus,

$$\langle\mathrm{d}\mathbf{F}\cdot\cdot\,\mathrm{d}^{\mathrm{p}}\mathbf{P}\rangle = \langle\mathrm{d}\mathbf{E} : \mathrm{d}^{\mathrm{p}}\mathbf{T}\rangle. \tag{13.15.10}$$

Furthermore, from Eq. (13.12.6), we have

$$\mathrm{d}\langle\mathbf{F}\rangle\cdot\cdot\,\mathrm{d}^{\mathrm{p}}\langle\mathbf{P}\rangle = \mathrm{d}\langle\mathbf{F}\rangle\cdot\cdot\,\langle\boldsymbol{\mathcal{K}}\rangle^{T} : \mathrm{d}^{\mathrm{p}}[\mathbf{T}] = \mathrm{d}[\mathbf{E}] : \mathrm{d}^{\mathrm{p}}[\mathbf{T}]. \tag{13.15.11}$$

Substitution of Eqs. (13.15.10) and (13.15.11) into Eq. (13.15.9) gives

$$\mathrm{d}[\mathbf{E}] : \mathrm{d}^{\mathrm{p}}[\mathbf{T}] = \langle\mathrm{d}\mathbf{E} : \mathrm{d}^{\mathrm{p}}\mathbf{T}\rangle + \langle\mathrm{d}^{\mathrm{s}}\mathbf{F}\cdot\cdot\,\boldsymbol{\Lambda}\cdot\cdot\,\mathrm{d}^{\mathrm{s}}\mathbf{F}\rangle. \tag{13.15.12}$$

The second-order quantity $\mathrm{d}\langle\mathbf{E}\rangle : \mathrm{d}^{\mathrm{p}}\langle\mathbf{T}\rangle$ is not equal to the volume average of the local quantity $\mathrm{d}\mathbf{E} : \mathrm{d}^{\mathrm{p}}\mathbf{T}$, because cycling $\mathrm{d}\langle\mathbf{E}\rangle$ macroscopically does not simultaneously cycle every $\mathrm{d}\mathbf{E}$ locally. We conclude that the macroscopic inequality $\mathrm{d}[\mathbf{E}] : \mathrm{d}^{\mathrm{p}}[\mathbf{T}] < 0$ is not guaranteed by the basic single crystal inequality at the local level $\mathrm{d}\mathbf{E} : \mathrm{d}^{\mathrm{p}}\mathbf{T} < 0$. However, since $\langle\mathrm{d}^{\mathrm{s}}\mathbf{F}\rangle = \mathbf{0}$, it is reasonable to expect that $\langle\mathrm{d}^{\mathrm{s}}\mathbf{F}\cdot\cdot\,\boldsymbol{\Lambda}\cdot\cdot\,\mathrm{d}^{\mathrm{s}}\mathbf{F}\rangle$ is small (being either positive or negative, since $\boldsymbol{\Lambda}$ is not necessarily positive definite); see Havner (1992).

In the infinitesimal strain theory, Eqs. (13.15.9) and (13.15.12) reduce to

$$\mathrm{d}\langle\boldsymbol{\varepsilon}\rangle : \mathrm{d}^{\mathrm{p}}\langle\boldsymbol{\sigma}\rangle = \langle\mathrm{d}\boldsymbol{\varepsilon} : \mathrm{d}^{\mathrm{p}}\boldsymbol{\sigma}\rangle - \langle\mathrm{d}^{\mathrm{s}}\boldsymbol{\varepsilon} : \boldsymbol{\Lambda} : \mathrm{d}^{\mathrm{s}}\boldsymbol{\varepsilon}\rangle. \tag{13.15.13}$$

Equation (13.15.13) was originally derived by Hill (1972). With the positive definite $\boldsymbol{\Lambda}$, it only implies that

$$\mathrm{d}\langle\boldsymbol{\varepsilon}\rangle : \mathrm{d}^{\mathrm{p}}\langle\boldsymbol{\sigma}\rangle > \langle\mathrm{d}\boldsymbol{\varepsilon} : \mathrm{d}^{\mathrm{p}}\boldsymbol{\sigma}\rangle. \tag{13.15.14}$$

Evidently, the stability at the microlevel, $\mathrm{d}\boldsymbol{\varepsilon} : \mathrm{d}^{\mathrm{p}}\boldsymbol{\sigma} < 0$, does not ensure the stability at the macrolevel, $\mathrm{d}\langle\boldsymbol{\varepsilon}\rangle : \mathrm{d}^{\mathrm{p}}\langle\boldsymbol{\sigma}\rangle < 0$.

It is noted that, dually to relation (13.14.22), we have

$$\langle \mathrm{d}\mathbf{F} \cdot\cdot\, \boldsymbol{\Lambda} \cdot\cdot\, \mathrm{d}\mathbf{F}\rangle - \langle \mathrm{d}^{\mathrm{s}}\mathbf{F} \cdot\cdot\, \boldsymbol{\Lambda} \cdot\cdot\, \mathrm{d}^{\mathrm{s}}\mathbf{F}\rangle = \mathrm{d}\langle\mathbf{F}\rangle \cdot\cdot\, [\boldsymbol{\Lambda}] \cdot\cdot\, \mathrm{d}\langle\mathbf{F}\rangle. \tag{13.15.15}$$

This can be verified by subtracting

$$\langle \mathrm{d}^{\mathrm{s}}\mathbf{F} \cdot\cdot\, \boldsymbol{\Lambda} \cdot\cdot\, \mathrm{d}^{\mathrm{s}}\mathbf{F}\rangle = \langle \mathrm{d}^{\mathrm{s}}\mathbf{F} \cdot\cdot\, \boldsymbol{\Lambda} \cdot\cdot\, (\mathrm{d}\mathbf{F} - \boldsymbol{\mathcal{F}} \cdot\cdot\, \mathrm{d}\langle\mathbf{F}\rangle)\rangle \tag{13.15.16}$$

from

$$\langle \mathrm{d}\mathbf{F} \cdot\cdot\, \boldsymbol{\Lambda} \cdot\cdot\, \mathrm{d}\mathbf{F}\rangle = \langle \big(\mathrm{d}^{\mathrm{s}}\mathbf{F} + \mathrm{d}\langle\mathbf{F}\rangle \cdot\cdot\, \boldsymbol{\mathcal{F}}^T\big) \cdot\cdot\, \boldsymbol{\Lambda} \cdot\cdot\, \mathrm{d}\mathbf{F}\rangle, \tag{13.15.17}$$

and by using the theorem on product averages for appropriate admissible fields. The results $\langle\boldsymbol{\mathcal{F}}^T \cdot\cdot\, \boldsymbol{\Lambda}\rangle = [\boldsymbol{\Lambda}]$ and $\langle \mathrm{d}^{\mathrm{s}}\mathbf{F}\rangle = \mathbf{0}$, from Eqs. (13.7.6) and (13.10.26), were also used.

We record an additional result. From Eq. (12.8.18) we have

$$\langle \mathbf{F} \cdot\cdot\, \mathrm{d}^{\mathrm{p}}\mathbf{P}\rangle = \langle \mathbf{C} : \mathrm{d}^{\mathrm{p}}\mathbf{T}\rangle, \tag{13.15.18}$$

where $\mathbf{C} = \mathbf{F}^T \cdot \mathbf{F}$ is the right Cauchy–Green deformation tensor. Thus, in conjunction with (13.3.4), we conclude that

$$[\mathbf{C}] : \mathrm{d}^{\mathrm{p}}[\mathbf{T}] = \langle \mathbf{C} : \mathrm{d}^{\mathrm{p}}\mathbf{T}\rangle + \langle \mathbf{F} \cdot\cdot\, \boldsymbol{\Lambda} \cdot\cdot\, \mathrm{d}^{\mathrm{s}}\mathbf{F}\rangle. \tag{13.15.19}$$

13.16. Analysis of Second-Order Work Quantities

Since $\mathrm{d}\mathbf{P}$ is statically and $\mathrm{d}\mathbf{F}$ is kinematically admissible, by the theorem on product averages, we can write for the volume average of the second-order work quantity

$$\langle \mathrm{d}\mathbf{P} \cdot\cdot\, \mathrm{d}\mathbf{F}\rangle = \langle \mathrm{d}\mathbf{P}\rangle \cdot\cdot\, \langle \mathrm{d}\mathbf{F}\rangle. \tag{13.16.1}$$

Recalling the definitions of plastic increments, we further have

$$\mathrm{d}\langle\mathbf{P}\rangle \cdot\cdot\, \mathrm{d}\langle\mathbf{F}\rangle = \mathrm{d}\langle\mathbf{P}\rangle \cdot\cdot\, \mathrm{d}^{\mathrm{p}}\langle\mathbf{F}\rangle + \mathrm{d}\langle\mathbf{P}\rangle \cdot\cdot\, [\mathbf{M}] \cdot\cdot\, \mathrm{d}\langle\mathbf{P}\rangle, \tag{13.16.2}$$

$$\langle \mathrm{d}\mathbf{P} \cdot\cdot\, \mathrm{d}\mathbf{F}\rangle = \langle \mathrm{d}^{\mathrm{p}}\mathbf{P} \cdot\cdot\, \mathrm{d}\mathbf{F}\rangle + \langle \mathrm{d}\mathbf{F} \cdot\cdot\, \boldsymbol{\Lambda} \cdot\cdot\, \mathrm{d}\mathbf{F}\rangle. \tag{13.16.3}$$

Since $\mathrm{d}\mathbf{F} = \mathrm{d}^{\mathrm{r}}\mathbf{F} + \mathbf{M} \cdot\cdot\, \boldsymbol{\mathcal{P}} \cdot\cdot\, \mathrm{d}\langle\mathbf{P}\rangle$, from Eq. (13.11.19), by expansion and the use of the product theorem, the last term on the right-hand side of Eq. (13.16.3) becomes

$$\langle \mathrm{d}\mathbf{F} \cdot\cdot\, \boldsymbol{\Lambda} \cdot\cdot\, \mathrm{d}\mathbf{F}\rangle = 2\,\mathrm{d}\langle\mathbf{P}\rangle \cdot\cdot\, \mathrm{d}^{\mathrm{p}}\langle\mathbf{F}\rangle + \mathrm{d}\langle\mathbf{P}\rangle \cdot\cdot\, [\mathbf{M}] \cdot\cdot\, \mathrm{d}\langle\mathbf{P}\rangle + \langle \mathrm{d}^{\mathrm{r}}\mathbf{F} \cdot\cdot\, \boldsymbol{\Lambda} \cdot\cdot\, \mathrm{d}^{\mathrm{r}}\mathbf{F}\rangle. \tag{13.16.4}$$

The relationship $\langle \boldsymbol{\mathcal{P}}^T \cdot\cdot \mathbf{M} \cdot\cdot \boldsymbol{\mathcal{P}} \rangle = [\mathbf{M}]$ from Eq. (13.8.12) was also used. The substitution of Eqs. (13.16.2)–(13.16.4) into Eq. (13.16.1) gives

$$\mathrm{d}\langle \mathbf{P} \rangle \cdot\cdot \mathrm{d}^{\mathrm{p}}\langle \mathbf{F} \rangle = -\langle \mathrm{d}^{\mathrm{p}}\mathbf{P} \cdot\cdot \mathrm{d}\mathbf{F} \rangle - \langle \mathrm{d}^{\mathrm{r}}\mathbf{F} \cdot\cdot \boldsymbol{\Lambda} \cdot\cdot \mathrm{d}^{\mathrm{r}}\mathbf{F} \rangle. \tag{13.16.5}$$

Furthermore, by summing the expressions in Eqs. (13.16.5) and (13.15.9), there follows

$$\mathrm{d}\langle \mathbf{P} \rangle \cdot\cdot \mathrm{d}^{\mathrm{p}}\langle \mathbf{F} \rangle + \mathrm{d}\langle \mathbf{F} \rangle \cdot\cdot \mathrm{d}^{\mathrm{p}}\langle \mathbf{P} \rangle = \langle \mathrm{d}^{\mathrm{s}}\mathbf{F} \cdot\cdot \boldsymbol{\Lambda} \cdot\cdot \mathrm{d}^{\mathrm{s}}\mathbf{F} \rangle - \langle \mathrm{d}^{\mathrm{r}}\mathbf{F} \cdot\cdot \boldsymbol{\Lambda} \cdot\cdot \mathrm{d}^{\mathrm{r}}\mathbf{F} \rangle. \tag{13.16.6}$$

The right-hand side can be recast as

$$\langle \mathrm{d}^{\mathrm{s}}\mathbf{F} \cdot\cdot \boldsymbol{\Lambda} \cdot\cdot \mathrm{d}^{\mathrm{p}}\mathbf{F} \rangle - \langle \mathrm{d}^{\mathrm{r}}\mathbf{F} \cdot\cdot \boldsymbol{\Lambda} \cdot\cdot \mathrm{d}^{\mathrm{p}}\mathbf{F} \rangle = \langle (\mathrm{d}^{\mathrm{r}}\mathbf{F} - \mathrm{d}^{\mathrm{s}}\mathbf{F}) \cdot\cdot \mathrm{d}^{\mathrm{p}}\mathbf{P} \rangle, \tag{13.16.7}$$

recalling Eqs. (13.10.33) and (13.11.32), and $\mathrm{d}^{\mathrm{p}}\mathbf{P} = -\boldsymbol{\Lambda} : \mathrm{d}^{\mathrm{p}}\mathbf{F}$.

Expressions dual to (13.16.5)–(13.16.7) can also be derived. We start from

$$\mathrm{d}\langle \mathbf{F} \rangle \cdot\cdot \mathrm{d}\langle \mathbf{P} \rangle = \mathrm{d}\langle \mathbf{F} \rangle \cdot\cdot \mathrm{d}^{\mathrm{p}}\langle \mathbf{P} \rangle + \mathrm{d}\langle \mathbf{F} \rangle \cdot\cdot [\boldsymbol{\Lambda}] \cdot\cdot \mathrm{d}\langle \mathbf{F} \rangle, \tag{13.16.8}$$

$$\langle \mathrm{d}\mathbf{F} \cdot\cdot \mathrm{d}\mathbf{P} \rangle = \langle \mathrm{d}^{\mathrm{p}}\mathbf{F} \cdot\cdot \mathrm{d}\mathbf{P} \rangle + \langle \mathrm{d}\mathbf{P} \cdot\cdot \mathbf{M} \cdot\cdot \mathrm{d}\mathbf{P} \rangle. \tag{13.16.9}$$

Since $\mathrm{d}\mathbf{P} = \mathrm{d}^{\mathrm{s}}\mathbf{P} + \boldsymbol{\Lambda} \cdot\cdot \boldsymbol{\mathcal{F}} \cdot\cdot \mathrm{d}\langle \mathbf{F} \rangle$, according to Eq. (13.10.20), by expansion and the use of the product theorem, the last term on the right-hand side of Eq. (13.16.9) becomes

$$\langle \mathrm{d}\mathbf{P} \cdot\cdot \mathbf{M} \cdot\cdot \mathrm{d}\mathbf{F} \rangle = 2\,\mathrm{d}\langle \mathbf{F} \rangle \cdot\cdot \mathrm{d}^{\mathrm{p}}\langle \mathbf{P} \rangle + \mathrm{d}\langle \mathbf{F} \rangle \cdot\cdot [\boldsymbol{\Lambda}] \cdot\cdot \mathrm{d}\langle \mathbf{F} \rangle + \langle \mathrm{d}^{\mathrm{s}}\mathbf{P} \cdot\cdot \mathbf{M} \cdot\cdot \mathrm{d}^{\mathrm{s}}\mathbf{P} \rangle. \tag{13.16.10}$$

The relationship $\langle \boldsymbol{\mathcal{F}}^T \cdot\cdot \boldsymbol{\Lambda} \cdot\cdot \boldsymbol{\mathcal{F}} \rangle = [\boldsymbol{\Lambda}]$ from (13.7.16) was used. Substituting Eqs. (13.16.8)–(13.16.10) into Eq. (13.16.1) then gives

$$\mathrm{d}\langle \mathbf{F} \rangle \cdot\cdot \mathrm{d}^{\mathrm{p}}\langle \mathbf{P} \rangle = -\langle \mathrm{d}^{\mathrm{p}}\mathbf{F} \cdot\cdot \mathrm{d}\mathbf{P} \rangle - \langle \mathrm{d}^{\mathrm{s}}\mathbf{P} \cdot\cdot \mathbf{M} \cdot\cdot \mathrm{d}^{\mathrm{s}}\mathbf{P} \rangle, \tag{13.16.11}$$

which is dual to Eq. (13.16.5).

On the other hand, by summing expressions in Eqs. (13.16.11) and (13.14.10), there follows

$$\mathrm{d}\langle \mathbf{F} \rangle \cdot\cdot \mathrm{d}^{\mathrm{p}}\langle \mathbf{P} \rangle + \mathrm{d}\langle \mathbf{P} \rangle \cdot\cdot \mathrm{d}^{\mathrm{p}}\langle \mathbf{F} \rangle = \langle \mathrm{d}^{\mathrm{r}}\mathbf{P} \cdot\cdot \mathbf{M} \cdot\cdot \mathrm{d}^{\mathrm{r}}\mathbf{P} \rangle - \langle \mathrm{d}^{\mathrm{s}}\mathbf{P} \cdot\cdot \mathbf{M} \cdot\cdot \mathrm{d}^{\mathrm{s}}\mathbf{P} \rangle. \tag{13.16.12}$$

The right-hand side is also equal to

$$\langle \mathrm{d}^{\mathrm{r}}\mathbf{P} \cdot\cdot \mathbf{M} \cdot\cdot \mathrm{d}^{\mathrm{p}}\mathbf{P} \rangle - \langle \mathrm{d}^{\mathrm{s}}\mathbf{P} \cdot\cdot \mathbf{M} \cdot\cdot \mathrm{d}^{\mathrm{p}}\mathbf{P} \rangle = \langle (\mathrm{d}^{\mathrm{s}}\mathbf{P} - \mathrm{d}^{\mathrm{r}}\mathbf{P}) \cdot\cdot \mathrm{d}^{\mathrm{p}}\mathbf{F} \rangle, \tag{13.16.13}$$

by Eqs. (13.10.34) and (13.11.33), and because $\mathrm{d^pF} = -\mathbf{M} : \mathrm{d^pM}$. It is easily verified that Eqs. (13.16.6) and (13.16.12) are in accord, since

$$\langle(\mathrm{d^r}\mathbf{F} - \mathrm{d^s}\mathbf{F}) \cdot\cdot\, \mathrm{d^p}\mathbf{P}\rangle = \langle(\mathrm{d^s}\mathbf{P} - \mathrm{d^r}\mathbf{P}) \cdot\cdot\, \mathrm{d^p}\mathbf{F}\rangle, \tag{13.16.14}$$

by Eq. (13.11.34).

We end this section by listing two additional identities. They are

$$\langle\, \mathrm{d^p}\mathbf{F} \cdot\cdot\, \boldsymbol{\Lambda} \cdot\cdot\, \mathrm{d^p}\mathbf{F}\,\rangle = \langle\, \mathrm{d^s}\mathbf{F} \cdot\cdot\, \boldsymbol{\Lambda} \cdot\cdot\, \mathrm{d^s}\mathbf{F}\,\rangle + \langle\, \mathrm{d^s}\mathbf{P} \cdot\cdot\, \mathbf{M} \cdot\cdot\, \mathrm{d^s}\mathbf{P}\,\rangle, \tag{13.16.15}$$

and

$$\langle\, \mathrm{d^p}\mathbf{P} \cdot\cdot\, \mathbf{M} \cdot\cdot\, \mathrm{d^p}\mathbf{P}\,\rangle = \langle\, \mathrm{d^r}\mathbf{F} \cdot\cdot\, \boldsymbol{\Lambda} \cdot\cdot\, \mathrm{d^r}\mathbf{F}\,\rangle + \langle\, \mathrm{d^r}\mathbf{P} \cdot\cdot\, \mathbf{M} \cdot\cdot\, \mathrm{d^r}\mathbf{P}\,\rangle. \tag{13.16.16}$$

For example, the first one follows from

$$\begin{aligned} \mathrm{d^p}\mathbf{F} \cdot\cdot\, \boldsymbol{\Lambda} \cdot\cdot\, \mathrm{d^p}\mathbf{F} &= (\mathrm{d^s}\mathbf{F} - \mathrm{d^s}\mathbf{P} \cdot\cdot\, \mathbf{M}) \cdot\cdot\, \boldsymbol{\Lambda} \cdot\cdot\, \mathrm{d^p}\mathbf{F} \\ &= \mathrm{d^s}\mathbf{F} \cdot\cdot\, \boldsymbol{\Lambda} \cdot\cdot\, \mathrm{d^p}\mathbf{F} + \mathrm{d^s}\mathbf{P} \cdot\cdot\, \mathbf{M} \cdot\cdot\, \mathrm{d^p}\mathbf{P}, \end{aligned} \tag{13.16.17}$$

by taking the volume average and by using Eqs. (13.10.33) and (13.10.34). Note that the left-hand sides in Eqs. (13.16.15) and (13.16.16) are actually equal to each other, both being equal to $-\langle\, \mathrm{d^p}\mathbf{P} \cdot\cdot\, \mathrm{d^p}\mathbf{F}\,\rangle$.

13.17. General Analysis of Macroscopic Plastic Potentials

A general study of the transmissibility of plastic potentials and normality rules from micro-to-macrolevel is presented in this section. The analysis is originally due to Hill and Rice (1973), who used the framework of general conjugate stress and strain measures in their formulation. Here, the formulation is conveniently cast by using the deformation gradient and the nominal stress. The plastic part of the free energy increment at the microlevel,

$$\mathrm{d^p}\Psi = \Psi\,(\mathbf{F}, \mathcal{H} + d\mathcal{H}) - \Psi\,(\mathbf{F}, \mathcal{H})\,, \tag{13.17.1}$$

is a potential for the plastic part of the nominal stress increment,

$$\mathrm{d^p}\mathbf{P} = \mathbf{P}\,(\mathbf{F}, \mathcal{H} + d\mathcal{H}) - \mathbf{P}\,(\mathbf{F}, \mathcal{H})\,, \tag{13.17.2}$$

such that

$$\mathrm{d^p}\mathbf{P} = \frac{\partial}{\partial \mathbf{F}}\,(\mathrm{d^p}\Psi)\,. \tag{13.17.3}$$

If the trace product of $\mathrm{d^p}\mathbf{P}$ with an elastic increment $\delta\mathbf{F}$ is positive,

$$\delta\mathbf{F} \cdot\cdot\, \mathrm{d^p}\mathbf{P} = \delta\mathbf{F} \cdot\cdot\, \frac{\partial}{\partial \mathbf{F}}\,(\mathrm{d^p}\Psi) = \delta\,(\mathrm{d^p}\Psi) > 0, \tag{13.17.4}$$

we say that the material response complies with the normality rule at microlevel in the deformation space.

Dually, the plastic part of the increment of complementary energy at the microlevel,

$$\mathrm{d}^{\mathrm{p}}\Phi = \Psi\left(\mathbf{P}, \mathcal{H} + d\mathcal{H}\right) - \Phi\left(\mathbf{P}, \mathcal{H}\right), \tag{13.17.5}$$

is a potential for the plastic part of the deformation gradient increment,

$$\mathrm{d}^{\mathrm{p}}\mathbf{F} = \mathbf{F}\left(\mathbf{P}, \mathcal{H} + d\mathcal{H}\right) - \mathbf{F}\left(\mathbf{P}, \mathcal{H}\right), \tag{13.17.6}$$

such that

$$\mathrm{d}^{\mathrm{p}}\mathbf{F} = \frac{\partial}{\partial \mathbf{P}}\left(\mathrm{d}^{\mathrm{p}}\Phi\right). \tag{13.17.7}$$

If the trace product of $\mathrm{d}^{\mathrm{p}}\mathbf{F}$ with an elastic increment $\delta\mathbf{P}$ is negative,

$$\delta\mathbf{P}\cdot\cdot\,\mathrm{d}^{\mathrm{p}}\mathbf{F} = \delta\mathbf{P}\cdot\cdot\,\frac{\partial}{\partial \mathbf{P}}\left(\mathrm{d}^{\mathrm{p}}\Phi\right) = \delta\left(\mathrm{d}^{\mathrm{p}}\Phi\right) < 0, \tag{13.17.8}$$

the material response complies with the normality rule at microlevel in the stress space. With these preliminaries from the microlevel, we now examine the macroscopic potentials and macroscopic normality rules.

13.17.1. Deformation Space Formulation

The plastic part of the increment of macroscopic free energy, associated with a cycle of the application and removal of an elastoplastic increment of the macroscopic deformation gradient $\mathrm{d}\langle\mathbf{F}\rangle$, is defined by

$$\mathrm{d}^{\mathrm{p}}\hat{\Psi} = \hat{\Psi}\left(\langle\mathbf{F}\rangle, \mathcal{H} + d\mathcal{H}\right) - \hat{\Psi}\left(\langle\mathbf{F}\rangle, \mathcal{H}\right). \tag{13.17.9}$$

The macroscopic free energy before the cycle is

$$\hat{\Psi}\left(\langle\mathbf{F}\rangle, \mathcal{H}\right) = \frac{1}{V^0}\int_{V^0} \Psi\left(\mathbf{F}, \mathcal{H}\right)\mathrm{d}V^0, \tag{13.17.10}$$

where $\mathbf{F}$ is the local deformation gradient field within the macroelement. After a cycle of $\mathrm{d}\langle\mathbf{F}\rangle$, the local deformation gradients within V^0 are in general not restored, so that

$$\begin{aligned}\hat{\Psi}\left(\langle\mathbf{F}\rangle, \mathcal{H} + d\mathcal{H}\right) &= \frac{1}{V^0}\int_{V^0} \Psi\left(\mathbf{F} + \mathrm{d}^{\mathrm{s}}\mathbf{F}, \mathcal{H} + d\mathcal{H}\right)\mathrm{d}V^0 \\ &= \frac{1}{V^0}\int_{V^0}\left[\Psi\left(\mathbf{F}, \mathcal{H} + d\mathcal{H}\right) + \frac{\partial\Psi}{\partial\mathbf{F}}\cdot\cdot\,\mathrm{d}^{\mathrm{s}}\mathbf{F}\right]\mathrm{d}V^0.\end{aligned} \tag{13.17.11}$$

Here, $\mathrm{d}^{\mathrm{s}}\mathbf{F}$ represents a residual increment of the deformation gradient that remains at the microlevel after macroscopic cycle of $\mathrm{d}\langle\mathbf{F}\rangle$. Upon substitution of Eqs. (13.17.10) and (13.17.11) into Eq. (13.17.9), there follows

$$\mathrm{d}^{\mathrm{p}}\hat{\Psi} = \frac{1}{V^0}\int_{V^0}\left[\Psi\left(\mathbf{F},\mathcal{H}+d\mathcal{H}\right)-\Psi\left(\mathbf{F},\mathcal{H}\right)\right]\mathrm{d}V^0 + \frac{1}{V^0}\int_{V^0}\mathbf{P}\cdot\cdot\,\mathrm{d}^{\mathrm{s}}\mathbf{F}\,\mathrm{d}V^0, \tag{13.17.12}$$

i.e.,

$$\mathrm{d}^{\mathrm{p}}\hat{\Psi} = \langle \mathrm{d}^{\mathrm{p}}\Psi\rangle + \langle \mathbf{P}\cdot\cdot\,\mathrm{d}^{\mathrm{s}}\mathbf{F}\rangle. \tag{13.17.13}$$

Recalling that $\mathbf{P}$ is statically admissible, while $\mathrm{d}^{\mathrm{s}}\mathbf{F}$ is kinematically admissible field, and since $\langle \mathrm{d}^{\mathrm{s}}\mathbf{F}\rangle = \mathbf{0}$ by Eq. (13.10.26), we have

$$\langle \mathbf{P}\cdot\cdot\,\mathrm{d}^{\mathrm{s}}\mathbf{F}\rangle = \langle \mathbf{P}\rangle\cdot\cdot\langle \mathrm{d}^{\mathrm{s}}\mathbf{F}\rangle = 0. \tag{13.17.14}$$

Equation (13.17.13) consequently reduces to

$$\mathrm{d}^{\mathrm{p}}\hat{\Psi} = \langle \mathrm{d}^{\mathrm{p}}\Psi\rangle. \tag{13.17.15}$$

Thus, the plastic increment of macroscopic free energy is a direct volume average of the plastic increment of microscopic free energy.

The potential property is established through

$$\begin{aligned}\frac{\partial}{\partial\langle\mathbf{F}\rangle}\left(\mathrm{d}^{\mathrm{p}}\hat{\Psi}\right) &= \frac{\partial}{\partial\langle\mathbf{F}\rangle}\langle \mathrm{d}^{\mathrm{p}}\Psi\rangle = \langle\,\frac{\partial\left(\mathrm{d}^{\mathrm{p}}\Psi\right)}{\partial\langle\mathbf{F}\rangle}\,\rangle \\ &= \langle\,\frac{\partial\left(\mathrm{d}^{\mathrm{p}}\Psi\right)}{\partial\mathbf{F}}\cdot\cdot\frac{\partial\mathbf{F}}{\partial\langle\mathbf{F}\rangle}\,\rangle = \langle \mathrm{d}^{\mathrm{p}}\mathbf{P}\cdot\cdot\boldsymbol{\mathcal{F}}\rangle.\end{aligned} \tag{13.17.16}$$

Since the plastic part of the increment of macroscopic nominal stress is a weighted volume average of the plastic part of the increment of local nominal stress, as seen from Eq. (13.10.14), we deduce that $\mathrm{d}^{\mathrm{p}}\hat{\Psi}$ is indeed a plastic potential for $\mathrm{d}^{\mathrm{p}}\langle\mathbf{P}\rangle$, i.e.,

$$\mathrm{d}^{\mathrm{p}}\langle\mathbf{P}\rangle = \frac{\partial}{\partial\langle\mathbf{F}\rangle}\left(\mathrm{d}^{\mathrm{p}}\hat{\Psi}\right). \tag{13.17.17}$$

If Eq. (13.17.17) is subjected to the trace product with an elastic increment $\delta\langle\mathbf{F}\rangle$, there follows

$$\delta\langle\mathbf{F}\rangle\cdot\cdot\,\mathrm{d}^{\mathrm{p}}\langle\mathbf{P}\rangle = \delta\langle\mathbf{F}\rangle\cdot\cdot\frac{\partial}{\partial\langle\mathbf{F}\rangle}\left(\mathrm{d}^{\mathrm{p}}\hat{\Psi}\right) = \delta(\mathrm{d}^{\mathrm{p}}\hat{\Psi}). \tag{13.17.18}$$

Substitution of (13.17.15) gives

$$\delta(\mathrm{d}^{\mathrm{p}}\hat{\Psi}) = \delta\,\langle \mathrm{d}^{\mathrm{p}}\Psi\,\rangle = \langle\,\delta(\mathrm{d}^{\mathrm{p}}\Psi)\,\rangle. \tag{13.17.19}$$

Thus, the normality at the microlevel ensures the normality at the macrolevel, i.e.,

$$\text{if}\quad \delta\left(\mathrm{d}^{\mathrm{p}}\Psi\right) > 0,\quad \text{then}\quad \delta\left(\mathrm{d}^{\mathrm{p}}\hat{\Psi}\right) > 0. \tag{13.17.20}$$

If the conjugate stress and strain measures $\mathbf{T}$ and $\mathbf{E}$ are utilized, Eq. (13.17.17) becomes

$$\mathrm{d^p}[\mathbf{T}] = \frac{\partial}{\partial[\mathbf{E}]}\,(\mathrm{d^p}\hat{\Psi}). \tag{13.17.21}$$

This follows because the relationships from Section 13.12 hold,

$$\mathrm{d^p}\langle\mathbf{P}\rangle = \langle\boldsymbol{\mathcal{K}}\rangle^T : \mathrm{d^p}[\mathbf{T}], \quad \frac{\partial}{\partial\langle\mathbf{F}\rangle} = \langle\boldsymbol{\mathcal{K}}\rangle^T : \frac{\partial}{\partial[\mathbf{E}]}\,. \tag{13.17.22}$$

13.17.2. Stress Space Formulation

In a dual analysis, we introduce the plastic part of the increment of macroscopic complementary energy, associated with a cycle of the application and removal of an elastoplastic increment of macroscopic stress $\mathrm{d}\langle\mathbf{P}\rangle$, such that

$$\mathrm{d^p}\hat{\Phi} = \hat{\Phi}\left(\langle\mathbf{P}\rangle, \mathcal{H}+d\mathcal{H}\right) - \hat{\Phi}\left(\langle\mathbf{P}\rangle, \mathcal{H}\right). \tag{13.17.23}$$

The macroscopic complementary energy before the cycle is

$$\hat{\Phi}\left(\langle\mathbf{P}\rangle, \mathcal{H}\right) = \frac{1}{V^0}\int_{V^0} \Phi\left(\mathbf{P}, \mathcal{H}\right)\mathrm{d}V^0, \tag{13.17.24}$$

where $\mathbf{P}$ is the local stress field within the macroelement. After a cycle of $\mathrm{d}\langle\mathbf{P}\rangle$, the local stresses within V^0 are in general not restored, so that

$$\begin{aligned}\hat{\Phi}\left(\langle\mathbf{P}\rangle, \mathcal{H}+d\mathcal{H}\right) &= \frac{1}{V^0}\int_{V^0} \Phi\left(\mathbf{P}+\mathrm{d^r}\mathbf{P}, \mathcal{H}+d\mathcal{H}\right)\mathrm{d}V^0 \\ &= \frac{1}{V^0}\int_{V^0} \left[\Phi\left(\mathbf{P}, \mathcal{H}+d\mathcal{H}\right) + \frac{\partial\Phi}{\partial\mathbf{P}}\cdot\cdot\,\mathrm{d^r}\mathbf{P}\right]\mathrm{d}V^0,\end{aligned} \tag{13.17.25}$$

where $\mathrm{d^r}\mathbf{P}$ represents a residual increment of stress that remains at the microlevel upon macroscopic cycle of $\mathrm{d}\langle\mathbf{P}\rangle$. Substitution of Eqs. (13.17.24) and (13.17.25) into Eq. (13.17.23) yields

$$\mathrm{d^p}\hat{\Phi} = \frac{1}{V^0}\int_{V^0}\left[\Phi\left(\mathbf{P}, \mathcal{H}+d\mathcal{H}\right) - \Phi\left(\mathbf{P}, \mathcal{H}\right)\right]\mathrm{d}V^0 + \frac{1}{V^0}\int_{V^0}\mathbf{F}\cdot\cdot\,\mathrm{d^r}\mathbf{P}\,\mathrm{d}V^0, \tag{13.17.26}$$

i.e.,

$$\mathrm{d^p}\hat{\Phi} = \langle\mathrm{d^p}\Phi\rangle + \langle\mathbf{F}\cdot\cdot\,\mathrm{d^r}\mathbf{P}\rangle. \tag{13.17.27}$$

Since $\mathbf{F}$ is kinematically admissible, while $\mathrm{d^r}\mathbf{P}$ is statically admissible field, and since $\langle\mathrm{d^r}\mathbf{P}\rangle = 0$ by Eq. (13.11.25), we have

$$\langle\mathbf{F}\cdot\cdot\,\mathrm{d^r}\mathbf{P}\rangle = \langle\mathbf{F}\rangle\cdot\cdot\,\langle\mathrm{d^r}\mathbf{P}\rangle = 0. \tag{13.17.28}$$

Consequently, Eq. (13.17.27) reduces to

$$\mathrm{d^p}\hat{\Phi} = \langle\mathrm{d^p}\Phi\rangle. \tag{13.17.29}$$

This shows that the plastic increment of macroscopic complementary energy is a direct volume average of the plastic increment of microscopic complementary energy.

The potential property follows from

$$\begin{aligned}\frac{\partial}{\partial\langle\mathbf{P}\rangle}(\mathrm{d}^{\mathrm{p}}\hat{\Phi}) &= \frac{\partial}{\partial\langle\mathbf{P}\rangle}\langle\mathrm{d}^{\mathrm{p}}\Phi\rangle = \langle\frac{\partial\,(\mathrm{d}^{\mathrm{p}}\Phi)}{\partial\langle\mathbf{P}\rangle}\rangle \\ &= \langle\frac{\partial\,(\mathrm{d}^{\mathrm{p}}\Phi)}{\partial\mathbf{P}}\cdot\cdot\frac{\partial\mathbf{P}}{\partial\langle\mathbf{P}\rangle}\rangle = \langle\mathrm{d}^{\mathrm{p}}\mathbf{F}\cdot\cdot\boldsymbol{\mathcal{P}}\rangle.\end{aligned} \tag{13.17.30}$$

Since the plastic part of the increment of macroscopic deformation gradient is a weighted volume average of the plastic part of the increment of local deformation gradient, as shown in Eq. (13.11.13), we deduce that $\mathrm{d}^{\mathrm{p}}\hat{\Phi}$ is indeed a plastic potential for $\mathrm{d}^{\mathrm{p}}\langle\mathbf{F}\rangle$, i.e.,

$$\mathrm{d}^{\mathrm{p}}\langle\mathbf{F}\rangle = \frac{\partial}{\partial\langle\mathbf{P}\rangle}(\mathrm{d}^{\mathrm{p}}\hat{\Phi}). \tag{13.17.31}$$

Furthermore, the trace product of Eq. (13.17.31) with an elastic increment $\delta\langle\mathbf{P}\rangle$ gives

$$\delta\langle\mathbf{P}\rangle\cdot\cdot\,\mathrm{d}^{\mathrm{p}}\langle\mathbf{F}\rangle = \delta\langle\mathbf{P}\rangle\cdot\cdot\frac{\partial}{\partial\langle\mathbf{P}\rangle}(\mathrm{d}^{\mathrm{p}}\hat{\Phi}) = \delta(\mathrm{d}^{\mathrm{p}}\hat{\Phi}). \tag{13.17.32}$$

In view of Eq. (13.17.29), therefore,

$$\delta(\mathrm{d}^{\mathrm{p}}\hat{\Phi}) = \delta\,\langle\mathrm{d}^{\mathrm{p}}\Phi\,\rangle = \langle\,\delta(\mathrm{d}^{\mathrm{p}}\Phi)\,\rangle. \tag{13.17.33}$$

From this we conclude that the normality at the microlevel, ensures the normality at the macrolevel, i.e.,

$$\text{if}\quad \delta\,(\mathrm{d}^{\mathrm{p}}\Phi) < 0,\quad \text{then}\quad \delta\,(\mathrm{d}^{\mathrm{p}}\hat{\Phi}) < 0. \tag{13.17.34}$$

It is observed that

$$\mathrm{d}^{\mathrm{p}}\hat{\Psi} + \mathrm{d}^{\mathrm{p}}\hat{\Phi} = 0, \tag{13.17.35}$$

since locally $\mathrm{d}^{\mathrm{p}}\Psi + \mathrm{d}^{\mathrm{p}}\Phi = 0$, as well. Thus, having in mind that

$$\frac{\partial}{\partial[\mathbf{E}]} = [\boldsymbol{\Lambda}_{(1)}] : \frac{\partial}{\partial[\mathbf{T}]}, \tag{13.17.36}$$

we can rewrite Eq. (13.17.21) as

$$\mathrm{d}^{\mathrm{p}}[\mathbf{T}] = [\boldsymbol{\Lambda}_{(1)}] : \frac{\partial}{\partial[\mathbf{T}]}(-\mathrm{d}^{\mathrm{p}}\hat{\Phi}). \tag{13.17.37}$$

Upon taking the trace product with $[\mathbf{M}_{(1)}] = [\boldsymbol{\Lambda}_{(1)}]^{-1}$, and recalling that $\mathrm{d}^{\mathrm{p}}[\mathbf{E}] = -[\mathbf{M}_{(1)}] : \mathrm{d}^{\mathrm{p}}[\mathbf{T}]$, Eq. (13.17.37) gives

$$\mathrm{d}^{\mathrm{p}}[\mathbf{E}] = \frac{\partial}{\partial[\mathbf{T}]}\,(\mathrm{d}^{\mathrm{p}}\hat{\Phi}). \tag{13.17.38}$$

This shows that $\mathrm{d}^{\mathrm{p}}\hat{\Phi}$, when expressed in terms of $[\mathbf{T}]$, is a potential for the plastic increment $\mathrm{d}^{\mathrm{p}}[\mathbf{E}]$.

13.18. Transmissibility of Ilyushin's Postulate

Suppose that the aggregate is taken through the deformation cycle which, at some stage, involves plastic deformation. Following an analogous analysis as in Section 8.5, the cycle emanates from the state $A^0\left(\langle\mathbf{F}\rangle^0, \mathcal{H}\right)$ within the macroscopic yield surface, it includes an elastic segment from A^0 to $A\left(\langle\mathbf{F}\rangle, \mathcal{H}\right)$ on the current yield surface, followed by an infinitesimal elastoplastic segment from A to $B\left(\langle\mathbf{F}\rangle + \mathrm{d}\langle\mathbf{F}\rangle, \mathcal{H} + d\mathcal{H}\right)$, and the elastic unloading segments from B to $C(\langle\mathbf{F}\rangle, \mathcal{H}+d\mathcal{H})$, and from C to $C^0\left(\langle\mathbf{F}\rangle^0, \mathcal{H} + d\mathcal{H}\right)$. The work done along the segments A^0A and CC^0 is

$$\begin{aligned}\int_{A^0}^{A} \langle\mathbf{P}\rangle \cdot\cdot\, \mathrm{d}\langle\mathbf{F}\rangle &= \int_{A^0}^{A} \frac{\partial\hat{\Psi}}{\partial\langle\mathbf{F}\rangle} \cdot\cdot\, \mathrm{d}\langle\mathbf{F}\rangle \\ &= \hat{\Psi}\left(\langle\mathbf{F}\rangle, \mathcal{H}\right) - \hat{\Psi}\left(\langle\mathbf{F}\rangle^0, \mathcal{H}\right),\end{aligned} \tag{13.18.1}$$

$$\begin{aligned}\int_{C}^{C^0} \langle\mathbf{P}\rangle \cdot\cdot\, \mathrm{d}\langle\mathbf{F}\rangle &= \int_{C}^{C^0} \frac{\partial\hat{\Psi}}{\partial\langle\mathbf{F}\rangle} \cdot\cdot\, \mathrm{d}\langle\mathbf{F}\rangle \\ &= \hat{\Psi}\left(\langle\mathbf{F}\rangle^0, \mathcal{H} + d\mathcal{H}\right) - \hat{\Psi}\left(\langle\mathbf{F}\rangle, \mathcal{H} + d\mathcal{H}\right).\end{aligned} \tag{13.18.2}$$

The work done along the segments AB and BC is, by the trapezoidal rule of quadrature,

$$\int_{A}^{B} \langle\mathbf{P}\rangle \cdot\cdot\, \mathrm{d}\langle\mathbf{F}\rangle = \langle\mathbf{P}\rangle \cdot\cdot\, \mathrm{d}\langle\mathbf{F}\rangle + \frac{1}{2}\,\mathrm{d}\langle\mathbf{P}\rangle : \mathrm{d}\langle\mathbf{F}\rangle, \tag{13.18.3}$$

$$\int_{B}^{C} \langle\mathbf{P}\rangle : \mathrm{d}\langle\mathbf{F}\rangle = -\langle\mathbf{P}\rangle \cdot\cdot\, \mathrm{d}\langle\mathbf{F}\rangle - \frac{1}{2}\left(\mathrm{d}\langle\mathbf{P}\rangle + \mathrm{d}^{\mathrm{p}}\langle\mathbf{P}\rangle\right) \cdot\cdot\, \mathrm{d}\langle\mathbf{F}\rangle, \tag{13.18.4}$$

to second-order terms. Consequently,

$$\oint_{\langle F\rangle} \langle\mathbf{P}\rangle \cdot\cdot\, \mathrm{d}\langle\mathbf{F}\rangle = -\frac{1}{2}\,\mathrm{d}^{\mathrm{p}}\langle\mathbf{P}\rangle \cdot\cdot\, \mathrm{d}\langle\mathbf{F}\rangle + (\mathrm{d}^{\mathrm{p}}\hat{\Psi})^0 - \mathrm{d}^{\mathrm{p}}\hat{\Psi}, \tag{13.18.5}$$

where

$$
\begin{aligned}
\mathrm{d}^{\mathrm{p}}\hat{\Psi} &= \hat{\Psi}\left(\langle\mathbf{F}\rangle, \mathcal{H}+d\mathcal{H}\right) - \hat{\Psi}\left(\langle\mathbf{F}\rangle, \mathcal{H}\right), \\
(\mathrm{d}^{\mathrm{p}}\hat{\Psi})^0 &= \hat{\Psi}\left(\langle\mathbf{F}\rangle^0, \mathcal{H}+d\mathcal{H}\right) - \hat{\Psi}\left(\langle\mathbf{F}\rangle^0, \mathcal{H}\right).
\end{aligned}
\tag{13.18.6}
$$

For the cycle with a sufficiently small segment along which the plastic deformation takes place, Eq. (13.18.5) becomes, to first order,

$$
\oint_{\langle F\rangle} \langle\mathbf{P}\rangle \cdot\cdot\, \mathrm{d}\langle\mathbf{F}\rangle = (\mathrm{d}^{\mathrm{p}}\hat{\Psi})^0 - \mathrm{d}^{\mathrm{p}}\hat{\Psi}. \tag{13.18.7}
$$

Since the plastic increment of macroscopic free energy is the volume average of the plastic increment of microscopic free energy, $\mathrm{d}^{\mathrm{p}}\hat{\Psi} = \langle \mathrm{d}^{\mathrm{p}}\Psi\rangle$, as shown in Eq. (13.17.15), we can rewrite Eq. (13.18.7) as

$$
\oint_{\langle F\rangle} \langle\mathbf{P}\rangle \cdot\cdot\, \mathrm{d}\langle\mathbf{F}\rangle = \langle\, (\mathrm{d}^{\mathrm{p}}\Psi)^0 - \mathrm{d}^{\mathrm{p}}\Psi\rangle. \tag{13.18.8}
$$

This holds even though the local $\mathbf{F}$ field is generally not restored in the macroscopic cycle of $\mathrm{d}\langle\mathbf{F}\rangle$. Equation (13.18.8) evidently implies, if

$$
(\mathrm{d}^{\mathrm{p}}\Psi)^0 - \mathrm{d}^{\mathrm{p}}\Psi > 0 \tag{13.18.9}
$$

at the microlevel, then

$$
\langle\, (\mathrm{d}^{\mathrm{p}}\Psi)^0 - \mathrm{d}^{\mathrm{p}}\Psi\,\rangle > 0 \tag{13.18.10}
$$

at the macrolevel. In other words, the restricted Ilyushin's postulate (for the specified deformation cycles with sufficiently small plastic segments) is transmitted from the microlevel to the macrolevel (Hill and Rice, 1973).

If the cycle begins from the point on the yield surface, i.e., if $A^0 = A$ and $\langle\mathbf{F}\rangle^0 = \langle\mathbf{F}\rangle$, Eq. (13.18.5) reduces to

$$
\oint_{\langle F\rangle} \langle\mathbf{P}\rangle \cdot\cdot\, \mathrm{d}\langle\mathbf{F}\rangle = -\frac{1}{2}\,\mathrm{d}^{\mathrm{p}}\langle\mathbf{P}\rangle \cdot\cdot\, \mathrm{d}\langle\mathbf{F}\rangle. \tag{13.18.11}
$$

On the other hand, from Eq. (13.15.9) we have

$$
\mathrm{d}\langle\mathbf{F}\rangle \cdot\cdot\, \mathrm{d}^{\mathrm{p}}\langle\mathbf{P}\rangle = \langle \mathrm{d}\mathbf{F} \cdot\cdot\, \mathrm{d}^{\mathrm{p}}\mathbf{P}\rangle + \langle \mathrm{d}^{\mathrm{s}}\mathbf{F} \cdot\cdot\, \mathbf{\Lambda} \cdot\cdot\, \mathrm{d}^{\mathrm{s}}\mathbf{F}\rangle. \tag{13.18.12}
$$

Since $\mathbf{\Lambda}$ is not necessarily positive definite, we conclude that the compliance with the restricted Ilyushin's postulate (for infinitesimal cycles emanating from the yield surface) at the microlevel,

$$
\oint_{F} \mathbf{P} \cdot\cdot\, \mathrm{d}\mathbf{F} = -\frac{1}{2}\,\mathrm{d}^{\mathrm{p}}\mathbf{P} \cdot\cdot\, \mathrm{d}\mathbf{F} > 0, \tag{13.18.13}
$$

is not necessarily transmitted to the macrolevel.

13.19. Aggregate Minimum Shear and Maximum Work Principle

Consider an aggregate macroelement in the deformed equilibrium configuration. The local deformation gradient and the nominal stress fields are $\mathbf{F}$ and $\mathbf{P}$. Let $\mathrm{d}\mathbf{F}$ be the actual increment of deformation gradient that physically occurs under prescribed increment of displacement $\mathrm{d}\mathbf{u}$ on the bounding surface S^0 of the aggregate macroelement. Furthermore, let $\mathrm{d}\bar{\mathbf{F}}$ be any kinematically admissible field of the increment of deformation gradient that is associated with the same prescribed increment of displacement $\mathrm{d}\mathbf{u}$ over S^0. By the Gauss divergence theorem, the volume averages of $\mathrm{d}\mathbf{F}$ and $\mathrm{d}\bar{\mathbf{F}}$, over the macroelement volume, are equal to each other,

$$\langle\, \mathrm{d}\mathbf{F}\,\rangle = \langle\, \mathrm{d}\bar{\mathbf{F}}\,\rangle = \int_{S^0} \mathrm{d}\mathbf{u} \otimes \mathbf{n}^0 \,\mathrm{d}S^0. \tag{13.19.1}$$

In addition, there is an equality

$$\langle\, \mathbf{P}\cdot\cdot\,\mathrm{d}\mathbf{F}\,\rangle = \langle\, \mathbf{P}\cdot\cdot\,\mathrm{d}\bar{\mathbf{F}}\,\rangle = \int_{S^0} \mathbf{p}_n \otimes \mathrm{d}\mathbf{u} \,\mathrm{d}S^0. \tag{13.19.2}$$

Suppose that simple shearing on active slip systems is the only mechanism of deformation in a rigid-plastic aggregate. Let n shears $\mathrm{d}\gamma^\alpha$ be a set of local slip increments which give rise to local strain increment $\mathrm{d}\mathbf{E}$. These are actual, physically operative slips, so that on each slip system of this set

$$|\tau^\alpha| = \tau^\alpha_{\mathrm{cr}}\,, \quad (\alpha = 1, 2, \ldots, n). \tag{13.19.3}$$

The slip in the opposite sense along the same slip direction is not considered as an independent slip system. The Bauschinger effect is assumed to be absent, so that $\tau^\alpha_{\mathrm{cr}}$ is equal in both senses along the same slip direction. In view of Eqs. (12.1.22) and (12.1.24), we can write

$$\mathrm{d}\mathbf{E} = \sum_{\alpha=1}^{n} \mathbf{P}^\alpha_0 \,\mathrm{d}\gamma^\alpha, \quad \mathbf{P}^\alpha_0 = \mathbf{F}^T\cdot\mathbf{P}^\alpha\cdot\mathbf{F} = \mathbf{F}^T\cdot(\mathbf{s}^\alpha\otimes\mathbf{m}^\alpha)_{\mathrm{s}}\cdot\mathbf{F}. \tag{13.19.4}$$

Further, let $\bar{n}$ shears $\mathrm{d}\bar{\gamma}^\alpha$ be a set of local slip increments which give rise to local strain increment $\mathrm{d}\bar{\mathbf{E}}$, but which are not necessarily physically operative, so that

$$|\bar{\tau}^\alpha| \le \bar{\tau}^\alpha_{\mathrm{cr}}\,, \quad (\alpha = 1, 2, \ldots, \bar{n}). \tag{13.19.5}$$

For this set we can write

$$\mathrm{d}\bar{\mathbf{E}} = \sum_{\alpha=1}^{\bar{n}} \bar{\mathbf{P}}^\alpha_0 \,\mathrm{d}\bar{\gamma}^\alpha, \quad \bar{\mathbf{P}}^\alpha_0 = \mathbf{F}^T\cdot\bar{\mathbf{P}}^\alpha\cdot\mathbf{F} = \mathbf{F}^T\cdot(\bar{\mathbf{s}}^\alpha\otimes\bar{\mathbf{m}}^\alpha)_{\mathrm{s}}\cdot\mathbf{F}. \tag{13.19.6}$$

The slip system vectors of the second set are denoted by $\bar{\mathbf{s}}^\alpha$ and $\bar{\mathbf{m}}^\alpha$. (Even if it happens that $\mathrm{d}\bar{\mathbf{E}} = \mathrm{d}\mathbf{E}$ at some point or the subelement, there still may be different sets of shears corresponding to that same $\mathrm{d}\mathbf{E}$. These are geometrically equivalent sets of shears, which were the main concern of the single crystal consideration in Section 12.19). Consequently,

$$\langle \mathbf{P} \cdot\cdot\, \mathrm{d}\mathbf{F} \rangle = \langle \mathbf{T} : \mathrm{d}\mathbf{E} \rangle = \Big\langle \sum_{\alpha=1}^{n} \tau^\alpha \, \mathrm{d}\gamma^\alpha \Big\rangle, \quad \tau^\alpha = \boldsymbol{\tau} : \mathbf{P}^\alpha, \tag{13.19.7}$$

$$\langle \mathbf{P} \cdot\cdot\, \mathrm{d}\bar{\mathbf{F}} \rangle = \langle \mathbf{T} : \mathrm{d}\bar{\mathbf{E}} \rangle = \Big\langle \sum_{\alpha=1}^{\bar{n}} \bar{\tau}^\alpha \, \mathrm{d}\bar{\gamma}^\alpha \Big\rangle, \quad \bar{\tau}^\alpha = \boldsymbol{\tau} : \bar{\mathbf{P}}^\alpha, \tag{13.19.8}$$

where $\boldsymbol{\tau} = \mathbf{F} \cdot \mathbf{P} = \mathbf{F} \cdot \mathbf{T} \cdot \mathbf{T}^T$ is the Kirchhoff stress (equal here to the Cauchy stress $\boldsymbol{\sigma}$, because the deformation of rigid-plastic polycrystalline aggregate is isochoric, $\det \mathbf{F} = 1$). Since slip in the opposite sense along the same slip direction is not considered as an independent slip system, $\mathrm{d}\gamma^\alpha < 0$ when $\tau^\alpha < 0$, and the above equations can be recast as

$$\Big\langle \sum_{\alpha=1}^{n} \tau^\alpha \, \mathrm{d}\gamma^\alpha \Big\rangle = \Big\langle \sum_{\alpha=1}^{n} |\tau^\alpha| \, |\mathrm{d}\gamma^\alpha| \Big\rangle = \Big\langle \sum_{\alpha=1}^{n} \tau_{\mathrm{cr}}^\alpha \, |\mathrm{d}\gamma^\alpha| \Big\rangle, \tag{13.19.9}$$

$$\Big\langle \sum_{\alpha=1}^{\bar{n}} \bar{\tau}^\alpha \, \mathrm{d}\bar{\gamma}^\alpha \Big\rangle = \Big\langle \sum_{\alpha=1}^{\bar{n}} |\bar{\tau}^\alpha| \, |\mathrm{d}\bar{\gamma}^\alpha| \Big\rangle \leq \Big\langle \sum_{\alpha=1}^{\bar{n}} \bar{\tau}_{\mathrm{cr}}^\alpha \, |\mathrm{d}\bar{\gamma}^\alpha| \Big\rangle. \tag{13.19.10}$$

Recall that $|\tau^\alpha| = \tau_{\mathrm{cr}}^\alpha$ and $|\bar{\tau}^\alpha| \leq \bar{\tau}_{\mathrm{cr}}^\alpha$. Thus, we conclude from Eqs. (13.19.2), (13.19.9), and (13.19.10) that

$$\Big\langle \sum_{\alpha=1}^{n} \tau_{\mathrm{cr}}^\alpha \, |\mathrm{d}\gamma^\alpha| \Big\rangle \leq \Big\langle \sum_{\alpha=1}^{\bar{n}} \bar{\tau}_{\mathrm{cr}}^\alpha \, |\mathrm{d}\bar{\gamma}^\alpha| \Big\rangle. \tag{13.19.11}$$

If the hardening in each grain is isotropic, we have

$$\Big\langle \tau_{\mathrm{cr}}^\alpha \sum_{\alpha=1}^{n} |\mathrm{d}\gamma^\alpha| \Big\rangle \leq \Big\langle \bar{\tau}_{\mathrm{cr}}^\alpha \sum_{\alpha=1}^{\bar{n}} |\mathrm{d}\bar{\gamma}^\alpha| \Big\rangle. \tag{13.19.12}$$

Assuming, in addition, that all grains harden equally, the critical resolved shear stress is uniform throughout the aggregate, and (13.19.12) reduces to

$$\Big\langle \sum_{\alpha=1}^{n} |\mathrm{d}\gamma^\alpha| \Big\rangle \leq \Big\langle \sum_{\alpha=1}^{\bar{n}} |\mathrm{d}\bar{\gamma}^\alpha| \Big\rangle. \tag{13.19.13}$$

This is the minimum shear principle for an aggregate macroelement. In the context of infinitesimal strain, the original proof was given by Bishop and Hill (1951a).

Bishop and Hill (*op. cit.*) also proved the maximum work principle for an aggregate of rigid-plastic crystals. Let $\dot{\mathbf{F}}$ be the rate of deformation gradient that takes place at the state of stress $\mathbf{P}$, and let $\mathbf{P}_*$ be any other state of stress which does not violate the yield condition on any slip system. The difference of the corresponding local rates of work per unit volume is, from Eq. (12.19.14),

$$(\boldsymbol{\tau} - \boldsymbol{\tau}_*) : \mathbf{D} = (\mathbf{P} - \mathbf{P}_*) \cdot\cdot \dot{\mathbf{F}} = (\mathbf{T} - \mathbf{T}_*) : \dot{\mathbf{E}} \geq 0. \tag{13.19.14}$$

Upon integration over the representative macroelement volume, there follows

$$\langle (\mathbf{P} - \mathbf{P}_*) \cdot\cdot \dot{\mathbf{F}} \rangle = (\langle \mathbf{P} \rangle - \langle \mathbf{P}_* \rangle) \cdot\cdot \langle \dot{\mathbf{F}} \rangle = ([\,\mathbf{T}\,] - [\,\mathbf{T}_*\,]) : [\dot{\mathbf{E}}] \geq 0. \tag{13.19.15}$$

If the current configuration is taken for the reference, we can write

$$(\{\boldsymbol{\sigma}\} - \{\boldsymbol{\sigma}_*\}) : \{\mathbf{D}\} \geq 0. \tag{13.19.16}$$

The last two expressions are the alternative statements of the maximum work principle for an aggregate.

13.20. Macroscopic Flow Potential for Rate-Dependent Plasticity

In a rate-dependent plastic aggregate, which exhibits the instantaneous elastic response to rapid loading or straining, the plastic part of the rate of macroscopic deformation gradient is defined by

$$\frac{\mathrm{d}^{\mathrm{p}} \langle \mathbf{F} \rangle}{\mathrm{d}t} = \frac{\mathrm{d} \langle \mathbf{F} \rangle}{\mathrm{d}t} - [\mathbf{M}] \cdot\cdot \frac{\mathrm{d} \langle \mathbf{P} \rangle}{\mathrm{d}t}, \tag{13.20.1}$$

where t stands for the physical time. By an analogous expression to (13.11.13), this is related to the local rate of deformation gradient by

$$\frac{\mathrm{d}^{\mathrm{p}} \langle \mathbf{F} \rangle}{\mathrm{d}t} = \langle \frac{\mathrm{d}^{\mathrm{p}} \mathbf{F}}{\mathrm{d}t} \cdot\cdot \boldsymbol{\mathcal{P}} \rangle. \tag{13.20.2}$$

The fourth-order tensor $\boldsymbol{\mathcal{P}}$ is the influence tensor of elastic heterogeneity, which relates the elastic increments of the local and macroscopic nominal stress, $\delta \mathbf{P} = \boldsymbol{\mathcal{P}} \cdot\cdot \delta \langle \mathbf{P} \rangle$.

Suppose that the flow potential exists at the microlevel, such that (see Section 8.4)

$$\frac{\mathrm{d}^{\mathrm{p}} \mathbf{F}}{\mathrm{d}t} = \frac{\partial \Omega\,(\mathbf{P}, \mathcal{H})}{\partial \mathbf{P}}. \tag{13.20.3}$$

Substitution of Eq. (13.20.3) into Eq. (13.20.2) gives

$$\frac{\mathrm{d}^{\mathrm{p}} \langle \mathbf{F} \rangle}{\mathrm{d}t} = \langle \frac{\partial \Omega}{\partial \mathbf{P}} \cdot\cdot \boldsymbol{\mathcal{P}} \rangle = \langle \frac{\partial \Omega}{\partial \langle \mathbf{P} \rangle} \rangle = \frac{\partial}{\partial \langle \mathbf{P} \rangle} \langle \Omega \rangle. \tag{13.20.4}$$

In the derivation, the partial differentiation enables the transition

$$\frac{\partial \Omega}{\partial \langle \mathbf{P} \rangle} = \frac{\partial \Omega}{\partial \mathbf{P}} \cdot\cdot \frac{\partial \mathbf{P}}{\partial \langle \mathbf{P} \rangle} = \frac{\partial \Omega}{\partial \mathbf{P}} \cdot\cdot \boldsymbol{\mathcal{P}}. \tag{13.20.5}$$

From Eq. (13.20.4) we conclude that the existence of the flow potential Ω at the microlevel implies the existence of the flow potential at the macrolevel. The macroscopic flow potential is equal to the volume average $\langle\, \Omega \,\rangle$ of the microscopic flow potentials.

Since

$$\frac{\mathrm{d}^{\mathrm{p}} \langle \mathbf{P} \rangle}{\mathrm{d}t} = -[\boldsymbol{\Lambda}] \cdot\cdot \frac{\mathrm{d}^{\mathrm{p}} \langle \mathbf{F} \rangle}{\mathrm{d}t}, \tag{13.20.6}$$

and since at fixed $\mathcal{H}$,

$$\frac{\partial}{\partial \langle \mathbf{F} \rangle} = [\boldsymbol{\Lambda}] \cdot\cdot \frac{\partial}{\partial \langle \mathbf{P} \rangle}, \tag{13.20.7}$$

we have, dually to Eq. (13.20.4),

$$\frac{\mathrm{d}^{\mathrm{p}} \langle \mathbf{P} \rangle}{\mathrm{d}t} = -\frac{\partial}{\partial \langle \mathbf{F} \rangle} \langle\, \Omega \,\rangle. \tag{13.20.8}$$

If the stress and strain measures $\mathbf{T}$ and $\mathbf{E}$ are used, there follows

$$\frac{\mathrm{d}^{\mathrm{p}} [\mathbf{E}]}{\mathrm{d}t} = \frac{\partial}{\partial [\mathbf{T}]} \langle\, \Omega \,\rangle, \tag{13.20.9}$$

$$\frac{\mathrm{d}^{\mathrm{p}} [\mathbf{T}]}{\mathrm{d}t} = -\frac{\partial}{\partial [\mathbf{E}]} \langle\, \Omega \,\rangle. \tag{13.20.10}$$

The original proof for the transmissibility of the flow potential from the local (subelement) to the macroscopic (aggregate) level is due to Hill and Rice (1973). See also Zarka (1972), Hutchinson (1976), and Ponter and Leckie (1976).

References

Bishop, J. F. W. and Hill, R. (1951a), A theory of plastic distortion of a polycrystalline aggregate under combined stresses, *Phil. Mag.*, Vol. 42, pp. 414–427.

Bishop, J. F. W. and Hill, R. (1951b), A theoretical derivation of the plastic properties of a polycrystalline face-centred metal, *Phil. Mag.*, Vol. 42, pp. 1298–1307.

Bui, H. D. (1970), Evolution de la frontière du domaine élastique des métaux avec l'écrouissage plastique et comportement élastoplastique d'un agregat de cristaux cubiques, *Mem. Artillerie Franç.: Sci. Tech. Armament*, Vol. 1, pp. 141–165.

Drugan, W. J. and Willis, J. R. (1996), A micromechanics-based nonlocal constitutive equation and estimates of representative volume element size for elastic composites, *J. Mech. Phys. Solids*, Vol. 44, pp. 497–524.

Hashin, Z. (1964), Theory of mechanical behavior of heterogeneous media, *Appl. Mech. Rev.*, Vol. 17, pp. 1–9.

Havner, K. S. (1973), An analytical model of large deformation effects in crystalline aggregates, in *Foundations of Plasticity*, ed. A. Sawczuk, pp. 93–106, Noordhoff, Leyden.

Havner, K. S. (1974), Aspects of theoretical plasticity at finite deformation and large pressure, *Z. angew. Math. Phys.*, Vol. 25, pp. 765–781.

Havner, K. S. (1986), Fundamental considerations in micromechanical modeling of polycrystalline metals at finite strain, in *Large Deformation of Solids: Physical Basis and Mathematical Modelling*, eds. J. Gittus, J. Zarka, and S. Nemat-Nasser, pp. 243–265, Elsevier, London.

Havner, K. S. (1992), *Finite Plastic Deformation of Crystalline Solids*, Cambridge University Press, Cambridge.

Hill, R. (1963), Elastic properties of reinforced solids: Some theoretical principles, *J. Mech. Phys. Solids*, Vol. 11, pp. 357–372.

Hill, R. (1967), The essential structure of constitutive laws for metal composites and polycrystals, *J. Mech. Phys. Solids*, Vol. 15, pp. 79–96.

Hill, R. (1972), On constitutive macro-variables for heterogeneous solids at finite strain, *Proc. Roy. Soc. Lond. A*, Vol. 326, pp. 131–147.

Hill, R. (1984), On macroscopic effects of heterogeneity in elastoplastic media at finite strain, *Math. Proc. Camb. Phil. Soc.*, Vol. 95, pp. 481–494.

Hill, R. (1985), On the micro-to-macro transition in constitutive analyses of elastoplastic response at finite strain, *Math. Proc. Camb. Phil. Soc.*, Vol. 98, pp. 579–590.

Hill, R. and Rice, J. R. (1973), Elastic potentials and the structure of inelastic constitutive laws, *SIAM J. Appl. Math.*, Vol. 25, pp. 448–461.

Hori, M. and Nemat-Nasser, S. (1999), On two micromechanics theories for determining micro-macro relations in heterogeneous solids, *Mech. Mater.*, Vol. 31, pp. 667–682.

Hutchinson, J. W. (1976), Bounds and self-consistent estimates for creep of polycrystalline materials, *Proc. Roy. Soc. Lond. A*, Vol. 348, pp. 101–127.

Kröner, E. (1972), *Statistical Continuum Mechanics*, CISM Lecture Notes – Udine, 1971, Springer-Verlag, Wien.

Kunin, I. A. (1982), *Elastic Media with Microstructure, I and II*, Springer-Verlag, Berlin.

Mandel, J. (1966), Contribution théorique à l'étude de l'écrouissage et des lois de l'écoulement plastique, in *Proc. 11th Int. Congr. Appl. Mech. (Munich 1964)*, eds. H. Görtler and P. Sorger, pp. 502–509, Springer-Verlag, Berlin.

Maugin, G. A. (1992), *The Thermomechanics of Plasticity and Fracture*, Cambridge University Press, Cambridge.

Mura, T. (1987), *Micromechanics of Defects in Solids*, Martinus Nijhoff, Dordrecht, The Netherlands.

Nemat-Nasser, S. (1999), Averaging theorems in finite deformation plasticity, *Mech. Mater.*, Vol. 31, pp. 493–523 (with Erratum, Vol. 32, 2000, p. 327).

Nemat-Nasser, S. and Hori, M. (1993), *Micromechanics: Overall Properties of Heterogeneous Materials*, North-Holland, Amsterdam.

Ortiz, M. (1987) A method of homogenization, *Int. J. Engng. Sci.*, Vol. 25, pp. 923–934.

Ortiz, M. and Phillips, R. (1999), Nanomechanics of defects in solids, *Adv. Appl. Mech.*, Vol. 36, pp. 1–79.

Petryk, H. (1998), Macroscopic rate-variables in solids undergoing phase transformation, *J. Mech. Phys. Solids*, Vol. 46, pp. 873–894.

Ponter, A. R. S. and Leckie, F. A. (1976), Constitutive relationships for the time-dependent deformation of metals, *J. Engng. Mater. Techn.*, Vol. 98, pp. 47–51.

Rice, J. R. (1970), On the structure of stress-strain relations for time-dependent plastic deformation in metals, *J. Appl. Mech.*, Vol. 37, pp. 728–737.

Rice, J. R. (1971), Inelastic constitutive relations for solids: An internal variable theory and its application to metal plasticity, *J. Mech. Phys. Solids*, Vol. 19, pp. 433–455.

Rice, J. R. (1975), Continuum mechanics and thermodynamics of plasticity in relation to microscale mechanisms, in *Constitutive Equations in Plasticity*, ed. A. S. Argon, pp. 23–79, MIT Press, Cambridge, Massachusetts.

Sanchez-Palencia, E. (1980), *Nonhomogeneous Media and Vibration Theory*, Lecture Notes in Physics, 127, Springer-Verlag, Berlin.

Sawicki, A. (1983), Engineering mechanics of elasto-plastic composites, *Mech. Mater.*, Vol. 2, pp. 217–231.

Stolz, C. (1997), Large plastic deformation of polycrystals, in *Large Plastic Deformation of Crystalline Aggregates*, ed. C. Teodosiu, pp. 81–108, Springer-Verlag, Wien.

Suquet, P. M. (1987), Elements of homogenization for inelastic solid mechanics, in *Homogenization Techniques for Composite Media*, eds. E. Sanchez-Palencia and A. Zaoui, pp. 193–278, Springer-Verlag, Berlin.

Torquato, S. (1991), Random heterogeneous media: Microstructure and improved bounds on effective properties, *Appl. Mech. Rev.*, Vol. 44, No. 2, pp. 37–76.

Willis, J. R. (1981), Variational and related methods for the overall properties of composites, *Adv. Appl. Mech.*, Vol. 21, pp. 1–78.

Yang, W. and Lee, W. B. (1993), *Mesoplasticity and its Applications*, Springer-Verlag, Berlin.

Zarka, J. (1972), Generalisation de la theorie du potentiel plastique multiple en viscoplasticite, *J. Mech. Phys. Solids*, Vol. 20, pp. 179–195.

CHAPTER 14

POLYCRYSTALLINE MODELS

The approximate models of the polycrystalline plastic response are discussed in this chapter. The objective is to correlate the polycrystalline to single crystal behavior and to derive the constitutive relation for a polycrystalline aggregate in terms of the known constitutive relations for single crystals and known (or assumed) distribution of crystalline grains within the aggregate. The classical model of Taylor (1938 a, b) and the analysis by Bishop and Hill (1951 a, b) are first presented. Determination of the polycrystalline axial stress-strain curve and the polycrystalline yield surface is considered. The main theme of the chapter is the self-consistent method, introduced in the polycrystalline plasticity by Kröner (1961), and Budiansky and Wu (1962). Hill's (1965 a) formulation and generalization of the method is followed in the presentation. The self-consistent calculations of elastic and elastoplastic moduli, the development of the crystallographic texture, and the effects of the grain-size on the aggregate response are then discussed.

14.1. Taylor-Bishop-Hill Analysis

The slip in an f.c.c. crystal occurs on the octahedral planes in the directions of the octahedron edges (Fig. 14.1). There are three possible slip directions in each of the four distinct slip planes, making a total of twelve slip systems (if counting both senses of a slip direction as one), or twenty four (if counting opposite directions separately). The positive senses of the slip directions are chosen as indicated in Table 14.1. The letters a, b, c, d refer to four slip planes. With attached indices 1, 2 and 3, they designate the slip rates in the respective positive slip directions.

If elastic (lattice) strains are disregarded, the components of the rate of deformation tensor $\mathbf{D}$, expressed on the cubic axes, due to simultaneous slip

Plane	(111)			$(\bar{1}\bar{1}1)$			$(\bar{1}11)$			$(1\bar{1}1)$		
Slip Rate	a_1	a_2	a_3	b_1	b_2	b_3	c_1	c_2	c_3	d_1	d_2	d_3
Slip Direction	$[0\bar{1}1]$	$[10\bar{1}]$	$[\bar{1}10]$	$[011]$	$[\bar{1}0\bar{1}]$	$[1\bar{1}0]$	$[0\bar{1}1]$	$[\bar{1}0\bar{1}]$	$[110]$	$[011]$	$[10\bar{1}]$	$[\bar{1}\bar{1}0]$

TABLE 14.1. Designation of slip systems in f.c.c. crystals

rates in twelve slip directions, are given by (Taylor, 1938a)

$$\sqrt{6}\, D_{11} = a_2 - a_3 + b_2 - b_3 + c_2 - c_3 + d_2 - d_3, \tag{14.1.1}$$

$$\sqrt{6}\, D_{22} = a_3 - a_1 + b_3 - b_1 + c_3 - c_1 + d_3 - d_1, \tag{14.1.2}$$

$$\sqrt{6}\, D_{33} = a_1 - a_2 + b_1 - b_2 + c_1 - c_2 + d_1 - d_2, \tag{14.1.3}$$

$$2\sqrt{6}\, D_{23} = -a_2 + a_3 + b_2 - b_3 - c_2 + c_3 + d_2 - d_3, \tag{14.1.4}$$

$$2\sqrt{6}\, D_{31} = -a_3 + a_1 + b_3 - b_1 + c_3 - c_1 - d_3 + d_1, \tag{14.1.5}$$

$$2\sqrt{6}\, D_{12} = -a_1 + a_2 - b_1 + b_2 + c_1 - c_2 + d_1 - d_2. \tag{14.1.6}$$

These are derived from the formulas in Section 12.17, i.e.,

$$\mathbf{D} = \sum_{\alpha=1}^{12} \mathbf{P}^\alpha \dot{\gamma}^\alpha = \sum_{\alpha=1}^{12} \frac{1}{2} (\mathbf{s}^\alpha \otimes \mathbf{m}^\alpha + \mathbf{m}^\alpha \otimes \mathbf{s}^\alpha)\, \dot{\gamma}^\alpha, \tag{14.1.7}$$

where $\mathbf{m}^\alpha$ is the unit slip plane normal, and $\mathbf{s}^\alpha$ is the slip direction. For example, the contribution from the slip rate $\dot{\gamma} = a_1$ is obtained by using

$$\mathbf{m} = \frac{1}{\sqrt{3}}(1,1,1), \quad \mathbf{s} = \frac{1}{\sqrt{2}}(0,-1,1), \tag{14.1.8}$$

which gives

$$\frac{1}{2}(\mathbf{s}^\alpha \otimes \mathbf{m}^\alpha + \mathbf{m}^\alpha \otimes \mathbf{s}^\alpha)\, a_1 = \frac{a_1}{2\sqrt{6}} \begin{pmatrix} 0 & -1 & 1 \\ -1 & -2 & 0 \\ 1 & 0 & 2 \end{pmatrix}. \tag{14.1.9}$$

An arbitrary rate of deformation tensor has five independent components ($\operatorname{tr}\mathbf{D} = 0$ for a rigid-plastic crystal), and therefore can only be produced by multiple slip over a group of slip systems containing an independent set of five. Of the $C_5^{12} = 792$ sets of five slips, only 384 are independent (Bishop

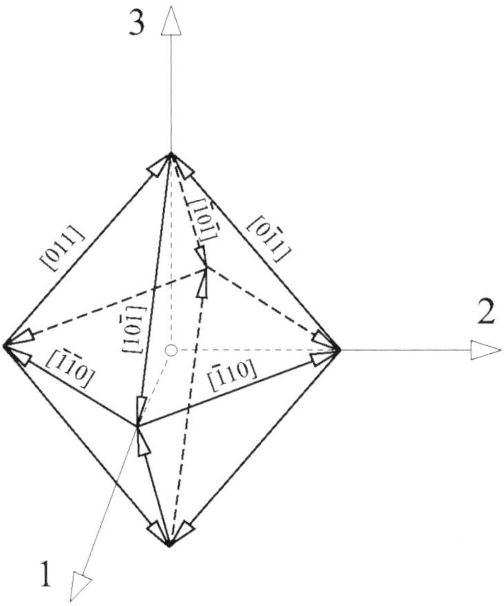

FIGURE 14.1. Twelve different slip directions in f.c.c. crystals (counting opposite directions as different) are the edges of the octahedron shown relative to principal cubic axes. Each slip direction is shared by two intersecting slip planes so that there is a total of 24 independent slip systems (12 if counting opposite slip directions as one).

and Hill, 1951b). The 408 dependent sets are identified as follows. First, as Taylor originally noted, only two of three slip systems in the same slip plane are independent. The unit slip rates along a_1, a_2 and a_3 directions together produce the zero resultant rate of deformation. The same applies to three slip directions in b, c and d slip planes. We write this symbolically as

$$a_1 + a_2 + a_3 = 0, \quad b_1 + b_2 + b_3 = 0, \quad c_1 + c_2 + c_3 = 0, \quad d_1 + d_2 + d_3 = 0. \tag{14.1.10}$$

Thus, if the set of five slip systems contains a_1, a_2 and a_3, there are $C_2^9 = 36$ possible combinations with the remaining nine slip systems. These 36 sets of five slips cannot produce an arbitrary $\mathbf{D}$, with five independent components, and are thus eliminated from 792 sets of five slips. Additional $3 \times 36 = 108$ sets, associated with dependent sets of three slips in b, c and d planes, can also be eliminated. This makes a total of 144 dependent sets corresponding to the constraints (14.1.10).

Of the remaining 648 sets of five slips, 324 involve two slips in each of two slip planes with one in a third ($6 \times 3^2 \times 6 = 324$), while 324 involve two slips in one slip plane and one in each of the other three slip planes ($4 \times 3^4 = 324$). In the latter group, there are $3 \times 8 = 24$ sets involving the combinations

$$a_1 - b_1 + c_1 - d_1 = 0, \quad a_2 - b_2 + c_2 - d_2 = 0, \quad a_3 - b_3 + c_3 - d_3 = 0. \quad (14.1.11)$$

These expressions can also be interpreted as meaning that such combinations of unit slips produce zero resultant rate of deformation. The 24 sets of five slips, involving four slip rates according to (14.1.11), can thus be eliminated (these sets necessarily consists of two slips in one plane and one slip in each of the remaining three slip planes). Additional 12 sets are eliminated, which correspond to conditions obtained from (14.1.11) by adding or subtracting $\sum a_i, \ldots, \sum d_i$, one at a time, to each of (14.1.11). A representative of these is $a_1 - b_1 + c_1 + d_2 + d_3 = 0$ (Havner, 1992).

There are $4 \times 33 = 132$ dependent sets associated with

$$a_1 + b_2 + d_3 = 0, \quad a_2 + b_1 + c_3 = 0, \quad a_3 + c_2 + d_1 = 0, \quad b_3 + c_1 + d_2 = 0. \quad (14.1.12)$$

Each group of 33 sets consists of 21 sets involving two slips in one plane and one slip in each of other three planes, and 12 sets involving two slips in two planes and one slip in one plane. Additional 84 sets can be eliminated by subtracting $\sum a_i$, $\sum b_i$ and $\sum d_i$, one at a time, from the first of (14.1.12), and similarly for the other three. This makes 12 groups of 7 sets. A representative group is associated with $a_1 + b_2 - d_1 - d_2 = 0$. Four of the 7 sets consist of two slips in two planes and one slip in one plane, while three sets consist of two slips in one plane and one slip in each of the other three planes. Finally, 12 more sets (making total of 228 dependent sets associated with (14.1.12) and their equivalents) can be eliminated by subtracting appropriate one of $\sum a_i, \ldots, \sum d_i$ from each of the 12 previous group equations. An example is $-a_1 + b_1 + b_3 + d_1 + d_2 = 0$. They all involve two slips in each of two planes and one slip in another plane.

In summary, there is a total of 408 dependent sets of five slips: 144 sets with three slips in the same plane, 108 sets with two slips in each of two planes and one in a third, and 156 sets with two slips in one plane and one slip in each of the other three planes. Taylor (1938a) originally considered only 216 sets as geometrically admissible (involving double slip in each of two planes), and did not observe 168 admissible sets with double slip in only one plane. These were originally identified by Bishop and Hill (1951b).

14.1.1. Polycrystalline Axial Stress-Strain Curve

In an early approach to predict the tensile yield stress of a polycrystalline aggregate, Sachs (1928) assumed that each grain is subjected to uniaxial stress parallel to the specimen axis and sufficient to initiate slip in the most critical slip system. Since each grain was assumed to deform only by a single slip, the deformations across the grain boundaries of differently oriented grains were incompatible. Furthermore, since the stress in each grain was assumed to be a simple tension, of the different amount from grain to grain, the equilibrium across the grain boundaries was not satisfied, either. Nonetheless, the obtained value for the aggregate tensile yield stress was about $2.2\,\tau$, where τ is the yield stress of a single crystal, which was not a very unsatisfactory estimate.

A more realistic model was proposed by Taylor (1938a), who assumed that every grain within a polycrystalline aggregate, subjected to macroscopically uniform deformation, sustains the same deformation (strain and rotation). This ensures compatibility, but not equilibrium, across the grain boundaries. As discussed below, the calculated value for the aggregate tensile yield stress is about $3.1\,\tau$. Taylor's assumption can be viewed as an extension of Voigt's (1889) uniform strain assumption for the elastic inhomogeneous bodies, as discussed later in Section 14.5.

Let ϕ, θ and ψ denote the Euler angles of the lattice axes of an arbitrary grain relative to the specimen axes. These can be defined as follows. Beginning with the coincident axes, imagine that the grain is first rotated by ϕ about [001] axis, then by θ about the current direction of the [010] axis, and finally by ψ about the new direction of the [001] axis. Counterclockwise rotations are positive. The corresponding orthogonal transformation defining the direction cosines of the crystal axes relative to the specimen axes is (Havner, 1992)

$$\mathbf{Q} = \begin{pmatrix} \cos\phi\cos\theta\cos\psi - \sin\phi\sin\psi & \sin\phi\cos\theta\cos\psi + \cos\phi\sin\psi & -\sin\theta\cos\psi \\ -\cos\phi\cos\theta\sin\psi - \sin\phi\cos\psi & -\sin\phi\cos\theta\sin\psi + \cos\phi\cos\psi & \sin\theta\sin\psi \\ \cos\phi\sin\theta & \sin\phi\sin\theta & \cos\theta \end{pmatrix}. \tag{14.1.13}$$

If the polycrystalline aggregate is subjected to uniform rate of deformation $\mathbf{D}_\infty$, the components of this tensor on the local crystal axes of an arbitrarily oriented grain are the components of the matrix $\mathbf{Q}\cdot\mathbf{D}_\infty\cdot\mathbf{Q}^T$. This in general has five independent components, and at least five independent slip systems must be active in the crystal to satisfy equations (14.1.1)–(14.1.6). Taylor assumed that only five systems will actually activate. As already discussed, there are 384 independent combinations of five slip rates that can produce a local rate of deformation with five independent components on the local crystal axes. Taylor (1938a, b) suggested, and Bishop and Hill (1951a) proved, that of all possible combinations of the slip rates, the actual one is characterized by the least sum of the absolute values of the slip rates. This was discussed in Section 12.7. From Eqs. (13.19.7) and (13.19.9), we can write

$$\{\boldsymbol{\sigma}:\mathbf{D}\} = \{\min\sum_\alpha \tau_{\rm cr}^\alpha\,|\dot\gamma^\alpha|\}, \tag{14.1.14}$$

where $\{\ \}$ denotes the orientation average. Assuming that the hardening of slip systems is isotropic, $\tau_{\rm cr}^\alpha = \tau_{\rm cr}$ for all slip systems within a grain, and since $\mathbf{D}$ is assumed to be equal to $\mathbf{D}_\infty$ in every grain, Eq. (14.1.14) becomes

$$\{\boldsymbol{\sigma}\}:\mathbf{D}_\infty = \{\tau_{\rm cr}\min\sum_\alpha|\dot\gamma^\alpha|\}. \tag{14.1.15}$$

The average critical resolved shear stress $\bar\tau_{\rm cr}$ of the aggregate can be defined by requiring that

$$\{\tau_{\rm cr}\min\sum_\alpha|\dot\gamma^\alpha|\} = \bar\tau_{\rm cr}\{\min\sum_\alpha|\dot\gamma^\alpha|\}, \tag{14.1.16}$$

so that

$$\{\boldsymbol{\sigma}\}:\mathbf{D}_\infty = \bar\tau_{\rm cr}\{\min\sum_\alpha|\dot\gamma^\alpha|\}. \tag{14.1.17}$$

If the macroscopic logarithmic strain in the direction of applied uniaxial tension σ is e (the lateral strain components of macroscopically isotropic specimen being equal to $-e/2$), the rate of work is

$$\sigma\,\dot e = \bar\tau_{\rm cr}\{\min\sum_\alpha|\dot\gamma^\alpha|\}. \tag{14.1.18}$$

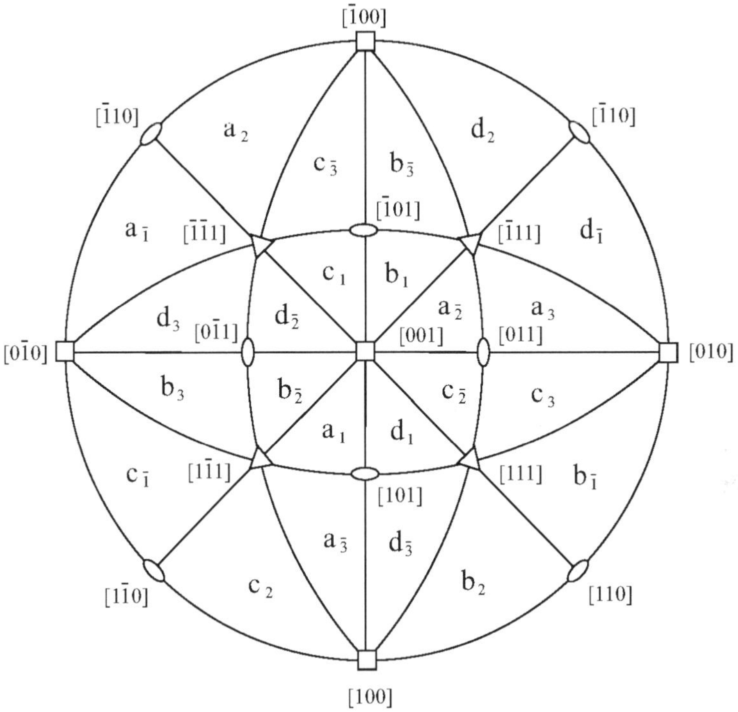

FIGURE 14.2. Standard [001] stereographic projection of cubic crystals. The 24 triangles represent regions in which a particular slip system operates. For f.c.c. crystals the letters a, b, c, d represent the four slip planes $\{111\}$, and the numbers (indices) $1, 2, 3$ designate the three slip directions $\langle 110\rangle$ (with attached bar, the number designates the opposite slip system). For b.c.c. crystals the letters represent the four slip directions $\langle 111\rangle$, and the numbers designate the three $\{110\}$ planes which contain each slip direction (from Havner, 1982; with permission from Elsevier Science).

The ratio

$$m = \frac{\sigma}{\bar{\tau}_{\text{cr}}} = \frac{\{\min\sum_{\alpha}|\mathrm{d}\gamma^{\alpha}|\}}{\mathrm{d}e} \tag{14.1.19}$$

is known as the Taylor orientation factor. Taylor (1938a) chose 44 initial orientations distributed uniformly over the spherical triangle [101][100][11$\bar{1}$] within a standard [001] stereographic projection ($d_{\bar{3}}$ in Fig. 14.2). The calculated value of m was $m = 3.10$. More accurate calculations of Bishop

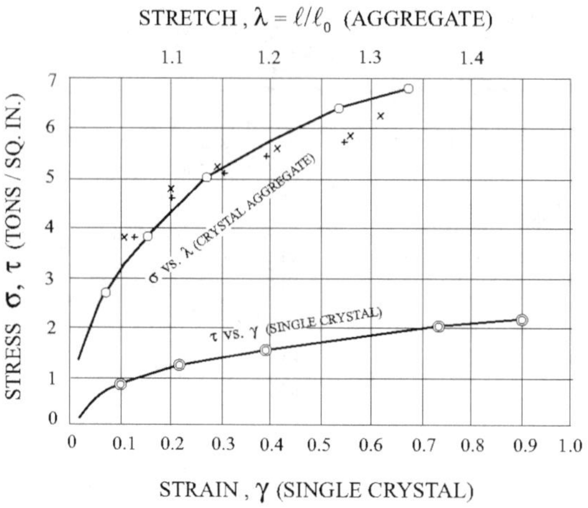

FIGURE 14.3. Single-crystal stress-strain curve $\tau = f(\gamma)$ and Taylor's prediction of the stress-stretch curve $\sigma = \sigma(\lambda)$ for the polycrystalline aggregate (from Taylor, 1938b; with permission from the Institute for Materials).

and Hill (1951b), accounting for all 384 geometrically admissible sets of five independent slip rates, resulted in an improved value of $m = 3.06$.

The polycrystalline stress-strain curve $\sigma = \sigma(e)$ can be deduced from Eq. (14.1.19) as

$$\sigma = m\,\bar{\tau}_{\rm cr} = m\,f(\bar{\gamma}) = m\,f\left(\int m\,{\rm d}e\right), \tag{14.1.20}$$

where

$$\bar{\gamma} = \int \{\min \sum_{\alpha} |{\rm d}\gamma^{\alpha}|\} = \int m\,{\rm d}e\,. \tag{14.1.21}$$

Here, it is assumed that the function f, relating $\bar{\tau}_{\rm cr}$ and $\bar{\gamma}$, is the same function that relates the shear stress and shear strain in a monocrystal under single slip, $\tau = f(\gamma)$. For aluminum crystals investigated by Taylor, this function was found to be nearly parabolic ($\sim \gamma^{1/2}$).

It is noted that the Taylor factor m depends on the strain level, because lattice rotations change the orientation of slip systems within grains relative to the specimen axes. The tensile stress-strain curve shown in Fig. 14.3 was obtained by Taylor using the constant value of m. Single crystal and polycrystalline data from uniaxial stress experiments can be found in Bell (1968).

14.1.2. Stresses in Grain

It is of interest to analyze the state of stress in an individual grain. Particularly important is to analyze whether there is a stress state, associated with a geometrically admissible set of slip rates, that is also physically admissible. For the physically admissible set, the resolved shear stress along inactive slip directions does not exceed the critical shear stress there. This problem was studied by Bishop and Hill (1951b). If the components of the uniform stress in the grain are σ_{ij}, relative to the local cubic axes, the resolved shear stresses (multiplied by $\sqrt{6}$) in the twelve f.c.c. slip systems are

$$\sqrt{6}\,\tau_{a_1} = -\sigma_{22} + \sigma_{33} + \sigma_{31} - \sigma_{12}, \quad \sqrt{6}\,\tau_{a_2} = -\sigma_{33} + \sigma_{11} - \sigma_{23} + \sigma_{12},$$

$$\sqrt{6}\,\tau_{a_3} = -\sigma_{11} + \sigma_{22} + \sigma_{23} - \sigma_{31}, \tag{14.1.23}$$

$$\sqrt{6}\,\tau_{b_1} = -\sigma_{22} + \sigma_{33} - \sigma_{31} - \sigma_{12}, \quad \sqrt{6}\,\tau_{b_2} = -\sigma_{33} + \sigma_{11} + \sigma_{23} + \sigma_{12},$$

$$\sqrt{6}\,\tau_{b_3} = -\sigma_{11} + \sigma_{22} - \sigma_{23} + \sigma_{31}, \tag{14.1.25}$$

$$\sqrt{6}\,\tau_{c_1} = -\sigma_{22} + \sigma_{33} - \sigma_{31} + \sigma_{12}, \quad \sqrt{6}\,\tau_{c_2} = -\sigma_{33} + \sigma_{11} - \sigma_{23} - \sigma_{12},$$

$$\sqrt{6}\,\tau_{c_3} = -\sigma_{11} + \sigma_{22} + \sigma_{23} + \sigma_{31}, \tag{14.1.27}$$

$$\sqrt{6}\,\tau_{d_1} = -\sigma_{22} + \sigma_{33} + \sigma_{31} + \sigma_{12}, \quad \sqrt{6}\,\tau_{d_2} = -\sigma_{33} + \sigma_{11} + \sigma_{23} - \sigma_{12},$$

$$\sqrt{6}\,\tau_{d_3} = -\sigma_{11} + \sigma_{22} - \sigma_{23} - \sigma_{31}. \tag{14.1.29}$$

The 12×6 matrix of the coefficients in these relations, between the 12 resolved shear stresses and 6 stress components, is the transpose of the 6×12 matrix of the coefficients relating the rate of deformation components to the slip rates in Eqs. (14.1.1)–(14.1.6). From the set of 12 equations (14.1.23)–(14.1.29) we can always find a stress state (apart from pressure) for which the resolved shear stress attains the critical value in five independent slip directions. The critical stress would usually be exceeded in one or more of the other seven slip directions. However, for any prescribed rate of deformation **D**, it is always possible to find at least one set of five slip rates, geometrically equivalent to $\mathbf{D}$, for which there exist a physically admissible stress state that does not violate the yield condition on other seven slip systems. Bishop (1953) actually proved that, for a given $\mathbf{D}$, a stress state determined by minimizing the rate of work $\dot{w} = \boldsymbol{\sigma} : \mathbf{D}$ will not exceed the critical shear stress in any other slip system. It is recalled that the work on physically operating slip rates is less than the work done on the slip rates that are only geometrically possible; see (12.19.8).

For example, a tension or compression of amount $\sqrt{6}\,\tau_{\rm cr}$ along a cubic axis is a stress state on an eightfold vertex of a polyhedral yield surface of the single crystal, since the substitution of $\sigma_{11} = \sigma_{22} = \sigma_{12} = \sigma_{23} = \sigma_{31} = 0$ and $\sigma_{33} = \sqrt{6}\,\tau_{\rm cr}$ into Eqs. (14.1.23)–(14.1.29) gives

$$\tau_{a_1} = -\tau_{a_2} = \tau_{b_1} = -\tau_{b_2} = \tau_{c_1} = -\tau_{c_2} = \tau_{d_1} = -\tau_{d_2} = \tau_{\rm cr}\,, \qquad (14.1.30)$$

and

$$\tau_{a_3} = \tau_{b_3} = \tau_{c_3} = \tau_{d_3} = 0. \qquad (14.1.31)$$

Differential hardening is assumed to be absent, so that all slip systems harden equally ($\tau_{\rm cr}$ equal on all slip systems). The microscopic Bauschinger effect is assumed to be absent, as well, so that the critical shear stress is equal in opposite senses along the same slip direction. A tension or compression of amount $\sqrt{6}\,\tau_{\rm cr}$ normal to an octahedral plane is a physically admissible stress state, too, being on a sixthfold vertex of the monocrystalline yield surface. Indeed, the substitution of $\sigma_{11} = \sigma_{22} = \sigma_{33} = 0$ and $\sigma_{12} = \sigma_{23} = \sigma_{31} = \sqrt{6}\,\tau_{\rm cr}/2$ into Eqs. (14.1.23)–(14.1.29) gives

$$\tau_{b_2} = -\tau_{b_1} = \tau_{c_3} = -\tau_{c_2} = \tau_{d_1} = -\tau_{d_3} = \tau_{\rm cr}\,, \qquad (14.1.32)$$

and

$$\tau_{a_1} = \tau_{a_2} = \tau_{a_3} = \tau_{b_3} = \tau_{c_1} = \tau_{d_2} = 0. \qquad (14.1.33)$$

The stresses in grains, associated with the assumption of equal deformation $\mathbf{D}_\infty$ in all grains, will not be in equilibrium across the grain boundaries. Denote this stress by $\hat{\boldsymbol{\sigma}}_{\rm c}$. Let $\boldsymbol{\sigma}_{\rm c}$ be the actual stress in the grain of a polycrystalline aggregate, corresponding to the actual rate of deformation $\mathbf{D}_{\rm c}$ that takes place in the grain. The fields $\boldsymbol{\sigma}_{\rm c}$ and $\mathbf{D}_{\rm c}$ are the true equilibrium and compatible fields of the polycrystalline aggregate. The orientation average

$$\{\boldsymbol{\sigma}_{\rm c}\} = \boldsymbol{\sigma}_\infty \qquad (14.1.34)$$

is the macroscopically uniform stress applied to the aggregate, and the average

$$\{\mathbf{D}_c\} = \mathbf{D}_\infty \tag{14.1.35}$$

is the corresponding macroscopically uniform (average) deformation rate in the aggregate. Since $\hat{\boldsymbol{\sigma}}_c$ is the stress state on the current yield surface of the grain, at which $\mathbf{D}_\infty$ would occur in the grain, from the maximum work principle (12.19.14) we can write

$$(\hat{\boldsymbol{\sigma}}_c - \boldsymbol{\sigma}_c) : \mathbf{D}_\infty \geq 0. \tag{14.1.36}$$

This holds because the stress $\boldsymbol{\sigma}_c$ does not violate the current yield condition for the grain, being the stress state at which the actual $\mathbf{D}_c$ takes place. Thus, upon averaging of (14.1.36), we obtain

$$\{\hat{\boldsymbol{\sigma}}_c\} : \mathbf{D}_\infty \geq \boldsymbol{\sigma}_\infty : \mathbf{D}_\infty. \tag{14.1.37}$$

This means that the actual rate of work done on a polycrystalline aggregate is not greater that the rate of work that would be done if all grains underwent the same (macroscopic) rate of deformation. Bishop and Hill (1951b) argued that the two rates of work are in fact nearly equal, and suggested an approximation

$$\{\hat{\boldsymbol{\sigma}}_c\} : \mathbf{D}_\infty \approx \boldsymbol{\sigma}_\infty : \mathbf{D}_\infty. \tag{14.1.38}$$

14.1.3. Calculation of Polycrystalline Yield Surface

The objective is now to calculate the polycrystalline yield surface in terms of the single crystal properties. First, since a superposed uniform hydrostatic stress throughout the aggregate does not affect the resolved shear stress on any slip system, and since slip is assumed to be governed by a pressure-independent Schmid law, the polycrystalline yield surface does not depend on the hydrostatic part of the applied stress. The surface is cylindrical, with its generator parallel to the hydrostatic stress axis. If there is no microscopic Bauschinger effect, the critical shear stress does not depend on the sense of slip along the slip direction, which implies that the polycrystalline yield surface is symmetric about the origin. Thus, if $\boldsymbol{\sigma}_\infty$ produces yielding of the aggregate, so does $-\boldsymbol{\sigma}_\infty$. When the aggregate is macroscopically isotropic, the corresponding yield surface possesses a sixfold symmetry in the deviatoric π plane of the principal stress space (e.g., Hill, 1950).

Bishop and Hill (1951 a) showed that, in the absence of the Bauschinger effect, the yield locus certainly lies between the two cylindrical surfaces. The inner locus is associated with the assumption that the stress state is uniform in all grains, but the displacement continuity is violated. The outer locus corresponds to deformation being considered uniform, and the equilibrium across the grain boundaries violated. Bishop and Hill (1951 b) subsequently introduced the following approximate method of calculating the shape of the yield surface. Equation (14.1.38) implies that the end point of the stress state $\{\hat{\boldsymbol{\sigma}}_{\rm c}\}$ lies on or very near the hyperplane in the macroscopic stress space that is orthogonal to $\mathbf{D}_\infty$ and tangent to the aggregate yield surface at the point $\boldsymbol{\sigma}_\infty$. The perpendicular distance from the stress origin to the yield hyperplane Σ, associated with $\mathbf{D}_\infty$, is

$$h_\Sigma = \frac{\boldsymbol{\sigma}_\infty : \mathbf{D}_\infty}{(\mathbf{D}_\infty : \mathbf{D}_\infty)^{1/2}} \approx \frac{\{\hat{\boldsymbol{\sigma}}_{\rm c}\} : \mathbf{D}_\infty}{(\mathbf{D}_\infty : \mathbf{D}_\infty)^{1/2}}. \tag{14.1.39}$$

The polycrystalline yield surface is then the envelope of all planes Σ for the complete range of the directions $\mathbf{D}_\infty$.

Rather than by a lengthy calculation of $\hat{\boldsymbol{\sigma}}_{\rm c}$ in each grain, corresponding to a prescribed $\mathbf{D}_\infty$, and the averaging procedure to find h_Σ, it is more convenient to use the maximum plastic work principle, i.e., to calculate the works done on $\mathbf{D}_\infty$ by the stress states that do not violate the crystalline yield conditions, and select from these the greatest. Bishop and Hill (1951 b) established that for an isotropic aggregate, in which all slip directions in every grain harden equally, it is only necessary to investigate 56 particular stress states, corresponding to the vertices of the polyhedral crystalline yield surface. Thirty-two of them correspond to a sixfold vertex (resolved shear stress attains the critical value in six different slip systems), and twenty-four stress states correspond to an eightfold vertex. These stress states can be recognized from Eqs. (14.1.23)–(14.1.29). In addition to the two types of stress state mentioned in the previous subsection, three more types of the stress states are: pure shear of amount $\sqrt{6}\,\tau_{\rm cr}$ in a cubic plane parallel to a cubic axis; pure shear of amount $\sqrt{3}\,\tau_{\rm cr}$ in a cubic plane and at $\pi/8$ to the cubic axes; and the stress state with the principal stresses $\pm\sqrt{6}\,\tau_{\rm cr}\,(1,0,-1/2)$, in which the zero principal stress is normal to an octahedral plane, and a $\sqrt{6}\,\tau_{\rm cr}/2$ principal stress is along a slip direction in that plane.

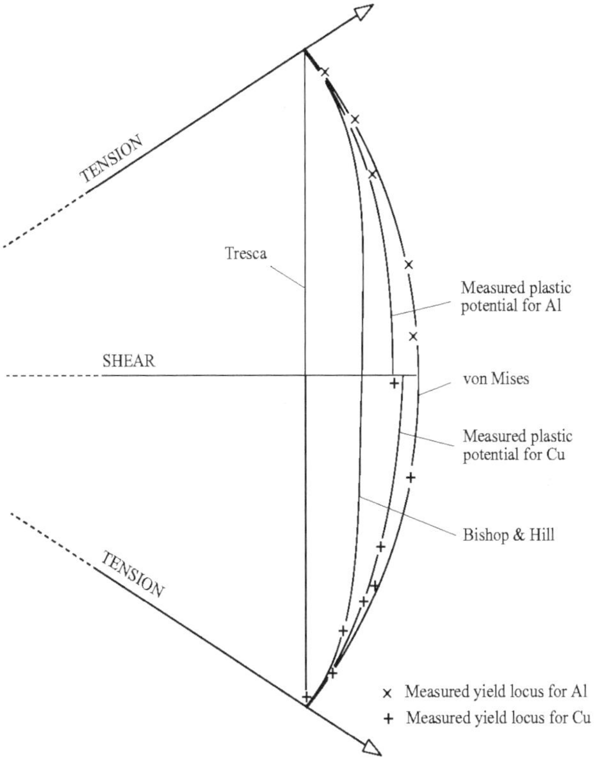

FIGURE 14.4. Polycrystalline yield loci for f.c.c. metals according to Bishop and Hill's theory, Tresca, and von Mises criteria. Indicated also are experimental data for aluminum and copper (from Bishop and Hill, 1951b; with permission from Taylor & Francis Ltd).

Because of the sixfold symmetry of the polycrystalline yield surface, only macroscopic rates of deformation $\mathbf{D}_{\infty}$ whose principal values are in the range

$$(1, -r, r-1)\, D_1^{\infty}\,, \quad \frac{1}{2} \leq r \leq 1 \tag{14.1.40}$$

need to be considered. The axis of the major rate of deformation is then restricted to one of the 48 identical spherical triangles in the standard stereographic projection, while other axes can rotate through half a revolution about the major axis. In calculations, Bishop and Hill took 5° intervals in θ and ϕ, and 18° intervals in ψ (these are the Euler angles of the principal axes relative to the cubic local axes). With an error estimated to be not more than one unit in the second decimal place, they obtained

$$\frac{h_\Sigma}{\tau_{\rm cr}} = \sqrt{\frac{2}{3}} \times 3.06 = 2.50 \qquad (14.1.41)$$

for an axisymmetric uniaxial tension ($r = 1/2$), and

$$\frac{h_\Sigma}{\tau_{\rm cr}} = \sqrt{\frac{2}{3}} \times 2.86 = 2.34 \qquad (14.1.42)$$

for pure shear ($r = 1$). Thus, the ratio of the yield stress in shear to that in tension is $2.86/(\sqrt{3} \times 3.06) = 0.54$, compared with 0.5 for the Tresca, and 0.577 for the von Mises criterion. A representative 60° sector of the calculated yield locus in the π plane is shown in Fig. 14.4. Also shown are the experimental data of Taylor and Quinney (1931), as well as the von Mises and Tresca yield loci. The calculated theoretical yield locus lies between the Tresca and von Mises loci. Since the value of h_Σ was obtained from the approximation given by the far right-hand side of (14.1.39), and since (14.1.37) actually holds, the calculated yield surface is an upper bound to the true yield surface. See, also, Hill (1967), Havner (1971), and Kocks (1970,1987). The development of the vertex at the loading point of the polycrystalline yield surface is discussed in Subsection 14.8.2, and the effects of the texture in Section 14.9.

14.2. Eshelby's Inclusion Problem of Linear Elasticity

An improved model of polycrystalline response can be constructed in which the interaction among grains is approximately taken into account by considering a grain to be embedded in the matrix with the overall aggregate properties, to be determined by the analysis. In this self-consistent method, discussed in detail in the subsequent sections, a prominent role plays the Eshelby inclusion problem. When an infinite elastic medium of the stiffness $\boldsymbol{\mathcal{L}}$, containing an ellipsoidal elastic inhomogeneity of the stiffness $\boldsymbol{\mathcal{L}}_{\rm c}$, is subjected to the far field uniform state of stress $\boldsymbol{\sigma}_\infty$, the state of stress $\boldsymbol{\sigma}_{\rm c}$ within the inhomogeneity is also uniform. This result was first obtained by Eshelby (1957, 1961), who derived it from the consideration of an auxiliary inclusion problem. Some aspects of that analysis are briefly reviewed in this section.

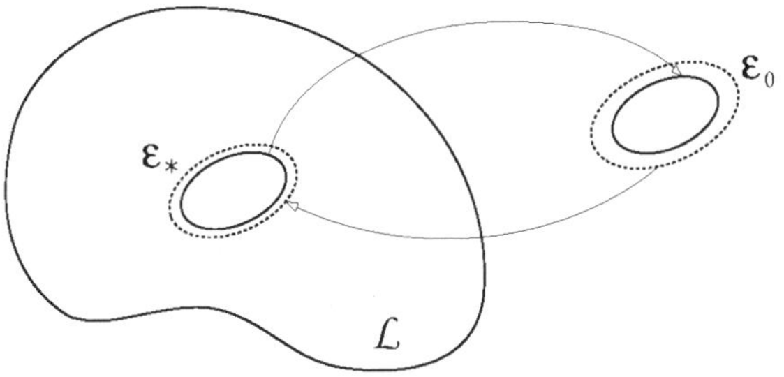

FIGURE 14.5. Schematics of Eshelby's inclusion problem. An ellipsoidal region, removed from an unstressed infinite medium, is subjected to an infinitesimal uniform eigenstrain ε_0 and inserted back into the medium. The state of strain in the inclusion after insertion is uniform and given by $\boldsymbol{\varepsilon}_* = \boldsymbol{S} : \boldsymbol{\varepsilon}_0$, where $\boldsymbol{S}$ is the Eshelby tensor.

14.2.1. Inclusion Problem

An ellipsoidal region of an unstressed infinitely extended homogeneous elastic medium is imagined to be removed from the medium and subjected to an infinitesimal uniform transformation strain (eigenstrain) $\boldsymbol{\varepsilon}_0$ (Fig. 14.5). When inserted back into the matrix material, the inclusion attains the strain

$$\boldsymbol{\varepsilon}_* = \boldsymbol{S} : \boldsymbol{\varepsilon}_0. \tag{14.2.1}$$

Eshelby (1957) has shown that the *in situ* strain $\boldsymbol{\varepsilon}_*$ is also uniform, by demonstrating that the components of the fourth-order nondimensional tensor $\boldsymbol{S}$ are functions of the elastic moduli ratios and the aspect ratios of the ellipsoid only. An arbitrary state of elastic anisotropy was assumed. The Eshelby tensor $\boldsymbol{S}$ is obviously symmetric with respect to the interchange of the leading pair of indices, and also of the terminal pair ($S_{ijkl} = S_{jikl} = S_{ijlk}$), but does not in general possess a reciprocal symmetry ($S_{ijkl} \neq S_{klij}$). Furthermore, since $\boldsymbol{\varepsilon}_0$ vanishes with $\boldsymbol{S} : \boldsymbol{\varepsilon}_0$, the tensor $\boldsymbol{S}$ has its inverse $\boldsymbol{S}^{-1}$. The rotation within the ellipsoidal inclusion is also uniform, and related to the prescribed eigenstrain by

$$\boldsymbol{\omega}_* = \Pi : \boldsymbol{\varepsilon}_0, \tag{14.2.2}$$

where Π is an appropriate fourth-order tensor (Eshelby, *op. cit.*). In the case of spherical inclusion, $\boldsymbol{\omega}_* = \mathbf{0}$.

If the material is isotropic, the components of $\boldsymbol{S}$ depend only on the Poisson ratio ν and the aspect ratios of the ellipsoid. Explicit formulae for S_{ijkl}, on the ellipsoidal axes, can be found in Eshelby's paper. In the case of spherical inclusion, $\boldsymbol{S}$ is an isotropic tensor,

$$\boldsymbol{S} = \alpha \boldsymbol{J} + \beta \boldsymbol{K}, \tag{14.2.3}$$

where

$$\beta = \frac{1+\nu}{3(1-\nu)}, \quad 5\alpha + \beta = 3, \tag{14.2.4}$$

and

$$K_{ijkl} = \frac{1}{3}\delta_{ij}\delta_{kl}, \quad J_{ijkl} = I_{ijkl} - K_{ijkl}. \tag{14.2.5}$$

The components of the fourth-order unit tensor are $I_{ijkl} = (\delta_{ik}\delta_{jl} + \delta_{il}\delta_{jk})/2$. Eshelby's tensor for anisotropic materials can be found in Mura's (1987) book, which contains the references to other related work. See, also, Willis (1964).

The state of stress within the inclusion is uniform and given by

$$\boldsymbol{\sigma}_* = \boldsymbol{\mathcal{L}} : (\boldsymbol{\varepsilon}_* - \boldsymbol{\varepsilon}_0) = \boldsymbol{\mathcal{L}} : (\boldsymbol{S} - \boldsymbol{I}) : \boldsymbol{\varepsilon}_0. \tag{14.2.6}$$

It is convenient to introduce the stress tensor $\boldsymbol{\sigma}_0$ that would be required to remove the eigenstrain $\boldsymbol{\varepsilon}_0$. This is

$$\boldsymbol{\sigma}_0 = -\boldsymbol{\mathcal{L}} : \boldsymbol{\varepsilon}_0, \tag{14.2.7}$$

so that

$$\boldsymbol{\sigma}_* = \boldsymbol{\mathcal{L}} : \boldsymbol{\varepsilon}_* + \boldsymbol{\sigma}_0. \tag{14.2.8}$$

The conjugate Eshelby tensor $\boldsymbol{T}$ is defined by

$$\boldsymbol{\sigma}_* = \boldsymbol{T} : \boldsymbol{\sigma}_0. \tag{14.2.9}$$

The relationship between $\boldsymbol{S}$ and $\boldsymbol{T}$ can be deduced from Eqs. (14.2.6) and (14.2.9), i.e.,

$$\boldsymbol{\mathcal{L}} : (\boldsymbol{S} - \boldsymbol{I}) : \boldsymbol{\varepsilon}_0 = -\boldsymbol{T} : \boldsymbol{\mathcal{L}} : \boldsymbol{\varepsilon}_0. \tag{14.2.10}$$

Since $\boldsymbol{\varepsilon}_0$ is an arbitrary uniform strain, this gives

$$\boldsymbol{\mathcal{L}} : (\boldsymbol{I} - \boldsymbol{S}) = \boldsymbol{T} : \boldsymbol{\mathcal{L}}, \quad (\boldsymbol{I} - \boldsymbol{T}) : \boldsymbol{\mathcal{L}} = \boldsymbol{\mathcal{L}} : \boldsymbol{S}. \tag{14.2.11}$$

An alternative derivation proceeds from

$$\boldsymbol{\varepsilon}_* = \boldsymbol{\mathcal{M}} : (\boldsymbol{\sigma}_* - \boldsymbol{\sigma}_0) = \boldsymbol{\mathcal{M}} : (\boldsymbol{T} - \boldsymbol{I}) : \boldsymbol{\sigma}_0, \tag{14.2.12}$$

where

$$\boldsymbol{\varepsilon}_0 = -\boldsymbol{\mathcal{M}} : \boldsymbol{\sigma}_0, \quad \boldsymbol{\mathcal{M}} = \boldsymbol{\mathcal{L}}^{-1}. \tag{14.2.13}$$

Thus

$$\boldsymbol{\mathcal{M}} : (\boldsymbol{T} - \boldsymbol{I}) : \boldsymbol{\sigma}_0 = -\boldsymbol{S} : \boldsymbol{\mathcal{M}} : \boldsymbol{\sigma}_0. \tag{14.2.14}$$

Since $\boldsymbol{\sigma}_0$ is an arbitrary uniform stress, there follows

$$\boldsymbol{\mathcal{M}} : (\boldsymbol{I} - \boldsymbol{T}) = \boldsymbol{S} : \boldsymbol{\mathcal{M}}, \quad (\boldsymbol{I} - \boldsymbol{S}) : \boldsymbol{\mathcal{M}} = \boldsymbol{\mathcal{M}} : \boldsymbol{T}. \tag{14.2.15}$$

If a far-field uniform state of stress $\boldsymbol{\sigma}_\infty = \boldsymbol{\mathcal{L}} : \boldsymbol{\varepsilon}_\infty$ is superposed to the matrix material, with an inserted inclusion, the states of stress and strain within the inclusion are, by superposition,

$$\boldsymbol{\sigma}_{\mathrm{i}} = \boldsymbol{\sigma}_* + \boldsymbol{\sigma}_\infty, \quad \boldsymbol{\varepsilon}_{\mathrm{i}} = \boldsymbol{\varepsilon}_* + \boldsymbol{\varepsilon}_\infty. \tag{14.2.16}$$

The inclusion stress and strain are related by

$$\boldsymbol{\sigma}_{\mathrm{i}} = \boldsymbol{\mathcal{L}} : (\boldsymbol{\varepsilon}_{\mathrm{i}} - \boldsymbol{\varepsilon}_0), \quad \boldsymbol{\varepsilon}_{\mathrm{i}} = \boldsymbol{\mathcal{M}} : (\boldsymbol{\sigma}_{\mathrm{i}} - \boldsymbol{\sigma}_0), \tag{14.2.17}$$

which follows from Eqs. (14.2.6) and (14.2.16). The states of stress and strain in the surrounding matrix are nonuniform and related by $\boldsymbol{\sigma}_{\mathrm{m}} = \boldsymbol{\mathcal{L}} : \boldsymbol{\varepsilon}_{\mathrm{m}}$. At infinity, $\boldsymbol{\sigma}_{\mathrm{m}}$ becomes $\boldsymbol{\sigma}_\infty$, and $\boldsymbol{\varepsilon}_{\mathrm{m}}$ becomes $\boldsymbol{\varepsilon}_\infty$.

14.2.2. Inhomogeneity Problem

Consider next an ellipsoidal inhomogeneity with elastic moduli $\boldsymbol{\mathcal{L}}_{\mathrm{c}}$, surrounded by an unstressed infinite medium with elastic moduli $\boldsymbol{\mathcal{L}}$. When subjected to the far field uniform state of stress and strain,

$$\boldsymbol{\sigma}_\infty = \boldsymbol{\mathcal{L}} : \boldsymbol{\varepsilon}_\infty, \tag{14.2.18}$$

the stress and strain in the inhomogeneity are also uniform and related by

$$\boldsymbol{\sigma}_{\mathrm{c}} = \boldsymbol{\mathcal{L}}_{\mathrm{c}} : \boldsymbol{\varepsilon}_{\mathrm{c}}. \tag{14.2.19}$$

Eshelby has shown that $\boldsymbol{\sigma}_{\mathrm{c}}$ and $\boldsymbol{\varepsilon}_{\mathrm{c}}$ can be calculated from the previously solved inclusion problem by specifying the inclusion eigenstrain $\boldsymbol{\varepsilon}_0$ such that

$$\boldsymbol{\sigma}_{\mathrm{c}} = \boldsymbol{\sigma}_{\mathrm{i}} \quad \text{and} \quad \boldsymbol{\varepsilon}_{\mathrm{c}} = \boldsymbol{\varepsilon}_{\mathrm{i}}\,. \tag{14.2.20}$$

The eigenstrain needed for this homogenization obeys, from Eqs. (14.2.17) and (14.2.19),

$$\boldsymbol{\mathcal{L}}_{\mathrm{c}} : \boldsymbol{\varepsilon}_{\mathrm{c}} = \boldsymbol{\mathcal{L}} : (\boldsymbol{\varepsilon}_{\mathrm{c}} - \boldsymbol{\varepsilon}_0). \tag{14.2.21}$$

In view of

$$\boldsymbol{\varepsilon}_{\rm c} = \boldsymbol{\varepsilon}_\infty + \boldsymbol{\varepsilon}_* = \boldsymbol{\varepsilon}_\infty + \boldsymbol{S} : \boldsymbol{\varepsilon}_0, \qquad (14.2.22)$$

the homogenization condition (14.2.21) becomes

$$(\boldsymbol{\mathcal{L}} - \boldsymbol{\mathcal{L}}_{\rm c}) : \boldsymbol{\varepsilon}_\infty = [\boldsymbol{\mathcal{L}} - (\boldsymbol{\mathcal{L}} - \boldsymbol{\mathcal{L}}_{\rm c}) : \boldsymbol{S}] : \boldsymbol{\varepsilon}_0. \qquad (14.2.23)$$

This specifies the homogenization eigenstrain,

$$\boldsymbol{\varepsilon}_0 = [\boldsymbol{\mathcal{L}} - (\boldsymbol{\mathcal{L}} - \boldsymbol{\mathcal{L}}_{\rm c}) : \boldsymbol{S}]^{-1} : (\boldsymbol{\mathcal{L}} - \boldsymbol{\mathcal{L}}_{\rm c}) : \boldsymbol{\varepsilon}_\infty, \qquad (14.2.24)$$

in terms of the known $\boldsymbol{\mathcal{L}}$, $\boldsymbol{\mathcal{L}}_{\rm c}$, $\boldsymbol{S}$, and $\boldsymbol{\varepsilon}_\infty$. Substituting Eq. (14.2.24) into Eq. (14.2.21), the strain in the inhomogeneity can be expressed as

$$\boldsymbol{\varepsilon}_{\rm c} = \boldsymbol{A}_{\rm c} : \boldsymbol{\varepsilon}_\infty, \qquad (14.2.25)$$

where $\boldsymbol{A}_{\rm c}$ is the concentration tensor

$$\boldsymbol{A}_{\rm c} = \boldsymbol{I} + \boldsymbol{S} : [\boldsymbol{\mathcal{L}} - (\boldsymbol{\mathcal{L}} - \boldsymbol{\mathcal{L}}_{\rm c}) : \boldsymbol{S}]^{-1} : (\boldsymbol{\mathcal{L}} - \boldsymbol{\mathcal{L}}_{\rm c}). \qquad (14.2.26)$$

Note also that

$$\boldsymbol{\sigma}_{\rm c} = \boldsymbol{\mathcal{L}} : \boldsymbol{\varepsilon}_{\rm c} + \boldsymbol{\sigma}_0, \qquad (14.2.27)$$

which can be compared with Eq. (14.2.7).

In a dual analysis, in place of Eq. (14.2.19), we have

$$\boldsymbol{\varepsilon}_{\rm c} = \boldsymbol{\mathcal{M}}_{\rm c} : \boldsymbol{\sigma}_{\rm c}. \qquad (14.2.28)$$

To find the homogenization stress $\boldsymbol{\sigma}_0$, in order that $\boldsymbol{\varepsilon}_{\rm c} = \boldsymbol{\varepsilon}_{\rm i}$ and $\boldsymbol{\sigma}_{\rm c} = \boldsymbol{\sigma}_{\rm i}$, we require that

$$\boldsymbol{\mathcal{M}}_{\rm c} : \boldsymbol{\sigma}_{\rm c} = \boldsymbol{\mathcal{M}} : (\boldsymbol{\sigma}_{\rm c} - \boldsymbol{\sigma}_0). \qquad (14.2.29)$$

Since

$$\boldsymbol{\sigma}_{\rm c} = \boldsymbol{\sigma}_\infty + \boldsymbol{\sigma}_* = \boldsymbol{\sigma}_\infty + \boldsymbol{T} : \boldsymbol{\sigma}_0, \qquad (14.2.30)$$

there follows

$$\boldsymbol{\sigma}_0 = [\boldsymbol{\mathcal{M}} - (\boldsymbol{\mathcal{M}} - \boldsymbol{\mathcal{M}}_{\rm c}) : \boldsymbol{T}]^{-1} : (\boldsymbol{\mathcal{M}} - \boldsymbol{\mathcal{M}}_{\rm c}) : \boldsymbol{\sigma}_\infty. \qquad (14.2.31)$$

Thus, the stress in the inhomogeneity can be expressed as

$$\boldsymbol{\sigma}_{\rm c} = \boldsymbol{B}_{\rm c} : \boldsymbol{\sigma}_\infty, \qquad (14.2.32)$$

where $\boldsymbol{B}_{\rm c}$ is a dual-concentration tensor

$$\boldsymbol{B}_{\rm c} = \boldsymbol{I} + \boldsymbol{T} : [\boldsymbol{\mathcal{M}} - (\boldsymbol{\mathcal{M}} - \boldsymbol{\mathcal{M}}_{\rm c}) : \boldsymbol{T}]^{-1} : (\boldsymbol{\mathcal{M}} - \boldsymbol{\mathcal{M}}_{\rm c}). \qquad (14.2.33)$$

We interpret $\boldsymbol{\varepsilon}_*$ and $\boldsymbol{\sigma}_*$ in

$$\boldsymbol{\varepsilon}_{\rm c} = \boldsymbol{\varepsilon}_\infty + \boldsymbol{\varepsilon}_*, \quad \boldsymbol{\sigma}_{\rm c} = \boldsymbol{\sigma}_\infty + \boldsymbol{\sigma}_* \qquad (14.2.34)$$

as the deviations of the strain and stress within the inhomogeneity from the applied remote fields, due to different elastic properties of the inhomogeneity and the surrounding matrix. Clearly, if $\boldsymbol{\mathcal{L}}_c = \boldsymbol{\mathcal{L}}$, then $\boldsymbol{A}_c = \boldsymbol{B}_c = \boldsymbol{I}$, i.e., $\boldsymbol{\varepsilon}_c = \boldsymbol{\varepsilon}_\infty$ and $\boldsymbol{\sigma}_c = \boldsymbol{\sigma}_\infty$.

The relationship between the concentration tensors $\boldsymbol{A}_c$ and $\boldsymbol{B}_c$ can be derived by either substituting Eqs. (14.2.25) and (14.2.32) into $\boldsymbol{\sigma}_c = \boldsymbol{\mathcal{L}}_c : \boldsymbol{\varepsilon}_c$, which gives

$$\boldsymbol{B}_c : \boldsymbol{\mathcal{L}} = \boldsymbol{\mathcal{L}}_c : \boldsymbol{A}_c, \tag{14.2.35}$$

or by substituting Eqs. (14.2.25) and (14.2.32) into $\boldsymbol{\varepsilon}_c = \boldsymbol{\mathcal{M}}_c : \boldsymbol{\sigma}_c$, which gives

$$\boldsymbol{A}_c : \boldsymbol{\mathcal{M}} = \boldsymbol{\mathcal{M}}_c : \boldsymbol{B}_c. \tag{14.2.36}$$

14.3. Inclusion Problem for Incrementally Linear Material

Consider an ellipsoidal grain (crystal) embedded in an infinite medium of a different (or differently oriented) material. Both materials are assumed to be incrementally linear, with fully symmetric tensors of instantaneous moduli $\boldsymbol{\mathcal{L}}_c$ and $\boldsymbol{\mathcal{L}}$. Superscript c stands for the crystalline grain. The instantaneous moduli relate the convected rate of the Kirchhoff stress and the rate of deformation tensor, such that

$$\overset{\triangle}{\underline{\boldsymbol{\tau}}} = \boldsymbol{\mathcal{L}} : \mathbf{D}. \tag{14.3.1}$$

Here, $\overset{\triangle}{\underline{\boldsymbol{\tau}}} = \overset{\triangle}{\boldsymbol{\sigma}} + \boldsymbol{\sigma}\,\mathrm{tr}\,\mathbf{D}$ is the rate of Kirchhoff stress with the current configuration as the reference. The instantaneous moduli tensor $\boldsymbol{\mathcal{L}}$ was denoted by $\underline{\boldsymbol{\mathcal{L}}}_{(1)}$ in earlier chapters, but for simplicity we omit in this chapter the underline symbol and the suffix (1). The same remark applies to $\boldsymbol{\mathcal{L}}_c$.

In the absence of body forces, the equations of continuing equilibrium require that the rate of nominal stress is divergence-free (see Section 3.11), i.e.,

$$\boldsymbol{\nabla} \cdot \dot{\underline{\mathbf{P}}} = \boldsymbol{\nabla} \cdot \left(\overset{\triangle}{\underline{\boldsymbol{\tau}}} + \boldsymbol{\sigma} \cdot \mathbf{L}^T \right) = \mathbf{0}. \tag{14.3.2}$$

The existing state of the Cauchy stress $\boldsymbol{\sigma}$ is in equilibrium, so that

$$\boldsymbol{\nabla} \cdot \boldsymbol{\sigma} = \mathbf{0}. \tag{14.3.3}$$

The term $\boldsymbol{\nabla} \cdot (\boldsymbol{\sigma} \cdot \mathbf{L}^T)$ in Eq. (14.3.2) would thus vanish identically in a field of uniform velocity gradient $\mathbf{L}$. In a nonuniform field of $\mathbf{L}$, the term will be disregarded presuming that the components of $\boldsymbol{\sigma}$ are small fractions of dominant instantaneous moduli, and that the spin components are not large compared to the rate of deformation components (Hill, 1965a). Thus, we take approximately

$$\boldsymbol{\nabla} \cdot \overset{\triangle}{\underline{\mathcal{T}}} = \mathbf{0}, \tag{14.3.4}$$

and for a prescribed $\mathbf{D}$ at infinity, the problem is analogous to Eshelby's problem of linear elasticity, considered in Section 14.2.

The rate of stress and strain are uniform within the ellipsoidal grain, and can be expressed as

$$\overset{\triangle}{\underline{\mathcal{T}}}_{\mathrm{c}} = \overset{\triangle}{\underline{\mathcal{T}}}_{\infty} + \overset{\triangle}{\underline{\mathcal{T}}}_{*}, \quad \mathbf{D}_{\mathrm{c}} = \mathbf{D}_{\infty} + \mathbf{D}_{*}, \tag{14.3.5}$$

with the connections

$$\overset{\triangle}{\underline{\mathcal{T}}}_{\mathrm{c}} = \boldsymbol{\mathcal{L}}_{\mathrm{c}} : \mathbf{D}_{\mathrm{c}}, \quad \overset{\triangle}{\underline{\mathcal{T}}}_{\infty} = \boldsymbol{\mathcal{L}} : \mathbf{D}_{\infty}. \tag{14.3.6}$$

Deviation from the far-field uniform rates $\overset{\triangle}{\underline{\mathcal{T}}}_{\infty}$ and $\mathbf{D}_{\infty}$ are denoted by $\overset{\triangle}{\underline{\mathcal{T}}}_{*}$ and $\mathbf{D}_{*}$. Note that during the deformation process an ellipsoidal crystal remains ellipsoidal, under the uniform deformation.

We retain the convected rate of stress in Eqs. (14.3.4)–(14.3.6) to preserve the objective structure of the rate-type constitutive relations. Also, the convected rate $\overset{\triangle}{\underline{\mathcal{T}}}_{\mathrm{c}}$ has a property that its average over a representative macroelement is an appropriate macrovariable in the constitutive analysis of the micro-to-macro transition, discussed in Section 13.4. In a truly infinitesimal formulation, we would simply proceed with the rates of the Cauchy stress $\dot{\boldsymbol{\sigma}}_{\mathrm{c}}$ and $\dot{\boldsymbol{\sigma}}_{\infty}$.

Hill (1965a) introduced a constrained tensor $\boldsymbol{\mathcal{L}}_{*}$ of the material surrounding the grain, such that

$$\overset{\triangle}{\underline{\mathcal{T}}}_{*} = -\boldsymbol{\mathcal{L}}_{*} : \mathbf{D}_{*}. \tag{14.3.7}$$

It will be shown in the sequel that $\boldsymbol{\mathcal{L}}_{*}$ depends only on $\boldsymbol{\mathcal{L}}$ and the aspect ratios of the ellipsoid, but not on $\boldsymbol{\mathcal{L}}_{\mathrm{c}}$. By substituting Eqs. (14.3.5) and (14.3.6) into Eq. (14.3.7), we obtain

$$\boldsymbol{\mathcal{L}}_{\mathrm{c}} : \mathbf{D}_{\mathrm{c}} - \boldsymbol{\mathcal{L}} : \mathbf{D}_{\infty} = -\boldsymbol{\mathcal{L}}_{*} : (\mathbf{D}_{\mathrm{c}} - \mathbf{D}_{\infty}), \tag{14.3.8}$$

i.e.,

$$(\boldsymbol{\mathcal{L}}_{\mathrm{c}} + \boldsymbol{\mathcal{L}}_{*}) : \mathbf{D}_{\mathrm{c}} = (\boldsymbol{\mathcal{L}} + \boldsymbol{\mathcal{L}}_{*}) : \mathbf{D}_{\infty}. \tag{14.3.9}$$

Thus,

$$\mathbf{D}_{\mathrm{c}} = \boldsymbol{A}_{\mathrm{c}} : \mathbf{D}_{\infty}, \tag{14.3.10}$$

where the concentration tensor $\boldsymbol{A}_{\mathrm{c}}$ is

$$\boldsymbol{A}_{\mathrm{c}} = (\boldsymbol{\mathcal{L}}_{\mathrm{c}} + \boldsymbol{\mathcal{L}}_*)^{-1} : (\boldsymbol{\mathcal{L}} + \boldsymbol{\mathcal{L}}_*). \tag{14.3.11}$$

To determine a constraint tensor $\boldsymbol{\mathcal{L}}_*$, we make use of Eshelby's inclusion problem and write, in analogy with Eq. (14.2.6),

$$\overset{\triangle}{\underline{\boldsymbol{T}}}_* = \boldsymbol{\mathcal{L}} : (\mathbf{D}_* - \mathbf{D}_0) = \boldsymbol{\mathcal{L}} : (\boldsymbol{S} - \boldsymbol{I}) : \mathbf{D}_0. \tag{14.3.12}$$

The homogenization rate of deformation is $\mathbf{D}_0$, and

$$\mathbf{D}_* = \boldsymbol{S} : \mathbf{D}_0. \tag{14.3.13}$$

The tensor $\boldsymbol{S}$ here depends on the instantaneous moduli ratios and the current aspect ratios of the deformed ellipsoid. In addition, from the Eshelby's inclusion problem we can express the spin tensor as $\mathbf{W}_* = \Pi : \mathbf{D}_0$. Comparing Eq. (14.3.12) with

$$\overset{\triangle}{\underline{\boldsymbol{T}}}_* = -\boldsymbol{\mathcal{L}}_* : \mathbf{D}_* = -\boldsymbol{\mathcal{L}}_* : \boldsymbol{S} : \mathbf{D}_0, \tag{14.3.14}$$

gives

$$\boldsymbol{\mathcal{L}}_* : \boldsymbol{S} = \boldsymbol{\mathcal{L}} : (\boldsymbol{I} - \boldsymbol{S}), \tag{14.3.15}$$

or

$$\boldsymbol{\mathcal{L}}_* = \boldsymbol{\mathcal{L}} : (\boldsymbol{S}^{-1} - \boldsymbol{I}). \tag{14.3.16}$$

If Eq. (14.3.15) is compared with Eq. (14.2.11), there follows

$$\boldsymbol{\mathcal{L}}_* : \boldsymbol{S} = \boldsymbol{T} : \boldsymbol{\mathcal{L}}. \tag{14.3.17}$$

Furthermore, from Eq. (14.3.15) we can write

$$\boldsymbol{S} = (\boldsymbol{\mathcal{L}} + \boldsymbol{\mathcal{L}}_*)^{-1} : \boldsymbol{\mathcal{L}}, \quad \boldsymbol{S}^{-1} = \boldsymbol{I} + \boldsymbol{\mathcal{M}} : \boldsymbol{\mathcal{L}}_*. \tag{14.3.18}$$

Alternatively, by taking a trace product of

$$\mathbf{D}_* = \boldsymbol{\mathcal{M}} : \overset{\triangle}{\underline{\boldsymbol{T}}}_* + \mathbf{D}_0 \tag{14.3.19}$$

with Eshelby's tensor $\boldsymbol{S}$ gives

$$(\boldsymbol{S} - \boldsymbol{I}) : \mathbf{D}_* = \boldsymbol{S} : \boldsymbol{\mathcal{M}} : \overset{\triangle}{\underline{\boldsymbol{T}}}_*. \tag{14.3.20}$$

The tensor of the instantaneous elastic compliances is $\boldsymbol{\mathcal{M}} = \boldsymbol{\mathcal{L}}^{-1}$. Since

$$\mathbf{D}_* = -\boldsymbol{\mathcal{M}}_* : \overset{\triangle}{\underline{\boldsymbol{T}}}_*, \quad \boldsymbol{\mathcal{M}}_* = \boldsymbol{\mathcal{L}}_*^{-1}, \tag{14.3.21}$$

we obtain

$$(\boldsymbol{I} - \boldsymbol{S}) : \boldsymbol{\mathcal{M}}_* = \boldsymbol{S} : \boldsymbol{\mathcal{M}}, \tag{14.3.22}$$

and

$$\boldsymbol{S} = \boldsymbol{\mathcal{M}}_* : (\boldsymbol{\mathcal{M}} + \boldsymbol{\mathcal{M}}_*)^{-1}. \tag{14.3.23}$$

The inverse of this is clearly in accord with Eq. (14.3.18).

14.3.1. Dual Formulation

In a dual approach, we use Eq. (14.3.21) and

$$\mathbf{D}_c = \boldsymbol{\mathcal{M}}_c : \overset{\triangle}{\underline{\boldsymbol{\tau}}}_c, \quad \mathbf{D}_\infty = \boldsymbol{\mathcal{M}} : \overset{\triangle}{\underline{\boldsymbol{\tau}}}_\infty, \tag{14.3.24}$$

where $\boldsymbol{\mathcal{M}}_c = \boldsymbol{\mathcal{L}}_c^{-1}$ is the crystalline instantaneous compliances tensor, to obtain

$$(\boldsymbol{\mathcal{M}}_c + \boldsymbol{\mathcal{M}}_*) : \overset{\triangle}{\underline{\boldsymbol{\tau}}}_c = (\boldsymbol{\mathcal{M}} + \boldsymbol{\mathcal{M}}_*) : \overset{\triangle}{\underline{\boldsymbol{\tau}}}_\infty. \tag{14.3.25}$$

Consequently

$$\overset{\triangle}{\underline{\boldsymbol{\tau}}}_c = \boldsymbol{B}_c : \overset{\triangle}{\underline{\boldsymbol{\tau}}}_\infty. \tag{14.3.26}$$

A dual-concentration tensor $\boldsymbol{B}_c$ is

$$\boldsymbol{B}_c = (\boldsymbol{\mathcal{M}}_c + \boldsymbol{\mathcal{M}}_*)^{-1} : (\boldsymbol{\mathcal{M}} + \boldsymbol{\mathcal{M}}_*). \tag{14.3.27}$$

To determine a constraint tensor $\boldsymbol{\mathcal{M}}_*$ in terms of $\boldsymbol{\mathcal{M}}$ and the conjugate Eshelby tensor $\boldsymbol{T}$, we write, in analogy with Eq. (14.2.12),

$$\mathbf{D}_* = \boldsymbol{\mathcal{M}} : (\overset{\triangle}{\underline{\boldsymbol{\tau}}}_* - \overset{\triangle}{\underline{\boldsymbol{\tau}}}_0) = \boldsymbol{\mathcal{M}} : (\boldsymbol{T} - \boldsymbol{I}) : \overset{\triangle}{\underline{\boldsymbol{\tau}}}_0. \tag{14.3.28}$$

The homogenization rate of stress is

$$\overset{\triangle}{\underline{\boldsymbol{\tau}}}_0 = -\boldsymbol{\mathcal{L}} : \mathbf{D}_0, \tag{14.3.29}$$

and

$$\overset{\triangle}{\underline{\boldsymbol{\tau}}}_* = \boldsymbol{T} : \overset{\triangle}{\underline{\boldsymbol{\tau}}}_0. \tag{14.3.30}$$

Note that $\mathbf{L}_0 = \mathbf{D}_0$, since $\mathbf{W}_0 = \mathbf{0}$, because only the rate of eigenstrain $\mathbf{D}_0$ gives rise to *in situ* stress and strain rates in the inclusion problem. Comparing Eq. (14.3.28) with

$$\mathbf{D}_* = -\boldsymbol{\mathcal{M}}_* : \overset{\triangle}{\underline{\boldsymbol{\tau}}}_* = -\boldsymbol{\mathcal{M}}_* : \boldsymbol{T} : \overset{\triangle}{\underline{\boldsymbol{\tau}}}_0, \tag{14.3.31}$$

gives

$$\boldsymbol{\mathcal{M}}_* : \boldsymbol{T} = \boldsymbol{\mathcal{M}} : (\boldsymbol{I} - \boldsymbol{T}), \tag{14.3.32}$$

or

$$\boldsymbol{\mathcal{M}}_* = \boldsymbol{\mathcal{M}} : (\boldsymbol{T}^{-1} - \boldsymbol{I}). \tag{14.3.33}$$

If Eq. (14.3.32) is compared with Eq. (14.2.15), there follows

$$\boldsymbol{\mathcal{M}}_* : \boldsymbol{T} = \boldsymbol{S} : \boldsymbol{\mathcal{M}}. \tag{14.3.34}$$

In addition, from Eq. (14.3.32) we can write

$$\boldsymbol{T} = (\boldsymbol{\mathcal{M}} + \boldsymbol{\mathcal{M}}_*)^{-1} : \boldsymbol{\mathcal{M}}, \quad \boldsymbol{T}^{-1} = \boldsymbol{I} + \boldsymbol{\mathcal{L}} : \boldsymbol{\mathcal{M}}_*. \tag{14.3.35}$$

Alternatively, by taking a trace product of

$$\overset{\triangle}{\underline{\boldsymbol{\tau}}}_* = \boldsymbol{\mathcal{L}} : \mathrm{D}_* + \overset{\triangle}{\underline{\boldsymbol{\tau}}}_0 \tag{14.3.36}$$

with the conjugate Eshelby tensor $\boldsymbol{T}$ gives

$$(\boldsymbol{T} - \boldsymbol{I}) : \overset{\triangle}{\underline{\boldsymbol{\tau}}}_* = \boldsymbol{T} : \boldsymbol{\mathcal{L}} : \mathrm{D}_*. \tag{14.3.37}$$

Having in mind Eq. (14.3.7), we arrive at

$$(\boldsymbol{I} - \boldsymbol{T}) : \boldsymbol{\mathcal{L}}_* = \boldsymbol{T} : \boldsymbol{\mathcal{L}}, \tag{14.3.38}$$

and

$$\boldsymbol{T} = \boldsymbol{\mathcal{L}}_* : (\boldsymbol{\mathcal{L}} + \boldsymbol{\mathcal{L}}_*)^{-1}. \tag{14.3.39}$$

The inverse of this is clearly in accord with Eq. (14.3.35).

14.3.2. Analysis of Concentration Tensors

It is first observed from Eqs. (14.3.18) and (14.3.39) that

$$\boldsymbol{S} : \boldsymbol{\mathcal{M}} = \boldsymbol{\mathcal{M}}_* : \boldsymbol{T} = (\boldsymbol{\mathcal{L}} + \boldsymbol{\mathcal{L}}_*)^{-1} = \boldsymbol{P}, \tag{14.3.40}$$

while, from Eqs. (14.3.23) and (14.3.35),

$$\boldsymbol{T} : \boldsymbol{\mathcal{L}} = \boldsymbol{\mathcal{L}}_* : \boldsymbol{S} = (\boldsymbol{\mathcal{M}} + \boldsymbol{\mathcal{M}}_*)^{-1} = \boldsymbol{Q}. \tag{14.3.41}$$

For convenience, the products that appear in Eqs. (14.3.40) and (14.3.41) are denoted by $\boldsymbol{P}$ and $\boldsymbol{Q}$ (Hill, 1965a). Evidently, since the instantaneous moduli and compliances possess the reciprocal symmetry, the tensors $\boldsymbol{P}$ and $\boldsymbol{Q}$ share the same symmetry, i.e.,

$$\boldsymbol{P}^T = \boldsymbol{P}, \quad \boldsymbol{Q}^T = \boldsymbol{Q}. \tag{14.3.42}$$

In view of Eqs. (14.2.11) and (14.2.15), we can write

$$\boldsymbol{P} = \boldsymbol{\mathcal{M}} : (\boldsymbol{I} - \boldsymbol{T}), \quad \boldsymbol{Q} = \boldsymbol{\mathcal{L}} : (\boldsymbol{I} - \boldsymbol{S}). \tag{14.3.43}$$

Furthermore, from Eqs. (14.3.18) and (14.3.35),

$$\boldsymbol{S} = \boldsymbol{P} : \boldsymbol{\mathcal{L}}, \quad \boldsymbol{T} = \boldsymbol{Q} : \boldsymbol{\mathcal{M}}. \tag{14.3.44}$$

A trace product of the second equation in (14.3.43) with $\boldsymbol{\mathcal{M}}$ from the left provides a connection between $\boldsymbol{P}$ and $\boldsymbol{Q}$,

$$\boldsymbol{P} : \boldsymbol{\mathcal{L}} + \boldsymbol{\mathcal{M}} : \boldsymbol{Q} = \boldsymbol{I}. \tag{14.3.45}$$

The concentration tensor $\boldsymbol{A}_{\mathrm{c}}$ can be expressed in terms of $\boldsymbol{P}$ as

$$\boldsymbol{A}_{\mathrm{c}} = (\boldsymbol{\mathcal{L}}_{\mathrm{c}} + \boldsymbol{\mathcal{L}}_{*})^{-1} : \boldsymbol{P}^{-1}, \tag{14.3.46}$$

which gives, by inversion,

$$\boldsymbol{A}_{\mathrm{c}}^{-1} = \boldsymbol{P} : (\boldsymbol{\mathcal{L}}_{\mathrm{c}} + \boldsymbol{\mathcal{L}}_{*}). \tag{14.3.47}$$

Since

$$\boldsymbol{P} : \boldsymbol{\mathcal{L}}_{*} = \boldsymbol{P} : (\boldsymbol{\mathcal{L}} + \boldsymbol{\mathcal{L}}_{*} - \boldsymbol{\mathcal{L}}) = \boldsymbol{I} - \boldsymbol{P} : \boldsymbol{\mathcal{L}}, \tag{14.3.48}$$

Equation (14.3.47) can be rewritten as

$$\boldsymbol{A}_{\mathrm{c}}^{-1} = \boldsymbol{I} + \boldsymbol{P} : (\boldsymbol{\mathcal{L}}_{\mathrm{c}} - \boldsymbol{\mathcal{L}}). \tag{14.3.49}$$

Similarly, the concentration tensor $\boldsymbol{B}_{\mathrm{c}}$ can be expressed in terms of $\boldsymbol{Q}$ as

$$\boldsymbol{B}_{\mathrm{c}} = (\boldsymbol{\mathcal{M}}_{\mathrm{c}} + \boldsymbol{\mathcal{M}}_{*})^{-1} : \boldsymbol{Q}^{-1}. \tag{14.3.50}$$

Upon inversion, this gives

$$\boldsymbol{B}_{\mathrm{c}}^{-1} = \boldsymbol{Q} : (\boldsymbol{\mathcal{M}}_{\mathrm{c}} + \boldsymbol{\mathcal{M}}_{*}). \tag{14.3.51}$$

Recalling that

$$\boldsymbol{Q} : \boldsymbol{\mathcal{M}}_{*} = \boldsymbol{Q} : (\boldsymbol{\mathcal{M}} + \boldsymbol{\mathcal{M}}_{*} - \boldsymbol{\mathcal{M}}) = \boldsymbol{I} - \boldsymbol{Q} : \boldsymbol{\mathcal{M}}, \tag{14.3.52}$$

Equation (14.3.51) can be recast as

$$\boldsymbol{B}_{\mathrm{c}}^{-1} = \boldsymbol{I} + \boldsymbol{Q} : (\boldsymbol{\mathcal{M}}_{\mathrm{c}} - \boldsymbol{\mathcal{M}}). \tag{14.3.53}$$

In addition, we recall from Section 14.2 that

$$\boldsymbol{B}_{\mathrm{c}} : \boldsymbol{\mathcal{L}} = \boldsymbol{\mathcal{L}}_{\mathrm{c}} : \boldsymbol{A}_{\mathrm{c}}, \quad \boldsymbol{A}_{\mathrm{c}} : \boldsymbol{\mathcal{M}} = \boldsymbol{\mathcal{M}}_{\mathrm{c}} : \boldsymbol{B}_{\mathrm{c}}. \tag{14.3.54}$$

14.3.3. Finite Deformation Formulation

To circumvent the approximation made in equilibrium equations (14.3.2), where the term $\boldsymbol{\nabla}\cdot(\boldsymbol{\sigma}\cdot\mathbf{L}^T)$ was neglected, based on an assumption that the stress components are small compared to dominant instantaneous moduli, we can consider an ellipsoidal grain in an infinitely extended matrix under the far-field uniform velocity gradient $\mathbf{L}_\infty$, and the corresponding rate of nominal stress

$$\underline{\dot{\mathbf{P}}}_\infty = \boldsymbol{\Lambda}\cdot\cdot\,\mathbf{L}_\infty. \tag{14.3.55}$$

The tensor of the instantaneous pseudomoduli for the matrix surrounding the crystalline grain is $\boldsymbol{\Lambda}$ (designated by $\underline{\boldsymbol{\Lambda}}$ in earlier chapters). The underline below $\dot{\mathbf{P}}$ is kept to indicate that the current configuration is taken for the reference. The problem was studied by Iwakuma and Nemat-Nasser (1984). As expected on physical grounds, the velocity gradient in the crystal must be uniform. Introducing the concentration tensor $\boldsymbol{A}_{\mathrm{c}}^0$, we write

$$\mathbf{L}_{\mathrm{c}} = \boldsymbol{A}_{\mathrm{c}}^0\cdot\cdot\,\mathbf{L}_\infty. \tag{14.3.56}$$

The velocity gradient $\mathbf{L}_{\mathrm{c}}$ can be represented as the sum of $\mathbf{L}_\infty$ and the deviation $\mathbf{L}_*$, caused by different pseudomoduli of the crystal and the surrounding medium. Thus,

$$\mathbf{L}_{\mathrm{c}} = \mathbf{L}_\infty + \mathbf{L}_* \quad \underline{\dot{\mathbf{P}}}_{\mathrm{c}} = \underline{\dot{\mathbf{P}}}_\infty + \underline{\dot{\mathbf{P}}}_*, \tag{14.3.57}$$

where

$$\underline{\dot{\mathbf{P}}}_{\mathrm{c}} = \boldsymbol{\Lambda}_{\mathrm{c}}\cdot\cdot\,\mathbf{L}_{\mathrm{c}}. \tag{14.3.58}$$

Introducing a constraint tensor $\boldsymbol{\Lambda}_*$ of the outer phase by

$$\underline{\dot{\mathbf{P}}}_* = -\boldsymbol{\Lambda}_*\cdot\cdot\,\mathbf{L}_*, \tag{14.3.59}$$

upon the substitution of (14.3.57) into (14.3.59), there follows

$$\underline{\dot{\mathbf{P}}}_{\mathrm{c}} - \underline{\dot{\mathbf{P}}}_\infty = -\boldsymbol{\Lambda}_*\cdot\cdot\,(\mathbf{L}_{\mathrm{c}} - \mathbf{L}_\infty), \tag{14.3.60}$$

i.e.,

$$(\boldsymbol{\Lambda}_{\mathrm{c}} + \boldsymbol{\Lambda}_*)\cdot\cdot\,\mathbf{L}_{\mathrm{c}} = (\boldsymbol{\Lambda} + \boldsymbol{\Lambda}_*)\cdot\cdot\,\mathbf{L}_\infty. \tag{14.3.61}$$

This defines the concentration tensor

$$\boldsymbol{A}_{\mathrm{c}}^0 = (\boldsymbol{\Lambda}_{\mathrm{c}} + \boldsymbol{\Lambda}_*)^{-1}\cdot\cdot\,(\boldsymbol{\Lambda} + \boldsymbol{\Lambda}_*). \tag{14.3.62}$$

Dually, we can start from

$$\mathbf{L}_* = -\mathbf{M}_* \cdot\cdot \underline{\dot{\mathbf{P}}}_*, \quad \mathbf{M}_* = \boldsymbol{\Lambda}_*^{-1}, \tag{14.3.63}$$

to obtain

$$\mathbf{L}_c - \mathbf{L}_\infty = -\mathbf{M}_* \cdot\cdot (\underline{\dot{\mathbf{P}}}_c - \underline{\dot{\mathbf{P}}}_\infty), \tag{14.3.64}$$

and

$$(\mathbf{M}_c + \mathbf{M}_*) \cdot\cdot \underline{\dot{\mathbf{P}}}_c = (\mathbf{M} + \mathbf{M}_*) \cdot\cdot \underline{\dot{\mathbf{P}}}_\infty. \tag{14.3.65}$$

This defines a dual-concentration tensor

$$\boldsymbol{B}_c^0 = (\mathbf{M}_c + \mathbf{M}_*)^{-1} \cdot\cdot (\mathbf{M} + \mathbf{M}_*), \tag{14.3.66}$$

such that

$$\underline{\dot{\mathbf{P}}}_c = \boldsymbol{B}_c^0 \cdot\cdot \underline{\dot{\mathbf{P}}}_\infty. \tag{14.3.67}$$

The connections between the two concentration tensors are easily established. They are

$$\boldsymbol{B}_c^0 : \boldsymbol{\Lambda} = \boldsymbol{\Lambda}_c \cdot\cdot \boldsymbol{A}_c^0, \quad \boldsymbol{A}_c^0 : \mathbf{M} = \mathbf{M}_c \cdot\cdot \boldsymbol{B}_c^0, \tag{14.3.68}$$

in line with Eqs. (14.2.35) and (14.2.36).

The analysis can be extended further by introducing the Eshelby-type tensor $\boldsymbol{H}$, and its conjugate tensor $\boldsymbol{G}$, which appear in the linear relationships

$$\mathbf{L}_* = \boldsymbol{H} : \mathbf{L}_0, \quad \underline{\dot{\mathbf{P}}}_* = \boldsymbol{G} \cdot\cdot \underline{\dot{\mathbf{P}}}_0. \tag{14.3.69}$$

Here, $\mathbf{L}_0$ is the eigenvelocity gradient in an Eshelby-type inclusion problem, cast with respect to $\mathbf{L}$ and $\underline{\dot{\mathbf{P}}}$ measures, while $\underline{\dot{\mathbf{P}}}_0 = -\boldsymbol{\Lambda} : \mathbf{L}_0$. These are such that

$$\underline{\dot{\mathbf{P}}}_* = \boldsymbol{\Lambda} \cdot\cdot (\mathbf{L}_* - \mathbf{L}_0), \quad \mathbf{L}_* = \mathbf{M} \cdot\cdot (\underline{\dot{\mathbf{P}}}_* - \underline{\dot{\mathbf{P}}}_0). \tag{14.3.70}$$

It follows

$$\boldsymbol{\Lambda} \cdot\cdot (\boldsymbol{I} - \boldsymbol{H}) = \boldsymbol{G} \cdot\cdot \boldsymbol{\Lambda}, \quad \mathbf{M} \cdot\cdot (\boldsymbol{I} - \boldsymbol{G}) = \boldsymbol{H} \cdot\cdot \mathbf{M}, \tag{14.3.71}$$

and

$$\boldsymbol{\Lambda}_* \cdot\cdot \boldsymbol{H} = \boldsymbol{\Lambda} \cdot\cdot (\boldsymbol{I} - \boldsymbol{H}), \quad (\boldsymbol{I} - \boldsymbol{H}) \cdot\cdot \mathbf{M}_* = \boldsymbol{H} \cdot\cdot \mathbf{M}, \tag{14.3.72}$$

$$\mathbf{M}_* \cdot\cdot \boldsymbol{G} = \mathbf{M} \cdot\cdot (\boldsymbol{I} - \boldsymbol{G}), \quad (\boldsymbol{I} - \boldsymbol{G}) \cdot\cdot \boldsymbol{\Lambda}_* = \boldsymbol{G} \cdot\cdot \boldsymbol{\Lambda}. \tag{14.3.73}$$

Evidently, by comparing Eqs. (14.3.71)–(14.3.73), we deduce that

$$\mathbf{M}_* \cdot\cdot \boldsymbol{G} = \boldsymbol{H} \cdot\cdot \mathbf{M}, \quad \boldsymbol{G} \cdot\cdot \boldsymbol{\Lambda} = \boldsymbol{\Lambda}_* \cdot\cdot \boldsymbol{H}. \tag{14.3.74}$$

Additional analysis can be found in the papers by Iwakuma and Nemat-Nasser (1984), Lipinski and Berveiller (1989), and Nemat-Nasser (1999). A construction of Green's functions needed for the calculation of the generalized Eshelby's tensor and the concentration tensors is there considered. The problem was also studied in connection with a possible loss of stability of the uniformly stressed homogeneous body at finite strain.

14.4. Self-Consistent Method

A self-consistent method was proposed in elasticity by Hershey (1954) and Kröner (1958) to determine the average elastic polycrystalline constants in terms of the single crystal constants. In this method, a single crystal is considered to be embedded in an infinite medium with the average polycrystalline moduli (homogeneous equivalent medium). The strain in the crystal is calculated in terms of the applied far-field strain by using the Eshelby inhomogeneity problem. It is then postulated that the average strain, over the relevant range of lattice orientations, is equal to the overall macroscopic strain applied to the polycrystalline aggregate (Fig. 14.6). The same results are obtained if it is required that the average stress over the relevant range of lattice orientations within crystalline grains is equal to the overall macroscopic stress applied to the polycrystalline aggregate. The method is in that respect self-consistent, thus the terminology. In contrast, the methods earlier suggested by Voigt (1889) and Reuss (1929), resulted in different estimates of the elastic polycrystalline constants (see Budiansky, 1965; Hill, 1965b, and the Subsection 14.5.1 of this chapter).

We proceed here with the rate-type formulation of the self-consistent method, following the presentation by Hill (1965a). Polycrystals are considered whose grains can be approximately treated as similar ellipsoids with their corresponding axes aligned (or as variously sized spheres). The lattice orientation, relative to the fixed frame of reference, may vary from grain to grain, either randomly or in the specified manner. The tensors $\boldsymbol{\mathcal{L}}_{\mathrm{c}}$ and $\boldsymbol{\mathcal{M}}_{\mathrm{c}}$ are the instantaneous moduli and compliances of a typical grain, and $\boldsymbol{\mathcal{L}}$ and $\boldsymbol{\mathcal{M}}$ are the overall tensors for the polycrystal itself. The tensors $\boldsymbol{\mathcal{L}}_*$ and $\boldsymbol{\mathcal{M}}_*$, as well as $\boldsymbol{S}$ and $\boldsymbol{T}$, correspond to an ellipsoid or sphere representing the average grain shape. The components of these tensors are constants, in the

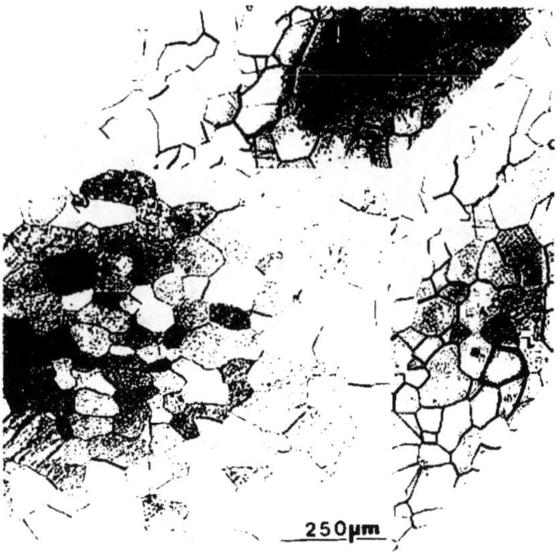

FIGURE 14.6. A micrograph of a polycrystalline sample of annealed tungsten (by courtesy of Professor M. A. Meyers).

fixed frame of reference, while the components of $\boldsymbol{\mathcal{L}}_{\rm c}$ and $\boldsymbol{\mathcal{M}}_{\rm c}$ depend on the local lattice orientation within the grain.

If the overall macroscopic rate of deformation $\mathbf{D}_\infty$, applied to the polycrystalline aggregate, is taken to be the orientation average of the crystalline rate of deformation $\mathbf{D}_{\rm c} = \boldsymbol{A}_{\rm c} : \mathbf{D}_\infty$, i.e.,

$$\{\mathbf{D}_{\rm c}\} = \mathbf{D}_\infty, \tag{14.4.1}$$

the orientation average of the concentration tensor $\boldsymbol{A}_{\rm c}$ is equal to the fourth-order unit tensor,

$$\{\boldsymbol{A}_{\rm c}\} = \boldsymbol{I}, \quad I_{ijkl} = \frac{1}{2}(\delta_{ik}\delta_{jl} + \delta_{il}\delta_{jk}). \tag{14.4.2}$$

In view of Eqs. (14.2.26) and (14.3.49), this implies that

$$\{\boldsymbol{I} + \boldsymbol{S} : [\boldsymbol{\mathcal{L}} - (\boldsymbol{\mathcal{L}} - \boldsymbol{\mathcal{L}}_{\rm c}) : \boldsymbol{S}]^{-1} : (\boldsymbol{\mathcal{L}} - \boldsymbol{\mathcal{L}}_{\rm c})\} = \boldsymbol{I}, \tag{14.4.3}$$

$$\{[\boldsymbol{I} + \boldsymbol{P} : (\boldsymbol{\mathcal{L}}_{\rm c} - \boldsymbol{\mathcal{L}})]^{-1}\} = \boldsymbol{I}. \tag{14.4.4}$$

In addition, since

$$\mathbf{D}_{\rm c} = \boldsymbol{\mathcal{M}}_{\rm c} : \overset{\triangle}{\underline{\boldsymbol{\tau}}}_{\rm c} = \boldsymbol{\mathcal{M}}_{\rm c} : \boldsymbol{B}_{\rm c} : \overset{\triangle}{\underline{\boldsymbol{\tau}}}_\infty, \quad \mathbf{D}_\infty = \boldsymbol{\mathcal{M}} : \overset{\triangle}{\underline{\boldsymbol{\tau}}}_\infty, \tag{14.4.5}$$

the substitution into Eq. (14.4.1) gives

$$\{\boldsymbol{\mathcal{M}}_{\rm c} : \boldsymbol{B}_{\rm c}\} = \boldsymbol{\mathcal{M}}. \tag{14.4.6}$$

Dually, if the macroscopic rate of stress $\overset{\triangle}{\underline{\boldsymbol{\tau}}}_{\infty}$ is taken to be the orientation average of the crystalline rate of stress $\overset{\triangle}{\underline{\boldsymbol{\tau}}}_{\rm c} = \boldsymbol{B}_{\rm c} : \overset{\triangle}{\underline{\boldsymbol{\tau}}}_{\infty}$, i.e.,

$$\{\overset{\triangle}{\underline{\boldsymbol{\tau}}}_{\rm c}\} = \overset{\triangle}{\underline{\boldsymbol{\tau}}}_{\infty}, \tag{14.4.7}$$

the orientation average of the concentration tensor $\boldsymbol{B}_{\rm c}$ is equal to the fourth-order unit tensor,

$$\{\boldsymbol{B}_{\rm c}\} = \boldsymbol{I}. \tag{14.4.8}$$

It is recalled from Section 13.4 that the macroscopic measure of the convected rate of Kirchhoff stress, with the current configuration as the reference, is indeed the volume (orientation) average of the local convected rate of the Kirchhoff stress. In view of Eqs. (14.2.33) and (14.3.51), Eq. (14.4.8) implies that

$$\{\boldsymbol{I} + \boldsymbol{T} : [\boldsymbol{\mathcal{M}} - (\boldsymbol{\mathcal{M}} - \boldsymbol{\mathcal{M}}_{\rm c}) : \boldsymbol{T}]^{-1} : (\boldsymbol{\mathcal{M}} - \boldsymbol{\mathcal{M}}_{\rm c})\} = \boldsymbol{I}, \tag{14.4.9}$$

$$\{[\boldsymbol{I} + \boldsymbol{Q} : (\boldsymbol{\mathcal{M}}_{\rm c} - \boldsymbol{\mathcal{M}})]^{-1}\} = \boldsymbol{I}. \tag{14.4.10}$$

In addition, since

$$\overset{\triangle}{\underline{\boldsymbol{\tau}}}_{\rm c} = \boldsymbol{\mathcal{L}}_{\rm c} : \mathbf{D}_{\rm c} = \boldsymbol{\mathcal{L}}_{\rm c} : \boldsymbol{A}_{\rm c} : \mathbf{D}_{\infty}, \quad \overset{\triangle}{\underline{\boldsymbol{\tau}}}_{\infty} = \boldsymbol{\mathcal{L}} : \mathbf{D}_{\infty}, \tag{14.4.11}$$

the substitution into Eq. (14.4.7) gives

$$\{\boldsymbol{\mathcal{L}}_{\rm c} : \boldsymbol{A}_{\rm c}\} = \boldsymbol{\mathcal{L}}. \tag{14.4.12}$$

This parallels the previously derived expression (14.4.6).

The self-consistency of the two approaches is easily established from the averaging of $\overset{\triangle}{\underline{\boldsymbol{\tau}}}_{*} = -\boldsymbol{\mathcal{L}}_{*} : \mathbf{D}_{*}$. This can be rewritten as

$$\overset{\triangle}{\underline{\boldsymbol{\tau}}}_{\rm c} - \overset{\triangle}{\underline{\boldsymbol{\tau}}}_{\infty} = -\boldsymbol{\mathcal{L}}_{*} : (\mathbf{D}_{\rm c} - \mathbf{D}_{\infty}), \tag{14.4.13}$$

and thus

$$\{\overset{\triangle}{\underline{\boldsymbol{\tau}}}_{\rm c}\} - \overset{\triangle}{\underline{\boldsymbol{\tau}}}_{\infty} = -\boldsymbol{\mathcal{L}}_{*} : (\{\mathbf{D}_{\rm c}\} - \mathbf{D}_{\infty}). \tag{14.4.14}$$

Recall that the components of the constraint tensor $\boldsymbol{\mathcal{L}}_{*}$ are constants in the fixed frame of reference. Since $\boldsymbol{\mathcal{L}}_{*}$ is nonsingular, we conclude from Eq. (14.4.14) that

$$\{\overset{\triangle}{\underline{\boldsymbol{\tau}}}_{\rm c}\} = \overset{\triangle}{\underline{\boldsymbol{\tau}}}_{\infty} \quad \text{whenever} \quad \{\mathbf{D}_{\rm c}\} = \mathbf{D}_{\infty}\,, \tag{14.4.15}$$

and *vice versa*, which establishes the self-consistency of the method.

14.4.1. Polarization Tensors

The rate of stress in the grain can be expressed as

$$\overset{\triangle}{\underline{\boldsymbol{\tau}}}_{\rm c} = \overset{\triangle}{\underline{\boldsymbol{\tau}}}_{\infty} + \overset{\triangle}{\underline{\boldsymbol{\tau}}}_{*} = \boldsymbol{\mathcal{L}} : \mathbf{D}_{\infty} - \boldsymbol{\mathcal{L}}_{*} : (\mathbf{D}_{\rm c} - \mathbf{D}_{\infty}). \tag{14.4.16}$$

Following Kröner's (1958) terminology, the polarization tensor is defined by

$$\overset{\triangle}{\underline{\boldsymbol{\tau}}}_{\rm c} - \boldsymbol{\mathcal{L}} : \mathbf{D}_{\rm c}\,, \tag{14.4.17}$$

so that

$$\overset{\triangle}{\underline{\boldsymbol{\tau}}}_{\rm c} - \boldsymbol{\mathcal{L}} : \mathbf{D}_{\rm c} = (\boldsymbol{\mathcal{L}} + \boldsymbol{\mathcal{L}}_{*}) : (\mathbf{D}_{\infty} - \mathbf{D}_{\rm c}). \tag{14.4.18}$$

The orientation average of the polarization tensor vanishes by Eq. (14.4.1), because $\boldsymbol{\mathcal{L}}$ and $\boldsymbol{\mathcal{L}}_{*}$ are constant tensors. Thus,

$$\{\overset{\triangle}{\underline{\boldsymbol{\tau}}}_{\rm c} - \boldsymbol{\mathcal{L}} : \mathbf{D}_{\rm c}\} = \mathbf{0}. \tag{14.4.19}$$

The polarization tensor can also be expressed as

$$\overset{\triangle}{\underline{\boldsymbol{\tau}}}_{\rm c} - \boldsymbol{\mathcal{L}} : \mathbf{D}_{\rm c} = (\boldsymbol{\mathcal{L}}_{\rm c} - \boldsymbol{\mathcal{L}}) : \mathbf{D}_{\rm c} = (\boldsymbol{\mathcal{L}}_{\rm c} - \boldsymbol{\mathcal{L}}) : \boldsymbol{A}_{\rm c} : \mathbf{D}_{\infty}. \tag{14.4.20}$$

The average of this vanishes for any applied $\mathbf{D}_{\infty}$, i.e.,

$$\{(\boldsymbol{\mathcal{L}}_{\rm c} - \boldsymbol{\mathcal{L}}) : \boldsymbol{A}_{\rm c}\} : \mathbf{D}_{\infty} = \mathbf{0}, \tag{14.4.21}$$

so that

$$\{\,(\boldsymbol{\mathcal{L}}_{\rm c} - \boldsymbol{\mathcal{L}}) : \boldsymbol{A}_{\rm c}\} = \mathbf{0}. \tag{14.4.22}$$

A condition of this type was employed by Eshelby (1961) to derive a cubic equation for the effective elastic shear modulus of an isotropic polycrystalline aggregate of cubic crystals. See also Hill (1965a).

Furthermore, from Eq. (14.3.49) we can write

$$\boldsymbol{A}_{\rm c}^{-1} = [(\boldsymbol{\mathcal{L}}_{\rm c} - \boldsymbol{\mathcal{L}})^{-1} + \boldsymbol{P}] : (\boldsymbol{\mathcal{L}}_{\rm c} - \boldsymbol{\mathcal{L}}), \tag{14.4.23}$$

and

$$(\boldsymbol{\mathcal{L}}_{\rm c} - \boldsymbol{\mathcal{L}}) : \boldsymbol{A}_{\rm c} = [(\boldsymbol{\mathcal{L}}_{\rm c} - \boldsymbol{\mathcal{L}})^{-1} + \boldsymbol{P}]^{-1}. \tag{14.4.24}$$

Consequently, by averaging and by using Eq. (14.4.22), there follows

$$\{[(\boldsymbol{\mathcal{L}}_{\rm c} - \boldsymbol{\mathcal{L}})^{-1} + \boldsymbol{P}]^{-1}\} = \mathbf{0}. \tag{14.4.25}$$

This condition was originally employed by Kröner (1958) in his derivation of the cubic equation for the effective shear modulus of an isotropic polycrystalline aggregate of cubic crystals.

A dual-polarization tensor is

$$\mathbf{D}_{\mathrm{c}} - \boldsymbol{\mathcal{M}} : \overset{\triangle}{\underline{\mathcal{T}}}_{\mathrm{c}} = (\boldsymbol{\mathcal{M}} + \boldsymbol{\mathcal{M}}_*) : (\overset{\triangle}{\underline{\mathcal{T}}}_{\infty} - \overset{\triangle}{\underline{\mathcal{T}}}_{\mathrm{c}}). \tag{14.4.26}$$

The orientation average of this also vanishes, in view of Eq. (14.4.7) and because $\boldsymbol{\mathcal{M}}$ and $\boldsymbol{\mathcal{M}}_*$ are constant tensors. Thus,

$$\{\mathbf{D}_{\mathrm{c}} - \boldsymbol{\mathcal{M}} : \overset{\triangle}{\underline{\mathcal{T}}}_{\mathrm{c}}\} = \mathbf{0}. \tag{14.4.27}$$

On the other hand, a dual-polarization tensor can be expressed as

$$\mathbf{D}_{\mathrm{c}} - \boldsymbol{\mathcal{M}} : \overset{\triangle}{\underline{\mathcal{T}}}_{\mathrm{c}} = (\boldsymbol{\mathcal{M}}_{\mathrm{c}} - \boldsymbol{\mathcal{M}}) : \overset{\triangle}{\underline{\mathcal{T}}}_{\mathrm{c}} = (\boldsymbol{\mathcal{M}}_{\mathrm{c}} - \boldsymbol{\mathcal{M}}) : \boldsymbol{B}_{\mathrm{c}} : \overset{\triangle}{\underline{\mathcal{T}}}_{\infty}. \tag{14.4.28}$$

The average here vanishes for any applied overall rate of stress $\overset{\triangle}{\underline{\mathcal{T}}}_{\infty}$, i.e.,

$$\{(\boldsymbol{\mathcal{M}}_{\mathrm{c}} - \boldsymbol{\mathcal{M}}) : \boldsymbol{B}_{\mathrm{c}}\} : \overset{\triangle}{\underline{\mathcal{T}}}_{\infty} = \mathbf{0}, \tag{14.4.29}$$

so that

$$\{(\boldsymbol{\mathcal{M}}_{\mathrm{c}} - \boldsymbol{\mathcal{M}}) : \boldsymbol{B}_{\mathrm{c}}\} = \mathbf{0}. \tag{14.4.30}$$

From Eq. (14.3.53) we further observe that

$$\boldsymbol{B}_{\mathrm{c}}^{-1} = [(\boldsymbol{\mathcal{M}}_{\mathrm{c}} - \boldsymbol{\mathcal{M}})^{-1} + \boldsymbol{Q}] : (\boldsymbol{\mathcal{M}}_{\mathrm{c}} - \boldsymbol{\mathcal{M}}), \tag{14.4.31}$$

and

$$(\boldsymbol{\mathcal{M}}_{\mathrm{c}} - \boldsymbol{\mathcal{M}}) : \boldsymbol{B}_{\mathrm{c}} = [(\boldsymbol{\mathcal{M}}_{\mathrm{c}} - \boldsymbol{\mathcal{M}})^{-1} + \boldsymbol{Q}]^{-1}. \tag{14.4.32}$$

Thus, by taking the average and by using Eq. (14.4.30), there follows

$$\{\,[(\boldsymbol{\mathcal{M}}_{\mathrm{c}} - \boldsymbol{\mathcal{M}})^{-1} + \boldsymbol{Q}]^{-1}\} = \mathbf{0}. \tag{14.4.33}$$

14.4.2. Alternative Expressions for Polycrystalline Moduli

The effective polycrystalline moduli can be expressed alternatively, in terms of $\boldsymbol{\mathcal{L}}_{\mathrm{c}}$ and the constraint tensor $\boldsymbol{\mathcal{L}}_*$, by taking the average of the concentration tensor $\boldsymbol{A}_{\mathrm{c}}$ in Eq. (14.3.11), which is

$$\{\boldsymbol{A}_{\mathrm{c}}\} = \{\,(\boldsymbol{\mathcal{L}}_{\mathrm{c}} + \boldsymbol{\mathcal{L}}_*)^{-1}\} : (\boldsymbol{\mathcal{L}} + \boldsymbol{\mathcal{L}}_*) = \boldsymbol{I}. \tag{14.4.34}$$

Therefore,

$$(\boldsymbol{\mathcal{L}} + \boldsymbol{\mathcal{L}}_*)^{-1} = \{\,(\boldsymbol{\mathcal{L}}_{\mathrm{c}} + \boldsymbol{\mathcal{L}}_*)^{-1}\}, \tag{14.4.35}$$

or

$$\boldsymbol{\mathcal{L}} = \{\,(\boldsymbol{\mathcal{L}}_{\mathrm{c}} + \boldsymbol{\mathcal{L}}_*)^{-1}\}^{-1} - \boldsymbol{\mathcal{L}}_*. \tag{14.4.36}$$

Dually, the effective polycrystalline compliances can be expressed in terms of $\mathcal{M}_c$ and the constraint tensor $\mathcal{M}_*$ by taking the average of the concentration tensor $\boldsymbol{B}_c$ in Eq. (14.3.27), which is

$$\{\boldsymbol{B}_c\} = \{\,(\mathcal{M}_c + \mathcal{M}_*)^{-1}\} : (\mathcal{M} + \mathcal{M}_*) = \boldsymbol{I}. \tag{14.4.37}$$

Thus,

$$(\mathcal{M} + \mathcal{M}_*)^{-1} = \{\,(\mathcal{M}_c + \mathcal{M}_*)^{-1}\}. \tag{14.4.38}$$

An equation of this type was used in the derivation of the effective polycrystalline compliances by Hershey (1954). It can be recast as

$$\mathcal{M} = \{\,(\mathcal{M}_c + \mathcal{M}_*)^{-1}\}^{-1} - \mathcal{M}_*. \tag{14.4.39}$$

In applications, either of equations (14.4.2), (14.4.8), (14.4.22), (14.4.25), (14.4.30), (14.4.33), (14.4.36), or (14.4.39) can be used to evaluate the overall (effective) instantaneous moduli or compliances of an incrementally linear polycrystalline aggregate.

14.4.3. Nonaligned Crystals

In the previous analysis it was assumed that the grains comprising a polycrystalline aggregate can be taken, on average, as spheres or aligned ellipsoids. A self-consistent generalization to nonaligned ellipsoidal crystals was suggested by Walpole (1969). In this generalization the local crystalline rate of deformation $\mathbf{D}_c$ is related to the average polycrystalline rate $\mathbf{D}_\infty$ by

$$\mathbf{D}_c = \boldsymbol{A}_c : \{\boldsymbol{A}_c\}^{-1} : \mathbf{D}_\infty. \tag{14.4.40}$$

This automatically satisfies $\{\mathbf{D}_c\} = \mathbf{D}_\infty$. The constraint tensor $\mathcal{L}_*$ depends on the grain orientation. Thus, upon averaging of

$$\overset{\triangle}{\underline{\boldsymbol{\tau}}}_c - \overset{\triangle}{\underline{\boldsymbol{\tau}}}_\infty = -\mathcal{L}_* : (\mathbf{D}_c - \mathbf{D}_\infty) = -\left(\mathcal{L}_* : \boldsymbol{A}_c : \{\boldsymbol{A}_c\}^{-1} - \mathcal{L}_*\right) : \mathbf{D}_\infty, \tag{14.4.41}$$

there follows

$$\{\overset{\triangle}{\underline{\boldsymbol{\tau}}}_c\} - \overset{\triangle}{\underline{\boldsymbol{\tau}}}_\infty = -\left(\{\mathcal{L}_* : \boldsymbol{A}_c\} : \{\boldsymbol{A}_c\}^{-1} - \{\mathcal{L}_*\}\right) : \mathbf{D}_\infty. \tag{14.4.42}$$

In order that $\{\overset{\triangle}{\underline{\boldsymbol{\tau}}}_c\} = \overset{\triangle}{\underline{\boldsymbol{\tau}}}_\infty$ for any $\mathbf{D}_\infty$, ensuring the self-consistency, it is required that

$$\{\mathcal{L}_* : \boldsymbol{A}_c\} : \{\boldsymbol{A}_c\}^{-1} - \{\mathcal{L}_*\} = \mathbf{0}, \tag{14.4.43}$$

i.e.,

$$\{\mathcal{L}_* : \boldsymbol{A}_c\} = \{\mathcal{L}_*\} : \{\boldsymbol{A}_c\}. \tag{14.4.44}$$

By substituting Eq. (14.4.40) into

$$\{\boldsymbol{\mathcal{L}}_{\mathrm{c}} : \mathbf{D}_{\mathrm{c}}\} = \boldsymbol{\mathcal{L}} : \mathbf{D}_{\infty}, \tag{14.4.45}$$

we obtain

$$\{\boldsymbol{\mathcal{L}}_{\mathrm{c}} : \boldsymbol{A}_{\mathrm{c}}\} = \boldsymbol{\mathcal{L}} : \{\boldsymbol{A}_{\mathrm{c}}\}. \tag{14.4.46}$$

In a dual formulation, let the local and average stress rates be related by

$$\overset{\triangle}{\underline{\boldsymbol{\tau}}}_{\mathrm{c}} = \boldsymbol{B}_{\mathrm{c}} : \{\boldsymbol{B}_{\mathrm{c}}\}^{-1} : \overset{\triangle}{\underline{\boldsymbol{\tau}}}_{\infty}, \tag{14.4.47}$$

which automatically satisfies $\{\overset{\triangle}{\underline{\boldsymbol{\tau}}}_{\mathrm{c}}\} = \overset{\triangle}{\underline{\boldsymbol{\tau}}}_{\infty}$. Upon averaging of

$$\mathbf{D}_{\mathrm{c}} - \mathbf{D}_{\infty} = -\boldsymbol{\mathcal{M}}_{*} : (\overset{\triangle}{\underline{\boldsymbol{\tau}}}_{\mathrm{c}} - \overset{\triangle}{\underline{\boldsymbol{\tau}}}_{\infty}) = -\left(\boldsymbol{\mathcal{M}}_{*} : \boldsymbol{B}_{\mathrm{c}} : \{\boldsymbol{B}_{\mathrm{c}}\}^{-1} - \boldsymbol{\mathcal{M}}_{*}\right) : \overset{\triangle}{\underline{\boldsymbol{\tau}}}_{\infty}, \tag{14.4.48}$$

there follows

$$\{\mathbf{D}_{\mathrm{c}}\} - \mathbf{D}_{\infty} = -\left(\{\boldsymbol{\mathcal{M}}_{*} : \boldsymbol{B}_{\mathrm{c}}\} : \{\boldsymbol{B}_{\mathrm{c}}\}^{-1} - \{\boldsymbol{\mathcal{M}}_{*}\}\right) : \overset{\triangle}{\underline{\boldsymbol{\tau}}}_{\infty}. \tag{14.4.49}$$

Thus, in order that $\{\mathbf{D}_{\mathrm{c}}\} = \mathbf{D}_{\infty}$ for any $\overset{\triangle}{\underline{\boldsymbol{\tau}}}_{\infty}$, which ensures the self-consistency, it is required that

$$\{\boldsymbol{\mathcal{M}}_{*} : \boldsymbol{B}_{\mathrm{c}}\} : \{\boldsymbol{B}_{\mathrm{c}}\}^{-1} - \{\boldsymbol{\mathcal{M}}_{*}\} = \mathbf{0}, \tag{14.4.50}$$

i.e.,

$$\{\boldsymbol{\mathcal{M}}_{*} : \boldsymbol{B}_{\mathrm{c}}\} = \{\boldsymbol{\mathcal{M}}_{*}\} : \{\boldsymbol{B}_{\mathrm{c}}\}. \tag{14.4.51}$$

In addition, the substitution of Eq. (14.4.47) into

$$\{\boldsymbol{\mathcal{M}}_{\mathrm{c}} : \overset{\triangle}{\underline{\boldsymbol{\tau}}}_{\mathrm{c}}\} = \boldsymbol{\mathcal{M}} : \overset{\triangle}{\underline{\boldsymbol{\tau}}}_{\infty}, \tag{14.4.52}$$

gives

$$\{\boldsymbol{\mathcal{M}}_{\mathrm{c}} : \boldsymbol{B}_{\mathrm{c}}\} = \boldsymbol{\mathcal{M}} : \{\boldsymbol{B}_{\mathrm{c}}\}. \tag{14.4.53}$$

14.4.4. Polycrystalline Pseudomoduli

If the macroscopic velocity gradient $\mathbf{L}_{\infty}$ is taken to be the orientation average of the crystalline velocity gradients $\mathbf{L}_{\mathrm{c}} = \boldsymbol{A}_{\mathrm{c}}^{0} : \mathbf{L}_{\infty}$, i.e., if

$$\{\mathbf{L}_{\mathrm{c}}\} = \mathbf{L}_{\infty}, \tag{14.4.54}$$

the orientation average of the concentration tensor $\boldsymbol{A}_{\mathrm{c}}^{0}$ is equal to the fourth-order unit tensor,

$$\{\boldsymbol{A}_{\mathrm{c}}^{0}\} = \boldsymbol{I}, \quad I_{ijkl} = \delta_{il}\delta_{jk}. \tag{14.4.55}$$

Since

$$\mathbf{L}_c = \mathbf{M}_c : \underline{\dot{\mathbf{P}}}_c = \mathbf{M}_c : \boldsymbol{B}_c^0 : \underline{\dot{\mathbf{P}}}_\infty, \quad \mathbf{L}_\infty = \mathbf{M} : \underline{\dot{\mathbf{P}}}_\infty, \tag{14.4.56}$$

the substitution into Eq. (14.4.54) gives

$$\{\mathbf{M}_c : \boldsymbol{B}_c^0\} = \mathbf{M}. \tag{14.4.57}$$

On the other hand, if the macroscopic rate of nominal stress $\underline{\dot{\mathbf{P}}}_\infty$ is taken to be the orientation average of the crystalline rate of nominal stress $\underline{\dot{\mathbf{P}}}_c = \boldsymbol{B}_c^0 : \underline{\dot{\mathbf{P}}}_\infty$, i.e., if

$$\{\underline{\dot{\mathbf{P}}}_c\} = \underline{\dot{\mathbf{P}}}_\infty, \tag{14.4.58}$$

the orientation average of the concentration tensor $\boldsymbol{B}_c^0$ is equal to the fourth-order unit tensor,

$$\{\boldsymbol{B}_c^0\} = \boldsymbol{I}. \tag{14.4.59}$$

Since

$$\underline{\dot{\mathbf{P}}}_c = \boldsymbol{\Lambda}_c : \mathbf{L}_c = \boldsymbol{\Lambda}_c : \boldsymbol{A}_c^0 : \mathbf{L}_\infty, \quad \underline{\dot{\mathbf{P}}}_\infty = \boldsymbol{\Lambda} : \mathbf{L}_\infty, \tag{14.4.60}$$

the substitution into Eq. (14.4.58) gives

$$\{\boldsymbol{\Lambda}_c : \boldsymbol{A}_c^0\} = \boldsymbol{\Lambda}. \tag{14.4.61}$$

The rate of nominal stress in the grain can be expressed as

$$\underline{\dot{\mathbf{P}}}_c = \underline{\dot{\mathbf{P}}}_\infty + \underline{\dot{\mathbf{P}}}_* = \boldsymbol{\Lambda} : \mathbf{L}_\infty - \boldsymbol{\Lambda}_* : (\mathbf{L}_c - \mathbf{L}_\infty). \tag{14.4.62}$$

The polarization-type tensor is defined by

$$\underline{\dot{\mathbf{P}}}_c - \boldsymbol{\Lambda} : \mathbf{L}_c = (\boldsymbol{\Lambda} + \boldsymbol{\Lambda}_*) : (\mathbf{L}_\infty - \mathbf{L}_c). \tag{14.4.63}$$

The orientation average of this vanishes by Eq. (14.4.54), because $\boldsymbol{\Lambda}$ and $\boldsymbol{\Lambda}_*$ are the constant tensors, so that

$$\{\underline{\dot{\mathbf{P}}}_c - \boldsymbol{\Lambda} : \mathbf{L}_c\} = \mathbf{0}. \tag{14.4.64}$$

The polarization tensor can also be expressed as

$$\underline{\dot{\mathbf{P}}}_c - \boldsymbol{\Lambda} : \mathbf{L}_c = (\boldsymbol{\Lambda}_c - \boldsymbol{\Lambda}) : \mathbf{L}_c = (\boldsymbol{\Lambda}_c - \boldsymbol{\Lambda}) : \boldsymbol{A}_c^0 : \mathbf{L}_\infty. \tag{14.4.65}$$

The average here vanishes for any applied $\mathbf{L}_\infty$, i.e.,

$$\{(\boldsymbol{\Lambda}_c - \boldsymbol{\Lambda}) : \boldsymbol{A}_c^0\} : \mathbf{L}_\infty = \mathbf{0}, \tag{14.4.66}$$

and

$$\{(\boldsymbol{\Lambda}_c - \boldsymbol{\Lambda}) : \boldsymbol{A}_c^0\} = \mathbf{0}. \tag{14.4.67}$$

A dual-polarization tensor is

$$\mathbf{L}_c - \mathbf{M} : \dot{\underline{\mathbf{P}}}_c = (\mathbf{M} + \mathbf{M}_*) : (\dot{\underline{\mathbf{P}}}_\infty - \dot{\underline{\mathbf{P}}}_c). \tag{14.4.68}$$

Its orientation average also vanishes, in view of Eq. (14.4.58) and because $\mathbf{M}$ and $\mathbf{M}_*$ are the constant tensors. Thus,

$$\{\mathbf{L}_c - \mathbf{M} : \dot{\underline{\mathbf{P}}}_c\} = \mathbf{0}. \tag{14.4.69}$$

A dual-polarization tensor can be alternatively expressed as

$$\mathbf{L}_c - \mathbf{M} : \dot{\underline{\mathbf{P}}}_c = (\mathbf{M}_c - \mathbf{M}) : \dot{\underline{\mathbf{P}}}_c = (\mathbf{M}_c - \mathbf{M}) : \boldsymbol{B}_c^0 : \dot{\underline{\mathbf{P}}}_\infty. \tag{14.4.70}$$

Its average vanishes for any applied $\dot{\underline{\mathbf{P}}}_\infty$, so that

$$\{(\mathbf{M}_c - \mathbf{M}) : \boldsymbol{B}_c^0\} : \dot{\underline{\mathbf{P}}}_\infty = \mathbf{0}, \tag{14.4.71}$$

and

$$\{(\mathbf{M}_c - \mathbf{M}) : \boldsymbol{B}_c^0\} = \mathbf{0}. \tag{14.4.72}$$

The effective polycrystalline pseudomoduli can be cast in terms of $\boldsymbol{\Lambda}_c$ and the constraint tensor $\boldsymbol{\Lambda}_*$ by taking the average of the concentration tensor $\boldsymbol{A}_c$ in Eq. (14.3.62), which is

$$\{\boldsymbol{A}_c^0\} = \{\, (\boldsymbol{\Lambda}_c + \boldsymbol{\Lambda}_*)^{-1}\} : (\boldsymbol{\Lambda} + \boldsymbol{\Lambda}_*) = \boldsymbol{I}. \tag{14.4.73}$$

Thus,

$$(\boldsymbol{\Lambda} + \boldsymbol{\Lambda}_*)^{-1} = \{\, (\boldsymbol{\Lambda}_c + \boldsymbol{\Lambda}_*)^{-1}\}, \tag{14.4.74}$$

and

$$\boldsymbol{\Lambda} = \{\, (\boldsymbol{\Lambda}_c + \boldsymbol{\Lambda}_*)^{-1}\}^{-1} - \boldsymbol{\Lambda}_*. \tag{14.4.75}$$

Alternatively, the effective polycrystalline pseudocompliances can be expressed in terms of $\mathbf{M}_c$ and the constraint tensor $\mathbf{M}_*$ by taking the average of the concentration tensor $\boldsymbol{B}_c^0$ in Eq. (14.3.66). This is

$$\{\boldsymbol{B}_c^0\} = \{\, (\mathbf{M}_c + \mathbf{M}_*)^{-1}\} : (\mathbf{M} + \mathbf{M}_*) = \boldsymbol{I}. \tag{14.4.76}$$

Therefore,

$$(\mathbf{M} + \mathbf{M}_*)^{-1} = \{\, (\mathbf{M}_c + \mathbf{M}_*)^{-1}\}, \tag{14.4.77}$$

or

$$\mathbf{M} = \{\, (\mathbf{M}_c + \mathbf{M}_*)^{-1}\}^{-1} - \mathbf{M}_*. \tag{14.4.78}$$

Nonaligned Crystals

In a self-consistent generalization to nonaligned ellipsoidal crystals, the local velocity gradient within a grain, $\mathbf{L}_c$, is related to the average polycrystalline velocity gradient, $\mathbf{L}_\infty$, by

$$\mathbf{L}_c = \boldsymbol{A}_c^0 : \{\boldsymbol{A}_c^0\}^{-1} : \mathbf{L}_\infty. \tag{14.4.79}$$

Thus,

$$\{\boldsymbol{\Lambda}_* : \boldsymbol{A}_c^0\} = \{\boldsymbol{\Lambda}_*\} : \{\boldsymbol{A}_c^0\}, \tag{14.4.80}$$

and

$$\{\boldsymbol{\Lambda}_c : \boldsymbol{A}_c^0\} = \boldsymbol{\Lambda} : \{\boldsymbol{A}_c^0\}. \tag{14.4.81}$$

Other normalizations for the nonaligned ellipsoidal grains were considered by Iwakuma and Nemat-Nasser (1984).

On the other hand, by defining

$$\underline{\dot{\mathbf{P}}}_c = \boldsymbol{B}_c^0 : \{\boldsymbol{B}_c^0\}^{-1} : \underline{\dot{\mathbf{P}}}_\infty, \tag{14.4.82}$$

we obtain

$$\{\mathbf{M}_* : \boldsymbol{B}_c^0\} = \{\mathbf{M}_*\} : \{\boldsymbol{B}_c^0\}, \tag{14.4.83}$$

and

$$\{\mathbf{M}_c : \boldsymbol{B}_c^0\} = \mathbf{M} : \{\boldsymbol{B}_c^0\}. \tag{14.4.84}$$

14.5. Isotropic Aggregates of Cubic Crystals

Consider a cubic crystal whose elastic moduli are defined by

$$\boldsymbol{\mathcal{L}}_c = 2c_{44}\boldsymbol{J} + (3c_{12} + 2c_{44})\boldsymbol{K} + (c_{11} - c_{12} - 2c_{44})\boldsymbol{Z}, \tag{14.5.1}$$

where $\boldsymbol{J}$ and $\boldsymbol{K}$ are defined by Eq. (14.2.5), and

$$\mathcal{Z}_{ijkl} = a_i a_j a_k a_l + b_i b_j b_k b_l + c_i c_j c_k c_l. \tag{14.5.2}$$

The vectors **a**, **b** and **c** are the orthogonal unit vectors along the principal cubic axes, and the usual notation for the elastic constants c_{11}, c_{12} and c_{44} is employed from Section 5.11. Two independent linear invariants of $\boldsymbol{\mathcal{L}}_c$ are

$$\mathcal{L}^c_{iijj} = 3(c_{11} + 2c_{12}), \quad \mathcal{L}^c_{ijij} = 3(c_{11} + 2c_{44}). \tag{14.5.3}$$

Denote by κ and μ the overall (effective) bulk and shear moduli of an isotropic aggregate of cubic crystals. The corresponding tensors of elastic moduli and compliances are

$$\boldsymbol{\mathcal{L}} = 2\mu\,\boldsymbol{J} + 3\kappa\,\boldsymbol{K}, \tag{14.5.4}$$

$$\boldsymbol{\mathcal{M}} = \frac{1}{2\mu}\boldsymbol{J} + \frac{1}{3\kappa}\boldsymbol{K}. \tag{14.5.5}$$

The Eshelby tensor for a spherical grain is

$$\boldsymbol{S} = \alpha \boldsymbol{J} + \beta \boldsymbol{K}, \quad \beta = 3 - 5\alpha = \frac{\kappa}{\kappa + 4\mu/3}. \tag{14.5.6}$$

Since the product of any pair of isotropic fourth-order tensors is isotropic and commutative, from (14.2.15) we deduce

$$\boldsymbol{T} = \boldsymbol{I} - \boldsymbol{S} = (1-\alpha)\boldsymbol{J} + (1-\beta)\boldsymbol{K}. \tag{14.5.7}$$

Thus,

$$\boldsymbol{P} = \boldsymbol{S} : \boldsymbol{\mathcal{M}} = \frac{\alpha}{2\mu}\boldsymbol{J} + \frac{\beta}{3\kappa}\boldsymbol{K}, \tag{14.5.8}$$

$$\boldsymbol{Q} = \boldsymbol{T} : \boldsymbol{\mathcal{L}} = 2\mu(1-\alpha)\boldsymbol{J} + 3\kappa(1-\beta)\boldsymbol{K}. \tag{14.5.9}$$

The constraint tensors are

$$\boldsymbol{\mathcal{L}}_* = \boldsymbol{\mathcal{L}} : (\boldsymbol{S}^{-1} - \boldsymbol{I}) = 2\mu\,\frac{1-\alpha}{\alpha}\,\boldsymbol{J} + 4\mu\boldsymbol{K}, \tag{14.5.10}$$

$$\boldsymbol{\mathcal{M}}_* = \frac{1}{2\mu}\frac{\alpha}{1-\alpha}\boldsymbol{J} + \frac{1}{4\mu}\boldsymbol{K}. \tag{14.5.11}$$

Upon substitution into Eq. (14.2.26) or (14.3.11), the concentration tensor becomes

$$\boldsymbol{A}_{\mathrm{c}} = \boldsymbol{I} + \alpha\left[a\boldsymbol{J} + (a+3b)(\boldsymbol{K} - \boldsymbol{Z})\right], \tag{14.5.12}$$

where

$$a = \frac{5(c_{11} + 2c_{12} + 4\mu)(\mu - c_{44})}{8\mu^2 + 3(c_{11} + 2c_{12} + 4c_{44})\mu + 2(c_{11} + 2c_{12})c_{44}}, \tag{14.5.13}$$

$$b = \frac{5(c_{11} + 2c_{12} + 4\mu)(c_{11} - c_{12} - 2\mu)}{6[8\mu^2 + 9c_{11}\mu + (c_{11} - c_{12})(c_{11} + 2c_{12})]}. \tag{14.5.14}$$

Since the cubic crystals under hydrostatic state of stress behave as isotropic materials, we have $D^{\mathrm{c}}_{ii} = D^{\infty}_{ii}$, which implies that $A^{\mathrm{c}}_{iikl} = \delta_{kl}$, as incorporated in Eq. (14.5.12). This also implies that $c_{11} + 2c_{12} = 3\kappa$.

The orientation average of the concentration tensors is

$$\{\boldsymbol{A}_{\mathrm{c}}\} = \boldsymbol{I} + \alpha\left[a\boldsymbol{J} + (a+3b)(\boldsymbol{K} - \{\boldsymbol{Z}\})\right]. \tag{14.5.15}$$

It can be shown by integration that

$$\{a_i a_j a_k a_l\} = \frac{1}{8\pi^2}\int_\Omega a_i a_j a_k a_l \,\mathrm{d}\Omega = \frac{1}{15}(\delta_{ij}\delta_{kl} + 2I_{ijkl}), \tag{14.5.16}$$

where $d\Omega = \sin\theta d\varphi d\theta d\psi$ is the solid angle, and φ, θ and ψ are the Euler angles. Thus,

$$\{\boldsymbol{Z}\} = \frac{2}{5}\boldsymbol{J} + \boldsymbol{K}, \tag{14.5.17}$$

and Eq. (14.5.15) becomes

$$\{\boldsymbol{A}_c\} = \boldsymbol{I} + \frac{3\alpha}{5}(a - 2b)\,\boldsymbol{J}. \tag{14.5.18}$$

From Eq. (14.4.2), this must be equal to the unit tensor $\boldsymbol{I}$, which requires that

$$a = 2b. \tag{14.5.19}$$

The substitution of expressions (14.5.14) and (14.5.13) into Eq. (14.5.19) yields a cubic equation for the effective shear modulus,

$$8\mu^3 + (5c_{11} + 4c_{12})\mu^2 - c_{44}(7c_{11} - 4c_{12})\mu - c_{44}(c_{11} - c_{12})(c_{11} + 2c_{12}) = 0. \tag{14.5.20}$$

This equation was originally derived by Kröner (1958). A quartic equation for μ, having the same single positive root, was previously derived by Hershey (1954). Willis (1981) showed that the cubic equation follows from an appropriate variational approach directly from the assumption of the aggregate isotropy, without commitment in the analysis to the spherical grain shape. The value of μ determined from the cubic equation is in-between upper and lower bounds provided by the Voigt and Reuss estimates (Hill, 1952). Closer bounds were derived by Hashin and Shtrikman (1962). See also Cleary, Chen, and Lee (1980), and Walpole (1981). The estimates of the higher order elastic constants were considered by Lubarda (1997), who also gives the reference to other related work.

14.5.1. Voigt and Reuss Estimates

According to the Voigt (1889) assumption, when a polycrystalline aggregate is subjected to the overall uniform strain, the individual crystals will all be in the same state of applied strain (which gives rise to stress discontinuities across the grain boundaries). Thus, by requiring that the overall stress is the average of the local stresses, there follows

$$\boldsymbol{\mathcal{L}} = \{\boldsymbol{\mathcal{L}}_c\}. \tag{14.5.21}$$

Instead of performing the integration

$$\mathcal{L}_{ijkl} = \frac{1}{8\pi^2} \int_\Omega \mathcal{L}^{\mathrm{c}}_{ijkl}\, d\Omega, \tag{14.5.22}$$

the effective polycrystalline constants can be obtained directly by observing that the linear invariants of $\boldsymbol{\mathcal{L}}$ and $\boldsymbol{\mathcal{L}}_{\mathrm{c}}$ must be equal. Thus, equating (14.5.3) to

$$\mathcal{L}_{iijj} = 9\kappa, \quad \mathcal{L}_{ijij} = 3\kappa + 10\mu, \tag{14.5.23}$$

we obtain the well-known Voigt estimates

$$\kappa = \frac{1}{3}(c_{11} + 2c_{12}), \quad \mu^{\mathrm{V}} = \frac{1}{5}(c_{11} - c_{12} + 3c_{44}). \tag{14.5.24}$$

According to the Reuss (1929) assumption, when a polycrystalline aggregate is subjected to the overall uniform stress, the individual crystals will all be in the same state of stress (which gives rise to incompatible deformations across the grain boundaries). Thus, by requiring that the overall strain is the average of the local strains, there follows

$$\boldsymbol{\mathcal{M}} = \{\boldsymbol{\mathcal{M}}_{\mathrm{c}}\}. \tag{14.5.25}$$

This gives the well-known Reuss estimates

$$\kappa = \frac{1}{3}(c_{11} + 2c_{12}), \quad \mu^{\mathrm{R}} = 5[4(c_{11} - c_{12})^{-1} + 3c_{44}^{-1}]^{-1}. \tag{14.5.26}$$

Hill (1952) proved that μ^{V} is the upper bound, and that μ^{R} is the lower bound on the true value of the effective shear modulus, i.e.,

$$\mu^{\mathrm{R}} \leq \mu \leq \mu^{\mathrm{V}}. \tag{14.5.27}$$

It can be easily shown that the effective Lamé constant is bounded such that $\lambda^{\mathrm{V}} \leq \lambda \leq \lambda^{\mathrm{R}}$; see Lubarda (1998).

14.6. Elastoplastic Crystal Embedded in Elastic Matrix

The analysis of the incrementally linear response presented in Section 14.4 is now extended to a piecewise linear elastoplastic response. We consider an elastoplastic ellipsoidal grain embedded in an elastic infinite medium, subjected to the far-field uniform rate of deformation $\mathbf{D}_\infty$. The crystalline rate of deformation $\mathbf{D}_{\mathrm{c}}$ is uniform within the ellipsoidal grain. Suppose that the plastic part of $\mathbf{D}_{\mathrm{c}}$, at the considered stress and deformation state involving n_0 potentially active (critical) slip systems, is produced by the crystallographic slip on a particular set of $n \leq n_0$ active slip systems. From Eq. (12.9.34), we can write

$$\overset{\triangle}{\boldsymbol{\mathcal{T}}}_{\mathrm{c}} = \boldsymbol{\mathcal{L}}_{\mathrm{c}}^{\mathrm{ep}} : \mathbf{D}_{\mathrm{c}}, \quad \boldsymbol{\mathcal{L}}_{\mathrm{c}}^{\mathrm{ep}} = \boldsymbol{\mathcal{L}}_{\mathrm{c}}^{\mathrm{e}} - \sum_{\alpha=1}^{n} \sum_{\beta=1}^{n} g_{\alpha\beta}^{\mathrm{c}\,-1}\, \mathbf{C}_{\mathrm{c}}^{\alpha} \otimes \mathbf{C}_{\mathrm{c}}^{\beta}. \tag{14.6.1}$$

The superscripts "e" and "ep" are added to indicate that $\boldsymbol{\mathcal{L}}_{\mathrm{c}}^{\mathrm{e}}$ and $\boldsymbol{\mathcal{L}}_{\mathrm{c}}^{\mathrm{ep}}$ are the instantaneous elastic and elastoplastic moduli of the crystal. Since the current state is used as the reference, the connections with the corresponding quantities used in Eq. (12.4.3) are

$$\mathbf{C}_{\mathrm{c}}^{\alpha} \leftrightarrow (\det \mathbf{F}_{\mathrm{c}})^{-1} \mathbf{C}_{\mathrm{c}}^{\alpha}, \quad g_{\alpha\beta}^{\mathrm{c}\,-1} \leftrightarrow (\det \mathbf{F}_{\mathrm{c}})\, g_{\alpha\beta}^{\mathrm{c}\,-1}. \tag{14.6.2}$$

The elastoplastic branch of the constitutive response given by Eq. (14.6.1) is associated with the crystallographic slip on a set of n active slip systems, so that the rate of deformation $\mathbf{D}_{\mathrm{c}}$ is directed within a pyramidal region defined by

$$\mathbf{C}_{\mathrm{c}}^{\beta} : \mathbf{D}_{\mathrm{c}} > 0, \quad \beta = 1, 2, \ldots, n. \tag{14.6.3}$$

Each $\mathbf{C}_{\mathrm{c}}^{\beta}$ is codirectional with the outward normal to the corresponding hyperplane of the local yield vertex in strain space.

If the prescribed $\mathbf{D}_{\infty}$ is such that the crystal is momentarily in the state of elastic unloading, then

$$\overset{\triangle}{\boldsymbol{\mathcal{T}}}_{\mathrm{c}} = \boldsymbol{\mathcal{L}}_{\mathrm{c}}^{\mathrm{e}} : \mathbf{D}_{\mathrm{c}}, \tag{14.6.4}$$

and

$$\mathbf{C}_{\mathrm{c}}^{\beta} : \mathbf{D}_{\mathrm{c}} \leq 0, \quad \beta = 1, 2, \ldots, n_0. \tag{14.6.5}$$

For other prescribed $\mathbf{D}_{\infty}$, the local $\mathbf{D}_{\mathrm{c}}$ may be directed within other pyramidal regions in the rate of deformation space, corresponding to other sets of active slip systems (from the set of all n_0 potentially active slip systems, which define the local vertex at a given state of stress and deformation). The whole rate of deformation space can thus be imagined as dissected into pyramidal regions by the set of hyperplanes $\mathbf{C}_{\mathrm{c}}^{\alpha} : \mathbf{D}_{\mathrm{c}} = 0$. The stress rate $\overset{\triangle}{\boldsymbol{\mathcal{T}}}_{\mathrm{c}}$ varies continuously with $\mathbf{D}_{\mathrm{c}}$ over the entire space. In each of the pyramidal regions, the instantaneous elastoplastic stiffness is constant, and the results from Section 14.3 can be accordingly applied, Hill (1965a).

14.6.1. Concentration Tensor

If the crystal is elastically unloading, the concentration tensor, appearing in the relationship $\mathbf{D}_{\mathrm{c}} = \boldsymbol{A}_{\mathrm{c}} : \mathbf{D}_{\infty}$, is

$$\boldsymbol{A}_{\mathrm{c}} = (\boldsymbol{\mathcal{L}}_{\mathrm{c}}^{\mathrm{e}} + \boldsymbol{\mathcal{L}}_{*})^{-1} : (\boldsymbol{\mathcal{L}}^{\mathrm{e}} + \boldsymbol{\mathcal{L}}_{*}^{\mathrm{e}}). \tag{14.6.6}$$

The instantaneous elastic stiffness tensor of the surrounding elastic matrix is $\boldsymbol{\mathcal{L}}^{\mathrm{e}}$, and $\boldsymbol{\mathcal{L}}_{*}^{\mathrm{e}}$ is the corresponding constraint tensor (independent of $\boldsymbol{\mathcal{L}}_{\mathrm{c}}^{\mathrm{e}}$ and the same for any constitutive branch of the crystalline response). The constraint tensor $\boldsymbol{\mathcal{L}}_{*}^{\mathrm{e}}$ of elastic matrix $\boldsymbol{\mathcal{L}}^{\mathrm{e}}$ is such that, from Eq. (14.3.15),

$$\boldsymbol{\mathcal{L}}_{*}^{\mathrm{e}} : \boldsymbol{S}^{\mathrm{e}} = \boldsymbol{\mathcal{L}}^{\mathrm{e}} : (\boldsymbol{I} - \boldsymbol{S}^{\mathrm{e}}). \tag{14.6.7}$$

The Eshelby tensor of the elastic matrix is denoted by $\boldsymbol{S}^{\mathrm{e}}$. We added the superscript "e" to $\boldsymbol{S}$ to indicate the elastic matrix. The concentration tensor in Eq. (14.6.6) applies in the elastic unloading range, which is defined by

$$\mathbf{C}_{\mathrm{c}}^{\beta} : (\boldsymbol{\mathcal{L}}_{\mathrm{c}}^{\mathrm{e}} + \boldsymbol{\mathcal{L}}_{*}^{\mathrm{e}})^{-1} : (\boldsymbol{\mathcal{L}}^{\mathrm{e}} + \boldsymbol{\mathcal{L}}_{*}^{\mathrm{e}}) : \mathbf{D}_{\infty} \leq 0, \quad \beta = 1, 2, \ldots, n_0, \tag{14.6.8}$$

from Eq. (14.6.5) and the relationship $\mathbf{D}_{\mathrm{c}} = \boldsymbol{A}_{\mathrm{c}} : \mathbf{D}_{\infty}$. The unloading condition can be rewritten as

$$\mathbf{C}_{\mathrm{c}}^{\beta} : (\boldsymbol{I} + \boldsymbol{\mathcal{M}}_{*}^{\mathrm{e}} : \boldsymbol{\mathcal{L}}_{\mathrm{c}}^{\mathrm{e}})^{-1} : (\boldsymbol{I} + \boldsymbol{\mathcal{M}}_{*}^{\mathrm{e}} : \boldsymbol{\mathcal{L}}^{\mathrm{e}}) : \mathbf{D}_{\infty} \leq 0, \quad \beta = 1, 2, \ldots, n_0. \tag{14.6.9}$$

If the crystal response is elastoplastic, with the crystallographic slip taking place over the set of n active slip systems, the concentration tensor becomes

$$\boldsymbol{A}_{\mathrm{c}} = \left(\boldsymbol{\mathcal{L}}_{\mathrm{c}}^{\mathrm{e}} + \boldsymbol{\mathcal{L}}_{*}^{\mathrm{e}} - \sum_{\alpha=1}^{n} \sum_{\beta=1}^{n} g_{\alpha\beta}^{\mathrm{c}\,-1} \, \mathbf{C}_{\mathrm{c}}^{\alpha} \otimes \mathbf{C}_{\mathrm{c}}^{\beta} \right)^{-1} : (\boldsymbol{\mathcal{L}}^{\mathrm{e}} + \boldsymbol{\mathcal{L}}_{*}^{\mathrm{e}}). \tag{14.6.10}$$

The inverse of the fourth-order tensor in Eq. (14.6.10) is given by Eq. (14.6.20) below. When this result is substituted into Eq. (14.6.10), there follows

$$\begin{aligned} \boldsymbol{A}_{\mathrm{c}} = [\boldsymbol{I} + \sum_{\alpha=1}^{n} \sum_{\beta=1}^{n} \hat{b}_{\alpha\beta}^{\mathrm{c}\,-1} \, (\boldsymbol{\mathcal{L}}_{\mathrm{c}}^{\mathrm{e}} + \boldsymbol{\mathcal{L}}_{*}^{\mathrm{e}})^{-1} : \mathbf{C}_{\mathrm{c}}^{\alpha} \otimes \mathbf{C}_{\mathrm{c}}^{\beta} \,] \\ : (\boldsymbol{\mathcal{L}}_{\mathrm{c}}^{\mathrm{e}} + \boldsymbol{\mathcal{L}}_{*}^{\mathrm{e}})^{-1} : (\boldsymbol{\mathcal{L}}^{\mathrm{e}} + \boldsymbol{\mathcal{L}}_{*}^{\mathrm{e}}), \end{aligned} \tag{14.6.11}$$

where

$$\hat{b}_{\alpha\beta}^{\mathrm{c}} = g_{\alpha\beta}^{\mathrm{c}} - \mathbf{C}_{\mathrm{c}}^{\alpha} : (\boldsymbol{\mathcal{L}}_{\mathrm{c}}^{\mathrm{e}} + \boldsymbol{\mathcal{L}}_{*}^{\mathrm{e}})^{-1} : \mathbf{C}_{\mathrm{c}}^{\beta}. \tag{14.6.12}$$

The corresponding plastic loading range is defined by

$$\mathbf{C}_{\mathrm{c}}^{\beta} : (\boldsymbol{\mathcal{L}}_{\mathrm{c}}^{\mathrm{e}} + \boldsymbol{\mathcal{L}}_{*}^{\mathrm{e}})^{-1} : (\boldsymbol{\mathcal{L}}^{\mathrm{e}} + \boldsymbol{\mathcal{L}}_{*}^{\mathrm{e}}) : \mathbf{D}_{\infty} > 0, \quad \beta = 1, 2, \ldots, n. \tag{14.6.13}$$

Derivation of the Inverse Tensor

We here derive a formula for the inverse of the fourth-order tensor used in the transition from (14.6.10) to (14.6.11). Consider first the constitutive structure in Eq. (14.6.1). A trace product with $\boldsymbol{\mathcal{L}}_{\mathrm{c}}^{\mathrm{e}\,-1}$ gives

$$\mathbf{D}_{\mathrm{c}} = \boldsymbol{\mathcal{L}}_{\mathrm{c}}^{\mathrm{e}\,-1} : \overset{\triangle}{\boldsymbol{\tau}}_{\mathrm{c}} + \sum_{\alpha=1}^{n} \sum_{\beta=1}^{n} g_{\alpha\beta}^{\mathrm{c}\,-1}\, \boldsymbol{\mathcal{L}}_{\mathrm{c}}^{\mathrm{e}\,-1} : \mathbf{C}_{\mathrm{c}}^{\alpha} \otimes \mathbf{C}_{\mathrm{c}}^{\beta} : \mathbf{D}_{\mathrm{c}}. \tag{14.6.14}$$

Upon application of the trace product with $\mathbf{C}_{\mathrm{c}}^{\gamma} : \boldsymbol{\mathcal{L}}_{\mathrm{c}}^{\mathrm{e}\,-1}$ to Eq. (14.6.1), we obtain

$$\mathbf{C}_{\mathrm{c}}^{\gamma} : \boldsymbol{\mathcal{L}}_{\mathrm{c}}^{\mathrm{e}\,-1} : \overset{\triangle}{\boldsymbol{\tau}}_{\mathrm{c}} = \sum_{\alpha=1}^{n} \sum_{\beta=1}^{n} b_{\gamma\alpha}^{\mathrm{c}}\, g_{\alpha\beta}^{\mathrm{c}\,-1}\, \mathbf{C}_{\mathrm{c}}^{\beta} : \mathbf{D}_{\mathrm{c}}, \tag{14.6.15}$$

where

$$b_{\gamma\alpha}^{\mathrm{c}} = g_{\gamma\alpha}^{\mathrm{c}} - \mathbf{C}_{\mathrm{c}}^{\gamma} : \boldsymbol{\mathcal{L}}_{\mathrm{c}}^{\mathrm{e}\,-1} : \mathbf{C}_{\mathrm{c}}^{\alpha}. \tag{14.6.16}$$

Suppose that the symmetric matrix with components $b_{\gamma\alpha}^{\mathrm{c}}$ is positive-definite. Then, by inversion, from Eq. (14.6.15),

$$\sum_{\beta=1}^{n} g_{\alpha\beta}^{-1}\, \mathbf{C}_{\mathrm{c}}^{\beta} : \mathbf{D}_{\mathrm{c}} = \sum_{\gamma=1}^{n} b_{\alpha\gamma}^{\mathrm{c}\,-1}\, \mathbf{C}_{\mathrm{c}}^{\gamma} : \boldsymbol{\mathcal{L}}_{\mathrm{c}}^{\mathrm{e}\,-1} : \overset{\triangle}{\boldsymbol{\tau}}_{\mathrm{c}}. \tag{14.6.17}$$

The substitution of (14.6.17) into (14.6.14) gives

$$\mathbf{D}_{\mathrm{c}} = \left(\boldsymbol{\mathcal{L}}_{\mathrm{c}}^{\mathrm{e}\,-1} + \sum_{\alpha=1}^{n} \sum_{\beta=1}^{n} b_{\alpha\beta}^{\mathrm{c}\,-1}\, \boldsymbol{\mathcal{L}}_{\mathrm{c}}^{\mathrm{e}\,-1} : \mathbf{C}_{\mathrm{c}}^{\alpha} \otimes \mathbf{C}_{\mathrm{c}}^{\beta} : \boldsymbol{\mathcal{L}}_{\mathrm{c}}^{\mathrm{e}\,-1} \right) : \overset{\triangle}{\boldsymbol{\tau}}_{\mathrm{c}}, \tag{14.6.18}$$

in agreement with the results from Section 12.11. The comparison of Eqs. (14.6.1) and (14.6.18) identifies the inverse tensor

$$\begin{gathered} \left(\boldsymbol{\mathcal{L}}_{\mathrm{c}}^{\mathrm{e}} - \sum_{\alpha=1}^{n} \sum_{\beta=1}^{n} g_{\alpha\beta}^{\mathrm{c}\,-1}\, \mathbf{C}_{\mathrm{c}}^{\alpha} \otimes \mathbf{C}_{\mathrm{c}}^{\beta} \right)^{-1} \\ = \boldsymbol{\mathcal{L}}_{\mathrm{c}}^{\mathrm{e}\,-1} + \sum_{\alpha=1}^{n} \sum_{\beta=1}^{n} b_{\alpha\beta}^{\mathrm{c}\,-1}\, \boldsymbol{\mathcal{L}}_{\mathrm{c}}^{\mathrm{e}\,-1} : \mathbf{C}_{\mathrm{c}}^{\alpha} \otimes \mathbf{C}_{\mathrm{c}}^{\beta} : \boldsymbol{\mathcal{L}}_{\mathrm{c}}^{\mathrm{e}\,-1}. \end{gathered} \tag{14.6.19}$$

When $\mathcal{L}_{\mathrm{c}}^{\mathrm{e}}$ in Eq. (14.6.19) is replaced with $\mathcal{L}_{\mathrm{c}}^{\mathrm{e}} + \mathcal{L}_{*}^{\mathrm{e}}$, we obtain

$$\begin{aligned} \left(\mathcal{L}_{\mathrm{c}}^{\mathrm{e}} + \mathcal{L}_{*}^{\mathrm{e}} - \sum_{\alpha=1}^{n} \sum_{\beta=1}^{n} g_{\alpha\beta}^{\mathrm{c}\,-1}\, \mathbf{C}_{\mathrm{c}}^{\alpha} \otimes \mathbf{C}_{\mathrm{c}}^{\beta} \right)^{-1} &= (\mathcal{L}_{\mathrm{c}}^{\mathrm{e}} + \mathcal{L}_{*}^{\mathrm{e}})^{-1} \\ + \sum_{\alpha=1}^{n} \sum_{\beta=1}^{n} \hat{b}_{\alpha\beta}^{\mathrm{c}\,-1} (\mathcal{L}_{\mathrm{c}}^{\mathrm{e}} + \mathcal{L}_{*}^{\mathrm{e}})^{-1} &: \mathbf{C}_{\mathrm{c}}^{\alpha} \otimes \mathbf{C}_{\mathrm{c}}^{\beta} : (\mathcal{L}_{\mathrm{c}}^{\mathrm{e}} + \mathcal{L}_{*}^{\mathrm{e}})^{-1}, \end{aligned} \tag{14.6.20}$$

which is a desired formula used in Eq. (14.6.11).

14.6.2. Dual-Concentration Tensor

Returning to Eq. (14.6.18), and introducing the tensor

$$\mathbf{G}_{\mathrm{c}}^{\alpha} = \mathcal{M}_{\mathrm{c}} : \mathbf{C}_{\mathrm{c}}^{\alpha}, \quad \mathcal{M}_{\mathrm{c}}^{\mathrm{e}} = \mathcal{L}_{\mathrm{c}}^{\mathrm{e}\,-1}, \tag{14.6.21}$$

the crystalline rate of deformation can be expressed in terms of $\overset{\triangle}{\underline{\tau}}_{\mathrm{c}}$ as

$$\mathbf{D}_{\mathrm{c}} = \mathcal{M}^{\mathrm{ep}} : \overset{\triangle}{\underline{\tau}}_{\mathrm{c}}, \quad \mathcal{M}^{\mathrm{ep}} = \mathcal{M}_{\mathrm{c}}^{\mathrm{e}} + \sum_{\alpha=1}^{n} \sum_{\beta=1}^{n} b_{\alpha\beta}^{\mathrm{c}\,-1}\, \mathbf{G}_{\mathrm{c}}^{\alpha} \otimes \mathbf{G}_{\mathrm{c}}^{\beta}. \tag{14.6.22}$$

The stress rate is here directed within the plastic loading range defined by

$$\mathbf{G}_{\mathrm{c}}^{\beta} : \overset{\triangle}{\underline{\tau}}_{\mathrm{c}} > 0, \quad \beta = 1, 2, \ldots, n. \tag{14.6.23}$$

A dual-concentration tensor, appearing in the transition $\overset{\triangle}{\underline{\tau}}_{\mathrm{c}} = \boldsymbol{B}_{\mathrm{c}} : \overset{\triangle}{\underline{\tau}}_{\infty}$, is

$$\boldsymbol{B}_{\mathrm{c}} = \left(\mathcal{M}_{\mathrm{c}}^{\mathrm{e}} + \mathcal{M}_{*}^{\mathrm{e}} + \sum_{\alpha=1}^{n} \sum_{\beta=1}^{n} b_{\alpha\beta}^{\mathrm{c}\,-1}\, \mathbf{G}_{\mathrm{c}}^{\alpha} \otimes \mathbf{G}_{\mathrm{c}}^{\beta} \right)^{-1} : (\mathcal{M}^{\mathrm{e}} + \mathcal{M}_{*}^{\mathrm{e}}), \tag{14.6.24}$$

from Eq. (14.3.27). A dual-constraint tensor $\mathcal{M}_{*}^{\mathrm{e}}$ of the elastic matrix $\mathcal{M}^{\mathrm{e}}$ obeys, from Eq. (14.3.22),

$$(\boldsymbol{I} - \boldsymbol{S}^{\mathrm{e}}) : \mathcal{M}_{*}^{\mathrm{e}} = \boldsymbol{S}^{\mathrm{e}} : \mathcal{M}^{\mathrm{e}}. \tag{14.6.25}$$

Upon inversion of the fourth-order tensor in Eq. (14.6.24), this becomes

$$\begin{aligned} \boldsymbol{B}_{\mathrm{c}} = [\boldsymbol{I} - \sum_{\alpha=1}^{n} \sum_{\beta=1}^{n} \hat{g}_{\alpha\beta}^{\mathrm{c}\,-1} (\mathcal{M}_{\mathrm{c}}^{\mathrm{e}} + \mathcal{M}_{*}^{\mathrm{e}})^{-1} : (\mathbf{G}_{\mathrm{c}}^{\alpha} \otimes \mathbf{G}_{\mathrm{c}}^{\beta}) \,] \\ : (\mathcal{M}_{\mathrm{c}}^{\mathrm{e}} + \mathcal{M}_{*}^{\mathrm{e}})^{-1} : (\mathcal{M}^{\mathrm{e}} + \mathcal{M}_{*}^{\mathrm{e}}), \end{aligned} \tag{14.6.26}$$

where

$$\hat{g}_{\alpha\beta}^{\mathrm{c}} = b_{\alpha\beta}^{\mathrm{c}} + \mathbf{G}_{\mathrm{c}}^{\alpha} : (\mathcal{M}_{\mathrm{c}}^{\mathrm{e}} + \mathcal{M}_{*}^{\mathrm{e}})^{-1} : \mathbf{G}_{\mathrm{c}}^{\beta}. \tag{14.6.27}$$

The above expression holds in the range of plastic loading,

$$\mathbf{G}_{\mathrm{c}}^{\beta} : (\mathcal{M}_{\mathrm{c}}^{\mathrm{e}} + \mathcal{M}_{*}^{\mathrm{e}})^{-1} : (\mathcal{M}^{\mathrm{e}} + \mathcal{M}_{*}^{\mathrm{e}}) : \overset{\triangle}{\underline{\tau}}_{\infty} > 0, \quad \beta = 1, 2, \ldots, n. \tag{14.6.28}$$

In the elastic unloading range, we have

$$\mathbf{G}_{\mathrm{c}}^{\beta} : (\boldsymbol{\mathcal{M}}_{\mathrm{c}}^{\mathrm{e}} + \boldsymbol{\mathcal{M}}_{*}^{\mathrm{e}})^{-1} : (\boldsymbol{\mathcal{M}}^{\mathrm{e}} + \boldsymbol{\mathcal{M}}_{*}^{\mathrm{e}}) : \overset{\triangle}{\underline{\boldsymbol{\tau}}}_{\infty} \leq 0, \quad \beta = 1, 2, \ldots, n_0, \tag{14.6.29}$$

with the concentration tensor

$$\boldsymbol{B}_{\mathrm{c}} = (\boldsymbol{\mathcal{M}}_{\mathrm{c}}^{\mathrm{e}} + \boldsymbol{\mathcal{M}}_{*}^{\mathrm{e}})^{-1} : (\boldsymbol{\mathcal{M}}^{\mathrm{e}} + \boldsymbol{\mathcal{M}}_{*}^{\mathrm{e}}). \tag{14.6.30}$$

The instantaneous compliances tensor of the surrounding matrix is $\boldsymbol{\mathcal{M}}^{\mathrm{e}}$, while $\boldsymbol{\mathcal{M}}_{*}^{\mathrm{e}}$ is the corresponding constraint tensor, which is the same for all constitutive branches of the crystalline response. The unloading condition (14.6.29) can also be expressed as

$$\mathbf{G}_{\mathrm{c}}^{\beta} : (\boldsymbol{I} + \boldsymbol{\mathcal{L}}_{*}^{\mathrm{e}} : \boldsymbol{\mathcal{M}}_{\mathrm{c}}^{\mathrm{e}})^{-1} : (\boldsymbol{I} + \boldsymbol{\mathcal{L}}_{*}^{\mathrm{e}} : \boldsymbol{\mathcal{M}}^{\mathrm{e}}) : \overset{\triangle}{\underline{\boldsymbol{\tau}}}_{\infty} \leq 0, \quad \beta = 1, 2, \ldots, n_0, \tag{14.6.31}$$

which is dual to (14.6.9).

14.6.3. Locally Smooth Yield Surface

When the yield surface is locally smooth, the elastoplastic branch of the crystalline response is

$$\overset{\triangle}{\underline{\boldsymbol{\tau}}}_{\mathrm{c}} = \left(\boldsymbol{\mathcal{L}}_{\mathrm{c}}^{\mathrm{e}} - \frac{1}{g_{\mathrm{c}}}\,\mathbf{C}_{\mathrm{c}} \otimes \mathbf{C}_{\mathrm{c}}\right) : \mathbf{D}_{\mathrm{c}}, \quad \mathbf{C}_{\mathrm{c}} : \mathbf{D}_{\mathrm{c}} > 0, \tag{14.6.32}$$

where $g_{\mathrm{c}} > 0$. The inverted form is

$$\mathbf{D}_{\mathrm{c}} = \left(\boldsymbol{\mathcal{M}}_{\mathrm{c}}^{\mathrm{e}} + \frac{1}{b_{\mathrm{c}}}\,\mathbf{G}_{\mathrm{c}} \otimes \mathbf{G}_{\mathrm{c}}\right) : \overset{\triangle}{\underline{\boldsymbol{\tau}}}_{\mathrm{c}}, \quad \mathbf{G}_{\mathrm{c}} : \overset{\triangle}{\underline{\boldsymbol{\tau}}}_{\mathrm{c}} > 0. \tag{14.6.33}$$

The relationships hold

$$\mathbf{G}_{\mathrm{c}} = \boldsymbol{\mathcal{M}}_{\mathrm{c}}^{\mathrm{e}} : \mathbf{C}_{\mathrm{c}}, \quad g_{\mathrm{c}} - b_{\mathrm{c}} = \mathbf{C}_{\mathrm{c}} : \boldsymbol{\mathcal{M}}_{\mathrm{c}}^{\mathrm{e}} : \mathbf{C}_{\mathrm{c}} = \mathbf{G}_{\mathrm{c}} : \boldsymbol{\mathcal{L}}_{\mathrm{c}}^{\mathrm{e}} : \mathbf{G}_{\mathrm{c}}. \tag{14.6.34}$$

The crystal is assumed to be in the hardening range, so that $b_{\mathrm{c}} > 0$ in Eq. (14.6.33). The corresponding concentration tensors are

$$\begin{aligned} \boldsymbol{A}_{\mathrm{c}} = [\boldsymbol{I} &+ \frac{1}{\hat{b}_{\mathrm{c}}}\,(\boldsymbol{\mathcal{L}}_{\mathrm{c}}^{\mathrm{e}} + \boldsymbol{\mathcal{L}}_{*}^{\mathrm{e}})^{-1} : (\mathbf{C}_{\mathrm{c}} \otimes \mathbf{C}_{\mathrm{c}})\,] \\ &: (\boldsymbol{\mathcal{L}}_{\mathrm{c}}^{\mathrm{e}} + \boldsymbol{\mathcal{L}}_{*}^{\mathrm{e}})^{-1} : (\boldsymbol{\mathcal{L}}^{\mathrm{e}} + \boldsymbol{\mathcal{L}}_{*}^{\mathrm{e}}), \end{aligned} \tag{14.6.35}$$

where

$$\hat{b}_{\mathrm{c}} = g_{\mathrm{c}} - \mathbf{C}_{\mathrm{c}} : (\boldsymbol{\mathcal{L}}_{\mathrm{c}}^{\mathrm{e}} + \boldsymbol{\mathcal{L}}_{*}^{\mathrm{e}})^{-1} : \mathbf{C}_{\mathrm{c}}, \tag{14.6.36}$$

and

$$\begin{aligned} \boldsymbol{B}_{\mathrm{c}} = [\boldsymbol{I} &- \frac{1}{\hat{g}}\,(\boldsymbol{\mathcal{M}}_{\mathrm{c}}^{\mathrm{e}} + \boldsymbol{\mathcal{M}}_{*}^{\mathrm{e}})^{-1} : (\mathbf{G}_{\mathrm{c}} \otimes \mathbf{G}_{\mathrm{c}})\,] \\ &: (\boldsymbol{\mathcal{M}}_{\mathrm{c}}^{\mathrm{e}} + \boldsymbol{\mathcal{M}}_{*}^{\mathrm{e}})^{-1} : (\boldsymbol{\mathcal{M}}^{\mathrm{e}} + \boldsymbol{\mathcal{M}}_{*}^{\mathrm{e}}), \end{aligned} \tag{14.6.37}$$

where

$$\hat{g}_{\mathrm{c}} = b_{\mathrm{c}} + \mathbf{G}_{\mathrm{c}} : (\boldsymbol{\mathcal{M}}_{\mathrm{c}}^{\mathrm{e}} + \boldsymbol{\mathcal{M}}_{*}^{\mathrm{e}})^{-1} : \mathbf{G}_{\mathrm{c}}. \tag{14.6.38}$$

It is noted that

$$\hat{b}_{\mathrm{c}} = \hat{g}_{\mathrm{c}}. \tag{14.6.39}$$

This can be verified by using the connection (14.6.34) and the relationships for the inverse tensors

$$(\boldsymbol{\mathcal{L}}_{\mathrm{c}} + \boldsymbol{\mathcal{L}}_{*})^{-1} = \boldsymbol{\mathcal{M}}_{\mathrm{c}}^{\mathrm{e}} - \boldsymbol{\mathcal{M}}_{\mathrm{c}}^{\mathrm{e}} : (\boldsymbol{\mathcal{M}}_{\mathrm{c}}^{\mathrm{e}} + \boldsymbol{\mathcal{M}}_{*}^{\mathrm{e}})^{-1} : \boldsymbol{\mathcal{M}}_{\mathrm{c}}^{\mathrm{e}}, \tag{14.6.40}$$

$$(\boldsymbol{\mathcal{M}}_{\mathrm{c}}^{\mathrm{e}} + \boldsymbol{\mathcal{M}}_{*}^{\mathrm{e}})^{-1} = \boldsymbol{\mathcal{L}}_{\mathrm{c}}^{\mathrm{e}} - \boldsymbol{\mathcal{L}}_{\mathrm{c}}^{\mathrm{e}} : (\boldsymbol{\mathcal{L}}_{\mathrm{c}}^{\mathrm{e}} + \boldsymbol{\mathcal{L}}_{*}^{\mathrm{e}})^{-1} : \boldsymbol{\mathcal{L}}_{\mathrm{c}}^{\mathrm{e}}. \tag{14.6.41}$$

The first of these follows because

$$\begin{aligned}
(\boldsymbol{\mathcal{L}}_{\mathrm{c}}^{\mathrm{e}} + \boldsymbol{\mathcal{L}}_{*}^{\mathrm{e}})^{-1} &= [\boldsymbol{\mathcal{L}}_{\mathrm{c}}^{\mathrm{e}} : (\boldsymbol{\mathcal{M}}_{\mathrm{c}}^{\mathrm{e}} + \boldsymbol{\mathcal{M}}_{*}^{\mathrm{e}}) : \boldsymbol{\mathcal{L}}_{*}^{\mathrm{e}}]^{-1} \\
&= \boldsymbol{\mathcal{M}}_{*} : (\boldsymbol{\mathcal{M}}_{\mathrm{c}}^{\mathrm{e}} + \boldsymbol{\mathcal{M}}_{*}^{\mathrm{e}})^{-1} : \boldsymbol{\mathcal{M}}_{\mathrm{c}}^{\mathrm{e}} \\
&= (\boldsymbol{\mathcal{M}}_{\mathrm{c}}^{\mathrm{e}} + \boldsymbol{\mathcal{M}}_{*}^{\mathrm{e}} - \boldsymbol{\mathcal{M}}_{\mathrm{c}}^{\mathrm{e}}) : (\boldsymbol{\mathcal{M}}_{\mathrm{c}}^{\mathrm{e}} + \boldsymbol{\mathcal{M}}_{*}^{\mathrm{e}})^{-1} : \boldsymbol{\mathcal{M}}_{\mathrm{c}}^{\mathrm{e}} \\
&= \boldsymbol{\mathcal{M}}_{\mathrm{c}}^{\mathrm{e}} - \boldsymbol{\mathcal{M}}_{\mathrm{c}}^{\mathrm{e}} : (\boldsymbol{\mathcal{M}}_{\mathrm{c}}^{\mathrm{e}} + \boldsymbol{\mathcal{M}}_{*}^{\mathrm{e}})^{-1} : \boldsymbol{\mathcal{M}}_{\mathrm{c}}^{\mathrm{e}},
\end{aligned} \tag{14.6.42}$$

and similarly for the second.

The plastic part of the crystalline stress rate is

$$\underline{\overset{\triangle}{\boldsymbol{\tau}}}{}_{\mathrm{c}}^{\mathrm{p}} = \underline{\overset{\triangle}{\boldsymbol{\tau}}}_{\mathrm{c}} - \boldsymbol{\mathcal{L}}_{\mathrm{c}}^{\mathrm{e}} : \mathbf{D}_{\mathrm{c}} = -\frac{1}{g_{\mathrm{c}}} (\mathbf{C}_{\mathrm{c}} \otimes \mathbf{C}_{\mathrm{c}}) : \mathbf{D}_{\mathrm{c}} = -\frac{1}{g_{\mathrm{c}}} (\mathbf{C}_{\mathrm{c}} \otimes \mathbf{C}_{\mathrm{c}}) : \boldsymbol{A}_{\mathrm{c}} : \mathbf{D}_{\infty}. \tag{14.6.43}$$

In view of (14.6.35), we can write

$$\mathbf{C}_{\mathrm{c}} : \boldsymbol{A}_{\mathrm{c}} = (g_{\mathrm{c}} / \hat{b}_{\mathrm{c}})\, \mathbf{C}_{\mathrm{c}} : (\boldsymbol{\mathcal{L}}_{\mathrm{c}}^{\mathrm{e}} + \boldsymbol{\mathcal{L}}_{*}^{\mathrm{e}})^{-1} : (\boldsymbol{\mathcal{L}}^{\mathrm{e}} + \boldsymbol{\mathcal{L}}_{*}^{\mathrm{e}}), \tag{14.6.44}$$

and the substitution into Eq. (14.6.43) gives

$$\underline{\overset{\triangle}{\boldsymbol{\tau}}}{}_{\mathrm{c}}^{\mathrm{p}} = -\frac{1}{\hat{b}_{\mathrm{c}}} (\mathbf{C}_{\mathrm{c}} \otimes \mathbf{C}_{\mathrm{c}}) : (\boldsymbol{\mathcal{L}}_{\mathrm{c}}^{\mathrm{e}} + \boldsymbol{\mathcal{L}}_{*}^{\mathrm{e}})^{-1} : (\boldsymbol{\mathcal{L}}^{\mathrm{e}} + \boldsymbol{\mathcal{L}}_{*}^{\mathrm{e}}) : \mathbf{D}_{\infty}. \tag{14.6.45}$$

Likewise, the plastic part of the crystalline rate of deformation is

$$\mathbf{D}_{\mathrm{c}}^{\mathrm{p}} = \mathbf{D}_{\mathrm{c}} - \boldsymbol{\mathcal{M}}_{\mathrm{c}}^{\mathrm{e}} : \underline{\overset{\triangle}{\boldsymbol{\tau}}}_{\mathrm{c}} = \frac{1}{b_{\mathrm{c}}} (\mathbf{G}_{\mathrm{c}} \otimes \mathbf{G}_{\mathrm{c}}) : \underline{\overset{\triangle}{\boldsymbol{\tau}}}_{\mathrm{c}} = \frac{1}{b_{\mathrm{c}}} (\mathbf{G}_{\mathrm{c}} \otimes \mathbf{G}_{\mathrm{c}}) : \boldsymbol{B}_{\mathrm{c}} : \underline{\overset{\triangle}{\boldsymbol{\tau}}}_{\infty}. \tag{14.6.46}$$

Since, from (14.6.37),

$$\mathbf{G}_{\mathrm{c}} : \boldsymbol{B}_{\mathrm{c}} = (b_{\mathrm{c}} / \hat{g}_{\mathrm{c}})\, \mathbf{G}_{\mathrm{c}} : (\boldsymbol{\mathcal{M}}_{\mathrm{c}}^{\mathrm{e}} + \boldsymbol{\mathcal{M}}_{*}^{\mathrm{e}})^{-1} : (\boldsymbol{\mathcal{M}}^{\mathrm{e}} + \boldsymbol{\mathcal{M}}_{*}^{\mathrm{e}}), \tag{14.6.47}$$

we obtain, upon substitution into Eq. (14.6.46),

$$\mathbf{D}_{\mathrm{c}}^{\mathrm{p}} = \frac{1}{\hat{g}_{\mathrm{c}}} (\mathbf{G}_{\mathrm{c}} \otimes \mathbf{G}_{\mathrm{c}}) : (\boldsymbol{\mathcal{M}}_{\mathrm{c}}^{\mathrm{e}} + \boldsymbol{\mathcal{M}}_{*}^{\mathrm{e}})^{-1} : (\boldsymbol{\mathcal{M}}^{\mathrm{e}} + \boldsymbol{\mathcal{M}}_{*}^{\mathrm{e}}) : \underline{\overset{\triangle}{\boldsymbol{\tau}}}_{\infty}. \tag{14.6.48}$$

Particular Cases

If the elastic properties of the grain and the surrounding matrix are identical, i.e., if

$$\mathcal{L}_{\mathrm{c}}^{\mathrm{e}} = \mathcal{L}^{\mathrm{e}}, \quad \mathcal{M}_{\mathrm{c}}^{\mathrm{e}} = \mathcal{M}^{\mathrm{e}}, \tag{14.6.49}$$

the preceding formulas simplify, and the concentration tensors become (Hill, 1965 a)

$$\boldsymbol{A}_{\mathrm{c}} = \boldsymbol{I} + \frac{(\mathcal{L}^{\mathrm{e}} + \mathcal{L}_{*}^{\mathrm{e}})^{-1} : (\mathbf{C}_{\mathrm{c}} \otimes \mathbf{C}_{\mathrm{c}})}{g_{\mathrm{c}} - \mathbf{C}_{\mathrm{c}} : (\mathcal{L}^{\mathrm{e}} + \mathcal{L}_{*}^{\mathrm{e}})^{-1} : \mathbf{C}_{\mathrm{c}}} = \boldsymbol{I} + \frac{\boldsymbol{P} : (\mathbf{C}_{\mathrm{c}} \otimes \mathbf{C}_{\mathrm{c}})}{g_{\mathrm{c}} - \mathbf{C}_{\mathrm{c}} : \boldsymbol{P} : \mathbf{C}_{\mathrm{c}}}, \tag{14.6.50}$$

$$\boldsymbol{B}_{\mathrm{c}} = \boldsymbol{I} - \frac{(\mathcal{M}^{\mathrm{e}} + \mathcal{M}_{*}^{\mathrm{e}})^{-1} : (\mathbf{G}_{\mathrm{c}} \otimes \mathbf{G}_{\mathrm{c}})}{b_{\mathrm{c}} + \mathbf{G}_{\mathrm{c}} : (\mathcal{M}^{\mathrm{e}} + \mathcal{M}_{*}^{\mathrm{e}})^{-1} : \mathbf{G}_{\mathrm{c}}} = \boldsymbol{I} - \frac{\boldsymbol{Q} : (\mathbf{G}_{\mathrm{c}} \otimes \mathbf{G}_{\mathrm{c}})}{b_{\mathrm{c}} + \mathbf{G}_{\mathrm{c}} : \boldsymbol{Q} : \mathbf{G}_{\mathrm{c}}}. \tag{14.6.51}$$

The plastic parts of the crystalline stress and strain rates are similarly

$$\overset{\triangle}{\underline{\boldsymbol{\tau}}}{}_{\mathrm{c}}^{\mathrm{p}} = -\frac{(\mathbf{C}_{\mathrm{c}} \otimes \mathbf{C}_{\mathrm{c}}) : \mathbf{D}_{\infty}}{g_{\mathrm{c}} - \mathbf{C}_{\mathrm{c}} : (\mathcal{L}^{\mathrm{e}} + \mathcal{L}_{*}^{\mathrm{e}})^{-1} : \mathbf{C}_{\mathrm{c}}} = -\frac{(\mathbf{C}_{\mathrm{c}} \otimes \mathbf{C}_{\mathrm{c}}) : \mathbf{D}_{\infty}}{g_{\mathrm{c}} - \mathbf{C}_{\mathrm{c}} : \boldsymbol{P} : \mathbf{C}_{\mathrm{c}}}, \tag{14.6.52}$$

$$\mathbf{D}_{\mathrm{c}}^{\mathrm{p}} = \frac{(\mathbf{G}_{\mathrm{c}} \otimes \mathbf{G}_{\mathrm{c}}) : \overset{\triangle}{\underline{\boldsymbol{\tau}}}_{\infty}}{b_{\mathrm{c}} + \mathbf{G}_{\mathrm{c}} : (\mathcal{M}^{\mathrm{e}} + \mathcal{M}_{*}^{\mathrm{e}})^{-1} : \mathbf{G}_{\mathrm{c}}} = \frac{(\mathbf{G}_{\mathrm{c}} \otimes \mathbf{G}_{\mathrm{c}}) : \overset{\triangle}{\underline{\boldsymbol{\tau}}}_{\infty}}{b_{\mathrm{c}} + \mathbf{G}_{\mathrm{c}} : \boldsymbol{Q} : \mathbf{G}_{\mathrm{c}}}. \tag{14.6.53}$$

These expressions can be further reduced if it is assumed that the elastic response is isotropic, and that the plastic response is incompressible ($\mathbf{G}_{\mathrm{c}}$ and $\mathbf{C}_{\mathrm{c}}$ deviatoric tensors). From Eqs. (14.5.8) and (14.5.9), we obtain in this case

$$\boldsymbol{P} = \frac{\alpha}{2\mu} \boldsymbol{J} + \frac{\beta}{3\kappa} \boldsymbol{K}, \quad \boldsymbol{Q} = 2\mu(1-\alpha)\boldsymbol{J} + 3\kappa(1-\beta)\boldsymbol{K}, \tag{14.6.54}$$

so that

$$\boldsymbol{P} : \mathbf{C}_{\mathrm{c}} = \frac{\alpha}{2\mu} \mathbf{C}_{\mathrm{c}}, \quad \boldsymbol{Q} : \mathbf{G}_{\mathrm{c}} = 2\mu(1-\alpha)\, \mathbf{G}_{\mathrm{c}}. \tag{14.6.55}$$

The components of the Eshelby tensor, α and β, are given in Eq. (14.5.6). Consequently,

$$\boldsymbol{A}_{\mathrm{c}} = \boldsymbol{I} + \frac{(\alpha/2\mu)\, \mathbf{C}_{\mathrm{c}} \otimes \mathbf{C}_{\mathrm{c}}}{g_{\mathrm{c}} - (\alpha/2\mu)\, \mathbf{C}_{\mathrm{c}} : \mathbf{C}_{\mathrm{c}}}, \tag{14.6.56}$$

$$\boldsymbol{B}_{\mathrm{c}} = \boldsymbol{I} - \frac{2\mu(1-\alpha)\, \mathbf{G}_{\mathrm{c}} \otimes \mathbf{G}_{\mathrm{c}}}{b_{\mathrm{c}} + 2\mu(1-\alpha)\, \mathbf{G}_{\mathrm{c}} : \mathbf{G}_{\mathrm{c}}}, \tag{14.6.57}$$

and

$$\overset{\triangle}{\underline{\boldsymbol{\tau}}}{}_{\mathrm{c}}^{\mathrm{p}} = -\frac{(\mathbf{C}_{\mathrm{c}} \otimes \mathbf{C}_{\mathrm{c}}) : \mathbf{D}_{\infty}}{g_{\mathrm{c}} - (\alpha/2\mu)\, \mathbf{C}_{\mathrm{c}} : \mathbf{C}_{\mathrm{c}}}, \tag{14.6.58}$$

$$\mathbf{D}_c^p = \frac{(\mathbf{G}_c \otimes \mathbf{G}_c) : \overset{\triangle}{\underline{\boldsymbol{\tau}}}_\infty}{b_c + 2\mu(1-\alpha)\,\mathbf{G}_c : \mathbf{G}_c}. \tag{14.6.59}$$

14.6.4. Rigid-Plastic Crystal in Elastic Matrix

Suppose that the crystal is rigid-plastic, i.e.,

$$\mathcal{M}_c^e = \mathbf{0}. \tag{14.6.60}$$

At the point where the yield surface is locally smooth, we have, from Eqs. (14.6.33) and (14.6.37),

$$\mathbf{D}_c = \mathcal{M}_c^p : \overset{\triangle}{\boldsymbol{\sigma}}_c, \quad \mathcal{M}_c^p = \frac{1}{b_c}(\mathbf{G}_c \otimes \mathbf{G}_c), \tag{14.6.61}$$

and

$$\boldsymbol{B}_c = \left[\boldsymbol{I} - \frac{\mathcal{L}_*^e : (\mathbf{G}_c \otimes \mathbf{G}_c)}{b_c + \mathbf{G}_c : \mathcal{L}_*^e : \mathbf{G}_c}\right] : (\boldsymbol{I} + \mathcal{L}_*^e : \mathcal{M}^e), \tag{14.6.62}$$

provided that

$$\mathbf{G}_c : (\boldsymbol{I} + \mathcal{L}_*^e : \mathcal{M}^e) : \overset{\triangle}{\underline{\boldsymbol{\tau}}}_\infty > 0. \tag{14.6.63}$$

Recall that for the rigid-plastic crystal

$$\overset{\triangle}{\underline{\boldsymbol{\tau}}}_c = \overset{\triangle}{\boldsymbol{\sigma}}_c. \tag{14.6.64}$$

Since, by Eq. (14.3.23),

$$\mathcal{M}_*^e = \mathcal{S}^e : (\mathcal{M}^e + \mathcal{M}_*^e) = (\mathcal{M}^e + \mathcal{M}_*^e) : \mathcal{S}^{eT}, \tag{14.6.65}$$

and since

$$\mathcal{S}^{e-1} = \boldsymbol{I} + \mathcal{M}^e : \mathcal{L}_*^e, \quad \mathcal{S}^{e-T} = \boldsymbol{I} + \mathcal{L}_*^e : \mathcal{M}^e, \tag{14.6.66}$$

the combination with Eq. (14.6.62) establishes

$$\boldsymbol{B}_c : \mathcal{S}^{eT} = \boldsymbol{I} - \frac{\mathcal{L}_*^e : \mathbf{G}_c \otimes \mathbf{G}_c}{b_c + \mathbf{G}_c : \mathcal{L}_*^e : \mathbf{G}_c}. \tag{14.6.67}$$

The plastic loading condition (14.6.63) can be expressed as

$$\mathbf{G}_c : \mathcal{S}^{e-T} : \overset{\triangle}{\underline{\boldsymbol{\tau}}}_\infty > 0. \tag{14.6.68}$$

Dually, in view of Eqs. (14.2.36) and (14.3.44), we have

$$\boldsymbol{A}_c : \mathcal{M}^e = \mathcal{M}_c^p : \boldsymbol{B}_c, \quad \boldsymbol{P} = \mathcal{S}^e : \mathcal{M}^e = \mathcal{M}^e : \mathcal{S}^{eT}, \tag{14.6.69}$$

and

$$\boldsymbol{A}_c : \boldsymbol{P} = \boldsymbol{A}_c : \mathcal{M}^e : \mathcal{S}^{eT} = \mathcal{M}_c^p : \boldsymbol{B}_c : \mathcal{S}^{eT}. \tag{14.6.70}$$

By substituting the expression (14.6.67) for $\boldsymbol{B}_{\mathrm{c}} : \boldsymbol{S}^{\mathrm{e}T}$ into Eq. (14.6.70), there follows

$$\boldsymbol{A}_{\mathrm{c}} : \boldsymbol{P} = \boldsymbol{\mathcal{M}}_{\mathrm{c}}^{\mathrm{p}} : \left(\boldsymbol{I} - \frac{\boldsymbol{\mathcal{L}}_{*}^{\mathrm{e}} : \mathbf{G}_{\mathrm{c}} \otimes \mathbf{G}_{\mathrm{c}}}{b_{\mathrm{c}} + \mathbf{G}_{\mathrm{c}} : \boldsymbol{\mathcal{L}}_{*}^{\mathrm{e}} : \mathbf{G}_{\mathrm{c}}} \right). \tag{14.6.71}$$

Upon using Eq. (14.6.61), this reduces to

$$\boldsymbol{A}_{\mathrm{c}} : \boldsymbol{P} = \frac{\mathbf{G}_{\mathrm{c}} \otimes \mathbf{G}_{\mathrm{c}}}{b_{\mathrm{c}} + \mathbf{G}_{\mathrm{c}} : \boldsymbol{\mathcal{L}}_{*}^{\mathrm{e}} : \mathbf{G}_{\mathrm{c}}}, \quad \mathbf{G}_{\mathrm{c}} : \boldsymbol{P}^{-1} : \mathbf{D}_{\infty} > 0. \tag{14.6.72}$$

Note the transition

$$\begin{aligned} \mathbf{G}_{\mathrm{c}} : \boldsymbol{S}^{\mathrm{e}-T} : \overset{\triangle}{\boldsymbol{\mathcal{T}}}_{\infty} &= \mathbf{G}_{\mathrm{c}} : \boldsymbol{S}^{\mathrm{e}-T} : \boldsymbol{\mathcal{L}}^{\mathrm{e}} : \mathbf{D}_{\infty} \\ &= \mathbf{G}_{\mathrm{c}} : (\boldsymbol{\mathcal{M}}^{\mathrm{e}} : \boldsymbol{S}^{\mathrm{e}T})^{-1} : \mathbf{D}_{\infty} = \mathbf{G}_{\mathrm{c}} : \boldsymbol{P}^{-1} : \mathbf{D}_{\infty}. \end{aligned} \tag{14.6.73}$$

14.7. Elastoplastic Crystal Embedded in Elastoplastic Matrix

The most general case in Hill's formulation of the self-consistent method is the consideration of an ellipsoidal elastoplastic crystal embedded in a homogeneous elastoplastic matrix. Suppose that the elastoplastic stiffness is uniform throughout the matrix, and given by (see Section 9.5)

$$\boldsymbol{\mathcal{L}}^{\mathrm{ep}} = \boldsymbol{\mathcal{L}}^{\mathrm{e}} - \sum_{\alpha=1}^{m} \sum_{\beta=1}^{m} g_{\alpha\beta}^{-1} \, \mathbf{C}^{\alpha} \otimes \mathbf{C}^{\beta}. \tag{14.7.1}$$

The tensor $\mathbf{C}^{\alpha}$ is codirectional with the outward normal to the corresponding hyperplane of the local yield vertex in strain space. The constitutive branch of the elastoplastic matrix response (14.7.1) is associated with m active yield segments at the vertex. It is assumed that these are activated when the applied $\mathbf{D}_{\infty}$ is such that

$$\mathbf{C}^{\beta} : \mathbf{D}_{\infty} > 0, \quad \beta = 1, 2, \ldots, m. \tag{14.7.2}$$

For other directions of the imposed $\mathbf{D}_{\infty}$, other constitutive branches at the yield vertex may apply, corresponding to other sets of active yield segments. In particular, the elastic unloading branch corresponds to $\mathbf{D}_{\infty}$ for which

$$\mathbf{C}^{\beta} : \mathbf{D}_{\infty} \leq 0, \quad \beta = 1, 2, \ldots, m_0, \tag{14.7.3}$$

where m_0 is the number of all yield segments forming a local vertex at the considered instant of deformation.

The concentration tensor associated with the elastoplastic matrix stiffness (14.7.1), and the elastoplastic crystalline stiffness (14.6.1), is

$$\boldsymbol{A}_{\mathrm{c}} = (\boldsymbol{\mathcal{L}}_{\mathrm{c}}^{\mathrm{ep}} + \boldsymbol{\mathcal{L}}_{*}^{\mathrm{ep}})^{-1} : (\boldsymbol{\mathcal{L}}^{\mathrm{ep}} + \boldsymbol{\mathcal{L}}_{*}^{\mathrm{ep}}). \tag{14.7.4}$$

The constraint tensor of the elastoplastic matrix $\mathcal{L}_*^{\rm ep}$ is defined such that

$$\mathcal{L}_*^{\rm ep} : \boldsymbol{S}^{\rm ep} = \mathcal{L}^{\rm ep} : (\boldsymbol{I} - \boldsymbol{S}^{\rm ep}). \tag{14.7.5}$$

The superscripts "ep" is added to $\boldsymbol{S}$ to indicate that $\boldsymbol{S}^{\rm ep}$ is the Eshelby tensor of the elastoplastic matrix. The branch of $\boldsymbol{S}^{\rm ep}$ corresponding to the elastoplastic matrix branch (14.7.1) is used in (14.7.5). We also note that the tensor $\boldsymbol{P}$, introduced in Subsection 14.3.2, is in this case

$$\boldsymbol{P} = (\mathcal{L}_*^{\rm ep} + \mathcal{L}^{\rm ep})^{-1} = \boldsymbol{S}^{\rm ep} : \mathcal{L}^{\rm ep\,-1}. \tag{14.7.6}$$

In an expanded form, the concentration tensor can be written as

$$\begin{aligned} \boldsymbol{A}_{\rm c} = & \left(\mathcal{L}_{\rm c}^{\rm e} + \mathcal{L}_*^{\rm ep} - \sum_{\alpha=1}^{n}\sum_{\beta=1}^{n} g_{\alpha\beta}^{{\rm c}\,-1}\, \mathbf{C}_{\rm c}^{\alpha} \otimes \mathbf{C}_{\rm c}^{\beta}\right)^{-1} \\ & : \left(\mathcal{L}^{\rm e} + \mathcal{L}_*^{\rm ep} - \sum_{\alpha=1}^{m}\sum_{\beta=1}^{m} g_{\alpha\beta}^{-1}\, \mathbf{C}^{\alpha} \otimes \mathbf{C}^{\beta}\right). \end{aligned} \tag{14.7.7}$$

Upon performing the required inversion in Eq. (14.7.7), this becomes

$$\begin{aligned} \boldsymbol{A}_{\rm c} = & \left[\boldsymbol{I} + \sum_{\alpha=1}^{n}\sum_{\beta=1}^{n} \hat{b}_{\alpha\beta}^{{\rm c}\,-1} (\mathcal{L}_{\rm c}^{\rm e} + \mathcal{L}_*^{\rm ep})^{-1} : (\mathbf{C}_{\rm c}^{\alpha} \otimes \mathbf{C}_{\rm c}^{\beta})\right] \\ & : (\mathcal{L}_{\rm c}^{\rm e} + \mathcal{L}_*^{\rm ep})^{-1} : \left(\mathcal{L}^{\rm e} + \mathcal{L}_*^{\rm ep} - \sum_{\alpha=1}^{m}\sum_{\beta=1}^{m} g_{\alpha\beta}^{-1}\, \mathbf{C}^{\alpha} \otimes \mathbf{C}^{\beta}\right), \end{aligned} \tag{14.7.8}$$

where

$$\hat{b}_{\alpha\beta}^{\rm c} = g_{\alpha\beta}^{\rm c} - \mathbf{C}_{\rm c}^{\alpha} : (\mathcal{L}_{\rm c}^{\rm e} + \mathcal{L}_*^{\rm ep})^{-1} : \mathbf{C}_{\rm c}^{\beta}. \tag{14.7.9}$$

The applied $\mathbf{D}_\infty$ is such that (14.7.2) holds, as well as

$$\mathbf{C}_{\rm c}^{\beta} : (\mathcal{L}_{\rm c}^{\rm e} + \mathcal{L}_*^{\rm ep})^{-1} : (\mathcal{L}^{\rm e} + \mathcal{L}_*^{\rm ep}) : \mathbf{D}_\infty > 0, \quad \beta = 1, 2, \ldots, n. \tag{14.7.10}$$

Formulation with Elastoplastic Compliances

In the formulation using the tensors of elastoplastic compliances, we have (see Section 9.6)

$$\mathcal{M}^{\rm ep} = \mathcal{M}^{\rm e} + \sum_{\alpha=1}^{m}\sum_{\beta=1}^{m} b_{\alpha\beta}^{-1}\, \mathbf{G}^{\alpha} \otimes \mathbf{G}^{\beta}, \tag{14.7.11}$$

where

$$\mathbf{G}^{\alpha} = \mathcal{M}^{\rm e} : \mathbf{C}^{\alpha}, \quad \mathcal{M}^{\rm e} = \mathcal{L}^{\rm e\,-1}, \tag{14.7.12}$$

and

$$b_{\alpha\beta} = g_{\alpha\beta} - \mathbf{C}^{\alpha} : \boldsymbol{\mathcal{M}}^{\mathrm{e}} : \mathbf{C}^{\beta}. \tag{14.7.13}$$

The tensor $\mathbf{G}^{\alpha}$ is codirectional with the outward normal to the corresponding hyperplane of the local yield vertex in stress space. The constitutive branch of the elastoplastic matrix response (14.7.11) is associated with m active yield segments of the vertex. It is assumed that, in the hardening range, these are activated when the applied $\overset{\triangle}{\underline{\boldsymbol{\tau}}}_{\infty}$ is such that

$$\mathbf{G}^{\beta} : \overset{\triangle}{\underline{\boldsymbol{\tau}}}_{\infty} > 0, \quad \beta = 1, 2, \ldots, m. \tag{14.7.14}$$

The elastic unloading branch corresponds to $\overset{\triangle}{\underline{\boldsymbol{\tau}}}_{\infty}$ for which

$$\mathbf{G}^{\beta} : \overset{\triangle}{\underline{\boldsymbol{\tau}}}_{\infty} \leq 0, \quad \beta = 1, 2, \ldots, m_0, \tag{14.7.15}$$

where m_0 is the number of all yield segments forming a local vertex at the considered state.

A dual-concentration tensor, associated with the elastoplastic matrix compliances (14.7.11) and the elastoplastic crystalline compliances (14.6.22), is

$$\boldsymbol{B}_{\mathrm{c}} = (\boldsymbol{\mathcal{M}}_{\mathrm{c}}^{\mathrm{ep}} + \boldsymbol{\mathcal{M}}_{*}^{\mathrm{ep}})^{-1} : (\boldsymbol{\mathcal{M}}^{\mathrm{ep}} + \boldsymbol{\mathcal{M}}_{*}^{\mathrm{ep}}). \tag{14.7.16}$$

A dual-constraint tensor of the elastoplastic matrix is $\boldsymbol{\mathcal{M}}_{*}^{\mathrm{ep}}$, such that

$$(\boldsymbol{I} - \boldsymbol{S}^{\mathrm{ep}}) : \boldsymbol{\mathcal{M}}_{*}^{\mathrm{ep}} = \boldsymbol{S}^{\mathrm{ep}} : \boldsymbol{\mathcal{M}}^{\mathrm{ep}}. \tag{14.7.17}$$

The tensor $\boldsymbol{Q}$, introduced in Subsection 14.3.2, is in this case

$$\boldsymbol{Q} = (\boldsymbol{\mathcal{M}}_{*}^{\mathrm{ep}} + \boldsymbol{\mathcal{M}}^{\mathrm{ep}})^{-1} = \boldsymbol{\mathcal{L}}_{*}^{\mathrm{ep}} : \boldsymbol{S}^{\mathrm{ep}}. \tag{14.7.18}$$

In an expanded form, a dual-concentration tensor is

$$\begin{aligned} \boldsymbol{B}_{\mathrm{c}} = & \left(\boldsymbol{\mathcal{M}}_{\mathrm{c}}^{\mathrm{e}} + \boldsymbol{\mathcal{M}}_{*}^{\mathrm{ep}} + \sum_{\alpha=1}^{n} \sum_{\beta=1}^{n} b_{\alpha\beta}^{\mathrm{c}\,-1}\, \mathbf{G}_{\mathrm{c}}^{\alpha} \otimes \mathbf{G}_{\mathrm{c}}^{\beta} \right)^{-1} \\ & : \left(\boldsymbol{\mathcal{M}}^{\mathrm{e}} + \boldsymbol{\mathcal{M}}_{*}^{\mathrm{ep}} + \sum_{\alpha=1}^{m} \sum_{\beta=1}^{m} b_{\alpha\beta}^{-1}\, \mathbf{G}^{\alpha} \otimes \mathbf{G}^{\beta} \right). \end{aligned} \tag{14.7.19}$$

Upon the required inversion, this becomes

$$\begin{aligned} \boldsymbol{B}_{\mathrm{c}} = & \left[\boldsymbol{I} - \sum_{\alpha=1}^{n} \sum_{\beta=1}^{n} \hat{g}_{\alpha\beta}^{\mathrm{c}\,-1} (\boldsymbol{\mathcal{M}}_{\mathrm{c}}^{\mathrm{e}} + \boldsymbol{\mathcal{M}}_{*}^{\mathrm{ep}})^{-1} : (\mathbf{G}_{\mathrm{c}}^{\alpha} \otimes \mathbf{G}_{\mathrm{c}}^{\beta}) \right] \\ & : (\boldsymbol{\mathcal{M}}_{\mathrm{c}}^{\mathrm{e}} + \boldsymbol{\mathcal{M}}_{*}^{\mathrm{ep}})^{-1} : (\boldsymbol{\mathcal{M}}^{\mathrm{e}} + \boldsymbol{\mathcal{M}}_{*}^{\mathrm{ep}} - \sum_{\alpha=1}^{m} \sum_{\beta=1}^{m} b_{\alpha\beta}^{-1}\, \mathbf{G}^{\alpha} \otimes \mathbf{G}^{\beta}), \end{aligned} \tag{14.7.20}$$

where

$$\hat{g}^{\rm c}_{\alpha\beta} = g^{\rm c}_{\alpha\beta} - \mathbf{G}^{\alpha}_{\rm c} : (\boldsymbol{\mathcal{M}}^{\rm e}_{\rm c} + \boldsymbol{\mathcal{M}}^{\rm ep}_{*})^{-1} : \mathbf{G}^{\beta}_{\rm c}. \tag{14.7.21}$$

The stress rate $\overset{\triangle}{\underline{\boldsymbol{\tau}}}_{\infty}$ is such that (14.7.14) holds, as well as

$$\mathbf{G}^{\beta}_{\rm c} : (\boldsymbol{\mathcal{M}}^{\rm e}_{\rm c} + \boldsymbol{\mathcal{M}}^{\rm ep}_{*})^{-1} : (\boldsymbol{\mathcal{M}}^{\rm e} + \boldsymbol{\mathcal{M}}^{\rm ep}_{*}) : \overset{\triangle}{\underline{\boldsymbol{\tau}}}_{\infty} > 0, \quad \beta = 1, 2, \ldots, n. \tag{14.7.22}$$

14.7.1. Locally Smooth Yield Surface

When the yield surfaces of the crystal and the matrix are both locally smooth, the corresponding elastoplastic stiffnesses are

$$\boldsymbol{\mathcal{L}}^{\rm ep}_{\rm c} = \boldsymbol{\mathcal{L}}^{\rm e}_{\rm c} - \frac{1}{g_{\rm c}}\, \mathbf{C}_{\rm c} \otimes \mathbf{C}_{\rm c}, \quad \mathbf{C}_{\rm c} : \mathbf{D}_{\rm c} > 0, \tag{14.7.23}$$

where $g_{\rm c} > 0$, and

$$\boldsymbol{\mathcal{L}}^{\rm ep} = \boldsymbol{\mathcal{L}}^{\rm e} - \frac{1}{g}\, \mathbf{C} \otimes \mathbf{C}, \quad \mathbf{C} : \mathbf{D}_{\infty} > 0, \tag{14.7.24}$$

where $g > 0$. The crystalline and matrix compliances are

$$\boldsymbol{\mathcal{M}}^{\rm ep}_{\rm c} = \boldsymbol{\mathcal{M}}^{\rm e}_{\rm c} + \frac{1}{b_{\rm c}}\, \mathbf{G}_{\rm c} \otimes \mathbf{G}_{\rm c}, \quad \mathbf{G}_{\rm c} : \overset{\triangle}{\underline{\boldsymbol{\tau}}}_{\rm c} > 0, \tag{14.7.25}$$

and

$$\boldsymbol{\mathcal{M}}^{\rm ep} = \boldsymbol{\mathcal{M}}^{\rm e} + \frac{1}{b}\, \mathbf{G} \otimes \mathbf{G}, \quad \mathbf{G} : \overset{\triangle}{\underline{\boldsymbol{\tau}}}_{\infty} > 0. \tag{14.7.26}$$

The connections hold

$$\mathbf{G}_{\rm c} = \boldsymbol{\mathcal{M}}^{\rm e}_{\rm c} : \mathbf{C}_{\rm c}, \quad g_{\rm c} - b_{\rm c} = \mathbf{C}_{\rm c} : \boldsymbol{\mathcal{M}}^{\rm e}_{\rm c} : \mathbf{C}_{\rm c} = \mathbf{G}_{\rm c} : \boldsymbol{\mathcal{L}}^{\rm e}_{\rm c} : \mathbf{G}_{\rm c}, \tag{14.7.27}$$

$$\mathbf{G} = \boldsymbol{\mathcal{M}}^{\rm e} : \mathbf{C}, \quad g - b = \mathbf{C} : \boldsymbol{\mathcal{M}}^{\rm e} : \mathbf{C} = \mathbf{G} : \boldsymbol{\mathcal{L}}^{\rm e} : \mathbf{G}. \tag{14.7.28}$$

The crystal and the matrix are both assumed to be in the hardening range, so that $b_{\rm c} > 0$ and $b > 0$ in Eqs. (14.7.25) and (14.7.26).

The corresponding concentration tensor is

$$\begin{aligned} \boldsymbol{A}_{\rm c} = &\left[\boldsymbol{I} + \frac{1}{\hat{b}_{\rm c}}\, (\boldsymbol{\mathcal{L}}^{\rm e}_{\rm c} + \boldsymbol{\mathcal{L}}^{\rm ep}_{*})^{-1} : (\mathbf{C}_{\rm c} \otimes \mathbf{C}_{\rm c})\right] \\ &: (\boldsymbol{\mathcal{L}}^{\rm e}_{\rm c} + \boldsymbol{\mathcal{L}}^{\rm ep}_{*})^{-1} : \left(\boldsymbol{\mathcal{L}}^{\rm e} + \boldsymbol{\mathcal{L}}^{\rm ep}_{*} - \frac{1}{g}\, \mathbf{C} \otimes \mathbf{C}\right), \end{aligned} \tag{14.7.29}$$

where

$$\hat{b}_{\rm c} = g_{\rm c} - \mathbf{C}_{\rm c} : (\boldsymbol{\mathcal{L}}^{\rm e}_{\rm c} + \boldsymbol{\mathcal{L}}^{\rm ep}_{*})^{-1} : \mathbf{C}_{\rm c}. \tag{14.7.30}$$

A dual-concentration tensor is similarly

$$\boldsymbol{B}_{\mathrm{c}} = \left[\boldsymbol{I} - \frac{1}{\hat{g}_{\mathrm{c}}}\,(\boldsymbol{\mathcal{M}}_{\mathrm{c}}^{\mathrm{e}} + \boldsymbol{\mathcal{M}}_{*}^{\mathrm{ep}})^{-1} : (\mathbf{G}_{\mathrm{c}} \otimes \mathbf{G}_{\mathrm{c}})\right] : (\boldsymbol{\mathcal{M}}_{\mathrm{c}}^{\mathrm{e}} + \boldsymbol{\mathcal{M}}_{*}^{\mathrm{ep}})^{-1} : \left(\boldsymbol{\mathcal{M}}^{\mathrm{e}} + \boldsymbol{\mathcal{M}}_{*}^{\mathrm{ep}} + \frac{1}{b}\,\mathbf{G} \otimes \mathbf{G}\right), \tag{14.7.31}$$

with

$$\hat{g}_{\mathrm{c}} = b_{\mathrm{c}} + \mathbf{G}_{\mathrm{c}} : (\boldsymbol{\mathcal{M}}_{\mathrm{c}}^{\mathrm{e}} + \boldsymbol{\mathcal{M}}_{*}^{\mathrm{ep}})^{-1} : \mathbf{G}_{\mathrm{c}}. \tag{14.7.32}$$

It is noted that $\hat{b}_{\mathrm{c}} = \hat{g}_{\mathrm{c}}$.

If the elastic properties of the crystal and the matrix are identical ($\boldsymbol{\mathcal{L}}_{\mathrm{c}}^{\mathrm{e}} = \boldsymbol{\mathcal{L}}^{\mathrm{e}}$, $\boldsymbol{\mathcal{M}}_{\mathrm{c}}^{\mathrm{e}} = \boldsymbol{\mathcal{M}}^{\mathrm{e}}$), the concentration tensors take on the simpler forms (Hill, *op. cit.*)

$$\boldsymbol{A}_{\mathrm{c}} = \left[\boldsymbol{I} + \frac{1}{\hat{b}_{\mathrm{c}}}\,(\boldsymbol{\mathcal{L}}^{\mathrm{e}} + \boldsymbol{\mathcal{L}}_{*}^{\mathrm{ep}})^{-1} : (\mathbf{C}_{\mathrm{c}} \otimes \mathbf{C}_{\mathrm{c}})\right] : \left[\boldsymbol{I} - \frac{1}{g}\,(\boldsymbol{\mathcal{L}}^{\mathrm{e}} + \boldsymbol{\mathcal{L}}_{*}^{\mathrm{ep}})^{-1} : (\mathbf{C} \otimes \mathbf{C})\right], \tag{14.7.33}$$

$$\boldsymbol{B}_{\mathrm{c}} = \left[\boldsymbol{I} - \frac{1}{\hat{g}_{\mathrm{c}}}\,(\boldsymbol{\mathcal{M}}^{\mathrm{e}} + \boldsymbol{\mathcal{M}}_{*}^{\mathrm{ep}})^{-1} : (\mathbf{G}_{\mathrm{c}} \otimes \mathbf{G}_{\mathrm{c}})\right] : \left[\boldsymbol{I} + \frac{1}{b}\,(\boldsymbol{\mathcal{M}}^{\mathrm{e}} + \boldsymbol{\mathcal{M}}_{*}^{\mathrm{ep}})^{-1} : (\mathbf{G} \otimes \mathbf{G})\right]. \tag{14.7.34}$$

14.7.2. Rigid-Plastic Crystal in Rigid-Plastic Matrix

The corresponding crystalline and matrix compliances are in this case

$$\boldsymbol{\mathcal{M}}_{\mathrm{c}}^{\mathrm{p}} = \frac{1}{b_{\mathrm{c}}}\,\mathbf{G}_{\mathrm{c}} \otimes \mathbf{G}_{\mathrm{c}}, \quad \mathbf{G}_{\mathrm{c}} : \overset{\triangle}{\boldsymbol{\sigma}}_{\mathrm{c}} > 0, \tag{14.7.35}$$

and

$$\boldsymbol{\mathcal{M}}^{\mathrm{p}} = \frac{1}{b}\,\mathbf{G} \otimes \mathbf{G}, \quad \mathbf{G} : \overset{\triangle}{\boldsymbol{\sigma}}_{\infty} > 0. \tag{14.7.36}$$

A dual-concentration tensor is

$$\boldsymbol{B}_{\mathrm{c}} = \left[\boldsymbol{I} - \frac{1}{\hat{g}_{\mathrm{c}}}\,\boldsymbol{\mathcal{L}}_{*}^{\mathrm{p}} : (\mathbf{G}_{\mathrm{c}} \otimes \mathbf{G}_{\mathrm{c}})\right] : \left[\boldsymbol{I} + \frac{1}{b}\,\boldsymbol{\mathcal{L}}_{*}^{\mathrm{p}} : (\mathbf{G} \otimes \mathbf{G})\right], \tag{14.7.37}$$

where

$$\hat{g}_{\mathrm{c}} = b_{\mathrm{c}} + \mathbf{G}_{\mathrm{c}} : \boldsymbol{\mathcal{L}}_{*}^{\mathrm{p}} : \mathbf{G}_{\mathrm{c}}. \tag{14.7.38}$$

The constraint tensors of the rigid-plastic matrix are $\boldsymbol{\mathcal{M}}_{*}^{\mathrm{p}}$ and $\boldsymbol{\mathcal{L}}_{*}^{\mathrm{p}} = \boldsymbol{\mathcal{M}}_{*}^{\mathrm{p}\,-1}$, such that

$$(\boldsymbol{I} - \boldsymbol{S}^{\mathrm{p}}) : \boldsymbol{\mathcal{M}}_{*}^{\mathrm{p}} = \boldsymbol{S}^{\mathrm{p}} : \boldsymbol{\mathcal{M}}^{\mathrm{p}}, \tag{14.7.39}$$

where $\boldsymbol{\mathcal{S}}^{\mathrm{p}}$ is the Eshelby tensor of the rigid-plastic matrix. The condition (14.7.22) becomes, for the rigid-plastic crystal and the rigid-plastic matrix,

$$\mathbf{G}_{\mathrm{c}} : \overset{\triangle}{\boldsymbol{\sigma}}_{\infty} > 0. \tag{14.7.40}$$

It is observed that

$$\boldsymbol{\mathcal{S}}^{\mathrm{p}-1} = \boldsymbol{I} + \boldsymbol{\mathcal{M}}^{\mathrm{p}} : \boldsymbol{\mathcal{L}}_*^{\mathrm{p}}, \quad \boldsymbol{\mathcal{S}}^{\mathrm{p}-T} = \boldsymbol{I} + \boldsymbol{\mathcal{L}}_*^{\mathrm{p}} : \boldsymbol{\mathcal{M}}^{\mathrm{p}}, \tag{14.7.41}$$

so that, from Eq. (14.7.37),

$$\boldsymbol{B}_{\mathrm{c}} : \boldsymbol{\mathcal{S}}^{\mathrm{p}T} = \boldsymbol{I} - \frac{\boldsymbol{\mathcal{L}}_*^{\mathrm{p}} : (\mathbf{G}_{\mathrm{c}} \otimes \mathbf{G}_{\mathrm{c}})}{b_{\mathrm{c}} + \mathbf{G}_{\mathrm{c}} : \boldsymbol{\mathcal{L}}_*^{\mathrm{p}} : \mathbf{G}_{\mathrm{c}}}, \tag{14.7.42}$$

in analogy with (14.6.67). The tensor $\boldsymbol{Q}$ is

$$\boldsymbol{Q} = (\boldsymbol{\mathcal{M}}^{\mathrm{p}} + \boldsymbol{\mathcal{M}}_*^{\mathrm{p}})^{-1} = \boldsymbol{\mathcal{L}}_*^{\mathrm{p}} : \boldsymbol{\mathcal{S}}^{\mathrm{p}}. \tag{14.7.43}$$

On the other hand, from Eqs. (14.2.36) and (14.3.44), we can write

$$\boldsymbol{A}_{\mathrm{c}} : \boldsymbol{\mathcal{M}}^{\mathrm{p}} = \boldsymbol{\mathcal{M}}_{\mathrm{c}}^{\mathrm{p}} : \boldsymbol{B}_{\mathrm{c}}, \quad \boldsymbol{P} = \boldsymbol{\mathcal{S}}^{\mathrm{p}} : \boldsymbol{\mathcal{M}}^{\mathrm{p}} = \boldsymbol{\mathcal{M}}^{\mathrm{p}} : \boldsymbol{\mathcal{S}}^{\mathrm{p}T}, \tag{14.7.44}$$

and

$$\boldsymbol{A}_{\mathrm{c}} : \boldsymbol{P} = \boldsymbol{A}_{\mathrm{c}} : \boldsymbol{\mathcal{M}}^{\mathrm{p}} : \boldsymbol{\mathcal{S}}^{\mathrm{p}T} = \boldsymbol{\mathcal{M}}_{\mathrm{c}}^{\mathrm{p}} : \boldsymbol{B}_{\mathrm{c}} : \boldsymbol{\mathcal{S}}^{\mathrm{p}T}. \tag{14.7.45}$$

By substituting the expression (14.7.42) for $\boldsymbol{B}_{\mathrm{c}} : \boldsymbol{\mathcal{S}}^{\mathrm{p}T}$ into Eq. (14.7.45), there follows

$$\boldsymbol{A}_{\mathrm{c}} : \boldsymbol{P} = \boldsymbol{\mathcal{M}}_{\mathrm{c}}^{\mathrm{p}} : \left(\boldsymbol{I} - \frac{\boldsymbol{\mathcal{L}}_*^{\mathrm{p}} : \mathbf{G}_{\mathrm{c}} \otimes \mathbf{G}_{\mathrm{c}}}{b_{\mathrm{c}} + \mathbf{G}_{\mathrm{c}} : \boldsymbol{\mathcal{L}}_*^{\mathrm{p}} : \mathbf{G}_{\mathrm{c}}} \right). \tag{14.7.46}$$

With the help of Eq. (14.7.36), this can be reduced to

$$\boldsymbol{A}_{\mathrm{c}} : \boldsymbol{P} = \frac{\mathbf{G}_{\mathrm{c}} \otimes \mathbf{G}_{\mathrm{c}}}{b_{\mathrm{c}} + \mathbf{G}_{\mathrm{c}} : \boldsymbol{\mathcal{L}}_*^{\mathrm{p}} : \mathbf{G}_{\mathrm{c}}}. \tag{14.7.47}$$

14.8. Self-Consistent Determination of Elastoplastic Moduli

Hill's general analysis presented in Section 14.7 can be applied to determine the polycrystalline elastoplastic moduli and compliances as follows. Assume that the constitutive branch of the elastoplastic response (set of active slip systems) is known for each grain of a polycrystalline aggregate subjected to the overall macroscopically uniform rate of deformation $\mathbf{D}_{\infty}$, so that $\boldsymbol{\mathcal{L}}_{\mathrm{c}}^{\mathrm{ep}}$ is known for each orientation of the grain relative to applied $\mathbf{D}_{\infty}$. The concentration tensor for a grain with the instantaneous stiffness $\boldsymbol{\mathcal{L}}_{\mathrm{c}}^{\mathrm{ep}}$, embedded in a matrix with the overall elastoplastic moduli $\boldsymbol{\mathcal{L}}^{\mathrm{ep}}$, is

$$\boldsymbol{A}_{\mathrm{c}} = (\boldsymbol{\mathcal{L}}_{\mathrm{c}}^{\mathrm{ep}} + \boldsymbol{\mathcal{L}}_*^{\mathrm{ep}})^{-1} : (\boldsymbol{\mathcal{L}}^{\mathrm{ep}} + \boldsymbol{\mathcal{L}}_*^{\mathrm{ep}}), \tag{14.8.1}$$

provided that

$$\mathbf{C}_{\mathrm{c}}^{\beta} : (\boldsymbol{\mathcal{L}}_{\mathrm{c}}^{\mathrm{e}} + \boldsymbol{\mathcal{L}}_{*}^{\mathrm{ep}})^{-1} : (\boldsymbol{\mathcal{L}}^{\mathrm{e}} + \boldsymbol{\mathcal{L}}_{*}^{\mathrm{ep}}) : \mathbf{D}_{\infty} > 0, \quad \beta = 1, 2, \ldots, n. \tag{14.8.2}$$

The corresponding constraint tensor $\boldsymbol{\mathcal{L}}_{*}^{\mathrm{ep}}$ is related to $\boldsymbol{\mathcal{L}}^{\mathrm{ep}}$ by

$$\boldsymbol{\mathcal{L}}_{*}^{\mathrm{ep}} : \boldsymbol{S}^{\mathrm{ep}} = \boldsymbol{\mathcal{L}}^{\mathrm{ep}} : (\boldsymbol{I} - \boldsymbol{S}^{\mathrm{ep}}). \tag{14.8.3}$$

The Eshelby tensor $\boldsymbol{S}^{\mathrm{ep}}$ is associated with the elastoplastic matrix with current (anisotropic) stiffness $\boldsymbol{\mathcal{L}}^{\mathrm{ep}}$.

According to the self-consistent method, an ellipsoidal elastoplastic grain is considered to be embedded in the elastoplastic matrix with the overall properties of the polycrystalline aggregate. It is required that the orientation average of the crystalline rate of deformation $\mathbf{D}_{\mathrm{c}} = \boldsymbol{A}_{\mathrm{c}} : \mathbf{D}_{\infty}$ is equal to applied $\mathbf{D}_{\infty}$. Thus,

$$\{\mathbf{D}_{\mathrm{c}}\} = \mathbf{D}_{\infty} \Rightarrow \{\boldsymbol{A}_{\mathrm{c}}\} = \boldsymbol{I}. \tag{14.8.4}$$

The brackets { } designate the appropriate orientation average. Furthermore, since $\boldsymbol{\mathcal{L}}^{\mathrm{ep}}$ is the overall instantaneous stiffness of the polycrystalline aggregate, we can write

$$\{\overset{\triangle}{\underline{\boldsymbol{\tau}}}_{\mathrm{c}}\} = \boldsymbol{\mathcal{L}}^{\mathrm{ep}} : \{\mathbf{D}_{\mathrm{c}}\} = \boldsymbol{\mathcal{L}}^{\mathrm{ep}} : \mathbf{D}_{\infty}. \tag{14.8.5}$$

Comparing this with

$$\{\overset{\triangle}{\underline{\boldsymbol{\tau}}}_{\mathrm{c}}\} = \{\boldsymbol{\mathcal{L}}_{\mathrm{c}}^{\mathrm{ep}} : \mathbf{D}_{\mathrm{c}}\} = \{\boldsymbol{\mathcal{L}}_{\mathrm{c}}^{\mathrm{ep}} : \boldsymbol{A}_{\mathrm{c}}\} : \mathbf{D}_{\infty}, \tag{14.8.6}$$

establishes

$$\boldsymbol{\mathcal{L}}^{\mathrm{ep}} = \{\boldsymbol{\mathcal{L}}_{\mathrm{c}}^{\mathrm{ep}} : \boldsymbol{A}_{\mathrm{c}}\}. \tag{14.8.7}$$

The substitution of Eq. (14.8.1), therefore, gives

$$\boldsymbol{\mathcal{L}}^{\mathrm{ep}} = \{\boldsymbol{\mathcal{L}}_{\mathrm{c}}^{\mathrm{ep}} : (\boldsymbol{\mathcal{L}}_{\mathrm{c}}^{\mathrm{ep}} + \boldsymbol{\mathcal{L}}_{*}^{\mathrm{ep}})^{-1} : (\boldsymbol{\mathcal{L}}^{\mathrm{ep}} + \boldsymbol{\mathcal{L}}_{*}^{\mathrm{ep}})\}. \tag{14.8.8}$$

This is a highly implicit equation for the polycrystalline moduli $\boldsymbol{\mathcal{L}}^{\mathrm{ep}}$. It involves the constraint tensor $\boldsymbol{\mathcal{L}}_{*}^{\mathrm{ep}}$, which itself depends on the polycrystalline moduli $\boldsymbol{\mathcal{L}}^{\mathrm{ep}}$, as seen from Eq. (14.8.3). Moreover, it is not known in advance which branch of $\boldsymbol{\mathcal{L}}^{\mathrm{ep}}$ and $\boldsymbol{\mathcal{L}}_{\mathrm{c}}^{\mathrm{ep}}$ is activated by a prescribed $\mathbf{D}_{\infty}$. The calculation requires an iterative procedure. It was originally devised by Hutchinson (1970). For a prescribed $\mathbf{D}_{\infty}$, a tentative guess is made for $\boldsymbol{\mathcal{L}}^{\mathrm{ep}}$, and $\boldsymbol{\mathcal{L}}_{*}^{\mathrm{ep}}$ is calculated from Eq. (14.8.3). The elastoplastic branch of the crystalline response (the set of active slip systems) is then assumed, the corresponding $\boldsymbol{\mathcal{L}}_{\mathrm{c}}^{\mathrm{ep}}$ calculated from (14.6.1), and the constraint tensor $\boldsymbol{A}_{\mathrm{c}}$ from (14.8.2). To ensure that the assumed set of active slip systems is indeed active, the condition (14.8.2) is verified. If it is not satisfied, a new set of active slip systems is selected until the correct $\boldsymbol{\mathcal{L}}_{\mathrm{c}}^{\mathrm{ep}}$ is found. This calculation is carried out for all grains and orientations. The results are substituted into (14.8.8) to find a new estimate for $\boldsymbol{\mathcal{L}}^{\mathrm{ep}}$. The whole procedure is repeated until a satisfactory convergence is obtained.

The calculation can also proceed by using the tensors of the instantaneous compliances $\boldsymbol{\mathcal{M}}_{\mathrm{c}}^{\mathrm{ep}}$ and $\boldsymbol{\mathcal{M}}^{\mathrm{ep}}$ (assuming that they exist). In this case we have

$$\boldsymbol{B}_{\mathrm{c}} = (\boldsymbol{\mathcal{M}}_{\mathrm{c}}^{\mathrm{ep}} + \boldsymbol{\mathcal{M}}_{*}^{\mathrm{ep}})^{-1} : (\boldsymbol{\mathcal{M}}^{\mathrm{ep}} + \boldsymbol{\mathcal{M}}_{*}^{\mathrm{ep}}), \tag{14.8.9}$$

provided that

$$\mathbf{G}_{\mathrm{c}}^{\beta} : (\boldsymbol{\mathcal{M}}_{\mathrm{c}}^{\mathrm{e}} + \boldsymbol{\mathcal{M}}_{*}^{\mathrm{ep}})^{-1} : (\boldsymbol{\mathcal{M}}^{\mathrm{e}} + \boldsymbol{\mathcal{M}}_{*}^{\mathrm{ep}}) : \overset{\triangle}{\underline{\boldsymbol{\tau}}}_{\infty} > 0, \quad \beta = 1, 2, \ldots, n. \tag{14.8.10}$$

The corresponding constraint tensor $\boldsymbol{\mathcal{M}}_{*}^{\mathrm{ep}}$ is related to $\boldsymbol{\mathcal{M}}^{\mathrm{ep}}$ via the Eshelby tensor $\boldsymbol{S}^{\mathrm{ep}}$ according to

$$(\boldsymbol{I} - \boldsymbol{S}^{\mathrm{ep}}) : \boldsymbol{\mathcal{M}}_{*}^{\mathrm{ep}} = \boldsymbol{S}^{\mathrm{ep}} : \boldsymbol{\mathcal{M}}^{\mathrm{ep}}. \tag{14.8.11}$$

The implicit equation for $\boldsymbol{\mathcal{M}}^{\mathrm{ep}}$ is thus

$$\boldsymbol{\mathcal{M}}^{\mathrm{ep}} = \{\boldsymbol{\mathcal{M}}_{\mathrm{c}}^{\mathrm{ep}} : \boldsymbol{B}_{\mathrm{c}}\}, \tag{14.8.12}$$

i.e.,

$$\boldsymbol{\mathcal{M}}^{\mathrm{ep}} = \{\boldsymbol{\mathcal{M}}_{\mathrm{c}}^{\mathrm{ep}} : (\boldsymbol{\mathcal{M}}_{\mathrm{c}}^{\mathrm{ep}} + \boldsymbol{\mathcal{M}}_{*}^{\mathrm{ep}})^{-1} : (\boldsymbol{\mathcal{M}}^{\mathrm{ep}} + \boldsymbol{\mathcal{M}}_{*}^{\mathrm{ep}})\}. \tag{14.8.13}$$

The calculation again requires an iterative procedure.

The elastic unloading branch can be determined more readily. It is associated with the pyramidal region defined by the inequalities

$$\mathbf{G}_{\mathrm{c}}^{\beta} : (\boldsymbol{\mathcal{M}}_{\mathrm{c}}^{\mathrm{e}} + \boldsymbol{\mathcal{M}}_{*}^{\mathrm{e}})^{-1} : (\boldsymbol{\mathcal{M}}^{\mathrm{e}} + \boldsymbol{\mathcal{M}}_{*}^{\mathrm{e}}) : \overset{\triangle}{\underline{\boldsymbol{\tau}}}_{\infty} < 0, \quad \beta = 1, 2, \ldots, n_0, \tag{14.8.14}$$

for all crystalline orientations. This can be rewritten as

$$\mathbf{G}_{\mathrm{c}}^{\beta} : (\boldsymbol{I} + \boldsymbol{\mathcal{L}}_{*}^{\mathrm{e}} : \boldsymbol{\mathcal{M}}_{\mathrm{c}}^{\mathrm{e}})^{-1} : (\boldsymbol{I} + \boldsymbol{\mathcal{L}}_{*}^{\mathrm{e}} : \boldsymbol{\mathcal{M}}^{\mathrm{e}}) : \overset{\triangle}{\underline{\boldsymbol{\tau}}}_{\infty} < 0, \quad \beta = 1, 2, \ldots, n_0. \tag{14.8.15}$$

The constraint tensor $\boldsymbol{\mathcal{M}}_{*}^{\mathrm{e}}$ is related to $\boldsymbol{\mathcal{M}}^{\mathrm{e}}$ by

$$(\boldsymbol{I} - \boldsymbol{S}^{\mathrm{e}}) : \boldsymbol{\mathcal{M}}_{*}^{\mathrm{e}} = \boldsymbol{S}^{\mathrm{e}} : \boldsymbol{\mathcal{M}}^{\mathrm{e}}. \tag{14.8.16}$$

The aggregate yield vertex is more or less pronounced depending on whether the directions

$$\mathbf{G}_{\mathrm{c}}^{\beta} : (\boldsymbol{I} + \boldsymbol{\mathcal{L}}_{*}^{\mathrm{e}} : \boldsymbol{\mathcal{M}}_{\mathrm{c}}^{\mathrm{e}})^{-1} = (\boldsymbol{I} + \boldsymbol{\mathcal{M}}_{\mathrm{c}}^{\mathrm{e}} : \boldsymbol{\mathcal{L}}_{*}^{\mathrm{e}})^{-1} : \mathbf{G}_{\mathrm{c}}^{\beta} \tag{14.8.17}$$

span large or small solid angle (Hill, 1965a). The overall elastic polycrystalline compliances are determined from

$$\boldsymbol{\mathcal{M}}^{\mathrm{e}} = \{\boldsymbol{\mathcal{M}}_{\mathrm{c}}^{\mathrm{e}} : (\boldsymbol{\mathcal{M}}_{\mathrm{c}}^{\mathrm{e}} + \boldsymbol{\mathcal{M}}_{*}^{\mathrm{e}})^{-1} : (\boldsymbol{\mathcal{M}}^{\mathrm{e}} + \boldsymbol{\mathcal{M}}_{*}^{\mathrm{e}})\}. \tag{14.8.18}$$

14.8.1. Kröner–Budiansky–Wu Method

In the original formulation of the self-consistent model of polycrystalline plasticity, Kröner (1961), and Budiansky and Wu (1962), in effect, suggested that the constraint tensor of the elastic matrix relates the differences between the local and overall stress and strain rates, even in the plastic range. Thus, it is assumed that

$$\overset{\triangle}{\underline{\tau}}_{\mathrm{c}} - \overset{\triangle}{\underline{\tau}}_{\infty} = -\boldsymbol{\mathcal{L}}_{*}^{\mathrm{e}} : (\mathbf{D}_{\mathrm{c}} - \mathbf{D}_{\infty}), \tag{14.8.19}$$

where

$$\overset{\triangle}{\underline{\tau}}_{\mathrm{c}} = \boldsymbol{\mathcal{L}}_{\mathrm{c}}^{\mathrm{ep}} : \mathbf{D}_{\mathrm{c}}, \quad \overset{\triangle}{\underline{\tau}}_{\infty} = \boldsymbol{\mathcal{L}}^{\mathrm{ep}} : \mathbf{D}_{\infty}, \tag{14.8.20}$$

and

$$\boldsymbol{\mathcal{L}}_{*}^{\mathrm{e}} : \boldsymbol{S}_{\mathrm{e}} = \boldsymbol{\mathcal{L}}^{\mathrm{e}} : (\boldsymbol{I} - \boldsymbol{S}^{\mathrm{e}}). \tag{14.8.21}$$

The tensor $\boldsymbol{\mathcal{L}}^{\mathrm{e}}$ is the overall elastic moduli tensor of the elastoplastic aggregate, and $\boldsymbol{S}^{\mathrm{e}}$ is the Eshelby tensor corresponding to $\boldsymbol{\mathcal{L}}^{\mathrm{e}}$. This leads to concentration tensors

$$\boldsymbol{A}_{\mathrm{c}} = (\boldsymbol{\mathcal{L}}_{\mathrm{c}}^{\mathrm{ep}} + \boldsymbol{\mathcal{L}}_{*}^{\mathrm{e}})^{-1} : (\boldsymbol{\mathcal{L}}^{\mathrm{ep}} + \boldsymbol{\mathcal{L}}_{*}^{\mathrm{e}}), \tag{14.8.22}$$

$$\boldsymbol{B}_{\mathrm{c}} = (\boldsymbol{\mathcal{M}}_{\mathrm{c}}^{\mathrm{ep}} + \boldsymbol{\mathcal{M}}_{*}^{\mathrm{e}})^{-1} : (\boldsymbol{\mathcal{M}}^{\mathrm{ep}} + \boldsymbol{\mathcal{M}}_{*}^{\mathrm{e}}). \tag{14.8.23}$$

The implicit equations for $\boldsymbol{\mathcal{L}}^{\mathrm{ep}}$ and $\boldsymbol{\mathcal{M}}^{\mathrm{ep}}$ are, thus,

$$\boldsymbol{\mathcal{L}}^{\mathrm{ep}} = \{\boldsymbol{\mathcal{L}}_{\mathrm{c}}^{\mathrm{ep}} : (\boldsymbol{\mathcal{L}}_{\mathrm{c}}^{\mathrm{ep}} + \boldsymbol{\mathcal{L}}_{*}^{\mathrm{e}})^{-1} : (\boldsymbol{\mathcal{L}}^{\mathrm{ep}} + \boldsymbol{\mathcal{L}}_{*}^{\mathrm{e}})\}, \tag{14.8.24}$$

$$\boldsymbol{\mathcal{M}}^{\mathrm{ep}} = \{\boldsymbol{\mathcal{M}}_{\mathrm{c}}^{\mathrm{ep}} : (\boldsymbol{\mathcal{M}}_{\mathrm{c}}^{\mathrm{ep}} + \boldsymbol{\mathcal{M}}_{*}^{\mathrm{e}})^{-1} : (\boldsymbol{\mathcal{M}}^{\mathrm{ep}} + \boldsymbol{\mathcal{M}}_{*}^{\mathrm{e}})\}. \tag{14.8.25}$$

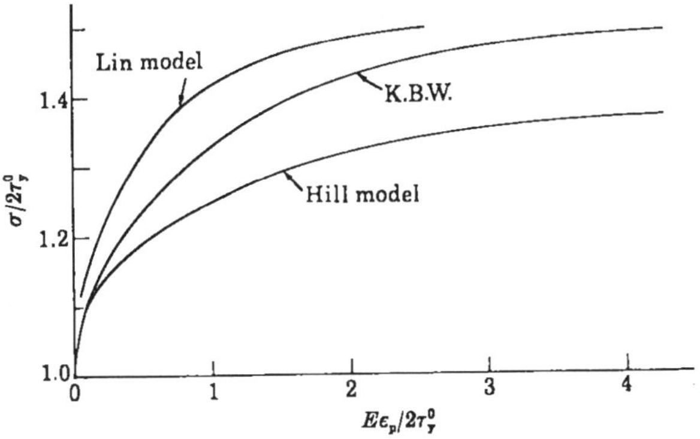

FIGURE 14.7. Polycrystalline stress-plastic strain curves for isotropic aggregate of isotropic ideally-plastic crystals (from Hutchinson, 1970; with permission from The Royal Society and the author).

14.8.2. Hutchinson's Calculations

Hutchinson's (1970) calculations of tensile stress-strain curves for polycrystals of spherical f.c.c. grains, with randomly oriented crystalline lattice, reveal that in the early stages of plastic deformation predictions based on Hill's and K.B.W. models are essentially identical, since $\mathcal{L}_*^{ep}$ is then approximately equal to $\mathcal{L}_*^{e}$. However, with progression of plastic deformation, the components of $\mathcal{L}^{ep}$ decrease, and so do the components of $\mathcal{L}_*^{ep}$, while the components of $\mathcal{L}_*^{e}$ remain constant. Consequently, the matrix constraint surrounding each grain is considerably weakened in Hill's model, and the stress required to produce a given amount of strain is lower in Hill's than in K.B.W. model (Fig. 14.7).

Hutchinson also calculated the initial and subsequent polycrystalline yield surfaces for the tensile deformation of an aggregate of isotropic non-hardening single crystals. The polycrystalline yield surface develops a corner after only a very small amount of plastic deformation. Figure 14.8 shows the traces of the yield surface on the two indicated planes in stress space. Since microscopic Bauschinger effect was not incorporated into calculations, the macroscopic Bauschinger type effect apparent in Fig. 14.8 is entirely due to grain interaction effects. The inclusion of crystal hardening will affect the yield surface evolution. The stronger (latent) hardening on inactive

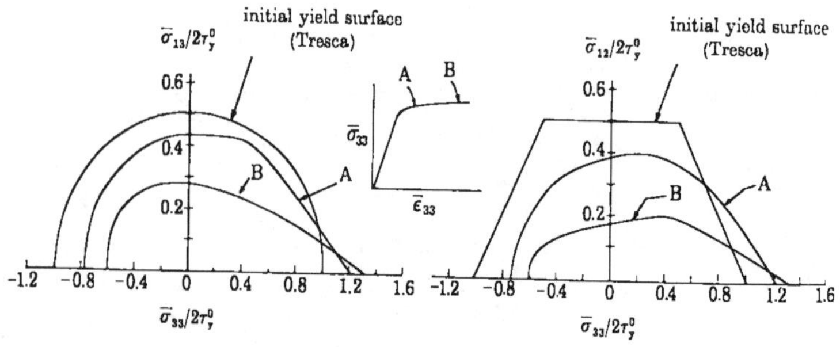

FIGURE 14.8. Evolution of the yield surface during tensile loading of an f.c.c. polycrystal comprised of isotropic ideally-plastic crystals (from Hutchinson, 1970; with permission from The Royal Society and the author).

than on active slip systems will cause the yield surface to contract less in the directions in stress space that are normal to the direction of the loading. The incorporation of the microscopic crystalline Bauschinger effect will cause the yield surface to contract more in the direction opposite to the loading direction.

The self-consistent calculations of the evolution of the yield surface were also performed by Iwakuma and Nemat-Nasser (1984), Berveiller and Zaoui (1986), Beradai, Berveiller, and Lipinski (1987). The studies of the rate-dependent polycrystalline response by the self-consistent method were done by Brown (1970), Hutchinson (1976), Weng (1981, 1982), Lin (1984), Nemat-Nasser and Obata (1986), Molinari, Canova, and Ahzi (1987), Harren (1989), Toth and Molinari (1994), Molinari (1997), Molinari, Ahzi, and Koddane (1997), Masson and Zaoui (1999), and others.

14.8.3. Berveiller and Zaoui Accommodation Function

The elastic moduli of the crystal and the aggregate are assumed to be identical in the K.B.W. model, both being given by the isotropic stiffness tensor $\boldsymbol{\mathcal{L}}^{\mathrm{e}} = 2\mu\,\boldsymbol{J} + 3\kappa\,\boldsymbol{K}$. Thus, in the case of spherical grain, the constraint tensor is

$$\boldsymbol{\mathcal{L}}^{\mathrm{e}}_{*} = 2\mu\left[\left(\frac{1}{\alpha} - 1\right)\boldsymbol{J} + 2\boldsymbol{K}\right], \tag{14.8.26}$$

and Eq. (14.8.19) becomes

$$\overset{\triangle}{\underline{\boldsymbol{\mathcal{T}}}}_{\mathrm{c}} - \overset{\triangle}{\underline{\boldsymbol{\mathcal{T}}}}_{\infty} = -2\mu\left(\frac{1}{\alpha} - 1\right)(\mathbf{D}_{\mathrm{c}} - \mathbf{D}_{\infty}), \quad \alpha = \frac{6(\kappa + 2\mu)}{5(3\kappa + 4\mu)}. \tag{14.8.27}$$

In an attempt to better represent the grain interaction and the matrix constraint, Berveiller and Zaoui (1979) suggested that the constraint tensor $\boldsymbol{\mathcal{L}}_*^{\mathrm{e}}$ in the K.B.W. model should be replaced by the constrained tensor corresponding to the elastoplastic stiffness of the polycrystal, which is approximated by an isotropic fourth-order tensor

$$\boldsymbol{\mathcal{L}}^{\mathrm{ep}} = 2\mu_{\mathrm{t}}\,\boldsymbol{J} + 3\kappa\,\boldsymbol{K}, \tag{14.8.28}$$

where μ_{t} is the tangent shear modulus of the polycrystal at the considered instant of elastoplastic deformation. For isochoric plastic deformation, $\kappa_{\mathrm{t}} = \kappa$. Thus, Eq. (14.8.26) is replaced with

$$\boldsymbol{\mathcal{L}}_*^{\mathrm{ep}} = 2\mu_{\mathrm{t}}\left[\left(\frac{1}{\alpha_{\mathrm{t}}} - 1\right)\boldsymbol{J} + 2\boldsymbol{K}\right], \quad \alpha_{\mathrm{t}} = \frac{6(\kappa + 2\mu_{\mathrm{t}})}{5(3\kappa + 4\mu_{\mathrm{t}})}, \tag{14.8.29}$$

and Eq. (14.8.27) with

$$\overset{\triangle}{\underline{\boldsymbol{\mathcal{T}}}}_{\mathrm{c}} - \overset{\triangle}{\underline{\boldsymbol{\mathcal{T}}}}_{\infty} = -2\mu_{\mathrm{t}}\left(\frac{1}{\alpha_{\mathrm{t}}} - 1\right)(\mathbf{D}_{\mathrm{c}} - \mathbf{D}_{\infty}). \tag{14.8.30}$$

If the elastoplastic partitions

$$\mathbf{D}_{\mathrm{c}} = \mathbf{D}_{\mathrm{c}}^{\mathrm{p}} + \frac{1}{2\mu}\overset{\triangle}{\underline{\boldsymbol{\mathcal{T}}}}_{\mathrm{c}}, \quad \mathbf{D}_{\infty} = \mathbf{D}_{\infty}^{\mathrm{p}} + \frac{1}{2\mu}\overset{\triangle}{\underline{\boldsymbol{\mathcal{T}}}}_{\infty} \tag{14.8.31}$$

are substituted into Eq. (14.8.30), there follows

$$\overset{\triangle}{\underline{\boldsymbol{\mathcal{T}}}}_{\mathrm{c}} - \overset{\triangle}{\underline{\boldsymbol{\mathcal{T}}}}_{\infty} = -2\varphi\mu\,(1-\alpha)(\mathbf{D}_{\mathrm{c}}^{\mathrm{p}} - \mathbf{D}_{\infty}^{\mathrm{p}}). \tag{14.8.32}$$

The parameter

$$\varphi = \frac{1-\alpha_{\mathrm{t}}}{1-\alpha}\,\frac{\mu_{\mathrm{t}}}{\alpha_{\mathrm{t}}\mu + (1-\alpha_{\mathrm{t}})\mu_{\mathrm{t}}} \tag{14.8.33}$$

is the so-called plastic accommodation function. The predicted stress strain curve falls between Hill's and K.B.W. curve in Fig. 14.7. When $\mu_{\mathrm{t}} = \mu$, it follows that $\alpha_{\mathrm{t}} = \alpha$ and $\varphi = 1$, so that Eq. (14.8.33) reduces to the original expression of the K.B.W. method.

14.8.4. Lin's Model

In an extension of Taylor's rigid-plastic model, Lin (1957) assumed that all grains in a polycrystalline aggregate deform equally ($\mathbf{D}_{\mathrm{c}} = \mathbf{D}_{\infty}$), even when elastic strains are not negligible. Thus, the concentration tensor is in this case $\boldsymbol{A}_{\mathrm{c}} = \boldsymbol{I}$, and Eq. (14.8.7) becomes

$$\mathcal{L}^{\mathrm{ep}} = \{\mathcal{L}^{\mathrm{ep}}_{\mathrm{c}}\}. \tag{14.8.34}$$

The prediction of the tensile stress-plastic strain curve from Lin's model is shown in Fig. 14.7. See also Hutchinson (1964a, b), Lin and Ito (1965, 1966), and Lin (1971). If the stresses in all grains are assumed to be equal, the tensor of the macroscopic aggregate compliances is

$$\mathcal{M}^{\mathrm{ep}} = \{\mathcal{M}^{\mathrm{ep}}_{\mathrm{c}}\}. \tag{14.8.35}$$

14.8.5. Rigid-Plastic Moduli

The rigid-plastic polycrystalline aggregates can be treated by considering the rigid-plastic crystals embedded in a rigid-plastic matrix. Suppose that all crystals deform by single slip, of different orientations in different grains. By averaging Eq. (14.7.47) we obtain an implicit equation for the compliances $\mathcal{M}^{\mathrm{p}}$,

$$\boldsymbol{P} = \{\eta \frac{\mathbf{G}_{\mathrm{c}} \otimes \mathbf{G}_{\mathrm{c}}}{b_{\mathrm{c}} + \mathbf{G}_{\mathrm{c}} : \mathcal{L}^{\mathrm{p}}_{*} : \mathbf{G}_{\mathrm{c}}}\}. \tag{14.8.36}$$

This was derived from $\{\boldsymbol{A}_{\mathrm{c}}\} = \boldsymbol{I}$, and the fact that $\boldsymbol{P} = \boldsymbol{\mathcal{S}}^{\mathrm{p}} : \mathcal{M}^{\mathrm{p}}$ is independent of the orientation of the crystalline lattice. The parameter η is equal to 1 or 0, depending on whether $\mathbf{G}_{\mathrm{c}} : \overset{\triangle}{\boldsymbol{\sigma}}_{\infty}$ is positive or negative.

If the slip mode $\mathbf{G}_{\mathrm{c}}$ is the same for all grains, then, for compatibility, the rate of deformation is necessarily uniform throughout the aggregate, so that

$$\mathbf{D}_{\infty} = \mathbf{D}_{\mathrm{c}}, \quad \mathbf{G} = \mathbf{G}_{\mathrm{c}}. \tag{14.8.37}$$

Recalling that for the rigid-plastic response

$$b_{\mathrm{c}}\mathbf{D}_{\mathrm{c}} = (\mathbf{G}_{\mathrm{c}} \otimes \mathbf{G}_{\mathrm{c}}) : \overset{\triangle}{\boldsymbol{\sigma}}_{\mathrm{c}}, \quad b\,\mathbf{D}_{\infty} = (\mathbf{G} \otimes \mathbf{G})\overset{\triangle}{\boldsymbol{\sigma}}_{\infty}, \tag{14.8.38}$$

and since $\{\overset{\triangle}{\boldsymbol{\sigma}}_{\mathrm{c}}\} = \overset{\triangle}{\boldsymbol{\sigma}}_{\infty}$, the averaging of Eq. (14.8.38) gives

$$b = \{b_{\mathrm{c}}\}. \tag{14.8.39}$$

Thus, in this particular case, the polycrystalline hardening rate is the average of the hardening rates in the individual crystals (Hill, 1965a).

14.9. Development of Crystallographic Texture

The formation of crystallographic texture is an important cause of anisotropy in polycrystalline materials. The texture has effects on macroscopic yield surface, the strain hardening characteristics (textural strengthening or softening effects), and may significantly affect the onset and the development of the localized modes of deformation. Some basic aspects of the texture analysis are discussed in this section. We restrict the consideration to crystallographic texture, although the development of morphological texture, due to the shape changes of the crystalline grains, may also be an important cause of the overall polycrystalline anisotropy at large strains.

In his treatment of axisymmetric tension of f.c.c. polycrystals, Taylor (1938a) observed that the crystallographic orientations of the grains in an initially isotropic aggregate tend toward the orientations with either (111) or (100) direction parallel to the direction of extension. His analysis was based on the rigid-plastic model considered in Section 14.1. The material spin tensor $\mathbf{W}_{\mathrm{c}}$ in each grain is caused by the lattice spin $\mathbf{W}_{\mathrm{c}}^{*}$ and by the slip induced spin, such that

$$\mathbf{W}_{\mathrm{c}} = \mathbf{W}_{\mathrm{c}}^{*} + \sum_{\alpha=1}^{12} \mathbf{Q}_{\mathrm{c}}^{\alpha}\, \dot{\gamma}^{\alpha}. \tag{14.9.1}$$

The components of the slip induced spin,

$$\boldsymbol{\Omega}_{\mathrm{c}} = \sum_{\alpha=1}^{12} \mathbf{Q}_{\mathrm{c}}^{\alpha}\, \dot{\gamma}^{\alpha} = \sum_{\alpha=1}^{12} \frac{1}{2} \left(\mathbf{s}^{\alpha} \otimes \mathbf{m}^{\alpha} - \mathbf{m}^{\alpha} \otimes \mathbf{s}^{\alpha}\right) \dot{\gamma}^{\alpha}, \tag{14.9.2}$$

expressed on the cubic axes, are

$$2\sqrt{6}\,\Omega_{12}^{\mathrm{c}} = a_1 + a_2 - 2a_3 + b_1 + b_2 - 2b_3 - c_1 - c_2 + 2c_3 - d_1 - d_2 + 2d_3, \tag{14.9.3}$$

$$2\sqrt{6}\,\Omega_{23}^{\mathrm{c}} = -2a_1 + a_2 + a_3 + 2b_1 - b_2 - b_3 - 2c_1 + c_2 + c_3 + 2d_1 - d_2 - d_3, \tag{14.9.4}$$

$$2\sqrt{6}\,\Omega_{31}^{\mathrm{c}} = a_1 - 2a_2 + a_3 - b_1 + 2b_2 - b_3 - c_1 + 2c_2 - c_3 + d_1 - 2d_2 + d_3. \tag{14.9.5}$$

The slip rates in the respective positive slip directions (see Table 14.1) are designated by a_i, b_i, c_i, d_i $(i = 1, 2, 3)$.

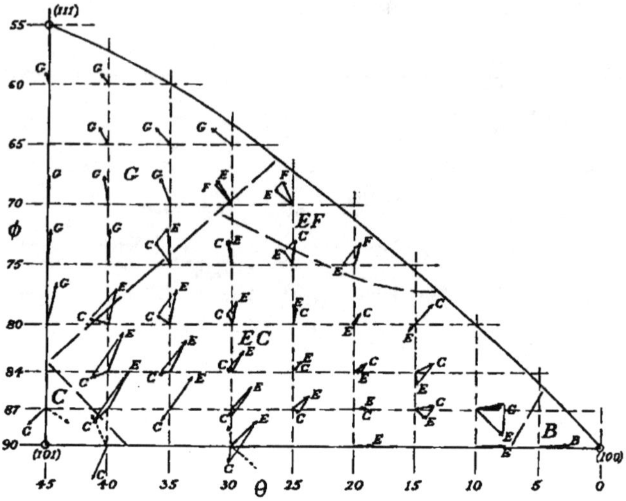

FIGURE 14.9. Taylor's prediction of the rotation of the specimen axis relative to the lattice axes of differently oriented grains in a polycrystalline aggregate at an extension of 2.37% (from Taylor, 1938b; with permission from the Institute for Materials).

According to Taylor's isostrain assumption, all grains are equally deformed, so that

$$\mathbf{D}_{\mathrm{c}} = \mathbf{D}_{\infty}, \quad \mathbf{W}_{\mathrm{c}} = \mathbf{W}_{\infty} = \mathbf{0}. \tag{14.9.6}$$

For a prescribed $\mathbf{D}_{\infty}$, a set of five independent slip rates can be found in each grain that is geometrically equivalent to this strain, and meets Taylor's minimum shear principle ($\min \sum_{\alpha} |\mathrm{d}\gamma^{\alpha}|$). The corresponding lattice spin in the grain is then

$$\mathbf{W}_{\mathrm{c}}^{*} = -\sum_{\alpha=1}^{5} \mathbf{Q}_{\mathrm{c}}^{\alpha}\,\dot{\gamma}^{\alpha}. \tag{14.9.7}$$

Since more than one set of five slip rates can be geometrically admissible and meet the minimum shear principle, the lattice spin $\mathbf{W}_{\mathrm{c}}^{*}$ is not necessarily uniquely determined in this model. Taylor plotted incremental rotation of the specimen axis relative to the lattice axes for selected 44 initial grain orientations in a polycrystalline bar extended 2.37%. The directions and relative magnitudes of the rotations are shown in Fig. 14.9. The angles ϕ and θ are defined in Fig. 14.10. Although the calculations were confined

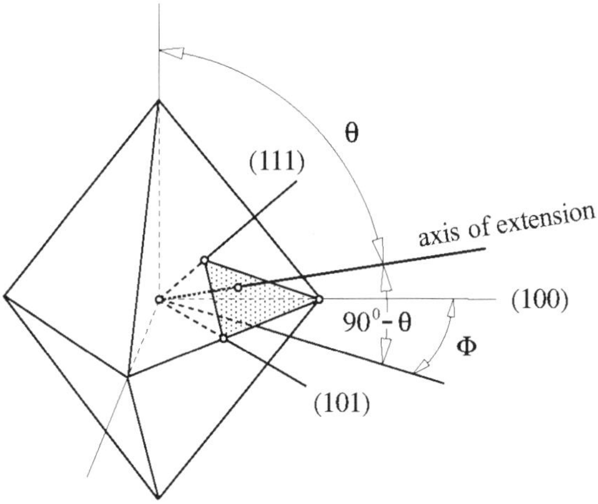

FIGURE 14.10. Definition of the angles ϕ and θ used in Fig. 14.9. The angles specify the orientation of the axis of specimen extension relative to local crystalline axes.

to a neighborhood of the initial yield, the initial trends of lattice rotations indicate a tendency toward a (111) – (100) texture development, as experimentally observed in stretched f.c.c. polycrystalline specimens. Since two different sets of five slips were geometrically equivalent and met the minimum shear principle for many of the initial grain orientations, two arrows emanate from the points corresponding to such orientations. For example, in the region EC either the set of five slips designated by E or C can occur. The angle between the two arrows then indicates the range of possible rotations of the specimen axis relative to the crystal axes.

Taylor's analysis motivated further experimental and theoretical studies of the texture in metal polycrystals. Bishop (1954) found an even more pronounced nonuniqueness of initial lattice rotations in the uniaxial compression of f.c.c. polycrystals. Chin and Mammel (1967) performed calculations for axisymmetric deformation of b.c.c. polycrystalline specimens. A significant amount of research was done to extend Taylor's analysis to large deformations. An early incremental application of Taylor's model to predict the evolving texture was presented by Kallend and Davies (1972), in the case of the plane strain idealization of cold rolling. Dillamore, Roberts and Bush (1979) examined the texture evolution in heavily rolled cubic metals in which shear bands become a dominant deformation mode. A method of the relaxed constraints was proposed by Honneff and Mecking (1978), and further developed by Canova, Kocks, and Jonas (1984), which includes the effects of the grain morphology and the changes in grain shape at large deformation. See also Van Houtte (1991). The texture evolution in plane strain compression and simple shear in the f.c.c. and h.c.p. aggregates was studied by finite elements and orientation distribution schemes by Prantil, Jenkins, and Dawson (1994), and Dawson and Kumar (1997). The calculations based on the self-consistent model were performed by Berveiller and Zaoui (1979, 1986), Molinari, Canova, and Ahzi (1987), Lipinski, Naddari, and Berveiller (1992), and Toth and Molinari (1994). The book by Yang and Lee (1993), and the reviews by Zaoui (1987) and Molinari (1997) can be consulted for additional references. Other aspects of the texture development are discussed in Gottstein and Lücke (1978), Bunge (1982, 1988), and Bunge and Nielsen (1997).

A large amount of research was devoted to deal with the nonuniqueness of lattice rotations due to the nonuniqueness of slip rates, and the resulting consequences on the texture predictions. Chin (1969) proposed that the operative set of slip rates is one with the maximum amount of the cross slip. Bunge (1970) used the average slips of all sets of admissible slip systems having the same minimum plastic work. Gil-Sevillano, Van Houtte, and Aernoudt (1975) selected the average of all admissible rotations in their calculations of texture, or randomly chose a set of slip rates from all admissible sets (Gil-Sevillano, Van Houtte, and Aernoudt, 1980). Lin and Havner (1994) adopted a minimum plastic spin postulate, introduced by Fuh and Havner (1989), according to which the operative set of slip rates minimizes the magnitude of the spin vector, associated with the components (14.9.3)–(14.9.5). The latter work provides a comprehensive analysis of the texture formation and the evolution of the macroscopic yield surface for f.c.c. polycrystalline metals in axisymmetric tension and compression, up to large strains. Taylor's model was incrementally used. In addition to Taylor's isotropic hardening, three other hardening rules were incorporated, accounting for the latent hardening on slip systems. The texture evolution in tension up to logarithmic strain $e_L = 1.61$ is depicted in Fig. 14.12. The

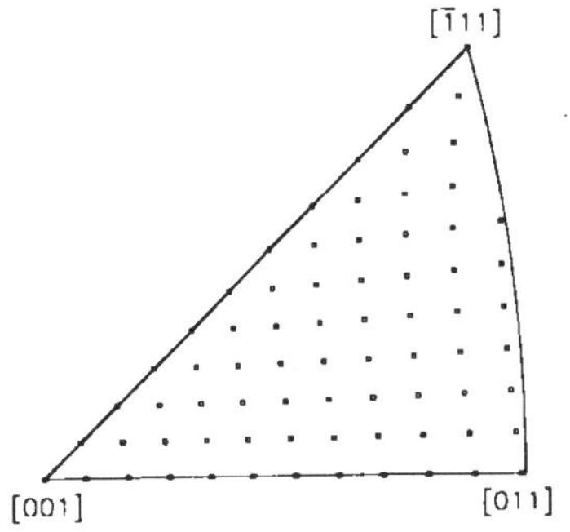

FIGURE 14.11. The initial distribution of the loading axis on the inverse pole figure, which is a [001] stereographic projection of the triangle [001][011][$\bar{1}$11] (from Lin and Havner, 1994; with permission from Elsevier Science).

distribution of the initial grain orientation is shown in Fig. 14.11. A comparative study of the hardening theories in torsion is given by Lin and Havner (1996). See also Wu, Neale, and Van der Giessen (1996).

Another approach used to resolve the nonuniqueness of lattice rotations, is to adopt a rate-dependent model of the crystallographic slip, in which the nonuniqueness of slip rates is eliminated altogether. This makes the lattice rotations and texture predictions unique. Using such an approach, Asaro and Needleman (1985) determined the texture evolution for the uniaxial and plane strain tensile and compressive loadings. Taylor's isostrain assumption, with the included elastic component of strain, was used in the large strain formulation of the model. Harren, Lowe, Asaro, and Needleman (1989) gave a comprehensive analysis of the shearing texture, with the stereographic pole and inverse pole figures corresponding to textures at various levels of finite shear strain. Anand and Kothari (1996) devised an iterative numerical procedure and a recipe based on the singular value decomposition to determine the unique set of active slip systems and slip increments in a rate-independent theory. The calculated stress-strain curves and the evolution of the crystallographic texture in simple compression were essentially indistinguishable from the corresponding calculations for a rate-dependent model (with a low value of the rate-sensitivity parameter), previously reported by

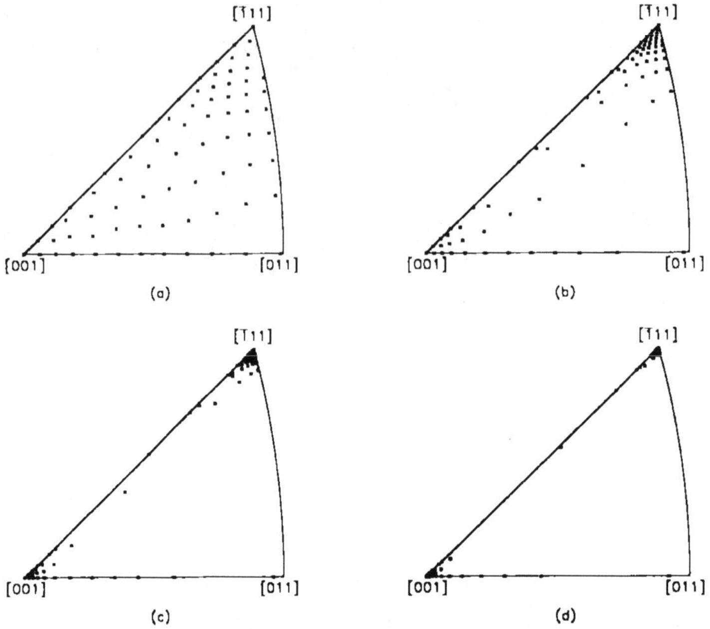

FIGURE 14.12. Inverse pole figures in tension for Taylor's hardening and Bunge's average slip method at the logarithmic strain levels of: (a) 0.23, (b) 0.69, (c) 1.15, and (d) 1.61 (from Lin and Havner, 1994; with permission from Elsevier Science).

Bronkhorst, Kalindini, and Anand (1992). They employed finite element calculations, as well as calculations based on Taylor's assumption of uniform deformation within each grain. The texture evolution in the aggregates of elastic-viscoplastic crystals with the low symmetry crystal lattices, lacking five independent slip systems, was studied by Parks and Ahzi (1990), Lee, Ahzi, and Asaro (1995), and Schoenfeld, Ahzi, and Asaro (1995). Further detailed analysis of various aspects of texture development can be found in a recent treatise by Kocks, Tomé, and Wenk (1998).

14.10. Grain Size Effects

The experimental evidence and the dislocation based models indicate that the macroscopic stress-strain response of a polycrystalline aggregate depends on the polycrystalline grain size. The well-known Hall–Petch relationship expresses the tensile yield stress of an aggregate, at a given amount of strain, as

$$\sigma = \sigma_0 + k\, l^{-1/2}, \tag{14.10.1}$$

where l is the average grain size (Fig. 14.13), and σ_0 and k are the appropriate constants (Hall, 1951; Petch, 1953). The constant k may be viewed as a measure of the average grain boundary resistance to slip propagation across the boundaries of differently oriented grains. Hall and Petch attributed the $l^{-1/2}$ dependence to stress acting on a dislocation pileup at the grain boundary. From an analytical solution derived by Eshelby, Frank, and Nabarro (1951), the stress exerted on the pinned dislocation at the boundary is equal to $\tau_{\text{pin}} = n\,\tau$, where τ is the applied shear stress on the pileup of n dislocations. For large n, the length of the pileup approaches $l = k_0\, n/\tau$, where $k_0 = \mu b/\pi(1-\nu)$ (b is the Burgers vector of edge dislocations in isotropic medium with the shear modulus μ and Poisson's ratio ν). By assuming that the length of the pileup is equal to the grain size, and by requiring that τ_{pin} is equal to the critical stress τ_* necessary to propagate the plastic deformation across the boundary, the Hall–Petch relation follows

$$\tau = \tau_0 + (k_0\tau_*)^{1/2}\, l^{-1/2}, \tag{14.10.2}$$

where τ_0 is the lattice friction stress. An alternative explanation of the $l^{-1/2}$ dependence is based on the measured dislocation density, which was found to be inversely proportional to the grain size, at a given amount of strain. If the flow stress increases in proportion to the square-root of the dislocation density, as suggested by Taylor's (1934) early dislocation model of strain hardening, the Hall–Petch relationship is again obtained.

Other micromechanical models were constructed to support the Hall–Petch relationship. Ashby (1970) suggested that geometrically necessary dislocations are generated in the vicinity of grain boundaries of the differently oriented grains, in order for them to fit together upon deformation under the applied stress. The density of these dislocations scales with the average strain in the grain divided by the grain size. Thus, the elevation in the yield stress scales with $l^{-1/2}$. Meyers and Ashworth (1982) proposed

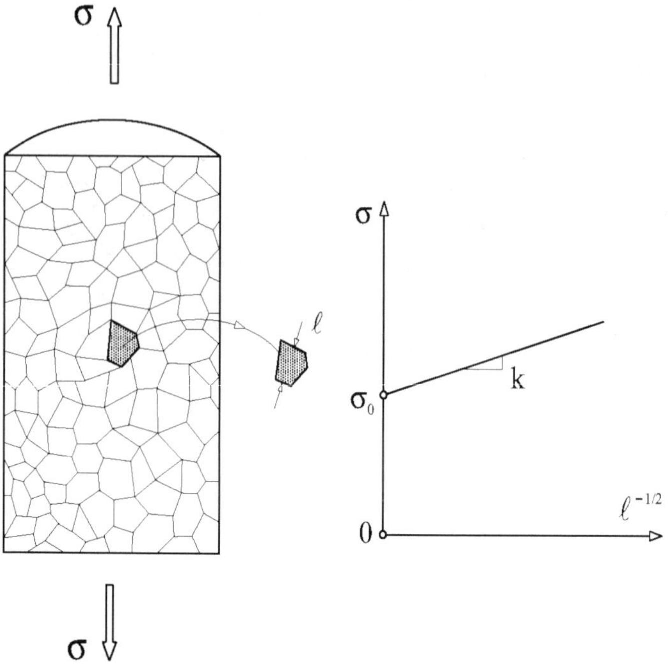

FIGURE 14.13. The yield stress σ of a polycrystalline aggregate as a function of the average grain size l, according to Hall–Petch inverse square-root relation.

that the grain size dependence of the yield stress is due to elastic incompatibility stresses at the grain boundaries. A work-hardened layer in the vicinity of grain boundaries, created by a network of geometrically necessary dislocations, acts as a reinforcement which elevates the yield stress. Modeling of the formation of organized dislocation structures by Lubarda, Blume, and Needleman (1993) can be employed to further study the microscopic structures causing the grain size effects. Related work includes Aifantis (1995), Van der Giessen and Needleman (1995), and Zbib, Rhee, and Hirth (1997).

The polycrystalline constitutive models considered in the previous sections of this chapter are unable to predict any grain size effect on the macroscopic response, because they were derived by the averaging schemes from the single crystal constitutive equations, which did not involve any length scale in their structure. In an approach toward a theoretical evaluation of the grain size effect on the overall behavior of polycrystals, Smyshlyaev and

Fleck (1996), and Shu and Fleck (1999) employed the strain gradient crystal plasticity theory of Fleck and Hutchinson (1997). The first-order gradients of the slip rates were included in this formulation. Since the nonlocal continuum theories are not considered in this book, we refer for the details of this approach to cited papers, and to Dillon and Kratochvil (1970), Zbib and Aifantis (1992), and Ning and Aifantis (1996a,b).

In the remainder of this section, we proceed in a simpler manner by partially addressing the grain size effects as follows. According to Armstrong, Codd, Douthwaite, and Petch (1962), and Armstrong (1970), it is assumed that the critical resolved shear stress of a single crystalline grain embedded in the surrounding polycrystal (effective medium) is grain size dependent, such that for an α slip system, at a given state of deformation,

$$\tau_{\mathrm{cr}}^{\alpha} = (\tau_{\mathrm{cr}}^{\alpha})^{\infty} + k_{\mathrm{c}}^{\alpha}\, l^{-1/2}. \tag{14.10.3}$$

Here, $(\tau_{\mathrm{cr}}^{\alpha})^{\infty}$ is the critical resolved shear stress in a free crystal (or in an infinite size crystal). The constant k_{c}^{α} (c stands for the crystal) reflects the fact that, when the grain is within a polycrystalline aggregate, dislocations arriving at the grain boundary cannot freely cross the boundary. This elevates the slip resistance and the required shear stress on the slip system. More generally, a hardening rule for the rate of the critical resolved shear stress could be specified, by extending (12.9.1), as

$$\dot{\tau}_{\mathrm{cr}}^{\alpha} = \sum_{\beta=1}^{n_0} \left(h_{\alpha\beta} + c_{\alpha\beta}\, l^{-1/2} \right) \dot{\gamma}^{\beta}, \quad \alpha = 1, 2, \ldots, N. \tag{14.10.4}$$

If Eq. (14.10.3) is adopted, the objective is to deduce the polycrystalline aggregate yield stress, for a given distribution of lattice orientations among the grains. A self-consistent calculation was presented by Weng (1983). We here employ a less involved analysis, based on Taylor's model of equal strain in all grains. Assuming that all slip systems within a grain harden equally, we write

$$(\tau_{\mathrm{cr}}^{\alpha})^{\infty} = \tau_{\mathrm{cr}}^{\infty}, \quad k_{\mathrm{c}}^{\alpha} = k_{\mathrm{c}}, \tag{14.10.5}$$

regardless of how much slip actually occurred on a particular slip system. From Eq. (14.1.16), the average values $\bar{\tau}_{\mathrm{cr}}^{\infty}$ and $\bar{k}$ for the aggregate are defined such that

$$\{ \left(\tau_{\mathrm{cr}}^{\infty} + k_{\mathrm{c}}\, l^{-1/2}\right) \min \sum_{\alpha} |\dot{\gamma}^{\alpha}| \} = \left(\bar{\tau}_{\mathrm{cr}}^{\infty} + \bar{k}\, l^{-1/2}\right) \{ \min \sum_{\alpha} |\dot{\gamma}^{\alpha}| \} . \tag{14.10.6}$$

In each grain it is assumed that

$$\tau_{\mathrm{cr}}^{\infty} = f\left(\int \min \sum_{\alpha} |\mathrm{d}\gamma^{\alpha}|\right), \quad k_{\mathrm{c}} = g\left(\int \min \sum_{\alpha} |\mathrm{d}\gamma^{\alpha}|\right), \tag{14.10.7}$$

and for the averages over all grains

$$\bar{\tau}_{\mathrm{cr}}^{\infty} = f(\bar{\gamma}), \quad \bar{k} = g(\bar{\gamma}), \quad \bar{\gamma} = \int \{ \min \sum_{\alpha} |\mathrm{d}\gamma^{\alpha}| \} . \tag{14.10.8}$$

Thus, extending Eq. (14.1.20), we have

$$\sigma = m\, \bar{\tau}_{\mathrm{cr}} = m \left(\bar{\tau}_{\mathrm{cr}}^{\infty} + \bar{k}\, l^{-1/2}\right) = m \left[f(\bar{\gamma}) + g(\bar{\gamma})\, l^{-1/2}\right], \tag{14.10.9}$$

i.e.,

$$\sigma = m \left[f\left(\int m\, \mathrm{de}\right) + g\left(\int m\, \mathrm{de}\right) l^{-1/2}\right], \tag{14.10.10}$$

since

$$\bar{\gamma} = \int \{ \min \sum_{\alpha} |\mathrm{d}\gamma^{\alpha}| \} = \int m\, \mathrm{de} . \tag{14.10.11}$$

Consequently, the Hall–Petch relation

$$\sigma = \sigma_0 + k\, l^{-1/2}, \tag{14.10.12}$$

with

$$\sigma_0 = m\, f\left(\int m\, \mathrm{de}\right), \quad k = m\, g\left(\int m\, \mathrm{de}\right). \tag{14.10.13}$$

As discussed earlier, the Taylor orientation factor m changes with the progression of deformation due to lattice rotation. For an initial random distribution of f.c.c. lattice orientation, $m = 3.06$, while for the random b.c.c. lattice orientation, $m = 2.83$.

References

Aifantis, E. C. (1995), Pattern formation in plasticity, *Int. J. Engng. Sci.*, Vol. 33, pp. 2161–2178.

Aifantis, E. C. (1995), From micro- to macro-plasticity: The scale invariance approach, *J. Engng. Mater. Techn.*, Vol. 117, pp. 352–355.

Anand, L. and Kothari, M. (1996), A computational procedure for rate-independent crystal plasticity, *J. Mech. Phys. Solids*, Vol. 44, pp. 525–558.

Armstrong, R. W., Codd, I., Douthwaite, R. M., and Petch, N. J. (1962), The plastic deformation of polycrystalline aggregates, *Phil. Mag.*, Vol. 7, pp. 45–58.

Armstrong, R. W. (1970), The influence of polycrystal grain size on several mechanical properties of materials, *Metall. Trans.*, Vol. 1, pp. 1169–1176.

Asaro, R. J. and Needleman, A. (1985), Texture development and strain hardening in rate dependent polycrystals, *Acta Metall.*, Vol. 33, pp. 923–953.

Ashby, M. F. (1970), The deformation of plastically non-homogeneous alloys, *Phil. Mag.*, Vol. 21, pp. 399–424.

Bell, J. F. (1968), The physics of large deformation of crystalline solids, *Springer Tracts in Natural Philosophy*, Vol. 14, Springer-Verlag, Berlin.

Beradai, Ch., Berveiller, M., and Lipinski, P. (1987), Plasticity of metallic polycrystals under complex loading paths, *Int. J. Plasticity*, Vol. 3, pp. 143–162.

Berveiller, M. and Zaoui, A. (1979), An extension of the self-consistent scheme to plastically-flowing polycrystals, *J. Mech. Phys. Solids*, Vol. 26, pp. 325–344.

Berveiller, M. and Zaoui, A. (1986), Some applications of the self-consistent scheme in the field of plasticity and texture of metallic polycrystals, in *Large Deformation of Solids: Physical Basis and Mathematical Modelling*, eds. J. Gittus, J. Zarka, and S. Nemat-Nasser, pp. 223–241, Elsevier, London.

Bishop, J. F. W. (1953), A theoretical examination of the plastic deformation of crystals by glide, *Phil. Mag.*, Vol. 44, pp. 51–64.

Bishop, J. F. W. (1954), A theory of the tensile and compressive textures of face-centred cubic metals, *J. Mech. Phys. Solids*, Vol. 3, pp. 130–142.

Bishop, J. F. W. and Hill, R. (1951a), A theory of plastic distortion of a polycrystalline aggregate under combined stresses, *Phil. Mag.*, Vol. 42, pp. 414–427.

Bishop, J. F. W. and Hill, R. (1951b), A theoretical derivation of the plastic properties of a polycrystalline face-centred metal, *Phil. Mag.*, Vol. 42, pp. 1298–1307.

Bronkhorst, C. A., Kalindini, S. R., and Anand, L. (1992), Polycrystal plasticity and the evolution of crystallographic texture in face-centred cubic metals, *Phil. Trans. Roy. Soc. Lond. A*, Vol. 341, pp. 443–477.

Brown, G. M. (1970), A self-consistent polycrystalline model for creep under combined stress states, *J. Mech. Phys. Solids*, Vol. 18, pp. 367–381.

Budiansky, B. (1965), On the elastic moduli of some heterogeneous materials, *J. Mech. Phys. Solids*, Vol. 13, pp. 223–227.

Budiansky, B. and Wu, T. Y. (1962), Theoretical prediction of plastic strains of polycrystals, in *Proc. 4th U.S. Nat. Congr. Appl. Mech.*, pp. 1175–1185.

Bunge, H.-J. (1970), Some applications of the Taylor theory of polycrystal plasticity, *Kristall und Technik*, Vol. 5, pp. 145–175.

Bunge, H.-J. (1982), *Texture Analysis in Materials Science – Mathematical Methods*, Butterworths, London.

Bunge, H.-J. (1988), Texture and directional properties of materials, in *Directional Properties of Materials*, ed. H.-J. Bunge, pp. 1–63, DGM Informationsgesellschaft-Verlag, Oberursel, FRG.

Bunge, H.-J. and Nielsen, I. (1997), Experimental determination of plastic spin in polycrystalline materials, *Int. J. Plasticity*, Vol. 13, pp. 435–446.

Canova, G. R., Kocks, U. F., and Jonas, J. J. (1984), Theory of torsion texture development, *Acta Metall.*, Vol. 32, pp. 211–226.

Chin, G. Y. (1969), Tension and compression textures, in *Textures in Research and Practice*, eds. J. Grewen and G. Wassermann, pp. 51–80, Springer-Verlag, Berlin.

Chin, G. Y. and Mammel, W. L. (1967), Computer solutions of the Taylor analysis for axisymmetric flow, *Trans. Met. Soc. AIME*, Vol. 239, pp. 1400–1405.

Cleary, M. P., Chen, I.-W., and Lee, S.-M. (1980), Self-consistent techniques for heterogeneous media, *J. Engrg. Mech., ASCE*, Vol. 106, pp. 861–887.

Dawson, P. R. and Kumar, A. (1997), Deformation process simulations using polycrystal plasticity, in *Large Plastic Deformation of Crystalline Aggregates*, ed. C. Teodosiu, pp. 173–246, Springer-Verlag, Wien.

Dillamore, I. L., Roberts, J. G., and Bush, A. C. (1979), Occurrence of shear bands in heavily rolled cubic metals, *Metal Sci.*, Vol. 13, pp. 73–77.

Dillon, O. W. and Kratochvil, J. (1970), A strain gradient theory of plasticity, *Int. J. Solids Struct.*, Vol. 6, pp. 1513–1533.

Eshelby, J. D., Frank, F. C., and Nabarro, F. R. N. (1951), The equilibrium of linear arrays of dislocations, *Phil. Mag.*, Vol. 42, pp. 351–364.

Eshelby, J. D. (1957), The determination of the elastic field of an ellipsoidal inclusion, and related problems, *Proc. Roy. Soc. Lond. A*, Vol. 241, pp. 376–396.

Eshelby, J. D. (1961), Elastic inclusions and inhomogeneities, in *Progress in Solid Mechanics*, Vol. 2, eds. I. N. Sneddon and R. Hill, pp. 87–140, North-Holland, Amsterdam.

Fleck, N. A. and Hutchinson, J. W. (1997), Strain gradient plasticity, *Adv. Appl. Mech.*, Vol. 33, pp. 295–362.

Fuh, S. and Havner, K. S. (1989), A theory of minimum plastic spin in crystal mechanics, *Proc. Roy. Soc. Lond. A*, Vol. 422, pp. 193–239.

Gil-Sevillano, J., Van Houtte, P., and Aernoudt, E. (1975), Deutung der Schertexturen mit Hilfe der Taylor-Analyse, *Z. Metallk.*, Vol. 66, pp. 367–373.

Gil-Sevillano, J., Van Houtte, P. and Aernoudt, E. (1980), Large strain work hardening and textures, *Progr. Mater. Sci.*, Vol. 25, pp. 69–412.

Gottstein, G. and Lücke, K., eds. (1978), *Textures of Materials*, Vol. I and II, Springer-Verlag, Berlin.

Hall, E. O. (1951), The deformation and ageing of mild steel, *Proc. Phys. Soc. Lond. B*, Vol. 64, pp. 747–753.

Harren, S. V. (1989), The finite deformation of rate-dependent polycrystals – I and II, *J. Mech. Phys. Solids*, Vol. 39, pp. 345–383.

Harren, S. V., Lowe, T. C., Asaro, R. J., and Needleman, A. (1989), Analysis of large-strain shear in rate-dependent face-centered cubic polycrystals: Correlation of micro- and macromechanics, *Phil. Trans. Roy. Soc. Lond. A*, Vol. 328, pp. 443–500.

Hashin, Z. and Shtrikman, S. (1962), A variational approach to the theory of the elastic behavior of polycrystals, *J. Mech. Phys. Solids*, Vol. 10, pp. 343–352.

Havner, K. S. (1971), A discrete model for the prediction of subsequent yield surface in polycrystalline plasticity, *Int. J. Solids Struct.*, Vol. 7, pp. 719–730.

Havner, K. S. (1982), The theory of finite plastic deformation of crystalline solids, in *Mechanics of Solids – The Rodney Hill 60th Anniversary Volume*, eds. H. G. Hopkins and M. J. Sewell, pp. 265–302, Pergamon Press, Oxford.

Havner, K. S. (1992), *Finite Plastic Deformation of Crystalline Solids*, Cambridge University Press, Cambridge.

Hershey, A. V. (1954), The elasticity of an isotropic aggregate of anisotropic cubic crystals, *J. Appl. Mech.* Vol. 21, pp. 236–240.

Hill, R. (1952) The elastic behavior of a crystalline aggregate, *Proc. Phys. Soc. London A*, Vol. 65, pp. 349–354.

Hill, R. (1965a), Continuum micro-mechanics of elastoplastic polycrystals, *J. Mech. Phys. Solids*, Vol. 13, pp. 89–101.

Hill, R. (1965b), A self-consistent mechanics of composite materials, *J. Mech. Phys. Solids*, Vol. 13, pp. 213–222.

Hill, R. (1967), The essential structure of constitutive laws for metal composites and polycrystals, *J. Mech. Phys. Solids*, Vol. 15, pp. 79–96.

Honneff, H. and Mecking, H. (1978), A method for the determination of the active slip systems and orientation changes during single crystal deformation, in *Texture of Materials*, eds. G. Gottstein and K. Lücke, pp. 265–275, Springer-Verlag, Berlin.

Hutchinson, J. W. (1964a), Plastic stress-strain relations of F.C.C. polycrystalline metals hardening according to Taylor rule, *J. Mech. Phys. Solids*, Vol. 12, pp. 11–24.

Hutchinson, J. W. (1964b), Plastic deformation of B.C.C. polycrystals, *J. Mech. Phys. Solids*, Vol. 12, pp. 25–33.

Hutchinson, J. W. (1970), Elastic-plastic behavior of polycrystalline metals and composites, *Proc. Roy. Soc. Lond. A*, Vol. 319, pp. 247–272.

Hutchinson, J. W. (1976), Bounds and self-consistent estimates for creep of polycrystalline materials, *Proc. Roy. Soc. Lond. A*, Vol. 348, pp. 101–127.

Iwakuma, T. and Nemat-Nasser, S. (1984), Finite elastic-plastic deformation of polycrystalline metals, *Proc. Roy. Soc. Lond. A*, Vol. 394, pp. 87–119.

Kallend, J. S. and Davies, G. J. (1972), A simulation of texture development in f.c.c. metals, *Phil. Mag.*, 8th Series, Vol. 25, pp. 471–490.

Kocks, U. F. (1970), The relation between polycrystalline deformation and single-crystal deformation, *Metall. Trans.*, Vol. 1, pp. 1121–1142.

Kocks, U. F. (1987), Constitutive behavior based on crystal plasticity, in *Unified Constitutive Equations for Creep and Plasticity*, ed. A. K. Miller, pp. 1–88, Elsevier Applied Science, London.

Kocks, U. F., Tomé, C. N., and Wenk, H.-R. (1998), *Texture and Anisotropy: Preferred Orientations in Polycrystals and their Effect on Materials Properties*, Cambridge University Press, Cambridge.

Kröner, E. (1958), Berechnung der elastischen Konstanten des Vielkristalls aus den Konstanten des Einkristalls, *Z. Physik*, Vol. 151, pp. 504–518.

Kröner, E. (1961), Zur plastichen Verformung des Vielkristalls, *Acta Metall.*, Vol. 9, pp. 155–161.

Lee, B. J., Ahzi, S., and Asaro, R. J. (1995), On the plasticity of low symmetry crystals lacking five independent slip systems, *Mech. Mater.*, Vol. 20, pp. 1–8.

Lin, G. and Havner, K. S. (1994), On the evolution of texture and yield loci in axisymmetric deformation of f.c.c. polycrystals, *Int. J. Plasticity*, Vol. 10, pp. 471–498.

Lin, G. and Havner, K. S. (1996), A comparative study of hardening theories in torsion using the Taylor polycrystal model, *Int. J. Plasticity*, Vol. 12, pp. 695–718.

Lin, T. H. (1957), Analysis of elastic and plastic strains of a face-centred cubic crystal, *J. Mech. Phys. Solids*, Vol. 5, pp. 143–149.

Lin, T. H. (1971), Physical theory of plasticity, *Adv. Appl. Mech.*, Vol. 11, pp. 255–311.

Lin, T. H. and Ito, Y. M. (1965), Theoretical plastic distortion of a polycrystalline aggregate under combined and reversed stresses, *J. Mech. Phys. Solids*, Vol. 13, pp. 103–115.

Lin, T. H. and Ito, Y. M. (1966), Theoretical plastic stress-strain relationships of a polycrystal and comparison with von Mises' and Tresca's plasticity theories, *Int. J. Engng. Sci.*, Vol. 4, pp. 543–561.

Lipinski, P. and Berveiller, M. (1989), Elastoplasticity of micro-inhomogeneous metals at large strains, *Int. J. Plasticity*, Vol. 5, pp. 149–172.

Lipinski, P., Naddari, A., and Berveiller, M. (1992), Recent results concerning the modelling of polycrystalline plasticity at large strains, *Int. J. Solids Struct.*, Vol. 29, pp. 1873–1881.

Lubarda, V. A. (1997), New estimates of the third-order elastic constants for isotropic aggregates of cubic crystals, *J. Mech. Phys. Solids*, Vol. 45, pp. 471–490.

Lubarda, V. A. (1998), A note on the effective Lamé constants of polycrystalline aggregates of cubic crystals, *J. Appl. Mech.*, Vol. 65, pp. 769–770.

Lubarda, V. A., Blume, J. A., and Needleman, A. (1993), An analysis of equilibrium dislocation distributions, *Acta Metall. Mater.*, Vol. 41, pp. 625–642.

Masson, R. and Zaoui, A. (1999), Self-consistent estimates for the rate-dependent elastoplastic behaviour of polycrystalline materials, *J. Mech. Phys. Solids*, Vol. 47, pp. 1543–1568.

Meyers, M. A. and Ashworth, E. (1982), A model for the effect of grain size on the yield stress of metals, *Phil. Mag.*, Vol. 46, pp. 737–759.

Molinari, A. (1997), Self-consistent modelling of plastic and viscoplastic polycrystalline materials, in *Large Plastic Deformation of Crystalline Aggregates*, ed. C. Teodosiu, pp. 173–246, Springer-Verlag, Wien.

Molinari, A., Ahzi, S., and Koddane, R. (1997), On the self-consistent modeling of elastic-plastic behavior of polycrystals, *Mech. Mater.*, Vol. 26, pp. 43–62.

Molinari, A., Canova, G. R., and Ahzi, S. (1987), A self-consistent approach of the large deformation polycrystal viscoplasticity, *Acta Metall.*, Vol. 35, pp. 2983–2994.

Mura, T. (1987), *Micromechanics of Defects in Solids*, Martinus Nijhoff Publishers, Dordrecht, The Netherlands.

Nemat-Nasser, S. (1999), Averaging theorems in finite deformation plasticity, *Mech. Mater.*, Vol. 31, pp. 493–523 (with Erratum, Vol. 32, 2000, p. 327).

Nemat-Nasser, S. and Obata, M. (1986), Rate-dependent, finite elasto-plastic deformation of polycrystals, *Proc. Roy. Soc. Lond. A*, Vol. 407, pp. 343–375.

Ning, J. and Aifantis, E. C. (1996a), Anisotropic yield and plastic flow of polycrystalline solids, *Int. J. Plasticity*, Vol. 12, pp. 1221–1240.

Ning, J. and Aifantis, E. C. (1996b), Anisotropic and inhomogeneous plastic deformation of polycrystalline solids, in *Unified Constitutive Laws of Plastic Deformation*, ed. A. K. Miller, pp. 319–341, Academic Press, San Diego.

Parks, D. M. and Ahzi, S. (1990), Polycrystalline plastic deformation and texture evolution for crystals lacking five independent slip systems, *J. Mech. Phys. Solids*, Vol. 38, pp. 701–724.

Petch, N. J. (1953), The cleavage strength of polycrystals, *J. Iron Steel Inst.*, Vol. 174, pp. 25–28.

Prantil, V. C., Jenkins, J. T., and Dawson, P. R., An analysis of texture and plastic spin for planar polycrystals, *J. Mech. Phys. Solids*, Vol. 41, pp. 1357–1382.

Reuss, A. (1929), Berechnung der Fliessgrenze von Mischkristallen auf Grund der Plastizitätsbedingung für Einkristalle, *Z. angew. Math. Mech.*, Vol. 9, pp. 49–58.

Sachs, G. (1928), Zur Ableitung einer Fleissbedingung, *Z. d. Ver. deut. Ing.*, Vol. 72, pp. 734–736.

Schoenfeld, S. E., Ahzi, S. and Asaro, R. J. (1995), Elastic-plastic crystal mechanics for low symmetry crystals, *J. Mech. Phys. Solids*, Vol. 43, pp. 415–446.

Shu, J. Y. and Fleck, N. A. (1999), Strain gradient crystal plasticity: Size-dependent deformation of bicrystals, *J. Mech. Phys. Solids*, Vol. 47, pp. 297–324.

Smyshlyaev, V. P. and Fleck, N. A. (1996), The role of strain gradients in the grain size effect for polycrystals, *J. Mech. Phys. Solids*, Vol. 44, pp. 465–495.

Taylor, G. I. (1934), The mechanism of plastic deformation of crystals, *Proc. Roy. Soc. Lond. A*, Vol. 145, pp. 362–387.

Taylor, G. I. (1938a), Plastic strain in metals, *J. Inst. Metals*, Vol. 62, pp. 307–324.

Taylor, G. I. (1938b), Analysis of plastic strain in a cubic crystal, in *Stephen Timoshenko 60th Anniversary Volume*, ed. J. M. Lessels, pp. 218–224, Macmillan, New York. (Reprinted with added note in *The Scientific Papers of Sir Geoffrey Ingram Taylor*, ed. G. K. Batchelor, Vol. I, pp. 439–445, Cambridge University Press, Cambridge, 1958).

Taylor, G. I. and Quinney, H. (1931), The plastic distortion of metals, *Phil. Trans. Roy. Soc. Lond. A*, Vol. 230, pp. 232–262.

Toth, L. and Molinari, A. (1994), Tuning a self-consistent viscoplastic model by finite-element results: Application to torsion textures, *Acta Metall. Mater.*, Vol. 42, pp. 2459–2466.

Van der Giessen, E. and Needleman, A. (1995), Discrete dislocation plasticity: A simple planar model, *Modelling Simul. Mater. Sci. Eng.*, Vol. 3, pp. 689–735.

Van Houtte, P. (1991), Models for the prediction of deformation textures, in *Advances and Applications of Quantitative Texture Analysis*, eds. H. J. Bunge and C. Esling, pp. 175–198, DGM Informationsgesellschaft-Verlag, Oberursel, FRG.

Voigt, W. (1889), Über die Beziehung zwischen den beiden Elastizitätskonst anten isotroper Körper, *Wied. Ann.*, Vol. 38, pp. 573–587.

Walpole, L. J. (1969), On the overall elastic moduli of composite materials, *J. Mech. Phys. Solids*, Vol. 17, pp. 235–251.

Walpole, L. J. (1981), Elastic behavior of composite materials: Theoretical foundations, *Adv. Appl. Mech.*, Vol. 21, pp. 169–242.

Weng, G. J. (1981), Self-consistent determination of time-dependent behavior of metals, *J. Appl. Mech.*, Vol. 48, pp. 41–46.

Weng, G. J. (1982), A unified, self-consistent theory for the plastic-creep deformation of metals, *J. Appl. Mech.*, Vol. 49, pp. 728–734.

Weng, G. J. (1983), A micromechanical theory of grain-size dependence in metal plasticity, *J. Mech. Phys. Solids*, Vol. 31, pp. 193–203.

Willis, J. R. (1964), Anisotropic elastic inclusion problems, *Quart. J. Mech. Appl. Math.*, Vol. 17, pp. 157–174.

Willis, J. R. (1981), Variational and related methods for the overall properties of composites, *Adv. Appl. Mech.*, Vol. 21, pp. 1–78.

Wu, P. D., Neale, K. W., and Van der Giessen, E. (1996), Simulation of the behaviour of FCC polycrystals during reversed torsion, *Int. J. Plasticity*, Vol. 12, pp. 1199–1219.

Yang, W. and Lee, W. B. (1993), *Mesoplasticity and its Applications*, Springer-Verlag, Berlin.

Zaoui, A. (1987), Approximate statistical modelling and applications, in *Homogenization Techniques for Composite Media*, eds. E. Sanchez-Palencia and A. Zaoui, pp. 338–397, Springer-Verlag, Berlin.

Zbib, H. M. and Aifantis, E. C. (1992), On the gradient-dependent theory of plasticity and shear banding, *Acta Mech.*, Vol. 92, pp. 209–225.

Zbib, H. M., Rhee, M., and Hirth, J. P. (1997), On plastic deformation and the dynamics of 3D dislocations, *Int. J. Mech. Sci.*, Vol. 40, pp. 113–127.

Author Index

Subject Index